Budding Yeast

A LABORATORY MANUAL

ALSO FROM COLD SPRING HARBOR LABORATORY PRESS

RELATED TITLES

The Early Days of Yeast Genetics

From **a** *to* α: *Yeast as a Model for Cellular Differentiation*

Human Fungal Pathogens

Landmark Papers in Yeast Biology

Methods in Yeast Genetics and Genomics: A Cold Spring Harbor Laboratory Course Manual, 2015 Edition

Mitosis

Navigating Metabolism

Yeast Intermediary Metabolism

OTHER LABORATORY MANUALS

Antibodies: A Laboratory Manual, Second Edition

Calcium Techniques: A Laboratory Manual

Cell Death Techniques: A Laboratory Manual

Manipulating the Mouse Embryo: A Laboratory Manual, Fourth Edition

Molecular Cloning: A Laboratory Manual, Fourth Edition

Molecular Neuroscience: A Laboratory Manual

Mouse Models of Cancer: A Laboratory Manual

Purifying and Culturing Neural Cells: A Laboratory Manual

Subcellular Fractionation: A Laboratory Manual

HANDBOOKS

At the Bench: A Laboratory Navigator, Updated Edition

At the Helm: Leading Your Laboratory, Second Edition

Experimental Design for Biologists, Second Edition

Lab Math: A Handbook of Measurements, Calculations, and Other Quantitative Skills for Use at the Bench

Lab Ref: A Handbook of Recipes, Reagents, and Other Reference Tools for Use at the Bench, Volume 1 and Volume 2

Next-Generation DNA Sequencing Informatics, Second Edition

Statistics at the Bench: A Step-by-Step Handbook for Biologists

Using R at the Bench: Step-by-Step Data Analytics for Biologists

WEBSITE

www.cshprotocols.org

Budding Yeast

A LABORATORY MANUAL

EDITED BY

Brenda Andrews
University of Toronto

Charles Boone
University of Toronto

Trisha N. Davis
University of Washington

Stanley Fields
*Howard Hughes Medical Institute
and University of Washington*

COLD SPRING HARBOR LABORATORY PRESS
Cold Spring Harbor, New York • www.cshlpress.org

BUDDING YEAST
A LABORATORY MANUAL

All rights reserved
© 2016 by Cold Spring Harbor Laboratory Press, Cold Spring Harbor, New York
Printed in the United States of America

Publisher	John Inglis
Acquisition Editor	Richard Sever
Managing Editor, *CSH Perspectives, CSH Protocols*	Maria Smit
Director of Editorial Services	Jan Argentine
Project Manager	Maryliz Dickerson
Production Editors, *CSH Protocols*	Joanne McFadden
Production Manager	Denise Weiss
Director of Product Development & Marketing	Wayne Manos

Cover art: A Pil1-GFP Hta2-mCherry-tagged yeast cell highlighting the eisosomes (green) and nuclei (red), imaged on the super-resolution confocal live imaging microscopy (SCLIM) system developed by the Akihiko Nakano laboratory at the RIKEN Center for Advanced Photonics in Wako, Japan (described in Kurokawa et al., *Methods in Cell Biology*, 2013). (Image courtesy of Erin B. Styles, Benjamin T. Grys, Kazuo Kurokawa, and Akihiko Nakano.)

Library of Congress Cataloging-in-Publication Data

Names: Andrews, Brenda, editor. | Boone, Charles, editor. | Davis, Trisha, 1954-,
    editor. | Fields, Stanley, editor. | Cold Spring Harbor
    Laboratory, issuing body.
Title: Budding yeast : a laboratory manual/edited by Brenda Andrews, Charles
    Boone, Trisha N. Davis, Stanley Fields.
Description: Cold Spring Harbor, New York : Cold Spring Harbor Laboratory
    Press, [2016] | Includes bibliographical references and index.
Identifiers: LCCN 2015039656| ISBN 9781621820550 | ISBN 9781621820567
Subjects: | MESH: Saccharomycetales–Laboratory Manuals. | Clinical
    Laboratory Techniques–methods–Laboratory Manuals.
Classification: LCC QK623.S23 | NLM QW 25 | DDC 579.5/63–dc23
LC record available at http://lccn.loc.gov/2015039656

2015020016

10 9 8 7 6 5 4 3 2 1

Students and researchers using the procedures in this manual do so at their own risk. Cold Spring Harbor Laboratory makes no representations or warranties with respect to the material set forth in this manual and has no liability in connection with the use of these materials. All registered trademarks, trade names, and brand names mentioned in this book are the property of the respective owners. Readers should please consult individual manufacturers and other resources for current and specific product information.

With the exception of those suppliers listed in the text with their addresses, all suppliers mentioned in this manual can be found on the BioSupplyNet website at www.biosupplynet.com.

All World Wide Web addresses are accurate to the best of our knowledge at the time of printing.

Procedures for the humane treatment of animals must be observed at all times. Check with the local animal facility for guidelines.

Certain experimental procedures in this manual may be the subject of national or local legislation or agency restrictions. Users of this manual are responsible for obtaining the relevant permissions, certificates, or licenses in these cases. Neither the authors of this manual nor Cold Spring Harbor Laboratory assume any responsibility for failure of a user to do so.

The materials and methods in this manual may infringe the patent and proprietary rights of other individuals, companies, or organizations. Users of this manual are responsible for obtaining any licenses necessary to use such materials and to practice such methods. COLD SPRING HARBOR LABORATORY MAKES NO WARRANTY OR REPRESENTATION THAT USE OF THE INFORMATION IN THIS MANUAL WILL NOT INFRINGE ANY PATENT OR OTHER PROPRIETARY RIGHT.

Authorization to photocopy items for internal or personal use, or the internal or personal use of specific clients, is granted by Cold Spring Harbor Laboratory Press, provided that the appropriate fee is paid directly to the Copyright Clearance Center (CCC). Write or call CCC at 222 Rosewood Drive, Danvers, MA 01923 (978-750-8400) for information about fees and regulations. Prior to photocopying items for educational classroom use, contact CCC at the above address. Additional information on CCC can be obtained at CCC Online at www.copyright.com.

For a complete catalog of all Cold Spring Harbor Laboratory Press publications, visit our website at www.cshlpress.org.

Contents

	Preface	xvii
CHAPTER 1	Historical Evolution of Laboratory Strains of *Saccharomyces cerevisiae*	

INTRODUCTION

Historical Evolution of Laboratory Strains of *Saccharomyces cerevisiae*   1
Edward J. Louis

CHAPTER 2	Classical Genetics with *Saccharomyces cerevisiae*	

INTRODUCTION

Classical Genetics with *Saccharomyces cerevisiae*   11
David C. Amberg and Daniel J. Burke

CHAPTER 3	Analysis of Recombination and Chromosome Structure during Yeast Meiosis	

INTRODUCTION

Analysis of Recombination and Chromosome Structure during Yeast Meiosis   21
G. Valentin Börner and Rita S. Cha

PROTOCOLS

1. Analysis of Yeast Sporulation Efficiency, Spore Viability, and Meiotic Recombination on Solid Medium   26
G. Valentin Börner and Rita S. Cha
2. Induction and Analysis of Synchronous Meiotic Yeast Cultures   32
G. Valentin Börner and Rita S. Cha
3. Analysis of Meiotic Recombination and Homolog Interaction during Yeast Meiosis   38
G. Valentin Börner and Rita S. Cha

CHAPTER 4	Investigating Filamentous Growth and Biofilm/Mat Formation in Budding Yeast	

INTRODUCTION

Investigating Filamentous Growth and Biofilm/Mat Formation in Budding Yeast   49
Paul J. Cullen

PROTOCOLS

1. The Plate-Washing Assay: A Simple Test for Filamentous Growth in Budding Yeast   53
Paul J. Cullen

2	Biofilm/Mat Assays for Budding Yeast *Paul J. Cullen*	57
3	Evaluating Filamentous Yeast Growth at the Single-Cell Level *Paul J. Cullen*	61
4	Evaluating the Activity of the Filamentous Growth Mitogen-Activated Protein Kinase Pathway in Yeast *Paul J. Cullen*	65

CHAPTER 5 Construction of Multifragment Plasmids by Homologous Recombination in Yeast

INTRODUCTION

Construction of Multifragment Plasmids by Homologous Recombination in Yeast *Jolanda van Leeuwen, Brenda Andrews, Charles Boone, and Guihong Tan*	73

PROTOCOL

1	Rapid and Efficient Plasmid Construction by Homologous Recombination in Yeast *Jolanda van Leeuwen, Brenda Andrews, Charles Boone, and Guihong Tan*	78

CHAPTER 6 Single-Molecule Analysis of Replicating Yeast Chromosomes

INTRODUCTION

Single-Molecule Analysis of Replicating Yeast Chromosomes *David Gallo, Gang Wang, Christopher M. Yip, and Grant W. Brown*	87

PROTOCOL

1	Analysis of Replicating Yeast Chromosomes by DNA Combing *David Gallo, Gang Wang, Christopher M. Yip, and Grant W. Brown*	90

CHAPTER 7 Measuring Chromatin Structure in Budding Yeast

INTRODUCTION

Measuring Chromatin Structure in Budding Yeast *Jon-Matthew Belton and Job Dekker*	103

PROTOCOLS

1	Chromosome Conformation Capture (3C) in Budding Yeast *Jon-Matthew Belton and Job Dekker*	108
2	Randomized Ligation Control for Chromosome Conformation Capture *Jon-Matthew Belton and Job Dekker*	115
3	Chromosome Conformation Capture Carbon Copy (5C) in Budding Yeast *Jon-Matthew Belton and Job Dekker*	121
4	Hi-C in Budding Yeast *Jon-Matthew Belton and Job Dekker*	127

CHAPTER 8 MX Cassettes for Knocking Out Genes in Yeast

INTRODUCTION

MX Cassettes for Knocking Out Genes in Yeast 141
John H. McCusker

PROTOCOLS

1 Introducing MX Cassettes into *Saccharomyces cerevisiae* 146
John H. McCusker

2 Popping Out MX Cassettes from *Saccharomyces cerevisiae* 153
John H. McCusker

CHAPTER 9 Multipurpose Transposon-Insertion Libraries in Yeast

INTRODUCTION

Multipurpose Transposon-Insertion Libraries in Yeast 161
Anuj Kumar

PROTOCOL

1 Using Yeast Transposon-Insertion Libraries for Phenotypic Screening and Protein Localization 165
Anuj Kumar

CHAPTER 10 Functional Genomics Using the *Saccharomyces cerevisiae* Yeast Deletion Collections

INTRODUCTION

Functional Genomics Using the *Saccharomyces cerevisiae* Yeast Deletion Collections 173
Corey Nislow, Lai Hong Wong, Amy Huei-Yi Lee, and Guri Giaever

PROTOCOL

1 Functional Profiling Using the Yeast Deletion Collections 179
Corey Nislow, Lai Hong Wong, Amy Huei-Yi Lee, and Guri Giaever

CHAPTER 11 Deep Mutational Scanning: A Highly Parallel Method to Measure the Effects of Mutation on Protein Function

INTRODUCTION

Deep Mutational Scanning: A Highly Parallel Method to Measure the Effects of Mutation on Protein Function 187
Lea M. Starita and Stanley Fields

PROTOCOLS

1 Deep Mutational Scanning: Library Construction, Functional Selection, and High-Throughput Sequencing 191
Lea M. Starita and Stanley Fields

2   Deep Mutational Scanning: Calculating Enrichment Scores for Protein Variants
    from DNA Sequencing Output Files   195
    Lea M. Starita and Stanley Fields

CHAPTER 12   Examination and Disruption of the Yeast Cell Wall

INTRODUCTION

Examination and Disruption of the Yeast Cell Wall   199
Hiroki Okada, Keiko Kono, Aaron M. Neiman, and Yoshikazu Ohya

PROTOCOLS

1   Fluorescent Labeling of Yeast Cell Wall Components   205
    Hiroki Okada and Yoshikazu Ohya

2   Assay for Spore Wall Integrity Using a Yeast Predator   209
    Hiroki Okada, Aaron M. Neiman, and Yoshikazu Ohya

3   Local and Acute Disruption of the Yeast Cell Surface   213
    Keiko Kono, Hiroki Okada, and Yoshikazu Ohya

CHAPTER 13   Analyzing and Understanding Lipids of Yeast: A Challenging Endeavor

INTRODUCTION

Analyzing and Understanding Lipids of Yeast: A Challenging Endeavor   217
Sepp D. Kohlwein

PROTOCOLS

1   Lipid Extraction from Yeast Cells   223
    Oskar L. Knittelfelder and Sepp D. Kohlwein

2   Thin Layer Chromatography to Separate Phospholipids and
    Neutral Lipids from Yeast   227
    Oskar L. Knittelfelder and Sepp D. Kohlwein

3   Derivatization and Gas Chromatography of Fatty Acids from Yeast   231
    Oskar L. Knittelfelder and Sepp D. Kohlwein

4   Quantitative Analysis of Yeast Phospholipids and Sterols by High-Performance
    Liquid Chromatography–Evaporative Light-Scattering Detection   235
    Oskar L. Knittelfelder and Sepp D. Kohlwein

CHAPTER 14   Methods for Synchronization and Analysis of the Budding Yeast Cell Cycle

INTRODUCTION

Methods for Synchronization and Analysis of the Budding Yeast Cell Cycle   239
Adam P. Rosebrock

PROTOCOLS

1   Synchronization and Arrest of the Budding Yeast Cell Cycle Using Chemical
    and Genetic Methods   242
    Adam P. Rosebrock

	2	Synchronization of Budding Yeast by Centrifugal Elutriation *Adam P. Rosebrock*	248
	3	Analysis of the Budding Yeast Cell Cycle by Flow Cytometry *Adam P. Rosebrock*	258

CHAPTER 15 High-Throughput Microscopy-Based Screening in *Saccharomyces cerevisiae*

INTRODUCTION

High-Throughput Microscopy-Based Screening in *Saccharomyces cerevisiae* 265
Erin B. Styles, Helena Friesen, Charles Boone, and Brenda Andrews

PROTOCOLS

1 Liquid Growth of Arrayed Fluorescently Tagged *Saccharomyces cerevisiae* Strains for Live-Cell High-Throughput Microscopy Screens 271
Michael J. Cox, Yolanda T. Chong, Charles Boone, and Brenda Andrews

2 Quantification of Cell, Actin, and Nuclear DNA Morphology with High-Throughput Microscopy and CalMorph 277
Hiroki Okada, Shinsuke Ohnuki, and Yoshikazu Ohya

CHAPTER 16 Single-Molecule Total Internal Reflection Fluorescence Microscopy

INTRODUCTION

Single-Molecule Total Internal Reflection Fluorescence Microscopy 283
Emily M. Kudalkar, Trisha N. Davis, and Charles L. Asbury

PROTOCOLS

1 Coverslip Cleaning and Functionalization for Total Internal Reflection Fluorescence Microscopy 287
Emily M. Kudalkar, Yi Deng, Trisha N. Davis, and Charles L. Asbury

2 Preparation of Reactions for Imaging with Total Internal Reflection Fluorescence Microscopy 294
Emily M. Kudalkar, Trisha N. Davis, and Charles L. Asbury

3 Data Analysis for Total Internal Reflection Fluorescence Microscopy 299
Charles L. Asbury

CHAPTER 17 Building Cell Structures in Three Dimensions: Electron Tomography Methods for Budding Yeast

INTRODUCTION

Building Cell Structures in Three Dimensions: Electron Tomography Methods for Budding Yeast 303
Eileen T. O'Toole, Thomas H. Giddings, Jr., and Mark Winey

PROTOCOL

1 Cryopreparation and Electron Tomography of Yeast Cells 308
Eileen T. O'Toole, Thomas H. Giddings, Jr., and Mark Winey

CHAPTER 18 The Yeast Two-Hybrid System: A Tool for Mapping Protein–Protein Interactions

INTRODUCTION

The Yeast Two-Hybrid System: A Tool for Mapping Protein–Protein Interactions 313
Jitender Mehla, J. Harry Caufield, and Peter Uetz

PROTOCOL

1 Mapping Protein–Protein Interactions Using Yeast Two-Hybrid Assays 319
Jitender Mehla, J. Harry Caufield, and Peter Uetz

CHAPTER 19 Membrane Yeast Two-Hybrid (MYTH) Mapping of Full-Length Membrane Protein Interactions

INTRODUCTION

Membrane Yeast Two-Hybrid (MYTH) Mapping of Full-Length Membrane Protein Interactions 331
Jamie Snider and Igor Stagljar

PROTOCOLS

1 Generation and Validation of MYTH Baits: iMYTH and tMYTH Variants 334
Jamie Snider and Igor Stagljar

2 MYTH Screening: iMYTH and tMYTH Variants 340
Jamie Snider and Igor Stagljar

CHAPTER 20 Protein-Fragment Complementation Assays for Large-Scale Analysis, Functional Dissection, and Spatiotemporal Dynamic Studies of Protein–Protein Interactions in Living Cells

INTRODUCTION

Protein-Fragment Complementation Assays for Large-Scale Analysis, Functional Dissection, and Spatiotemporal Dynamic Studies of Protein–Protein Interactions in Living Cells 347
Stephen W. Michnick, Christian R. Landry, Emmanuel D. Levy, Guillaume Diss, Po Hien Ear, Jacqueline Kowarzyk, Mohan K. Malleshaiah, Vincent Messier, and Emmanuelle Tchekanda

PROTOCOLS

1 The Dihydrofolate Reductase Protein-Fragment Complementation Assay: A Survival-Selection Assay for Large-Scale Analysis of Protein–Protein Interactions 350
Stephen W. Michnick, Emmanuel D. Levy, Christian R. Landry, Jacqueline Kowarzyk, and Vincent Messier

2 Combining the Dihydrofolate Reductase Protein-Fragment Complementation Assay with Gene Deletions to Establish Genotype-to-Phenotype Maps of Protein Complexes and Interaction Networks 359
Guillaume Diss and Christian R. Landry

3 Dissecting the Contingent Interactions of Protein Complexes with the Optimized Yeast Cytosine Deaminase Protein-Fragment Complementation Assay 366
Po Hien Ear, Jacqueline Kowarzyk, and Stephen W. Michnick

4 Real-Time Protein-Fragment Complementation Assays for Studying Temporal,
   Spatial, and Spatiotemporal Dynamics of Protein–Protein Interactions in
   Living Cells   373
      Mohan Malleshaiah, Emmanuelle Tchekanda, and Stephen W. Michnick

CHAPTER 21   Protein Complex Purification by Affinity Capture

INTRODUCTION

Protein Complex Purification by Affinity Capture   383
   John LaCava, Javier Fernandez-Martinez, Zhanna Hakhverdyan, and Michael P. Rout

PROTOCOLS

1 Optimized Affinity Capture of Yeast Protein Complexes   388
   John LaCava, Javier Fernandez-Martinez, Zhanna Hakhverdyan, and Michael P. Rout

2 Native Elution of Yeast Protein Complexes Obtained by Affinity Capture   393
   John LaCava, Javier Fernandez-Martinez, and Michael P. Rout

3 Density Gradient Ultracentrifugation to Isolate Endogenous Protein
   Complexes after Affinity Capture   397
   Javier Fernandez-Martinez, John LaCava, and Michael P. Rout

CHAPTER 22   Proteomic Analysis of Protein Posttranslational Modifications by Mass Spectrometry

INTRODUCTION

Proteomic Analysis of Protein Posttranslational Modifications by Mass Spectrometry 401
   Danielle L. Swaney and Judit Villén

PROTOCOLS

1 Enrichment of Phosphopeptides via Immobilized Metal Affinity Chromatography   404
   Danielle L. Swaney and Judit Villén

2 Enrichment of Modified Peptides via Immunoaffinity Precipitation with
   Modification-Specific Antibodies   409
   Danielle L. Swaney and Judit Villén

CHAPTER 23   Protein Microarrays: Flexible Tools for Scientific Innovation

INTRODUCTION

Protein Microarrays: Flexible Tools for Scientific Innovation   415
   *Johnathan Neiswinger, Ijeoma Uzoma, Eric Cox, HeeSool Rho, Guang Song, Corry Paul,
   Jun Seop Jeong, Kuan-Yi Lu, Chien-Sheng Chen, and Heng Zhu*

PROTOCOLS

1 Characterization of Protein–Protein Interactions Using Protein Microarrays   418
   Corry Paul, HeeSool Rho, Johnathan Neiswinger, and Heng Zhu

2 Characterization of RNA-Binding Proteins Using Protein Microarrays   422
   Guang Song, Johnathan Neiswinger, and Heng Zhu

3 Characterization of Lipid–Protein Interactions Using Nonquenched Fluorescent
   Liposomal Nanovesicles and Yeast Proteome Microarrays   428
   Kuan-Yi Lu, Chien-Sheng Chen, Johnathan Neiswinger, and Heng Zhu

4 Posttranslational Modification Assays on Functional Protein Microarrays   433
   Johnathan Neiswinger, Ijeoma Uzoma, Eric Cox, HeeSool Rho, Jun Seop Jeong, and Heng Zhu

CHAPTER 24 Synthetic Genetic Arrays: Automation of Yeast Genetics

INTRODUCTION

Synthetic Genetic Arrays: Automation of Yeast Genetics   441
Elena Kuzmin, Michael Costanzo, Brenda Andrews, and Charles Boone

PROTOCOL

1 Synthetic Genetic Array Analysis   448
   Elena Kuzmin, Michael Costanzo, Brenda Andrews, and Charles Boone

CHAPTER 25 Systematic Mapping of Chemical–Genetic Interactions in *Saccharomyces cerevisiae*

INTRODUCTION

Systematic Mapping of Chemical–Genetic Interactions in *Saccharomyces cerevisiae*   463
*Sundari Suresh, Ulrich Schlecht, Weihong Xu, Walter Bray, Molly Miranda, Ronald W. Davis,
Corey Nislow, Guri Giaever, R. Scott Lokey, and Robert P. St.Onge*

PROTOCOLS

1 A High-Throughput Yeast Halo Assay for Bioactive Compounds   468
   Walter Bray and R. Scott Lokey

2 Identification of Chemical–Genetic Interactions via Parallel Analysis
   of Barcoded Yeast Strains   471
   *Sundari Suresh, Ulrich Schlecht, Weihong Xu, Molly Miranda, Ronald W. Davis,
   Corey Nislow, Guri Giaever, and Robert P. St.Onge*

CHAPTER 26 Prions

INTRODUCTION

Prions   481
   Dmitry Kryndushkin, Herman K. Edskes, Frank P. Shewmaker, and Reed B. Wickner

PROTOCOLS

1 Genetic Methods for Studying Yeast Prions   488
   Reed B. Wickner, Herman K. Edskes, Dmitry Kryndushkin, and Frank P. Shewmaker

2 Prion Transfection of Yeast   495
   Herman K. Edskes, Dmitry Kryndushkin, Frank P. Shewmaker, and Reed B. Wickner

3 Isolation and Analysis of Prion and Amyloid Aggregates from Yeast Cells   501
   Dmitry Kryndushkin, Natalia Pripuzova, and Frank P. Shewmaker

CHAPTER 27 Gene-Centered Yeast One-Hybrid Assays

INTRODUCTION

Gene-Centered Yeast One-Hybrid Assays 509
  Juan I. Fuxman Bass, John S. Reece-Hoyes, and Albertha J.M. Walhout

PROTOCOLS

1 Generating Bait Strains for Yeast One-Hybrid Assays 514
  Juan I. Fuxman Bass, John S. Reece-Hoyes, and Albertha J.M. Walhout

2 Performing Yeast One-Hybrid Library Screens 521
  Juan I. Fuxman Bass, John S. Reece-Hoyes, and Albertha J.M. Walhout

3 Colony Lift Colorimetric Assay for β-Galactosidase Activity 527
  Juan I. Fuxman Bass, John S. Reece-Hoyes, and Albertha J.M. Walhout

4 Zymolase-Treatment and Polymerase Chain Reaction Amplification from Genomic and Plasmid Templates from Yeast 530
  Juan I. Fuxman Bass, John S. Reece-Hoyes, and Albertha J.M. Walhout

CHAPTER 28 Transposon Calling Cards

INTRODUCTION

Transposon Calling Cards 533
  David Mayhew and Robi D. Mitra

PROTOCOL

1 Calling Card Analysis in Budding Yeast 536
  David Mayhew and Robi D. Mitra

CHAPTER 29 Transcription Factor–DNA Binding Motifs in *Saccharomyces cerevisiae*: Tools and Resources

INTRODUCTION

Transcription Factor–DNA Binding Motifs in *Saccharomyces cerevisiae*: Tools and Resources 547
  Joshua L. Schipper and Raluca M. Gordân

CHAPTER 30 The *Saccharomyces* Genome Database: A Tool for Discovery

INTRODUCTION

The *Saccharomyces* Genome Database: A Tool for Discovery 557
  J. Michael Cherry

PROTOCOLS

1 The *Saccharomyces* Genome Database: Exploring Biochemical Pathways and Mutant Phenotypes 561
  J. Michael Cherry

2 The *Saccharomyces* Genome Database: Advanced Searching Methods and
Data Mining 566
J. Michael Cherry

3 The *Saccharomyces* Genome Database: Gene Product Annotation of Function,
Process, and Component 570
J. Michael Cherry

4 The *Saccharomyces* Genome Database: Exploring Genome Features and Their
Annotations 574
J. Michael Cherry

CHAPTER 31   BioGRID: A Resource for Studying Biological Interactions in Yeast

INTRODUCTION

BioGRID: A Resource for Studying Biological Interactions in Yeast 577
Rose Oughtred, Andrew Chatr-aryamontri, Bobby-Joe Breitkreutz, Christie S. Chang, Jennifer M. Rust, Chandra L. Theesfeld, Sven Heinicke, Ashton Breitkreutz, Daici Chen, Jodi Hirschman, Nadine Kolas, Michael S. Livstone, Julie Nixon, Lara O'Donnell, Lindsay Ramage, Andrew Winter, Teresa Reguly, Adnane Sellam, Chris Stark, Lorrie Boucher, Kara Dolinski, and Mike Tyers

PROTOCOL

1 Use of the BioGRID Database for Analysis of Yeast Protein and Genetic
Interactions 582
Rose Oughtred, Andrew Chatr-aryamontri, Bobby-Joe Breitkreutz, Christie S. Chang, Jennifer M. Rust, Chandra L. Theesfeld, Sven Heinicke, Ashton Breitkreutz, Daici Chen, Jodi Hirschman, Nadine Kolas, Michael S. Livstone, Julie Nixon, Lara O'Donnell, Lindsay Ramage, Andrew Winter, Teresa Reguly, Adnane Sellam, Chris Stark, Lorrie Boucher, Kara Dolinski, and Mike Tyers

CHAPTER 32   Exploratory Analysis of Biological Networks through Visualization, Clustering, and Functional Annotation in Cytoscape

PROTOCOL

1 Exploratory Analysis of Biological Networks through Visualization, Clustering,
and Functional Annotation in Cytoscape 591
Anastasia Baryshnikova

CHAPTER 33   Metabolomics in Yeast

INTRODUCTION

Metabolomics in Yeast 599
Amy A. Caudy, Michael Mülleder, and Markus Ralser

PROTOCOLS

1 Metabolite Extraction from *Saccharomyces cerevisiae* for Liquid
Chromatography–Mass Spectrometry 603
Adam P. Rosebrock and Amy A. Caudy

2 A High-Throughput Method for the Quantitative Determination of Free Amino Acids in *Saccharomyces cerevisiae* by Hydrophilic Interaction Chromatography–Tandem Mass Spectrometry 608
Michael Mülleder, Katharina Bluemlein, and Markus Ralser

3 Spectrophotometric Analysis of Ethanol and Glucose Concentrations in Yeast Culture Media 614
Amy A. Caudy

CHAPTER 34 High-Throughput Yeast Strain Sequencing

INTRODUCTION

High-Throughput Yeast Strain Sequencing 621
Katja Schwartz and Gavin Sherlock

PROTOCOL

1 Preparation of Yeast DNA Sequencing Libraries 625
Katja Schwartz and Gavin Sherlock

CHAPTER 35 Budding Yeast Strains and Genotype–Phenotype Mapping

INTRODUCTION

Budding Yeast Strains and Genotype–Phenotype Mapping 631
Gianni Liti, Jonas Warringer, and Anders Blomberg

PROTOCOLS

1 Isolation and Laboratory Domestication of Natural Yeast Strains 636
Gianni Liti, Jonas Warringer, and Anders Blomberg

2 Mapping Quantitative Trait Loci in Yeast 641
Gianni Liti, Jonas Warringer, and Anders Blomberg

3 Yeast Reciprocal Hemizygosity to Confirm the Causality of a Quantitative Trait Loci–Associated Gene 646
Jonas Warringer, Gianni Liti, and Anders Blomberg

CHAPTER 36 Genetic Analysis of Complex Traits in *Saccharomyces cerevisiae*

INTRODUCTION

Genetic Analysis of Complex Traits in *Saccharomyces cerevisiae* 651
Ian M. Ehrenreich and Paul M. Magwene

PROTOCOL

1 Genetic Dissection of Heritable Traits in Yeast Using Bulk Segregant Analysis 656
Ian M. Ehrenreich and Paul M. Magwene

CHAPTER 37 Chemostat Culture for Yeast Physiology and Experimental Evolution

INTRODUCTION

Chemostat Culture for Yeast Physiology and Experimental Evolution — 661
Maitreya J. Dunham, Emily O. Kerr, Aaron W. Miller, and Celia Payen

PROTOCOLS

1 Assembly of a Mini-Chemostat Array — 665
Aaron W. Miller, Emily O. Kerr, and Maitreya J. Dunham

2 Chemostat Culture for Yeast Physiology — 673
Emily O. Kerr and Maitreya J. Dunham

3 Chemostat Culture for Yeast Experimental Evolution — 679
Celia Payen and Maitreya J. Dunham

CHAPTER 38 Methods to Synthesize Large DNA Fragments for a Synthetic Yeast Genome

PROTOCOL

1 Methods to Synthesize Large DNA Fragments for a Synthetic Yeast Genome — 685
Yizhi Cai and Junbiao Dai

APPENDIX General Safety and Hazardous Material Information — 693

INDEX — 699

Preface

In this laboratory manual, we have tried to capture basic techniques in yeast molecular genetics and cell biology, methodologies for functional genomics, some approaches to computational analysis, and the use of relevant databases. Our hope is that this information will serve as a starting point for the next phase of yeast analysis, which will include such areas as the emerging field of synthetic biology. We anticipate that the combination of the methods described here and new methods undoubtedly to come will allow the yeast community to assemble the first comprehensive and working model of a cell.

It is fitting that this book is published by Cold Spring Harbor Laboratory Press, given the critical role that Cold Spring Harbor Laboratory has long played in the yeast community. In particular, CSHL established the Yeast Genetics Course in 1970, which continues to this day. Many leaders in the field embarked on their careers as yeast geneticists after taking this course, catalyzing the growth of a remarkable community of researchers who focused on using yeast as a model system to understand the eukaryotic cell. Over the years, CSHL Press has produced several notable publications on budding yeast. The two-volume monograph *The Molecular Biology of the Yeast* Saccharomyces, published in 1981 and 1982, highlighted research advances in genome dynamics, metabolism, and gene expression—many of which were catalyzed by the then-recent development of robust methods for yeast transformation and genome manipulation. In the following decade, the monograph was updated with *The Molecular and Cellular Biology of the Yeast* Saccharomyces, a three-volume series published in 1991, 1992, and 1997 that provided an encyclopedia of fundamental knowledge of yeast biology.

Given the experimental tractability of budding yeast and the collaborative nature of the yeast community, *Saccharomyces cerevisiae* was unsurprisingly the first eukaryotic organism to have its genome sequenced, and that drove the development of numerous functional genomics methods and resources, including DNA microarrays, systematic reverse genetics with genome-wide mutant collections, and technologies for the analysis of the proteome. Throughout, the *Saccharomyces* Genome Database (SGD) led the effort to annotate the yeast genome and continues today to curate all aspects of its postgenome biology. With this history in mind, we hope that the experimental and computational methods described here will assist another generation of yeast researchers to make further progress on our understanding of this wonderful organism.

Brenda Andrews
Charles Boone
Trisha N. Davis
Stanley Fields

General Safety and Hazardous Material Information

This manual should be used by laboratory personnel with experience in laboratory and chemical safety or students under the supervision of such trained personnel. The procedures, chemicals, and equipment referenced in this manual are hazardous and can cause serious injury unless performed, handled, and used with care and in a manner consistent with safe laboratory practices. Students and researchers using the procedures in this manual do so at their own risk. It is essential for your safety that you consult the appropriate Material Safety Data Sheets, the manufacturers' manuals accompanying equipment, and your institution's Environmental Health and Safety Office, as well as the General Safety and Disposal Cautions in the Appendix for proper handling of hazardous materials in this manual. Cold Spring Harbor Laboratory makes no representations or warranties with respect to the material set forth in this manual and has no liability in connection with the use of these materials.

All registered trademarks, trade names, and brand names mentioned in this book are the property of the respective owners. Readers should please consult individual manufacturers and other resources for current and specific product information.

Appropriate sources for obtaining safety information and general guidelines for laboratory safety are provided in the General Safety and Hazardous Material Information Appendix.

CHAPTER 1

Historical Evolution of Laboratory Strains of *Saccharomyces cerevisiae*

Edward J. Louis[1]

Centre for Genetic Architecture of Complex Traits, Department of Genetics, University of Leicester, Leicester LE1 7RH, United Kingdom

Budding yeast strains used in the laboratory have had a checkered past. Historically, the choice of strain for any particular experiment depended on the suitability of the strain for the topic of study (e.g., cell cycle vs. meiosis). Many laboratory strains had poor fermentation properties and were not representative of the robust strains used for domestic purposes. Most strains were related to each other, but investigators usually had only vague notions about the extent of their relationships. Isogenicity was difficult to confirm before the advent of molecular genetic techniques. However, their ease of growth and manipulation in laboratory conditions made them "the model" model organism, and they still provided a great deal of fundamental knowledge. Indeed, more than one Nobel Prize has been won using them. Most of these strains continue to be powerful tools, and isogenic derivatives of many of them—including entire collections of deletions, overexpression constructs, and tagged gene products—are now available. Furthermore, many of these strains are now sequenced, providing intimate knowledge of their relationships. Recent collections, new isolates, and the creation of genetically tractable derivatives have expanded the available strains for experiments. But even still, these laboratory strains represent a small fraction of the diversity of yeast. The continued development of new laboratory strains will broaden the potential questions that can be posed. We are now poised to take advantage of this diversity, rather than viewing it as a detriment to controlled experiments.

CONGENIC VERSUS ISOGENIC STRAINS

By far, the most widely used laboratory strain of budding yeast has been S288c (Mortimer and Johnston 1986; Olson et al. 1986; Winston et al. 1995; Brachmann et al. 1998). The history of S288c and its derivatives has been discussed by Mortimer and Johnston (1986); they describe its complicated pedigree, involving the use of various sources of yeast in crosses that were not well-controlled. As we will see below, one of the reasons for its wide use, in addition to the generosity of Mortimer and other yeast researchers (including Fred Sherman, David Botstein, Gerry Fink, Phil Heiter, Francois Lacroute, Fred Winston, and many others), was the fact that isogenic *MAT*a and *MAT*α versions were available through a spontaneous rare MAT switching in the original S288cα (ho) strain. Two spores from a spontaneous diploid (X2180-1A and X2180-1B) were widely distributed and used, and these allowed the development of isogenic series of strains through crossing (Mortimer and Johnston 1986). Until molecular genetic tools became available, most researchers working with other strains had to deal with varying levels of congenicity through backcrossing to one or another parental input (usually S288c or a congenic derivative). This invariably led to interesting observations as well as

[1]Correspondence: ejl21@le.ac.uk

Copyright © Cold Spring Harbor Laboratory Press; all rights reserved
Cite this introduction as *Cold Spring Harb Protoc*; doi:10.1101/pdb.top077750

problems with controls. One series of congenic strains is YNN216 and derivatives YPH499, YPH500, and YPH501 (Johnston and Davis 1984; Sikorski and Hieter 1989), which have been widely used and are isogenic within the set. Another strain developed in this way was W303, derived from crossing a set of desired markers using S288c and several other strains for several generations (Rodney Rothstein as described at the *Saccharomyces* Genome Database [SGD]). Although highly related to S288c, 15% of its genetic material from elsewhere, it is phenotypically different from S288c in several respects as will be seen below. Once molecular genetics and the cloning of genes came to exist, the HO endonuclease could be used to create isogenic derivatives of the opposite mating type, as was done for W303, overcoming the issue of variation in congenic derivatives. Genetic combinations of mutants in W303, for example, could be compared with W303 rather than S288c once these isogenic series were available.

The S288c pedigree also gave rise to A364a, another widely used strain, particularly for cell-cycle studies. Although obviously related, these two strains display quite different properties and were used in different types of experiments (Hartwell 1967; Kaiser and Schekman 1990) and use of both has led to Nobel Prizes (2001 and 2013). Table 1 lists some of the popular laboratory strains during the rise of yeast genetics and biology. There are many more described at SGD (see Box 1). In addition to S288c (and derivatives), A364a, and W303, these include FL100, CEN.PK, YNN216, YPH499, YPH500, YPH501, and Sigma1278b. All of these are related in some way to S288c. Other strains such as SK1 and Y55 were derived from independently isolated yeast and have been developed for specific studies, including sporulation for both (Tauro and Halvorson 1966; Kane and Roth 1974; Bishop et al. 1992) and various mutant screens for Y55 (McCusker et al. 1987; McCusker and Haber 1988a,b). More recently, there has been an explosion of independent yeast isolates that have been made genetically tractable for various studies including RM11-1A derived from a vineyard yeast (Brem et al. 2002), YJM789, a clinical isolate (Wei et al. 2007), and the *Saccharomyces* Genome Resequencing Project (SGRP) clean lineages from Wine/European (WE), North American (NA), West African (WA), Sake (SA), and Malaysian (MA) populations (Liti et al. 2009). Other isolates have been sequenced, such as EC1118 from the wine industry (Novo et al. 2009) and PE-2 from the bioethanol industry (Argueso et al. 2009), but they have yet to be adapted to easy laboratory use. Most recently, there has been an expansion of *Saccharomyces cerevisiae* diversity with the discovery of new populations from China (Wang et al. 2012). Although not entirely sequenced, the relationships of these new populations to the other known populations of yeast can be seen in Figure 1. Each laboratory strain has an interesting story behind their derivation and use and a few will illustrate important lessons.

CONDITIONAL EFFECTS AND THE SCOURGE OF CONGENICITY

In the 1980s, there were many debates over the requirement of certain genes for certain functions, and one memorable one was whether or not *FUS3* (an MAPK kinase) was required for pheromone-mediated signal transduction with an East coast versus West coast of the United States difference in outcome associated with a *fus3* mutation. The resolution came down to a difference between W303 and S288c. S288c has a mutation in *KSS1*, another MAPK kinase with overlapping function, while it is intact in W303 (Elion et al. 1991). The synthetic effect on S288c, in this case, we could say conditional effect, was due to the genetic background difference. Here there is only one relevant difference between the strains. There are many other phenotypic differences between these two strains as well as other related strains, which are due to genetic differences between the backgrounds. At the time of the *FUS3/KSS1* story, it was used as a lesson for proper controlled experiments with isogenic strains as well as making sure the genetic context of an experiment was clear. Now we take advantage of such differences to map other genes involved in particular traits. An early example of this was the rediscovery of filamentous growth in yeast, not seen in S288c, due to three genetic differences between S288c and Sigma1278b-related strains (Liu et al. 1993). The difference in the mating pathway between the two strains has recently been elucidated (Chin et al. 2012).

TABLE 1. Budding yeast laboratory strains: old and new

Strain	Genotype	Relationship	Sequenced	Reference
S288C	MATα SUC2 gal2 mal mel flo1 flo8-1 hap1 ho bio1 bio6	Mosaic	Yes	Goffeau et al. 1996; Mortimer and Johnston 1986
X2180-1A	MATa SUC2 mal mel gal2 CUP1	Isogenic with S288c	No	Mortimer and Johnston 1986
X2180-1B	MATα SUC2 mal mel gal2 CUP1	Isogenic with S288c	No	Mortimer and Johnston 1986
BY4730	MATa leu2Δ0 met15Δ0 ura3Δ0	Isogenic with S288c	Yes	Brachmann et al. 1998
BY4739	MATα leu2Δ0 lys2Δ0 ura3Δ0	Isogenic with S288c	Yes	Brachmann et al. 1998
BY4741	MATa his3Δ1 leu2Δ0 met15Δ0 ura3Δ0	Isogenic with S288c	Yes	Brachmann et al. 1998
BY4742	MATα his3Δ1 leu2Δ0 lys2Δ0 ura3Δ0	Isogenic with S288c	Yes	Brachmann et al. 1998
BY4743	MATa/α his3Δ1/his3Δ1 leu2Δ0/leu2Δ0 LYS2/lys2Δ0met15Δ0/MET15 ura3Δ0/ura3Δ0	Isogenic with S288c	Yes	Brachmann et al. 1998
FY4	MATa	Isogenic with S288c	No	Winston et al. 1995
F1679	MATa/α ura3-52/ura3-52 trp1Δ63/TRP1 leu2Δ1/LEU2his3Δ200/HIS3 GAL2/GAL	Isogenic with S288c	Yes	Winston et al. 1995
AB792	MATα X2180-1B trp10 [rho 0]	Isogenic with S288c	Yes	Olson et al. 1986
A364a	MATa ade1 ade2 ura1 his7 lys2 tyr1 gal1 SUC mal cup BIO	Related to S288c	Some	Mortimer and Johnston 1986
W303	MATa or MATα leu2-3,112 trp1-1 can1-100 ura3-1 ade2-1his3-11,15	Congenic with S288c	Yes	Liti et al. 2009; Ralser et al. 2012
FL100	MATa	Congenic with S288c	Yes	Lacroute 1968
YNN216	MATa/α ura3-52/ura3-52 lys2-801amber/lys2-801amberade2-101ochre/ade2-101ochre	Congenic with S288c	No	Johnston and Davis 1984; Sikorski and Hieter 1989
YPH499	MATa ura3-52 lys2-801_amber ade2-101_ochre trp1-Δ63his3-Δ200 leu2-Δ1	Isogenic with YNN216	No	Johnston and Davis 1984; Sikorski and Hieter 1989
YPH500	MATα ura3-52 lys2-801_amber ade2-101_ochre trp1-Δ63his3-Δ200 leu2-Δ1	Isogenic with YNN216	No	Johnston and Davis 1984; Sikorski and Hieter 1989
Sigma1278b	Prototrophic with congenic series of markers available	Related with S288c	Yes	Dowell et al. 2010
CEN.PK	MATa/α ura3-52/ura3-52 trp1-289/trp1-289 leu2-3_112/leu2-3_112his3 Δ1/his3 Δ1 MAL2-8C/MAL2-8C SUC2/SUC2	Congenic with S288c	Yes	van Dijken et al. 2000
RM11-1A	MATa leu2Δ0 ura3-Δ0 HO::kanMX	Wine/EU population	Yes	Brem et al. 2002
SK1	MATa/α HO gal2 cupS can1R BIO (many markers and ho available)	Mosaic/West African	Yes	Bishop et al. 1992; Kane and Roth 1974
Y55	MATa /MATα HO/HO (many markers and ho available)	Mosaic/West African	Yes	McCusker and Haber 1988a; McCusker and Haber 1988b; Tauro and Halvorson 1966
DVBPG6765	MATa ura3::kanMX HO::hygMXMATαura3::kanMX HO::hygMX	Wine/EU	Yes	Liti et al. 2009
DVBPG6044	MATa ura3::kanMX HO::hygMX	West African	Yes	Liti et al. 2009
YPS128	MATa ura3::kanMX HO::hygMXMATα ura3::kanMX HO::hygMX	North American	Yes	Liti et al. 2009
Y12	MATa ura3::kanMX HO::hygMXMATα ura3::kanMX HO::hygMX	Sake	Yes	Liti et al. 2009
UWOPS03-461.4	MATa ura3::kanMX HO::hygMXMATα ura3::kanMX HO::hygMX	Malaysian	Yes	Liti et al. 2009

The importance of this conditional effect of mutations dependent on genetic background can be seen best in the comparison of gene knockouts in S288c Sigma1278b, which is closely related to S288c with at least half of the genome identical. When only essential genes are considered, Boone and colleagues (Dowell et al. 2010) found several genes that were essential in one background and not in the other. These conditional essential genes were then found to have more than one modifying genetic difference between the strains underlying the essentiality (Dowell et al. 2010). Individual cases of conditional essential genes were seen in the past such as with the clathrin heavy chain, first reported to be nonessential (Payne and Schekman 1985) and then found to be essential once an unlinked segregating suppressor was found (Lemmon and Jones 1987). Life versus death is the most extreme phenotype, and this observation is only the tip of the iceberg of genetic background effects. Synthetic lethal screen exploits the simplest of these genetic background effects with only two differences to

Chapter 1

> **BOX 1. BUDDING YEAST RESOURCES**
>
> *Saccharomyces* Genome Database (SGD) contains almost all you need to know about *S. cerevisiae*, with connections to other data and resources. www.yeastgenome.org
>
> **Strain/reagent collections/repositories**:
>
> American Type Culture Collection (ATCC) maintains yeast stocks and clones. www.lgcstandards-atcc.org
>
> EUROSCARF, the EUROpean *Saccharomyces Cerevisiae* ARchive for Functional analysis, maintains a collection of systematic deletion strains searchable by gene name. web.uni-frankfurt.de/fb15/mikro/euroscarf/
>
> National Collection of Yeast Cultures (NCYC) maintains over 3100 nonpathogenic yeasts, including type strains, strains of general interest for education and research, strains of industrial importance, and genetically marked strains. www.ncyc.co.uk
>
> Common Access to Biological Resources and Information (CABRI) includes catalogs from European culture collections for yeast and other organisms. www.cabri.org
>
> Yeast Genetic Resource Center (YGRC) maintains over 4800 *S. pombe* strains and over 9000 *S. cerevisiae* strains. yeast.lab.nig.ac.jp/nig/index_en.html
>
> Industrial Yeasts Collection DBVPG is an academic biological resource center that specializes in yeasts and yeast-like microorganisms. www.agr.unipg.it/dbvpg
>
> The collections at Centraalbureau voor Schimmelcultures (CBS) offer comprehensive coverage of the culturable biodiversity of the fungal kingdom. www.cbs.knaw.nl
>
> Addgene is a nonprofit plasmid repository that distributes many plasmids for yeast research. It includes a collection of Yeast Advanced Gateway Destination Vectors created by Dr Susan Lindquist's laboratory. www.addgene.org/yeast_gateway
>
> Yeast-GFP Clone Collection from Dr Erin O'Shea and Dr Jonathan Weissman at UCSF consists of carboxy-terminal tagged open reading frames (ORFs). clones.lifetechnologies.com/cloneinfo.php?clone=yeastgfp
>
> Yeast GST-Tagged Collection for inducible overexpression of yeast ORFs was developed in the Andrews laboratory at the University of Toronto. www.thermoscientificbio.com/non-mammalian-cdna-and-orf/yeast-gst-tagged-orfs
>
> Yeast Knockout (YKO) Collection is available from the *Saccharomyces* Genome Deletion Consortium. www-sequence.stanford.edu/group/yeast_deletion_project/deletions3.html
>
> Yeast-TAP Fusion Library from Dr Erin O'Shea and Dr Jonathan Weissman at UCSF contains open reading frames (ORFs) that are tagged with high-affinity epitopes and are expressed from their natural chromosomal locations. www.thermoscientificbio.com/non-mammalian-cdna-and-orf/yeast-tap-tagged-orfs
>
> Yeast Tet-promoters Hughes Collection contains 800 essential yeast genes for which expression is regulated by doxycycline. www.thermoscientificbio.com/non-mammalian-cdna-and-orf/yeast-tet-promoters-collection/?redirect=true

consider and other synthetic interactions between two gene knockouts are the basis of large genetic interactions studies (Tong et al. 2001; Jorgensen et al. 2002; Tong et al. 2004).

EXPANSION OF AVAILABLE STRAINS AND THE EMBRACEMENT OF DIVERSITY

As seen above, the majority of laboratory strains in early use were all related to each other. The exceptions were the independently isolated SK1 (Kane and Roth 1974; Bishop et al. 1992) and Y55 (Tauro and Halvorson 1966; McCusker and Haber 1988a,b). Until recently, it was thought that these isolates, both used in studies of sporulation and meiotic recombination, were very different from each other genetically, as well as from S288c, as they behaved differently phenotypically. We now know through sequencing that these two isolates are indeed related to each other and have segments of their genomes from other populations (Liti et al. 2009). One could speculate that early studies with these strains, Y55 was used for sporulation studies in the Halvorson laboratory (Tauro and Halvorson 1966) and Kane was used in Halvorson's laboratory before starting his own studies of sporulation using SK1 (Kane and Roth 1974), led to their inadvertent interbreeding.

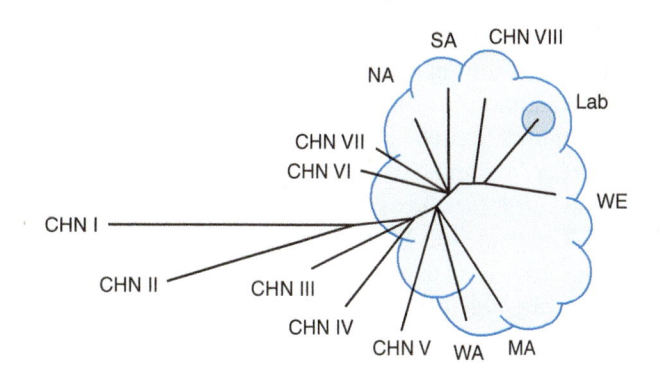

FIGURE 1. Topology of the phylogenetic relationship among yeast isolates. The relationships of the main lineages of *S. cerevisiae* isolates are displayed without all strains indicated. The topology shown is adapted from the SGRP analysis of whole-genome sequences (Liti et al. 2009) and the analysis of nine concatenated gene sequences for the Chinese isolates (Wang et al. 2012). The main lineages from the SGRP analysis are North American (NA), Wine/European (WE), Sake (SA), West African (WA), and Malaysian (MA). The Chinese analysis revealed eight new populations (CHN I through VIII), some of which are closely related to the previously known lineages, whereas some are more diverse. The cloud represents the known whole-genome sequence space of *S. cerevisiae* as a species. The majority of laboratory strains fall into the small circular area within the cloud. The exceptions being SK1 and Y55, which fall close the WA population.

Through the efforts of many people, there are a great many diverse strains available for use, opening the door to new areas of inquiry. This has been particularly evident in complex trait analysis where the standing genetic variation between strains and their phenotypic effects is embraced as a source of data (Brem et al. 2002; Steinmetz et al. 2002; Sinha et al. 2006; Cubillos et al. 2011; Parts et al. 2011), rather than treated as a source of noise if not worse. As with the above example of the conditional essential genes between two related laboratory strains, most phenotypes are due to complex interactions between more than two genes as well as with the environment.

S288c was the first eukaryotic organism sequenced (Goffeau et al. 1996) enhancing its status as the most widely used yeast strain. This sequence is the reference for many studies including the assessment of genetic variation in other isolates. The genetic diversity and relationships between strains were first gleaned from microarray data (Winzeler et al. 2003; Schacherer et al. 2007; Schacherer et al. 2009). This revealed that there were two possible domesticated populations, for Wine and Sake use (Fay and Benavides 2005), and that the laboratory strains were all related to each other. As sequencing became more efficient, many more strains have been sequenced. At the end of the first-generation sequence era, 36 isolates of *S. cerevisiae* from a wide variety of sources were sequenced (Liti et al. 2009) revealing that there are at least five "clean" lineages of yeast with laboratory strains and many others being recent mosaic outbreds of two or more of these populations. Second-generation technology has led to the genome sequences of many other strains, and new populations are being discovered that will add to the sequence space of the species (Wang et al. 2012). In Figure 1, we can see the global phylogenetic relationship between the current known groups of *S. cerevisiae*. Those within the cloud have genetically tractable derivatives (McCusker and Haber 1988a,b; Bishop et al. 1992; Winston et al. 1995; Brachmann et al. 1998; Brem et al. 2002; Cubillos et al. 2009). The new diverse population will likely be amenable to genetic manipulations in the near future. It is clear that the classical laboratory strains represent a small fraction of the available genetic diversity. Future studies may benefit from taking advantage of this increased diversity.

RESOURCES AVAILABLE

This expansion of available strains as well as additional tools to facilitate research has led to a plethora of resources, both in terms of strain collections and in plasmid collections, vectors, and drug markers. In Box 1, the current URLs are provided for a variety of these resources, many of which are also available through the authors of the work or through members of the community using these resources. The font of most knowledge about yeast is of course the *Saccharomyces* Genome Database, which in addition to containing detailed information about genes, genomes, and phenotypes, has a large section dedicated to community information and resources.

There are a number of culture collections around the world with various yeast strains from diverse origins available (see Box 1 for details). These include the ATCC in the United States, the NCYC in the United Kingdom, the YGRC in Japan, the DBVPG in Italy, and the CBS in The Netherlands. In addition, there is a collection of culture collection catalogs available at CABRI.

There are large collections of yeast strains for systematic studies. Perhaps the most widely used now is the yeast knockout collection, now with more than 20,000 strains available as *MAT*a and *MAT*α haploids, heterozygous knockout diploids (these contain the essential gene deletions, whereas the other sets do not), and homozygous knockout diploids. Created with unique barcodes, flanking the KANMX disruption cassette, these collections are useful for both specific gene studies and global studies. They are all created in one or another of the BY series, isogenic with S288c (Table 1) and have useful auxotrophies for systematic analyses (Shoemaker et al. 1996; Winzeler et al. 1999). They are available from the *Saccharomyces* Genome Deletion Consortium as well as EUROSCARF and can also be obtained from members of the community. Complementing this collection are over 800 essential genes expressed under a regulatable promoter (also in the BY strain background) (Mnaimneh et al. 2004; Davierwala et al. 2005) (Box 1), which facilitates the study of loss of function of these essential genes. There are also ways to generate temperature-sensitive alleles of essential genes for functional studies (Ben-Aroya et al. 2010; Li et al. 2011).

Complementing the deletion collections are various sets of tagged proteins for a variety of studies. Over 4150 proteins have been tagged with GFP in BY4741 for localization studies (Box 1) (Huh et al. 2003). A similar collection of TAP fusions of over 4250 proteins in BY4741 is also available for high-affinity purification and protein complex studies (Ghaemmaghami et al. 2003). Over 5000 of the genes have been GST-tagged for overexpression studies as well, also in BY4741 (Sopko et al. 2006).

Other resources include the collection of each gene under their native promoter on CEN plasmids, each with a molecular barcode (the MoBY collection) (Ho et al. 2009). This is useful for both individual gene studies and systematic complementation studies to identify mutations. Functional studies that go beyond use of the collections already described are facilitated by various vector collections including the 288 yeast advanced gateway vectors (Box 1) (Alberti et al. 2007).

WHAT STRAIN(S) TO USE

In the past, the choice of strain depended in large part on the markers available (Barnett 2007). Early genetic markers were auxotrophies generated through spontaneous or induced mutagenesis. The advent of molecular techniques allowed the creation of specific targeted mutations and deletions, and in recent years, the use of dominant drug-resistant markers adapted for use in yeast has made virtually any strain amenable to genetic manipulation by creation of appropriate gene knockouts (*HO*, for example, to create stable haploids, or specific auxotrophies such as *ura3*). Auxotrophies that continue to be used for numerous experiments are *ura3*, *lys2*, *trp1*, *his3*, and *leu2*, but there are many others used for various purposes. The major drug marker used confers G418 resistance and is generally taken from the KANMX cassette created for gene knockouts in yeast (Wach et al. 1994). There are several other marker cassettes available as well (Goldstein and McCusker 1999; Goldstein et al. 1999). The problem of limited marker availability, when multiple knockouts are required, has been overcome with various recycling schemes. A recent advance is the use of modified loxP sites and dominant drug markers such that any strain can be genetically modified (Carter and Delneri 2010), which opens the door to the diverse wild populations that may harbor interesting genetic and phenotypic variation.

The choice of strain also depends on the questions being addressed (Barnett 2007). The ease of genetic and molecular manipulation today makes it feasible to do experiments, even high-throughput global experiments, in virtually any and perhaps in multiple strain backgrounds. We now know that rather than being a problem, differences in outcome between different genetic backgrounds can be highly informative. For many questions, the workhorse of yeast genetics, S288c, and all the collections derived from it, may be the best solution. It is usually the most efficient solution as all the hard work and infrastructure are already in place. However, S288c is not the best representative of the species and other isolates may be worth the effort to work up for some experiments. For example, there are at least six well-studied gene families not present in S288c and 38 new gene families not present in S288c were found in a population genomics survey (Liti et al. 2009).

For many studies, the deletion collection and resources available for S288c are probably the first choice as there is a tremendous amount of information and experimental data available for help in particular studies as well as for comparison. However, this workhorse is not always the best for all studies and in particular studies of variation such as quantitative genetics of complex traits or in functional analysis of genes not present in its genome. Some studies would be better using more industrially relevant strains such as the commonly used Ethanol Red (available in many places and free to use without restrictions) or the wine strain EC1118 (Novo et al. 2009) which has much better fermentation properties than S288c. Perhaps the most appropriate approach to strain choice is to screen a variety of strains first for the particular study of interest. Choose ones that behave best under the conditions of the experiment before investing a great deal of time. Simple things like growth rates under various conditions, flocculation, temperature sensitivity (cold and hot), mating ability, etc., can be done easily before undertaking an experiment. As it is the differences between individuals that makes biology interesting, it may be wise to choose several strains in a comparative study as these differences should be embraced (Liti and Schacherer 2011; Nieduszynski and Liti 2011).

FUTURE PROSPECTS

Saccharomyces cerevisiae remains at the forefront of biology as a model organism. This is in large part because of the tools and resources that facilitate and enable sophisticated experiments. As functional studies go beyond the S288c laboratory strain, new tools and collections will become useful. Entire deletion collections in other strain backgrounds will become useful, facilitating the study of conditional phenotypes (Dowell et al. 2010) as well as in reciprocal hemizygosity confirmation of candidate QTLs (Cubillos et al. 2011). There is even an argument for deletion collections in the closely related species. Other tools such as tagged proteins and ORF collections on plasmids in other strain backgrounds will also be useful to the community. Finally, it is clear that a great deal of phenotypic variation comes from presence/absence of genes and gene families as well as copy-number variation (Bergstrom et al. 2014), and this genomic variation, mostly in the subtelomeres, is underexplored. As sequencing becomes less expensive (Wilkening et al. 2013) and new strains are identified (Wang et al. 2012), the choice of strains to use will become both more informed and more difficult due to the huge number of possibilities. In contrast, the prospects of the synthetic yeast genome may remove the need for many of these tools and resources as specific strains for specific studies could be synthesized as needed (Annaluru et al. 2014).

REFERENCES

Alberti S, Gitler AD, Lindquist S. 2007. A suite of Gateway cloning vectors for high-throughput genetic analysis in *Saccharomyces cerevisiae*. *Yeast* 24: 913–919.

Annaluru N, Muller H, Mitchell LA, Ramalingam S, Stracquadanio G, Richardson SM, Dymond JS, Kuang Z, Scheifele LZ, Cooper EM, et al. 2014. Total synthesis of a functional designer eukaryotic chromosome. *Science* 344: 55–58.

Argueso JL, Carazzolle MF, Mieczkowski PA, Duarte FM, Netto OV, Missawa SK, Galzerani F, Costa GG, Vidal RO, Noronha MF, et al. 2009. Genome structure of a *Saccharomyces cerevisiae* strain widely used in bioethanol production. *Genome Res* 19: 2258–2270.

Barnett JA. 2007. A history of research on yeasts 10: Foundations of yeast genetics. *Yeast* 24: 799–845.

Ben-Aroya S, Pan X, Boeke JD, Hieter P. 2010. Making temperature-sensitive mutants. *Methods Enzymol* **470**: 181–204.

Bergstrom A, Simpson JT, Salinas F, Barre B, Parts L, Zia A, Nguyen Ba AN, Moses AM, Louis EJ, Mustonen V, et al. 2014. A high-definition view of functional genetic variation from natural yeast genomes. *Mol Biol Evol* **31**: 872–888.

Bishop DK, Park D, Xu L, Kleckner N. 1992. DMC1: A meiosis-specific yeast homolog of *E. coli* recA required for recombination, synaptonemal complex formation, and cell cycle progression. *Cell* **69**: 439–456.

Brachmann CB, Davies A, Cost GJ, Caputo E, Li J, Hieter P, Boeke JD. 1998. Designer deletion strains derived from *Saccharomyces cerevisiae* S288C: A useful set of strains and plasmids for PCR-mediated gene disruption and other applications. *Yeast* **14**: 115–132.

Brem RB, Yvert G, Clinton R, Kruglyak L. 2002. Genetic dissection of transcriptional regulation in budding yeast. *Science* **296**: 752–755.

Carter Z, Delneri D. 2010. New generation of loxP-mutated deletion cassettes for the genetic manipulation of yeast natural isolates. *Yeast* **27**: 765–775.

Chin BL, Ryan O, Lewitter F, Boone C, Fink GR. 2012. Genetic variation in *Saccharomyces cerevisiae*: Circuit diversification in a signal transduction network. *Genetics* **192**: 1523–1532.

Cubillos FA, Billi E, Zorgo E, Parts L, Fargier P, Omholt S, Blomberg A, Warringer J, Louis EJ, Liti G. 2011. Assessing the complex architecture of polygenic traits in diverged yeast populations. *Mol Ecol* **20**: 1401–1413.

Cubillos FA, Louis EJ, Liti G. 2009. Generation of a large set of genetically tractable haploid and diploid *Saccharomyces* strains. *FEMS Yeast Res* **9**: 1217–1225.

Davierwala AP, Haynes J, Li Z, Brost RL, Robinson MD, Yu L, Mnaimneh S, Ding H, Zhu H, Chen Y, et al. 2005. The synthetic genetic interaction spectrum of essential genes. *Nature Genetics* **37**: 1147–1152.

Dowell RD, Ryan O, Jansen A, Cheung D, Agarwala R, Danford T, Bernstein DA, Rolfe PA, Heisler LE, Chin B, et al. 2010. Genotype to phenotype: A complex problem. *Science* **328**: 469.

Elion EA, Brill JA, Fink GR. 1991. FUS3 represses CLN1 and CLN2 and in concert with KSS1 promotes signal transduction. *Proc Natl Acad Sci* **88**: 9392–9396.

Fay JC, Benavides JA. 2005. Evidence for domesticated and wild populations of *Saccharomyces cerevisiae*. *PLoS Genet* **1**: 66–71.

Ghaemmaghami S, Huh WK, Bower K, Howson RW, Belle A, Dephoure N, O'Shea EK, Weissman JS. 2003. Global analysis of protein expression in yeast. *Nature* **425**: 737–741.

Goffeau A, Barrell BG, Bussey H, Davis RW, Dujon B, Feldmann H, Galibert F, Hoheisel JD, Jacq C, Johnston M, et al. 1996. Life with 6000 genes. *Science* **274**: 546, 563–567.

Goldstein AL, McCusker JH. 1999. Three new dominant drug resistance cassettes for gene disruption in *Saccharomyces cerevisiae*. *Yeast* **15**: 1541–1553.

Goldstein AL, Pan X, McCusker JH. 1999. Heterologous URA3MX cassettes for gene replacement in *Saccharomyces cerevisiae*. *Yeast* **15**: 507–511.

Hartwell LH. 1967. Macromolecule synthesis in temperature-sensitive mutants of yeast. *J Bacteriol* **93**: 1662–1670.

Ho CH, Magtanong L, Barker SL, Gresham D, Nishimura S, Natarajan P, Koh JL, Porter J, Gray CA, Andersen RJ, et al. 2009. A molecular barcoded yeast ORF library enables mode-of-action analysis of bioactive compounds. *Nat Biotechnol* **27**: 369–377.

Huh WK, Falvo JV, Gerke LC, Carroll AS, Howson RW, Weissman JS, O'Shea EK. 2003. Global analysis of protein localization in budding yeast. *Nature* **425**: 686–691.

Johnston M, Davis RW. 1984. Sequences that regulate the divergent GAL1-GAL10 promoter in *Saccharomyces cerevisiae*. *Mol Cell Biol* **4**: 1440–1448.

Jorgensen P, Nelson B, Robinson MD, Chen Y, Andrews B, Tyers M, Boone C. 2002. High-resolution genetic mapping with ordered arrays of *Saccharomyces cerevisiae* deletion mutants. *Genetics* **162**: 1091–1099.

Kaiser CA, Schekman R. 1990. Distinct sets of SEC genes govern transport vesicle formation and fusion early in the secretory pathway. *Cell* **61**: 723–733.

Kane SM, Roth R. 1974. Carbohydrate metabolism during ascospore development in yeast. *J Bacteriol* **118**: 8–14.

Lacroute F. 1968. Regulation of pyrimidine biosynthesis in *Saccharomyces cerevisiae*. *J Bacteriol* **95**: 824–832.

Lemmon SK, Jones EW. 1987. Clathrin requirement for normal growth of yeast. *Science* **238**: 504–509.

Li Z, Vizeacoumar FJ, Bahr S, Li J, Warringer J, Vizeacoumar FS, Min R, Vandersluis B, Bellay J, Devit M, et al. 2011. Systematic exploration of essential yeast gene function with temperature-sensitive mutants. *Nat Biotechnol* **29**: 361–367.

Liti G, Carter DM, Moses AM, Warringer J, Parts L, James SA, Davey RP, Roberts IN, Burt A, Koufopanou V, et al. 2009. Population genomics of domestic and wild yeasts. *Nature* **458**: 337–341.

Liti G, Schacherer J. 2011. The rise of yeast population genomics. *C R Biol* **334**: 612–619.

Liu H, Styles CA, Fink GR. 1993. Elements of the yeast pheromone response pathway required for filamentous growth of diploids. *Science* **262**: 1741–1744.

McCusker JH, Haber JE. 1988a. crl mutants of *Saccharomyces cerevisiae* resemble both mutants affecting general control of amino acid biosynthesis and omnipotent translational suppressor mutants. *Genetics* **119**: 317–327.

McCusker JH, Haber JE. 1988b. Cycloheximide-resistant temperature-sensitive lethal mutations of *Saccharomyces cerevisiae*. *Genetics* **119**: 303–315.

McCusker JH, Perlin DS, Haber JE. 1987. Pleiotropic plasma membrane ATPase mutations of *Saccharomyces cerevisiae*. *Mol Cell Biol* **7**: 4082–4088.

Mnaimneh S, Davierwala AP, Haynes J, Moffat J, Peng WT, Zhang W, Yang X, Pootoolal J, Chua G, Lopez A, et al. 2004. Exploration of essential gene functions via titratable promoter alleles. *Cell* **118**: 31–44.

Mortimer RK, Johnston JR. 1986. Genealogy of principal strains of the yeast genetic stock center. *Genetics* **113**: 35–43.

Nieduszynski CA, Liti G. 2011. From sequence to function: Insights from natural variation in budding yeasts. *Biochim Biophys Acta* **1810**: 959–966.

Novo M, Bigey F, Beyne E, Galeote V, Gavory F, Mallet S, Cambon B, Legras JL, Wincker P, Casaregola S, et al. 2009. Eukaryote-to-eukaryote gene transfer events revealed by the genome sequence of the wine yeast *Saccharomyces cerevisiae* EC1118. *Proc Natl Acad Sci* **106**: 16333–16338.

Olson MV, Dutchik JE, Graham MY, Brodeur GM, Helms C, Frank M, MacCollin M, Scheinman R, Frank T. 1986. Random-clone strategy for genomic restriction mapping in yeast. *Proc Natl Acad Sci* **83**: 7826–7830.

Parts L, Cubillos FA, Warringer J, Jain K, Salinas F, Bumpstead SJ, Molin M, Zia A, Simpson JT, Quail MA, et al. 2011. Revealing the genetic structure of a trait by sequencing a population under selection. *Genome Res* **21**: 1131–1138.

Payne GS, Schekman R. 1985. A test of clathrin function in protein secretion and cell growth. *Science* **230**: 1009–1014.

Ralser M, Kuhl H, Werber M, Lehrach H, Breitenbach M, Timmermann B. 2012. The *Saccharomyces cerevisiae* W303-K6001 cross-platform genome sequence: Insights into ancestry and physiology of a laboratory mutt. *Open Biol* **2**: 120093.

Schacherer J, Ruderfer DM, Gresham D, Dolinski K, Botstein D, Kruglyak L. 2007. Genome-wide analysis of nucleotide-level variation in commonly used *Saccharomyces cerevisiae* strains. *PLoS One* **2**: e322.

Schacherer J, Shapiro JA, Ruderfer DM, Kruglyak L. 2009. Comprehensive polymorphism survey elucidates population structure of *Saccharomyces cerevisiae*. *Nature* **458**: 342–345.

Shoemaker DD, Lashkari DA, Morris D, Mittmann M, Davis RW. 1996. Quantitative phenotypic analysis of yeast deletion mutants using a highly parallel molecular bar-coding strategy. *Nat Genet* **14**: 450–456.

Sikorski RS, Hieter P. 1989. A system of shuttle vectors and yeast host strains designed for efficient manipulation of DNA in *Saccharomyces cerevisiae*. *Genetics* **122**: 19–27.

Sinha H, Nicholson BP, Steinmetz LM, McCusker JH. 2006. Complex genetic interactions in a quantitative trait locus. *PLoS Genet* **2**: e13.

Sopko R, Huang D, Preston N, Chua G, Papp B, Kafadar K, Snyder M, Oliver SG, Cyert M, Hughes TR, et al. 2006. Mapping pathways and phenotypes by systematic gene overexpression. *Mol Cell* **21**: 319–330.

Steinmetz LM, Sinha H, Richards DR, Spiegelman JI, Oefner PJ, McCusker JH, Davis RW. 2002. Dissecting the architecture of a quantitative trait locus in yeast. *Nature* **416**: 326–330.

Tauro P, Halvorson HO. 1966. Effect of gene position on the timing of enzyme synthesis in synchronous cultures of yeast. *J Bacteriol* **92**: 652–661.

Tong AH, Evangelista M, Parsons AB, Xu H, Bader GD, Page N, Robinson M, Raghibizadeh S, Hogue CW, Bussey H, et al. 2001. Systematic genetic analysis with ordered arrays of yeast deletion mutants. *Science* **294**: 2364–2368.

Tong AH, Lesage G, Bader GD, Ding H, Xu H, Xin X, Young J, Berriz GF, Brost RL, Chang M, et al. 2004. Global mapping of the yeast genetic interaction network. *Science* **303**: 808–813.

van Dijken JP, Bauer J, Brambilla L, Duboc P, Francois JM, Gancedo C, Giuseppin ML, Heijnen JJ, Hoare M, Lange HC, et al. 2000. An interlaboratory comparison of physiological and genetic properties of four *Saccharomyces cerevisiae* strains. *Enzyme Microb Technol* **26**: 706–714.

Wach A, Brachat A, Pohlmann R, Philippsen P. 1994. New heterologous modules for classical or PCR-based gene disruptions in *Saccharomyces cerevisiae*. *Yeast* **10**: 1793–1808.

Wang QM, Liu WQ, Liti G, Wang SA, Bai FY. 2012. Surprisingly diverged populations of *Saccharomyces cerevisiae* in natural environments remote from human activity. *Mol Ecol* **21**: 5404–5417.

Wei W, McCusker JH, Hyman RW, Jones T, Ning Y, Cao Z, Gu Z, Bruno D, Miranda M, Nguyen M, et al. 2007. Genome sequencing and comparative analysis of *Saccharomyces cerevisiae* strain YJM789. *Proc Natl Acad Sci* **104**: 12825–12830.

Wilkening S, Tekkedil MM, Lin G, Fritsch ES, Wei W, Gagneur J, Lazinski DW, Camilli A, Steinmetz LM. 2013. Genotyping 1000 yeast strains by next-generation sequencing. *BMC genomics* **14**: 90.

Winston F, Dollard C, Ricupero-Hovasse SL. 1995. Construction of a set of convenient *Saccharomyces cerevisiae* strains that are isogenic to S288C. *Yeast* **11**: 53–55.

Winzeler EA, Castillo-Davis CI, Oshiro G, Liang D, Richards DR, Zhou Y, Hartl DL. 2003. Genetic diversity in yeast assessed with whole-genome oligonucleotide arrays. *Genetics* **163**: 79–89.

Winzeler EA, Shoemaker DD, Astromoff A, Liang H, Anderson K, Andre B, Bangham R, Benito R, Boeke JD, Bussey H, et al. 1999. Functional characterization of the *S. cerevisiae* genome by gene deletion and parallel analysis. *Science* **285**: 901–906.

CHAPTER 2

Classical Genetics with *Saccharomyces cerevisiae*

David C. Amberg[1,3] and Daniel J. Burke[2,3]

[1]*Department of Biochemistry and Molecular Biology, SUNY Upstate Medical University, Syracuse, New York 13210;* [2]*Department of Biochemistry and Molecular Genetics, University of Virginia School of Medicine, Charlottesville, Virginia 22908-0733*

The budding yeast *Saccharomyces cerevisiae* is an outstanding experimental model organism that has been exploited since the early part of the twentieth century for studies in biochemistry and genetics. It has been the premiere experimental system for modern functional genomics and continues to make important contributions to many areas of biology. Here we discuss its many virtues as an organism for classical genetic research.

THE VIRTUE OF BUDDING YEAST

The budding yeast *S. cerevisiae* (hereafter referred to as "yeast") is an outstanding model organism for use in a wide variety of disciplines including classical and molecular genetics. There are several excellent "how-to" resources that detail how to grow and manipulate yeast for all of the manipulations we discuss here, and we refer the reader to these resources for specifics on media preparation, sources for supplies, resources, and so forth (Amberg et al. 2005). Our goal is to describe some of the many virtues of using budding yeast for classical genetic analysis and some of the issues that should be considered when using classical yeast genetics to dissect complex biological problems.

GROWTH CONDITIONS

Yeast is a microorganism that is simple to propagate, grows in large numbers in small volumes, and is relatively cheap to grow and maintain. Yeast has a short doubling time that is temperature-dependent, and cells are routinely cultivated at temperatures between 12°C and 37°C. Healthy wild-type cells divide approximately every 90 min at the preferred temperature of 30°C and grow to high-cell densities of $\sim 10^7$–10^8 per mL depending on the type of growth media. Yeast can be propagated in rich media that contains yeast extract, peptone, and dextrose (YPD), and this medium routinely supports growth to the highest cell densities ($\sim 2 \times 10^8$ cells per mL at saturation). Yeast cells prefer glucose as their carbon source and prefer to use the glucose via fermentation as opposed to respiration. Therefore, they use very little oxygen during batch culture until the dextrose is used up (the diauxic shift) when cells depend on the nonfermentable carbon source (ethanol) to reach saturation. Therefore, even though the cells do not respire during the majority of growth, it is advisable to have a well-aerated culture to achieve maximal cell densities. Yeast prefer to grow at temperatures close to 30°C, and heat-sensitive mutants (often referred to as temperature-sensitive; Ts$^-$) are often used to study

---

[3]Correspondence: ambergd@upstate.edu; dburke@virginia.edu

Copyright © Cold Spring Harbor Laboratory Press; all rights reserved
Cite this introduction as *Cold Spring Harb Protoc*; doi:10.1101/pdb.top077628

essential genes. Ts⁻ mutants are typically propagated at 23°C (permissive temperature), and the phenotype is scored at 37°C (restrictive temperature). Growing yeast at 37°C induces a heat shock (stress) response and therefore phenotypes of temperature-sensitive mutants are potentially complicated by the fact that they are being assessed in the background of a stress response; therefore, Ts⁻ mutants should be carefully compared with wild-type cells grown at the same temperature (Morano et al. 2012). Cold-sensitive mutants are used less often but are an excellent option, and they typically have a permissive temperature of 23°C and a restrictive temperature of 12°C (Weems et al. 2014). Mutants may not follow these temperature guidelines, and therefore permissive and nonpermissive conditions may need to be operationally defined. Yeast can also be grown in a synthetic media (referred to as synthetic complete or SC) that contain a mixture of vitamins (Yeast Nitrogen Base), amino acids, uracil, adenine, ammonium acetate (nitrogen source), and glucose (carbon source). Many variations of SC are possible by simply omitting one of the ingredients and are referred to by indicating the omitted ingredient (such as SC-ura) and are typically referred to as "drop-out medium." Prototrophic yeast can grow in a minimal medium (SD) containing only the vitamins (Yeast Nitrogen Base), a carbon source (typically glucose), and a nitrogen source (typically ammonium) because prototrophs are able to synthesize all amino acids, purines, and pyrimidines de novo.

AUXOTROPHIC MUTANTS

Auxotrophic mutants harbor mutations in genes that encode enzymes that are essential for a de novo synthesis pathway (such as histidine, uracil etc.) and require that the medium be supplemented. For example, *URA3* encodes orotidine-5′-phosphate (OMP) decarboxylase, which catalyzes the sixth enzymatic step in the de novo biosynthesis of pyrimidines. A *ura3* mutant strain cannot grow on SC-ura or on SD medium because uracil must be supplemented for growth (referred to as SC-ura "dropout" medium). Most laboratory strains have multiple auxotrophies that are indicated in the genotype, and the strain should be tested for the predicted phenotype before starting experiments. The phenotype of a strain (with respect to auxotrophies) is determined by the medium on which it will grow. Testing the phenotype on SD medium is the most stringent test of the phenotype because a strain with two mutations (say *his3* and *ura3*) will not grow on SD + ura medium, indicating that there is at least one more auxotrophic mutation in the strain. Yeast genotypes follow a specific convention for naming genes. Recessive mutants (the most common kind of mutant) are lowercase and italicized (*ura3*). Alleles are indicated by dashes (*ura3–52*), by semicolons, or a Δ sign if a portion of the gene is deleted (*ura3::KanMx, ura3Δ*). Uppercase designates the dominant allele (*URA3*), which is usually normal wild type except for the rare case of gain-of-function mutations. A wild-type strain is rarely a strain with no mutations, but it is typically a well-established, and well-behaved, laboratory strain with a specific genotype that is the starting point of the genetic experiment.

PROPAGATING YEAST CULTURES

Yeast can be grown to high cell density in both liquid and agar containing (solid) YPD medium. Cell growth is more variable on synthetic medium (SD and SC), and cell densities at saturation are typically ∼10⁷ per mL. Cell number can be determined directly using a hemocytometer or by using an electronic counting device such as a Coulter counter. Cell number can also be estimated by measuring the extent of light scattering with a spectrophotometer. The latter method is convenient, but it has limitations especially if the mutant changes the cell size or shape and care must be taken to calibrate the optical density (at 600 nm) with cell number. Budding yeast is a true eukaryotic organism and is typically propagated as either a haploid or a diploid. A single cell from a healthy culture plated onto a YPD plus agar-containing Petri plate (YPD plate) will grow into a visible colony in 2 d at 30°C. Cells are transferred from one plate to the other either using sterile toothpicks to transfer one colony at a time or using sterile velveteen cloths (replica plating) to transfer the entire content of one Petri plate to the other while preserving the location of each colony relative to the others. All experiments are initiated by streaking cells onto YPD plates, under conditions where cells form individual colonies. A single, well-isolated colony is a population of clonally derived cells and is used as the starting point in any experiment. To not follow this golden rule could mean that your experiments are complicated by unrealized genetic heterogeneity.

FORWARD GENETICS: MUTANTS TO GENES

Forward genetics is the process of isolating and characterizing mutants that define the genes that contribute to a biological process. The goal is to isolate mutants in every gene required for the process and thereby determine the genetic complexity of that process. The starting strain is the wild type, usually one of several well-characterized laboratory strains, and should be chosen carefully. Haploids are useful for forward genetics because all mutants, dominant and recessive, display their phenotype. Therefore, most forward genetic screens begin with haploid strains. The haploids have two complementary mating types (*MAT*a and *MAT*α) that are determined by alternative alleles of the *MAT* locus. Mixing cells of opposite mating type is used to generate diploids. The mating type of cells is determined by "mating-type testers" that are often strains with a single auxotrophic mutation that is uncommonly used in your laboratory strains (e.g., *MAT*a *ade1* and *MAT*α *ade1*; *ade1* mutants are particularly useful in that they are red and mating can be scored by the formation of white colonies). Cells only mate in the G_1 phase of the cell cycle and cells of each mating type secrete small peptide-based pheromones that cause the cells of the opposite cell type to arrest in G_1 and prepare to mate. The mating haploids must be grown together for several hours on solid medium to allow the cells to communicate via the pheromones and orchestrate the mating process, which involves a complex pathway of cell communication followed by cell and nuclear fusion. Diploids can be selected using complementing recessive auxotrophic mutants with mutations in different genes (such as *ura3* and *his3*). The recessive auxotrophic mutants will complement, and diploids can be selected on SC-ura-his or supplemented SD. The mating-type testers are used in a similar way to determining if cells mate. The *MAT*a tester will only mate with *MAT*α cells and will not mate with *MAT*α or diploid (*MAT*a /*MAT*α) cells. *MAT*α cells only mate with *MAT*a cells and will not mate with *MAT*α or diploid (*MAT*a /*MAT*α) cells. Mating type can be scored by replica plating to SD medium if all auxotrophies are expected to complement or to the appropriate SC drop-out medium. Alternatively, diploids can be selected after mating haploids harboring different dominant drug resistance markers and selecting on medium that contains both drugs. Zygotes can be isolated manually from a mixture of mating cells with a micromanipulator if the diploid strain cannot be selected. Typically, a small amount of a fresh grown colony is mixed with similarly fresh grown cells of opposite mating type on YPD plates for 3 h. A small drop of sterile water is added to dilute the cells, and the diluted cells are spread out in a small area of the YPD plate. Zygotes are identified by a characteristic "Mickey Mouse-like" appearance and are isolated using a micromanipulator. Putative diploid colonies are recovered after 2 d and tested for the ability to mate to mating-type testers. Sterile colonies (unable to mate to either tester) are diploids.

Diploids are used to test if the mutants are dominant or recessive, and for complementation tests to test for allelism. The basis for the complementation test is to cross two recessive haploid mutants with identical phenotypes and determine if the diploid has the mutant or the wild-type phenotype. For example, suppose two mutants are histidine auxotrophs and if the mutations are allelic, then a diploid derived from crossing the two mutants would be homozygous and unable to grow on SC-his medium; the mutants are noncomplementing. If the diploid grows on SC-his, then the mutants are complementing and likely define different genes. The complementation test is really a complementation and mating test. Testing a pair of mated haploids for a phenotype in the presumed diploid assumes that one or both haploid mutants is not sterile (unable to mate). Therefore, it is important to either test all mutants for their ability to mate (using mating type testers) or first select for diploids using some other markers before testing for complementation of the desired phenotype. In some cases, mutants in the same gene will complement and this is called intragenic complementation

(Shaffer et al. 1969; Ohya and Botstein 1994). Such mutants are extremely useful as they suggest that the gene is multifunctional and that the complementing alleles are defective in distinct and separable functions of that gene. To prove allelism in such cases requires tetrad dissection of the double-mutant diploids (see below).

THE AWESOME POWER OF TETRAD GENETICS

Diploids are also used to induce meiosis as a way to generate new combinations of mutations. Meiosis occurs during sporulation, which is specific to *MATa/MATα* diploids and can be induced by growing diploid cells in acetate-containing medium (YPA) that is nonfermentable. Cells are washed from the YPD and resuspended at 2×10^7 cells per mL in 1%–3% potassium acetate and incubated with aeration for several days (23°C or 30°C). Diploid cells sporulate and undergo meiosis to produce the four products of a single meiosis inside a single-walled ascus, which is commonly referred to as a "tetrad." The ascus wall can be removed enzymatically with Zymolyase, and the four spores can be isolated by micromanipulation and allowed to grow on Petri plates (in groups of four). Such "tetrad dissections" are a very powerful genetic tool that is the most important procedure used in any yeast genetics laboratory. Tetrad dissection allows you to determine if the phenotype that you are studying is due to a mutation in a single gene, which is instantly identified as $2^+:2^-$ segregation of the phenotype (referred to as 2:2 segregation) in a cross with a wild-type strain. This is a direct demonstration of Mendel's first law. Furthermore, tetrad dissection allows you to determine if different aspects of the phenotype are linked, and therefore cosegregate. The phenotype is the sum of all things you can measure and if you find that a mutant has two attributes to its phenotype that are different from wild type, you can determine if they result from a mutation in a single gene. Both attributes will segregate 2:2, and if every spore that has attribute #1 also has attribute #2, then they are both due to a mutation in the same gene. If the attributes are due to mutations in different genes, then the attributes will segregate independently and you would expect that only half of the colonies with attribute #1 would also have attribute #2 after tetrad dissection. 2:2 cosegregation is definitive evidence that two attributes are due to a mutation in a single gene. Tetrad dissection is also used to construct double mutants when two mutations are unlinked. This is easiest to imagine if the phenotypes of the single mutants are distinguishable (e.g., *MATa his3* and *MATα ura3*) and the double mutant is *MATa* or *MATα his3 ura3* which is identified as the colony that cannot grow on SC-his or on SC-ura after replica plating. Double mutants can be identified even if the phenotypes of the single mutants are indistinguishable (for example, *MATa his3* and *MATα his7* with mutations in genes encoding two different enzymes required for histidine synthesis). The *his* markers are on different chromosomes (unlinked) and will segregate 2his$^+$:2his$^-$. If the spores that inherit the *his3* allele also inherits the *his7* allele, then the tetrad has 2 his$^+$ spores (both wild-type alleles) and 2 his$^-$ spores (both *his3 his7* double mutants). If the spores that inherit the two *his3* allele also inherit the wild-type *HIS7* allele, then the other two spores must have the two *his7* alleles. The latter tetrad has 4 his$^-$ spores and the tetrad is said to be "Parental Ditype" or PD, meaning that there are only two types of spores in the tetrad (either *his3* or *his7*) and each is like one of the parents. The tetrad with 2 his$^+$ spores (both wild-type alleles) and 2 his$^-$ spores (both *his3 his7* double mutants) is said to be "NonParental Ditype" or NPD, meaning that there are only two types of spores in the tetrads and neither is like the parents. For unlinked genes, PD and NPD tetrads are equally frequent. If the original mutants are indistinguishable in phenotype, the double mutants are obtained after tetrad dissection in the NPD tetrads. The virtue of tetrads cannot be overstated as the analysis will produce 2:2 segregation of all heterozygous mutations, and the pattern of segregation for two independent markers follows precise rules and produce expected patterns (called Parental Ditype or PD, NonParental Ditype or NPD, and Tetratype or T, the pattern of which arises when there is a crossover between one of the markers and its centromere). In most crosses, the number of PD tetrads equals the number of NPD tetrads, and the proportion of T tetrads depends on the distance between the markers and their respective centromeres, but the ratios do not exceed 1PD:1NPD:4T. Yeast has a very high frequency of recombination in meiosis (crossovers per

Kb) so that genes that are on the same chromosome often behave as if they are unlinked. Segregation of independent markers in tetrads is a direct demonstration of Mendel's Second Law. The precise segregation of the markers also assures you that the chromosomes containing the markers are behaving normally in the haploid parents. Poor spore viability in a tetrad dissection can be a clue that there are issues with the chromosomes such as aneuploidy or chromosome translocations (Kotval et al. 1983).

Forward genetics generates mutants and by definition, mutant cells that are not working properly. Microorgansims that grow rapidly can adapt quickly to deleterious effects of mutations, often by a compensatory mutation in a second unlinked gene. Long-term propagation of any mutant will often result in such compensatory mutations, and therefore great care must be taken to minimize serial propagation, and tetrad analysis must be used regularly to assure that the phenotype you are studying is due to the single mutation that you believe is responsible. Fortunately, yeast cultures can be easily frozen at −70°C by simply adding an equal volume of 25% glycerol to a saturated culture grown in liquid medium and stored in cryotubes. The cells are revived by scraping the frozen culture and streaking cells onto a YPD plate. Care must be taken to return to frozen cultures and to minimize serial propagating of mutant strains.

ISOGENIC STRAINS

Not all yeast strains are the same, and great care should be taken to study isogenic strains (i.e., strains that are genetically identical and differ in defined ways). In the perfect genetic experiment, the experimental strain differs from the control by a single mutation, and it is a very powerful statement when you can say that a phenotype is due to single nucleotide difference between the wild-type and the mutant strains. The common laboratory yeast strains can differ dramatically from each other, and many common strains that are used are not interchangeable (Schacherer et al. 2007). For example, the strain SK1 is commonly used to study meiosis because the diploids sporulate efficiently and synchronously (Ben-Ari et al. 2006). Most other strains sporulate with lower efficiencies and varying synchrony and are therefore less desirable for studying meiosis. Some yeast strains can be induced to go through a reversible developmental transition from a single cell to a multicellular filamentous form. This switch causes haploid cells to invade agar in response to carbon starvation and causes diploid cells to form elongated cells (pseudohyphae) in response to nitrogen starvation. This morphological switch is similar to that seen in certain pathogenic fungi for which yeast has served as a model (Berman 2006; Arkowitz and Bassilana 2011). The switch happens efficiently in the strain Σ1278b, but it cannot be induced in most other common laboratory strains (Ryan et al. 2012). Therefore, the wild-type strain is chosen based partly on the process that is being studied. The yeast community is large with a history of collegial cooperation. It is often convenient to get strains from other laboratories, but you must be careful to use the same genetic background. When requesting mutants, you should always also obtain the congenic wild-type strain as a control. The genetic differences between laboratory strains are mostly due to differences in the genomic DNA, and the DNA sequences of most common laboratory strains has been determined (see the *Saccharomyces* Genome Database; http://www.yeastgenome.org). DNA-mediated transformation allows you to transfer mutations from one genetic background to another, but you must be careful to maintain your strains as isogenic. Any strain that is made by DNA-mediated transformation should not be used until it is crossed to the isogenic wild type, sporulated, and tetrads dissected to be sure that the expected phenotype segregates 2:2 as the products of transformation often have additional unexpected mutations (Yuen et al. 2007).

Auxotrophic mutants such as *ura3* and *his3* result from mutations in the structural genes encoding enzymes for specific steps in the biochemical pathway (uracil or histidine), and the wild-type alleles for some are used as dominant selection markers for DNA-mediated molecular genetic manipulations (transformation). If using auxotrophic genes as dominant selectable markers, then it is best to use nonrevertable alleles (such as deletions) to assure transformation success and reduce the number of false positives.

Chapter 2

FORWARD GENETIC APPROACHES

There are two general ways in which mutants are identified and collected using forward genetics. A selection is an experimental condition where wild-type cells die and only the mutants of interest survive. A screen is an experimental condition to identify mutants based on visual inspection where the mutants look different than the wild type, for example, growth arrest at a nonpermissive temperature at a specific step in the cell cycle (Hartwell et al. 1974). Selections have the virtue that you can use very large numbers of cells to identify rare events. If the selection is stringent (low background of false positives), then it can be accomplished without the need for chemical mutagens, and rare spontaneous mutants are isolated. The spontaneous mutation frequency varies slightly over the genome, but an average gene has a spontaneous mutation frequency of $\sim 10^{-7}$ (Lang and Murray 2011). Therefore, with a stringent selection, it is possible to plate 10^8 cells from a liquid culture onto a Petri plate under selective conditions, and the colonies that survive are candidate mutants of interest. The number of colonies per plate will depend on when the mutation arose in the culture. If the mutation arose early, there will be a larger number of mutants than if the mutation arose late in the growth of the culture. If there are multiple colonies on the plate after selection, then they could be derived from the same mutational event, and if so, they would be called siblings. It is desirable to have multiple independently derived mutations (independent alleles), and therefore, only one mutant should be selected from a given culture. The experimental design is to streak the wild-type strains out on YPD plate and to isolate single colonies, because each colony is a population of genetically identical cells that are clonally derived. Independent cultures for multiple colonies are established in liquid YPD medium. The cultures are subjected to the selection, and a single candidate is isolated from a given culture after the selection. This protocol assures that all mutants are independently derived. Visual screens are more laborious, and individual colonies have to be inspected one at a time. This is usually done after plating ~ 200 colonies per plate after replica plating from YPD medium to some test medium or test condition that exposes the phenotype of interest. It is impractical to perform a visual screen to identify spontaneous mutants because it would require an overwhelming number of Petri plates. Therefore, mutagens such as ethylmethanesulfonate (EMS) or ultraviolet light are used to increase the mutation frequency at least 100-fold above the spontaneous rate. Protocols for mutagenesis were determined empirically so that the visual screens could be done on reasonable numbers of colonies with a high success rate. Haploid cells are mutagenized and screened by visual inspection and the extent of mutagenesis can be determined by selecting for canavanine-resistant mutants. Canavanine is an α-amino acid found in certain plants and is structurally related to arginine. Yeast can use canavanine instead of arginine in protein synthesis, resulting in cell death. Yeast becomes resistant to 60 µg/mL canavanine by mutations in a single gene (*CAN1*) that encodes the arginine permease. The proportion of canavanine-resistant mutants per viable cell should be determined before and after mutagenesis to determine if the mutagen has increased the mutation frequency 100-fold. Canavanine-resistant mutants result from inactivating mutations and are therefore indicative of the most common type of recessive (loss-of-function) mutations. In either strategy (screen or selection), it is advisable to have a secondary test that is independent of the primary screen or selection to reduce the number of false positives that are identified. Standard EMS mutagenesis generates ~ 20 mutations per genome so that there is a large background of mutations relative to the mutation that causes the phenotype of interest (Timmermann et al. 2010). Once a mutant is identified, it is crossed to a wild-type strain (backcross) to determine if it is dominant or recessive and the diploid is sporulated and tetrads are dissected. Only mutants that segregate the phenotype 2:2 (and thus are due to a mutation in a single gene) are considered further. The first backcross will produce mutants in both mating types, and the recessive mutants can be tested for complementation. The backcross also removes additional mutations that are introduced during mutagenesis, and several backcrosses are needed to remove all additional mutations. It is best to have complementing auxotrophic mutations in the two mating types so that diploids can be selected first and then tested for complementation of the desired phenotype. The goal of a genetic screen is to identify all of the genes that function in the process. This is determined empirically by isolating a large number of independently derived mutants and determining allelism by complementation. The screen or selection is considered saturated when no new complementation groups are discovered. Most genetic screens fail to identify all of the genes that function in a biological process. This can be due to many reasons, but the most obvious is that the phenotype that you choose imposes limitations on the mutants that you isolate. Usually, an alternative screen or selection that is directed toward the same biological process will identify new alleles of already identified genes and new genes. There is great value in having multiple alleles as some mutants may retain part of the function of the gene and this can be very informative. Identifying the gene that is mutated can be done a number of ways. One is to use libraries or cloned yeast genes to transform by DNA-mediated transformation into your mutant with the goal of complementing the mutant phenotype. Once the complementing DNA is isolated, the cognate DNA from the mutant should be recovered by polymerase chain reaction (PCR) amplification and the DNA sequence should be determined. Finally, the mutant sequence should be used in a one-step gene replacement to reconstruct the mutation in wild-type cells and confirm the phenotype. Alternatively, the gene could be identified by whole-genome sequencing of several mutants with different alleles looking for the gene in common that is mutated in the strains. Again, the mutant sequence should be used in a one-step gene replacement to reconstruct the mutation in wild-type cells.

Finally, using mutants that completely eliminate the function of the gene is the best route for determining the function of the gene. Genetic analysis works best when you can say that the phenotype is due to something being missing. Interpreting the function of the gene is less certain if the gene activity is reduced but not eliminated. The collection of ~5000 deletion mutants that completely eliminate the function of all nonessential genes is an invaluable resource. If one or more of your genes is among them, you should transfer the allele into your genetic background by PCR amplification and one-step gene replacement and determine the phenotype of the null allele. Essential genes are more problematic. Temperature-sensitive mutants usually define them, and the best approach is to have multiple alleles and determine which are the most severe. It is also advisable to epitope tag the mutant proteins and determine their stability at the restrictive temperature. Temperature-sensitive mutants sometimes encode labile proteins, and if you have an allele where the protein is eliminated at the restrictive temperature, then you have identified a conditional, null allele.

A robust mutant collection has many uses because the mutants become substrates for a variety of experiments. The mutants often reveal unexpected connections between biological processes and provide important materials for biochemical analysis. They are the starting point for applying genome-wide analysis such as Synthetic Genetic Array or Complex Haplo-Insufficiency and are critical to our future understanding of how networks of genes function and how cells have evolved to respond to internal and external perturbations.

STRUCTURE–FUNCTION STUDIES BY SITE-DIRECTED MUTAGENESIS: GENES TO MUTANTS

The yeast system is extremely powerful for doing structure/function studies with site-directed mutants of a gene or genes of interest. Because it is relatively easy to purify proteins from large cultures of yeast, defined mutants can be combined with biochemical approaches to identify mutants with highly specific defects that can be mapped to the structure of the protein. Typical random forward genetics approaches may not generate the diversity of mutants required for this type of analysis, and the mutants identified may have gross structural defects lending them less useful for structure/function studies. Site-directed mutagenesis strategies can target residues likely to be on the surface of the protein and therefore less likely to grossly affect structure and more likely to be involved in protein–protein interactions. One strategy that has been particularly fruitful is to construct sets of alanine-scan mutants whereby groups of charged residues in the primary sequence of the protein are simultaneously mutated to alanine. The thinking behind this approach is that protein–protein interactions are stabilized by many electrostatic interactions and that disruption of a protein's interactions requires mutating multiple residues in the binding site. Furthermore, targeting clusters of charged residues is more likely to affect surface exposed regions of the protein. One of the first targets for this approach was yeast actin (Wertman et al. 1992), and the resulting collection of 36 mutants have been extremely useful for structure function studies of actin-binding proteins (Amberg et al. 1995) and for genetic interaction studies (Haarer et al. 2007), as well as elucidating the diversity of actin functions through phenotype analysis of these mutants (Drubin et al. 1993). To identify the actin alanine-scan alleles, the primary sequence was scanned with a window of five amino acids looking for windows that contained two or more charged residues.

Construction of alanine-scan mutants (or any other site-directed mutation of interest) can be done using fusion PCR with overlapping mutagenic primers (Ho et al. 1989). It is best to first construct a template for these PCRs that has a selectable marker linked to (usually in the 3′ end) the gene of interest to select for transformants of the mutants and to follow the mutants in genetic crosses. Transformation is best done into a diploid in which one copy of the gene to be mutated is wild type and the second copy has a complete deletion and replacement of the open reading frame with a selectable marker (called a Δ0 allele). The fusion PCR product should be designed such that there are regions of homology of ∼200 bp beyond the ends of the deletion to facilitate homologous recombination and gene replacement at the Δ0 locus. In the transformation, first select for the linked marker to the mutant allele and then screen (usually by replica plating) the positive transformants for loss of the selectable marker that was at the Δ0 locus. Only in this manner can you be sure that the recombination integrated the desired mutation into the gene of interest. The mutants can then be isolated as haploids by tetrad dissection and their phenotypes analyzed. Note that it is best to backcross the mutants to be confident that the phenotype is linked to the alanine-scan mutant allele.

The modern era of functional genomics and the extensive number of genomic tools that are available to yeast geneticists have kept yeast as one of the premier experimental organisms for modern genetics research (Rine 2014). Ira Herskowitz coined the term "the awesome power of yeast genetics" in the 1980s, and it has become an integral phrase in the yeast scientific community (Botstein 2004). The awesome power lies in the simple beauty of the yeast life cycle and the ability to switch between haploids and diploids and in the ability to do tetrad genetics. Classical genetics coupled with forward genetic screens and site-directed mutagenesis remains an integral part of basic science and will continue to provide one of the best avenues for discovery-based science for generations to come.

REFERENCES

Amberg DC, Basart E, Botstein D. 1995. Defining protein interactions with yeast actin in vivo. *Nat Struct Biol* **2:** 28–35.

Amberg DC, Burke DJ, Strathern JN. 2005. *Methods in yeast genetics: A Cold Spring Harbor Laboratory course manual*. Cold Spring Harbor Laboratory Press, Cold Spring Harbor, NY.

Arkowitz RA, Bassilana M. 2011. Polarized growth in fungi: Symmetry breaking and hyphal formation. *Semin Cell Dev Biol* **22:** 806–815.

Ben-Ari G, Zenvirth D, Sherman A, David L, Klutstein M, Lavi U, Hillel J, Simchen G. 2006. Four linked genes participate in controlling sporulation efficiency in budding yeast. *PLoS Genet* **2:** e195.

Berman J. 2006. Morphogenesis and cell cycle progression in *Candida albicans*. *Curr Opin Microbiol* **9:** 595–601.

Botstein D. 2004. Ira Herskowitz: 1946–2003. *Genetics* **166:** 653–660.

Drubin DG, Jones HD, Wertman KF. 1993. Actin structure and function: Roles in mitochondrial organization and morphogenesis in budding yeast and identification of the phalloidin-binding site. *Mol Biol Cell* **4:** 1277–1294.

Haarer B, Viggiano S, Hibbs MA, Troyanskaya OG, Amberg DC. 2007. Modeling complex genetic interactions in a simple eukaryotic genome: Actin displays a rich spectrum of complex haploinsufficiencies. *Genes Dev* **21:** 148–159.

Hartwell LH, Culottie J, Pringle JR, Reid BJ. 1974. Genetic control of cell division in yeast. *Science* **183:** 46–51.

Ho SN, Hunt HD, Horton RM, Pullen JK, Pease LR. 1989. Site-directed mutagenesis by overlap extension using the polymerase chain reaction. *Gene* **77:** 51–59.

Kotval J, Zaret KS, Consaul S, Sherman F. 1983. Revertants of a transcription termination mutant of yeast contain diverse genetic alterations. *Genetics* **103:** 367–388.

Lang GI, Murray AW. 2011. Mutation rates across budding yeast chromosome VI are correlated with replication timing. *Genome Biol Evol* **3:** 799–811.

Morano KA, Grant CM, Moye-Rowley WS. 2012. The response to heat shock and oxidative stress in *Saccharomyces cerevisiae*. *Genetics* **190:** 1157–1195.

Ohya Y, Botstein D. 1994. Diverse essential functions revealed by complementing yeast calmodulin mutants. *Science* **263:** 963–966.

Rine J. 2014. A future of the model organism model. *Mol Biol Cell* **25:** 549–553.

Ryan O, Shapiro RS, Kurat CF, Mayhew D, Baryshnikova A, Chin B, Lin ZY, Cox MJ, Vizeacoumar F, Cheung D, et al. 2012. Global gene deletion analysis exploring yeast filamentous growth. *Science* **337:** 1353–1356.

Schacherer J, Ruderfer DM, Gresham D, Dolinski K, Botstein D, Kruglyak L. 2007. Genome-wide analysis of nucleotide-level variation in commonly used *Saccharomyces cerevisiae* strains. *PLoS One* **2:** e322.

Shaffer B, Rytka J, Fink GR. 1969. Nonsense mutations affecting the his4 enzyme complex of yeast. *Proc Natl Acad Sci* **63:** 1198–1205.

Timmermann B, Jarolim S, Russmayer H, Kerick M, Michel S, Kruger A, Bluemlein K, Laun P, Grillari J, Lehrach H, et al. 2010. A new dominant peroxiredoxin allele identified by whole-genome re-sequencing of random mutagenized yeast causes oxidant-resistance and premature aging. *Aging* **2**: 475–486.

Weems AD, Johnson CR, Argueso JL, McMurray MA. 2014. Higher-order septin assembly is driven by GTP-promoted conformational changes: Evidence from unbiased mutational analysis in *Saccharomyces cerevisiae*. *Genetics* **196**: 711–727.

Wertman KF, Drubin DG, Botstein D. 1992. Systematic mutational analysis of the yeast *ACT1* gene. *Genetics* **132**: 337–350.

Yuen KW, Warren CD, Chen O, Kwok T, Hieter P, Spencer FA. 2007. Systematic genome instability screens in yeast and their potential relevance to cancer. *Proc Natl Acad Sci* **104**: 3925–3930.

CHAPTER 3

Analysis of Recombination and Chromosome Structure during Yeast Meiosis

G. Valentin Börner[1,3] and Rita S. Cha[2,3]

[1]*Center for Gene Regulation in Health and Disease, Department of Biological, Geological and Environmental Sciences, Cleveland State University, 2121 Euclid Avenue, Cleveland, Ohio 44115-2214;* [2]*North West Cancer Research Institute, School of Medical Sciences, Bangor University, Bangor LL57 2UW, United Kingdom*

Meiosis is a diploid-specific differentiation program that consists of a single round of genome duplication followed by two rounds of chromosome segregation. These events result in halving of the genetic complement, which is a requirement for formation of haploid reproductive cells (i.e., spores in yeast and gametes in animals and plants). During meiosis I, homologous maternal and paternal chromosomes (homologs) pair and separate, whereas sister chromatids remain connected at the centromeres and separate during the second meiotic division. In most organisms, accurate homolog disjunction requires crossovers, which are formed as products of meiotic recombination. For the past two decades, studies of yeast meiosis have provided invaluable insights into evolutionarily conserved mechanisms of meiosis.

BIOLOGY OF YEAST MEIOSIS

Meiosis in budding yeast is an integral part of the differentiation program that results in formation of spores, a dormant cell type that provides resistance against many types of environmental stress. Yeast meiosis is induced by deprivation of nutrients, in particular nitrogen and a fermentable carbon source. This condition relieves glucose-dependent repression of Ime1, which promotes transcription of early meiosis-specific genes (Kassir et al. 1988). Despite its specialized biological function, evolutionary conservation of the meiotic program makes yeast an attractive model system (e.g., Caryl et al. 2000; Kumar et al. 2010).

SPECIAL FEATURES OF MEIOSIS

Chromosome Segregation

During meiosis, one round of genome duplication is followed by two rounds of chromosome segregation. During the first segregation or meiosis I (MI), homologous chromosomes separate, resulting in halving of the ploidy from diploid to haploid (Fig. 1A). The second meiotic division or meiosis II (MII) is equational, where sister chromatids separate. As a result, four haploid nuclei are generated from a single diploid precursor cell. In yeast, each nucleus becomes engulfed by a separate spore wall during sporogenesis. All four haploid spores, the products of a single meiotic event, remain contained as a tetrad within a common cell wall structure called ascus.

[3]Correspondence: g.boerner@csuohio.edu; r.cha@bangor.ac.uk

Copyright © Cold Spring Harbor Laboratory Press; all rights reserved
Cite this introduction as *Cold Spring Harb Protoc*; doi:10.1101/pdb.top077636

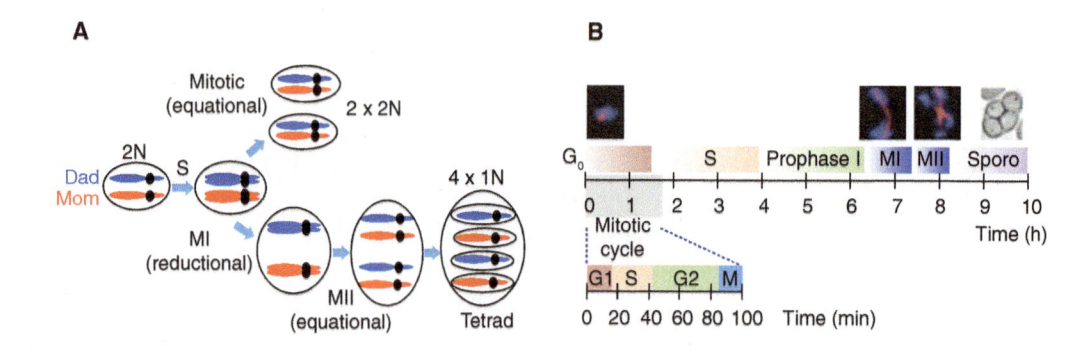

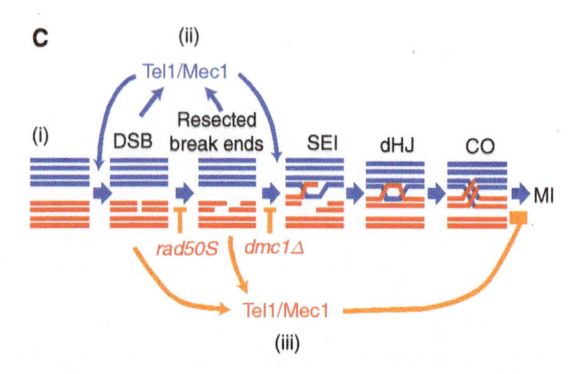

FIGURE 1. (A) Mitotic (equational) versus meiotic divisions. (S) Mitotic or meiotic replication; (MI) meiosis I; (MII) meiosis II. (B) Timeline of key meiotic events in wild-type SK1 compared to the corresponding events in vegetative cells (Sporo) Sporogenesis. (S) Mitotic or Meiotic replication; Meiotic recombination and homolog synapsis occur during prophase I. Images above the timeline are nuclear morphologies of mono-, di-, and tetra-nucleate cells, and an ascus-engulfed tetrad, observable before MI, post MI, post MII, and post sporogenesis, respectively. DNA and tubulin are stained in blue and red, respectively. (C) Key intermediates leading to crossover (CO) formation in meiotic recombination. (i) DSB: double strand break; SEI: single-end invasions; dHJ: double Holliday junctions (Schwacha and Kleckner 1997; Hunter and Kleckner 2001). (ii) Tel1 and Mec1 are activated by DSBs and regulate Spo11-activity to maintain DSB-homeostasis and ensure IH-bias (Carballo et al. 2008, 2013). (iii) Accumulation of recombination intermediates in rad50S and dmc1Δ background, respectively, trigger Tel1/Mec1-dependent meiotic arrest and/or delay (Lydall et al. 1996; Usui et al. 2001).

Meiotic S Phase

In all organisms examined to date, several stages of meiosis occupy substantially longer time intervals than the corresponding mitotic counterparts (Fig. 1B). For example, meiotic and mitotic S-phases in budding yeast last 60–80 min and ∼15–20 min, respectively (Williamson et al. 1983; Cha et al. 2000). Although the rate of replication fork progression and the number of replication origins used are comparable in meiotic versus mitotic replication, delayed firing of many replication origins appears to contribute to the prolongation of meiotic S-phase (Collins and Newlon 1994; Blitzblau et al. 2012).

Meiotic Recombination

In most organisms, meiotic recombination is initiated after S-phase by Spo11 catalysis of DNA double-strand breaks (DSBs) (Bergerat et al. 1997; Keeney et al. 1997). Recombinational repair of meiotic DSBs proceeds with an interhomolog (IH) bias, whereby the majority of breaks are repaired using an intact nonsister homologous chromosome as a template. IH-bias ensures formation of crossovers (COs) essential for homolog segregation (e.g., Schwacha and Kleckner 1997) (Fig. 1C).

Meiotic Chromosome Structure

Meiotic recombination takes place in the context of a tripartite proteinaceous structure, called the synaptonemal complex (SC), which intimately juxtaposes homologous chromosomes (Sym et al. 1993). The extent of SC development defines substages of meiotic prophase, where leptotene, zygotene, pachytene, and diplotene refer, respectively, to stages of minimal, partial, and full synapsis as well as the stage at which the SC has disassembled.

Regulation of Meiotic Progression

Meiotic progression is regulated by factors known to also control the mitotic cell cycle (e.g., cyclin-dependent kinase [Cdc28] and polo kinase [Cdc5]) (e.g., Wan et al. 2008; Sourirajan and Lichten 2008). In addition, meiosis-specific proteins such as the transcription factor Ndt80 regulate meiotic progression (Xu et al. 1995). Initiation of meiotic recombination activates the conserved ATM/ATR kinases (i.e., Tel1/Mec1 in budding yeast) (Fig. 1C). During unperturbed meiosis, Tel1/Mec1 function to control the extent of Spo11-catalysis and to promote IH-bias (Carballo et al. 2008, 2013). Tel1/Mec1 also function as bona fide checkpoint regulators, triggering meiotic arrest in response to meiotic recombination defects (Fig. 1C) (Lydall et al. 1996; Usui et al. 2001).

ANALYSIS OF MEIOSIS IN YEAST

In accompanying protocols, we present selected methods for analyzing meiotic products in budding yeast following sporulation on solid medium (Protocol 1: Analysis of Yeast Sporulation Efficiency, Spore Viability, and Meiotic Recombination on Solid Medium [Börner and Cha 2015a]), preparing highly synchronous meiotic cultures in liquid medium (Protocol 2: Induction and Analysis of Synchronous Meiotic Yeast Cultures [Börner and Cha 2015b]), and analyzing meiotic recombination and chromosome dynamics in samples collected from synchronous meiotic cultures (Protocol 3: Analysis of Meiotic Recombination and Homolog Interaction during Yeast Meiosis [Börner and Cha 2015c]). These protocols provide a practical introduction for the novice to the study of yeast meiosis. Additional yeast meiosis methods (e.g., screening for and isolating meiotic mutants) as well as methods for investigating meiosis in other organisms have been published elsewhere (Guthrie and Fink 1991; Keeney 2009a,b).

Choice of Strain Background

Several strain backgrounds have been used for meiotic analysis (e.g., Cotton et al. 2009; Elrod et al. 2009). However, the methods described in the above-mentioned protocols are optimized for SK1, which is a readily sporulating strain of *Saccharomyces cerevisiae*. The SK1 strain completes both meiotic divisions by 10 h after transfer to sporulation medium (SPM) at 30°C (Padmore et al. 1991). Spore formation is observed in >90% of cells within 24 h and the spores typically show >90% viability (e.g., Padmore et al. 1991).

Synchronous Meiotic Cultures

Induction of synchronous meiosis entails a presporulation growth protocol that generates a large number of G_0-arrested cells, which are subsequently transferred to liquid SPM (see Protocol 2: Induction and Analysis of Synchronous Meiotic Yeast Cultures [Börner and Cha 2015b]). In a typical synchronous culture at 30°C, 50% of active cells have entered meiotic S-phase by ~2.5 h, followed by completion of S-phase ~1 h later (Padmore et al. 1991; Cha et al. 2000). Meiotic recombination is initiated after S-phase, with 50% of DSBs formed at 4 h and exit from this stage at ~5 h. The 50% entry time for appearance of COs and NCOs is ~6 h, and the first meiotic division ensues at ~7 h (Fig. 1B) (Padmore et al. 1991). Some optimization may be required to obtain a highly synchronous SK1 meiotic culture, with variability attributable to factors such as water quality, temperature, pH, aeration, and pre-SPM growth conditions (e.g., Cha et al. 2000; Börner et al. 2004).

Chapter 3

Spore Formation and Spore Viability

Most mutations that affect meiosis result in either reduced spore formation and/or reduced spore viability. Reduced sporulation efficiency (see Protocol 1: Analysis of Yeast Sporulation Efficiency, Spore Viability, and Meiotic Recombination on Solid Medium [Börner and Cha 2015a]) may result from defects in positive drivers of meiotic progression, for example, Ndt80 or Cdc5 (see above). Alternatively, failure to complete meiosis could be due to a Tel1/Mec1-mediated recombination checkpoint response. In the latter case, introduction of mutations that prevent initiation of meiotic recombination (e.g., *spo11*) or abolish checkpoint response (e.g., *MEC1*-activating protein Rad24) restores sporulation efficiency (Lydall et al. 1996).

Effects of mutations on spore viability can be substantial (<5%) or moderate (~30%–60%). Mutants that sporulate efficiently but generate inviable spores often harbor defects in genes required for initiation of meiotic recombination (e.g., *spo11Δ*) or IH-bias (e.g., *mek1Δ*), resulting in a lack of COs (e.g., Malone et al. 1991; Schwacha and Kleckner 1997; Carballo et al. 2008). Spore inviability in these cases can be restored by introducing *spo13Δ*, a mutation that leads to a single equational division, thus bypassing the requirement for CO formation (Malone and Esposito 1981). Intermediate spore viability can result from several causes including reduced levels of meiotic DSBs (e.g., Martini et al. 2006; Carballo et al. 2013).

Recombination Analysis by Genetic and Physical Assays

The yeast system is amenable to detailed genetic analysis of meiotic recombination at both the local and genome-wide levels (e.g., Chen et al. 2008; Mancera et al. 2008). However, the reliance of these analyses on the formation of viable spores in substantial numbers of cells renders them unsuitable for mutants with defects in spore formation and/or low spore viability. In mutants that fail to generate viable spores, recombination intermediates and products can be investigated using physical analysis in synchronized meiotic cultures (Schwacha and Kleckner 1997; Allers and Lichten 2001; Börner et al. 2004). This technique is based on hetero-allele systems that allow quantitative assessment of recombination intermediates and products at various hotspots of recombination and has provided important mechanistic insights into meiotic recombination. A widely used assay system, in which high levels of recombination occur at the *HIS4* locus due to presence of a *LEU2*-marked DSB hotspot (*HIS4::LEU2*), carries restriction length polymorphisms in the "Mom" and "Dad" versions of this recombination hotspot, resulting in distinct gel electrophoretic mobility of recombination intermediates and products (e.g., Börner et al. 2004) (see Protocol 3: Analysis of Meiotic Recombination and Homolog Interaction during Yeast Meiosis [Börner and Cha 2015c]).

Visualization of Meiotic Chromosomes

The combination of DAPI (4′,6-diamidino-2-phenylindole) and antibodies to specific structural components of meiotic chromosomes provides an effective approach for studying meiotic chromosome dynamics (see Protocol 3: Analysis of Meiotic Recombination and Homolog Interaction during Yeast Meiosis [Börner and Cha 2015c]). Widely used markers of meiotic chromosome axes include Rec8 (a meiosis-specific component of cohesin; Klein et al. 1999), Red1 and Hop1 (meiotic chromosome axis proteins; Hollingsworth et al. 1990; Smith and Roeder 1997), and Zip1 (a component of the SC central element; Sym et al. 1993). In addition, tethering of fluorescent marker proteins via a cognate DNA-binding domain (e.g., TetR-GFP) to a DNA array (e.g., *tetOx112*) provides a positional marker and/or a tool for monitoring the pairing status of homologous chromosomes (Klein et al. 1999).

CONCLUDING REMARKS

Yeast is an excellent model system for studying meiosis owing to the high degree of evolutionary conservation and the relative ease with which essential meiotic processes can be analyzed in a large number of cells undergoing synchronous meiosis. These features have significantly contributed to the current understanding of the genetic and molecular basis of meiosis. Work in the G.V.B. lab is supported by National Institutes of Health/National Institute of General Medical Sciences (NIH-NIGMS) Grant GM099056-01.

REFERENCES

Allers T, Lichten M. 2001. Differential timing and control of noncrossover and crossover recombination during meiosis. *Cell* **106:** 47–57.

Bergerat A, de Massy B, Gadelle D, Varoutas P-C, Nicolas A, Forterre P. 1997. An atypical topoisomerase II from archaea with implications for meiotic recombination. *Nature* **386:** 414–417.

Blitzblau H, Chan C, Hochwagen A, Bell S. 2012. Separation of DNA replication from the assembly of break-competent meiotic chromosomes. *PLoS Genet* **8:** e1002643.

Börner GV, Cha RS. 2015a. Analysis of yeast sporulation efficiency, spore viability, and meiotic recombination on solid medium. *Cold Spring Harb Protoc* doi: 10.1101/pdb.prot085027.

Börner GV, Cha RS. 2015b. Induction and analysis of synchronous meiotic yeast cultures. *Cold Spring Harb Protoc* doi: 10.1101/pdb.prot085035.

Börner GV, Cha RS. 2015c. Analysis of meiotic recombination and homolog interaction during yeast meiosis. *Cold Spring Harb Protoc* doi: 10.1101/pdb.prot085050.

Börner GV, Kleckner N, Hunter N. 2004. Crossover/noncrossover differentiation, synaptonemal complex formation, and regulatory surveillance at the leptotene/zygotene transition of meiosis. *Cell* **117:** 29–45.

Carballo JA, Johnson AL, Sedgwick SG, Cha RS. 2008. Phosphorylation of the axial element protein Hop1 by Mec1/Tel1 ensures meiotic interhomolog recombination. *Cell* **132:** 758–770.

Carballo JA, Panizza S, Serrentino ME, Johnson AL, Geymonat M, Borde V, Klein F, Cha RS. 2013. Budding yeast ATM/ATR control meiotic double-strand break (DSB) levels by down-regulating Rec114, an essential component of the DSB-machinery. *PLoS Genet* **9:** e1003545.

Caryl AP, Armstrong SJ, Jones GH, Franklin FC. 2000. A homologue of the yeast *HOP1* gene is inactivated in the *Arabidopsis* meiotic mutant *asy1*. *Chromosoma* **109:** 62–71.

Cha RS, Weiner BM, Keeney S, Dekker J, Kleckner N. 2000. Progression of meiotic DNA replication is modulated by interchromosomal interaction proteins, negatively by Spo11p and positively by Rec8p. *Genes Dev* **14:** 493–503.

Chen SY, Tsubouchi T, Rockmill B, Sandler JS, Richards DR, Vader G, Hochwagen A, Roeder GS, Fung JC. 2008. Global analysis of the meiotic crossover landscape. *Dev Cell* **15:** 401–415.

Collins I, Newlon CS. 1994. Chromosomal DNA replication initiates at the same origins in meiosis and mitosis. *Mol Cell Biol* **14:** 3524–3534.

Cotton VE, Hoffmann ER, Abdullah MF, Borts RH. 2009. Interaction of genetic and environmental factors in *Saccharomyces cerevisiae* meiosis: The devil is in the details. *Methods Mol Biol* **557:** 3–20.

Elrod SL, Chen SM, Schwartz K, Shuster EO. 2009. Optimizing sporulation conditions for different *Saccharomyces cerevisiae* strain backgrounds. *Methods Mol Biol* **557:** 21–26.

Guthrie C, Fink G. ed. 1991. *Guide to yeast genetics and molecular biology*. Academic Press, San Diego.

Hollingsworth NM, Goetsch L, Byers B. 1990. The *HOP1* gene encodes a meiosis-specific component of yeast chromosomes. *Cell* **61:** 73–84.

Hunter N, Kleckner N. 2001. The single-end invasion: An asymmetric intermediate at the double-strand break to double-holliday junction transition of meiotic recombination. *Cell* **106:** 59–70.

Kassir Y, Granot D, Simchen G. 1988. *IME1*, a positive regulator gene of meiosis in *S cerevisiae*. *Cell* **52:** 853–862.

Keeney S, ed. 2009a. *Meiosis*. Volume 1. *Molecular and genetic methods*. Methods in molecular biology 557. Humana Press/Springer, New York.

Keeney S, ed. 2009b. *Meiosis*. Volume 2. *Cytological methods*. Methods in molecular biology 558. Humana Press/Springer, New York.

Keeney S, Giroux CN, Kleckner N. 1997. Meiosis-specific DNA double-strand breaks are catalyzed by Spo11, a member of a widely conserved protein family. *Cell* **88:** 375–384.

Klein F, Mahr P, Galova M, Buonomo SB, Michaelis C, Nairz K, Nasmyth K. 1999. A central role for cohesins in sister chromatid cohesion, formation of axial elements, and recombination during yeast meiosis. *Cell* **98:** 91–103.

Kumar R, Bourbon HM, de Massy B. 2010. Functional conservation of Mei4 for meiotic DNA double-strand break formation from yeasts to mice. *Genes Dev* **24:** 1266–1280.

Lydall D, Nikolsky Y, Bishop DK, Weinert T. 1996. A meiotic recombination checkpoint controlled by mitotic checkpoint genes. *Nature* **383:** 840–843.

Malone RE, Bullard S, Hermiston M, Rieger R, Cool M, Galbraith A. 1991. Isolation of mutants defective in early steps of meiotic recombination in the yeast *Saccharomyces cerevisiae*. *Genetics* **128:** 79–88.

Malone RE, Esposito RE. 1981. Recombinationless meiosis in *Saccharomyces cerevisiae*. *Mol Cell Biol* **1:** 891–901.

Mancera E, Bourgon R, Brozzi A, Huber W, Steinmetz LM. 2008. High-resolution mapping of meiotic crossovers and non-crossovers in yeast. *Nature* **454:** 479–485.

Martini E, Diaz RL, Hunter N, Keeney S. 2006. Crossover homeostasis in yeast meiosis. *Cell* **126:** 285–295.

Padmore R, Cao L, Kleckner N. 1991. Temporal comparison of recombination and synaptonemal complex formation during meiosis in *S. cerevisiae*. *Cell* **66:** 1239–1256.

Schwacha A, Kleckner N. 1997. Interhomolog bias during meiotic recombination: Meiotic functions promote a highly differentiated interhomolog-only pathway. *Cell* **90:** 1123–1135.

Smith AV, Roeder GS. 1997. The yeast Red1 protein localizes to the cores of meiotic chromosomes. *J Cell Biol* **136:** 957–967.

Sourirajan A, Lichten M. 2008. Polo-like kinase Cdc5 drives exit from pachytene during budding yeast meiosis. *Genes Dev* **22:** 2627–2632.

Sym M, Engebrecht JA, Roeder GS. 1993. Zip1 is a synaptonemal complex protein required for meiotic chromosome synapsis. *Cell* **72:** 365–378.

Usui T, Ogawa H, Petrini JH. 2001. A DNA damage response pathway controlled by Tel1 and the Mre11 complex. *Mol Cell* **7:** 1255–1266.

Wan L, Niu H, Futcher B, Zhang C, Shokat KM, Boulton SJ, Hollingsworth NM. 2008. Cdc28-Clb5 (CDK-S) and Cdc7-Dbf4 (DDK) collaborate to initiate meiotic recombination in yeast. *Genes Dev* **22:** 386–397.

Williamson DH, Johnston LH, Fennell DJ, Simchen G. 1983. The timing of the S phase and other nuclear events in yeast meiosis. *Exp Cell Res* **145:** 209–217.

Xu L, Ajimura M, Padmore R, Klein C, Kleckner N. 1995. *NDT80*, a meiosis-specific gene required for exit from pachytene in *Saccharomyces cerevisiae*. *Mol Cell Biol* **15:** 6572–6581.

Protocol 1

Analysis of Yeast Sporulation Efficiency, Spore Viability, and Meiotic Recombination on Solid Medium

G. Valentin Börner[1,3] and Rita S. Cha[2,3]

[1]Center for Gene Regulation in Health and Disease, Department of Biological, Geological and Environmental Sciences, Cleveland State University, 2121 Euclid Avenue, Cleveland, Ohio 44115-2214; [2]North West Cancer Research Institute, School of Medical Sciences, Bangor University, Bangor LL57 2UW, United Kingdom

Under conditions of nutrient deprivation, yeast cells initiate a differentiation program in which meiosis is induced and spores are formed. During meiosis, one round of genome duplication is followed by two rounds of chromosome segregation (meiosis I and meiosis II) to generate four haploid nuclei. Meiotic recombination occurs during prophase I. During sporogenesis, each nucleus becomes surrounded by an individual spore wall, and all four haploid spores become contained as a tetrad within an ascus. Important insights into the meiotic function(s) of a gene of interest can be gained by observing the effects of gene mutations on spore viability and viability patterns among tetrads. Moreover, recombination frequencies among viable spores can reveal potential involvement of the gene during meiotic exchange between homologous chromosomes. Here, we describe methods for inducing spore formation on solid medium, determining spore viability, and measuring, via tetrad analysis, frequencies of crossing over and gene conversion as indicators of meiotic chromosome exchange.

MATERIALS

It is essential that you consult the appropriate Material Safety Data Sheets and your institution's Environmental Health and Safety Office for proper handling of equipment and hazardous material used in this protocol.

RECIPES: Please see the end of this protocol for recipes indicated by <R>. Additional recipes can be found online at http://cshprotocols.cshlp.org/site/recipes.

Reagents

DAPI (4'6-diamidino-2-phenylindole) solution (0.1 mg/mL)
Ethanol (40%, v/v)
Selection plates (for genotyping spore colonies following tetrad dissection)
Spheroplast solution <R>
SPM plates <R>
Water (deionized and sterile)
Yeast strains of interest (stored at −80°C)

This protocol was optimized for SK1, which is a readily sporulating yeast strain that completes nuclear divisions and spore formation in >90% of cells and typically shows spore viabilities of >90% at 30°C in liquid medium. In >70% of cells, both meiotic divisions are completed by 10 h after transfer to sporulation medium (SPM) (Padmore et al. 1991).

---

[3]Correspondence: g.boerner@csuohio.edu; r.cha@bangor.ac.uk

Copyright © Cold Spring Harbor Laboratory Press; all rights reserved
Cite this protocol as Cold Spring Harb Protoc; doi:10.1101/pdb.prot085027

YPD plates <R>
YPG plates <R>

Equipment

Coverslips for microscope slides (No. 1.5; 20 × 20 mm)
Dissection needles (Singer Instruments)
Fluorescence microscope (for DAPI analysis)
Microcentrifuge tubes (1.5-mL; sterile)
Microscope slides
Phase-contrast light microscope (for assessing sporulation efficiency)
Phase-contrast light microscope with micromanipulater (for tetrad dissection) (Singer Instruments)
Replicating block
Replicating velvet cloths (sterile; ∼13 cm × 13 cm)
Toothpicks (flat; sterile)

METHOD

Analysis of Sporulation Efficiency and Spore Viability from Tetrad Dissection

1. Thaw mutant and wild-type yeast strains of interest from stocks stored at −80°C and transfer drop-sized samples onto an YPG agar plate. Incubate overnight (∼16 h) at 30°C.

   Overnight incubation on YPG medium, which contains glycerol as a nonfermentable carbon source, selects against spontaneously arising petite mutants that lack functional mitochondria. Petite mutants are defective for meiosis. For temperature-sensitive strains, perform all presporulation incubations at the permissive temperature. Perform all cell transfers with the wide end of a sterile flat toothpick.

2. Streak each strain for single colonies on separate YPD plates, and incubate at 30°C until ∼1.5–2-mm-diameter colonies form.

   For a wild-type SK1 strain, incubation for 2 d at 30°C leads to single colonies ∼1.5–2 mm in diameter. Prolonged incubation under limiting nutrient conditions may trigger some cells to initiate meiosis and commit to meiotic recombination, potentially resulting in mutation-induced chromosome imbalances. For strains with mitotic growth defects, including temperature-sensitive mutants, incubate at the permissive temperature until the colony size reaches ∼1.5–2 mm in diameter.

3. Transfer several freshly isolated single colonies from YPD to an SPM agar plate. Incubate for 2 d at 30°C (or at the restrictive temperature if the strain is temperature sensitive).

   If incubation is performed at a temperature greater than 30°C, optimize the conditions first to ensure adequate sporulation of the wild-type strain because budding yeast meiosis is temperature sensitive (e.g., Börner et al. 2004).

Analysis of Sporulation Efficiency

4. Remove a ∼1-mm^2 patch of sporulated cells and suspend in 100 µL of sterile water in a 1.5-mL microcentrifuge tube. Vortex to mix.

5. Place ∼4 µL of the suspension on a microscope slide.

6. Under a phase-contrast microscope, count the number of asci containing 1–4 spores and the unsporulated cells (Fig. 1A). Count a total of ≥100 cells. Use these numbers to calculate overall sporulation efficiency and the fraction of monads or dyads versus tetrads.

   Asci containing three spores are counted as tetrads on the assumption that the fourth spore is in a different optical plane. Always assay an isogenic wild-type strain on the same SPM plate to account for subtle differences in experimental conditions that can affect sporulation efficiency and viability (e.g., incubator temperature and the exact composition of plate media).

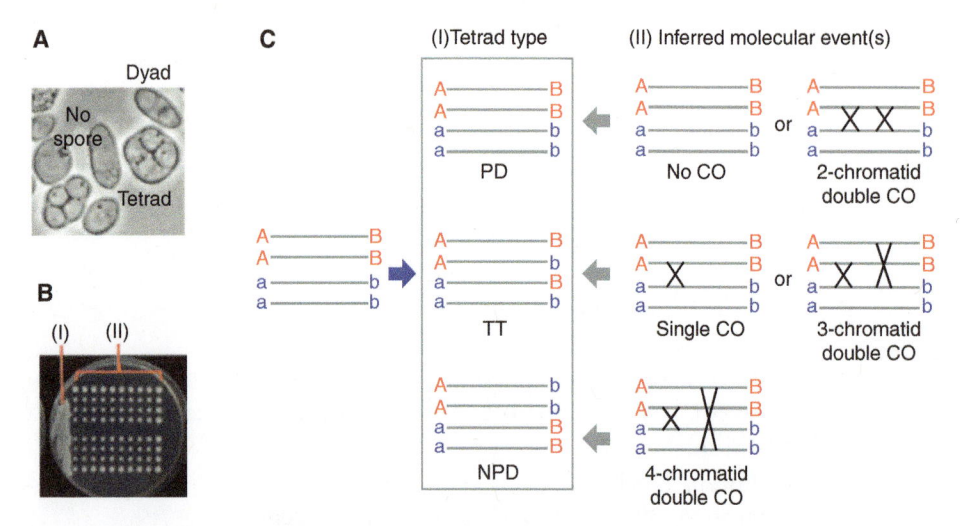

FIGURE 1. (A) Light microscopy image showing tetrad and dyad asci, as well as unsporulated cells. (B) A tetrad-dissection plate of wild-type SK1 after 2-d incubation at 30°C. (I) An area where spheroplasts were spread using a sterile toothpick. (II) All 80 spores from 20 tetrads (10 above and 10 below) gave rise to a colony. (C) Tetrad analysis of two linked loci. (I) Three different tetrad types are observed for an AB/ab diploid: parental ditype (PD), tetratype (TT), and nonparental ditype (NPD). (II) Inferred recombination events leading to the observed tetrad types. Most PD and TT tetrads result from no crossover (CO) or a single CO, respectively. NPD arises from double COs involving all four chromatids. Minorities of PD and TT result from double COs involving two and three chromatids, respectively. The numbers of PD, NPD, and TT tetrads are used for determining genetic distance between two loci in centimorgans (cM); see text.

7. (Optional) If there are no spores, examine the cells for nuclear divisions to determine whether the absence of spores is caused by a defect in meiotic entry or a failure to complete meiosis.

    i. Remove a 1-mm^2 patch of cells and suspend in 100 μL of 40% ethanol in a 1.5-mL microcentifuge tube. Vortex to mix.

    ii. Place ~2 μL of the cell suspension on a microscope slide. Mix with 2 μL of 0.1 mg/mL DAPI solution and immediately cover with a coverslip.

    iii. Analyze under a fluorescence microscope equipped with a DAPI filter (see Protocol 2: Induction and Analysis of Synchronous Meiotic Yeast Cultures [Börner and Cha 2015]).

Analysis of Spore Viability

Tetrad analysis is recommended for samples containing >5% tetrads. It will be difficult to find a sufficient number of suitable tetrads for dissection if the fraction becomes too small.

8. Transfer a ~1-mm^2 patch of cells from an SPM plate (after incubation in Step 3) to 100 μL of spheroplast solution in a sterile 1.5-mL microcentrifuge tube. Vortex to make the suspension homogeneous.

9. Incubate the suspension for 30 min at 37°C, and then place it on ice.

    Spheroplasted samples can be stored for ≤10 d at 4°C.

10. With a sterile toothpick, gently spread spheroplasts on the edge of a YPD plate (Fig. 1B,I). Using a phase-contrast microscope equipped with a micromanipulator and a glass dissection needle, pick a tetrad, transfer to a cell-free area on the same YPD plate, and transfer each of the spores to four well-defined positions separated by about 7 mm (Fig. 1B,II). Dissect at least 60 tetrads to obtain reliable estimates of spore viabilities.

    Typically, 20 tetrads can be dissected on each 100-mm plate.

11. Incubate the plate for 2–4 d at 30°C or for a longer period at a permissive temperature if the strain is temperature sensitive.

12. Calculate the spore viability by dividing the number of visible spore colonies by the number of spores placed on the plate and comparing this to the viability of the wild-type strain sporulated in parallel.

   Spore viabilities of <5% are suggestive of defective recombination initiation (in mutants such as spo11, rec114, rec102) or severe defects in processing recombination intermediates, either due to a failure to repair double-strand breaks (e.g., rad51) or a high degree of exchange between sister chromatids (e.g., mek1). Spore viability patterns with an increased incidence of two- and zero- versus three- and one-viable spore tetrads are an indication for homolog nondisjunction during meiosis I.

Tetrad Analysis for Crossing Over and Gene Conversion (Non-Mendelian Marker Segregation)

13. Using a strain carrying multiple, appropriately spaced heterozygous markers along one or several chromosomes (e.g., NHY1848; Oh et al. 2007), follow Steps 1–12 above. Use standard replica plating techniques to transfer spore colonies from the YPD master plate to selective medium and incubate until the majority of tetrads containing four viable spores have grown with the expected Mendelian (2^+:2^-) marker segregation pattern.

   Whereas the majority of tetrads show Mendelian (2^+:2^-) marker segregation, some may show aberrant, non-Mendelian segregation patterns (3:1 or 4:0), indicating gene conversion of that particular marker. Segmentation of spore colonies, with only half of a given colony scoring positive for a given marker, is indicative of postmeiotic segregation (PMS). Tetrads showing apparent non-Mendelian segregation for multiple markers are likely derived from association of unrelated spores and should be excluded from further analysis.

14. For tetrads showing four viable spores, score each marker. Use the YPD master plate as a reference for colony sizes because colonies with multiple auxotrophic markers tend to be smaller and might result in poor cell transfer. Use the YPD master plate to confirm PMS as well.

15. Determine frequencies of gene conversion and PMS for a particular marker by dividing the number of tetrads with a gene conversion or PMS event for that marker by the total number of four-viable-spore tetrads.

16. For crossover analyses, include only linked marker pairs, where both markers show Mendelian segregation (2^+:2^-). Score tetrads as parental ditypes (PD; each spore shows the parental marker configuration for that particular marker pair), tetratypes (TT; two spores are parental and two are recombinant for the marker pair), or nonparental ditypes (NPD; all four spores show reciprocal recombinant marker combinations for a given marker pair) (Fig. 1C,I). Then use the Perkins equation $[100 (6NPD + TT)/2(PD + NPD + TT)]$ to determine the recombination frequency or genetic map distance in cM for each marker pair (Perkins 1949).

   Most TTs indicate a single crossover event. Minorities of PDs and TTs are derived from double crossovers involving the same pair of chromatids or three chromatids, respectively (Fig. 1C,II). In the absence of double crossovers, 1 cM corresponds to one crossover (i.e., one tetratype) in 50 tetrads. The proportion of undetected double crossovers (PDs or TTs) increases when genetic intervals >25 cM are included in the analysis, resulting in an underestimation of genetic distances. The Perkins equation is used to take into account for genetic distance calculations the number of undetected double crossovers from observed NPDs.

17. Statistically evaluate differences in genetic distances between mutant and wild type for each marker pair. Refer to instructions on the Stahl Laboratory website (http://molbio.uoregon.edu/~fstahl/EquationsMapDistance.html) as necessary.

    i. Determine the absolute value of the difference between the two map distances ($|X1 - X2|$, where X is the map distance).

    ii. Using frequencies of tetratypes (fTT) and nonparental ditypes (fNPD) for a given interval and genotype, determine the sampling variance of the Perkins equation. Note that this sampling variance is given by

$$\text{Var}[X] = 0.25\,\text{var}[f\text{TT}] + 9\,\text{var}[f\text{NPD}] + 3\,\text{covar}[f\text{TT}, f\text{NPD}]$$
$$= 0.25\,[(f\text{TT})(1 - f\text{TT})/n] + 9\,[(f\text{NPD})(1 - f\text{NPD})/n] + 3\,[-(f\text{TT})(f\text{NPD})/n],$$

where f denotes the frequency of each type of tetrad (TT/n, NPD/n, PD/n) and n denotes the total number of four viable spore tetrads showing Mendelian marker segregation for the particular marker pair.

iii. Compare the value obtained in Step 17.i with the standard error (SE) of the difference of the two map distances. The difference in genetic distances between two genotypes is significant if the absolute value of the difference between the two map distances is greater than twice the standard error of the difference in map distances (i.e., $|X1 - X2| > 2 \times \text{SE}$, where SE = square root $(\text{Var}[X1] + \text{Var}[X2])$).

Although assaying genetic distances via tetrad analysis allows concurrent recombination analysis of multiple chromosome regions in a single experiment, it frequently requires ≥800 four viable spore tetrads for detection of statistically significant changes, especially when small intervals or small differences are examined.

RECIPES

Spheroplast Solution

Reagent	Final concentration
Potassium phosphate (pH 7.5)	50 mM
EDTA	10 mM
Sorbitol	1.2 M
Zymolyase 100T (e.g., MP Biomedicals)	0.25 mg/mL
2-Mercaptoethanol	1%

SPM Plates

Reagent	Final concentration
Potassium acetate	0.3%
Raffinose	0.02%
Bacto agar	2%

YPD Plates

Reagent	Final concentration
Bacto yeast extract	1%
Bacto peptone	2%
Glucose	2%
Bacto agar	2%

YPG Plates

Reagent	Final concentration
Bacto yeast extract	1%
Bacto peptone	2%
Glycerol	2%
Bacto agar	2%

REFERENCES

Börner GV, Cha RS. 2015. Induction and analysis of synchronous meiotic yeast cultures. *Cold Spring Harb Protoc* doi: 10.1101/pdb.prot085035.

Börner G, Kleckner N, Hunter N. 2004. Crossover/noncrossover differentiation, synaptonemal complex formation, and regulatory surveillance at the leptotene/zygotene transition of meiosis. *Cell* 117: 29–45.

Oh S, Lao J, Hwang P, Taylor A, Smith G, Hunter N. 2007. BLM ortholog, Sgs1, prevents aberrant crossing-over by suppressing formation of multichromatid joing molecules. *Cell* 130: 259–272.

Padmore R, Cao L, Kleckner N. 1991. Temporal comparison of recombination and synaptonemal complex formation during meiosis in *S. cerevisiae*. *Cell* 66: 1239–1256.

Perkins DD. 1949. Biochemical mutants in the smut fungus *Ustilago maydis*. *Genetics* 34: 607–626.

WWW RESOURCES

Stahl Laboratory website http://molbio.uoregon.edu/~fstahl/EquationsMapDistance.html

Protocol 2

Induction and Analysis of Synchronous Meiotic Yeast Cultures

G. Valentin Börner[1,3] and Rita S. Cha[2,3]

[1]Center for Gene Regulation in Health and Disease, Department of Biological, Geological and Environmental Sciences, Cleveland State University, Cleveland, Ohio 44115-2214; [2]North West Cancer Research Institute, School of Medical Sciences, Bangor University, Bangor LL57 2UW, United Kingdom

Meiosis in *Saccharomyces cerevisiae* can be induced by deprivation of nutrients. Here, we present a protocol for inducing synchronous meiosis in SK1, the most efficient and synchronous yeast strain for meiosis, by exposing SK1 cells to liquid medium that contains potassium acetate as a nonfermentable carbon source and lacks nitrogen. These synchronous meiotic yeast cultures can be subjected to a range of molecular and cytological analyses, making them useful for investigating the genetic and molecular determinants of meiosis.

MATERIALS

It is essential that you consult the appropriate Material Safety Data Sheets and your institution's Environmental Health and Safety Office for proper handling of equipment and hazardous material used in this protocol.

RECIPES: Please see the end of this protocol for recipes indicated by <R>. Additional recipes can be found online at http://cshprotocols.cshlp.org/site/recipes.

Reagents

DAPI (4′,6-diamidino-2-phenylindole) stock solution (0.1 mg/mL)
Ethanol (40%, v/v)
Spheroplast storage buffer (SSB) <R>
Water (deionized and sterilized)
Yeast presporulation medium (SPS) <R>
Yeast sporulation medium (SPM) <R>
Yeast strains of interest (stored at −80°C)

This protocol was optimized for SK1, a readily sporulating yeast strain. In a typical wild-type culture undergoing synchronous meiosis, >70% of cells have completed both meiotic divisions by 10 h after transfer to liquid SPM at 30°C (Padmore et al. 1991). Sporulation efficiency and spore viability usually exceed 90%.

YPD plates <R>
YPG plates <R>

Equipment

Centrifuge (with rotors and sterile centrifuge tubes ranging in sizes from 15 to 500 mL depending on the culture volume)
Coverslips for microscope slides (No. 1.5; 20 × 20 mm)

---

[3]Correspondence: g.boerner@csuohio.edu; r.cha@bangor.ac.uk

Copyright © Cold Spring Harbor Laboratory Press; all rights reserved
Cite this protocol as *Cold Spring Harb Protoc*; doi:10.1101/pdb.prot085035

Falcon tubes (15 or 50 mL depending on sample volume)
Flasks

Use flasks that are 10 times the culture volume (e.g., 1-L flasks for 100-mL cultures). For cultures with volumes >300 mL, use 2.8-L triple-baffled Fernbach flasks (e.g., from Bellco Glass).

Fluorescence microscope (for DAPI analysis)
Incubators (stationary and shaking)
Microcentrifuge
Microcentrifuge tubes (1.5-mL; sterile)
Microscope slides
Pipette tips (filtered)
Roller drum
Spectrophotometer (OD_{600}) and plastic cuvettes

METHOD

Induction of Synchronous Meiosis in Liquid Medium

This procedure was modified from Padmore et al. (1991) and Keeney (2009).

1. Thaw mutant and wild-type yeast strains of interest from stocks stored in 25% glycerol at −80°C. Transfer drop-sized samples onto an YPG agar plate and incubate overnight (∼16 h) at 30°C.

   Overnight incubation on YPG medium, which contains glycerol as a nonfermentable carbon source, selects against spontaneously arising petite mutants that lack functional mitochondria. Petite mutants are defective for meiosis. For temperature-sensitive strains, perform all presporulation incubations at the permissive temperature.

2. Streak each strain for single colonies on separate YPD plates and incubate at 30°C until the colonies are ∼1.5–2 mm in diameter.

   For a wild-type SK1 strain, incubation for 2 d at 30°C leads to single colonies ∼1.5–2 mm in diameter. Prolonged incubation under limiting nutrient conditions may trigger some cells to initiate meiosis and commit to meiotic recombination, potentially resulting in mutation-induced chromosome imbalances. For strains with mitotic growth defects, including temperature-sensitive mutants, incubate at the permissive temperature until the colony size reaches ∼1.5–2 mm in diameter.

3. Use a single colony to inoculate 4–5 mL of liquid YPD. Incubate on a roller drum with aeration for 24–26 h at 30°C.

4. Approximately 16–18 h before the desired time of meiosis induction, inoculate 100 mL of SPS in a 1-L flask with the YPD culture aiming for an OD_{600} of 0.01. Start with a 1/200 dilution (i.e., 0.5 mL into 100 mL of SPS) and adjust accordingly to obtain the correct OD by adding either more YPD culture or more SPS medium. Incubate with agitation at 200–300 rpm overnight at 30°C.

   The volume of SPS culture can be scaled up to 450 mL depending on the nature of meiotic events to be analyzed. For example, use 100 mL for DAPI and FACS analyses and use 300–450 mL for physical analysis of recombination (see below). In all cases, the ratio between the culture volume and the flask size should be ∼1:10 to ensure adequate aeration. For culture volumes >200 mL, use a baffled Fernbach flask to enhance aeration.

5. When the OD_{600} of the culture reaches 1.2–1.4 (∼2×10^7 cells/mL after 16–18 h incubation), harvest the cells by centrifuging at 3000g for 3 min at room temperature.

6. Wash the pellet once with one volume of prewarmed (30°C) SPM. Resuspend in 100 mL (i.e., the same volume as the SPS culture) of prewarmed SPM.

7. Start the synchronous meiotic time-course by incubating the culture in a shaker incubator at 30°C (or at a desired temperature) with vigorous agitation (≥350 rpm). Collect and store

Chapter 3

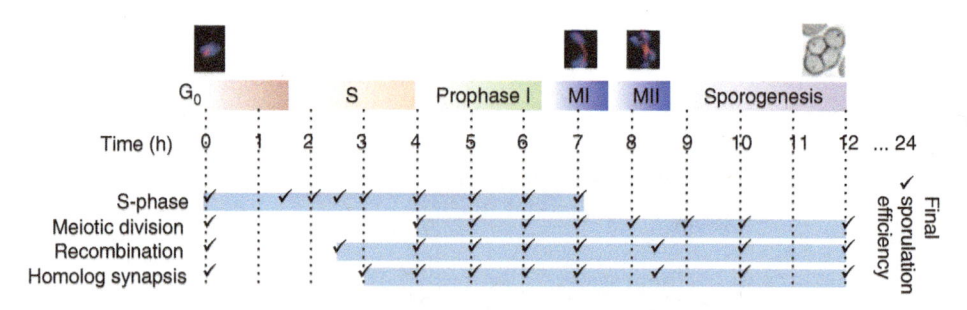

FIGURE 1. Timeline of key meiotic stages in wild-type SK1. Meiosis in budding yeast is induced by placing G₀ cells into liquid sporulation medium (SPM), which lacks nitrogen and a fermentable carbon source. *Above* the timeline (data from Padmore et al. 1991) are nuclear morphologies of mono-, di-, and tetra-nucleate species, and a four-spore ascus observed before meiosis I (MI), after meiosis I (MI), after meiosis II (MII), and after sporogenesis, respectively. Blue indicates DNA and red indicates tubulin. *Below* the timeline are the durations of key meiotic events and the recommended time points of sample collection for analysis of a specific meiotic event.

sample(s) of recommended volume from the flask at the appropriate time points, including $t = 0$ (see below and Fig. 1).

To determine the extent of synchrony by DAPI analysis, follow Steps 8–11. To collect and store samples for physical analysis of meiotic recombination by Southern analysis, follow Steps 12–16. To collect and store samples for cytological analysis of chromosomes, follow Steps 17–18. To collect and store samples for western blots, follow Steps 19–22. Typically, samples collected during synchronous meiosis are analyzed within a few days after the time course.

Assessment of Efficiency and Synchrony of Meiotic Culture by DAPI Analysis of Nuclear Divisions

Recommended sample collection times for analysis of synchrony in a wild-type culture are 0, 4, 5, 6, 7, 8, 9, 10, 12, and 24 h (see Fig. 1).

8. Collect 1-mL samples of culture (from Step 7) and harvest the cells by centrifuging at 13,000g for 15 sec. Resuspend the cells in 250 μL of 40% ethanol.

    The fixed cells can be stored for up to 1 wk at 4°C or for up to several months at −20°C.

9. Place 3 μL of DAPI solution on a microscope slide.

10. Vortex the fixed cells briefly to ensure that they are in suspension. Mix 3 μL of cells with the DAPI solution on the slide. Cover the cells immediately with a coverslip.

11. Under a fluorescence microscope equipped with a 100× oil-immersion objective and a DAPI filter, count the number of cells showing one, two, or three/four nuclei at each time point (Fig. 1).

    We recommend starting with the 7-h sample of the wild-type culture, which should show 30%–60% of cells with ≥2 nuclei, indicative of completion of meiosis I (MI). An efficient wild-type culture typically contains ~ 80% cells that have undergone MI by $t = 10$–12 h, with divided nuclei first appearing at ~$t = 5$ h. Further analysis of transient meiotic events, including physical analysis of meiotic recombination and chromosome synapsis, should be limited to cultures showing good synchrony as indicated by occurrence of >70% of meiotic divisions within a 3-h time window.

Collection and Storage of Samples for Physical Analysis of Meiotic Recombination by Southern Analysis

Recommended sample collection times for analysis of meiotic recombination in a wild-type culture are 0, 2.5, 4, 5, 6, 7, 8.5, 10, 12, and optionally 24 h (see Fig. 1).

12. Harvest 9–25 mL of culture (from Step 7) by centrifuging at 4000g for 2 min at 4°C.

13. Wash with one volume of SSB at 4°C.

14. Resuspend in 1 mL of SSB at 4°C and transfer to a 1.5-mL microcentrifuge tube.
15. Centrifuge at 14,000g for 10 sec and remove the supernatant.
16. Freeze the cell pellet on dry ice and store at −80°C.

    The experimental procedure for using these samples is described in Protocol 3: Analysis of Meiotic Recombination and Homolog Interaction during Yeast Meiosis (Börner and Cha 2015a).

Collection and Storage of Samples for Surface Spreading of Nuclei for Immunofluorescence Analysis

Recommended sample collection times for a wild-type culture are 0, 3, 4, 5, 6, 7, 8.5, 10, and 12 h (see Fig. 1).

17. Collect 1–10 mL of culture (from Step 7) and store on ice.

    1 mL gives sufficient cell material for three microscope slides; the remaining sample volume can be used for additional analysis, such as for physical analysis of meiotic recombination.

18. Process samples for spreading within 3 d of collection (see Protocol 3: Analysis of Meiotic Recombination and Homolog Interaction during Yeast Meiosis [Börner and Cha 2015a]).

Collection and Storage of Samples for Western Blot Analysis

For recommended sample collection times for the process of interest, see Figure 1.

19. Collect 12 mL of culture (from Step 7), which should be sufficient for three western blot experiments. Harvest the cells by centrifuging at 3000g for 3 min at room temperature.
20. Discard the supernatant, and resuspend the cells in 1.2 mL of 30% glycerol.
21. Transfer 0.4-mL aliquots into three 1.5-mL microcentrifuge tubes.
22. Freeze the mixture on dry ice, and store at −80°C.

    These samples are suitable for trichloroacetic acid whole-cell extract preparation following the method of Foiani et al. (1994).

DISCUSSION

Sporulation efficiency and synchrony of meiotic cultures are affected by both pre-growth and sporulation conditions. Notably, small fluctuations in the experimental conditions such as extent of aeration and temperature of a shaker incubator can often lead to substantial variations (e.g., Börner et al. 2004). It is not uncommon to observe such variations in meiotic progression in parallel cultures of the same genetic background, including wild-type cultures (Cha et al. 2000). As such, quantitative comparison between different experiments should be avoided unless the respective wild-type controls behaved comparably. We also recommend analyzing no more than four cultures at a time to minimize general sloppiness in sample handling, which tends to compromise efficiency and/or synchrony of meiosis.

RECIPES

Spheroplast Storage Buffer

Reagent	Final concentration
Potassium phosphate (pH 7.5)	50 mM
EDTA	10 mM
Sorbitol	1.2 M
Glycerol	20% (v/v)

Yeast Presporulation Medium

Reagent	Final concentration
Bacto yeast extract	0.5% (w/v)
Bacto peptone	1% (w/v)
Potassium acetate	1% (w/v)
Yeast nitrogen base without amino acids and ammonium sulfate	0.17% (w/v)
Ammonium sulfate	0.5% (w/v)
Potassium hydrogen phthalate	0.05 M

Adjust the pH to 5.5 with 10 N KOH, add 20 µL of antifoam agent (polypropyleneglycol 2000) per liter of solution, and autoclave.

Yeast Sporulation Medium

Reagent	Final concentration
Potassium acetate	1% (w/v)
Raffinose	0.02% (w/v)

Prepare sporulation medium (SPM) by combining the ingredients above, adjusting the pH to 7 with 10 N KOH, adding 20 µL of antifoam agent (polypropylene glycol 2000) per liter of solution, and aotoclaving. Supplement SPM with nutrients below according to the auxotrophic requirements of the yeast strain of interest prior to autoclaving.

Nutrient	Stock concentration (in 0.05 M HCl except for uracil)	Final concentration
Adenine	0.5%	1.6 mL/L
Arginine	2.0%	0.2 mL/L
Histidine	2.0%	0.2 mL/L
Isolucine	1.0%	0.6 mL/L
Leucine	1.0%	1.2 mL/L
Lysine	1.5%	0.4 mL/L
Methionine	2.0%	0.2 mL/L
Threonine	6.0%	1.0 mL/L
Tryptophan	1.0%	0.4 mL/L
Tyrosine	0.25%	2.0 mL/L
Valine	3.0%	1.0 mL/L
Uracil	0.2% in 1% Na_2CO_3	1.0 mL/L

YPD Plates

Reagent	Final concentration
Bacto yeast extract	1%
Bacto peptone	2%
Glucose	2%
Bacto agar	2%

YPG Plates

Reagent	Final concentration
Bacto yeast extract	1%
Bacto peptone	2%
Glycerol	2%
Bacto agar	2%

REFERENCES

Börner GV, Cha RS. 2015. Analysis of meiotic recombination and homolog interaction during yeast meiosis. *Cold Spring Harb Protoc* doi: 10.1101/pdb.prot085050.

Börner GV, Kleckner N, Hunter N. 2004. Crossover/noncrossover differentiation, synaptonemal complex formation, and regulatory surveillance at the leptotene/zygotene transition of meiosis. *Cell* 117: 29–45.

Cha RS, Weiner BM, Keeney S, Dekker J, Kleckner N. 2000. Progression of meiotic DNA replication is modulated by interchromosomal interaction proteins, negatively by Spo11p and positively by Rec8p. *Genes Dev* 14: 493–503.

Foiani M, Marini F, Gamba D, Lucchini G, Plevani P. 1994. The B subunit of the DNA polymerase α-primase complex in *Saccharomyces cerevisiae* executes an essential function at the initial stage of DNA replication. *Mol Cell Biol* 14: 923–933.

Keeney S, ed. 2009. *Meiosis. Volume 1, Molecular and Genetic Methods.* Methods in Molecular Biology series 557. Humana Press/Springer, NY.

Padmore R, Cao L, Kleckner N. 1991. Temporal comparison of recombination and synaptonemal complex formation during meiosis in *S. cerevisiae*. *Cell* 66: 1239–1256.

Protocol 3

Analysis of Meiotic Recombination and Homolog Interaction during Yeast Meiosis

G. Valentin Börner[1,3] and Rita S. Cha[2,3]

[1]*Center for Gene Regulation in Health and Disease, Department of Biological, Geological and Environmental Sciences, Cleveland State University, Cleveland, Ohio 44115-2214;* [2]*North West Cancer Research Institute, School of Medical Sciences, Bangor University, Bangor LL57 2UW, United Kingdom*

During meiosis, one round of genome duplication is followed by two rounds of chromosome segregation, resulting in the halving of the genetic complement and the formation of haploid reproductive cells. In most organisms, intimate juxtaposition of homologous chromosomes and homologous recombination during meiotic prophase are required for meiotic success. Here we present a general protocol for visualizing chromosomal proteins and homolog interaction on surface-spread nuclei and a widely used protocol for analyzing meiotic recombination based on an engineered hotspot referred to as *HIS4::LEU2*.

MATERIALS

It is essential that you consult the appropriate Material Safety Data Sheets and your institution's Environmental Health and Safety Office for proper handling of equipment and hazardous material used in this protocol.

RECIPES: Please see the end of this protocol for recipes indicated by <R>. Additional recipes can be found online at http://cshprotocols.cshlp.org/site/recipes.

Reagents

[α-^{32}P]-dCTP (6000 Ci/mmol)
Antifade reagent (e.g., ProLong Gold from Life Technologies)
Bovine serum albumin (BSA; fraction V; 10 mg/mL)
DAPI (4',6-diamidino-2-phenylindole) solution (0.1 mg/mL)
Depurination solution (0.25 N HCl)
Dithiothreitol (DTT; 1 M stock solution)
EDTA
Ethanol (100% and 70%)
Ethidium bromide stock solution (10 mg/mL)
Fixative solution for spheroplasts <R>
Glycerol (30%, v/v)
Hybridization solution
   Prepare by adding denatured probe to prehybridization solution (see Step 84).

---

[3]Correspondence: g.boerner@csuohio.edu; r.cha@bangor.ac.uk

Copyright © Cold Spring Harbor Laboratory Press; all rights reserved
Cite this protocol as *Cold Spring Harb Protoc*; doi:10.1101/pdb.prot085050

Isopropanol

Lipsol detergent (1%, v/v)

> Obtain this reagent from a meiosis laboratory. (Available on request from the authors.)

Loading dye (6×) <R>

Lysis buffer for yeast spheroplasts <R>

MES wash buffer <R>

Phenol:chloroform:isoamylalcohol (25:24:1)

> Buffer at pH 8.0 with Tris–HCl and stabilize with 0.1% (w/v) 8-hydroxyquinoline as antioxidant/coloring agent. Store at −20°C and thaw as needed.

Photo-Flo 200 solution (Kodak)

Potassium acetate (5 M; without pH adjustment)

Prehybridization solution <R>

Proteinase K stock solution

> Prepare a solution of 10 mg/mL proteinase K in 100 mM Tris–HCl (pH 8.0) and 50 mM EDTA (pH 8.0). Make fresh on the day of DNA extraction. Store on ice.

Restriction enzymes (XhoI, BamHI, and NgoMIV) (New England BioLabs)

RNase A work solution (10 µg/mL in TE buffer)

Primary antibodies

> Validated antibodies include rat anti-HA 3F10 monoclonal antibody (Roche), rabbit anti-GFP polyclonal antibody (Life Technologies A-11122), and goat anti-Zip1 polyclonal antibody (Santa Cruz Biotechnology yC-19).

Prime-It RmT random primer labeling kit (Agilent)

Probe 4

> This probe is for the hotspot HIS4::LEU2 and corresponds to nucleotides 63086–63640 of Saccharomyces cerevisiae chromosome III excised from a plasmid (Ahuja and Börner 2011).

Secondary antibodies

> For single staining, FITC-conjugated, secondary antibodies (e.g., from Life Technologies) are preferred because they provide superior resolution compared to red fluorophores. For double staining with different fluorophores (e.g., FITC and Rhodamine), avoid interactions between secondary antibodies by using only secondary antibodies raised in different animals than the primary antibodies.

Sodium acetate

Sodium dodecyl sulfate (SDS; 20% [w/v] in deionized water)

Sodium phosphate (50 mM; pH 7.2)

Southern wash buffer (0.1% SDS in 0.1× SSC)

SSC (10×, pH 7.0) <R>

TBS for immunostaining <R>

TE buffer <R>

Tris-HCL (200 mM; pH 7.5)

Yeast strains of interest

ZK buffer (50 mM Tris [pH7.5] containing 0.5 M KCl)

Zymolyase buffer <R>

Zymolyase stock solution

> Prepare a solution of 20 mg/mL Zymolyase 100T (e.g., MP Biomedicals) in 50 mM Tris (pH 7.5) containing 2% glucose.

Zymolyase work solution

> Immediately before use, prepare a solution of 100 µg/mL Zymolyase 100T (e.g., MP Biomedicals) and 1% (v/v) β-mercaptoethanol in Zymolyase buffer. Vortex for 5 min to dissolve the Zymolyase.

Equipment

Agarose gel electrophoresis equipment (gel box and power supply)

Aluminum foil

Coplin staining jar
Coverslips for microscope slides (No. 1.5; 20 × 50 mm)
Fluorescence microscope (equipped with 100×, 1.3–1.4 NA oil-immersion objective, CCD camera, and appropriate fluorescence filters [e.g., DAPI, fluorescein, or rhodamine])
Heating blocks (set at 42°C and 100°C)
Hybridization bottles
Hybridization oven with rotisserie (set at 65°C)
Light microscope with 20× objective (to monitor status of spheroplast formation)
Marker pen (with water-resistant ink)
Microcentrifuge
Microscope slides (frosted)
Microscope slide storage box (capacity 100 slides)
Moist chamber (e.g., a microslide storage box containing a wet paper towel)
Nail polish (clear)
Nylon membrane
Paper towels
Pasteur pipettes (1 mL; glass)
Phosphorimager
Pipette tips (filtered)
Plastic container (2 L)
Saran Wrap
Shaking incubator (set at 65°C)
Spin column (GE Healthcare Life Sciences)
Timer
UV crosslinker
Water bath (set at 65°C)
Whatman 3MM paper

METHOD

Visualization of Meiotic Chromosomes

Surface Spreading of Nuclei

This procedure is modified from Loidl (1995).

1. Set up synchronous meiotic yeast cultures and collect samples as described in Protocol 2: Induction and Analysis of Synchronous Meiotic Yeast Cultures (Börner and Cha 2015). For each experiment, include a control strain that lacks the epitope of interest.

2. Soak frosted microscope slides overnight in a Coplin jar containing 100% ethanol.

3. Pellet the cells from ≥1 mL of meiotic culture by spinning in a microcentrifuge at 5000 rpm for 2 min.

4. Discard the supernatant and add 0.25 mL of 200 mM Tris-HCL (pH 7.5) containing 20 mM DTT.

5. Resuspend the cell pellet and incubate for 2 min at room temperature.

6. Pellet the cells by spinning in a microcentrifuge at 5000 rpm for 2 min.

7. Discard the supernatant and ensure it is all removed by inverting the tube on a paper towel.

8. Resuspend the cell pellet in 250 µL of ZK buffer.

9. Add 2 µL of Zymolyase stock solution.

10. Incubate for 25 min at 30°C. Invert the tubes four times at regular intervals during the incubation. The cells will clump together and then unclump. Check the progress of spheroplast formation by placing 1 µL of the cell suspension in a drop of deionized water and observe under a light microscope. Spheroplasts without cell walls will explode, leaving behind wrinkled-appearing "ghosts."

11. Pellet the spheroplasts by spinning in a microcentrifuge at 5000 rpm for 3 min.

12. Using filter tips, resuspend the spheroplasts in 1 mL of MES wash buffer. Use only plastic pipettes because spheroplasts stick to glass.

13. Pellet the spheroplasts by spinning in a microcentrifuge at 5000 rpm for 3 min.

14. Using filter tips, carefully resuspend the spheroplasts in ~80 µL of MES wash buffer.

    As an indication of the appropriate resuspension volume, it should be just barely possible to see through the suspension. For example, it should be possible to detect black printed letters, but not read them.

    At this point in the experiment, the spheroplasts can be placed on ice for up to 24 h.

15. Take the microscope slides out of 100% ethanol and lean them upright against a solid support to air dry with the frosted area facing upward.

16. Using a marker pen with water-resistant ink, label the dried slides on the frosted part of slide. Prepare three slides per sample.

17. Prepare three adjustable micropipettes set to 20, 40, and 80 µL.

18. Find the focal plane of the spheroplasts under a light microscope equipped with a 20× objective.

19. Line up three slides per sample on the edge of the bench and transfer 20 µL of spheroplasts to the nonfrosted end of each slide.

20. Add 40 µL of fixative solution for spheroplasts to each slide.

21. Quickly add 80 µL of 1% lipsol to each slide.

    Lipsol acts as a spreading agent by solubilizing cell and nuclear membranes. Addition of fixative before the addition of detergent prevents overspreading.

22. Mix and swirl gently by picking up the frosted end of each slide and rolling the drop from side to side.

23. Incubate for 1 min.

24. Observe under the light microscope and monitor until ~80% of the spheroplasts have lysed.

    Focus on a group of 10 spheroplasts; lysis is indicated by sudden retraction of the cell body from the focal plane. With fresh cell samples and adequate Zymolyase treatment, ~80% lysis is usually achieved within 1 min. Protoplasts stored on ice for 2 or 3 d may take >2 min to lyse. Time for lysis can be determined for the first few samples and maintained for subsequent samples without microscopic monitoring.

25. Add 80 µL of fixative solution for spheroplasts to each slide.

26. Swirl the preparation gently by picking up the frosted end of the slide and rolling the drop from side to side.

27. Pull a clean 1-mL glass Pasteur pipette across the group of three parallel slides, spreading cells toward the frosted part. Slide the pipette along the top of the liquid without rolling it on the slide.

28. Dry the slides horizontally in a fume hood overnight.

29. Store indefinitely in a 100-slide storage box at −80°C.

Immunodecoration of Surface-Spread Nuclei

30. Dip the slides in 0.2% (v/v) Photo-Flo 200 solution for 30 sec.

31. Air dry the slides in an upright orientation.

32. Incubate the slides in TBS for immunostaining with mild agitation in a vertical Coplin jar for 15 min.

33. Drain off the excess liquid on a paper towel for 1 min. Do not allow to dry out.
34. Place the slides horizontally in a moist chamber. Add 200 µL of TBS for immunostaining containing 1% BSA. Cover with a coverslip and incubate for 10 min at room temperature.
35. Gently shake off the coverslip to avoid excessive shearing. Place the slide in a moist chamber and cover the entire nonfrosted area with 200 µL of primary antibody solution in TBS for immunostaining containing 1% BSA (try 1:100–1:1000 dilutions for typical primary antibodies) and cover with a coverslip.
36. Incubate in a moist chamber overnight at 4°C.
37. Shake off the coverslip. Place the slide in a moist chamber and cover the entire nonfrosted area with 200 µL of secondary antibody in TBS for immunostaining containing 1% BSA (try 1:500–1:5000 dilutions for typical secondary antibodies) and cover with a coverslip.

    Perform all procedures with fluorescent secondary antibodies under subdued light.

38. Incubate in moist chamber in the dark for 1–3 h at room temperature.
39. Shake off the coverslip. Gently agitate in TBS for immunostaining containing 1 µg/mL DAPI in a Coplin jar covered with aluminum foil.
40. Drain the slides, dry in the dark, and add 1 drop of antifade. Allow the antifade to cure as specified by the manufacturer, cover with a coverslip, and seal with clear nail polish.
41. Store for 1 mo at −20°C or for up to 2 yr at −80°C.

Fluorescence Microscopy

42. Take images with a 100× 1.4 NA oil-immersion objective on two or three channels individually.
43. Score stages of Zip1 assembly according to the following criteria: no staining (pre-meiotic), a few focus-like signals (leptotene; foci correspond to 16 centromeric signals [Tsubouchi and Roeder 2005]), foci with some lines (zygotene), and mostly lines with a few foci (pachytene) (Fig. 1A and B).

    In synchronous wild-type meiotic cultures incubated at 30°C, the pachytene peak usually occurs at the time when 10%–20% of cells have completed the first meiotic division.

Physical Analysis of Meiotic Recombination

For an illustration of physical analysis of meiotic recombination at HIS4::LEU2 locus, see Figure 1C.

DNA Extraction and Purification

44. Set up synchronous meiotic yeast cultures and collect samples as described in Protocol 2: Induction and Analysis of Synchronous Meiotic Yeast Cultures (Börner and Cha 2015).

    Only cultures showing good synchrony are processed further for genomic DNA extraction because detection of transient and low-abundant recombination intermediates depends on a high degree of synchrony. Cohorts of no more than eight samples should be processed to ensure consistent DNA extraction efficiencies.

45. Using a P1000 filtered tip, resuspend a cell pellet in 0.5 mL of Zymolyase work solution by stirring and slowly pipetting up and down. Transfer the cell suspension to a 1.5-mL microcentrifuge tube and incubate for 30 min at 37°C, inverting the tubes at least three times during the incubation.

    Throughout the protocol, avoid vortexing and minimize pipetting so as to reduce mechanical stress that might affect the integrity of genomic DNA.

46. Spin the spheroplasted cells in a microcentrifuge at 7000 rpm for 5 min. Remove the supernatant with a pipette, leaving as little liquid as possible.
47. For every cohort of eight tubes, prepare a master mix of lysis work solution with proteinase K by premixing 5.5 mL of lysis buffer for yeast spheroplasts with 200 µL of 10 mg/mL proteinase

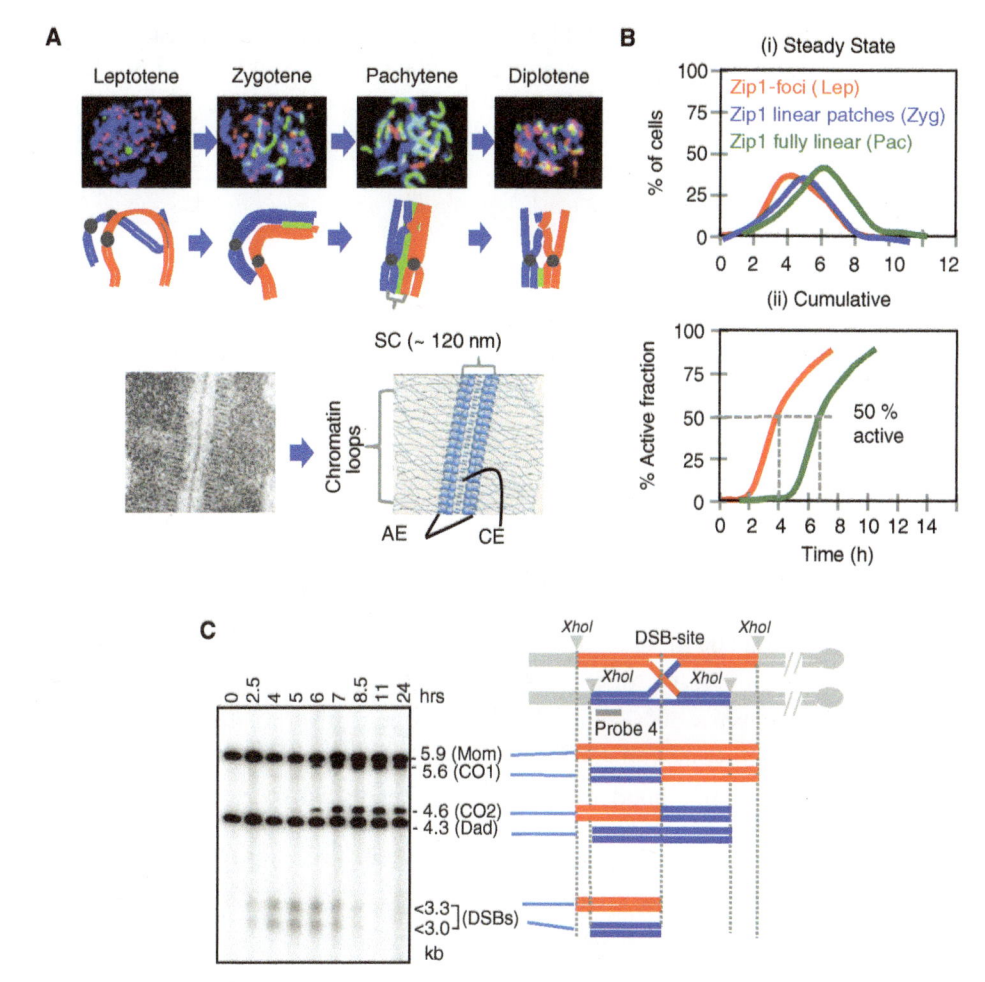

FIGURE 1. (*A*) Physical interactions between homologs during prophase I. The synaptonemal complex (SC) has three parts: the central element (CE) and two axial elements (AEs), which, following CE assembly, are referred to as lateral elements (LEs). The extent of the SC is used to define substages of meiotic prophase I as leptotene, zygotene, pachytene, and diplotene, which correspond to stages showing minimal, partial, full synapsis, and complete disassembly of the SC, respectively. Hop1/Red1 and Zip1 are components of axial elements and central elements, respectively (Hollingsworth et al. 1990; Sym et al. 1993; Smith and Roeder 1997). Blue represents DNA stained with DAPI, green represents Zip1-GFP, and red represents Mec1/Tel1 targets visualized using antibodies against Mec1/Tel1-specific phosphorylation (e.g., Carballo et al. 2008). The electron microscopy image is of pachytene chromosomes in *Neotallia* (Westergaard and von Wettstein 1972; reproduced with permission of Annual Review of Genetics, copyright © 1972, Annual Reviews, http://www.annualreviews.org). (*B*) Quantitative analysis of homolog synapsis. (i) At each time point, the fraction of nuclei showing Zip1 in foci, patches, and (semi-)continuous lines corresponding to leptotene, zygotene, and pachytene stages, respectively, are scored and plotted as a function of time to determine the steady-state levels. (ii) The same data set can be used to calculate cumulative fractions, that is, the fraction of the active cells that have reached the respective state by any given time point (Padmore et al. 1991). Cumulative analysis provides information regarding temporal relationship among different events. (*C*) Physical analysis of meiotic recombination at the *HIS4::LEU2* DSB hotspot. The diagram depicts relevant XhoI restriction sites, the location of the double-strand-break (DSB) site, and the probe used for Southern analysis (Probe 4). Parental "Dad" and "Mom" homologs are distinguished via restriction site polymorphism. Shown is an example of Southern blot analysis of a one-dimensional gel from a meiotic time course with the sizes and identities of each species (e.g., Börner et al. 2004).

K stock solution. Add 570 μL of lysis work solution to each cell pellet. Resuspend the viscous spheroplast pellet by setting a P1000 pipette to 400 μL and pipetting up and down slowly with a filtered tip. Incubate in a water bath for 45 min at 65°C. Vigorously flick the tubes during the first 10 min of incubation until the pellets are completely resuspended. Continue flicking tubes during the incubation.

> *The solution will remain opaque, but no particles or streaks should be visible at the end of incubation. Flicking is done by plucking individual tubes like a guitar string.*

48. Let the samples cool on ice. Add 150 μL of 5 M potassium acetate. Mix by repeated inverting, keep on ice for 15 min, and spin in a microcentrifuge at maximum speed for 20 min at 4°C.

49. Transfer 650 μL of the supernatant into a tube containing 700 μL of 100% ethanol. During transfer, avoid disturbing the pellet. If the pellet becomes loose, centrifuge again, and process fewer samples at a time. Invert the mixture of supernatant and ethanol more than five times to mix.

   > *A sizable white precipitate should be visible in all samples except those collected before genome duplication, typically t <3 h.*

50. Spin in a microcentrifuge at maximum speed for 5 min at room temperature. Discard the supernatant completely, but do not dry the pellet.

51. Add 500 μL of RNase A work solution to the pellet and store overnight at 4°C.

52. The next day, flick the tubes until the pellet separates from the bottom of the tube. Incubate for 10 min at 50°C and flick the tubes frequently until the pellet is completely dissolved. Incubate for another 30 min at 37°C, flicking each tube at least three times.

53. In a fume hood, add 500 μL of phenol:chloroform:isoamylalcohol (25:24:1). Invert the tubes 10 times to mix and spin in a tabletop centrifuge at 13,500 rpm for 20 min. Remove the tubes promptly and transfer the aqueous phase to tubes containing 600 μL of isopropanol. Avoid disturbing the interphase.

54. Invert the tubes several times. Spin in a microcentrifuge at 13,500 rpm.

   > *The precipitate will be smaller than that produced by the ethanol precipitation.*

55. Discard the supernatant. Rinse the pellet with ∼100 μL of 70% (v/v) ethanol. Place the tubes in a heating block set at 42°C and cover them loosely with Saran Wrap.

56. After the pellet has dried completely, add 40 μL of TE buffer and allow the genomic DNA to dissolve overnight at 4°C.

57. The next day, flick the tubes vigorously until the pellet is completely in solution. Collect the genomic DNA solution by spinning in a microcentrifuge at maximum speed for 1 min.

58. To assess the consistency of DNA extraction, analyze 1–2 μL of undigested DNA on a 0.6% agarose gel.

   > *Typical yields are 0.5–1 mg of DNA per mL of meiotic culture.*

Restriction Enzyme Digestion

59. Digest 10–15 μL of genomic DNA from a meiotic culture (Step 57) with 30 U of XhoI in a reaction volume of 80 μL. Incubate for 16 h at 37°C.

   > *For crossover/noncrossover analysis, add 30 U of BamHI or NgoMIV for an additional 6 h at 37°C (Martini et al. 2006; Ahuja and Börner 2011).*

60. Stop the digestion by adding 3 volumes (240 μL) of 96% ethanol/150 mM sodium acetate. Invert the tube and centrifuge for 10 min at 4°C. Discard the supernatant. Wash the pellet with 70% ethanol. Dry completely in a heating block set at 42°C. Add 36 μL of 50 mM Tris, 1 mM EDTA (pH 8.0) and allow to dissolve for >5 h at room temperature.

61. Mix by flicking the tube. Add 8 μL of 6X loading dye and mix again thoroughly.

Electrophoresis and Southern Transfer

The protocol for Southern analysis described here uses transfer to neutral nylon membrane with 10× SSC as transfer buffer. Alkaline transfer to positively charged nylon membrane somewhat reduces the sensitivity in our hands.

62. For one-dimensional gel analysis at the *HIS4::LEU2* hotspot, separate XhoI- or XhoI/BamHI-restricted DNA fragments on a 26-cm 0.6% agarose gel at 1.6 V/cm for 25 h at room temperature. Flick the genomic digest and spin in a tabletop centrifuge at maximum speed for 2 min immediately before loading the gel. Load 18 µL of the digest on to the gel.

    Ethidium bromide in the agarose gel changes the mobility of resected double-strand breaks and should be avoided.

63. Following electrophoresis, submerse the gel together with the gel tray into a large Pyrex pan of deionized water containing 0.3 µg/mL ethidium bromide and agitate gently for 45 min at room temperature. Use gel tray for all subsequent transfers.

64. Transfer the gel into a tray containing 1 L of deionized water.

    Washing gels in a large excess volume of water improves depurination efficiency and ensures quantitative transfer of large DNA molecules as required for quantitative analysis of, for example, double-strand-break bands.

65. Take an ultraviolet (UV) image of the gel to confirm that comparable amounts of DNA were loaded in each lane and that genomic DNA is completed digested.

66. Transfer the gel into a tray containing depurination solution and agitate gently for 20 min or until the bromophenol blue turns yellow.

67. Rinse the gel by submersing it briefly in deionized water. Transfer the gel into neutralization buffer and agitate gently for 30 min.

68. Rinse the gel briefly in deionized water, transfer it into 10× SSC (nucleic acid transfer buffer), and agitate gently for 30 min.

69. Pour new 10× SSC into a large Pyrex pan and place a glass plate across the middle of the pan. Prewet a piece of Whatman 3MM paper (23 × 46 cm) by holding it at diagonally opposite corners and lowering it into 10× SSC. Place the paper across the glass plate. Similarly, prewet a small (gel-sized) piece of Whatman 3MM paper and place it on top of the bridge. Repeat this step with a second small piece of 3MM paper.

    At this and subsequent steps, roll a 5-mL serological pipette across the paper to remove trapped air bubbles.

70. To invert the gel, slide it from the gel tray to a similarly sized Plexiglas plate. Cover with second Plexiglas plate. Reach underneath the lower plate with one hand (thumb pointing away from you), and reach with the other hand across the gel (thumb pointing toward you), grip firmly, and invert.

71. Slide the gel from the Plexiglas onto the pre-wetted small piece of Whatman 3MM paper.

72. Wearing gloves, cut nylon membrane to the size of the gel. Label the back of the membrane in the top left-hand corner using a sharp pencil, which will make the label visible following hybridization and blot exposure.

73. Prewet the nylon membrane in 10× SSC and place it on the gel, starting at one corner of gel. If you have to shift the membrane after placing it, start over with a new membrane.

74. Cover the entire blotting prewetted setup with Saran Wrap to minimize evaporation. Cut along the edge of the nylon membrane with a razor blade. Remove the cut-out center piece of Saran Wrap.

75. Cover the nylon membrane with two pieces of prewetted Whatman 3MM paper.

76. Pile 5 cm of dry paper towels on top of the small pieces of Whatman 3MM paper. Layer a glass plate on top of the paper towels and distribute four bottles with a total weight of ∼0.5 kg evenly on the glass plate.

77. Allow transfer to occur overnight.

78. Following transfer, dismantle the blot, remove the membrane, and neutralize by incubating briefly in 50 mM sodium phosphate (pH 7.2). UV-crosslink the membrane in a UV crosslinker set at 120,000 µJ/cm². Store the membrane between two pieces of Whatman 3MM paper in a resealable freezer bag at room temperature or for long-term storage at −20°C.

Hybridization and Exposure

Use approved procedures while working with radioactivity. Always handle membranes with gloves and avoid creasing or bending (which will result in streaks that obscure the real signal) as well as stretching (which will distort signals).

79. Preheat prehybridization solution and hybridization bottles to 65°C.

80. Pour 30 mL of prehybridization solution into a large hybridization bottle. Roll up the UV-crosslinked nylon membrane and transfer it into the hybridization bottle. Rotate in the hybridization oven rotisserie for at least 30 min at 65°C.

81. Following the supplier's instructions, label 30 ng of Probe 4 using [α-^{32}P]-dCTP (6000 Ci/mmol) and the Prime-It RmT random primer labeling kit.

82. To reduce unspecific background, remove unincorporated nucleotides using a spin column following the manufacturer's instructions.

83. Collect the probe from the spin column and transfer into a 1.5-mL screw-cap tube. Incubate in a heating block for 5 min at 100°C.

    If a microcentrifuge tube with a flip-open cap is used, poke a hole in the lid using a needle to prevent the cap from popping open during the incubation.

84. Prepare the hybridization solution by adding the denatured probe to 30 mL of prehybridization solution preheated to 65°C in a 50-mL Falcon tube. Do not attempt to mix this reagent—it is important to maintain the temperature.

85. Discard the prehybridization solution from the bottle containing the nylon membrane.

86. Pour the hybridization solution into the bottle containing the membrane. Incubate in the rotisserie overnight at 65°C.

87. Discard the hybridization solution in the radioactive waste. Rinse the membrane with 60 mL of Southern wash buffer preheated to 65°C. Discard the wash buffer. Transfer the membrane into a 2-L plastic container and unfold completely.

88. Add 1 L of wash buffer preheated to 65°C and wash in a shaking incubator for 20 min at 65°C. Discard the wash buffer and perform two more identical washes.

89. Remove the nylon membrane from the wash buffer. Pack it between two layers of Saran Wrap, squeeze out excess wash solution by rolling a serological pipette across the membrane, and fold in the edges of the Saran Wrap.

90. Tape the membrane into an exposure cassette. Expose to an erased imaging plate for >30 h. Scan the imaging plate with a phosphorimager at 100 µm/pixel resolution (files are ~20 Mbytes).

91. Using quantitation software such as QuantityOne or ImageJ, quantitate double-strand-break and crossover bands as percent of total DNA, subtracting band-sized background volumes in the same lane from each signal.

RELATED INFORMATION

Tethering epitope-tagged marker proteins via a DNA-binding domain (e.g., TetR-GFP) to an array of cognate binding sites (e.g., *tetOx112*) provides a tool for determining whether chromosomes are homologously paired (Klein et al. 1999).

RECIPES

Fixative Solution for Spheroplasts

Reagent	Final concentration
Paraformaldehyde	3% (w/v)
Sucrose	3.4% (w/v)

Heat 25 mL of deionized water to 60°C. Add 0.75 g of paraformaldehyde and 2 drops of 1 N NaOH, and stir with a magnetic stir bar for 5–10 min. Filter sterilize and adjust the pH to 6.5 using pH paper and 1 M HCl. Add 0.85 g of sucrose. Store for no longer than 1 mo at 4°C.

Loading Dye (6×)

Reagent	Final concentration
Bromophenol blue	0.25% (w/v)
Xylene cyanol	0.25% (w/v)
Ficoll 400	15% (w/v)

Dissolve the reagents in water. Mix thoroughly, filter sterilize, and store at 4°C.

Lysis Buffer for Yeast Spheroplasts

Reagent	Final concentration
Tris–HCl (pH 8.0)	100 mM
EDTA (pH 8.0)	50 mM
SDS	0.5% (w/v)

Make fresh on day of DNA isolation. Store at room temperature.

MES Wash Buffer

Reagent	Final concentration
Sorbitol	1 M
EDTA	1 mM
MgCl$_2$	0.5 mM
MES	0.1 M

Adjust the pH to 6.5 with NaOH.

Prehybridization Solution

Reagent	Final concentration
Sodium phosphate (pH 7.2)	0.25 M
NaCl	0.25 M
EDTA	1 mM
SDS	7%

SSC (10×, pH 7.0)

Reagent	Final concentration
Sodium citrate	300 mM
Sodium chloride	1 M

Adjust pH to 7.0 with HCl. Store at room temperature.

TBS for Immunostaining

Reagent	Quantity (for 1 L)	Final concentration
NaCl	8 g	136 mM
KCl	0.2 g	3 mM
Tris base	3 g	25 mM

Adjust the volume to 1 L and the pH to 8.0. Store at room temperature.

TE Buffer

Reagent	Quantity (for 100 mL)	Final concentration
EDTA (0.5 M, pH 8.0)	0.2 mL	1 mM
Tris-Cl (1 M, pH 8.0)	1 mL	10 mM
H_2O	to 100 mL	

Zymolyase Buffer

Reagent	Quantity (for 100 mL)	Final concentration
Sorbitol (2 M)	50 mL	1 M
Potassium phosphate (pH 7.5)	4.2 mL of 1 M K_2HPO_4	50 mM
	0.8 mL of 1 M KH_2PO_4	
EDTA (0.5 M, pH 8.0)	1 mL	5 mM

Bring to volume with deionized water and filter sterilize. Prepare aliquots and store at −20°C.

REFERENCES

Ahuja JS, Börner GV. 2011. Analysis of meiotic recombination intermediates by two-dimensional gel electrophoresis. *Methods Mol Biol* **745:** 99–116.

Börner GV, Cha RS. 2015. Induction and analysis of synchronous meiotic yeast cultures. *Cold Spring Harb Protoc* doi: 10.1101/pdb.prot085035.

Börner GV, Kleckner N, Hunter N. 2004. Crossover/noncrossover differentiation, synaptonemal complex formation, and regulatory surveillance at the leptotene/zygotene transition of meiosis. *Cell* **117:** 29–45.

Carballo JA, Johnson AL, Sedgwick SG, Cha RS. 2008. Phosphorylation of the axial element protein Hop1 by Mec1/Tel1 ensures meiotic interhomolog recombination. *Cell* **132:** 758–770.

Hollingsworth NM, Goetsch L, Byers B. 1990. The *HOP1* gene encodes a meiosis-specific component of yeast chromosomes. *Cell* **61:** 73–84.

Klein F, Mahr P, Galova M, Buonomo SB, Michaelis C, Nairz K, Nasmyth K. 1999. A central role for cohesins in sister chromatid cohesion, formation of axial elements, and recombination during yeast meiosis. *Cell* **98:** 91–103.

Loidl J. 1995. Meiotic chromosome pairing in triploid and tetraploid *Saccharomyces cerevisiae*. *Genetics* **139:** 1511–1520.

Martini E, Diaz RL, Hunter N, Keeney S. 2006. Crossover homeostasis in yeast meiosis. *Cell* **126:** 285–295.

Padmore R, Cao L, Kleckner N. 1991. Temporal comparison of recombination and synaptonemal complex formation during meiosis in *S. cerevisiae*. *Cell* **66:** 1239–1256.

Smith AV, Roeder GS. 1997. The yeast Red1 protein localizes to the cores of meiotic chromosomes. *J Cell Biol* **136:** 957–967.

Sym M, Engebrecht JA, Roeder GS. 1993. *ZIP1* is a synaptonemal complex protein required for meiotic chromosome synapsis. *Cell* **72:** 365–378.

Tsubouchi T, Roeder GS. 2005. A synaptonemal complex protein promotes homology-independent centromere coupling. *Science* **308:** 870–873.

Westergaard M, von Wettstein D. 1972. The synaptinemal complex. *Annu Rev Genet* **6:** 71–110.

CHAPTER 4

Investigating Filamentous Growth and Biofilm/Mat Formation in Budding Yeast

Paul J. Cullen[1]

Department of Biological Sciences, State University of New York at Buffalo, Buffalo, New York 14260

In response to nutrient limitation, budding yeast can undergo filamentous growth by differentiating into elongated chains of interconnected cells. Filamentous growth is regulated by signal transduction pathways that oversee the reorganization of cell polarity, changes to the cell cycle, and an increase in cell adhesion that occur in response to nutrient limitation. Each of these changes can be easily measured. Yeast can also grow colonially atop surfaces in a biofilm or mat of connected cells. Filamentous growth and biofilm/mat formation require cooperation among individuals; therefore, studying these responses can shed light on the origin and genetic basis of multicellular behaviors. The assays introduced here can be used to study analogous behaviors in other fungal species, including pathogens, which require filamentous growth and biofilm/mat formation for virulence.

INTRODUCTION

Microbial species use diverse strategies to compete for nutrients. Being nonmotile, fungal microorganisms have developed a unique behavior, called filamentous growth, in which cells change their shape and band together in chains or filaments to scavenge for nutrients. Many fungal species can also grow in interconnected mats of cells called biofilms. The budding yeast *Saccharomyces cerevisiae* shows these behaviors, providing a genetically tractable system to study the pathways that control nutrient-dependent foraging. Studies on filamentous growth have provided insights into how eukaryotic cells differentiate and cooperate with each other, and how genetic pathways control fungal pathogenesis. Fungal pathogens require filamentous growth and biofilm formation for virulence.

Filamentous Growth

In budding yeast, filamentous growth is triggered by nutrient limitation (Gimeno et al. 1992; Cullen and Sprague 2012). In particular, depletion of glucose or fixed nitrogen induces filamentous growth in both haploid and diploid cells (Cullen and Sprague 2002). The balance of the cell's nutrient levels is critical for commitment to the filamentous growth program: complete removal of nutrients triggers entry into stationary phase (G_0) in both haploids and diploids, and in diploids, depletion of both carbon and nitrogen sources induces sporulation (Neiman 2011). The decision of whether or not to undergo filamentous growth, and the coordination of the response itself, is regulated by signal transduction pathways. Among the pathways that regulate filamentous growth are the RAS protein kinase A (PKA) pathway (Gimeno et al. 1992) and a mitogen-activated protein kinase (MAPK) pathway called the filamentous growth pathway (Roberts and Fink 1994). These pathways regulate

[1]Correspondence: pjcullen@buffalo.edu

Copyright © Cold Spring Harbor Laboratory Press; all rights reserved
Cite this introduction as *Cold Spring Harb Protoc*; doi:10.1101/pdb.top077495

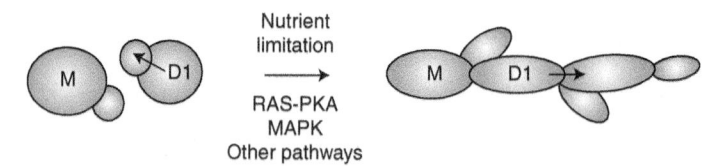

		Nutrient limitation → RAS-PKA MAPK Other pathways		
Cell shape:	Round		Elongated (G$_2$ extension)
Polarity:	Axial budding		Distal-unipolar budding (Bud8)
Adhesion:	Cells separate		Cells remain attached (Flo11)

FIGURE 1. Morphological changes that occur during filamentous growth in yeast. Under nutrient-rich conditions (*left*), yeast-form cells are round in shape and produce daughters that fully separate from their mothers. In haploid cells (shown), daughter cells (D1) bud back toward the mother cell (M) by axial budding (arrow toward *left*). Under nutrient-limiting conditions (*right*), cells become elongated and remain attached through Flo11. Daughter cells bud away from the mother cell (arrow toward *right*) by distal-unipolar budding by using the distal landmark Bud8. Signal transduction pathways (including RAS-PKA and MAPK) regulate these changes.

changes in gene expression and alter the activity of target proteins, leading to the construction of a new cell type.

Although filamentous growth is thought to be a complex differentiation response, it includes three easily observable changes in cells (Fig. 1). First, cell polarity is altered. Cell polarity is determined by bud-site-selection proteins, which mark the different poles of the cell (Park and Bi 2007) and direct growth in different directions. Haploid cells bud in an axial pattern, and diploid cells in a bipolar pattern (Chant and Pringle 1995). During filamentous growth, both haploid and diploid cells switch to a distal-unipolar pattern [Fig. 1; (Gimeno et al. 1992; Roberts and Fink 1994; Cullen and Sprague 2002)] by using the distal-pole landmark Bud8 (Harkins et al. 2001). Second, an increase cell length occurs, resulting from a delay in the G$_2$ phase of the cell cycle [Fig. 1 (Kron et al. 1994)]. Finally, cells remain attached to each other (Fig. 1). Unlike yeast-form cells that fully separate after cytokinesis, filamentous cells retain connections between proteins and carbohydrates on the cell wall. One such protein, Flo11, is a mucin-like flocculin and the major cell adhesion molecule that regulates cell–cell adherence (Lambrechts et al. 1996; Lo and Dranginis 1998; Guo et al. 2000; Halme et al. 2004). Together, these changes contribute to the filamentous growth response.

Biofilm/Mat Formation

Budding yeast can also undergo biofilm/mat formation (Reynolds and Fink 2001; Vachova et al. 2011), an ancient microbial response that involves regulating colonial growth at the level of cellular connectivity. In addition to its role in regulating filamentous growth, Flo11 is also required for biofilm/mat formation. Flo11 specifically regulates the complex colony morphology of biofilm/mats (Granek and Magwene 2010), their rim-and-spoke pattern (Reynolds and Fink 2001), and the expansion of mats across surfaces (Reynolds and Fink 2001). Flo11 mediates cellular "sliding" in part because the protein is shed from cells, thereby attenuating cell adhesion and potentially conferring a cellular lubrication property (Karunanithi et al. 2010). Filamentous growth and biofilm formation have distinct regulatory features (Sarode et al. 2011; Ryan et al. 2012), yet are related in that both occur in response to nutrient limitation and require overlapping signaling pathways and target proteins. Under some conditions, filamentous growth and biofilm formation occur in concert, which indicates that these behaviors may represent aspects of a global foraging response (Karunanithi et al. 2012).

TECHNICAL APPROACHES

The associated protocols describe assays to measure and quantitate the changes that occur during filamentous growth and biofilm formation in yeast. The assays are designed to distinguish between phenotypes that occur in high- and low-nutrient environments and between wild-type and mutants strains. A key feature of several of these assays is their simplicity. The plate-washing assay (see Protocol 1: The Plate-Washing Assay: A Simple Test for Filamentous Growth in Budding Yeast [Cullen 2015a]) and biofilm/mat assay (see Protocol 2: Biofilm/Mat Assays for Budding Yeast [Cullen 2015b]) require minimal reagents and measure changes in colony patterns that are visible to the naked eye and are interpretable without specialized equipment. In the single-cell invasive growth assay and pseudohyphal growth assay (both presented in Protocol 3: Evaluating Yeast Filamentous Growth at the Single-Cell Level [Cullen 2015c]), microscopic examination of cells allows quantitation of changes in budding pattern and cell length that occur during filamentous growth.

Three related assays measure the activity of the MAPK pathway that controls filamentous growth (all are presented in Protocol 4: Evaluating the Activity of the Filamentous Growth Mitogen-Activated Protein Kinase Pathway in Yeast [Cullen 2015d]). First, detecting phosphorylated MAPKs in yeast by western blotting using commercially available antibodies provides a direct measure of MAPK activity. Second, the pectinase assay measures the enzymatic activity of a secreted pectinase that is a target of the filamentous growth pathway. Finally, during filamentous growth, Flo11 (mentioned above) and other mucin-like proteins are shed from cells. Measuring Flo11 shedding provides information about protein levels and biofilm/mat patterning. Another mucin-like protein, Msb2, is the signaling glycoprotein that regulates the filamentous growth pathway (Cullen et al. 2004). Cleavage and release of the extracellular inhibitory domain of Msb2 is required for MAPK activity (Vadaie et al. 2008), which also corresponds to filamentous growth MAPK activity. Secretion profiling of yeast mucin-like proteins provides information about the role of MAPKs in the regulation of filamentous growth.

Finally, the *FLO11* gene is regulated by a large and complex promoter where multiple signals converge (Rupp et al. 1999). Measuring changes the expression of *FLO11* (using techniques not described here) can provide a diagnostic readout of changes in the filamentous growth response.

Most yeast strains used in the laboratory do not show filamentous growth because they have acquired mutations as a result of genetic manipulation (Liu et al. 1996). The filamentous (Σ1278b) background is typically used to study filamentous growth (Gimeno et al. 1992). The genome sequence of the Σ1278b background is available (Dowell et al. 2010) as it is a collection of ordered deletion mutants (Ryan et al. 2012). These tools facilitate the genetic analysis of this growth response.

CONCLUSIONS

The current picture of filamentous growth is a complex one, in which multiple pathways and hundreds of targets coordinate a highly integrated response that we are only beginning to understand. Future studies of filamentous growth will aid in understanding of the genetic basis of cell differentiation, development, and the regulation of multicellularity in eukaryotes. The assays described in the associated protocols are attractive in terms of their simplicity and potential use as teaching tools. Their versatility furthermore allows analysis of filamentous growth and biofilm formation in diverse fungal species including pathogens.

ACKNOWLEDGMENTS

P.J.C. is supported by a U.S. Public Health Service grant (GM098629).

REFERENCES

Chant J, Pringle JR. 1995. Patterns of bud-site selection in the yeast *Saccharomyces cerevisiae*. *J Cell Biol* 129: 751–765.

Cullen PJ. 2015a. The plate-washing assay: A simple test for filamentous growth in budding yeast. *Cold Spring Harb Protoc* doi: 10.1101/pdb.prot085068.

Cullen PJ. 2015b. Biofilm/mat assays for budding yeast. *Cold Spring Harb Protoc* doi: 10.1101/pdb.prot085076.

Cullen PJ. 2015c. Evaluating yeast filamentous growth at the single-cell level. *Cold Spring Harb Protoc* doi: 10.1101/pdb.prot085084.

Cullen PJ. 2015d. Evaluating the activity of the filamentous growth mitogen-activated protein kinase pathway in yeast. *Cold Spring Harb Protoc* doi: 10.1101/pdb.prot085092.

Cullen PJ, Sabbagh W Jr, Graham E, Irick MM, van Olden EK, Neal C, Delrow J, Bardwell L, Sprague GF Jr. 2004. A signaling mucin at the head of the Cdc42- and MAPK-dependent filamentous growth pathway in yeast. *Genes Dev* **18**: 1695–1708.

Cullen PJ, Sprague GF Jr. 2002. The roles of bud-site-selection proteins during haploid invasive growth in yeast. *Mol Biol Cell* **13**: 2990–3004.

Cullen PJ, Sprague GF Jr. 2012. The regulation of filamentous growth in yeast. *Genetics* **190**: 23–49.

Dowell RD, Ryan O, Jansen A, Cheung D, Agarwala S, Danford T, Bernstein DA, Rolfe PA, Heisler LE, Chin B, et al. 2010. Genotype to phenotype: A complex problem. *Science* **328**: 469.

Gimeno CJ, Ljungdahl PO, Styles CA, Fink GR. 1992. Unipolar cell divisions in the yeast *S. cerevisiae* lead to filamentous growth: Regulation by starvation and RAS. *Cell* **68**: 1077–1090.

Granek JA, Magwene PM. 2010. Environmental and genetic determinants of colony morphology in yeast. *PLoS Genet* **6**: e1000823.

Guo B, Styles CA, Feng Q, Fink GR. 2000. A *Saccharomyces* gene family involved in invasive growth, cell-cell adhesion, and mating. *Proc Natl Acad Sci* **97**: 12158–12163.

Halme A, Bumgarner S, Styles C, Fink GR. 2004. Genetic and epigenetic regulation of the *FLO* gene family generates cell-surface variation in yeast. *Cell* **116**: 405–415.

Harkins HA, Page N, Schenkman LR, De Virgilio C, Shaw S, Bussey H, Pringle JR. 2001. Bud8p and Bud9p, proteins that may mark the sites for bipolar budding in yeast. *Mol Biol Cell* **12**: 2497–2518.

Karunanithi S, Joshi J, Chavel C, Birkaya B, Grell L, Cullen PJ. 2012. Regulation of mat responses by a differentiation MAPK pathway in *Saccharomyces cerevisiae*. *PLoS ONE* **7**: e32294.

Karunanithi S, Vadaie N, Chavel CA, Birkaya B, Joshi J, Grell L, Cullen PJ. 2010. Shedding of the mucin-like Flocculin Flo11p reveals a new aspect of fungal adhesion regulation. *Curr Biol* **20**: 1389–1395.

Kron SJ, Styles CA, Fink GR. 1994. Symmetric cell division in pseudohyphae of the yeast *Saccharomyces cerevisiae*. *Mol Biol Cell* **5**: 1003–1022.

Lambrechts MG, Bauer FF, Marmur J, Pretorius IS. 1996. Muc1, a mucin-like protein that is regulated by Mss10, is critical for pseudohyphal differentiation in yeast. *Proc Natl Acad Sci* **93**: 8419–8424.

Liu H, Styles CA, Fink GR. 1996. *Saccharomyces cerevisiae* S288C has a mutation in *FLO8*, a gene required for filamentous growth. *Genetics* **144**: 967–978.

Lo WS, Dranginis AM. 1998. The cell surface flocculin Flo11 is required for pseudohyphae formation and invasion by *Saccharomyces cerevisiae*. *Mol Biol Cell* **9**: 161–171.

Neiman AM. 2011. Sporulation in the budding yeast *Saccharomyces cerevisiae*. *Genetics* **189**: 737–765.

Park HO, Bi E. 2007. Central roles of small GTPases in the development of cell polarity in yeast and beyond. *Microbiol Mol Biol Rev* **71**: 48–96.

Reynolds TB, Fink GR. 2001. Bakers' yeast, a model for fungal biofilm formation. *Science* **291**: 878–881.

Roberts RL, Fink GR. 1994. Elements of a single MAP kinase cascade in *Saccharomyces cerevisiae* mediate two developmental programs in the same cell type: Mating and invasive growth. *Genes Dev* **8**: 2974–2985.

Rupp S, Summers E, Lo HJ, Madhani H, Fink G. 1999. MAP kinase and cAMP filamentation signaling pathways converge on the unusually large promoter of the yeast FLO11 gene. *Embo J* **18**: 1257–1269.

Ryan O, Shapiro RS, Kurat CF, Mayhew D, Baryshnikova A, Chin B, Lin ZY, Cox MJ, Vizeacoumar F, Cheung D, et al. 2012. Global gene deletion analysis exploring yeast filamentous growth. *Science* **337**: 1353–1356.

Sarode N, Miracle B, Peng X, Ryan O, Reynolds TB. 2011. Vacuolar protein sorting genes regulate mat formation in *Saccharomyces cerevisiae* by Flo11p-dependent and -independent mechanisms. *Eukaryot Cell* **10**: 1516–1526.

Vachova L, Stovicek V, Hlavacek O, Chernyavskiy O, Stepanek L, Kubinova L, Palkova Z. 2011. Flo11p, drug efflux pumps, and the extracellular matrix cooperate to form biofilm yeast colonies. *J Cell Biol* **194**: 679–687.

Vadaie N, Dionne H, Akajagbor DS, Nickerson SR, Krysan DJ, Cullen PJ. 2008. Cleavage of the signaling mucin Msb2 by the aspartyl protease Yps1 is required for MAPK activation in yeast. *J Cell Biol* **181**: 1073–1081.

Protocol 1

The Plate-Washing Assay: A Simple Test for Filamentous Growth in Budding Yeast

Paul J. Cullen[1]

Department of Biological Sciences, State University of New York at Buffalo, Buffalo, New York 14260

Filamentous growth is a foraging response that occurs in fungal species. It allows fungal pathogens to invade cells and tissues of a host organism. Budding yeast undergoes filamentous growth and can invade semisolid agar plates, penetrating the agar surface. These cells cannot be removed by rinsing with water and form an invasive scar. The plate-washing assay is an easy first test for filamentous growth and is performed at low cost with minimal reagents. The assay is versatile: It can be used as a teaching tool, is amenable to high-throughput genetic analysis, and is used to evaluate filamentous growth in different fungal species, including pathogens like *Candida albicans*.

MATERIALS

It is essential that you consult the appropriate Material Safety Data Sheets and your institution's Environmental Health and Safety Office for proper handling of equipment and hazardous materials used in this protocol.

RECIPES: Please see the end of this protocol for recipes indicated by <R>. Additional recipes can be found online at http://cshprotocols.cshlp.org/site/recipes.

Reagents

Distilled water, sterile
Synthetic defined (SD) medium <R>
Yeast strains of interest

The Σ1278b background undergoes filamentous growth (Gimeno et al. 1992). Commonly used laboratory strains have lost the ability to undergo filamentous growth (Liu et al. 1996). A mutant defective for agar invasion (e.g., flo11Δ) should be included in the assay as a negative control. Maintain stocks on SD plates (not YEPD).

YEPD agar plates <R>

Equipment

Digital camera
Flat toothpicks, sterile
Glass tubes with metal tops, sterile
ImageJ software (http://imagej.nih.gov; Schneider et al. 2012)
Incubator set at 30°C
Inoculation loop or long wooden toothpick, sterile
Light box

---

[1]Correspondence: pjcullen@buffalo.edu

Copyright © Cold Spring Harbor Laboratory Press; all rights reserved
Cite this protocol as *Cold Spring Harb Protoc*; doi:10.1101/pdb.prot085068

Light microscope with differential interference contrast (DIC) optics and 100× objective
Microcentrifuge
Microcentrifuge tubes, sterile
Microscope oil
Microscope slides and coverslips
Pipette tips, sterile
Shaking incubator or a shaker in a 30°C room

METHOD

1. Using a long wooden toothpick or inoculation loop, inoculate yeast from freshly patched synthetic defined (SD) plates (from freezer stocks or a plate stored at 4°C) into 5 mL of SD medium in a glass tube with a metal top and grow until saturation.

   We typically grow cultures for 16 h (overnight) at 30°C with shaking at 5g. SD medium reduces clumpiness by minimizing expression of FLO11. Cell aggregation complicates quantitating cell density and culture manipulation.

2. Remove 1 mL of cells from the overnight culture. Pellet the cells by centrifugation at 16,000g for 1 min at 20°C. Resuspend the cells in 1 mL of sterile distilled H_2O.

3. Spot 10 µL of cells directly onto YEPD agar plates. If testing multiple strains (and/or serial dilutions from tubes) on a single plate, spot cells equidistant from each other and the edges of the plate.

   Colonies that have more access to nutrients (e.g., along an edge) will invade better than colonies in the center of the plate. Invasive growth occurs more robustly when fewer colonies are spotted.

   A quick and commonly used alternative to spotting is to patch cells in quadrants on YEPD plates using a toothpick. Patching is quicker, whereas spotting is more quantitative.

4. Incubate plates for 2–10 d at 30°C until invasive growth occurs.

   See Discussion.

5. Photograph the colonies.

   See Figure 1.

6. Wash the plates in a stream of water. Use the same flow rate and water temperature for all plates. Wash each colony for the same period of time; do not wash more invasive colonies longer.

   This step can be separated into a "soft" and "hard" wash. For the soft wash, use a stream of water. For the hard wash, rub your finger over the colonies to remove cells from the plate. Perform the soft wash first. If informative, photograph the plate after the soft wash.

7. Photograph the plates by transmitted light, using a light box, to reveal details of the invasive pattern.

   The invasive scar will be at a different focal plane than the colony. Change the focus of the camera accordingly.

   See Figure 1.

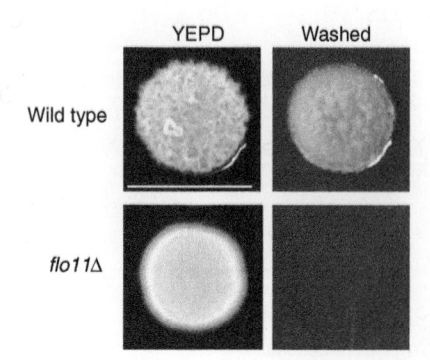

FIGURE 1. Sample results from a plate-washing assay. Colonies of the Σ1278b background grown on YEPD have a ruffled appearance (*top left*). Washing the plate in a stream of water reveals an invasive scar (*top right*). A mutant lacking the *FLO11* gene (*flo11Δ*) is defective for filamentous growth. It has a smooth pattern (*bottom left*) and is defective for agar invasion (*bottom right*). Bar, 1 cm.

8. Examine the invaded cells by microscopy. Gently scrape the invasive scar with a toothpick to excise invaded cells. Place the cells onto a slide spotted with 7 µL of H_2O and add a coverslip. Examine the cells by DIC at 100× magnification by oil immersion microscopy.

> Cells undergoing filamentous growth form chains of elongated and connected cells. The localization of proteins in filamentous cells can be determined with GFP fusion proteins and fluorescent microscopy. Additionally, excised cells can be evaluated by molecular techniques such as RNA-seq, DNA microarray analysis, or real-time polymerase chain reaction (RT-PCR).

9. Quantitate agar invasion.

   i. Analyze photographs of colonies and washed plates by densitometry using ImageJ software. Use the invert function and subtract background values.

   ii. Examine multiple replicates and perform statistical analysis to assess whether differences in agar invasion are statistically significant.

   > See Zupan and Raspor (2008).

DISCUSSION

Agar invasion occurs after 2–10 d depending on the strain (different Σ1278b derivatives show different degrees of agar invasion) and the number of colonies spotted per plate. Washing multiple plates over a time series will establish the optimal incubation time. Some mutants/strains show hyperinvasive growth, which can be measured by washing plates at early time points (1–2 d), before the invasive growth seen in wild-type cells. Even cells that are defective for agar invasion (e.g., cells lacking a MAPK pathway) show agar invasion over an extended time period.

Colonies become ruffled over time because of cell adhesion mediated by *FLO11* (Granek and Magwene 2010). Colony ruffling typically correlates with invasive growth (Fig. 1). Longer incubation (1–2 wk) results in exaggerated colony ruffling, robust invasive growth, and an increase in cell length, which can facilitate quantitation of these phenotypes.

RELATED INFORMATION

This assay has been adapted for high-throughput genetic analysis (Ryan et al. 2012) and for measuring filamentous growth in *Candida albicans* (Warenda et al. 2003).

RECIPES

Synthetic Defined (SD) Medium

2% glucose
6.7 g/L yeast nitrogen base without amino acids
Amino acids:
    20 mg/L histidine
    120 mg/L leucine
    60 mg/L lysine
    20 mg/L arginine
    20 mg/L tryptophan
    20 mg/L tyrosine
    40 mg/L threonine
    20 mg/L methionine
    50 mg/L phenylalanine
20 mg/L uracil
20 mg/L adenine

20 g/L agar (omit for liquid medium)

Include additional amino acids that may be required for the growth of an auxotrophic strain. Prepare in H$_2$O and sterilize by autoclaving. For agar plates, fill sterile Petri dishes with ~25 mL of autoclaved medium.

Yeast Extract-Peptone-Dextrose Growth Medium (YEPD)

Reagent	Quantity (for 1 L)	Final concentration (w/v)
Bacto peptone	20 g	2%
Yeast extract	10 g	1%
Dextrose	20 g	2%
H$_2$O	to 1 L	

Sterilize by autoclaving.

YEPD Agar Plates

Reagent	Quantity
Bacto-agar (2%)	20 g
YEPD liquid medium	1 L

Add Bacto-agar to YEPD liquid medium in a 2-L flask and autoclave. Fill sterile Petri dishes with 30–40 mL of autoclaved medium.

ACKNOWLEDGMENTS

P.J.C. is supported by a U.S. Public Health Service grant (GM098629).

REFERENCES

Gimeno CJ, Ljungdahl PO, Styles CA, Fink GR. 1992. Unipolar cell divisions in the yeast *S. cerevisiae* lead to filamentous growth: Regulation by starvation and RAS. *Cell* 68: 1077–1090.

Granek JA, Magwene PM. 2010. Environmental and genetic determinants of colony morphology in yeast. *PLoS Genet* 6: e1000823.

Liu H, Styles CA, Fink GR. 1996. *Saccharomyces cerevisiae* S288C has a mutation in *FLO8*, a gene required for filamentous growth. *Genetics* 144: 967–978.

Ryan O, Shapiro RS, Kurat CF, Mayhew D, Baryshnikova A, Chin B, Lin ZY, Cox MJ, Vizeacoumar F, Cheung D, et al. 2012. Global gene deletion analysis exploring yeast filamentous growth. *Science* 337: 1353–1356.

Schneider CA, Rasband WS, Eliceiri KW. 2012. NIH Image to ImageJ: 25 years of image analysis. *Nat Methods* 9: 671–675.

Warenda AJ, Kauffman S, Sherrill TP, Becker JM, Konopka JB. 2003. *Candida albicans* septin mutants are defective for invasive growth and virulence. *Infect Immun* 71: 4045–4051.

Zupan J, Raspor P. 2008. Quantitative agar-invasion assay. *J Microbiol Methods* 73: 100–104.

Protocol 2

Biofilm/Mat Assays for Budding Yeast

Paul J. Cullen[1]

Department of Biological Sciences, State University of New York at Buffalo, Buffalo, New York 14260

Many microbial species form biofilms/mats under nutrient-limiting conditions, and fungal pathogens rely on this social behavior for virulence. In budding yeast, mat formation is dependent on the mucin-like flocculin Flo11, which promotes cell-to-cell and cell-to-substrate adhesion in mats. The biofilm/mat assays described here allow the evaluation of the role of Flo11 in the formation of mats. Cells are grown on surfaces with different degrees of rigidity to assess their expansion and three-dimensional architecture, and the cells are also exposed to plastic surfaces to quantify their adherence. These assays are broadly applicable to studying biofilm/mat formation in microbial species.

MATERIALS

It is essential that you consult the appropriate Material Safety Data Sheets and your institution's Environmental Health and Safety Office for proper handling of equipment and hazardous materials used in this protocol.

RECIPES: Please see the end of this protocol for recipes indicated by <R>. Additional recipes can be found online at http://cshprotocols.cshlp.org/site/recipes.

Reagents

Crystal violet (1% w/v in H_2O)
Distilled water, sterile
Yeast strains of interest
 The Σ1278b background undergoes biofilm/mat formation (Gimeno et al. 1992). Commonly used laboratory strains have lost the ability to undergo biofilm/mat formation (Liu et al. 1996). A mutant defective for biofilm/mat formation (e.g., flo11Δ) should be included in the assay as a negative control.

YEPD agar plates with varying agar concentrations (e.g., 0.3%, 2%, and 4%) <R>
 Do not invert 0.3% agar plates.

Equipment

Digital camera
Flat-end toothpicks, sterile
ImageJ software (http://imagej.nih.gov; Schneider et al. 2012)
Incubator set at 30°C
Light microscope with 100× objective
Nitrocellulose filters (circular)

---

[1]Correspondence: pjcullen@buffalo.edu

Plastic wrap (e.g., Saran Wrap)
Polystyrene or polypropylene plate (96-well)
Spectrophotometer

METHOD

1. Using a sterile toothpick, transfer cells of the yeast strain of interest to each of the plates listed below. Gently touch the toothpick containing cells to the center of each plate. On a separate set of plates, transfer cells of mutant strain defective for mat formation (a negative control; e.g., *flo11Δ*).

   - YEPD plate with 0.3% agar

     0.3% agar is optimal to observe colony spreading. On these plates, colonies can show a radial spoke pattern.

   - YEPD plate with 2% agar

     2% agar is optimal to observe mat architecture. It also allows assessment of invasive growth by the plate-washing assay (see Protocol 1: The Plate-Washing Assay: A Simple Test for Filamentous Growth in Budding Yeast [Cullen 2015]).

   - YEPD plate with 4% agar

     4% agar is optimal to observe colony ruffling and z-axis growth (see Fig. 1). The high surface rigidity reduces expansion in the plane of the xy-axis and promotes formation of dense architecturally complex mats that grow upward in the plane of the z-axis.

   - YEPD plate with 4% agar plate with a nitrocellulose filter placed on top

     This plate maximizes complex colony morphology.

2. Maintain the plates in a 30°C incubator in a location where vibrations are minimized. Examine mat expansion visually starting at 24 h and continuing over the course of several weeks.

   Mats can be defined by their FLO11-dependent colony architecture and degree of expansion in the x-, y-, and z-axes by visual inspection (e.g., see Fig. 1).

3. Photograph the biofilms.

   i. Measure the mat areas by photographing the plates and analyzing the photos with ImageJ software (http://imagej.nih.gov; Schneider et al. 2012).

   ii. Examine the ruffled morphology characteristic of mats by photography with a digital camera at 1× to 5× magnification.

4. To separate adherent from nonadherent cells, use an overlay adhesion assay (Reynolds et al. 2008).

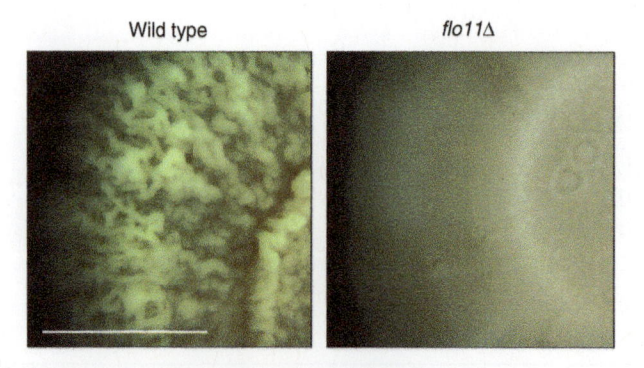

FIGURE 1. Sample results from a biofilm/mat assay. When grown on medium containing a high agar concentration (YEPD medium containing 4% agar), wild-type cells (*left*) produce a ruffled pattern, whereas *flo11Δ* mutant cells (*right*) do not. Bar, 1 cm. Magnification, 3×.

i. Place plastic wrap over the agar plate, and gently pull the plastic wrap off the plate.
   When performing this action, the mat may or may not be removed from the agar. This distinguishes cells that adhere to the plastic from cells that adhere to the agar.

ii. Photograph both the plate and the wrap.

iii. (Optional) Remove cells for additional analysis.

5. Quantify the degree of cell adhesion by measuring the adhesion of yeast cells to a plastic surface.

   i. At various times after mat formation has occurred, remove cells using a toothpick. Resuspend in 500 μL of H_2O.
      Always compare the cells of interest with flo11Δ cells (a negative control).

   ii. Adjust the optical density of the cells to A_{600} = 2.0.

   iii. Add 100 μL of the cell suspension to a 96-well polystyrene or polypropylene plate.

   iv. Incubate the cells for 4 h at 25°C to allow the cells to settle to the bottom of the wells.

   v. Add 100 μL of 1% crystal violet to each well. Incubate for 20 min at 25°C.

   vi. Wash wells five times with water and photograph adherent cells with a digital camera or by microscopy at 10×.
       Wild-type cells will adhere to the plastic surface and will be violet in color. In contrast, flo11Δ cells will not adhere, leaving a transparent plastic surface that is relatively free of cells.

RELATED INFORMATION

For more discussion on the biology of biofilm/mat formation and applications of the assays described here, see Reynolds and Fink (2001), Blankenship and Mitchell (2006), and Karunanithi et al. (2012).

RECIPES

YEPD Agar Plates

Reagent	Quantity
Bacto agar (2%)	20 g
YEPD liquid medium	1 L

Add Bacto agar to YEPD liquid medium in a 2-L flask and autoclave. Fill sterile Petri dishes with 30–40 mL of autoclaved medium.

Yeast Extract-Peptone-Dextrose Growth Medium (YEPD)

Reagent	Quantity (for 1 L)	Final concentration (w/v)
Bacto peptone	20 g	2%
Yeast extract	10 g	1%
Dextrose	20 g	2%
H_2O	to 1 L	

Sterilize by autoclaving.

ACKNOWLEDGMENTS

The author thanks T. Reynolds for engaging discussions about biofilm/mat form growth. P.J.C. is supported by a U.S. Public Health Service grant (GM098629).

REFERENCES

Blankenship JR, Mitchell AP. 2006. How to build a biofilm: A fungal perspective. *Curr Opin Microbiol* **9**: 588–594.

Cullen PJ. 2015. The plate-washing assay: A simple test for filamentous growth in budding yeast. *Cold Spring Harb Protoc* doi: 10.1101/pdb.prot085068.

Gimeno CJ, Ljungdahl PO, Styles CA, Fink GR. 1992. Unipolar cell divisions in the yeast *S. cerevisiae* lead to filamentous growth: Regulation by starvation and RAS. *Cell* **68**: 1077–1090.

Karunanithi S, Joshi J, Chavel C, Birkaya B, Grell L, Cullen PJ. 2012. Regulation of mat responses by a differentiation MAPK pathway in *Saccharomyces cerevisiae*. *PLoS ONE* **7**: e32294.

Liu H, Styles CA, Fink GR. 1996. *Saccharomyces cerevisiae* S288C has a mutation in *FLO8*, a gene required for filamentous growth. *Genetics* **144**: 967–978.

Reynolds TB, Fink GR. 2001. Bakers' yeast, a model for fungal biofilm formation. *Science* **291**: 878–881.

Reynolds TB, Jansen A, Peng X, Fink GR. 2008. Mat formation in *Saccharomyces cerevisiae* requires nutrient and pH gradients. *Eukaryot Cell* **7**: 122–130.

Schneider CA, Rasband WS, Eliceiri KW. 2012. NIH image to ImageJ: 25 years of image analysis. *Nat Methods* **9**: 671–675.

Protocol 3

Evaluating Yeast Filamentous Growth at the Single-Cell Level

Paul J. Cullen[1]

Department of Biological Sciences, State University of New York at Buffalo, Buffalo, New York 14260

Budding yeast can undergo filamentous growth in response to nutrient limitation. Filament formation can be examined under glucose-limiting conditions by the single-cell invasive growth assay, or under nitrogen-limiting conditions with the pseudohyphal growth assay, both described here. The single-cell assay allows robust quantitation of changes in budding pattern and cell length, and most cells in the population show the response. The pseudohyphal growth assay reveals filamentous patterns in larger microcolonies and adjoining subpopulations of cells. Historically, the single-cell assay has been used to study filamentous growth in haploid cells and the pseudohyphal growth assay in diploid cells. However, both assays can be used in either cell type.

MATERIALS

It is essential that you consult the appropriate Material Safety Data Sheets and your institution's Environmental Health and Safety Office for proper handling of equipment and hazardous materials used in this protocol.

RECIPES: Please see the end of this protocol for recipes indicated by <R>. Additional recipes can be found online at http://cshprotocols.cshlp.org/site/recipes.

Reagents

Distilled H_2O (dH_2O), sterile
Glucose (50%) (optional; see Step 5)
Plates for single-cell invasive growth assay or pseudohyphal growth assay (see Step 5)
 Isobutanol agar plates

  These are SD agar plates with 1% isobutanol or isoamyl alcohol added after autoclaving.

 S-GLU agar plates

  These are SD agar plates with no glucose added.

 SLAHD agar plates <R>

  Prepare all plates within 24 h of the experiment, as the level of moisture in the plates is critical for optimal filamentous growth.

Synthetic defined (SD) medium <R>
Yeast strains of interest

 The Σ1278b background undergoes filamentous growth (Gimeno et al. 1992). Commonly used laboratory strains have lost the ability to undergo filamentous growth (Liu et al. 1996). A MAPK pathway mutant (e.g., ste11Δ) should be included in the assay as a negative control. Maintain stocks on synthetic defined (SD) plates (not YEPD).

---

[1]Correspondence: pjcullen@buffalo.edu

Copyright © Cold Spring Harbor Laboratory Press; all rights reserved
Cite this protocol as *Cold Spring Harb Protoc*; doi:10.1101/pdb.prot085084

Chapter 4

Equipment

These assays require the limitation of glucose or nitrogen. Use clean dishware and glass spreaders, as trace glucose or nitrogen contamination suppresses the filamentous morphology.

Coverslips
Flat toothpicks, sterile
Glass plate spreader, sterile
Glass tubes with metal tops, sterile
Incubator set at 30°C
Inoculation loop or long wooden toothpick, sterile
Light microscope with 100× objective
Manual dissection microscope (e.g., SPOREPLAY, Singer Instruments) (optional; see Step 5)
Microcentrifuge
Microcentrifuge tubes, sterile
Microscope oil (optional; see Step 7)
Parafilm
Pipette tips, sterile
Shaking incubator or a shaker in a 30°C room

METHOD

1. Using a long wooden toothpick or inoculation loop, inoculate yeast from freshly patched SD plates (from freezer stocks or a plate stored at 4°C) into 5 mL of SD medium in a glass tube with a metal top and grow until saturation is reached.

   We typically grow cultures for 16 h (overnight) at 30°C with shaking at 5g. Synthetic (SD) medium reduces clumpiness, and clumps of cells are a poor starting material for the single-cell assay.

2. Remove 1 mL of cells from the overnight culture. Pellet the cells by centrifugation at 16,000g for 1 min at 20°C.

3. Wash the cells in 1 mL of sterile dH$_2$O. Centrifuge. Repeat. Resuspend the cells in 1 mL of sterile dH$_2$O.

4. Dilute 50 µL of cells in 1 mL dH$_2$O.

5. Spread the cells on plates as follows. Alternatively, place individual cells onto the appropriate plates under a dissection microscope.

   - For the single-cell invasive growth assay, spread 50 µL of cells from the dilution onto S-GLU agar plates using a sterile glass spreader. Allow the plates to dry for 5 min, and then proceed to Step 6.

     Isobutanol can been used to stimulate filamentous growth (Lorenz et al. 2000; Chen and Fink 2006). Grow cells on isobutanol agar plates prepared with and without glucose.

     To visualize yeast-form morphology (within the region of diffused glucose) and filamentous-form morphology (outside the region of diffused glucose) (Fig. 1), spot 10 µL of 50% glucose at the center of the plate after the cells are spread.

   - For the pseudohyphal growth assay, spread 50 µL of cells from the dilution onto SLAHD agar plates using a glass spreader. Allow the plates to dry for 5 min, and then proceed to Step 6.

6. Wrap agar plates with lids in Parafilm and incubate them without inversion for 16 h at 30°C for the single-cell invasive growth assay or for 48 h for the pseudohyphal growth assay.

7. Remove the plates from the incubator and allow them to equilibrate to room temperature. Place the plates on the microscope stage and examine at 20×. Place a coverslip on the plates, directly over the cells, and observe at 40×, or at 100× using oil immersion.

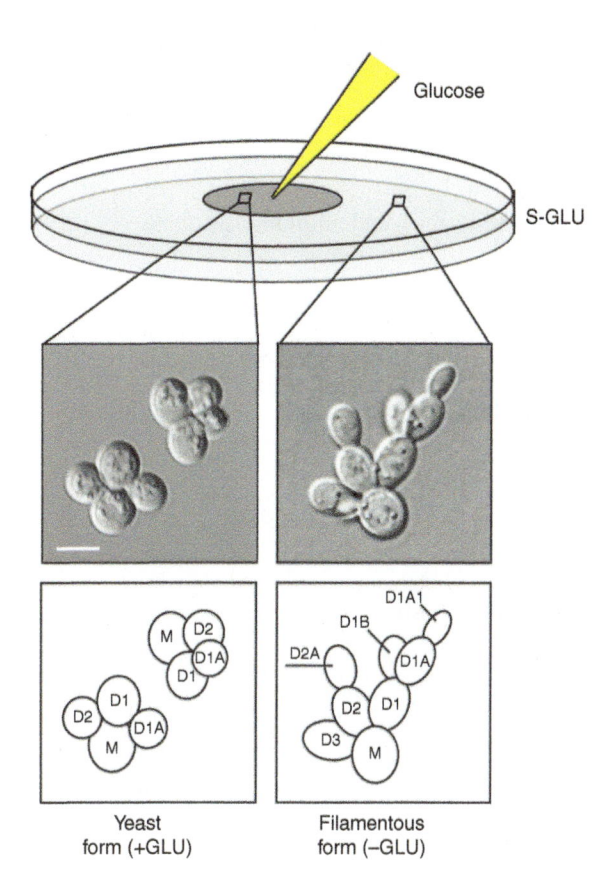

FIGURE 1. The single-cell invasive growth assay. Cells grown on synthetic media lacking glucose undergo filamentous growth within the first few cell divisions. Round mother cells (M) produce elongated progeny (D1, D2, and D3) that propagate away from the mother (D2A, D1A, D1B, and D1A1). Bar, 5 μm. Spotting glucose (GLU) at the center of the plate rescues the round cell shape and axial budding pattern.

Be careful when manipulating plates on a microscope stage. Microscope presets are designed for slides. Objectives can be damaged if the objective runs into the plate.

DISCUSSION

In the single-cell invasive growth assay (Fig. 1; Cullen and Sprague 2000), cells within the region of diffused glucose should be in the yeast form, appear round, and show an axial budding pattern. Cells outside the region of diffused glucose should be in the filamentous form, appear elongated, and show a distal-unipolar budding pattern (e.g., D1A and D1A1). If desired, microcolony development can be visualized by photographing the cells over a time series starting immediately after they have been spread.

In the pseudohyphal growth assay (Fig. 2; Gimeno et al. 1992), wild-type cells should produce filaments, whereas MAPK pathway mutant cells should not.

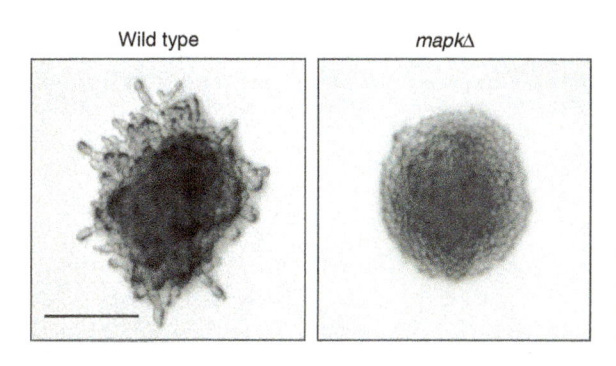

FIGURE 2. Results from a pseudohyphal growth assay. Wild-type cells grown on limiting nitrogen produce filaments (*left*), whereas a MAPK mutant (*right*) does not. Bar, 100 μm.

RECIPES

SLAHD Agar Plates

2% glucose
6.7 g/L yeast nitrogen base without amino acids and ammonium sulfate
20 mg/L uracil
20 mg/L histidine or 500 mg/L proline
20 g/L agar

Histidine or proline can be used interchangeably, depending on the genotype of the yeast strain used. Include any additional amino acids that may be required for the growth of an auxotrophic strain. Prepare in H_2O and sterilize by autoclaving. Fill sterile Petri dishes with ~25 mL of autoclaved medium.

Synthetic Defined (SD) Medium

2% glucose
6.7 g/L yeast nitrogen base without amino acids
Amino acids:
  20 mg/L histidine
  120 mg/L leucine
  60 mg/L lysine
  20 mg/L arginine
  20 mg/L tryptophan
  20 mg/L tyrosine
  40 mg/L threonine
  20 mg/L methionine
  50 mg/L phenylalanine
20 mg/L uracil
20 mg/L adenine
20 g/L agar (omit for liquid medium)

Include additional amino acids that may be required for the growth of an auxotrophic strain. Prepare in H_2O and sterilize by autoclaving. For agar plates, fill sterile Petri dishes with ~25 mL of autoclaved medium.

ACKNOWLEDGMENTS

P.J.C. is supported by a U.S. Public Health Service grant (GM098629).

REFERENCES

Chen H, Fink GR. 2006. Feedback control of morphogenesis in fungi by aromatic alcohols. *Genes Dev* **20**: 1150–1161.

Cullen PJ, Sprague GF Jr. 2000. Glucose depletion causes haploid invasive growth in yeast. *Proc Natl Acad Sci* **97**: 13619–13624.

Gimeno CJ, Ljungdahl PO, Styles CA, Fink GR. 1992. Unipolar cell divisions in the yeast *S. cerevisiae* lead to filamentous growth: Regulation by starvation and RAS. *Cell* **68**: 1077–1090.

Liu H, Styles CA, Fink GR. 1996. *Saccharomyces cerevisiae* S288C has a mutation in FLO8, a gene required for filamentous growth. *Genetics* **144**: 967–978.

Lorenz MC, Cutler NS, Heitman J. 2000. Characterization of alcohol-induced filamentous growth in *Saccharomyces cerevisiae*. *Mol Biol Cell* **11**: 183–199.

Protocol 4

Evaluating the Activity of the Filamentous Growth Mitogen-Activated Protein Kinase Pathway in Yeast

Paul J. Cullen[1]

Department of Biological Sciences, State University of New York at Buffalo, Buffalo, New York 14260

Mitogen-activated protein kinase (MAPK) pathways are evolutionarily conserved signaling pathways that regulate diverse processes in eukaryotes. One such pathway regulates filamentous growth, a nutrient limitation response in budding yeast and other fungal species. This protocol describes three assays used to measure the activity of the filamentous growth pathway. First, western blotting for phosphorylated (activated) MAPKs (P~MAPKs; Slt2p, Kss1p, Fus3p, and Hog1p) provides a measure of MAPK activity in yeast and other fungal species. Second, the *PGU1* gene is a transcriptional target of the filamentous growth pathway. Cells that undergo filamentous growth secrete Pgu1p, an endopolygalacturonase that degrades the plant-specific polysaccharide pectin. We describe an assay that measures secreted pectinase activity, which reflects an active filamentous growth pathway. Finally, in yeast, two mucin-like glycoproteins, Msb2 and Flo11, regulate filamentous growth. Secretion of the processed and shed glycodomain of Msb2 is an indicator of MAPK activity. Flo11, the major adhesion molecule that controls filamentous growth and biofilm/mat formation, is also shed from cells. Detecting shed mucins with epitope-tagged versions of the proteins (secretion profiling) provides information about the regulation of filamentous growth across fungal species.

MATERIALS

It is essential that you consult the appropriate Material Safety Data Sheets and your institution's Environmental Health and Safety Office for proper handling of equipment and hazardous materials used in this protocol.

RECIPES: Please see the end of this protocol for recipes indicated by <R>. Additional recipes can be found online at http://cshprotocols.cshlp.org/site/recipes.

Reagents

α-Factor mating pheromone
Anti-HA antibody (Roche 12CA5)
β-Mercaptoethanol
Bovine serum albumin (BSA)
Distilled H_2O, sterile
Fus3p antibody (Santa Cruz Sc-6773)
Goat α-mouse IgG–HRP (Bio-Rad 170-6516)
Goat α-rabbit IgG-HRP (Jackson ImmunoResearch Laboratories 111-035-144)
HCl (1 N)

---

[1]Correspondence: pjcullen@buffalo.edu

Copyright © Cold Spring Harbor Laboratory Press; all rights reserved
Cite this protocol as *Cold Spring Harb Protoc*; doi:10.1101/pdb.prot085092

Immunoblot detection kit
KCl (5 M)
Kss1p antibody (Santa Cruz Sc-6775-R)
Liquid nitrogen (optional; see Step 3)
Nonfat dry milk
Pectinase agar plates <R>
Pgk1p antibody (Life Technologies 459250)
Phospho-p38 MAPK antibody (Cell Signaling Technology 9211)

This antibody recognizes Hog1, the high osmolarity glycerol response (HOG) MAPK.

Phospho-p44/42 MAPK antibody (Erk1/2) (D13.14.4E) (Cell Signaling Technology 4370)

This antibody recognizes the phosphorylated forms of the MAPKs Kss1 (filamentous growth), Fus3 (mating or pheromone response), and Slt2 (protein kinase C [PKC]).

Resuspension buffer <R>
Ruthenium red (1 mg/mL; Sigma-Aldrich)
SDS-PAGE gel (12%)
SDS gel-loading buffer (2×), prepared with freshly added 200 mM β-mercaptoethanol <R>
TBST <R>
TCA buffer <R>
Yeast extract-peptone-dextrose growth medium (YEPD) <R>
Yeast strains of interest

The Σ1278b background undergoes filamentous growth (Gimeno et al. 1992). Commonly used laboratory strains have lost the ability to undergo filamentous growth (Liu et al. 1996). Use appropriate negative controls (for western blotting, see Step 2; for the pectinase assay, see Step 13). Strains with epitope-tagged versions of mucins can be used for secretion profiling (see Step 18).

YEPD agar plates <R>
YEP-GAL medium <R>

Equipment

Digital camera
Forceps, sterile
Glass beads (Sigma-Aldrich G9143)

Wash beads in 1 N HCl for 5 min, rinse in saturating H₂O to pH 5, and dry.

Glass tubes with metal tops, sterile
Glassware, sterile
Heat block set to 100°C
ImageJ software (http://imagej.nih.gov; Schneider et al. 2012)
Incubator set at 30°C
Inoculation loop or long wooden toothpick, sterile
Microcentrifuge
Microcentrifuge tubes, sterile
Microscope
Multitube vortexer
Nitrocellulose filters, round
Nitrocellulose membrane (Protran BA85)
Protein transfer blot apparatus
SDS-PAGE apparatus
Shaking incubator or a shaker in a 30°C room and in a 37°C room (see Step 2)
Spectrophotometer
Toothpicks, sterile

METHOD

Western Blotting to Detect P~MAPKs

1. Using a long wooden toothpick or inoculation loop, inoculate yeast (from a fresh plate) into 5 mL of YEPD medium in a glass tube with a metal top and grow until saturation is reached.

   We typically grow cultures for 16 h (overnight) at 30°C with shaking at 225 rpm.

2. Dilute aliquots of the saturated culture in the control and inducing conditions listed below.

   - Add 400 µL of saturated cell culture to 10 mL of YEPD. Incubate cells at 30°C with shaking for 6 h or until they reach mid-log phase (1×10^8 cells/mL; determined by spectrophotometry, $A_{600} = 1.0$).

     This is the control condition. Grow control cells and induced cells to the same cell number, not necessarily for the same amount of time.

   - Add 750 µL of saturated cell culture to 10 mL of YEP-GAL. Incubate cells at 30°C with shaking for 6 h or until they reach mid-log phase (1×10^8 cells/mL; determined by spectrophotometry, $A_{600} = 1.0$).

     Σ1278b cells undergo filamentous growth in YEP-GAL liquid medium, which can be confirmed by morphological examination (remove an aliquot and examine by microscopy at 100×). This condition can be used to assess P~Kss1.

   - Add 400 µL of saturated cell culture to 10 mL of YEPD. Grow cells to mid-log phase as described above. Wash cells twice in sterile distilled H_2O and resuspend in YEPD ± 100 nM α-factor mating pheromone. Incubate at 30°C with shaking for 30 min.

     Shmoo response can be confirmed by microscopy (at 100×). This condition can be used to assess P~Fus3.

   - Add 400 µL of saturated cell culture to 10 mL of YEPD. Grow cells to mid-log phase as described above. Transfer cells to 37°C with shaking for 30 min.

     This condition can be used to assess P~Slt2.

   - Add 400 µL of saturated cell culture to 10 mL of YEPD. Grow cells to mid-log phase as described above. Add KCl to the medium to a final concentration of 0.4 M. Incubate at 30°C with shaking for 5 min.

     This condition can be used to assess P~Hog1.

     MAPK mutant yeast strains can be used as negative controls. Basal levels of P~Fus3, P~Kss1, and P~Slt2 are visible under noninducing conditions (Fig. 1). Basal P~Hog1 is not detectable under noninducing conditions.

3. Centrifuge the cells at 16,000g for 2 min at 20°C. Discard the supernatant. Snap-freeze the cell pellets in liquid nitrogen or in an ultracold freezer (at −80°C).

4. Thaw the cell pellets by adding 300 µL of TCA buffer.

5. Add ~0.2 mL of acid-washed glass beads to each sample. Apply five 1-min pulses at full speed in a multitube vortexer at 25°C. Place samples on ice for 3 min between each cycle.

6. Transfer cell lysates to sterile 1.5-mL tubes and centrifuge at 16,000g for 10 min at 4°C.

7. Discard the supernatant. Resuspend each pellet in 150 µL of resuspension buffer.

8. Boil the samples for 5 min at 100°C and centrifuge 16,000g for 30 sec at 25°C. Collect the supernatant. Add an equal volume of 2× SDS-PAGE loading dye.

9. Load the samples onto a 12% SDS-PAGE gel and run the gel using standard SDS gel techniques.

   We generally load 40 µL (10 µg of total protein) per lane.

   Load two SDS-PAGE gels, one to assess phosphorylation and one as a control for overall protein levels (using antibodies to a control protein, such as Pgk1).

10. Transfer the proteins to nitrocellulose membranes using standard techniques.

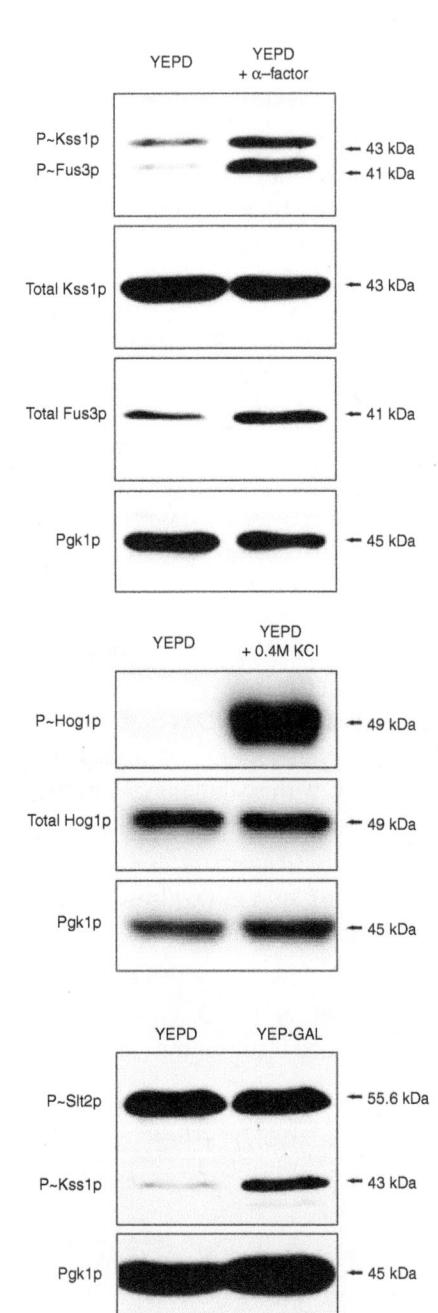

FIGURE 1. Detecting phosphorylated (active) MAP kinases in yeast. Immunoblot analysis of phosphorylated MAP kinases. P~Slt2p, P~Kss1p, and P~Fus3p are detected by p42/p44 (ERK-type MAP kinase) antibodies. P~Hog1p is detected by anti-phospho-p38 antibodies. Blots also show total MAPKs and Pgk1 as a control for protein levels.

11. Perform standard immunoblot analysis with the antibodies of interest.

    We use the following antibodies at the concentrations indicated.

    - *Fus3p antibody: 1:5000 dilution in 5% nonfat dried milk in TBST*
    - *Kss1p antibody: 1:5000 dilution in 5% nonfat dried milk in TBST*
    - *Pgk1p antibody: 1:20,000 dilution in 5% nonfat dried milk in TBST*
    - *Phospho-p38 MAPK antibody: 1:5000 dilution in 5% BSA in TBST*
    - *Phospho-p44/42 MAPK antibody: 1:1000 dilution in 5% BSA in TBST*
    - *Goat α-mouse IgG–HRP: 1:3000 dilution in 5% nonfat dried milk in TBST*

    We typically perform primary antibody incubations for 16 h at 4°C and secondary antibody incubations for 1 h at 25°C with rocking.

12. Detect antibodies using an immunoblot detection kit.

    See Figure 1 for a representative analysis and Lee and Dohlman (2008) and Roman et al. (2005) for sample experiments.

Pectinase Assay

13. Using a sterile toothpick, take cells of the yeast strain of interest from a fresh stock plate (1–2 d old) and make a circular patch on a pectinase plate.

    The pgu1Δ mutant strain can be used as a negative control.

14. Incubate the plates for 2 d at 30°C.

15. Pour freshly prepared ruthenium red over the plate containing the patched cells. Incubate for 8 h at 25°C.

16. Remove excess ruthenium red with a paper towel and rinse the plate under a stream of water.

17. Examine the stained pectin haloes by photography. Quantitate haloes with ImageJ software.

    Haloes are indicators of secreted pectinase (see Fig. 2). For more information, see Gainvors et al. (1994), Blanco et al. (1998), Gognies et al. (1999), and Madhani et al. (1999).

Secretion Profiling of Yeast Mucins

18. Using a long wooden toothpick or inoculation loop, inoculate yeast (from a fresh plate) into 5 mL of YEPD medium and grow until saturation is reached.

    We typically grow cultures for 16 h (overnight) at 30°C with shaking at 5g. Strains carrying functional epitope-tagged versions of mucins (e.g., Msb2 [Cullen et al. 2004; Vadaie et al. 2008; Chavel et al. 2010] and Flo11 [Karunanithi et al. 2010]) are required for this procedure. Alternatively, untagged strains can be used if there is an antibody available to the protein of interest.

19. Remove 0.5 mL of cells from the overnight culture. Pellet the cells by centrifugation at 16,000g for 1 min at 20°C. Wash the cells twice by adding 0.5 mL of sterile distilled H_2O and centrifuging as above. Resuspend the cells in 0.5 mL of sterile distilled H_2O.

20. Using sterile forceps, place a round nitrocellulose filter on a YEPD plate.

21. Spot 10 μL of the resuspended cells onto the nitrocellulose filter.

22. Incubate the plate for 48 h at 30°C.

23. Photograph the colonies with a digital camera.

24. Rub the cells off of the filter in a stream of water.

25. Remove the filter from the YEPD plate.

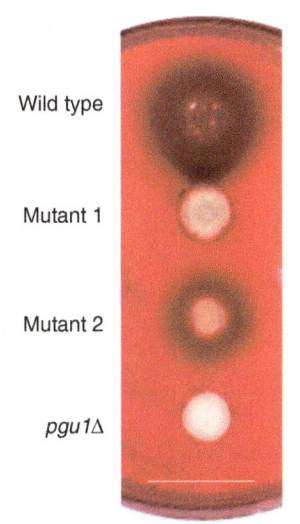

FIGURE 2. Detecting halos (an indicator of secreted pectinase) in the pectinase assay. The size of the halo is reduced in certain mutants (mutant 1 and mutant 2, *middle*) and absent in other mutants (e.g., pgu1Δ mutant, *bottom*). Bar, 1 cm.

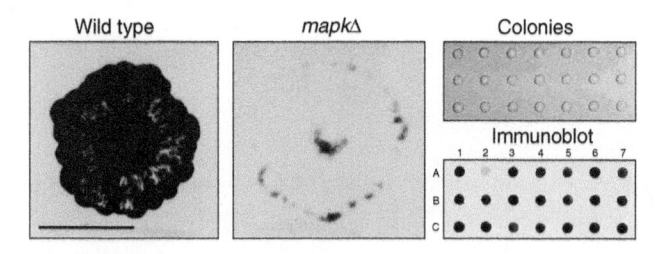

FIGURE 3. Secretion profiling of yeast mucins. Colony immunoblot of hemagglutinin (HA)-tagged Flo11 shed from a biofilm in wild-type cells (*left*) compared to a MAPK pathway mutant (*middle*) defective for *FLO11* expression. (*Right*) Secretion profiling of yeast deletion mutant transformed with a plasmid carrying HA-Msb2 (*top* panel, colonies; *bottom* panel, anti-HA immunoblot). The mutant at position A2 is defective for Msb2 shedding. Bar, 1 cm.

26. Allow the filter to air dry for several minutes.

27. Perform standard immunoblot analysis with the nitrocellulose membrane containing shed epitope-tagged proteins.

    We use the following antibodies at the concentrations indicated.
    - *Anti-HA antibody: 1:2000 dilution in 5% nonfat dry milk in TBST*
    - *Goat α-rabbit IgG-HRP: 1:5000 dilution in 5% nonfat dry milk in TBST*

    We typically perform primary antibody incubations for 16 h at 4°C and secondary antibody incubations for 1 h at 25°C with rocking.

28. Detect antibodies using an immunoblot detection kit.

    See Figure 3 for a representative assay. This approach can be adapted for use in genomic collections (e.g., Fig. 3, right panel). Secretion profiling in various fungal species has been described by Perez-Nadales and Di Pietro (2011), Puri et al. (2012), and Szafranski-Schneider et al. (2012).

RECIPES

Pectinase Agar Plates

1% polygalacturonic acid (Fluka)
6.7 g/L yeast nitrogen base without amino acids
2 g/L ammonium sulfate
2% glucose (or 2% galactose; see below)
2% agar
Amino acids:
   20 mg/L histidine
   120 mg/L leucine
   60 mg/L lysine
   20 mg/L arginine
   20 mg/L tryptophan
   20 mg/L tyrosine
   40 mg/L threonine
   20 mg/L methionine
   50 mg/L phenylalanine
20 mg/L uracil
20 mg/L adenine
50 mM potassium phosphate buffered solution (pH 8)

Use 2% galactose instead of 2% glucose to measure the induction of the filamentous growth pathway in response to glucose limitation. Include any additional amino acids that may be required for the growth of an auxotrophic strain. Prepare in H_2O. Heat the solution to 70°C with stirring for 30 min before autoclaving (polygalacturonic acid has low solubility in H_2O). Fill sterile Petri dishes with ~25 mL of autoclaved medium.

Potassium Phosphate-Buffered Solution (1 M, pH 8)

94 mL K$_2$HPO$_4$ (1 M, pH 8)
6 mL KH$_2$PO$_4$ (1 M, pH 8)

Resuspension Buffer

0.1 M Tris–HCl (pH 11.0)
3% SDS

SDS Gel-Loading Buffer (2×)

100 mM Tris-Cl (pH 6.8)
4% (w/v) SDS (sodium dodecyl sulfate; electrophoresis grade)
0.2% (w/v) bromophenol blue
20% (v/v) glycerol
200 mM DTT (dithiothreitol)

Store the SDS gel-loading buffer without DTT at room temperature. Add DTT from a 1 M stock just before the buffer is used. 200 mM β-mercaptoethanol can be used instead of DTT.

TBST

10 mM Tris–HCl (pH 8)
150 mM NaCl
0.05% Tween 20

TCA Buffer

10 mM Tris–HCl (pH 8.0)
10% trichloroacetic acid
25 mM NH$_4$OAc
1 mM Na$_2$EDTA

Yeast Extract-Peptone-Dextrose Growth Medium (YEPD)

Reagent	Quantity (for 1 L)	Final concentration (w/v)
Bacto peptone	20 g	2%
Yeast extract	10 g	1%
Dextrose	20 g	2%
H$_2$O	to 1 L	

Sterilize by autoclaving.

YEPD Agar Plates

Reagent	Quantity
Bacto-agar (2%)	20 g
YEPD liquid medium	1 L

Add Bacto-agar to YEPD liquid medium in a 2-L flask and autoclave. Fill sterile Petri dishes with 30–40 mL of autoclaved medium.

YEP-GAL Medium

Reagent	Quantity (for 1 L)	Final concentration (w/v)
Bacto peptone	20 g	2%
Yeast extract	10 g	1%
Galactose	20 g	2%
H_2O	to 1 L	

Sterilize by autoclaving.

ACKNOWLEDGMENTS

Thanks to P. Pryciak and H. Dohlman for suggestions about the phosphoblot analysis. Thanks to G. Fink and T. Reynolds for discussions about mucin shedding. P.J.C. is supported by a U.S. Public Health Service grant (GM098629).

REFERENCES

Blanco P, Sieiro C, Reboredo NM, Villa TG. 1998. Cloning, molecular characterization, and expression of an endo-polygalacturonase-encoding gene from *Saccharomyces cerevisiae* IM1-8b. *FEMS Microbiol Lett* 164: 249–255.

Chavel CA, Dionne HM, Birkaya B, Joshi J, Cullen PJ. 2010. Multiple signals converge on a differentiation MAPK pathway. *PLoS Genet* 6: e1000883.

Cullen PJ, Sabbagh W Jr, Graham E, Irick MM, van Olden EK, Neal C, Delrow J, Bardwell L, Sprague GF Jr. 2004. A signaling mucin at the head of the Cdc42- and MAPK-dependent filamentous growth pathway in yeast. *Genes Dev* 18: 1695–1708.

Gainvors A, Frezier V, Lemaresquier H, Lequart C, Aigle M, Belarbi A. 1994. Detection of polygalacturonase, pectin-lyase and pectinesterase activities in a *Saccharomyces cerevisiae* strain. *Yeast* 10: 1311–1319.

Gimeno CJ, Ljungdahl PO, Styles CA, Fink GR. 1992. Unipolar cell divisions in the yeast *S. cerevisiae* lead to filamentous growth: Regulation by starvation and RAS. *Cell* 68: 1077–1090.

Gognies S, Gainvors A, Aigle M, Belarbi A. 1999. Cloning, sequence analysis and overexpression of a *Saccharomyces cerevisiae* endopolygalacturonase-encoding gene (PGL1). *Yeast* 15: 11–22.

Karunanithi S, Vadaie N, Chavel CA, Birkaya B, Joshi J, Grell L, Cullen PJ. 2010. Shedding of the mucin-like Flocculin Flo11p reveals a new aspect of fungal adhesion regulation. *Curr Biol* 20: 1389–1395.

Lee MJ, Dohlman HG. 2008. Coactivation of G protein signaling by cell-surface receptors and an intracellular exchange factor. *Curr Biol* 18: 211–215.

Liu H, Styles CA, Fink GR. 1996. *Saccharomyces cerevisiae* S288C has a mutation in *FLO8*, a gene required for filamentous growth. *Genetics* 144: 967–978.

Madhani HD, Galitski T, Lander ES, Fink GR. 1999. Effectors of a developmental mitogen-activated protein kinase cascade revealed by expression signatures of signaling mutants. *Proc Natl Acad Sci* 96: 12530–12535.

Perez-Nadales E, Di Pietro A. 2011. The membrane Mucin Msb2 regulates invasive growth and plant infection in *Fusarium oxysporum*. *Plant Cell* 23: 1171–1185.

Puri S, Kumar R, Chadha S, Tati S, Conti HR, Hube B, Cullen PJ, Edgerton M. 2012. Secreted aspartic protease cleavage of *Candida albicans* Msb2 activates Cek1 MAPK signaling affecting biofilm formation and oropharyngeal candidiasis. *PLoS ONE* 7: e46020.

Roman E, Nombela C, Pla J. 2005. The Sho1 adaptor protein links oxidative stress to morphogenesis and cell wall biosynthesis in the fungal pathogen *Candida albicans*. *Mol Cell Biol* 25: 10611–10627.

Schneider CA, Rasband WS, Eliceiri KW. 2012. NIH Image to ImageJ: 25 years of image analysis. *Nat Methods* 9: 671–675.

Szafranski-Schneider E, Swidergall M, Cottier F, Tielker D, Roman E, Pla J, Ernst JF. 2012. Msb2 shedding protects *Candida albicans* against antimicrobial peptides. *PLoS Pathog* 8: e1002501.

Vadaie N, Dionne H, Akajagbor DS, Nickerson SR, Krysan DJ, Cullen PJ. 2008. Cleavage of the signaling mucin Msb2 by the aspartyl protease Yps1 is required for MAPK activation in yeast. *J Cell Biol* 181: 1073–1081.

CHAPTER 5

Construction of Multifragment Plasmids by Homologous Recombination in Yeast

Jolanda van Leeuwen,[1,3] Brenda Andrews,[1,2] Charles Boone,[1,2] and Guihong Tan[1,3]

[1]*Donnelly Centre for Cellular and Biomolecular Research, University of Toronto, Toronto, Ontario M5S 3E1, Canada;* [2]*Department of Molecular Genetics, University of Toronto, Toronto, Ontario M5S 3E1, Canada*

Over the past decade, the focus of cloning has shifted from constructing plasmids that express a single gene of interest to creating multigenic constructs that contain entire pathways or even whole genomes. Traditional cloning methods that rely on restriction digestion and ligation are limited by the number and size of fragments that can efficiently be combined. Here, we focus on the use of homologous-recombination-based DNA manipulation in the yeast *Saccharomyces cerevisiae* for the construction of plasmids from multiple DNA fragments. Owing to its simplicity and high efficiency, cloning by homologous recombination in yeast is very accessible and can be applied to high-throughput construction procedures. Its applications extend beyond yeast-centered purposes and include the cloning of large mammalian DNA sequences and entire bacterial genomes.

INTRODUCTION

DNA cloning has been a fundamental aspect of molecular biology for decades. However, with the widespread availability of whole-genome sequences and the increased interest in synthetic and systems biology, the focus of cloning has evolved from inserting a single gene of interest into a vector to creating large, multigenic constructs that express entire pathways. With this new focus comes the need for cloning strategies that allow for efficient multifragment assembly.

Traditionally, cloning was based on the introduction of restriction sites onto the ends of a DNA sequence of interest via the tails of a primer used in the polymerase chain reaction (PCR). After digestion of the PCR product and the vector with the proper restriction enzyme(s), ligation and transformation into *Escherichia coli*, the resulting plasmid could be obtained (Cohen et al. 1973). As the used restriction sites cannot be present in the sequence of interest and as the cloning efficiency drops as the number of fragments increases, it quickly becomes problematic to clone large or multiple DNA fragments using this method.

In homologous-recombination-based cloning techniques, the primers add a sequence to the ends of the amplified product that is homologous to the incorporation site in the vector. The homologous sequences can recombine into a single DNA molecule either in vitro using specific enzymes (Aslanidis and de Jong 1990; Gibson et al. 2009) or in vivo using homologous-recombination-competent microorganisms such as the yeast *Saccharomyces cerevisiae* (Ma et al. 1987; Raymond et al. 1999; Gibson 2009). Homologous-recombination-based cloning strategies do not rely on the availability of restriction sites, and so any set of DNA fragments can be joined together independent of their sequence. This

[3]Correspondence: guihong.tan@utoronto.ca; jolanda.vanleeuwen@utoronto.ca

Copyright © Cold Spring Harbor Laboratory Press; all rights reserved
Cite this introduction as *Cold Spring Harb Protoc*; doi:10.1101/pdb.top084111

Chapter 5

eliminates both the main disadvantage and various time-consuming steps of ligase-based cloning and increases the overall efficiency of the process.

IN VITRO AND *ESCHERICHIA COLI*–BASED RECOMBINATION METHODS

In vitro recombination relies on the creation of single-stranded DNA overhangs by an enzyme with exonuclease activity such as T4 DNA polymerase or T5 exonuclease (Aslanidis and de Jong 1990; Gibson et al. 2009) or by commercially available kits such as In-Fusion by Clontech or Gateway Cloning by Life Technologies (Benoit et al. 2006; Liang et al. 2013). Complementary single-stranded sequences can then anneal in vitro, followed by transformation into *E. coli* to seal the remaining nicks in the DNA (Fig. 1A). Alternatively, the commonly used bacterium *E. coli* can be used as a host for in vivo recombination reactions. As homologous recombination is not very proficient in wild-type *E. coli* strains, several mutant strains such as *recBC sbcA* strains and strains overexpressing Redαβ or RecET prophage proteins have been used to increase the recombination efficiency (Zhang et al. 1998; Wang 2000; Fu et al. 2012). Although these in vitro and *E. coli*

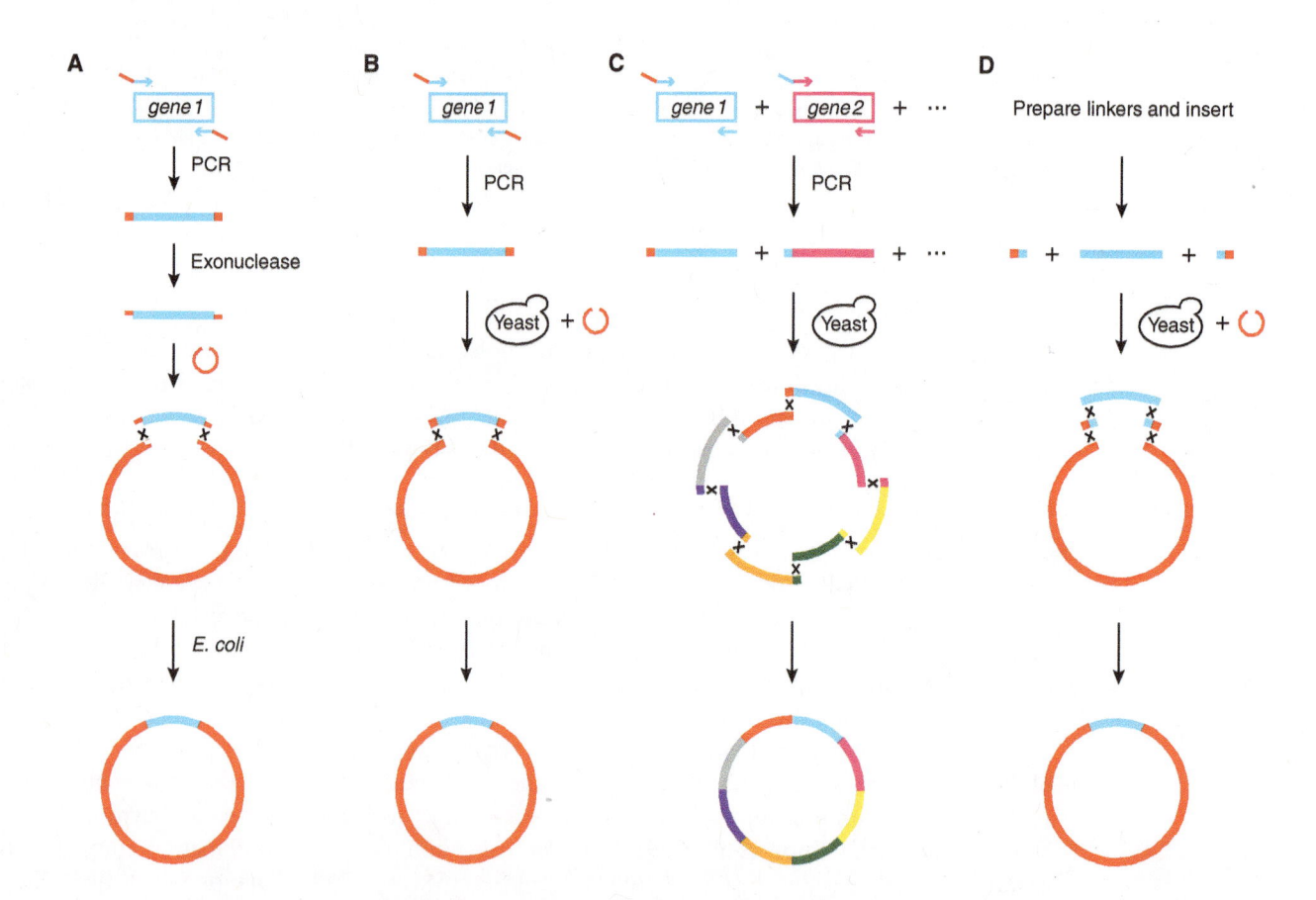

FIGURE 1. Cloning by homologous recombination. (*A*) In vitro homologous recombination of a linearized yeast plasmid and an insert through the creation of complementary single-stranded DNA overhangs by means of an exonuclease. (*B*) In vivo recombination of a linearized yeast plasmid and an insert in budding yeast using homologous double-stranded DNA sequences introduced by PCR. (*C*) In vivo recombination of multiple fragments in budding yeast using homologous double-stranded DNA sequences introduced by PCR. Instead of using a linearized yeast plasmid as one fragment, the selection marker(s) and replication origin(s) can be introduced on different DNA fragments. (*D*) In vivo recombination in budding yeast using homologous sequences present on separate DNA-linker fragments. In this case, the DNA linkers have homology regions to both the acceptor vector and the insert and are useful for cloning inserts from different sources when PCR amplification is difficult or undesirable.

methods are very straightforward for regular two-fragment cloning, the overall efficiency of these recombination strategies in multifragment reactions remains low, and there are limitations to the size of the fragments that can be cloned.

RECOMBINATIONAL CLONING USING *S. Cerevisiae*

The homologous recombination pathway in *S. cerevisiae* efficiently repairs double-strand DNA breaks and can be exploited to join multiple DNA fragments together (Fig. 1B–D). These DNA fragments can be oligonucleotides, PCR products, synthetic DNA, or parts of plasmids or chromosomes. Although homologous recombination in yeast is extremely practical when the downstream applications are also in yeast, its application extends beyond yeast-specific research questions. For example, yeast can be used to assemble constructs that, owing to their size, the number of DNA fragments or the nature of the sequence, are difficult to clone by other methods (Kouprina et al. 2003; Gibson et al. 2008a,b; Vieira et al. 2010; Noskov et al. 2011).

One of the first applications of homologous recombination in yeast for the construction of plasmids from DNA fragments was described by the Botstein laboratory (Ma et al. 1987). In these early experiments, large homology regions of several-hundred base pairs were used. However, it quickly became clear that as little as 30- to 50-bp homology regions are sufficient to induce recombination, which opened the door for PCR-mediated strategies in which homology regions are introduced in the primers (Fig. 1B) (Baudin et al. 1993; Manivasakam et al. 1995; Hua et al. 1997; Oldenburg et al. 1997).

As recombinational repair is far more prominent in yeast than ligation or nonhomologous end-joining of a linearized plasmid, the background of nonrecombinant vector is generally very low (Ma et al. 1987; Hudson et al. 1997; Raymond et al. 1999; Tsvetanova et al. 2011). Up to 25 large (17–35 kb) DNA fragments have been assembled efficiently into one vector after cotransformation in yeast without direct selection for any of the fragments except the vector backbone (Gibson et al. 2008b). If a low recombination rate is anticipated, for example owing to the low concentration or large size of an insert, loss of a (counter)selectable marker in the vector can be used to screen for recombinants among the resulting transformants (Ma et al. 1987; Gunyuzlu et al. 2001; Noskov et al. 2002; Raymond et al. 2002; Kitazono 2009). Alternatively, the yeast selection marker and replication origin can be introduced on different DNA fragments, thereby separating the elements necessary for survival and eliminating background transformants caused by self-closure of a vector backbone (Fig. 1C) (Kuijpers et al. 2013).

PCR-FREE RECOMBINATION IN YEAST

Several recombination-based methods have been described for cloning DNA fragments from one vector into another (Ma et al. 1987; Erickson and Johnston 1993; Prado and Aguilera 1994; Gunyuzlu et al. 2001; Iizasa and Nagano 2006). These methods rely on homology regions between the two used plasmids that can either recombine directly to yield the plasmid of interest (Ma et al. 1987; Erickson and Johnston 1993; Iizasa and Nagano 2006) or recombination can be directed by the use of double-stranded DNA linker molecules (Gunyuzlu et al. 2001). These DNA-linkers have homology regions to both the acceptor vector and the insert and can also be used to clone inserts from different sources when PCR amplification is difficult or undesirable (Fig. 1D) (Raymond et al. 1999; DeMarini et al. 2001; Gunyuzlu et al. 2001; Tsvetanova et al. 2011). Furthermore, single-stranded DNA can serve as a DNA-linker (DeMarini et al. 2001; Raymond et al. 2002) or even as the insert itself, as was shown by the assembly of 38 overlapping single-stranded oligonucleotides into one plasmid after cotransformation with linearized vector in yeast (Gibson 2009).

Chapter 5

APPLICATIONS OF RECOMBINATIONAL CLONING IN YEAST

Large genomic fragments from either yeast or more-complex genomes can be captured in a circular vector or on yeast artificial chromosomes (YACs) by cotransforming yeast cells with linearized vector and the genomic DNA of the organism of interest (Orr-Weaver and Szostak 1983; Fairhead et al. 1996; Noskov et al. 2002). Clones that carry up to 2 Mb of human DNA sequences have been reported (Marschall et al. 1999). An example of the use of homologous recombination in yeast for multifragment heterologous cloning is the construction of a single vector that expresses the complete zeaxanthin biosynthetic pathway by assembling cassettes of eight genes from other microorganisms combined with *S. cerevisiae* promoters and terminators (Shao et al. 2009). Another highlight of multifragment heterologous cloning by homologous recombination in yeast is the assembly of the full ~1.1-Mb *Mycoplasma mycoides* genome in three stages from 1-kb DNA fragments (Gibson et al. 2010). The *M. mycoides* genome was stable in yeast as a centromeric plasmid and was fully functional after transferring to a bacterial host cell (Gibson et al. 2010).

Yeast homologous recombination can easily be adapted to introduce point mutations or partial deletions, add a carboxy- or amino-terminal tag, or create chimeric genes using either yeast genes or DNA sequences from other organisms (Pompon and Nicolas 1989; Muhlrad et al. 1992; Marykwas and Passmore 1995; Storck et al. 1996; Oldenburg et al. 1997; Raymond et al. 1999). For example, Hudson and others cloned fusions of nearly all yeast open-reading frames (ORFs) to the activation domain of the transcriptional regulator Gal4p, which can be used in large-scale yeast two-hybrid analysis (Hudson et al. 1997). Additionally, yeast homologous recombination can be used to create mutants in genomes of other species that do not have the genetic tools available for such a modification directly. This was shown by the creation of a deletion allele of a nonessential restriction endonuclease gene in the complete *M. mycoides* genome by homologous recombination in yeast (Lartigue et al. 2009).

CONCLUDING REMARKS

Over the past few decades, yeast-based recombinational cloning has evolved from inserting just a single fragment into a yeast–bacterial shuttle vector using homology regions of several-hundred basepairs, to the impressive creation of large plasmid-free assemblies of many fragments with short homologous sequences. As plasmid construction by homologous recombination in yeast is straightforward and highly efficient, we expect its use and applications to further expand in the future. In the accompanying protocol, we present a method using homologous recombination in the budding yeast to assemble multiple DNA fragments into one construct; see Protocol 1: Rapid and Efficient Plasmid Construction by Homologous Recombination in Yeast (van Leeuwen et al. 2015).

ACKNOWLEDGMENTS

This work was supported by grants HHMI 55007643, CIHR MOP-130358, and Ministry of Research and Innovation GL-01-022 to B.A. and C.B. and a CIHR fellowship held by J.v.L.

REFERENCES

Aslanidis C, de Jong PJ. 1990. Ligation-independent cloning of PCR products (LIC-PCR). *Nucleic Acids Res* 18: 6069–6074.

Baudin A, Ozier-Kalogeropoulos O, Denouel A, Lacroute F, Cullin C. 1993. A simple and efficient method for direct gene deletion in *Saccharomyces cerevisiae*. *Nucleic Acids Res* 21: 3329–3330.

Benoit RM, Wilhelm RN, Scherer-Becker D, Ostermeier C. 2006. An improved method for fast, robust, and seamless integration of DNA fragments into multiple plasmids. *Protein Expr Purif* 45: 66–71.

Cohen SN, Chang AC, Boyer HW, Helling RB. 1973. Construction of biologically functional bacterial plasmids in vitro. *Proc Natl Acad Sci* 70: 3240–3244.

DeMarini DJ, Creasy CL, Lu Q, Mao J, Sheardown SA, Sathe GM, Livi GP. 2001. Oligonucleotide-mediated, PCR-independent cloning by homologous recombination. *BioTechniques* 30: 520–523.

Erickson JR, Johnston M. 1993. Direct cloning of yeast genes from an ordered set of lambda clones in *Saccharomyces cerevisiae* by recombination in vivo. *Genetics* 134: 151–157.

Fairhead C, Llorente B, Denis F, Soler M, Dujon B. 1996. New vectors for combinatorial deletions in yeast chromosomes and for gap-repair cloning using "split-marker" recombination. *Yeast* **12**: 1439–1457.

Fu J, Bian X, Hu S, Wang H, Huang F, Seibert PM, Plaza A, Xia L, Muller R, Stewart AF, et al. 2012. Full-length RecE enhances linear-linear homologous recombination and facilitates direct cloning for bioprospecting. *Nat Biotechnol* **30**: 440–446.

Gibson DG. 2009. Synthesis of DNA fragments in yeast by one-step assembly of overlapping oligonucleotides. *Nucleic Acids Res* **37**: 6984–6990.

Gibson DG, Benders GA, Andrews-Pfannkoch C, Denisova EA, Baden-Tillson H, Zaveri J, Stockwell TB, Brownley A, Thomas DW, Algire MA, et al. 2008a. Complete chemical synthesis, assembly, and cloning of a *Mycoplasma genitalium* genome. *Science* **319**: 1215–1220.

Gibson DG, Benders GA, Axelrod KC, Zaveri J, Algire MA, Moodie M, Montague MG, Venter JC, Smith HO, Hutchison CA 3rd. 2008b. One-step assembly in yeast of 25 overlapping DNA fragments to form a complete synthetic *Mycoplasma genitalium* genome. *Proc Natl Acad Sci* **105**: 20404–20409.

Gibson DG, Young L, Chuang RY, Venter JC, Hutchison CA 3rd, Smith HO. 2009. Enzymatic assembly of DNA molecules up to several hundred kilobases. *Nat Methods* **6**: 343–345.

Gibson DG, Glass JI, Lartigue C, Noskov VN, Chuang RY, Algire MA, Benders GA, Montague MG, Ma L, Moodie MM, et al. 2010. Creation of a bacterial cell controlled by a chemically synthesized genome. *Science* **329**: 52–56.

Gunyuzlu PL, Hollis GF, Toyn JH. 2001. Plasmid construction by linker-assisted homologous recombination in yeast. *BioTechniques* **31**: 1246, 1248, 1250.

Hua SB, Qiu M, Chan E, Zhu L, Luo Y. 1997. Minimum length of sequence homology required for in vivo cloning by homologous recombination in yeast. *Plasmid* **38**: 91–96.

Hudson JR Jr, Dawson EP, Rushing KL, Jackson CH, Lockshon D, Conover D, Lanciault C, Harris JR, Simmons SJ, Rothstein R, et al. 1997. The complete set of predicted genes from *Saccharomyces cerevisiae* in a readily usable form. *Genome Res* **7**: 1169–1173.

Iizasa E, Nagano Y. 2006. Highly efficient yeast-based in vivo DNA cloning of multiple DNA fragments and the simultaneous construction of yeast/Escherichia coli shuttle vectors. *BioTechniques* **40**: 79–83.

Kitazono AA. 2009. Improved gap-repair cloning method that uses oligonucleotides to target cognate sequences. *Yeast* **26**: 497–505.

Kouprina N, Leem SH, Solomon G, Ly A, Koriabine M, Otstot J, Pak E, Dutra A, Zhao S, Barrett JC, et al. 2003. Segments missing from the draft human genome sequence can be isolated by transformation-associated recombination cloning in yeast. *EMBO Rep* **4**: 257–262.

Kuijpers NG, Solis-Escalante D, Bosman L, van den Broek M, Pronk JT, Daran JM, Daran-Lapujade P. 2013. A versatile, efficient strategy for assembly of multi-fragment expression vectors in *Saccharomyces cerevisiae* using 60 bp synthetic recombination sequences. *Microb Cell Fact* **12**: 47.

Lartigue C, Vashee S, Algire MA, Chuang RY, Benders GA, Ma L, Noskov VN, Denisova EA, Gibson DG, Assad-Garcia N, et al. 2009. Creating bacterial strains from genomes that have been cloned and engineered in yeast. *Science* **325**: 1693–1696.

Liang X, Peng L, Baek CH, Katzen F. 2013. Single step BP/LR combined Gateway reactions. *BioTechniques* **55**: 265–268.

Ma H, Kunes S, Schatz PJ, Botstein D. 1987. Plasmid construction by homologous recombination in yeast. *Gene* **58**: 201–216.

Manivasakam P, Weber SC, McElver J, Schiestl RH. 1995. Micro-homology mediated PCR targeting in *Saccharomyces cerevisiae*. *Nucleic Acids Res* **23**: 2799–2800.

Marschall P, Malik N, Larin Z. 1999. Transfer of YACs up to 2.3 Mb intact into human cells with polyethylenimine. *Gene Ther* **6**: 1634–1637.

Marykwas DL, Passmore SE. 1995. Mapping by multifragment cloning in vivo. *Proc Natl Acad Sci* **92**: 11701–11705.

Muhlrad D, Hunter R, Parker R. 1992. A rapid method for localized mutagenesis of yeast genes. *Yeast* **8**: 79–82.

Noskov V, Kouprina N, Leem SH, Koriabine M, Barrett JC, Larionov V. 2002. A genetic system for direct selection of gene-positive clones during recombinational cloning in yeast. *Nucleic Acids Res* **30**: E8.

Noskov VN, Lee NC, Larionov V, Kouprina N. 2011. Rapid generation of long tandem DNA repeat arrays by homologous recombination in yeast to study their function in mammalian genomes. *Biol Proced Online* **13**: 8.

Oldenburg KR, Vo KT, Michaelis S, Paddon C. 1997. Recombination-mediated PCR-directed plasmid construction in vivo in yeast. *Nucleic Acids Res* **25**: 451–452.

Orr-Weaver TL, Szostak JW. 1983. Yeast recombination: The association between double-strand gap repair and crossing-over. *Proc Natl Acad Sci* **80**: 4417–4421.

Pompon D, Nicolas A. 1989. Protein engineering by cDNA recombination in yeasts: Shuffling of mammalian cytochrome P-450 functions. *Gene* **83**: 15–24.

Prado F, Aguilera A. 1994. New in-vivo cloning methods by homologous recombination in yeast. *Curr Genet* **25**: 180–183.

Raymond CK, Pownder TA, Sexson SL. 1999. General method for plasmid construction using homologous recombination. *BioTechniques* **26**: 134–138, 140–131.

Raymond CK, Sims EH, Olson MV. 2002. Linker-mediated recombinational subcloning of large DNA fragments using yeast. *Genome Res* **12**: 190–197.

Shao Z, Zhao H, Zhao H. 2009. DNA assembler, an in vivo genetic method for rapid construction of biochemical pathways. *Nucleic Acids Res* **37**: e16.

Storck T, Kruth U, Kolhekar R, Sprengel R, Seeburg PH. 1996. Rapid construction in yeast of complex targeting vectors for gene manipulation in the mouse. *Nucleic Acids Res* **24**: 4594–4596.

Tsvetanova B, Peng L, Liang X, Li K, Yang JP, Ho T, Shirley J, Xu L, Potter J, Kudlicki W, et al. 2011. Genetic assembly tools for synthetic biology. *Methods Enzymol* **498**: 327–348.

van Leeuwen JS, Andrews BJ, Boone C, Tan G. 2015. Rapid and efficient plasmid construction by homologous recombination in yeast. *Cold Spring Harb Protoc* doi: 10.1101/pdb.prot085100.

Vieira N, Pereira F, Casal M, Brown AJ, Paiva S, Johansson B. 2010. Plasmids for in vivo construction of integrative *Candida albicans* vectors in *Saccharomyces cerevisiae*. *Yeast* **27**: 933–939.

Wang PL. 2000. Creating hybrid genes by homologous recombination. *Dis Markers* **16**: 3–13.

Zhang Y, Buchholz F, Muyrers JP, Stewart AF. 1998. A new logic for DNA engineering using recombination in *Escherichia coli*. *Nat Genet* **20**: 123–128.

Protocol 1

Rapid and Efficient Plasmid Construction by Homologous Recombination in Yeast

Jolanda van Leeuwen,[1,3] Brenda Andrews,[1,2] Charles Boone,[1,2] and Guihong Tan[1,3]

[1]Donnelly Centre for Cellular and Biomolecular Research, University of Toronto, Toronto, Ontario M5S 3E1, Canada; [2]Department of Molecular Genetics, University of Toronto, Toronto, Ontario M5S 3E1, Canada

The cloning of DNA fragments is a fundamental aspect of molecular biology. Traditional DNA cloning techniques rely on the ligation of an insert and a linearized plasmid that have been digested with restriction enzymes and the subsequent introduction of the ligated DNA into *Escherichia coli* for propagation. However, this method is limited by the availability of restriction sites, which often becomes problematic when cloning multiple or large DNA fragments. Furthermore, using traditional methods to clone multiple DNA fragments requires experience and multiple laborious steps. In this protocol, we describe a simple and efficient cloning method that relies on homologous recombination in the yeast *Saccharomyces cerevisiae* to assemble multiple DNA fragments, with 30-bp homology regions between the fragments, into one sophisticated construct. This method can easily be extended to clone plasmids for other organisms, such as bacteria, plants, and mammalian cells.

MATERIALS

It is essential that you consult the appropriate Material Safety Data Sheets and your institution's Environmental Health and Safety Office for proper handling of equipment and hazardous material used in this protocol.

RECIPES: Please see the end of this protocol for recipes indicated by <R>. Additional recipes can be found online at http://cshprotocols.cshlp.org/site/recipes.

Reagents

Agarose gel (1.2%) and electrophoresis reagents
Bacterial strains
  Any standard Escherichia coli *strain such as Top10 will suffice.*
Deionized water (sterilized)
Dimethyl sulfoxide (DMSO)
dNTPs (10 mM)
Ethanol (70%)
Glass beads (0.4–0.6 mm)
Glycerol (10%)
Isopropanol
Lithium acetate (LiAc)

[3]Correspondence: guihong.tan@utoronto.ca; jolanda.vanleeuwen@utoronto.ca

Copyright © Cold Spring Harbor Laboratory Press; all rights reserved
Cite this protocol as *Cold Spring Harb Protoc*; doi:10.1101/pdb.prot085100

TABLE 1. Sequence of the oligonucleotides used in this protocol

Name	Sequence (5'–3')
ori-F	GATACTAACGCCGCCATCCAGTTT<u>CCCGGG</u>aaaggcggtaatacggtta
ori-R	<u>CCCGGG</u>ttgataatctcatgaccaaaatcc
ampR-F	TGGTCATGAGATTATCAA<u>CCCGGG</u>aaaggatcttcacctagatcct
ampR-R	<u>GGG</u>cacttttcggggaaatgtgcg
CEN-F	GTTCCGCGCACATTTCCCCGAAAAGTG<u>CCCGGG</u>tccttttcatcacgtgc
CEN-R	<u>GGG</u>cttaggacggatcgcttgc
LEU2-F	AGTTACAGGCAAGCGATCCGTCCTAAG<u>CCCGGG</u>aactgtgggaatactcaggt
LEU2-R	Cgtgtcgtttctattatgaatttc
LYS2-F	TTTATAAATGAAATTCATAATAGAAACGA<u>CCCGGG</u>cttcaatagttttgccagcg
LYS2-R	GCT<u>CCCGGG</u>catatcatacgtaatgctca
URA3-F	TTGAGCATTACGTATGATATG<u>CCCGGG</u>agcttttcaattcatctttttttttttgttc
URA3-R	<u>GGG</u>taataactgatataattaaattgaagc
HIS3-F	GCTTCAATTTAATTATATCAGTTATTA<u>CCCGGG</u>cttcattcaacgtttcccatt
HIS3-R	<u>GGG</u>tgatgcattaccttgtcatc
kanR-F	TACTGAAGATGACAAGGTAATGCATCA<u>CCCGGG</u>tagcccatacatccccatgt
kanR-R	<u>CCCGGG</u>TAAATCACGCTAACATTTGA

All oligonucleotides are desalted and <60 bp in length. The sequences shown in lower case are homologous to sequences upstream (F) or downstream (R) of the template gene or origin; the sequences shown in upper case are homologous to one of the other fragments; the SmaI sites are underlined.

> To prepare a 1 M solution, dissolve 10.2 g of LiAc in 100 mL of deionized water, filter-sterilize, and store at room temperature.

Luria–Bertani (LB) medium plus ampicillin <R>
Lysis buffer for yeast <R>
Milli-Q water
Oligonucleotides (desalted)
> For a list of all the oligonucleotides used in this protocol, see Table 1.

Phenol (saturated)
Phusion high-fidelity DNA polymerase and buffer (5×)
> For example, consider using Phusion Hot Start II DNA Polymerase (Thermo Scientific); kit includes bespoke buffers.

Plasmid miniprep kit
Polyethylene glycol (PEG) 3350
> Prepare 100 mL of a 50% solution by dissolving 50 g of PEG3350 in 50 mL deionized water, adjusting the volume to 100 mL with deionized water, followed by filter sterilization.

Single-stranded DNA (ssDNA; e.g., Sigma-Aldrich D8899)
SmaI restriction endonuclease
Synthetic amino-acid-dropout medium (SD-all) <R>
Synthetic lysine-dropout medium (SD-lys) <R>
Terrific broth (TB) medium <R>
Tris–EDTA (TE) buffer (10×)
> Add 0.2 mL of EDTA (0.5 M, pH 8.0) and 1 mL of Tris–Cl (1 M, pH 8.0) to 99 mL of deionized water. Filter-sterilize. Store at room temperature.

Yeast extract-peptone-dextrose (YEPD) <R>
Yeast strains
> This protocol uses BY4709 MATα ura3Δ0 (ATCC200872), BY4712 MATα leu2Δ0 (ATCC200875), and BY4742 MATα his3Δ1 leu20 lys2Δ0 met15Δ0 ura3Δ0 (ATCC201389), but many other standard laboratory yeast strains can also be used.

Equipment

Agarose gel electrophoresis materials
Benchtop centrifuge/microcentrifuge

Chapter 5

Electroporation system (e.g., MicroPulser by Bio-Rad)
Ice
Incubators (30°C and 37°C)
Microcentrifuge tubes (1.5-mL)
Orbital shaker
PCR tubes
Pipettes
Plates
Thermal cycler (PCR machine)
Vortex
Water bath (42°C)

METHOD

To illustrate the cloning of multiple DNA fragments into one construct by homologous recombination, we describe the assembly of five different yeast selection markers (HIS3, LEU2, LYS2, URA3, and kanMX6) together with origins of replication for both yeast and E. coli (CEN6/ARS4 and ori) and an E. coli selection marker (ampR) into one plasmid (Fig. 1A).

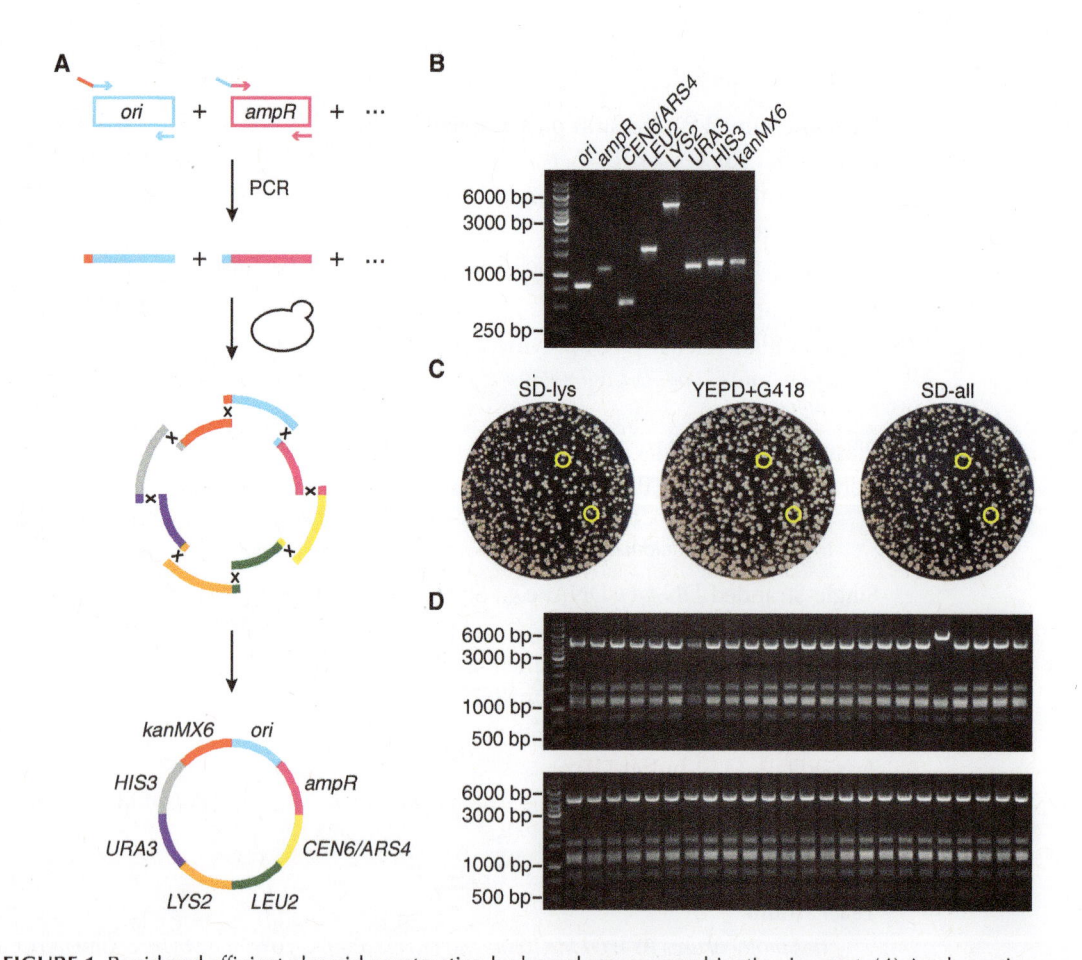

FIGURE 1. Rapid and efficient plasmid construction by homologous recombination in yeast. (*A*) A schematic representation of the described assembly of eight DNA fragments in yeast using 30-bp recombination sequences. (*B*) The PCR fragments used in the featured assembly. (*C*) The selection of yeast transformants on synthetic lysine-dropout medium (SD-lys) and subsequent confirmation of the presence of the six fragments containing yeast sequences by replica plating on YEPD + G418 and synthetic-dropout medium without amino acids (SD-all). Only two transformants fail to grow after replica plating (yellow circles). (*D*) The SmaI profiles of the final plasmids isolated from 24 yeast colonies (*upper* panel) or 24 E. coli colonies (*lower* panel).

Cite this protocol as *Cold Spring Harb Protoc*; doi:10.1101/pdb.prot085100

Preparation of DNA Fragments

This procedure should take 3–4 h on day 1.

1. Prepare genomic DNA from a mixture of the yeast strains BY4709 and BY4712 to serve as a template in the polymerase chain reaction (PCR) (see Steps 23–35).

2. Prepare a 20-µL PCR.

Sterile deionized water	13.4 µL
Buffer (5×)	4 µL
dNTP mixture (10 mM)	0.4 µL
Forward primer (10 µM)	0.5 µL
Reverse primer (10 µM)	0.5 µL
Template DNA	1 µL
Phusion DNA polymerase	0.2 µL

3. Use the following PCR cycling conditions.

Initial denaturation	60 sec at 98°C
5 Cycles	10 sec at 98°C, 20 sec at 55°C, and 90 sec at 72°C
25 Cycles	10 sec at 98°C, 20 sec at 62°C, and 90 sec at 72°C

   HIS3, LEU2, LYS2, and URA3 can be amplified from ~100 ng of genomic DNA of a mixture of the yeast strains BY4709 and BY4712. CEN6/ARS4, ampR, and ori can be amplified from 1 pg of pRS416 (Sikorski and Hieter 1989), and kanMX6 can be amplified from 1 pg of pFA6-kanMX6 (Bahler et al. 1998).

4. Run 2 µL of each PCR product on a 1.2% agarose gel (Fig. 1B).

   If a PCR product contains multiple fragments, all fragments can potentially be assembled in the final construct. In this case, purification of the correct PCR product from the gel might be necessary. Alternatively, the unpurified PCR product with multiple fragments can be used, in which case the number of colonies that have to be screened for the expected construct has to be increased.

   The PCR products can be stored at 4°C or −20°C.

Preparation of Competent Yeast Cells

The following procedure is a modified version of Gietz's method (Gietz and Woods 2002) and should take 5 min on day 1 and 10–15 min on day 2.

5. Patch the yeast strain BY4742 on a 2-cm² area on YEPD agar and incubate overnight at 30°C.

   Alternatively, the yeast strain can be used to inoculate a 5-mL culture of liquid YEPD medium. Incubate overnight at 30°C while shaking on an orbital shaker at 200 rpm. Any standard laboratory yeast strain, such as BY4741 or W303, can be used instead of BY4742.

6. Scrape a 50-µL portion of yeast cells from the YEPD plate and resuspend the cells in 1 mL of sterile deionized water in a 1.5-mL microcentrifuge tube.

   Alternatively, harvest 1–1.5 mL of a liquid culture.

7. Pellet the cells by centrifugation at 3000 rpm (800 rcf) for 1 min at room temperature in a microcentrifuge and discard the supernatant.

8. Wash the pellet once with 1 mL of 0.1 M LiAc.

9. Estimate the volume of the cell pellet and resuspend the cells in an equal volume of 0.1 M LiAc supplemented with 10% glycerol.

   By using larger volumes of liquid culture in Steps 5 and 6, a large amount of competent cells can be made at once. These competent cells can be stored at −80°C in 0.1 M LiAc supplemented with 10% glycerol.

Yeast Transformation

This procedure is a modified version of Gietz's method (Gietz and Woods 2002) and should take 3 h (not including the time needed to grow the yeast cells) on day 2.

10. Prepare the transformation buffer by mixing 800 µL of 50% PEG3350, 100 µL of 1 M LiAc, 100 µL of 10× TE, and 50 µL of DMSO.
11. Boil the ssDNA (10 mg/mL) for 5 min and place it on ice.
12. Add ~100 ng of each DNA fragment, 2 µL of ssDNA, and 12 µL of competent yeast cells into a 1.5-mL microcentrifuge tube; mix gently by pipetting up and down.

    If the total volume of all the DNA fragments together is >20 µL, the overall transformation efficiency will decrease. The volume can be reduced by air-drying the DNA overnight at room temperature.

13. Add 100 µL of the transformation buffer from Step 10 and vortex for ~10 sec.
14. Incubate for 30 min at room temperature.
15. Incubate for 15 min in a 42°C water bath.
16. Incubate on ice for 5 min.
17. Pellet the cells by centrifugation at 3000 rpm (800 rcf) for 1 min and remove the supernatant.
18. Resuspend the cells in 1 mL of YEPD.
19. (Optional) Incubate the cells for 2 h at 30°C.

    This step allows the cells to produce the antibiotic resistance and/or auxotrophic marker proteins before applying selection and thereby increases the overall transformation efficiency.

20. Plate 250 µL of the cell suspension on SD-lys plates.

    Other media that select for one of the other cloned genes can also be used.

21. Incubate the plates for 2–3 d at 30°C.
22. Replica-plate the colonies on SD-all and YEPD + G418 plates (Fig. 1C).

    The SD-lys plates select only for one of the yeast markers (LYS2) and for presence of the CEN6/ARS4. This replica-plating step tests whether the other yeast selection markers are present in the construct and functional. We have obtained 470 colonies on the SD-lys plate, only two of which failed to grow on the YEPD + G418 (which selects for kanMX6) and SD-all (which selects for HIS3, LEU2, and URA3) plates.

Preparation of Yeast Genomic DNA

This procedure should take 1 h on day 4.

23. Wash all the colonies from the SD-lys plate from Step 21 using 5 mL of sterile water and transfer 1 mL of cells to a 1.5-mL microcentrifuge tube.
24. Pellet the cells by centrifugation at 3000 rpm (800 rcf) for 1 min and remove the supernatant.
25. Resuspend the pellet in 250 µL of lysis buffer for yeast.
26. Add 250 µL of saturated phenol and ~200 µL (~200 mg) of glass beads.
27. Close the cap tightly and vortex for >2 min at room temperature.
28. Centrifuge at 13,000 rpm (15,700 rcf) for 5 min.
29. Transfer 150 µL of the aqueous top layer to a new 1.5-mL microcentrifuge tube and centrifuge again at 13,000 rpm (15,700 rcf) for 5 min.
30. Transfer 100 µL of the top layer to a new 1.5 mL microcentrifuge tube, add 100 µL of 100% isopropanol and mix thoroughly by inversion.
31. Centrifuge at 13,000 rpm (15,700 rcf) for 10 min and remove the supernatant.
32. Wash the pellet once with 500 µL of 70% ethanol.
33. Briefly centrifuge at 13,000 rpm (15,700 rcf) for 10 sec and remove the remaining ethanol.
34. Dry the pellet at room temperature for ~10 min.
35. Dissolve the pellet in 40 µL of sterile water.

Preparation of Competent *E. coli* Cells

This procedure should take 10 min on day 4.

36. Inoculate 2 mL of TB using one bacterial colony and incubate overnight at 37°C while shaking at 250 rpm.

    In the example described here, we use Top10 cells.

37. Add 2 mL of TB and incubate for another 25 min at 37°C while shaking at 250 rpm in an orbital shaker.

38. Divide the culture into aliquots of 1 mL.

    This is approximately the amount of culture needed for one transformation.

39. Collect the cells by centrifuging at 13,000 rpm (15,700 rcf) for 1 min.

40. Wash the cells four times with Milli-Q water.

41. Estimate the volume of the cell pellet and resuspend the cells in an equal volume of Milli-Q water.

    More water (up to five times the volume of the cell pellet) can be used to obtain "more" competent cells. However, the transformation efficiency will slightly decrease because the cells are more diluted. By using greater volumes of liquid culture in Steps 36–41, a large number of competent cells can be made at once. These competent cells can be stored at −80°C in 10% glycerol.

    Other methods can be used to make competent E. coli cells, but these will require longer preparation times. Also, with this method, all the steps after growing the culture can be performed at room temperature, which makes it more straightforward and robust.

Plasmid Recovery from *Escherichia coli*

This procedure should take ~5 h on day 5 and day 6.

42. Add 20 µL of competent *E. coli* cells to a 200-µL PCR tube and add 0.5 µL of yeast genomic DNA.

43. Mix the DNA with the competent cells by pipetting up and down.

44. Transfer the mixture to an electroporation cuvette (1 mm gap).

45. Electroporate the cells according to the manufacturer's manual.

46. Add 250 µL of LB and plate all the cells on an LB plate containing ampicillin.

    Note that kanMX6 is functional in both S. cerevisiae and E. coli. Therefore, transformants can also be selected on LB media containing kanamycin.

47. Incubate the plates overnight at 37°C.

48. Purify the plasmid DNA with a plasmid miniprep kit.

49. Confirm the correct assembly of the fragments by a SmaI digestion (Fig. 1D).

    In the example described here, SmaI restriction sites were added to each fragment by means of the primers to show the accuracy of homologous recombination in yeast. However, the addition of restriction sites is not necessary for this method.

DISCUSSION

Homologous recombination in yeast has been successfully used in the cloning of both natural and synthetic DNA fragments (Ma et al. 1987; Wang 2000; Chen et al. 2005; Gibson et al. 2008a,b; Gibson 2009; Shao and Zhao 2009; Tan and Tan 2010; Tsvetanova et al. 2011; Liang et al. 2012; Shao and Zhao 2013). However, many of these methods are based on yeast–bacterial shuttle plasmids, which rather restricted the technology to a small niche within molecular biology. For an introduction to these technologies, see Introduction: Construction of Multifragment Plasmids by Homologous Recombination in Yeast (van Leeuwen et al. 2015). Recently, however, Kuijpers et al. (2013) have presented an efficient strategy for assembling a plasmid from mutiple DNA fragments by using overlapping homology regions of 60 bp. As this method does not make use of a vector backbone, it can easily be extended to construct plasmids for other organisms.

In this protocol, we have described the use of yeast as a host to assemble eight DNA fragments, with short homology regions (30 bp) between the fragments, into one plasmid (Fig. 1A). Equal amounts of PCR products were cotransformed into yeast, and transformants were selected on synthetic media lacking lysine (SD-lys), which confirms the presence and functionality of *LYS2* and the *CEN6/ARS4* origin of replication. The presence and activity of *HIS3*, *LEU2*, *URA3*, and *kanMX6* were confirmed by replica plating (Step 22). Only two out of 470 transformants selected on SD-lys failed to grow after replica plating. The genomic DNA was individually purified from 24 randomly picked yeast colonies from the SD-lys plate and from 24 randomly picked E. coli colonies from Steps 46–48, followed by recovery of the plasmids. SmaI sites were introduced between all the fragments when the primers were designed, which enabled confirmation of the correct assembly of the construct by a SmaI digestion. In one out of the 48 isolated plasmids, a SmaI site was missing, which could be caused by impurities in the primers used, whereas the other 47 plasmids all showed the expected DNA bands (Fig. 1D). Based on the growth phenotypes on the different media and the SmaI digestion profiles, we can conclude that the overall assembly efficiency is >95%.

This protocol uses short (<60 bp), desalted oligonucleotides and unpurified PCR products. This makes the method highly cost- and time-efficient, especially when cloning multiple fragments. We modified the yeast and bacterial transformation methods, which facilitates high-throughput cloning. However, this method has the common disadvantage that inserts that are toxic to either yeast or bacteria cannot be cloned. Also, because homologous recombination is so efficient in yeast, the cloning of inverted or repetitive sequences can be problematic.

Interestingly, we have found that the *kanMX6* selector module is functional both in yeast and bacteria, which reduces the number of required selection markers from two (one for yeast and one for E. coli) to one. Although, in the example described here, separate fragments are used for the E. coli selection marker and origin of replication (*ampR* and *ori*), we have routinely used the shortest backbone of pBlueScript II (Alting-Mees and Short 1989), which contains both *ori* and *ampR*, as one fragment in our assemblies. Finally, this method can be extended to clone constructs for other organisms by combining one or more fragment(s) with the genes or sequences of interest with a fragment containing a linearized nonyeast plasmid, and another fragment containing a *CEN/ARS* origin and yeast selection marker.

RECIPES

Luria–Bertani (LB) Medium Plus Ampicillin

Reagent	Quantity
Agar	20 g
NaCl	10 g
Tryptone	10 g
Yeast extract	5 g

Prepare the above-listed ingredients in 1 L of deionized water. Adjust the pH to 7.0 with 5 N NaOH. Autoclave for 20 min at 15 psi (1.05 kg/cm^2). Cool to ~60°C and add ampicillin (final concentration 120 µg/mL). Pour the medium into Petri dishes (~25 mL per 100-mm plate). Store the LB plates at 4°C; they will keep for at least 4 mo.

Lysis Buffer for Yeast

Reagent	Quantity	Final concentration
Triton X-100	10 mL	2% (v/v)
SDS (10%)	50 mL	1% (w/v)
NaCl (5 M)	10 mL	100 mM
Tris-Cl (1 M, pH 8.0)	5 mL	10 mM
EDTA (0.5 M, pH 8.0)	1 mL	1 mM

Prepare in 500 mL of deionized water. Autoclave for 20 min at 15 psi (1.05 kg/cm^2). Store at room temperature; it will keep for at least 1 yr.

Supplements for SD-Lys

Reagent	Quantity
L-Leucine	0.84 g
L-Histidine HCl	0.42 g
L-Methionine	0.42 g
Uracil	0.25 g

In separate tubes, prepare 100× stocks of each of the four reagents by dissolving the quantities indicated in 100 mL of deionized water and then filter-sterilizing. Store all stocks at room temperature; they will keep for at least 6 mo.

Synthetic Amino-Acid-Dropout Medium (SD-All)

Reagent	Quantity	Final concentration
Difco yeast nitrogen base without amino acids	6.7 g	6.7 g/L
Agar	20 g	20 g/L
Dextrose (40%)	50 mL	20 g/L

Add 950 mL of deionized water to 6.7 g Difco yeast nitrogen base without amino acids and 20 g of agar. Autoclave for 20 min at 15 psi (1.05 kg/cm^2). After autoclaving, add 50 mL of a 40% dextrose solution. Cool the medium to ~60°C and pour into Petri dishes (~25 mL per 100-mm plate). Store the SD-all plates at 4°C; they will keep for at least 6 mo.

Synthetic Lysine-Dropout Medium (SD-Lys)

Reagent	Quantity	Final concentration
Difco yeast nitrogen base without amino acids	6.7 g	6.7 g/L
Agar	20 g	20 g/L
Supplements for SD-lys:		
L-Leucine (100×)	10 mL	8.4 mg/L
L-Histidine HCl (100×)	10 mL	4.2 mg/L
L-Methionine (100×)	10 mL	4.2 mg/L
Uracil (100×)	10 mL	2.5 mg/L
Dextrose (40%)	50 mL	20 g/L

Combine 10 mL of each of the SD-lys supplements (100× L-leucine, 100× L-histidine HCl, 100× L-methionine, 100× uracil) with 6.7 g of Difco yeast nitrogen base without amino acids and 20 g agar, and add 950 mL of deionized water. Autoclave for 20 min at 15 psi (1.05 kg/cm^2). After autoclaving, add 50 mL of a 40% dextrose solution. Cool the medium to ~60°C and pour into Petri dishes (~25 mL per 100-mm plate). Store the SD-lys plates at 4°C; they will keep for at least 6 mo. (Note that this recipe is optimized for BY4742 strains. It should be adjusted if another yeast strain is used as a recombination host.)

Terrific Broth (TB) Medium

Reagent	Quantity	Final concentration
Yeast extract	24 g	24 g/L
Tryptone	20 g	20 g/L
Glycerol	4 mL	4 mL/L
Phosphate buffer (0.17 M KH$_2$PO$_4$, 0.72 M K$_2$HPO$_4$)	100 mL	0.017 M KH$_2$PO$_4$, 0.072 M K$_2$HPO$_4$

Add 900 mL of deionized water to 24 g of yeast extract, 20 g of tryptone, and 4 mL of glycerol. Shake or stir until the solutes have dissolved and sterilize by autoclaving for 20 min at 15 psi (1.05 kg/cm^2). Allow the solution to cool to ~60°C and add 100 mL of sterile phosphate buffer. Store TB at room temperature; it will keep for at least 1 yr.

Yeast Extract-Peptone-Dextrose (YEPD)

Reagent	Quantity	Final concentration
Bacto peptone	20 g	2% (w/v)
Yeast extract	10 g	1% (w/v)
Dextrose	20 g	2% (w/v)
Agar (optional)	20 g	2% (w/v)
G418 (200 mg/mL; optional)	1 mL	200 mg/L

Add 1 L of deionized water to 20 g bacto peptone, 10 g yeast extract, and 20 g dextrose (and 20 g of agar for YEPD plates). Sterilize by autoclaving for 20 min at 15 psi (1.05 kg/cm^2). To prepare YEPD plus G418 plates, allow the solution to cool to ~60°C, add 1 mL of G418 stock solution, and pour into Petri dishes (~25 mL per 100-mm plate). Store YEPD medium without G418 at room temperature, and store YEPD containing medium G418 at 4°C.

ACKNOWLEDGMENTS

This work was supported by grants HHMI 55007643, CIHR MOP-130358, and Ministry of Research and Innovation GL-01-022 to B.A. and C.B. and a CIHR fellowship held by J.v.L.

REFERENCES

Alting-Mees MA, Short JM. 1989. pBluescript II: Gene mapping vectors. *Nucleic Acids Res* 17: 9494.

Bahler J, Wu JQ, Longtine MS, Shah NG, McKenzie A 3rd, Steever AB, Wach A, Philippsen P, Pringle JR. 1998. Heterologous modules for efficient and versatile PCR-based gene targeting in *Schizosaccharomyces pombe*. *Yeast* 14: 943–951.

Chen X, Yuan H, He W, Hu X, Lu H, Li Y. 2005. Construction of a novel kind of expression plasmid by homologous recombination in *Saccharomyces cerevisiae*. *Sci China C Life Sci* 48: 330–336.

Gibson DG. 2009. Synthesis of DNA fragments in yeast by one-step assembly of overlapping oligonucleotides. *Nucleic Acids Res* 37: 6984–6990.

Gibson DG, Benders GA, Andrews-Pfannkoch C, Denisova EA, Baden-Tillson H, Zaveri J, Stockwell TB, Brownley A, Thomas DW, Algire MA, et al. 2008a. Complete chemical synthesis, assembly, and cloning of a *Mycoplasma genitalium* genome. *Science* 319: 1215–1220.

Gibson DG, Benders GA, Axelrod KC, Zaveri J, Algire MA, Moodie M, Montague MG, Venter JC, Smith HO, Hutchison CA 3rd. 2008b. One-step assembly in yeast of 25 overlapping DNA fragments to form a complete synthetic *Mycoplasma genitalium* genome. *Proc Natl Acad Sci* 105: 20404–20409.

Gietz RD, Woods RA. 2002. Transformation of yeast by lithium acetate/single-stranded carrier DNA/polyethylene glycol method. *Methods Enzymol* 350: 87–96.

Kuijpers NG, Solis-Escalante D, Bosman L, van den Broek M, Pronk JT, Daran JM, Daran-Lapujade P. 2013. A versatile, efficient strategy for assembly of multi-fragment expression vectors in *Saccharomyces cerevisiae* using 60 bp synthetic recombination sequences. *Microb Cell Fact* 12: 47.

Liang X, Peng L, Tsvetanova B, Li K, Yang JP, Ho T, Shirley J, Xu L, Potter J, Kudlicki W, et al. 2012. Recombination-based DNA assembly and mutagenesis methods for metabolic engineering. *Methods Mol Biol* 834: 93–109.

Ma H, Kunes S, Schatz PJ, Botstein D. 1987. Plasmid construction by homologous recombination in yeast. *Gene* 58: 201–216.

Shao Z, Zhao H. 2009. DNA assembler, an in vivo genetic method for rapid construction of biochemical pathways. *Nucleic Acids Res* 37: e16.

Shao Z, Zhao H. 2013. Construction and engineering of large biochemical pathways via DNA assembler. *Methods Mol Biol* 1073: 85–106.

Sikorski RS, Hieter P. 1989. A system of shuttle vectors and yeast host strains designed for efficient manipulation of DNA in *Saccharomyces cerevisiae*. *Genetics* 122: 19–27.

Tan G, Tan C. 2010. SMC, a simple method to rapidly assemble multiple fragments into one construct. *Front Biosci (Elite Ed)* 2: 1105–1114.

Tsvetanova B, Peng L, Liang X, Li K, Yang JP, Ho T, Shirley J, Xu L, Potter J, Kudlicki W, et al. 2011. Genetic assembly tools for synthetic biology. *Methods Enzymol* 498: 327–348.

van Leeuwen JS, Andrews BJ, Boone C, Tan G. 2015. Construction of multi-fragment plasmids by homologous recombination in yeast. *Cold Spring Harb Protoc* doi: 10.1101/pdb.top084111.

Wang PL. 2000. Creating hybrid genes by homologous recombination. *Dis Markers* 16: 3–13.

CHAPTER 6

Single-Molecule Analysis of Replicating Yeast Chromosomes

David Gallo,[1,3] Gang Wang,[1,3] Christopher M. Yip,[1,2,3,4] and Grant W. Brown[1,3,4]

[1]*Department of Biochemistry, University of Toronto, Toronto, Ontario M5S 1A8, Canada;* [2]*Department of Chemical Engineering and Applied Chemistry, University of Toronto, Toronto, Ontario M5S 3E5, Canada;* [3]*Donnelly Centre, University of Toronto, Toronto, Ontario M5S 3E1, Canada*

The faithful replication of eukaryotic chromosomal DNA occurs during S phase once per cell cycle. Replication is highly regulated and is initiated at special structures, termed origins, from which replication forks move out bidirectionally. A wide variety of techniques have been developed to study the features and kinetics of replication. Many of these, such as those based on flow cytometry and two-dimensional and pulsed-field gel electrophoresis, give a population-level view of replication. However, an alternative approach, DNA fiber analysis, which was originally developed more than 50 years ago, has the advantage of revealing features of replication at the level of individual DNA fibers. Initially based on autoradiography, this technique has been superseded by immunofluorescence-based detection of incorporated halogenated thymidine analogs. Furthermore, derivations of this technique have been developed to distribute and stretch the labeled DNA fibers uniformly on optically clear surfaces. As described here, one such technique—DNA combing, in which DNA is combed onto silanized coverslips—has been used successfully to monitor replication fork progression and origin usage in budding yeast.

FUNDAMENTALS OF YEAST CHROMOSOMAL REPLICATION

Faithful transmission of chromosomes to daughter cells during cell division is a highly coordinated process in the budding yeast *Saccharomyces cerevisiae*. An integral part of this process is accurate and complete DNA replication during S phase of the cell cycle (reviewed in Kelly and Brown 2000; Bell and Dutta 2002; Remus and Diffley 2009; Araki 2010). DNA replication is initiated from origins marked by the autonomously replicating sequence (ARS) consensus sequence and bound by a multisubunit protein complex called the origin recognition complex.

In G_1 phase, proteins collectively known as the pre-RC are assembled onto origins. Activation of pre-RCs is induced at the start of S phase by the cyclin-dependent and Dbf4-dependent kinases, resulting in activation of the replicative DNA helicase and recruitment of the DNA replication machinery. Double-stranded DNA is unwound, forming a replication "bubble," with two replication forks moving bidirectionally away from the origin.

DNA replication kinetics can be altered by changes in the efficiency of replication origin firing, changes in the rates at which the replication forks move and by changes in the processivity of the individual forks. These types of alterations can be caused by genetic mutations, chemical or physical agents that damage DNA, or inhibitors of nucleotide metabolism.

---

[4]Correspondence: grant.brown@utoronto.ca; christopher.yip@utoronto.ca

Copyright © Cold Spring Harbor Laboratory Press; all rights reserved
Cite this introduction as *Cold Spring Harb Protoc*; doi:10.1101/pdb.top077784

Chapter 6

TECHNIQUES TO STUDY CHROMOSOMAL REPLICATION

Population-Level Technologies

A number of techniques are available to study DNA replication in yeast that give a population-level view of chromosome replication. These include flow cytometry (Tercero and Diffley 2001), pulsed-field gel electrophoresis (Versini et al. 2003), two-dimensional (2D) gel electrophoresis (Friedman and Brewer 1995), density transfer (Raghuraman et al. 2001; McCune et al. 2008), immunoprecipitation of incorporated thymidine analogs (Katou et al. 2003; Poli et al. 2012), and copy-number variation (Raveendranathan et al. 2006; Davidson et al. 2012).

Single-Fiber-Level Methods

As an alternative to the population-level approaches, the technique of DNA fiber analysis reveals the replication profiles of individual replicating chromosomes. The oldest of these techniques is DNA fiber autoradiography (Cairns 1963; Huberman and Riggs 1966, 1968), in which tritiated thymidine that has been incorporated into nascent DNA is detected on spread DNA fibers by autoradiography.

More recently, this has been refined by replacing autoradiography with immunofluorescence detection by incorporation of halogenated thymidine analogs, allowing for relatively fast analysis of samples by microscopy (Gratzner 1982). Furthermore, techniques have been developed to distribute and stretch DNA fibers uniformly on optically clear surfaces, such that fiber length (measured in microns) is uniformly proportional to DNA length (in base pairs). These techniques take two forms, using either microfluidic capillary channels (Sidorova et al. 2009) or an air–water interface to stretch the DNA fibers (Bensimon et al. 1994; Michalet et al. 1997; Yokota et al. 1997). In the latter method, high-molecular-mass DNA binds to a glass surface by means of the DNA ends, and the glass is then pulled out of the DNA solution at a constant rate. The meniscus formed at the air–water interface uniformly stretches the DNA fibers along the entire length of the surface.

DNA combing, combined with labeling of the nascent strand DNA with halogenated thymidine analogs, has been used successfully to monitor replication fork progression and origin usage in budding yeast during normal conditions and when under replication stress (Lengronne et al. 2001; Versini et al. 2003; Tourriere et al. 2005; Luke et al. 2006; Semple et al. 2006; Czajkowsky et al. 2008; Falbo et al. 2009; Crabbe et al. 2010; Cheung-Ong et al. 2012; Ma et al. 2012; Tittel-Elmer et al. 2012; Sheu et al. 2014). In an accompanying protocol, we present a detailed method for this procedure, including preparation of the combing surfaces and the design of a simple machine to prepare the DNA samples (see Protocol 1: Analysis of Replicating Yeast Chromosomes by DNA Combing [Gallo et al. 2015]).

ACKNOWLEDGMENTS

The authors' laboratories are supported by the Canadian Institutes of Health Research, the Natural Sciences and Engineering Research Council of Canada, and the Canadian Cancer Society.

REFERENCES

Araki H. 2010. Cyclin-dependent kinase-dependent initiation of chromosomal DNA replication. *Curr Opin Cell Biol* 22: 766–771.

Bell SP, Dutta A. 2002. DNA replication in eukaryotic cells. *Annu Rev Biochem* 71: 333–374.

Bensimon A, Simon A, Chiffaudel A, Croquette V, Heslot F, Bensimon D. 1994. Alignment and sensitive detection of DNA by a moving interface. *Science* 265: 2096–2098.

Cairns J. 1963. The bacterial chromosome and its manner of replication as seen by autoradiography. *J Mol Biol* 6: 208–213.

Cheung-Ong K, Song KT, Ma Z, Shabtai D, Lee AY, Gallo D, Heisler LE, Brown GW, Bierbach U, Giaever G, et al. 2012. Comparative chemogenomics to examine the mechanism of action of DNA-targeted platinum-acridine anticancer agents. *ACS Chem Biol* 7: 1892–1901.

Crabbe L, Thomas A, Pantesco V, De Vos J, Pasero P, Lengronne A. 2010. Analysis of replication profiles reveals key role of RFC-Ctf18 in yeast replication stress response. *Nat Struct Mol Biol* 17: 1391–1397.

Czajkowsky DM, Liu J, Hamlin JL, Shao Z. 2008. DNA combing reveals intrinsic temporal disorder in the replication of yeast chromosome VI. *J Mol Biol* 375: 12–19.

Davidson MB, Katou Y, Keszthelyi A, Sing TL, Xia T, Ou J, Vaisica JA, Thevakumaran N, Marjavaara L, Myers CL, et al. 2012. Endogenous

DNA replication stress results in expansion of dNTP pools and a mutator phenotype. *EMBO J* **31:** 895–907.

Falbo KB, Alabert C, Katou Y, Wu S, Han J, Wehr T, Xiao J, He X, Zhang Z, Shi Y, et al. 2009. Involvement of a chromatin remodeling complex in damage tolerance during DNA replication. *Nat Struct Mol Biol* **16:** 1167–1172.

Friedman KL, Brewer BJ. 1995. Analysis of replication intermediates by two-dimensional agarose gel electrophoresis. *Methods Enzymol* **262:** 613–627.

Gallo D, Wang G, Yip CM, Brown GW. 2015. Analysis of replicating yeast chromosomes by DNA combing. *Cold Spring Harb Protoc* doi: 10.1101/pdb.prot085118.

Gratzner HG. 1982. Monoclonal antibody to 5-bromo- and 5-iododeoxyuridine: A new reagent for detection of DNA replication. *Science* **218:** 474–475.

Huberman JA, Riggs AD. 1966. Autoradiography of chromosomal DNA fibers from Chinese hamster cells. *Proc Natl Acad Sci* **55:** 599–606.

Huberman JA, Riggs AD. 1968. On the mechanism of DNA replication in mammalian chromosomes. *J Mol Biol* **32:** 327–341.

Katou Y, Kanoh Y, Bando M, Noguchi H, Tanaka H, Ashikari T, Sugimoto K, Shirahige K. 2003. S-phase checkpoint proteins Tof1 and Mrc1 form a stable replication-pausing complex. *Nature* **424:** 1078–1083.

Kelly TJ, Brown GW. 2000. Regulation of chromosome replication. *Annu Rev Biochem* **69:** 829–880.

Lengronne A, Pasero P, Bensimon A, Schwob E. 2001. Monitoring S phase progression globally and locally using BrdU incorporation in TK$^+$ yeast strains. *Nucleic Acids Res* **29:** 1433–1442.

Luke B, Versini G, Jaquenoud M, Zaidi IW, Kurz T, Pintard L, Pasero P, Peter M. 2006. The cullin Rtt101p promotes replication fork progression through damaged DNA and natural pause sites. *Curr Biol* **16:** 786–792.

Ma E, Hyrien O, Goldar A. 2012. Do replication forks control late origin firing in *Saccharomyces cerevisiae*? *Nucleic Acids Res* **40:** 2010–2019.

McCune HJ, Danielson LS, Alvino GM, Collingwood D, Delrow JJ, Fangman WL, Brewer BJ, Raghuraman MK. 2008. The temporal program of chromosome replication: Genomewide replication in clb5 *Saccharomyces cerevisiae*. *Genetics* **180:** 1833–1847.

Michalet X, Ekong R, Fougerousse F, Rousseaux S, Schurra C, Hornigold N, van Slegtenhorst M, Wolfe J, Povey S, Beckmann JS, et al. 1997. Dynamic molecular combing: Stretching the whole human genome for high-resolution studies. *Science* **277:** 1518–1523.

Poli J, Tsaponina O, Crabbe L, Keszthelyi A, Pantesco V, Chabes A, Lengronne A, Pasero P. 2012. dNTP pools determine fork progression and origin usage under replication stress. *EMBO J* **31:** 883–894.

Raghuraman MK, Winzeler EA, Collingwood D, Hunt S, Wodicka L, Conway A, Lockhart DJ, Davis RW, Brewer BJ, Fangman WL. 2001. Replication dynamics of the yeast genome. *Science* **294:** 115–121.

Raveendranathan M, Chattopadhyay S, Bolon YT, Haworth J, Clarke DJ, Bielinsky AK. 2006. Genome-wide replication profiles of S-phase checkpoint mutants reveal fragile sites in yeast. *EMBO J* **25:** 3627–3639.

Remus D, Diffley JF. 2009. Eukaryotic DNA replication control: Lock and load, then fire. *Curr Opin Cell Biol* **21:** 771–777.

Semple JW, Da-Silva LF, Jervis EJ, Ah-Kee J, Al-Attar H, Kummer L, Heikkila JJ, Pasero P, Duncker BP. 2006. An essential role for Orc6 in DNA replication through maintenance of pre-replicative complexes. *EMBO J* **25:** 5150–5158.

Sheu YJ, Kinney JB, Lengronne A, Pasero P, Stillman B. 2014. Domain within the helicase subunit Mcm4 integrates multiple kinase signals to control DNA replication initiation and fork progression. *Proc Natl Acad Sci* **18:** 1899–1908.

Sidorova JM, Li N, Schwartz DC, Folch A, Monnat RJ Jr. 2009. Microfluidic-assisted analysis of replicating DNA molecules. *Nat Protoc* **4:** 849–861.

Tercero JA, Diffley JF. 2001. Regulation of DNA replication fork progression through damaged DNA by the Mec1/Rad53 checkpoint. *Nature* **412:** 553–557.

Tittel-Elmer M, Lengronne A, Davidson MB, Bacal J, Francois P, Hohl M, Petrini JH, Pasero P, Cobb JA. 2012. Cohesin association to replication sites depends on rad50 and promotes fork restart. *Mol Cell* **48:** 98–108.

Tourriere H, Versini G, Cordon-Preciado V, Alabert C, Pasero P. 2005. Mrc1 and tof1 promote replication fork progression and recovery independently of rad53. *Mol Cell* **19:** 699–706.

Versini G, Comet I, Wu M, Hoopes L, Schwob E, Pasero P. 2003. The yeast Sgs1 helicase is differentially required for genomic and ribosomal DNA replication. *EMBO J* **22:** 1939–1949.

Yokota H, Johnson F, Lu H, Robinson RM, Belu AM, Garrison MD, Ratner BD, Trask BJ, Miller DL. 1997. A new method for straightening DNA molecules for optical restriction mapping. *Nucleic Acids Res* **25:** 1064–1070.

Protocol 1

Analysis of Replicating Yeast Chromosomes by DNA Combing

David Gallo,[1,3] Gang Wang,[1,3] Christopher M. Yip,[1,2,3,4] and Grant W. Brown[1,3,4]

[1]Department of Biochemistry, University of Toronto, Toronto, Ontario M5S 1A8, Canada; [2]Department of Chemical Engineering and Applied Chemistry, University of Toronto, Toronto, Ontario M5S 3E5, Canada; [3]Donnelly Centre, University of Toronto, Toronto, Ontario M5S 3E1, Canada

Molecular combing of DNA fibers is a powerful technique to monitor origin usage and DNA replication fork progression in the budding yeast *Saccharomyces cerevisiae*. In contrast to traditional flow cytometry, microarray, or sequencing techniques, which provide population-level data, DNA combing provides DNA replication profiles of individual molecules. DNA combing uses yeast strains that express human thymidine kinase, which facilitates the incorporation of thymidine analogs into nascent DNA. First, DNA is isolated and stretched uniformly onto silanized glass coverslips. Following immunodetection with antibodies that recognize the thymidine analog and the DNA, the DNA fibers are imaged using a fluorescence microscope. Finally, the lengths of newly replicated DNA tracks are measured and converted to base pairs, allowing calculations of the speed of the replication fork and of interorigin distances. DNA combing can be applied to monitor replication defects caused by gene mutations or by chemical agents that induce replication stress. Here, we present a methodology for studying replicating yeast chromosomes by molecular DNA combing. We begin with procedures for the preparation of silanized coverslips and for assembly of a DNA combing machine (DCM) and conclude by presenting a detailed protocol for molecular DNA combing in yeast.

MATERIALS

It is essential that you consult the appropriate Material Safety Data Sheets and your institution's Environmental Health and Safety Office for proper handling of equipment and hazardous materials used in this protocol.

RECIPES: Please see the end of this protocol for recipes indicated by <R>. Additional recipes can be found online at http://cshprotocols.cshlp.org/site/recipes.

Reagents

2-(*N*-morpholino)ethanesulfonic acid (MES) buffer (7:3 [v:v] of MES hydrate:MES sodium salt [50 mM, pH 5.7])
Acetone
α-factor (5 mg/mL in 95% ethanol; stored at −20°C)
Anhydrous ethanol
Anti-BrdU solution (BrdU antibody [AbD Serotec MCA2060], freshly diluted 1:40 in blocking buffer)
Anti-DNA solution (DNA antibody [Millipore MAB3034], freshly diluted 1:50 in blocking buffer)
Antisecondary solution <R>
Argon
β-Agarase I (New England Biolabs M0392)

[4]Correspondence: grant.brown@utoronto.ca; christopher.yip@utoronto.ca

Copyright © Cold Spring Harbor Laboratory Press; all rights reserved
Cite this protocol as *Cold Spring Harb Protoc*; doi:10.1101/pdb.prot085118

Blocking buffer (PBS-T containing 10% [w/v] BSA; freshly prepared and sterilized with a 0.22-μm syringe filter)

BrdU (Sigma-Aldrich B5002; freshly prepared at 10 mg/mL in double-distilled H_2O and filter-sterilized with a 0.22-μm syringe filter)

Chloroform

Cyanoacrylate glue

Double-distilled H_2O (ddH_2O), filtered

EDTA (0.5 M) (optional; see Step 35)

Heptane (anhydrous, 99%; Sigma-Aldrich)

Lambda DNA (Sigma-Aldrich D9768-5U)

Low-melting-point (LMP) agarose (Bioshop AGA101; freshly prepared at 1% [w/v] in 50 mM EDTA [pH 8.0])

Methanol

NaOH (1 M; filtered)

Octenyltrichlorosilane (mixture of isomers, 96% purity; Sigma-Aldrich 539279)

PBS (2 mM KH_2PO_4, 10 mM $NaHPO_4$, 2.7 mM KCl, 137 mM NaCl [pH 7.4])

PBS-T (PBS containing 0.05% [v/v] Tween-20)

Prolong Gold antifade reagent (Molecular Probes 36930)

Pronase (10 mg/mL in double-distilled H_2O; freshly prepared)

Proteinase K solution <R>

SCE buffer <R>

Sodium azide (10% [w/v] in double-distilled H_2O)

TE_{50} buffer (10 mM Tris–HCl [pH 7.0], 50 mM EDTA)

TE buffer (10 mM Tris–HCl [pH 7.0], 1 mM EDTA)

Yeast cultures (see Step 23)

YOYO-1 solution (Molecular Probes Y3601, diluted 1:150 in TE_{50} buffer)

YPD <R>

Equipment

Arduino Uno microcontroller board (www.arduino.cc)

Beakers (500-mL and 100-mL)

Bulldog clips

Cardboard box

Centrifuges (clinical centrifuge and microcentrifuge)

Coplin jar

Coverslip mini-racks (Molecular Probes C14784)

Coverslip staining rack, stainless steel

Desiccation chamber

Drierite desiccant

Drying oven

Dual H-bridge motor driver chip (SN754410 or L293D)

Enclosure (Nalgene 6740-1101 Acrylic Beta Box #3283-9aO)

Filter, sterile 0.22-μm

Flow cytometer

Fluorescence microscope, equipped with a 63× oil-immersion objective, FITC and CY3 filter sets, and a charge-coupled device (CCD) camera

Forceps

Fume hood

Gas regulator

Glass microscope coverslips (22 × 22-mm)

Chapter 6

Glass microscope slides (76 × 26-mm, with frosted end)
Heating block with fittings for 1.5- and 14-mL tubes
Humidity chamber
Hybridization oven
Kimwipes
Liquid water bath shaker
M3 10-mm screws, nuts
M3 25-mm supporting posts
Mini-gel comb
Motor and chassis (Sanyo Denki 103H548-0498 stepping motor)
Pasteur pipette bulb
Pasteur pipette, 9-inch (heated at the end and formed into a U-shaped scoop)
Pencils and waterproof markers
Phase-contrast microscope, equipped with a 40× air objective
PLA filament (1.75-mm-diameter; Solidoodle)
Plasma cleaner (Harrick Plasma, PDC-32G, 115V)
Plug mold (Biorad 170-3713)
Polycarbonate tubes (14-mL round bottom)
Polypropylene centrifuge tubes (50-mL conical)
Razor blade
Retort stand
Retort stand clamp
Rotary cutting tools
Screw cap tubes (2-mL)
Snap action (travel-limit) switches (COM-00098; www.sparkfun.com)
Solidoodle 2 3D-printer
Spectrophotometer
SPST momentary normally open (N.O.) pushbutton (COM-11992; www.sparkfun.com)
Syringe, sterile
Touch screen (Adafruit 2.8″ TFT Touch Shield for Arduino; www.adafruit.com/products/376)
Vacuum pump
Water bath sonicator
Whatman paper

METHOD

Generating Glass Surfaces Suitable for DNA Combing

1. Place eight coverslips into a Teflon mini-rack and then, using forceps to hold the rack, completely submerge the coverslips in a 100-mL beaker filled with acetone to rinse them.

   It is important not to touch the coverslips with anything except forceps during all manipulations.

2. Using forceps, transfer the racks to a 500-mL beaker with 250 mL of 50% methanol in double-distilled water (ddH$_2$O).

   One 500-mL beaker can fit up to four coverslip racks.

3. Secure the beaker in a retort stand clamp and position the retort stand beside the water bath sonicator. Lower the beaker into the bath until the total volume of liquid in the beaker is submerged. Sonicate for 20 min.

   The sonication steps must be carried out in a fume hood. It is important that coverslips remain separated during sonication to ensure uniform cleaning/coating.

4. While in the fume hood, remove racks from the methanol beaker and rinse in a 100-mL beaker filled with chloroform. Place racks in a 500-mL beaker with 250 mL of chloroform and sonicate for 20 min.

5. Remove the racks from the chloroform and transfer the coverslips to a coverslip staining rack. Let residual chloroform evaporate in the fume hood at room temperature (RT).

6. While the coverslips are drying, prepare the plasma cleaner.

    i. Turn on the vacuum pump and attach the front cover with needle valve fully closed.
    ii. Set the RF level to MED and bleed in some air by slightly opening the needle valve for 3–4 sec; if successful, there will be a purple glow visible through the holes on top of the instrument.
    iii. Run with no sample for 10 min. When finished, set RF to OFF and turn off vacuum.
    iv. Fully open the needle valve to allow air into the chamber.
    v. Once atmospheric pressure is reached, remove the front cover.

7. Remove the rack from the fume hood and place into the plasma cleaner. Run plasma cleaner with RF level set to LOW for 10 min.

    Transport rack covered in a plastic box outside of the fume hood to minimize exposure to dust particles.

8. Remove rack from the plasma cleaner and bake in a drying oven for 1 h at 100°C.

    The stainless steel rack is hot after plasma cleaning and drying—use appropriate personal protection equipment.

9. Return the coverslip staining rack to the fume hood and move the coverslips back to the Teflon mini-racks. Place the racks in a 500-mL beaker with 250 mL of heptane. Add 250 μL of octenyltrichlorosilane and swirl gently to mix.

    Once opened, store octenyltrichlorosilane under argon gas in a desiccator with drierite to minimize oxidation and polymerization. Discard opened containers of octenyltrichlorosilane after 3 mo.

10. Place beaker into desiccation chamber with drierite in the fume hood and incubate overnight.

11. Transfer racks to a fresh 500-mL beaker with 250 mL of heptane and sonicate for 5 min.

12. Remove the racks from the heptane and rinse in a 100-mL beaker of ddH$_2$O. Place racks in a 500-mL beaker with 250 mL of ddH$_2$O. Sonicate for 5 min.

    Be careful when transferring racks to ddH$_2$O because the nonpolar heptane can cause the coverslips to stick together.

13. Transfer the racks to a 500-mL beaker with 250 mL of chloroform and sonicate for 5 min.

14. Remove the racks from chloroform and transfer the coverslips back to the coverslip staining rack. Allow excess chloroform to evaporate in the fume hood.

    See Troubleshooting.

15. Store the coverslips in the coverslip staining rack, protected from light and dust, at RT.

Building a Simple Machine Suitable for DNA Combing

16. Assemble the motor shield (Fig. 1C). (The 5-V voltage source and the ground connection are all provided by the Arduino Uno.)

17. Upload dcm_firmware.ino (available as a supplementary file online at http://cshprotocols.cshlp.org/; also available at http://bigten.med.utoronto.ca/tools/open-source_resources/dna-combing-machine) onto the Arduino Uno.

18. Use a rotary cutting tool with a cutting guide to make holes on the enclosure box, as indicated in 3D printing file enclosure.skp (available as a supplementary file online at http://cshprotocols.cshlp.org/; also available at http://bigten.med.utoronto.ca/tools/open-source_resources/dna-combing-machine).

Chapter 6

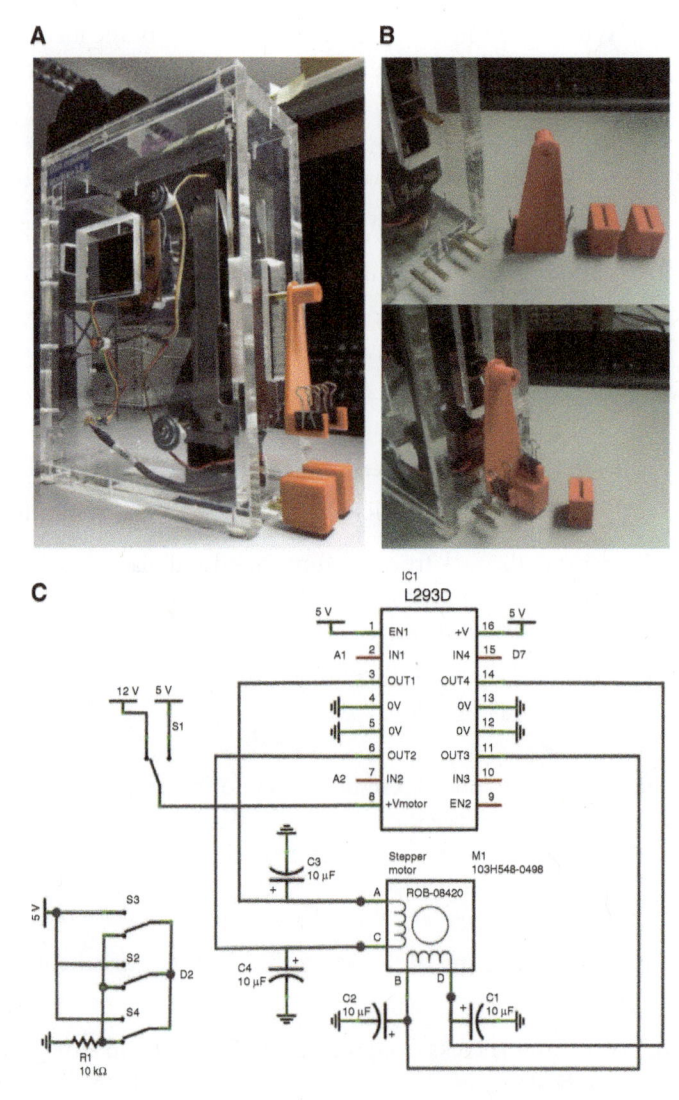

FIGURE 1. A simple machine for analyzing chromosome replication by DNA combing. (*A*) The assembled combing machine. (*B*) Installation of coverslip holder and sample reservoirs on M3 support posts. (*C*) Motor shield circuit diagram. A, Arduino Uno analog pins; D, Arduino Uno digital pins; C, capacitor; EN, enable pin; IC, integrated circuit (L293D); IN, input pin; kΩ, kilohm; M, motor; OUT, output pin; R, resistor; S1, motor power level switch; S2, lower travel-limit switch; S3, upper travel-limit switch; S4, push button; V, volt; Vmotor, voltage input for motor; +V, voltage supply for the integrated circuit; µF, microfarad.

19. 3D-print the coverslip holder and sample reservoir using holders.skp (available as a supplementary file online at http://cshprotocols.cshlp.org/; also available at http://bigten.med.utoronto.ca/tools/open-source_resources/dna-combing-machine).

20. Connect electrical components with jumper wires.

21. Install switches, Arduino Uno, and motor chassis in the enclosure with M3 screws and supporting posts (enclosure.skp; Fig. 1A). Install bulldog clips on coverslip holder. Install coverslip holder and sample reservoir on the motor chassis and enclosure box, using M3 supporting posts (Fig. 1B).

22. Set the operating parameters via the touch screen. Note that in the left column, there are icons for position, speed, incubation timer, type of travel, and calibration. The middle column displays the parameter values. In the right column, there are two icons to increase or decrease the value of each parameter; a third triangular icon modifies the increments for the "increase" and "decrease"

buttons. On selecting a parameter, it is highlighted on the screen and its value is instantly updated as it is modified by means of the "increase" or "decrease" buttons.

  i. Position. The position of the combing stage (in µm) is determined from the stepper motor by the microcontroller. As such, the position is constantly updated on the display during motion. With the "position" button selected, push the "increase" button to raise the DCM stage to a user-defined maximum height (see Step 22.v); conversely, push the "decrease" button to lower the DCM stage to its minimum position (0 µm).

  ii. Speed. The combing speed can be set to between 0 and 1 cm/sec. We use 710 µm/sec.

  iii. Type of travel. The travel button switches between "one-way" and "two-way" travel for the stage; two-way travel returns the stage to the starting position after a user-defined incubation time. The motor can be stopped midway by either the pushbutton or travel-limit switches located above and below the moving stage.

  iv. Incubation timer. For programmed sample incubation, a user can define a time period for which the machine, on reaching its bottom-most position, will pause before initiating withdrawal of the slide.

  v. Range and calibration. The "Range" button allows a user to change the maximum stage height (in µm). This sets the upper travel-limit relative to 0 µm, which is the lower travel-limit. The "Calibration" button enables users to modify the current position thus shifting the entire travel range up or down to accommodate different coverslip dimensions or reservoir depths.

Molecular Combing of DNA Fibers

An overview of the procedure is shown in Figure 2.

Cell Synchronization and BrdU Labeling

23. Grow yeast cultures in YPD liquid broth at 30°C to early logarithmic phase (OD_{600} = 0.20–0.30) in a water bath shaker. Allow at least two cell doublings when diluting from a saturated culture. Remove an aliquot for flow cytometry.

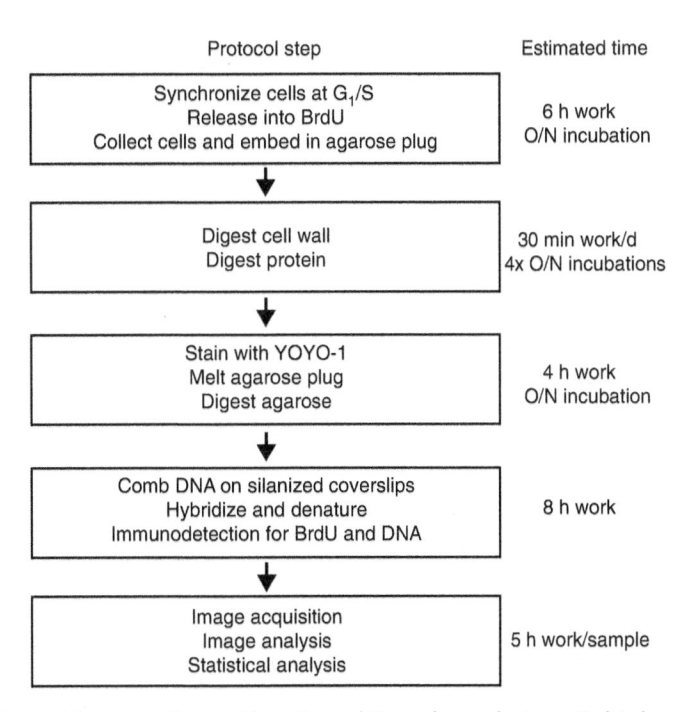

FIGURE 2. The DNA combing procedure, with estimated times for each stage. BrdU, bromodeoxyuridine; O/N, overnight.

S. cerevisiae cells are unable to incorporate BrdU because they lack a thymidine kinase. Strains used for molecular combing must ectopically express the herpes simplex virus thymidine kinase (HSV-TK). Ectopic expression of the human equilibrative nucleoside transporter 1 (hENT1) improves BrdU uptake from the media but is not mandatory for incorporation into DNA.

24. To arrest cells in G_1 phase, add 5 mg/mL α-factor to 2.5 μM final (0.83 μL/mL) and continue growth for 75 min. Add an additional 0.33 μL/mL α-factor (1 μM final) and continue growth for 45 min.

25. Add 40 μL/mL BrdU (400 μg/mL final) and continue growing for 30 min.

    Addition of BrdU while cells are arrested in G_1 phase facilitates uptake into the cells.

26. Inspect culture under a phase-contrast microscope. Approximately >90% of cells should have the "shmoo" morphology—if not, grow culture for an additional 30 min and check again. Remove an aliquot for flow cytometry before proceeding.

27. To release cells into S phase, add pronase to a final concentration of 100 μg/mL (10 μL of stock solution per milliliter of culture) directly to the culture and continue growth. The duration of S phase labeling can be varied, but 30 min is typical.

28. To harvest cells, transfer 20 mL of culture to a precooled 50 mL conical polypropylene centrifuge tube containing 10 μL sodium azide stock per milliliter of culture (0.1% w/v final), mix by inverting several times and incubate on ice for 15 min. If multiple samples are being collected, samples can be left on ice for up to 2 h. Remove an aliquot for flow cytometry.

    If desired, the cell cycle arrest and release can be confirmed by analyzing cellular DNA contents by flow cytometry, using the aliquots collected at Steps 23, 26, and 28.

Agarose Plug Preparation and Digestion

29. Determine the cell density (expressed as cells/mL) and centrifuge 1.2×10^8 cells in prechilled centrifuge tubes at 800g for 3 min at 4°C.

    The density of cells is an important step in plug preparation. If the density is too low, the concentration of DNA fibers will be too low, making the microscopy difficult and time consuming; conversely, if the density is too high, the subsequent plug digestion and melting steps will not be efficient, resulting in clumped and tangled DNA fibers. The cell density used here is a midpoint, and optimal density can vary from this by as much as 10-fold. Density should be optimized for each strain and experimental setup. We analyze samples as indicated by flow cytometry to ensure that the G_1 arrest and release into S phase are as expected.

30. Aspirate the supernatant, wash the pellet in ice-cold TE_{50} buffer, transfer to a prechilled microcentrifuge tube, and centrifuge at 16,000g for 1 min at 4°C.

    Steps 31–33 should be carried out in succession for each sample so the low-melting-point (LMP) agarose does not solidify. Ensure you have enough dissolved 1% LMP agarose solution at 68°C and SCE buffer prepared before continuing.

31. Resuspend pellet in SCE buffer at RT such that the total volume of cells plus buffer is 200 μL. For the density used here, 160 μL of SCE is sufficient.

32. Add 200 μL of 1% LMP agarose and mix by gentle pipetting. Avoid air bubbles.

33. Transfer 100 μL into the plug mold by pipetting down the side to avoid air bubbles or voids. Cast three plugs per sample.

    BrdU is light sensitive, and precautions should be taken from this step forward to minimize light exposure.

34. Incubate plugs in molds for 45 min at 4°C to allow agarose to solidify.

35. Using a Pasteur pipette bulb, eject plugs into 14-mL round-bottom polycarbonate tubes and add 0.5 mL of SCE buffer per plug. Incubate overnight at 37°C.

    Alternatively, plugs can be stored in 0.5 M EDTA in 2-mL screw cap tubes until digestion. This is not optimal but can be done if the samples need to prepared and then shipped to another location.

36. Remove old SCE buffer and replace with the same volume of fresh SCE buffer. Incubate again at 37°C overnight.

    A 15-well SDS-PAGE mini-gel comb can be placed over the top of the tube, allowing the old solution to be poured out while retaining the plugs.

37. Remove SCE buffer and rinse plugs three times in 1 mL of TE$_{50}$ buffer. Add 0.5 mL of prewarmed proteinase K solution per plug and incubate overnight at 50°C.

38. Remove old proteinase K solution and replace with fresh prewarmed proteinase K solution. Incubate overnight at 50°C. Repeat this step once more—making a total of three overnight incubations with proteinase K solution.

39. Remove the last proteinase K solution and wash plugs five times for 10 min in TE$_{50}$ at RT.
    i. Transfer plugs, using a 9-inch Pasteur pipette formed into a U-shaped scoop, to 2-mL screwcap tubes with 1 mL TE$_{50}$.
    ii. Store at 4°C, protected from light.

    Plugs are extremely fragile and should be handled with care; they are stable in TE$_{50}$ at 4°C for many months.

Plug Melting and DNA Combing

40. Remove one plug and transfer to a round-bottom polycarbonate tube. Add 150 µL of YOYO-1 solution and incubate at RT for 30 min.

41. Remove YOYO-1 solution and wash plugs three times for 5 min in 10 mL of TE buffer.

42. Remove last TE wash and incubate in 2 mL of MES buffer for 5 min at RT.

43. Remove MES buffer and replace with 2 mL of fresh MES buffer. Incubate for 10–15 min at 72°C. Gently rock tube horizontally once to disperse agarose and incubate for an additional 10 min at 72°C.

    It is crucial that the agarose plug is completely melted into the MES solution or DNA fibers will appear clumped during analysis. From this point on, the DNA fibers in solution are extremely fragile and must be handled gently to avoid mechanical shearing, which will result in short fibers.

44. Transfer the DNA fiber solution to 42°C and equilibrate for 15 min. Add 3 units of β-agarase I and incubate overnight at 42°C.

    Do not mix the fiber solution.

45. Heat the DNA fiber solution for 10 min to 72°C and cool to RT.

46. Carefully pour the fiber solution into the reservoir of the combing machine (see Steps 16–22).

47. Mount the silanized coverslip (prepared in Steps 1–15), lower it into the solution, and incubate for 5 min.

    The incubation time can be increased to up to 20 min to facilitate more DNA fiber binding to the coverslip.

48. Pull the coverslip out of the solution at a constant speed of 710 µm/sec.

49. Place coverslips on Whatman paper in a cardboard box and bake in a hybridization oven for 90 min at 60°C.

    Be careful to note the orientation of DNA fibers on the coverslip and maintain the direction of the fibers.

50. Mount the coverslip on a glass slide by placing a small drop of cyanoacrylate glue ~1 cm from the clear end. Carefully mount the combed coverslip, centered on the drop of glue, with the end of the coverslip that was clamped pointing toward the frosted end. This orients the DNA fibers parallel to the long side of the glass slide. Label the frosted end of the slide with pencil. Leave to dry in a cardboard box for 5 min at RT.

    See Troubleshooting.

    For optimal immunodetection of BrdU, proceed directly to immunostaining, but, if necessary, mounted coverslips can be stored overnight at −20°C.

Immunodetection

51. Place slides in Coplin jars and dehydrate by incubating sequentially in 70%, 90%, and anhydrous ethanol for 5 min each at RT.

Make sure there is enough volume to completely submerge the coverslip during incubations. Dilute anhydrous ethanol in filtered ddH$_2$O.

52. Remove slides and wipe excess ethanol with a Kimwipe, being careful not to touch the coverslip. Place slides in covered cardboard box and let air dry for 5 min at RT.

    Slides can now be labeled with a waterproof marker.

53. Place slides into a clean Coplin jar and denature DNA in 1 M NaOH for 25 min at RT.

54. Remove NaOH and wash five times for 1 min in PBS and then incubate for 5 min in PBS-T in the Coplin jar.

55. Remove slides from jar, wipe excess PBS-T from around the edges and place in a humidity chamber. Add 21 µL of blocking buffer on the coverslip and place a loose coverslip on top to evenly disperse the liquid. Incubate in the humidity chamber for 30 min at 37°C.

    See Troubleshooting.

56. Dip slide into Coplin jar containing PBS-T to remove coverslip and place back into humid chamber. Add 21 µL of anti-BrdU solution and incubate in humidity chamber for 1 h at 37°C.

57. Remove coverslips and wash three times for 5 min in PBS-T. Add 21 µL of anti-DNA solution and incubate in humidity chamber for 1 h at 37°C.

58. Remove coverslip and wash three times for 5 min in PBS-T. Add 21 µL of antisecondary solution and incubate in humidity chamber for 1 h at 37°C.

59. Remove coverslip and wash three times for 5 min in PBS-T.

60. Wipe excess PBS-T from slide. Add 10 µL of ProLong Gold antifade reagent, cover with a fresh coverslip, and leave to dry in cardboard box overnight at RT.

    Slides can be stored long term at −20°C.

Image Acquisition and Analysis

61. Perform fiber visualization with an appropriate fluorescence microscope equipped with a CCD camera for image acquisition. Acquire images under a 63× oil-immersion objective lens with CY3 and FITC filter sets for ssDNA and BrdU, respectively.

    An image of combed DNA fibers is shown in Figure 3.

62. Analyze images using the open-source software ImageJ (http://rsb.info.nih.gov/ij).

63. Measure lengths in pixels and convert to base pairs using a conversion factor. This factor depends on the magnification of the objective, the pixel size of the CCD camera, and the stretching of DNA fibers. DNA fibers of known length, such as bacteriophage λ DNA, can be combed to determine the conversion factor, as summarized below.

    i. Prepare a 2-mL solution of λ DNA at 250 ng/mL in MES buffer.

       Avoid pipetting or vortexing, which can shear the DNA.

    ii. Heat for 10 min at 65°C (to increase the fraction of monomeric DNA molecules) and then transfer to ice for 10 min.

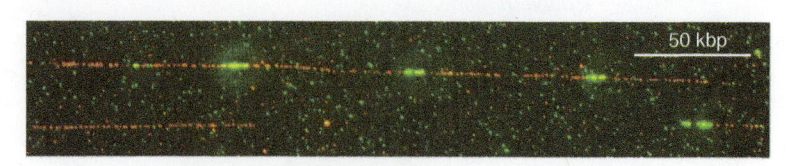

FIGURE 3. Raw merged image of combed DNA fibers. AlexaFluor 546 (red) marks the DNA and AlexaFluor 488 (green) marks the BrdU incorporated into replicating DNA.

iii. Pour the DNA solution into the reservoir of the combing machine (Step 46) and complete Steps 47 through 62 (omitting the steps and reagents used to detect BrdU) to detect the DNA fibers.

iv. Measure the lengths of the DNA fibers (expressed as pixels) and plot a histogram to determine the mode of the main peak (which is the 48502-bp monomer).

v. Use the mode to calculate a pixel per base-pair conversion factor.

> It is routine to observe a sharp peak of very short DNA molecules that represent molecules that have been sheared before the combing steps; smaller peaks at multiples of the mode are concatamers of lambda DNA. As the length of the lambda DNA in pixels depends on the imaging system used, we recommend calibrating the microscope using a micrometer. We find that combed λ DNA is 20–22 µm in length.

64. Depict track lengths and interorigin distance values graphically as box plots. The distributions of these values are non-normal and thus a Mann–Whitney U-test should be performed to determine the statistical significance of differences between sample distributions.

    > BrdU track lengths represent bidirectional replication forks progressing from a single origin.

65. To provide an estimate of replication fork rate, divide the median track length in half and then divide by the labeling time.

66. Calculate the median distance between labeled tracks on the same fiber (interorigin distance) to provide a measure of origin firing efficiency. Alternatively, express origin usage as the number of active origins per mega-base-pair of total DNA.

    > Note that, for accurate interorigin distance (IOD) comparisons between samples, it is important that DNA fibers be consistently of similar lengths and approximately four times longer than the average IOD (Tuduri et al. 2010).

TROUBLESHOOTING

Problem (Step 14): There is a white residue present after the chloroform evaporates on the coverslips.
Solution: The coverslips should be clear with no residue—if the white residue persists, discard the coverslips.

Problem (Step 50): Excess glue seeps from the edge and dries on the coverslip.
Solution: Do not put too much glue on the slide when mounting as it will interfere with subsequent immunodetection steps—if some glue does seep out, use a razor blade to carefully scrape the coverslip clean.

Problem (Step 55): Air bubbles appear under the coverslip.
Solution: To avoid air bubbles, place one edge of the coverslip down and use a pipette tip to help lower the other edge down—all antibody stages (Steps 56–60) are performed in the same manner.

DISCUSSION

In this protocol, we began with procedures for preparation of silanized coverslips (Steps 1–15), then described the assembly of a machine suitable for DNA combing (Steps 16–22), and concluded by presenting a method for molecular DNA combing in yeast (Steps 23–66; Fig. 2). Steps 1–15 featured an adaptation of the liquid-phase silanization procedure described previously by Labit et al. (2008). Alternative procedures for silanization in the vapor phase can also be considered (Schwob et al. 2009). Note too that suitable surfaces are also available commercially (http://www.genomicvision.com).

We next presented a methodology (Steps 16–22) for a robust, easy-to-use, and cost-efficient combing machine to pull coverslips from a reservoir of DNA solution at a constant speed. In this machine (Fig. 1), which can be built for ~$150, a touch screen and a pushbutton act as input devices for a microcontroller, which communicates with a motor shield to drive a stepper motor to control the DCM stage. Note too that suitable combing machines are also available commercially (http://www.genomicvision.com).

In the final part of this protocol, we have outlined our standard DNA combing protocol for detecting incorporation of the halogenated thymidine analog BrdU into newly replicated DNA isolated from budding yeast. It is optimized for use with the E1670 yeast strain that lacks an endogenous thymidine kinase but expresses seven copies of the human thymidine kinase to allow incorporation of halogenated thymidine analogs into nascent DNA (Lengronne et al. 2001). Following pulse-labeling with BrdU, the cells are then embedded into agarose, where the cell wall and protein components are digested. The plug is melted and the DNA is combed onto silanized coverslips, where it is denatured and subjected to immunodetection for BrdU and DNA. The coverslips are then imaged using fluorescence microscopy (Fig. 3) and the images are analyzed using computer software to measure nascent DNA track lengths and distances between replication origins. This procedure is suitable for measuring replication fork rates and replication origin usage. The protocol (Steps 23–66) is adapted from the procedures of the Schwob and Pasero laboratories (Lengronne et al. 2001; Versini et al. 2003; Schwob et al. 2009; Bianco et al. 2012). It is also amenable to more complicated double-labeling procedures involving the sequential addition of IdU and CldU to measure replication fork stalling and fork asymmetry.

For an overview of replication analysis techniques suitable for yeasts, see Introduction: Single-Molecule Analysis of Replicating Yeast Chromosomes (Gallo et al. 2015).

RECIPES

Antisecondary Solution

Prepare blocking buffer by adding bovine serum albumin (BSA) to 10% (w/v) in PBS-T (phosphate-buffered saline with 0.05% [v/v] Tween-20). Then, add Alexa Fluor anti-rat 488 (Molecular probes A11006) at a 1:75 dilution and Alexa Fluor anti-mouse 546 (Molecular probes A11030) at a 1:50 dilution into blocking buffer. Prepare fresh before use.

Proteinase K Solution

1 mg/mL proteinase K
1% (w/v) sarkosyl
10 mM Tris–HCl (pH 7.0)
50 mM EDTA

Prepare fresh. Preheat to 50°C for 30 min before use.

SCE Buffer

1 M sorbitol
100 mM sodium citrate
10 mM EDTA (pH 8.0)
0.125% (v/v) β-mercaptoethanol
10 U/mL zymolyase (Bioshop ZYM001.1)

Add β-mercaptoethanol and zymolyase fresh before use.

YPD

Peptone, 20 g
Glucose, 20 g
Yeast extract, 10 g
H$_2$O to 1000 mL

YPD (YEPD medium) is a complex medium for routine growth of yeast. To prepare plates, add 20 g of Bacto Agar (2%) before autoclaving.

ACKNOWLEDGMENTS

We thank Philippe Pasero and Etienne Schwob for introducing us to the DNA combing procedure. We also extend thanks to Michael Chang, Fred Dong, Johnny Tkach, and Jay Yang for modifications to the procedure and for helpful discussions, and to Michael Lee for help developing the machine. The authors' laboratories are supported by the Canadian Institutes of Health Research, the Natural Sciences and Engineering Research Council of Canada, and the Canadian Cancer Society.

REFERENCES

Bianco JN, Poli J, Saksouk J, Bacal J, Silva MJ, Yoshida K, Lin YL, Tourriere H, Lengronne A, Pasero P. 2012. Analysis of DNA replication profiles in budding yeast and mammalian cells using DNA combing. *Methods* **57**: 149–157.

Gallo D, Wang G, Yip CM, Brown GW. 2015. Single-molecule analysis of replicating yeast chromosomes. *Cold Spring Harb Protoc* doi: 10.1101/pdb.top077784.

Labit H, Goldar A, Guilbaud G, Douarche C, Hyrien O, Marheineke K. 2008. A simple and optimized method of producing silanized surfaces for FISH and replication mapping on combed DNA fibers. *BioTechniques* **45**: 649–652, 654, 656–648.

Lengronne A, Pasero P, Bensimon A, Schwob E. 2001. Monitoring S phase progression globally and locally using BrdU incorporation in TK$^+$ yeast strains. *Nucleic Acids Res* **29**: 1433–1442.

Schwob E, de Renty C, Coulon V, Gostan T, Boyer C, Camet-Gabut L, Amato C. 2009. Use of DNA combing for studying DNA replication in vivo in yeast and mammalian cells. *Methods Mol Biol* **521**: 673–687.

Tuduri S, Tourriere H, Pasero P. 2010. Defining replication origin efficiency using DNA fiber assays. *Chromosome Res* **18**: 91–102.

Versini G, Comet I, Wu M, Hoopes L, Schwob E, Pasero P. 2003. The yeast Sgs1 helicase is differentially required for genomic and ribosomal DNA replication. *EMBO J* **22**: 1939–1949.

CHAPTER 7

Measuring Chromatin Structure in Budding Yeast

Jon-Matthew Belton and Job Dekker[1]

Program in Systems Biology, University of Massachusetts Medical School, Worcester, Massachusetts 01605

Chromosome conformation capture (3C) has revolutionized the ways in which the conformation of chromatin and its relationship to other molecular functions can be studied. 3C-based techniques are used to determine the spatial arrangement of chromosomes in organisms ranging from bacteria to humans. In particular, they can be applied to the study of chromosome folding and organization in model organisms with small genomes and for which powerful genetic tools exist, such as budding yeast. Studies in yeast allow the mechanisms that establish or maintain chromatin structure to be analyzed at very high resolution with relatively low cost, and further our understanding of these fundamental processes in higher eukaryotes as well. Here we provide an overview of chromatin structure and introduce methods for performing 3C, with a focus on studies in budding yeast. Variations of the basic 3C approach (e.g., 3C-PCR, 5C, and Hi-C) can be used according to the scope and goals of a given experiment.

INTRODUCTION

Since chromosomes were first described, understanding the structure of these enormous molecules has been a fundamental topic in biology. For decades, light and electron microscopy have been instrumental in studying the relevance of chromosome structure to genome functions such as gene expression, DNA replication, and chromosome transmission. In 2002, a molecular approach called chromosome conformation capture (3C) (Dekker et al. 2002) was developed to complement these imaging-based methods. 3C relies on formaldehyde cross-linking to detect the physical association of genomic loci in three-dimensional (3D) space (Fig. 1A). Briefly, chromatin is first cross-linked with formaldehyde and then digested with a restriction enzyme. Cross-linked pairs of restriction fragments are ligated together and purified, resulting in a library of chimeric DNA molecules which represent 3D spatial interactions between the fragments. The library can be analyzed in a variety of ways, including via polymerase chain reaction (PCR) and deep sequencing. The frequency with which a pair of restriction fragments is found cross-linked and ligated indicates the likelihood that the two fragments of chromatin interact in the cell. Detailed methods for performing 3C and its related techniques are provided in the protocols that accompany this introduction (see below).

3C-based studies of chromatin conformation have revealed many aspects of chromatin behavior. It is clear that chromatin is not randomly dispersed in the eukaryotic nucleus. In higher eukaryotes (e.g., mouse and human), the genome is organized hierarchically. The highest level of organization is that of chromosome territories, which are on the order of tens to hundreds of megabases in size and are formed because the chromosomes only rarely intermingle with other chromosomes. The next level of chromatin hierarchy is compartments, which are formed by the association of large regions (one to tens of megabases) that have similar properties (active and inactive) and are spatially separated from

---

[1]Correspondence: job.dekker@umassmed.edu

Copyright © Cold Spring Harbor Laboratory Press; all rights reserved
Cite this introduction as *Cold Spring Harb Protoc*; doi:10.1101/pdb.top077552

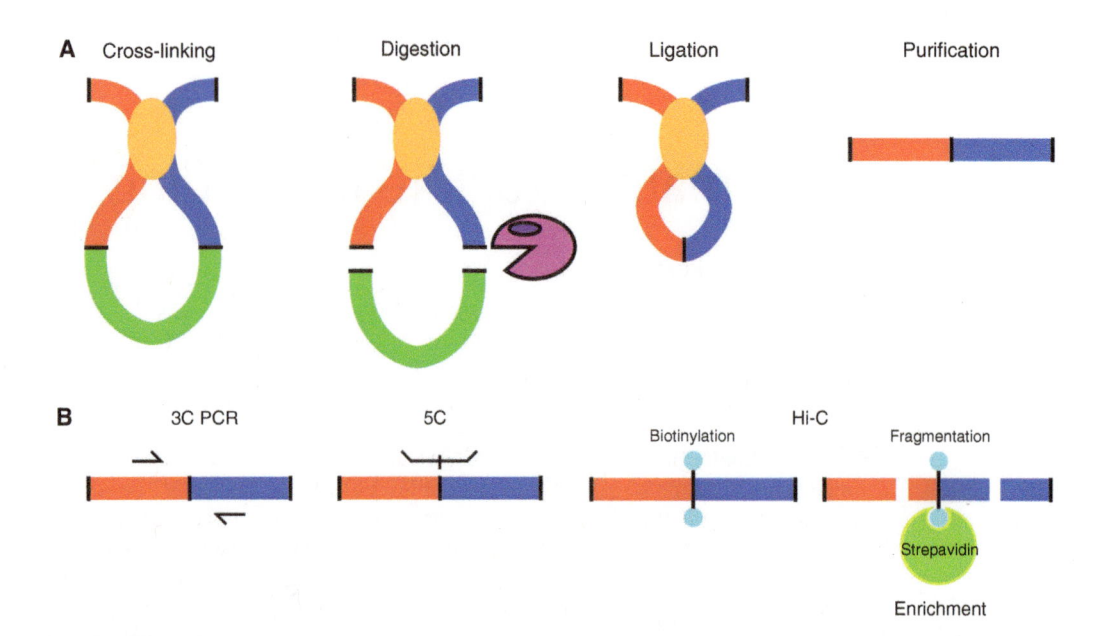

FIGURE 1. Schematic overview of the 3C, 5C, and Hi-C techniques. (*A*) Common steps in 3C-based techniques. First, cells are cross-linked with formaldehyde. This process covalently links protein–DNA complexes in the cell nucleus. The DNA is then digested with the restriction enzyme of choice, breaking up the cross-linked chromatin and freeing covalently linked protein–DNA complexes into solution. These complexes are diluted in a ligation reaction. The dilution favors ligation between two chromatin fragments that are attached to the same complex. After ligation the DNA is purified. (*B*) There are different ways to detect 3C ligation products. Locus-specific PCR primers can be designed to detect specific interactions of interest. To detect many interactions at once, 5C uses pools of hundreds to millions of short oligonucleotides which hybridize to ligation products of interest and are then ligated together. The "carbon-copies" are analyzed by high-throughput sequencing to quantify interaction frequencies. For unbiased and genome-wide analysis using Hi-C, ends of chromatin fragments are biotinylated after digestion but before ligation to mark the ligation junctions. After ligation, DNA is fragmented by sonication and the molecules that contain a biotin at the ligation junction are enriched by pulling down with streptavidin beads. The resulting short molecules are analyzed directly by high-throughput DNA sequencing.

each other in the genome (Lieberman-Aiden et al. 2009; Zhang et al. 2012). Compartments are further subdivided into topologically associating domains (TADs), which are regions (100 kb to >1 Mb) in which chromatin is found preferentially interacting with itself but not with neighboring regions (Dixon et al. 2012; Nora et al. 2012). These hierarchical levels of chromatin structure play a role in many biological functions, such as gene expression (Bau et al. 2011; Sanyal et al. 2012; Jin et al. 2013; Symmons et al. 2014) and DNA repair (Zhang et al. 2012). Defects in this organization have also been implicated in human pathology (e.g., in the premature aging disease Hutchinson–Gilford progeria syndrome [McCord et al. 2013]).

CHROMATIN ORGANIZATION IN YEAST

Interestingly, budding yeast (*Saccharomyces cerevisiae*) chromosomes, which range from hundreds of kilobases to >1 Mb in size, have a very different chromatin topological landscape than mammalian chromosomes. Studies employing 3C-based techniques, live cell imaging, and polymer simulations have revealed global and some local features of yeast nuclear organization. For instance, very strong interactions between pairs of centromeres have been observed (Berger et al. 2008; Duan et al. 2010), initially by microscopy (Jin et al. 1998). These interactions are caused by clustering of centromeres at one pole of the nucleus adjacent to the spindle pole body (SPB). Yeast nuclei display prominent centromere clustering that is reduced in nondividing cells and in meiotic prophase (Jin et al. 1998, 2000). Centromere clustering is also a major determinant of yeast interphase nuclear organization (Jin et al. 2000). Strong interactions between pairs of telomeres located on separate chromosomes have also been observed (Berger et al. 2008; Duan et al. 2010). This phenomenon is consistent with observations that telomeres are tethered to the nuclear periphery (Gotta et al. 1996). The clustering of telomeres coincides with colocalization with Rap1, Sir3, and Sir4 proteins in wild-type *S. cerevisiae* (Gotta et al. 1996; Trelles-Sticken et al. 2000). Meiotic telomere protein Ndj1p is required for meiosis-specific telomere distribution, bouquet formation and efficient homolog pairing (Trelles-Sticken et al. 2000). A limiting of mobility from a 3D volume to a 2D surface, or direct interactions between telomeres, may explain their colocalization with one another.

The conformational biology of yeast Chromosome XII is also unique. The rDNA array is located on the right arm of Chromosome XII. Usually one or two copies of the rDNA locus is included in most assemblies, but in reality there are approximately 150 copies per cell (Kobayashi et al. 1998). This massive array of rDNA repeats forms a crescent-shaped nucleolus which is located at the pole opposite of the SPB (Berger et al. 2008). Therefore, unique constraints are placed on the two portions of the right arm of Chromosome XII. The portion of the right arm upstream of the rDNA array is tethered to both the SPB and the nucleolus, making it the only portion of a chromosome which is tethered to both poles of the nucleus. The portion of the right arm downstream from the rDNA array is the only chromosome section which is anchored at the opposite pole from the SPB and thus projects into the nucleus from a different pole than all of the other chromosome arms.

It has also been observed that there is somewhat of a difference in the behavior of short arms compared to long arms of yeast chromosomes (Tjong et al. 2012). Short arms seem to occupy a volume very close to the SPB, whereas long arms stretch out in the rest of the nucleus. Polymer simulations suggest this is an effect of volume exclusion near the SPB (Tjong et al. 2012).

Some aspects of yeast nuclear organization are not constitutive but rather are acquired, depending on specific cell conditions. For example, SAGA-dependent genes are confined to the nuclear periphery when activated (Cabal et al. 2006). Likewise, it has been shown that a double-stranded break at the MAT locus on Chromosome III, which initiates mating type switching, also becomes confined to the nuclear periphery (Oza et al. 2009). In addition, the hidden MAT left (HML) and hidden MAT right (HMR) loci on the ends of Chromosome III have been shown to not only be confined to the nuclear periphery but to colocalize with one another (Miele et al. 2009).

The properties of chromatin at finer scales (e.g., the sub-TAD scale in mammals) are less well characterized, in part because it is costly to perform comprehensive chromatin interaction studies at the resolution of single restriction fragments in organisms with large genomes. Because the number of possible pairwise chromatin interactions scales with the square of the number of restriction fragments in a genome, organisms with small genomes present a unique opportunity to study the conformation of chromatin at high resolution in a cost-effective manner. *S. cerevisiae* has a genome size of ~12 Mb distributed over 16 chromosomes, it is genetically tractable, and its genome can easily be manipulated. Coupled with the vast depth of knowledge about the physiology of budding yeast, these features allow for the analysis of chromatin interactions in contexts that are not readily available in higher eukaryotes, including an array of unique environmental conditions. Further, the roles of specific proteins suspected to be involved in chromatin organization can be straightforwardly studied. In addition, yeast cultures are easily synchronized to study changes in chromosome conformation during the cell cycle.

3C-BASED TECHNIQUES

3C-based techniques have revolutionized the way we study and think about the structure of chromosomes. The type of technique that should be used for a given study depends on the question and scope of the experiment. All rely on formaldehyde cross-linking of chromatin and subsequent proximity ligation of cross-linked restriction fragments to detect chromatin interactions; the only difference is the manner in which the products are detected (Fig. 1B). In 3C-PCR, locus-specific PCR primers designed to target chromatin fragments of interest are used to detect and quantify specific interactions one at a time. 3C-PCR is applicable when there are only a few (key) regions/loci of interest for which one would like to study physical associations. Typically, experiments with less than 50 restriction fragments to be analyzed can be readily performed using 3C-PCR. Classical 3C-PCR is described in Protocol 1: Chromosome Conformation Capture (3C) in Budding Yeast (Belton and Dekker 2015a).

High-throughput versions of 3C have also been described. 3C-on-ChIP/Circular 3C (4C) (Simonis et al. 2006; Zhao et al. 2006) employs inverse PCR to detect interactions for single loci genome-wide. 4C is suitable for studies of the genome-wide interactions of a single locus of interest, e.g., a centromere. For more comprehensive studies of target regions (e.g., whole chromosomes) that are not genome-wide, chromosome conformation capture carbon copy (5C) is appropriate; see Protocol 3: Chromosome Conformation Capture Carbon Copy (5C) in Budding Yeast (Belton and Dekker 2015b). 5C allows for the high-throughput detection of many chromatin interactions in a single reaction (Dostie et al. 2006). In brief, short oligonucleotide probes are first hybridized to ligation junctions of interest in a 3C library. The probes are then ligated together by *Taq* ligase, in effect making "carbon-copies" of the 3C ligation products. The 5C products are detected by high-throughput sequencing. The types of questions that can be addressed with 5C depend on how the probes are designed. The probes can be arranged for all restriction fragments along a chromosome; this yields comprehensive information on the spatial distribution of the entire chromosome. Such information can be used to generate 3D models of the chromosome (Bau and Marti-Renom 2011; Bau et al. 2011). The probes can also be designed so that one set hybridizes to restriction fragments containing regulatory elements and the other set hybridizes to fragments which overlap genes. This design allows for a comprehensive analysis of the spatial interactions between genes and their regulatory elements (Sanyal et al. 2012). Many other variations of these two generic probe designs are possible. 5C experiments typically use 50–5000 probes to detect hundreds to millions of chromatin interactions in parallel. Note that for both 3C and 5C, production of a random control library is essential to account for intrinsic biases in restriction fragments, probes, and primers; see Protocol 2: Randomized Ligation Control for Chromosome Conformation Capture (Belton and Dekker 2015c).

When the goal of an experiment is to obtain information regarding the spatial organization of a complete genome, Hi-C is the technique of choice; see Protocol 4: Hi-C in Budding Yeast (Belton and Dekker 2015d). Hi-C (Lieberman-Aiden et al. 2009) is an unbiased whole-genome conformation capture assay which uses biotin incorporation to enrich for ligation junctions. In brief, the ends left after restriction digestion are filled in with biotinylated nucleotides before ligation. As a result, ligation junctions are labeled with biotin and can be purified using streptavidin-coated agarose beads. Purified ligation junctions obtained using Hi-C can be analyzed and quantified using high-throughput sequencing to generate a genome-wide interaction map.

REFERENCES

Bau D, Marti-Renom MA. 2011. Structure determination of genomic domains by satisfaction of spatial restraints. *Chromosome Res* 19: 25–35.

Bau D, Sanyal A, Lajoie BR, Capriotti E, Byron M, Lawrence JB, Dekker J, Marti-Renom MA. 2011. The three-dimensional folding of the α-globin gene domain reveals formation of chromatin globules. *Nat Struct Mol Biol* 18: 107–114.

Belton J-M, Dekker J. 2015a. Chromosome conformation capture (3C) in budding yeast. *Cold Spring Harb Protoc* doi: 10.1101/pdb.prot085175.

Belton J-M, Dekker J. 2015b. Chromosome conformation capture carbon copy (5C) in budding yeast. *Cold Spring Harb Protoc* doi: 10.1101/pdb.prot085191.

Belton J-M, Dekker J. 2015c. Randomized ligation control for chromosome conformation capture. *Cold Spring Harb Protoc* doi: 10.1101/pdb.prot085183.

Belton J-M, Dekker J. 2015d. Hi-C in budding yeast. *Cold Spring Harb Protoc* doi: 10.1101/pdb.prot085209.

Berger AB, Cabal GG, Fabre E, Duong T, Buc H, Nehrbass U, Olivo-Marin J-C, Gadal O, Zimmer C. 2008. High-resolution statistical mapping reveals gene territories in live yeast. *Nat Meth* 5: 1031–1037.

Cabal GG, Genovesio A, Rodriguez-Navarro S, Zimmer C, Gadal O, Lesne A, Buc H, Feuerbach-Fournier F, Olivo-Marin J-C, Hurt EC, et al. 2006. SAGA interacting factors confine sub-diffusion of transcribed genes to the nuclear envelope. *Nature* 441: 770–773.

Dekker J, Rippe K, Dekker M, Kleckner N. 2002. Capturing chromosome conformation. *Science* 295: 1306–1311.

Dixon JR, Selvaraj S, Yue F, Kim A, Li Y, Shen Y, Hu M, Liu JS, Ren B. 2012. Topological domains in mammalian genomes identified by analysis of chromatin interactions. *Nature* 485: 376–380.

Dostie J, Richmond TA, Arnaout RA, Selzer RR, Lee WL, Honan TA, Rubio ED, Krumm A, Lamb J, Nusbaum C, et al. 2006. Chromosome Conformation Capture Carbon Copy (5C): A massively parallel solution for mapping interactions between genomic elements. *Genome Res* 16: 1299–1309.

Duan Z, Andronescu M, Schutz K, McIlwain S, Kim YJ, Lee C, Shendure J, Fields S, Blau CA, Noble WS. 2010. A three-dimensional model of the yeast genome. *Nature* 465: 363–367.

Gotta M, Laroche T, Formenton A, Maillet L, Scherthan H, Gasser SM. 1996. The clustering of telomeres and colocalization with Rap1, Sir3, and Sir4 proteins in wild-type *Saccharomyces cerevisiae*. *J Cell Biol* **134**: 1349–1363.

Jin Q-W, Trelles-Sticken E, Scherthan H, Loidl J. 1998. Yeast nuclei display prominent centromere clustering that is reduced in nondividing cells and in meiotic prophase. *J Cell Biol* **141**: 21–29.

Jin QW, Fuchs J, Loidl J. 2000. Centromere clustering is a major determinant of yeast interphase nuclear organization. *J Cell Sci* **113**: 1903–1912.

Jin F, Li Y, Dixon JR, Selvaraj S, Ye Z, Lee AY, Yen C-A, Schmitt AD, Espinoza CA, Ren B. 2013. A high-resolution map of the three-dimensional chromatin interactome in human cells. *Nature* **503**: 290–294.

Kobayashi T, Heck DJ, Nomura M, Horiuchi T. 1998. Expansion and contraction of ribosomal DNA repeats in *Saccharomyces cerevisiae*: Requirement of replication fork blocking (Fob1) protein and the role of RNA polymerase I. *Genes Dev* **12**: 3821–3830.

Lieberman-Aiden E, van Berkum NL, Williams L, Imakaev M, Ragoczy T, Telling A, Amit I, Lajoie BR, Sabo PJ, Dorschner MO, et al. 2009. Comprehensive mapping of long-range interactions reveals folding principles of the human genome. *Science* **326**: 289–293.

McCord RP, Nazario-Toole A, Zhang H, Chines PS, Zhan Y, Erdos MR, Collins FS, Dekker J, Cao K. 2013. Correlated alterations in genome organization, histone methylation, and DNA–lamin A/C interactions in Hutchinson-Gilford progeria syndrome. *Genome Res* **23**: 260–269.

Miele A, Bystricky K, Dekker J. 2009. Yeast silent mating type loci form heterochromatic clusters through silencer protein-dependent long-range interactions. *Plos Genet* **5**: e1000478. doi: 10.1371/journal.pgen.1000478. Epub 2009 May 8.

Nora EP, Lajoie BR, Schulz EG, Giorgetti L, Okamoto I, Servant N, Piolot T, van Berkum NL, Meisig J, Sedat J, et al. 2012. Spatial partitioning of the regulatory landscape of the X-inactivation centre. *Nature* **485**: 381–385.

Oza P, Jaspersen SL, Miele A, Dekker J, Peterson CL. 2009. Mechanisms that regulate localization of a DNA double-strand break to the nuclear periphery. *Genes Dev* **23**: 912–927.

Sanyal A, Lajoie BR, Jain G, Dekker J. 2012. The long-range interaction landscape of gene promoters. *Nature* **489**: 109–113.

Simonis M, Klous P, Splinter E, Moshkin Y, Willemsen R, de Wit E, van Steensel B, de Laat W. 2006. Nuclear organization of active and inactive chromatin domains uncovered by chromosome conformation capture-on-chip (4C). *Nat Genet* **38**: 1348–1354.

Symmons O, Uslu VV, Tsujimura T, Ruf S, Nassari S, Schwarzer W, Ettwiller L, Spitz F. 2014. Functional and topological characteristics of mammalian regulatory domains. *Genome Res* **24**: 390–400.

Tjong H, Gong K, Chen L, Alber F. 2012. Physical tethering and volume exclusion determine higher-order genome organization in budding yeast. *Genome Res* **22**: 1295–1305.

Trelles-Sticken E, Dresser ME, Scherthan H. 2000. Meiotic telomere protein Ndj1p is required for meiosis-specific telomere distribution, bouquet formation and efficient homologue pairing. *J Cell Biol* **151**: 95–106.

Zhang Y, McCord Rachel P, Ho Y-J, Lajoie Bryan R, Hildebrand Dominic G, Simon Aline C, Becker Michael S, Alt Frederick W, Dekker J. 2012. Spatial organization of the mouse genome and its role in recurrent chromosomal translocations. *Cell* **148**: 908–921.

Zhao Z, Tavoosidana G, Sjolinder M, Gondor A, Mariano P, Wang S, Kanduri C, Lezcano M, Singh Sandhu K, Singh U, et al. 2006. Circular chromosome conformation capture (4C) uncovers extensive networks of epigenetically regulated intra- and interchromosomal interactions. *Nat Genet* **38**: 1341–1347.

Protocol 1

Chromosome Conformation Capture (3C) in Budding Yeast

Jon-Matthew Belton and Job Dekker[1]

Program in Systems Biology, University of Massachusetts Medical School, Worcester, Massachusetts 01605

Chromosome conformation capture (3C) is a method for studying chromosomal organization that takes advantage of formaldehyde cross-linking to measure the spatial association of two pieces of chromatin. The 3C method begins with whole-cell formaldehyde fixation of chromatin. After cell lysis, solubilized chromatin is digested with a type II restriction endonuclease, and cross-linked DNA fragments are ligated together. Cross-links are reversed by degradation with proteinase K, and chimeric DNA molecules are purified by standard phenol:chloroform extraction. The resulting 3C library represents chromatin fragments that may be separated by large genomic distances or located on different chromosomes, but are close enough in three-dimensional space for cross-linking. Locus-specific oligonucleotide primers are used to detect interactions of interest in the 3C library using end-point polymerase chain reaction (PCR).

MATERIALS

It is essential that you consult the appropriate Material Safety Data Sheets and your institution's Environmental Health and Safety Office for proper handling of equipment and hazardous materials used in this protocol.

RECIPES: Please see the end of this protocol for recipes indicated by <R>. Additional recipes can be found online at http://cshprotocols.cshlp.org/site/recipes.

Reagents

Agarose
ATP (100 mM)
Bovine serum albumin (BSA) (10 mg/mL)
DNA ladder (1 kb) (New England Biolabs N3232S)
DNA quantification reagents for the method of choice (e.g., Bioanalyzer, qPCR, or fluorometry)
Ethanol (100%)
Formaldehyde (37%) (Fisher Scientific BP531-500)
Glycine (2.5 M, filter-sterilized)
Ligation buffer for 3C (10×) <R>
PCR reagents, including locus-specific oligonucleotide primers for end-point PCR

    3C PCR primers should be designed 100–200 bp upstream of the restriction cut site on the fragments of interest (Fig. 1B). All 3C primers should hybridize to the "−" strand. This ensures that amplification occurs only when the fragments have been digested and then ligated in the opposite (head-to-head) orientation, and eliminates false-positive signals that can result from partial digestion products. Further discussion on 3C primer design and experimental setup can be found in Dekker (2006) and Naumova et al. (2012).

Phase Lock Gel Light tubes (15-mL) (5-Prime 2302840)

[1]Correspondence: job.dekker@umassmed.edu

Copyright © Cold Spring Harbor Laboratory Press; all rights reserved
Cite this protocol as Cold Spring Harb Protoc; doi:10.1101/pdb.prot085175

Measuring Chromatin Structure in Budding Yeast

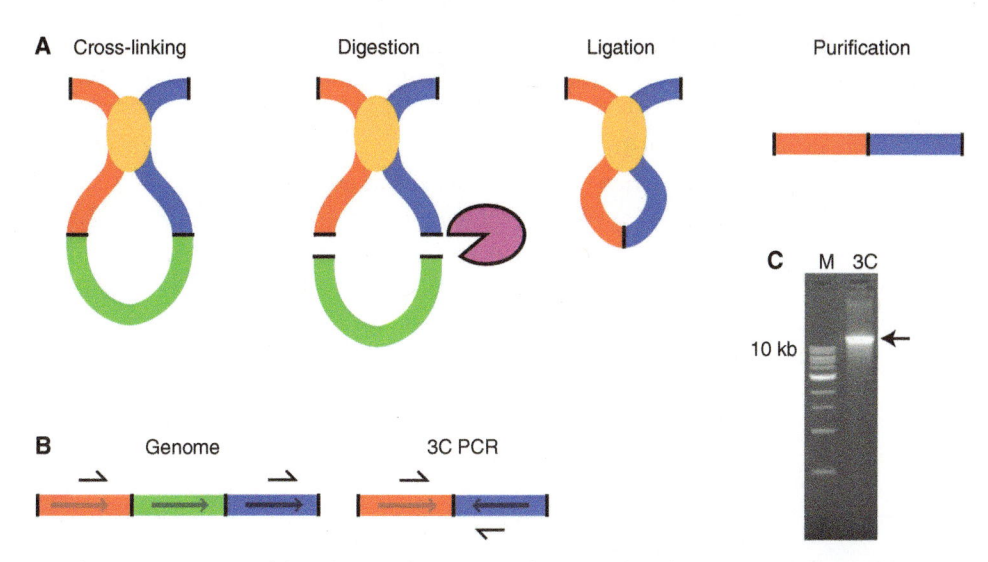

FIGURE 1. Schematic overview of the 3C procedure, primer design, and quality control. (A) A 3C library is generated by cross-linking chromatin with formaldehyde, digesting the chromatin with a restriction enzyme of choice, ligating cross-linked chromatin fragments together, and purifying the chimeric ligation products (which represent spatial interactions between pairs of restriction fragments). (B) Locus-specific 3C primers are designed 100–200 bp upstream of genomic restriction fragments. Each primer is designed to hybridize to the "–" strand so that PCR amplicons will only be formed when the two fragments of interest have been digested. This ensures that only digested ligation products are detected. (C) The 3C library runs at ~12 kb on a 0.8% agarose gel, indicated by the arrow. M, 1-kb DNA ladder.

Phase Lock Gel Light tubes (50-mL) (5-Prime 2302860)
Phenol:chloroform (1:1) <R>
Proteinase K (Life Technologies 25530-015)
  Prepare a solution of 10 mg/mL in TE buffer. <R>
Restriction enzyme, type II (20,000 U/mL), with the appropriate reaction buffer
  Enzymes that rely on NEBuffer 4 should be avoided, as it contains acetate ions that precipitate SDS.
RNase A, DNase-free (10 mg/mL)
Saccharomyces cerevisiae
  Grow yeast cells (using growth medium and conditions appropriate for the strain and experiment) for 2–3 doubling times to obtain mid-log phase cells. Usually, a 200-mL culture produces enough cells to generate two or three 3C libraries.
Sodium acetate (3 M, pH 5.2) <R>
Sodium dodecyl sulfate (SDS) (10% and 1% [w/v])
T4 DNA ligase (1 U/μL) (Life Technologies 15224-017)
TE buffer (1×) <R>
Triton X-100 (10% [v/v])

Equipment

Centrifugal filters (Amicon 15-mL, 30 kDa) (Millipore UFC903024)
Centrifuge (high-speed, refrigerated, for up to 18,000g)
Centrifuge (tabletop, refrigerated, for up to 3100g)
DNA quantification equipment for the method of choice (e.g., Bioanalyzer, qPCR, fluorometry)
Dry ice
Flask (200-mL)
Gel electrophoresis apparatus
Incubator/shaker at 25°C
Liquid nitrogen
Microcentrifuge tubes (1.7-mL)

Chapter 7

Mortar and pestle
> *Immediately before use, precool a mortar and pestle by placing them on dry ice and adding enough liquid nitrogen to the mortar to cover the head of the pestle. Let the liquid nitrogen evaporate before use (Step 9).*

PCR plate (96-well)
Spectrometer to measure optical density at 600 nm
Thermal cycler
Thermomixer or water baths at 16°C, 37°C, and 65°C
Tubes, conical (15-, 50-, and 200-mL)
Tubes, screw cap (35- and 250-mL, suitable for high-speed centrifugation)
Vacuum aspirator
Vortex

METHOD

An overview of the 3C protocol is provided in Figure 1.

Cross-Linking Chromatin

Because the output of the 3C method is the rate of cross-linking, it is essential to standardize cross-linking across all samples. Fixation of whole cells followed by nitrogen grinding, described here, is preferable to fixation of spheroplasts, because the spheroplasting process may affect cell integrity and nuclear organization.

1. Add 37% formaldehyde to the cultured yeast cells to a final concentration of 3% of the culture medium.

2. Shake the culture at 200 rpm for 20 min at 25°C.

3. Quench the cross-linking by adding 2.5 M glycine at 2× the volume of formaldehyde used in Step 1.

4. Shake the culture at 200 rpm for 5 min at 25°C.

5. Transfer the cells to a 200-mL tube and centrifuge the cells at 1800g for 5 min in a tabletop centrifuge.

6. Pour off the medium and wash the cells in 100 mL of sterile H_2O by pipetting up and down until the cell pellet is resuspended.

7. Centrifuge the cells at 1800g for 5 min in a tabletop centrifuge.

8. Pour off the supernatant and resuspend the cells in 5 mL of 1× restriction enzyme buffer by pipetting up and down.

9. Add liquid nitrogen to a prechilled mortar and pour the sample into the liquid nitrogen. Once the sample has frozen, begin to crush it with the pestle. When the sample is broken into little pieces, begin to grind it with the pestle. Grind the sample for 10 min, adding liquid nitrogen as necessary (approximately every 3 min).

10. Scrape the sample into a 50-mL tube on ice. Add 45 mL of ice-cold 1× restriction enzyme buffer to the lysate.

11. Centrifuge the lysate at 1800g for 5 min at 4°C in a tabletop centrifuge.

12. Pour off the supernatant and resuspend the cells in 1× restriction enzyme buffer to an OD_{600} of 10.0.
    The lysate can be stored in 5-mL aliquots at −80°C for several years.

Digesting Cross-Linked Chromatin

It is critical to place the samples on ice after the incubations in Steps 18 and 23, as high temperature reverses formaldehyde cross-links.

13. If frozen, thaw one 5-mL aliquot of lysate on ice. Transfer the lysate to a 50-mL tube and wash with 45 mL of ice-cold 1× restriction enzyme buffer by inverting the tube several times.

14. Centrifuge the lysate at 3100g for 10 min at 4°C in a tabletop centrifuge.
15. Aspirate the supernatant with a vacuum. Resuspend the pellet in 3.8 mL of 1× restriction enzyme buffer.
16. Distribute 38 µL of cells into 96 wells across one 96-well PCR plate.
17. Solubilize the chromatin by adding 3.8 µL of 1% SDS per well. Mix by pipetting up and down but avoid making bubbles.
18. Incubate the plate in a thermal cycler for 10 min at 65°C. Place the plate on ice immediately after the incubation.
19. Quench the SDS by adding 4.4 µL of 10% Triton X-100 per well. Mix by pipetting up and down but avoid making bubbles.
20. Digest the chromatin by adding 5 µL of restriction enzyme (20,000 U/mL) per well. Mix by pipetting up and down but avoid making bubbles.
21. Incubate the plate in a thermal cycler overnight at 37°C.
22. Denature the restriction enzyme by adding 8.6 µL of 10% SDS per well. Mix by pipetting up and down but avoid making bubbles.
23. Incubate the plate in a thermal cycler for 20 min at 65°C. Place the plate on ice immediately after the incubation.

Ligating Cross-Linked Chromatin Fragments

Below, chromatin fragments are ligated in dilute conditions to minimize intermolecular ligation. Because variations in ligation conditions can contribute to variation in background ligation (intermolecular ligation) and reduce the signal-to-noise ratio, it is essential to standardize intramolecular ligation across all samples.

24. Assemble the ligation reaction by adding the following to a 200-mL flask on ice, pooling together the 96 digestion reactions from Step 23. Mix the reaction by gently swirling but avoid making bubbles.

10% Triton X-100	7.15 mL
10× Ligation buffer	7.15 mL
BSA (10 mg/mL)	768 µL
ATP (100 mM)	768 µL
H$_2$O (sterile)	57.2 mL
3C digestion reactions (combined)	5.74 mL
T4 DNA ligase (1 U/µL)	384 µL

25. Split the ligation reaction into eight 15-mL tubes (~9.6 mL per tube) on ice. Incubate all ligation reactions for 4 h at 16°C.

Reverse Cross-Links and Purifying Ligation Products

26. Add 60 µL of proteinase K (10 mg/mL) to each ligation reaction from Step 25. Mix by inverting. Incubate reactions for 4 h at 65°C.
27. Add another 60 µL of proteinase K (10 mg/mL) to each reaction. Mix by inverting. Incubate reactions overnight at 65°C.
28. Transfer each reaction to a 50-mL tube. Add 19.8 mL of phenol:chloroform (1:1) to each tube and vortex each tube for 30 sec.
29. Pour each vortexed sample into a prespun 50-mL tube of Phase Lock Gel Light. Centrifuge at 1500g for 10 min.
30. Pour the aqueous phase of each sample into a fresh 50-mL tube. Add 19.8 mL of phenol:chloroform (1:1) to each tube and vortex each tube for 30 sec.

31. Pour each vortexed sample into a prespun 50-mL tube of Phase Lock Gel Light. Centrifuge at 1500g for 10 min.
32. Pool the aqueous phases from all samples into two 250-mL tubes (~39.6 mL/tube).
33. Precipitate the 3C ligation products by adding 1/10th volumes (3.96 mL) of 3 M sodium acetate (pH 5.2) and 2.5× volumes (99.0 mL) of 100% ethanol to each tube.
34. Incubate the tubes on dry ice for 30–45 min. Make sure the liquid is very cold and thick but not frozen.
35. Centrifuge the tubes at 10,000g for 20 min at 4°C.
36. Carefully pour off the supernatant and discard. Centrifuge at 10,000g for 30 sec at 4°C to collect the remaining drops of liquid at the bottom of the tube.
37. Aspirate the remaining alcohol using a vacuum.
38. Resuspend both 3C DNA pellets in 8 mL total of 1× TE and transfer the sample to a 50-mL tube.
39. Add 2 volumes (16 mL) of phenol:chloroform (1:1) to the sample and vortex for 30 sec.
40. Transfer the sample to a prespun 50-mL tube of Phase Lock Gel Light. Centrifuge in a tabletop centrifuge at 1500g for 10 min.
41. Pour the aqueous phase into a fresh 50-mL tube. Add 2 volumes (16 mL) of phenol:chloroform (1:1) and vortex the tube for 30 sec.
42. Transfer the sample to a prespun 50-mL tube of Phase Lock Gel Light. Centrifuge in a tabletop centrifuge at 1500g for 10 min.
43. Transfer the aqueous phase to a 35-mL tube. Precipitate the 3C DNA with 1/10th volume (800 µL) of sodium acetate (pH 5.2) and 2.5× volumes (20 mL) of 100% ethanol.
44. Incubate the tube on dry ice until the precipitation mixture is thick but not frozen.
45. Centrifuge the tube at 18,000g for 20 min at 4°C.
46. Decant the supernatant and let the 3C DNA pellet air-dry at room temperature.
47. Resuspend the sample in 15 mL of 1× TE.
48. Desalt the sample by adding it to a 15-mL Amicon 30 kDa filter column.
49. Centrifuge the sample in a tabletop centrifuge at 3100g for 15 min.
50. Discard the flowthrough and wash the sample once with 15 mL of 1× TE.
51. Centrifuge the sample in a tabletop centrifuge at 3100g for 15 min.
52. Pipet the 3C DNA sample out of the filter with a 200-µL micropipette and transfer it to a clean 1.7-mL tube.
53. Degrade any co-precipitated RNA by adding 1/10th volume of RNase A (10 mg/mL). Incubate for 1 h at 37°C.
54. Quantify the DNA using the method of choice and electrophorese the sample on a 0.8% agarose gel alongside the 1-kb DNA ladder to verify the 3C products (Fig. 1C).

> The 3C products should appear as a tight band ~12 kb on the gel. Any lower molecular mass smearing indicates a poor quality library and is probably caused by overvigorous lysis.
>
> In addition to 3C-PCR (below), the resulting library alternatively can be used for 5C; see Protocol 3: Chromosome Conformation Capture Carbon Copy (5C) in Budding Yeast (Belton and Dekker 2015a).

Performing End-Point 3C-PCR

55. Titrate all libraries to be used in a given experiment to determine the amount needed to obtain quantitative PCR signals as follows.

    i. Begin with 10 ng of the 3C library. Prepare 8–10 twofold serial dilutions.

ii. Perform PCR with each dilution using several sets of primers to detect interactions from a variety of genomic distances. For instance, use one primer pair to detect a long-range interaction (100–150 kb), another to detect a medium-range interaction (50–70 kb), and third to detect a close-range interaction (10–20 kb).

iii. Quantify the resulting products via gel electrophoresis and plot the quantities graphically.

iv. Choose a DNA amount which is in the linear range of detection for all three distances in all 3C libraries of interest, and use this amount for PCR with all primer pairs in the 3C experiment.

56. Conduct end-point PCR using locus-specific oligonucleotide primers to detect specific interactions of interest.

RELATED INFORMATION

When conducting 3C-PCR, it is essential to account for intrinsic biases in restriction fragments and primers. This can be accomplished by generating a random control library; see Protocol 2: Randomized Ligation Control for Chromosome Conformation Capture (Belton and Dekker 2015b).

RECIPES

Ligation Buffer for 3C (10×)

Reagent	Amount to add (for 1 L)	Final concentration (10×)
Tris–HCl (1 M, pH 7.5)	500 mL	500 mM
MgCl$_2$ (1 M)	100 mL	100 mM
Dithiothreitol (DTT) (1 M)	100 mL	100 mM

Prepare in deionized H$_2$O. Store at −20°C in 15-mL aliquots.

Phenol:Chloroform (1:1)

In a chemical fume hood, adjust the pH of the phenol to 8.0 with Tris buffer. Shake vigorously. Mix 500 mL of phenol (pH 8.0) and 500 mL of chloroform in a 1-L glass bottle with a lid. Shake the mixture vigorously and let it separate overnight at 4°C. Store at 4°C for up to 1 mo.

Sodium Acetate (3 M, pH 5.2)

Dissolve 246.1 g of sodium acetate in 500 mL of deionized H$_2$O. Adjust the pH to 5.2 with glacial acetic acid. Allow the solution to cool overnight. Adjust the pH once more to 5.2 with glacial acetic acid. Adjust the final volume to 1 L with deionized H$_2$O and filter-sterilize.

TE Buffer

Reagent	Quantity (for 100 mL)	Final concentration
EDTA (0.5 M, pH 8.0)	0.2 mL	1 mM
Tris-Cl (1 M, pH 8.0)	1 mL	10 mM
H$_2$O	to 100 mL	

REFERENCES

Belton J-M, Dekker J. 2015a. Chromosome conformation capture carbon copy (5C) in budding yeast. *Cold Spring Harb Protoc* doi: 10.1101/pdb.prot085191.

Belton J-M, Dekker J. 2015b. Randomized ligation control for chromosome conformation capture. *Cold Spring Harb Protoc* doi: 10.1101/pdb.prot085183.

Dekker J. 2006. The three "C"s of chromosome conformation capture: Controls, controls, controls. *Nat Methods* 3: 17–21.

Naumova N, Smith EM, Zhan Y, Dekker J. 2012. Analysis of long-range chromatin interactions using Chromosome Conformation Capture. *Methods* 58: 192–203.

Protocol 2

Randomized Ligation Control for Chromosome Conformation Capture

Jon-Matthew Belton and Job Dekker[1]

Program in Systems Biology, University of Massachusetts Medical School, Worcester, Massachusetts 01605

In experiments using chromosome conformation capture followed by PCR (3C-PCR) or chromosome conformation capture carbon copy (5C), it is critical to control for intrinsic biases in the restriction fragments of interest and the probes or primers used for detection. Characteristics such as GC%, annealing temperature, efficiency of 3C primers or 5C probes, and length of restriction fragment can cause variations in primer or probe performance and fragment ligation efficiency. Bias can be measured empirically by production of a random control library, as described here, to be used with the 3C library of interest.

MATERIALS

It is essential that you consult the appropriate Material Safety Data Sheets and your institution's Environmental Health and Safety Office for proper handling of equipment and hazardous materials used in this protocol.

RECIPES: Please see the end of this protocol for recipes indicated by <R>. Additional recipes can be found online at http://cshprotocols.cshlp.org/site/recipes.

Reagents

3C library from Protocol 1: Chromosome Conformation Capture (3C) in Budding Yeast (Belton and Dekker 2015a)

3C-PCR and/or 5C reagents, including oligonucleotide primers and/or probes from Protocol 1: Chromosome Conformation Capture (3C) in Budding Yeast (Belton and Dekker 2015a) or Protocol 3: Chromosome Conformation Capture Carbon Copy (5C) in Budding Yeast (Belton and Dekker 2015b)

Agarose
ATP (10 mM)
Bovine serum albumin (BSA) (1 mg/mL)
DNA quantification reagents for the method of choice (e.g., Bioanalyzer, qPCR, or fluorometry)
EDTA (0.5 M, pH 8.0)
Ethanol (100% and 80%)
Isopropanol (100%)
Ligation buffer for 3C (10×) <R>
Lysing buffer I, freshly prepared <R>

---

[1]Correspondence: job.dekker@umassmed.edu

Copyright © Cold Spring Harbor Laboratory Press; all rights reserved
Cite this protocol as *Cold Spring Harb Protoc*; doi:10.1101/pdb.prot085183

Chapter 7

    Phase Lock Gel Light (2-mL tubes) (5-Prime 2302820)
    Phase Lock Gel Light (50-mL tubes) (5-Prime 2302860)
    Phenol:chloroform (1:1) <R>
    Potassium acetate (5 M)
    Proteinase K
        Prepare a solution of 20 mg/mL in TE buffer.
    Restriction enzyme used to generate the 3C library in Protocol 1: Chromosome Conformation Capture (3C) in Budding Yeast (Belton and Dekker 2015a), with the appropriate 10× reaction buffer
    RNase A, DNase-free (10 mg/mL)
    Saccharomyces cerevisiae (200-mL overnight culture, grown in appropriate medium and conditions)
    Sodium acetate (3 M, pH 5.2) <R>
    Spheroplasting buffer II, freshly prepared <R>
    T4 DNA ligase (1 U/μL) (Life Technologies 15224-017)
    TE buffer <R>

Equipment

    Centrifugal filters (Amicon 15-mL, 30 kDa) (Millipore UFC903024)
    Centrifugal filters (Amicon 500-μL, 30 kDa) (Millipore UFC5030BK)
    Centrifuge (high-speed, refrigerated, for up to 18,000g)
    Centrifuge (tabletop, for up to 3100g)
    DNA quantification equipment for the method of choice (e.g., Bioanalyzer, qPCR, or fluorometry)
    DNA sequencer for 5C
    Gel electrophoresis apparatus
    Microcentrifuge (refrigerated)
    Microcentrifuge tubes (1.7- and 2.0-mL)
    PCR plate (96-well)
    Spectrometer to measure optical density at 600 nm
    Thermal cycler
    Thermomixer or water baths at 16°C, 37°C, and 65°C
    Tubes, conical (50- and 200-mL)
    Tubes, screw cap (35-mL, suitable for high-speed centrifugation)
    Vortex

METHOD

Isolating Yeast Chromosomal DNA

1. Transfer the cultured yeast cells to a 200-mL tube. Centrifuge the cells at 3100g for 10 min in a tabletop centrifuge. Remove the supernatant promptly.

2. Resuspend the cells in 20 mL of spheroplasting buffer II and transfer to a 35-mL tube. Incubate the cells for 40 min at 37°C.

    The solution should appear stringy.

3. Add 4 mL of lysing buffer I.

4. Add 400 μL of proteinase K (20 mg/mL) and incubate the tube for 30 min at 65°C.

    The solution should now appear clearer.

5. Add 4 mL of 5 M potassium acetate and incubate for 10 min in ice water.

6. Pellet the cell debris by centrifuging the tube at 18,000g for 20 min at 4°C.

7. Divide the supernatant into four fresh 35-mL tubes (7.1 mL per tube). To each tube, add 16.4 mL of ice-cold 100% ethanol. Invert the tubes five times to mix.
8. Precipitate the DNA by centrifuging the samples at 18,000g for 10 min at 4°C.
9. Remove the supernatant carefully and let the DNA pellets dry completely.
10. Resuspend all four DNA pellets in a total of 20 mL of 1× TE buffer containing 10 µg/mL RNase A and transfer to one 50-mL tube.
11. Incubate the sample for 30 min at 37°C to degrade RNA. Occasionally tap the tube gently to mix.
12. Add 20.0 mL (1 volume) of phenol:chloroform (1:1) to the sample. Vortex the mixture for 30 sec and then transfer the sample to a prespun 50-mL tube of Phase Lock Gel Light.
13. Centrifuge the sample at 1500g for 10 min in a tabletop centrifuge to separate the organic and aqueous phases.
14. Transfer the aqueous (upper) phase to two fresh 35-mL tubes (10.0 mL per tube).
15. Precipitate the DNA by adding 10 mL of isopropanol to each tube and invert the tubes five times to mix.
    DNA should be visible as a string-like ball after isopropanol is added.
16. Centrifuge the sample at 18,000g for 10 min at room temperature. Remove the supernatant from the DNA pellet.
17. Wash the DNA pellet by resuspending with 20 mL of 80% ethanol with mild agitation.
18. Centrifuge the sample at 18,000g for 10 min at room temperature. Remove the supernatant from the DNA pellet and allow it to dry at room temperature.
19. Resuspend the DNA pellet in 2 mL of 1× TE buffer.
20. Determine the DNA concentration using absorption spectroscopy.

Digesting Genomic DNA

21. Prepare 20 digests in 1.7-mL tubes containing the following reagents per tube.

Genomic DNA from Step 20	10.0 µg
10× restriction enzyme buffer	40.0 µL
Restriction enzyme	60 U
H$_2$O (sterile)	to 400 µL

22. Incubate the reactions for 3 h at 37°C.
23. Pool all 20 reactions in a 50-mL tube.
    The total volume will be 8.0 mL.
24. Extract the restriction enzyme from the reaction by adding 16.0 mL of phenol:chloroform (1:1). Vortex the extraction mixture for 30 sec.
25. Transfer the vortexed mixture to a prespun 50-mL tube of Phase Lock Gel Light.
26. Centrifuge the sample at 1500g for 10 min in a tabletop centrifuge.
27. Pour the upper aqueous phase into a fresh 35-mL tube. Precipitate the DNA by adding 800.0 µL (1/10th volume) of 3 M sodium acetate (pH 5.2) and 20.0 mL (2.5 volumes) of 100% ethanol. Mix by inversion.
28. Incubate the sample at −80°C until the solution becomes thick but not frozen (~1 h).
29. Centrifuge the sample for 10 min at 18,000g at 4°C. Pour off the supernatant and let the DNA pellet dry at room temperature.
30. Resuspend the DNA pellet in 15 mL of sterile H$_2$O. Vortex to dissolve the pellet.

31. Add the sample to a 15-mL Amicon filter column and centrifuge at 3100g in a tabletop centrifuge until the sample volume is concentrated to 400 µL.

Ligating Digested Genomic Restriction Fragments

32. Prepare 20 reactions containing the following reagents in 1.7-mL tubes.

Digested DNA from Step 31	20.0 µL
Ligation buffer (10×)	3.0 µL
BSA (1 mg/mL)	3.0 µL
ATP (10 mM)	3.0 µL
T4 DNA ligase (1 U/µL)	1.0 µL

33. Incubate the reactions for 1 h at 16°C.

34. Pool all 20 reactions in a 2-mL tube.
    The total volume will be 600 µL.

35. Stop the reaction by adding 12.0 µL of 0.5 M EDTA (pH 8.0).

36. Add 1.2 mL (2 volumes) of phenol:chloroform (1:1) and vortex for 30 sec.

37. Divide the extraction mixture into two prespun 2-mL tubes of Phase Lock Gel Light (900 µL per tube). Centrifuge the tubes in a microcentrifuge at 18,000g for 5 min.

38. Transfer the upper aqueous phase from each tube to a fresh 1.7-mL tube with a micropipette.
    The volume should be 300 µL per tube.

39. Precipitate the DNA by adding 30.0 µL of 3 M sodium acetate (pH 5.2) and 750 µL of 100% ethanol to each tube. Mix by inversion.

40. Centrifuge the tubes in a microcentrifuge at 18,000g for 10 min at 4°C.

41. Aspirate the supernatant with a vacuum or pipette and let the DNA pellet air dry at room temperature.

42. Resuspend both pellets in a total volume of 500 µL of 1× TE.

43. Wash out any excess salt as follows.

    i. Transfer the sample to a 500-µL Amicon filter column.

    ii. Centrifuge in a microcentrifuge at 18,000g for 5 min.

    iii. Pour out the flowthrough in the bottom of the collection tube.

    iv. Bring the volume up to 500 µL with 1× TE and repeat Steps 43.ii–43.iii.

    v. Repeat Step 43.iv for a total of two washes.

44. Pipette the sample out of the filter with a 200-µL micropipette and transfer it to a fresh collection tube. Adjust the final volume to 400 µL with 1× TE. Quantify the DNA using the method of choice.

45. Titrate the library using 3C-PCR primers as follows.

    i. Begin with 100 ng of the control library. Prepare 8–10 twofold serial dilutions.

    ii. Perform PCR with each dilution using primers to detect interactions from a variety of genomic distances. For instance, use one primer pair to detect a long-range interaction (100–150 kb), another to detect a medium-range interaction (50–70 kb), and third to detect a close-range interaction (10–20 kb).

    iii. Quantify the resulting products via agarose gel electrophoresis and plot the quantities graphically.
        This step verifies that the library is a true random ligation control. All of the titration curves should be very similar since there is no longer any biology represented in this control library.

iv. Choose a DNA amount which is in the linear range of detection for all three distances, and use this amount of control library for PCR with all primer pairs in the experiment.

46. Conduct 3C-PCR or 5C as described in Protocol 1: Chromosome Conformation Capture (3C) in Budding Yeast (Belton and Dekker 2015a) or Protocol 3: Chromosome Conformation Capture Carbon Copy (5C) in Budding Yeast (Belton and Dekker 2015b) using the control and experimental libraries.

- To normalize an experimental 3C library using 3C-PCR, amplify the interactions of interest and quantify the gel band signals for each primer pair using both the experiment and the control libraries. Calculate the ratio of the experimental signal to the control signal.

- To correct 5C data, prepare a 5C library from the control library as described in Protocol 3: Chromosome Conformation Capture Carbon Copy (5C) in Budding Yeast (Belton and Dekker 2015b). Sequence this library and use the interaction frequencies to estimate the inherent bias of each pair of interacting restriction fragments.

DISCUSSION

In this protocol, a random control library is produced by digesting naked genomic DNA with the same restriction enzyme used in the 3C experiment of interest (Dekker 2006; Naumova et al. 2012). Contrary to the standard 3C protocol, ligation of these restriction fragments is performed in concentrated conditions to maximize the formation of random inter-molecular combinations of chimeric ligation products. The resulting ligation product library is used for either 3C-PCR or 5C as normal. The signal produced from any combination of 3C primers or 5C probes is assumed to be the same since all biological signal has been removed from the data; therefore, any deviation in the signal can be attributed to the inherent biases of the fragments being interrogated and the probes and primers (Imakaev et al. 2012; Yaffe and Tanay 2011). The experimental 3C/5C data can be corrected for these biases using the empirically measured deviations.

RECIPES

Ligation Buffer for 3C (10×)

Reagent	Amount to add (for 1 L)	Final concentration (10×)
Tris–HCl (1 M, pH 7.5)	500 mL	500 mM
MgCl$_2$ (1 M)	100 mL	100 mM
Dithiothreitol (DTT) (1 M)	100 mL	100 mM

Prepare in deionized H$_2$O. Store at −20°C in 15-mL aliquots.

Lysing Buffer I

Reagent	Amount to add (for 100 mL)	Final concentration
EDTA (0.5 M, pH 8.0)	50 mL	0.25 M
Tris base (1.5 M)	33.3 mL	0.5 M
SDS (15%, v/v)	16.7 mL	2.5% (v/v)

Prepare fresh before each experiment.

Phenol:Chloroform (1:1)

In a chemical fume hood, adjust the pH of the phenol to 8.0 with Tris buffer. Shake vigorously. Mix 500 mL of phenol (pH 8.0) and 500 mL of chloroform in a 1-L glass bottle with a lid. Shake the mixture vigorously and let it separate overnight at 4°C. Store at 4°C for up to 1 mo.

Sodium Acetate (3 M, pH 5.2)

Dissolve 246.1 g of sodium acetate in 500 mL of deionized H$_2$O. Adjust the pH to 5.2 with glacial acetic acid. Allow the solution to cool overnight. Adjust the pH once more to 5.2 with glacial acetic acid. Adjust the final volume to 1 L with deionized H$_2$O and filter-sterilize.

Spheroplasting Buffer II

Reagents	Amount to add (for 100 mL)	Final concentration
Sodium phosphate buffer (100 mM, pH 7.2)	10 mL	10 mM
EDTA (500 mM)	2.0 µL	10 mM
β-Mercaptoethanol	1 mL	1% (v/v)
Zymolyase 100-T (20 mg/mL)	0.5 mL	100 µg/mL

Prepare fresh before each experiment.

TE Buffer

Reagent	Quantity (for 100 mL)	Final concentration
EDTA (0.5 M, pH 8.0)	0.2 mL	1 mM
Tris-Cl (1 M, pH 8.0)	1 mL	10 mM
H$_2$O	to 100 mL	

Zymolyase 100-T (20 mg/mL)

Reagent	Amount to add (for 1 mL)	Final concentration
Zymolyase 100-T	20 mg	20 mg/mL
Glucose (40%, w/v)	50.0 µL	2% (w/v)
Tris (1 M, pH 7.5)	50.0 µL	50 mM

Prepare in sterile H$_2$O. Store for up to 1 month at 4°C.

REFERENCES

Belton J-M, Dekker J. 2015a. Chromosome conformation capture (3C) in budding yeast. *Cold Spring Harb Protoc* doi: 10.1101/pdb.prot085175.

Belton J-M, Dekker J. 2015b. Chromosome conformation capture carbon copy (5C) in budding yeast. *Cold Spring Harb Protoc* doi: 10.1101/pdb.prot085191.

Dekker J. 2006. The three 'C's of chromosome conformation capture: Controls, controls, controls. *Nat Methods* 3: 17–21.

Imakaev M, Fudenberg G, McCord RP, Naumova N, Goloborodko A, Lajoie BR, Dekker J, Mirny LA 2012. Iterative correction of Hi-C data reveals hallmarks of chromosome organization. *Nat Methods* 9: 999–1003.

Naumova N, Smith EM, Zhan Y, Dekker J. 2012. Analysis of long-range chromatin interactions using Chromosome Conformation Capture. *Methods* 58: 192–203.

Yaffe E, Tanay A. 2011. Probabilistic modeling of Hi-C contact maps eliminates systematic biases to characterize global chromosomal architecture. *Nat Genet* 43: 1059–1065.

Protocol 3

Chromosome Conformation Capture Carbon Copy (5C) in Budding Yeast

Jon-Matthew Belton and Job Dekker[1]

Program in Systems Biology, University of Massachusetts Medical School, Worcester, Massachusetts 01605

Chromosome conformation capture carbon copy (5C) is a high-throughput method for detecting ligation products of interest in a chromosome conformation capture (3C) library. 5C uses ligation-mediated amplification (LMA) to generate carbon copies of 3C ligation product junctions using single-stranded oligonucleotide probes. This procedure produces a 5C library of short DNA molecules which represent the interactions between the corresponding restriction fragments. The 5C library can be amplified using universal primers containing the Illumina paired-end adaptor sequences for subsequent high-throughput sequencing.

MATERIALS

It is essential that you consult the appropriate Material Safety Data Sheets and your institution's Environmental Health and Safety Office for proper handling of equipment and hazardous materials used in this protocol.

Reagents

3C library DNA (5.33 ng/µL) from Protocol 1: Chromosome Conformation Capture (3C) in Budding Yeast (Belton and Dekker 2015)

5C probes, forward and reverse
  5C experiments typically use 50–5000 probes, which should be designed according to the experiment. See Discussion.

Agarose

ATP (10 mM)

DNA purification reagents for gel extraction

DNA quantification reagents for the method of choice (e.g., Bioanalyzer, qPCR, or fluorometry)

dNTPs (100 mM)

Low-molecular-weight DNA ladder (New England BioLabs N3233S)

NEBuffer 4 (10×) (New England BioLabs B7004S)

PfuUltra II Fusion HS DNA polymerase with 10× reaction buffer (Agilent Technologies 600850)

T4 polynucleotide kinase (10 U/µL) with 10× PNK reaction buffer (New England BioLabs M0201S)

Taq DNA ligase (40 U/µL) with 10× reaction buffer (New England BioLabs M0208S)

---

[1]Correspondence: job.dekker@umassmed.edu

Copyright © Cold Spring Harbor Laboratory Press; all rights reserved
Cite this protocol as *Cold Spring Harb Protoc*; doi:10.1101/pdb.prot085191

Universal oligonucleotide primers containing Illumina paired-end sequences (HPLC-purified)
Universal Primer 1: 5′-AATGATACGGCGACCACCGAGATCTACACTCTTTCCCTACACGAC GCTCTTCCGATCTCCTCTCTATGGGCAGTCGGTGAT-3′ (80 µM)
Universal Primer 2: 5′-CAAGCAGAAGACGGCATACGAGATCGGTCTCGGCATTCCTGCT GAACCGCTCTTCCGATCTCTGCCCCGGGTTCCTCATTCTCT-3′ (80 µM)

Universal primers for 5C should contain adapter sequences for a specific sequencing platform of choice. This saves time and money when preparing the library for sequencing. The above primers contain overhanging Illumina paired-end sequences, which are incorporated into the molecule after amplification.

Equipment

DNA quantification equipment for the method of choice (e.g., Bioanalyzer, qPCR, or fluorometry)
Gel electrophoresis apparatus (for long gels; see Step 18)
Microcentrifuge tubes (1.7-mL)
PCR plates (96-well)
PCR tube strips, eight-well (100 µL)
Scalpel
Thermal cycler
Thermomixer or water baths at 37°C and 65°C
Tubes, conical (15-mL)
UV light box and face shield

METHOD

An overview of the 5C protocol is provided in Figure 1.

Preparing 5C Probes

1. In a 1.7-mL tube, prepare an equimolar mixture of all 5C forward probes. Dilute the mixture to a final concentration of 0.2 fmol per probe per µL.

2. In a separate 1.7-mL tube, prepare an equimolar mixture of all 5C reverse probes.

3. Phosphorylate the reverse probes.

    i. Combine the following reagents in a 1.7-mL tube.

Reverse probe mix from Step 2	300 pmol
T4 polynucleotide kinase (10 U/µL)	1.0 µL
ATP (10 mM)	5.0 µL
PNK reaction buffer (10×)	5.0 µL
H$_2$O (sterile)	to 50.0 µL

    ii. Incubate the reaction for 45 min at 37°C.

    iii. Incubate the reaction for 20 min at 65°C to inactivate the enzyme.

4. Dilute the reverse probe mixture to a final concentration of 0.2 fmol per probe per µL.

5. Mix equal proportions of the forward and reverse probe mixtures to a final concentration of 0.1 fmol per probe per µL.

    Large batches of this primer pool can be prepared in advance, divided into aliquots and stored at −20°C for several years.

Measuring Chromatin Structure in Budding Yeast

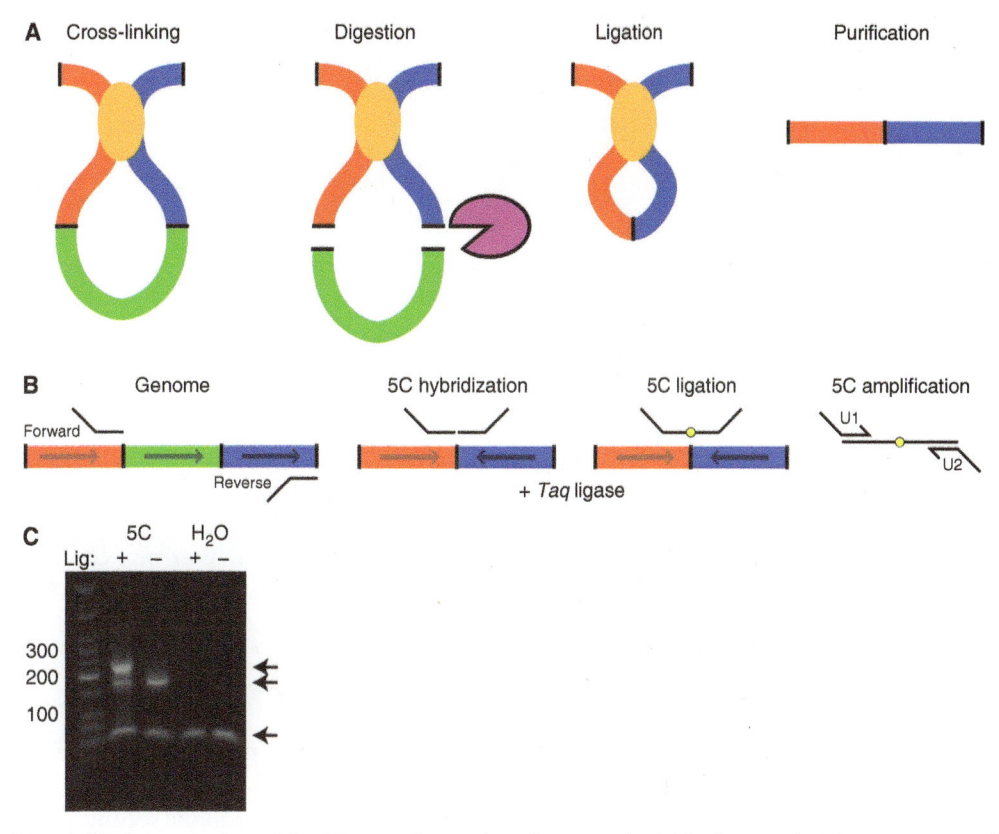

FIGURE 1. Schematic overview of the 5C procedure and quality control. (*A*) The first step in 5C is generation of a 3C library to reflect the relative interaction frequencies between pairs of restriction fragments. In 3C, the chromatin is cross-linked with formaldehyde, digested with a restriction enzyme, and cross-linked fragments are ligated in an intra-molecular fashion; see Protocol 1: Chromosome Conformation Capture (3C) in Budding Yeast (Belton and Dekker 2015). After DNA purification, quantifying the resulting library of chimeric molecules provides a measurement of how frequently pairs of fragments interact with each other spatially. (*B*) 5C uses LMA to make copies of 3C ligation junctions. This process can be done at a high level of multiplexing, with hundreds to thousands of 5C probes, so that thousands to millions of chromatin interactions can be quantified in a single reaction. The 5C probes are designed to the genome directly adjacent to restriction fragment cut sites on the fragments of interest. There are two types of probes, forward (hybridizes to the "−" strand) and reverse (hybridizes to the "+" strand). This arrangement prevents the detection of ligation products formed by self-ligation of partial digestion products. In the LMA reaction, the 3C library is denatured and the 5C probes are allowed to hybridize to 3C molecules. In cases where two restriction fragments of interest are ligated together in the 3C library, the two respective 5C probes anneal adjacent to each other. *Taq* ligase is used to repair the nick between the 5C probes, creating a continuous molecule. Universal primers are then used to amplify the 5C ligation products and to incorporate sequences at the tails for deep-sequencing. The molecules can be sequenced with the platform of choice following the manufacturer's instructions. (*C*) Electrophoresis of a low molecular mass DNA ladder, 5C products and controls. Two bands are visible in the 5C+*Taq* ligase reaction. The correct 5C product is the top band, indicated by the *upper* arrow. The *middle* arrow indicates the extended reverse probes which are present irrespective of the *Taq* ligase. The *bottom* arrow indicates the remaining universal primers from the PCR reaction. The exact size of the 5C library will depend on the size of the probes designed.

Annealing 5C Probes

6. In an eight-well PCR tube strip, prepare six 5C reactions containing the following.

NEBuffer 4 (10×)	4.0 µL
3C library DNA (5.33 ng/µL)	10.0 µL (4 million genome copies)
5C probe mixture (0.1 fmol/probe/µL)	10.0 µL
H$_2$O (sterile)	16.0 µL

7. Prepare two water-only controls by combining 4.0 µL of 10× NEB4 buffer with 36.0 µL of sterile H$_2$O per reaction.

Chapter 7

8. In a thermal cycler, perform the following program on all reactions.

   i. Incubate for 9 min at 95°C to denature the 3C library DNA.

   ii. Slowly decrease the temperature to 55°C at a 0.1°C/sec ramp.
   Annealing temperature should be determined separately for any probes that were not designed following the recommendations below (see Discussion).

   iii. Incubate overnight at 55°C.
   Reactions should remain at 55°C until ready to proceed with Step 11.

Ligating Fragments

9. In a 1.7-mL tube, prepare a "+ ligase" master mix containing the following reagents. Vortex to mix.

Taq DNA ligase (40 U/µL)	3.0 µL
Taq DNA ligase reaction buffer (10×)	24.0 µL
H$_2$O (sterile)	213.0 µL

10. In a separate 1.7-mL tube, prepare a "− ligase" master mix containing the following reagents. Vortex to mix.

Taq DNA ligase reaction buffer (10×)	8.0 µL
H$_2$O (sterile)	72.0 µL

11. While the 5C reactions are still at 55°C, add 40 µL of the "+ ligase" master mix to five of the 5C reactions ("5C + ligase") and one water control ("water + ligase"). Add 40 µL of the "− ligase" master mix to the last 5C reaction ("5C − ligase") and the second water control ("water − ligase").

12. Incubate all reactions for 1 h at 55°C.

Amplifying the 5C Products

13. Combine the following reagents in a 1.7-mL tube to prepare a PCR master mix.

PfuUltra II reaction buffer (10×)	200.0 µL
Universal Primer 1 (80 µM)	40.0 µL
Universal Primer 2 (80 µM)	40.0 µL
dNTPs (100 mM)	16.0 µL
H$_2$O (sterile)	1184.0 µL
PfuUltra II	40.0 µL

14. In a PCR plate, combine 12.0 µL of each "5C + ligase" reaction with 38.0 µL of PCR master mix per well. Repeat using the entire volume of each of the five "5C + ligase" reactions to prepare ∼30 reactions. Prepare only one PCR for each of the three controls ("5C − ligase," "water + ligase," and "water − ligase").

15. In a thermal cycler, perform the following program for all reactions: 9 min at 95°C; 30 cycles of 30 sec at 95°C, 30 sec at 65°C, and 45 sec at 72°C; 7 min at 72°C; hold at 4°C.

16. Pool all of the "5C + ligase" reactions in one 15.0-mL tube.

17. On a 2% agarose gel, perform gel electrophoresis using three dilutions of the pooled "5C + ligase," 10 µL of each control, and the low molecular mass DNA standard. Quantify the 5C bands compared with the standard to obtain an estimate of the concentration of the 5C library.

    Two bands will be visible in the "5C + ligase" lane (Fig. 1C). The lower band represents double-stranded probes that result from reverse probes binding to their universal primer. This band will also be present in the "5C − ligase" control. The other, larger molecular mass band contains the 5C library. This band will not be present in the "5C − ligase" control, indicating that it is dependent on Taq ligase. The size of this band should be the size of 2 probes plus the Illumina PE adaptors. Only the full-length product should be quantified. No product should be observed in either of the water controls.

See Troubleshooting.

18. Perform gel electrophoresis of the remainder of the 5C + ligase reaction on a long 2% agarose gel for 3 h at 4°C.

    This ensures that bands are spaced adequately for easy excision of the full-length 5C product.

19. Place the gel on a UV light box. Using a scalpel and wearing a face shield, excise the larger, full-length band from the gel.

20. Purify the 5C DNA product using the preferred gel extraction procedure.

21. Quantify the 5C product using the method of choice before DNA sequencing.

TROUBLESHOOTING

Problem (Step 17): The five 5C reactions do not produce enough DNA for downstream processing.
Solution: The yield of a 5C reaction will depend on how the 5C probes are arranged in the genome. For instance, a probe arrangement which detects only interchromosomal interactions will yield fewer products compared to a similar design which detects only intrachromosomal interactions. The ratio of 4 million genome copies to 1.0 fmol of each primer per 5C reaction is optimized for detecting a mixture of both interchromosomal and intrachromosomal interactions. If the yield from the 5C reactions is too low, simply prepare more 5C reactions using the same conditions and pool all completed reactions at Step 16.

DISCUSSION

The 5C copying process preserves the information contained in the junctions between pairs of restriction fragments in a 3C library, but reduces the size of the molecules and modifies their tails to facilitate high-throughput DNA sequencing. 5C can be performed with a high degree of multiplexing, which allows for the simultaneous detection of hundreds of thousands of ligation products at once in a single tube.

Advantages of 5C

There are several advantages to using 5C to detect 3C ligation products of interest: (1) 5C increases the speed of data production compared to 3C-PCR for the same number of interactions; (2) 5C can be used to assay many more interactions of interest than is practical with 3C-PCR; (3) the dynamic range of quantification of the 5C products is much longer than that of semiquantitative 3C-PCR; and (4) for very large and comprehensive studies, the cost per interaction is much lower than that of 3C-PCR.

5C Probe Design

5C employs two types of probes: forward and reverse (Dostie et al. 2006). For each restriction fragment, either a forward or a reverse 5C probe is designed adjacent to the cut site. Forward probes hybridize to the "−" strand and contain the universal sequence at the 5′ end of the molecule. Reverse probes hybridize to the "+" strand and contain the universal sequence at the 3′ end of the molecule. This eliminates false-positive signal arising from self-ligation of partial digestion products, as 5C can only detect interactions between forward and reverse probes.

There are many ways 5C probes can be positioned in the genome. Probes can be positioned such that forward and reverse probes alternate at consecutive restriction fragments. This layout will give the most density of structural information and therefore is ideal for structural/modeling studies of chromatin (Bau and Marti-Renom 2011; Bau et al. 2011). Probes can also be placed such that the two types represent different types of chromatin classes. For instance, one type of probe can represent gene promoters while the other represents enhancers (Sanyal et al. 2012).

5C probes should be designed with a length of 40 nucleotides (excluding universal tail sequences) and an optimal T_m of 65°C. The probes should hybridize to unique sequences in the genome to avoid off-target effects. The web-based tool http://my5c.umassmed.edu can be used to design 5C probes (Lajoie et al. 2009).

REFERENCES

Bau D, Marti-Renom MA. 2011. Structure determination of genomic domains by satisfaction of spatial restraints. *Chromosome Res* **19**: 25–35.

Bau D, Sanyal A, Lajoie BR, Capriotti E, Byron M, Lawrence JB, Dekker J, Marti-Renom MA. 2011. The three-dimensional folding of the α-globin gene domain reveals formation of chromatin globules. *Nat Struct Mol Biol* **18**: 107–114.

Belton J-M, Dekker J. 2015. Chromosome conformation capture (3C) in budding yeast. *Cold Spring Harb Protoc* doi: 10.1101/pdb.prot085175.

Dostie J, Richmond TA, Arnaout RA, Selzer RR, Lee WL, Honan TA, Rubio ED, Krumm A, Lamb J, Nusbaum C, et al. 2006. Chromosome Conformation Capture Carbon Copy (5C): A massively parallel solution for mapping interactions between genomic elements. *Genome Res* **16**: 1299–1309.

Lajoie BR, van Berkum NL, Sanyal A, Dekker J. 2009. My5C: Web tools for chromosome conformation capture studies. *Nat Methods* **6**: 690–691.

Sanyal A, Lajoie BR, Jain G, Dekker J. 2012. The long-range interaction landscape of gene promoters. *Nature* **489**: 109–113.

Protocol 4

Hi-C in Budding Yeast

Jon-Matthew Belton and Job Dekker[1]

Program in Systems Biology, University of Massachusetts Medical School, Worcester, Massachusetts 01605

Hi-C enables simultaneous detection of interaction frequencies between all possible pairs of restriction fragments in the genome. The Hi-C method is based on chromosome conformation capture (3C), which uses formaldehyde cross-linking to fix chromatin regions that interact in three-dimensional space, irrespective of their genomic locations. In the Hi-C protocol described here, cross-linked chromatin is digested with HindIII and the ends are filled in with a nucleotide mix containing biotinylated dCTP. These fragments are ligated together, and the resulting chimeric molecules are purified and sheared to reduce length. Finally, biotinylated ligation junctions are pulled down with streptavidin-coated beads, linked to high-throughput sequencing adaptors, and amplified via polymerase chain reaction (PCR). The resolution of the Hi-C data set will depend on the depth of sequencing and choice of restriction enzyme. When sufficient sequence reads are obtained, information on chromatin interactions and chromosome conformation can be derived at single restriction fragment resolution for complete genomes.

MATERIALS

It is essential that you consult the appropriate Material Safety Data Sheets and your institution's Environmental Health and Safety Office for proper handling of equipment and hazardous materials used in this protocol.

RECIPES: Please see the end of this protocol for recipes indicated by <R>. Additional recipes can be found online at http://cshprotocols.cshlp.org/site/recipes.

Reagents

Agarose

AmPure XP magnetic bead-based purification system (Beckman Coulter)

The AmPure XP mixture is used to fractionate the library into 100- to 300-bp fragments for high-throughput sequencing.

ATP (100 mM)

Binding buffer (BB) (1× and 2×) <R>

Bovine serum albumin (BSA) (10 mg/mL)

Buffer EB, prewarmed to 65°C (QIAgen 19086)

DNA polymerase I, large (Klenow) fragment (5 U/μL) (New England Biolabs M0210S)

DNA quantification reagents for the method of choice (e.g., Bioanalyzer, qPCR, or fluorometry)

Dynabeads MyOne Streptavidin C1 beads (Life Technologies 65001)

Ethanol (100% and 70%)

Formaldehyde (37%) (Fisher Scientific BP531-500)

---

[1]Correspondence: job.dekker@umassmed.edu

Copyright © Cold Spring Harbor Laboratory Press; all rights reserved
Cite this protocol as *Cold Spring Harb Protoc*; doi:10.1101/pdb.prot085209

Glycine (2.5 M, filter-sterilized)
Klenow Fragment (3'→5' exo-) (5 U/μL) (New England Biolabs M0212S)
Ligation buffer for 3C (10×) <R>
NEBuffer 2 (New England Biolabs B7002S) (10× and 1×)
Nucleotides
　　Biotin-14-dCTP (0.4 mM) (Life Technologies 19518-018)
　　dATP (100 mM and 1 mM)
　　dGTP (100 mM)
　　dTTP (100 mM)
　　Nucleotide (dNTP) mix (25 mM per nucleotide; 100 mM total)
　　Nucleotide (dNTP) mix (1.25 mM per nucleotide; 5 mM total)
Oligonucleotide primers
　　Hi5: GTTTCCGAAAATCCACGACGAACCAG (80 μM)
　　Hi6: ATATTTTCGCCGGAGGTGCTGGAAAT (80 μM)
　　PE1.0:
　　AATGATACGGCGACCACCGAGATCTACACTCTTTCCCTACACGACGCTCTTCCGATCT
　　(HPLC-purified) (25 μM)
　　PE2.0:
　　CAAGCAGAAGACGGCATACGAGATCGGTCTCGGCATTCCTGCTGAACCGCTCTTCC
　　GATCT (HPLC-purified) (25 μM)
Paired-end adapters compatible with the Illumina HiSeq or GAII platforms (available from Illumina, Kapa Biosystems, or Bioo Scientific)
PfuUltra II Fusion HS polymerase with 10× reaction buffer (Agilent Technologies 600850)
Phase Lock Gel Light tubes (15-mL) (5-Prime 2302840)
Phase Lock Gel Light tubes (50-mL) (5-Prime 2302860)
Phenol:chloroform (1:1) <R>
Proteinase K (Life Technologies 25530-015)
　　Prepare a solution of 10 mg/mL in TE buffer <R>.

Quick Ligation Kit (Quick T4 DNA ligase and 2× Quick Ligation Reaction Buffer) (New England Biolabs M2200S)
Restriction enzymes
　　HindIII (20,000 U/mL) (New England Biolabs R0104S)
　　NheI (10,000 U/mL) (New England Biolabs R0131S)

RNase A, DNase-free (10 mg/mL)
Saccharomyces cerevisiae
　　Grow yeast cells (using growth medium and conditions appropriate for the strain and experiment) for 2–3 doubling times to obtain mid-log phase cells. Usually, a 100-mL culture produces enough cells to generate one Hi-C library.

Sodium acetate (3 M, pH 5.2) <R>
Sodium dodecyl sulfate (SDS) (10% and 1% [w/v])
Standards for gel electrophoresis (e.g., 1-kb ladder and low molecular mass DNA ladder [New England Biolabs N3232S and N3233S])
T4 DNA ligase (1 U/μL) (Life Technologies 15224-017)
T4 DNA ligase reaction buffer (10×) (New England Biolabs B0202S)
T4 DNA polymerase (3 U/μL) (New England Biolabs M0203L)
T4 polynucleotide kinase (10 U/μL) (New England Biolabs M0201S)
TBE buffer <R>
TE buffer <R>

TLE buffer <R>
Triton X-100 (10% [v/v])
Tween wash buffer (TWB) <R>

Equipment

AFA microTUBES with snap caps (Covaris 520045)
Centrifugal filters (Amicon 15-mL, 30 kDa) (Millipore UFC903024)
Centrifugal filters (Amicon 500-µL, 30 kDa) (Millipore UFC5030BK)
Centrifuge (high-speed, refrigerated, for up to 18,000g)
Centrifuge (tabletop, refrigerated, for up to 3100g)
Densitometer
DNA LoBind tubes (1.7-mL)
DNA quantification equipment for the method of choice (e.g., Bioanalyzer, qPCR, or fluorometry)
Dry ice
Flask (250-mL)
Gel electrophoresis apparatus
Incubator at 37°C, with rotating platform
Incubator/shaker at 25°C
Liquid nitrogen
Magnetic particle separator (MPS)
Microcentrifuge tubes (1.7-mL)
MinElute columns (QIAgen 28004)
Mortar and pestle

Immediately before use, precool a mortar and pestle by placing them on dry ice and adding enough liquid nitrogen to the mortar to cover the head of the pestle. Let the liquid nitrogen evaporate before use (Step 9).

PCR tubes (0.2-mL)
Rotator
Sonicator (Covaris S2)

The Covaris S2 is the sonicator of choice since it shears the DNA to 50–700 bp fragments in only 4 min. (Any other sonicator will also work, but the exact conditions and setting must be determined empirically for naked DNA, and it may not produce as tight of a smear.) Fill the water chamber of the sonicator with deionized water and allow it to cool down and degas for at least 30 min before sonication (Step 73).

Spectrometer to measure optical density at 600 nm
Thermal cycler
Thermomixer or water baths at 16°C, 20°C, 37°C, 65°C, and 75°C
Tubes, conical (15- and 50-mL)
Tubes, screw cap (35- and 250-mL, suitable for high-speed centrifugation)
Vacuum aspirator
Vortex

METHOD

An overview of Hi-C is provided in Figure 1.

Cross-Linking Chromatin

Because the output of Hi-C is the rate of cross-linking, it is essential to standardize cross-linking across all samples.

1. Add 37% formaldehyde to the cultured yeast cells to a final concentration of 3% of the culture medium (i.e., 8.82 mL of 37% formaldehyde per 100-mL culture).

2. Shake the culture at 200 rpm for 20 min at 25°C.

3. Quench the cross-linking by adding 2.5 M glycine at 2× the volume of formaldehyde used in Step 1 (i.e., 17.64 mL of 2.5 M glycine per 100 mL of culture).

Chapter 7

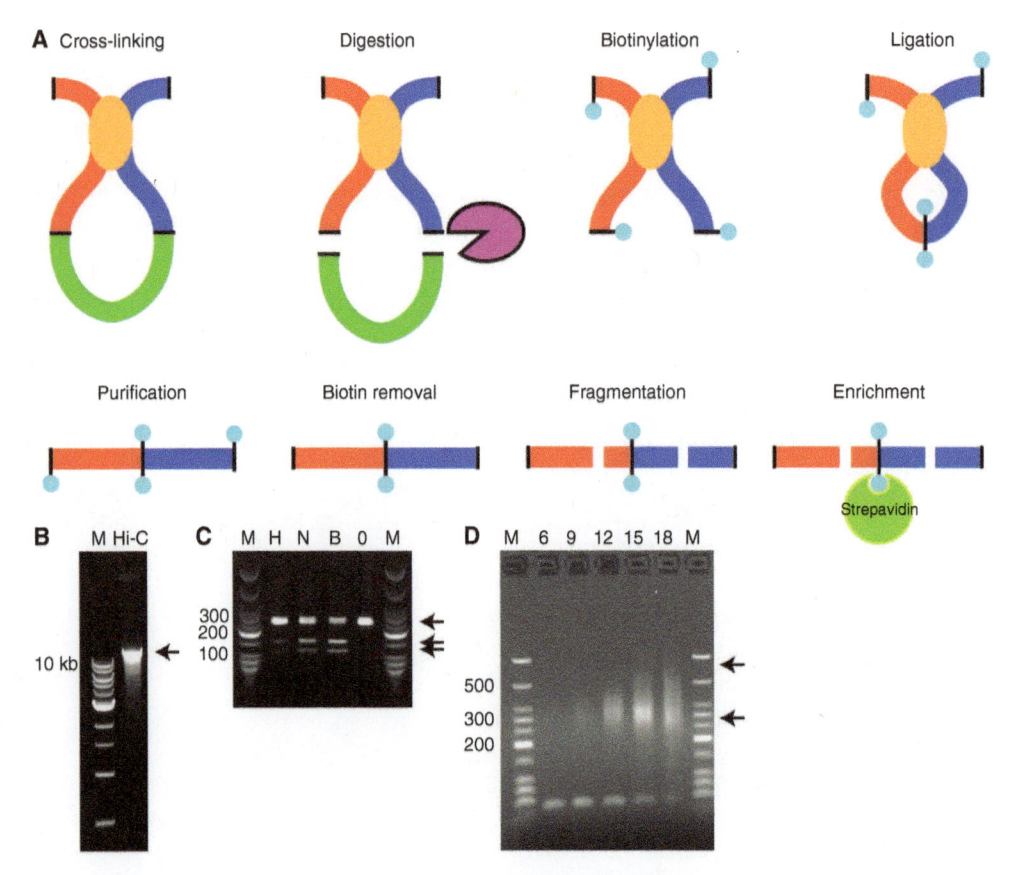

FIGURE 1. Schematic overview of the Hi-C procedure and quality control. (A). Hi-C is based on 3C, in which the chromatin is cross-linked using formaldehyde and digested with a restriction enzyme (see Protocol 1: Chromosome Conformation Capture (3C) in Budding Yeast [Belton and Dekker 2015]). Hi-C is unique in that the overhangs of the restriction cut site are filled in with a biotinylated nucleotide before ligation. Intra-molecular ligation and purification are very similar to the 3C protocol; however, the Hi-C method contains a few additional steps to enrich for ligation products. First, biotin incorporated into free ends (not at a ligation junction of two different fragments) is removed using the exonuclease activity of T4 DNA polymerase. The library is then fragmented by sonication and true ligation products are enriched by pull-down using streptavidin-coated beads. (B) The purified Hi-C library runs at ~12 kb on a 0.8% agarose gel (Step 65), indicated by the arrow. M, 1-kb DNA ladder. (C) Agarose gel for estimation of Hi-C efficiency (Step 68). H, HindIII digested; N, NheI digested; B, both HindIII and NheI digested; 0, no digestion of the PCR product; M, low molecular mass marker. The *upper* arrow indicates the undigested PCR product and the two *lower* arrows indicate the digestion products. (D) Electrophoresis of PCR cycle titration (Step 110). M, low molecular mass marker; "6," "9," "12," "15," "18" are the number of cycles used to amplify the library. The *bottom* arrow indicates the correct size of the Hi-C library and the *upper* arrow indicates high-molecular-mass species that arise from overamplifying the library.

4. Shake the culture at 200 rpm for 5 min at 25°C.

5. Transfer the cells to a 250-mL tube and centrifuge the cells at 1800g for 5 min in a tabletop centrifuge.

6. Pour off the medium and wash the cells in 100 mL of sterile H_2O by pipetting up and down until the cell pellet is resuspended.

7. Centrifuge the cells at 1800g for 5 min in a tabletop centrifuge.

8. Pour off the supernatant and resuspend the cells in 5 mL of 1× NEBuffer 2 by pipetting up and down.

9. Add liquid nitrogen to a prechilled mortar and pour the sample into the liquid nitrogen. Once the sample has frozen, begin to crush it with the pestle. When the sample is broken into little pieces, begin to grind it with the pestle. Grind the sample for 10 min, adding liquid nitrogen as necessary (approximately every 3 min).

10. Scrape the sample into a 50-mL tube on ice. Add 45 mL of ice-cold 1× NEBuffer 2 to the lysate.
11. Centrifuge the lysate at 1800g for 5 min at 4°C in a tabletop centrifuge.
12. Pour off the supernatant and resuspend the cells in 1× NEBuffer 2 to an OD$_{600}$ of 10.0.
    The lysate can be stored in 7.2-mL aliquots at −80°C for several years.

Digesting Cross-Linked Chromatin with HindIII

13. If frozen, thaw one 7.2-mL aliquot of lysate on ice. Transfer the lysate to a 50-mL tube and wash with 50 mL of ice-cold 1× NEBuffer 2 by inverting the tube several times.
14. Centrifuge the lysate at 3100g for 10 min at 4°C in a tabletop centrifuge.
15. Aspirate the supernatant with a vacuum. Resuspend the pellet in 5.5 mL of 1× NEBuffer 2.
16. Distribute 456 µL of cell lysate into each of twelve 1.7-mL tubes.
17. Solubilize the chromatin by adding 45.6 µL of 1% SDS per tube. Mix by pipetting up and down. Resuspend any precipitated cellular debris, but avoid making bubbles.
18. Incubate the tubes for 10 min at 65°C. Place the tubes on ice immediately after the incubation.
    It is critical to place the samples on ice after this incubation, as high temperature reverses formaldehyde cross-links.
19. Centrifuge the tubes for 10 sec to remove liquid from the tube caps.
20. Quench the SDS by adding 52.8 µL of 10% Triton X-100 per tube. Mix by pipetting up and down. Resuspend any precipitated cellular debris, but avoid making bubbles.
21. Digest the chromatin by adding 60 µL of HindIII (20,000 units/mL) per tube. Mix by pipetting up and down. Resuspend any precipitated cellular debris, but avoid making bubbles.
22. Incubate the tubes overnight at 37°C.
    It is preferable to agitate the sample during incubation, so a rotating platform in a 37°C incubator is recommended.

Incorporating Biotin at Digested Ends

23. Centrifuge the tubes for 10 sec to remove liquid from the tube caps.
24. Add the following reagents to each tube.

NEBuffer 2 (10×)	6.4 µL
dATP (100 mM)	0.18 µL
dGTP (100 mM)	0.18 µL
dTTP (100 mM)	0.18 µL
Biotin-14-dCTP (0.4 mM)	45.0 µL
DNA polymerase I (Klenow) (5 U/µL)	12.0 µL

25. Incubate all reactions for 2 h at 37°C.
    It is preferable to agitate the sample during the incubation, so a rotating platform in a 37°C incubator is recommended.
26. Centrifuge the tubes for 10 sec to remove liquid from the tube caps.
27. Denature the enzyme by adding 115.2 µL of 10% SDS to each tube. Mix by pipetting up and down but avoid making bubbles.
28. Incubate the tubes for 20 min at 65°C. Place the tubes on ice immediately after the incubation.

Ligating Cross-Linked Chromatin Fragments

Below, chromatin fragments are ligated in a dilute reaction to favor intramolecular ligation. Because variations in ligation conditions can contribute to variation in background ligation (intermolecular ligation) and reduce the signal-to-noise ratio, it is essential to standardize intra-molecular ligation across all samples.

29. Assemble the ligation master mix by adding the following to a 250-mL flask on ice. Mix the reaction by gently swirling but avoid making bubbles.

Triton X-100 (10%)	13.0 mL
Ligation buffer for 3C (10×)	13.0 mL
BSA (10 mg/mL)	1.4 mL
ATP (100 mM)	1.4 mL
H$_2$O (sterile)	104 mL
T4 DNA ligase (1 U/µL)	3.51 mL

30. Add 10.5 mL of ligation master mix to each of twelve 15-mL tubes on ice.

31. Centrifuge the tubes for 10 sec to remove the liquid from the caps.

32. Using a 1-mL micropipette, transfer each Hi-C reaction from Step 28 to a 15-mL tube containing ligation master mix. Mix the tubes gently by inverting.

33. Incubate the ligation reactions for 8 h at 16°C. Invert the tubes every hour.

Reversing Cross-Links and Purifying Ligation Products

34. Add 72 µL of proteinase K (10 mg/mL) to each ligation reaction from Step 33. Mix by inverting. Incubate reactions at 65°C overnight.

35. Add another 72 µL of proteinase K (10 mg/mL) to each reaction. Mix by inverting. Incubate reactions for 2 h at 65°C.

36. Transfer each reaction to a 50-mL tube. Add 23.0 mL of phenol:chloroform (1:1) to each tube and vortex each tube for 30 sec.

37. Pour each vortexed sample into a prespun 50-mL tube of Phase Lock Gel Light. Centrifuge at 1500g for 10 min.

38. Pour the aqueous phase of each sample into a fresh 50-mL tube. Add 23.0 mL of phenol:chloroform (1:1) to each tube and vortex each tube for 30 sec.

39. Pour each vortexed sample into a prespun 50-mL tube of Phase Lock Gel Light. Centrifuge at 1500g for 10 min.

40. Pool the aqueous phases from all samples into two 250-mL tubes (∼67.8 mL/tube).

41. Precipitate the 3C ligation products by adding 1/10th volumes (6.78 mL) of 3 M sodium acetate (pH 5.2) and 2.5× volumes (169.5 mL) of 100% ethanol to each tube.

42. Incubate the tubes on dry ice for 30–45 min. Make sure the liquid is very cold and thick but not frozen.

43. Centrifuge the tubes at 10,000g for 20 min at 4°C.

44. Carefully pour off the supernatant and discard. Centrifuge at 10,000g for 30 sec at 4°C to collect the remaining drops of liquid at the bottom of the tube.

45. Aspirate the remaining alcohol using a vacuum.

46. Resuspend both Hi-C DNA pellets in 4 mL total of TE buffer and transfer the sample to a 15-mL tube.

47. Add 8.0 mL of phenol:chloroform (1:1) to the Hi-C sample and vortex for 30 sec.

48. Transfer the sample to a prespun 15-mL tube of Phase Lock Gel Light. Centrifuge in a tabletop centrifuge at 1500g for 10 min.

49. Transfer the aqueous phase to a fresh 15-mL tube. Add 8 mL of phenol:chloroform (1:1) and vortex the tube for 30 sec.

50. Transfer the sample to a prespun 15-mL tube of Phase Lock Gel Light. Centrifuge in a tabletop centrifuge at 1500g for 10 min.

51. Transfer the aqueous phase to a 35-mL tube. Precipitate the 3C DNA with 400 µL of sodium acetate (pH 5.2) and 10.0 mL of 100% ethanol.
52. Centrifuge the sample at 18,000g for 20 min.
53. Decant the supernatant and centrifuge the sample at 10,000g for 30 sec at 4°C to collect remaining liquid.
54. Aspirate the remaining alcohol using a vacuum and allow the DNA pellet to air dry.
55. Resuspend the Hi-C sample in 15 mL of TE and vortex for 60 sec.
56. Transfer the sample to a 15-mL Amicon 30-kDa filter column for desalting.
57. Centrifuge the sample in a tabletop centrifuge at 3100g for 15 min.
58. Discard the flowthrough and wash the sample with 15 mL of TE. Repeat the centrifugation in Step 57.
59. Elute the Hi-C DNA from the 15-mL filter by pipetting the remaining sample out of the filter with a 200-µL micropipette. Transfer the sample to a 500-µL Amicon 30 kDa filter column.
60. Wash the surface of the 15-mL filter with 200 µL of TE. Pipette the 200 µL off the filter and add this to the rest of the sample in the 500-µL filter column.
61. Centrifuge the sample in the 500-µL filter column at 18,000g for 20 min. If the volume in the column is >20 µL, centrifuge repeatedly for 5 min at a time until it is <20 µL.
62. Elute the Hi-C sample by inverting the column and placing it in a fresh collection tube. Centrifuge the tube at 18,000g for 20 min.
63. Adjust the sample volume to 20 µL with TE.
64. Degrade any co-precipitated RNA by adding 2 µL of RNase A (10 mg/mL). Incubate the sample for 1 h at 37°C.
65. Perform gel electrophoresis of dilutions of the library DNA (0.25 to 2.0 µL) on a 0.8% agarose gel in 0.5× TBE. Quantify the DNA by comparison to known concentration standards (500 to 100 ng).

    The Hi-C library DNA should appear as a tight band ~12 kb on the gel (Fig. 1B). Any smearing of the sample may be caused by overly vigorous lysis or the presence of endogenous nucleases in the strain.

    The typical yield of a Hi-C library is 3–10 µg, with an average yield of ~5 µg.

Estimating Hi-C Efficiency

Hi-C efficiency is the percentage of ligation products for a single, neighboring interaction (an interaction between restriction fragments that are adjacent to one another in the genome) that are biotinylated. This value is used to standardize the amount of streptavidin beads used to enrich the library for biotinylated ligation products and gives a measure of the efficiency of the biotin fill-in.

66. Use 40 ng of the Hi-C library to PCR-amplify a neighboring interaction as follows. Place the rest of the Hi-C library DNA at 4°C until ready to proceed with Step 69.

    i. Prepare the reaction by combining the following reagents in a 0.2-mL PCR tube.

Hi-C library DNA from Step 65	40 ng
PfuUltra II reaction buffer (10×)	10 µL
dNTPs (100 mM; 25 mM each nucleotide)	0.8 µL
Primer Hi5 (80 µM)	0.5 µL
Primer Hi6 (80 µM)	0.5 µL
PfuUltra II polymerase	2.0 µL
H$_2$O (sterile)	to 100 µL

    ii. Amplify the ligation product of interest using the following program on a thermal cycler: 5 min at 95°C; 35 cycles of 30 sec at 95°C, 30 sec at 65°C, and 30 sec at 72°C; and 8 min at 72°C.

67. Digest the amplicon with HindIII, NheI, and HindIII/NheI as follows.

    i. Prepare the HindIII digestion containing 15 µL of PCR product, 1.9 µL of 10× NEBuffer 2, 0.19 µL of BSA, 0.95 µL of HindIII (20,000 units/mL), and 0.95 µL of sterile H$_2$O.

    ii. Prepare the NheI digestion containing 15 µL of PCR product, 1.9 µL of 10× NEBuffer 2, 0.19 µL of BSA, 0.95 µL of NheI, and 0.95 µL of sterile H$_2$O.

    iii. Prepare the double-digest containing 15 µL of PCR product, 1.9 µL of 10× NEBuffer 2, 0.19 µL of BSA, 0.95 µL of HindIII, and 0.95 µL of NheI.

    iv. Prepare a control containing 15 µL of PCR product, 1.9 µL of 10× NEBuffer 2, 0.19 µL of BSA, and 1.9 µL of sterile H$_2$O.

    v. Incubate the digestion reactions overnight at 37°C.

68. Perform gel electrophoresis of the digested and undigested bands on a 2% agarose gel in 0.5× TBE (Fig. 1C). Quantify the bands by densitometry, and estimate the Hi-C efficiency.

    Not all PCR products will be cut in the NheI and HindIII combined reaction. This is likely due to inefficiency of digestion and point mutations introduced during amplification, and may also be caused by DNA breakage during the procedure leading to noncanonical ligation products. Thus, the Hi-C efficiency, which is the percentage of digested products in the NheI digestion, should be corrected by the percent cleavable, which is determined from the combined reaction. The final Hi-C efficiency is therefore calculated as the percent digested with NheI divided by the percent digested with NheI and HindIII. Generally, 30%–60% of the amplicon will digest with NheI.

    See Troubleshooting.

Removing Biotin from Free Ends

Proceed with the entire Hi-C sample (set aside in Step 66) throughout the rest of the protocol.

69. For each 1 µg of Hi-C library DNA, add the following reagents to the tube of DNA: 0.1 µL of BSA (10 mg/mL), 1.0 µL of 10× NEBuffer 2, 0.1 µL of 5 mM dNTPs (1.25 mM per nucleotide), and 1.0 µL of T4 DNA polymerase (3 U/µL). Adjust the final volume to 10.0 µL with sterile H$_2$O.

70. Incubate the reaction for 4 h at 20°C and then for 20 min at 75°C to inactivate the enzyme.

71. Adjust the volume to 102 µL by adding sterile H$_2$O or concentrating the sample using a 500-µL 30-kDa Amicon filter as appropriate.

Shearing the Hi-C Library

72. Load 101 µL of the Hi-C sample into an AFA microTUBE. Set aside 1.0 µL of unsonicated DNA for electrophoresis analysis in Step 95.

73. Place the tube into the Covaris-approved S2 holder and run the following program: Duty Cycle, 10%; Intensity, 5; Cycles per Burst, 200; Mode, Frequency sweeping. Continue to degas during the sonication. Run the program for a total of 4 min. Set aside 1.0 µL of sonicated DNA for electrophoresis analysis in Step 95.

74. Purify the DNA using one MinElute column for every 5 µg of Hi-C reaction. Elute the DNA from each column using 31 µL of prewarmed (65°C) Buffer EB.

Repairing DNA Ends and A-Tailing

75. Add the following reagents to each of the sonicated elutions from Step 74.

T4 DNA Ligase Reaction Buffer (10×)	14.0 µL
dNTPs (100 mM; 25 mM each)	1.4 µL
T4 DNA polymerase (3 U/µL)	5.0 µL
T4 polynucleotide kinase (10 U/µL)	5.0 µL
DNA polymerase I (Klenow) (5 U/µL)	1.0 µL
H$_2$O (sterile)	13.6 µL

76. Incubate the reactions for 30 min at 20°C.
77. Purify the DNA using one MinElute column per reaction. Elute each sample with 31 µL of prewarmed (65°C) Buffer EB.
78. Add the following reagents to each of the end-repaired samples.

NEBuffer 2 (10×)	5.0 µL
dATP (1.0 mM)	10.0 µL
Klenow fragment (*exo-*) (5 U/µL)	3.0 µL
H₂O (sterile)	2.0 µL

79. Incubate the reactions for 30 min at 37°C.
80. Inactivate the enzyme by incubating the reactions for 20 min at 65°C.

Fractionating the Sonicated Hi-C Library

81. Pool all end-repaired samples from Step 80 and adjust the volume to 500 µL with Buffer EB.
82. Add 450 µL of AMPure XP to the Hi-C sample. Label this sample "0.9X."
83. Vortex the mixture briefly and then quickly centrifuge to remove liquid from the tube cap. Incubate at room temperature for 10 min.

    Proceed to Step 84 during this incubation.

84. While the "0.9X" sample is incubating, prepare another tube ("1.1X") containing AmPure XP as follows.

    i. Add 500 µL of AmPure XP to a new 1.7-mL tube and label this tube "1.1X".
    ii. Place the "1.1X" tube on an MPS for 5 min to collect the AmPure XP beads to the side of the tube.
    iii. While the tube is still on the MPS, remove and discard the AmPure solution using a 1-mL micropipette.
    iv. Resuspend the collected beads in 100 µL of Ampure XP.

        This process increases the number of beads in the Ampure XP mixture and ensures there is enough binding capacity to collect all of the DNA in the Hi-C sample.

85. Place the "0.9X" sample on an MPS for 5 min to collect the AmPure XP beads to the side of the tube.
86. While still on the MPS, transfer the supernatant from the "0.9X" sample to the "1.1X" tube using a 1-mL micropipette.

    Retain the "0.9X" beads for washing in Step 90.

87. Vortex the "1.1X" sample briefly and then quickly centrifuge to remove liquid from the cap. Incubate the mixture at room temperature for 10 min.
88. Place the "1.1X" sample on an MPS for 5 min to collect the Ampure XP beads to the side of the tube.
89. While still on the MPS, remove and discard the AmPure solution from the sample using a 1-mL micropipette.
90. Wash the "0.9X" and "1.1X" beads twice with 200 µL of 70% ethanol per wash as follows.

    i. Resuspend the beads in 70% ethanol and centrifuge the tube quickly.
    ii. Collect the beads on an MPS.
    iii. Discard the ethanol.

91. After the second ethanol wash, centrifuge the sample again and collect any remaining ethanol. Air-dry the beads at room temperature.

92. Resuspend the "0.9X" and the "1.1X" beads in 30 µL of prewarmed (65°C) Buffer EB per tube. Elute the DNA at room temperature for 10 min.

93. Place the "0.9X" and the "1.1X" tubes on an MPS for 5 min to collect the AmPure XP beads to the side of each tube.

94. While still on the MPS, transfer the AmPure solution from each tube to a separate 1.7-mL tubes using a 1-mL micropipette.

    The "0.9X" sample contains DNA fragments >300 bp and the "1.1X" sample contains DNA fragments between 100 bp and 300 bp. The "1.1X" sample, which contains the portion of the library that is compatible with the Illumina platform, will be used for all subsequent steps.

95. Perform gel electrophoresis of 1.0 µL of each of the unsonicated (Step 72), sonicated (Step 73), "0.9X" and "1.1X" samples on a short, 2% agarose gel in 0.5X TBE at 250 volts for 30 min. Verify the size of the 1.1X fraction and quantify it by comparison to a known concentration standard (10.0 to 100 ng).

Enriching Biotinylated Ligation Products

96. Calculate the quantity of biotinylated ligation product in the library as follows.

    i. Multiply the amount of DNA (ng) in the 1.1X fraction, measured in Step 95, by 200 (the average length of the sonicated Hi-C library). Divide this amount by 8000 (the average length of the presonication library).

       This quantity is an estimate of the amount of the sample that contains a ligation product.

    ii. Multiply the quantity from Step 96.i by the proportion of biotinylated ligation products (Hi-C efficiency) calculated in Step 68.

        This is the quantity of biotinylated ligation product in the library. This is used to calculate the amount of Dynabeads MyOne Strepavidin C1 beads to use for the pull down.

97. For each 1.0 ng of biotinylated ligation product calculated in Step 96, add 1.0 µL of Dynabeads MyOne Strepavidin C1 beads to a 1.7-mL LoBind tube.

    LoBind tubes are used in this step to reduce nonspecific pull-down of DNA.

98. Wash the beads twice with TWB as follows.

    i. Resuspend the beads in the wash buffer using a micropipette.

    ii. Transfer the beads to a new LoBind tube and incubate for 3 min with rotation.

    iii. Quickly centrifuge the beads to the bottom of the tube.

    iv. Reclaim the beads on an MPS for 1 min.

99. Resuspend the beads in 30 µL of 2× BB.

100. Add the 1.1X fraction from Step 94 to the beads and incubate for 1 h at room temperature with rotation.

101. Reclaim the beads on an MPS for 1 min.

     The supernatant should be retained as insurance until the end of the protocol.

102. Wash the beads once with 200 µL of 1× BB as described in Steps 98.i–98.iv.

103. Wash the beads once with 50 µL of 1× NEB Quick Ligation Buffer as described in Steps 98.i–98.iv.

104. Resuspend the beads in 10 µL of sterile H_2O.

105. Ligate the Illumina PE adapters to the biotinylated products by adding 13 µL of 2× Quick Ligation Reaction Buffer, 2 µL of Illumina PE adapters and 1 µL of Quick T4 DNA Ligase to the beads. Incubate the ligation reaction at room temperature for 15 min.

106. Reclaim the beads on an MPS for 1 min.

107. Conduct the following washes as described in Step 98.i–98.iv.

    i. Wash the beads twice with 200 µL of TWB.

    ii. Wash the beads once with 100 µL of 1× BB.

    iii. Wash the beads once with 100 µL of 1× NEBuffer 2.

    iv. Wash the beads once with 25 µL of 1× NEBuffer 2.

108. Resuspend the beads in the amount of 1× NEBuffer 2 equivalent to the starting volume of beads used in Step 97.

    For example, if 10 µL of Dynabeads MyOne Strepavidin C1 beads were used to pull down the library, then resuspend the library in 10 µL of 1X NEBuffer 2.

Amplifying the Adapter-Modified Hi-C Library Using Paired-End PCR

109. Perform a PCR to titrate the number of cycles for library amplification.

    i. Combine the following reagents in a 0.2-mL PCR tube.

Hi-C-coated streptavidin beads	3 µL
PfuUltra II buffer (10×)	5 µL
dNTPs (100 mM; 25 mM each)	0.4 µL
PE1.0 primer (25 µM)	0.7 µL
PE2.0 primer (25 µM)	0.7 µL
PfuUltra II	1 µL
H₂O (sterile)	39.2 µL

    ii. Amplify the library using the following program on a thermal cycler: 30 sec at 98°C; 6 cycles of 10 sec at 98°C, 30 sec at 65°C, and 30 sec at 72°C; and 2 min at 72°C.

    iii. Vortex the PCR reaction and remove 2.0 µL for gel electrophoresis.

    iv. Quickly centrifuge the reaction and repeat Steps 109.ii and 109.iii, except run the PCR program for only 3 cycles.

    v. Repeat Step 109.iv until a total of 18 cycles has been completed.

110. Perform gel electrophoresis of all reactions on a short 2% agarose gel in 0.5× TBE (Fig. 1D). Quantify the amount of DNA product for each cycle number by comparison to a known concentration standard.

111. Choose a number of cycles for amplification that will yield at least 100 fmol of final Hi-C library when all of the beads are amplified, but that does not produce higher molecular mass artifacts which arise from overcycling. Choose the least number of cycles possible.

    All libraries in a given experiment should be amplified using the same number of cycles.

112. Using the reaction setup described in Step 109.i, prepare enough reactions to amplify all of the remaining beads (3 µL of beads/reaction).

113. Amplify the reactions using the thermal cycler program described in Step 109.ii with the optimal number of cycles determined in Step 111.

114. After amplification, pool all reactions together in a 1.7-mL tube. Add 1.8 volumes of AmPure XP beads to the pooled reactions.

115. Place the tube on an MPS for 1 min to collect the AmPure XP beads to the side of the tube.

116. While still on the MPS, remove the supernatant and discard.

117. While still on the MPS, wash the AmPure XP beads twice with 1 mL of 70% ethanol per wash as described in Steps 90–91. Air-dry the beads on the magnet.

118. Elute the DNA by resuspending the beads in 14 µL of TLE buffer.

119. Place the tube on an MPS for 1 min to collect the beads to the side of the tube.
120. While still on the MPS, transfer the supernatant to a fresh 1.7-mL tube.
121. Quantify the Hi-C library DNA using the method of choice.

TROUBLESHOOTING

Problem (Step 68): The Hi-C efficiency is low.
Solution: This can be caused by a poor biotinylation reaction or a poor ligation reaction, or both. Repeat these reactions as needed.

RELATED INFORMATION

Hi-C provides the most comprehensive genome-wide analysis of any of the 3C-based techniques for studying chromatin conformation (Lieberman-Aiden et al. 2009; Belton et al. 2012). Classic 3C (Dekker et al. 2002) is described in Protocol 1: Chromosome Conformation Capture (3C) in Budding Yeast (Belton and Dekker 2015).

RECIPES

Binding Buffer (BB)

Reagent	Amount to add (for 100 mL)	Final concentration (1×)
Tris–HCl (1 M, pH 8.0)	500 µL	5.0 mM
EDTA (500 mM)	100 µL	0.5 mM
NaCl (5 M)	20 mL	1 M

For 2× BB, double the volume of each reagent added. Prepare in deionized H_2O. Store for up to 1 yr at room temperature.

Ligation Buffer for 3C (10×)

Reagent	Amount to add (for 1 L)	Final concentration (10×)
Tris–HCl (1 M, pH 7.5)	500 mL	500 mM
$MgCl_2$ (1 M)	100 mL	100 mM
Dithiothreitol (DTT) (1 M)	100 mL	100 mM

Prepare in deionized H_2O. Store at −20°C in 15-mL aliquots.

Phenol:Chloroform (1:1)

In a chemical fume hood, adjust the pH of the phenol to 8.0 with Tris buffer. Shake vigorously. Mix 500 mL of phenol (pH 8.0) and 500 mL of chloroform in a 1-L glass bottle with a lid. Shake the mixture vigorously and let it separate overnight at 4°C. Store for up to 1 mo at 4°C.

Sodium Acetate (3 M, pH 5.2)

Dissolve 246.1 g of sodium acetate in 500 mL of deionized H_2O. Adjust the pH to 5.2 with glacial acetic acid. Allow the solution to cool overnight. Adjust the pH once more to 5.2 with glacial acetic acid. Adjust the final volume to 1 L with deionized H_2O and filter-sterilize.

TE Buffer

Reagent	Quantity (for 100 mL)	Final concentration
EDTA (0.5 M, pH 8.0)	0.2 mL	1 mM
Tris-Cl (1 M, pH 8.0)	1 mL	10 mM
H$_2$O	to 100 mL	

TLE Buffer

Reagent	Amount to add (for 1 L)	Final concentration
Tris–HCl (1 M, pH 8.0)	10 mL	10 mM
EDTA (0.5 M, pH 8.0)	200 µL	0.1 mM

Prepare in deionized H$_2$O. Store for up to 1 yr at room temperature.

Tween Wash Buffer (TWB)

Reagent	Amount to add (for 100 mL)	Final concentration
Tris–HCl (1 M, pH 8.0)	500 µL	5.0 mM
EDTA (500 mM)	100 µL	0.5 mM
NaCl (5 M)	20 mL	1 M
Tween 20 (100%)	50.0 µL	0.05%

Prepare in deionized H$_2$O. Store for up to 1 yr at room temperature.

TBE Buffer

Prepare a 5× stock solution in 1 L of H$_2$O:
- 54 g of Tris base
- 27.5 g of boric acid
- 20 mL of 0.5 M EDTA (pH 8.0)

The 0.5× working solution is 45 mM Tris-borate/1 mM EDTA. TBE is usually made and stored as a 5× or 10× stock solution. The pH of the concentrated stock buffer should be ~8.3. Dilute the concentrated stock buffer just before use and make the gel solution and the electrophoresis buffer from the same concentrated stock solution. Some investigators prefer to use more concentrated stock solutions of TBE (10× as opposed to 5×). However, 5× stock solution is more stable because the solutes do not precipitate during storage. Passing the 5× or 10× buffer stocks through a 0.22-µm filter can prevent or delay formation of precipitates.

REFERENCES

Belton J-M, Dekker J. 2015. Chromosome conformation capture (3C) in budding yeast. *Cold Spring Harb Protoc* doi: 10.1101/pdb.prot085175.

Belton J-M, McCord RP, Gibcus JH, Naumova N, Zhan Y, Dekker J. 2012. Hi–C: A comprehensive technique to capture the conformation of genomes. *Methods* **58:** 268–276.

Dekker J, Rippe K, Dekker M, Kleckner N. 2002. Capturing chromosome conformation. *Science* **295:** 1306–1311.

Lieberman-Aiden E, van Berkum NL, Williams L, Imakaev M, Ragoczy T, Telling A, Amit I, Lajoie BR, Sabo PJ, Dorschner MO, et al. 2009. Comprehensive mapping of long-range interactions reveals folding principles of the human genome. *Science* **326:** 289–293.

CHAPTER 8

MX Cassettes for Knocking Out Genes in Yeast

John H. McCusker[1]

Department of Molecular Genetics and Microbiology, Duke University School of Medicine, Durham, North Carolina 27710

Precise modifications of the *Saccharomyces cerevisiae* genome use marker cassettes, most often in the form of "knockout" (KO) marker cassettes, to delete genes. Many different KO marker cassettes exist, some of which require strains with specific genotypes, such as auxotrophic mutations, and others that have no strain genotype requirements, such as selections for drug resistance and one of two selections for nitrogen source utilization. This introduction focuses on the most frequently used family of KO cassettes—the MX cassettes. In particular, we focus on and describe the different types of MX cassettes and selections; specifically, selections for prototrophy; selections for utilization of cytosine or acetamide as sole nitrogen sources; and selections for resistance to six different drugs. The use of cassettes to place genes under regulated control is briefly discussed. Also discussed are strain genotype requirements (where applicable); media requirements; how to "recycle" or "pop out" cassettes; and counterselections against specific KO cassettes.

INTRODUCTION

One of the major advantages of the *Saccharomyces cerevisiae* model system is the ease of site specifically integrating cassettes into the genome, most often to "knockout" (KO) specific genes. Although there are many different cassettes, the MX cassettes are by far the most widely used family of KO marker cassettes and are the subject of this introduction and accompanying protocol (see Protocol 1: Introducing MX Cassettes into *Saccharomyces cerevisiae* [McCusker 2015a]). The MX cassettes use the heterologous promoter and terminator of the *Ashbya gossyppi* TEF gene (Wach et al. 1994) to express mostly heterologous open reading frames (ORFs) that provide selection for integration into the genome. MX cassette heterology eliminates homologous recombination with the *S. cerevisiae* genome, which greatly facilitates gene targeting of the MX cassettes. In the cases where they have been examined, single copy MX cassettes integrated at *ho*, a neutral location in the genome, are themselves phenotypically neutral (Baganz et al. 1997, 1998; Goldstein and McCusker 1999), making such MX cassettes suitable markers for small-scale competition experiments that look for subtle effects of mutations.

MX CASSETTE AMPLIFICATION

MX cassettes are most often polymerase chain reaction (PCR) amplified with primers that have short 3′ tails (homologous to all MX cassettes) and 5′ tails (usually ≥40 bp) homologous to Your Favorite Gene (YFG1) (Fig. 1). Both the short primer lengths and the ability to use such primers to amplify all MX cassettes are advantageous. The precise PCR amplification conditions are often primer-specific

[1]Correspondence: john.mccusker@duke.edu

Copyright © Cold Spring Harbor Laboratory Press; all rights reserved
Cite this introduction as *Cold Spring Harb Protoc*; doi:10.1101/pdb.top080689

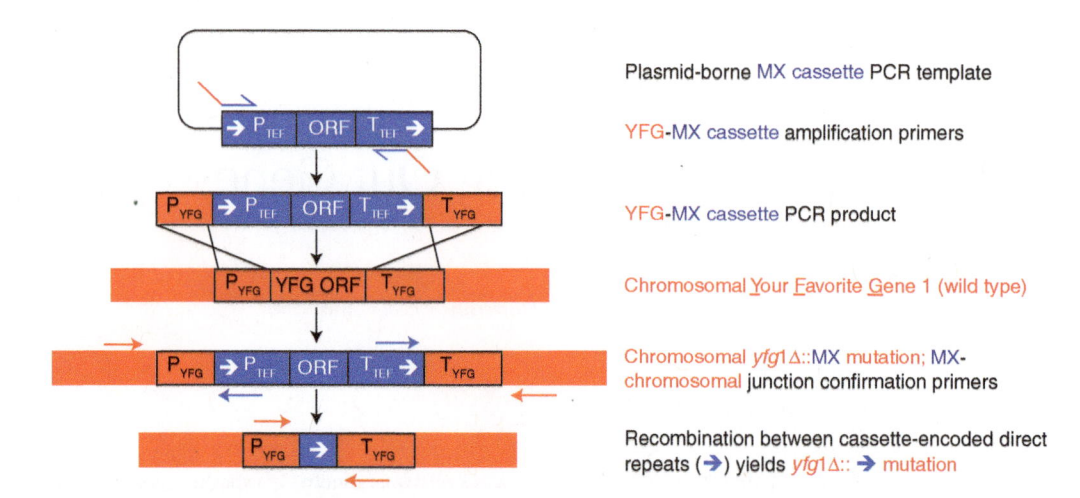

FIGURE 1. MX cassette PCR amplification, integration into the genome, and confirmation. The plasmid-borne PCR template consists of an MX cassette (P_{TEF}-ORF for selectable marker-T_{TEF}; → denotes a cassette-encoded direct repeat [i.e., MX3, PR, or loxP] that is present in some but not all MX cassettes). Ideally, the PCR template should be a nonreplicating (in yeast) plasmid; otherwise, it is necessary to purify the PCR product. The cassette amplification primers consist of ≥35 bases with TEF homology (universal for MX cassettes) at the 3′ ends and ≥40 bases of gene-specific targeting homology at the 5′ ends as shown. After the introduction of an MX PCR product (ideally, purified) into the host to recombine with a chromosomal (wild-type) gene (your favorite gene; YFG), the chromosomal *yfg1Δ*::MX mutation must be confirmed by PCR of both junctions. For MX cassettes flanked by cassette-encoded direct repeats, after recombination, test for chromosomal *yfg1Δ*::→ mutation by PCR.

(e.g., melting temperature, length, structure) and MX cassette-specific; for example, because of the high GC contents of their ORFs, successful PCR amplification of patMX and natMX cassettes requires the addition of 5% DMSO (Goldstein and McCusker 1999). Ideally, MX cassettes should be amplified from nonreplicating (in yeast) plasmid templates. If amplifying an MX cassette from a plasmid that replicates in yeast, the PCR product must always be purified before introduction into yeast; failure to purify the PCR product before introduction into yeast will result in a high number of transformant colonies because of the presence of the replicating plasmid. The presence, amount, and size of MX cassette PCR products should always be confirmed before introduction into yeast.

INTRODUCTION OF MX CASSETTES INTO *S. cerevisiae*

MX cassette PCR products are introduced into *S. cerevisiae* using a LiAc/PEG protocol, variants of which have been optimized for efficiency, speed, and throughput. One key factor in LiAc/PEG transformation efficiency is that cells must be freshly grown and growing well. LiAc/PEG protocols have been extensively and thoroughly described and reviewed (Gietz and Woods 2002; Gietz and Schiestl 2007a,b,c) elsewhere and are thus not discussed further here. PCR of both genome-cassette junctions, using template DNA generated by colony PCR (Amberg et al. 2006; Bergkessel and Guthrie 2013) or "smash and grab" (Hoffman 2001), is required to confirm successful introduction of MX cassettes into the genome. Successful introduction of MX cassette PCR products into *S. cerevisiae* strains results in insertion of the MX cassette into the genome by homologous recombination. Depending on primer design, MX cassette integration can be used to generate KO (most often, deletions of entire ORFs, *yfg1Δ*::MX), other insertion mutations, or point mutations (Horecka and Davis 2014).

GENE REGULATION CASSETTES

In addition to their frequent use in generating KO mutations, MX cassettes can be the selectable components of larger gene regulation cassettes. Briefly, cassette-borne gene regulatory elements include auxin-inducible degrons that regulate gene product stability (Nishimura et al. 2009; Morawska and Ulrich 2013); multiple *S. cerevisiae* promoters that regulate transcription (Kaufmann and Knop 2011); and heterologous tetracycline/doxycycline responsive aptamers (Suess et al. 2003; Kotter et al. 2009) and transcription factors/promoters (Gari et al. 1997; Nagahashi et al. 1997; Belli et al. 1998a, 1998b) that regulate translation and transcription, respectively. Finally, there is cassette-borne, estradiol-mediated transcriptional regulation (McIsaac et al. 2011, 2013). Because estradiol has no obvious off target effects and estradiol-mediated transcriptional regulation is tunable and rapid (McIsaac et al. 2011, 2013), the estradiol system is highly advantageous. Gene regulation cassettes are not further described here.

MULTIPLE MX CASSETTES AND MULTIPLE MX CASSETTE SELECTIONS

There are multiple MX cassettes, each with a different selection, as described in the accompanying protocol (see Protocol 1: Introducing MX Cassettes into *Saccharomyces cerevisiae* [McCusker 2015a]). If for a given strain, one MX cassette or selection does not work or cannot be used, another MX cassette is likely to work. Based on their selections, there are three general types of MX cassettes. First, there are MX cassettes that complement mutations, ideally nonreverting deletions, that confer auxotrophy; that is, integration of these MX cassettes converts auxotrophs to prototrophs. There are MX cassettes that complement *his3* (Wach et al. 1997; Longtine et al. 1998; Gueldener et al. 2002), *ura3* (Goldstein et al. 1999), *lys5* (Ito-Harashima and McCusker 2004), and (in *ura* mutant backgrounds) *fcy1* (Hartzog et al. 2005) mutations. MX cassette selections for prototrophy are straightforward but the requirement for specific strain mutations limits strains and cassette usage as well as environments.

Second, there are MX cassettes that select for utilization of specific compounds as sole nitrogen sources. There are MX cassettes that select for utilization of cytosine (Hartzog et al. 2005), which requires a specific strain genotype (*fcy1*), and acetamide (Solis-Escalante et al. 2013), which has no strain genotype requirements, as sole nitrogen sources. MX cassette sole nitrogen source selections are straightforward but require selection media that lack other nitrogen sources, which includes not only ammonium sulfate, which is present in the most commonly used form of Yeast Nitrogen Base, but also most amino acids (Ljungdahl and Daignan-Fornier 2012). Thus, these sole nitrogen source MX cassette selections must use Yeast Nitrogen Base *without* ammonium sulfate and without amino acids and should use strains with no amino acid auxotrophic mutations; however, because histidine and lysine are not used as nitrogen sources by *S. cerevisiae* (Ljungdahl and Daignan-Fornier 2012), strains may include *his* or *lys* auxotrophic mutations.

Finally, there are MX cassettes that select for drug resistance, which requires no specific strain genotypes. There are MX cassettes that select for resistance to the DNA damaging agent phleomycin (Gueldener et al. 2002), D-serine (Vorachek-Warren and McCusker 2004), bialaphos (a tri-peptide that contains phosphinothricin, a glutamine synthase inhibitor) (Goldstein and McCusker 1999), and the translation inhibiting aminoglycosides G418 (Wach et al. 1994), hygromycin B, and ClonNat (nourseothricin) (Goldstein and McCusker 1999). Of the three general types of MX cassette selections, MX cassette drug selections have the fewest limitations and are thus the most versatile.

RECYCLING MX CASSETTES

Specific MX cassettes can be recycled or "popped out" of the genome by recombination between cassette-encoded direct repeats, such as the large MX3 (Wach et al. 1994) and PR (Hartzog et al. 2005) direct repeats as well as site-specific recombination between loxP direct repeats induced by a plasmid-borne, galactose-regulated Cre site-specific recombinase (Güldener et al. 1996; Gueldener et al. 2002) (Fig. 1). Recycling cassettes allows the same cassettes to be reused in a strain to generate additional KO mutations. Recycling cassettes also eliminates MX homology in a strain, which facilitates the subsequent generation of additional KO mutations with MX cassettes. Methods of recycling, or popping out, MX cassettes, are described in the accompanying protocol (see Protocol 2: Popping Out MX Cassettes from *Saccharomyces cerevisiae* [McCusker 2015b]). Although all MX cassettes have positive selections, there are counterselections for LYS5MX (Ito-Harashima and McCusker 2004), FCY1/FCA1MX (Hartzog et al. 2005), URA3MX (Goldstein et al. 1999), and amdSYM (Solis-Escalante et al. 2013) cassettes. In addition to facilitating cassette recycling, MX cassette counterselections, coupled with appropriate primer design, allow the generation of scar-free mutations (Solis-Escalante et al. 2013; Horecka and Davis 2014). MX cassettes with counter selections are also described in the accompanying protocol (see Protocol 2: Popping Out MX Cassettes from *Saccharomyces cerevisiae* [McCusker 2015b]).

OBTAINING MX CASSETTES

Most, if not all, MX (and other) cassettes have been deposited into plasmid collections, such as Addgene (http://www.addgene.org) and EUROSCARF (http://web.uni-frankfurt.de/fb15/mikro/euroscarf/data/Del_plas.html; http://web.uni-frankfurt.de/fb15/mikro/euroscarf/plasmid.html). Although there may be exceptions, most people who have constructed MX (and other) cassettes greatly prefer that they be requested from such collections.

ACKNOWLEDGMENTS

J.H.M. is supported by a U.S. Public Health Service grant (GM098287).

REFERENCES

Amberg DC, Burke DJ, Strathern JN. 2006. Yeast colony PCR. *CSH Protoc* doi: 10.1101/pdb.prot4170.

Baganz F, Hayes A, Farquhar R, Butler PR, Gardner DC, Oliver SG. 1998. Quantitative analysis of yeast gene function using competition experiments in continuous culture. *Yeast* 14: 1417–1427.

Baganz F, Hayes A, Marren D, Gardner DCJ, Oliver SG. 1997. Suitability of replacement markers for functional analysis studies in *Saccharomyces cerevisiae*. *Yeast* 13: 1563–1573.

Belli G, Gari E, Aldea M, Herrero E. 1998a. Functional analysis of yeast essential genes using a promoter-substitution cassette and the tetracycline-regulatable dual expression system. *Yeast* 14: 1127–1138.

Belli G, Gari E, Piedrafita L, Aldea M, Herrero E. 1998b. An activator/repressor dual system allows tight tetracycline-regulated gene expression in budding yeast. *Nucleic Acids Res* 26: 942–947.

Bergkessel M, Guthrie C. 2013. Colony PCR. *Methods Enzymol* 529: 299–309.

Gari E, Piedrafita L, Aldea M, Herrero E. 1997. A set of vectors with a tetracycline-regulatable promoter system for modulated gene expression in *Saccharomyces cerevisiae*. *Yeast* 13: 837–848.

Gietz RD, Schiestl RH. 2007a. Microtiter plate transformation using the LiAc/SS carrier DNA/PEG method. *Nat Protoc* 2: 5–8.

Gietz RD, Schiestl RH. 2007b. High-efficiency yeast transformation using the LiAc/SS carrier DNA/PEG method. *Nat Protoc* 2: 31–34.

Gietz RD, Schiestl RH. 2007c. Quick and easy yeast transformation using the LiAc/SS carrier DNA/PEG method. *Nat Protoc* 2: 35–37.

Gietz RD, Woods RA. 2002. Transformation of yeast by lithium acetate/single-stranded carrier DNA/polyethylene glycol method. *Methods Enzymol* 350: 87–96.

Goldstein AL, McCusker JH. 1999. Three new dominant drug resistance cassettes for gene disruption in *Saccharomyces cerevisiae*. *Yeast* 15: 1541–1553.

Goldstein AL, Pan X, McCusker JH. 1999. Heterologous URA3MX cassettes for gene replacement in *Saccharomyces cerevisiae*. *Yeast* 15: 507–511.

Gueldener U, Heinisch J, Koehler GJ, Voss D, Hegemann JH. 2002. A second set of loxP marker cassettes for Cre-mediated multiple gene knockouts in budding yeast. *Nucleic Acids Res* 30: e23.

Güldener U, Heck S, Fielder T, Beinhauer J, Hegemann JH. 1996. A new efficient gene disruption cassette for repeated use in budding yeast. *Nucleic Acids Res* 24: 2519–2524.

Hartzog PE, Nicholson BP, McCusker JH. 2005. Cytosine deaminase MX cassettes as positive/negative selectable markers in *Saccharomyces cerevisiae*. *Yeast* 22: 789–798.

Hoffman CS. 2001. Preparation of yeast DNA. *Curr Protoc Mol Biol* 2001 May; Chapter 13:Unit13.11. doi: 10.1002/0471142727.mb1311s39.

Horecka J, Davis RW. 2014. The 50:50 method for PCR-based seamless genome editing in yeast. *Yeast* 31: 103–112.

Ito-Harashima S, McCusker JH. 2004. Positive and negative selection LYS5MX gene replacement cassettes for use in *Saccharomyces cerevisiae*. *Yeast* 21: 53–61.

Kaufmann A, Knop M. 2011. Genomic promoter replacement cassettes to alter gene expression in the yeast *Saccharomyces cerevisiae*. *Methods Mol Biol* 765: 275–294.

Kotter P, Weigand JE, Meyer B, Entian KD, Suess B. 2009. A fast and efficient translational control system for conditional expression of yeast genes. *Nucleic Acids Res* 37: e120.

Ljungdahl PO, Daignan-Fornier B. 2012. Regulation of amino acid, nucleotide, and phosphate metabolism in *Saccharomyces cerevisiae*. *Genetics* 190: 885–929.

Longtine MS, McKenzie A 3rd, Demarini DJ, Shah NG, Wach A, Brachat A, Philippsen P, Pringle JR. 1998. Additional modules for versatile and economical PCR-based gene deletion and modification in *Saccharomyces cerevisiae*. *Yeast* 14: 953–961.

McCusker JH. 2015a. Introducing MX cassettes into *Saccharomyces cerevisiae*. *Cold Spring Harb Protoc* doi: 10.1101/pdb.prot088104.

McCusker JH. 2015b. Popping out MX cassettes from *Saccharomyces cerevisiae*. *Cold Spring Harb Protoc* doi: 10.1101/pdb.prot088120.

McIsaac RS, Oakes BL, Wang X, Dummit KA, Botstein D, Noyes MB. 2013. Synthetic gene expression perturbation systems with rapid, tunable, single-gene specificity in yeast. *Nucleic Acids Res* 41: e57.

McIsaac RS, Silverman SJ, McClean MN, Gibney PA, Macinskas J, Hickman MJ, Petti AA, Botstein D. 2011. Fast-acting and nearly gratuitous induction of gene expression and protein depletion in *Saccharomyces cerevisiae*. *Mol Biol Cell* 22: 4447–4459.

Morawska M, Ulrich HD. 2013. An expanded tool kit for the auxin-inducible degron system in budding yeast. *Yeast* 30: 341–351.

Nagahashi S, Nakayama H, Hamada K, Yang H, Arisawa M, Kitada K. 1997. Regulation by tetracycline of gene expression in *Saccharomyces cerevisiae*. *Mol Gen Genet* 255: 372–375.

Nishimura K, Fukagawa T, Takisawa H, Kakimoto T, Kanemaki M. 2009. An auxin-based degron system for the rapid depletion of proteins in nonplant cells. *Nat Methods* 6: 917–922.

Solis-Escalante D, Kuijpers NG, Bongaerts N, Bolat I, Bosman L, Pronk JT, Daran JM, Daran-Lapujade P. 2013. amdSYM, a new dominant recyclable marker cassette for *Saccharomyces cerevisiae*. *FEMS Yeast Res* 13: 126–139.

Suess B, Hanson S, Berens C, Fink B, Schroeder R, Hillen W. 2003. Conditional gene expression by controlling translation with tetracycline-binding aptamers. *Nucleic Acids Res* 31: 1853–1858.

Vorachek-Warren MK, McCusker JH. 2004. DsdA (D-serine deaminase): A new heterologous MX cassette for gene disruption and selection in *Saccharomyces cerevisiae*. *Yeast* 21: 163–171.

Wach A, Brachat A, Alberti-Segui C, Rebischung C, Philippsen P. 1997. Heterologous HIS3 marker and GFP reporter modules for PCR-targeting in *Saccharomyces cerevisiae*. *Yeast* 13: 1065–1075.

Wach A, Brachat A, Pohlmann R, Philippsen P. 1994. New heterologous modules for classical or PCR-based gene disruptions in *Saccharomyces cerevisiae*. *Yeast* 10: 1793–1808.

Protocol 1

Introducing MX Cassettes into *Saccharomyces cerevisiae*

John H. McCusker[1]

Department of Molecular Genetics and Microbiology, Duke University School of Medicine, Durham, North Carolina 27710

The *Saccharomyces cerevisiae* genome can be readily and precisely modified with the use of knock out (KO) marker cassettes to delete genes. The most frequently used family of KO cassettes is the MX cassettes. This protocol describes how to use the different types of MX cassettes by selecting for prototrophy, utilization of cytosine or acetamide as a sole nitrogen source, or resistance to one of six different drugs.

MATERIALS

It is essential that you consult the appropriate Material Safety Data Sheets and your institution's Environmental Health and Safety Office for proper handling of equipment and hazardous materials used in this protocol.

RECIPES: Please see the end of this protocol for recipes indicated by <R>. Additional recipes can be found online at http://cshprotocols.cshlp.org/site/recipes.

Reagents

Agar plates appropriate for the selection of the MX cassette(s) of interest (see Table 1 and Step 3)

Synthetic dextrose (SD) plates (with or without cytosine) <R>

SD + cytosine medium must have no uracil because FCY/FCAMX cassette selection requires that cytosine be the sole source of uracil. Prepare cytosine solutions and cytosine-containing medium fresh to limit spontaneous deamination to uracil and NH_4^+, which reduces the effectiveness of the positive selection for utilization of cytosine as a pyrimidine source.

Synthetic dextrose glutamate (SDE) plates with drugs <R>

Basal level resistance/sensitivity to the drugs used in this medium and those listed below can vary among different yeast strains. Therefore, before transformation, determine the levels of resistance to single drugs on plates. If demanding simultaneous resistance to two drugs, particularly for two aminoglycosides (i.e., G418, hygromycin B, and clonNAT), empirically determine the levels of both drugs required to inhibit growth on plates before selecting for transformants. Relative to selection medium containing one aminoglycoside, selection medium containing two aminoglycosides will typically require ≥50% higher levels of both aminoglycosides.

Synthetic dextrose (SD) plates without ammonium sulfate or amino acids (with or without acetamide or cytosine) <R>

For sole nitrogen source selections, it is essential that other compounds that can serve as a nitrogen source, including ammonium sulfate and most amino acids, be absent from the selection media. This requires the use of yeast nitrogen base without ammonium sulfate and without amino acids, and the avoidance of most and preferably all amino acid auxotrophic mutations and corresponding nutritional supplements. As above, prepare cytosine solutions and cytosine-containing medium fresh.

---

[1]Correspondence: john.mccusker@duke.edu

Copyright © Cold Spring Harbor Laboratory Press; all rights reserved
Cite this protocol as *Cold Spring Harb Protoc*; doi:10.1101/pdb.prot088104

TABLE 1. Summary of MX cassettes, strain genotypes, positive selections, and media

MX cassette	*S. cerevisiae* strain	Selection	Type of selection
HIS3MX (Wach et al. 1997) SpHIS5MX (Longtine et al. 1998; Gueldener et al. 2002)	*his3* strain	SD plates	Prototrophy
URA3MX (Goldstein et al. 1999)	*ura3* strain	SD plates	Prototrophy
LYS5MX (Ito-Harashima and McCusker 2004)	*lys5* strain	SD plates	Prototrophy
FCY1MX (Hartzog et al. 2005) FCA1MX (Hartzog et al. 2005)	*fcy1 ura* Ade$^+$ strain	SD + cytosine plates (to select for cytosine as the pyrimidine source)	Prototrophy
amdSYM (Solis-Escalante et al. 2013)	Avoid strains with most amino acid auxotrophic mutations (*his* or *lys* mutations are okay)	SD plates without ammonium sulfate and without amino acids + acetamide	Nitrogen source utilization
FCY1MX (Hartzog et al. 2005) FCA1MX (Hartzog et al. 2005)	*fcy1* Ade$^+$ strain Avoid strains with most amino acid auxotrophic mutations (*his* or *lys* mutations are okay)	SD without ammonium sulfate and without amino acids + cytosine (to select for cytosine as the sole nitrogen source).	Nitrogen source utilization
bleMX (Gueldener et al. 2002)		YPD + phleomycin	Drug resistance on YPD
kanMX (Wach et al. 1994)		YPD + G418	Drug resistance on YPD
hphMX (Goldstein and McCusker 1999)		YPD + hygromycin B	Drug resistance on YPD
natMX (Goldstein and McCusker 1999)		YPD + ClonNat (nourseothricin)	Drug resistance on YPD
kanMX (cassette: Wach et al. 1994; medium: Cheng et al. 2000)		SDE without ammonium sulfate and without amino acids + G418	Drug resistance on SD
hphMX (cassette: Goldstein and McCusker 1999; medium (with drug substitution): Cheng et al. 2000)		SDE without ammonium sulfate and without amino acids + hygromycin B	Drug resistance on SD
natMX (cassette: Goldstein and McCusker 1999; medium (with drug substitution): Cheng et al. 2000)		SDE without ammonium sulfate and without amino acids + clonNAT (nourseothricin)	Drug resistance on SD
dsdAMX (Vorachek-Warren and McCusker 2004)	Avoid strains with most amino acid auxotrophic mutations	SDP without ammonium sulfate and without amino acids + D-serine	Drug resistance on SDP
patMX (Goldstein and McCusker 1999).	Avoid strains with most amino acid auxotrophic mutations	SDP without ammonium sulfate and without amino acids + bialaphos or phosphinothricin	Drug resistance on SDP

Synthetic dextrose proline (SDP) plates with drugs <R>
Yeast extract peptone dextrose (YPD) liquid medium and plates with drugs <R>
Agarose gel and buffer of choice
Plasmid templates for MX cassette(s) of interest (see Table 1 and Step 3)
> Most, if not all, MX (and other) cassettes have been deposited into plasmid collections, such as Addgene (http://www.addgene.org) and EUROSCARF (http://web.uni-frankfurt.de/fb15/mikro/euroscarf/data/Del-plas.html; http://web.uni-frankfurt.de/fb15/mikro/euroscarf/plasmid.html).

Polymerase chain reaction (PCR) reagents to amplify MX cassette(s) of interest (see Step 1)
Reagents to perform colony PCR or the "smash and grab" technique (see Step 6)
Reagents to perform an LiAc/PEG transformation (see Step 2)
Yeast strain of interest with the required genotypes (if applicable; see Table 1 and Step 3), freshly grown in YPD

Equipment

Agarose gel apparatus
Incubator
PCR tubes
Petri dishes (100 mm × 15 mm)
Thermocycler

Chapter 8

METHOD

1. Using published PCR conditions for the MX cassette of interest, PCR-amplify MX cassette(s) from the appropriate plasmid template. Confirm the presence, amount, and size of MX cassette PCR products via gel electrophoresis.

    Ideally, MX cassettes should be amplified from a plasmid that does not replicate in yeast. If a plasmid that replicates in yeast is used, the PCR product must be purified before introduction into yeast; failure to purify the PCR product before introduction into yeast will result in a high number of transformant colonies due to the presence of the replicating plasmid.

    The precise PCR amplification conditions are often primer-specific (e.g., melting temperature, length, structure) and MX cassette-specific; for example, because of the high GC contents of their open reading frames (ORFs), successful PCR amplification of patMX and natMX cassettes requires the addition of 5% DMSO (Goldstein and McCusker 1999). Therefore, PCR conditions for MX cassettes are not described here; users should instead consult the original papers and may need to further optimize PCR conditions.

    PCR primer design depends on whether the desired mutation is a simple yfg1Δ::MX mutation (see Fig. 1 in Introduction: MX Cassettes for Knocking Out Genes in Yeast [McCusker 2015a]); a seamless/scar-free yfg1Δ mutation (Solis-Escalante et al. 2013; Horecka and Davis 2014) (see Fig. 1 in Protocol 2: Popping Out MX Cassettes from Saccharomyces cerevisiae [McCusker 2015b]); or scar-free indel or point mutations (Horecka and Davis 2014) (see Fig. 2 in Protocol 2: Popping Out MX Cassettes from Saccharomyces cerevisiae [McCusker 2015b]).

    See Troubleshooting.

2. Use an LiAc/PEG transformation protocol (Gietz and Woods 2002; Gietz and Schiestl 2007a; Gietz and Schiestl 2007c; Gietz and Schiestl 2007b) to introduce the MX cassette PCR products into freshly growing yeast.

3. Plate $\sim10^6$ yeast cells and, separately, $\sim10^6$ mock-transformed yeast cells onto the appropriate type of plate.

    - If selecting for cassettes that are selected based on prototrophy (HIS3MX, SpHIS5MX, URA3MX, LYS5MX), plate cells on SD plates.

        Heterologous Saccharomyces kluyveri HIS3 (HIS3MX) (Wach et al. 1997) and heterologous Schizosaccharomyces pombe HIS5 (SpHIS5MX) (Longtine et al. 1998; Gueldener et al. 2002) MX cassettes complement S. cerevisiae his3 mutations. Heterologous (Candida albicans ORF) URA3MX cassettes complement S. cerevisiae ura3 mutations (Goldstein et al. 1999). Homologous (S. cerevisiae ORF) and heterologous (C. albicans ORF) LYS5MX cassettes complement S. cerevisiae lys5 mutations (Ito-Harashima and McCusker 2004).

    - If selecting for cytosine as the pyrimidine source using FCY1/FCA1MX cassettes in an *fcy1* strain, plate cells onto SD + cytosine plates.

        Homologous FCY1MX (S. cerevisiae ORF) and heterologous FCA1MX (C. albicans ORF) cassettes encode cytosine deaminase, which deaminates cytosine to uracil and NH_4^+. Thus, in an fcy1 ura strain, FCY1MX and FCA1MX cassettes allow positive selection—in the absence of exogenous uracil—for the use of cytosine (25 μM) as the pyrimidine source (Hartzog et al. 2005). Cytosine transport requires Fcy2p (Pantazopoulou and Diallinas 2007). Because cytosine and adenine compete for uptake by Fcy2p, fcy1 ura strains should be Ade^+ to avoid having to supplement the cytosine-containing media with adenine and thus likely reducing the effectiveness of the FCY1/FCA1 cassette selection.

    - If selecting for the amdSYM cassette, plate cells onto SD plates without ammonium sulfate and without amino acids + acetamide.

        The heterologous amdSYM cassette encodes acetamidase, which deaminates acetamide to acetate and NH_4^+, with the NH_4^+ able to serve as the sole nitrogen source (Solis-Escalante et al. 2013). Fps1p transports acetamide (Shepherd and Piper 2010); thus, strains should be FPS1.

    - If selecting for cytosine as the sole nitrogen source using FCY1/FCA1MX cassettes in an *fcy1* strain, plate cells onto SD without ammonium sulfate and without amino acids + cytosine.

        The homologous FCY1MX (S. cerevisiae ORF) and heterologous FCA1MX (C. albicans ORF) cassettes encode cytosine deaminase, which deaminates cytosine to uracil and NH_4^+, with the NH_4^+ able to serve as the sole nitrogen source. In an fcy1 strain, FCY1MX and FCA1MX cassettes allow positive selection for the use of cytosine (1 mM) as the sole nitrogen source (Hartzog et al. 2005). Cytosine transport requires Fcy2p (Pantazopoulou and Diallinas 2007). Because cytosine and adenine compete for uptake by

Fcy2p, fcy1 strains should be Ade⁺ to avoid having to supplement the cytosine-containing media with adenine and thus likely reducing the effectiveness of the FCY1/FCA1 cassette selection.

- If selecting for bleMX, kanMX, hphMX, or natMX cassettes, plate cells on YPD containing phleomycin, G418, hygromycin B, or ClonNat, respectively.

    In general, kanMX, hphMX, or natMX can be selected for on YPD + drug. Use SDE without ammonium sulfate and without amino acids + drug if a second selection requires the use of defined medium.

    There are four heterologous MX cassettes that confer drug resistance in YPD: bleMX, kanMX, hphMX, and natMX. The bleMX cassette encodes a protein that tightly binds and confers resistance to the DNA damaging agent phleomycin (Gueldener et al. 2002). Phleomycin, bleomycin, and zeocin are closely related DNA damaging agents. The kanMX, hphMX, and natMX cassettes encode enzymes that specifically modify and confer resistance to the translation inhibiting aminoglycosides G418, hygromycin B, and ClonNat (nourseothricin), respectively (Wach et al. 1994; Goldstein and McCusker 1999).

- If selecting for the dsdAMX cassette, plate cells on SDP without ammonium sulfate and without amino acids + D-serine. If selecting for the patMX cassette, plate cells on SDP without ammonium sulfate and without amino acids + bialaphos.

    The dsdAMX cassette encodes D-serine deaminase, which confers resistance to toxic D-serine via deamination to nontoxic pyruvate and NH_4^+ (Vorachek-Warren and McCusker 2004). The patMX cassette encodes phosphinothricin acetyl transferase, which confers resistance to both the tri-peptide bialaphos (Ala-Ala-phosphinothricin) and to the glutamine synthase-inhibiting glutamate analog phosphinothricin (Goldstein and McCusker 1999). Uptake of and sensitivity to D-serine requires the GAP1-encoded general amino acid permease (Rytka 1975); uptake of bialaphos likely requires the PTR2-encoded peptide transporter. The activity and/or expression of GAP1 and PTR2 requires the absence of good nitrogen sources, such as ammonium sulfate and most amino acids, and the presence of a poor nitrogen source, such as 0.5 g proline/L. This requires the use of yeast nitrogen base without ammonium sulfate and without amino acids and the avoidance of most amino acid auxotrophic mutations and corresponding nutritional supplements that may affect the activity and/or expression of GAP1 and PTR2. Instead of ammonium sulfate, the nitrogen source is proline (0.5 mg/mL) (Goldstein and McCusker 1999; Vorachek-Warren and McCusker 2004).

    There are five heterologous MX cassettes that confer drug resistance in SD: dsdAMX, patMX, kanMX, hphMX, and natMX. For all of these drug resistance selections in SD, use of yeast nitrogen base without ammonium sulfate and without amino acids is required to reduce the salt concentration; instead of ammonium sulfate, the nitrogen source can be monosodium glutamate or proline (Cheng et al. 2000).

4. Incubate the plates at 30°C as follows.
    - Incubate SD plates, SD + cytosine plates, or YPD plates with drugs for ≥2 d.
    - Incubate SD plates without ammonium sulfate and without amino acids containing cytosine or acetamide as the sole nitrogen source for 4–6 d.
    - Incubate SDE or SDP plates without ammonium sulfate and without amino acids with drugs for 3–7 d.

5. Pick transformant colonies.

    See Troubleshooting.

6. Using colony PCR (Amberg et al. 2006; Bergkessel and Guthrie 2013) or "smash and grab" (Hoffman 2001) to isolate genomic DNA, test both junctions to confirm successful introduction of MX cassettes into the genome (see Fig. 1 in Introduction: MX Cassettes for Knocking Out Genes in Yeast [McCusker 2015a]).

7. Test the phenotype(s) of transformants.

TROUBLESHOOTING

Problem (Step 1): An appropriately sized PCR product corresponding to the MX cassette PCR product was not present on an agarose gel.
Solution: Consider the following.

- For natMX and patMX cassettes that have high GC contents, amplify in the presence of DMSO.
- If using a plasmid-encoded patMX cassette as a template, use instead as a template a patMX cassette integrated into the genome of an *S. cerevisiae* strain. patMX may be deleterious in *Escherichia coli*; therefore, if PCR using a plasmid-encoded patMX does not yield transformants, use genomic DNA from an ho::patMX strain (obtainable from the Fungal Genetics Stock Center [http://www.fgsc.net]) as a template.

Problem (Step 5): No or few transformants were recovered on the selection medium.
Solution: Consider the following.

- Check the LiAc/PEG reagents (Gietz and Woods 2002; Gietz and Schiestl 2007a; Gietz and Schiestl 2007c; Gietz and Schiestl 2007b).
- Check the strain genotype and phenotype as well as the selection medium.
- Repeat transformation using PCR product formation verified for amount and correct size, freshly made LiAc/PEG reagents, freshly grown cells, plating more cells, and/or remade medium.
- If applicable, reduce the drug concentration.
- If using translation-inhibiting aminoglycosides, increase the recovery time following LiAc/PEG transformation (minimum of 2 h) in nonselective medium (YPD) before plating onto selective medium. In extreme cases, recovery time can be increased to an overnight incubation to ensure expression of the cassette-encoded, drug-inactivating enzyme.
- Confirm the presence of the PCR product on a gel before transformation (as described in Step 1).

Problem (Step 5): Too many colonies or confluent growth was recovered on selection media.
Solution: Consider the following.

- The auxotrophic mutation in the strain may have reverted or the strain culture may be contaminated. Check the strain genotype and phenotype.
- The selection medium may be incorrect. In particular, when noted, the selection medium must use yeast nitrogen base *without ammonium sulfate and without amino acids* and should not contain any amino acids that can serve as nitrogen sources. In the case of cytosine selections, use freshly made cytosine solution and freshly made cytosine-containing selection medium.
- If using a yeast strain with a point or unknown mutation in *his3*, *fcy1*, *ura3*, or *lys5*, use instead strains with *his3Δ*, *fcy1Δ*, *ura3Δ*, or *lys5Δ* mutation(s).
- If introducing homologous FCY1MX or LYS5MX cassettes, use only strains with *fcy1Δ* and *lys5Δ* mutations, respectively. Alternatively, introduce heterologous FCA1MX or LYS5MX cassettes.
- Plate fewer cells onto selective medium.
- If the strain already contains an MX cassette, "pop-out" the existing integrated MX cassette (see Protocol 2: Popping Out MX Cassettes from *Saccharomyces cerevisiae* [McCusker 2015b]); select for both the existing integrated MX cassette marker and the newly introduced MX cassette marker; or introduce the MX cassette into an MX cassette–free strain.
- Amplify the MX cassette using a nonreplicating plasmid as a template. Alternatively, purify the MX cassette PCR product away from the template.
- If applicable, increase the drug concentration(s). If the selection medium contains two aminoglycosides, the concentrations of both aminoglycosides must be increased. Empirically determine increased drug concentrations.
- Replica plate the initial drug selection plate(s) containing too many colonies to fresh drug plates. This frequently eliminates many false positive transformants.

- For other D-serine-resistant mutations, distinguish between spontaneous *gap1* mutations (which also confer D-serine resistance) and desired *yfg1Δ::dsdMX* mutations by the inability of *gap1* mutants to use L-citrulline as a sole nitrogen source (Rytka 1975; Vorachek-Warren and McCusker 2004).

DISCUSSION

The introduction of an MX cassette PCR product with short (~40-bp) regions of homology results in successful transformation in most cases. An important exception, however, is if a strain already contains an integrated MX cassette. In this case, the existing integrated MX cassette has very large regions of homology with the incoming MX PCR product, which will result in no or very few of the desired *yfg1Δ::MX* transformants. Two simple solutions are to either "pop-out" the existing integrated MX cassette (see Protocol 2: Popping Out MX Cassettes from *Saccharomyces cerevisiae* [McCusker 2015b]), thereby eliminating the large MX recombination target, or to select for both the existing integrated MX cassette marker and the newly introduced MX cassette marker.

Even in the absence of an already integrated MX cassette, the introduction of an MX cassette PCR product with short (~40-bp) regions of homology may produce no transformants in a specific strain and/or in a specific gene. In such cases, amplify the MX cassette with a different primer pair and/or primer pairs with longer (≥60-bp) regions of genomic homology. Alternatively, if the MX cassette will integrate into another strain, PCR-amplify using primers ≥500 bp flanking the integrated MX cassette and introduce the resulting PCR product, which has much longer regions of homology, into the desired strain.

RECIPES

Synthetic Dextrose Glutamate (SDE) Plates with Drugs

Mix yeast nitrogen base without ammonium sulfate and without amino acids (1.7 g/L), L-glutamate (monosodium salt; 1 g/L), dextrose (20 g/L), and Bacto agar (20 g/L) in 1 L of (distilled or deionized) H_2O, and autoclave. After autoclaving, add one of the following filter-sterilized drugs, as required, to the appropriate final concentration: G418 (200 µg/mL), hygromycin B (300 µg/mL), or clonNAT (100 µg/mL). After cooling to 55°C–60°C, pour medium into Petri dishes. If using glutamic acid instead of the suggested monosodium glutamate, adjust medium pH to 5.5 before autoclaving.

Synthetic Dextrose (SD) Plates (with or without Cytosine)

Mix yeast nitrogen base with ammonium sulfate and without amino acids (6.7 g/L), dextrose (20 g/L), and Bacto agar (20 g/L) in 1 L of (deionized or distilled) H_2O. (Other nutritional supplements may be required depending on strain genotypes; if necessary, dropout mixtures may be made as described [Rose et al. 1990] or obtained from commercial sources such as ClonTech.) Autoclave. If necessary, add filter-sterilized cytosine to a final concentration of 25 µM (Hartzog et al. 2005) after autoclaving. After cooling to 55°C–60°C, pour into Petri dishes.

Synthetic Dextrose (SD) Plates without Ammonium Sulfate or Amino Acids (with or without Acetamide or Cytosine)

Mix yeast nitrogen base without ammonium sulfate and without amino acids (1.7 g/L), dextrose (20 g/L), and Bacto agar (20 g/L) in 1 L of (deionized or distilled) H_2O, and autoclave. If necessary, add filter-sterilized acetamide to a final concentration of 600 mg/L (Solis-Escalante et al. 2013) or filter-sterilized cytosine to a final concentration of 1 mM (Hartzog et al. 2005) after autoclaving. After the medium has cooled to 55°C–60°C, pour into Petri dishes.

Synthetic Dextrose Proline (SDP) Plates with Drugs

Mix yeast nitrogen base without ammonium sulfate and without amino acids (1.7 g/L), L-proline (0.5 g/L), dextrose (20 g/L), and Bacto agar (20 g/L) in 1 L of (distilled or deionized) H_2O, and autoclave. After autoclaving, add one of the following filter-sterilized drugs, as required, to the appropriate final concentration: bialaphos (200 µg/mL), phosphinothricin (600–800 µg/mL), or D-serine (2 mg/mL). After cooling to 55°C–60°C, pour medium into Petri dishes.

Yeast Extract Peptone Dextrose (YPD) Liquid Medium and Plates with Drugs

For YPD liquid medium, mix yeast extract (10 g/L), Bacto peptone (20 g/L), and dextrose (20 g/L) in 1 L of (deionized or distilled) H_2O. Adjust the pH to 5.5 with 1 M HCl, and autoclave.

For YPD plates with drugs, prepare liquid medium as above, but add Bacto Agar (20 g/L) before autoclaving. After autoclaving, add one of the following filter-sterilized drugs, as required, to the appropriate final concentration: clonNAT (100 µg/mL), G418 (200 µg/mL), hygromycin B (300 µg/mL), or phleomycin (7.5 µg/mL). After cooling to 55°C–60°C, pour liquid into Petri dishes.

ACKNOWLEDGMENTS

J.H.M. is supported from a U.S. Public Health Service grant (GM098287).

REFERENCES

Amberg DC, Burke DJ, Strathern JN. 2006. Yeast colony PCR. *CSH Protoc* doi: 10.1101/pdb.prot4170.

Bergkessel M, Guthrie C. 2013. Colony PCR. *Methods Enzymol* 529: 299–309.

Cheng TH, Chang CR, Joy P, Yablok S, Gartenberg MR. 2000. Controlling gene expression in yeast by inducible site-specific recombination. *Nucleic acids research* 28: E108.

Gietz RD, Schiestl RH. 2007a. Microtiter plate transformation using the LiAc/SS carrier DNA/PEG method. *Nat Protoc* 2: 5–8.

Gietz RD, Schiestl RH. 2007b. High-efficiency yeast transformation using the LiAc/SS carrier DNA/PEG method. *Nat Protoc* 2: 31–34.

Gietz RD, Schiestl RH. 2007c. Quick and easy yeast transformation using the LiAc/SS carrier DNA/PEG method. *Nat Protoc* 2: 35–37.

Gietz RD, Woods RA. 2002. Transformation of yeast by lithium acetate/single-stranded carrier DNA/polyethylene glycol method. *Methods Enzymol* 350: 87–96.

Goldstein AL, McCusker JH. 1999. Three new dominant drug resistance cassettes for gene disruption in *Saccharomyces cerevisiae*. *Yeast* 15: 1541–1553.

Goldstein AL, Pan X, McCusker JH. 1999. Heterologous URA3MX cassettes for gene replacement in *Saccharomyces cerevisiae*. *Yeast* 15: 507–511.

Gueldener U, Heinisch J, Koehler GJ, Voss D, Hegemann JH. 2002. A second set of loxP marker cassettes for Cre-mediated multiple gene knockouts in budding yeast. *Nucleic Acids Res* 30: e23.

Hartzog PE, Nicholson BP, McCusker JH. 2005. Cytosine deaminase MX cassettes as positive/negative selectable markers in *Saccharomyces cerevisiae*. *Yeast* 22: 789–798.

Hoffman CS. 2001. Preparation of yeast DNA. *Curr Protoc Mol Biol*. 2001 May; Chapter 13:Unit13.11. doi: 10.1002/0471142727.mb1311s39.

Horecka J, Davis RW. 2014. The 50:50 method for PCR-based seamless genome editing in yeast. *Yeast* 31: 103–112.

Ito-Harashima S, McCusker JH. 2004. Positive and negative selection LYS5MX gene replacement cassettes for use in *Saccharomyces cerevisiae*. *Yeast* 21: 53–61.

Longtine MS, McKenzie A 3rd, Demarini DJ, Shah NG, Wach A, Brachat A, Philippsen P, Pringle JR. 1998. Additional modules for versatile and economical PCR-based gene deletion and modification in *Saccharomyces cerevisiae*. *Yeast* 14: 953–961.

McCusker JH. 2015a. MX cassettes for knocking out genes in yeast. *Cold Spring Harb Protoc* doi: 10.1101/pdb.top080689.

McCusker JH. 2015b. Popping out MX cassettes from *Saccharomyces cerevisiae*. *Cold Spring Harb Protoc* doi: 10.1101/pdb.prot088120.

Pantazopoulou A, Diallinas G. 2007. Fungal nucleobase transporters. *FEMS Microbiol Rev* 31: 657–675.

Rose MD, Winston F, Hieter P. 1990. *Methods in yeast genetics: A laboratory course manual*. Cold Spring Harbor Laboratory Press, Cold Spring Harbor, NY.

Rytka J. 1975. Positive selection of general amino acid permease mutants in *Saccharomyces cerevisiae*. *J Bacteriol* 121: 562–570.

Shepherd A, Piper PW. 2010. The Fps1p aquaglyceroporin facilitates the use of small aliphatic amides as a nitrogen source by amidase-expressing yeasts. *FEMS Yeast Res* 10: 527–534.

Solis-Escalante D, Kuijpers NG, Bongaerts N, Bolat I, Bosman L, Pronk JT, Daran JM, Daran-Lapujade P. 2013. amdSYM, a new dominant recyclable marker cassette for *Saccharomyces cerevisiae*. *FEMS Yeast Res* 13: 126–139.

Vorachek-Warren MK, McCusker JH. 2004. DsdA (D-serine deaminase): A new heterologous MX cassette for gene disruption and selection in *Saccharomyces cerevisiae*. *Yeast* 21: 163–171.

Wach A, Brachat A, Alberti-Segui C, Rebischung C, Philippsen P. 1997. Heterologous HIS3 marker and GFP reporter modules for PCR-targeting in *Saccharomyces cerevisiae*. *Yeast* 13: 1065–1075.

Wach A, Brachat A, Pohlmann R, Philippsen P. 1994. New heterologous modules for classical or PCR-based gene disruptions in *Saccharomyces cerevisiae*. *Yeast* 10: 1793–1808.

Protocol 2

Popping Out MX Cassettes from *Saccharomyces cerevisiae*

John H. McCusker[1]

Department of Molecular Genetics and Microbiology, Duke University School of Medicine, Durham, North Carolina 27710

MX cassettes are frequently used to generate knockout (KO) mutations in *Saccharomyces cerevisiae*. The recycling or "popping out" of an MX cassette flanked by direct repeats allows the same cassette to be reused in a strain to generate additional KO mutations. Popping out MX cassettes also eliminates MX homology in a strain, which facilitates the subsequent generation of additional KO mutations with other MX cassettes. MX cassettes can be recycled or "popped out" of the genome by spontaneous recombination between large, cassette-borne MX3 or PR direct repeats and by Cre-mediated, site-specific recombination between small, cassette-borne loxP direct repeats. Both of these techniques leave a mutation with a cassette-encoded "scar." For the URA3MX, LYS5MX, FCA1/FCY1MX, and amdSYM cassettes, there are counterselections. Counterselections are extremely useful as they allow for positive selection for plasmid shuffling, transplacement of mutant alleles into the genome, and recycling or popping out cassettes flanked by cassette-encoded direct repeats to yield mutations with a cassette-encoded scar. Finally, after amplifying with the appropriately designed primers, integrated counterselectable MX cassettes can be popped out to generate seamless or "scar-free" deletion mutations, as well as indel and point mutations.

MATERIALS

It is essential that you consult the appropriate Material Safety Data Sheets and your institution's Environmental Health and Safety Office for proper handling of equipment and hazardous materials used in this protocol.

RECIPES: Please see the end of this protocol for recipes indicated by <R>. Additional recipes can be found online at http://cshprotocols.cshlp.org/site/recipes.

Reagents

Agar plate(s) to detect the loss of the MX cassette using cassettes with counterselection (see Steps 2 and 3):

SD + 5-fluorocytosine medium <R>
SD + fluoroacetamide medium <R>
SD + LYS + α-aminoadipate medium <R>
SD + URA + 5-FOA medium <R>
SDP + URA + 5-FOA medium <R>
YPD + 5-fluorocytosine medium <R>

---

[1]Correspondence: john.mccusker@duke.edu

Copyright © Cold Spring Harbor Laboratory Press; all rights reserved
Cite this protocol as *Cold Spring Harb Protoc*; doi:10.1101/pdb.prot088120

Chapter 8

Agar plates to detect the loss of the MX cassette using cassettes without counterselection (see Step 1)
> This will be the same used to select for the presence of the MX cassette. See Protocol 1: Introducing MX Cassettes into Saccharomyces cerevisiae (McCusker 2015b).

Reagents needed only for popping out MX cassettes with Cre-mediated recombination (see Step 4):
Agar plates to detect the loss of the MX cassette
> This will be the same used to select for the presence of the MX cassette. See Protocol 1: Introducing MX Cassettes into Saccharomyces cerevisiae (McCusker 2015b).

Dextrose-containing yeast agar plates appropriate to select for only the plasmid-borne selectable marker following transformation in Step 4.i

Plasmid containing the galactose-inducible, Cre recombinase
> Many yeast Cre-containing plasmids have been deposited into plasmid collections, such as Addgene (http://www.addgene.org) and EUROSCARF (http://web.uni-frankfurt.de/fb15/mikro/euroscarf/data/Del_plas.html; http://web.uni-frankfurt.de/fb15/mikro/euroscarf/plasmid.html).

Reagents to perform a LiAc/PEG transformation (see Step 4.i)
YPGal medium <R>

Reagents to perform colony PCR or the "smash and grab" technique (see Step 5)
YPD liquid medium and plates with drugs <R>
Yeast strain containing MX cassette of interest, freshly grown in YPD medium without selection for the MX cassette marker

Equipment

Incubator
Petri dishes (100-mm × 15-mm)
Polypropylene round-bottom tubes or flasks (see Step 4)
Replica plating tool (optional; see Step 4)
Roller drum or tube shaker (see Step 4)
Thermocycler
Velvets (sterile, for replica plating) (optional; see Step 4)

METHOD

Select the appropriate MX cassette pop-out strategy from Steps 1–4, and then proceed to Step 5. Mechanisms to "pop out" MX cassettes from the genome by spontaneous recombination between large, cassette-borne MX3 are discussed in Wach et al. (1994), by spontaneous recombination between PR direct repeats are discussed in Hartzog et al. (2005), and by Cre-mediated, site-specific recombination between small, cassette-borne loxP direct repeats are discussed in Güldener et al. (1996) and Gueldener et al. (2002). These techniques leave a mutation with a cassette-encoded "scar" (see Fig. 1 from Introduction: MX Cassettes for Knocking Out Genes in Yeast [McCusker 2015a]). Solis-Escalante et al. (2013) and Horecka and Davis (2014) discuss how MX cassettes can be popped out to generate seamless or "scar-free" deletion mutations (Fig. 1), and Horecka and Davis (2014) discuss how MX cassettes can be popped out to create indel and point mutations (Fig. 2).

Pop Out Integrated MX Cassettes (with No Counterselection) Flanked by Large MX3 or PR Direct Repeats

1. Plate $\geq 10^2$ yeast cells (which have been grown in YPD without cassette selection) onto each of ≥ 10 YPD plates and, after 2–3 d growth at 30°C, replica plate to medium that will detect the loss of the MX cassette.

   > The selection medium used for replica plating here will be the same selection used to select for the presence of the MX cassette. See Protocol 1: Introducing MX Cassettes into Saccharomyces cerevisiae (McCusker 2015b).
   >
   > For MX cassettes flanked by large MX3 (Wach et al. 1994) or PR repeats (Hartzog et al. 2005), the spontaneous frequencies of cassette loss will be 10^{-3} to 10^{-4}.
   >
   > See Troubleshooting.

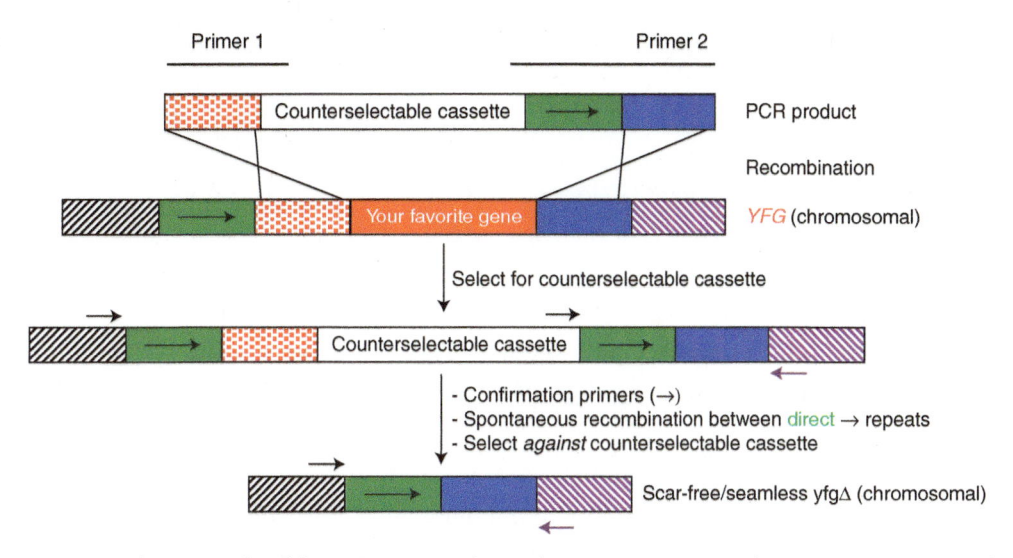

FIGURE 1. Use of counterselectable cassettes, together with appropriate primer design, to generate scar-free or seamless deletion mutations (Solis-Escalante et al. 2013; Horecka and Davis 2014).

Pop Out Integrated MX Cassettes (with Counterselection) Flanked by Large MX3 or PR Direct Repeats

2. Plate 10^3–10^4 yeast cells (which have been grown in YPD without cassette selection) onto medium that selects against the MX cassette (see below) and, depending on the counter selection medium, incubate for 3–7 d at 30°C.

   - For LYS5MX cassettes, use SD + LYS + α-aminoadipate medium.
   - For FCA1/FCY1MX cassettes, use SD + 5-fluorocytosine medium or YPD + 5-fluorocytosine medium.

     Use YPD + 5-fluorocytosine medium unless the background of the strain or a plasmid in the strain indicates a need for selection on SD.

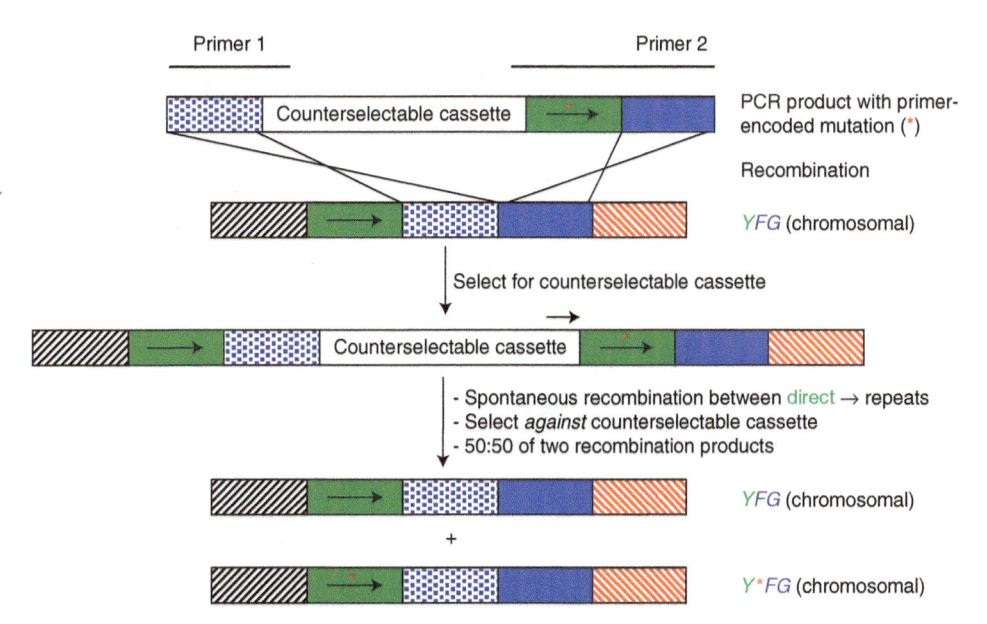

FIGURE 2. Use of counterselectable cassettes, together with appropriate primer design, to generate scar-free indel and point mutations (Horecka and Davis 2014).

- For URA3MX cassettes, use SD + URA + 5-FOA medium or SDP + URA + 5-FOA medium.

  The use of proline as the sole nitrogen source (in place of ammonium sulfate), together with lower amounts of uracil, allows much lower concentrations of 5-FOA (an expensive reagent) to be used. However, not all genetic backgrounds are sensitive to SDP + URA + 5-FOA medium and the addition of some nutritional supplements abolishes sensitivity to 5-FOA (McCusker and Davis 1991).

- For amdSYM cassettes, use SD + fluoroacetamide medium.

  See Troubleshooting.

Pop Out Integrated MX Cassettes (with Counterselection) Flanked, After Integration, by Primer-Encoded Direct Repeats

See Solis-Escalante et al. (2013) and Horecka and Davis (2014).

3. Follow the instructions in Step 2, but plate 10^7 yeast cells (which have been grown in YPD without cassette selection) onto multiple counterselection plates.

    More cells must be plated here because the primer-encoded direct repeats are much shorter than the cassette-encoded direct repeats (e.g., MX3), and there is a relationship between direct repeat lengths and recombination frequencies.

    See Troubleshooting.

Pop Out Integrated MX Cassettes Flanked by loxP Direct Repeats

4. Proceed as follows.

    i. Use a LiAc/PEG transformation protocol (Gietz and Woods 2002; Gietz and Schiestl 2007a,b,c) to introduce a plasmid containing the galactose-inducible Cre recombinase (Güldener et al. 1996; Gueldener et al. 2002) into yeast that contain an integrated MX cassette flanked by loxP sites, freshly grown in YPD without cassette selection. Following transformation, select for only the plasmid-borne selectable marker on dextrose containing medium.

        The type of medium used to select for the Cre plasmid will depend on the type of plasmid used. For example, if Cre on a URA3 plasmid is introduced into a ura3 strain, the plasmid should be selected using SD plates.

    ii. Grow yeast strains containing the integrated MX cassette and the galactose-regulated, Cre recombinase plasmid in YPGal medium for ≥ 2 h without selection for the integrated MX cassette marker.

        Grow strains using standard yeast techniques. As an example, grow strains in 1–5 mL of YPGal in polypropylene round-bottom tubes at 30°C (the temperature can be changed if the strain or experimental design indicates otherwise). Place the tubes in a roller drum to incubate. Alternatively, grow cells in YPGal in a flask, using a shaker. The integrated MX cassette will be lost in 80%–90% of the cells (Güldener et al. 1996). Because of the high efficiency of cassette loss, all loxP flanked MX cassettes can be used. If using counterselectable MX cassettes, counterselection should not be necessary.

        Alternatively, grow yeast strains containing the integrated MX cassette and the Cre recombinase plasmid in YPD medium overnight without selection for the integrated MX cassette marker. Even in YPD, the integrated MX cassette will be lost in ∼5% of the cells (likely because the Cre promoter is leaky) (Gueldener et al. 2002). Because of the high efficiency of cassette loss, all loxP-flanked MX cassettes can be used. If using counterselectable MX cassettes, counterselection should not be necessary.

    iii. Plate $\geq 10^2$ yeast cells onto YPD plates and, after 2–3 d growth at 30°C, replica plate to medium that will detect the loss of the MX cassette.

        The selection medium used for replica plating here will be the same selection used to select for the presence of the MX cassette. See Protocol 1: Introducing MX Cassettes into Saccharomyces cerevisiae (McCusker 2015b).

        See Troubleshooting.

Confirmation Successful Loss of MX Cassettes

5. After performing the appropriate MX cassette pop-out method above, test the phenotype(s) of putative MX-free colonies. Use colony PCR (Amberg et al. 2006; Bergkessel and Guthrie 2013) or "smash and grab" (Hoffman 2001) to confirm successful loss of MX cassettes from the genome.

    See Troubleshooting.

TROUBLESHOOTING

Problem (Steps 1–4): No colonies were recovered in final plating step.
Solution: Consider the following.

- Check and remake the counterselection medium (Steps 2 and 3).
- Determine if the MX cassette is indeed flanked by direct repeats. If not, remake the mutation with an MX cassette that is flanked by direct repeats.
- For galactose-induced, Cre-mediated recombination between loxP sites, check and remake the YPGal induction medium. Determine if the strain is Gal$^+$.
- Plate more cells.

Problem (Steps 2 and 3): Too many colonies or confluent growth was observed on counterselection media.
Solution: Consider the following.

- Determine if the strain is contaminated.
- Check and remake the counterselection medium.

Problem (Steps 2 and 3): No colonies, too few colonies, or only petite colonies were recovered on 5-fluoroorotate- and/or 5-fluorocytosine-containing selection media.
Solution: High plating density of URA3MX and FCA1/FCY1 cassette-containing strains on 5-FOA- and 5-fluorocytosine-containing media, respectively, may result in reduced recovery of resistant colonies and/or a high incidence of resistant but ρ$^-$ (petite, respiration deficient) colonies due to fratricide. That is, the large number of URA3MX- and FCA1/FCY1-cassette-containing sensitive cells may produce sufficient 5-fluorouracil to affect neighboring cells. Plate fewer cells.

Problem (Step 5): The MX cassette is still present in colonies growing on the counterselection medium.
Solution: Consider the following.

- Cassette-borne forward mutations, such as *URA3* to *ura3*, will allow growth on the counterselection media. Specific chromosomal forward mutations will also allow growth on the counterselection media. For example, mutation of *LYS2* to *lys2* will allow growth on α-aminoadipate medium (Chattoo et al. 1979). In addition, FPS1 is required for uptake of acetamide (Shepherd and Piper 2010) and presumably fluoroacetamide.
- Test additional colonies selected for growth on counterselection media.
- Remake the strain and select for growth on counterselection media.

Chapter 8

RECIPES

SD + LYS + α-Aminoadipate Medium

1. Dissolve 1.7 g of yeast nitrogen base without ammonium sulfate and without amino acids and 20 g of dextrose in 100 mL of dH$_2$O. Add 2 mL of 1.5% L-lysine. Filter-sterilize.

2. Dissolve 2 g of DL-α-aminoadipate in 100 mL of dH$_2$O. Add pellets of KOH until all of the DL-α-aminoadipate goes into solution. Then adjust the pH to 5.5 with glacial acetic acid. Filter-sterilize.

3. Add 20 g of Bacto agar to 800 mL of dH$_2$O, and autoclave.

4. Add the filter-sterilized solutions to the autoclaved agar, mix thoroughly, and pour the medium into Petri dishes.

    Although it is possible to add some nutritional supplements to α-aminoadipate medium, the best results are obtained with either no supplements (using a strain containing only the required lys5 mutation) or supplements that either cannot be utilized as nitrogen sources (e.g., adenine, histidine, uracil) or are poorly utilized (e.g., tryptophan).

    This recipe was adapted from Chattoo et al. (1979) and Ito-Harashima and McCusker (2004).

SD + URA + 5-FOA Medium

1. Dissolve 6.7 g of yeast nitrogen base with ammonium sulfate and without amino acids, 1 g of 5-fluoroortic acid (5-FOA), 20 g of dextrose, and 50 mg of uracil in 500 mL of dH$_2$O. If necessary, add other nutritional supplements, such as drop-out mixtures. Heat to 65°C to dissolve all components; adjust to pH 3.5–4; and filter-sterilize.

2. Add 20 g of Bacto agar to 500 mL of dH$_2$O, and autoclave in a flask with a volume ≥2 L.

3. Add the filter-sterilized solution to autoclaved agar, mix thoroughly, and pour the medium into Petri dishes.

    This recipe was adapted from Boeke et al. (1987).

SDP + URA + 5-FOA Medium

1. Dissolve 1.7 g of yeast nitrogen base without ammonium sulfate and without amino acids, 1 g of L-proline, ≥25 mg of 5-fluoroortic acid (5-FOA), 10 mg of uracil, and 20 g of dextrose in 100 mL of dH$_2$O. Heat to 65°C to dissolve all components. Adjust the pH to 3.5–4. Filter-sterilize.

    The use of proline as the sole nitrogen source (in place of ammonium sulfate), together with lower amounts of uracil, allows much lower concentrations of 5-FOA to be used. However, not all genetic backgrounds are sensitive to SDP + URA + 5-FOA medium and the addition of some nutritional supplements abolishes sensitivity to 5-FOA (McCusker and Davis 1991).

2. Add 20 g of Bacto agar to 900 mL of dH$_2$O, and autoclave in a 2-L flask.

3. Add the filter-sterilized solution to the autoclaved agar, mix thoroughly, and pour the medium into Petri dishes.

    This recipe is adapted from McCusker and Davis (1991).

SD + 5-Fluorocytosine Medium

1. Dissolve 6.7 g of yeast nitrogen base with ammonium sulfate and without amino acids, 20 g of dextrose, and 5-fluorocytosine (5FC) in 100 mL of deionized water. Filter-sterilize.

   In the absence of adenine and uracil, the final concentration of 5FC should be 100 µM. In the presence of adenine and uracil (based on the genotype of the yeast strain being plated), the final concentration of 5FC should be 1 mM. 5FC solutions should be made up fresh to limit spontaneous deamination to 5FC, which reduces the effectiveness of the counter selection for resistance to 5FC.

2. Autoclave 20 g of Bacto agar in 900 mL of deionized water.

3. Add the filter-sterilized solution to the autoclaved agar; mix thoroughly, and pour the medium into Petri dishes.

   5FC-containing media are best used fresh but, if necessary, can be stored for several months at 4 °C in the dark.

   This recipe is adapted from Hartzog et al. (2005).

SD + Fluoroacetamide Medium

1. Dissolve 6.7 g of yeast nitrogen base with ammonium sulfate and without amino acids, 20 g of dextrose, and 2.3 g of fluoroacetamide in 100 mL of deionized water. Filter-sterilize.

2. Autoclave 20 g of Bacto agar in 900 mL of deionized water.

3. Add the filter-sterilized solution to autoclaved agar, mix thoroughly, and pour the medium into Petri dishes.

   This recipe is adapted from Solis-Escalante et al. (2013).

YPD + 5-Fluorocytosine Medium

1. Autoclave YPD (1% yeast extract, 2% Bacto peptone, 2% dextrose, and 2% Bacto agar).

2. After autoclaving, add filter-sterilized 5-fluorocytosine to a final concentration of 1 mM, mix thoroughly, and pour the medium into Petri dishes.

   5-Fluorocytosine solutions should be made up fresh to limit spontaneous deamination to 5-fluorouracil, which reduces the effectiveness of the counter selection for resistance to 5-fluorocytosine. 5-Fluorocytosine-containing media are best used fresh but, if necessary, can be stored for several months at 4°C in the dark.

   This recipe is adapted from Hartzog et al. (2005).

YPD Liquid Medium and Plates with Drugs

For YPD liquid medium, mix yeast extract (10 g/L), Bacto peptone (20 g/L), and dextrose (20 g/L) in 1 L of (deionized or distilled) H_2O. Adjust the pH to 5.5 with 1 M HCl, and autoclave.

For YPD plates with drugs, prepare liquid medium as above, but add Bacto Agar (20 g/L) before autoclaving. After autoclaving, add one of the following filter-sterilized drugs, as required, to the appropriate final concentration: clonNAT (100 µg/mL), G418 (200 µg/mL), hygromycin B (300 µg/mL), or phleomycin (7.5 µg/mL). After cooling to 55°C–60°C, pour liquid into Petri dishes.

YPGal Medium

Mix yeast extract (10 g/L) and Bacto peptone (20 g/L) in 900 mL of (deionized or distilled) H_2O, and autoclave. Add 100 mL of filter-sterilized 20% galactose solution to the autoclaved medium.

ACKNOWLEDGMENTS

J.H.M. is supported from a U.S. Public Health Service grant (GM098287).

REFERENCES

Amberg DC, Burke DJ, Strathern JN. 2006. Yeast colony PCR. *CSH Protoc* doi: 10.1101/pdb.prot4170.

Bergkessel M, Guthrie C. 2013. Colony PCR. *Methods Enzymol* **529**: 299–309.

Boeke JD, Trueheart J, Natsoulis G, Fink GR. 1987. 5-Fluoroorotic acid as a selective agent in yeast molecular genetics. *Methods Enzymol* **154**: 164–175.

Chattoo BB, Sherman F, Azubalis DA, Fjellstedt TA, Mehnert D, Ogur M. 1979. Selection of *lys2* mutants of the yeast *Saccharomyces cerevisiae* by the utilization of α-aminoadipate. *Genetics* **93**: 51–65.

Gietz RD, Schiestl RH. 2007a. Microtiter plate transformation using the LiAc/SS carrier DNA/PEG method. *Nat Protoc* **2**: 5–8.

Gietz RD, Schiestl RH. 2007b. High-efficiency yeast transformation using the LiAc/SS carrier DNA/PEG method. *Nat Protoc* **2**: 31–34.

Gietz RD, Schiestl RH. 2007c. Quick and easy yeast transformation using the LiAc/SS carrier DNA/PEG method. *Nat Protoc* **2**: 35–37.

Gietz RD, Woods RA. 2002. Transformation of yeast by lithium acetate/single-stranded carrier DNA/polyethylene glycol method. *Methods Enzymol* **350**: 87–96.

Gueldener U, Heinisch J, Koehler GJ, Voss D, Hegemann JH. 2002. A second set of loxP marker cassettes for Cre-mediated multiple gene knockouts in budding yeast. *Nucleic Acids Res* **30**: e23.

Güldener U, Heck S, Fielder T, Beinhauer J, Hegemann JH. 1996. A new efficient gene disruption cassette for repeated use in budding yeast. *Nucleic Acids Res* **24**: 2519–2524.

Hartzog PE, Nicholson BP, McCusker JH. 2005. Cytosine deaminase MX cassettes as positive/negative selectable markers in *Saccharomyces cerevisiae*. *Yeast* **22**: 789–798.

Hoffman CS. 2001. Preparation of yeast DNA. *Curr Protoc Mol Biol* 2001 May; Chapter 13:Unit13.11. doi: 10.1002/0471142727.mb1311s39.

Horecka J, Davis RW. 2014. The 50:50 method for PCR-based seamless genome editing in yeast. *Yeast* **31**: 103–112.

Ito-Harashima S, McCusker JH. 2004. Positive and negative selection LYS5MX gene replacement cassettes for use in *Saccharomyces cerevisiae*. *Yeast* **21**: 53–61.

McCusker JH. 2015a. MX cassettes for knocking out genes in yeast. *Cold Spring Harb Protoc* doi: 10.1101/pdb.top080689.

McCusker JH. 2015b. Introducing MX cassettes into *Saccharomyces cerevisiae*. *Cold Spring Harb Protoc* doi: 10.1101/pdb.prot088104.

McCusker JH, Davis RW. 1991. The use of proline as a nitrogen source causes hypersensitivity to, and allows more economical use of 5FOA in *Saccharomyces cerevisiae*. *Yeast* **7**: 607–608.

Shepherd A, Piper PW. 2010. The Fps1p aquaglyceroporin facilitates the use of small aliphatic amides as a nitrogen source by amidase-expressing yeasts. *FEMS Yeast Res* **10**: 527–534.

Solis-Escalante D, Kuijpers NG, Bongaerts N, Bolat I, Bosman L, Pronk JT, Daran JM, Daran-Lapujade P. 2013. amdSYM, a new dominant recyclable marker cassette for *Saccharomyces cerevisiae*. *FEMS Yeast Res* **13**: 126–139.

Wach A, Brachat A, Pohlmann R, Philippsen P. 1994. New heterologous modules for classical or PCR-based gene disruptions in *Saccharomyces cerevisiae*. *Yeast* **10**: 1793–1808.

CHAPTER 9

Multipurpose Transposon-Insertion Libraries in Yeast

Anuj Kumar[1]

Department of Molecular, Cellular, and Developmental Biology, University of Michigan, Ann Arbor, Michigan 48109-1048

Libraries of transposon-insertion alleles constitute powerful and versatile tools for large-scale analysis of yeast gene function. Transposon-insertion libraries are constructed most simply through mutagenesis of a plasmid-based genomic DNA library; modification of the mutagenizing transposon by incorporation of yeast selectable markers, recombination sites, and an epitope tag enables the application of insertion alleles for phenotypic screening and protein localization. In particular, yeast genomic DNA libraries have been mutagenized with modified bacterial transposons carrying the *URA3* marker, *lox* recombination sites, and sequence encoding multiple copies of the hemagglutinin (HA) epitope. Mutagenesis with these transposons has yielded a large resource of insertion alleles affecting nearly 4000 yeast genes in total. Through well-established protocols, these insertion libraries can be introduced into the desired strain backgrounds and the resulting insertional mutants can be screened or systematically analyzed. Relative to alternative methods of UV irradiation or chemical mutagenesis, transposon-insertion alleles can be easily identified by PCR-based approaches or high-throughput sequencing. Transposon-insertion libraries also provide a cost-effective alternative to targeted deletion approaches, although, in contrast to start-codon to stop-codon deletions, insertion alleles might not represent true null-mutants. For protein-localization studies, transposon-insertion alleles can provide encoded epitope tags in-frame with internal codons; in many cases, these transposon-encoded epitope tags can provide a more accurate localization for proteins in which terminal sequences are crucial for intracellular targeting. Thus, overall, transposon-insertion libraries can be used quickly and economically and have a particular utility in screening for desired phenotypes and localization patterns in nonstandard genetic backgrounds.

INTRODUCTION

The budding yeast *Saccharomyces cerevisiae* has long served as a workhorse for genetics, and numerous approaches have been used to successfully mutagenize yeast DNA. Few of these approaches, however, have proven to be more successful for functional genomics than the methods of transposon mutagenesis. The availability of transposon-insertion libraries consisting of thousands of plasmid-based transposon-insertion alleles has greatly facilitated the systematic and large-scale analysis of gene function, particularly in nonstandard genetic backgrounds (e.g., strains other than the wild-type S288c or backgrounds already carrying a mutation of interest). This introduction reviews the available resources for the application of transposon-insertion alleles toward phenotypic screening and protein localization. It highlights the advantages and limitations associated with the use of transposon-insertion libraries relative to other approaches for the generation of mutations and epitope/fluorescent-protein fusion-tagged alleles. Finally, it also summarizes several applications of transposon-insertion libraries for phenotypic screening and protein localization.

[1]Correspondence: anujk@umich.edu

Copyright © Cold Spring Harbor Laboratory Press; all rights reserved
Cite this introduction as *Cold Spring Harb Protoc*; doi:10.1101/pdb.top080259

Chapter 9

RESOURCES AND FEATURES

Michael Snyder's laboratory pioneered the large-scale application of transposon mutagenesis in yeast for the construction of transposon-insertion libraries (Burns et al. 1994; Ross-Macdonald et al. 1997). These insertion libraries were generated by mutagenesis of a plasmid-based yeast genomic DNA library by using modified prokaryotic transposons (Ross-Macdonald et al. 1999; Kumar et al. 2004). The transposons were derived from either Tn3 or Tn7 and have been used for in vivo and in vitro mutagenesis, respectively; the Tn7 transposon was developed as an in vitro mutagenesis system in Nancy Craig's laboratory (Stellwagen and Craig 1997; Biery et al. 2000). For mutagenesis of yeast genomic DNA, the yeast selectable marker URA3 was incorporated in the transposon. Both the Tn3- and Tn7-derived transposons were designed to facilitate multiple functional analyses, as illustrated in Figure 1A. Insertion of the full-length transposon within a promoter sequence or gene coding sequence is likely to generate a loss-of-function mutation. The transposons carry a promoterless and 5′-truncated *lacZ* reporter, with production of β-galactosidase being indicative of in-frame insertion within the coding sequence (Fig. 1B). Consequently, the transposon can be used as a reporter (Kumar et al. 2002b). In addition, the transposons carry sequence encoding three copies of the HA epitope and *lox* sites for Cre-mediated recombination. The *lox* sites are positioned such that recombination will excise most of the transposon sequence, leaving behind a residual insertion element encoding 99 amino acids; this recombined insertion element consists of the transposon terminal sequence, a single *lox* site and HA epitope sequences and can be used to generate an epitope-tagged allele of the host gene for subsequent studies of protein localization (Kumar et al. 2002a). In some instances, the truncated insertion results in a hypomorphic allele, as it might not completely disrupt gene function. We estimate that 20% of truncated insertions yield hypomorphic alleles. This type of mutation is particularly useful in the analysis of essential genes. Thus, the engineered transposons can facilitate a variety of studies.

The Tn3- and Tn7-derived transposon-insertion libraries each encompass in excess of 300,000 insertions. The insertion alleles can be introduced into a desired yeast strain through straightforward methods of DNA transformation. Each plasmid-born fragment of genomic DNA carrying a single transposon insertion will integrate at its corresponding genomic locus by homologous recombination. The collection of resulting yeast transformants can be screened for the desired phenotypes. Subsequently, the site of transposon insertion can be identified in the strains of interest by inverse or vectorette PCR-based approaches (Kumar et al. 2002c; Xu et al. 2011). Direct Sanger sequencing with yeast genomic DNA as the template has also been used (Horecka and Jigami 2000). Alternatively,

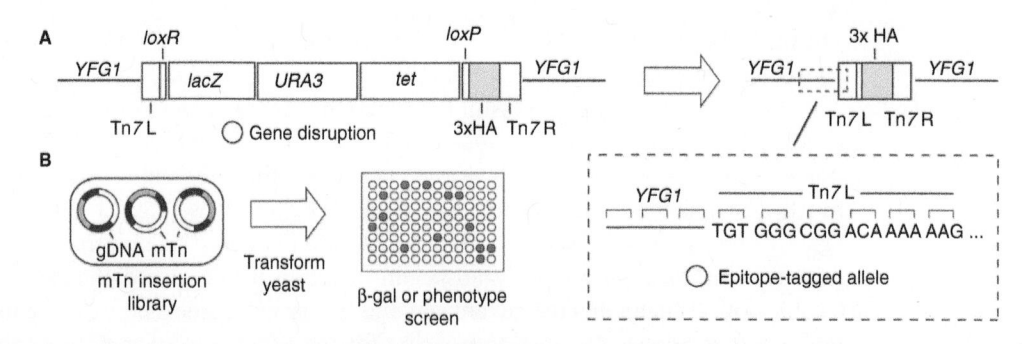

FIGURE 1. Multipurpose transposon-insertion libraries. (*A*) The Tn7-derived transposon illustrated here can be used to disrupt gene function, and, provided the insertion is in-frame with the host coding sequence, the insertion can be truncated by Cre-*lox* recombination to yield an epitope-tagged allele for the analysis of protein localization. (*B*) The transposon-insertion library can be introduced into a desired strain of yeast to screen for the phenotypes of interest; the insertion mutants can also be screened for β-galactosidase activity to identify in-frame insertions within coding sequence. β-gal, β-galactosidase; gDNA, genomic DNA; 3xHA, three copies of HA tag; *lacZ*, β-galactosidase gene; *loxP*, DNA sequence recognized by cre enzyme; *loxR*, second DNA sequence recognized by cre enzyme; mTn, mini-transposon; *tet*, tetracycline-resistance gene; tn7, transposon Tn7; URA3, gene encoding orotidine 5′-phosphate decarboxylase; YFG1, "your favorite gene 1."

next-generation sequencing is a viable means to identify an insertion site. Insertion mutants can be analyzed by using an enrichment strategy to capture DNA fragments containing the terminal transposon sequence; the captured DNA fragments can be sequenced to identify the junction between the transposon sequence and the native yeast genomic DNA.

ADVANTAGES AND LIMITATIONS IN USING TRANSPOSON-INSERTION LIBRARIES

Relative to targeted approaches for loss-of-function phenotypic screening and/or protein-localization studies, transposon mutagenesis offers two advantages. First, the application of transposon-insertion libraries is economical. The transformation methods and media for screening integrated transposon-insertion libraries typically require inexpensive reagents, and procedures for the identification of transposon insertions are straightforward (Coelho et al. 2000). Second, transposon-insertion alleles can be highly informative; by using insertional libraries with the transposons described here, a single insertion allele can facilitate more studies than can a targeted full-length deletion allele. A given insertion allele can enable analyses of loss-of-function phenotypes, and, by Cre-mediated recombination within the transposon sequence, the resulting transposon-encoded epitope-tagged allele can also be used to facilitate protein-localization studies, provided that the insertion is in-frame with the host-gene coding sequence. If the truncated epitope-tagging insertion is too large to accommodate proper folding of the host protein, it is possible that the allele might generate a hypomorphic or partial loss-of-function mutation that can be very valuable in the analysis of essential gene function. Furthermore, transposon insertions can cluster to varying degrees within certain sequences, and the availability in the insertional library of full-length or shortened transposon insertions at multiple sites within a single gene can often be advantageous in defining domains and other functional elements within the host coding sequence.

Two principal limitations exist in applying transposon-insertion libraries for large-scale studies. First, insertion alleles can be less effective for their respective purposes than start-codon to stop-codon gene-deletion alleles or fluorescent-protein fusions targeted to 5′- or 3′-gene termini. For loss-of-function phenotypic screens, a given transposon insertion can occur within a region of gene coding sequence where gene function is not compromised by the interrupting transposon sequence. For example, an insertion toward the 5′-end of a given gene might not yield a true null allele. Similarly, a truncated transposon insertion for epitope-tagging might not result in a properly folded and localized protein, depending upon the specific location of the insertion relative to the host protein domains. Fluorescent protein fusions to either 5′- or 3′-gene coding sequence might be more likely to yield accurate localization data, at least in some cases. Second, transposons are subject to some bias in insertion-site preference, decreasing the effective coverage of genes over the genome as a whole. From our analysis, Tn7-derived transposons show less insertional bias than do other bacterial transposons, and, in general, bacterial transposons are much more nonspecific than Ty1 transposable elements or those from higher eukaryotes. Thus, utilization of the Tn7 insertional library does mitigate this concern somewhat as insertions affecting 2613 genes have been identified from the analysis of Tn7-mutagenized yeast genomic DNA.

Thus, in summary, transposon-insertion libraries are optimal tools for mutagenesis if, first, a large number of mutant alleles is desired, and, second, a genetic background is needed for which a large collection of targeted alleles is unavailable.

APPLICATIONS OF TRANSPOSON-INSERTION LIBRARIES

The transposon-insertion libraries described here have been used for several large-scale studies. In a study by Ross-Macdonald and colleagues, the authors analyzed the Tn3 insertional library extensively for 20 phenotypes associated with cell growth under various stress conditions and drug treatments, identifying insertion alleles in 407 genes yielding phenotypes distinct from that of the wild type (Ross-Macdonald et al. 1999). The Tn3 library has also been used for a large-scale analysis of protein localization, enabling the determination of subcellular localization patterns for 2744 proteins and localization to nine subcellular structures or compartments (Kumar et al. 2002a).

Mosch and Fink applied the Tn3-derived insertional library to screen for mutants defective in pseudohyphal growth and identified insertion alleles in 16 genes that blocked filamentation (Mosch and Fink 1997). More recently, a collection of 3627 insertion alleles from the Tn3- and Tn7-based insertional libraries has been introduced into the pseudohyphal strain background Σ1278b for analysis of filamentation defects (Jin et al. 2008). This work identified 309 genes yielding defects in pseudohyphal growth upon integration of the transposon-insertion alleles—the collection of defined and sequenced insertion alleles used in that study is available from Open Biosystems/Thermo Scientific (www.thermoscientificbio.com).

Collectively, these studies provide a brief overview of experimental strategies using transposon-insertion libraries for phenotypic screening and protein localization, particularly for the analysis of gene function in nonstandard genetic backgrounds that complement applications of the targeted gene deletion and fluorescent protein-fusion resources in S288c.

In an accompanying protocol, I present a methodology for the use of a transposon-insertion library for phenotypic screening and for large-scale protein localization, providing a relatively straightforward alternative to time-consuming targeted approaches for generating deletion alleles and fluorescent protein fusions (see Protocol 1: Using Yeast Transposon-Insertion Libraries for Phenotypic Screening and Protein Localization [Kumar 2015]).

ACKNOWLEDGMENTS

The transposon-insertion libraries described here were constructed in Michael Snyder's laboratory, and Tn7 mutagenesis was accomplished with generous assistance from Nancy Craig's laboratory. Research in the Kumar laboratory is supported by grants 1R01-A1098450-01A1 from the National Institutes of Health and 1-FY11-403 from the March of Dimes.

REFERENCES

Biery M, Stewart F, Stellwagen A, Raleigh E, Craig N. 2000. A simple *in vitro* Tn7-based transposition system with low target site selectivity for genome and gene analysis. *Nucleic Acids Res* 28: 1067–1077.

Burns N, Grimwade B, Ross-Macdonald PB, Choi E-Y, Finberg K, Roeder GS, Snyder M. 1994. Large-scale characterization of gene expression, protein localization and gene disruption in *Saccharomyces cerevisiae*. *Genes Dev* 8: 1087–1105.

Coelho PS, Kumar A, Snyder M. 2000. Genome-wide mutant collections: Toolboxes for functional genomics. *Curr Opin Microbiol* 3: 309–315.

Horecka J, Jigami Y. 2000. Identifying tagged transposon insertion sites in yeast by direct genomic sequencing. *Yeast* 16: 967–970.

Jin R, Dobry CJ, McCown PJ, Kumar A. 2008. Large-scale analysis of yeast filamentous growth by systematic gene disruption and overexpression. *Mol Biol Cell* 19: 284–296.

Kumar A. 2015. Using yeast transposon-insertion libraries for phenotypic screening and protein localization. *Cold Spring Harb Protoc* doi: 10.1101/pdb.prot085217.

Kumar A, Agarwal S, Heyman JA, Matson S, Heidtman M, Piccirillo S, Umansky L, Drawid A, Jansen R, Liu Y, et al. 2002a. Subcellular localization of the yeast proteome. *Genes Dev* 16: 707–719.

Kumar A, Harrison PM, Cheung KH, Lan N, Echols N, Bertone P, Miller P, Gerstein MB, Snyder M. 2002b. An integrated approach for finding overlooked genes in yeast. *Nat Biotechnol* 20: 58–63.

Kumar A, Vidan S, Snyder M. 2002c. Insertional mutagenesis: Transposon-insertion libraries as mutagens in yeast. *Methods Enzymol* 350: 219–229.

Kumar A, Seringhaus M, Biery M, Sarnovsky RJ, Umansky L, Piccirillo S, Heidtman M, Cheung K-H, Dobry CJ, Gerstein M, et al. 2004. Large-scale mutagenesis of the yeast genome using a Tn7-derived multipurpose transposon. *Genome Res* 14: 1975–1986.

Mosch HU, Fink GR. 1997. Dissection of filamentous growth by transposon mutagenesis in *Saccharomyces cerevisiae*. *Genetics* 145: 671–684.

Ross-Macdonald P, Coelho PS, Roemer T, Agarwal S, Kumar A, Jansen R, Cheung KH, Sheehan A, Symoniatis D, Umansky L, et al. 1999. Large-scale analysis of the yeast genome by transposon tagging and gene disruption. *Nature* 402: 413–418.

Ross-Macdonald P, Sheehan A, Roeder GS, Snyder M. 1997. A multipurpose transposon system for analyzing protein production, localization, and function in *Saccharomyces cerevisiae*. *Proc Natl Acad Sci* 94: 190–195.

Stellwagen A, Craig N. 1997. Gain-of-function mutations in TnsC, an ATP-dependent transposition protein that activates the bacterial transposon Tn7. *Genetics* 145: 573–585.

Xu T, Bharucha N, Kumar A. 2011. Genome-wide transposon mutagenesis in *Saccharomyces cerevisiae* and *Candida albicans*. *Methods Mol Biol* 765: 207–224.

Protocol 1

Using Yeast Transposon-Insertion Libraries for Phenotypic Screening and Protein Localization

Anuj Kumar[1]

Department of Molecular, Cellular, and Developmental Biology, University of Michigan, Ann Arbor, Michigan 48109-1048

This protocol details how to use a transposon-insertion library for phenotypic screening and protein localization. The insertion library was generated by mutagenesis of a plasmid-based yeast genomic DNA library by using a multipurpose transposon; the transposon produces gene disruptions, and, by Cre-mediated recombination at *lox* sites incorporated within the transposon, alleles with an in-frame insertion can be truncated to a residual transposon encoding multiple copies of the hemagglutinin epitope. Insertions are generated in yeast by shuttle mutagenesis. Yeast genomic DNA containing a transposon insertion is released from the library, and the mutagenized DNA sequences are introduced into a desired strain of yeast, where the insertion alleles replace native loci by homologous recombination. The insertion mutants can be screened for phenotypes, and the site of transposon insertion can subsequently be identified in selected mutants by inverse polymerase chain reaction (PCR). In-frame insertions within genes of interest can be truncated to an epitope-tagged allele by Cre-*lox* recombination, and the subcellular localization of the encoded protein product can be identified by standard methods of indirect immunofluorescence. In summary, the transposon-insertion libraries represent an informative resource for large-scale mutagenesis, presenting a straightforward alternative to labor-intensive targeted approaches for the construction of deletion alleles and fluorescent protein fusions.

MATERIALS

It is essential that you consult the appropriate Material Safety Data Sheets and your institution's Environmental Health and Safety Office for proper handling of equipment and hazardous material used in this protocol.

RECIPES: Please see the end of this protocol for recipes indicated by <R>. Additional recipes can be found online at http://cshprotocols.cshlp.org/site/recipes.

Reagents

5-Fluoroorotic acid (5-FOA) plates <R>
AluI or DraI restriction endonuclease
ATP (5 mM)
Chloroform
Complete minimal (CM) or synthetic complete (SC) and drop-out media (e.g., SC–Ura) <R>
  For plates, add agar to 2%.
dNTP mix (2.5 mM of each dNTP)

---

[1]Correspondence: anujk@umich.edu

Copyright © Cold Spring Harbor Laboratory Press; all rights reserved
Cite this protocol as *Cold Spring Harb Protoc*; doi:10.1101/pdb.prot085217

Glycerol (sterile)
MgCl$_2$ (1 M)
NotI restriction endonuclease
One-step buffer <R>
pGAL-cre plasmid

The CEN-containing pGAL-cre plasmid is ampicillin resistant and carries the LEU2 marker.

Primers for inverse PCR

ABP1 primer	5′-GAAGGAGAGGACGCTGTCTGTCGAAGGTAAGG AACGGACGA-GAGAAGGGAGAG-3′
ABP2 primer	5′-GACTCTCCCTTCTCGAATCGTAACCGTTCGT ACGAGAATCGCTGT-CCTCTCCTTC-3′
UV primer	5′-CGAATCGTAACCGTTCGTACGAGAATCGCT-3′
mTn primer	5′-CGCCAGGGTTTTCCCAGTCACGAC-3′

Primers ABP1 and ABP2 are used to form an anchor bubble; the third primer required is a universal vectorette (UV) primer, and the fourth is a primer that is complementary to sequence in the transposon used to generate the insertion library (the mini-transposon [mTn] primer).

Raffinose
Salmon sperm DNA, denatured
Taq DNA polymerase and PCR buffer (New England BioLabs)
TE buffer <R>
T4 DNA ligase and buffer
Transposon-mutagenized library DNA

Tn7 insertion library DNA (available upon request from Anuj Kumar, University of Michigan)
Tn3 insertion library DNA (available upon request from Michael Snyder, Stanford University)

To assist the research community, several micrograms of plasmid DNA from individual pools of the transposon-mutagenized library will be distributed to users upon request. This DNA can be amplified in Escherichia coli or is suitable for direct introduction into yeast cells by the transformation methods described in this protocol.

Water, sterile
X-gal plates <R>
YPD medium <R>

Equipment

Agarose gel electrophoresis apparatus
Bench-top microcentrifuge
Clinical tabletop centrifuge
Filter paper
Heat block
Petri dishes, glass (9- and 15-cm)
Thermal cycler
Water bath, heated

METHOD

Transformation of Yeast with Transposon-Insertion Library DNA

1. Digest ~1–2 µg of insertion library DNA with the restriction endonuclease NotI, and then store the reaction mixture at 4°C for use in Step 4.

2. Grow a 10-mL culture of a Ura$^-$ yeast strain to mid-log phase (density of 10^7 cells/mL or OD$_{600}$ of ~1) with appropriate selection.

3. Centrifuge cells in a clinical tabletop centrifuge at room temperature for 5 min at 1100g, then wash the resulting pellet once with five volumes of one-step buffer.

4. Suspend the washed cells in 1 mL of one-step buffer supplemented with 1 mg of denatured salmon sperm DNA; add 100-µL aliquots from this suspension to 0.1–1 µg of NotI-digested plasmid DNA from Step 1, and mix contents thoroughly by vortexing. Incubate the transformation mixture for 30 min at 45°C.

5. Pellet the cells by centrifugation for 5 sec at maximum speed at room temperature in a benchtop microcentrifuge. Suspend the resulting pellet in 400 µL of SC–Ura dropout medium, spread aliquots from this mixture onto SC–Ura plates, and then incubate for 3–4 d at 30°C.

    Up to 1000 transformants can be recovered per microgram of library DNA.

Screening Transformants for β-Galactosidase Activity

6. Transfer transformant colonies onto YPD plates with a sterile toothpick, forming small patches or spotted cultures.

7. Place a sterile disc of filter paper onto a SC–Ura plate and replicate transformant cells onto filter-covered plates.

    By this approach, corresponding colonies can be easily identified on the YPD plate.

8. Incubate overnight at 30°C.

9. Lift filters from the plates and place in the lid of a 9-cm glass Petri dish.

10. Place the lid inside a closed 15-cm glass Petri dish containing chloroform and incubate for ~10–30 min at room temperature.

11. Place filters with the colony side facing up onto X-gal plates, and incubate inverted for up to 2 d at 30°C.

12. Select transformants from the regrown YPD plates corresponding to those exhibiting blue staining on the X-gal plates.

    According to needs, strains of interest can be stored long-term in 15% (v/v) glycerol at –70°C.

Identification of Insertion Sites by Inverse PCR

13. Prepare four primers for inverse PCR (e.g., ABP1, ABP2, UV, and mTn; see Reagents).

    i. Prepare a solution containing 2–4 mM each of primers ABP1 and ABP2.

    ii. Prepare separate 20-µM solutions of UV primer and of mTn primer.

14. Denature the ABP1 and ABP2 primers by incubating for 5 min at 95°C in a heat block.

15. Add 1 M $MgCl_2$ to the ABP1 and ABP2 primer mix to a final concentration of 2 mM.

16. Anneal the ABP1 and ABP2 primers by removing the sample from the heating block and placing it onto a bench-top until it cools to room temperature.

17. Prepare genomic DNA from the transposon-insertion mutants of interest (Guthrie and Fink 1991), and then digest 1–3 µg of genomic DNA with 10 units of either AluI or DraI in a total reaction volume of 20 µL. Incubate overnight at 37°C.

18. On the next day, inactivate the restriction enzyme by incubating for 20 min at 65°C.

19. Add the following to the reaction mixture from Step 18:

10× T4 DNA ligase buffer	5 µL
Sterile water	22.5 µL
Annealed anchor bubble (generated in Steps 14–16)	1 µL
5 mM ATP	0.5 µL
T4 DNA ligase	1 µL (400 U)

20. Incubate the ligation reaction mixture for 9–24 h at 16°C.
21. Prepare a 100-µL PCR mix by withdrawing 5 µL from the ligation reaction and adding this 5-µL aliquot to the following:

10× *Taq* PCR buffer	10 µL
Sterile water	71 µL
2.5 mM dNTP mix	8 µL
20 µM mTn primer	2.5 µL
20 µM UV primer	2.5 µL
Taq DNA polymerase	1 µL (5U)

22. Transfer to a thermal cycler and incubate for 2 min at 92°C.
23. Perform PCR as follows:

    i. Perform 35 cycles of 20 sec at 92°C (denaturation), 30 sec at 67°C (annealing), and 45 sec at 72°C (extension).

    ii. Perform 1 cycle of 90 sec at 72°C (extension).

24. Analyze 80 µL of the PCR reaction mixture by agarose gel electrophoresis.
    A single band containing ~200–400 ng of DNA should be visible.
25. Excise the band and recover the DNA in 12 µL of TE buffer.
26. Analyze 4–6 µL of the recovered product by DNA sequencing to identify the precise site of transposon insertion.

Cre-*lox* Recombination to Generate Epitope-Tagged Alleles

27. Introduce pGAL-cre into the insertion mutants of interest by standard methods of yeast transformation (Gietz and Schiestl 2007); note that, if transposon mutagenesis is to be used for large-scale localization studies, introduce the pGAL-cre plasmid into the background yeast strain before transformation with the insertion library.
28. Inoculate transformants into 2 mL SC–Ura–Leu media with 2% raffinose.
29. Incubate at 30°C with shaking at 120 rpm until the culture has grown to saturation.
30. Dilute cultures 100-fold in 2 mL of fresh SC–Leu media containing 2% galactose; also dilute an aliquot of this same culture 100-fold into 2 mL of fresh SC–Leu media containing 2% glucose for use as a control.
31. Incubate cultures for 2 d at 30°C with shaking at 120 rpm.
32. Proceed as follows:

    i. If visible growth is evident, dilute cultures 100-fold in sterile water and withdraw a 10-µL aliquot. If no growth is evident, withdraw a 10-µL aliquot from the undiluted culture.

    ii. Spot and streak aliquots from both cultures onto 5-FOA plates.

    iii. Incubate 5-FOA plates at 30°C until growth is visible on those plates inoculated with strains grown in galactose (alternatively, cultures can be plated onto SC medium and replicated onto SC–Ura medium).

    iv. Incubate for 2 d at 30°C.
    Cultures grown in galactose should yield ~100-fold more cells on SC–Ura medium than identical cultures grown in glucose.

33. Single colonies from strains that have lost *URA3* can be stored indefinitely in 15% (v/v) glycerol at –70°C.

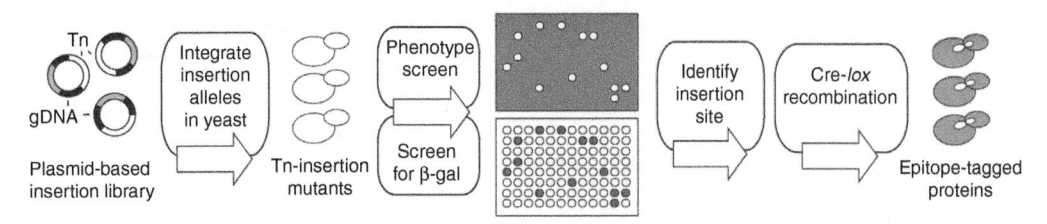

FIGURE 1. Flowchart of major steps in the application of transposon-insertion libraries for phenotypic screening and protein localization. Methodologies described in this protocol are indicated with arrows. β-gal, β-galactosidase; gDNA, genomic DNA; tn, transposon.

DISCUSSION

It is well established that transposon-insertion libraries can be used for phenotypic screening and protein localization in yeast (Ross-Macdonald et al. 1999; Kumar et al. 2004; Kumar 2008). The protocol given here first describes the transformation of yeast cells with DNA from a transposon-insertion library, then goes on to detail how to screen the transformants for β-galactosidase activity before summarizing the procedure for identifying the sites of insertion by inverse PCR. It concludes by summarizing how to use Cre-*lox* recombination to generate epitope-tagged alleles. The major steps in the procedures are diagrammed in Figure 1.

When generating epitope-tagged alleles, the protocol takes advantage of the fact that the Cre recombinase can be conditionally expressed by using the pGAL-cre vector in which a galactose-inducible promoter drives *cre* transcription (Ross-Macdonald et al. 1997; Ross-Macdonald et al. 1999). In the presented procedure, transposon insertions that have undergone loss of the transposon-encoded *URA3* marker by Cre-*lox* recombination are selected on medium containing 5-FOA, a fluorinated derivative of the pyrimidine precursor orotic acid. Typically, *URA3* is lost by this strategy in over 90% of colonies (Kumar et al. 2002; Kumar et al. 2004; Xu et al. 2011). In summary, the gene disruption alleles and epitope-tagged alleles are particularly useful for the analysis of gene function in nonstandard genetic backgrounds, and as such, complement existing reagent collections constructed in the S288C strain.

RELATED INFORMATION

Accompanying this protocol is an introduction to the use of transposon-insertion libraries, summarizing their advantages and limitations, compared with other established approaches, and showing how they can be used efficiently and cheaply, particularly for screening for desired phenotypes and for elucidating protein-localization patterns in nonstandard genetic backgrounds (see Introduction: Multipurpose Transposon-Insertion Libraries in Yeast [Kumar 2015]).

RECIPES

Complete Minimal (CM) or Synthetic Complete (SC) and Drop-Out Media

Bact-yeast nitrogen base without amino acids[a], 6.7 g
Glucose, 20 g
Bacto Agar, 20 g
Drop-out mix, 2 g <R>
H$_2$O, to 1000 mL

To test the growth requirements of strains, it is useful to have media in which each of the commonly encountered auxotrophies is supplemented except the one of interest (dropout media). Dry growth supplements are stored premixed. CM (or SC) is a medium in which the drop-out mix contains all possible supplements (i.e., nothing is "dropped out").

[a]Yeast nitrogen base without amino acids (YNB) is sold either with or without ammonium sulfate. This recipe is for YNB with ammonium sulfate. If the bottle of YNB is lacking ammonium sulfate, add 5 g of ammonium sulfate and only 1.7 g of YNB.

Dropout Mix

Reagent	Amount to add (g)
Adenine	0.5
Alanine	2.0
Arginine	2.0
Asparagine	2.0
Aspartic acid	2.0
Cysteine	2.0
Glutamine	2.0
Glutamic acid	2.0
Glycine	2.0
Histidine	2.0
Inositol	2.0
Isoleucine	2.0
Leucine	10.0
Lysine	2.0
Methionine	2.0
para-Aminobenzoic acid	0.2
Phenylalanine	2.0
Proline	2.0
Serine	2.0
Threonine	2.0
Tryptophan	2.0
Tyrosine	2.0
Uracil	2.0
Valine	2.0

Combine the appropriate ingredients, minus the relevant supplements, and mix in a sealed container. Turn the container end-over-end for at least 15 min; add several clean marbles to help mix the solids.

5-Fluoroorotic Acid (5-FOA) Plates

0.67% yeast nitrogen base
0.2% SC–Ura drop-out mix
2% glucose
50 µg/mL uracil
0.1% 5-FOA

Combine the ingredients above, and add distilled water to give a total volume of 500 mL. Filter-sterilize using a 0.2-µm filter. Dissolve 10 g of bacto-agar in 500 mL of distilled water, autoclave, and cool to ~80°C. Mix the two solutions, and pour the plates while the solution is still warm. Store the plates for up to 3 mo at 4°C.

One-Step Buffer

0.2 M lithium acetate
40% (w/v) PEG 4000
100 mM 2-mercaptoethanol

Store for up to 1 wk at 4°C in a light-protected bottle.

TE Buffer

Reagent	Quantity (for 100 mL)	Final concentration
EDTA (0.5 M, pH 8.0)	0.2 mL	1 mM
Tris-Cl (1 M, pH 8.0)	1 mL	10 mM
H$_2$O	to 100 mL	

X-Gal Plates

Reagent	Quantity
Yeast nitrogen base without amino acids/ammonium sulfate	1.7 g
Ammonium sulfate	5.0 g
Dextrose	20.0 g
Agar	20.0 g
Dropout powder	0.8 g
NaOH	one pellet

Combine ingredients, add water to 898 mL, and autoclave. Cool to 45°C–50°C. Add 100 mL of 0.7 M potassium phosphate (pH 7.0). Add 2 mL of X-gal solution (20 mg/mL in 100% *N,N*-dimethylformamide). Pour the plates while the solution is still warm.

YPD

Peptone, 20 g
Glucose, 20 g
Yeast extract, 10 g
H$_2$O to 1000 mL

YPD (YEPD medium) is a complex medium for routine growth of yeast.
To prepare plates, add 20 g of Bacto Agar (2%) before autoclaving.

ACKNOWLEDGMENTS

Research in the Kumar laboratory is supported by grants 1R01-A1098450-01A1 from the National Institutes of Health and 1-FY11-403 from the March of Dimes.

REFERENCES

Gietz RD, Schiestl RH. 2007. High-efficiency yeast transformation using the LiAc/SS carrier DNA/PEG method. *Nat Protoc* **2:** 31–34.

Guthrie C, Fink G. 1991. *Guide to yeast genetics and molecular biology.* Academic, San Diego.

Kumar A. 2008. Multipurpose transposon insertion libraries for large-scale analysis of gene function in yeast. *Methods Mol Biol* **416:** 117–129.

Kumar A. 2015. Multi-purpose transposon-insertion libraries in yeast. *Cold Spring Harb Protoc* doi: 10.1101/pdb.top080259.

Kumar A, Seringhaus M, Biery M, Sarnovsky RJ, Umansky L, Piccirillo S, Heidtman M, Cheung K-H, Dobry CJ, Gerstein M, et al. 2004. Large-scale mutagenesis of the yeast genome using a Tn7-derived multipurpose transposon. *Genome Res* **14:** 1975–1986.

Kumar A, Vidan S, Snyder M. 2002. Insertional mutagenesis: Transposon-insertion libraries as mutagens in yeast. *Methods Enzymol* **350:** 219–229.

Ross-Macdonald P, Coelho PS, Roemer T, Agarwal S, Kumar A, Jansen R, Cheung KH, Sheehan A, Symoniatis D, Umansky L, et al. 1999. Large-scale analysis of the yeast genome by transposon tagging and gene disruption. *Nature* **402:** 413–418.

Ross-Macdonald P, Sheehan A, Roeder GS, Snyder M. 1997. A multipurpose transposon system for analyzing protein production, localization, and function in *Saccharomyces cerevisiae*. *Proc Natl Acad Sci* **94:** 190–195.

Xu T, Bharucha N, Kumar A. 2011. Genome-wide transposon mutagenesis in *Saccharomyces cerevisiae* and *Candida albicans*. *Methods Mol Biol* **765:** 207–224.

CHAPTER 10

Functional Genomics Using the *Saccharomyces cerevisiae* Yeast Deletion Collections

Corey Nislow, Lai Hong Wong, Amy Huei-Yi Lee, and Guri Giaever[1]

Department of Pharmaceutical Sciences, University of British Columbia, Vancouver, British Columbia V6T 1Z3, Canada

Constructed by a consortium of 16 laboratories, the *Saccharomyces* genome-wide deletion collections have, for the past decade, provided a powerful, rapid, and inexpensive approach for functional profiling of the yeast genome. Loss-of-function deletion mutants were systematically created using a polymerase chain reaction (PCR)-based gene deletion strategy to generate a start-to-stop codon replacement of each open reading frame by homologous recombination. Each strain carries two molecular barcodes that serve as unique strain identifiers, enabling their growth to be analyzed in parallel and the fitness contribution of each gene to be quantitatively assessed by hybridization to high-density oligonucleotide arrays or through the use of next-generation sequencing technologies. Functional profiling of the deletion collections, using either strain-by-strain or parallel assays, provides an unbiased approach to systematically survey the yeast genome. The *Saccharomyces* yeast deletion collections have proved immensely powerful in contributing to the understanding of gene function, including functional relationships between genes and genetic pathways in response to diverse genetic and environmental perturbations.

THE *Saccharomyces* GENOME DELETION PROJECT

Completion of the *Saccharomyces cerevisiae* genome sequencing project in 1996 identified more than 6000 open reading frames (ORFs) that were predicted to encode proteins (Goffeau et al. 1996). This number was astonishing considering that in the previous century, genetic studies had attributed function to only approximately 1000 yeast genes, meaning that 80% of the yeast genome was uncharacterized (Goffeau et al. 1996). Following completion of the sequencing project, bioinformatic analyses were able to further classify ~50% of the yeast proteome based on sequence similarity to other proteins, including 11% predicted to be involved in cellular metabolism; 7% in transcription; 6% in translation; and 3% in DNA replication, recombination, and repair (Dujon 1996; Goffeau et al. 1996). However, sequence homology allows only general descriptions and predictions of biochemical activities, with direct experimentation being necessary to confirm functions of the proteins encoded by these putative ORFs.

A long-standing and powerful approach to elucidate gene function is to create loss-of-function mutants and to observe the resulting phenotypes. With the completed genome sequence, combined with the introduction of microhomology-mediated recombination and dominant selectable genetic markers, a comprehensive deletion project became possible (Orr-Weaver et al. 1981; Wach et al. 1994; Manivasakam et al. 1995). In 1998, the *Saccharomyces* Genome Deletion Project, an international

---

[1]Correspondence: g.giaever@ubc.ca; ggiaever@gmail.com

Copyright © Cold Spring Harbor Laboratory Press; all rights reserved
Cite this introduction as *Cold Spring Harb Protoc*; doi:10.1101/pdb.top080945

consortium comprised of 16 laboratories, initiated an effort to create genome-wide deletion collections (YKOv1, Yeast Knockout version 1) in the auxotrophic parental yeast strain BY4743 (Winzeler et al. 1999). Each ORF in the *S. cerevisiae* genome was replaced with a *KanMX* module to generate a start-to-stop codon deletion with two unique 20-mer molecular barcodes flanking the deletion (Fig. 1) (Baudin et al. 1993; Wach et al. 1994; Winzeler et al. 1999) (http://www-sequence.stanford.edu/group/yeast_deletion_project/deletions3.html). *KanMX* expression in yeast allows the selection of mutants with geneticin (G418) (Davies and Jimenez 1980). In total, four yeast deletion strain collections were generated by the *Saccharomyces* Genome Deletion Project: (1) MATa, (2) MATα haploid strains, (3) homozygous diploid deletion strains each carrying a deletion in one of approximately 5000 nonessential genes, and (4) heterozygous diploid deletion strains each with a single-copy deletion of approximately 1000 essential or approximately 5000 nonessential genes. These deletion collections provide a powerful resource for numerous applications—for example, to identify the relative requirement of all genes for growth under specific environmental conditions. Furthermore, such genotype–phenotype associations can be performed in a rapid and inexpensive manner when

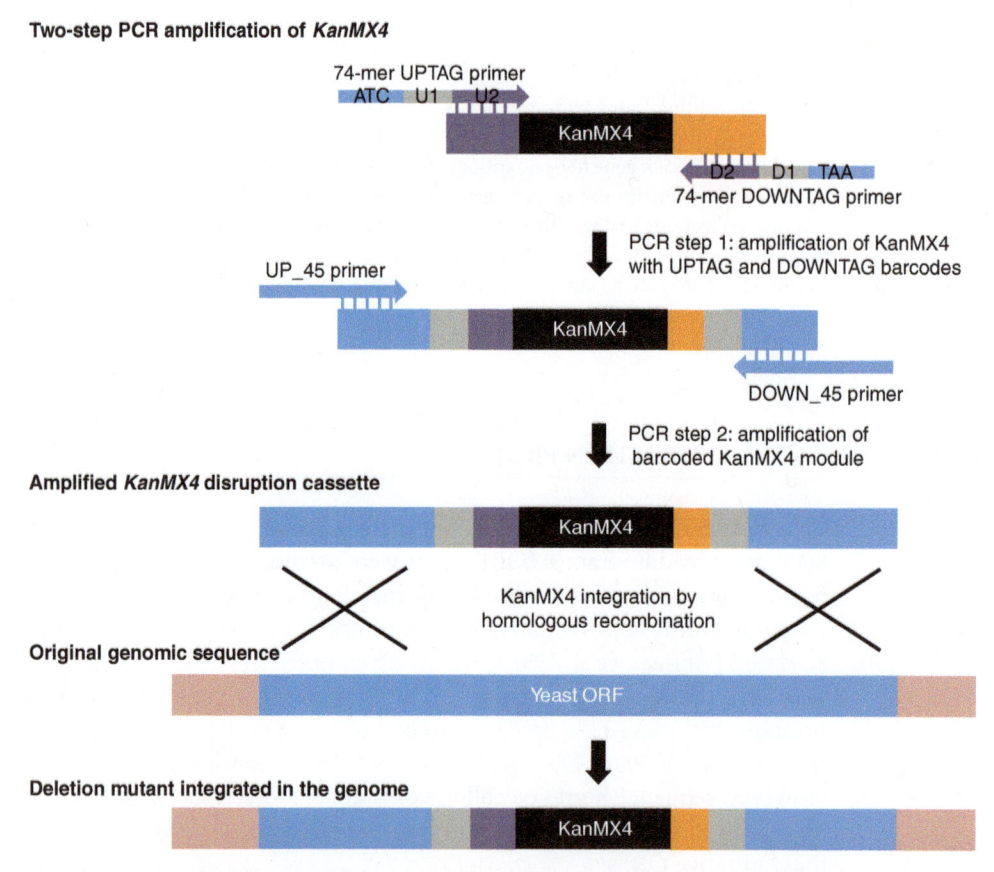

FIGURE 1. Construction of single-gene deletions in *S. cerevisiae*. The gene disruption *KanMX* cassette was constructed using two-step PCR. In the first PCR, 74 bp UPTAG and DOWNTAG primers, which flank the upstream 5′ and downstream 3′ end of the *KanMX* ORF, respectively, were used to amplify the *KanMX* gene from pFA6-kanMX4 template. UPTAG (U) and DOWNTAG (D) are primers with a sequence common to all open reading frame (ORF) disruptions (U1: 5′-GATGTCCACGAGGTCTCT-3′ or D1: 5′-CGGTGTCGGTCTCGTAG-3′), a 20-bp unique "molecular bar-code" TAG sequence, and a sequence homologous to the *KanMX4* cassette (U2: 5′-CGTACGCTGCAGGTCGAC-3′ or D2: 5′-ATCGATGAATTCGAGCTCG-3′). In the second PCR, amplification was performed with two 45-mer primers that have specific homology with 45 bp upstream and downstream from each ORF start and stop codon (UP_45 and DOWN_45, respectively). This extended ORF-specific homology increases specificity during mitotic recombination of the gene disruption cassette. For detailed descriptions of primers, go to: http://www-sequence.stanford.edu/group/yeast_deletion_project/Primer_specs.html#uptagpri. The two-step PCR was designed to accommodate the costs and accuracy of chemical synthesis that were prevalent in 1997. Presently, a single 100–120-nt primer is typically used to create such barcoded deletions.

competitive fitness assays are used (Giaever et al. 2002). This functional genome-wide profiling of the yeast genome provides a systematic and unbiased approach to understand functional relationships among genes, pathways, and drug–gene interactions in response to diverse genetic, chemical, and physical perturbations.

Use of the auxotrophic strain BY4743 provides a genetic background that allows traditional genetic manipulation, whereas the unique molecular barcodes associated with each deletion strain serve both as strain identifiers and as a means for the development of parallel fitness assays. Parallel screening of pooled mutants, pioneered in *Escherichia coli* (Hensel et al. 1995) and refined in yeast (Shoemaker et al. 1996), allows for the entire yeast deletion collection to be grown in a single mixed culture under diverse conditions. The abundance of each strain is then independently assessed during the course of the experiment and barcode abundance is quantified by amplification of the barcodes (using common primers) followed by microarray hybridization. The greater the requirement of each gene for growth the more rapidly the associated barcodes became depleted, providing a direct readout for relative sensitivity (or resistance) to a condition of choice. The initial deletion collection comprised 5916 mutants, representing 96.5% of all ORFs of $\geq$100 amino acids identified at the project's inception. An update to the yeast genome collection, YKOv2, has been released with more than 300 ORF annotation changes, including the addition of nondubious small ORFs and the substitution of poorly performing deletion strains (Giaever et al. 2002; Chu and Davis 2008). It is important to recognize that as computational tools improve, and our definition of a gene evolves, additional strains may be added to the collection. Furthermore, some ORFs will prove refractory to "start-to-stop" deletion because of their genomic context (e.g., they overlap adjacent gene regulatory elements). The unique barcodes in each deletion strain have recently been resequenced at a high sequencing depth, confirming that 99.5% of barcodes map to the correct loci in the yeast genome (Eason et al. 2004; Smith et al. 2009).

GENOME-WIDE FUNCTIONAL PROFILING OF THE DELETION COLLECTIONS

The most common phenotype used as a readout for screens of the deletion collections is fitness (i.e., growth over time). Fitness can be monitored individually either by pinning mutants on solid media and measuring colony size or by measuring growth rate in liquid media (Giaever et al. 2002; Costanzo et al. 2010). Colony size on solid media can be quantified using a variety of metrics such as the "S score" (Collins et al. 2006) or the "Synthetic Genetic Array (SGA) epsilon score" (Baryshnikova et al. 2010a). These metrics normalize colony size by taking into account various technical effects (such as media batch or plate positions) that can otherwise overshadow experimental effects. Parallel fitness assays allow the deletion strain collection to be grown competitively and relative fitness to be quantified using the barcodes. Parallel strain analysis is generally more sensitive than fitness measurements based on colony size (Giaever et al. 2002; Costanzo et al. 2010). Costanzo et al. (2010) directly compared fitness on solid and in liquid media and found that in highly controlled experimental conditions, the two approaches yield highly correlated fitness data. The choice of assay format chosen is ultimately based on technical and experimental constraints, in addition to the particular biological question being addressed. For a review of the diverse ways in which the deletion collection has been used, see Giaever and Nislow (2014).

A number of different algorithms have been developed to assess deletion strain fitness, either as individual colonies or in pools (Giaever et al. 2002; St.Onge et al. 2007; Baryshnikova et al. 2010b; Li et al. 2011; Wagih et al. 2013; see also Introduction of Chapter 25: Systematic Mapping of Chemical–Genetic Interactions in *Saccharomyces cerevisiae* [Suresh et al. 2015]). In brief, barcode abundance can be determined using traditional microarray readouts that rely on normalized fluorescence intensity from the Affymetrix Tag4 array (Giaever et al. 2004; Pierce et al. 2006, 2007). More recently, "next-generation" sequencing technologies have been applied to fitness profiling, providing another powerful tool to determine barcode abundance based on the frequency with which each barcode is sequenced (Smith et al. 2009). Genome-wide screens that use barcode analysis by sequencing

(BAR-Seq) deliver comparable results to data generated by barcode microarray hybridization, if not outperforming it with higher sensitivity, larger dynamic range, and more reliable detection.

Interrogation of the yeast deletion collection has been performed by assessing phenotypic changes other than fitness, including cell morphology, cell size, colony color, bud site selection, metabolite synthesis, mating, germination, and sporulation (Ni and Snyder 2001; Deutschbauer et al. 2002; Jorgensen et al. 2002; Zhang et al. 2002; Holland et al. 2007; Kloimwieder and Winston 2011). For example, cell size measurements and bud site location of deletion mutants provide information on cell cycle and division regulation (Ni and Snyder 2001). Many of these screens required considerable effort (e.g., manual inspection of each mutant strain) but nonetheless would have been impossible without the yeast deletion collections. More recently, a number of "designer" screens, in which reporter genes or exogenous genes are introduced into the yeast deletion collections, have been reported (Boone et al. 2007), with synthetic dosage lethality screening as a noteworthy example (Sopko et al. 2006).

FURTHER APPLICATIONS OF THE DELETION COLLECTIONS

As is the case for most genome-wide experimental assays, the data resulting from any one particular screen is valuable on its own, but the combination of multiple screens can allow the detection of trends and similarities in the phenotypes resulting from the ablation of single genes or groups of genes. For example, comprehensive analyses of genes that respond to environmental stress conditions, such as heat shock, oxidative stress, weak acid, and ionic stress, or osmotic shock and of genes involved in the mechanisms of drug action have been accomplished by two assays that use the diploid deletion collections in parallel assays known as haploinsufficiency profiling and homozygous deletion profiling (HIPHOP) assays (see Introduction of Chapter 25: Systematic Mapping of Chemical–Genetic Interactions in *Saccharomyces cerevisiae* [Suresh et al. 2015]; Giaever et al. 1999, 2002, 2004). Using the heterozygous diploid deletion in essential genes collection, drug target candidates can be identified by determining heterozygous deletion strains with the greatest drug sensitivity due to decreased dosage of the gene encoding the drug target. In the complementary HOP assay, homozygous deletion strains in nonessential genes can be used to reveal drug target pathways that buffer the drug-induced growth sensitivity. The deletion collections have also been widely applied to the study of digenic interactions. This methodology, known as synthetic genetic array analysis (SGA) (Tong et al. 2001), elucidates the phenotypic consequences of double mutants by crossing a query strain carrying a mutant gene of interest to the haploid deletion collection. Further applications can be achieved by modifications to the deletion collections, such as substituting the *KanMX* cassette with fluorescent tags or any other reporter system (Kainth et al. 2009; Sassi et al. 2009). Yet other modifications to the collections aim to address some of their initial limitations, such as restoring prototrophy to the BY4743 genetic background to allow characterization of deletion collections in conditions that mimic a "wild-type" strain (Mulleder et al. 2012; Gibney et al. 2013).

The yeast deletion project has revolutionized yeast genetics by providing a systematic and global approach to analyzing gene functions. It has also served as a technology test bed for genome-wide assays in other organisms; for example, screening of interfering RNAs in mammalian cells borrows many of its design principles from yeast pooled screening (Ketela et al. 2011). The yeast deletion collections can be considered as the founding members of genome-wide deletion collections and have inspired various genome-wide deletion collections in other organisms, such as *Schizosaccharomyces pombe* (Baryshnikova et al. 2010a; Kim et al. 2010), *Arabidopsis thaliana* (Alonso et al. 2003), and *E. coli* (Baba et al. 2006).

ACKNOWLEDGMENTS

Work in the laboratories of C.N. and G.G. has been supported by grants from the National Institutes of Health Research (NHGRI), the Canadian Institutes for Health Research (CIHR), and the Canadian

Cancer Society Research Institute (CCSRI). The authors thank all 16 yeast deletion collection consortium laboratories and especially Ron Davis and Mark Johnston for spearheading the effort.

REFERENCES

Alonso JM, Stepanova AN, Leisse TJ, Kim CJ, Chen H, Shinn P, Stevenson DK, Zimmerman J, Barajas P, Cheuk R, et al. 2003. Genome-wide insertional mutagenesis of *Arabidopsis thaliana*. *Science* 301: 653–657.

Baba T, Ara T, Hasegawa M, Takai Y, Okumura Y, Baba M, Datsenko KA, Tomita M, Wanner BL, Mori H. 2006. Construction of *Escherichia coli* K-12 in-frame, single-gene knockout mutants: The Keio collection. *Mol Syst Biol* 2: 2006.00081–11.

Baryshnikova A, Costanzo M, Dixon S, Vizeacoumar FJ, Myers CL, Andrews B, Boone C. 2010a. Synthetic genetic array (SGA) analysis in *Saccharomyces cerevisiae* and *Schizosaccharomyces pombe*. *Methods Enzymol* 470: 145–179.

Baryshnikova A, Costanzo M, Kim Y, Ding H, Koh J, Toufighi K, Youn JY, Ou J, San Luis BJ, Bandyopadhyay S, et al. 2010b. Quantitative analysis of fitness and genetic interactions in yeast on a genome scale. *Nat Methods* 7: 1017–1024.

Baudin A, Ozier-Kalogeropoulos O, Denouel A, Lacroute F, Cullin C. 1993. A simple and efficient method for direct gene deletion in *Saccharomyces cerevisiae*. *Nucleic Acids Res* 21: 3329–3330.

Boone C, Bussey H, Andrews BJ. 2007. Exploring genetic interactions and networks with yeast. *Nat Rev Genet* 8: 437–449.

Chu AM, Davis RW. 2008. High-throughput creation of a whole-genome collection of yeast knockout strains. *Methods Mol Biol* 416: 205–220.

Collins SR, Schuldiner M, Krogan NJ, Weissman JS. 2006. A strategy for extracting and analyzing large-scale quantitative epistatic interaction data. *Genome Biol* 7: R63.

Costanzo M, Baryshnikova A, Bellay J, Kim Y, Spear ED, Sevier CS, Ding H, Koh JL, Toufighi K, Mostafavi S, et al. 2010. The genetic landscape of a cell. *Science* 327: 425–431.

Davies J, Jimenez A. 1980. A new selective agent for eukaryotic cloning vectors. *Am J Trop Med Hyg* 29: 1089–1092.

Deutschbauer AM, Williams RM, Chu AM, Davis RW. 2002. Parallel phenotypic analysis of sporulation and postgermination growth in *Saccharomyces cerevisiae*. *Proc Natl Acad Sci* 99: 15530–15535.

Dujon B. 1996. The yeast genome project: What did we learn? *Trends Genet* 12: 263–270.

Eason RG, Pourmand N, Tongprasit W, Herman ZS, Anthony K, Jejelowo O, Davis RW, Stolc V. 2004. Characterization of synthetic DNA bar codes in *Saccharomyces cerevisiae* gene-deletion strains. *Proc Natl Acad Sci* 101: 11046–11051.

Giaever G, Chu AM, Ni L, Connelly C, Riles L, Veronneau S, Dow S, Lucau-Danila A, Anderson K, Andre B, et al. 2002. Functional profiling of the *Saccharomyces cerevisiae* genome. *Nature* 418: 387–391.

Giaever G, Flaherty P, Kumm J, Proctor M, Nislow C, Jaramillo DF, Chu AM, Jordan MI, Arkin AP, Davis RW. 2004. Chemogenomic profiling: Identifying the functional interactions of small molecules in yeast. *Proc Natl Acad Sci* 101: 793–798.

Giaever G, Nislow C. 2014. The yeast deletion collection: A decade of functional genomics. *Genetics* 197: 451–465.

Giaever G, Shoemaker DD, Jones TW, Liang H, Winzeler EA, Astromoff A, Davis RW. 1999. Genomic profiling of drug sensitivities via induced haploinsufficiency. *Nat Genet* 21: 278–283.

Gibney PA, Lu C, Caudy AA, Hess DC, Botstein D. 2013. Yeast metabolic and signaling genes are required for heat-shock survival and have little overlap with the heat-induced genes. *Proc Natl Acad Sci* 110: E4393–E4402.

Goffeau A, Barrell BG, Bussey H, Davis RW, Dujon B, Feldmann H, Galibert F, Hoheisel JD, Jacq C, Johnston M, et al. 1996. Life with 6000 genes. *Science* 274: 546, 563–567.

Hensel M, Shea JE, Gleeson C, Jones MD, Dalton E, Holden DW. 1995. Simultaneous identification of bacterial virulence genes by negative selection. *Science* 269: 400–403.

Holland S, Lodwig E, Sideri T, Reader T, Clarke I, Gkargkas K, Hoyle DC, Delneri D, Oliver SG, Avery SV. 2007. Application of the comprehensive set of heterozygous yeast deletion mutants to elucidate the molecular basis of cellular chromium toxicity. *Genome Biol* 8: R268.

Jorgensen P, Nishikawa JL, Breitkreutz BJ, Tyers M. 2002. Systematic identification of pathways that couple cell growth and division in yeast. *Science* 297: 395–400.

Kainth P, Sassi HE, Pena-Castillo L, Chua G, Hughes TR, Andrews B. 2009. Comprehensive genetic analysis of transcription factor pathways using a dual reporter gene system in budding yeast. *Methods* 48: 258–264.

Ketela T, Heisler LE, Brown KR, Ammar R, Kasimer D, Surendra A, Ericson E, Blakely K, Karamboulas D, Smith AM, et al. 2011. A comprehensive platform for highly multiplexed mammalian functional genetic screens. *BMC Genomics* 12: 213.

Kim DU, Hayles J, Kim D, Wood V, Park HO, Won M, Yoo HS, Duhig T, Nam M, Palmer G, et al. 2010. Analysis of a genome-wide set of gene deletions in the fission yeast *Schizosaccharomyces pombe*. *Nat Biotechnol* 28: 617–623.

Kloimwieder A, Winston F. 2011. A screen for germination mutants in *Saccharomyces cerevisiae*. *G3 (Bethesda)* 1: 143–149.

Li Z, Vizeacoumar FJ, Bahr S, Li J, Warringer J, Vizeacoumar FS, Min R, Vandersluis B, Bellay J, Devit M, et al. 2011. Systematic exploration of essential yeast gene function with temperature-sensitive mutants. *Nat Biotechnol* 29: 361–367.

Manivasakam P, Weber SC, McElver J, Schiestl RH. 1995. Micro-homology mediated PCR targeting in *Saccharomyces cerevisiae*. *Nucleic Acids Res* 23: 2799–2800.

Mulleder M, Capuano F, Pir P, Christen S, Sauer U, Oliver SG, Ralser M. 2012. A prototrophic deletion mutant collection for yeast metabolomics and systems biology. *Nat Biotechnol* 30: 1176–1178.

Ni L, Snyder M. 2001. A genomic study of the bipolar bud site selection pattern in *Saccharomyces cerevisiae*. *Mol Biol Cell* 12: 2147–2170.

Orr-Weaver TL, Szostak JW, Rothstein RJ. 1981. Yeast transformation: A model system for the study of recombination. *Proc Natl Acad Sci* 78: 6354–6358.

Pierce SE, Davis RW, Nislow C, Giaever G. 2007. Genome-wide analysis of barcoded *Saccharomyces cerevisiae* gene-deletion mutants in pooled cultures. *Nat Protoc* 2: 2958–2974.

Pierce SE, Fung EL, Jaramillo DF, Chu AM, Davis RW, Nislow C, Giaever G. 2006. A unique and universal molecular barcode array. *Nat Methods* 3: 601–603.

Sassi HE, Bastajian N, Kainth P, Andrews BJ. 2009. Reporter-based synthetic genetic array analysis: A functional genomics approach for investigating the cell cycle in *Saccharomyces cerevisiae*. *Methods Mol Biol* 548: 55–73.

Shoemaker DD, Lashkari DA, Morris D, Mittmann M, Davis RW. 1996. Quantitative phenotypic analysis of yeast deletion mutants using a highly parallel molecular bar-coding strategy. *Nat Genet* 14: 450–456.

Smith AM, Heisler LE, Mellor J, Kaper F, Thompson MJ, Chee M, Roth FP, Giaever G, Nislow C. 2009. Quantitative phenotyping via deep barcode sequencing. *Genome Res* 19: 1836–1842.

Sopko R, Papp B, Oliver SG, Andrews BJ. 2006. Phenotypic activation to discover biological pathways and kinase substrates. *Cell Cycle* 5: 1397–1402.

St.Onge RP, Mani R, Oh J, Proctor M, Fung E, Davis RW, Nislow C, Roth FP, Giaever G. 2007. Systematic pathway analysis using high-resolution fitness profiling of combinatorial gene deletions. *Nat Genet* 39: 199–206.

Suresh S, Schlecht U, Xu W, Bray W, Miranda M, Davis RW, Nislow C, Giaever G, Lokey S, St.Onge RP. 2015. Systematic mapping of chemical–genetic interactions in *Saccharomyces cerevisiae*. *Cold Spring Harb Protoc* doi: 10.1101/pdb.top077701.

Tong AH, Evangelista M, Parsons AB, Xu H, Bader GD, Page N, Robinson M, Raghibizadeh S, Hogue CW, Bussey H, et al. 2001. Systematic genetic analysis with ordered arrays of yeast deletion mutants. *Science* 294: 2364–2368.

Wach A, Brachat A, Pohlmann R, Philippsen P. 1994. New heterologous modules for classical or PCR-based gene disruptions in *Saccharomyces cerevisiae*. *Yeast* **10**: 1793–1808.

Wagih O, Usaj M, Baryshnikova A, VanderSluis B, Kuzmin E, Costanzo M, Myers CL, Andrews BJ, Boone CM, Parts L. 2013. SGAtools: One-stop analysis and visualization of array-based genetic interaction screens. *Nucleic Acids Res* **41**: W591–596.

Winzeler EA, Shoemaker DD, Astromoff A, Liang H, Anderson K, Andre B, Bangham R, Benito R, Boeke JD, Bussey H, et al. 1999. Functional characterization of the *S. cerevisiae* genome by gene deletion and parallel analysis. *Science* **285**: 901–906.

Zhang J, Schneider C, Ottmers L, Rodriguez R, Day A, Markwardt J, Schneider BL. 2002. Genomic scale mutant hunt identifies cell size homeostasis genes in *S. cerevisiae*. *Curr Biol* **12**: 1992–2001.

Protocol 1

Functional Profiling Using the *Saccharomyces* Genome Deletion Project Collections

Corey Nislow, Lai Hong Wong, Amy Huei-Yi Lee, and Guri Giaever[1]

Department of Pharmaceutical Sciences, University of British Columbia, Vancouver, British Columbia V6T 1Z3, Canada

The ability to measure and quantify the fitness of an entire organism requires considerably more complex approaches than simply using traditional "omic" methods that examine, for example, the abundance of RNA transcripts, proteins, or metabolites. The yeast deletion collections represent the only systematic, comprehensive set of null alleles for any organism in which such fitness measurements can be assayed. Generated by the *Saccharomyces* Genome Deletion Project, these collections allow the systematic and parallel analysis of gene functions using any measurable phenotype. The unique 20-bp molecular barcodes engineered into the genome of each deletion strain facilitate the massively parallel analysis of individual fitness. Here, we present functional genomic protocols for use with the yeast deletion collections. We describe how to maintain, propagate, and store the deletion collections and how to perform growth fitness assays on single and parallel screening platforms. Phenotypic fitness analyses of the yeast mutants, described in brief here, provide important insights into biological functions, mechanisms of drug action, and response to environmental stresses. It is important to bear in mind that the specific assays described in this protocol represent some of the many ways in which these collections can be assayed, and in this description particular attention is paid to maximizing throughput using growth as the phenotypic measure.

MATERIALS

It is essential that you consult the appropriate Material Safety Data Sheets and your institution's Environmental Health and Safety Office for proper handling of equipment and hazardous materials used in this protocol.

RECIPES: Please see the end of this protocol for recipes indicated by <R>. Additional recipes can be found online at http://cshprotocols.cshlp.org/site/recipes.

Reagents

Compounds appropriate for experimental goals (optional; see Steps 7, 11, and 18)
Dimethyl sulfoxide (DMSO)
Distilled water, sterile
Ethanol (70% v/v)
Frozen glycerol (or DMSO) stocks of the yeast deletion collections in 96-well microtiter plates

Stocks can be obtained from EUROSCARF (European Saccharomyces cerevisiae Archive for Functional Analysis) (http://web.uni-frankfurt.de/fb15/mikro/euroscarf/yeast.html), ATCC (http://www.atcc.org/), or Thermo Scientific (http://www.lifetechnologies.com/ca/en/home.html).

---

[1]Correspondence: ggiaever@gmail.com; g.giaever@ubc.ca

Copyright © Cold Spring Harbor Laboratory Press; all rights reserved
Cite this protocol as *Cold Spring Harb Protoc*; doi:10.1101/pdb.prot088039

Geneticin (G418)
Liquid medium of choice appropriate for experimental goals (optional; see Steps 11 and 18)
Solid agar medium of choice appropriate for experimental goals (optional; see Step 7)
YPD media (for liquid medium and agar plates) <R>

Equipment

96-well pin tool (V&P Scientific)
384 pin replicator (V&P Scientific)
384-well plates (Thermo Scientific Nunc 242757)
Adhesive plate seal (ABgene AB-0580) (optional; see Step 18)
Aluminum 96-well plate-sealer
Bunsen burner
Capped PCR (polymerase chain reaction) tubes (eight- or 12-strip)
Cell culture plate (single-well, e.g., NUNC OmniTray)
Cell culture plate (48-well; Greiner 677102) (optional; see Step 18)
Cell spreader
Colony manipulation robot (RoTor, Singer, or Biomatrix colony arrayer, S&P Robotics) (optional; see Step 8 and Step 11)
Library Copier (V&P Scientific, VP381)
Microcentrifuge tubes (1.5-mL)
Microplate reader (Tecan, Sunrise)
Multichannel dispenser (e.g., Rainin or Mettler Toledo)
Polypropylene tubes (50-mL) (optional; see Step 14)
Programmable liquid handler (e.g., Janus, PerkinElmer, or Evo, Tecan) (optional; see Step 18)
UV light (optional; Step 13)

METHOD

Inoculation of Yeast Deletion Collections

1. Completely thaw yeast deletion collection frozen glycerol stocks in 96-well microtiter plates at room temperature.

   Yeast cells may have settled to the bottom of the wells before plates were frozen; therefore, ensure plates are completely thawed. Remove plates from the freezer in manageable numbers (e.g., sets of 20) to avoid leaving thawed cells at room temperature for >2 h.

2. Sterilize a 96-well pin tool by dipping the pin tool in water to rinse away any cells, followed by two dips in a 70% ethanol (v/v) bath and flame sterilization.

   The water should be replaced every four to six pinnings and the pin tool sterilized in between each pinning.

3. Insert the sterile 96-well pin tool into the thawed 96-well glycerol stocks so that it rests on the bottom of the plate. Swirl gently and transfer to a NUNC OmniTray containing 50 mL of YPD agar with 200 µg/mL G418. Allow the pin tool to sit on the agar for 5–10 sec.

4. Reseal the 96-well glycerol stock plate with new aluminum plate-sealer and return it to the −80°C freezer.

5. Repeat Steps 2–4 for remaining plates.

6. Grow colonies until they are at maximal size (~3.0 mm) by incubating the agar plates for 2–3 d at 30°C.

Phenotype Characterization or Fitness Measurements of the Deletion Collections

Phenotype or fitness can be characterized by different approaches after growth of individual strains on solid medium (Steps 7–10) or in liquid medium (Steps 11–12) or pooled in liquid medium (Steps 13–20) as described below and illustrated in Figure 1.

Growth of Yeast Deletion Strains on Solid Medium

7. Pour YPD agar or solid agar medium of choice containing compound(s) appropriate to experimental goals into NUNC OmniTrays. Dry plates for at least 1 d at room temperature before pinning.

   We recommend 50 mL of agar medium per OmniTray but 40 mL will suffice if precious compounds are used.

8. Replica-pin deletion collections onto plates either manually or robotically directly from thawed–frozen stocks in 96-well microtiter plates (see Step 1) or from agar plates (see Step 6).

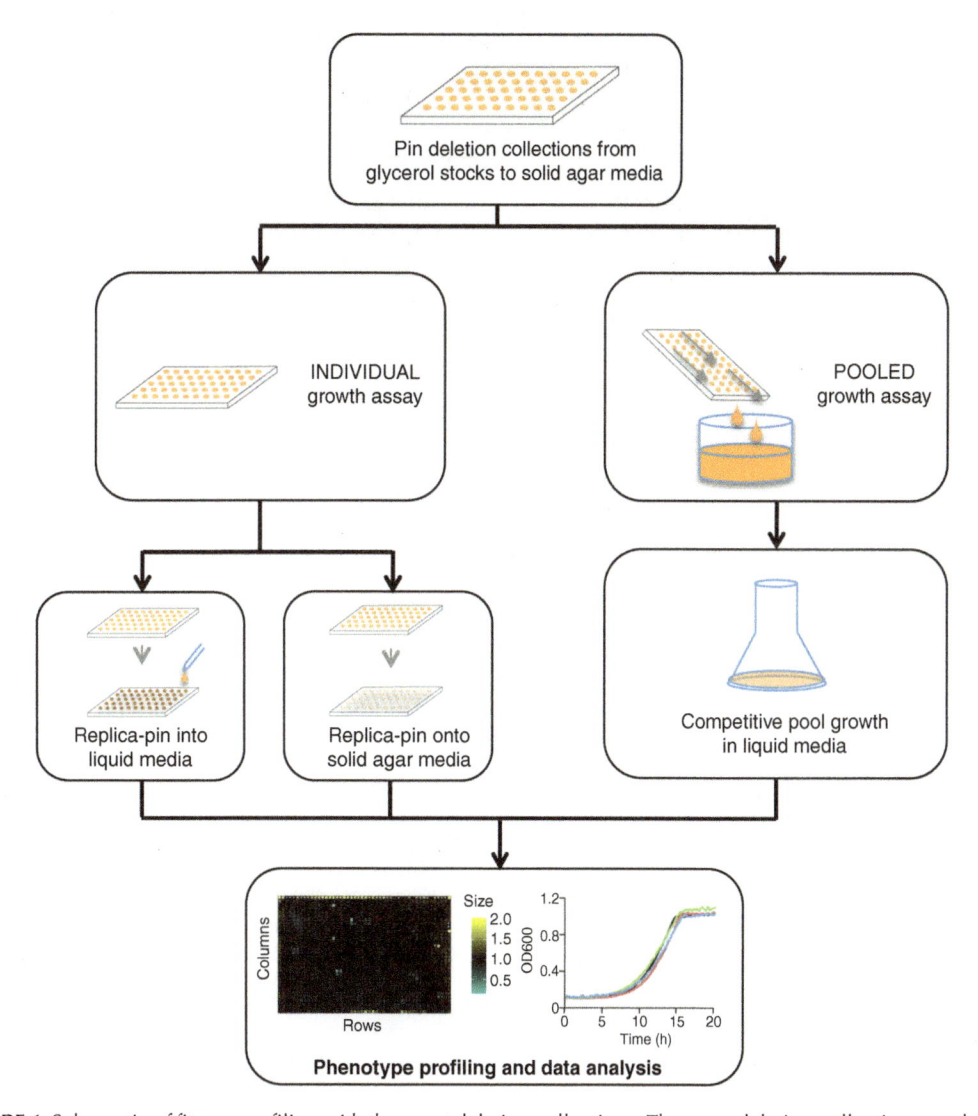

FIGURE 1. Schematic of fitness profiling with the yeast deletion collections. The yeast deletion collections can be used for individual growth or pooled growth assays. To perform individual growth assays, glycerol stocks of deletion collections are replica-pinned into individual wells of 96-microtiter plates containing liquid medium, or onto solid agar in OmniTrays. To perform a pooled growth assay, each strain in a deletion collection is pooled together and the pool is grown in liquid medium. Phenotype profiling and data analysis pipelines discussed in this protocol can be used to analyze the growth assay results.

- For manual pinning, flame sterilize a 96-well pin tool and transfer deletion strain colonies onto OmniTrays containing solid medium.
- For robotic pinning, plate colonies robotically onto OmniTrays containing solid medium using either a Singer RoTor or S&P Biomatrix robot. Follow the manufacturer's instructions.

  Be sure to use the latest software: https://www.google.ca/webhp?sourceid=chrome-instant&ion=1&espv=2&ie=UTF-8q=singer+instruments+rotator or http://www.sprobotics.com/.

9. Incubate plates at the appropriate temperature for 2–3 d (30°C is typical).

   Growth conditions for the experimental assay and appropriate controls should be defined by the user according to experimental goals.

10. Assess fitness measurements using web-based SGAtools (Wagih et al. 2013).

Growth of Individual Yeast Deletion Strains in Liquid Medium

11. Pin deletion strains manually or robotically directly from thawed–frozen stocks in 96-well microtiter plates (see Step 1) or from agar plates (see Step 6) into 96-well plates containing 100 µL per well of YPD with 200 µg/mL G418 or medium of choice containing compound(s) appropriate to experimental goals.

    - For manual pinning, flame sterilize a 96-well pin tool and transfer deletion strain colonies into 96-well plates containing liquid medium, using gentle swirling for optimum resuspension.
    - For automated pinning, use an S&P Biomatrix or Singer RoTor robot to pin mutants into 96-well plates containing liquid medium. Program the pinhead to move up and down at least three times to ensure proper mixing.

      Growth conditions for the experimental assay and appropriate controls should be defined by the user according to experimental goals.

12. Measure fitness of individual deletion strains using a spectrophotometer microtiter plate reader to read OD_{600} over time as a proxy for cell growth.

    Relative growth rate or fitness can be calculated by a variety of growth metrics, such as average rate or logarithmic rate, depending on the experimental goal. To eliminate potential plate effects, normalize data against the control on each plate and across all plates in the screen.

Yeast Deletion Strain Pool Construction and Pooled Growth in Liquid Medium

13. Generate a pool of yeast deletion strains using a 96-well pin tool to create a set of 384-well arrayed strains on YPD-agar containing 200 µg/mL G418 by serially pinning four 96-well plates onto one OmniTray as a 384 (16 rows × 24 columns) colony array using a Library Copier register. Allow the colonies to grow for 2 d at 30°C, and then repin three copies using a 384 pin replicator and regrow for 2 d at 30°C.

    Disposable pin tools can be bleached, washed, and sterilized by UV light to enable reuse.

14. Collect colonies from the triplicate plates by flooding each plate with 5 mL of medium, wait for 30 min, scrape with a cell spreader, and decant into a beaker. Rinse each plate with an additional 5 mL of medium.

    For slow-growing strains, add to the pool additional or an overabundance of cells equivalent in volume to the colony size of normal growing cells to ensure equal strain representation. Typically, 250 mL of cell suspension is collected for one pool.

    See Troubleshooting.

15. Measure the OD_{600} of the pool and adjust the pool to a final OD_{600} of 50.

16. Add DMSO to a final concentration of 7% (v/v), mix well, and aliquot three-quarters of the pooled cells into several 50 mL tubes, and the remaining one-quarter into several 1.5 mL tubes and hundreds of eight- or 12-strip capped PCR tubes. Store at −80°C.

    An aliquot can be used directly in Step 18 without freezing.

17. Thaw a frozen aliquot of pooled cells on ice.

18. Use pool immediately or allow recovery at 30°C with shaking for one to two generations. Grow cultures under manual or automated conditions appropriate for experimental goals for a desired number of generations (typically five to 20).

   - For manual cell growth, inoculate cells in medium containing appropriate drugs for experimental goals at a starting OD_{600} of 0.002 in 50 mL total volume in a 250 mL culture flask and grow at 30°C at 250 rpm. In parallel, prepare appropriate control culture(s).

     Drug dosage should be predetermined using wild-type cells. Cells with a final OD_{600} of 2 will have undergone 10 generations of growth.

   - For automated cell growth, inoculate cells in medium containing appropriate drugs for experimental goals or appropriate solvent control(s) at a starting OD_{600} of 0.0625 in a total volume of 700 µL in a 48-well plate (any programmable liquid handler can be used), and seal with an adhesive plate seal. Similarly, prepare wells as above but without cells to be used by the robot for inoculation at five-generation intervals. If the conditions require optimal aerobic growth (e.g., nonfermentable carbon sources), poke a small hole in the center of the membrane seal of each well. Grow in a spectrophotometer at 30°C with a predetermined optimal shaking regimen.

     The growth is divided into five-generation intervals to allow for continuous culture in the log phase for 20 generations. Drug dosage should be predetermined using wild-type cells. Cells grown to a final OD_{600} of 2 have typically gone through five generations of growth. A fraction of the cell suspension can be saved on a cold plate of the robotic deck to define generation times (for details of this procedure, see Proctor et al. 2011; http://med.stanford.edu/sgtc/technology/access.html; or contact C. Nislow or G. Giaever).

     See Troubleshooting.

19. Collect cells at a final $OD_{600} \sim 2$ in 1.5-mL microcentrifuge tubes. Centrifuge cells and remove medium. Store cell pellets at −20°C.

    The OD_{600} of cells grown in culture flasks needs to be normalized between all samples before pellet collection.

20. Analyze pooled strains by microarray or sequencing methods to determine barcode abundance (see Giaever et al. 2004; Smith et al. 2009; or Protocol 2 of Chapter 25: Identification of Chemical–Genetic Interactions via Parallel Analysis of Barcoded Yeast Strains [Suresh et al. 2015]).

TROUBLESHOOTING

Problem (Step 14): Slow-growing strains are observed during pool construction.
Solution: Known slow-growing strains, listed at http://www.yeastgenome.org/, can be supplemented at three times more biomass compared with the other strains in the collection (i.e., three normal colony equivalents). Alternatively, slow-growing strains can be screened separately.

Problem (Step 18): Slow growth and long lag time of the pool due to lack of pool recovery before experimental perturbation are observed.
Solution: In our initial studies (Giaever et al. 2002), we inoculated a pool aliquot (the volume of the aliquot depends on the OD_{600} of the aliquot and the desired starting concentration) into 200 mL of medium and allowed the cells to recover for 16 h at 30°C before harvesting them in mid-logarithmic growth (OD_{600} 1) and then diluting to a starting OD_{600} of 0.06. More recently, we have diluted thawed pool cells directly into the medium of choice to achieve the desired starting OD_{600}, typically 0.0625. The latter procedure results in an additional 60 min of lag time in the first generation of growth.

DISCUSSION

The yeast deletion collections offer a set of single-gene deletion strains in *S. cerevisiae*, which enables unbiased analysis of all the approximately 6000 growth phenotypes of individual strains in a single assay. These collections can be obtained from EUROSCARF (http://web.uni-frankfurt.de/fb15/mikro/euroscarf/yeast.html), Thermo Scientific (formerly Invitrogen) (http://www.lifetechnologies.com/ca/en/home.html), or the ATCC (http://www.atcc.org/). It is important to appreciate certain caveats with the yeast deletion collections when designing experiments. For example, some deletion mutants in the homozygous, nonessential gene deletion collection show sick or slower growth phenotypes, which pose a problem during pooled competitive growth assays, where slower growing mutants can be depleted before the start of a screen. Hence, screens of this yeast deletion collection are typically assayed for a limited number (e.g., five) of generations. The number of generations can also increase the number of second site mutations in single-gene deletion mutants (Hughes et al. 2000; Lehner et al. 2007). Furthermore, evidence suggests that ~5% of haploid mutants have the capacity to diploidize and show aneuploidy, which suggests that the single-gene deletions in the haploid yeast collections may show higher genomic instability compared to their diploid counterparts (Hughes et al. 2000). Several studies also suggest that auxotrophic markers in yeast deletion collections can induce cross-feeding or other unanticipated effects that can have an impact on the experimental growth conditions used in screens (Pronk 2002; Bauer et al. 2003; Canelas et al. 2010; Corbacho et al. 2011; Hanscho et al. 2012; Heavner et al. 2012; Hueso et al. 2012; Mulleder et al. 2012).

When analyzing gene function using drugs, other considerations should be taken into account. For example, the complete loss of function of a gene may not phenocopy the action of a drug on the gene's product, and by extension, the absence of a gene product precludes identification of the drug target. Hence, for all screens, additional assays should be used to validate a candidate gene's response to perturbation. Also, caution is needed when single-gene deletions are used to study signaling pathways because of the ability of mutants to rewire and adapt; protein dynamics can be modified such that they do not reflect normal interactions. Potential neighboring gene effects can also confound gene function characterization (Ben-Shitrit et al. 2012). Notwithstanding these caveats, fitness-profiling data from the yeast deletion collections enable an excellent proof-of-principle phase in the comprehensive analyses of genes' contributions to diverse biological activities in an in vivo context.

RECIPE

YPD Media (for Liquid Medium and Agar Plates)

Reagent	Quantity
Yeast extract	10 g
Peptone	20 g
Agar (for plates only)	20 g
H_2O	950 mL
Glucose (40%, w/v; sterile)	50 mL

Combine the first four ingredients, and autoclave. Add the glucose, and mix well.

REFERENCES

Bauer BE, Rossington D, Mollapour M, Mamnun Y, Kuchler K, Piper PW. 2003. Weak organic acid stress inhibits aromatic amino acid uptake by yeast, causing a strong influence of amino acid auxotrophies on the phenotypes of membrane transporter mutants. *Eur J Biochem* 270: 3189–3195.

Ben-Shitrit T, Yosef N, Shemesh K, Sharan R, Ruppin E, Kupiec M. 2012. Systematic identification of gene annotation errors in the widely used yeast mutation collections. *Nat Methods* 9: 373–378.

Canelas AB, Harrison N, Fazio A, Zhang J, Pitkanen JP, van den Brink J, Bakker BM, Bogner L, Bouwman J, Castrillo JI, et al. 2010. Integrated multilaboratory systems biology reveals differences in protein metabolism between two reference yeast strains. *Nat Commun* 1: 145.

Corbacho I, Teixido F, Velazquez R, Hernandez LM, Olivero I. 2011. Standard YPD, even supplemented with extra nutrients, does not always compensate growth defects of *Saccharomyces cerevisiae* auxotrophic strains. *Antonie Van Leeuwenhoek* 99: 591–600.

Giaever G, Chu AM, Ni L, Connelly C, Riles L, Veronneau S, Dow S, Lucau-Danila A, Anderson K, Andre B, et al. 2002. Functional profiling of the *Saccharomyces cerevisiae* genome. *Nature* 418: 387–391.

Giaever G, Flaherty P, Kumm J, Proctor M, Nislow C, Jaramillo DF, Chu AM, Jordan MI, Arkin AP, Davis RW. 2004. Chemogenomic profiling: Identifying the functional interactions of small molecules in yeast. *Proc Natl Acad Sci* **101**: 793–798.

Hanscho M, Ruckerbauer DE, Chauhan N, Hofbauer HF, Krahulec S, Nidetzky B, Kohlwein SD, Zanghellini J, Natter K. 2012. Nutritional requirements of the BY series of *Saccharomyces cerevisiae* strains for optimum growth. *FEMS Yeast Res* **12**: 796–808.

Heavner BD, Smallbone K, Barker B, Mendes P, Walker LP. 2012. Yeast 5—An expanded reconstruction of the *Saccharomyces cerevisiae* metabolic network. *BMC Syst Biol* **6**: 55.

Hueso G, Aparicio-Sanchis R, Montesinos C, Lorenz S, Murguia JR, Serrano R. 2012. A novel role for protein kinase Gcn2 in yeast tolerance to intracellular acid stress. *Biochem J* **441**: 255–264.

Hughes TR, Roberts CJ, Dai H, Jones AR, Meyer MR, Slade D, Burchard J, Dow S, Ward TR, Kidd MJ, et al. 2000. Widespread aneuploidy revealed by DNA microarray expression profiling. *Nat Genet* **25**: 333–337.

Lehner KR, Stone MM, Farber RA, Petes TD. 2007. Ninety-six haploid yeast strains with individual disruptions of open reading frames between YOR097C and YOR192C, constructed for the *Saccharomyces* genome deletion project, have an additional mutation in the mismatch repair gene MSH3. *Genetics* **177**: 1951–1953.

Mulleder M, Capuano F, Pir P, Christen S, Sauer U, Oliver SG, Ralser M. 2012. A prototrophic deletion mutant collection for yeast metabolomics and systems biology. *Nat Biotechnol* **30**: 1176–1178.

Proctor M, Urbanus ML, Fung EL, Jaramillo DF, Davis RW, Nislow C, Giaever G. 2011. The automated cell: Compound and environment screening system (ACCESS) for chemogenomic screening. *Methods Mol Biol* **759**: 239–269.

Pronk JT. 2002. Auxotrophic yeast strains in fundamental and applied research. *Appl Environ Microbiol* **68**: 2095–2100.

Smith AM, Heisler LE, Mellor J, Kaper F, Thompson MJ, Chee M, Roth FP, Giaever G, Nislow C. 2009. Quantitative phenotyping via deep barcode sequencing. *Genome Res* **19**: 1836–1842.

Suresh S, Schlecht U, Xu W, Miranda M, Davis RW, Nislow C, Giaever G, St Onge RP. 2015. Identification of chemical–genetic interactions via parallel analysis of barcoded yeast strains. *Cold Spring Harb Protoc* doi: 10.1101/pdb.prot088054.

Wagih O, Usaj M, Baryshnikova A, VanderSluis B, Kuzmin E, Costanzo M, Myers CL, Andrews BJ, Boone CM, Parts L. 2013. SGAtools: One-stop analysis and visualization of array-based genetic interaction screens. *Nucleic Acids Res* **41**: W591–W596.

CHAPTER 11

Deep Mutational Scanning: A Highly Parallel Method to Measure the Effects of Mutation on Protein Function

Lea M. Starita[1,4] and Stanley Fields[1,2,3,4]

[1]Department of Genome Sciences, University of Washington, Seattle, Washington 98195; [2]Department of Medicine, University of Washington, Seattle, Washington 98195; [3]Howard Hughes Medical Institute, Seattle, Washington 98195

Deep mutational scanning is a method that makes use of next-generation sequencing technology to measure in a single experiment the activity of 10^5 or more unique variants of a protein. Because of this depth of mutational coverage, this strategy provides data that can be analyzed to reveal many protein properties. Deep mutational scanning approaches are particularly amenable to being performed in *Saccharomyces cerevisiae*, given the extensive toolkit of reagents and technologies available for this organism.

INTRODUCTION

Mutational analyses have long been critical for determining the role of individual amino acids in proteins. Single amino acid changes can affect folding, thermodynamic stability, enzymatic activity, interactions, posttranslational modifications, and other properties. Previous efforts to determine the consequences of these changes typically required many steps, including mutagenesis of the gene encoding the protein, protein expression and purification, and cellular or in vitro assays. For example, an alanine scan is a method devised to assess the changes of each amino acid in a given protein to alanine (Cunningham and Wells 1989). Applying this approach to a protein of 100 amino acids would generate 1900 missense variants (100 amino acids × 19 changes). The overall analysis would require months to carry out the mutagenesis step at each of the 100 codons and the subsequent individual assay of each variant. Clearly, determining the consequence of all possible variants is not tenable here. With the advent and combination of developing technologies, however, >10^5 variants of a protein now can be scored for function in parallel in a single experiment (Fowler et al. 2010; Hietpas et al. 2011). Because of the depth of mutational coverage and the analyses they enable, these experiments yield a wealth of knowledge, providing insight into the in vivo and in vitro properties of the proteins under study. These high-throughput approaches are particularly powerful in *Saccharomyces cerevisiae*, given the multitude of genetic selections available for this organism.

DEEP MUTATIONAL SCANNING

Deep mutational scanning combines a genotype–phenotype system with high-throughput DNA sequencing (Araya and Fowler 2011). Variants of a protein expressed in such a coupled system can be

[4]Correspondence: fields@uw.edu; lstarita@uw.edu

Copyright © Cold Spring Harbor Laboratory Press; all rights reserved
Cite this introduction as *Cold Spring Harb Protoc*; doi:10.1101/pdb.top077503

subjected to a selection dependent on the protein's function, such that variants that perform well increase in abundance and those that perform poorly decrease. These changes in frequency can be assessed by sequencing the DNA that encodes the variants, both before and after the selection step (simplified cartoon, Fig. 1A). A score, called the enrichment ratio, provides a measurement of each variant's function. The ratio for each variant is calculated by dividing the frequency of a variant in the selected population by its frequency in the starting or input population. An enrichment ratio of 1, representing no change in a variant's abundance during the selection, indicates that a mutation is neutral; an enrichment ratio of >1, representing an increase in abundance, indicates that a mutation is beneficial; and an enrichment ratio of <1, representing a decrease in abundance, indicates that a mutation is deleterious. These scores can be visualized as a sequence–function map (Fig. 1B).

Assays that couple genotype to phenotype and are therefore amenable to this approach are diverse. For example, a protein essential for yeast growth under a specified condition can be expressed from a plasmid-encoded gene (Hietpas et al. 2011; Kim et al. 2013; Melamed et al. 2013; Roscoe et al. 2013). A library of mutated versions of this gene is generated and transformed into yeast, under a selection condition that demands the transformants express the activity of the protein. Many other assays that allow the separation of functional from nonfunctional variants can be experimentally incorporated. This partitioning can be achieved by the use, for example, of fluorescent reporters followed by fluorescence-activated cell sorting or of yeast protein display and selection of functional variants by their binding to an immobilized or fluorescent ligand (Whitehead et al. 2012).

Human proteins expressed in yeast also can be subjected to deep mutational scanning assays to reveal the functional consequences of amino acid substitutions that arise from human genetic variation. The human protein cystathionine-β-synthase (CBS), deleterious mutants of which cause homocystinuria, was used in one such study. Because human CBS and the *S. cerevisiae* cystathionine-β-synthase (Cys4) perform the same function—synthesis of cysteine and glutathione—expression of human CBS in yeast can complement the deletion of yeast *CYS4* (Kruger and Cox 1994). In a cross-species complementation experiment, 84 alleles of human *CBS* were tested for their ability to support the growth of yeast on media lacking glutathione (Mayfield et al. 2012). Extending this approach to a deep mutational scan of CBS would entail scoring every possible missense mutation for function, resulting in a prediction for function of mutant alleles before they are found in the human population. Even human proteins that do not have orthologous proteins in yeast may be amenable to many yeast-based technologies, including one-hybrid assays to test for transcriptional activity, two-hybrid assays to detect protein–protein interaction, and three-hybrid assays to analyze protein–RNA interaction.

Carrying out a mutational scan under differing conditions can reveal various features of a protein's interaction network. The interactions between a protein of interest and other genes or pathways may

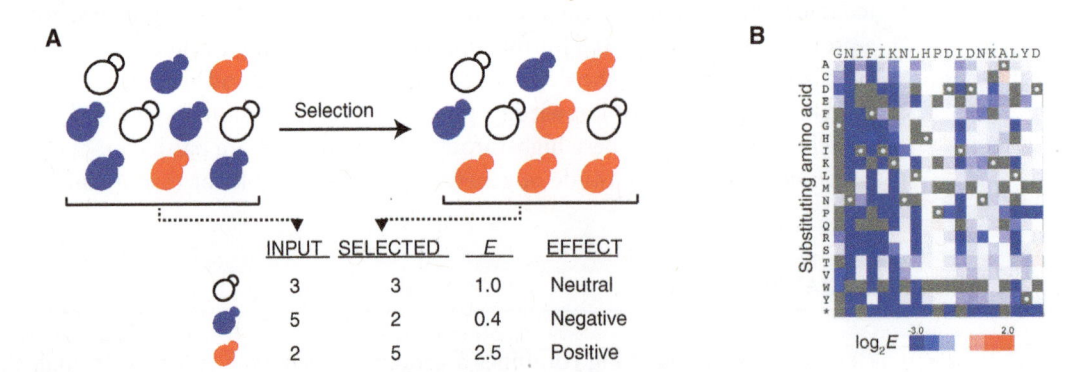

FIGURE 1. Deep mutational scanning. (A) Schematic of a deep mutational scanning experiment. Change in the abundance of variants is tracked by next-generation sequencing of DNA variants before and after selection for the function of the protein. The enrichment ratio (E) is a measure of the performance of each variant. (B) A sequence–function map of log$_2$ enrichment ratios (E) for variants with a single amino acid change in a portion of the Pab1 protein. Blue, white, and red boxes represent variants that were depleted, neutral, or enriched, respectively, during the selection process; gray represents that no data passed quality filters; boxed with white dots represent the wild-type residue.

be identified by performing selections in different genetic backgrounds. For example, a deep mutational scan of a protein performed both in wild-type yeast and in a strain deleted for a specific chaperone protein may reveal mutations that affect the client (partner) status of the chaperone. An increase in temperature or the addition of a denaturant during selection can exacerbate protein misfolding, and thus be used to identify mutations that affect folding (Traxlmayr et al. 2012). Chemical interactions can also be probed by the addition of small molecules to the selection. In this case, a protein's resistance profile to a specific drug may be determined by performing a deep mutational scan in the presence and absence of the drug.

Protein engineering is another field in which deep mutational scanning has provided insights. For example, in an experiment to analyze binding affinity, a protein designed to bind to the influenza virus hemagglutinin was scanned using a yeast display selection. The protein variants were analyzed to determine the positions of amino acid changes that resulted in increased binding to the viral protein (Whitehead et al. 2012). The results of this experiment were used to identify a new protein having changes at five positions with a concomitant 25-fold increase in binding affinity.

INFORMATION FOUND IN SEQUENCE–FUNCTION MAPS

Aspects of the structure and function of a protein also can be gleaned from the sequence–function maps generated by deep mutational scanning. In the most basic analysis, regions of the protein that are critical for function will be revealed as the most intolerant to substitution. For example, the catalytically important positions in an enzyme typically are intolerant to many changes. In contrast, rare substitutions that enhance catalytic activity may be identified in a deep mutational scan. These variants can be used to probe the biochemical mechanisms of catalysis (Starita et al. 2013) and may be industrially relevant. There are other regions of a given protein—outside the catalytic core—that will tolerate only a subset of amino acid substitutions. For example, a protein may tolerate the substitution of amino acids that make up its hydrophobic core for other hydrophobic amino acids, whereas changes in these positions to charged amino acids may cause the protein to become unstable (Fowler et al. 2010; Melamed et al. 2013; Roscoe et al. 2013).

Enrichment ratios of variants containing more than one amino acid substitution can reveal mutations that show epistasis in the presence of other mutations. For example, an amino acid substitution that thermodynamically stabilizes a protein can be identified as a mutation that rescues another deleterious mutation when the two mutations are paired in a double mutant (Araya et al. 2012).

Yeast should continue to be a proving ground for new uses of deep mutational scanning approaches. The combination of the many genetic strategies and resources that are available in yeast, combined with the ingenuity of yeast researchers, are likely to ensure that this organism remains the model in deciphering protein function at high throughput. In the associated protocols, we discuss the unique challenges of performing deep mutational scanning experiments in yeast (see Protocol 1: Deep Mutational Scanning: Library Construction, Functional Selection, and High-Throughput Sequencing [Starita and Fields 2015a] and Protocol 2: Deep Mutational Scanning: Calculating Enrichment Scores for Protein Variants from DNA Sequencing Output Files [Starita and Fields 2015b]).

REFERENCES

Araya CL, Fowler DM. 2011. Deep mutational scanning: Assessing protein function on a massive scale. *Trends Biotech* 29: 435–442.

Araya CL, Fowler DM, Chen W, Muniez I, Kelly JW, Fields S. 2012. A fundamental protein property, thermodynamic stability, revealed solely from large-scale measurements of protein function. *Proc Natl Acad Sci* 109: 16858–16863.

Cunningham BC, Wells JA. 1989. High-resolution epitope mapping of hGH-receptor interactions by alanine-scanning mutagenesis. *Science* 244: 1081–1085.

Fowler DM, Araya CL, Fleishman SJ, Kellogg EH, Stephany JJ, Baker D, Fields S. 2010. High-resolution mapping of protein sequence–function relationships. *Nat Meth* 7: 741–746.

Hietpas RT, Jensen JD, Bolon DNA. 2011. Experimental illumination of a fitness landscape. *Proc Natl Acad Sci* 108: 7896–7901.

Kim I, Miller CR, Young DL, Fields S. 2013. High-throughput analysis of in vivo protein stability. *Mol Cell Prot* 12: 3370–3378.

Kruger WD, Cox DR. 1994. A yeast system for expression of human cystathionine beta-synthase: Structural and functional conservation of the human and yeast genes. *Proc Natl Acad Sci* **91**: 6614–6618.

Mayfield JA, Davies MW, Dimster-Denk D, Pleskac N, McCarthy S, Boydston EA, Fink L, Lin AS, Meighan M, Rine J. 2012. Surrogate genetics and metabolic profiling for characterization of human disease alleles. *Genetics* **190**: 1309–1323.

Melamed D, Young DL, Gamble CE, Miller CR, Fields S. 2013. Deep mutational scanning of an RRM domain of the *Saccharomyces cerevisiae* Poly(A)-binding protein. *RNA* **19**: 1537–1551.

Roscoe BP, Thayer KM, Zeldovich KB, Fushman D, Bolon DNA. 2013. Analyses of the effects of all ubiquitin point mutants on yeast growth rate. *J Mol Biol* **425**: 1363–1377.

Starita LM, Fields S. 2015a. Deep mutational scanning: Library construction, functional selection, and high-throughput sequencing. *Cold Spring Harb Protoc* doi: 10.1101/pdb.prot085225.

Starita LM, Fields S. 2015b. Deep mutational scanning: Calculating enrichment scores for protein variants from DNA sequencing output files. *Cold Spring Harb Protoc* doi: 10.1101/pdb.prot085233.

Starita LM, Pruneda JN, Lo RS, Fowler DM, Kim HJ, Hiatt JB, Shendure J, Brzovic PS, Fields S, Klevit RE. 2013. Activity-enhancing mutations in an E3 ubiquitin ligase identified by high-throughput mutagenesis. *Proc Natl Acad Sci* **110**: E1263–E1272.

Traxlmayr MW, Hasenhindl C, Hackl M, Stadlmayr G, Rybka JD, Borth N, Grillari J, Rüker F, Obinger C. 2012. Construction of a stability landscape of the CH3 domain of human IgG1 by combining directed evolution with high throughput sequencing. *J Mol Biol* **423**: 397–412.

Whitehead TA, Chevalier A, Song Y, Dreyfus C, Fleishman SJ, De Mattos C, Myers CA, Kamisetty H, Blair P, Wilson IA, Baker D. 2012. Optimization of affinity, specificity and function of designed influenza inhibitors using deep sequencing. *Nat Biotech* **30**: 543–548.

Protocol 1

Deep Mutational Scanning: Library Construction, Functional Selection, and High-Throughput Sequencing

Lea M. Starita[1,4] and Stanley Fields[1,2,3,4]

[1]Department of Genome Sciences, University of Washington, Seattle, Washington 98195; [2]Department of Medicine, University of Washington, Seattle, Washington 98195; [3]Howard Hughes Medical Institute, Seattle, Washington 98195

Deep mutational scanning is a highly parallel method that uses high-throughput sequencing to track changes in >10^5 protein variants before and after selection to measure the effects of mutations on protein function. Here we outline the stages of a deep mutational scanning experiment, focusing on the construction of libraries of protein sequence variants and the preparation of Illumina sequencing libraries.

MATERIALS

It is essential that you consult the appropriate Material Safety Data Sheets and your institution's Environmental Health and Safety Office for proper handling of equipment and hazardous material used in this protocol.

Reagents

KAPA Illumina Library Quantification Kit (KAPA Biosystems)

Reagents for selection experiment (see Step 5), polymerase chain reaction (PCR) (see Step 7), and sequencing (see Step 9)

Yeast strain to be transformed, low-copy-number expression plasmid, and reagents for ligation and high-efficiency transformation (see Steps 3–4)

Zymoprep Yeast Plasmid Miniprep II (Zymo Research)

Equipment

Bioanalyzer (Agilent) or Qubit (Invitrogen/Life Technologies)

High-throughput sequencing apparatus (e.g., Illumina MiSeq or HiSeq)

We use the Illumina high-throughput DNA sequencing platform; thus, this protocol is based on this platform.

Polyacrylamide gel electrophoresis system

Thermocycler or qPCR machine (e.g., MiniOpticon from Bio-Rad)

---

[4]Correspondence: fields@uw.edu; lstarita@uw.edu

Copyright © Cold Spring Harbor Laboratory Press; all rights reserved
Cite this protocol as *Cold Spring Harb Protoc*; doi:10.1101/pdb.prot085225

Chapter 11

METHOD

A deep mutational scan is dependent on high-throughput DNA sequencing; therefore, an assay must be developed in which the number of sequence reads of the DNA encoding a protein variant depends on the activity of that variant. The parameters of the assay should be optimized to maximize the separation of a wild-type protein from nonfunctional variants, with partially functional variants falling in between the two classes.

1. Identify the region of the protein on which to perform the deep mutational scan. Then select the corresponding DNA sequence that encodes the protein region of interest.

    There are multiple parameters to consider when choosing the size of the mutated region. Most importantly, consider the sequencing platform to be used. The maximum number of bases that can be sequenced by using an Illumina MiSeq (the platform that we use) is currently 500. Because the error rate of Illumina technology is ~1%, each base must be read twice using paired-end sequencing to reduce the error rate to ~0.01%. This constraint therefore limits the read length to 250 bp or 83 amino acids. See Troubleshooting.

2. Order (e.g., from IDT or TriLink) a collection of synthetic oligonucleotides for the library that have a specific error rate, in which "incorrect" nucleotides are "doped" (randomly inserted) into the nucleotide mix during synthesis. Use the following formula to determine the frequency of random insertions in the library,

$$([\text{error rate} \div \text{number of nucleotides}] \times 100) \div 3,$$

    where 3 is the number of possible incorrect bases.

    For example, for a region of 250 nt (83 amino acids) with an error rate of 2%, request that each incorrect base be doped in at 0.27%. Each correct base should be present at 99.19% [99.99 = 100−(0.27 × 3)].
    See Discussion.

3. Clone the variant library into a low-copy-number yeast expression plasmid.

    We find it advantageous to obtain as many variants as possible (10^5–10^6) during this cloning step. Using a low-copy-number plasmid ensures that each cell will have one to two copies of the gene and that the copy number does not vary widely among cells, as is the case for a high-copy-number vector. Gene variants can also be integrated into the yeast genome.

4. Transform the selected yeast strain with the variant library using a high-efficiency plasmid transformation protocol.

    Even though high-throughput sequencing is massively parallel, it is desirable to limit the diversity of the library because increasing the read counts per variant produces higher-quality data and better reproducibility between replicates. Therefore, at this step, it is important to consider the total number of reads expected for the input variant library and to limit the diversity of the library to 1% of that number. For example, if 10 million reads are required, the variant library should consist of a maximum of 1×10^5 unique clones, allowing for 100 reads per variant. If too many transformants are produced, the population can be put through a bottleneck by using only a subset of the total transformants in the following steps.

5. Carry out the appropriate selection experiment. Separate functional from nonfunctional variants as determined by preliminary experiments. Include biological replicates in the experimental design. Collect samples of 16 OD units (1 OD unit is the number of cells in 1 mL of culture at $OD_{600} = 1$) at time = 0 and at one or more additional time points. Split each 16 OD-unit sample into four aliquots of 4 OD units to use as technical replicates or as backups in the case of sample loss in the next steps.

    Yeast populations separated by flow sorting into bins based on fluorescent signal or another property can be plated to increase cell number (Whitehead et al. 2012).

6. Purify plasmid DNA from the 4-OD-unit aliquots using the Zymoprep Yeast Plasmid Miniprep II kit according to the manufacturer's instructions. Include a freeze/thaw cycle after the zymolyase step and elute the DNA in 10 μL of EB (included with the miniprep kit).

7. Create the Illumina sequencing libraries by amplifying the variable regions of the cloned DNA by PCR. Use primers that add cluster-generating sequences; if the libraries will be multiplexed, also include a different 4–9-nt index sequence for each sample. To minimize PCR bias, maximize the

amount of template DNA in the reaction. (Typically half of the purified plasmid DNA from Step 6 is used per 50 µL PCR.) Use a high-fidelity polymerase and optimize conditions so that a single robust band can be seen using <20 cycles. Include SYBR Green in the reaction to monitor the progress of the reaction by qPCR. Stop the reaction during exponential amplification and remove samples before they reach saturation.

8. Measure the concentration of purified double-stranded DNA product by using a Qubit, Bioanalyzer, or the qPCR-based KAPA Illumina Library Quantification Kit (we find this kit to be the most accurate).

    Illumina sequencers are sensitive to DNA concentration. The DNA concentration must be >2 nM.

9. Multiplex and sequence the Illumina libraries according to the manufacturer's instructions.

10. Analyze the resulting sequence data.

    The data can be used to generate a sequence–function map of the protein under study. This involves the calculation of enrichment scores for the variants (see Protocol 2: Deep Mutational Scanning: Calculating Enrichment Scores for Protein Variants from DNA Sequencing Output Files [Starita and Fields 2015]).
    See Troubleshooting.

TROUBLESHOOTING

Problem (Step 1): Short read sequencing technology limits the size of the variable region to 250 nt.
Solution: To increase the size of the variable region, build separate libraries and perform experiments in parallel (Melamed et al. 2013). Alternatively, use barcoding and subassembly strategies to exceed the sequence length of sequencing technologies. A barcode is a short (10–20-bp) degenerate sequence cloned near the mutated gene, which acts as a molecular tag to enable the assembly of the full-length variable region (Hiatt et al. 2010; Patwardhan et al. 2012). The advantages of barcoding and assembly include reduced sequencing costs, as each variable region is sequenced and assigned to a barcode only once. Subsequent sequencing after experimental selection is limited to the barcode only. Reduced sequencing costs and using shorter sequencing runs allow multiple replicates to be included and the use of different experimental conditions. Moreover, counting statistics can be applied to the data because each variant is represented by multiple barcodes.

Problem (Step 10): The sequencing reads are of low quality or otherwise unclear.
Solution: Illumina libraries are considered to be biased if most of the reads have the same sequence, as is the case in a deep mutational scanning library. Library bias causes problems for Illumina sequencers in cluster calling and focus, and results in poor performance. The manufacturer recommends, therefore, a spike-in of 5% random PhiX genomic DNA (from Illumina) to remedy these problems. We have found, however, that a spike-in of at least 20% PhiX DNA is necessary. In these cases, standard Illumina sequencing primers are mixed with the custom primers for sequencing the variable region, and care must be taken to avoid the formation of primer dimers. Sequencing primers longer than 60 nt should be purified by polyacrylamide gel electrophoresis.

DISCUSSION

In a deep mutational scan, different strategies may be used to create a library of variants. There is a trade-off between minimizing the amount of wild-type DNA in the library and including so many mutations per DNA molecule that it is difficult to parse the functional consequence of each mutation. We have found that a variant library consisting of partially randomized (or "doped") synthetic oligonucleotides with a 2%–4% error rate provides good coverage of most possible single mutations and a substantial fraction of double mutations, with a tolerable proportion of 10%–20% of wild-type sequences in the library (Fowler et al. 2010).

The use of doped synthetic oligonucleotides has several limitations—for example, amino acid substitutions that require 3-nt changes are rare, and, even though 10%–20% of the variants may be wild-type, another 10%–20% of the variants may contain multiple changes (four or more) that are difficult to track and waste valuable sequencing capacity. Other problems are concerned with length limitations. The maximum length of doped oligonucleotides, at the time of this writing, is typically 200 bases. If a longer length is required, two or more oligonucleotides can be used to encode the variable region. These can then be ligated using a bridge oligonucleotide(s), and the resulting product can be "stitched" (incorporated) into a larger sequence by annealing and polymerase extension or by Gibson assembly (Gibson et al. 2009; Melamed et al. 2013; Starita et al. 2013). Another consideration is that oligonucleotides that are close to the 200-base limit often have deletions, even after purification by gel electrophoresis.

Alternative methods for creating libraries that have one codon change per variant include site-saturation mutagenesis by methods such as overlap extension PCR (McLaughlin et al. 2012; Whitehead et al. 2012) or PFunkel (Firnberg and Ostermeier 2012). These methods are more costly and more laborious, but the libraries contain only single-codon changes, which allows shotgun sequencing to be used and increases the likelihood of sampling every possible single amino acid substitution.

REFERENCES

Firnberg E, Ostermeier M. 2012. PFunkel: Efficient, expansive, user-defined mutagenesis. *PLoS ONE* 7: e52031.

Fowler DM, Araya CL, Fleishman SJ, Kellogg EH, Stephany D, Fields S. 2010. High-resolution mapping of protein sequence–function relationships. *Nat Meth* 7: 741–746.

Gibson DG, Young L, Chuang R-Y, Venter JC, Hutchison HO. 2009. Enzymatic assembly of DNA molecules up to several hundred kilobases. *Nat Meth* 6: 343–345.

Hiatt JB, Patwardhan RP, Turner EH, Lee C, Shendure J. 2010. Parallel, tag-directed assembly of locally derived short sequence reads. *Nat Meth* 7: 119–122.

McLaughlin RN, Poelwijk FJ, Raman A, Gosal WS, Ranganathan R. 2012. The spatial architecture of protein function and adaptation. *Nature* 491: 138–142.

Melamed D, Young DL, Gamble CE, Miller CR, Fields S. 2013. Deep mutational scanning of an RRM domain of the *Saccharomyces cerevisiae* Poly(A)-binding protein. *RNA* 19: 1537–1551.

Patwardhan RP, Hiatt JB, Witten DM, Kim MJ, Smith D, Lee JM, Lee GM, Ahituv N, Pennacchio LA, Shendure J. 2012. Massively parallel functional dissection of mammalian enhancers in vivo. *Nat Biotech* 30: 265–270.

Starita LM, Fields S. 2015. Deep mutational scanning: Calculating enrichment scores for protein variants from DNA sequencing output files. *Cold Spring Harb Protoc* doi: 10.1101/pdb.prot085233.

Starita LM, Pruneda JN, Lo RS, Fowler HJ, Hiatt JB, Shendure J, Brzovic PS, Fields S, Klevit RE. 2013. Activity-enhancing mutations in an E3 ubiquitin ligase identified by high-throughput mutagenesis. *Proc Natl Acad Sci* 110: E1263–E1272.

Whitehead TA, Chevalier A, Song Y, Dreyfus C, Fleishman SJ, De Mattos C, Myers CA, Kamisetty H, Blair P, Wilson IA, Baker D. 2012. Optimization of affinity, specificity and function of designed influenza inhibitors using deep sequencing. *Nat Biotech* 30: 543–548.

Protocol 2

Deep Mutational Scanning: Calculating Enrichment Scores for Protein Variants from DNA Sequencing Output Files

Lea M. Starita[1,4] and Stanley Fields[1,2,3,4]

[1]*Department of Genome Sciences, University of Washington, Seattle, Washington 98195;* [2]*Department of Medicine, University of Washington, Seattle, Washington 98195;* [3]*Howard Hughes Medical Institute, Seattle, Washington 98195*

During a deep mutational scanning experiment, a collection of variants of a given protein is subjected to high-throughput sequencing before and after selection. The variants that perform well during selection will increase in abundance, whereas those that perform poorly will decrease. Generating a sequence–function map of a protein from a deep mutational scan requires the calculation and comparison of the enrichment scores for each protein variant, based on the results of high-throughput DNA sequencing output files. Here we describe the use of the software program Enrich, which was written specifically for the data analysis phase of a deep mutational scanning experiment.

MATERIALS

It is essential that you consult the appropriate Material Safety Data Sheets and your institution's Environmental Health and Safety Office for proper handling of equipment and hazardous materials used in this protocol.

Equipment

DNA sequencing files (FASTQ-formatted files [Cock et al. 2010]) from a deep mutational scanning experiment (see Protocol 1: Deep Mutational Scanning: Library Construction, Functional Selection, and High-Throughput Sequencing [Starita and Fields 2015a])
Enrich (http://depts.washington.edu/sfields/software/enrich/; Fowler et al. 2011)
Java Tree View (http://jtreeview.sourceforge.net/)
matplotlib (http://matplotlib.org/)
NumPy (http://www.numpy.org/)
PyMOL (http://www.pymol.org/)
Python (http://www.python.org/)
R (http://www.r-project.org/)
SciPy (http://www.scipy.org/)
Sequencing Analysis Viewer (SAV) (Illumina)
SolexaQA (http://solexaqa.sourceforge.net/)

---

[4]Correspondence: fields@uw.edu; lstarita@uw.edu

Copyright © Cold Spring Harbor Laboratory Press; all rights reserved
Cite this protocol as *Cold Spring Harb Protoc*; doi:10.1101/pdb.prot085233

Chapter 11

METHOD

The Enrich software will split the FASTQ-formatted DNA files by index, fuse each forward and reverse read into a single consensus read, align each consensus read to the wild-type sequence, identify mutations in the DNA and translated protein sequence, and tally the number of all unique variants. For variants present in both the input and in selected libraries, the program calculates an enrichment ratio.

1. Retrieve the FASTQ-formatted sequencing files from the Illumina sequencing run of the deep mutational scanning experiment (see Protocol 1: Deep Mutational Scanning: Library Construction, Functional Selection, and High-Throughput Sequencing [Starita and Fields 2015a]). Assess the quality of the sequencing run by using Illumina SAV or software such as SolexaQA (Cox et al. 2010).

   See Troubleshooting.

2. Use the Enrich software package to calculate an enrichment ratio for each variant found in the input and in the selected library.

   See Troubleshooting.

3. Visualize the data in plots generated automatically by Enrich.

   The plots are designed to give a broad overview of the data, including library diversity and the performance of variants at each position of the protein sequence. A detailed description of each plot is provided in the Enrich documentation.

4. Use the Enrich output files (Table 1) in the data/output directory for appropriate downstream analyses.

   - Use a statistical software package (e.g., R) to test for the correlation of enrichment ratios between replicates.

     See Troubleshooting.

   - If selected populations were sampled at multiple time points, calculate slopes from a line fit through the read frequencies at each point.

     The slopes can replace the enrichment ratio for a data set with reduced noise (Araya et al. 2012).

   - Use the "unlink_" matrices to make sequence–function maps using Java Tree View or similar software.

     For an example of a sequence-function map, see Figure 1B in Introduction: Deep Mutational Scanning: A Highly Parallel Method to Measure the Effects of Mutation on Protein Function (Starita and Fields 2015b).

   - Use PyMOL to map position-averaged enrichment ratios onto the structure of the protein.

     Every protein has its own role within the cell, requiring specialized downstream analyses of the mutational data. These analyses can be the most time-consuming part of this approach, and they can demand particular creativity to tease out all the insight that the mutational data can provide.

TABLE 1. Enrich output files

File types	Contents
"counts_" files	Tally and frequency information for each sequence from each sample
"ratios_" files	Enrichment ratio for each variant as well as the enrichment ratio normalized to the enrichment ratio for the wild-type sequence
"unlink_" files	Enrichment scores[a] of each amino acid change at each position in matrix format

[a]A two-sided Poisson exact test is used to calculate a P-value for the significance of the enrichment. The P-value is corrected for multiple testing; the resulting q-value is used to determine which variants are significantly enriched or depleted during selection.

TROUBLESHOOTING

Problem (Step 1): The software reveals low quality sequencing reads or run/lane failure.
Solution: Review the run with Illumina technical support. Problems with read bias or over/underloading can be remedied before the next sequencing attempt. Problems with sequencing primer design can also be remedied by testing the primers in Sanger sequencing reactions to make sure that they bind to the amplicons correctly. Removal of ∼10% of the reads in the next steps due to poor quality is not uncommon.

Problem (Step 2): The Enrich software fails to run.
Solution: Contact your local information technology technician for help installing Enrich and its dependencies (NumPy, SciPy, and matplotlib).

Problem (Step 2): Enrich appears to run without errors, but the output files are empty.
Solution: Run the "Example" data set supplied with Enrich. If this data set works as designed, the parameters for Enrich have not been set correctly in the configuration file for your own data. Use only a small subset of your data to rapidly locate where the analysis pipeline is failing and correct the parameters.

Problem (Step 2): Enrich runs slowly.
Solution: Enrich was designed to run on a personal computer and over multiple nodes on a clustered computer network. Ask your local information technology technician about interfacing Enrich with your department's computer cluster.

Problem (Step 4): Low reproducibility is seen between biological replicates.
Solution: A source of noise in deep mutational scanning data arises from variants with low input read counts. Filter for high-quality data by removing variants with low read counts from the input sample and recalculate the enrichment ratios. The lower limit can be set arbitrarily or synonymous mutations can be used to determine the read count cutoff where the variation between synonymous codons is close to zero. Another approach is to increase the average read quality parameter in the Enrich configuration file and reanalyze the data.

REFERENCES

Araya CL, Fowler DM, Chen W, Muniez I, Kelly JW, Fields S. 2012. A fundamental protein property, thermodynamic stability, revealed solely from large-scale measurements of protein function. *Proc Natl Acad Sci* 109: 16858–16863.

Cock PJA, Fields CJ, Goto N, Heuer ML, Rice PM. 2010. The Sanger FASTQ file format for sequences with quality scores, and the Solexa/Illumina FASTQ variants. *Nucl Acids Res* 38: 1767–1771.

Cox MP, Peterson DA, Biggs PJ. 2010. SolexaQA: At-a-glance quality assessment of Illumina second-generation sequencing data. *BMC Bioinform* 11: 485.

Fowler DM, Araya CL, Gerard W, Fields S. 2011. Enrich: Software for analysis of protein function by enrichment and depletion of variants. *Bioinformatics* 27: 3430–3431.

Starita LM, Fields S. 2015a. Deep mutational scanning: Library construction, functional selection, and high-throughput sequencing. *Cold Spring Harb Protoc* doi: 10.1101/pdb.prot085225.

Starita LM, Fields S. 2015b. Deep mutational scanning: A highly parallel method to measure the effects of mutation on protein function. *Cold Spring Harb Protoc* doi: 10.1101/pdb.top077503.

CHAPTER 12

Examination and Disruption of the Yeast Cell Wall

Hiroki Okada,[1] Keiko Kono,[2] Aaron M. Neiman,[3] and Yoshikazu Ohya[1,4]

[1]Department of Integrated Biosciences, Graduate School of Frontier Sciences, The University of Tokyo, 5-1-5 Kashiwanoha, Kashiwa, Chiba Prefecture 277-8562, Japan; [2]Department of Cell Biology, Graduate School of Medical Sciences, Nagoya City University, 1, Kawasumi, Mizuho-cho, Mizuho-ku, Nagoya, Aichi Prefecture 467-8601, Japan; [3]Department of Biochemistry and Cell Biology, Stony Brook University, Stony Brook, New York 11794-5215

The cell wall of *Saccharomyces cerevisiae* is a complicated extracellular organelle. Although the barrier may seem like a technical nuisance for researchers studying intracellular biomolecules or conditions, the rigid wall is an essential aspect of the yeast cell. Without it, yeast cells are unable to proliferate or carry out their life cycle. The chemical composition of the cell wall and the biosynthetic pathways and signal transduction mechanisms involved in cell wall remodeling have been studied extensively, but many unanswered questions remain. This introduction describes techniques for investigating abnormalities in the cell and spore walls and performing cell wall disruption.

THE YEAST CELL WALL

Cell Wall Function

The yeast cell wall is a rigid structure that serves as a barrier to the extracellular environment. Cell wall functions include providing osmotic integrity to the cell, maintaining mechanical strength, defining cell shape, and providing a scaffold for presentation of adhesive glycoproteins to other yeast cells (Klis et al. 2006). A recent study indicated that the cell wall is important for phenotypic robustness during vegetative growth (Okada et al. 2014). Moreover, the spore wall, the specialized wall which covers the spore after meiosis, is responsible for spore resistance to severe environmental conditions (Neiman 2011). A strategy for studying yeast spore survival under natural environmental stress is described in Protocol 2: Assay for Spore Wall Integrity Using a Yeast Predator (Okada et al. 2015).

Simple experiments can provide information about the important functions of the cell wall (Fig. 1). For example, after shaking intact cells with glass beads, the empty shell maintains an oval shape even after cell breakage, indicating that cell shape is dictated by the cell wall. Spheroplasts, in which the cell wall has been almost completely removed after treatment with lytic enzymes such as Zymolyase, burst upon exposure to hypo-osmotic conditions.

Components of the Cell Wall

The cell wall is composed of filamentous polysaccharides and nonfilamentous glycoproteins (Orlean 2012). The most abundant filamentous component is 1,3-β-glucan, which accounts for 30%–60% of the cell wall mass (Aguilar-Uscanga and François 2003). There are several minor filamentous

---

[4]Correspondence: ohya@k.u-tokyo.ac.jp

Copyright © Cold Spring Harbor Laboratory Press; all rights reserved
Cite this introduction as *Cold Spring Harb Protoc*; doi:10.1101/pdb.top078659

Chapter 12

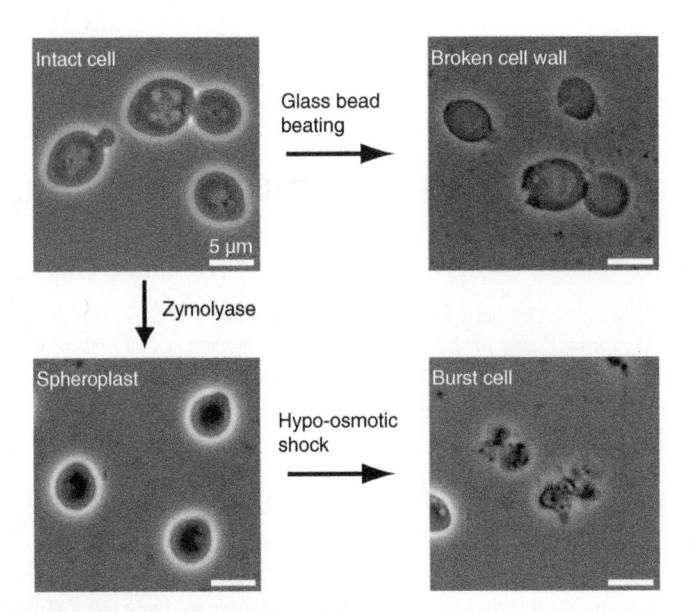

FIGURE 1. Phase contrast images of an intact cell, a broken cell wall, a spheroplast, and a burst cell. Intact cells were broken with glass beads. Spheroplasts were formed by treatment with the lytic enzyme Zymolyase. Spheroplasts burst easily under hypo-osmotic conditions.

components, including the relatively short 1,6-β-glucan and chitin. Chitin is localized in specific areas in the cell wall; it is normally deposited in a ring at the neck between a mother cell and its emerging bud, in the primary septum during division, and, to a lesser extent, in the lateral walls of newly separated daughter cells (Cabib et al. 2001). Other major components in the cell wall include the nonfilamentous mannoproteins, which are heavily glycosylated mannose-containing glycoproteins. 1,3-β-Glucan, chitin, and mannoproteins can be visualized using Protocol 1: Fluorescent Labeling of Yeast Cell Wall Components (Okada and Ohya 2015).

Studies using electron microscopy and specific antibodies have showed that 1,3-β-glucan is located in the interior of the cell wall during vegetative growth, whereas the exterior portion contains mannoproteins (Fig. 2A). Accordingly, the exterior wall surface presents to the environment mannoproteins with specialized activity, such as agglutinins and flocculins. In the spore wall, however, dityrosine and chitosan form an external layer outside of the mannoprotein and glucan layers (Fig. 2B). Dityrosine fills the outermost pores of the mesh made by internal filamentous components (Suda et al. 2009) and is likely responsible for the barrier against environmental stress. The order of wall layers is inverted during germination: The first budded cells from the spore harbor the mannoprotein layer in the exterior of the new vegetative cell wall and in the interior of the old spore wall (Fig. 2C).

CONSTRUCTION OF THE CELL WALL

Of the main cell wall components, 1,3-β-glucan can be synthesized in vitro from UDP-glucose to make a polymer with a chain length of 60–80 glucose molecules (Shematek et al. 1980). The multisubunit enzyme 1,3-β-glucan synthase is localized at sites of polarized growth and cell wall modeling on the plasma membrane (Utsugi et al. 2002). It is composed of a catalytic subunit (Fks1 or Fks2) and a regulatory subunit (Rho1), and its activity is stimulated by guanosine triphosphate (Douglas 1994; Drgonová et al. 1996; Qadota et al. 1996).

Mannoproteins are synthesized in the lumen of the endoplasmic reticulum (ER) and modified with N-linked glycans through the secretory pathway (Helenius and Aebi 2001; Larkin and Imperiali 2011; Orlean 2012). N-glycosylation requires preassembly of a branched 14-sugar oligosaccharide on the carrier dolichol-pyrophosphate in the ER membrane followed by transfer of the oligosaccharide to selected asparagine residues of the cell wall proteins in the ER lumen. N-linked glycans on proteins are

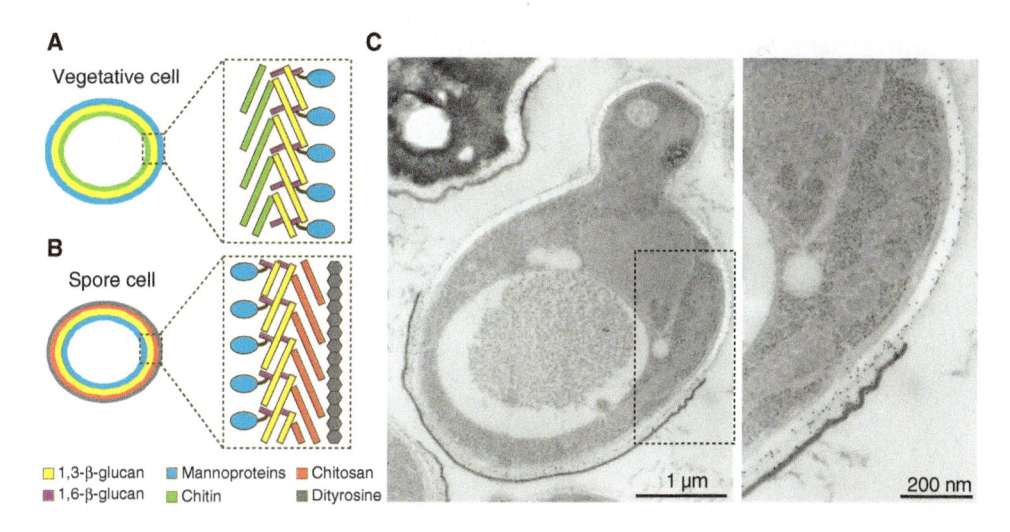

FIGURE 2. Organization of the cell wall and spore wall. (*A*) Model for the vegetative cell wall representing the interior and exterior layers. (*B*) Model for the layered organization of the spore wall. (*C*) Electron microscope picture of a germinating cell. Immunoelectron microscopy reveals the localization of 1,3-β-glucan (represented by dots).

further extended in the Golgi with a Man10–14 core-type structure or with mannan outer chains containing up to 150–200 mannose molecules (Jigami 2008). Many cell wall proteins are also modified with the addition of up to five linear mannose oligosaccharides to Ser or Thr residues (O-glycosylation) in the ER and in the Golgi (Schuldiner et al. 2013).

Chitin is synthesized by chitin synthase I, II, and III, which require the catalytic proteins Chs1, Chs2, and Chs3, respectively (Cabib 2004; Orlean 2012). All chitin synthases are activated on the plasma membrane. Chitin synthase II and III are involved in septation, and additional proteins are required to ensure the correct spatial and temporal localization of chitin synthesis.

Although many genes are involved in 1,6-β-glucan synthesis, the enzyme responsible for 1,6-β-glucan synthesis in vitro remains unclear (Orlean 2012). Experiments measuring the formation of 1,6-β-glucan in cells permeabilized by osmotic shock and incubated with radiolabeled UDP-glucose suggested that 1,6-β-glucan is synthesized intracellularly (Aimanianda et al. 2009); however, immune staining failed to detect 1,6-β-glucan intracellularly (Montijn et al. 1999).

Cell wall components are cross-linked to each other, building a highly organized dynamic network structure in the cell wall (Cabib and Arroyo 2013). The reducing ends of 1,6-β-glucan chains can be attached to 1,3-β-glucan (Kollár et al. 1997). The reducing end of a chitin chain is connected to either the nonreducing end of a 1,3-β-glucan or the nonreducing end of a 1,3-β-glucose that branches off 1,6-β-glucan (Kollár et al. 1995; Cabib et al. 2007; Cabib 2009). Many mannoproteins are glycosylphosphatidylinositol (GPI)-anchored proteins connected to 1,6-β-glucan chains through the sugar moiety of their GPI anchors (Kollár et al. 1997; Kapteyn et al. 1995). Additionally, other cell wall proteins undergo crosslinking reactions that incorporate them into the wall. Although the local and overall architecture of the cell wall remain undefined, covalent linkages between the various components of the cell wall give rise to its elastic and plastic properties, providing a strong, continuous fabric.

CELL WALL DISRUPTORS

That a long list of genes affects the structure and function of the cell wall is not surprising; ~180 genes encode proteins directly involved in the biosynthesis or remodeling of the cell wall (Orlean 2012). A survey of deletion strains revealed that up to a quarter of the genes in *S. cerevisiae* affect cell wall functions (de Groot et al. 2001). Several researchers have studied easily scored phenotypes to identify mutants with abnormal cell walls. Cell-wall-defective mutants show altered sensitivity or resistance to a

TABLE 1. List of cell wall disruptors

Category	Disruptor	Description	References
Chemical disruptor	Caffeine	Cell wall stress agent	Martin 2000
	Calcofluor white	Disrupting cell wall assembly by binding to chitin	Herth 1980
	Congo red	Disrupting cell wall assembly by binding to chitin	Herth 1980
	Demethylallosamidin	Inhibitor of chitinase	Sakuda et al. 1990
	D75-4590	Inhibitor of 1,6-β-glucan synthesis	Kitamura et al. 2009
	Echinocandins (caspofungin, etc.)	Inhibitor of 1,3-β-glucan synthase	Douglas 2001
	HM-1 killer toxin	Inhibitor of 1,3-β-glucan synthesis	Yamamoto et al. 1986
	K1 killer toxin	Fungicidal pore-forming protein that targets 1,6-β-glucan as a primary receptor	Hutchins and Bussey 1983
	Nikkomycin Z	Inhibitor of chitin synthase	Cabib 1991
	Polyoxin D	Inhibitor of chitin synthase	Cabib 1991
	Rhodamine-3-acetic acid derivatives (OGT2468)	Inhibitor of O-glycosylation	Orchard et al.2004; Arroyo et al. 2011
	SDS	Detergent to destabilize the cell wall	Shimizu et al. 1994
	Triacetylchitotriose	Inhibitor of crosslinking between chitin and 1,3-β-glucan	Blanco et al. 2012
	Tunicamycin	Inhibitor of N-glycosylation	Barnes et al. 1984
	Zymolyase	Lytic enzyme of 1,3-β-glucan	Kitamura et al. 1974; Ovalle et al. 1999
	2-Deoxyglucose	Inhibitor of 1,3-β-glucan synthesis	Johnson 1968
Physical disruptor	Glass bead beating	Inducing mechanical crack of cell wall	Ranhand 1974
	Low osmotic pressure	Bursting the cell by inner turgor pressure	Stateva et al. 1991
	Laser damage	Inducing optical damage to the cell wall; see Protocol 3: Local and Acute Disruption of the Yeast Cell Surface	Kono et al. 2012; Kono et al. 2015
	Sonication	Disruption of the cell wall by shear forces and cavitation	Ruiz et al. 1999

variety of chemical disruptors (Table 1), including caffeine, echinocandins, nikkomycin Z, Zymolyase, SDS, K1 killer toxin, calcofluor white, and Congo red; the latter two compounds bind to filamentous components of the yeast cell wall (Ram and Klis 2006). In the class of physical disruptors (Table 1), the most commonly used treatment is glass bead beating. The newly developed technique of physical disruption using laser-equipped microscopy is described in Protocol 3: Local and Acute Disruption of the Yeast Cell Surface (Kono et al. 2015).

Remodeling of the cell wall occurs under the control of the cell wall integrity signal transduction pathway in response to environmental stress (Levin 2011). This pathway transmits wall stress signals from the cell surface, activates the Slt2 MAP kinase, and regulates the production and polarized delivery of various components to the site of cell wall remodeling. Cell wall stresses include exposure to elevated temperatures, mating pheromone, hypo-osmotic shock, cell wall stress agents, cell wall biogenesis mutations, actin cytoskeleton depolarization, ER stress, turgor pressure, and plasma membrane stretch (Levin 2011). A second MAP kinase pathway, the high osmolarity glycerol response pathway, is also involved in response to cell wall damage such as digestion of the 1,3-β-glucan fiber by treatment with Zymolyase (Bermejo et al. 2008).

As an essential fungal structure that is absent in mammalian cells, the cell wall is an obvious target for antifungal drugs (Munro 2013; Table 1). Echinocandins such as echinocandin B and micafungin target the catalytic subunit of 1,3-β-glucan synthase and inhibit its activity (Kurtz and Douglas 1997; Perlin 2007). Nikkomycin Z binds to chitin synthase and inhibits its activity (Cabib 1991). Additionally, several reagents specifically inhibit synthesis of the cell wall components. Tunicamycin inhibits GlcNAc phosphotransferase, which plays a role in the initial step of N-glycosylation of cell wall mannoproteins (Barnes et al. 1984). The rhodamine-3-acetic acid derivative OGT2468 decreases protein O-mannosyltransferase activity (Arroyo et al. 2011; Orchard et al. 2004). These reagents are useful for investigating the physiological functions of the cell wall and studying the effects of stress on the cell wall.

REFERENCES

Aguilar-Uscanga B, François JM. 2003. A study of the yeast cell wall composition and structure in response to growth conditions and mode of cultivation. *Lett Appl Microbiol* 37: 268–274.

Aimanianda V, Clavaud C, Simenel C, Fontaine T, Delepierre M, Latgé J-P. 2009. Cell wall β-(1,6)-glucan of *Saccharomyces cerevisiae*: Structural characterization and in situ synthesis. *J Biol Chem* 284: 13401–13412.

Arroyo J, Hutzler J, Bermejo C, Ragni E, García-Cantalejo J, Botías P, Piberger H, Schott A, Sanz AB, Strahl S. 2011. Functional and genomic analyses of blocked protein O-mannosylation in baker's yeast. *Mol Microbiol* 79: 1529–1546.

Barnes G, Hansen WJ, Holcomb CL, Rine J. 1984. Asparagine-linked glycosylation in *Saccharomyces cerevisiae*: Genetic analysis of an early step. *Mol Cell Biol* 4: 2381–2388.

Bermejo C, Rodríguez E, García R, Rodríguez-Peña JM, Rodríguez de la Concepción ML, Rivas C, Arias P, Nombela C, Posas F, Arroyo J. 2008. The sequential activation of the yeast HOG and SLT2 pathways is required for cell survival to cell wall stress. *Mol Biol Cell* 19: 1113–1124.

Blanco N, Reidy M, Arroyo J, Cabib E. 2012. Crosslinks in the cell wall of budding yeast control morphogenesis at the mother-bud neck. *J Cell Sci* 125: 5781–5789.

Cabib E. 1991. Differential inhibition of chitin synthetases 1 and 2 from *Saccharomyces cerevisiae* by polyoxin D and nikkomycins. *Antimicrob Agents Chemother* 35: 170–173.

Cabib E. 2004. The septation apparatus, a chitin-requiring machine in budding yeast. *Arch Biochem Biophys* 426: 201–207.

Cabib E. 2009. Two novel techniques for determination of polysaccharide cross-links show that Crh1p and Crh2p attach chitin to both β(1–6)- and β(1–3)glucan in the *Saccharomyces cerevisiae* cell wall. *Eukaryot Cell* 8: 1626–1636.

Cabib E, Arroyo J. 2013. How carbohydrates sculpt cells: Chemical control of morphogenesis in the yeast cell wall. *Nat Rev Microbiol* 11: 648–655.

Cabib E, Blanco N, Grau C, Rodríguez-Peña JM, Arroyo J. 2007. Crh1p and Crh2p are required for the cross-linking of chitin to β(1–6) glucan in the *Saccharomyces cerevisiae* cell wall. *Mol Microbiol* 63: 921–935.

Cabib E, Roh DH, Schmidt M, Crotti LB, Varma A. 2001. The yeast cell wall and septum as paradigms of cell growth and morphogenesis. *J Biol Chem* 276: 19679–19682.

Douglas CM. 2001. Fungal β(1,3)-D-glucan synthesis. *Med Mycol* 39: 55–66.

Douglas CM. 1994. The *Saccharomyces cerevisiae FKS1 (ETG1)* gene encodes an integral membrane protein which is a subunit of 1,3-β-D-glucan synthase. *Proc Natl Acad Sci* 91: 12907–12911.

Drgonová J, Drgon T, Tanaka K, Kollár R, Chen GC, Ford RA, Chan CS, Takai Y, Cabib E. 1996. Rho1p, a yeast protein at the interface between cell polarization and morphogenesis. *Science* 272: 277–279.

de Groot PW, Ruiz C, Vázquez de Aldana CR, Duenas E, Cid VJ, Del Rey F, Rodríquez-Peña JM, Pérez P, Andel A, Caubín J, et al. 2001. A genomic approach for the identification and classification of genes involved in cell wall formation and its regulation in *Saccharomyces cerevisiae*. *Comp Funct Genomics* 2: 124–142.

Helenius A, Aebi M. 2001. Intracellular functions of N-linked glycans. *Science* 291: 2364–2369.

Herth W. 1980. Calcofluor white and Congo red inhibit chitin microfibril assembly of Poterioochromonas: Evidence for a gap between polymerization and microfibril formation. *J Cell Biol* 87: 442–450.

Hutchins K, Bussey H. 1983. Cell wall receptor for yeast killer toxin: Involvement of (1 leads to 6)-β-D-glucan. *J Bacteriol* 154: 161–169.

Jigami Y. 2008. Yeast glycobiology and its application. *Biosci Biotechnol Biochem* 72: 637–648.

Johnson BF. 1968. Lysis of yeast cell walls induced by 2-deoxyglucose at their sites of glucan synthesis. *J Bacteriol* 95: 1169–1172.

Kapteyn JC, Montijn RC, Dijkgraaf GJ, Van den Ende H, Klis FM. 1995. Covalent association of beta-1,3-glucan with β-1,6-glucosylated mannoproteins in cell walls of *Candida albicans*. *J Bacteriol* 177: 3788–3792.

Kitamura A, Someya K, Hata M, Nakajima R, Takemura M. 2009. Discovery of a small-molecule inhibitor of β-1,6-glucan synthesis. *Antimicrob Agents Chemother* 53: 670–677.

Kitamura K, Kaneko T, Yamamoto Y. 1974. Lysis of viable yeast cells by enzymes of *Arthrobacter luteus* II. Purification and properties of an enzyme, zymolyase, which lyses viable yeast cells. *J Gen Appl Microbiol* 20: 323–344.

Klis FM, Boorsma A, De Groot PW. 2006. Cell wall construction in *Saccharomyces cerevisiae*. *Yeast* 23: 185–202.

Kollár R, Petráková E, Ashwell G, Robbins PW, Cabib E. 1995. Architecture of the yeast cell wall. The linkage between chitin and β(1→3)-glucan. *J Biol Chem* 270: 1170–1178.

Kollár R, Reinhold BB, Petráková E, Yeh HJ, Ashwell G, Drgonová J, Kapteyn JC, Klis FM, Cabib E. 1997. Architecture of the yeast cell wall. β(1→6)-glucan interconnects mannoprotein, β(1→3)-glucan, and chitin. *J Biol Chem* 272: 17762–17775.

Kono K, Okada H, Ohya Y. 2015. Local and acute disruption of the yeast cell surface. *Cold Spring Harb Protoc* doi: 10.1101/pdb.prot085266.

Kono K, Saeki Y, Yoshida S, Tanaka K, Pellman D. 2012. Proteasomal degradation resolves competition between cell polarization and cellular wound healing. *Cell* 150: 151–164.

Kurtz MB, Douglas CM. 1997. Lipopeptide inhibitors of fungal glucan synthase. *J Med Vet Mycol* 35: 79–86.

Larkin A, Imperiali B. 2011. The expanding horizons of asparagine-linked glycosylation. *Biochemistry* 50: 4411–4426.

Levin DE. 2011. Regulation of cell wall biogenesis in *Saccharomyces cerevisiae*: The cell wall integrity signaling pathway. *Genetics* 189: 1145–1175.

Martin H. 2000. Regulatory mechanisms for modulation of signaling through the cell integrity Slt2-mediated pathway in *Saccharomyces cerevisiae*. *J Biol Chem* 275: 1511–1519.

Montijn RC, Vink E, Müller WH, Verkleij AJ, Van Den Ende H, Henrissat B, Klis FM. 1999. Localization of synthesis of β1,6-glucan in *Saccharomyces cerevisiae*. *J Bacteriol* 181: 7414–7420.

Munro CA. 2013. Chitin and glucan, the yin and yang of the fungal cell wall, implications for antifungal drug discovery and therapy. *Adv Appl Microbiol* 83: 145–172.

Neiman AM. 2011. Sporulation in the budding yeast *Saccharomyces cerevisiae*. *Genetics* 189: 737–765.

Okada H, Ohya Y. 2015. Fluorescent labeling of yeast cell wall components. *Cold Spring Harb Protoc* doi:10.1101/pdb.prot085241.

Okada H, Neiman A, Ohya Y. 2015. Assay for spore wall integrity using a yeast predator. *Cold Spring Harb Protoc* doi:10.1101/pdb.prot085258.

Okada H, Ohnuki S, Roncero C, Konopka JB, Ohya Y. 2014. Distinct roles of cell wall biogenesis in yeast morphogenesis as revealed by multivariate analysis of high-dimensional morphometric data. *Mol Biol Cell* 25: 222–233.

Orchard MG, Neuss JC, Galley CM, Carr A, Porter DW, Smith P, Scopes DI, Haydon D, Vousden K, Stubberfield CR, et al. 2004. Rhodanine-3-acetic acid derivatives as inhibitors of fungal protein mannosyl transferase 1 (PMT1). *Bioorg Med Chem Lett* 14: 3975–3978.

Orlean P. 2012. Architecture and biosynthesis of the *Saccharomyces cerevisiae* cell wall. *Genetics* 192: 775–818.

Ovalle R, Spencer M, Thiwanont M, Lipke PN. 1999. The spheroplast lysis assay for yeast in microtiter plate format. *Appl Environ Microbiol* 65: 3325–3327.

Perlin DS. 2007. Resistance to echinocandin-class antifungal drugs. *Drug Resist Updat* 10: 121–130.

Qadota H, Python CP, Inoue SB, Arisawa M, Anraku Y, Zheng Y, Watanabe T, Levin DE, Ohya Y. 1996. Identification of yeast Rho1p GTPase as a regulatory subunit of 1,3-β-glucan synthase. *Science* 272: 279–281.

Ram AFJ, Klis FM. 2006. Identification of fungal cell wall mutants using susceptibility assays based on Calcofluor white and Congo red. *Nat Protoc* 1: 2253–2256.

Ranhand JM. 1974. Simple, inexpensive procedure for the disruption of bacteria. *Appl Microbiol* 28: 66–69.

Ruiz C, Cid VJ, Lussier M, Molina M, Nombela C. 1999. A large-scale sonication assay for cell wall mutant analysis in yeast. *Yeast* 15: 1001–1008.

Sakuda S, Nishimoto Y, Ohi M, Watanabe M, Takayama S, Isogai A, Yamada Y. 1990. Effects of demethylallosamidin, a potent yeast chitinase inhibitor, on the cell division of yeast. *Agric Biol Chem* **54**: 1333–1335.

Schuldiner M, Schwappach B, Loibl M, Strahl S. 2013. Protein O-mannosylation: What we have learned from baker's yeast. *Biochim Biophys Acta–Mol Cell Res* **1833**: 2438–2446.

Shematek E, Braatz J, Cabib E. 1980. Biosynthesis of the yeast cell wall. I. Preparation and properties of β-(1 leads to 3)glucan synthetase. *J Biol Chem* **255**: 888–894.

Shimizu J, Yoda K, Yamasaki M. 1994. The hypo-osmolarity-sensitive phenotype of the *Saccharomyces cerevisiae hpo2* mutant is due to a mutation in *PKC1*, which regulates expression of β-glucanase. *Mol Gen Genet* **242**: 641–648.

Stateva LI, Oliver SG, Trueman LJ, Venkov PV. 1991. Cloning and characterization of a gene which determines osmotic stability in *Saccharomyces cerevisiae*. *Mol Cell Biol* **11**: 4235–4243.

Suda Y, Rodriguez RK, Coluccio AE, Neiman AM. 2009. A screen for spore wall permeability mutants identifies a secreted protease required for proper spore wall assembly. *PLoS One* **4**: e7184.

Utsugi T, Minemura M, Hirata A, Abe M, Watanabe D, Ohya Y. 2002. Movement of yeast 1,3-β-glucan synthase is essential for uniform cell wall synthesis. *Genes Cells* **7**: 1–9.

Yamamoto T, Hiratani T, Hirata H, Imai M, Yamaguchi H. 1986. Killer toxin from *Hansenula mrakii* selectively inhibits cell wall synthesis in a sensitive yeast. *FEBS Lett* **197**: 50–54.

Protocol 1

Fluorescent Labeling of Yeast Cell Wall Components

Hiroki Okada and Yoshikazu Ohya[1]

Department of Integrated Biosciences, Graduate School of Frontier Sciences, The University of Tokyo, 5-1-5 Kashiwanoha, Kashiwa, Chiba Prefecture 277-8562, Japan

Yeast cells stained with a fluorescent dye that specifically binds to one of the cell wall components can be observed under a fluorescent microscope. Visualization of the components 1,3-β-glucan, mannoproteins, and/or chitin not only provides information concerning the cell wall, but also reveals clues about various cellular activities such as cell polarity, vesicular transport, establishment of budding pattern, apical and isotropic bud growth, and replicative cell age. This protocol describes a standard method for visualizing different components of the yeast cell wall.

MATERIALS

It is essential that you consult the appropriate Material Safety Data Sheets and your institution's Environmental Health and Safety Office for proper handling of equipment and hazardous materials used in this protocol.

Reagents

Fluorescent dye(s)

Aniline blue, water-soluble (Wako 016-21302) (for staining 1,3-β-glucan)

Prepare a solution of 5 mg/mL in PBS.

Calcofluor white M2R/Fluorescent Brightener 28 (Sigma-Aldrich F3543) (for staining chitin)

Prepare a solution of 1 mg/mL in deionized H_2O.

Concanavalin A, fluorescein isothiocyanate-conjugated (FITC-ConA) (Sigma-Aldrich C7642) (for staining mannoproteins)

Prepare a solution of 1 mg/mL in PBS.

All dye solutions should be stored at 4°C in the dark and can be kept for several weeks.

Formaldehyde solution (37%) (Wako 064-00406) (optional; see Step 1)
Phosphate-buffered saline (PBS) (TaKaRa T900)
Saccharomyces cerevisiae, exponentially growing (~1×10^7 cells/mL)

Equipment

Coverslips (22 × 22-mm)
Incubator/shaker at temperature for yeast cell culture
Microcentrifuge (Eppendorf Minispin 5452)

---

[1]Correspondence: ohya@k.u-tokyo.ac.jp

Copyright © Cold Spring Harbor Laboratory Press; all rights reserved
Cite this protocol as *Cold Spring Harb Protoc*; doi:10.1101/pdb.prot085241

Chapter 12

Microcentrifuge tubes (1.5-mL)
Microscope outfitted with a camera, appropriate light source, appropriate filter sets, and image analysis software

The following list contains one example of an acceptable microscope setup.
   Axio Imager M1 fluorescence microscope with an EC Plan-Neofluar 100 ×/1.30 oil objective (Carl Zeiss)
   CoolSNAP HQ charge-coupled device (CCD) camera (Roper Scientific)
   HBO 100 Microscope Illuminating System (Carl Zeiss)
   FITC/4',6-diamidino-2-phenylindole (DAPI) filter sets (Carl Zeiss) and/or XF09 filter set (Opto Science, Inc.) (see Step 8)
   AxioVision ver. 4.5 software with multidimensional acquisition/viewer (Carl Zeiss)

Microscope slides, glass (76 × 26-mm)
Sonicator (TITEC VP-5S)

METHOD

Preparing for Specimen Staining

To stain live cells, skip Step 1.

1. (Optional) Add formaldehyde solution to a log-phase culture to a final concentration of 3.7%. Agitate the cells for 30 min at the culturing temperature.

2. Transfer a small volume of culture (~10^7 cells) to a microcentrifuge tube for harvest and staining of 1,3-β-glucan (Steps 3–8), mannoproteins (Steps 9–16), or chitin (Steps 17–27).

Staining Cell Wall Components

Images of stained cell walls are provided in Figure 1A.

Staining 1,3-β-Glucan

3. Collect the cells by centrifugation at 10,000 rpm for 30 sec at room temperature. Discard the supernatant.

4. Add 1 mL of PBS and disperse the cells by sonication for 5 sec at level 3 to 4.

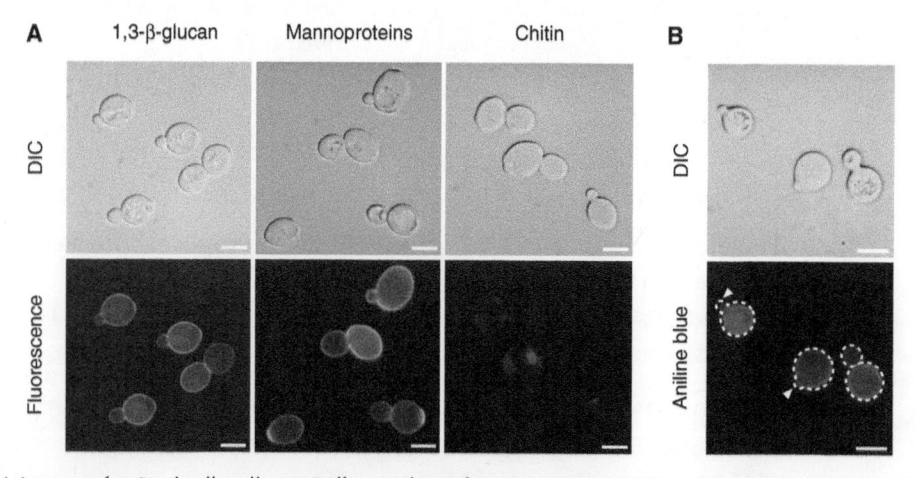

FIGURE 1. Images of stained cell walls. (*A*) Differential interference contrast (DIC) and fluorescent images of wild-type (BY4743) cells were captured after staining for the three cell wall components 1,3-β-glucan, mannoproteins, and chitin using the fluorescent dyes aniline blue, FITC-ConA, and calcofluor white, respectively. (*B*) Cells of a temperature-sensitive 1,3-β-glucan synthase mutant (*fks1-1154 fks2Δ*) were cultured to the log phase at 25°C followed by additional incubation for 2 h at 37°C. Cells were harvested and then observed after staining with aniline blue. Arrowheads denote fewer stained buds. Scale bar, 5 μm.

5. Repeat Step 3.
6. Resuspend the cells in 90 µL of PBS and 10 µL of aniline blue solution.
7. Incubate the cell suspension for 5 min at room temperature.
8. Mount the cell suspension on a glass slide and observe the cells using a DAPI filter set.

    Observation using a common DAPI filter set tends to interfere with the staining signal from the cytosol, particularly in fixed cells. To focus on the signal from the cell wall, the XF09 filter set can be used for observation of 1,3-β-glucan (excitation wavelength, 340–390 nm; emission wavelength, 517.5–552.5 nm).

Staining Mannoproteins

9. Collect the cells by centrifugation at 10,000 rpm for 30 sec at room temperature. Discard the supernatant.
10. Add 1 mL of PBS and disperse the cells by sonication for 5 sec at level 3 to 4.
11. Repeat Step 9.
12. Quickly resuspend the cells in 450 µL of PBS and 50 µL of FITC-ConA solution.

    Quick resuspension is essential for uniform staining.

13. Incubate the cell suspension for 10 min at room temperature.
14. Repeat Step 9.

    Supernatant should be removed completely to reduce background noise during microscopy.

15. Resuspend the cells in 100 µL of PBS.
16. Mount the cell suspension on a glass slide and observe the cells using an FITC filter set.

Staining Chitin

17. Collect the cells by centrifugation at 10,000 rpm for 30 sec at room temperature. Discard the supernatant.
18. Add 1 mL of deionized H_2O and disperse the cells by sonication for 5 sec at level 3 to 4.
19. Repeat Step 17.
20. Resuspend the cells in 100 µL of calcofluor white M2R solution.
21. Incubate the cell suspension for 1 min at room temperature.
22. Repeat Step 17.
23. Resuspend the cells in 1 mL of deionized H_2O.
24. Repeat Step 17.
25. Repeat Steps 23–24.

    Supernatant should be removed completely to reduce background noise during microscopy.

26. Resuspend the cells in 100 µL of deionized H_2O.
27. Mount the cell suspension on a glass slide and observe the cells using a DAPI filter set.

DISCUSSION

This simple staining method is frequently used to investigate abnormalities of the yeast cell wall. Failure to synthesize cell wall components decreases the staining signals of the components (Fig. 1B; Watanabe et al. 2001; Sanz et al. 2004). Conversely, a strong calcofluor white M2R signal suggests upregulation of chitin synthesis via the cell wall integrity pathway activated by various cell wall stresses (de Nobel et al. 2000). Loss of cell polarity or abnormal intracellular vesicular transport also results in an altered staining pattern (Schmidt et al. 2003). Because one function of the cell wall is to dictate cell shape, aberrant cell morphology is observed upon perturbation of synthesis or crosslinking of the components (de Groot et al. 2001; Blanco et al. 2012). Treatment with drugs that affect cell walls, such as echinocandin B, tunicamycin, and nikkomycin Z, results in cells with broader necks and more morphological variation (Okada et al. 2014).

Because cell wall dyes have a high specificity for each cell wall component, they can also be used for quantification. Aniline blue (Shedletzky et al. 1997) and calcofluor white M2R (Costa-de-Oliveira et al. 2013) have been used to quantify the amounts of 1,3-β-glucan and chitin, respectively. Some cell wall dyes, such as calcofluor white M2R, act as cell wall disruptors, inhibiting cell growth. Therefore, sensitivity to dyes that interfere with cell wall components can be used to identify genes involved in cell wall biogenesis.

Chitin staining with calcofluor white M2R is used to monitor budding pattern and aging. For example, during the cell division of S. cerevisiae, a ring-structured bud scar composed of chitin appears on the surface of the mother cell at the point of cytokinesis. "Axial" and "bipolar" patterns are predominantly observed in haploid and diploid cells of wild-type strains, respectively, and therefore are used as markers of cell polarity establishment (Chant and Pringle 1995). Because the number of bud scars increases in every division, numeration of the bud scars provides information regarding replicative cell age (Molin et al. 2011).

Mannoprotein staining with FITC-ConA is used in time lapse experiments because it has no obvious inhibitory activity on S. cerevisiae cell growth. Once cells are stained with FITC-ConA, the newly synthesized portion of the bud loses the staining signal and can be discriminated from the old cell wall. Apical bud growth in the early cell cycle stage is distinguished from later isotropic bud growth in this way (Watanabe et al. 2009). Staining with AlexaFluor488 5-TFP has similarly been used as a marker in growing cell populations (Váchová and Palková 2005). Unlike S. cerevisiae, Schizosaccharomyces pombe is not sensitive to calcofluor white M2R. Therefore, visualization of the S. pombe septum to reveal in situ development is possible using time-lapse imaging (Muñoz et al. 2013).

REFERENCES

Blanco N, Reidy M, Arroyo J, Cabib E. 2012. Crosslinks in the cell wall of budding yeast control morphogenesis at the mother-bud neck. *J Cell Sci* **125:** 5781–5789.

Chant J, Pringle JR. 1995. Patterns of bud-site selection in the yeast *Saccharomyces cerevisiae*. *J Cell Biol* **129:** 751–765.

Costa-de-Oliveira S, Silva AP, Miranda IM, Salvador A, Azevedo MM, Munro CA, Rodrigues AG, Pina-Vaz C. 2013. Determination of chitin content in fungal cell wall: An alternative flow cytometric method. *Cytometry A* **83:** 324–328.

de Groot PW, Ruiz C, Vázquez de Aldana CR, Duenas E, Cid VJ, Del Rey F, Rodríquez-Peña JM, Pérez P, Andel A, Caubín J, et al. 2001. A genomic approach for the identification and classification of genes involved in cell wall formation and its regulation in *Saccharomyces cerevisiae*. *Comp Funct Genomics* **2:** 324–342.

de Nobel H, Ruiz C, Martin H, Morris W, Brul S, Molina M, Klis FM. 2000. Cell wall perturbation in yeast results in dual phosphorylation of the Slt2/Mpk1 MAP kinase and in an Slt2-mediated increase in *FKS2-lacZ* expression, glucanase resistance and thermotolerance. *Microbiology* **146:** 2121–2132.

Molin M, Yang J, Hanzén S, Toledano MB, Labarre J, Nyström T. 2011. Life span extension and H(2)O(2) resistance elicited by caloric restriction require the peroxiredoxin Tsa1 in *Saccharomyces cerevisiae*. *Mol Cell* **43:** 823–833.

Muñoz J, Cortés JCG, Sipiczki M, Ramos M, Clemente-Ramos JA, Moreno MB, Martins IM, Pérez P, Ribas JC. 2013. Extracellular cell wall β(1,3) glucan is required to couple septation to actomyosin ring contraction. *J Cell Biol* **203:** 265–282.

Okada H, Ohnuki S, Roncero C, Konopka JB, Ohya Y. 2014. Distinct roles of cell wall biogenesis in yeast morphogenesis as revealed by multivariate analysis of high-dimensional morphometric data. *Mol Biol Cell* **25:** 222–233.

Schmidt M, Varma A, Drgon T, Bowers B, Cabib E. 2003. Septins, under Cla4p regulation, and the chitin ring are required for neck integrity in budding yeast. *Mol Biol Cell* **14:** 2128–2141.

Sanz M, Castrejón F, Durán A, Roncero C. 2004. *Saccharomyces cerevisiae* Bni4p directs the formation of the chitin ring and also participates in the correct assembly of the septum structure. *Microbiology* **150:** 2129–2141.

Shedletzky E, Unger C, Delmer DP. 1997. A microtiter-based fluorescence assay for (1,3)-β-glucan synthases. *Anal Biochem* **249:** 88–93.

Váchová L, Palková Z. 2005. Physiological regulation of yeast cell death in multicellular colonies is triggered by ammonia. *J Cell Biol* **169:** 711–717.

Watanabe D, Abe M, Ohya Y. 2001. Yeast Lrg1p acts as a specialized RhoGAP regulating 1,3-β-glucan synthesis. *Yeast* **18:** 943–951.

Watanabe M, Watanabe D, Nogami S, Morishita S, Ohya Y. 2009. Comprehensive and quantitative analysis of yeast deletion mutants defective in apical and isotropic bud growth. *Curr Genet* **55:** 365–380.

Protocol 2

Assay for Spore Wall Integrity Using a Yeast Predator

Hiroki Okada,[1] Aaron M. Neiman,[2,3] and Yoshikazu Ohya[1,3]

[1]Department of Integrated Biosciences, Graduate School of Frontier Sciences, The University of Tokyo, 5-1-5 Kashiwanoha, Kashiwa, Chiba Prefecture 277-8562, Japan; [2]Department of Biochemistry and Cell Biology, Stony Brook University, Stony Brook, New York 11794-5215

During the budding yeast life cycle, a starved diploid cell undergoes meiosis followed by production of four haploid spores, each surrounded by a spore wall. The wall allows the spores to survive in harsh environments until conditions improve. Spores are also more resistant than vegetative cells to treatments such as ether vapor, glucanases, heat shock, high salt concentrations, and exposure to high or low pH, but the relevance of these treatments to natural environmental stresses remains unclear. This protocol describes a method for assaying the yeast spore wall under natural environmental conditions by quantifying the survival of yeast spores that have passed through the digestive system of a yeast predator, the fruit fly.

MATERIALS

It is essential that you consult the appropriate Material Safety Data Sheets and your institution's Environmental Health and Safety Office for proper handling of equipment and hazardous materials used in this protocol.

RECIPES: Please see the end of this protocol for recipes indicated by <R>. Additional recipes can be found online at http://cshprotocols.cshlp.org/site/recipes.

Reagents

Drosophila melanogaster wild-type strain (e.g., Canton-R), 12–15 per dish, anesthetized at 4°C

To avoid contamination from yeast present in standard fly food, flies must be starved in a humidity chamber (e.g., a 50-mL conical-bottom centrifuge tube containing a small piece of wet filter paper) for over 6 h to empty their gut contents before use in the assay. Just before placing the flies in the dish containing yeast (Step 10), transfer the flies to a 4°C chamber for anesthesia.

Saccharomyces cerevisiae strain(s) of interest, growing as single colonies

Cells should be labeled with a fluorescent protein (e.g., TEF2-GFP) to distinguish intact cells in flyspecks.
A sporulation efficiency over 80% is desirable for this experiment; therefore, use of a high-efficiency and rapid-sporulation strain (e.g., SK-1 derivative [Kane and Roth 1974]) is recommended. Prior to the experiment, verify the high sporulation efficiency of the testing strain using an established sporulation protocol.

Solid media, freshly prepared:

Agar medium (2%) <R>

SPO agar <R>

---

[3]Correspondence: ohya@k.u-tokyo.ac.jp

Copyright © Cold Spring Harbor Laboratory Press; all rights reserved
Cite this protocol as *Cold Spring Harb Protoc*; doi:10.1101/pdb.prot085258

Chapter 12

>YPA agar <R>
>YPD agar <R>

Equipment

>Coverslips (22 × 22-mm)
>Incubator at 25°C
>Microscope outfitted with a camera, appropriate light source, appropriate filter sets, and image analysis software
>>*The following list contains one example of an acceptable microscope setup.*
>>>*Axio Imager M1 fluorescence microscope with an EC Plan-Neofluar 100 × /1.30 oil objective (Carl Zeiss)*
>>>*CoolSNAP HQ charge-coupled device (CCD) camera (Roper Scientific)*
>>>*HBO 100 Microscope Illuminating System (Carl Zeiss)*
>>>*FITC filter set (Carl Zeiss)*
>>>*AxioVision ver. 4.5 software with multidimensional acquisition/viewer (Carl Zeiss)*
>
>Microscope slides, glass (76 × 26-mm)
>Toothpicks (sterile)
>Tweezers

METHOD

Preparing Prey Yeast Cells

1. Using a sterile toothpick, isolate a single yeast colony and create a yeast patch on the agar surface of an YPD agar plate.
2. Incubate the cells overnight at 25°C.
3. Re-patch the cells on an YPA agar plate.
4. Incubate the cells overnight at 25°C.
5. Re-patch the cells on an SPO agar plate.
6. Incubate the cells for 1–5 d at 25°C.
7. Determine the sporulation efficiency under a microscope.

Fly Feeding Using Prey Yeast Cells

8. Using a sterile toothpick, transfer a yeast patch from the SPO agar plate to the agar surface of a 2% agar plate.
   The agar surface should be well-dried to avoid trapping flies on a wet surface.
9. Attach two coverslips to the underside of the lid of the Petri dish using 1 µL of deionized H_2O as "glue." Place the Petri dish right-side-up with the coverslips on the lid to avoid contamination from falling exterior fly components.
10. Place 12 to 15 anesthetized flies in each Petri dish containing yeast.
11. Store the Petri dish right-side-up overnight at room temperature (Fig. 1A).

Observing Yeast in Fly Feces

12. Remove the flies from the dish and remove the coverslips from the lid using tweezers.
13. Place a coverslip on a microscope slide and observe individual excreta directly using light and fluorescence microscopy.

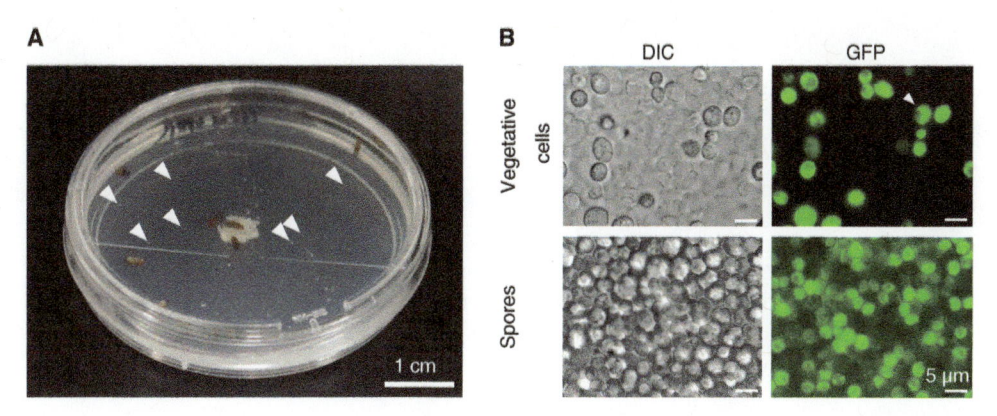

FIGURE 1. Yeast survival assay based on passage through the fly digestive system. (*A*) Image of the fly feeding experiment. Yeast specimen (patched on the center of the agar surface) is subjected to predation by flies. Arrowheads denote flyspecks on the lid-attached coverslips. (*B*) Differential interference contrast (DIC) and fluorescent images of yeast cells in flyspecks. Arrowhead shows example of intact cells.

14. Score over 100 yeast cells in each flyspeck as either intact or dead.

    Intact cells appear healthy based on differential interference (DIC) microscopy and are fluorescently active based on fluorescence microscopy (Fig. 1B). Dead cells appear as ghosts (empty cells) and lose fluorescent signal.

DISCUSSION

The current protocol is based on the prey–predator interaction between two well-known model organisms—namely, yeast and fruit fly. Drosophilid species tend to use yeast as a food source in nature, and thus yeasts must deal with predation in their natural life cycle. Because the midgut of the fruit fly contains regions of both high and low pH for digestion, vegetative yeast cells are killed by passage through the gut. However, spores show enhanced survival relative to vegetative cells.

This method was used to screen for yeast genes essential to spore survival during passage through the fly gut (Coluccio et al. 2008). The outermost layers of the spore wall, composed of chitosan and dityrosine, are the most important components for protection against the outside environment. Spores of mutants defective in the biosynthesis of these components are highly sensitive to the fly digestive system (Coluccio et al. 2008). Interestingly, predation by the fly may be beneficial for yeast by providing an opportunity for transmission to a new location using a "flying" vector. In addition, spore predation from *Drosophila* may increase yeast outbreeding as a result of ascus sac digestion (Reuter et al. 2007). This would contribute to genetic mixing of yeast populations, potentially enhancing fitness during colonization of diverse environments. Spores are exclusively distributed to the outer layer of a yeast colony (Piccirillo and Honigberg 2010), and this distribution bias may be a strategy for spores to take advantage of predation.

Spore cells of the fission yeast *Schizosaccharomyces pombe* are more sensitive to the *Drosophila* digestive system than those of *S. cerevisiae*. Considering the natural phenomenon of symbiosis, it is possible that an unknown preferential partner for *S. pombe* exists. This protocol facilitates the study of symbiotic relationships as well as the physiological analysis of a single organism.

RECIPES

Agar Medium (2%)

1. Prepare a solution of 2% agar in deionized H_2O.
2. Sterilize by autoclaving for 20 min at 15 psi.
3. Pour 5–10 mL of agar medium per 52 × 10-mm Petri dish.

SPO Agar

1. Combine the following reagents in deionized H$_2$O.

Reagent	Final concentration
Agar	2%
Potassium acetate	2%

2. Sterilize by autoclaving for 20 min at 15 psi.
3. Add trace nutrients appropriate for the yeast background of interest.
4. Pour 20–25 mL of agar medium per 90 × 15-mm Petri dish.

YPA Agar

1. Combine the following reagents in deionized H$_2$O.

Reagent	Final concentration
Adenine sulfate	20 µg/mL
Agar	2%
Potassium acetate	2%
Peptone (BD 211677)	2%
Yeast extract (BD 212750)	1%

2. Sterilize by autoclaving for 20 min at 15 psi.
3. Pour 20–25 mL of agar medium per 90 × 15-mm Petri dish.

YPD Agar

1. Combine the following reagents in deionized H$_2$O.

Reagent	Final concentration
Agar	2%
Peptone (BD 211677)	2%
Yeast extract (BD 212750)	1%

2. Prepare a solution of 20% glucose in deionized H$_2$O.
3. Sterilize both solutions by autoclaving for 20 min at 15 psi.
3. Add the sterile 20% glucose solution to the medium from Step 1 to a final concentration of 2% glucose.
4. Pour 20–25 mL of agar medium per 90 × 15-mm Petri dish.

REFERENCES

Coluccio AE, Rodriguez RK, Kernan MJ, Neiman AM. 2008. The yeast spore wall enables spores to survive passage through the digestive tract of *Drosophila*. *PLoS One* **3:** e2873.

Kane SM, Roth R. 1974. Carbohydrate metabolism during ascospore development in yeast. *J Bacteriol* **118:** 8–14.

Piccirillo S, Honigberg SM. 2010. Sporulation patterning and invasive growth in wild and domesticated yeast colonies. *Res Microbiol* **161:** 390–398.

Reuter M, Bell G, Greig D. 2007. Increased outbreeding in yeast in response to dispersal by an insect vector. *Curr Biol* **17:** R81–R83.

Protocol 3

Local and Acute Disruption of the Yeast Cell Surface

Keiko Kono,[1,3] Hiroki Okada,[2] and Yoshikazu Ohya[2]

[1]Department of Cell Biology, Graduate School of Medical Sciences, Nagoya City University, 1, Kawasumi, Mizuho-cho, Mizuho-ku, Nagoya, Aichi Prefecture 467-8601, Japan; [2]Department of Integrated Biosciences, Graduate School of Frontier Sciences, The University of Tokyo, 5-1-5 Kashiwanoha, Kashiwa, Chiba Prefecture 277-8562, Japan

In nature, the yeast cell barrier encounters threats ranging from physical impact to abrupt changes in osmolality after rainfall. Genetic materials are protected from these environmental attacks by the rigid cell wall. Laboratory methods for challenging cell wall integrity have made an enormous contribution to the study of the yeast cell surface, but most have targeted whole-cell populations in place of single-cell analysis. This protocol describes pulse-laser-based acute disruption of the yeast cell surface, which enables the observation of single-cell response to submicron-scale damage.

MATERIALS

It is essential that you consult the appropriate Material Safety Data Sheets and your institution's Environmental Health and Safety Office for proper handling of equipment and hazardous materials used in this protocol.

RECIPES: Please see the end of this protocol for recipes indicated by <R>. Additional recipes can be found online at http://cshprotocols.cshlp.org/site/recipes.

Reagents

Yeast strain of interest with a biomarker of cell wounding introduced (e.g., Pkc1-GFP), grown overnight in an appropriate medium (e.g., complete minimal [CM] or synthetic complete [SC] and drop-out media <R>)

Dilute the overnight culture in fresh medium and incubate to a density of $5 \times 10^6 – 1 \times 10^7$ cells/mL.

Equipment

MicroPoint Galvo pulse laser illumination and ablation system (Andor Technology), with an NL100 nitrogen laser (Stanford Research Systems)

Use a Coumarin 440 laser dye cell.

Microscope outfitted with a camera, appropriate light source, appropriate filter sets, and image analysis software

The following list contains one example of an acceptable microscope setup.

    Axiovert 200M motorized inverted microscope (Carl Zeiss) with a 63× Plan-Apochromat objective and an MS-2000 stage (Applied Scientific Instrumentation)

    CoolSNAP HQ charge-coupled device (CCD) camera (Roper Scientific)

    Lambda LS 175-watt Xenon light source (Sutter Instruments)

    BrightLine GFP filter set (Semrock)

    Dichroic beam splitter with an 85/15 transmission-reflection ratio

    SlideBook software (Intelligent Imaging Innovations)

---

[3]Correspondence: konoamed.nagoya-cu.ac.jp

Copyright © Cold Spring Harbor Laboratory Press; all rights reserved
Cite this protocol as *Cold Spring Harb Protoc*; doi:10.1101/pdb.prot085266

Chapter 12

ONIX Microfluidic Platform (CellASIC)
Alternatively, use an agarose bed (Rines et al. 2011).

METHOD

1. Load yeast cells (5.0×10^6–1.0×10^7 cells/mL) in the microfluidic platform or place the cells on an agarose bed.

2. Set the laser conditions as follows.

   i. Tune the laser to 440 nm using a Coumarin 440 dye cell.

   ii. Place a dichroic beam splitter with an 85/15 transmission-reflection ratio.

   iii. Optimize the laser power depending on the laser system and background of the yeast strain.
   Optimal laser power is typically 48%–58%, empirically determined based on the laser power sufficient for Pkc1-GFP recruitment without resulting in cell lysis.

3. Acquire images at appropriate intervals (e.g., 30 sec). Acquire at least three frames before laser damage (Step 4) to allow averaging of the starting signal intensity.

4. Choose the region of interest, and irradiate with the laser.

5. Continue imaging at appropriate intervals until there are no more responses.

DISCUSSION

Prior work suggests that the cell wall integrity (CWI) pathway and downstream chitin synthesis are the key modes of yeast cell wall repair (Levin 2011). Because commonly used cell-wall-damaging reagents attack the whole cell cortex, it has been unclear whether cell polarity is disorganized or tightly targeted to the local wound. The laser damage assay described here was established to address whether budding yeast can repair local damage to cell walls and/or the plasma membrane (Kono et al. 2012). Data from the assay showed that a chitin synthase, Chs3, accumulated around the damage site, staunching the wound with newly synthesized chitin (Fig. 1). In addition to Chs3, type V myosin Myo2 and exocyst component Exo70 were recruited to the wound. These results show convincingly that cell polarity is established toward the local damage.

In higher eukaryotes, the scar at the wounded plasma membrane is cleared by endocytosis and exocytosis (Sonnemann and Bement 2011). Perhaps in yeast, chitin at the staunched wound is replaced by more common cell wall components such as mannoprotein or 1,3-β-glucan. Alternatively,

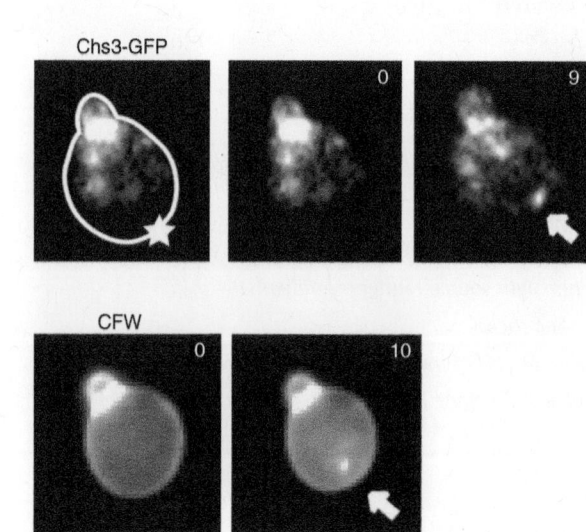

FIGURE 1. A local cell wall construction during the wound healing response in budding yeast. Local activation of the chitin emergency response was assayed by live cell imaging of Chs3-GFP localization and calcofluor white (CFW) staining of chitin after damage induction. The star denotes the site of laser damage. Numbers in the upper-right corner of each panel indicate the time after damage (min). Arrows show recruitment of Chs3-GFP or newly synthesized chitin. (Reprinted from Kono et al. 2012; © 2012 Elsevier, Inc.)

it may remain at the damaged site, as with bud scars. The laser damage assay should provide a useful tool for addressing whether the repaired cell wall can perform critical functions, and if not, how the yeast cell manages the deficiency.

RECIPES

Complete Minimal (CM) or Synthetic Complete (SC) and Drop-Out Media

Bacto-yeast nitrogen base without amino acids*, 6.7 g
Glucose, 20 g
Bacto agar, 20 g
Drop-out mix, 2 g <R>
H$_2$O, to 1000 mL

To test the growth requirements of strains, it is useful to have media in which each of the commonly encountered auxotrophies is supplemented except the one of interest (drop-out media). Dry growth supplements are stored premixed.

CM (or SC) is a medium in which the drop-out mix contains all possible supplements (i.e., nothing is "dropped out").

*Yeast nitrogen base without amino acids (YNB) is sold either with or without ammonium sulfate. This recipe is for YNB with ammonium sulfate. If the bottle of YNB is lacking ammonium sulfate, add 5 g of ammonium sulfate and only 1.7 g of YNB.

Drop-Out Mix

Reagent	Amount to add (g)
Adenine	0.5
Alanine	2.0
Arginine	2.0
Asparagine	2.0
Aspartic acid	2.0
Cysteine	2.0
Glutamine	2.0
Glutamic acid	2.0
Glycine	2.0
Histidine	2.0
Inositol	2.0
Isoleucine	2.0
Leucine	10.0
Lysine	2.0
Methionine	2.0
para-Aminobenzoic acid	0.2
Phenylalanine	2.0
Proline	2.0
Serine	2.0
Threonine	2.0
Tryptophan	2.0
Tyrosine	2.0
Uracil	2.0
Valine	2.0

Combine the appropriate ingredients, minus the relevant supplements, and mix in a sealed container. Turn the container end-over-end for at least 15 min; add several clean marbles to help mix the solids.

REFERENCES

Kono K, Saeki Y, Yoshida S, Tanaka K, Pellman D. 2012. Proteasomal degradation resolves competition between cell polarization and cellular wound healing. *Cell* **150:** 151–164.

Levin DE. 2011. Regulation of cell wall biogenesis in *Saccharomyces cerevisiae*: The cell wall integrity signaling pathway. *Genetics* **189:** 1145–1175.

Rines DR, Thomann D, Dorn JF, Goodwin P, Sorger PK. 2011. Live cell imaging of yeast. *Cold Spring Harb Protoc* **2011:** top065482.

Sonnemann KJ, Bement WM. 2011. Wound repair: Toward understanding and integration of single-cell and multicellular wound responses. *Annu Rev Cell Dev Biol* **27:** 237–263.

CHAPTER 13

Analyzing and Understanding Lipids of Yeast: A Challenging Endeavor

Sepp D. Kohlwein[1]

Institute of Molecular Biosciences, University of Graz, BioTechMed Graz, 8010 Graz, Austria

Lipids are essential biomolecules with diverse biological functions, ranging from building blocks for all biological membranes to energy substrates, signaling molecules, and protein modifiers. Despite advances in lipid analytics by mass spectrometry, the extraction and quantitative analysis of the diverse classes of lipids are still an experimental challenge. Yeast is a model organism that provides several advantages for studying lipid metabolism, because most biosynthetic pathways are well described and a great deal of information is available on the regulatory mechanisms that control lipid homeostasis. In addition, the composition of yeast lipids is much less complex than that of mammalian lipids, making yeast an excellent reference system for studying lipid-associated cell functions.

INTRODUCTION

Lipid research has gained great momentum in recent years, fueled not only by the pandemic dimensions of lipid-associated disorders in humans, but also by the potential for exploiting single-cell lipids as a source for biofuel ("biodiesel"). In addition, a major limitation in lipid research—the analytical part— has recently been overcome by the implementation of mass spectrometry techniques that provide high-resolution insight into lipid molecular species content and metabolism in cells and tissues. The yeast *Saccharomyces cerevisiae* represents an excellent experimental reference organism to study physiological and pathophysiological roles of lipids at the cellular level. Lipid metabolic pathways have been well elucidated and a great deal of information is available about the regulatory processes controlling lipid homeostasis in yeast (Nohturfft and Zhang 2009; Carman and Han 2011; Henry et al. 2012; Kohlwein et al. 2013). Most metabolic pathways are conserved, and also the regulatory circuitry that links lipid metabolism to energy homeostasis, cell growth, and development shares several levels of homology with mammalian cells (Nohturfft and Zhang 2009; Natter and Kohlwein 2013). Despite this remarkable level of functional conservation, yeast lipid composition is rather simple compared to that of mammalian cells (Ejsing et al. 2009; Shui et al. 2010; Klose et al. 2012), facilitating experimental setups and interpretation of the complex dynamic processes that govern lipid homeostasis.

Nevertheless, work with lipids is still a challenging task. First, lipids have highly diverse chemical structures, although they all have a rather hydrophobic or amphipathic nature and are defined by their "solubility in organic solvents." Second, various lipid classes differ dramatically in their abundance, ranging from "bulk" lipids such as storage triacylglycerols and membrane phospholipids to less abundant sphingolipids, which are lipid-signaling molecules present only in trace amounts. Third, despite the relative simple lipid composition in yeast (see below), there is a great deal of combinatorial chemistry in

---

[1]Correspondence: sepp.kohlwein@uni-graz.at

Copyright © Cold Spring Harbor Laboratory Press; all rights reserved
Cite this introduction as *Cold Spring Harb Protoc*; doi:10.1101/pdb.top078956

terms of variations in acyl chain composition in various phospholipid classes, giving rise to a quite complex set of lipid molecular species even in the "simple" yeast cell (Guan and Wenk 2006; Ejsing et al. 2009; Shui et al. 2010; Knittelfelder et al. 2014; Casanovas et al. 2015). Fourth, lipid metabolism is a highly dynamic process, which correlates with cell growth, nutritional supply, and even the cell cycle (Kurat et al. 2006, 2009; Zanghellini et al. 2008; Klose et al. 2012; Chauhan et al. 2015). Thus, comparing lipid data between cells grown under different conditions or between mutants can be very misleading, making functional and mechanistic conclusions from lipidomic data typically quite difficult.

LIPIDS OF YEAST: COMPOSITION, METABOLISM, AND CELLULAR FUNCTION

The lipid composition of yeast is rather simple and comprises some 150 distinct molecular species that have been identified to date (Guan and Wenk 2006; Ejsing et al. 2009; Guan et al. 2010; Klose et al. 2012; Casanovas et al. 2015). Table 1 summarizes the total lipid composition of various yeast wild-type strains grown in minimal medium supplemented with inositol (Daum et al. 1999).

Fatty Acids

Fatty acids (FAs) are the essential building blocks of all glycerolipids and sphingolipids (see below) and they are either synthesized de novo from acetyl-CoA units that are assembled in a cyclic series of

TABLE 1. Lipid composition of various yeast wild-type strains: (A) phospholipid content; (B) sterol and triacylglycerol content; and (C) fatty acid (FA) content and ratios of C16 to C18 and unsaturated to saturated FAs

A

| Yeast strain | PL* | Percentage of total phospholipids ||||||||
		PA	PS	PE	PC	PI	CL	PDME	Lyso
JL 20	31.5	4.0	5.0	24.5	40.6	18.9	2.1	2.3	2.3
YJ3C 366	28.0	2.7	6.9	21.0	41.8	21.6	1.9	3.3	0.6
JS91.13-23	30.2	7.4	3.3	13.6	45.9	23.1	3.1	2.5	1.1
W303-1A	26.4	1.9	5.5	21.0	46.2	18.9	3.1	2.7	0.9
FY1679	31.5	2.0	9.2	20.4	40.0	19.1	2.9	4.0	2.0
CEN.PK2-1C	31.4	2.2	5.4	19.0	49.4	16.7	2.9	2.3	0.6

B

| Yeast strain | Ergosterol* | Sterol and triacylglycerol content |||||| |
		SE*	TS*	E (%)	L (%)	Sq (%)	TAG*
JL 20	5.7	9.1	14.9	76.5	6.2	16.8	4.7
YJ3C 366	4.7	7.8	12.5	94.1	3.9	2.0	8.1
JS91.13-23	6.0	8.7	14.8	73.8	17.5	8.7	10.0
W303-1A	4.1	14.1	18.2	85.4	9.7	4.9	7.8
FY1679	6.0	2.6	8.5	75.3	17.0	7.7	2.4
CEN.PK2-1C	4.0	16.0	20.0	78.9	14.3	6.8	15.2

C

| Yeast strain | Percentage of total fatty acids |||| Ratios C16/C18 | Unsat/Sat |
	C16:0	C16:1	C18:0	C18:1		
JL 20	12.0	42.7	5.3	40.0	1.2	4.8
YJ3C 366	6.3	38.3	3.6	51.8	0.8	9.1
W303-1A	6.1	24.6	5.8	63.5	0.44	7.4
FY1679	20.8	45.1	4.8	27.1	2.1	2.8
CEN.PK2-1C	4.5	40.0	2.7	52.0	0.8	12.8

Modified, with permission, from Daum et al. 1999. Copyright © 1999 John Wiley & Sons, Ltd.

Note that values are quite comparable between wild-type strains, with some marked differences in sterol and triacylglycerol as well as FA composition, which may be the result of different growth characteristics. Asterisks indicate values are milligrams lipid per gram cell dry weight.

PL, total phospholipids; PA, phosphatidic acid; PS, phosphatidylserine; PE, phosphatidylethanolamine; PC, phosphatidylcholine; PI, phosphatidylinositol; CL, cardiolipin; PDME, phosphatidyldimethylethanolamine; Lyso, lysophospholipids; SE, steryl esters; TS, total sterol; E, ergosterol; L, lanosterol; Sq, squalene; TAG, triacylglycerol; C16:0, palmitic acid; C16:1, palmitoleic acid; 18:0, stearic acid; C18:1, oleic acid.

reactions by the fatty acid synthase complex, or they may be taken up by yeast cells if they are supplied in the culture medium (Tehlivets et al. 2007). In yeast, de novo synthesis delivers CoA-activated FAs (acyl-CoAs), whereas exogenously added FAs require activation with CoA before their incorporation into lipids, or further modification, such as elongation or desaturation (Black and DiRusso 2007). Yeast contains mostly FAs with 16 or 18 carbon atoms, which are saturated (~20%) or mono-unsaturated (~80%) (see Table 1). Minor FA species are C14 and very long chain FAs, up to C26, which are mostly present in sphingolipids and glycosylphosphatidylinositol anchors (Tehlivets et al. 2007; Ejsing et al. 2009). In contrast to mammalian cells, yeast contains only mono-unsaturated FAs harboring a double bond between C9 and C10, which drastically reduces the combinatorial acyl-chain complexity in the yeast lipids; the double bond is introduced into the acyl chain by the essential Ole1 Δ9 FA desaturase (Martin et al. 2007).

FA analysis is typically performed by gas–liquid chromatography after transmethylation of lipid-bound FAs to their respective volatile methylesters. Since yeast does not contain polyunsaturated FAs, which are susceptible to (per)oxidation during the isolation procedure, no specific precautions need to be taken.

Glycerophospholipids

Glycerolipids are the most abundant class of lipids in yeast. They are derived from glycerol and harbor one, two, or three acyl chains esterified to the hydroxyl groups of the glycerol moiety. In triacylglycerols (triglycerides; TAG), all three hydroxyl groups are esterified with an FA, and these molecules are "neutral lipids," which are uncharged and have a highly apolar structure. TAG are typically stored in cytosolic lipid droplets and provide a most efficient means to store FAs, which may serve as a source of energy by β-oxidation on demand, for example, in the absence of other carbon sources. In addition, FAs released by lipolysis may also serve as precursors for membrane lipids. *sn*-1,2-diacylglycerol (DAG) is only a minor neutral lipid intermediate and is a precursor for the synthesis of TAG or glycerophospholipids. In addition, DAG is generated in the course of signaling processes (e.g., degradation of phosphatidylinositol phosphates) and may activate protein kinase C in the plasma membrane. The main route of DAG synthesis is via the dephosphorylation of phosphatidic acid (PA), which is also the central precursor for all glycerophospholipids. PA (also a minor lipid class) harbors two acyl chains in positions *sn*-1 and *sn*-2 and a phosphate group at position *sn*-3, which is negatively charged at intracellular pH. In addition to its metabolic role as a key intermediate in DAG, TAG, and phospholipid synthesis, PA also has signaling function, which operates through its specific interaction with the transcriptional repressor Opi1. This binding of Opi1 to PA regulates the subcellular distribution of the repressor and hence its nuclear repressing activity (Loewen et al. 2004; Henry et al. 2012; Hofbauer et al. 2014).

PA is synthesized by two sequential acylation reactions of glycerol-3-phosphate, each of which is catalyzed by two redundant acyl-CoA-dependent acyltransferases. Subsequently, in de novo phospholipid synthesis, PA is activated with CTP to CDP-DAG (a very minor lipid class), which is the substrate for several transesterification reactions: (i) CDP-DAG and inositol give rise to phosphatidylinositol (PI); (ii) CDP-DAG and glycerol-3-phosphate are precursors of phosphatidylglycerol phosphate (PGP), which is further processed to the mitochondria-specific lipid, cardiolipin (CL); and (iii) transesterification of CDP-DAG with serine generates phosphatidylserine (PS). PGP, CL, PI, and PS are all negatively charged phospholipids. PS is converted by decarboxylation to phosphatidylethanolamine (PE), which is methylated in a three-step reaction to phosphatidylcholine (PC), the major glycerophospholipid in yeast. In an alternative pathway, PE and PC can be generated by transesterification of CDP–ethanolamine or CDP–choline, and DAG, respectively (Henry et al. 2012). This so-called CDP–choline or Kennedy pathway is essential in the absence of the de novo methylation pathway (i.e., in respective mutants) and requires supplementation of cells with choline. However, this pathway is also highly active in the absence of exogenous ethanolamine or choline to recycle precursors derived from lipid turnover. PE and PC both carry positive and negative charges at cellular pH, but appear overall as neutral. Intermediates of glycerophospholipids during synthesis or degradation are so-called lysophospholipids (i.e., with the structure *sn*-1-acyl-glycerol-3-phosphate), which are typically present only in very low amounts, and may have detrimental, detergent-like effects on cell membranes. The observation of elevated lysolipids in cellular extracts indicates some level of degradation during the isolation procedure.

All glycerophospholipids have an amphipathic structure, which is the basis for their spontaneous assembly into bilayers in which the hydrophobic acyl chains interact with each other to exclude water (the hydrophobic effect) and the highly polar, charged, or uncharged head groups are oriented toward the aqueous phase, where they interact with ions (to balance the charges), proteins, and metabolites. Most biosynthetic pathways of glycerolipid synthesis are associated with the endoplasmic reticulum, but some reactions take place in the cytosol or mitochondria. The occurrence of specific lipid profiles in subcellular membranes (Schneiter et al. 1999) requires extensive lipid trafficking and remodeling, and also implies that various subcellular lipid (precursor) pools must exist, dependent on the localization and activity of the anabolic or catabolic enzymes. Nevertheless, mutant analyses have shown that yeast cells are able to tolerate dramatic alterations in their FA and lipid composition (Henry et al. 2012; Klose et al. 2012).

It should also be noted that lateral movement of lipid molecules and proteins is a highly dynamic process, whereas transversal movement of lipids (flip–flop) from one monolayer to the opposite side is kinetically extremely unfavored and requires enzymes for catalysis. As a consequence, lipid composition of both leaflets of the membrane is typically asymmetric, which has profound biological consequences for membrane function. These subcellular differences have to be taken into account for the interpretation of lipidomics data, when analyzing total lipid extracts!

Sphingolipids

In addition to glycerophospholipids, yeast cells contain a second class of phospholipids termed sphingolipids (Schneiter 1999; Guan and Wenk 2006; Dickson 2010). These lipids are derived from sphingosine or phytosphingosine, which are produced in a condensation reaction between serine and palmitoyl-CoA. Further addition of a very long chain FA (up to C26) in an amide linkage generates ceramide. Ceramide is modified with inositol phosphate (PI is the inositol phosphate donor), giving rise to the essential inositol phosphoryl ceramide (IPC), which is further modified with sugar (mannose) to form mannosyl inositol phosphoryl ceramide (MIPC) and with one additional inositol phosphate to form mannosyl di-inositol phosphate ceramide (M$[IP]_2$C). These complex sphingolipids are not essential under normal growth conditions, but lack of these lipids renders cells highly temperature and ion (e.g., calcium) sensitive. Sphingolipids are enriched in the plasma membrane and contribute to membrane stability; they also serve as membrane anchors for proteins.

Sterols

Sterols are chemically and structurally quite distinct from other classes of lipids. They are membrane-active compounds, but do not form bilayers on their own. The main yeast sterol is ergosterol, which is synthesized in a complex series of condensation, cyclization, and redox reactions from C5 (isoprene) building blocks (Espenshade and Hughes 2007). Fungal sterol biosynthesis follows steps similar to mammalian cholesterol synthesis, with some distinct differences, which make sterol biosynthesis (an essential pathway) a prime target for antifungal drugs (Lupetti et al. 2002). Free sterols modulate membrane fluidity and are enriched in the yeast plasma membrane, where their content is regulated not only by the activity of key biosynthetic enzymes, but also by their partial sequestration by acylation and storage as steryl esters, together with TAG, in cytosolic lipid droplets (Czabany et al. 2007; Kohlwein et al. 2013).

Minor Lipid Components in Yeast

In addition to lipid intermediates, such as lysolipids, PA, CDP-DAG, and DAG, numerous other classes of lipids are present only in minor amounts in yeast lipid extracts: These include phosphatidylinositol phosphate (PIP) and its higher phosphorylated derivatives, PI(3)P, PI(4)P, PI(3,5)P_2, and

PI(4,5)P$_2$, and sphingosine derivatives, some of which are present only in trace amounts, but nevertheless play essential roles as signaling molecules.

YEAST LIPID ANALYTICS

Advances in lipid mass spectrometry have significantly facilitated lipid analysis in yeast and other cell types and have unveiled unprecedented insight into the molecular species composition of yeast lipids (Guan and Wenk 2006; Ejsing et al. 2009; Guan et al. 2010; Shui et al. 2010; Klose et al. 2012; Casanovas et al. 2015). Because lipid molecular masses are typically <1000 Da, they are excellent targets for mass spectrometry approaches. Despite the relatively simple composition of the yeast lipidome, major differences in chemical nature and abundance make a total lipid analysis from yeast cells a challenging task, starting with cell breakage and extraction, and various levels of qualitative and quantitative analysis. It should be noted that not all lipids are extracted with the same efficacy in a given solvent system; thus, depending on the type of lipid under analysis, alternative extraction procedures may need to be considered (e.g., for sphingolipids or PIPs). For a standard protocol for the isolation of lipids from yeast, see Protocol 1: Lipid Extraction from Yeast Cells (Knittelfelder and Kohlwein 2015a). For protocols describing the quantification of FAs, the major glycerophospholipids, and sterols/steryl esters from yeast, see Protocol 2: Thin-Layer Chromatography to Separate Phospholipids and Neutral Lipids from Yeast (Knittelfelder and Kohlwein 2015b), Protocol 3: Derivatization and Gas Chromatography of Fatty Acids from Yeast (Knittelfelder and Kohlwein 2015c), and Protocol 4: Quantitative Analysis of Yeast Phospholipids and Sterols by High-Performance Liquid Chromatography–Evaporative Light-Scattering Detection (Knittelfelder and Kohlwein 2015d).

ACKNOWLEDGMENTS

I would like to thank the members of my laboratory for helpful suggestions. Research in my laboratory is supported by the PhD program "Molecular Enzymology" (W901-B05) and project F3005-B12 (SFB LIPOTOX), which are funded by the Austrian Science Funds (FWF). Additional support NAWI Graz and BioTechMed Graz is gratefully acknowledged.

REFERENCES

Black PN, DiRusso CC. 2007. Yeast acyl-CoA synthetases at the crossroads of fatty acid metabolism and regulation. *Biochim Biophys Acta* 1771: 286–298.

Carman GM, Han GS. 2011. Regulation of phospholipid synthesis in the yeast *Saccharomyces cerevisiae*. *Annu Rev Biochem* 80: 859–883.

Casanovas A, Sprenger RR, Tarasov K, Ruckerbauer DE, Hannibal-Bach HK, Zanghellini J, Jensen ON, Ejsing CS. 2015. Quantitative analysis of proteome and lipidome dynamics reveals functional regulation of global lipid metabolism. *Chem Biol* 22: 412–425.

Chauhan N, Visram M, Cristobal-Sarramian A, Sarkleti F, Kohlwein SD. 2015. Morphogenesis checkpoint kinase Swe1 is the executor of lipolysis-dependent cell-cycle progression. *Proc Natl Acad Sci* 112: E1077–E1085.

Czabany T, Athenstaedt K, Daum G. 2007. Synthesis, storage and degradation of neutral lipids in yeast. *Biochim Biophys Acta* 1771: 299–309.

Daum G, Tuller G, Nemec T, Hrastnik C, Balliano G, Cattel L, Milla P, Rocco F, Conzelmann A, Vionnet C, et al. 1999. Systematic analysis of yeast strains with possible defects in lipid metabolism. *Yeast* 15: 601–614.

Dickson RC. 2010. Roles for sphingolipids in *Saccharomyces cerevisiae*. In *Advances in experimental medicine and biology* (eds. Chalfant C, Del Poeta M), pp. 217–231, Kluwer Academic, New York.

Ejsing CS, Sampaio JL, Surendranath V, Duchoslav E, Ekroos K, Klemm RW, Simons K, Shevchenko A. 2009. Global analysis of the yeast lipidome by quantitative shotgun mass spectrometry. *Proc Natl Acad Sci* 106: 2136–2141.

Espenshade PJ, Hughes AL. 2007. Regulation of sterol synthesis in eukaryotes. *Annu Rev Genetics* 41: 401–427.

Guan XL, Wenk MR. 2006. Mass spectrometry-based profiling of phospholipids and sphingolipids in extracts from *Saccharomyces cerevisiae*. *Yeast* 23: 465–477.

Guan XL, Riezman I, Wenk MR, Riezman H. 2010. Yeast lipid analysis and quantification by mass spectrometry. *Methods Enzymol* 470: 369–391.

Henry SA, Kohlwein SD, Carman GM. 2012. Metabolism and regulation of glycerolipids in the yeast *Saccharomyces cerevisiae*. *Genetics* 190: 317–349.

Hofbauer HF, Schopf FH, Schleifer H, Knittelfelder OL, Pieber B, Rechberger GN, Wolinski H, Gaspar ML, Kappe CO, Stadlmann J, et al. 2014. Regulation of gene expression through a transcriptional repressor that senses acyl-chain length in membrane phospholipids. *Dev Cell* 29: 729–739.

Klose C, Surma MA, Gerl MJ, Meyenhofer F, Shevchenko A, Simons K. 2012. Flexibility of a eukaryotic lipidome—Insights from yeast lipidomics. *PLoS One* 7: e35063.

Knittelfelder OL, Kohlwein SD. 2015a. Lipid extraction from yeast cells. *Cold Spring Harb Protoc* doi: 10.1101/pdb.prot085449.

Knittelfelder OL, Kohlwein SD. 2015b. Thin-layer chromatography to separate phospholipids and neutral lipids from yeast. *Cold Spring Harb Protoc* doi: 10.1101/pdb.prot085456.

Knittelfelder OL, Kohlwein SD. 2015c. Derivatization and gas chromatography of fatty acids from yeast. *Cold Spring Harb Protoc* doi: 10.1101/pdb.prot085464.

Knittelfelder OL, Kohlwein SD. 2015d. Quantitative analysis of yeast phospholipids and sterols by high-performance liquid chromatography—Evaporative light-scattering detection. *Cold Spring Harb Protoc* doi: 10.1101/pdb.prot085472.

Knittelfelder OL, Weberhofer BP, Eichmann TO, Kohlwein SD, Rechberger GN. 2014. A versatile ultra-high performance LC-MS method for lipid profiling. *J Chromatogr B Analyt Technol Biomed Life Sci* **951–952**: 119–128.

Kohlwein SD, Veenhuis M, van der Klei IJ. 2013. Lipid droplets and peroxisomes: Key players in cellular lipid homeostasis or a matter of fat—Store 'em up or burn 'em down. *Genetics* **193**: 1–50.

Kurat CF, Natter K, Petschnigg J, Wolinski H, Scheuringer K, Scholz H, Zimmermann R, Leber R, Zechner R, Kohlwein SD. 2006. Obese yeast: Triglyceride lipolysis is functionally conserved from mammals to yeast. *J Biol Chem* **281**: 491–500.

Kurat CF, Wolinski H, Petschnigg J, Kaluarachchi S, Andrews B, Natter K, Kohlwein SD. 2009. Cdk1/Cdc28-dependent activation of the major triacylglycerol lipase Tgl4 in yeast links lipolysis to cell-cycle progression. *Mol Cell* **33**: 53–63.

Loewen CJ, Gaspar ML, Jesch SA, Delon C, Ktistakis NT, Henry SA, Levine TP. 2004. Phospholipid metabolism regulated by a transcription factor sensing phosphatidic acid. *Science* **304**: 1644–1647.

Lupetti A, Danesi R, Campa M, Del Tacca M, Kelly S. 2002. Molecular basis of resistance to azole antifungals. *Trends Mol Med* **8**: 76–81.

Martin CE, Oh CS, Jiang Y. 2007. Regulation of long chain unsaturated fatty acid synthesis in yeast. *Biochim Biophys Acta* **1771**: 271–285.

Natter K, Kohlwein SD. 2013. Yeast and cancer cells—Common principles in lipid metabolism. *Biochim Biophys Acta* **1831**: 314–326.

Nohturfft A, Zhang SC. 2009. Coordination of lipid metabolism in membrane biogenesis. *Annu Rev Cell Dev Biol* **25**: 539–566.

Schneiter R. 1999. Brave little yeast, please guide us to Thebes: Sphingolipid function in *S. cerevisiae*. *BioEssays* **21**: 1004–1010.

Schneiter R, Brugger B, Sandhoff R, Zellnig G, Leber A, Lampl M, Athenstaedt K, Hrastnik C, Eder S, Daum G, et al. 1999. Electrospray ionization tandem mass spectrometry (ESI-MS/MS) analysis of the lipid molecular species composition of yeast subcellular membranes reveals acyl chain-based sorting/remodeling of distinct molecular species en route to the plasma membrane. *J Cell Biol* **146**: 741–754.

Shui G, Guan XL, Low CP, Chua GH, Goh JS, Yang H, Wenk MR. 2010. Toward one step analysis of cellular lipidomes using liquid chromatography coupled with mass spectrometry: Application to *Saccharomyces cerevisiae* and *Schizosaccharomyces pombe* lipidomics. *Mol Biosyst* **6**: 1008–1017.

Tehlivets O, Scheuringer K, Kohlwein SD. 2007. Fatty acid synthesis and elongation in yeast. *Biochim Biophys Acta* **1771**: 255–270.

Zanghellini J, Natter K, Jungreuthmayer C, Thalhammer A, Kurat CF, Gogg-Fassolter G, Kohlwein SD, Von Grünberg HH. 2008. Quantitative modeling of triacylglycerol homeostasis in yeast—Metabolic requirement for lipolysis to promote membrane lipid synthesis and cellular growth. *FEBS J* **275**: 5552–5563.

Protocol 1

Lipid Extraction from Yeast Cells

Oskar L. Knittelfelder and Sepp D. Kohlwein[1]

Institute of Molecular Biosciences, University of Graz, BioTechMed Graz, 8010 Graz, Austria

The diversity of lipid molecules in biological tissues makes their analysis an experimental challenge. Not only do lipids differ greatly in their chemical structures and biophysical properties, but they also occur in greatly varying concentrations in living cells. Accordingly, even for an organism with a relatively simple lipidome such as yeast, multiple extraction and analysis protocols have been developed because none of them allows comprehensive and quantitative determination of all the diverse molecular lipid species. Here we describe an extraction procedure that results in good yields of neutral lipids and glycerophospholipids from yeast. The resulting samples are suitable for analysis by thin-layer chromatography, gas chromatography/mass spectrometry, or high-performance liquid chromatography.

MATERIALS

It is essential that you consult the appropriate Material Safety Data Sheets and your institution's Environmental Health and Safety Office for proper handling of equipment and hazardous material used in this protocol.

Reagents

Chloroform ($CHCl_3$) (Sigma-Aldrich 34854)

The chloroform used in this protocol is stabilized with amylene (pentene). Other stabilizers can have a negative impact on lipid extraction and can lead to undesired side reactions.

All solvents used for extraction and lipid analysis are at least gradient grade.

Growth medium (appropriate for yeast strain of interest)
Liquid nitrogen
Methanol (MeOH) (Carl Roth 4627.2)
$MgCl_2$ (0.034% in H_2O) (Merck 172571)
Yeast strain of interest

Equipment

Centrifuge (tabletop)
Glass beads (Sartorius BBI-8541701)
Sample vials (glass; 1.5-mL and 200-µL)
Shaker (Multi Reax from Heidolph)
Test tubes (Pyrex; with caps with Teflon seals; 25-mL and 15-mL)

---

[1]Correspondence: sepp.kohlwein@uni-graz.at

Copyright © Cold Spring Harbor Laboratory Press; all rights reserved
Cite this protocol as *Cold Spring Harb Protoc*; doi:10.1101/pdb.prot085449

Chapter 13

METHOD

Cell Growth and Harvest

1. Grow the yeast cells in an appropriate medium until the desired cell density is attained.

    1×10^9 cells are typically sufficient for a complete fatty acid and phospholipid analysis (with samples in triplicate).

2. Harvest the cells by centrifuging at 1000g for 3 min at room temperature. Remove the supernatant completely.

3. Shock-freeze the cells in liquid nitrogen and store them at −75°C.

Lipid Extraction

This method is based on the extraction procedure described by Folch et al. (1957) and has been adapted for yeast cells (Schneiter and Daum 2006; Deranieh et al. 2013). It results in good yields of neutral and glycerophospholipids. Sphingolipids are also extracted, but only partially, and the quantitative extraction of these lipids requires modifications to the protocol (see Angus and Lester 1972; Guan and Wenk 2006), which are not covered here.

4. Put ∼1 mL of washed glass beads in a 25-mL Pyrex tube.

    Use a microcentrifuge tube to scoop the glass beads.

5. Using a glass pipette, add 5 mL of cold (4°C) $CHCl_3$/MeOH (2:1) to the frozen cell pellet. Vortex briefly and transfer the cell suspension quantitatively with the pipette to the tube containing the glass beads.

    The solvent is very volatile and tends to drip from the pipette tip. Prerinse the pipette to saturate the tip.

6. Screw the Teflon-sealed cap tightly on the tube. Place it in the Multi Reax shaker and shake at maximum speed for 30 min at 4°C.

7. Add 1 mL of a cold (4°C) 0.034% $MgCl_2$ solution to the suspension.

    This step leads to a phase separation.

8. Close the tube tightly and shake again at maximum speed for 10 min at 4°C.

9. Centrifuge the sample in a tabletop centrifuge at 1000g for 3 min at room temperature.

    The milky suspension should separate into two clear phases, and the organic lower phase contains the extracted lipids. The interphase contains precipitated proteins. See Troubleshooting.

10. Remove and discard the aqueous upper phase.

    Make sure that the aqueous phase, which also contains methanol and $CHCl_3$, is disposed of properly.

11. Add 2 mL of artificial upper phase MeOH/H_2O/$CHCl_3$ (48:47:3) to the extract and vortex extensively.

    This step extracts some of the proteins from the organic phase.

12. Centrifuge the sample in a tabletop centrifuge at 1000g for 3 min at room temperature.

13. Remove and discard the aqueous upper phase.

    Make sure that the aqueous phase, which also contains methanol and $CHCl_3$, is disposed of properly.

14. Transfer the organic lower phase to a 15-mL Pyrex tube.

    Use a pulled-out glass pipette to avoid contamination with plastic compounds. Avoid the transfer of protein precipitate from the interphase.

15. Wash the remaining glass beads with 2 mL of $CHCl_3$/MeOH (2:1) and 1 mL of artificial upper phase MeOH/H_2O/$CHCl_3$ (48:47:3). Close the tube and vortex vigorously.

16. Centrifuge the sample in a tabletop centrifuge at 1000g for 3 min at room temperature.

17. Remove and discard the aqueous upper phase.

    Make sure that the aqueous phase, which also contains methanol and $CHCl_3$, is disposed of properly.

18. Transfer the remaining organic lower phase to the 15-mL Pyrex tube containing the extract from Step 14.

    The organic phase can turn to a white emulsion indicating the presence of a small amount of water, which typically is not a problem. It will be removed by the drying step.

19. Repeat Steps 15–18 once.

20. Evaporate the solvent under a stream of nitrogen in a fume hood.

    To accelerate this process, heat the Pyrex tubes in a sample evaporator.

21. Dissolve the dried lipids in 1 mL of $CHCl_3$/MeOH (2:1), and transfer the extract into a glass sample vial.

22. Evaporate the solvent under a stream of nitrogen in a fume hood.

23. Store the dried lipid extracts at −75°C.

    The resulting extracts can be analyzed by thin-layer chromatography, gas chromatography/mass spectrometry, or high-performance liquid chromatography (see Protocol 2: Thin-Layer Chromatography to Separate Phospholipids and Neutral Lipids from Yeast [Knittelfelder and Kohlwein 2015a], Protocol 3: Derivatization and Gas Chromatography of Fatty Acids from Yeast [Knittelfelder and Kohlwein 2015b], and Protocol 4: Quantitative Analysis of Yeast Phospholipids and Sterols by High-Performance Liquid Chromatography– Evaporative Light-Scattering Detection [Knittelfelder and Kohlwein 2015c]). Samples can be stored for several months without noticeable degradation. See Troubleshooting.

TROUBLESHOOTING

Problem (Step 9): No phase separation occurs after the addition of aqueous $MgCl_2$ solution.
Solution: Owing to the different volatilities of $CHCl_3$ and MeOH in the extraction solvents, the ratio of these components may change upon extended storage. Prepare all solvent mixtures freshly before starting a new extraction. Do not store solvent mixtures for >1 wk.

Problem (Step 23): Lipids are degraded.
Solution: The glass beads have reactive surfaces that may lead to lipid degradation. It is therefore important to wash the glass beads carefully before use. Zirconia beads (0.5 mm) (e.g., from BioSpec Products) may be used instead of glass beads. Use of Zymolyase, which degrades the cell wall, also facilitates homogenization and improves lipid yield. However, Zymolyase incubation may take 10–30 min at room temperature or at 30°C, which may lead to alterations in the lipid profile. Old and stationary phase cells are less susceptible to Zymolyase treatment. Alternatives are freeze-drying of cells and lipid extraction in the presence of glass beads under vigorous shaking overnight at 4°C. The latter procedure can be adapted for multiple samples (e.g., in a 48-well Eppendorf shaker in the cold room).

Problem (Step 23): Samples are contaminated with plastic materials. Plastic pipette tips and tubes contain plasticizer that may be extracted by solvents and interfere with subsequent analytical steps.
Solution: Avoid the use of plastic materials (tips and tubes) in all steps involving organic solvents—use glassware. Rinse new and recycled or washed Pyrex tubes twice with solvent before use. Cap seals should be made of Teflon to avoid possible contamination and should be rinsed with solvent before use.

ACKNOWLEDGMENTS

We thank members of our laboratory for helpful suggestions. O.L.K. is a member of the PhD program "Molecular Enzymology" funded by the Austrian Science Funds (FWF), which also supported project F3005-B12 (SFB LIPOTOX) to S.D.K. Additional support by NAWI Graz and BioTechMed Graz is gratefully acknowledged.

Chapter 13

REFERENCES

Angus WW, Lester RL. 1972. Turnover of inositol- and phosphorus-containing lipids in *Saccharomyces cerevisiae*: Extracellular accumulation of glycerophosphorylinositol derived from phosphatidylinositol. *Arch Biochem Biophys* **151**: 483–495.

Deranieh RM, Joshi AS, Greenberg ML. 2013. Thin-layer chromatography of phospholipids. In *Methods in molecular biology: Membrane biogenesis*, 1st ed. (ed. Rapaport D, Herrmann JM), Vol. **1033**, pp. 21–27. Humana Press, Totowa, NJ.

Folch J, Lees M, Sloane GH. 1957. A simple method for the isolation and purification of total lipides from animal tissues. *J Biol Chem* **226**: 497–509.

Guan XL, Wenk MR. 2006. Mass spectrometry-based profiling of phospholipids and sphingolipids in extracts from *Saccharomyces cerevisiae*. *Yeast* **23**: 465–477.

Knittelfelder OL, Kohlwein SD. 2015a. Thin-layer chromatography to separate phospholipids and neutral lipids from yeast. *Cold Spring Harb Protoc* doi: 10.1101/pdb.prot085456.

Knittelfelder OL, Kohlwein SD. 2015b. Derivatization and gas chromatography of fatty acids from yeast. *Cold Spring Harb Protoc* doi: 10.1101/pdb.prot085464.

Knittelfelder OL, Kohlwein SD. 2015c. Quantitative analysis of yeast phospholipids and sterols by high-performance liquid chromatography—Evaporative light-scattering detection. *Cold Spring Harb Protoc* doi: 10.1101/pdb.prot085472.

Schneiter R, Daum G. 2006. Extraction of yeast lipids. In *Methods in molecular biology: Yeast protocols*, 2nd ed. (ed. Wei X), Vol. **313**, pp. 41–45. Humana Press, Totowa, NJ.

Protocol 2

Thin-Layer Chromatography to Separate Phospholipids and Neutral Lipids from Yeast

Oskar L. Knittelfelder and Sepp D. Kohlwein[1]

Institute of Molecular Biosciences, University of Graz, BioTechMed Graz, 8010 Graz, Austria

Thin-layer chromatography (TLC) is a versatile technique for the separation of lipid classes. It provides excellent resolution, can be used for both preparative and analytical applications, and does not require expensive equipment. Here we describe the use of different solvent systems to separate yeast phospholipids and neutral lipids by TLC in one dimension. Resolved lipid species are visualized by iodine vapor or by charring after treatment with sulfuric acid and manganese chloride. Neither of these staining techniques yields a quantitative readout because the mixture of various lipids in yeast affects iodine absorption and charring efficiency; standard curves are required to obtain semiquantitative estimates of the relative lipid composition.

MATERIALS

It is essential that you consult the appropriate Material Safety Data Sheets and your institution's Environmental Health and Safety Office for proper handling of equipment and hazardous material used in this protocol.

Reagents

Acetic acid (glacial) (EMD Millipore 1000632511)
  All solvents used for extraction and lipid analysis are at least gradient grade.

Chloroform ($CHCl_3$) (Sigma-Aldrich 34854)
Diethylether (EMD Millipore 1009211000)
Ergosterol (Acros Organics 117810050)
Ethanol (EMD Millipore 1009832500)
Iodine crystals (EMD Millipore 1047610100)
L-α-phosphatidylinositol (Avanti Polar Lipids 840042P)
Lipid extracts (dried) (see Protocol 1: Lipid Extraction from Yeast Cells [Knittelfelder and Kohlwein 2015])
Manganese chloride ($MnCl_2$) (Sigma-Aldrich M3634)
Methanol (MeOH) (Carl Roth 4627.2)
Petroleum ether (Carl Roth T173.2)
Phosphatidylcholine 34:1 (Avanti Polar Lipids 850457P)
Phosphatidylethanolamine 34:1 (Avanti Polar Lipids 850757P)
Phosphatidylserine 34:1 (Avanti Polar Lipids 840034P)

---

[1]Correspondence: sepp.kohlwein@uni-graz.at

Copyright © Cold Spring Harbor Laboratory Press; all rights reserved
Cite this protocol as *Cold Spring Harb Protoc*; doi:10.1101/pdb.prot085456

Sulfuric acid (H$_2$SO$_4$) (EMD Millipore 1007311000)
Triacylglycerol 48:3 (Larodan 33-1610-12)

Equipment

Automated sample applicator (Automatic TLC Sampler 4 from CAMAG Scientific)
Drying oven (set at 100°C for activating plates and 120°C for charring plates)
Filter paper (Whatman)
Glass thin-layer chromatography (TLC) developing chambers
TLC silica gel 60 plates (EMD Millipore 1055530001)

METHOD

1. Prepare the solvent mixtures.

   - For separation of phospholipids, prepare a CHCl$_3$/MeOH/water (32.5:12.5:2) mixture (Mangold 1961).

   - For separation of neutral lipids, prepare a petroleum ether/diethylether/acetic acid (32:8:0.8) mixture.

2. Add the solvent mixtures from Step 1 to separate glass TLC developing chambers.

   Instead of running two separate TLC plates for neutral lipid and phospholipid separation, both steps can be combined by successive development of plates in two different solvent mixtures (Deranieh et al. 2013).

3. Saturate the atmosphere in the chambers for at least 30–60 min.

   To facilitate saturation with solvent, cover the inside of the chamber with filter paper and seal the chamber tightly with a glass lid. Avoid frequent opening of the lid.

4. Activate two TLC silica gel 60 plates (20 × 10 cm; one for neutral lipids and one for phospholipids) in a drying oven for 30–60 min at 100°C.

5. Dissolve the dried lipid extracts in 1 mL of CHCl$_3$/MeOH (2:1) and the lipid standards at a concentration of 0.5 mg/mL in CHCl$_3$/MeOH (2:1).

6. Spot 10 μL of each sample and the lipid standards onto the plates.

   Use an automated sample applicator to facilitate spotting and to improve the resolution of separation of lipid species. If the samples are applied manually, spot in a small volume with a glass syringe.

7. Develop the plates in the appropriate solvent system until the solvent front reaches the top of the plate.

   To maintain a solvent-saturated atmosphere in the chamber, do not open the lid fully while the plates are developing.

8. Dry the plates under a fume hood.

   Because of the water content of the solvent, this step may take several minutes and can be accelerated by using a hair dryer at very low heat. Cool the plates to room temperature before visualization.

9. Visualize the lipids resolved by TLC using either of the following methods.

   - Place a few crystals of iodine inside a glass chamber. Insert the plate and close the lid.

     Iodine binds reversibly to lipids that contain double bonds (e.g., fatty acids and sterols). Iodine staining is not quantitative, because the staining intensity depends on the number of double bonds in the lipid species. However, it is a reversible procedure that may provide a quick estimate about relative changes in the lipid content between extracts.

   - Dip the plate for 10 sec into a solution containing 50% ethanol in water, 3.2% H$_2$SO$_4$, and 0.5% MnCl$_2$. Char the plate for 30 min at 120°C (Fig. 1).

     Be aware that iodine and sulfuric acid visualize all organic compounds present on the TLC plate. To monitor possible contaminations, include solvent blanks.
     See Troubleshooting.

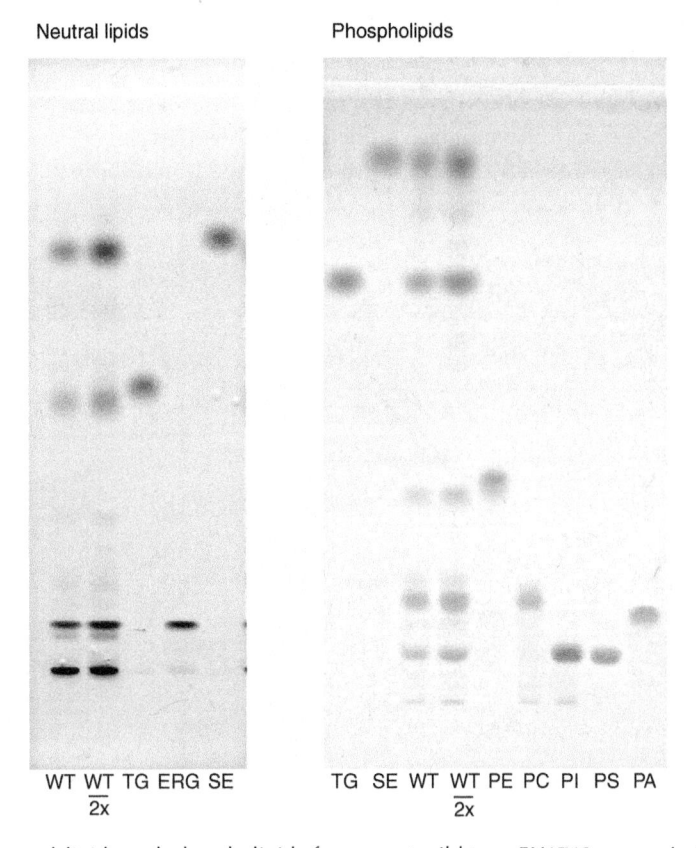

FIGURE 1. TLC of neutral lipids and phospholipids from yeast wild-type BY4742 grown in YPD medium to the stationary phase (i.e., for ~48 h). Lipid extracts were prepared from ~3 × 10^8 cells and dissolved in 1 mL of CHCl$_3$/MeOH 2:1 (v/v). 20-μL samples were applied to TLC plates using an automated applicator. 10-μL samples of appropriate lipid standards (0.5 mg/mL in CHCl$_3$/MeOH 2:1 [v/v]) were also applied. Lipid spots were visualized by charring after treatment with sulfuric acid and MnCl$_2$. WT, wild-type lipid extract; TG, triacyglycerol; ERG, ergosterol; SE, steryl esters; PE, phosphatidylethanolamine; PC, phosphatidylcholine; PI, phosphatidylinositol; PS, phosphatidylserine; PA, phosphatidic acid.

TROUBLESHOOTING

Problem (Step 9): The lipid classes are poorly resolved.

Solution: Use an automated sample applicator to improve the separation of the different lipid species. To achieve optimal saturation of the chamber with solvent vapor, insert sheets of filter paper to cover the chamber walls, close the chambers tightly with a lid, and open only very briefly to insert the plate.

RELATED INFORMATION

As an alternative for quantification after chromatographic separation, phospholipids may be radiolabeled by incubating yeast cultures to steady state with [^{32}P] orthophosphate, which facilitates quantification by autoradiography, PhosphorImager, or scintillation counting (Deranieh et al. 2013). Total lipids, including sterols, may be quantified by incubating yeast cultures to steady state with [2-^{14}C] acetate (Morlock et al. 1988). The use of radiolabeled substrates requires appropriate facilities, safety precautions, and special care with TLC plates to avoid release of radioactive dust particles.

ACKNOWLEDGMENTS

We thank members of our laboratory for helpful suggestions. O.L.K. is a member of the PhD program "Molecular Enzymology" funded by the Austrian Science Funds (FWF), which also supported project F3005-B12 (SFB LIPOTOX) to S.D.K. Additional support by NAWI Graz and BioTechMed Graz is gratefully acknowledged.

REFERENCES

Deranieh RM, Joshi AS, Greenberg ML. 2013. Thin-layer chromatography of phospholipids. In *Methods in molecular biology: Membrane biogenesis*, 1st ed. (ed. Rapaport D, Herrmann JM), Vol. 1033, pp. 21–27. Humana Press, Totowa, NJ.

Knittelfelder OL, Kohlwein SD. 2015. Lipid extraction from yeast cells. *Cold Spring Harb Protoc* doi: 10.1101/pdb.prot085449.

Mangold HK. 1961. Thin-layer chromatography of lipids. *J Am Oil Chem Soc* 38: 708–721.

Morlock KR, Lin YP, Carman GM. 1988. Regulation of phosphatidate phosphatase activity by inositol in *Saccharomyces cerevisiae*. *J Bacteriol* 170: 3561–3566.

Protocol 3

Derivatization and Gas Chromatography of Fatty Acids from Yeast

Oskar L. Knittelfelder and Sepp D. Kohlwein[1]

Institute of Molecular Biosciences, University of Graz, BioTechMed Graz, 8010 Graz, Austria

Analysis of fatty acids by gas chromatography (GC) is facilitated by the generation of volatile fatty acid methyl ester (FAME) derivatives. Here we describe the esterification procedure and a typical program for separating FAMEs using a gas chromatograph equipped with a flame ionization detector or a dual stage quadrupole-mass spectrometry detector. GC is a rather simple technique that provides quantitative information on cellular fatty acid content and composition by use of an internal fatty acid standard.

MATERIALS

It is essential that you consult the appropriate Material Safety Data Sheets and your institution's Environmental Health and Safety Office for proper handling of equipment and hazardous material used in this protocol.

Reagents

BF_3 in MeOH (Sigma-Aldrich B1252)

BF_3 is toxic and may deteriorate over time. For best results, purchase small quantities (100 mL) and store at 4°C. It has a limited shelf life; if any precipitate is visible, discard the solution and order a new one.

C14:0 fatty acid (Sigma-Aldrich 70079)
C16:0 fatty acid (Larodan 10-1600-13)
C16:1 fatty acid (Larodan 10-1601-30)
C17:0 fatty acid (Sigma-Aldrich H3500)
C18:0 fatty acid (Larodan 10-1800-13)
C18:1 fatty acid (Larodan 10-1801-17)
Chloroform ($CHCl_3$) (Sigma-Aldrich 34854)
Hexane (J.T. Baker 9262)
Lipid extracts (dried) (see Protocol 1: Lipid Extraction from Yeast Cells [Knittelfelder and Kohlwein 2015])
Methanol (MeOH) (Carl Roth 4627.2)
Toluene (EMD Millipore 1083252500)

Equipment

Centrifuge (tabletop)
Gas chromatograph (e.g., Trace GC Ultra from Thermo Scientific) equipped with an FID or DSQ-MS detector

---

[1]Correspondence: sepp.kohlwein@uni-graz.at

Copyright © Cold Spring Harbor Laboratory Press; all rights reserved
Cite this protocol as *Cold Spring Harb Protoc*; doi:10.1101/pdb.prot085464

Chapter 13

GC column for FID (e.g., a type MH 30-15 fused silica 25-m column from Varian)
GC column for MS (e.g., a type DB-5 60-m column from Agilent Technologies)
Incubation oven (set at 100°C)
Overhead shaker
Test tubes (Pyrex; with caps with Teflon seals; 15 mL)

METHOD

Generation of FAME Derivatives

1. Prepare samples in triplicate for a calibration curve. Dissolve desired fatty acids (e.g., C14:0, C16:0, C16:1, C18:0, and C18:1) each at a concentration of 0.4 mg/mL in methanol and prepare serial dilutions (up to 10 concentrations).

   Store fatty acid stock solutions at −20°C.

2. Dissolve dried lipid extracts in 1 mL of $CHCl_3$/MeOH (2:1).

3. Transfer 200 µL of the lipid extract or 100 µL of the desired fatty acid stock solutions into individual 15-mL Pyrex tubes.

4. Add 100 µL of internal standard solution (i.e., 250 µM C17:0 in methanol) to each tube.

5. Evaporate the solvent under a stream of nitrogen in a fume hood.

   To accelerate this process, heat the Pyrex tubes in a sample evaporator.

6. Add 0.5 mL of toluene to each tube.

   Use dry toluene to avoid side reactions with water and the resultant decreased recovery of FAMEs.

7. Add 2 mL of BF_3 in methanol to each tube.

   Instead of BF_3 in methanol, HCl, or H_2SO_4 dissolved in methanol can be used for the esterification of FAs (Christie 1993). Substitute the 2 mL of BF_3 in methanol in Step 7 above, with 3 mL of 2% HCl in methanol and proceed with the same protocol.

8. Vortex briefly and screw the Teflon-sealed caps tightly on the tubes.

9. Incubate in an oven for 1 h at 100°C.

10. Cool the samples on ice.

11. Add 1 mL of ice-cold distilled/deionized water to each tube.

12. Add 2 mL of hexane/$CHCl_3$ (4:1) to each tube.

    Prepare solvent mixtures immediately before use.

13. Vortex the samples vigorously and shake on an overhead shaker for 15 min.

14. Centrifuge in a tabletop centrifuge at 2000 rpm for 5 min at room temperature.

15. Transfer the upper organic phase from each sample into a clean Pyrex tube.

16. Repeat Steps 11–14 once.

17. Combine the upper phases from each sample and dry under a stream of nitrogen in the fume hood.

18. Add 500 µL of hexane to each tube and vortex briefly to dissolve and collect the FAMEs at the bottom of the tube.

    Carefully rinse the inside of the Pyrex tubes with the hexane to collect all of each FAME sample.

19. Dry the samples under a stream of nitrogen.

20. Dissolve each sample in 100 µL of hexane, vortex briefly, and transfer the FAMEs into sample vials.

    The samples can be stored up to several months at −20°C. Please note that hexane is very volatile and may need to be refilled to dissolve the FAME.

Separation of FAME Derivatives by GC

21. Analyze the samples on a gas chromatograph equipped with an FID or DSQ-MS detector according to the following guidelines (Fig. 1).

    FID

    - Injection: Sample volume, 1–2 µL; injector temperature, 230°C; column, 25 m fused silica. Apply temperature gradient from 150°C (hold for 0 min) to 250°C (hold for 2 min) with 5°C/min. Apply a second ramp to increase the temperature to 260°C (hold for 5 min) with 10°C/min.
    - Detector settings: Base temperature, 200°C; ignition threshold, 0.2 pA; air flow rate, 200 mL/min; H_2 flow rate, 30 mL/min; makeup, 20 mL/min.

    DSQ-MS

    - Injection: Sample volume, 1–2 µL; injector temperature, 250°C; column, 60 m DB-5. Apply temperature gradient from 110°C (hold for 4 min) to 300°C (hold for 10 min) with 20°C/min.
    - DSQ-MS settings: Electron impact ionization (EI); ion source temperature, 280°C; start time, 8.5 min; total scan time, 0.25 sec; scan mode, full scan; mass range, *m/z* 50–700.

      See Troubleshooting.

      Numerous other GC columns are available for the separation of FAMEs. The DB-5 column represents a standard column and is suitable for separating simple (yeast) FAME mixtures. Note that the retention time (which is used to identify a particular FAME) varies according to the polarity of the column. Thus, for instance, saturated and unsaturated FAMEs may change the order in which they elute (Fig. 1). To

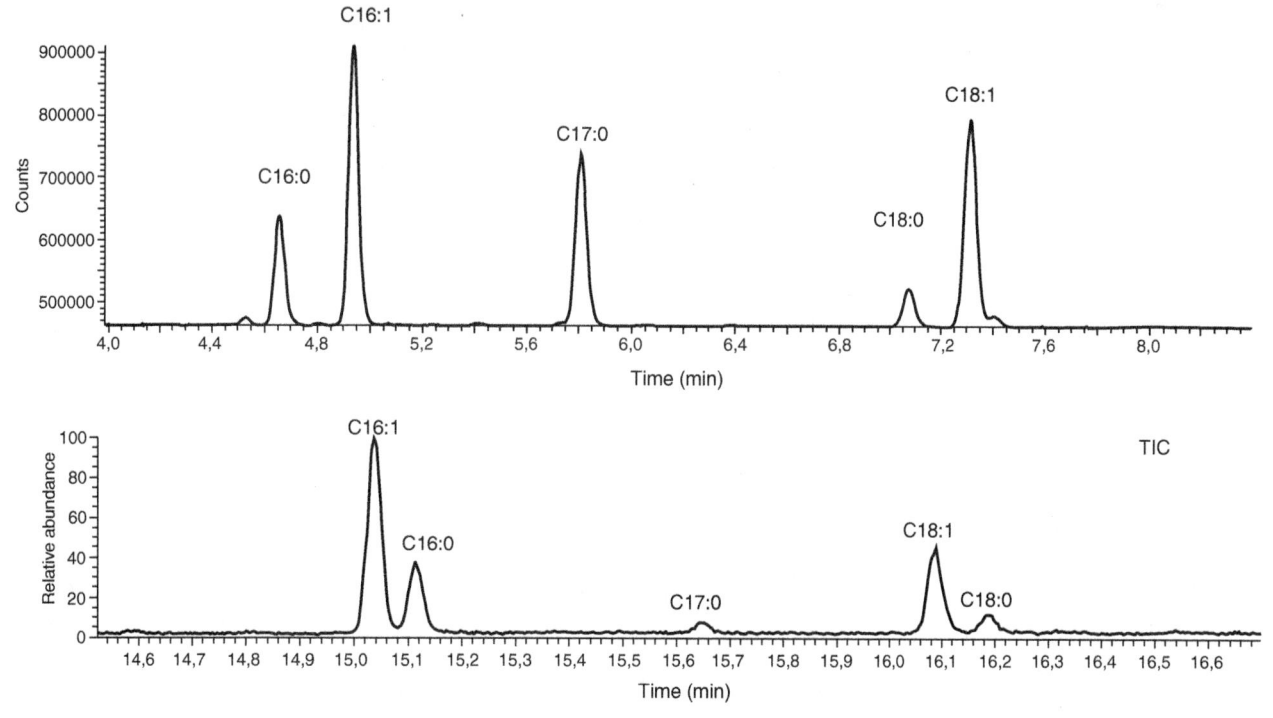

FIGURE 1. Fatty acid analysis from yeast extracts. (*Upper* panel) GC-FID chromatogram of FAMEs from wild-type yeast strain BY4742 grown to stationary phase in YPD. (*Lower* panel) GC-MS analysis of FAMEs from wild-type yeast grown in minimal medium and harvested in the logarithmic growth phase. Note that the retention times may vary dependent on the type of column used for separation. C17:0 (heptadecanoic acid) does not occur in yeast lipids and is used as an internal standard.

verify the retention times with an FID, it is necessary to measure single standards first. MS analysis defines FAMEs based on their mass. However, the fragmentation efficiency differs for different FAMEs, and the signal intensity (total ion current) does not accurately reflect the actual amounts. Rigorous standardization is required to obtain dose-response curves for individual FAMEs.

TROUBLESHOOTING

Problem (Step 21): Samples are contaminated with plasticizer or other volatile compounds.
Solution: Include blanks (i.e., CHCl$_3$/MeOH 2:1 solution without lipids) to check for possible contamination of internal stock solution, solvents, and glass tubes. Prepare calibration curves and solvent mixtures immediately before use.

Problem (Step 21): Only baseline signals are obtained.
Solution: Hexane can evaporate during measurements when sample vials are not tightly closed. Close sample vials as tightly as possible.

ACKNOWLEDGMENTS

We thank members of our laboratory for helpful suggestions. O.L.K. is a member of the PhD program "Molecular Enzymology" funded by the Austrian Science Funds (FWF), which also supported project F3005-B12 (SFB LIPOTOX) to S.D.K. Additional support by NAWI Graz and BioTechMed Graz is gratefully acknowledged.

REFERENCES

Christie WW. 1993. Preparation of ester derivatives of fatty acids for chromatographic analysis. In *Advances in lipid methodology-two*, 1st ed. (ed. Christie WW), pp. 69–111. Oily Press, Dundee, United Kingdom.

Knittelfelder OL, Kohlwein SD. 2015. Lipid extraction from yeast cells. *Cold Spring Harb Protoc* doi: 10.1101/pdb.prot085449.

Protocol 4

Quantitative Analysis of Yeast Phospholipids and Sterols by High-Performance Liquid Chromatography–Evaporative Light-Scattering Detection

Oskar L. Knittelfelder and Sepp D. Kohlwein[1]

Institute of Molecular Biosciences, University of Graz, BioTechMed Graz, 8010 Graz, Austria

Normal-phase high-performance liquid chromatography (HPLC) is a standard method for separating the major lipid classes in an extract. Owing to the absence of a common property like light absorbance in the various lipid classes, evaporative light-scattering detection (ELSD) is the method of choice for qualitative and quantitative lipid detection. In most cases, neutral lipids and polar lipids are separated by different solvent systems, making it necessary to perform multiple analyses. Compared with other techniques like thin-layer chromatography, normal-phase HPLC–ELSD has better reproducibility and allows a higher degree of automation. Here we describe a method for separating and quantifying yeast neutral lipids and glycerophospholipids in one analytical run.

MATERIALS

It is essential that you consult the appropriate Material Safety Data Sheets and your institution's Environmental Health and Safety Office for proper handling of equipment and hazardous material used in this protocol.

Reagents

Acetic acid (EMD Millipore 1000632511)
Acetone (Carl Roth 7328.2)
Ammonium acetate (J.T. Baker 0599-08)
Chloroform (CHCl$_3$) (Sigma-Aldrich 34854)
Cholesteryl palmitate (Sigma-Aldrich C6072)
Ergosterol (ACROS Organics 117810050)
Ethylacetate (EMD Millipore 1109721000)
Isooctane (Carl Roth 7340.1)
Isopropanol (Carl Roth 7343.1)
L-α-phosphatidylinositol (Avanti Polar Lipids 840042P)
Lipid extracts (dried) (see Protocol 1: Lipid Extraction from Yeast Cells [Knittelfelder and Kohlwein 2015])
Methanol (MeOH) (Carl Roth 4627.2)

---

[1]Correspondence: sepp.kohlwein@uni-graz.at

Copyright © Cold Spring Harbor Laboratory Press; all rights reserved
Cite this protocol as *Cold Spring Harb Protoc*; doi:10.1101/pdb.prot085472

Chapter 13

Phosphatidylcholine 34:1 (Avanti Polar Lipids 850457P)
Phosphatidylethanolamine 34:1 (Avanti Polar Lipids 850757P)
Phosphatidylserine 34:1 (Avanti Polar Lipids 840034P)
Triacylglycerol 48:3 (Larodan 33-1610-12)
Water for high-performance liquid chromatography (HPLC; LC–MS grade) (VWR 83645.320)

Equipment

Data acquisition and analysis software (Agilent ChemStation B.04.01)
Evaporative light-scattering detector (ELSD) (Sedex 85, SEDERE)
HPLC column (BETASIL Diol, 100 × 4.6-mm, 5-μm, Thermo Scientific)
HPLC system (Agilent 1100 comprising pump, injector, precooled sample manager, and column oven)
Test tubes (Pyrex; with caps with Teflon seals; 15-mL)

METHOD

1. Prepare 1-mg/mL stock solutions of triacylglycerol 48:3, ergosterol, cholesteryl palmitate, phosphatidylcholine 34:1, phosphatidylethanolamine 34:1, phosphatidylserine 34:1, and L-α-phosphatidylinositol in $CHCl_3$/MeOH 2:1. Use 15-mL Pyrex tubes and store stock solutions at −20°C.

2. Combine all the standards at a starting concentration of 0.35 mg/mL each in $CHCl_3$/MeOH (2:1). Dilute the standards with $CHCl_3$/MeOH (2:1) in successive 1:1 steps to obtain a calibration curve (seven to eight dilutions). Prepare calibration curve samples in triplicate.

3. Dissolve dried lipid extracts in 1 mL of $CHCl_3$/MeOH (2:1).

   If necessary, concentrate the samples by dissolving them in a lower volume.

4. Set up the HPLC system comprising pump, injector, precooled sample manager (set at 4°C), and column oven (set at 40°C). Use a ternary gradient (see Table 1) with a BETASIL Diol column to

TABLE 1. Evaporative light-scattering detector (ELSD) settings and solvent gradient

Temperature			60.0°C	
Gain			5 (0–15.6 min; 48–49 min) 7 (15.6–48 min)	
			Autozero when gain is changed	
Pressure			3.5 bar	

Time (min)	A (%)	B (%)	C (%)	Flow rate (mL/min)
0	100	0	0	0.5
3.5	100	0	0	0.5
5.5	97	3	0	0.5
9	94	6	0	0.5
11	35	50	15	0.5
13	34	39	27	0.5
17	43	30	27	0.5
18	43	30	27	0.5
21	40	0	60	0.5
25	40	0	60	0.5
35	0	100	0	0.5
40	0	100	0	0.5
42	100	0	0	0.5
49	100	0	0	0.5

Modified from Graeve and Janssen (2009).

The ternary gradient system consists of eluent A, which is isooctane:ethylacetate 99.8:0.2 (v/v), eluent B, which is acetone:ethylacetate 2:1 (v/v) containing 0.02% (v/v) acetic acid, and eluent C, which is isopropanol:water 85:15 (v/v) containing 0.05% (v/v) acetic acid and 3 mM ammonium acetate.

separate the various lipid classes. To detect lipids, use a Sedex 85 ELSD with parameters as specified in Table 1. Perform data acquisition using ChemStation B.04.01 software.

5. Inject 10 µL of sample or calibration standard.

   To establish the calibration curve, start with the lowest concentration of the dilution series. Verify the retention times of the analytes by injecting single standards.

6. Analyze the resulting chromatograms with suitable software (e.g., ChemStation B.04.01) (see Fig. 1).

   ELSD calibration curves are not linear. Calculation of absolute amounts can be simplified by using a logarithmic (log 10) correlation between peak area and concentration.

   See Troubleshooting.

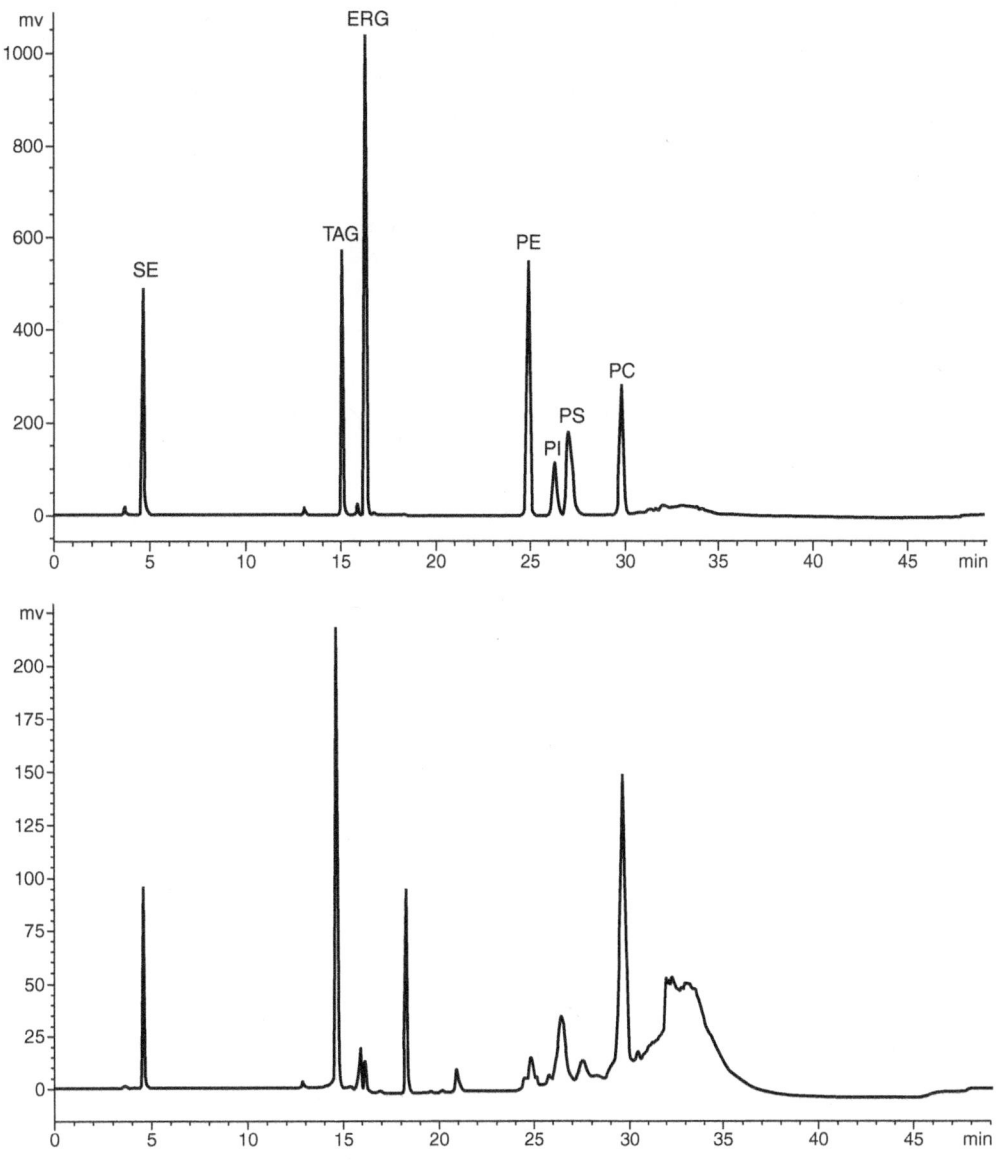

FIGURE 1. HPLC–ELSD chromatogram of 10 µL of a lipid standard mix (*upper* chromatogram; neutral lipids 0.3 mg/mL, phospholipids 0.35 mg/mL) and 10 µL of lipid extract from wild-type BY4742 (*lower* chromatogram). Note the different sensitivities for the various lipids in the ELSD, which requires standardization for all lipid classes. TAG, triacyglycerol; ERG, ergosterol; SE, steryl esters; PC, phosphatidylcholine, PS, phosphatidylserine; PI, phosphatidylinositol; PE, phosphatidylethanolamine.

TROUBLESHOOTING

Problem (Step 6): Signals are too low.

Solution: This may happen when the concentration of samples is low. Evaporate the sample and dissolve it in a smaller volume of $CHCl_3$/MeOH 2:1. If the signals of the calibration curve are too low, check the stock solutions, prepare fresh ones if necessary, and clean the nebulizer of the detector. The signal intensities can be further improved by increasing the gain of the instrument. Batyl alcohol may be added to the yeast cells before extraction as an internal standard, which elutes between neutral lipids (triacylglycerol and steryl esters) and phospholipids. This standardization can improve the results by accounting for different extraction efficacies.

ACKNOWLEDGMENTS

We thank members of our laboratory for helpful suggestions. O.L.K. is a member of the PhD program "Molecular Enzymology" funded by the Austrian Science Funds (FWF), which also supported project F3005-B12 (SFB LIPOTOX) to S.D.K. Additional support by NAWI Graz and BioTechMed Graz is gratefully acknowledged.

REFERENCES

Graeve M, Janssen D. 2009. Improved separation and quantification of neutral and polar lipid classes by HPLC-ELSD using a monolithic silica phase: Application to exceptional marine lipids. *J Chromatogr B Analyt Technol Biomed Life Sci* **877**: 1815–1819.

Knittelfelder OL, Kohlwein SD. 2015. Lipid extraction from yeast cells. *Cold Spring Harb Protoc* doi: 10.1101/pdb.prot085449.

CHAPTER 14

Methods for Synchronization and Analysis of the Budding Yeast Cell Cycle

Adam P. Rosebrock[1]

Donnelly Centre for Cellular and Biomolecular Research, University of Toronto, Toronto, Ontario M5S 3E1, Canada

Like other eukaryotes, budding yeast temporally separate cell growth and division. DNA synthesis is distinct from chromosome segregation. Storage carbohydrates are accumulated slowly and then rapidly liquidated once per cycle. Cyclin-dependent kinase associates with multiple different transcriptionally and posttranslationally regulated cyclins to drive the cell cycle. These and other crucial events of cellular growth and division are limited to narrow windows of the cell cycle. Many experiments in the yeast laboratory treat a culture of cells as a homogeneous mixture. Measurements of asynchronous cultures are, however, confounded by the presence of cells in various cell cycle stages; measuring a population average in unsynchronized cells provides at best a decreased signal and at worst an artifactual result. A number of experimentally tractable methods have been developed to generate populations of yeast cells that are synchronized with respect to cell cycle phase. Robust methods for determining cell cycle position have also been developed. These methods are introduced here.

SYNCHRONIZING CELLS

While growing in most common nutrient conditions, budding yeast cells spend time in each stage of the cell cycle: gap 1 (G_1), synthesis (S), gap 2 (G_2), and mitosis (M) (Fig. 1). Different methods have been developed to synchronize a population of cells at particular stages of this cycle. For some purposes, it is sufficient to block the cell cycle progression of a culture, permitting all cells to continue through to the blocked position, and then to examine the resulting culture of cells at a single cell cycle phase. Arrests in early S phase and in G_2/M by using hydroxyurea and nocodazole, respectively, are provided in an accompanying protocol (see Protocol 1: Synchronization and Arrest of the Budding Yeast Cell Cycle Using Chemical and Genetic Methods [Rosebrock 2015c]). These approaches are technically simple, scalable, and afford quantitative synchronization: The appropriate drug is added to cells, and, after more than one doubling time, a fully arrested culture of cells is harvested for analysis.

When studying dynamic behaviors, examining cells blocked at one or more cell cycle positions may be inadequate. Synchronously cycling *Saccharomyces cerevisiae* cultures can be generated by genetic or chemical block-release methods, such as the reversible G_1 and M/G_1 arrests using α-factor mating pheromone and the temperature-sensitive *cdc15-2* allele, respectively (see Protocol 1: Synchronization and Arrest of the Budding Yeast Cell Cycle Using Chemical and Genetic Methods [Rosebrock 2015a]), or by elutriation, which is a physical method to isolate cells of a narrow size, and therefore cell cycle window (see Protocol 2: Synchronization of Budding Yeast by Centrifugal Elutriation [Rosebrock 2015b]). Block-release methods enable use of all cells in the culture—as with other cell cycle

---

[1]Correspondence: adam.rosebrock@utoronto.ca

Copyright © Cold Spring Harbor Laboratory Press; all rights reserved
Cite this introduction as *Cold Spring Harb Protoc*; doi:10.1101/pdb.top080630

Chapter 14

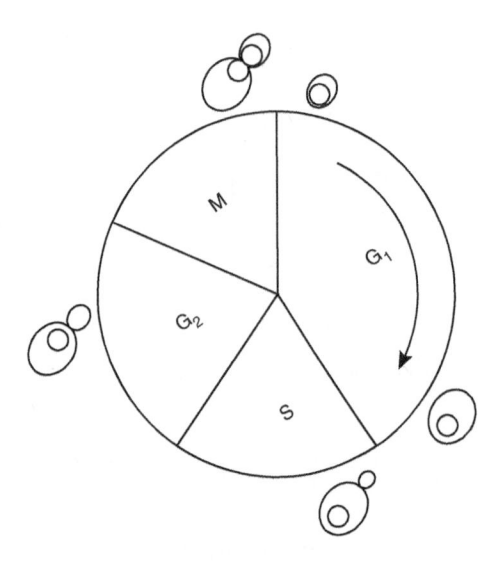

FIGURE 1. The budding yeast cell cycle is separated into cytologically distinct G_1, S, G_2, and M phases. Cells are born as small G_1 (1N DNA content) daughters that grow in size to reach critical size and pass the G_1/S restriction point, known as START. During S phase (>1N, <2N DNA content), cells bud and replicate their genomes. Bud size increases during G_2 (2N DNA content), and, at the G_2/M transition, the nucleus is brought to the bud neck for partitioning into mother and daughter cells.

arrests, the culture is blocked for a time sufficient to ensure that all cells will have "piled up" at the block. Elutriation, in contrast, separates a mixed-size asynchronous culture into one or more fractions of similarly sized cells. Generating 1 L of mid-log cells by block-release requires <1 L of cells to start—an equivalent yield from elutriation requires 10–20 L of input!

During normal growth, the cell cycle is linked to cell size. Cell cycle arrests, including block-release synchronizations, decouple growth from cell cycle progression: Blocked cells continue to synthesize protein, build biomass, and grow in size despite not progressing through the cell cycle. Consequently, blocked cells are not of a homogeneous size: Cells that were *just past* the point of arrest at the start of an experiment will be nearly "normal" in size at release, whereas cells that were arrested *just before* the arrest point will be released at a size far larger than normal. *One must interpret data from experiments based on cell cycle blocks in light of increased and heterogeneous cell size.*

The mother–daughter asymmetry of budding yeast implicitly limits the length of time that synchrony can be maintained. In block-release experiments using cells grown in rich media, synchrony is maintained for two or three cell cycles following release. However, cells grown in relatively "slow" carbon sources maintain synchrony only until the first division, after which the culture becomes an admixture of subpopulations of synchronized mother and daughter cells, the latter subject to delay before reaching the critical size.

Techniques used for cell cycle synchronization, to a greater degree than most experimental manipulations, are susceptible to observer effects: The methods used alter the biology being studied. It is essential to corroborate experimental findings using orthogonal synchronization techniques. For example, heat-shock effects from a *cdc15-2* synchronization are not present in an α-factor block release, and the confound of large cell size of block-release experiments is absent from elutriation. Elutriation, however, exposes cells to altered media conditions and cell-wall stress from prolonged centrifugation not experienced in chemical or genetic methods.

Note that the accompanying protocols have been optimized for the S288c strain background, one of the most commonly used laboratory strains and the basis for the original yeast deletion collection (Giaever et al. 2002). Irrespective of strain background and genotype, it is essential that synchrony be empirically assessed and experiments interpreted as a function of both cell cycle phase and quality of synchronization.

DETERMINING CELL CYCLE POSITION

Different methods for determining cell cycle position suit various experimental designs. Transmitted light microscopy can be used to count cells with and without buds and to provide a crude measure of mother and bud size. Epifluorescent microscopy of a DNA-binding dye or intercalating dyes such as 4′,6-diamidino-2-phenylindole (DAPI) can be used to monitor nuclear migration and chromosome segregation late in the cell cycle. Measurements of cell size provide a window into cell cycle position and are crucial to successful elutriation; an introduction to Coulter counting in the context of elutriation is available (see Protocol 2: Synchronization of Budding Yeast by Centrifugal Elutriation [Rosebrock 2015b]).

The most direct, universally applicable readout of cell cycle position is the measurement of DNA content by flow cytometry. A robust, high-throughput protocol to prepare budding yeast cells for flow cytometric measurement of DNA content is provided as a separate method (see Protocol 3: Analysis of the Budding Yeast Cell Cycle by Flow Cytometry [Rosebrock 2015c]).

REFERENCES

Giaever G, Chu AM, Ni L, Connelly C, Riles L, Véronneau S, Dow S, Lucau-Danila A, Anderson K, André B, et al. 2002. Functional profiling of the *Saccharomyces cerevisiae* genome. *Nature* 418: 387–391.

Rosebrock AP. 2015a. Synchronization and arrest of the budding yeast cell cycle using chemical and genetic methods. *Cold Spring Harb Protoc* doi: 10.1101/pdb.prot088724.

Rosebrock AP. 2015b. Synchronization of budding yeast by centrifugal elutriation. *Cold Spring Harb Protoc* doi: 10.1101/pdb.prot088732.

Rosebrock AP. 2015c. Analysis of the budding yeast cell cycle by flow cytometry. *Cold Spring Harb Protoc* doi: 10.1101/pdb.prot088740.

Protocol 1

Synchronization and Arrest of the Budding Yeast Cell Cycle Using Chemical and Genetic Methods

Adam P. Rosebrock[1]

Donnelly Centre for Cellular and Biomolecular Research, University of Toronto, Toronto, Ontario M5S 3E1, Canada

The cell cycle of budding yeast can be arrested at specific positions by different genetic and chemical methods. These arrests enable study of cell cycle phase–specific phenotypes that would be missed during examination of asynchronous cultures. Some methods for arrest are reversible, with kinetics that enable release of cells back into a synchronous cycling state. Benefits of chemical and genetic methods include scalability across a large range of culture sizes from a few milliliters to many liters, ease of execution, the absence of specific equipment requirements, and synchronization and release of the entire culture. Of note, cell growth and division are decoupled during arrest and block-release experiments. Cells will continue transcription, translation, and accumulation of protein while arrested. If allowed to reenter the cell cycle, cells will do so as a population of mixed, larger-than-normal cells. Despite this important caveat, many aspects of budding yeast physiology are accessible using these simple chemical and genetic tools. Described here are methods for the block and release of cells in G_1 phase and at the M/G_1 transition using α-factor mating pheromone and the temperature-sensitive *cdc15-2* allele, respectively, in addition to methods for arresting the cell cycle in early S phase and at G_2/M by using hydroxyurea and nocodazole, respectively.

MATERIALS

It is essential that you consult the appropriate Material Safety Data Sheets and your institution's Environmental Health and Safety Office for proper handling of equipment and hazardous material used in this protocol.

Reagents

α-Factor mating pheromone, WHWLQLKPGQPMY (5 mg/mL stock in ethanol or methanol, stored at −20°C) (for synchronization in G_1 [Steps 1–7] only)

Where small quantities of α-factor are required, it can be purchased from commercial suppliers, including Sigma-Aldrich, Zymo Research, and US Biological. For large or numerous experiments (or to share among multiple laboratories), it is cost effective to synthesize and purify α-factor. At a 1-g scale, outsourced peptide synthesis can produce α-factor for ~1% of the cost of "catalog" reagents of equal purity. Custom-synthesized pheromone must be purified by one or more rounds of reverse-phase high-performance liquid chromatography before use—truncations and branched peptide by-products generated during solid-phase synthesis can act as dominant negatives to α-factor signaling if not removed.

Culture of yeast to be synchronized

Select either a BAR1 or bar1 strain for synchronization in G_1 (Steps 1–7). Use a cdc15-2 strain for synchronization at the M/G_1 transition (Steps 8–17).

---

[1]Correspondence: adam.rosebrock@utoronto.ca

Copyright © Cold Spring Harbor Laboratory Press; all rights reserved
Cite this protocol as *Cold Spring Harb Protoc*; doi:10.1101/pdb.prot088724

Strains can be made bar1 by homologous-recombination-based gene disruption using bar1::KanMX or a similar construct. bar1 strains mate at extremely low efficiency and should not be carried through high-throughput mating schemes including synthetic genetic arrays (SGAs). bar1 cells are, by their nature, extremely sensitive to α-factor. Conversely, arrest of BAR1 strains requires 200- to 400-fold more pheromone than isogenic bar1 strains. BAR1 strains constitutively degrade α-factor; Bar1p activity continues unabated during arrest, resulting in the eventual spontaneous release into S phase.

Reagent for cell cycle arrest (optional; for Steps 18–22 only)
    Hydroxyurea (2 M stock in culture medium, freshly prepared)
    Nocodazole (7.5 mg/mL stock in DMSO, prepared in advance and stored at −20°C)

Wet ice (for synchronization at the M/G$_1$ transition [Steps 8–17 only])
Yeast culture media

YEPD (rich) medium is used in this protocol. α-Factor will quantitatively arrest cells grown in either rich or synthetic media. For S288c and W303 (and likely other backgrounds), cells grown in YEP-based media have better postrelease synchrony than cells grown on identical carbon sources in synthetic media. Other protocols (Futcher 1999; Amberg et al. 2006) have recommended reduction of media pH to anywhere from 5.0 to 3.5 to reduce the activity of Bar1p. Although these methods can decrease the amount of α-factor required for an experiment, comparison of cells grown in different media contexts must be undertaken carefully. Quantitative arrest for >2 h can be achieved in unbuffered YEPD media, as described in Steps 1–7. We find that cdc15-2 synchronization and release are more effective for cultures grown in YEPD or similar media, but either rich or synthetic media can be used.

Equipment

Calibrated thermometer or thermocouple (accurate to at least 0.5°C resolution) (only needed for Steps 8–17)
Centrifuge and bottles (rotor appropriately sized for culture volumes) (optional; for Steps 6.v–6.ix only)
Filters (0.45-μm; nylon, cellulose acetate, or nitrocellulose), corresponding fritted filter holder (47- or 90-mm), and vacuum pump or house vacuum connection with inline trap (optional; for Steps 6.i–6.iv only)
Heat-block shaker (e.g., Eppendorf Thermomixer) and conical tubes (50-mL, with vented lids) (optional; see Steps 15–17)
Microscope, transmitted light (with differential interference contrast or phase-contrast optics suitable for bud-index determination)
Shaking incubator or roller drum, set at 30°C
Shaking incubators, set at 23°C and at 37°C (only needed for Steps 8–17; see Step 9)
Sonicator with microtip probe
Spectrophotometer (for measurement of OD$_{600}$) or Coulter counter

METHOD

The methods described in this protocol exploit various aspects of the biology of replicating yeast cells (see Fig. 1 and Discussion). Select the series of steps appropriate for the experimental goals (see Introduction: Methods for Synchronization and Analysis of the Budding Yeast Cell Cycle [Rosebrock 2015]). Steps 1–7 describe synchronization in G$_1$ using α-factor mating pheromone; Steps 8–17 describe synchronization at the M/G$_1$ transition using the temperature-sensitive cdc15-2 allele; and Steps 18–22 describe arrest in early S phase and at G$_2$/M by using hydroxyurea and nocodazole, respectively.

Synchronization in G$_1$ Using α-Factor Mating Pheromone

In this series of steps, α-factor mating pheromone is used to enact a reversible arrest of MATa haploids in G$_1$.

Chapter 14

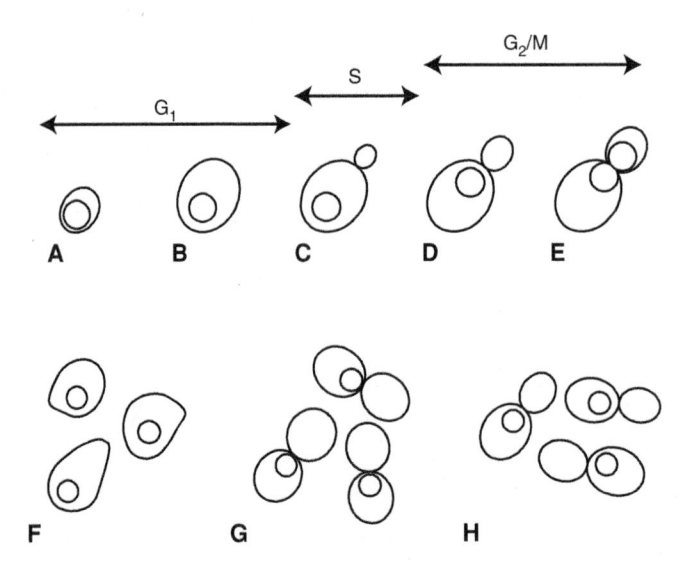

FIGURE 1. Cytological features of cell cycle progression and arrest. The budding yeast cell cycle, like that of higher eukaryotes, consists of G_1, S, G_2, and M phases. In contrast to the symmetric division of most higher eukaryotic cells and fission yeast, cells of the budding yeast Saccharomyces cerevisiae bud asymmetrically: Mother cells give birth to smaller daughters. The mother–daughter size difference is affected by growth rate and nutrient source. Small G_1 cells grow (A) in volume until they reach a critical size at START (B). At the G_1–S transition, cells begin to bud (C). The bud, and to a lesser extent the mother cell, increases in size throughout S and G_2 (D) until M phase (E). Mother cells have implicitly passed START and, under many growth conditions, rebud with little delay; mothers begin the cycle around B. Cell cycle arrest can be monitored by microscopy. Cells arrested with α-factor will be large and oblong, with a characteristic "schmoo" morphology (F). cdc15-2 and nocodazole arrested cultures will be large "dumbbells" consisting of two rounded cell bodies (G). Hydroxyurea-treated cells will arrest with large buds of varying size (H).

Cell Cycle Block

1. Grow an appropriate volume (e.g., 1 L) of a culture of yeast cells to a density of 2–5×10^6 cells/mL, as measured by a spectrophotometer for OD_{600} or by means of a Coulter counter.

    Growth at temperatures lower than 30°C will result in slower growth and might be experimentally useful to extend the cell cycle.

2. Add α-factor to a final concentration of 5 μg/mL (for *BAR1* strains) or 25 ng/mL (for *bar1* strains), corresponding to a 1:1000 or 1:200,000 dilution of a 5 mg/mL stock, respectively. Continue to incubate the culture with shaking at the appropriate growth temperature (usually 30°C).

3. Prewarm an equal volume of yeast culture medium to the appropriate growth temperature (usually 30°C); this will be used to resuspend the released cells in Step 6. In addition, prewarm a sufficient volume of medium to be used for the release procedure chosen in Step 6.

4. After 1.5 h of shaking at 30°C, separate cells with a sonicator with a microtip probe, and then examine the cells with a light microscope. Collect and fix cells for quantitative bud counting and DNA content measurement by flow cytometry, if desired.

    Sonication settings should be determined empirically as the minimum time required to separate mother–daughter pairs. Excess sonication can result in the formation of ruptured "ghost" cells. No small buds should be visible. Keep in mind that S288c-derived cultures will be primarily large single cells with an oblong "schmoo" morphology.

5. After ~2 h total incubation (i.e., when the cells have quantitatively arrested), collect a final prerelease fraction (if appropriate for the experiment). Release the cells from arrest by removing the medium containing α-factor (see Steps 6 and 7).

    Arresting cells for 1.5 cell cycles is a convenient rule of thumb. Cells that are released after insufficient arrest will vary greatly in cell size and can show poor synchrony.

    BAR1 cells will spontaneously release from arrest and should be monitored from 1.5 to 2 h.

Cell Cycle Release

6. Collect the cells by either filtration or centrifugation, as follows. Filtration is generally preferred owing to the inevitable thermal, nutrient, and physical stress responses induced by centrifugation.

Filtration

   i. Assemble the filter assembly and membrane, and prewet it with fresh medium from Step 3.
   ii. Filter the cells onto the filter membrane.

   Do not allow the membrane to dry out. Cells will starve within seconds to the removal of media.

   iii. Wash the membrane twice with 25 mL of prewarmed medium from Step 3. Allow nearly all of the medium to be pulled through the filter between each of the two washes.

   bar1 strains might require additional washes to release; they are extremely sensitive to α-factor. Conversely, BAR1 strains constitutively degrade α-factor; Bar1p activity continues unabated during arrest, resulting in the eventual spontaneous release into S phase. Bar1p obviates extensive washing of BAR1 cultures for release—residual α-factor will be rapidly degraded.

   iv. Remove the filter from the manifold, and place it into a flask of prewarmed medium (use the same volume of medium as in Step 2). Swirl vigorously to resuspend cells. Immediately collect a postrelease fraction, and proceed to Step 7.

Centrifugation

   v. Transfer the culture from Step 5 to balanced centrifuge bottles.
   vi. Harvest the cells by centrifugation at 4000g for 5–10 min at room temperature.
   vii. Discard the supernatant, and resuspend the pellet with 25 mL of prewarmed fresh medium from Step 3 (pooling the pellets from several bottles, if applicable).
   viii. Repellet the cells by centrifugation at 4000g for 5 min, and discard the wash medium.
   ix. Resuspend the cells in fresh medium from Step 3 (use the same volume of medium as in Step 2). Immediately collect a postrelease fraction, and proceed to Step 7.

7. Return the culture to a shaking incubator at 30°C. Sample the cells with a 5- to 10-min sampling interval as desired.

   Wild-type cells grown in YEPD medium will retain synchrony for two or more cycles.

Synchronization at M/G$_1$ Using *cdc15-2*

In the following steps, cells harboring a temperature-sensitive mutation in the CDC15 kinase are shifted to a restrictive temperature, inducing a reversible arrest at the M/G$_1$ transition.

Steps 8–14 use larger volumes (e.g., 1 L) of cells; modifications for small volumes (15 mL or less) are outlined in Steps 15–17.

Cell Cycle Block

8. Grow an appropriate volume (e.g., 1 L) of a yeast strain harboring *cdc15-2* at 23°C (use a calibrated thermometer or thermocouple for accurate temperature control) to a concentration of 2×10^6 cells/mL, as measured by OD$_{600}$ or a Coulter counter.

9. Transfer the culture to an air incubator, not a water bath shaker, set at 37°C to inhibit the CDC15 kinase.

   A rapid increase in temperature (e.g., caused by placement into a water bath incubator) results in a greater stress response and poor release from arrest. Use an air-heated incubator and relatively large culture volumes (0.75 L or more in a 2-L flask) to ensure slow equilibration to 37°C. Alternatively, a single incubator set to 23°C can be reset to warm to 37°C to slow the rate of temperature increase.

10. Continue to incubate the cells at 37°C. During this incubation, chill 0.6 culture volumes of medium on wet ice in preparation for the release step.

11. Starting at 2 h after the temperature shift, regularly assay the efficiency of arrest by examining a sonicated aliquot of cells by microscopy.

    Arrested cells should appear as large "dumbbell"-shaped pairs (see Fig. 1G).

12. Continue incubation at 37°C for as long as needed (usually ~3 h) until nearly all the cells are large dumbbells.

    The culture volume, initial cell density, and temperature ramp rates will affect the experimental kinetics.

Release of Cultures into a Synchronous Cell Cycle

13. Release the culture by adding 0.6 volumes of ice-cold medium while mixing, and then return the culture to a shaker set at 23°C.

    The addition of 0.6 volumes of cold medium will immediately drop the temperature of the culture to 23°C. Shifting the cells to a water bath shaker set to 23°C is an alternative, but this results in slightly poorer synchrony and is difficult to reproducibly scale up or down.

14. Take samples of the released culture as required.

    Budding will occur within 45 min to 1 h. Cells grown in YEPD will remain synchronous for one or two generations.

Modifications for Small Volumes

When only small volumes of cells are needed, we recommend the following modifications to the procedure described in Steps 8–14.

15. Grow the cells in 50-mL vented-lid conical tubes (15-mL maximum culture volume).

16. Enact arrest by increasing the block temperature on a heat-block shaker (e.g., an Eppendorf Thermomixer) from 23°C to 37°C.

    The thermal mass of the block will result in a desirable slow increase in temperature.

17. Release as described in Step 13, and then return the cells to a room-temperature shaker or roller for outgrowth.

Drug-Induced Arrest in Early S Phase or at G$_2$/M

In the following steps, cell cycle arrest is induced in early S phase using hydroxyurea or at the G$_2$/M transition using nocodazole.

18. Grow an appropriate volume (e.g., 1 L) of cells to a density of 2–5 × 10^6 cells/mL, as measured by OD$_{600}$ or a Coulter counter.

19. Collect a prearrested control fraction.

20. Initiate arrest by adding the drug of choice, as follows.
    - To achieve arrest in early S phase, add hydroxyurea to a final concentration of 200 mM (i.e., use a 1:10 dilution of a 2 M stock hydroxyurea solution that has been freshly prepared in prewarmed media).
    - To achieve a G$_2$/M arrest, add nocodazole to give a final concentration of 15 µg/mL (i.e., use a 1:500 dilution of a 7.5 mg/mL stock nocodazole solution).

21. Expose the cells to the drug of choice for more than one full cell cycle by incubating them as appropriate (e.g., in YEPD medium for 2 h at 30°C).

22. At end of the drug incubation, immediately harvest the cells as appropriate.

    For example, collect fractions for staining for DNA content, determining the bud index by microscopy, and size distribution by use of a Coulter counter.

DISCUSSION

The technique for enacting G_1 arrest (Steps 1–7) takes advantage of the fact that budding yeast cells secrete mating-type-specific pheromones that permit directional polarized growth—a process known as "schmooing"—when cells of opposite mating types are in close proximity (Fig. 1F). *MAT*a cells arrest in response to α-factor only during G_1; in an asynchronous culture, it will take up to one full cell cycle for the entire population to have traversed to the arrest point.

The method for synchronization at M/G_1 (Steps 8–17) relies on selective manipulation of the CDC15 kinase, which is essential for activating the protein kinase–activator complex DBF2–MOB1 to enact mitotic exit. When shifted to a restrictive temperature of 37°C, cells harboring *cdc15-2* fail to exit anaphase and arrest at the M/G_1 transition; a subsequent return to a permissive temperature results in synchronous reentry into the cell cycle, but without complete separation of the dumbbell-shaped cells that are characteristic of this arrest (Fig. 1G). Cell cycle progression can be monitored by fluorescence microscopy to examine nuclear location and morphology, western blotting for cell cycle–regulated proteins, or measurement of cell cycle–regulated transcripts; flow-cytometric and cell-size analyses are unsuitable because of incomplete cell separation following release of *cdc15-2*-arrested cells.

Finally, cultures can be used for measurement of biochemical, cytological, or transcriptional state in G_2/M and early S phase by inducing arrest through application to the culture of specific chemical compounds, including hydroxyurea and nocodazole (described in Steps 18–22). Nocodazole-arrested cells will have a dumbbell morphology similar to the *cdc15-2* arrest with 2N DNA content (Fig. 1G); hydroxyurea-arrested cells have large buds and a DNA content between 1N and 2N, as determined by flow cytometry (Fig. 1H). Note that although these compounds enable collection of large numbers of cells at a specific cell cycle position, the drug wash-out will fail to generate synchrony comparable to that of other methods.

REFERENCES

Amberg DC, Burke DJ, Strathern JN. 2006. Inducing yeast cell synchrony: α-Factor arrest using low pH. *Cold Spring Harb Protoc* doi: 10.1101/pdb.prot4172.

Futcher B. 1999. Cell cycle synchronization. *Methods Cell Sci* 21: 79–86.

Rosebrock AP. 2015. Methods for synchronization and analysis of the budding yeast cell cycle. *Cold Spring Harb Protoc* doi: 10.1101/pdb.top080630.

Protocol 2

Synchronization of Budding Yeast by Centrifugal Elutriation

Adam P. Rosebrock[1]

Donnelly Centre for Cellular and Biomolecular Research, University of Toronto, Toronto, Ontario M5S 3E1, Canada

In yeast, cell size is normally tightly linked to cell cycle progression. Centrifugal elutriation is a method that fractionates cells based on the physical properties of cell size—fluid drag and buoyant density. Using a specially modified centrifuge and rotor system, cells can be physically separated into one or more cohorts of similar size and therefore cell cycle position. Small G_1 daughters are collected first, followed by successively larger cells. Elutriated populations can be analyzed immediately or can be returned to medium and permitted to synchronously progress through the cell cycle. This protocol describes two different elutriation methods. In the first, one or more fractions of synchronized cells are obtained from an asynchronous starting population, reincubated, and followed prospectively across a time series. In the second, an asynchronous starting population is separated into multiple fractions of similarly sized cells, and each cohort of similarly sized cells can be analyzed separately without further growth.

MATERIALS

It is essential that you consult the appropriate Material Safety Data Sheets and your institution's Environmental Health and Safety Office for proper handling of equipment and hazardous materials used in this protocol.

Reagents

Cultures of yeast for experimentation
  The protocol has been optimized for S288c- and W303-derived strains.

DAPI (4′,6-diamidino-2-phenylindole; 2.5 mg/mL 1000× stock in deionized water; store at −20°C in the dark)

Ethanol (70% in sterile deionized water)

Ethanol (95%)
  Prealiquot volumes of 1 mL into 1.7-mL microcentrifuge tubes.

Formaldehyde (37% in deionized water)
  Prealiquot volumes of 120 µL into 1.7-mL microcentrifuge tubes.

Hydrogen peroxide (5% w/v in sterile deionized water) (optional; see Step 5)
  Peroxide is recommended as an orthogonal sterilization method to be used in addition to 70% ethanol. However, for elutriation systems dedicated to yeast, it is sufficient to use exclusively ethanol-based decontamination and sterilization.

---

[1]Correspondence: adam.rosebrock@utoronto.ca

Copyright © Cold Spring Harbor Laboratory Press; all rights reserved
Cite this protocol as *Cold Spring Harb Protoc*; doi:10.1101/pdb.prot088732

Isoton II buffer (Beckman Coulter)
Sodium azide (NaN$_3$) (optional; see Step 54)
Wet ice, in a large basin (optional; see Step 56)
Yeast medium, filter-sterilized (1–12 L)

Autoclaved media, especially those including yeast extract/peptone, frequently contain particulate matter that can interfere with achieving stable flow in the elutriation chamber.

Equipment

Centrifuge, equipped with elutriation rotor, tether cable, and control system (e.g., Beckman J6-MI or Avanti J-26 XP equipped with JE-5.0 Elutriation system)

At the time of writing, Beckman Coulter is the only vendor selling research (nonclinical) centrifugal elutriators. Numerous models and versions of the Beckman elutriation system exist. The principles from this protocol can be transferred to different chambers or instruments after optimization of flow rates and rotor speeds.

Centrifuge equipped with rotor and bottles suitable for harvest of multiliter cultures
Conical-bottom (or flat-bottom) centrifuge bottles, polypropylene (225 or 250 mL)
Cotton-tipped wooden applicators
Coulter Z2/Channelizer/Multisizer or equivalent cell-volume counter (50-μm aperture)
Cuvettes
Elutriation chambers (40-mL) horizontal (see Fig. 1)

Elutriation can be performed with one or two size-matched chambers. The use of tandem 40-mL chambers will provide moderately increased capacity and a large improvement in fraction resolution because it adds a second elutriation boundary past which all separated cells must flow. In practice, tandem-chamber elutriation enables collection of more-synchronous fractions from a larger total number of cells loaded. Note that tandem operation requires a specific "A" and "B" chamber and that altered rotor plumbing is necessary, as specified in the system manual.

Flasks (2-L), sterile, for culture growth and to contain clarified medium
Flasks (500-mL), sterile, for culture outgrowth
High-vacuum grease
Microscope (transmitted light, with differential interference contrast (DIC) or phase-contrast optics for bud-index determination)
Pipettor and tips (P200)
Shaking incubator, set at 30°C
Sonicator, with standard (0.5-in.) and microtip probes

A microtip can be used exclusively, but this will require increased processing time before sample loading.

Spectrophotometer, for measurement of OD$_{600}$
Tandem peristaltic pump, capable of total flow rates of 6–40 mL/min

By their nature, peristaltic pumps generate pulsatile flow. By stacking multiple pump heads in an offset (45°) configuration and teeing their input and output connections, pulsation can be reduced by nearly 95%. Elutriation with stacked pump heads reduces the likelihood of chamber inlet clogs, significantly improves the stability of the elutriation front, and leads to tighter size distributions of elutriated cells.

Tubing and fittings, from sample or medium input through pumps and output (Masterflex size 13 and 14 recommended)

METHOD

The following series of steps describe first how to prepare cultures of cells suitable for elutriation (Steps 1–4), and then how to set up and sterilize an elutriator (Steps 5–21). Sample preparation and monitoring by Coulter counter are described in Steps 22–33. The preferred methodology for collection of small G$_1$ cells for prospective sampling is given in Steps 34–53; Steps 54–58 give an alternative strategy for directly acquiring multiple fractions of increasing cell size.

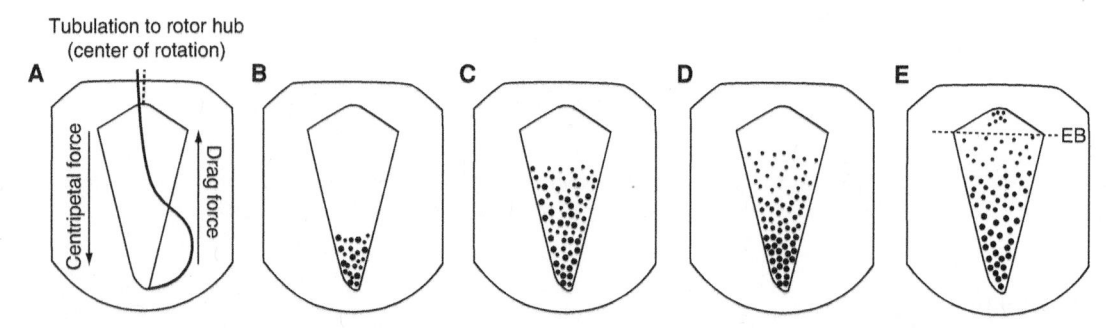

FIGURE 1. Elutriator chamber design and schematized operation. Elutriation operates on the balance of centripetal force generated by rotation of the centrifuge, pushing cells toward the chamber bottom, and fluid drag generated by the flow of medium toward the top of the chamber, pushing cells upward toward the rotor hub (A). Before loading, the elutriation chamber(s) is filled with conditioned medium. Cells are loaded into the chamber from the bottom as a mixed-size population (B). Loading of additional cells moves the sample "front" of cells higher in the chamber and the cells begin stratification by size, with the largest at the bottom (C). Loading of new cells is stopped at this point. With continued flow of medium, cells stratify in the chamber; the smallest cells are pushed to the top because of their relatively high drag (D). With increased flow rate, the cell front is pushed further up the chamber. When cells move past the elutriation boundary (EB), the linear flow rate increases rapidly and the cells are swept from the chamber through the upper port (E).

Culture Pregrowth and Calculation of Inoculation Volumes

1. Two days before elutriation, inoculate a 5-mL culture of yeast for overnight growth in a shaking incubator at 30°C, using the same yeast medium chosen for elutriation.

2. On the day before elutriation, subculture the 5-mL overnight culture into 100 mL of fresh yeast medium. With a spectrophotometer, measure OD_{600} at regular (30-min) intervals for at least two generations.

3. Determine the doubling time of the culture either by calculating a logarithmic best-fit across multiple measurements or by using the start and final OD_{600} measurements in Equations 1 and 2. Solving for doubling time using base$_{10}$ logarithms yields the following:

$$OD_{600}\text{start}/OD_{600}\text{final} = 2^{(\text{Elapsed Time})/(\text{Doubling Time})}, \quad (1)$$

$$\text{Doubling time} = \frac{\text{Elapsed time}}{\frac{\log \frac{OD_{600}\text{final}}{OD_{600}\text{start}}}{\log 10(2)}}. \quad (2)$$

Accurate doubling-time measurement is crucial to set up large-volume cultures required for elutriation and is best calculated by a logarithmic best-fit across multiple measurements. Growth should be measured over at least two generations in mid-log phase.

4. Inoculate large-scale cultures to be elutriated. Use Equations 3–5 to determine the appropriate volume. Aim to acquire an OD_{600} of 0.3 at the desired elutriation start time. Rearranging for amount to subculture yields the following:

$$\text{Generations} = \text{Doubling time}/\text{Time until elutriation}, \quad (3)$$

$$\text{Generations} = \log_2([OD_{\text{final}} * \text{Volume}_{\text{final}}]/[OD_{\text{subculture}} * \text{Volume}_{\text{subculture}}]), \quad (4)$$

$$\text{Volume}_{\text{subculture}} = (OD_{\text{final}} * \text{Volume}_{\text{final}}/OD_{\text{subculture}})/2^{(\text{Time to elutriation}/\text{Doubling time})}. \quad (5)$$

For example, wild-type S288c yeast cells in synthetic-complete (SC) medium with a doubling time of 2 h will grow eight generations during a 16-h period. To have 8 L of SC-grown wild-type cells at OD_{600} = 0.3 after 16 h while starting from a culture at OD_{600} = 0.6, the investigator will need

$$\text{Volume}_{\text{subculture}} = (0.3 \times 8000/0.6)/2^{(16/2)} = 15.625 \text{ mL}.$$

Elutriator Setup and Sterilization

Comprehensive assembly and maintenance instructions are available from Beckman Coulter (e.g., pn JE5-IM-13AA). Perform the following steps on the day intended for elutriation.

5. Use 70% ethanol in water (and/or hydrogen peroxide) and cotton-tipped applicators to clean the ports of the elutriation chambers and centrifuge rotor assembly. Carefully remove any grease remaining in the ports to avoid clogging internal passages.

6. Use dry applicators to remove cleaning solvent from the ports.

7. Inspect silicon seal between top and bottom halves of chamber and then assemble elutriation chamber.

    A small amount of torque is sufficient to seal the assembly; do not overtighten.

8. Apply a thin film of high-vacuum grease to the O-rings on the ends of each connecting tube.

    Do not overlubricate. Excess grease will lead to system blockage and loss of sample.

9. Assemble fluidics by attaching chambers to the hub using greased connection tubes.

10. Secure chambers to hub assembly; turn each screw one-half turn counterclockwise (loosen) after hand-tightening to enable radial expansion of the assembly.

11. Connect sample inlet and outlet tubing and fittings to the hub.

12. Place sample and medium lines into 70% ethanol, set peristaltic pump to dispense 20 mL/min, and fill the system and chamber(s).

    Tilting the fluidics assembly is necessary to remove air. When using a tandem-chamber system, the "A" chamber is first in the liquid path and should be oriented downward until filled. Higher flow rates can be used, but these can lead to leakage caused by radial expansion of the chambers when the fluidics assembly is not mounted in the centrifuge rotor.

13. After the system is filled, place the effluent line into the 70% ethanol reservoir.

14. Continue pumping for 15 min to recirculate ethanol. Periodically switch the medium/loading stopcock to clean both input lines. Briefly lift each input line to introduce ~1 cm worth of air bubbles; these act to "squeegee" the walls of the tubing and enact a thorough cleaning.

15. Remove the medium and sample input lines from ethanol, and then pump air into the system to drain the rotor.

16. Refill rotor with sterile water to wash. Next drain the rotor and repeat the rinsing twice.

17. Ensure that all bubbles have been cleared from chamber assembly. Forcefully bump the chamber-end of the fluidics assembly using the heel of the hand to dislodge any bubbles.

    Air bubbles must be avoided. At best, bubbles disrupt the elutriation front and lead to poor fraction resolution, and, at worst, bubbles can block flow, create high back-pressure, and lead to sample loss.

18. Mount the completed fluidics assembly to the centrifuge rotor.

    The chamber assembly is designed to expand laterally under centripetal force. It might be necessary to manually squeeze the two chambers together toward the hub during mounting.

19. Secure the rotor hub to the chamber wall using the provided tether cable.

    Note that failure to secure the hub will invariably result in loss of sample and can cause damage to the tubing and rotor assembly.

20. Wet-test rotor assembly by running at 2400 rpm with 20 mL/min flow for several minutes. Examine the fluidics hub and centrifuge chamber for signs of leakage.

21. Pump air into the system to drain water from the chamber and lines.

    The sterilized elutriator can be stored in this run-ready state for hours to days.

Chapter 14

Sample Preparation

22. Measure the OD_{600} of overnight yeast cultures (from Step 4) using a spectrophotometer.

    It is advisable to start monitoring the OD_{600} one doubling time before the calculated value. If cells have grown faster than expected, an earlier harvest can then be performed.

23. Collect a sample before culture processing.

24. When OD_{600} reaches 0.2 to 0.5, harvest mid-log yeast cultures by centrifugation in a large-volume centrifuge.

    Centrifugation at 3000g–4000g for 5–10 min is sufficient for most strains.

25. Decant conditioned medium (supernatant) into sterile 2-L flasks and set these aside.

26. Repeat centrifugation (Steps 24–25), as necessary, to harvest all input cells. Reserve all supernatant.

27. Resuspend and pool collected cells into 50–100 mL of conditioned medium at room temperature. Use a plastic centrifuge bottle for the final resuspension.

    A 225- or 250-mL conical-bottom centrifuge bottle is ideal, but a flat-bottom centrifuge bottle can be used.

28. Sonicate the pooled yeast sample using a standard-size sonicator tip to liberate daughter cells. Use a duty cycle of no greater than 50% and swirl the culture to mix between cycles.

    Power levels and cycle times will need to be determined empirically by microscopy of processed cells. Sufficient sonication will produce well-separated daughters, but oversonication will result in a large number of lysed "ghost" cells. Using a Misonix 3000 sonicator and 0.5-in. probe, S288c- and W303-derived strains can be reliably sonicated in 2 min at 40% power (10 sec on, 10 sec off).

29. Collect a sample of concentrated, asynchronous pre-elutriated cells at this step for bud index, Coulter sizing, and DNA content, if desired. Asynchronous pre-elutriated cells may also be needed as controls for other downstream steps (e.g., western blotting).

Determination of Cell Size by Coulter Counter

Before use, the Coulter Z2 Counter must be primed with fresh Isoton II buffer. Other routine maintenance is necessary, including regular cleaning of the system and routine calibration with yeast-sized beads (size L3). Acquisition of a background reading from pure Isoton II is a necessary control to enable accurate cell counts.

The following method (Steps 30–33) can be used for counting (e.g., as required after Step 29 and during the procedure described in Steps 45–53).

30. Dilute yeast culture in Isoton II buffer to a total volume of 10 mL in a clean cuvette.

    During elutriation and when following culture outgrowth in 500-mL flasks, a 1/100 dilution (100 µL culture in 9.9 mL buffer) provides a suitable working concentration of cells. When measuring the concentrated input, a 1/1000 or 1/10,000 dilution might be necessary to achieve an appropriate event rate for the instrument. Use serial dilution to minimize error.

31. Mix and briefly sonicate the diluted cell mixture.

    Ten seconds with a tuned microtip at full power is generally sufficient.

32. Count 1 mL of diluted culture across an appropriate size range for the strain being measured.

    S288c- and W303-derived haploids can be accurately measured across a 5–125-fL window. Diploids will require a larger upper threshold of at least 150 fL. Altering instrument parameters might be necessary to include the full range of sizes present in a culture. Instrument precision is inversely related to window width—avoid unnecessarily wide acquisition settings.

33. Note the cell count, apply a dilution correction as necessary, and save size and count data to disk for further analysis and comparison of all collected fractions.

    Proceed to either Steps 34–53 or Steps 54–58.

Collection of Small G_1 Cells for Prospective Sampling

The following method (Steps 34–53) is preferred for most applications. One or more fractions of small G_1 daughter cells are collected and subsequently sampled as they proceed synchronously through the cell cycle. A complementary procedure is given in Steps 54–58.

Methods for Synchronization and Analysis of the Budding Yeast Cell Cycle

Common elutriation centrifuges are not actively warmed, although the system will warm during use. Ensure that chamber temperatures do not rise past 30°C. Above this temperature, cells will begin to exhibit heat-shock responses.

Loading the Elutriator

34. Set pump to a flow rate of 15 mL/min.

    For a tandem pump arrangement, flow rate is twice the displayed setting. For example, for 15 mL/min, set the pump displayed rate to 7.5 mL/min.

35. Purge air from both the medium and sample lines, and then set the stopcock to allow flow from the medium line. Place the sample line into the tube of sonicated yeast cells (from Step 28), and then put the effluent tube into a fresh 500-mL flask.

    Avoid introducing a bubble at the end of the sample tube.

36. Fill fluidics with conditioned medium, carefully dislodging all air bubbles.

37. Mount fluidics to centrifuge rotor, reconnecting tether cable to the chamber wall.

38. Close the centrifuge chamber; observe effluent flow and ensure that the tubing is not pinched by the chamber lid.

    A slightly constricted flow can cause the inlet tubing to "throb" and the outlet flow to become pulsatile. Kinked or pinched tubing will lead to poor fraction resolution.

39. Set centrifuge to 2400 rpm with no time limit ("hold"), then start the centrifuge.

40. Observe effluent flow during centrifuge acceleration. If the flow slows or ceases, immediately turn off pump, stop centrifuge, and debubble fluidics.

    Note that air will be pushed toward the center of rotation and will block flow in a tandem-chamber system. With the pump running, pressure will increase until the system leaks.

41. While looking into the viewing port on the chamber lid, adjust the strobe delay potentiometer to move the "A" chamber into field.

    Both the A and B chambers can be visualized by adjusting the strobe delay. Cells are first introduced into the "A" chamber.

42. When the centrifuge reaches full speed, switch the stopcock from medium to sample input.

    Cells will be pumped into the system and will become visible as a darkening at the bottom edge of the chamber (see Fig. 1B).

43. Observe the "front" of cells that forms in the first chamber. When the "A" chamber is filled, turn the strobe delay potentiometer to move the "B" chamber into view.

    See Troubleshooting.

44. When the cell front is approximately two-thirds into the "B" chamber, switch the stopcock from sample to medium inlet.

    It is crucial that fluid constantly flows into the chamber. Be careful not to switch the stopcock into a position where neither medium nor sample is being pumped as this will lead to pelleting of cells in the chambers and clogging of the system. Depending on cell size, concentration, and amount of input, not all of the sample might be required to load the chamber; record the volume of sample remaining in the sample tube.

Collection and Monitoring of Fractions

45. Move the effluent tube to a fresh 500-mL flask and continue pumping fresh medium at a rate of 15 mL/min; observe the region above the sample front and collect both a 100-µL sample of effluent for Coulter counting and several-hundred microliters for microscopy.

    A P200 pipette tip can be positioned just past the end of the effluent line to collect a fresh sample. Do not occlude the effluent tube with the pipette tip during sampling. Briefly centrifuge the sample for microscopy, decant supernatant, leaving ~10 µL of supernatant into which the sample is resuspended and visualized.

46. Measure the fraction size and count by Coulter counter (Steps 30–33). Examine cells by light microscopy.

    At this stage, only debris, cell ghosts, and a few small cells should be visible.

47. Increase the pump speed in 0.5 mL/min increments until the sample front nears the elutriation boundary (see Fig. 1C,D). Collect fractions for Coulter sizing (Steps 30–33) and microscopy to monitor contents of the effluent.

48. When the sample front reaches the elutriation boundary, run the pump with speed unchanged for 5–10 min. During this time, debris will be flushed from the system.

49. Place the effluent line in a fresh 500-mL flask. Increase the pump speed by 1 mL/min and monitor output by Coulter counting (Steps 30–33) and light microscopy.

50. Immediately after switching to a new 500-mL flask, note the collected volume and pump speed. Collect a Coulter sample and transfer 1 mL of culture to a microcentrifuge tube containing 150 μL of 37% formaldehyde to fix for bud-index determination. Collect 500 μL of culture into 1 mL of 95% ethanol for DNA content staining. Use DAPI (2.5 μg/mL) for fluorescence microscopy or staining suitable for flow cytometry (see Protocol 3: Analysis of the Budding Yeast Cell Cycle by Flow Cytometry [Rosebrock 2015a]).

    See Troubleshooting.

51. When the eluted cell concentration drops, switch to a fresh 500-mL flask, increase pump speed, monitor effluent (as in Step 47), and repeat as needed.

    Cells are elutriated in order of increasing size. The first few fractions contain extremely small cells that make up a minority of the culture. For applications where collection of a large number of relatively small cells is desired, the smallest cells can be "washed out" by increasing the pump speed by several milliliters per minute from the point of initial elution, waiting for the eluent to clear, then increasing the speed by 1 mL/min for the next fraction. See Figure 1E.

    See Troubleshooting.

52. If the OD_{600} of the collected culture is >0.2, dilute appropriately using conditioned medium.

53. Return collected fractions to shaking incubator (30°C) and sample periodically by Coulter counter (Steps 30–33), as required.

    Cells will synchronously proceed through one cell cycle. After the first division, the mother and daughter cell cycles will differ in length, as daughter cells spend longer in G_1 before reaching the critical size.

Exhaustive Fractionation for Immediate Collection

In the following method, cells can be separated into multiple fractions of increasing cell size. This is a complementary procedure to Steps 34–53. Fractions collected in this variation have broader size distributions than those of the method listed above. More of the input cells are recovered, however, making this a method of value when downstream techniques require large numbers of cells. In addition, cells can be elutriated at 4°C using this method, reducing confounding effects of cell growth during elutriation.

54. Load the elutriator as described above in Steps 34–45.

    The sample and centrifuge can be prechilled to 4°C if desired. The cells can also be treated with sodium azide (NaN_3) before harvesting to inhibit the electron transport chain and deplete cellular ATP stores to block cell growth and ATP-dependent biochemical processes.

55. Collect the first fraction, as described above in Steps 46–49.

    See Troubleshooting.

56. Increase pump speed in increments of 1 mL/min, collecting 250–500 mL per fraction, until the cell concentration of the eluent decreases. Collect the fractions on wet ice.

    See Troubleshooting.

57. Repeat fraction collection for each increase in pump speed. Refill the medium reservoir periodically, as required.

    Carefully decant additional medium into the input reservoir. Beware that any bubbles formed might be aspirated into the pump tubing and cause a fluidics blockage.

58. After collecting the final desired fraction, decelerate the centrifuge and collect the chamber retentate.

    Successful elutriation will result in depletion of all cells up to a given size in the chamber. Coulter sizing of the remaining material is an important measurement of experimental success.

Cleanup and Preparation for Storage

59. Decelerate centrifuge before turning off the pump.

    This is desirable as stopping the flow while the centrifuge is rotating will result in pelleting of the remaining cells and clogging of the chamber inlet.

60. Remove the fluidics assembly from the rotor, leaving the tubing connected.

61. Remove sample and medium lines from the liquid to pump air into the system; discard the remaining cells by tilting chambers to allow cells and medium to flow toward the center (outlet) position.

62. Place sample and medium lines into sterile water. Alternate aspiration of water and air to clean the input lines.

63. Flush several liters of water through the system, alternating between chamber filling and draining with water and air, respectively.

64. Flush the system with 70% ethanol, as described in Steps 12–15, and then drain the system.

    The system can be stored in this state for several days. For longer-term storage, disassembly of the fluidics harness and chambers is recommended.

65. Release tubing from the peristaltic pump cams.

    This is necessary as long-term compression of peristaltic tubing leads to distortion or blockage.

TROUBLESHOOTING

Problem (Steps 43 and 55): Cells accumulate at the bottom edge (inlet) of the chamber.
Solution: This phenomenon is normal and is of no concern unless flow is being blocked (see below). Large cells and aggregates that persist following sonication accumulate at the rear of the chamber. With a tandem-pump arrangement, the relatively constant flow of fluid will preclude formation of a solid plug of cells. Observe the inlet and outlet lines. If effluent flow is continuous (not pulsatile) and the inlet tubing is not throbbing, elutriation can proceed as usual. Plugging of the chamber inlet generates a feed-forward loop—lack of flow into the chamber results in pelleting of cells, exacerbating the problem. Increased backpressure on the inlet tubing will eventually result in the pump stalling or leaking.

Problem (Steps 43 and 55): A solid clog forms at the chamber inlet.
Solution: There are several possible causes and remedies.

- There may be insufficient countercurrent flow. Increase the pump speed.
- The input cells may have been poorly sonicated. Increase the sonication time or power; verify disaggregation of input by microscopy and Coulter sizing.
- A flocculent yeast strain may have been used. Knock out *flo* genes, add EDTA, or supplement with appropriate flocculation-blocking sugars.
- The ceramic seal in the rotor hub may not be tight, causing the fluid to bypass the chamber. Disassemble and clean the rotating seal assembly; relap the seal as necessary.

Problem (Steps 50 and 56): The size distribution of fractions is too broad or the adjacent fractions largely overlap.

Solution: Collection of narrow-distribution fractions relies on, first, depleting the chamber of small cells and, second, changing the flow rate by a small increment. It is necessary to collect 5–10 chamber volumes of effluent to exhaust the system of cells that will elute at a given flow rate. The change in flow rate is proportional to the width of the fraction and number of cells that will be eluted—small increases in flow rate (0.5–1.0 mL/min for early fractions) will generate tight size distributions.

Problem (Step 51): Large-cell fractions are contaminated with smaller cells.
Solution: When elutriating live cultures at ambient temperatures, cells will continue to grow and divide while in the chamber. Even without sonication, daughter cells will spontaneously separate from mothers and will be quickly eluted from the system. This problem is exacerbated by elutriation of fast-growing cells. As elutriation separates all cells below a given ratio of surface area to density, small cells will be continually eluted. Where multiple elutriated fractions of a wide size range are desired, elutriation of cold, azide-treated cells is recommended.

Problem: Medium and cells leak into the centrifuge chamber.
Solution: The fluidics harness contains numerous connections and fittings, including a seal assembly that withstands moderate pressure while rotating at >2000 rpm. Increased backpressure from a clogged chamber or pinched tubing will invariably lead to leakage or tubing blow-out. Proper preventative maintenance and cleaning are imperative. Additional common sources of leakage include the O-ring that seals the output tube to the rotating seal and the pair of O-rings that seal the input line to the rotating seal. These should be regularly inspected and replaced as necessary.

DISCUSSION

In contrast to chemical methods and approaches based on genetic or temperature-sensitive (*ts*) mutations (see Protocol 1: Synchronization and Arrest of the Budding Yeast Cell Cycle Using Chemical and Genetic Methods [Rosebrock 2015b]), elutriation can generate synchronized cells of any genotype without confounding heat-shock or chemical stress responses. However, unlike "simple" chemical or genetic *ts* methods, elutriation requires specific equipment, is limited to a relatively narrow range of input cell quantities, and entails hours of preparation, execution, and instrument maintenance. There is no need for cells to be growing—or even alive—during separation.

Elutriation is based on the balance of centripetal force and fluid drag, which is related to linear flow rate. Simplified models of elutriation often reference Stokes' law, with a velocity term equal to the volumetric flow rate divided by the cross-sectional area of the chamber at a given point. In reality, flow rate is inversely related to unoccupied volume in the system. As the chambers are loaded with cells, the effective volume of the system decreases and the effective flow rate increases. This protocol directs starting with an excess of cells that are loaded until the system is filled. This method enables reproducible starting set-points for flow rate and centrifuge speed and generates excellent run-to-run reproducibility, with minimum optimization. Fewer starting cells can be used, with a (potentially significant) increase in initial pump speed and different incremental flow-rate increases.

Effective synchronization by elutriation is predicated upon size diversity in asynchronous cells and the ability to maintain this diversity in the elutriator long enough to load and elute one or more fractions. Rapidly growing yeast cultures—for example, wild-type cells grown in YEPD medium—pose multiple challenges. After loading with live cells at ambient temperatures, cell growth and division continues. The smallest cells in the population grow alongside their sister and mother cells. Daughter cells are born in the chamber, but they are not quantitatively released from their mothers (necessitating sonication, see above). The relatively long collection times required to collect a given size population—10–15 min for early fractions—are a significant percentage, 20% or more, of the cell cycle for rapidly growing cells. Note that elutriation of cultures grown in fermentable carbon sources results in the evolution of CO_2 within the chamber. These bubbles disrupt the elutriation front and lead to poor fraction resolution.

Decreasing the growth rate will increase the amount of time cells spend at a given size, and significantly improve resolution of cell cycle events by elutriation. The growth rate can be modulated by temperature (e.g., decreasing from 30°C to 23°C) and changes in medium (e.g., ethanol, instead of glucose, as a carbon source). Changes in nutrient conditions and, to a lesser extent, temperature, affect many aspects of cell physiology. It follows that it is important to consider the effects of experimental context when interpreting data, even from synchronized cells.

REFERENCES

Rosebrock AP. 2015a. Analysis of the budding yeast cell cycle by flow cytometry. *Cold Spring Harb Protoc* doi: 10.1101/pdb.prot088740.

Rosebrock AP. 2015b. Synchronization and arrest of the budding yeast cell cycle using chemical and genetic methods. *Cold Spring Harb Protoc* doi: 10.1101/pdb.prot088724.

Protocol 3

Analysis of the Budding Yeast Cell Cycle by Flow Cytometry

Adam P. Rosebrock[1]

Donnelly Centre for Cellular and Biomolecular Research, University of Toronto, Toronto, Ontario M5S 3E1, Canada

DNA synthesis is one of the landmark events in the cell cycle: G_1 cells have one copy of the genome, S phase cells are actively engaged in DNA synthesis, and G_2 cells have twice as much nuclear DNA as G_1 cells. Cellular DNA content can be measured by staining with a fluorescent dye followed by a flow-cytometric readout. This method provides a quantitative measurement of cell cycle position on a cell-by-cell basis at high speed. Using flow cytometry, tens of thousands of single-cell measurements can be generated in a few seconds. This protocol details staining of cells of the budding yeast *Saccharomyces cerevisiae* for flow cytometry using Sytox Green dye in a method that can be scaled widely—from one sample to many thousands and operating on inputs ranging from 1 million to more than 100 million cells. Flow cytometry is preferred over light microscopy or Coulter analyses for the analysis of the cell cycle as DNA content and cell cycle position are being directly measured.

MATERIALS

It is essential that you consult the appropriate Material Safety Data Sheets and your institution's Environmental Health and Safety Office for proper handling of equipment and hazardous material used in this protocol.

RECIPES: Please see the end of this protocol for recipes indicated by <R>. Additional recipes can be found online at http://cshprotocols.cshlp.org/site/recipes.

Reagents

Cultures of yeast to be stained
   The protocol has been optimized for the S288c strain background.

Deionized water, autoclaved or filter-sterilized
Ethanol (95%)
Ethylenediaminetetraacetic acid (EDTA) (0.1 M, pH 7.5) (optional; see Step 19)
Proteinase K (20 mg/mL) <R>
RNase A (DNase-free [i.e., boiled]; 10 mg/mL) <R>
Sodium citrate buffer (50 mM; filter-sterilized, and with pH adjusted to 7.2 with citric acid)
Sytox Green (5 mM stock solution in DMSO; Molecular Probes S7020)

Equipment

Aluminum foil
Centrifuge with buckets for microtiter plates (for high-throughput staining only; see Step 26)

[1]Correspondence: adam.rosebrock@utoronto.ca

Copyright © Cold Spring Harbor Laboratory Press; all rights reserved
Cite this protocol as *Cold Spring Harb Protoc*; doi:10.1101/pdb.prot088740

Flow cytometer with a 488-nm excitation line and emission filter appropriate for Sytox Green
Sytox Green stain is readily excited by the 488 nm "blue" laser line present on most cytometers. Peak emission is at ~525 nm, with a profile similar to that of fluorescein isothiocyanate (FITC) and yellow fluorescent protein (YFP). Note that the emission spectrum of green fluorescent protein (GFP) is ~20 nm blue-shifted from that of Sytox Green, making use of narrow-bandwidth GFP-specific filters a suboptimal choice.

Freezer, set at −20°C
Incubators, set at 37°C and 55°C
Microcentrifuge tubes
Microscope, equipped with filters suitable for epifluorescent detection of Sytox Green (optional; see Step 19)
Multichannel pipettors (P20, P200) (for high-throughput staining only; see Steps 24–32)
Reagent troughs (for high-throughput staining only; see Step 28)
Round-bottom polypropylene plates (96-well) (for high-throughput staining only; see Step 24)
Round-bottom polystyrene plates (96-well) (for high-throughput staining only; see Step 31)
Sonicator, with microplate feed horn (QSonica 431MPX) (for high-throughput staining only; see Step 32)
Sonicator, with microtip horn (e.g., Qsonica 4417 microtip and Q700 power supply)
Tubes (polystyrene), for loading sample to cytometer (instrument-specific)
v-bottom polypropylene plates (96-well) (for high-throughput staining only; see Step 28)

METHODS

Ethanol Fixation of Budding Yeast

This protocol has been successfully used for a wide range of culture volumes and densities (e.g., 0.03–5.0 OD_{600} or 1×10^6–1.5×10^8 cells).

1. Add two volumes of 95% ethanol to cultures of yeast in polypropylene microcentrifuge tubes.

   For example, add 1 mL of 95% ethanol to 0.5 mL of culture.

2. Store cells at −20°C in a freezer, for at least overnight.

   Overnight "incubation" at −20°C is crucial for reliable permeabilization of yeast cells. Place 96-well plates of cells at −20°C overnight with nonsealing covers. Longer-term storage will lead to evaporative loss of ethanol from plates without seals. Properly sealed, cells can be left indefinitely at −20°C before processing.

Rehydration of Fixed Cells

3. Pellet fixed cells in centrifuge at 5000g for 20 min at 4°C. Alternatively, use room-temperature centrifugation at higher speeds (>10,000g).

   The cells will form a loose pellet or smear on the side of the tubes; when working with small quantities of cells, rotate tubes 180° and recentrifuge to improve pelleting efficiency.

4. Remove ethanolic supernatant by decanting or pipetting.

   Work gently and be careful to avoid aspiration or dislodging of the loose pellet. Quantitative removal of supernatant is unnecessary as the remaining ethanol will be diluted and removed in subsequent washes. When working with small numbers of cells, it might be difficult to visualize a pellet.

5. Resuspend cells in 800 μL of 50 mM sodium citrate buffer (pH 7.2), and then vortex to mix.

6. Incubate for 10 min at room temperature.

   Complete rehydration of cells and removal of ethanol is crucial for downstream enzymatic reactions and low coefficients of variation (CVs) during subsequent flow cytometry.

7. Collect cells by centrifugation at 5000g for 5 min at room temperature, and then remove and discard the supernatant.

8. Repeat Steps 4–7 for a second wash.

Chapter 14

Enzymatic Digestion and DNA Staining

9. Resuspend pelleted cells in 500 µL of 50 mM sodium citrate buffer (pH 7.2) containing 20 µg/mL RNase A and 2.5 µM Sytox Green. Vortex to mix.

   Excess Sytox Green is deliberately used to provide saturating conditions across varied input cell numbers. Titration of Sytox Green to a lower concentration is possible, but will result in varying fluorescence intensities between samples containing different concentrations of cells and confounding batch analysis.

10. Incubate RNase digestion for at least 1 h at 37°C in the dark.

    Sytox Green is relatively photostable in ambient light but reasonable precautions should be taken to minimize light exposure. Some loosely tented aluminum foil works well as a light shield.

11. Add 10 µL of 20 mg/mL proteinase K, and then briefly vortex to mix.

12. Incubate the proteinase K digest for at least 1 h at 55°C in the dark.

    Complete proteinase digestion of the sample is essential to generate uniform optical scatter measurements and reduce the frequency of multicell aggregates.

13. Incubate the cells at 4°C overnight.

    Stained cells are stable for weeks to months at 4°C in the dark.

Flow-Cytometric Analysis of Sytox-Stained Budding Yeast

14. Transfer stained cells to a tube appropriate for loading onto the flow cytometer being used. Sonicate for 30 sec using a microtip probe set to full power.

    Sonication is necessary to separate mothers from daughters and disassociate aggregates of cells; the quality of sonication can be assessed by light microscopy. The cells can be sonicated at any time after staining and before cytometry; store at 4°C after sonication.

15. Configure the flow cytometer to acquire:

    - Forward-scatter (FSC—narrow-angle) area and width, on a linear scale;
    - Side-scatter (SSC—orthogonal light scatter) area and width, on a linear scale; and
    - Sytox Green fluorescence (often labeled FITC, GFP, or YFP) area and width, on a linear scale.

    Modern instruments are able to acquire pulse area and width for multiple parameters, whereas older instruments, including the widely available BD Biosciences FACSCalibur, can only acquire width measurements for a single parameter. In such a case, the pulse width of the Sytox Green signal should be collected.

16. Using the lowest sample flow rate possible, acquire a test sample of data to check the event rate and optimize photomultiplier tube (PMT) voltages and instrument gain. See Figure 1 for example data and gating strategy.

    The optimal event rate will vary based on instrumentation. Samples must be run at low flow rates to minimize the fluid core stream diameter and provide minimal CVs. If samples are too concentrated, dilute with sodium citrate buffer (pH 7.2) containing 2.5 mM Sytox Green. Overly dilute samples can be concentrated by centrifugation and resuspension in a fraction of the supernatant.

    The PMT voltage and/or gain should be adjusted to place the haploid G_1 peak at 20% of full scale, or diploid G_1/haploid G_2 at 40%. This configuration enables acquisition of 1N through 4N events, spanning haploid G_1 through diploid G_2 (see Fig. 1E).

    Because of the characteristic pattern of 1C, 2C, and 4C peaks from Sytox staining, running unstained cells as a control is not absolutely necessary. Inclusion of at least one tube of a wild-type mid-log culture of identical ploidy is strongly recommended.

17. Set up acquisition gates in the flow cytometer software (Fig. 1A–D). Use FSC area and SSC area to distinguish cells from debris, and use pulse width (*y*-axis) by pulse area (*x*-axis) to distinguish single cells from doublets.

18. Acquire data for each sample.

    The number of events collected will depend on sample concentration, instrument availability, and downstream analytical goals. As a general rule, we collect 50,000 gated events for each sample, although far fewer events are needed to establish G_1 versus G_2 distributions.

    See Troubleshooting.

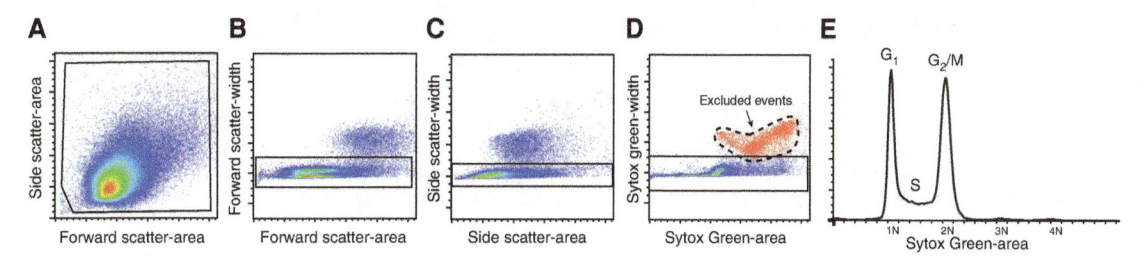

FIGURE 1. Flow-cytometric gating of budding yeast cells and a representative result of the cell cycle distribution. Flow cytometry data must be gated to exclude events that confound accurate cell cycle phenotyping. (*A*) Individual budding yeast cells can be distinguished from debris and aggregates by examining forward and side-scatter pulse areas (FSC-A and SSC-A, respectively). Small debris has low scatter (*bottom left*), whereas large debris and aggregates are above the quantitative range of either or both parameters. A polygonal gate, as shown (black line), excludes both of these confounding subpopulations. (*B,C*) Plotting pulse area [x] versus pulse width [y] permits exclusion of overlapping "doublet" events. Events corresponding to single cells have a continuous distribution along both pulse width and area measurements. A rectangular gate excludes doublets (*upper right*) that appear as a freestanding cluster. Doublet gating should be performed on both scatter parameters as well as Sytox Green, using either Boolean AND gating or a suitable hierarchy. (*D*) The effect of pulse-width gating is evident when comparing cells that were included in all gates (small, multicolored points) with those excluded by one or more gates (red points, dotted line). The combination of width parameters excludes events that would be incorrectly passed by a single gate. (*E*) Asynchronous budding yeast grown in YEPD media have well separated G_1 and G_2 peaks. Cytometer PMT voltages were set to place haploid G_1 (1N) cells at 20% of maximum signal. Cells in G_2 (2N) are located at approximately twice the G_1 peak. Cells in S phase have an intermediate DNA content and are located between the G_1 and G_2 peaks. Even after gating to exclude doublets, as in *D*, minor 3N and 4N peaks are just visible—these represent a combination of unseparated mother–daughter pairs and cells that simultaneously traversed the excitation laser during cytometry. By placing 1N cells at 20% of full scale, diploid cells will show G_1 and G_2 peaks at ~40% and 80% of full scale, respectively.

19. Export data, perform any postrun cleaning, and return the cells to 4°C if future acquisition is desired. Assess cell morphology by epifluorescence microscopy if desired.

    We have successfully analyzed the same sample, stored at 4°C in the dark, over >6 mo with no change in profile. Bacterial contamination of stored samples is a concern with long-term storage and can be mitigated by addition of EDTA to 1 mM final concentration after RNase digestion.

    Sytox Green photobleaches rapidly under high-intensity illumination. This is of no consequence for flow cytometry, where cells are measured in a "one-shot" fashion. When examining populations by epifluorescence microscopy, it might be necessary to traverse different visual fields to observe unbleached areas. The nuclei and mitochondria will appear bright green against a dark cellular background.

Data Analysis

Many packages are available for analysis of flow cytometry data. Software used for data acquisition can provide first-pass qualitative analysis. Details of software operation vary and are beyond the scope of this protocol. A general gating scheme for identifying single cells is presented in Figure 1.

20. As necessary, repeat gating (Step 17) to exclude debris, aggregates, and cell doublets.

21. Export flow-cytometry standard format (FCS) files of gated data for downstream analysis.

    Alternatively, skip Step 20, export all events, and repeat gating in the subsequent software tool.

22. Examine DNA content peaks.

    The signal ratio for 1C:2C cells should be between 1.9 and 2.1. Rapidly dividing cultures (e.g., wild-type cells grown to mid-log in YEPD media) will have a significant fraction of cells in S phase, located between the G_1 and G_2 peaks.

    See Troubleshooting.

23. For quantitative analysis of G_1, S, and G_2 compartments, use the Watson model (Watson et al. 1987), fitting Gaussian peaks to G_1 and G_2 and with S phase defined as cells in between.

    FlowJo (TreeStar Inc) and ModFit LT (Verity Software House) provide built-in functionality for model-based cell cycle analysis.

Chapter 14

Modifications for High-Throughput Staining

Many experimental designs require staining of multiple samples. The manual protocol above is labor intensive and scales poorly past a full microcentrifuge of samples. We use a similar approach listed below to process dozens to thousands of samples in parallel using 96- and 384-well plates. This high-throughput protocol can be performed manually using multichannel pipettes and has been adapted in our laboratory for fully automated operation using a Biomek automation platform. Following high-throughput staining, samples can either be transferred to tubes for analysis on a standard flow cytometer or be directly analyzed on an instrument equipped with a plate loading autosampler.

24. Fix 100 µL of culture with 180 µL of 95% ethanol in a round-bottom 96-well polypropylene plate.

25. Increase the duration of fixation to a minimum of 24 h at −20°C in a freezer.

26. Centrifuge cells at >2000g for 20 min at 4°C in a swing-bucket rotor.

27. Rapidly decant ethanolic supernatant from the plate by a swift "flick" of the plates, without impact, over a suitable vessel.

    The goal is to not disrupt the loose film of cells while removing the majority of ethanolic fixative.

28. Rehydrate the cells with 200 µL of sodium citrate buffer (pH 7.2) at room temperature, and then transfer to a 96-well v-bottom polypropylene plate following resuspension.

    Use of v-bottom plates facilitates pelleting of the small quantity of cells used. Multichannel pipettors greatly speed processing of samples in microplates. Use reagent troughs to efficiently hold liquid reagents.

29. Spin resuspended cells at the maximum speed permissible by the plate and centrifuge (>3000g).

30. Discard the supernatant while carefully avoiding the pellet and repeat wash (Steps 28–29) with sodium citrate buffer (pH 7.2).

31. Resuspend cells in 50 mM sodium citrate buffer (pH 7.2) containing 20 µg/mL RNase A and 2.5 µM of Sytox Green at room temperature and transfer to a 96-well round-bottom polystyrene plate for incubation.

    Rigid polystyrene plates are necessary for efficient sonication.

32. Sonicate using a microplate feedhorn, ensuring that sufficient deionized water is present to acoustically couple the feedhorn to the plate.

    Refer to the sonicator operating manual for further detail.

TROUBLESHOOTING

Problem (Step 18): A low event rate is observed during flow cytometry data collection.
Solution: Consider the following possibilities.

- Ethanol-fixed cells do not pellet well before rehydration and can be lost during processing. It is preferable to carry over a small amount of supernatant between washes than to lose cells by overzealous aspiration; use of careful manual aspiration is recommended for tube-based processing. For high-throughput processing, it is preferable to leave behind some of the ethanolic fixative after the first spin instead of trying to quantitatively remove liquid. Overzealous removal of fixative from plates is a frequent cause of partial or total sample loss.

- Check for instrument clogs. Note that samples are not filtered during this protocol. Flow operators routinely pass samples through a 70-µm mesh filter before loading—we have found this unnecessary across hundreds of thousands of samples. Nevertheless, particulate matter, especially in yeast extract or peptone-containing media, can clog the sample-introduction tube. Use check beads or a "known-good" sample to check for fluidic clogs.

Problem (Step 22): Distinct G_1 and G_2 peaks are not observed.
Solution: The presence of a broad, diffuse signal across a wide range of fluorescence intensities suggests the presence of residual RNA—Sytox Green is not DNA-specific. This can be confirmed through microscopy: The cytosol of RNA-containing cells will fluoresce bright green, compared

with the punctate staining of nuclei and mitochondria in properly digested cells. Ensure that residual ethanol is being removed during washing, and check the activity of the RNase on purified RNA.

Problem (Step 22): Broad peaks in fluorescence intensity (high CVs) are observed.
Solution: High CVs can arise from biological, instrument, or protocol issues.

- Biological: In contrast to mammalian cells, where size varies twofold between G_1 and M-phase cells, asymmetrically dividing *S. cerevisiae* cells can have a 10-fold difference in cell volume between virgin daughters and old mothers. Mitochondrial DNA content is correlated with cell size; old mothers in G_1 have significantly more DNA than daughters. Strains with defects in cell cycle progression will have significantly higher CVs owing to decoupling of the regulation of cell size from cell cycle phase.

- Instrument: Improperly aligned optics is a common cause of high CVs. Use of check-beads is highly recommended to monitor instrument performance. Sample flow rates must be kept low to minimize the core-stream diameter and ensure that cells pass through a narrow laser intercept point. For many instruments, it is advisable to let the sample stream stabilize for 10–60 sec after tube loading before beginning data acquisition.

- Protocol: The presence of broad but resolvable peaks, in conjunction with higher-than-anticipated scatter values, suggests failure to completely fix and permeabilize the cells. Longer fixation at −20°C might be necessary. Alternatively, a second, longer, digestion with proteinase K (Steps 11–13) might "rescue" these samples.

RECIPES

Boiled RNase A

Dissolve RNase A at a concentration of 10 mg/mL in 0.01 M sodium acetate (pH 5.2); heat the solution to 100°C for 15 min. Cool slowly to room temperature and adjust the pH by addition of 0.1 volume of 1 M Tris-Cl (pH 7.4). Aliquot and store at −20°C to prevent microbial growth.

If not used immediately, samples should be stored and protected from light at 4°C.

Proteinase K (20 mg/mL)

Purchase as a lyophilized powder and dissolve at a concentration of 20 mg/mL in sterile 50 mM Tris (pH 8.0), 1.5 mM calcium acetate. Divide the stock solution into small aliquots and store at −20°C. Each aliquot can be thawed and refrozen several times but should then be discarded. Unlike much cruder preparations of protease (e.g., pronase), proteinase K need not be self-digested before use.

REFERENCES

Watson JV, Chambers SH, Smith PJ. 1987. A pragmatic approach to the analysis of DNA histograms with a definable G1 peak. *Cytometry* **8**: 1–8.

CHAPTER 15

High-Throughput Microscopy-Based Screening in *Saccharomyces cerevisiae*

Erin B. Styles, Helena Friesen, Charles Boone,[1] and Brenda Andrews[1]

The Donnelly Centre, University of Toronto, Toronto, Ontario M5S 3E1, Canada

The budding yeast *Saccharomyces cerevisiae* has served as the pioneer model organism for virtually all genome-scale methods, including genome sequencing, DNA microarrays, gene deletion collections, and a variety of proteomic platforms. Yeast has also provided a test-bed for the development of systematic fluorescence-based imaging screens to enable the analysis of protein localization and abundance in vivo. Especially important has been the integration of high-throughput microscopy with automated image-processing methods, which has allowed researchers to overcome issues associated with manual image analysis and acquire unbiased, quantitative data. Here we provide an introduction to automated imaging in budding yeast.

INTRODUCTION

Saccharomyces cerevisiae is the premier model organism for high-throughput (HTP) genomic assays, and it has played a central role in the development of major tools and methods in the field of functional genomics (Botstein and Fink 2011). Fluorescence-based HTP microscopy assays are particularly useful in yeast, because they provide the spatiotemporal resolution required to dissect the dynamics of biological pathways and networks with unprecedented sensitivity. In the past decade or so, numerous qualitative and quantitative cellular imaging and analysis techniques have been developed to complement other HTP assays commonly used in eukaryotic cell biology. New imaging and analysis pipelines have been designed to track the movement of fluorescent markers through time, identify protein and compartment morphologies in three-dimensional space, and image dynamic changes in the subcellular distribution of the proteome in response to a variety of conditions at the single-cell level (reviewed by Liberali et al. [2015]).

REAGENTS AND METHODS FOR HTP IMAGING OF YEAST

Although yeast cells are smaller than most eukaryotic cells, ranging from ∼5 to 10 μM in diameter (30–40 fL in volume), they feature most of the same organelles and many similar subcellular structures (Feldmann 2010). However, unlike for other model systems including mammalian cell lines, strain collections are available for proteome-scale analysis of localization, abundance, and turnover of yeast proteins using fluorescence microscopy. The "ORF-GFP" collection provided the first proteome-scale view of protein localization in a eukaryotic cell (Fig. 1; Huh et al. 2003). This collection contains

---

[1]Correspondence: charlie.boone@utoronto.ca; brenda.andrews@utoronto.ca

Copyright © Cold Spring Harbor Laboratory Press; all rights reserved
Cite this introduction as *Cold Spring Harb Protoc*; doi:10.1101/pdb.top087593

Chapter 15

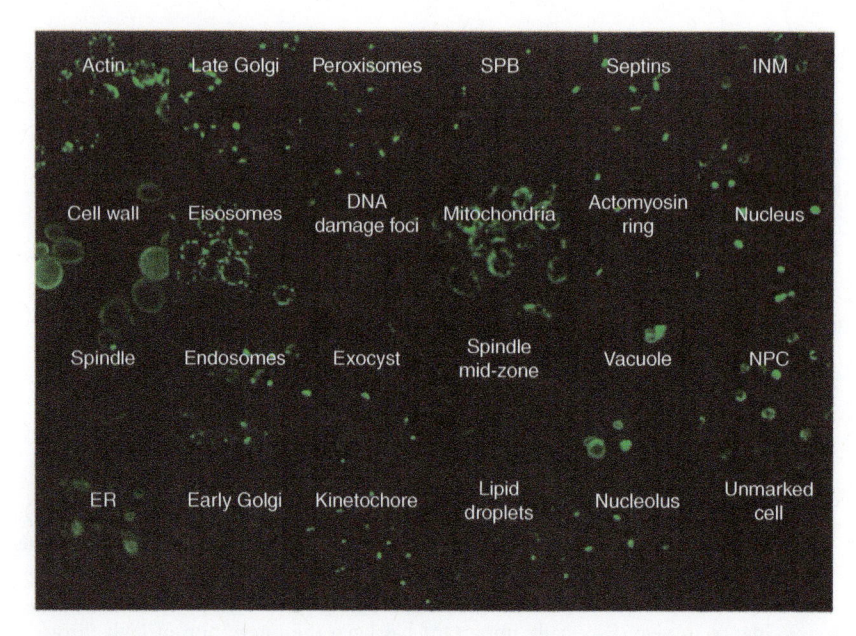

FIGURE 1. A subset of visible GFP-labeled subcellular compartments in *S. cerevisiae*. SPB, spindle pole body; INM, inner nuclear membrane; NPC, nuclear pore complex; ER, endoplasmic reticulum.

~4100 yeast strains, each with an open-reading frame (ORF) uniquely tagged with the green fluorescent protein (GFP) moiety to generate a full-length protein with a carboxy-terminus GFP fusion, the expression of which is driven by the endogenous ORF promoter (Huh et al. 2003). Manual inspection of the strains in the collection using wide-field microscopy assigned most proteins to 22 subcellular localizations. More recently, genome-scale arrays of strains expressing genes tagged with a tandem fluorescent timer (tFT) have been constructed, facilitating systematic analysis of protein turnover and mobility in vivo (Khmelinskii et al. 2012, 2014). These unique strain collections as well as other arrayed mutant collections (Table 1), combined with methods for automated yeast genetics, make the budding yeast system highly valuable for development and implementation of HTP cell-imaging methods. In particular, the synthetic genetic array (SGA) technique provides a platform for automating the introduction of any mutant or tagged allele into any set of arrayed yeast strains, and can be used to rapidly construct arrays of strains suitable for HTP imaging experiments (Tong et al. 2001). The SGA technique is introduced in Introduction of Chapter 24: Synthetic Genetic Arrays: Automation of Yeast Genetics (Kuzmin et al. 2015a), and a protocol for SGA is provided in Protocol 1 of Chapter 24: Synthetic Genetic Array Analysis (Kuzmin et al. 2015b).

HTP IMAGING PIPELINES

In the past few years, yeast researchers have taken advantage of the methods and reagents summarized above, along with advances in automated image analysis, to implement a number of HTP imaging protocols, some of which we highlight below.

- Automated imaging can be used to discover genes that regulate the localization and/or abundance of a protein of interest. In this type of experiment, SGA is used to introduce fluorescent markers of key cellular compartments, along with sensitizing mutations, into yeast mutant collections. Live cell imaging of the mutant arrays allows quantitative assessment of the abundance and localization of the fluorescent reporters, providing cell biological readouts of specific pathways and cellular structures in response to thousands of genetic perturbations (e.g., Vizeacoumar et al. 2010; Witkin et al. 2012; Neumuller et al. 2013). The effects of genetic perturbation on a fluorescent protein fused to a reporter sequence or localization signal can also be assessed (e.g., GFP fused to a

TABLE 1. Summary of arrayed mutant collections amenable to analysis by HTP microscopy

Genetic perturbation	Collection	Description	Selectable marker	Reference
Loss of function	Haploid deletion	Precise start-to-stop-codon deletions of each of the ~5000 nonessential ORFs	KanMX + Barcode	Winzeler et al. 1999; Giaever et al. 2002
Loss of function	Tet	602 promoter shutoff (TetO$_7$-promoter) alleles of essential genes	KanMX	Mnainmeh et al. 2004
Loss of function	DAmP	Disruption of 3′-UTR for 842 essential genes	KanMX	Breslow et al. 2008
Loss of function	TS	250 strains carrying temperature-sensitive alleles in essential genes	URA3MX + Barcode	Ben-Aroya et al. 2008
Loss of function	TS	787 strains carrying temperature-sensitive alleles of essential genes, covering 497 of the 1101 essential genes	KanMX	Li et al. 2011
Overexpression	MORF (moveable ORF)	5854 strains containing high copy plasmids, each expressing a sequence-verified ORF with carboxy-terminal His6-HA-ZZ tag under control of the GAL promoter	URA3	Gelperin et al. 2005
Overexpression	GST	5280 strains containing high copy plasmids, each expressing an ORF with amino-terminal GST-His tag under control of the GAL promoter	URA3	Sopko et al. 2006
Overexpression	FLEX (Full Length EXpression ready)	5192 strains containing CEN plasmids, each expressing an untagged ORF under control of the GAL promoter	URA3	Hu et al. 2007; Douglas et al. 2012
GFP fusion	GFP	4156 strains expressing ORFs under control of endogenous promoter with carboxy-terminal GFP fusion	HIS3	Huh et al. 2003
mCherry/superfolder-GFP fusion	tFT (tandem Fluorescent Timer)	4044 strains expressing ORFs under the control of under endogenous promoter with carboxy-terminal tFT tag	Excisable URA3	Khmelinskii et al. 2014

peroxisome targeting sequence [Wolinski et al. 2009]). Most screens of this type have made use of the collection of viable haploid deletion mutants, but some screens assessing the cell biological phenotypes caused by mutation of essential genes have been performed using collections of strains carrying either Tet-repressible (Singh and Tyers 2009) or temperature-sensitive (ts) alleles of essential genes (Li et al. 2011; Neumuller et al. 2013). Screens for genes whose overexpression affects the localization or abundance of the protein of interest have also been performed using an array of strains that inducibly overexpress GST-tagged alleles of yeast genes (Singh and Tyers 2009).

- The ORF–GFP array (see above) can be used to systematically identify proteins that change localization or abundance in response to a given environmental, chemical, or genetic perturbation. The array has been assessed directly for protein localization changes after pheromone treatment by creation of cell microarrays from cultures of fixed cells, followed by quantitative image analysis (Narayanaswamy et al. 2009). Alternatively, SGA can be used to introduce fluorescent markers that facilitate computational identification of yeast cells within an image, along with any mutation of interest, into the ORF–GFP array, enabling the combination of HTP genetics and imaging (e.g., Kaluarachchi Duffy et al. 2012; Tkach et al. 2012; Breker et al. 2013; Handfield et al. 2013; Chong et al. 2015).

- Quantitative morphological phenotypes can be used to assess phenotypic variation between mutants in available strain collections, in genetically diverse natural isolates, or in strains treated with chemicals or drugs. To generate a morphological profile, cell morphology parameters of a mutant or drug-treated cell are compared with profiles of other strains in a mutant collection (e.g., Ohya et al. 2005; Li et al. 2011; Okada et al. 2014). This comparative approach, which requires careful quantitative measures of multiple morphological features, has yielded new insight into gene function and cellular pathways (Ohya et al. 2005; Okada et al. 2010; Li et al. 2011), enabled prediction of drug targets (Okada et al. 2014), and shed light on the phenotypic diversity present in yeast (Skelly et al. 2013; Yang et al. 2014). For further detail, see Protocol 2: Quantification of Cell, Actin, and Nuclear DNA Morphology with High-Throughput Microscopy and CalMorph (Okada et al. 2015).

Chapter 15

AUTOMATED IMAGE ANALYSIS

The execution of useful HTP microscopy-based assays that produce high-quality phenotypic data requires the integration of a pipeline for automated image analysis. Manual image analysis is technically challenging and suboptimal, because biases between individuals can lead to discrepancies in data. Also, manual analysis limits the possible scale of the experiments, and is not quantitative (reviewed by Chong et al. [2012]). The development of high content screening (HCS), which integrates HTP microscopy with quantitative, computerized image-processing methods, allows for unbiased data analysis that delivers standardized morphological feature measurements amenable to further statistical analysis (Walter et al. 2010). In addition, automated image analysis addresses the data-generation bottleneck associated with HTP imaging, which acquires images much faster than they can be assessed by manual inspection.

Quantitative analysis of cell images typically begins with segmentation to identify the objects of interest, such as cells or cell nuclei, within an image. A number of free software packages for automatic cell segmentation, including CellProfiler (Carpenter et al. 2006), EBImage (Pau et al. 2010), and ImageJ (http://rsb.info.nih.gov/ij/), are available and have been used to segment yeast images. The incorporation of cell segmentation and downstream computational analysis into an HTP-screening protocol requires careful consideration of the biological questions to be addressed using the data and the overall goals of the screen. For example, for live cell imaging using a confocal microscope, cell markers must be chosen taking into account possible effects of markers on cell growth, as well as the impact of the cell cycle, genetic background, and environmental conditions on the localization and efficiency of the markers. Markers may include fluorescent proteins that mark the cytosol (e.g., Witkin et al. 2012; Breker et al. 2013; Chong et al. 2015) or the nucleus (e.g., Tkach et al. 2012; Mazumder et al. 2013), or dyes that provide the appropriate contrast to allow cell segmentation (e.g., da Silva Pedrini et al. 2014). It is also important to carefully consider the optical channels, lasers, and filter sets available in the screening microscope to ensure that the chosen fluorescent proteins are compatible and spectrally distinct (Lee et al. 2013). It is typically advantageous to fluorescently tag proteins of high abundance and choose bright fluorescent moieties, although the potential toxicity or effects on cell physiology must also be carefully considered.

CONCLUSION

This overview of HTP imaging experiments in yeast emphasizes available reagents and the general approaches used to set up a successful integrated screening and automated image analysis pipeline. An effective integrated screening platform is built on a reproducible, technically robust assay that includes sample preparation, image acquisition, data storage and handling, image analysis, and statistical and bioinformatics-based data analysis. Two related protocols provide examples of experimental designs that have been used successfully to perform HCS assays of yeast cells in response to perturbations. Our Protocol 1: Liquid Growth of Arrayed Fluorescently Tagged *Saccharomyces cerevisiae* Strains for Live-Cell High-Throughput Microscopy Screens (Cox et al. 2015) details a robust method for semiautomated preparation of live cells for HCS and a pipeline for systematically imaging these cells on an HTP confocal microscope to generate micrographs suitable for automated image analysis. The Protocol 2: Quantification of Cell, Actin, and Nuclear DNA Morphology with High-Throughput Microscopy and CalMorph (Okada et al. 2015) describes generation of fluorescent images of fixed cell populations that are compatible with quantitative morphological analysis using CalMorph software.

ACKNOWLEDGMENTS

This work was supported by Grant MOP-97939 from the Canadian Institutes for Health Research (CIHR) to B.A. and C.B. and a CIHR Fellowship held by E.S.

REFERENCES

Ben-Aroya S, Coombes C, Kwok T, O'Donnell KA, Boeke JD, Hieter P. 2008. Toward a comprehensive temperature-sensitive mutant repository of the essential genes of *Saccharomyces cerevisiae*. *Mol Cell* 30: 248–258.

Botstein D, Fink GR. 2011. Yeast: An experimental organism for 21st century biology. *Genetics* 189: 695–704.

Breker M, Gymrek M, Schuldiner M. 2013. A novel single-cell screening platform reveals proteome plasticity during yeast stress responses. *J Cell Biol* 200: 839–850.

Breslow DK, Cameron DM, Collins SR, Schuldiner M, Stewart-Ornstein J, Newman HW, Braun S, Madhani HD, Krogan NJ, Weissman JS. 2008. A comprehensive strategy enabling high-resolution functional analysis of the yeast genome. *Nat Methods* 5: 711–718.

Carpenter AE, Jones TR, Lamprecht MR, Clarke C, Kang IH, Friman O, Guertin DA, Chang JH, Lindquist RA, Moffat J, et al. 2006. CellProfiler: Image analysis software for identifying and quantifying cell phenotypes. *Genome Biol* 7: R100.

Chong YT, Cox MJ, Andrews B. 2012. Proteome-wide screens in *Saccharomyces cerevisiae* using the yeast GFP collection. *Adv Exp Med Biol* 736: 169–178.

Chong YT, Koh JLY, Friesen H, Kaluarachchi-Duffy S, Cox MJ, Moses AM, Moffat J, Boone C, Andrews BJ. 2015. Yeast proteome dynamics from single cell imaging and automated analysis. *Cell* 161: 1413–1424.

Cox MJ, Chong YT, Boone C, Andrews B. 2015. Liquid growth of arrayed fluorescently tagged *Saccharomyces cerevisiae* strains for live-cell high-throughput microscopy screens. *Cold Spring Harb Protoc* doi: 10.1101/pdb.prot088799.

da Silva Pedrini MR, Dupont S, de Anchieta Camara A Jr, Beney L, Gervais P. 2014. Osmoporation: A simple way to internalize hydrophilic molecules into yeast. *Appl Microbiol Biotechnol* 98: 1271–1280.

Douglas AC, Smith AM, Sharifpoor S, Yan Z, Durbic T, Heisler LE, Lee AY, Ryan O, Gottert H, Surendra A, et al. 2012. Functional analysis with a barcoder yeast gene overexpression system. *G3 (Bethesda, MD)* 2: 1279–1289.

Feldmann H. 2012. *Yeast: Molecular and cell biology*. Wiley-Blackwell, Weinheim, Germany.

Gelperin DM, White MA, Wilkinson ML, Kon Y, Kung LA, Wise KJ, Lopez-Hoyo N, Jiang L, Piccirillo S, Yu H, et al. 2005. Biochemical and genetic analysis of the yeast proteome with a movable ORF collection. *Genes Dev* 19: 2816–2826.

Giaever G, Chu AM, Ni L, Connelly C, Riles L, Veronneau S, Dow S, Lucau-Danila A, Anderson K, Andre B, et al. 2002. Functional profiling of the *Saccharomyces cerevisiae* genome. *Nature* 418: 387–391.

Handfield LF, Chong YT, Simmons J, Andrews BJ, Moses AM. 2013. Unsupervised clustering of subcellular protein expression patterns in high-throughput microscopy images reveals protein complexes and functional relationships between proteins. *PLoS Comput Biol* 9: e1003085.

Hu Y, Rolfs A, Bhullar B, Murthy TVS, Zhu C, Berger MF, Camargo AA, Kelley F, McCarron S, Jepson D, et al. 2007. Approaching a complete repository of sequence-verified protein-encoding clones for *Saccharomyces cerevisiae*. *Genome Res* 17: 536–543.

Huh WK, Falvo JV, Gerke LC, Carroll AS, Howson RW, Weissman JS, O'Shea EK. 2003. Global analysis of protein localization in budding yeast. *Nature* 425: 686–691.

Kaluarachchi Duffy S, Friesen H, Baryshnikova A, Lambert JP, Chong YT, Figeys D, Andrews B. 2012. Exploring the yeast acetylome using functional genomics. *Cell* 149: 936–948.

Khmelinskii A, Keller PJ, Bartosik A, Meurer M, Barry JD, Mardin BR, Kaufmann A, Trautmann S, Wachsmuth M, Pereira G, et al. 2012. Tandem fluorescent protein timers for in vivo analysis of protein dynamics. *Nat Biotechnol* 30: 708–714.

Khmelinskii A, Blaszczak E, Pantazopoulou M, Fischer B, Omnus DJ, Le Dez G, Brossard A, Gunnarsson A, Barry JD, Meurer M, et al. 2014. Protein quality control at the inner nuclear membrane. *Nature* 516: 410–413.

Kuzmin E, Costanzo M, Andrews B, Boone C. 2015a. Synthetic genetic arrays: Automation of yeast genetics. *Cold Spring Harb Protoc* doi: 10.1101/pdb.top086652.

Kuzmin E, Costanzo M, Andrews B, Boone C. 2015b. Synthetic genetic arrays analysis. *Cold Spring Harb Protoc* doi: 10.1101/pdb.prot088807.

Lee S, Lim WA, Thorn KS. 2013. Improved blue, green, and red fluorescent protein tagging vectors for *S. cerevisiae*. *PLoS One* 8: e67902.

Li Z, Vizeacoumar FJ, Bahr S, Li J, Warringer J, Vizeacoumar FS, Min R, Vandersluis B, Bellay J, Devit M, et al. 2011. Systematic exploration of essential yeast gene function with temperature-sensitive mutants. *Nat Biotechnol* 29: 361–367.

Liberali P, Snijder B, Pelkmans L. 2015. Single-cell and multivariate approaches in genetic perturbation screens. *Nat Rev Genet* 16: 18–32.

Mazumder A, Pesudo LQ, McRee S, Bathe M, Samson LD. 2013. Genome-wide single-cell-level screen for protein abundance and localization changes in response to DNA damage in *S. cerevisiae*. *Nucleic Acids Res* 41: 9310–9324.

Mnainmeh S, Davierwala AP, Haynes J, Moffat J, Peng W-T, Zhang W, Yang X, Pootoolal J, Chua G, Lopez A, et al. 2004. Exploration of essential gene functions via titrable promoter alleles. *Cell* 118: 31–44.

Narayanaswamy R, Moradi EK, Niu W, Hart GT, Davis M, McGary KL, Ellington AD, Marcotte EM. 2009. Systematic definition of protein constituents along the major polarization axis reveals an adaptive reuse of the polarization machinery in pheromone-treated budding yeast. *J Proteome Res* 8: 6–19.

Neumuller RA, Gross T, Samsonova AA, Vinayagam A, Buckner M, Founk K, Hu Y, Sharifpoor S, Rosebrock AP, Andrews B, et al. 2013. Conserved regulators of nucleolar size revealed by global phenotypic analyses. *Sci Signal* 6: ra70.

Ohya Y, Sese J, Yukawa M, Sano F, Nakatani Y, Saito TL, Saka A, Fukuda T, Ishihara S, Oka S, et al. 2005. High-dimensional and large-scale phenotyping of yeast mutants. *Proc Natl Acad Sci* 102: 19015–19020.

Okada H, Ohnuki S, Ohya Y. 2015. Quantification of cell, actin and nuclear DNA morphology with high-throughput microscopy and CalMorph. *Cold Spring Harb Protoc* doi: 10.1101/pdb.prot078667.

Okada H, Abe M, Asakawa-Minemura M, Hirata A, Qadota H, Morishita K, Ohnuki S, Nogami S, Ohya Y. 2010. Multiple functional domains of the yeast l,3-β-glucan synthase subunit Fks1p revealed by quantitative phenotypic analysis of temperature-sensitive mutants. *Genetics* 184: 1013–1024.

Okada H, Ohnuki S, Roncero C, Konopka JB, Ohya Y. 2014. Distinct roles of cell wall biogenesis in yeast morphogenesis as revealed by multivariate analysis of high-dimensional morphometric data. *Mol Biol Cell* 25: 222–233.

Pau G, Fuchs F, Sklyar O, Boutros M, Huber W. 2010. EBImage—An R package for image processing with applications to cellular phenotypes. *Bioinformatics* 26: 979–981.

Singh J, Tyers M. 2009. A Rab escort protein integrates the secretion system with TOR signaling and ribosome biogenesis. *Genes Dev* 23: 1944–1958.

Skelly DA, Merrihew GE, Riffle M, Connelly CF, Kerr EO, Johansson M, Jaschob D, Graczyk B, Shulman NJ, Wakefield J, et al. 2013. Integrative phenomics reveals insight into the structure of phenotypic diversity in budding yeast. *Genome Res* 23: 1496–1504.

Sopko R, Huang D, Preston N, Chua G, Papp B, Kafadar K, Snyder M, Oliver SG, Cyert MS, Hughes TR, et al. 2006. Mapping pathways and phenotypes by systematic gene overexpression. *Mol Cell* 21: 319–330.

Tkach JM, Yimit A, Lee AY, Riffle M, Costanzo M, Jaschob D, Hendry JA, Ou J, Moffat J, Boone C, et al. 2012. Dissecting DNA damage response pathways by analysing protein localization and abundance changes during DNA replication stress. *Nat Cell Biol* 14: 966–976.

Tong AH, Evangelista M, Parsons AB, Xu H, Bader GD, Page N, Robinson M, Raghibizadeh S, Hogue CW, Bussey H, et al. 2001. Systematic genetic analysis with ordered arrays of yeast deletion mutants. *Science* 294: 2364–2368.

Vizeacoumar FJ, van Dyk N, S Vizeacoumar F, Cheung V, Li J, Sydorskyy Y, Case N, Li Z, Datti A, Nislow C, et al. 2010. Integrating High-Throughput genetic interaction mapping and high-content screening to explore yeast spindle morphogenesis. *J Cell Biol* 188: 69–81.

Walter T, Shattuck DW, Baldock R, Bastin ME, Carpenter AE, Duce S, Ellenberg J, Fraser A, Hamilton N, Pieper S, et al. 2010. Visualization of image data from cells to organisms. *Nat Methods* 7: S26–S41.

Winzeler EA, Shoemaker DD, Astromoff A, Liang H, Anderson K, Andre B, Bangham R, Benito R, Boeke JD, Bussey H, et al. 1999. Functional characterization of the *S. cerevisiae* genome by gene deletion and parallel analysis. *Science* **285:** 901–906.

Witkin KL, Chong YT, Shao S, Webster MT, Lahiri S, Walters AD, Lee B, Koh JL, Prinz WA, Andrews BJ, et al. 2012. The budding yeast nuclear envelope adjacent to the nucleolus serves as a membrane sink during mitotic delay. *Curr Biol* **22:** 1128–1133.

Wolinski H, Petrovic U, Mattiazzi M, Petschnigg J, Heise B, Natter K, Kohlwein SD. 2009. Imaging-based live cell yeast screen identifies novel factors involved in peroxisome assembly. *J Proteome Res* **8:** 20–27.

Yang M, Ohnuki S, Ohya Y. 2014. Unveiling nonessential gene deletions that confer significant morphological phenotypes beyond natural yeast strains. *BMC Genomics* **15:** 932.

Protocol 1

Liquid Growth of Arrayed Fluorescently Tagged *Saccharomyces cerevisiae* Strains for Live-Cell High-Throughput Microscopy Screens

Michael J. Cox,[1] Yolanda T. Chong,[2] Charles Boone,[1] and Brenda Andrews[1,3]

[1]The Donnelly Centre, University of Toronto, Toronto, Ontario, M5S 3E1 Canada; [2]Janssen Pharmaceutica, Johnson & Johnson, Beerse, Belgium 2340

This protocol describes culturing arrays of fluorescently tagged yeast strains to early log-phase in a 96-well format for imaging on a high-throughput (HTP) microscope. The method assumes the use of the synthetic genetic array (SGA) technique to create the array of marked strains. When this approach is coupled with automated image analysis, the subcellular distribution and abundance of tagged proteins can be systematically and quantitatively examined in different genetic backgrounds and/or under different growth regimes.

MATERIALS

It is essential that you consult the appropriate Material Safety Data Sheets and your institution's Environmental Health and Safety Office for proper handling of equipment and hazardous materials used in this protocol.

RECIPES: Please see the end of this protocol for recipes indicated by <R>. Additional recipes can be found online at http://cshprotocols.cshlp.org/site/recipes.

Reagents

Drug or compound for treatment (optional; Step 7)

Prepare a 10× stock of the desired drug in the vehicle specified by the manufacturer. Store per manufacturer's instructions.

Medium for yeast cell culture

Alternative medium for medium switch (optional; Step 8)

SD_{MSG} medium (standard and low-fluorescent) <R>

Yeast strains expressing fluorescently tagged proteins arrayed in 96- or 384-colony format on an SGA haploid selection plate

This protocol assumes that the SGA technology has been used to create the yeast strain array, and that strains have undergone at least two rounds of haploid selection as described in Protocol 1 of Chapter 24: Synthetic Genetic Array Analysis (Kuzmin et al. 2015).

When designing an array for screening by the HTP microscopy, it is recommended that each subarray of 96 colonies is given unique identifying marks, which allow it to be distinguished from other subarrays. This can be done by leaving unique positions on each subarray devoid of colonies so that they appear as empty wells on the 384-well microscope slide and unambiguously identify the parent subarray. Alternatively, strains that express a

---

[3]Correspondence: brenda.andrews@utoronto.ca

Copyright © Cold Spring Harbor Laboratory Press; all rights reserved
Cite this protocol as *Cold Spring Harb Protoc*; doi:10.1101/pdb.prot088799

Chapter 15

fluorescent marker that can be easily distinguished from all others on the array can be included at positions which uniquely identify each subarray of 96.

Equipment

Aluminum foil
Bottles (1-L disposable, sterile with 2-μm filter)
Cap mats for sealing 96-well plates
Centrifuge (microplate-compatible) (if required for medium switch)
Deep-well blocks, polypropylene (2-mL, 96-well, with lids)
Glass beads (3-mm)
High-throughput microscope with 60× objective lens (Perkin Elmer Opera)
   See Discussion.
Manual pipetting machine or liquid-handling robot (96-tip)
Microplate lids, polypropylene (Evergreen Scientific 290-8020-03L)
Microplates, polypropylene (96-well, round-bottomed) (Evergreen Scientific 290-8353-03R)
Microscope slides (384-well)
Optical microplates, polystyrene (96-well, flat-bottomed)
Orbital shaker at 30°C; with racks for deep-well blocks
Pipette tips (20-μL [if required] and 200-μL; for 96-well liquid-handling system)
Replica plating pads (RePads) (96-long-pin) (Singer Instruments RP-MP-2L)
Reservoirs, polypropylene (96-well, pyramid bottom)
Sealing membranes (Axygen BF-400)
Spectrophotometer (microplate-compatible)

METHOD

Use sterile medium and plasticware for all growth steps. (Polypropylene deep-well blocks, microplates, and pin RePads can be washed after use and sterilized by autoclaving.) Perform all pipetting using a 96-tip liquid handler.

Preparing Culture Materials

To ensure uniform growth of yeast cultures, a single 3-mm glass bead must be added to each well of the microplates and deep-well blocks used for the culturing steps in this procedure.

1. Prepare beaded microplates and deep-well culture blocks as follows.
   i. Cover a 96-well cap mat with 3-mm glass beads. Tilt to remove excess beads.
   ii. Invert the 96-well polypropylene microplate or deep-well block over the cap mat. Invert again to fill the wells with beads.
   iii. Repeat Steps 1.i-1.ii for each microplate or deep-well block.
   iv. Place lids on the beaded microplates and blocks. Wrap the microplates in aluminum foil and sterilize in an autoclave at 121°C and 30 psi for 30 min.
2. Fill a reservoir with SD_{MSG} and aliquot 200 μL/well to a beaded microplate with lid. Fill one microplate for each 96 colonies to be cultured.
3. Prepare inoculation microplates as follows.
   i. Push a sterile 96-long-pin RePad gently into yeast colonies on a 96- or 384-format SGA haploid selection plate.

ii. Using the RePad, transfer the yeast cells to a beaded microplate containing SD$_{MSG}$. Rotate the RePad in the medium to dislodge the yeast cells.

iii. Seal the microplate with a breathable sealing membrane and replace the lid. Grow the yeast cells to saturation (~16 h at 30°C) with agitation in an orbital shaker.

> The use of saturated cultures minimizes the well-to-well variation in the OD$_{600}$ values of the 96 deep-well block subcultures and ensures that each well in the 384-well microscope slide contains a similar number of cells.

Preparing Subcultures

The volumes transferred and the incubation times used for subculturing will be dependent on the growth rate of the yeast cultures, which is influenced by the fitness of the yeast strains, the composition of the culture medium, and the temperature of incubation. A pilot experiment should be conducted before screening to determine the optimal volumes for a particular experimental design. Table 1 shows a typical set of dilutions used to grow haploid strains in low fluorescent SD$_{MSG}$ + hygromycin + nourseothricin + methionine at 30°C to OD$_{600}$ ~0.2.

4. Prepare materials for subculturing as follows.

    i. Fill a beaded deep-well block with 600 µL/well of low fluorescent SD$_{MSG}$.

    ii. Fill an unbeaded microplate with 230 µL/well of low fluorescent SD$_{MSG}$.

    This is the dilution microplate.

5. Subculture the cells as follows.

    i. Transfer 5–20 µL/well of saturated culture from the inoculation microplate (Step 3.iii) to the dilution microplate. Mix the contents of the wells thoroughly by repeatedly pipetting a 150-µL volume.

    ii. Transfer 5–20 µL of culture from the dilution microplate to the beaded deep-well block containing low fluorescent SD$_{MSG}$. Grow the subcultures to an OD$_{600}$ of ~0.2 (14–20 h) in an orbital shaker at 30°C.

6. Transfer 100 µL/well of subculture from the deep-well block to a flat-bottomed optical microplate and measure the OD$_{600}$. If the average OD$_{600}$/well is ~0.2, transfer the cells to a microscope slide for imaging (Step 9) or proceed to optional drug treatment (Step 7) or medium switch (Step 8).

(Optional) Treating Cells

If a screen is designed to monitor localization/abundance changes in fluorescently tagged proteins following acute exposure to drugs or other compounds, these can be added to cultures before they are transferred to a 384-well slide as described in Step 7. If a screen is designed to monitor localization/abundance changes in response to acute changes in nutrient availability, the medium can be switched before imaging as described in Step 8.

TABLE 1. Example dilution volumes for subcultures of haploid strains in low fluorescent SD$_{MSG}$ + hygromycin + nourseothricin + methionine at 30°C

Volume of saturated culture (µL)	Volume in dilution plate (µL)	Volume transferred to deep well block (µL)	Time to reach OD$_{600}$ ~ 0.2 (h)
20	230	20	14
20	230	10	17
10	230	10	20
10	230	5	23
5	230	5	26

Chapter 15

Drug Treatment

7. Treat cells with the drug or compound of interest as follows.

    i. Transfer 100–200 µL/well of 10× drug/compound stock solution to an empty sterile microplate for use in Step 7.iii.

    ii. Transfer 180 µL/well of early log-phase culture from the deep-well block to an empty sterile microplate.

    If prolonged incubation is required, wells can contain 3-mm glass beads.

    iii. Transfer 20 µL/well of 10× drug/compound stock solution to the microplate containing cells. Mix thoroughly.

    iv. Proceed directly to Step 9, or seal microplate with a breathable membrane and incubate in an orbital shaker for the desired time.

Medium Switch

8. Perform a medium switch as follows

    i. Fill two sterile microplates with 250 µL/well of alternative medium for use in Step iii.

    ii. Transfer 200 µL/well of early log-phase culture from the deep-well block to an empty sterile microplate. Centrifuge at 960 rcf for 3 min.

    iii. Remove the supernatant and wash the cell pellets in 200 µL/well of alternative medium.

    iv. Repeat washing as needed.

    When removing liquid from a microplate, most liquid handlers leave some residual volume. If removal of all residual SD_{MSG} is critical, the number of wash steps can be increased.

Imaging Cells

9. Transfer 30–100 µL/well of early log-phase culture to a 384-well slide. Repeat for each quadrant of the slide. Leave the slide undisturbed for ~10 min to allow the cells to settle to the bottom of the wells.

    The precise volume transferred will depend on the average OD_{600} of each 96-well plate and the average cell density per field of view desired for the screen. In screens that require accurate bud size information to monitor cell cycle position, it may be desirable to take many images of relatively dispersed cell populations. Using our equipment, we find that transferring a volume equal to 8/Plate average (OD_{600}) µL/well of early log-phase culture results in densities of ~100 cells/field when images are taken using a 60× objective. Transferring less than 20 µL/well to a dry 384-well slide can be inaccurate and can also result in incomplete coverage of the well. These problems can be avoided by adding 30 µL of medium to each well before the addition of low volumes of yeast culture.

10. Acquire images of the cells on an HTP microscope.

    Optical settings and exposure times are determined by the spectral characteristics and brightness of the fluorophores and capabilities of the microscope.

DISCUSSION

The protocol described here was designed for imaging on a PerkinElmer Opera HTP confocal microscope equipped with four cameras and a quadruple bandpass dichroic mirror, which allows images in multiple fluorescent channels to be acquired simultaneously. Because of this, there is no need to anchor cells to the bottom of the microscope slide. On microscopes equipped with only a single camera, cells may move between sequential exposures of the same field of view. This should be avoided, as it will prevent accurate computational analysis of these images. To prevent cell movement, slides can be coated with concanavalin A before culture transfer; see Protocol 2: Quantification of Cell, Actin, and Nuclear DNA Morphology with High-Throughput Microscopy and CalMorph (Okada et al. 2015).

RECIPES

Dropout/Add-Back Solution (10×)

L-isoleucine (300 mg/L)
L-valine (1.5 g/L)
L-adenine (400 mg/L)
L-arginine (200 mg/L)
L-histidine (200 mg/L)
L-leucine (1 g/L)
L-lysine (300 mg/L)
L-methionine (1.5 g/L)
L-phenylalanine (500 mg/L)
L-threonine (2 g/L)
L-tryptophan (400 mg/L)
L-tyrosine (300 mg/L)
L-uracil (200 mg/L)

Prepare in dH$_2$O. Autoclave at 121°C, 30 psi for 30 min. For dropout or add-back solutions, omit or include, respectively, the desired amino acids.

SD$_{MSG}$ Medium

850 mL yeast nitrogen base + monosodium glutamate (standard <R> or low fluorescent <R>)
100 mL dropout/add-back solution (10×) <R>
50 mL glucose solution (40%, sterile [i.e., autoclaved at 121°C, 30 psi for 30 min])
1 mL each of G418, hygromycin B, and/or nourseothricin

The medium must be tailored to suit the yeast strains created using the SGA technology. To allow for an effective selection of strains expressing drug resistance cassettes, monosodium glutamate is used as a nitrogen source rather than ammonium sulfate, and the required drugs are included in the medium. Drugs are prepared in advance as 1000× stocks in dH$_2$O (200 mg/mL G418 sulfate, 300 mg/mL hygromycin B, or 100 mg/mL nourseothricin), filter-sterilized, and stored at 4°C. To select for the growth of only those cells in a colony that express specific genetically encoded auxotrophic marker cassettes, omit the corresponding amino acids from the dropout/add-back solution included in the recipes. L-Canavanine sulfate and thialysine (*S*-[2-aminoethyl]-L-cysteine hydrochloride) are used in the SGA technique to select against diploid cell growth in colonies (see Protocol: Synthetic Genetic Array Analysis [Kuzmin et al. 2015]). If strains have been subjected to two rounds of haploid selection on solid medium in the presence of these drugs, and these haploids express the *STE2pr-LEU2* mating-type reporter gene, L-canavanine and thialysine can be omitted from the liquid medium described here without the significant growth of diploids in the resulting cell cultures. For low-fluorescent SD$_{MSG}$ medium, use 850 mL of yeast nitrogen base + monosodium glutamate (low fluorescent) instead of 850 mL of yeast nitrogen base + monosodium glutamate (standard) and filter-sterilize the solution to remove particulate matter that may be visualized during microscopy.

Yeast Nitrogen Base + Monosodium Glutamate
(Low Fluorescent)

1 g monosodium glutamate
1.7 g yeast nitrogen base without ammonium sulfate, without folic acid, without riboflavin (MP Biomedicals 4030-512)
850 mL dH$_2$O

Autoclave at 121°C, 30 psi for 30 min. Final volume should be 850 mL.

Yeast Nitrogen Base + Monosodium Glutamate (Standard)

1 g monosodium glutamate
1.7 g yeast nitrogen base without ammonium sulfate (BD 233520)
850 mL dH$_2$O

Autoclave at 121°C, 30 psi for 30 min. Final volume should be 850 mL.

REFERENCES

Kuzmin E, Costanzo M, Andrews B, Boone C. 2015. Synthetic genetic arrays analysis. *Cold Spring Harb Protoc* doi: 10.1101/pdb.prot088807.

Okada H, Ohnuki S, Ohya Y. 2015. Quantification of cell, actin and nuclear DNA morphology with high-throughput microscopy and CalMorph. *Cold Spring Harb Protoc* doi: 10.1101/pdb.prot078667.

Protocol 2

Quantification of Cell, Actin, and Nuclear DNA Morphology with High-Throughput Microscopy and CalMorph

Hiroki Okada, Shinsuke Ohnuki, and Yoshikazu Ohya[1]

Department of Integrated Biosciences, Graduate School of Frontier Sciences, The University of Tokyo, 5-1-5 Kashiwanoha, Kashiwa, Chiba Prefecture 277-8562, Japan

Automated image acquisition and processing systems have been developed to quantitatively describe yeast cell morphology. These systems are superior to the preceding qualitative methods in terms of reproducibility, as they completely avoid subjective recognition of images. Because high-throughput microscopy has enabled rapid production of numerous cellular images, reinforcement of high-performance and high-throughput automated image-processing techniques has been in increasing demand in the field of biology. This protocol describes how to use a high-throughput microscope in conjunction with the image-processing software CalMorph, which outputs more than 500 morphological parameters, for quantification of cell, actin, and nuclear DNA morphology.

MATERIALS

It is essential that you consult the appropriate Material Safety Data Sheets and your institution's Environmental Health and Safety Office for proper handling of equipment and hazardous materials used in this protocol.

RECIPES: Please see the end of this protocol for recipes indicated by <R>. Additional recipes can be found online at http://cshprotocols.cshlp.org/site/recipes.

Reagents

4′,6-Diamidino-2-phenylindole (DAPI) solution (Wako 049-18801)
Prepare a solution of 1 µg/mL in deionized H_2O. Store in 1-mL aliquots in the dark for up to ~1 yr at 4°C.

Concanavalin A (ConA) (Wako 037-08771)
Prepare a solution of 1 mg/mL in deionized H_2O. Frozen aliquots may be stored for up to ~1 yr.

Concanavalin A, fluorescein isothiocyanate-conjugated (FITC-ConA) (Sigma-Aldrich C7642)
Prepare a solution of 1 mg/mL in P buffer. Store in 1-mL aliquots in the dark at 4°C. Storage longer than 1 wk may cause uneven staining.

Glycerol solution <R>
P buffer <R>
Phosphate-buffered saline (PBS) (TaKaRa T900)
Rhodamine phalloidin (Rh-ph) solution (Molecular Probes R415)
Prepare a solution of 200 U/mL in methanol. Store in 1-mL aliquots in the dark for up to several months at −80°C.

Saccharomyces cerevisiae haploid strain(s) of interest, exponentially growing in 20-mL culture (4×10^6–1×10^7 cells/mL)

---

[1]Correspondence: ohya@k.u-tokyo.ac.jp

Copyright © Cold Spring Harbor Laboratory Press; all rights reserved
Cite this protocol as *Cold Spring Harb Protoc*; doi:10.1101/pdb.prot078667

Triton X-100 (10% [v/v] solution in deionized H$_2$O)
Yeast fixation solution <R>

Equipment

CalMorph (v 1.1)
CalMorph (v 1.1) can be downloaded from the Saccharomyces cerevisiae Morphological Database (SCMD) site at http://scmd.gi.k.u-tokyo.ac.jp/datamine/. Java Runtime Environment version 1.4.2 or later is required.

Centrifuge, with holders for 50-mL conical-bottom tubes and 384-well microplates (Hitachi Koki CF7D2)
High-throughput microscope (IN Cell Analyzer 2000 [GE Healthcare] with 100× objective lens [Nikon])
Microcentrifuge (Eppendorf Minispin 5452)
Microcentrifuge tubes (1.5-mL)
Microplates (384-well flat-bottom, black, with lid and microclear bottom [Greiner 781091])
Sonicator (TITEC VP-5S)
Tubes (15- and 50-mL, conical-bottom)
Water bath shaker at 25°C

METHOD

Fixing Yeast Cells

1. Transfer 20 mL of log-phase culture (4×10^6–1×10^7 cells/mL) to a 50-mL tube containing 5 mL of yeast fixation solution. Close the tube.

   The culture volume can be reduced to 0.5 mL as described in Ohnuki et al. (2012).

2. Agitate the closed tube for 30 min at 25°C in a water bath shaker.

3. Collect the cells by centrifugation at 3000 rpm for 5 min at 25°C. Discard the supernatant.

4. Resuspend the cells in a mixture of 2 mL of fixation solution and 8 mL of deionized H$_2$O.

5. Incubate the sample for 45 min at 25°C.

6. Repeat Step 3.

7. Resuspend the cells in 1 mL of PBS.

   Although it is possible to store fixed cells at 4°C for several days, immediate staining of cells is recommended.

Staining Yeast Cells

Keep samples on ice throughout the following section.

8. Transfer the suspended cells to a microcentrifuge tube on ice.

9. Collect the cells by centrifugation at 10,000 rpm for 30 sec at room temperature. Discard the supernatant.

10. Resuspend the cells in 600 µL of PBS.

11. Collect the cells by centrifugation at 10,000 rpm for 30 sec at room temperature. Discard the supernatant.

12. Resuspend the pellet in a mixture of 90 µL of PBS, 10 µL of Rh-ph solution and 1 µL of Triton X-100 solution. Incubate overnight in the dark at 4°C.

13. Repeat Step 11.

14. Wash the cells as described in Steps 10–11 using 600 µL of PBS.

15. Wash the cells as described in Steps 10–11 using 600 µL of P buffer.

16. Resuspend the cells in 488 µL of P buffer and 12 µL of FITC-ConA solution. Incubate the mixture for 10 min in the dark at 4°C.

17. Wash the cells as described in Steps 10–11 using 600 µL of P buffer. Keep the cells on ice in the dark until ready to proceed with Step 24.

Coating the Microplate with ConA

18. Transfer 50 µL of ConA solution to each well of a microplate.

19. Remove the solution from the well.

20. Repeat Steps 18 and 19.

21. Air-dry the plate for 30 min.

Preparing Specimens for High-Throughput Microscopy

22. Transfer 2.8 mL of glycerol solution to a 15-mL tube.

23. Add 105 µL of DAPI solution to the tube from Step 22. Mix well and keep the tube on ice in the dark until ready to proceed with Step 26.

24. Add 100 µL of PBS to the cells from Step 17. Disperse the cells by sonication for 5 s at levels 3–4.

25. Transfer 35 µL of the suspended cells to a microcentrifuge tube.

26. Add 415 µL of glycerol solution with DAPI from Step 23 to the cell sample.

    The final concentration of cells is 2×10^6–3×10^6 cells/mL. This can be adjusted as needed in Step 25.

27. Transfer 50 µL of the suspended cells into each ConA-coated well from Step 21.

28. Retain the cells at the bottom of the well by centrifugation at 1500 rpm for 5 min.

Acquiring Images and Processing Data

29. Using the IN Cell Analyzer 2000 according to the manufacturer's instructions, acquire images of the cell wall, actin and nuclear DNA in the same field of view.

    We used the following settings:
    Excitation/Emission filters: DAPI, FITC and Cy3
    Binning: 2 × 2
    Image: 2-D Deconvolution
    Laser power: 30%

30. Save the images in 8-bit grayscale JPEG format (image size: 520 × 696 pixels) in the directory.

    Files should be named according to the following (Fig. 1A).
    [folder name]-C[number].jpg for FITC-ConA images
    [folder name]-D[number].jpg for DAPI images
    [folder name]-A[number].jpg for Rh-ph images

31. Run CalMorph to quantify cell morphology.

    Five hundred and one yeast morphological parameters have been defined using CalMorph (Ohya et al. 2005). CalMorph input and output are illustrated in Figure 1B. Protocols for manual operation of a fluorescent microscope, helpful tools and original data can be downloaded from the SCMD site and http://www.yeast.ib.k.u-tokyo.ac.jp/CalMorph.

DISCUSSION

When we used the IN Cell Analyzer 2000 to take pictures with 20 fields of view in each well, we were able to recognize more than 4000 cells in 1 h—approximately seven times faster than manual operation of a fluorescent microscope. Quantification of cell, actin, and nuclear DNA morphology by image processing software is highly sensitive to image quality. For this reason, it is important to

Chapter 15

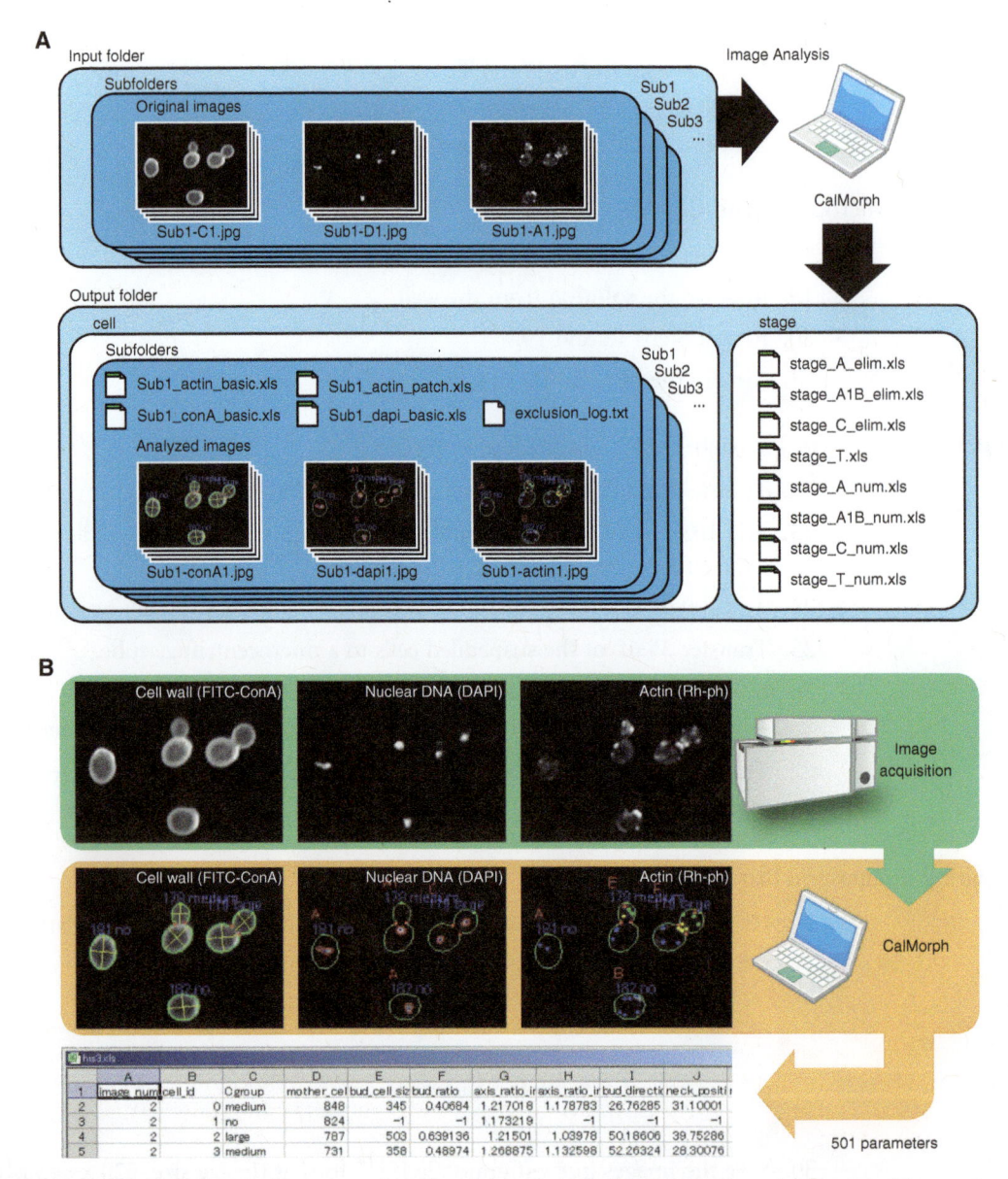

FIGURE 1. (A) Storage of original images and analyzed data. Rounded squares indicate folders stored in a computer. Input and output folders specify a folder containing original images and analyzed data (images and text files), respectively. (B) Schematic drawing of the analytical flow. After *his3* cells (EUROSCARF acc. no. Y02458) were stained for the cell wall (FITC-ConA), nuclear DNA (DAPI), and actin (Rh-ph), images of triple-stained cells were acquired in the same field of view (green region) using a high-throughput microscope. In the CalMorph-processed images, an ellipse is fitted to each mother cell and bud to measure morphological parameters (orange region). The yellow lines show the long and short axes of the fitted ellipses. In the processed nuclear DNA images, light blue bullets indicate recognized nuclei. In the processed actin images, colored dots indicate recognized actin dots.

confirm that the auto-focusing function of the microscope is effective in every well. We do not recommend using dark and foggy images that are easily recognized as low-quality images by eye. It is also important to ensure that there is no reduction in signal intensity during microscope operation. Quantitative data cannot be obtained from clumping cells, cells with unevenly stained cell walls, cells touching an external margin of the image, or cells that move during a filter change. CalMorph (v 1.1) sometimes fails to correctly recognize haploid cells with an extraordinary cell shape. To quantify diploid cells, we recommend CalMorph (v 1.3) (Yvert et al. 2013).

A well-designed experiment followed by appropriate statistical analyses can provide a wealth of biologically meaningful information on cell morphology. Quantification of morphological features by

CalMorph was completed for the cataloged haploid deletion mutant collection (Ohya et al. 2005) as well as a set of isolated wild yeast strains (Yvert et al. 2013). With quantitative and high-dimensional morphological data sets, yeast mutants were classified according to their morphology (Okada et al. 2010). It was also possible to search for a set of mutants with similar morphology to drug-treated cells, thus predicting drug targets in the cell (Iwaki et al. 2013). Although high-dimensional data are useful for describing various aspects of cellular activities, identifying primary features can be problematic due to the complexity of the data. Multivariate analyses such as principal component analysis and support vector machines can be applied to define the biologically meaningful phenotype for effective analyses (Ohnuki et al. 2012).

RECIPES

Glycerol Solution

Reagent	Amount to add
Phosphate-buffered saline (PBS) (10×; TaKaRa T900)	600 µL
Glycerol for fluorescence microscopy (Merck 104095)	5400 µL

Mix gently with agitation.

P Buffer

Reagent	Final concentration
Sodium phosphate (pH 7.2)	10 mM
NaCl	150 mM

Yeast Fixation Solution

Formaldehyde solution (37%; Wako 064-00406)
Potassium phosphate buffer (1 M, pH 6.5)

Combine equal volumes of these reagents immediately before use.

REFERENCES

Iwaki A, Ohnuki S, Suga Y, Izawa S, Ohya Y. 2013. Vanillin inhibits translation and induces messenger ribonucleoprotein (mRNP) granule formation in *Saccharomyces cerevisiae*: Application and validation of high-content, image-based profiling. *PLoS ONE* **8**: e61748.

Ohnuki S, Kobayashi T, Ogawa H, Kozone I, Ueda JY, Takagi M, Shin-Ya K, Hirata D, Nogami S, Ohya Y. 2012. Analysis of the biological activity of a novel 24-membered macrolide JBIR-19 in *Saccharomyces cerevisiae* by the morphological imaging program CalMorph. *FEMS Yeast Res* **12**: 293–304.

Ohya Y, Sese J, Yukawa M, Sano F, Nakatani Y, Saito TL, Saka A, Fukuda T, Ishihara S, Oka S, et al. 2005. High-dimensional and large-scale phenotyping of yeast mutants. *Proc Natl Acad Sci* **102**: 19015–19020.

Okada H, Abe M, Asakawa-Minemura M, Hirata A, Qadota H, Morishita K, Ohnuki S, Nogami S, Ohya Y. 2010. Multiple functional domains of the yeast 1,3-β-glucan synthase subunit Fks1p revealed by quantitative phenotypic analysis of temperature-sensitive mutants. *Genetics* **184**: 1013–1024.

Yvert G, Ohnuki S, Nogami S, Imanaga Y, Fehrmann S, Schacherer J, Ohya Y. 2013. Single-cell phenomics reveals intra-species variation of phenotypic noise in yeast. *BMC Syst Biol* **7**: 54.

CHAPTER 16

Single-Molecule Total Internal Reflection Fluorescence Microscopy

Emily M. Kudalkar,[1] Trisha N. Davis,[1,3] and Charles L. Asbury[2,3]

[1]*Department of Biochemistry, University of Washington, Seattle, Washington 98195;* [2]*Department of Physiology and Biophysics, University of Washington, Seattle, Washington 98195*

The advent of total internal reflection fluorescence (TIRF) microscopy has permitted visualization of biological events on an unprecedented scale: the single-molecule level. Using TIRF, it is now possible to view complex biological interactions such as cargo transport by a single molecular motor or DNA replication in real time. TIRF allows for visualization of single molecules by eliminating out-of-focus fluorescence and enhancing the signal-to-noise ratio. TIRF has been instrumental for studying in vitro interactions and has also been successfully implemented in live-cell imaging. Visualization of cytoskeletal structures and dynamics at the plasma membrane, such as endocytosis, exocytosis, and adhesion, has become much clearer using TIRF microscopy. Thanks to recent advances in optics and commercial availability, TIRF microscopy is becoming an increasingly popular and user-friendly technique. In this introduction, we describe the fundamental properties of TIRF microscopy and the advantages of using TIRF for single-molecule investigation.

SINGLE-MOLECULE VISUALIZATION USING TIRF MICROSCOPY

Since the development of fluorescence microscopy, a long-standing problem has been the limitation of visualizing single fluorescent molecules. Standard microscopes have a resolution limit near 0.2 µm, much too large to spatially distinguish individual protein complexes. Recent techniques have been developed that greatly enhance resolution, such as deconvolution and confocal microscopy, but these are still confounded by out-of-focus light or rapid photobleaching. One major problem with standard fluorescence imaging is limiting the excitation of fluorescent molecules to a precise focal plane. Out-of-focus fluorescence increases the background noise and detracts from the intensity of true signal, making spatial resolution difficult, if not impossible. The development of TIRF microscopy has effectively eliminated out-of-focus fluorescence by restricting excitation to a very thin section near the coverslip, making it possible to achieve single-molecule or -particle detection (Axelrod et al. 1984). This selective excitation also reduces photobleaching of fluorophores in solution and prevents harmful light damage when imaging live cells.

BASIC PHYSICS OF TIRF MICROSCOPY

The basic concept of TIRF microscopy applies the fundamental properties of optical physics to generate an evanescent field to excite fluorophores instead of using direct illumination. In a typical

---

[3]Correspondence: tdavis@uw.edu; casbury@uw.edu

Copyright © Cold Spring Harbor Laboratory Press; all rights reserved
Cite this introduction as *Cold Spring Harb Protoc;* doi:10.1101/pdb.top077800

Chapter 16

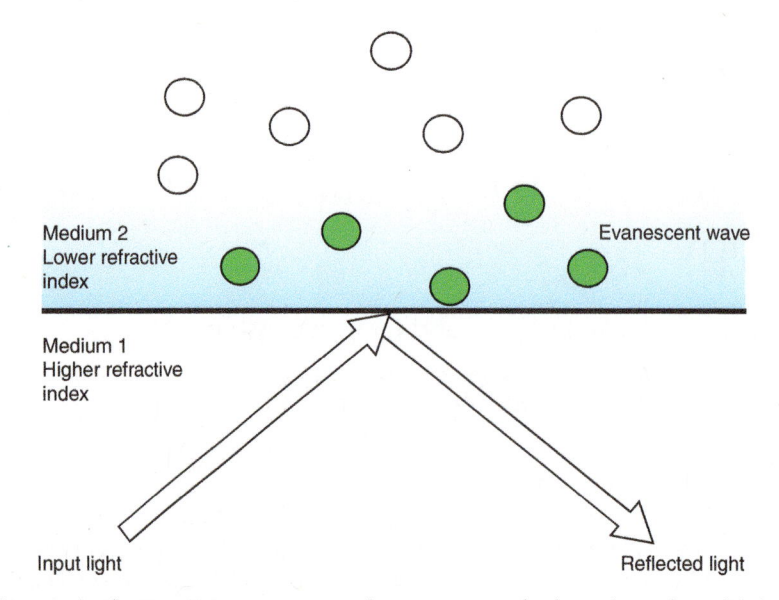

FIGURE 1. Total internal reflection (TIR) microscopy. The proportion of refracted to reflected light can be changed when passing light through two different materials: one of higher refractive index and the other of lower refractive index. A critical angle can be reached that defines the point where all the light is refracted parallel to the boundary between the two mediums. Once this critical angle is passed, light is totally internally reflected and creates an evanescent wave that transmits into the second medium and decays exponentially. In total internal reflection fluorescence (TIRF) microscopy, this phenomenon is exploited to restrict illumination of fluorophores to only those within ~100 nm of the coverslip (green spheres) while eliminating background excitation of fluorophores in solution (white spheres).

TIRF setup, light is transmitted through two adjacent materials, one of a higher refractive index (such as a glass coverslip), followed by another with a lower refractive index (such as an aqueous solution). As light passes through the media, it is partially reflected and refracted depending on the incident angle. By using two materials with different refractive indices, a critical angle can be reached: the point where all the light is refracted parallel to the interface of the two materials. Once this critical angle is passed, the phenomenon of total internal reflection (TIR) occurs (Fig. 1). Although light no longer passes through the second medium, the reflected light creates an electromagnetic field that penetrates into the less refractive material. This electromagnetic field is evanescent and decays exponentially with the depth of penetration. TIRF microscopy exploits this evanescent field, which is typically ~100 nm thick, to exclusively excite fluorophores at the boundary of the two media, that is, very near to the coverslip. This restriction eliminates excitation of molecules away from the coverslip, thereby significantly reducing out-of-focus background fluorescence. Until recently, reaching the proper incident angle to induce TIR was rather challenging due to the limitations of commercially available microscope objectives. However, the development of new objectives with sufficiently high numerical aperture (typically >1.45 Å) to reach the proper incident angle has made TIRF microscopy much more accessible to researchers in recent years (Axelrod 2001). This advancement combined with the increase in commercial availability of TIRF microscopes has greatly enhanced the popularity and usage of TIRF microscopy in today's research.

APPLICATIONS OF TIRF MICROSCOPY

TIRF microscopy is especially useful for studying protein–protein and protein–nucleic acid biochemical interactions. TIRF permits direct visualization of binding events and quantification of kinetic on and off rates. Standard bulk biochemical assays report on the average properties of a population and often miss individual variability and stochasticity that can only be appreciated at the single-molecule level. Visualizing single complexes can reveal sample heterogeneity such as the presence of different oligomeric states of a complex. TIRF can be used for stoichiometric quantification of proteins within complexes that are tagged with different fluorescent probes or by quantifying photobleach steps using single probes (Ulbrich and Isacoff 2007). Because the concentration range used in TIRF assays is often on the pM–nM scale, experiments can be performed using minute amounts of protein or DNA. This becomes extremely useful when visualizing native complexes purified from cells because often very little material is recovered. TIRF also makes it possible to watch assembly and disassembly of macromolecular complexes and can reveal hierarchical orders that may be obscured using other techniques (Hoskins et al. 2011). Using TIRF, it is now possible to view complex biological interactions such as cargo transport by a single molecular motor or DNA replication in real time (Axelrod et al. 1983; Vale et al. 1996; Ha et al. 1999; Tanner et al. 2009; Yardimci et al. 2010; Hoskins et al. 2011).

Single-molecule imaging has been vital to the characterization of molecular movements along a substrate, especially polymers such as actin, microtubules, or DNA (Funatsu et al. 1995; Vale et al. 1996; Harada et al. 1999). In addition, visualization of cytoskeletal structures and dynamics at the plasma membrane, such as endocytosis, exocytosis, and adhesion, has become much clearer using TIRF microscopy (Mashanov et al. 2003; Cai et al. 2007). Molecular diffusion and movement can be difficult to see within cells due to the complexity of structures within the cytoplasm. Recapitulating these events in vitro is a powerful method used for understanding biochemical function. Visualizing molecular events in real time provides key information about their temporal regulation, such as the stalling and reinitiation that helicases undergo while unwinding DNA (Ha et al. 2002). Single-molecule analysis is an excellent method for elucidating the function of enzymes and characterizing the timing of individual steps in catalytic reactions. TIRF has been instrumental in understanding the precise function of molecular machines such as the hand-over-hand motion of myosin V along microtubules (Yildiz et al. 2003), GroEL-assisted protein folding (Yamasaki et al. 1999), the dynamics of actin polymerization (Amann and Pollard 2001), and the analysis of the dynamic assembly of the spliceosome (Hoskins et al. 2011).

Coupling the properties of TIRF microscopy with other optical and biophysical techniques has been quite successful for the development of new combinatorial technologies. TIRF has been used to develop single-pair fluorescence resonance energy transfer (spFRET), giving even higher spatial resolution to single molecular colocalization (Weiss 1999). spFRET can reveal conformational changes within single enzymes during their interactions with substrates. TIRF combined with super-resolution methods, such as stochastic optical reconstruction microscopy (STORM), has been useful for refining molecular localization to within a few nanometers. Coupling TIRF with optical tweezer microscopy has also been tremendously powerful. Simultaneous use of both technologies allows for direct correlation of mechanical movement with structural changes (Ishijima et al. 1998; Lang et al. 2003). These techniques have lead to important and sometimes surprising findings, such as the demonstration that ATP hydrolysis and mechanical movement are not always simultaneous (Ishijima et al. 1998).

PROTOCOLS FOR TIRF MICROSCOPY

In the accompanying protocols, we provide step-by-step procedures for preparing and imaging samples for TIRF and for analyzing the resulting data. See Protocol 1: Coverslip Cleaning and Functionalization for Total Internal Reflection Fluorescence Microscopy (Kudalkar et al. 2015a), Protocol 2: Preparation of Reactions for Imaging with Total Internal Reflection Fluorescence Microscopy (Kudalkar et al. 2015b), and Protocol 3: Data Analysis for Total Internal Reflection Fluorescence Microscopy (Asbury 2015).

ACKNOWLEDGMENTS

The authors would like to thank Neil Umbreit and Erik Yusko for their help with developing experimental protocols. E.M.K., T.N.D., Y.D., and C.L.A. are supported by the National Institutes of Health

(F32GM099223 to E.M.K., R01GM040506 to T.N.D., R01GM079373 and S10RR026406 to C.L.A.) and Y.D. and C.L.A. are also supported by the Packard Fellowship (2006-30521 to C.L.A.).

REFERENCES

Amann KJ, Pollard TD. 2001. Direct real-time observation of actin filament branching mediated by Arp2/3 complex using total internal reflection fluorescence microscopy. *Proc Natl Acad Sci* 98: 15009–15013.

Asbury CL. 2015. Data analysis for total internal reflection fluorescence microscopy. *Cold Spring Harb Protoc* doi: 10.1101/pdb.prot085571.

Axelrod D. 2001. Selective imaging of surface fluorescence with very high aperture microscope objectives. *J Biomed Opt* 6: 6–13.

Axelrod D, Thompson NL, Burghardt TP. 1983. Total internal inflection fluorescent microscopy. *J Microsc* 129: 19–28.

Axelrod D, Burghardt TP, Thompson NL. 1984. Total internal reflection fluorescence. *Annu Rev Biophys Bioeng* 13: 247–268.

Cai D, Verhey KJ, Meyhofer E. 2007. Tracking single Kinesin molecules in the cytoplasm of mammalian cells. *Biophys J* 92: 4137–4144.

Funatsu T, Harada Y, Tokunaga M, Saito K, Yanagida T. 1995. Imaging of single fluorescent molecules and individual ATP turnovers by single myosin molecules in aqueous solution. *Nature* 374: 555–559.

Ha T, Ting AY, Liang J, Caldwell WB, Deniz AA, Chemla DS, Schultz PG, Weiss S. 1999. Single-molecule fluorescence spectroscopy of enzyme conformational dynamics and cleavage mechanism. *Proc Natl Acad Sci* 96: 893–898.

Ha T, Rasnik I, Cheng W, Babcock HP, Gauss GH, Lohman TM, Chu S. 2002. Initiation and re-initiation of DNA unwinding by the *Escherichia coli* Rep helicase. *Nature* 419: 638–641.

Harada Y, Funatsu T, Murakami K, Nonoyama Y, Ishihama A, Yanagida T. 1999. Single-molecule imaging of RNA polymerase–DNA interactions in real time. *Biophys J* 76: 709–715.

Hoskins AA, Friedman LJ, Gallagher SS, Crawford DJ, Anderson EG, Wombacher R, Ramirez N, Cornish VW, Gelles J, Moore MJ. 2011. Ordered and dynamic assembly of single spliceosomes. *Science* 331: 1289–1295.

Ishijima A, Kojima H, Funatsu T, Tokunaga M, Higuchi H, Tanaka H, Yanagida T. 1998. Simultaneous observation of individual ATPase and mechanical events by a single myosin molecule during interaction with actin. *Cell* 92: 161–171.

Kudalkar EM, Deng Y, Davis TN, Asbury CL. 2015a. Coverslip cleaning and functionalization for total internal reflection fluorescence microscopy. *Cold Spring Harb Protoc* doi: 10.1101/pdb.prot085548.

Kudalkar EM, Davis TN, Asbury CL. 2015b. Preparation of reactions for imaging with total internal reflection fluorescence microscopy. *Cold Spring Harb Protoc* doi: 10.1101/pdb.prot085563.

Lang MJ, Fordyce PM, Block SM. 2003. Combined optical trapping and single-molecule fluorescence. *J Biol* 2: 6.

Mashanov GI, Tacon D, Knight AE, Peckham M, Molloy JE. 2003. Visualizing single molecules inside living cells using total internal reflection fluorescence microscopy. *Methods* 29: 142–152.

Tanner NA, Loparo JJ, Hamdan SM, Jergic S, Dixon NE, van Oijen AM. 2009. Real-time single-molecule observation of rolling-circle DNA replication. *Nucleic Acids Res* 37: e27.

Ulbrich MH, Isacoff EY. 2007. Subunit counting in membrane-bound proteins. *Nat Methods* 4: 319–321.

Vale RD, Funatsu T, Pierce DW, Romberg L, Harada Y, Yanagida T. 1996. Direct observation of single kinesin molecules moving along microtubules. *Nature* 380: 451–453.

Weiss S. 1999. Fluorescence spectroscopy of single biomolecules. *Science* 283: 1676–1683.

Yamasaki R, Hoshino M, Wazawa T, Ishii Y, Yanagida T, Kawata Y, Higurashi T, Sakai K, Nagai J, Goto Y. 1999. Single molecular observation of the interaction of GroEL with substrate proteins. *J Mol Biol* 292: 965–972.

Yardimci H, Loveland AB, Habuchi S, van Oijen AM, Walter JC. 2010. Uncoupling of sister replisomes during eukaryotic DNA replication. *Mol Cell* 40: 834–840.

Yildiz A, Forkey JN, McKinney SA, Ha T, Goldman YE, Selvin PR. 2003. Myosin V walks hand-over-hand: Single fluorophore imaging with 1.5-nm localization. *Science* 300: 2061–2065.

Protocol 1

Coverslip Cleaning and Functionalization for Total Internal Reflection Fluorescence Microscopy

Emily M. Kudalkar,[1] Yi Deng,[2] Trisha N. Davis,[1,3] and Charles L. Asbury[2,3]

[1]Department of Biochemistry, University of Washington, Seattle, Washington 98195; [2]Department of Physiology and Biophysics, University of Washington, Seattle, Washington 98195

Total internal reflection fluorescence (TIRF) microscopy allows visualization of biological events at the single-molecule level by restricting excitation to a precise focal plane near the coverslip and eliminating out-of-focus fluorescence. The quality of TIRF imaging relies on a high signal-to-noise ratio and therefore it is imperative to prevent adherence of molecules to the glass coverslip. Nonspecific interactions can make it difficult to distinguish true binding events and may also interfere with accurate quantification of background noise. In addition, nonspecific binding of the fluorescently tagged protein will lower the effective working concentration, thereby altering values used to calculate affinity constants. To prevent spurious interactions, we thoroughly clean the surface of the coverslip and then functionalize the glass either by applying a layer of silane or by coating with a lipid bilayer.

MATERIALS

It is essential that you consult the appropriate Material Safety Data Sheets and your institution's Environmental Health and Safety Office for proper handling of equipment and hazardous material used in this protocol.

RECIPES: Please see the end of this protocol for recipes indicated by <R>. Additional recipes can be found online at http://cshprotocols.cshlp.org/site/recipes.

Reagents

Biotinylated antibody (200 μM stock solution)
Biotinyl Cap PE (0.1 mg/mL in 2:1 chloroform/methanol) (Avanti Polar Lipids 870277X)
BRB80 (or any buffer with pH 5–9 and <200 mM ionic strength) <R>
Butylamine
Chloroform
Concentrated HCl (37%)
Concentrated sulfuric acid (95%–98%)
1,2-Dipalmitoyl-sn-glycero-3-phosphocholine (DPPC; 10 mg/mL in chloroform) (Avanti Polar Lipids)
Ethanol
Hydrogen peroxide (30%)
Imaging buffer <R>

[3]Correspondence: tdavis@uw.edu; casbury@uw.edu

Chapter 16

κ-casein (5 mg/mL stock solution; filter-sterilized)
Methanol
2-Methoxy(polyethyleneoxy)propyltrimethoxysilane (Gelest SIM-6492.7)
Nitrogen gas (dry)
Purified protein of interest
Streptavidin (1 mg/mL stock solution)
Toluene
Wash buffer (BRB80 containing 1 mg/mL κ-casein)

Equipment

Beaker (2-L, with Teflon lid and gas inlets)
Beakers (glass; 1- and 2-L)
Coverslips (glass; 22 × 60 mm)
Double-sided tape
Drying oven (set at 50°C)
Fume hood
Hot plate (set at 50°C)
Mercury lamp
Metal bar to hold Teflon racks
Nail polish
pH strips (range 2.0–7.0)
Pipettes (glass)
Plasma cleaner
Quartz cuvette
Razor blade
Safety gear (laboratory coat, goggles, gloves, and UV eye protection)
Slides (glass; 3 × 1 in)
Sonicating water bath
Sonifier (with microtip probe)
Spray bottle
Syringes (glass; 10-µL and 100-µL)
Teflon racks (custom-fabricated; see Fig. 1)
Teflon straw
Test tubes (glass)
Vacuum pump and desiccator
Vials (glass; for storing lipids)

Use Teflon tape for sealing the vials. Do not use rubber cap liners.

FIGURE 1. Custom-fabricated Teflon coverslip rack. Our coverslip racks hold 13 glass coverslips (22 × 60 mm) with minimal contact and four units can be linked together to clean and silanize 52 coverslips in one preparation. Teflon can withstand all the chemicals used in the protocol and the racks can be reused for several years.

Vortexer
Water bath (set at 50°C)

METHOD

Coverslips are thoroughly cleaned and the glass surface is functionalized either by applying a layer of silane (Steps 1–21) or by coating with a lipid bilayer (Steps 22–42).

Functionalizing Glass Coverslips by Silanization

The silanization method (adapted from Cras et al. 1999 and Walba et al. 2004) involves a harsh acid wash to clean the glass followed by application of a functionalized layer of silane, which adheres to the glass and creates a uniform, covalently linked layer on the coverslip surface. The silane is linked to a short polyethyleneglycol molecule at one end, creating a hydrophilic layer pointing away from the surface. The hydrophilic layer attracts water and helps prevent proteins from adhering to the coverslip.

Acid-Washing the Coverslips

1. Load glass coverslips (22 × 60 mm) into four custom-fabricated Teflon racks (Fig. 1).

   A typical rack contains 13 slots and four units can be utilized together to clean and silanize 52 coverslips in one preparation. The racks are handled by inserting a horizontal metal bar into the top loops of each rack. All subsequent manipulations are performed with the coverslips loaded into the racks.

2. Holding the racks by the metal bar, briefly rinse the coverslips by dipping the racks into a 1-L beaker filled with distilled water.

3. Mix 500 mL of methanol with 500 mL of concentrated HCl in a 2-L beaker in the fume hood. Cool on ice for ~10 min (the reaction is exothermic) to allow time for the fuming to stop. Place the methanol/HCl beaker into a sonicating water bath.

4. Remove the coverslips from the distilled water and gently shake off excess water.

5. Carefully lower the coverslips into the methanol/HCl solution and gently dip up and down several times. Sonicate for 30 min in the sonicating water bath.

6. Prepare five wash beakers by filling five 2-L beakers with 1 L of distilled water each.

7. After sonication, gently shake off excess methanol/HCl solution and dip the coverslips several times into the first wash beaker. Repeat with the second wash beaker.

8. Place the second wash beaker into the sonicating water bath and sonicate for 3 min.

9. Thoroughly rinse the coverslips sequentially in the remaining wash beakers. Carefully dip the racks up and down several times in each wash solution. Check the pH of the water in each wash beaker using pH strips to ensure that the HCl is being washed away. (By the fourth or fifth beaker, the pH should equal to that of distilled water.) Thoroughly, but gently, shake off excess water after the final wash.

10. Pour 1 L of concentrated sulfuric acid into a clean 2-L beaker in the fume hood. Place the beaker into the sonicating water bath.

11. Lower the coverslips into the sulfuric acid, gently dipping up and down several times as in Step 5. Sonicate for 30 min.

12. Prepare seven wash beakers by filling seven 2-L beakers with 1 L of distilled water each.

    These additional washes are required to fully rinse off the sulfuric acid.

13. Rinse the coverslips sequentially in the wash beakers and check the pH of the rinse water. Sonicate the wash beakers for 10 min during the second and third washes. (By the seventh wash, the sulfuric acid should be removed.) Shake off excess water following the final wash.

14. Fill a spray bottle with 300 mL of ethanol. Spray the coverslips extensively with ethanol and place them in the drying oven for 30 min at 50°C.

15. When the coverslips are thoroughly dry, continue with silanization (Steps 16–21).

Silanization

16. Mix the silanization solution in a 2-L beaker by combining 960 mL of toluene, 16.8 mL of 2-methoxy(polyethyleneoxy)propyltrimethoxysilane, and 6 mL of butylamine.

17. Transfer the acid-cleaned, dry coverslips into the silanization solution and cap with a Teflon lid with N_2 gas inlets. Incubate for 90 min under N_2 gas at a flow rate of 2–3 mL/min.

18. Prepare two wash beakers by filling two 1-L beakers with 500 mL of toluene each. After incubating for 90-min, remove the coverslips from the silanization solution and gently shake to remove excess liquid. Dip the coverslips sequentially into the toluene washes, thoroughly shaking off the liquid between each wash.

19. Place the coverslips into a clean 2-L beaker with a Teflon lid and gas inlets. Fit two Teflon straws through the gas inlets and cover the beaker with the lid. The straws should reach the bottom of the beaker to achieve the best results. Dry thoroughly with N_2 gas at a flow rate of 5 mL/min for 90 min.

20. Cure overnight with N_2 gas at a flow rate of 1 mL/min.

21. Turn off the N_2 gas the following morning and cap the inlets. Store the coverslips at room temperature under N_2.

    To use silanized coverslips for TIRF microscopy experiments, see Protocol 2: Preparation of Reactions for Imaging with Total Internal Reflection Fluorescence Microscopy (Kudalkar et al. 2015).

Functionalizing Glass Coverslips by Lipid Passivation

Besides silanization, coating surfaces with a lipid bilayer provides an alternative approach for specifically immobilizing molecules of interest, while eliminating nonspecific interactions between the surface and background molecules. In general, lipid bilayers appear to be inert to most biomolecules, including nucleic acids and most soluble proteins. For the purpose of TIRF imaging, lipid bilayers can be created by the spontaneous deposition of small unilamellar vesicles (SUVs) on a flat hydrophilic surface such as a clean glass coverslip (Sackmann 1996; Cremer and Boxer 1999). Lipids that carry modified head groups can be mixed with regular phospholipids to introduce specific immobilization of the molecule of interest on the surface. We use biotinylated phosphoethanolamine (Biotinyl Cap PE) as the affinity anchor. The density of the active linkers can be conveniently adjusted by altering the fractions of modified and regular lipids. In addition, the extensive variety of lipid species allows one to tune the physical properties of the lipid bilayer such as its diffusivity and phase.

Preparation of Small Unilamellar Lipid Vesicles

22. In the fume hood, transfer 1 mL of chloroform into three glass test tubes for washing the glass syringes.

23. Wash a fourth test tube briefly with chloroform and drain. This tube will be the dry lipid test tube.

24. Wash a 100-µL glass syringe three times with the washing chloroform in each of the three test tubes before use. Transfer 70 µL of chloroform to the lipid test tube.

25. Transfer 30 µL of the DPPC lipid solution to the lipid test tube. Wash the 100-µL glass syringe nine times with the washing chloroform.

26. Wash the 10-µL glass syringe nine times with the washing chloroform. Transfer 4 µL of Biotinyl Cap PE to the lipid test tube. Wash the 10-µL syringe nine times.

27. Vortex to mix the lipid.

28. Evaporate the chloroform from the mixture in the lipid test tube by gently blowing N_2 gas above the mixture for at least 5 min while slowly rotating the tube.

    The dried lipid mixture will form a white patch at the bottom of the test tube.

29. Place the lipid test tube in the desiccator and further dry under vacuum for at least 15 min.

    A plasma cleaner can also be used to dry the lipid. Keep the plasma off if drying with plasma cleaner.

30. Add 200 µL of BRB80 (or other desired buffer) to the dried lipid and vortex briefly. Place the test tube in a water bath set at 50°C. Vortex every 5 min until the lipid patch at the bottom of the tube is completely rehydrated into giant unilamellar vesicles (GUVs), which appear as a white opaque suspension.

    This step typically takes <15 min.

31. Transfer the rehydrated lipid to a 1.5-mL microcentrifuge tube. Immerse the sonifier microtip probe into the suspension and sonicate for 3 min, while keeping the test tube in a water bath set >50°C. To prevent spurious bubbling, lower the sonication duty cycle and the output power.

    The white GUV suspension should turn into a clear SUV suspension.

32. (Optional) If low fluorescence background is required, add 100 µL of warm 30% hydrogen peroxide to the liposome suspension and transfer the mixture to a quartz cuvette. Place the cuvette under a mercury lamp and irradiate for 30 min to bleach the fluorophores in the liposome.

    This step can be performed at room temperature. Proper eye protection is required to prevent UV damage.

    UV irradiation and peroxide treatment have no observable effect on the passivation ability of saturated phospholipid and the biotin functional group, but they significantly lower the fluorescence of the contaminants in the liposome.

33. Keep the SUV preparation on a hotplate or in a water bath set at 50°C. Use the liposomes within 1 d.

Coating Glass Coverslips with a Lipid Bilayer

34. Clean coverslips and glass slides with a plasma cleaner.

35. Apply several pieces of double-sided tape crosswise to a plasma-cleaned glass slide, leaving ~1–3 mm gaps in between the pieces to create chambers.

36. Place a plasma-cleaned coverslip on the slide. Gently press down to ensure a tight and even seal between the coverslip and the tape. Place on a hot plate set for 5 min at 50°C and periodically press down on the coverslip to secure sealing.

37. Using a razor blade, trim off excess tape from either side of the coverslip so it is flush with the edge.

38. Transfer the slide to a warm humidity chamber, such as a pipette tip box half-filled with hot water. Flow in BRB80 and prewet the chamber to prevent lifting of the double-sided tape. Flowthrough SUV in warm buffer (from Step 33). Incubate for 5 min to allow the lipid bilayer to form. Before or during the incubation, prepare a 0.25 mg/mL solution of streptavidin in wash buffer.

39. Wash the chamber with 70 µL of wash buffer. Move the slide to the humid chamber at room temperature. Flowthrough the streptavidin solution and incubate for 5 min. Before or during the incubation, prepare a 20 µM solution of biotinylated antibody solution in wash buffer.

40. Wash the chamber with 70 µL of wash buffer. Flow through the antibody solution. Incubate for 5–15 min. During the incubation, prepare the protein solution in wash buffer.

41. Wash the chamber with 70 µL of wash buffer. Flowthrough the purified protein solution and incubate for 5 min. Before or during the incubation, prepare the imaging buffer.

42. Wash the chamber with 70 µL of wash buffer. Flowthrough the imaging buffer, seal the chamber with nail polish. Observe the reaction under the microscope (see Protocol 3: Data Analysis for Total Internal Reflection Fluorescence Microscopy [Asbury 2015]).

DISCUSSION

As with silanization, the lipid coating method provides a way to block nonspecific interactions between the glass surface of the coverslip and the solute while keeping a particular molecule of interest attached to the surface. We find the lipid coating procedure to be effective against nonspecific adsorption of DNA oligonucleotides, single protein molecules such as green fluorescent protein (GFP), and large protein complexes such as purified kinetochore particles. However, we do not recommend the lipid coating method when using microtubules because the lipid bilayer has an extremely high affinity for microtubules. (See Protocol 2: Preparation of Reactions for Imaging with Total Internal Reflection Fluorescence Microscopy [Kudalkar et al. 2015] for further details on using silanized coverslips in TIRF microscopy experiments for studying the binding of kinetochore proteins to microtubules.) Fine-tuning the ratio between functionalized and regular lipids can accurately control the density of the active functional groups on the bilayer surface. Each DPPC lipid occupies 50 Å^2 and 0.1% of the lipid carries a biotin group according to the ratio used in the protocol. Thus, the density of biotin on the bilayer is 2000 µm^{-2}.

Lipid bilayers have complex phase behavior according to temperature and composition. Above the transition temperature, the bilayer forms a two-dimensional fluid and the lipid molecules are diffusive. For TIRF imaging applications, the diffusion will cause motion blur, and the severity of this depends on the exposure time. This can be avoided by using lipids with high phase-transition temperature (such as DPPC) that can form a two-dimensional gel on the surface. However, the passivation efficiency is impaired if the lipid is in the gel phase. Supplementing the incubation buffer with κ-casein restores the passivation, presumably by blocking flaws in the gel-phase bilayer. When lateral constraint is not required, we recommend using low transition temperature lipids such as κ-POPC. In the latter case, all procedures can be performed at room temperature without κ-casein. The passivation efficiency of POPC is higher, and the POPC liposomes can be stored for 3–5 d at 4°C.

RECIPES

BRB80 (5×)

Reagent	Quantity (for 100 mL)	Final concentration (1×)
K-PIPES	12.1 g	80 mM
MgCl$_2$ (1 M)	0.5 mL	1 mM
EGTA (0.5 M)	1 mL	1 mM

Dissolve the K-PIPES and 2.7 g of KOH pellets in 85 mL of Milli-Q purified water. Add the MgCl$_2$ and EGTA. Stir until all the K-PIPES has dissolved. If necessary, add the KOH pellets one at a time until the K-PIPES goes into solution, but be careful not to exceed pH 6.8. Do not put a pH probe into the solution until all the PIPES is dissolved. Measure the pH and bring up to pH 6.8 using 5 M KOH. Adjust the volume to 100 mL with Milli-Q water and divide into 15-mL aliquots. Store the aliquots at −20°C. A working aliquot may be stored for up to 1 mo at 4°C. Prepare fresh 1× BRB80 (diluted in Milli-Q water) each day.

Imaging Buffer

Reagent	Final concentration
Glucose oxidase (oxygen scavenger; 10 mg/mL)	200 mg/mL
Catalase (oxygen scavenger; 1.75 mg/mL)	35 mg/mL
Glucose (1.25 M)	25 mM
Dithiothreitol (DTT; 250 mM)	5 mM
κ-casein (5 mg/mL)	1 mg/mL

REFERENCES

Asbury CL. 2015. Data analysis for total internal reflection fluorescence microscopy. *Cold Spring Harb Protoc* doi: 10.1101/pdb.prot085571.

Cras JJ, Rowe-Taitt CA, Nivens D, Ligler FS. 1999. Comparison of chemical cleaning methods of glass in preparation for silanization. *Biosens Bioelectron* **14:** 683–688.

Cremer PS, Boxer SG. 1999. Formation and spreading of lipid bilayers on planar glass supports. *J Phys Chem B* **103:** 2554–2559.

Kudalkar EM, Davis TN, Asbury CL. 2015. Preparation of reactions for imaging with total internal reflection fluorescence microscopy. *Cold Spring Harb Protoc* doi: 10.1101/pdb.prot085563.

Sackmann E. 1996. Supported membranes: Scientific and practical applications. *Science* **271:** 43–48.

Walba DMLC, Korblova E, Farrow M, Furtak T, Chow BC, Schwartz DK, Freeman AS, Douglas K, Williams SD, Klittnick AF, Clark NA. 2004. Self-assembled monolayers for liquid crystal alignment: Simple preparation on glass using alkyltrialkoxysilanes. *Liquid Crystals* **31:** 481–489.

Protocol 2

Preparation of Reactions for Imaging with Total Internal Reflection Fluorescence Microscopy

Emily M. Kudalkar,[1] Trisha N. Davis,[1,3] and Charles L. Asbury[2,3]

[1]Department of Biochemistry, University of Washington, Seattle, Washington 98195; [2]Department of Physiology and Biophysics, University of Washington, Seattle, Washington 98195

Here we present our standard protocol for studying the binding of kinetochore proteins to microtubules as a paradigm for designing single-molecule total internal reflection fluorescence (TIRF) microscopy experiments. Several aspects of this protocol require empirical optimization, including the method for anchoring the polymer or substrate to the coverslip, the type and amount of blocking protein to prevent nonspecific protein adsorption to the glass, the appropriate protein concentration, the laser power, and the duration of imaging. Our method uses bovine serum albumin and κ-casein as blocking agents to coat any imperfections in the coverslip silanization and thereby prevent protein adsorption to the coverslip. Protein concentration and duration of imaging must be optimized for each experiment and protein of interest. Ideally, a range is determined that allows for resolution of single complexes binding to microtubules to ensure proper measurement of kinetic off rates and diffusion along microtubules. Excessively high concentrations may lead to overlapping binding of proteins on microtubules, making it impossible to resolve single binding events. The duration of imaging must be long enough to capture very low off rates (long residence time on microtubules) and we typically image at 10 frames/sec for 200 sec. The laser power can be adjusted to prevent photobleaching, but must be high enough to achieve a sufficient signal/noise ratio.

MATERIALS

It is essential that you consult the appropriate Material Safety Data Sheets and your institution's Environmental Health and Safety Office for proper handling of equipment and hazardous material used in this protocol.

RECIPES: Please see the end of this protocol for recipes indicated by <R>. Additional recipes can be found online at http://cshprotocols.cshlp.org/site/recipes.

Reagents

Alexa-568-labeled GMPCPP seeds (Hyman et al. 1991)
Alexa-568-labeled, paclitaxel-stabilized microtubules (Hyman et al. 1991)
BB80 (BRB80 <R> containing 8 mg/mL BSA)
BB80T (BB80 containing 10 μM taxol)
Bovine serum albumin (BSA; 40 mg/mL stock solution; filter-sterilized)
Catalase (1.75 mg/mL stock solution)
Dithiothreitol (DTT; 250 mM stock solution)
Ethanol

[3]Correspondence: tdavis@uw.edu; casbury@uw.edu

Copyright © Cold Spring Harbor Laboratory Press; all rights reserved
Cite this protocol as *Cold Spring Harb Protoc*; doi:10.1101/pdb.prot085563

Glucose (1.25 M stock solution)
Glucose oxidase (10 mg/mL stock solution)
GTP (100 mM stock solution, pH 7.0)
κ-casein (5 mg/mL stock solution; filter sterilized)
Microtubule growth buffer (GB; BB80 containing 1 mM GTP)
Purified protein of interest
Rigor kinesin (purified in our laboratory)
Tubulin (bovine; 1:100 Alexa-568-labeled to unlabeled) (purified and labeled in our laboratory)

Equipment

Adaptors for peristaltic pump (custom-made)
Adhesive transfer tape (3M F9473PC)
Coverslips (glass; 22 × 60 mm; silanized) (see Protocol 1: Coverslip Cleaning and Functionalization for Total Internal Reflection Fluorescence Microscopy [Kudalkar et al. 2015])
Drill (with diamond bit)
Double-sided tape
Nail polish
Razor blade
Slides (glass; 3 × 1 in)
Vacuum grease

METHOD

This protocol for using TIRF microscopy to study the binding of kinetochore proteins to microtubules at the single-molecule level is adapted from Gestaut et al. (2010).

Assembly of Flow Chamber

The design of the flow chamber depends on whether the experiment is to be performed with the peristaltic pump method (see Steps 1–8) or the sealed slide method (see Steps 9–12). The peristaltic pump method is used to introduce reactants to the flow chamber during imaging on the microscope and the sealed slide method is used to introduce reactants before imaging. Both methods use the same procedure for washes, adherence of rigor kinesin, and introduction of microtubules and proteins. If the reactants are introduced before imaging, the entire protocol is performed at the bench using an aspirator to draw liquid through the chambers and then the slide is sealed with either clear nail polish or vacuum grease.

Flow Chamber for Peristaltic Pump Method

1. Drill eight holes into a 3 × 1 in glass slide using a diamond bit—two sets of four holes on the long axis of the slide, directly across from one another (Fig. 1A).

2. Clean the drilled glass slide with ethanol. Place double-sided tape crosswise between the drilled holes leaving ∼3 mm between each piece of tape to create four equivalently sized flow chambers.

3. Place a silanized coverslip lengthwise over the middle of the slide. Ensure all eight holes are covered to create four perfusion chambers. Gently press down on the surface of the coverslip to ensure a tight seal (Fig. 1B).

4. Remove the excess tape on the sides of the coverslip using a razor blade.

5. Using a cotton-tipped applicator, seal the chambers by gently pressing vacuum grease into each opening on the outer edges of the slide until it reaches the holes. Wipe off excess grease with ethanol.

6. Flip the glass slide over so the coverslip is underneath. Apply adhesive tape to four of the holes on this side of the slide and, using forceps, remove circles of tape that cover each hole (Fig. 1C).

Chapter 16

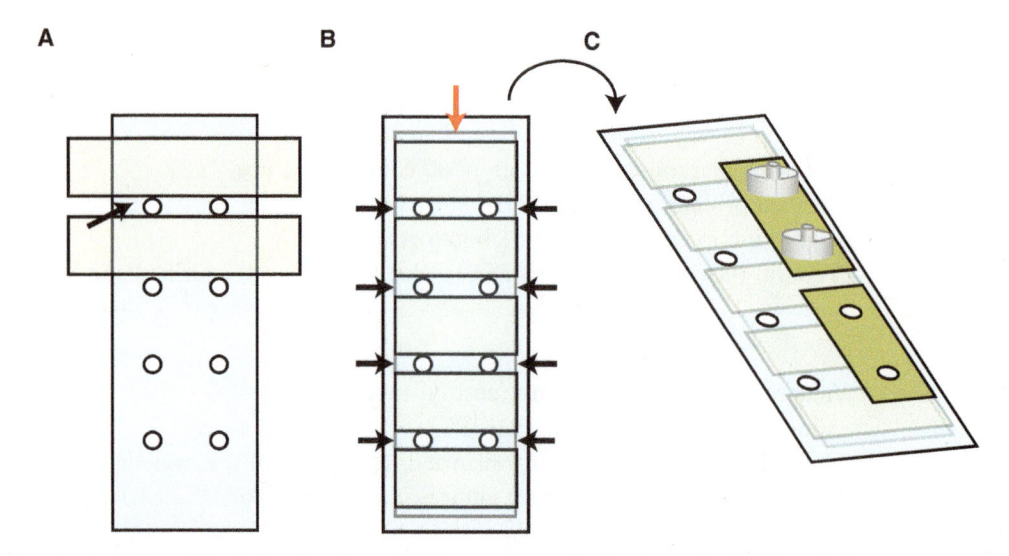

FIGURE 1. Assembly of flow chamber for peristaltic pump method. (*A*) Apply double-sided tape to a glass slide predrilled with two sets of four holes in a lengthwise arrangement. Position the tape to create four 3-mm-wide chambers (black arrows). (*B*) Position a cleaned coverslip (red arrow) in the *middle* of the slide and press firmly to seal against the tape. Trim the tape on both sides using a razor blade so it is flush with the coverslip edge. Push vacuum grease into the sides of chambers (black arrows) to seal. Do not cover the holes. (*C*) Flip the slide over so coverslip is face down. Apply 3M adhesive tape to cover the holes on one edge of slide (yellow). Using forceps, remove the tape covering each hole to create an open channel. Center custom-made flow adaptors (gray) on top of each hole and press firmly to seal.

7. Place custom-made adaptors on the tape, centering each one with the holes in the slide. Press gently to ensure a good seal (Fig. 1C).
8. Make a small circle around each remaining hole using vacuum grease to create a pool for buffers.

Flow Chamber for Sealed Slide Method

9. Clean a 3 × 1 in glass slide thoroughly with ethanol.
10. Apply several pieces of double-sided tape crosswise to the slide, leaving a gap of ∼3 mm in between the pieces to create chambers.
11. Place a silanized coverslip crosswise on the slide leaving an equal-sized ledge on either side of the slide. Gently press down to ensure a tight and even seal between the coverslip and tape.
12. Using a razor blade, trim off excess tape from either side of the coverslip so it is flush with the edge.

Binding Interactions Using Paclitaxel-Stabilized Microtubules

Direct adsorption of microtubules to the coverslip can interfere with their ability to grow and shorten (if using non-stabilized microtubules) and could potentially hinder protein binding. To attach microtubules to the coverslip without direct adherence to the glass, we use an established method that uses a mutated kinesin ("rigor kinesin") that is competent to bind, but not release, microtubules (Rice et al. 1999). We optimize the concentration of rigor kinesin to ensure that microtubules are stably tethered to the coverslip, but the kinesin does not interfere with experimental protein binding.

13. Flow 100 μL of Milli-Q purified water through the chamber twice.
14. Flow 25 μL of rigor kinesin (diluted in BB80T) through the chamber and incubate for 5 min. Determine the dilution factor for rigor kinesin empirically for each coverslip preparation to ensure proper anchoring and coverage of microtubules. Apply 50 μL of BB80T to the edge or hole of the flow chamber during incubation to prevent the chamber from drying.
15. Flow 50 μL of BB80T through the chamber.

16. Flow 15 µL of Alexa-568-labeled, paclitaxel-stabilized microtubules diluted in BB80T through the chamber. Determine the dilution factor empirically to achieve an appropriate amount of microtubule coverage (typically about three to seven nonoverlapping microtubules per field is optimal for analysis). Incubate for 5 min. Apply 50 µL of BB80T to the edge or hole of the flow chamber during incubation to prevent the chamber from drying.

17. During the microtubule incubation, prepare the experimental reaction, which typically contains 10–100 pM protein, 25 mM glucose, 5 mM DTT, and oxygen scavengers 200 µg/mL glucose oxidase and 35 µg/mL catalase. Adjust the volume to 50 µL with BB80T.

    To prevent protein loss to the tube (a problem that can occur when working in the pM concentration range), dilute the protein of interest immediately before adding to the reaction mixture.

18. Flow 50 µL of BB80T through the chamber and then 50 µL of reaction mixture. Image immediately on both 488 and 561 nm channels (see Protocol 3: Data Analysis for Total Internal Reflection Fluorescence Microscopy [Asbury 2015]).

Binding Interactions Using Dynamic Microtubules

19. Follow Steps 13 and 14 of the protocol using stabilized microtubules.

    The concentration of rigor kinesin may be increased to ensure proper anchoring of growing microtubule extensions.

20. Flow 50 µL of GB through the chamber.

21. Flow 15 µL of Alexa-568-labeled GMPCPP seeds diluted in BB80 through the chamber. Incubate for 1 min. Optimize the concentration of seeds to ensure proper coverslip coverage.

22. Flow 50 µL of GB through the chamber.

23. Prepare the tubulin mix, which contains 2 mg/mL bovine tubulin (1:100 Alexa-568-labeled to unlabeled), 25 mM glucose, 5 mM DTT, and oxygen scavengers 200 µg/mL glucose oxidase and 35 µg/mL catalase. Adjust the volume to 50 µL with BB80. Flow 50 µL of tubulin mix through the chamber. Incubate for 15 min to allow microtubules to extend off the Alexa-568 GMPCPP seeds.

24. During the incubation, focus the microscope on the channel.

    Once the reaction mix is added (Step 25), the microtubules will immediately begin to depolymerize. Therefore, it is essential to have the field of view already in focus so as to begin imaging promptly on addition of the reaction mix.

25. Prepare the reaction mix containing 10 pM to 1 nM protein, 200 µg/mL glucose oxidase, 35 µg/mL catalase, 25 mM glucose, and 5 mM DTT. Adjust the volume to 50 µL with BB80. Flow 50 µL of reaction mix through the chamber. Begin imaging immediately (see Protocol 3: Data Analysis for Total Internal Reflection Fluorescence Microscopy [Asbury 2015]).

RECIPE

BRB80 (5×)

Reagent	Quantity (for 100 mL)	Final concentration (1×)
K-PIPES	12.1 g	80 mM
MgCl$_2$ (1 M)	0.5 mL	1 mM
EGTA (0.5 M)	1 mL	1 mM

Dissolve the K-PIPES and 2.7 g of KOH pellets in 85 mL of Milli-Q purified water. Add the MgCl$_2$ and EGTA. Stir until all the K-PIPES has dissolved. If necessary, add the KOH pellets one at a time until the K-PIPES goes into solution, but be careful not to exceed pH 6.8. Do not put a pH probe into the solution until all the PIPES is dissolved. Measure the pH and bring up to pH 6.8 using 5 M KOH. Adjust the volume to 100 mL with Milli-Q water and divide into 15-mL aliquots. Store the aliquots at −20°C. A working aliquot may be stored for up to 1 mo at 4°C. Prepare fresh 1× BRB80 (diluted in Milli-Q water) each day.

REFERENCES

Asbury CL. 2015. Data analysis for total internal reflection fluorescence microscopy. *Cold Spring Harb Protoc* doi: 10.1101/pdb.prot085571.

Gestaut DR, Cooper J, Asbury CL, Davis TN, Wordeman L. 2010. Reconstitution and functional analysis of kinetochore subcomplexes. *Methods Cell Biol* 95: 641–656.

Hyman A, Drechsel D, Kellogg D, Salser S, Sawin K, Steffen P, Wordeman L, Mitchison T. 1991. Preparation of modified tubulins. *Methods Enzymol* 196: 478–485.

Kudalkar EM, Deng Y, Davis TN, Asbury CL. 2015. Coverslip cleaning and functionalization for total internal reflection fluorescence microscopy. *Cold Spring Harb Protoc* doi: 10.1101/pdb.prot085548.

Rice S, Lin AW, Safer D, Hart CL, Naber N, Carragher BO, Cain SM, Pechatnikova E, Wilson-Kubalek EM, Whittaker M, et al. 1999. A structural change in the kinesin motor protein that drives motility. *Nature* 402: 778–784.

Protocol 3

Data Analysis for Total Internal Reflection Fluorescence Microscopy

Charles L. Asbury[1]

Department of Physiology and Biophysics, University of Washington, Seattle, Washington 98195

In the microscopes we use to analyze total internal reflection fluorescence (TIRF), the emitted fluorescence is split chromatically, using dichroic filters, into either two or three different colors ("channels"). In our two-color instrument, the green emission wavelengths (405–488 nm; for imaging green fluorescent protein [GFP]-tagged proteins) and far-red emission wavelengths (650–800 nm; for imaging Alexa-647-labeled microtubules) are projected onto the upper and lower halves, respectively, of a single camera. A single filter can be swapped to collect near-red wavelengths (561–640 nm; for imaging mCherry, or Alexa-568-labeled microtubules) instead of far-red. Our three-color instrument is very similar except that the green, near-red, and far-red color ranges are projected onto three separate cameras. In either case, the different colors can be imaged simultaneously. Typically, we collect images at 10 frames/sec for ∼200 sec. We have developed a series of semiautomated image analysis programs, written in LabView, to obtain the brightness, residence time, and mobility of individual particles bound to single microtubules. The basic analysis steps are straightforward and could also be implemented using ImageJ or Matlab. For convenience, this protocol describes the analysis of a single microtubule. Data from many microtubules across many experimental trials are needed to obtain robust conclusions that are independent of stochastic and trial-to-trial variability.

MATERIALS

Equipment

Data analysis program (e.g., Igor Pro, Matlab, or Microsoft Excel; see Step 7)
EMCCD camera (iXon DV887 from Andor)
Image analysis programs custom written in LabView (see Gestaut et al. 2010)
   Copies of our programs are available for free upon request.
Microscope (Ti-U from Nikon or equivalent)

METHOD

For appropriate sample preparation procedures, see Protocol 1: Coverslip Cleaning and Functionalization for Total Internal Reflection Fluorescence Microscopy (Kudalkar et al. 2015a) and Protocol 2: Preparation of Reactions for Imaging with Total Internal Reflection Fluorescence Microscopy (Kudalkar et al. 2015b).

1. Trace the contour of a microtubule by hand using a multisegment line tool (Fig. 1A).

---

[1]Correspondence: casbury@uw.edu

Copyright © Cold Spring Harbor Laboratory Press; all rights reserved
Cite this protocol as *Cold Spring Harb Protoc*; doi:10.1101/pdb.prot085571

Chapter 16

The filaments in the experiments described here should remain stationary on the coverslips. Assuming negligible drift, the contours traced in an early image will apply to the entire duration of the time-lapse experiment.

2. Map the microtubule contour automatically onto the other color channel(s), and use it to generate a kymograph(s) showing the arrival of particles on the microtubule, their movement along the filament, and their release (Fig. 1B).

   Accurate mapping from one color channel to another usually requires coordinates to be translated (i.e., moved along x and y directions) and also slightly rotated. Depending on the microscope setup, slight magnification changes might also be needed. The required mapping operation can be determined empirically using reference particles that fluoresce in all the channels.

3. For each particle visible on the kymograph, estimate its location along the microtubule by finding the brightest pixel at each time point.

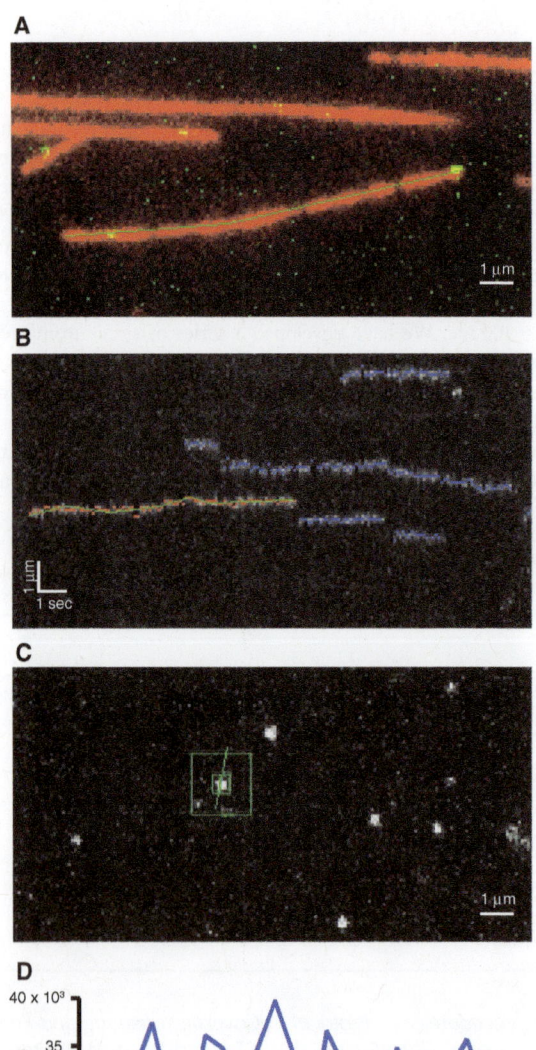

FIGURE 1. TIRF data analysis. (*A*) Overlapping green and red channels depict a microtubule (red) that was traced by hand using a multisegmented tool. Yellow spots show GFP-tagged proteins binding to the microtubule. (*B*) Kymograph created from a traced microtubule as in *A*. Red dots and green line indicate the currently selected binding event, blue lines show previously selected events, and untraced events are shown in gray. (*C*) Particle brightness is computed at each time point of event from *B*. Integrated brightness is measured using small green box surrounding particle and larger green box is used to calculate background intensity. (*D*) Example plot of pixel brightness (blue trace) and background level (red trace) over time generated from *C*.

4. Map the one-dimensional estimate of location versus time from Step 3 back onto the original, two-dimensional particle image using the contour from Step 2.

5. Obtain the pixel coordinates of the particle in each image by using a two-dimensional search for the brightest pixel within a small square (7 × 7 pixels) centered on the estimated position from Step 4.

   Alternatively, fit a two-dimensional Gaussian function to the intensity distribution within the small square. A Gaussian fit is more time-consuming and generally does not improve localization accuracy for single GFP molecules imaged at 10 frames/sec in our microscopes. However, in cases where more photons are collected per frame, Gaussian fitting may improve localization accuracy.

6. After coordinates are obtained in Step 5, compute particle brightness at each time point by integrating pixel intensities over a small square, centered on the particle position. Estimate background levels by integrating over a larger concentric square area, excluding the central small square (Fig. 1C).

7. For every particle, save a file containing the pixel coordinates, integrated brightness, and background level at each time point (Fig. 1D).

   Once the particle data are saved, we use another graphing and data analysis program, Igor Pro, to carry out the subsequent analysis steps. These steps could also be performed using Matlab or Microsoft Excel.

8. Examine plots of brightness versus time for each event to confirm the arrival and release times, and to identify photobleaching steps.

9. Compute the residence times for each event from the start and end times obtained in Steps 7 and 8. Plot a distribution of residence times, either in the form of a histogram (e.g., see Gestaut et al. 2008 or Powers et al. 2009) or as a cumulative survival probability versus time (e.g., see Sarangapani et al. 2013).

   Histograms are more intuitive, but cumulative distributions avoid the need for binning and facilitate comparisons because many can be overlayed onto a single graph without loss of clarity.

10. Note that the residence time distribution is often (but not always) well described by a single exponential decay, except that the lowest bins (corresponding to the shortest residence times) may be underpopulated due to the finite time resolution of the instrument. Invert the mean from the best-fit exponential (excluding the lowest bins) or, equivalently, its time constant, τ, to give the off rate, $k_{off} = \tau^{-1}$ (e.g., in units of sec^{-1}).

11. For each event, also calculate a mean squared displacement along the microtubule long axis, $<x^2>$, for every possible time lag, Δt. Average the $<x^2>$ values across many events to generate a plot of $<x^2>$ versus Δt for a population (each individual binding event contributing equally to the population average).

12. Compute the one-dimensional diffusion coefficient, D, from the slope, m, of a linear fit to the plot according to $D = (1/2)m = (1/2) <x^2> \Delta t^{-1}$.

REFERENCES

Gestaut DR, Graczyk B, Cooper J, Widlund PO, Zelter A, Wordeman L, Asbury CL, Davis TN. 2008. Phosphoregulation and depolymerization-driven movement of the Dam1 complex do not require ring formation. *Nat Cell Biol* 10: 407–414.

Gestaut DR, Cooper J, Asbury CL, Davis TN, Wordeman L. 2010. Reconstitution and functional analysis of kinetochore subcomplexes. *Methods Cell Biol* 95: 641–656.

Kudalkar EM, Deng Y, Davis TN, Asbury CL. 2015a. Coverslip cleaning and functionalization for total internal reflection fluorescence microscopy. *Cold Spring Harb Protoc* doi: 10.1101/pdb.prot085548.

Kudalkar EM, Davis TN, Asbury CL. 2015b. Preparation of reactions for imaging with total internal reflection fluorescence microscopy. *Cold Spring Harb Protoc* doi: 10.1101/pdb.prot085563.

Powers AF, Franck AD, Gestaut DR, Cooper J, Gracyzk B, Wei RR, Wordeman L, Davis TN, Asbury CL. 2009. The Ndc80 kinetochore complex forms load-bearing attachments to dynamic microtubule tips via biased diffusion. *Cell* 136: 865–875.

Sarangapani KK, Akiyoshi B, Duggan NM, Biggins S, Asbury CL. 2013. Phosphoregulation promotes release of kinetochores from dynamic microtubules via multiple mechanisms. *Proc Natl Acad Sci* 110: 7282–7287.

CHAPTER 17

Building Cell Structures in Three Dimensions: Electron Tomography Methods for Budding Yeast

Eileen T. O'Toole,[1] Thomas H. Giddings, Jr., and Mark Winey[1]

Boulder Laboratory for 3-D Electron Microscopy of Cells, Department of Molecular, Cellular, and Developmental Biology, University of Colorado, Boulder, Colorado 80309-0347

Saccharomyces cerevisiae has been an important model system for numerous cellular, genetic, and molecular studies. However, this small eukaryote presents a challenge for imaging at the electron microscope level. Preparation of yeast using high-pressure freezing followed by freeze-substitution (HPF/FS) results in excellent preservation of cell structure in these difficult-to-fix samples. In particular, cells prepared by HPF/FS can be used for 3D electron tomography (ET) studies where optimum cell preservation is critical. Here, we discuss the advantages of using HPF/FS for ET and show examples of the utility of this method for building yeast cell structures in three dimensions.

INTRODUCTION

The use of budding yeast as a powerful model system has advanced our understanding of cell and molecular biology. Its importance is without question, yet yeast is considered difficult to prepare for cellular studies using transmission electron microscopy (TEM). These studies are hampered by the fact that the yeast cell wall acts as a barrier to the diffusion of chemical fixatives into the cell; therefore, its removal is required for conventional chemical fixation (Byers and Goetsch 1991). Despite this rather invasive preparation, important landmark studies have used conventional chemical fixation to describe the cellular features of this organism and to compare ultrastructural defects that result from mutations in key genes (Byers and Goetsch 1974, 1975; Winey et al. 1991; Novick et al. 1980). Since the time of those studies, methods using cryoimmobilization followed by freeze substitution have been developed to provide excellent preservation of intact cells (Baba and Osumi 1987; Giddings et al. 2001; McDonald and Müller-Reichert 2002). These approaches involve rapid freezing of the sample with subsequent substitution treatment to replace frozen water in the sample with an organic solvent and fixatives. Yeast prepared with these methods are used in three-dimensional (3D) analyses for which sampling of the cell is performed at unprecedented resolution; for these studies, optimal preservation is therefore key. In this introduction, we summarize the common methods for preparing budding yeast for EM and discuss the advantages and disadvantages of each with an emphasis on the utility of the method for 3D electron tomography (ET) studies.

PREPARATION OF YEAST FOR TEM

Several methods are currently available for the preparation of budding yeast cells for TEM, including conventional room temperature fixation and various freezing techniques (Table 1). Conventional

[1]Correspondence: eileen.otoole@colorado.edu; mark.winey@colorado.edu

Copyright © Cold Spring Harbor Laboratory Press; all rights reserved
Cite this introduction as *Cold Spring Harb Protoc*; doi:10.1101/pdb.top077685

TABLE 1. Specimen preparation techniques for TEM of budding yeast

Fixation	Advantages	Disadvantages	References
Conventional room temperature chemical fixation and resin embedding	Reagents and equipment readily available, high yield of cells	Cell wall presents a diffusion barrier, enzymatic digestion recommended	Byers and Goetsch 1991
High-pressure freezing followed by low-temperature freeze substitution and resin embedding	Excellent preservation of cell structure Resin embedding allows room temperature sectioning and ease of microscopy High yield of cells	Specialized, expensive equipment	Giddings et al. 2001; McDonald and Müller-Reichert 2002; McDonald 2007
High-pressure freezing followed by vitreous sectioning and cryo-EM	Cells are in as close to native state—no fixative or stain	Technically challenging Specialized equipment necessary, low contrast	Pierson et al. 2010
Plunge freezing-whole cells followed by low-temperature freeze substitution and embedding	Freezing equipment not as expensive	Low yield of cells	Baba and Osumi 1987
Plunge freezing-isolated organelles, cryoimaging	Organelles imaged without fixative or stain, close to native state	Organelles isolated from cellular context Challenging technique	Bullitt et al. 1997; Kollman et al. 2010

room temperature fixation methods have the advantage that the reagents and equipment for these preparations are readily available in a typical EM facility (Byers and Goetsch 1991). These fixation techniques involve the diffusion of chemical fixatives into the cell followed by dehydration and embedding in plastic resins (Fig. 1, blue boxes and related image; see also Hayat 2000). No one chemical will fix all cellular components, and typically two chemical fixatives are used: glutaraldehyde to cross-link proteins followed by osmium tetroxide to fix membranes and proteins. The yeast cell wall limits diffusion of these chemical fixatives, however, and its removal by enzymatic digestion is therefore recommended (Byers and Goetsch 1991). Because the removal of the cell wall can result in a change in the 3D organization of the cell, cells prepared by conventional chemical fixation are not well suited for 3D studies using ET. Nevertheless, conventional chemical fixation is a useful method for standard, thin-section TEM studies both because of its ease and because of the availability of standard equipment for EM.

Cryoimmobilization techniques rely on an initial fixation by rapid freezing of essentially all components of the cell within milliseconds. The advantage of cryoimmobilization is that all molecules in the intact cell are frozen almost instantaneously, thereby preserving the cell in its most native state (reviewed in Gilkey and Staehelin 1986). Cells in the frozen hydrated state can be imaged directly in the EM or further subjected to freeze substitution at low temperature and embedded with resin (Table 1; Fig. 1, red boxes in top panel). The common cryoimmobilization techniques used to prepare yeast samples for EM are high-pressure freezing and plunge freezing.

Plunge freezing has been used in conjunction with cryo-EM and cryo-ET to advance the field of molecular microscopy. In this method of cryopreservation, isolated protein complexes, organelles, or cells grown on polycarbonate membranes are plunged into a vessel of ethane that has been cooled by liquid nitrogen. Most isolated protein complexes and organelles are small enough to be imaged directly in the frozen-hydrated state. Indeed, high-resolution detail has been obtained for the structural arrangements of the yeast gamma tubulin complex and organelles such as the yeast spindle pole body (SPB; Kollman et al. 2010; Bullitt et al. 1997). However, intact yeast cells prepared by plunge freezing are too large to image directly in the frozen-hydrated state. For successful analysis by EM, frozen cells must be further treated by freeze substitution and then sectioned (Baba and Osumi 1987). This method results in excellent preservation of the cellular ultrastructure, yet has the disadvantage that only a small sample size is obtained from cells grown on the polycarbonate membranes. In the associated protocol, we describe our method for cryopreparation of yeast cells for 3D studies using tomography (see Protocol 1: Cryopreparation and Electron Tomography of Yeast Cells [O'Toole et al. 2015]).

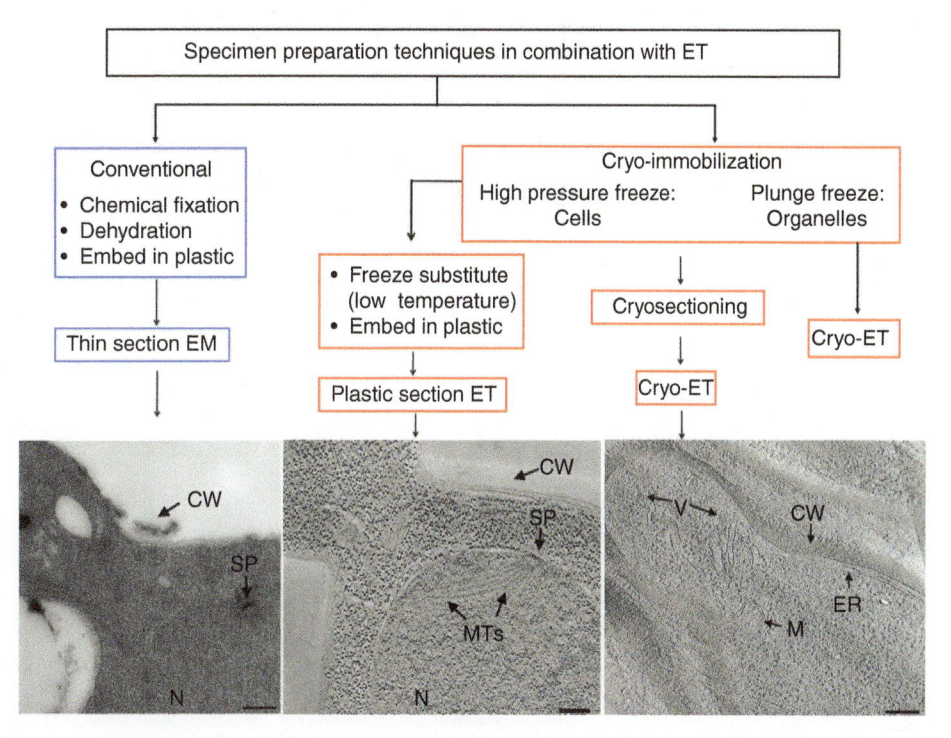

FIGURE 1. Specimen preparation techniques used with ET studies. Conventional chemical fixation (blue boxes) involves the use of digestive enzymes to remove the cell wall and thus is not well suited for ET. Cells prepared by cryoimmobilization (high-pressure freezing or plunge freezing) can be freeze substituted and sectioned for room temperature ET or imaged directly in the frozen-hydrated state using cryo-ET (red boxes). CW, cell wall; SP, spindle pole; N, nucleus; MTs, microtubules; V, vesicles; M, mitochondria; ER, endoplasmic reticulum. Scale bar, 200 nm.

HIGH-PRESSURE FREEZING AND FREEZE SUBSTITUTION FOR 3D STUDIES

The preparation of cells by high-pressure freezing followed by freeze substitution (HPF/FS) and resin embedding has the advantage of excellent preservation of cell ultrastructure (Fig. 1, middle image of lower panel). In this approach, cells are prepared by high-pressure freezing at ~2045 bar, a pressure that slows the formation of damaging ice crystals (comprehensively reviewed in Steinbrecht and Zierold 1987). The frozen cells are then subjected to freeze substitution at low temperature in an organic solvent and secondary fixative. The samples are infiltrated with resin and the polymerized blocks are sectioned and imaged at room temperature (Giddings et al. 2001; McDonald and Müller-Reichert 2002; McDonald 2007). Currently, HPF/FS is the method of choice for preparing cells for ET and has the advantage that a large pellet of well-preserved cells can be processed. In addition, room temperature serial sectioning facilitates the ability to track and model complete cellular assemblies for reconstruction (O'Toole et al. 1999; West et al. 2011). Cells prepared by HPF can also be sectioned; a technique commonly referred to as vitreous sectioning (Fig. 1, right image of lower panel). Imaging frozen sections enables the visualization of cell structure in as close to the native state as possible. The disadvantages of this emerging technology are that it is technically challenging and requires specialized, expensive equipment.

3D RECONSTRUCTION USING ET

In recent years, ET has emerged as a method for reconstructing cellular structures at unprecedented resolution (Hoenger and McIntosh 2009). Numerous published studies have described yeast cell structures in 3D and the structural changes that occur in mutant strains. Because the 3D volumes

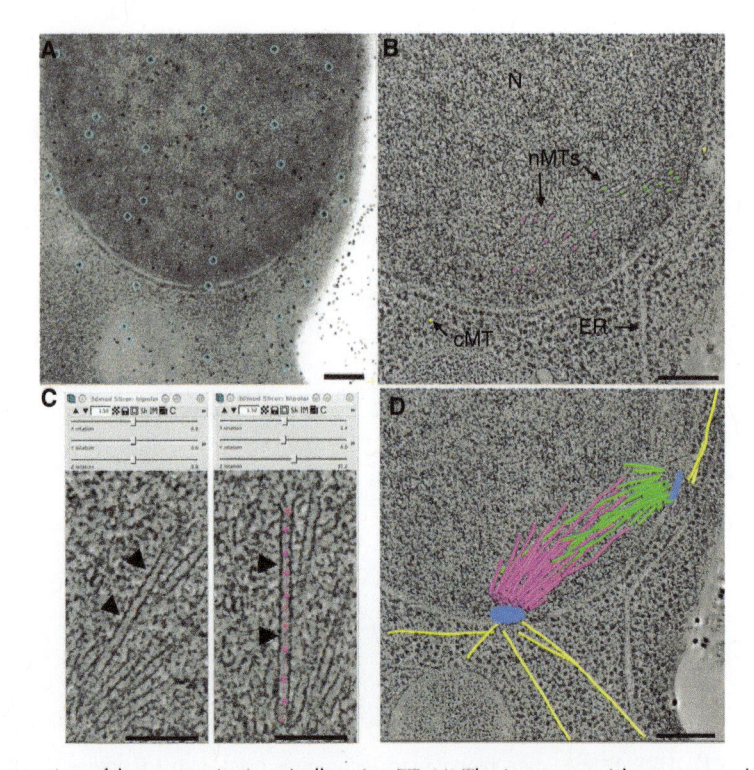

FIGURE 2. Reconstruction of the yeast mitotic spindle using ET. (*A*) The image provides an example of a thick section used for a tilt series collection. Gold particles (blue circles) are used as alignment markers. (*B*) A selected tomographic slice shows model points on spindle microtubules in the nucleus (nMTs; green and pink circles) and on cytoplasmic MTs (cMTs; yellow circles). The nucleus (N) is round and organelles such as the endoplasmic reticulum (ER) are well preserved. (*C*) Spindle MTs can be unambiguously modeled using the slicer tool in 3dmod (a program for 3D reconstruction and modeling). The *left* panel shows a portion of an oblique MT (arrowheads). In the *right* panel, the slice of image data is rotated to view the complete MT along its length (arrowheads). Model points can then be deposited along the length of this MT (pink). (*D*) The complete 3D model is projected on a selected tomographic slice. Spindle MTs (green and pink lines) can be seen arising from each SPB (denoted as blue discs). Numerous cytoplasmic MTs are displayed as yellow lines. Scale bar, 200 nm (*A,B,D*); 100 nm (*C*). (See online Movie 1 at cshprotocols.cshlp.org.) Aligned, tilted views of a 300-nm-thick section. (See online Movie 2 at cshprotocols.cshlp.org.) Serial, tomographic slices through the complete volume. The movie first displays the slices and then shows model points marking the positions of the spindle MT from either spindle pole (green and pink circles, respectively). (See online Movie 3 at cshprotocols.cshlp.org.) The complete 3D model projected onto serial, tomographic slices and then rotated to view the geometry of the yeast mitotic spindle. Spindle pole bodies (spb) are represented as blue discs, green and pink lines represent the spindle MT from each spindle pole. Cytoplasmic MTs are shown in yellow.

allow sampling of cell structure at 4–6-nm resolution, the use of superior and most precise structural preservation methods are essential.

Figure 2 shows the overall reconstruction of a mitotic spindle as an example of the steps involved in tomography. The raw data from which a tomogram is computed is a tilt series collected from a thick section of 200–300 nm (see Fig. 2A and online Movie 1 at cshprotocols.cshlp.org). Images are collected every degree over a ±60° or 70° range; gold particles on the top and bottom surface of the section are used as fiducial markers for alignment of the serial, tilted images (shown as blue circles in Fig. 2A). The tomographic reconstruction is computed from the aligned tilt series using weighted back projection algorithms. An example of a selected, 1 nm tomographic slice can be seen in Figure 2B; a movie through the complete volume is shown in online Movie 2 at cshprotocols.cshlp.org. The tomographic slices show exquisite detail of the yeast cell, including the spindle microtubules in the nucleus (nMTs), cytoplasmic microtubules (cMT), and the endoplasmic reticulum (ER). The complex arrangement of spindle MTs can be modeled with the IMOD software program, using the slicer window, where a slice through the 3D volume can be rotated to get the best view to unambiguously model a given structure (Fig. 2C). The complete model is projected in 3D to study the geometry of the mitotic spindle (Fig. 2D). A moving sequence showing the 3D model emerging from serial, tomographic slices is shown in online Movie 3 at cshprotocols.cshlp.org. By modeling the 3D volume, spindle MTs (seen as green and pink lines in Fig. 2D) arising from each spindle pole SPB as blue discs in Fig. 2D) can be identified as well as the numerous cytoplasmic MTs at the SPB (cMTs are seen as yellow lines). The panels of Figure 2 illustrate the significance of ET for understanding the 3D organization of complex, cellular structures.

SUMMARY AND FUTURE DIRECTIONS

ET of well-preserved cells has contributed to our understanding of the complex 3D arrangements of cell structures and reveals the alterations that can occur in mutant strains. However, future work is still needed to develop EM labels that will allow protein identification within the volume of a well-preserved cell. In addition, the future development of cryo-ET, including volume averaging of structures in 3D volumes, will provide enhanced molecular detail within the cell.

ACKNOWLEDGMENTS

The authors wish to thank Kent McDonald and Mary Morphew for the numerous contributions they have made in HPF/FS technology, and Mark Ladinsky and Jason Pierson for the yeast cryo-ET. The Boulder Laboratory for 3D Electron Microscopy of Cells by grant P41GM103431 from the National Institute of General Medical Sciences to Andreas Hoenger.

REFERENCES

Baba M, Osumi M. 1987. Transmission and scanning electron microscopic examination of intracellular organelles in freeze-substituted *Kloeckera* and *Saccharomyces cerevisiae* yeast cells. *J Electron Microsc Tech* 5: 249–261.

Bullitt E, Rout MP, Kilmartin JV, Akey CW. 1997. The yeast spindle pole body is assembled around a central crystal of Spc42p. *Cell* 89: 1077–1086.

Byers B, Goetsch L. 1975. Behavior of spindles and spindle plaques in the cell cycle and conjugation of *Saccharomyces cerevisiae. J Bacteriol* 124: 511–523.

Byers B, Goetsch L. 1974. Duplication of spindle plaques and integration of the yeast cell cycle. *Cold Spring Harb Symp Quant Biol* 38: 123–131.

Byers B, Goetsch L. 1991. Preparation of yeast cells for thin-section electron microscopy. *Methods Enzymol* 194: 602–608.

Giddings TH, O'Toole ET, Morphew M, Mastronarde DN, McIntosh JR, Winey M. 2001. Using rapid freeze and freeze-substitution for the preparation of yeast cells for electron microscopy and three-dimensional analysis. *Methods Cell Biol* 67: 27–42.

Gilkey JC, Staehelin LA. 1986. Advances in ultrarapid freezing for the preservation of cellular ultrastructure. *J Electron Microsc Tech* 3: 177–210.

Hayat MA. 2000. *Principles and techniques of electron microscopy: Biological applications.* Cambridge University Press, Cambridge.

Hoenger A, McIntosh JR. 2009. Probing the macromolecular organization of cells by electron tomography. *Curr Opin Cell Biol* 21: 89–96.

Kollman JM, Polka JK, Zelter A, Davis TN, Agard DA. 2010. Microtubule nucleating γ-TuSC assembles structures with 13-fold microtubule-like symmetry. *Nature* 466: 879–882.

McDonald K. 2007. Cryopreparation methods for electron microscopy of selected model systems. *Methods Cell Biol* 79: 23–56.

McDonald K, Müller-Reichert T. 2002. Cryomethods for thin section electron microscopy. *Methods Enzymol* 351: 96–123.

Novick P, Field C, Schekman R. 1980. Identification of 23 complementation groups required for post-translational events in the yeast secretory pathway. *Cell* 21: 205–215.

O'Toole ET, Giddings TH, Winey M. 2015. Cryopreparation and electron tomography of yeast cells. *Cold Spring Harb Protoc* doi: 10.1101/pdb.prot085589.

O'Toole ET, Winey M, McIntosh JR. 1999. High-voltage electron tomography of spindle pole bodies and early mitotic spindles in the yeast *Saccharomyces cerevisiae. Mol Biol Cell* 10: 2017–2031.

Pierson J, Fernández JJ, Bos E, Amini S, Gnaegi H, Vos M, Bel B, Adolfsen F, Carrascosa JL, Peters PJ. 2010. Improving the technique of vitreous cryo-sectioning for cryo-electron tomography: Electrostatic charging for section attachment and implementation of an anti-contamination glove box. *J Struct Biol* 169: 219–225.

Steinbrecht RA, Zierold K. eds. 1987. *Cryotechniques in biological electron microscopy.* Springer, Berlin, Heidelberg.

West M, Zurek N, Hoenger A, Voeltz GK. 2011. A 3D analysis of yeast ER structure reveals how ER domains are organized by membrane curvature. *J Cell Biol* 193: 333–346.

Winey M, Goetsch L, Baum P, Byers B. 1991. MPS1 and MPS2: Novel yeast genes defining distinct steps of spindle pole body duplication. *J Cell Biol* 114: 745–754.

Protocol 1

Cryopreparation and Electron Tomography of Yeast Cells

Eileen T. O'Toole,[1] Thomas H. Giddings, Jr., and Mark Winey[1]

Boulder Laboratory for 3-D Electron Microscopy of Cells, Department of Molecular, Cellular, and Developmental Biology, University of Colorado, Boulder, Colorado 80309-0347

Three-dimensional imaging of cells using electron tomography enables analysis of cell structure at unprecedented resolution. The preparation of cells for tomography using rapid freezing followed by freeze-substitution is an essential first step to ensure the optimal preservation of the cell structure for 3D studies. This protocol outlines a method for obtaining well-preserved cells using high-pressure freezing followed by freeze-substitution. We have found that this method is particularly well suited for electron tomography studies and has the added bonus of preserving antigenicity for immuno-electron microscopy. The steps involved in imaging cells and performing tomographic analysis of cellular structures are also outlined.

MATERIALS

It is essential that you consult the appropriate Material Safety Data Sheets and your institution's Environmental Health and Safety Office for the proper handling of equipment and hazardous materials used in this protocol.

RECIPES: Please see the end of this protocol for recipes indicated by <R>. Additional recipes can be found online at http://cshprotocols.cshlp.org/site/recipes.

Reagents

Acetone (anhydrous; EM grade)
Colloidal gold (10- or 15-nm; undiluted) to serve as fiducial markers
Freeze-substitution fixative for yeast <R>
Lead citrate for postsection stain (Reynold's lead citrate) <R>
Liquid nitrogen
Lowicryl HM20 embedding resin (e.g., from Ted Pella, Inc.)
  Prepare the Lowicryl HM20 embedding resin according to the manufacturer's instructions. Store the resin in an amber bottle at −20°C until use.

Uranyl acetate (2%, prepared in H_2O)
Yeast cultures

Equipment

BEEM capsules
Cryovials (Nalgene)
Dissecting microscope

---

[1]Correspondence: eileen.otoole@colorado.edu; mark.winey@colorado.edu

Copyright © Cold Spring Harbor Laboratory Press; all rights reserved
Cite this protocol as *Cold Spring Harb Protoc*; doi:10.1101/pdb.prot085589

Forceps (fine)

Freeze-substitution device with an ultraviolet source for polymerizing Lowicryl resins (e.g., Leica Automatic Freeze Substitution [AFS] Chamber)

If a commercial device is not available, a simple, homemade box with dry ice and a cold metal block can be used (McDonald and Müller-Reichert 2002).

High-pressure freezer

Microtome and diamond knives for cutting thin (50-nm) or semi-thick (200- to 300-nm) sections

Millipore filter (0.45-μm; Fisher)

Parafilm

Sample carriers for high-pressure freezing (100- or 200-μm wells) (Ted Pella or Leica Microsystems)

Shaker/bath for yeast cultures (controlled temperature)

Slot grids coated with 0.7% formvar

We use copper/rhodium grids with a 2 mm × 1 mm slot.

Software for computing tomographic reconstructions

Tomography packages are both commercially and freely available. Our lab has developed the IMOD software package (http://bio3d.colorado.edu/imod) that contains a suite of programs for computation, display, and analysis of tomographic reconstructions.

Software for image acquisition

Automated image acquisition packages are both commercially and freely available. Our lab has developed the SerialEM image acquisition package that can be downloaded for free from http://bio3d.colorado.edu/SerialEM.

Tomography-capable electron microscope

Toothpicks and flat spatula

Vacuum filtration apparatus (15-mL) (Millipore/Fisher)

Vacuum source (house line)

Vials (~20-mL)

METHOD

High-Pressure Freezing

1. Grow liquid cultures of yeast cells to early log phase (0.3–0.5 OD$_{600}$) on a shaker bath, maintaining them at an appropriate growth temperature.

2. Collect ~15 mL of cells onto a 0.45-μm Millipore filter using the vacuum filtration apparatus.

   Be careful not to dry the cells during the process. The cells should form a wet slurry, with a consistency similar to that of apple sauce.

3. Remove the filter from the apparatus. Under a dissecting scope, scrape the cells from the filter using a flattened spatula or toothpick. Transfer the cells with a sharpened toothpick to fill the well of a sample carrier.

4. Freeze the cells in a high-pressure freezer, and immediately transfer the frozen cells to liquid nitrogen.

   The cells can be stored in liquid nitrogen until they are processed by freeze substitution.

Freeze Substitution

5. Under liquid nitrogen, transfer the sample carriers with the frozen cells into cryovials containing 1 mL of freeze substitution fixative.

6. Set up the AFS device by filling the chamber with liquid nitrogen and setting the temperature to −90°C. Alternatively, if using a box setup, fill the box with dry ice.

7. Freeze substitute the sample for 2–3 days at −90°C in an AFS or at −80°C on dry ice.

8. Place the four glass-capped vials listed below in the AFS and allow them to cool to −35°C:
   i. One vial with pure acetone;
   ii. one vial with 100% Lowicryl HM20 embedding resin;
   iii. one vial for mixing resin and acetone; and
   iv. one vial for waste.

9. Warm the frozen samples to −35°C. Using the tips of a fine forceps, remove the cryovials containing the yeast pellets from the metal specimen holders. Allow the yeast to remain in the freeze substitution mixture for another hour.

10. Rinse the samples twice in pure acetone at −35°C.

11. Slowly infiltrate the samples with a graded series of acetone/Lowicryl HM20 at −35°C using the following schedule.

    i. Immerse the samples overnight in 25% Lowicryl HM20 and 75% acetone.
    ii. Immerse the samples for 8 h in 50% Lowicryl HM20 and 50% acetone.
    iii. Immerse the samples overnight in 75% Lowicryl HM20 and 25% acetone.
    iv. Transfer the samples through three changes of 100% Lowicryl HM20 (1 h each).
    v. Transfer the samples through a final change of 100% Lowicryl HM20. Leave the samples in this solution overnight at −35°C until ready to proceed to Step 12.

12. Transfer the samples to BEEM capsules, return them to the AFS, and polymerize the resin by UV irradiation for 2 d at −35°C. Store the samples at room temperature until ready to proceed to Step 13.

Preparation of Grids for Electron Tomography

This series of steps should be performed at room temperature.

13. Using a standard microtome, cut a series of 200- to 300-nm-thick sections of the samples. Collect the sections onto formvar-coated slot grids.

14. Poststain the sections in 2% uranyl acetate for 8 min and then in Reynold's lead citrate for 5 min.

15. Affix 15-nm gold particles to each side of the sections on the grid.

    i. Place a ~20-µL drop of colloidal gold solution onto a small square of Parafilm, and float the grid on top of the drop for 5 min.
    ii. Remove excess gold from the grid by using a triangle of filter paper at the metal edge of the grid, taking care not to touch or damage the formvar film.
    iii. Turn the grid over, and repeat Steps 15.i–15.ii for the other side of the grid.

Tilt Series Acquisition and Tomographic Reconstruction

16. Place the grids into a tomography-capable electron microscope with a high tilt holder. Find the cell profiles of the organelle of interest. For larger organelles and structures (e.g., a mitotic spindle) that span several serial sections, map their locations in serial sections.

17. Select an appropriate imaging magnification (commonly 1- to 1.2-nm pixel), and preirradiate the specimen for several minutes or for a total electron dose of ~2000 electrons/Å^2.

18. Use image acquisition software to collect a tilt series of images every degree from +60° to −60°. After the first tilt series is acquired, rotate the grid 90° and collect another tilt series about the orthogonal axis.

19. Use an appropriate tomography software package to align the serial, tilted views and compute the tomographic volume using weighted back-projection.
20. Join the serial tomograms to increase the volume of the reconstructed cell. Display the model in 3D, and analyze the features of interest.
    See Troubleshooting.

TROUBLESHOOTING

Problem (Step 20): The cells appear to be dense and contracted, with little fine structure preserved.
Solution: The yeast may have dried before freezing. Take care not to dry the yeast during vacuum filtration in Step 2. Apply sufficient suction to produce a wet slurry of yeast onto the filter. Then work quickly when loading the yeast into the wells of the sample carriers in Step 3.

Problem (Step 20): The cell structures appear to be damaged or distorted.
Solution: The samples may have been improperly loaded into the sample carriers such that air became trapped. When loading the sample carriers in Step 3, overfill the wells slightly so that there is no space for air bubbles. Additional details on sample loading can be found in McDonald and Müller-Reichert (2002).

Problem (Step 20): There appear to be large, open areas in the specimen that have the appearance of "chicken wire"; elsewhere, halos are seen around organelles.
Solution: These features are indicative of ice crystal damage. Gross ice crystal damage can be recognized by the large open areas, whereas more subtle ice damage results in the "halo" effect. The samples may have been warmed unintentionally. Keep material under liquid nitrogen after freezing to avoid warming and precool forceps before transferring frozen samples to freeze-substitution vials.

RELATED INFORMATION

Additional protocols on the preparation of yeast using high-pressure freezing and freeze-substitution can be found in Giddings et al. (2001), McDonald and Müller-Reichert (2002), and McDonald (2007). Detailed information on tomographic reconstruction can be found in O'Toole et al. (2002).

RECIPES

Freeze-Substitution Fixative for Yeast

Carry out the following steps wearing protective gloves and working in a fume hood.
1. In a 5-mL conical tube, prepare a stock solution of 5% uranyl acetate by dissolving 0.1 g of uranyl acetate crystals in 2 mL of methanol. Cover the tube with foil and place it on a rocker for ~10 min at room temperature to dissolve the crystals.
2. Prepare a stock solution of 10% glutaraldehyde (anhydrous EM grade) in acetone (anhydrous EM grade).
3. Transfer 1.25 mL of the 10% glutaraldehyde solution from Step 2 into a 50-mL conical screw-cap tube, and add 1 mL of the 5% uranyl acetate solution from Step 1.
4. Fill the tube to 50 mL with acetone; the final concentrations of the components are 0.25% glutaraldehyde and 0.1% uranyl acetate.
5. Distribute 1-mL aliquots of the fixative into cryovials and freeze them in liquid nitrogen. Store the aliquots in liquid nitrogen.

Lead Citrate for EM

1. Dissolve 1.33 g of lead nitrate in 30 mL of distilled water.
2. Add 1.76 g of sodium citrate (e.g., Calbiochem/EMD Millipore 567446) and shake well, until solution is milky.
3. Add 8 mL of 1 m sodium hydroxide and shake well.
4. Make up the volume to 50 mL with distilled water.

ACKNOWLEDGMENTS

The Boulder Laboratory for 3D Electron Microscopy of Cells by grant P41GM103431 from National Institute of General Medical Sciences to Andreas Hoenger.

REFERENCES

Giddings TH, O'Toole ET, Morphew M, Mastronarde DN, McIntosh JR, Winey M. 2001. Using rapid freeze and freeze-substitution for the preparation of yeast cells for electron microscopy and three-dimensional analysis. *Methods Cell Biol* **67**: 27–42.

McDonald K. 2007. Cryopreparation methods for electron microscopy of selected model systems. *Methods Cell Biol* **79**: 23–56. doi: 10.1016/S0091-679X(06)79002-1.

McDonald K, Müller-Reichert T. 2002. Cryomethods for thin section electron microscopy. *Methods Enzymol* **351**: 96–123.

O'Toole ET, Winey M, McIntosh JR, Mastronarde DN. 2002. Electron tomography of yeast cells. *Methods Enzymol* **351**: 81–95.

CHAPTER 18

The Yeast Two-Hybrid System: A Tool for Mapping Protein–Protein Interactions

Jitender Mehla, J. Harry Caufield, and Peter Uetz[1]

Center for the Study of Biological Complexity, Virginia Commonwealth University, Richmond, Virginia 23284

Virtually all processes in living cells are dependent on protein–protein interactions (PPIs). Understanding PPI networks is thus essential for molecular biology and disease research. One powerful genetic system for mapping PPIs both at a small scale and in a high-throughput manner is the yeast two-hybrid (Y2H) screen. In Y2H screening, PPIs are detected through the activation of reporter genes responding to a reconstituted transcription factor. In this introduction, we describe library- and array-based Y2H methods and explain their basic theory. We also include the rationale behind different Y2H approaches and strategies for optimizing results.

INTRODUCTION

Proteins are responsible for all major biological processes inside cells, either individually or in close coordination with other macromolecules. Outside cells, proteins and their interactions help to mediate cell–cell communication. Millions of micro- and macromolecular interactions are required to maintain the normal functional and structural architecture of a cell at any given time.

Comprehensive analysis of specific protein–protein interactions (PPIs) on a genome-wide scale is a challenging task of proteomics and has been best explored in organisms such as budding yeast, *Escherichia coli*, and human. However, major portions of most genomes remain uncharacterized, including 20%–50% of the predicted human genes (Pawłowski 2008) and 745 predicted genes of *Saccharomyces cerevisiae* (http://www.yeastgenome.org/cache/genomeSnapshot.html). High-throughput approaches have been developed to characterize unknown proteins and determine inter- and intramolecular interactions and networks in many organisms. Affinity purification and subsequent analysis by mass spectrometry (AP/MS) and the yeast two-hybrid (Y2H) system are the preferred methods for mapping protein–protein networks on a large scale. The Y2H assay is a genetic method that employs in vivo screening for binary PPIs (Fields and Song 1989). A two-hybrid screen may make use of random libraries (from genomic DNA or cDNA) or defined clone sets called ORFeomes. The latter may be used in array-based Y2H assays and are the preferred choice if such clones are available, because they are easily scalable from a few proteins to whole genomes. Most importantly, arrays are more efficient at avoiding false positives, although the number of false negatives may be also higher than in random library screens.

RATIONALE FOR THE YEAST TWO-HYBRID SYSTEM

The original concept of the Y2H system is based on detection of an interaction between two proteins via the reconstitution of a transcription factor that activates one or more reporter genes (Fields and

---

[1]Correspondence: phu878@gmail.com

Copyright © Cold Spring Harbor Laboratory Press; all rights reserved
Cite this introduction as *Cold Spring Harb Protoc*; doi:10.1101/pdb.top083345

Song 1989). The system is enabled by the fact that protein domains of a fully functional protein can be separated and recombined to reconstitute the fully functional protein. Specifically, a transcription factor can be split into a DNA-binding domain (DBD) and an activation domain (AD). In a typical Y2H system, a protein "X" for which one wants to explore physically interacting partners is fused to the DBD of the yeast transcription factor Gal4, though other DBDs have been used. The DBD-X fusion is generally referred to as a *bait*. The DBD can bind to the upstream activating sequence (UAS) of a promoter but can only activate transcription if an activation domain is present. Simultaneously, open reading frames (ORFs) or random protein fragments are fused to the AD of Gal4. These fusions are called *preys*. The AD can activate transcription but cannot bind to the promoter of a reporter gene without a DBD. Thus, only if both the bait and an interacting prey are coexpressed will a fully functional transcription factor be reconstituted and the reporter gene transcribed (Fig. 1). Note that the terms bait and prey are sometimes used in the literature for both the fusion proteins and the proteins fused to the DBD and AD.

Both bait and prey fusion proteins must be coexpressed in the same yeast cell, although they may be expressed first in different yeast strains of opposite mating types (e.g., the strains AH109 and Y187). In the latter case, the two strains must be mated to form diploid cells containing both bait and prey constructs. However, only the cells in which the reporter genes are expressed will grow on selective media. For example, the *HIS3* reporter gene encodes the histidine biosynthetic enzyme imidazole-glycerol-phosphate (IGP) dehydratase. The Y2H strains lack an endogenous copy of *HIS3*, rendering them auxotrophic for histidine. Interacting bait and prey pairs provide functional Gal4, enabling expression of the *HIS3* reporter gene, histidine synthesis, and growth on medium lacking histidine. For noninteracting baits and preys, the *HIS3* reporter gene is not expressed and thus these cells cannot grow (Fig. 2).

Note that alternative Y2H systems have been devised that are not based on transcription factors, for example, the split ubiquitin system (Obrdlik et al. 2004). Yet other systems use alternative reporter genes such as luciferase or certain metabolic markers (Ozawa et al. 2001; Remy and Michnick 2006).

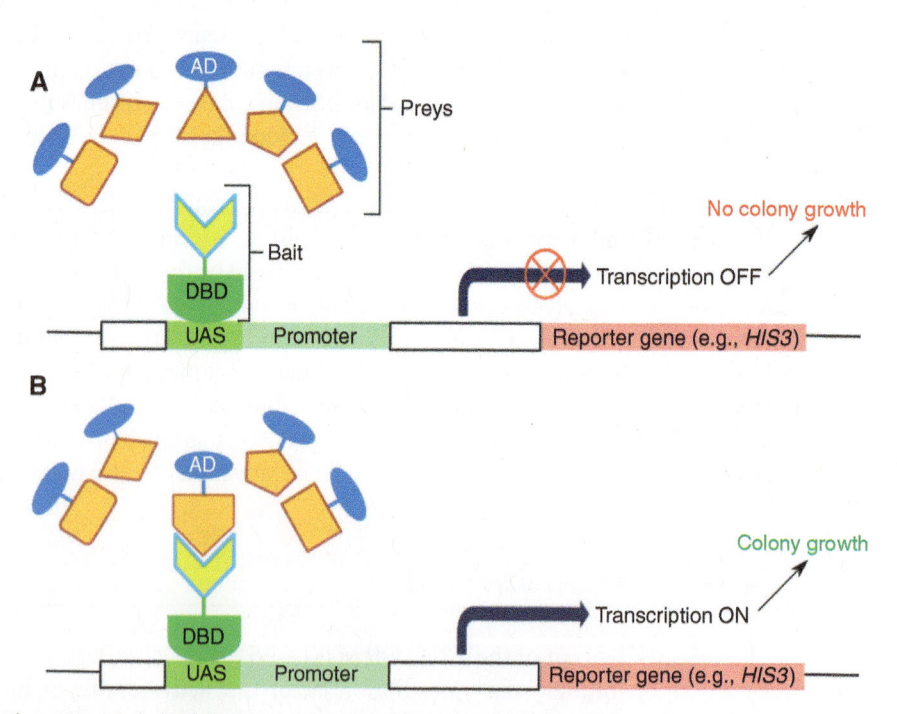

FIGURE 1. The yeast two-hybrid system is a genetic tool used to detect interactions between two proteins. (*A*) A lack of physical interaction between bait and prey prevents transcription of the reporter gene, preventing cell growth. (*B*) When bait and prey proteins interact, a functional transcription factor is reconstituted from the Gal4 DNA binding domain (DBD) and activation domain (AD). Under these conditions, a reporter gene (i.e., *HIS3*) is transcribed, enabling yeast cell growth.

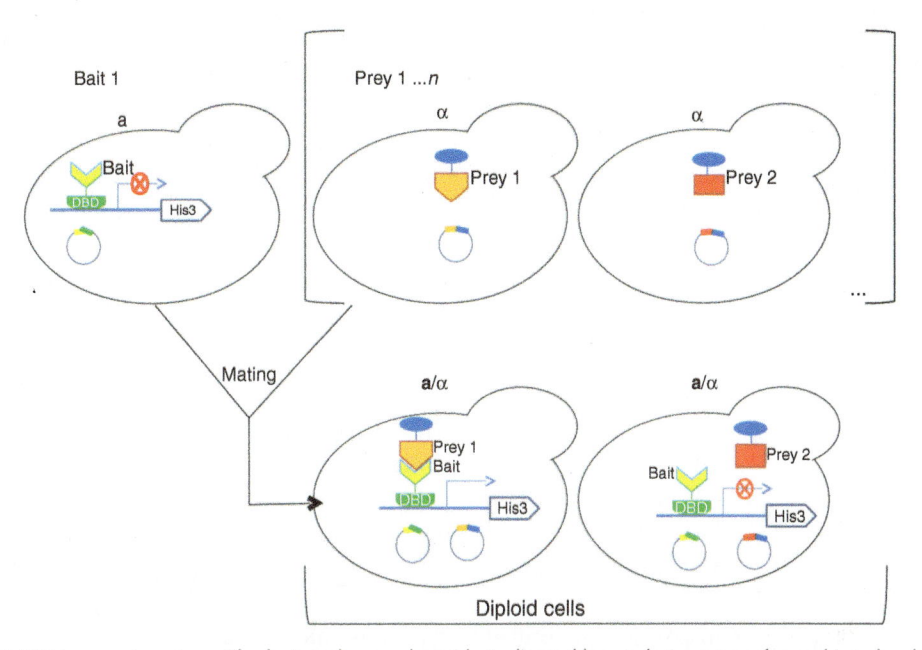

FIGURE 2. Y2H screening steps. The bait and prey plasmids (indicated by circles) are transformed into haploid yeast cells of mating type "**a**" and "α", respectively. A single bait strain (bait 1) is mated with a prey library (Prey 1...n). The resulting diploids (**a**/α) carry both bait and prey vectors leading to expression of both bait and prey in the same cell. The interacting pair of bait and prey activates expression of the *HIS3* reporter gene and the cells grow on minimal medium lacking histidine (diploid on the *left*). The diploid cells with noninteracting bait and prey cannot grow on selective medium as they cannot synthesize histidine (diploid on the *right*).

An array-based Y2H screen can be completed within ∼2 wk, starting from the initial mating to scoring and analyzing the interactions. With automated systems, 20 or more screens can be initiated per day, depending on the size of the library and the level of automation. Screening a random library takes about the same time as an array-based screen, but after the initial screen, positives should be verified using additional tests, which may take another 2 wk.

CRITICAL PARAMETERS IN Y2H

Choosing a Vector

The selection of vectors in a two-hybrid assay will determine its outcome. A number of different DBD- and AD-containing vectors are available either commercially or from academic laboratories. Using a combination of different vectors will likely produce different results. We have used four different vectors (pGBGT7g, pGBKCg, pGADT7g, and pGADCg) in combination for mapping interactomes, as they allow for both amino-terminal and carboxy-terminal fusion proteins.

Testing Baits for Self-Activation

Ten to 20% of all baits can activate transcription of the reporter gene in the absence of a bait/prey interaction, leading to false positives. The activation of reporter genes by the bait protein in the absence of a prey is termed *self-activation*. Y2H experiments should therefore begin with checking baits for their potential for self-activation. Different baits can have different levels of auto-activation. 3-amino-triazole (3-AT), a competitive inhibitor of the His3 protein, may be used to suppress background reporter activity and reduce false positives when *HIS3* is used as a reporter gene; see Protocol 1: Mapping Protein–Protein Interactions Using Yeast Two-Hybrid Assays (Mehla et al. 2015). The concentration of 3-AT in selective media should be optimized for each screen.

Chapter 18

Selecting Host Strains

The strains used for transformation of bait and prey plasmids must be compatible for mating. For example, if the bait plasmid is in an "a" strain (e.g., AH109), then the prey plasmid must be transformed into a strain of mating type "α" (e.g., Y187). Different yeast strains have different transformation efficiencies (Dohmen et al. 1991; Hayama et al. 2002). Because most Y2H screens involve only a few bait strains but require many prey strains, the prey should be transformed into a strain with high transformation efficiency before mating.

GENOMIC OR cDNA LIBRARY SCREENING

Systematic cloning of every prey construct is not required for screening a library of preys. However, the prey library must be generated before performing any actual screening. Prey libraries can be prepared using either genomic DNA or cDNA. A strategy for the construction of a genomic library has been described by Rain et al. (2001). A cDNA library can be either made by RT-PCR of mRNA (from specific cell types or whole organisms) or obtained from commercial sources. In a library screening, a specific bait or a pool of baits is screened against a library of prey proteins; see Protocol 1: Mapping Protein–Protein Interactions Using Yeast Two-Hybrid Assays (Mehla et al. 2015). The prey proteins are then identified by PCR amplification and sequencing of the insert.

Library screens offer several advantages:

1. A library screen using a cDNA or genomic library can detect not only full-length interacting proteins, but also interacting domains or subdomains of a given protein. Detection of fragment interactions increases the sensitivity of a screen.
2. Random library screens allow exhaustive screening of large libraries. Libraries of random fragments, for example, may contain >10^6 different clones. There are more random libraries available than ORFeome clone sets.
3. Library screens yield a higher number of interactions than array-based screens. The reason for this phenomenon is most likely the fact that random libraries contain a collection of fragments, one of which may interact whereas a full-length protein may not (e.g., proteins with transmembrane domains may yield false positives or no positives in array-based Y2H).
4. Library screens are less time-consuming and resource-intensive than array-based screens. After mating, cells can be plated directly onto selective dropout medium to select for interactions. Randomly generated prey plasmid libraries can also be transformed directly into haploid yeast prey strains to save time. Library screens can be performed without specialized equipment and with minimal labware.

Library screens also have some disadvantages. Interactions of protein fragments may be nonspecific and thus represent false positives. A library screen may also be difficult to reproduce.

ARRAY-BASED SCREENING

Array-based Y2H screens are commonly used to test a number of defined prey proteins for interactions with a single bait or a pool of baits. The prey proteins are tested individually for interactions with the bait proteins. In a typical Y2H screen (Fig. 2), the baits are expressed in one yeast strain (e.g., AH109) and the preys are expressed in another yeast strain of opposite mating type (e.g., Y187). The two strains are then mated to produce both bait and prey in diploid cells as described in Protocol 1: Mapping Protein–Protein Interactions Using Yeast Two-Hybrid Assays (Mehla et al. 2015).

Array screens offer several advantages:

1. Interacting proteins can be easily identified without further downstream experiments like sequencing because the identity of the interacting proteins is already known and does not require further confirmation.

2. Array-based screens are adaptable to size demands. A few proteins may be screened or the method may be scaled up to include all ORFs within a genome (e.g., in Gateway vectors).
3. Because assays are performed in parallel, we can immediately see if one interaction is stronger than another. Using 3-AT to titrate HIS3 activity renders array screens semiquantitative.
4. True interacting signals can be differentiated from background (false positive) signals because most positions are expected to be negative.

Array screens have one significant disadvantage: they require upfront construction of all bait and prey clones. Until the advent of Y2H arrays, constructing and screening a random (genomic or cDNA) library was the standard approach to interaction screening. An intermediary approach is to pool multiple clones in an array format for mixed library/array-based screens (see below).

POOLED ARRAY SCREENING

Both library- and array-based screens have advantages and disadvantages to keep in mind before planning a project. In the absence of automation, array-based screens may be time-consuming and tedious. Some screens may fail to yield a single interaction. For a large set of clones, a pooling strategy may be more efficient but may also produce more false negatives. For instance, pooled screens in *Campylobacter jejuni* led to more false negatives than 1:1 testing screens of *Treponema pallidum* (Parrish et al. 2007; Rajagopala et al. 2007). If an initial pool screen is performed, the interacting pair needs to be identified either by sequencing or 1:1 retesting of all baits or subsets thereof. In fact, there may be a five- to sixfold reduction in yield when pooling is used (J. Mehla, unpubl.).

EVALUATING Y2H SCREEN RESULTS

Ideally, no growth should occur with noninteracting pairs; however, some baits may yield extensive background growth. Thus, the colony size of interacting pairs must be significantly larger than that of noninteracting pairs, rather than simply a matter of growth versus lack of growth.

Some "sticky" preys seem to interact nonspecifically with a large number of baits and thus may or may not have any biological relevance. Such signals may appear reproducible but lack specificity and can be easily identified. These artifacts can be excluded from screening data by filtering results on the basis of interactions per bait or prey. For instance, if a bait or prey is involved in 20 or more interactions, many of these are likely unspecific. However, the threshold is somewhat arbitrary, depending on the size of the library and the nature of the bait or prey protein. Some proteins, such as chaperones, have many bona fide interactions (Rajagopala and Uetz 2011). Several protocols have been developed to filter raw interaction data to improve the credibility and quality of data, including logistic regression, which uses positive and negative training sets of interactions (Bader et al. 2004; von Mering et al. 2007; Titz et al. 2008).

Previous studies have shown that the **false-negative rate** in typical array-based screens is on the order of 75% (Rajagopala et al. 2007). That is, only ~25% of all "true" interactions are detected. Possible reasons for missed interactions may be steric hindrance, degradation and instability of proteins, failure of nuclear localization and posttranslational modifications. The rate of false negatives in Y2H can be reduced by either taking a combinatorial approach or choosing a different experimental approach. For example, using combinations of different Y2H vectors can reduce false negative rates to 20%–25% (Chen et al. 2010).

Y2H methods have historically been plagued by two distinct types of false positive results. In a **technical false-positive** interaction, the reporter gene is activated without an interaction between bait and prey. This effect may be the result of baits acting as transcriptional activators. Some bait or prey proteins may fail to interact but may still enhance yeast viability on selective media. The second type of effect, a **biological false positive**, is an actual two-hybrid interaction that has no biological relevance.

For instance, protein pairs may interact in the Y2H assay despite being usually expressed in different tissues or developmental stages. Y2H interactions should therefore be validated using additional experiments or other existing knowledge to enhance the credibility of data. For example, Uetz et al. (2006) validated the Y2H interactions of Kaposi's sarcoma-associated herpesvirus (KSHV) by CoIP and found that ∼50% of interactions could be confirmed. Simply retesting all positive interactions can also help to reduce false positives (Serebriiskii et al. 2000; Serebriiskii and Golemis 2001; Koegl and Uetz 2007).

REFERENCES

Bader JS, Chaudhuri A, Rothberg JM, Chant J. 2004. Gaining confidence in high-throughput protein interaction networks. *Nat Biotechnol* **22**: 78–85.

Chen YC, Rajagopala SV, Stellberger T, Uetz P. 2010. Exhaustive benchmarking of the yeast two-hybrid system. *Nat Methods* **7**: 667–668.

Dohmen RJ, Strasser AW, Höner CB, Hollenberg CP. 1991. An efficient transformation procedure enabling long-term storage of competent cells of various yeast genera. *Yeast* **7**: 691–692.

Fields S, Song O. 1989. A novel genetic system to detect protein–protein interactions. *Nature* **340**: 245–246.

Hayama Y, Fukuda Y, Kawai S, Hashimoto W, Murata K. 2002. Extremely simple, rapid and highly efficient transformation method for the yeast *Saccharomyces cerevisiae* using glutathione and early log phase cells. *J Biosci Bioeng* **94**: 166–171.

Koegl M, Uetz P. 2007. Improving yeast two hybrid screening systems. *Brief Funct Genomic Proteomic* **6**: 302–312.

Mehla J, Caufield JH, Uetz P. 2015. Mapping protein–protein interactions using yeast two-hybrid assays. *Cold Spring Harb Protoc* doi: 10.1101/pdb.prot086157.

Obrdlik P, El-Bakkoury M, Hamacher T, Cappellaro C, Vilarino C, Fleischer C, Ellerbrok H, Kamuzinzi R, Ledent V, Blaudez D, et al. 2004. K$^+$ channel interactions detected by a genetic system optimized for systematic studies of membrane protein interactions. *Proc Natl Acad Sci* **101**: 12242–12247.

Ozawa T, Kaihara A, Sato M, Tachihara K, Umezawa Y. 2001. Split luciferase as an optical probe for detecting protein–protein interactions in mammalian cells based on protein splicing. *Anal Chem* **73**: 2516–2521.

Parrish JR, Yu J, Liu G, Hines JA, Chan JE, Mangiola BA, Zhang H, Pacifico S, Fotouhi F, DiRita VJ, et al. 2007. A proteome-wide protein interaction map for *Campylobacter jejuni*. *Genome Biol* **8**: R130.

Pawłowski K. 2008. Uncharacterized/hypothetical proteins in biomedical "omics" experiments: Is novelty being swept under the carpet? *Brief Funct Genomic Proteomic* **7**: 283–290.

Rain JC, Selig L, De Reuse H, Battaglia V, Reverdy C, Simon S, Lenzen G, Petel F, Wojcik J, Schachter V, et al. 2001. The protein–protein interaction map of *Helicobacter pylori*. *Nature* **409**: 211–215.

Rajagopala SV, Uetz P. 2011. Analysis of protein–protein interactions using high-throughput yeast two hybrid screens. *Methods Mol Biol* **781**: 1–29.

Rajagopala SV, Titz B, Goll J, Parrish JR, Wohlbold K, McKevitt MT, Palzkill T, Mori H, Finley RL Jr, Uetz P. 2007. The protein network of bacterial motility. *Mol Syst Biol* **3**: 128.

Remy I, Michnick SW. 2006. A highly sensitive protein–protein interaction assay based on *Gaussia* luciferase. *Nat Methods* **3**: 977–979.

Serebriiskii IG, Golemis EA. 2001. Two hybrid system and false positives. Approaches to detection and elimination. *Methods Mol Biol* **177**: 123–134.

Serebriiskii I, Estojak J, Berman M, Golemis EA. 2000. Approaches to detecting false positives in yeast two hybrid systems. *Biotechniques* **28**: 328–330, 332–336.

Titz B, Rajagopala SV, Goll J, Häuser R, McKevitt MT, Palzkill T, Uetz P. 2008. The binary protein interactome of *Treponema pallidum*—The syphilis spirochete. *PloS ONE* **3**: e2292.

Uetz P, Dong YA, Zeretzke C, Atzler C, Baiker A, Berger B, Rajagopala SV, Roupelieva M, Rose D, Fossum E, et al. 2006. Herpesviral protein networks and their interaction with the human proteome. *Science* **311**: 239–242.

von Mering C, Jensen LJ, Kuhn M, Chaffron S, Doerks T, Kruger B, Snel B, Bork P. 2007. STRING 7-recent developments in the integration and prediction of protein interactions. *Nucleic Acids Res* **35**: D358–D362.

Protocol 1

Mapping Protein–Protein Interactions Using Yeast Two-Hybrid Assays

Jitender Mehla, J. Harry Caufield, and Peter Uetz[1]

Center for the Study of Biological Complexity, Virginia Commonwealth University, Richmond, Virginia 23284

Yeast two-hybrid (Y2H) screens are an efficient system for mapping protein–protein interactions and whole interactomes. The screens can be performed using random libraries or collections of defined open reading frames (ORFs) called ORFeomes. This protocol describes both library and array-based Y2H screening, with an emphasis on array-based assays. Array-based Y2H is commonly used to test a number of "prey" proteins for interactions with a single "bait" (target) protein or pool of proteins. The advantage of this approach is the direct identification of interacting protein pairs without further downstream experiments: The identity of the preys is known and does not require further confirmation. In contrast, constructing and screening a random prey library requires identification of individual prey clones and systematic retesting. Retesting is typically performed in an array format.

MATERIALS

It is essential that you consult the appropriate Material Safety Data Sheets and your institution's Environmental Health and Safety Office for proper handling of equipment and hazardous materials used in this protocol.

RECIPES: Please see the end of this protocol for recipes indicated by <R>. Additional recipes can be found online at http://cshprotocols.cshlp.org/site/recipes.

Reagents

Adenine (optional, for mating; see Step 41)
Carrier (sheared salmon sperm) DNA (10 mg/mL) (Life Technologies)
  Store carrier DNA at −20°C. Boil for 10–12 min and cool on ice 3–4 min before use.

Ethanol (≥95% v/v)
Glycerol
Lithium acetate (LiAc) (0.1 M in TE), freshly prepared
  Immediately before use, combine 1 mL of 1.0 M LiAc with 1 mL of 10× TE buffer (0.1 M Tris–HCl, 10 mM EDTA [pH 7.5]) and 8 mL of dH$_2$O.

PCR and DNA sequencing reagents (for identification of interacting preys in library screens)
PEG (40%)
  Prepare a 44% PEG solution in advance by dissolving 44.0 g of polyethylene glycol 3000 (PEG) (Sigma-Aldrich) in dH$_2$O for a total volume of 100 mL, and then sterilize by autoclaving. Prepare 40% PEG immediately before use (Step 7) by combining 10 mL of 44% PEG with 1 mL of freshly prepared 0.1 M LiAc (prepared in TE).

[1]Correspondence: phu878@gmail.com

Copyright © Cold Spring Harbor Laboratory Press; all rights reserved
Cite this protocol as *Cold Spring Harb Protoc*; doi:10.1101/pdb.prot086157

Chapter 18

TABLE 1. Y2H bait and prey vectors and their properties (see also Fig. 1)

	Vector	Promoter[a]	Gal4-Fusion DBD	Gal4-Fusion AD	Selection Yeast	Selection Bacterial	Ori	Source
Baits	pGBKCg	t-ADH1	Carboxy-terminal	–	Trp1	Kanamycin	2μ	Stellberger et al. 2010
	pGBGT7g	t-ADH1	Amino-terminal	–	Trp1	Gentamycin	2μ	Rajagopala et al. 2014
Preys	pGADCg	fl-ADH1	–	Carboxy-terminal	Leu2	Ampicillin	2μ	Stellberger et al. 2010
	pGADT7g	fl-ADH1	–	Amino-terminal	Leu2	Ampicillin	2μ	Chen et al. 2010

[a]t-ADH1, fl-ADH1 = truncated or full-length ADH1 promoter.

Plasmid DNA constructs containing bait protein(s) of interest, plus empty vector control

Gateway cloning has become the method of choice for vector assembly; this allows bait constructs (DNA-binding domain [DBD]-ORF fusions) to be generated by transfer of ORFs from entry vectors into specific bait vectors via recombination. We use the bait vectors pGBGT7g (a gentamycin-resistance-encoding version of pGBKT7g) and pGBKCg (Stellberger et al. 2010); see Table 1 and Figure 1. See Landy (1989) and Walhout et al. (2000) for more details regarding Gateway cloning and vectors.

Ensure that bait plasmids are compatible with prey plasmids, especially with respect to markers.

Plasmid DNA constructs containing prey proteins of interest (e.g., prey libraries or ORFeomes), plus empty vector control

Prey constructs (activation domain [AD]-ORF fusions) must be cloned or obtained in appropriate prey vectors. As with bait vectors, there are several Gateway-compatible prey vectors available, for example, pGADT7g and pGADCg (Stellberger et al. 2010); see Table 1 and Figure 1.

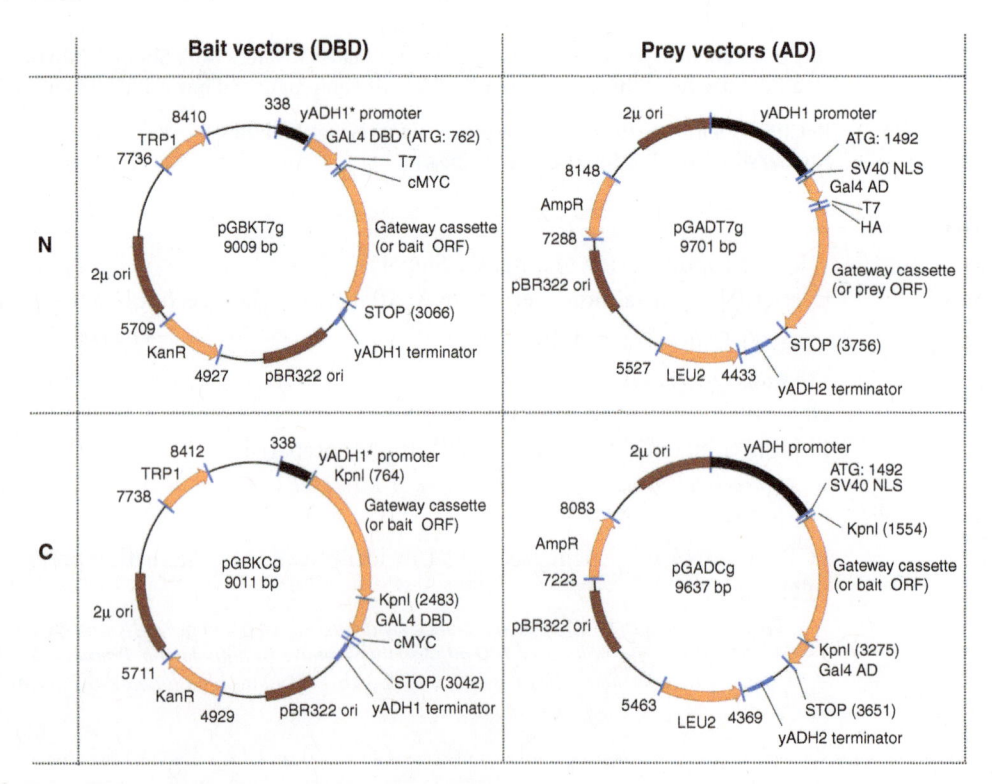

FIGURE 1. Selected bait and prey vectors. The pGBKT7g and pGADT7g generate amino-terminal DBD and AD fusions, respectively. pGBKCg and pGADCg fuse the DBD or AD at the carboxyl terminus of inserted ORF. (Reprinted from Stellberger et al. 2010.)

Y2H prey libraries for many species, tissues, and cell types are available from several commercial sources or can be constructed in the laboratory using available protocols (Rain et al. 2001). ORFeomes are available for genomes of various sizes, from small viral genomes (e.g., KSHV and VZV [Uetz et al. 2006]) to the bacterial genomes of Escherichia coli, Bacillus anthracis and Yersinia pestis (Rajagopala et al. 2010). Bacterial ORFeomes can be obtained from BEI Resources (https://www.beiresources.org) and academic laboratories. Clone sets from multicellular eukaryotes, such as (Lamesch et al. 2004), human (Rual et al. 2004), and Arabidopsis (Gong et al. 2004) have also been described, though not all genes of interest are yet available in convenient vectors. Cloning and Y2H procedures may be automated using 96-well plates such that entire ORFeomes can be processed in parallel.

Selective liquid medium

Follow the recipe for selective solid medium for Y2H, but omit the agar.

Selective solid medium for Y2H (including −L, −T, −LT, and −LTH, with or without 3-amino-1,2,4-triazole [3-AT]), prepared in single-well Omnitray plates <R>

3-AT (Sigma-Aldrich) is added to selective medium at varying concentrations (0–100 mM) during the bait self-activation test (Steps 18–23). The 3-AT concentration in the selective (−LTH) medium used for protein interaction screening (Steps 33 and 43) will depend on the results of the self-activation test.

Selective solid medium can also be prepared in Qtray plates (Molecular Devices); see Step 15. If plates containing solid medium are wet, allow them to dry for a few hours or overnight at 30°C before plating cells.

Yeast strains (haploid, mating-competent)

AH109 (for baits)

The AH109 genotype is MATa, trp1-901, leu2-3, 112, ura3-52, his3-200, gal4Δ, gal80Δ, LYS2::GAL1$_{UAS}$-GAL1$_{TATA}$-HIS3, GAL2$_{UAS}$-GAL2$_{TATA}$-ADE2, URA3::MEL1$_{UAS}$-MEL1$_{TATA}$-lacZ (after James et al. 1996).

Y187 (for preys)

The Y187 genotype is MATα, ura3-52, his3-200, ade2-101, trp1-901, leu2-3, 112, gal4Δ, met⁻, gal80Δ, URA3::GAL1$_{UAS}$-GAL1$_{TATA}$-lacZ (after Harper et al. 1993).

YEPAD liquid medium

Follow the recipe for YEPAD solid medium for Y2H, but omit the agar.

YEPAD solid medium for Y2H, prepared in single-well Omnitray plates <R>

If plates containing solid medium are wet, allow them to dry for a few hours or overnight at 30°C before plating cells.

Equipment

Aluminum foil

Centrifuge

Centrifuge tubes (15- and 50-mL)

High-density plate replicator (96-pin and/or 384-pin) (Beckman Coulter [Biomek FX 96/384 HDR Tool Body] or V&P Scientific)

Solid pins of 0.4–0.5 mm or 1.1–1.6 mm ("thin" or "thick" pins, respectively) for use with the plate replicator are also available from Beckman Coulter or V&P Scientific.

Before use and between all transfers, sterilize the plate replicator by dipping the pins into the following series: 20% bleach for 60 sec, sterile H$_2$O for 20 sec, 95% ethanol for 60 sec, and sterile H$_2$O again for 20 sec.

Incubator at 30°C

Incubator/shaker at 30°C, for aerating liquid cultures

A benchtop shaker placed inside a standard incubator may also be used.

Laboratory automation robot (Beckman Coulter Biomek 2000, FX or FXP) (optional)

A robotic workstation may be used to speed up the screening procedures and maximize reproducibility. Automation is strongly recommended if many clones need to be tested and for screening large numbers of baits, given that manual work is not only boring, straining, and error prone, but also time consuming and tedious. However, maintaining and operating robots can also be expensive and may justify collaboration with a laboratory already in possession of automation equipment.

Microcentrifuge

Microcentrifuge tubes (1.5-mL)

Microtiter plates (96-well, flat-bottom) (Corning)
Omnitray plates (single-well, 86 × 128-mm, with lids) (Nunc)
Parafilm
PCR and DNA sequencing equipment (for identification of interacting preys in library screens)
Spectrophotometer
Vortex mixer
Water bath at 42°C

METHOD

A rationale for the selection of library- or array-based Y2H is provided in Introduction: The Yeast Two-Hybrid System: A Tool for Mapping Protein–Protein Interactions (Mehla et al. 2015).

Transforming Yeast

This method is suitable for transformation of bait or prey plasmid clones into their respective mating-competent, haploid yeast strains: Prey vectors are transformed into Y187, mating type "α," while bait vectors are transformed into AH109, mating type "a." The following is optimized for 100 transformations; a success rate of 100% per transformation should be expected. It also can be scaled down or up as required for high-throughput transformation, and most steps can be automated. As with all high-throughput procedures, small discrepancies can have noticeable downstream consequences. Verify the sterility and pH of reagents and the phenotype of yeast strains before transformation.

1. Inoculate 10–20 mL of YEPAD liquid medium with the appropriate haploid yeast strain. Grow overnight at 30°C in a shaking incubator (∼200 rpm, depending on model). After 12–16 h, measure the absorbance (OD_{600}) of the culture.

   The OD_{600} should be >1.5.

2. Dilute the cells in 100 mL of fresh liquid YEPAD medium to an OD_{600} of ∼0.2–0.3. Grow the cells for ∼3–4 h (depending on the generation time of the strain) at 30°C in a shaking incubator (∼200 rpm).

   The OD_{600} should be between 0.5 and 1.0 (at least one doubling), indicating that the cells are in log phase of growth.

3. Transfer the culture to two 50-mL tubes. Collect the cells by brief centrifugation (∼700g for up to 5 min at room temperature) and discard the supernatant.

4. Wash the cells once with 25–50 mL sterile dH_2O. Collect the cells by brief centrifugation (∼700g for up to 5 min at room temperature) and discard the supernatant.

5. Resuspend the cells in 1 mL of 0.1 M LiAc (prepared in TE) and incubate on ice for 15 min.

6. Transfer the suspension to 1.5-mL microcentrifuge tubes and briefly centrifuge (∼350g for 30–60 sec). Discard the supernatant. Resuspend the cells in 1 mL of 0.1 M LiAc (prepared in TE) and incubate them on ice for another 15–20 min.

7. Prepare a transformation mix for ∼100 transformation reactions as follows.

   i. Add 250 µL of carrier DNA (10 mg/mL) to 10 mL of 40% PEG in a sterile 50-mL tube. Vortex for at least 30 sec.

   ii. Add 1–2 mL of freshly prepared competent yeast cells from Step 6. Vortex for an additional 60 sec.

8. Transfer 100 µL of reaction mix to each well of a 96-well plate using a multichannel pipette or robotic liquid handler.

   Depending on the number of samples, the final volume can be divided such that each volume does not exceed 120 µL per well.

   For fewer samples, 1.5-mL tubes may be used.

9. Add at least 100 ng of the desired plasmid DNA construct to each well and mix well with the pipette tip. Include one negative control containing carrier DNA but no plasmid DNA and at least one positive control containing the empty vector construct.

10. Seal the 96-well plate with adhesive aluminum foil and Parafilm to secure the edges.

    All exposed plate surfaces should be sealed to protect against contamination.

11. Vortex the plate gently for 3–4 min, ensuring that the wells do not cross-contaminate.

12. Place the plate for 45 min at 30°C with gentle shaking (~60 rpm).

13. Transfer the plate to a water bath at 42°C and incubate for 30 min. Ensure the plate is floating and water is not entering the wells.

    For small scale transformation using microcentrifuge tubes, this incubation may be reduced to 15 min.

14. After 30 min, centrifuge the plate at ~700g for 8–10 min at room temperature. Discard the supernatant by multichannel pipette and rinse the plate surface with 95% ethanol to avoid cross-contamination between wells.

15. Resuspend the contents of each well in 50–100 µL of sterile dH$_2$O. Spread each volume on a separate Omnitray plate (or distinct sections of a Qtray) containing selective solid medium.

16. Incubate the plates for at least 2–3 d at 30°C.

    Some strains, such as Y187, may require 3–4 d to yield noticeable transformant growth.

    Plates may be stored for several months at 4°C as long as the medium does not dry out or become contaminated. For long-term storage, transformants may be cultured in rich (YEPAD) or selective liquid medium overnight, resuspended in 20% glycerol, and transferred to a freezer at −80°C.

17. After preparation of the desired bait transformants, proceed to self-activation testing of baits (Steps 18–23) followed by library-based (Steps 24–35) and/or array-based (Steps 36–44) screening.

Self-Activation Testing of Bait Transformants Using 3-AT

Performing self-activation testing before Y2H screening helps optimize screen conditions, saving time and resources. 3-AT is a competitive inhibitor of the HIS3 gene product and is frequently used in Y2H screening to increase the stringency of selection when using the HIS3 reporter gene. The following assay measures the background activity of the HIS3 reporter in baits in the presence of empty prey vector.

18. Prepare 5-mL cultures in YEPAD liquid medium using empty-vector transformants in the prey strain (i.e., the Y187 strain containing the pGADCg empty vector). Grow the cultures for 12–16 h at 30°C in a shaking incubator (~200 rpm).

19. Plate the bait clones on YEPAD using a sterilized plate replicator.

    Either the standard 96-spot or 384-spot format can be used, depending on the number of clones.

20. Perform yeast mating between the bait clones and the prey strain carrying empty vector as follows.
    i. Transfer the empty-vector prey culture from Step 18 to an empty Omnitray plate.
    ii. Touch the liquid culture with the pins of a sterilized plate replicator, then touch the pins directly onto the plated bait spots from Step 19.
    iii. Incubate the cells for 36–48 h at 30°C to allow mating.

21. To select for diploid colonies, transfer the colonies to plates containing the appropriate double-dropout selective medium (−LT) using a sterilized plate replicator. Grow the cells for 2–3 d at 30°C until the colonies are 1 mm in diameter.

22. To measure *HIS3* reporter gene activity, transfer the diploid colonies to plates containing the appropriate triple-dropout selective medium (−LTH) plus 3-AT at the following concentrations: 0, 1, and 3 mM.

    The initial 3-AT concentrations may be increased to 10, 25, 50, or 100 mM if necessary. Higher 3-AT concentrations are inadvisable as they will halt all growth.

Chapter 18

23. Incubate the selective plates for ~1 wk at 30°C. Record the lowest 3-AT concentration at which background colony growth is completely prevented.

The recorded concentration of 3-AT should be added to selective plates in the subsequent interaction screens to suppress reporter activation in the absence of an interacting prey.

Screening for Protein Interactions Using Genomic/cDNA Libraries/ORFeomes

In random library screening (Fig. 2), a single bait of interest is screened against a full library of prey proteins in a single screen. The interacting prey partner is identified by sequencing.

Preparing the Prey and Bait Cultures (Day 1)

24. Prepare an activated prey library as follows.

    i. Inoculate 200 mL of selective liquid medium with an aliquot of the yeast strain containing the prey library at 30°C.

    ii. Incubate the culture at 30°C overnight with shaking (180 rpm). After ~16 h (or the next day), measure the absorbance (OD_{600}) of the culture.

    The OD_{600} should reach 0.9–1.0 before mating.

25. Prepare a bait liquid culture and negative control as follows.

    i. Inoculate ~10 mL of selective liquid medium with the yeast strain containing the bait construct at 30°C.

    ii. In parallel, start a negative control culture by inoculating identical selective medium with yeast carrying empty prey vector.

    iii. Incubate overnight at 30°C with shaking (180 rpm).

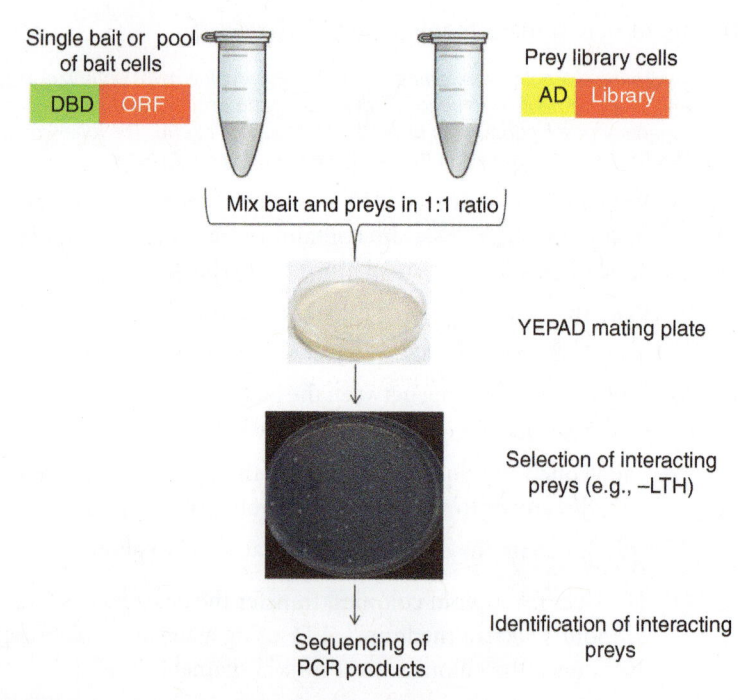

FIGURE 2. Y2H random library screen. The bait construct (DBD fusion protein) and the prey library (AD fusions with a cDNA or random genomic library) in haploid yeast strains are mixed in a 1:1 ratio. The mating culture is plated on a YEPAD agar plate. After incubation of mating plates for 6 h at 30°C, the diploid cells are transferred to –LTH plates (with or without 3-AT) for selection. The interacting preys are then identified by PCR amplification and sequencing of the vector inserts.

Mating Cells (Day 2)

26. Mix the bait and prey construct-containing cultures in equal proportion (1:1 ratio by volume) in 15-mL tubes. For every bait of interest, include a negative control by mating the bait with the empty-vector prey transformant at a 1:1 ratio.
27. Pellet the cells by centrifugation at ∼3000g for 2 min. Discard the medium.
28. Resuspend the cells in 500 μL of fresh YEPAD liquid medium.
29. Plate the cells on YEPAD solid medium. Incubate the plates for 6 h at 30°C or at room temperature overnight.
30. Wash the plate with 2 mL of sterile dH$_2$O to collect the cells. Pellet the cells as in Step 27.
31. Resuspend the cells in 2 mL of sterile dH$_2$O. Pellet the cells as in Step 27, and resuspend in 2 mL of selective medium (−LTH).
32. Check the mating efficiency as follows.

    i. Prepare a 1:100 dilution in liquid medium and plate the cells on −L, −T, and −LT plates.

    ii. Incubate the plates for 48–72 h at 30°C. Count the total number of resulting colonies on each type of selective medium.

    For both diploid (those on −LT) and haploid (those on −L or −T) colonies, viability is the colony count (cfu) per ml (adjusted by the 1:100 dilution factor). Mating efficiency can be calculated by dividing the diploid viability by the lower of the two haploid viability results.

    Viability of diploids is generally at least 95% that of haploid bait or prey cultures but varies by strain.

33. Plate 100–150 μL of cells on plates containing selective solid medium (−LTH with or without 3-AT).

    The appropriate concentration of 3-AT in selective medium should be determined using the bait self-activation test (Steps 18–23).

34. Incubate the plates for 6–8 d at 30°C or until the colonies are ∼1 mm in diameter.
35. Select Y2H-positive colonies. Identify the interacting prey partner by amplifying the insert in the prey vector using colony PCR (Rajagopala and Uetz 2011) followed by sequencing.

    Library screens usually yield dozens of positives that should be retested using array-based Y2H.

Screening for Protein Interactions Using Arrays

After baits are screened for potential self-activation (Steps 18–23), they may be tested for interactions versus the prey array as follows. Arrays are used as a primary screen or to validate the positives resulting from a library screen. Given the built-in controls present in array-based screens, they are the best choice for differentiating strong and weak interactions from background and efficiently reducing false positives. Usually, preys rather than baits are arrayed in a defined order, as described here, due to the high self-activation potential among baits; the activation domain (AD) used with prey proteins does not generally produce self-activation. An overview of array-based screening is provided in Figure 3.

Whether screening is manual or automated, a high-density (96-pin or 384-pin) replicating tool is used for routine transfer of colonies from one plate to another. In the procedure described here, the prey array is gridded on single-well (86 × 128-mm) Omnitray plates in a 96- or 384-colony format.

36. After preparation of the desired prey transformants (each carrying one specific AD-containing plasmid), array the preys on plates containing selective solid medium (−L) in either a 96- or 384- format in duplicates or quadruplicates.

    Baits can be arrayed likewise, as needed. Other array formats such as 768 or 1536 colonies per plate can be used, but the higher densities may render scoring positives more difficult.

    For long-term storage, all arrays may be cultured in rich (YEPAD) or selective liquid medium overnight, resuspended in 20% glycerol, and transferred to a freezer at −80°C. For day-to-day use, the arrays can be stored on solid selective medium (−L or −T) for up to 3 mo at 4°C. Fresh array plates may be produced using these storage plates before each screen. Copying the array onto fresh selective plates every 2 wk should protect against plasmid loss.

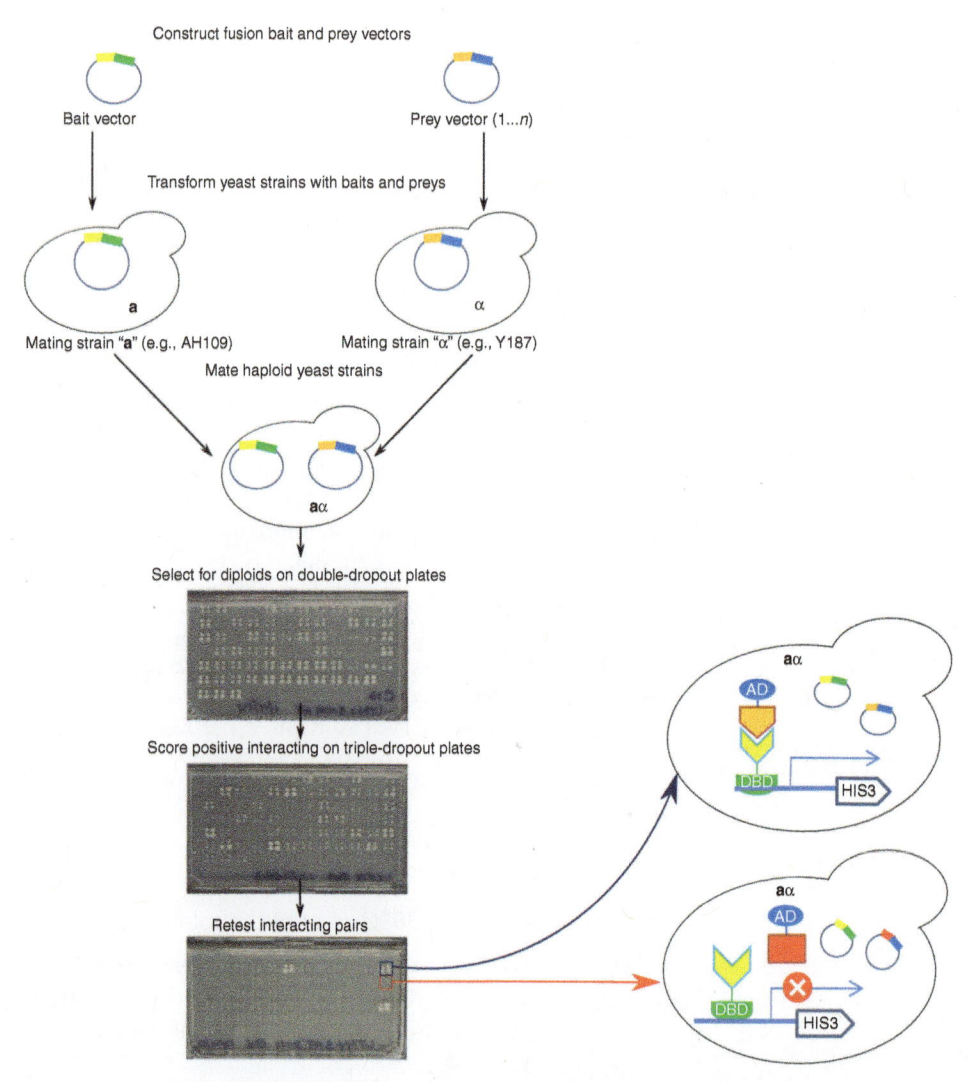

FIGURE 3. Array-based two-hybrid assays. Bait (pGBKCg, pGBGT7g) and prey (pGADCg, pGADT7g) constructs are transformed into haploid yeast cells of opposite mating type (**a** and α). Preys are arrayed on selective solid medium (−L). Baits should be tested for self-activation. The Y2H screening begins with mating, in array format, of two haploid strains carrying bait and prey plasmids. Diploid cells are selected by transferring the mated array to double-dropout (−LT) followed by triple-dropout (−LTH) plates. The rectangles on the selective plate (*bottom* panel) mark negative and positive interactions between the bait and the prey at these specific positions of the array.

37. Use the sterile plate replicator to transfer the yeast prey array to plates containing YEPAD solid medium. Grow the array overnight at 30°C.

    Alternatively, a fresh prey array can be prepared in YEPAD liquid medium in a 96-well plate. Note that only the master prey array (Step 36) should be used to make fresh copies for mating.

38. Transfer 15–20 mL of liquid YEPAD medium to a 50-mL tube. Inoculate the medium with the bait strain (or pool of baits) and grow overnight at 30°C in a shaking incubator (at ∼200 rpm).

    A bait pooling strategy may be used to reduce the time of Y2H screening. The pool of baits may be grown together in 15–20 ml YEPAD broth overnight.

    If the bait strains are frozen, they should be streaked or pinned on selective solid medium plates and grown for 1–2 d at 30°C before any experiment. A fresh bait culture should always be prepared for Y2H mating, as extended culturing in rich media may result in plasmid loss.

39. Dip the sterile plate replicator in the bait liquid culture and transfer the cells onto a plate containing YEPAD solid medium. Replicate as needed. Allow the spots to dry for 15–20 min.

    Ensure YEPAD plates are free of residual moisture before pinning to ensure consistent colony formation.

40. Touch the fresh prey array cultures from Step 37 with the sterilized plate replicator and transfer them directly onto the plated bait spots. Ensure that each of the bait spots receives different prey cells (i.e., a different AD-fusion protein).

41. Incubate the cells for 36–48 h at 30°C to allow mating.

    Supplemental adenine in the bait culture medium may increase the mating efficiency of some baits. Optimization of this step is subject to yeast strains, bait and prey vectors, and laboratory conditions.

42. To select diploid cells, transfer the colonies from the mating plates to plates containing double-dropout selective solid medium (–LT) using the sterilized plate replicator. Grow the cells for 2–3 d at 30°C until the colonies are 1 mm in diameter.

    This step ensures successful mating, as only diploid cells containing the selection markers in the prey and bait vectors will grow. Low mating efficiency will noticeably reduce the quality of screen results by obscuring signal and noise.

43. Transfer the colonies from –LT plates to plates containing triple-dropout selective solid medium (–LTH, with or without 3-AT) using the sterilized plate replicator. Incubate the plates for 6–8 d at 30°C.

    The choice of 3-AT concentrations will depend on the results of the bait self-activation test (Steps 18–23).

44. Examine the plates and score the positive interactions by counting colonies that are significantly large in size compared with background. If colonies have been plated in duplicate or quadruplicate, at least half of the colonies in each set should be significantly larger than background growth to indicate a positive result.

    Plates should be examined and monitored every day for positive colonies and contamination. Most Y2H positive colonies should appear within 3–5 d but some positive interactions may require more time. For most screens, observing >10 positive results per set of 96 different preys constitutes a suspicious result and may indicate false positives.

    See Troubleshooting.

TROUBLESHOOTING

Problem (Step 44): There are high background growth levels or false positives due to self-activation of baits.

Solution: Some baits can activate transcription of the reporter gene (e.g., *HIS3*) in the absence of any prey partner. Bait self-activation of the *HIS3* reporter gene can be suppressed by adding 3-AT to the selective medium. A concentration of 1–10 mM is usually sufficient to alleviate most self-activation, though up to 100 mM may be necessary in some cases.

Cloning the bait gene into a different vector can also reduce self-activation. Self-activating baits expressed as amino-terminal fusion products (i.e., in pGBGT7g) may show no self-activation at all when expressed as carboxy-terminal fusions (i.e., in pGBKCg). Alternatively, a bait construct can be modified to remove the region responsible for the self-activation. This step may be necessary for baits which strongly activate transcription.

Problem (Step 44): There are high background growth levels or false positives due to nonreproducible, seemingly random sets of positive results.

Solution: The main limiting factor associated with the Y2H assay in general is the potential for a large number of false positives. Without proper controls, >90% of all interactions in a Y2H screen can be the result of nonreproducible background (Uetz 2002). The following can be used to increase reliability and confidence in Y2H interaction data sets.

- Retesting all positive interactions by repeated mating usually reduces the number of false positives and provides confirmation of positive interactions. Interactions may be repeated in quadruplicate or by mating the interacting pair and scoring its signal strength with reference to empty prey vector.

- The use of different vector combinations can provide additional evidence for positive interactions in multiple screens, especially when protein interactions are expected to be inhibited by certain conditions (i.e., when a binding domain is only exposed in a particular binding conformation). This protocol was designed for use with combinations of the vectors used by Stellberger et al. (2010) (Fig. 1).

- Potential false-positives also can be retested by employing a full set of amino-terminal and carboxy-terminal protein fusions (Stellberger et al. 2010).

- Alternative methods, such as β-galactosidase assays, are also available for further confirmation of Y2H interactions; see the protocols by Serebriiskii et al. (2005) and Rajagopala et al. (2007).

RECIPES

Amino Acid Dropout Mixture for Y2H

Reagent	Quantity (for 1 L of medium)
Adenine	10 mg
L-Arginine	50 mg
L-Aspartic acid	80 mg
L-Histidine	20 mg
L-Isoleucine	50 mg
L-Leucine	100 mg
L-Lysine	50 mg
L-Methionine	20 mg
L-Phenylalanine	50 mg
L-Threonine	100 mg
L-Tryptophan	50 mg
L-Tyrosine	50 mg
L-Valine	140 mg
Uracil	20 mg

Combine the appropriate ingredients, and mix in a sealed container. Omit the desired dropout component. Haploid yeast requires a single dropout (i.e., −Trp=−T or −Leu= −L); diploid yeast requires double dropouts (i.e., −Leu −Trp=−LT); and selecting for interactions requires triple dropouts (i.e., −Leu−Trp −His=−LTH, plus 3-AT). Premixed amino acid mixtures are available as complete supplement mixtures (CSM) from commercial suppliers (e.g., Sunrise Science Products) and ensure consistency in amino acid concentrations.

Selective Solid Medium for Y2H

Reagent	Amount to add (for 1 L)	Final concentration (w/v)
Yeast nitrogen base without amino acids	1.7 g	0.17%
Ammonium sulfate	5 g	0.5%
Dextrose	20 g	2%
Agar	15 g	1.5%
Amino acid dropout mixture for Y2H <R>	0.69 g	Varies by amino acid
3-Amino-1,2,4-triazole (3-AT)	as needed	0–100 mM
dH$_2$O	to 1 L	

The choice of 3-AT concentration depends on the results of the bait self-activation test. After combining the ingredients, sterilize by autoclaving not more than 15 min/L. Pour 40–50 mL of sterilized medium per plate into single-well Omnitray microtiter plates (86 × 128 mm, with lids; Nunc) and allow to solidify.

YEPAD Solid Medium for Y2H

Reagent	Amount to add (for 1 L)	Final concentration (w/v)
Yeast extract	10 g	1%
Peptone	20 g	2%
Dextrose	20 g	2%
Adenine hemisulfate salt	100 mg	0.01%
Agar	16 g	1.6%
dH$_2$O	to 1 L	

Sterilize by autoclaving. Pour 40–50 mL of sterilized medium per plate into single-well Omnitray microtiter plates (86 × 128 mm, with lids; Nunc) and allow to solidify.

REFERENCES

Chen YC, Rajagopala SV, Stellberger T, Uetz P. 2010. Exhaustive benchmarking of the yeast two-hybrid system. *Nat Methods* **7:** 667–668.

Gong W, Shen YP, Ma LG, Pan Y, Du YL, Wang DH, Yang JY, Hu LD, Liu XF, Dong CX, et al. 2004. Genome-wide ORFeome cloning and analysis of *Arabidopsis* transcription factor genes. *Plant Physiol* **135:** 773–782.

Harper JW, Adami GR, Wei N, Keyomarsi K, Elledge SJ. 1993. The p21 Cdk-interacting protein Cip1 is a potent inhibitor of G1 cyclin-dependent kinases. *Cell* **75:** 805–816.

James P, Halladay J, Craig EA. 1996. Genomic libraries and a host strain designed for highly efficient two-hybrid selection in yeast. *Genetics* **144:** 1425–1436.

Lamesch P, Milstein S, Hao T, Rosenberg J, Li N, Sequerra R, Bosak S, Doucette-Stamm L, Vandenhaute J, Hill DE, et al. 2004. *C. elegans* ORFeome version 3.1: Increasing the coverage of ORFeome resources with improved gene predictions. *Genome Res* **14:** 2064–2069.

Landy A. 1989. Dynamic, structural, and regulatory aspects of λ site-specific recombination. *Annu Rev Biochem* **58:** 913–949.

Mehla J, Caufield JH, Uetz P. 2015. The yeast two-hybrid system: A tool for mapping protein–protein interactions. *Cold Spring Harb Protoc* doi: 10.1101/pdb.top083345.

Rain JC, Selig L, De Reuse H, Battaglia V, Reverdy C, Simon S, Lenzen G, Petel F, Wojcik J, Schachter V, et al. 2001. The protein–protein interaction map of *Helicobacter pylori*. *Nature* **409:** 211–215.

Rajagopala SV, Uetz P. 2011. Analysis of protein–protein interactions using high-throughput yeast two-hybrid screens. *Methods Mol Biol* **781:** 1–29.

Rajagopala SV, Titz B, Uetz P. 2007. Array-based yeast two-hybrid screening for protein–protein interactions. In *Yeast gene analysis*, 2nd ed. (Methods in Microbiology Series) (Stansfield I, Stark MJ), Vol. 36, pp. 139–163. Academic, London.

Rajagopala SV, Yamamoto N, Zweifel AE, Nakamichi T, Huang HK, Mendez-Rios JD, Franca-Koh J, Boorgula MP, Fujita K, Suzuki K, et al. 2010. The *Escherichia coli* K-12 ORFeome: A resource for comparative molecular microbiology. *BMC Genomics* **11:** 470.

Rajagopala SV, Sikorski P, Kumar A, Mosca R, Vlasblom J, Arnold R, Franca-Koh J, Pakala SB, Phanse S, Ceol A, et al. 2014. The binary protein–protein interaction landscape of *Escherichia coli*. *Nat Biotechnol* **32:** 285–290.

Rual JF, Hirozane-Kishikawa T, Hao T, Bertin N, Li S, Dricot A, Li N, Rosenberg J, Lamesch P, Vidalain PO, et al. 2004. Human ORFeome version 1.1: A platform for reverse proteomics. *Genome Res* **14:** 2128–2135.

Serebriiskii I, Golemis E, Uetz P. 2005. The yeast two-hybrid system for detecting interacting proteins. In *The proteomics handbook* (Walker JM), pp. 653–682. Humana, Totowa, NJ.

Stellberger T, Hauser R, Baiker A, Pothineni VR, Haas J, Uetz P. 2010. Improving the yeast two-hybrid system with permutated fusions proteins: The *Varicella Zoster* virus interactome. *Proteome Sci* **8:** 8.

Uetz P. 2002. Two-hybrid arrays. *Curr Opin Chem Biol* **6:** 57–62.

Uetz P, Dong YA, Zeretzke C, Atzler C, Baiker A, Berger B, Rajagopala SV, Roupelieva M, Rose D, Fossum E, et al. 2006. Herpesviral protein networks and their interaction with the human proteome. *Science* **311:** 239–242.

Walhout AJ, Temple GF, Brasch MA, Hartley JL, Lorson MA, van den Heuvel S, Vidal M. 2000. GATEWAY recombinational cloning: Application to the cloning of large numbers of open reading frames or ORFeomes. *Methods Enzymol* **328:** 575–592.

CHAPTER 19

Membrane Yeast Two-Hybrid (MYTH) Mapping of Full-Length Membrane Protein Interactions

Jamie Snider[1,4] and Igor Stagljar[1,2,3,4]

[1]*Donnelly Centre, University of Toronto, Toronto, Ontario M5S 3E1, Canada;* [2]*Department of Molecular Genetics, University of Toronto, Toronto, Ontario M5S 1A8, Canada;* [3]*Department of Biochemistry, University of Toronto, Toronto, Ontario M5S 1A8, Canada*

Mapping of protein interaction networks is a major strategy for obtaining a global understanding of protein function in cells and represents one of the primary goals of proteomics research. Membrane proteins, which play key roles in human disease and as drug targets, are of considerable interest; however, because of their hydrophobic nature, mapping their interactions presents significant technical challenges and requires the use of special methodological approaches. One powerful approach is the membrane yeast two-hybrid (MYTH) assay, a split-ubiquitin-based system specifically suited to the study of full-length membrane protein interactions in vivo using the yeast *Saccharomyces cerevisiae* as a host. The system can be used in both low- and high-throughput formats to study proteins from a wide range of different organisms. There are two primary variants of MYTH: integrated (iMYTH), which involves endogenous expression and tagging of baits and is suitable for studying native yeast membrane proteins, and traditional (tMYTH), which involves ectopic plasmid-based expression of tagged baits and is suitable for studying membrane proteins from other organisms. Here we provide an introduction to the MYTH assay, including both the iMYTH and tMYTH variants. MYTH can be set up in almost any laboratory environment, with results typically obtainable within 4 to 6 wk.

THE MYTH ASSAY

Integral membrane proteins are of incredible biological importance, playing roles in a diverse range of cellular processes (Engel and Gaub 2008). The functional association of many integral membrane proteins with disease has also made them popular drug targets, with a significant percentage of modern therapeutics directed toward the modulation of their function (Overington et al. 2006). Acquiring a greater understanding of membrane protein mechanism and function is therefore of great clinical interest.

A useful approach for increasing our understanding of protein function is through the identification of protein–protein interactions. Current high-throughput proteomics technologies allow for the generation of protein interaction maps (or "interactomes") on a previously unprecedented scale, providing us with a global view of the dynamic interplay between cellular systems (Petschnigg et al. 2011). One particularly powerful technology is the membrane yeast two-hybrid (MYTH) system, which enables high-throughput identification of the interactors of full-length integral membrane proteins in an in vivo environment (Iyer et al. 2005; Stagljar et al. 1998). The MYTH system makes use of the yeast *Saccharomyces cerevisiae* as a host and is suitable for studying proteins from many different organisms. The only requirements for candidate proteins are that (1) they can be effectively

---

[4]Correspondence: igor.stagljar@utoronto.ca; jamie.snider@utoronto.ca

Copyright © Cold Spring Harbor Laboratory Press; all rights reserved
Cite this introduction as *Cold Spring Harb Protoc*; doi:10.1101/pdb.top077560

expressed in yeast and (2) they have cytosolic amino- or carboxy-termini available for MYTH tagging. Unlike the conventional yeast two-hybrid assay, MYTH does not require the localization of proteins to the nucleus of the cell, and it is thus ideally suited for the study of full-length membrane proteins in the context of the membrane environment.

MYTH is a "split ubiquitin"–based protein complementation technology. Ubiquitin, a highly conserved 76-amino acid protein involved in 26S proteasome-mediated degradation of protein targets, can be split into two stable "moieties" referred to as NubI (the amino-terminal fragment) and Cub (the carboxy-terminal fragment). MYTH exploits the ability of these two stably divided moieties to functionally reassociate into a full-sized "pseudoubiquitin" molecule for use as a sensor of protein–protein interactions (Johnsson and Varshavsky 1994; Stagljar et al. 1998).

In MYTH, membrane protein "baits" are terminally fused to a tag consisting of the Cub fragment linked to a transcription factor (TF) composed of the LexA DNA-binding domain from *Escherichia coli* and the VP16 transcriptional activation domain of the herpes simplex virus. Preys, which can be either membrane bound or soluble, are terminally fused to a mutant version of the NubI fragment ("NubG") containing an isoleucine 13 to glycine mutation. This mutation prevents Nub and Cub from spontaneously reassociating, thereby necessitating an interaction between the bait and prey proteins to bring the Nub and Cub into close enough proximity to form pseudoubiquitin. Once pseudoubiquitin is formed, it is recognized by cytosolic deubiquitinating enzymes (DUBs), which cleave off the TF. The newly freed TF is directed to the nucleus of the cell, where it activates a reporter system, allowing for selection of cells in which a bait–prey interaction has occurred (Stagljar et al. 1998). A schematic diagram outlining the MYTH system is provided in Figure 1.

iMYTH AND tMYTH

There are currently two primary forms of MYTH—integrated and traditional. Integrated MYTH (iMYTH) involves endogenous, 3′ tagging of genes within the yeast chromosome. It is the preferred method for use with native yeast proteins possessing cytosolic carboxyl-terminus, as it leaves bait expression under the control of the corresponding natural promoter. This helps reduce problems sometimes associated with bait overexpression, such as a high false-positive rate and failure to

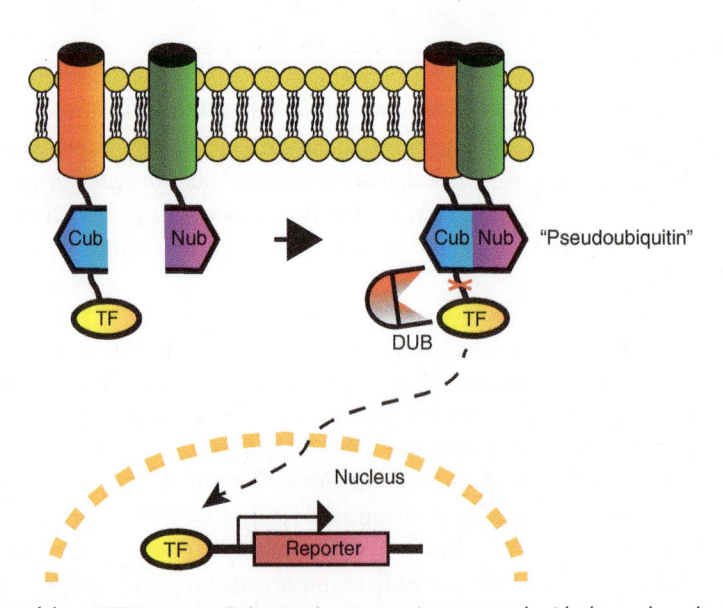

FIGURE 1. Principles of the MYTH system. Bait proteins (orange) are tagged with the carboxyl terminus of ubiquitin (Cub, blue) fused to a transcription factor (TF, yellow). Prey proteins (green) can be soluble or membrane-bound and are fused to the amino terminus of ubiquitin (Nub, purple). Interaction of bait and prey proteins allows for association of Cub and Nub into a full-size "pseudoubiquitin" molecule, which is recognized by cytosolic deubiquitinating enzymes (DUB, red). Subsequent DUB cleavage leads to release of the TF, which enters the nucleus and activates a reporter system, allowing for selection of cells in which a bait–prey interaction has occurred.

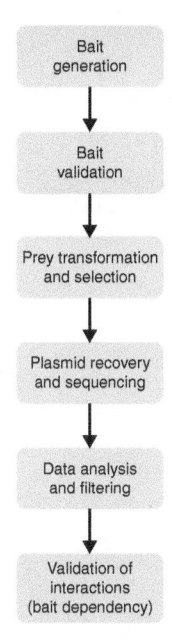

FIGURE 2. Major steps involved in MYTH screening.

properly localize to the membrane. Traditional MYTH (tMYTH) involves ectopic expression of tagged baits from plasmids and is used with proteins not native to yeast, as well as native yeast proteins with only cytosolic amino termini. It allows for both amino- and carboxy-terminal tagging of baits.

The MYTH system is readily scalable, and can be easily used in both high- and low-throughput applications. Examples of lower-throughput applications include validating known interactions, checking for the effects of mutations on protein binding, and screening for interactors within small protein subsets. High-throughput MYTH library screening studies, on the other hand, are invaluable for generating comprehensive interactome maps on a large scale (Snider et al. 2013; Gulati et al. 2015). A general overview of the major steps involved in a typical MYTH screening is shown in Figure 2.

We have provided two protocols for carrying out both variants of MYTH. Protocol 1: Generation and Validation of MYTH Baits: iMYTH and tMYTH Variants (Snider and Stagljar 2015a) describes the steps involved in the generation and validation of MYTH baits. Protocol 2: MYTH Screening: iMYTH and tMYTH Variants (Snider and Stagljar 2015b) provides detailed instructions on how to carry out high-throughput MYTH library screening. Note that the principles described for high-throughput analysis can be easily adapted for use in lower-throughput formats. The results from MYTH screening provide valuable information about how specific membrane proteins fit into cellular systems, and should serve as an excellent first step toward more detailed studies of protein mechanism and function.

REFERENCES

Engel A, Gaub HE. 2008. Structure and mechanics of membrane proteins. *Annu Rev Biochem* 77: 127–148. doi:10.1146/annurev.biochem.77.062706.154450.

Gulati S, Balderes D, Kim C, Guo ZA, Wilcox L, Area-Gomez E, Snider J, Wolinski H, Stagljar I, Granato JT, et al. 2015. ATP-binding cassette transporters and sterol O-acyltransferases interact at membrane microdomains to modulate sterol uptake and esterification. *FASEB J* doi:10.1096/fj.14-264796.

Iyer K, Bürkle L, Auerbach D, Thaminy S, Dinkel M, Engels K, Stagljar I. 2005. Utilizing the split-ubiquitin membrane yeast two-hybrid system to identify protein-protein interactions of integral membrane proteins. *Sci STKE* 2005 Mar 15; **2005**: pl3. doi:10.1126/stke.2752005pl3.

Johnsson N, Varshavsky A. 1994. Split ubiquitin as a sensor of protein interactions in vivo. *Proc Natl Acad Sci* 91: 10340–10344.

Overington JP, Al-Lazikani B, Hopkins AL. 2006. How many drug targets are there? *Nat Rev Drug Discov* 5: 993–996. doi:10.1038/nrd2199.

Petschnigg J, Snider J, Stagljar I. 2011. Interactive proteomics research technologies: Recent applications and advances. *Curr Opin Biotechnol* 22: 50–58. doi:10.1016/j.copbio.2010.09.001.

Snider S, Stagljar I. 2015a. Generation and validation of MYTH baits: iMYTH and tMYTH variants. *Cold Spring Harb Protoc* doi:10.1101/pdb.prot087817.

Snider S, Stagljar I. 2015b. MYTH screening: iMYTH and tMYTH variants. *Cold Spring Harb Protoc* doi:10.1101/pdb.prot087825.

Snider J, Hanif A, Lee ME, Jin K, Yu AR, Graham C, Chuk M, Damjanovic D, Wierzbicka M, Tang P, et al. 2013. Mapping the functional yeast ABC transporter interactome. *Nat Chem Biol* 9: 565–572.

Stagljar I, Korostensky C, Johnsson N, te Heesen S. 1998. A genetic system based on split-ubiquitin for the analysis of interactions between membrane proteins in vivo. *Proc Natl Acad Sci* 95: 5187–5192.

Protocol 1

Generation and Validation of MYTH Baits: iMYTH and tMYTH Variants

Jamie Snider[1,4] and Igor Stagljar[1,2,3,4]

[1]*Donnelly Centre, University of Toronto, Toronto, Ontario M5S 3E1, Canada;* [2]*Department of Molecular Genetics, University of Toronto, Toronto, Ontario M5S 1A8, Canada;* [3]*Department of Biochemistry, University of Toronto, Toronto, Ontario M5S 1A8, Canada*

Generation of baits for membrane yeast two-hybrid (MYTH) screening differs depending on the nature of the protein(s) being studied. When using native yeast proteins with cytoplasmic carboxyl termini, the integrated form of MYTH (iMYTH) is the method of choice. iMYTH involves endogenous carboxy-terminal tagging of the gene of interest within the yeast chromosome, leaving the gene under the control of its natural promoter. When studying proteins not native to yeast, or native yeast proteins with only cytoplasmic amino termini, traditional MYTH (tMYTH) must be used. In the tMYTH approach, amino- or carboxy-terminally tagged proteins are expressed ectopically from a plasmid. In this protocol, we describe the generation and validation of iMYTH and tMYTH baits. MYTH bait generation can typically be completed in ∼1–2 wk.

MATERIALS

It is essential that you consult the appropriate Material Safety Data Sheets and your institution's Environmental Health and Safety Office for proper handling of equipment and hazardous materials used in this protocol.

RECIPES: Please see the end of this protocol for recipes indicated by <R>. Additional recipes can be found online at http://cshprotocols.cshlp.org/site/recipes.

Reagents

α-LexA/α-VP16 antibody and reagents for immunofluorescence (as needed; see Step 18)
Glycerol
NubI and NubG control prey plasmid DNA (available from Dr. Igor Stagljar)
PCR reagents
Reagents for iMYTH

    G418 (200 mg/mL stock solution in sterile ddH$_2$O)
    KanMX reverse primer (5′-GAGCGTTTCCCTGCTCGCAG-3′)
    L2 or L3 integration cassette plasmid DNA (available from Dr. Igor Stagljar)
        Both the L2 and L3 plasmids encode integration cassettes consisting of a MYTH tag followed by the KanMX gene (for selection of transformants). The L3 MYTH tag also includes YFP.
    L2 or L3 sequencing primers
        L2 reverse primer (5′-GCCGTTAACGCTTTCATGC-3′)

[4]Correspondence: igor.stagljar@utoronto.ca; jamie.snider@utoronto.ca

L3 reverse primer (5′-TTGTGCCCATTAACATCACC-3′)

Oligonucleotide primer for bait gene of interest (forward, internal)
A suitable forward internal primer is typically 18–20 bases in length and 100- to 200-bp upstream of the end of the open reading frame (ORF) of the gene of interest.

Oligonucleotide primers for cassette amplification
Instructions for primer design are provided in Figure 1B.

Yeast genomic DNA extraction kit
YPAD liquid medium <R>

Reagents for tMYTH
    DNA plasmid miniprep kit

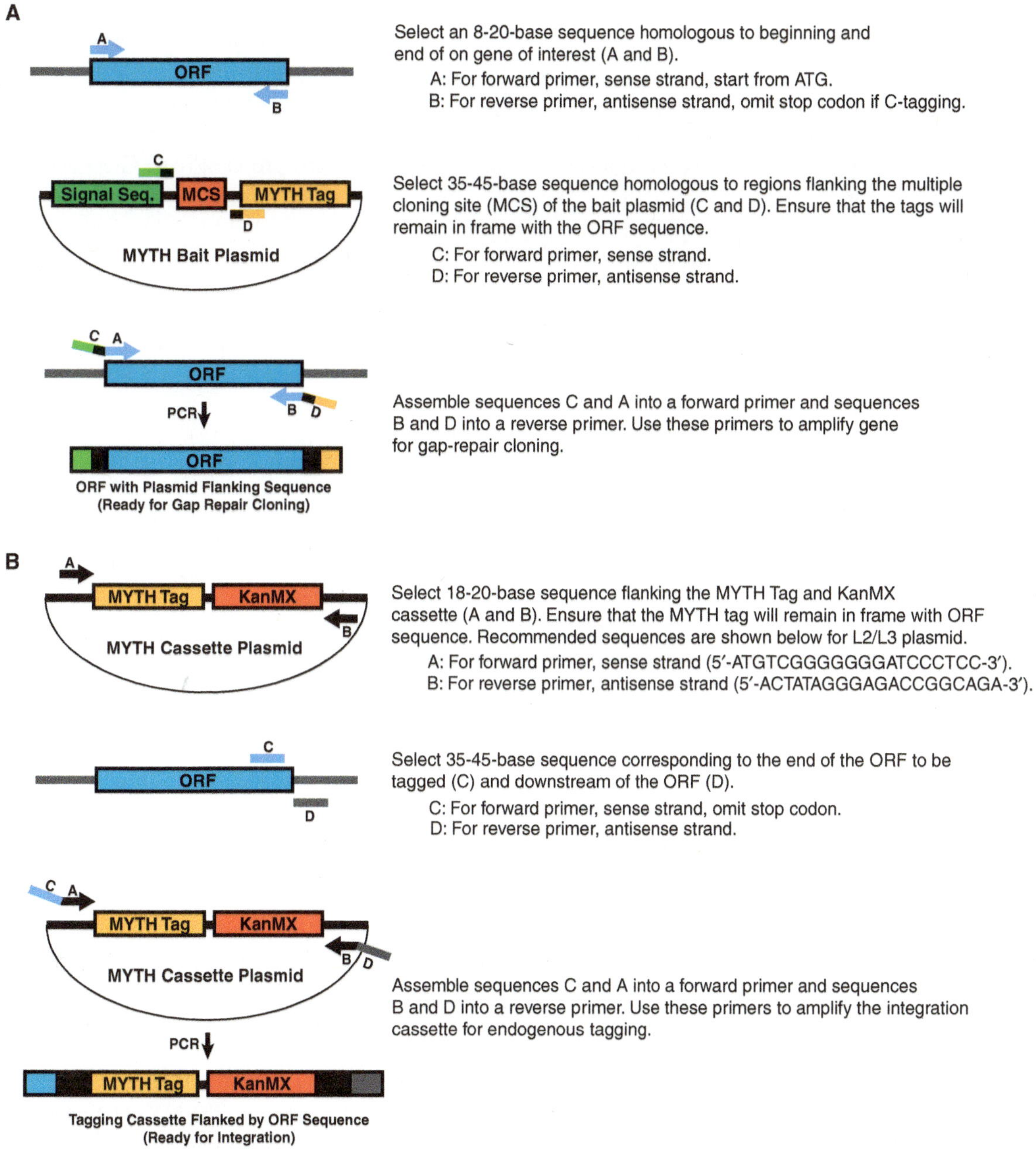

FIGURE 1. Primer design for MYTH bait generation. (*A*) tMYTH primer design. (*B*) iMYTH primer design.

Chapter 19

Escherichia coli (competent, for plasmid propagation)

Kanamycin (50 mg/mL stock solution in ddH$_2$O, sterilized by filtration)

LB solid or liquid medium <R>

MYTH bait vector DNA (available from Dr. Igor Stagljar)
> *Either an amino- or carboxy-terminal tagging vector can be selected. The MYTH tag must be cytosolic. Note that bait performance can vary considerably with expression level and is sometimes improved by including specific yeast signal or leader sequences. As such, it may be necessary to test various vector backbones to find the optimal configuration for use in the screening of your particular bait.*

Oligonucleotide primers for gap-repair cloning of the tMYTH bait ORF
> *Instructions for primer design are provided in Figure 1A.*

Oligonucleotide primers for sequence verification of cloned baits

Restriction enzyme for bait vector digestion

Source DNA for gene of interest, for generating bait protein

Saccharomyces cerevisiae NMY51 MYTH reporter strain (*MATa his3delta200 trp1-901 leu2-3,112 ade2 LYS2::(lexAop)$_4$-HIS3 ura3::(lexAop)$_8$-lacZ (lexAop)$_8$-ADE2 GAL4*)

Synthetic drop-out (SD) media, prepared with 3-AT (3-amino-1,2,4-triazole) as needed (see Step 17) <R>

Yeast DNA transformation reagents for a standard protocol (e.g., Gietz and Woods 2006)

Equipment

Fluorescence microscope

Shaking and standing incubators at 30°C and 37°C

Soda lime glass beads (0.5 mm) and vortex (for tMYTH)

Thermocycler for PCR

METHOD

For background on MYTH, see Introduction: Membrane Yeast Two-Hybrid (MYTH) Mapping of Full-Length Membrane Protein Interactions (Snider and Stagljar 2015a). To perform the MYTH assay, you MUST select a bait protein with its amino and/or carboxyl terminus (i.e., the site of MYTH-tagging) in the cytosol to ensure that the MYTH tag is accessible to cytosolic deubiquitinating enzymes. If a membrane protein does not meet this criterion it cannot be screened in its full-length form using the MYTH system.

Once a bait protein has been chosen, the suitable MYTH variant (tMYTH or iMYTH) can be selected. iMYTH involves endogenous carboxy-terminal tagging of baits in the yeast chromosome and is the method of choice for proteins native to yeast that have a cytosolic carboxyl terminus. The key advantage of iMYTH is that bait expression remains under the control of its native promoter, helping to reduce problems sometimes associated with overexpression, such as improper localization and a higher false-positive rate. tMYTH involves ectopic, plasmid-based expression of baits and is used with nonyeast proteins, or native yeast proteins with only a cytosolic amino terminus. The key advantages of tMYTH are that it allows both amino- and carboxy-terminal tagging and can be used to screen a much wider range of proteins.

Generating a Bait Protein

If performing tMYTH, follow Steps 1–9 for bait generation. If performing iMYTH, follow Steps 10–15.

tMYTH Bait Generation

1. PCR-amplify the target gene DNA using primers suitable for gap-repair homologous recombination into the selected MYTH bait vector (Ma et al. 1987) (Fig. 1A).

2. Digest the MYTH bait vector DNA with an appropriate restriction enzyme.

3. Transform the PCR product and digested vector into the MYTH NMY51 reporter strain following a standard protocol. Grow the cells on plates containing SD-Leu solid medium for 2–3 d at 30°C in a standing incubator.

4. Grow a single colony in SD-Leu liquid medium overnight at 30°C in a shaking incubator. Use this culture to isolate plasmid DNA using a commercial miniprep kit with the following modification.

   i. To help disrupt the yeast cell wall and ensure efficient cell lysis, add a small volume of 0.5-mm soda lime glass beads to the cells in the initial miniprep resuspension buffer and vortex vigorously for 5–10 min.

   ii. Proceed with the miniprep protocol as normal.

5. Transform the yeast miniprep DNA into a competent *E. coli* strain suitable for plasmid propagation. Grow the cells on plates containing LB solid medium with kanamycin (50 µg/mL) overnight at 37°C in a standing incubator.

   Passage through E. coli *substantially increases DNA yield and purity.*

6. Grow a single colony from Step 5 in LB liquid medium with kanamycin (50 µg/mL) overnight at 37°C in a shaking incubator.

7. Using the culture from Step 6, generate a glycerol stock and isolate the plasmid DNA using a commercial miniprep kit.

8. Sequence the purified plasmid to verify proper bait construction.

9. Transform the sequenced bait plasmid back into the NMY51 yeast reporter strain for bait validation (Steps 16–18) and screening (see Protocol 2: MYTH Screening: iMYTH and tMYTH Variants [Snider and Stagljar 2015b]).

iMYTH Bait Generation

10. PCR-amplify the tagging cassette from L2 or L3 plasmid DNA using appropriate primers (Fig. 1B).

11. Transform the PCR product into the NMY51 MYTH reporter strain using a standard protocol. Grow cells on plates containing YPAD solid medium with G418 (200 µg/mL) for 2–3 d at 30°C in a standing incubator.

12. Grow a single colony from Step 11 in YPAD liquid medium overnight at 30°C in a shaking incubator.

13. Use the culture from Step 12 to generate a glycerol stock and isolate genomic DNA using a standard protocol or commercial kit.

14. PCR-amplify the tagging site from the genomic DNA using a forward internal primer specific to your bait gene of interest and the KanMX reverse primer.

15. Sequence the PCR product using the L2 or L3 reverse sequencing primer to verify proper tagging before proceeding to bait validation (Steps 16–18) and screening (see Protocol 2: MYTH Screening: iMYTH and tMYTH Variants [Snider and Stagljar 2015b]).

Validating the Bait Protein

Once the bait strain (i.e., the NMY51 MYTH reporter strain expressing the tagged bait protein) has been shown to both pass the NubGI test and display proper membrane localization as described below, it is suitable for use in MYTH screening as described in Protocol 2: MYTH Screening: iMYTH and tMYTH Variants (Snider and Stagljar 2015b).

Performing NubGI Testing

16. Transform the bait strain with 100–200 ng of NubI "positive" and NubG "negative" control prey plasmid DNA using a standard protocol. Plate the cells onto SD-Trp-Leu-Ade-His (tMYTH) or SD-Trp-Ade-His (iMYTH) solid medium. Grow for 2–3 d at 30°C in a standing incubator.

    To be suitable for screening, bait strains must grow on SD-Trp-Leu-Ade-His or SD-Trp-Ade-His medium only when transformed with the NubI, but not the NubG, control prey.

17. If growth occurs in the NubG strain, repeat the test using medium containing 3-AT. Only proceed with screening if NubG growth can be eliminated using 75 mM 3-AT or less.

   3-AT inhibits the activity of the His3p reporter enzyme and is useful for titrating out background growth resulting from basal expression of the HIS3 gene.

Verifying Subcellular Localization

18. Verify the bait is properly localized to the membrane using fluorescence microscopy.

   If using a fluorescent MYTH tag (e.g., L3 cassette), live cells can be visualized directly. Note that the fluorescent MYTH tag is bulkier than the standard tag, and may prevent detection of certain interactions due to increased steric hindrance. If not using a fluorescent tag, a standard immunofluorescence staining protocol (e.g., Hasek 2006) with a primary antibody against the LexA or VP16 components of the MYTH tag can be used.

RECIPES

Drop-Out Solution (10×)

300 mg	isoleucine
1500 mg	valine
400 mg	adenine sulfate dihydrate
200 mg	histidine monohydrochloride
1000 mg	leucine
300 mg	lysine
1500 mg	methionine
500 mg	phenylalanine
2000 mg	threonine
400 mg	tryptophan
300 mg	tyrosine
200 mg	uracil
200 mg	arginine hydrochloride

Combine the ingredients, omitting components as required. Bring to 1 L with ddH$_2$O. Autoclave and store at 4°C.

LB Solid or Liquid Medium

10 g	tryptone
5 g	yeast extract
5 g	NaCl
20 g	agar (for solid medium only)

Combine the ingredients and bring to 1 L with ddH$_2$O. Autoclave and store at room temperature or 4°C.

Sodium Phosphate Solution

70 g	Na$_2$HPO$_4$
30 g	NaH$_2$PO$_4$ · H$_2$O

Combine the ingredients and bring to 1 L with ddH$_2$O. Autoclave and store at room temperature.

Synthetic Drop-Out (SD) Media

100 mL	drop-out solution (10×) <R>
6.7 g	yeast nitrogen base

20 g	D-glucose
20 g	agar (solid medium only)
100 mL	sodium phosphate solution <R>
0.8 mL	X-gal (5-bromo-4-chloro-3-indolyl-β-D-galactopyranoside) solution (100 mg/mL in N,N-dimethyl formamide)
x mL	3-AT (3-amino-1,2,4-triazole) solution (1 M in ddH$_2$O, sterilized by filtration)

To prepare 1 L of SD medium, combine the first four ingredients, omitting components from the drop-out solution as required. Bring to 1 L with ddH$_2$O. Autoclave, and store at room temperature or 4°C.

To prepare SD medium with X-gal, combine the first four ingredients and bring to only 900 mL with ddH$_2$O. Autoclave. After the medium has cooled to ~50°C, add the sodium phosphate solution and X-gal. Store for up to 2 wk at 4°C.

To prepare SD medium with 3-AT, combine the first four ingredients in a volume of ddH$_2$O that has been reduced by the volume of 3-AT solution needed to obtain the desired final concentration. Autoclave. Add 3-AT after medium has cooled to ~50°C. Store for up to 2 wk at 4°C.

YPAD Liquid Medium

10 g	Yeast extract
20 g	Peptone
20 g	Glucose
40 mg	Adenine sulfate dihydrate

Combine the ingredients and bring to 1 L with ddH$_2$O. Autoclave and store at room temperature or 4°C.

REFERENCES

Gietz RD, Woods RA. 2006. Yeast transformation by the LiAc/SS Carrier DNA/PEG method. *Methods Mol Biol* **313**: 107–120. doi: 10.1385/1-59259-958-3:107.

Hasek J. 2006. Yeast fluorescence microscopy. *Methods Mol Biol* **313**: 85–96. doi: 10.1385/1-59259-958-3:085.

Ma H, Kunes S, Schatz PJ, Botstein D. 1987. Plasmid construction by homologous recombination in yeast. *Gene* **58**: 201–216.

Snider S, Stagljar I. 2015a. Membrane yeast two-hybrid (MYTH) mapping of full-length membrane protein interactions. *Cold Spring Harb Protoc* doi: 10.1101/pdb.top077560.

Snider S, Stagljar I. 2015b. MYTH screening: iMYTH and tMYTH variants. *Cold Spring Harb Protoc* doi: 10.1101/pdb.prot087825.

Protocol 2

MYTH Screening: iMYTH and tMYTH Variants

Jamie Snider[1,4] and Igor Stagljar[1,2,3,4]

[1]Donnelly Centre, University of Toronto, Toronto, Ontario M5S 3E1, Canada; [2]Department of Molecular Genetics, University of Toronto, Toronto, Ontario M5S 1A8, Canada; [3]Department of Biochemistry, University of Toronto, Toronto, Ontario M5S 1A8, Canada

Once a bait has been generated and validated for the membrane yeast two-hybrid (MYTH) assay, it can be used for either high-throughput screening to generate a detailed interaction map (interactome) or in low-throughput experiments to examine interactions with specific targets. Here we describe how to carry out high-throughput MYTH library screening of a validated bait generated using integrated or traditional MYTH. The principles herein can be easily adapted for use in a smaller-scale format if required. A typical MYTH library screen can be completed in ~3–4 wk.

MATERIALS

It is essential that you consult the appropriate Material Safety Data Sheets and your institution's Environmental Health and Safety Office for proper handling of equipment and hazardous materials used in this protocol.

RECIPES: Please see the end of this protocol for recipes indicated by <R>. Additional recipes can be found online at http://cshprotocols.cshlp.org/site/recipes.

Reagents

Ampicillin (100 mg/mL stock solution in ddH$_2$O, sterilized by filtration)
Dimethyl sulfoxide (DMSO) (>99%)
DNA plasmid miniprep kit (96-well capacity)
Escherichia coli (competent, for plasmid propagation)
LB solid or liquid medium <R>
MYTH prey library DNA (available from Dr. Igor Stagljar)
NaCl (0.9% in ddH$_2$O, sterilized by autoclaving)
NubG sequencing primers

> Forward primer (5′-CCGATACCATCGACAACGTTAAGTCG-3′)
>
> Reverse primer (5′-CGACTTAACGTTGTCGATGGTATCGG-3′)
> *When using libraries expressing protein with the NubG tag at the amino terminus, the forward sequencing primer should be used. For libraries expressing protein with the NubG tag at the carboxyl terminus, the reverse sequencing primer should be used.*

Saccharomyces cerevisiae NMY51 MYTH reporter strain expressing the tagged bait protein, prepared as described in Protocol 1: Generation and Validation of MYTH Baits: iMYTH and tMYTH Variants (Snider and Stagljar 2015a)
> *In addition, express an unrelated tagged control bait in the NMY51 MYTH reporter strain.*

---

[4]Correspondence: igor.stagljar@utoronto.ca; jamie.snider@utoronto.ca

Copyright © Cold Spring Harbor Laboratory Press; all rights reserved
Cite this protocol as *Cold Spring Harb Protoc*; doi:10.1101/pdb.prot087825

Salmon sperm DNA solution <R>

Synthetic drop-out (SD) medium, prepared with X-gal or 3-amino-1,2,4-triazole (3-AT) as needed <R>

If 3-AT was needed for your bait to pass the NubGI test (see Protocol 1: Generation and Validation of MYTH Baits: iMYTH and tMYTH Variants [Snider and Stagljar 2015a]), include it in SD medium at the same concentration where indicated.

Transformation solution I, freshly prepared <R>

Transformation solution II, freshly prepared <R>

Yeast DNA transformation reagents for a standard protocol (e.g., Gietz and Woods 2006)

YPAD medium (for bait strains generated using iMYTH) <R>

YPAD medium (2×) <R>

Zymolyase

Equipment

Centrifuge tubes, screw-cap (15- and 50-mL)

Centrifuges (clinical and tabletop) at 4°C

Culture blocks (96-well) for cell growth (sterile)

Microcentrifuge tubes (1.5-mL, sterile)

Robotic setup for automated colony picking (optional; see Step 21)

Shaking and standing incubators at 30°C and 37°C

Spectrophotometer

Standing water baths at 30°C and 42°C

Vortex

METHOD

The following protocol can be used to screen baits generated using integrated MYTH (iMYTH) or traditional MYTH (tMYTH); see Introduction: Membrane Yeast Two-Hybrid (MYTH) Mapping of Full-Length Membrane Protein Interactions (Snider and Stagljar 2015b).

Transforming the Library

1. Grow a 4-mL overnight culture of the bait strain (i.e., the NMY51 MYTH reporter strain expressing the tagged bait protein) in YPAD (iMYTH) or SD-Leu (tMYTH) medium at 30°C in a shaking incubator.

2. Using this overnight culture, inoculate 200 mL of YPAD (iMYTH) or SD-Leu (tMYTH) medium at an initial OD_{600} of 0.15. Grow at 30°C in a shaking incubator until an OD_{600} of 0.6–0.8 is reached (∼4–5 h).

3. Divide the 200 mL culture into 4 × 50-mL screw-cap tubes and centrifuge at 700g for 5 min at 4°C. Discard the supernatant.

4. Resuspend each pellet in 40 mL of sterile ddH_2O.

   This wash step can help improve transformation efficiency.

5. Centrifuge at 700g for 5 min at 4°C. Discard the supernatant.

6. Resuspend each cell pellet in 1 mL of transformation solution I and transfer the samples to sterile 1.5 mL tubes.

7. Centrifuge at 700g for 5 min at 4°C. Discard the supernatant.

8. Resuspend each cell pellet in 600 µL of transformation solution I and transfer the samples to 15-mL screw-cap tubes.

Chapter 19

9. To each tube, add 2.5 mL of transformation solution II, 100 µL of salmon sperm DNA solution, and 10 µg of MYTH prey library DNA. Vortex vigorously for 30–60 sec.
10. Incubate the tubes in a standing water bath for 45 min at 30°C. Vortex briefly every 15 min during incubation.
11. Add 160 µL of DMSO to each tube and mix immediately by inversion.
12. Incubate the tubes in a standing water bath for 20 min at 42°C.
13. Centrifuge at 1500g for 5 min at 4°C. Discard the supernatant.
14. Resuspend each cell pellet in 3 mL of 2× YPAD medium. Pool the samples together in a single 50-mL screw-cap tube.
15. Incubate the tube for 90 min at 30°C in a shaking incubator.
16. Centrifuge at 700g for 5 min at 4°C. Discard the supernatant.
17. Resuspend the cell pellet in 4.9 mL of 0.9% NaCl solution.
18. Prepare 10×, 100×, and 1000× serial dilutions of cells in 0.9% NaCl. Plate 100 µL volumes onto SD-Trp (iMYTH) or SD-Trp-Leu (tMYTH) solid medium. Grow overnight at 30°C and count the number of transformants obtained.

    The total number of transformants is a measure of the "coverage" of your screen. Coverage should equal, and preferably exceed, the complexity of the library. The more the coverage exceeds the library complexity, the greater the likelihood that all unique prey constructs within the library have been screened.

    $$\text{Total Transformants} = \#\text{Colonies on Plate} \times \text{Dilution Factor} \times 49$$

19. Divide the remaining cell suspension (from Step 17) equally and plate onto SD-Trp-Ade-His (iMYTH) or SD-Trp-Leu-Ade-His (tMYTH) solid medium, including 3-AT if used previously for the NubGI test. Grow the transformants for 3–4 d at 30°C in a standing incubator.

    Typically, ten 150-mm plates are sufficient for obtaining isolated colonies.

Performing Secondary Screening/Prey Identification

20. Resuspend single colonies from Step 19 in 150 µL volumes of sterile 0.9% NaCl solution and spot onto plates containing SD-Trp-Ade-His + X-gal (iMYTH) or SD-Trp-Leu-Ade-His + X-gal (tMYTH) solid medium, including 3-AT if used previously. Grow the cells for 3–4 d at 30°C in a standing incubator.

    This step is a secondary validation using the β-galactosidase reporter.

21. Select the colonies that turn blue AND display robust growth. Grow these up overnight in 96-well blocks containing 1.2 mL of SD-Trp medium per well at 30°C in a shaking incubator.

    If available, the use of robotics to assist in colony picking is helpful if a large number of transformants are obtained.

22. Isolate plasmid DNA from the overnight cultures using a commercial 96-well miniprep kit.

    To help disrupt the yeast cell wall, it is advisable to treat cells with Zymolyase before proceeding with the miniprep.

23. Transform the yeast miniprep DNA into a competent *E. coli* strain suitable for plasmid propagation. Grow the cells on LB solid medium with ampicillin (100 µg/mL) overnight at 37°C in a standing incubator.

    Passage through E. coli *substantially increases DNA yield and purity.*

    It is recommended to carry out transformations in a 96-well format to maximize efficiency.

24. Grow up single colonies from Step 23 in 96-well blocks containing 1.2 mL of LB medium with ampicillin (100 µg/mL) per well overnight at 37°C in a shaking incubator.
25. Isolate the plasmid DNA from the overnight cultures using a commercial 96-well miniprep kit.

26. Sequence the plasmid DNA using the NubG forward or reverse sequencing primer as appropriate for your library.
27. Identify preys using BLAST analysis.

    Be careful to verify that detected sequences are properly in frame with the NubG sequence. This ensures that the correct gene is identified and eliminates spurious interactions with short peptide sequences.

Bait-Dependency Testing

28. Retransform the interactor DNA into the original bait strain and an unrelated control bait strain using a standard protocol. Spot onto SD-Trp-Ade-His + XGal (iMYTH) or SD-Trp-Leu-Ade-His + XGal (tMYTH) solid medium as in Step 20, including 3-AT if used previously. Grow the cells for 3–4 d at 30°C.
29. Remove spurious preys that produce robust growth and blue color in the unrelated control bait strain. In addition, eliminate any preys that fail to interact on retransformation into the original bait strain.
30. Assemble the remaining preys into your interactome.

    This interactome represents a powerful tool for identifying the interaction partners of your bait of interest. Interactome quality should be further assessed using bioinformatics analysis and orthogonal interaction assays (e.g., co-immunoprecipitation, other protein complementation approaches). The final interactome should serve as a useful starting point for in-depth functional follow-up studies.

RECIPES

Drop-Out Solution (10×)

300 mg	isoleucine
1500 mg	valine
400 mg	adenine sulfate dihydrate
200 mg	histidine monohydrochloride
1000 mg	leucine
300 mg	lysine
1500 mg	methionine
500 mg	phenylalanine
2000 mg	threonine
400 mg	tryptophan
300 mg	tyrosine
200 mg	uracil
200 mg	arginine hydrochloride

Combine the ingredients, omitting components as required. Bring to 1 L with ddH$_2$O. Autoclave and store at 4°C.

LB Solid or Liquid Medium

10 g	tryptone
5 g	yeast extract
5 g	NaCl
20 g	agar (for solid medium only)

Combine the ingredients and bring to 1 L with ddH$_2$O. Autoclave and store at room temperature or 4°C.

Salmon Sperm DNA Solution

Dissolve 200 mg of salmon sperm DNA in 100 mL of sterile ddH$_2$O. Prepare aliquots, boil for 5 min at 100°C, and transfer to ice. Store aliquots at −20°C.

Chapter 19

Sodium Phosphate Solution

70 g	Na$_2$HPO$_4$
30 g	NaH$_2$PO$_4 \cdot$ H$_2$O

Combine the ingredients and bring to 1 L with ddH$_2$O. Autoclave and store at room temperature.

Synthetic Drop-Out (SD) Media

100 mL	drop-out solution (10×) <R>
6.7 g	yeast nitrogen base
20 g	D-glucose
20 g	agar (For solid medium only)
100 mL	sodium phosphate solution <R>
0.8 mL	X-gal (5-bromo-4-chloro-3-indolyl-β-D-galactopyranoside) solution (100 mg/mL in N,N-dimethyl formamide)
x mL	3-AT (3-amino-1,2,4-triazole) solution (1 M in ddH$_2$O, sterilized by filtration)

To prepare 1 L of SD medium, combine the first four ingredients, omitting components from the drop-out solution as required. Bring to 1 L with ddH$_2$O. Autoclave, and store at room temperature or 4°C.

To prepare SD medium with X-gal, combine the first four ingredients and bring to only 900 mL with ddH$_2$O. Autoclave. After the medium has cooled to ~50°C, add the sodium phosphate solution and X-gal. Store for up to 2 wk at 4°C.

To prepare SD medium with 3-AT, combine the first four ingredients in a volume of ddH$_2$O that has been reduced by the volume of 3-AT solution needed to obtain the desired final concentration. Autoclave. Add 3-AT after medium has cooled to ~50°C. Store for up to 2 wk at 4°C.

Transformation Solution I

To prepare 10 mL, combine the following. Use sterile components.

1.1 mL	lithium acetate (1 M)
1.1 mL	TE (100 mM Tris, 10 mM EDTA, pH 7.5)
7.8 mL	ddH$_2$O

Transformation Solution II

To prepare 15 mL, combine the following. Use sterile components.

1.5 mL	lithium acetate (1 M)
1.5 mL	TE (100 mM Tris, 10 mM EDTA, pH 7.5)
12 mL	Polyethylene glycol 3350 (50%) prepared in ddH$_2$O and sterilized by autoclaving.

YPAD Liquid Medium

10 g	yeast extract
20 g	peptone
20 g	glucose
40 mg	adenine sulfate dihydrate

Combine the ingredients and bring to 1 L with ddH$_2$O. Autoclave and store at room temperature or 4°C.

YPAD Liquid Medium (2×)

20 g	yeast extract
40 g	peptone
40 g	glucose
40 mg	adenine sulfate dihydrate

Combine the ingredients and bring to 1 L with ddH$_2$O. Autoclave and store at room temperature or 4°C.

REFERENCES

Gietz RD, Woods RA. 2006. Yeast transformation by the LiAc/SS Carrier DNA/PEG method. *Methods Mol Biol* **313**: 107–120. doi:10.1385/1-59259-958-3:107.

Snider S, Stagljar I. 2015a. Generation and validation of MYTH baits: iMYTH and tMYTH variants. *Cold Spring Harb Protoc* doi:10.1101/pdb.prot087817.

Snider S, Stagljar I. 2015b. Membrane yeast two-hybrid (MYTH) mapping of full-length membrane protein interactions. *Cold Spring Harb Protoc* doi: 10.1101/pdb.top077560.

CHAPTER 20

Protein-Fragment Complementation Assays for Large-Scale Analysis, Functional Dissection, and Spatiotemporal Dynamic Studies of Protein–Protein Interactions in Living Cells

Stephen W. Michnick,[1,7] Christian R. Landry,[2,7] Emmanuel D. Levy,[3,7] Guillaume Diss,[2] Po Hien Ear,[1,4] Jacqueline Kowarzyk,[1] Mohan K. Malleshaiah,[1,5] Vincent Messier,[1,6] and Emmanuelle Tchekanda[1]

[1]*Département de Biochimie et Médecine Moléculaire, Université de Montréal, Montréal, Québec H3C 3J7, Canada;* [2]*Département de Biologie, Institut de Biologie Intégrative et des Systèmes, PROTEO-Québec Research Network on Protein Function, Structure and Engineering, Université Laval, Québec, Québec G1V 0A6, Canada;* [3]*Department of Structural Biology, Weizmann Institute of Science, Rehovot 76100, Israel;* [4]*Department of Genetics, Harvard Medical School, Boston, Massachusetts 02115;* [5]*Department of Systems Biology, Harvard Medical School, Boston, Massachusetts 02115;* [6]*Department of Medical Research, University of Toronto, Toronto, Ontario M5S 3E1, Canada*

Protein-fragment complementation assays (PCAs) comprise a family of assays that can be used to study protein–protein interactions (PPIs), conformation changes, and protein complex dimensions. We developed PCAs to provide simple and direct methods for the study of PPIs in any living cell, subcellular compartments or membranes, multicellular organisms, or in vitro. Because they are complete assays, requiring no cell-specific components other than reporter fragments, they can be applied in any context. PCAs provide a general strategy for the detection of proteins expressed at endogenous levels within appropriate subcellular compartments and with normal posttranslational modifications, in virtually any cell type or organism under any conditions. Here we introduce a number of applications of PCAs in budding yeast, *Saccharomyces cerevisiae*. These applications represent the full range of PPI characteristics that might be studied, from simple detection on a large scale to visualization of spatiotemporal dynamics.

PRINCIPLES OF PCAs

The general principle of PCA is the re-creation of the spontaneous folding of a protein from two complementary peptide amino- and carboxy-terminal fragments brought together in space by the interaction of two proteins to which the fragments are, respectively, fused (Fig. 1; Pelletier and Michnick 1997; Pelletier et al. 1998; Michnick et al. 2000). An important consequence of PCAs as a folding phenomenon (vs. protein-docking) is that the assays are generally reversible. That is, when associated protein complexes fall apart, the PCA reporter protein spontaneously unfolds (Michnick et al. 2007). Thus, the PCA reporter does not contribute to rates of association of a complex subunit, and PCAs can therefore be used to study the dynamics of PPIs.

This general principle of PCAs has been applied to generate assays using reporter proteins, whose activities are measured in a number of different ways, depending on the desired applications (reviewed by Michnick et al. [2007]). Furthermore, PCAs are highly sensitive to the topological details and

---

[7]Correspondence: stephen.michnick@umontreal.ca; christian.landry@bio.ulaval.ca; emmanuel.levy@weizmann.ac.il

Copyright © Cold Spring Harbor Laboratory Press; all rights reserved
Cite this introduction as *Cold Spring Harb Protoc*; doi:10.1101/pdb.top083543

Chapter 20

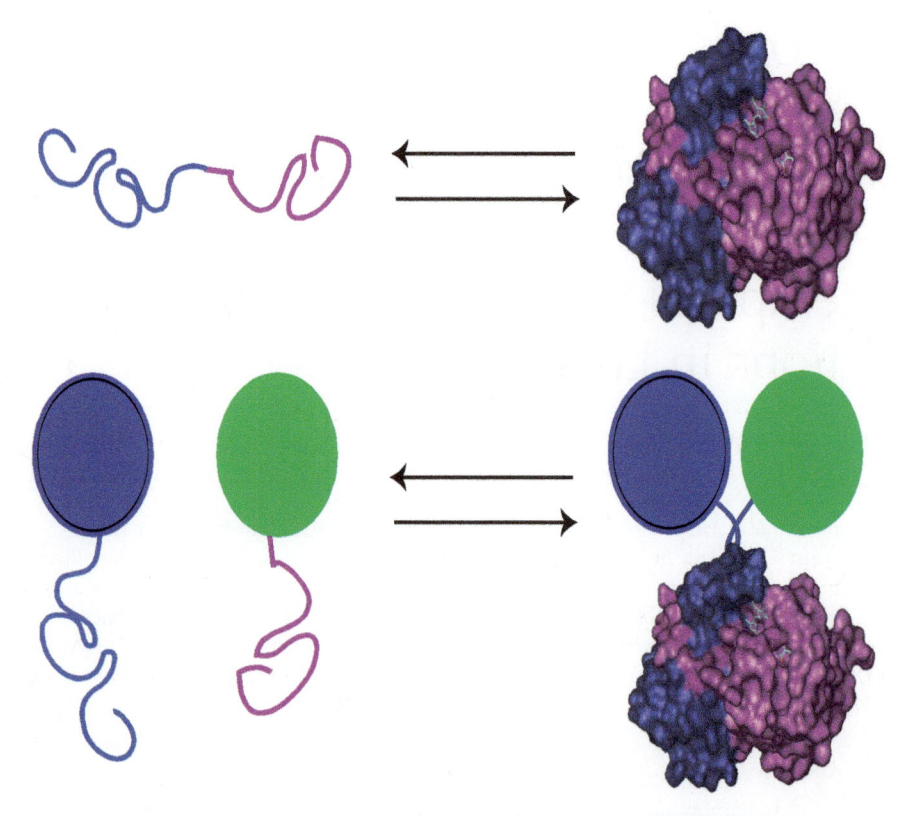

FIGURE 1. Principles of PCAs. Protein folding occurs spontaneously for a complete polypeptide chain (*upper*), but can also be driven by PPI-mediated localization of amino- and carboxy-terminal complementary fragments (*lower*). The folding of the protein from the fragments, with the reconstitution of enzyme activity, becomes a reporter for the PPI that drove it. Whether a PCA successfully occurs depends on the distance between the points of attachment of the interacting proteins to the reporter protein fragments; this, in turn, can be changed by integrating linker peptides of different lengths between proteins and fragments. We generally insert linker peptides of 5–15 amino acids between proteins and reporter protein fragments.

conformation changes of protein complexes; they have been used to detect topological features in not only individual protein complexes but also an entire protein interactome (Remy et al. 1999; Tarassov et al. 2008).

GENERAL CONSIDERATIONS FOR PCAs

We should not be surprised that what is measured by a PCA in vivo might be quite different than what one measures using in vitro methods, such as affinity or immuno-purification or two-hybrid assays. Because of the steric requirement for detection of PPIs via PCAs, a PCA is only possible if proteins make direct or indirect contacts that allow the PCA fragments to fold together into a functional reporter protein. This condition is met if the termini of the proteins to which the PCA fragments are attached are within a distance determined by the length of the polypeptide linkers separating the proteins and fragments (Fig. 1) (Remy et al. 1999; Tarassov et al. 2008).

There are also practical considerations to consider when applying any PCA to study PPIs. The sensitivity of PCAs, like other reporter assays, depends on the type of signal the reporter produces and the background signal of the cells. For example, in the DHFR PCA (see Protocol 1: The Dihydrofolate Reductase Protein-Fragment Complementation Assay: A Survival-Selection Assay for Large-Scale Analysis of Protein–Protein Interactions [Michnick et al. 2015]), we estimate that a mere 25 complexes per cell of a complex with a dissociation constant of ~1 nM is sufficient for detection. We have more generally shown that we can detect complexes of proteins expressed endogenously at fewer than

100 copies per cell (Remy and Michnick 1999; Tarassov et al. 2008). Further, we have estimated an upper limit on detection of PPIs with dissociation complexes in the range of 10–100 μM, well within the range of the highest dissociation complexes one would expect to observe for protein complexes in vivo (Campbell-Valois et al. 2005; Ear and Michnick 2009).

Detectability of a PPI by a PCA will also depend on whether the signal is enzymatically amplified and on the levels of interfering background cell signal that might need to be overcome. For example, fluorescent protein PCA reporters are less sensitive and produce higher cell background than luciferase-based PCAs, which benefit both from enzymatic amplification and virtually no cell background (MacDonald et al. 2006; Remy and Michnick 2006; Stefan et al. 2007, 2011; Malleshaiah et al. 2010).

Finally, as in any experiment in which proteins of interest are fused to reporter proteins, it is essential in PCAs to determine whether fusions to amino- or carboxy-termini affect protein stability or localization (Remy and Michnick 2001; Breitkreutz et al. 2010). In the case of PCAs, there are potentially eight different configurations in which the assay can be performed. We have found that when all eight of these combinations are tested, we observe a range of results from no observed interaction for any combination to observed interactions with any combination (SW Michnick, unpubl.). Therefore, for small-scale studies, it is desirable to test all possible configurations before implementing a PCA.

We have provided detailed protocols for PCAs based on distinct reporters with different readouts and for different applications; see Protocol 1: The Dihydrofolate Reductase Protein-Fragment Complementation Assay: A Survival-Selection Assay for Large-Scale Analysis of Protein–Protein Interactions (Michnick et al. 2015), Protocol 2: Combining the Dihydrofolate Reductase Protein-Fragment Complementation Assay with Gene Deletions to Establish Genotype-to-Phenotype Maps of Protein Complexes and Interaction Networks (Diss and Landry 2015), Protocol 3: Dissecting the Contingent Interactions of Protein Complexes with the Optimized Yeast Cytosine Deaminase Protein-Fragment Complementation Assay (Ear et al. 2015), and Protocol 4: Real-Time Protein-Fragment Complementation Assays for Studying Temporal, Spatial, and Spatiotemporal Dynamics of Protein–Protein Interactions in Living Cells (Malleshaiah et al. 2015).

REFERENCES

Breitkreutz A, Choi H, Sharom JR, Boucher L, Neduva V, Larsen B, Lin ZY, Breitkreutz BJ, Stark C, Liu G, et al. 2010. A global protein kinase and phosphatase interaction network in yeast. *Science* **328:** 1043–1046.

Campbell-Valois FX, Tarassov K, Michnick SW. 2005. Massive sequence perturbation of a small protein. *Proc Natl Acad Sci* **102:** 14988–14993.

Diss G, Landry CR. 2015. Combining the dihydrofolate reductase protein-fragment complementation assay with gene deletions to establish genotype-to-phenotype maps of protein complexes and interaction networks. *Cold Spring Harb Protoc*. doi: 10.1101/pdb.prot090035.

Ear PH, Michnick SW. 2009. A general life-death selection strategy for dissecting protein functions. *Nat Methods* **6:** 813–816.

Ear PH, Kowarzyk J, Michnick SW. 2015. Dissecting the contingent interactions of protein complexes with the optimized yeast cytosine deaminase protein-fragment complementation assay. *Cold Spring Harb Protoc*. doi: 10.1101/pdb.prot090043.

MacDonald ML, Lamerdin J, Owens S, Keon BH, Bilter GK, Shang Z, Huang Z, Yu H, Dias J, Minami T, et al. 2006. Identifying off-target effects and hidden phenotypes of drugs in human cells. *Nat Chem Biol* **2:** 329–337.

Malleshaiah MK, Shahrezaei V, Swain PS, Michnick SW. 2010. The scaffold protein Ste5 directly controls a switch-like mating decision in yeast. *Nature* **465:** 101–105.

Malleshaiah M, Tchekanda E, Michnick SW. 2015. Real-time protein-fragment complementation assays for studying temporal, spatial, and spatiotemporal dynamics of protein–protein interactions in living cells. *Cold Spring Harb Protoc*. doi: 10.1101/pdb.prot090068.

Michnick SW, Remy I, Campbell-Valois FX, Vallee-Belisle A, Pelletier JN. 2000. Detection of protein–protein interactions by protein fragment complementation strategies. *Methods Enzymol* **328:** 208–230.

Michnick SW, Ear PH, Manderson EN, Remy I, Stefan E. 2007. Universal strategies in research and drug discovery based on protein-fragment complementation assays. *Nat Rev* **6:** 569–582.

Michnick SW, Levy ED, Landry CR, Kowarzyk J, Messier V. 2015. The dihydrofolate reductase protein-fragment complementation assay: A survival-selection assay for large-scale analysis of protein–protein interactions. *Cold Spring Harb Protoc*. doi: 10.1101/pdb.prot090027.

Pelletier JN, Michnick SW. 1997. A protein complementation assay for detection of protein–protein interactions in vivo. *Protein Eng* **10:** 89.

Pelletier JN, Campbell-Valois FX, Michnick SW. 1998. Oligomerization domain-directed reassembly of active dihydrofolate reductase from rationally designed fragments. *Proc Natl Acad Sci* **95:** 12141–12146.

Remy I, Michnick SW. 1999. Clonal selection and in vivo quantitation of protein interactions with protein-fragment complementation assays. *Proc Natl Acad Sci* **96:** 5394–5399.

Remy I, Michnick SW. 2001. Visualization of biochemical networks in living cells. *Proc Natl Acad Sci* **98:** 7678–7683.

Remy I, Michnick SW. 2006. A highly sensitive protein–protein interaction assay based on *Gaussia* luciferase. *Nat Methods* **3:** 977–979.

Remy I, Wilson IA, Michnick SW. 1999. Erythropoietin receptor activation by a ligand-induced conformation change. *Science* **283:** 990–993.

Stefan E, Aquin S, Berger N, Landry CR, Nyfeler B, Bouvier M, Michnick SW. 2007. Quantification of dynamic protein complexes using *Renilla* luciferase fragment complementation applied to protein kinase A activities in vivo. *Proc Natl Acad Sci* **104:** 16916–16921.

Stefan E, Malleshaiah MK, Breton B, Ear PH, Bachmann V, Beyermann M, Bouvier M, Michnick SW. 2011. PKA regulatory subunits mediate synergy among conserved G-protein-coupled receptor cascades. *Nat Commun* **2:** 598.

Tarassov K, Messier V, Landry CR, Radinovic S, Serna Molina MM, Shames I, Malitskaya Y, Vogel J, Bussey H, Michnick SW. 2008. An in vivo map of the yeast protein interactome. *Science* **320:** 1465–1470.

Protocol 1

The Dihydrofolate Reductase Protein-Fragment Complementation Assay: A Survival-Selection Assay for Large-Scale Analysis of Protein–Protein Interactions

Stephen W. Michnick,[1,5] Emmanuel D. Levy,[2,5] Christian R. Landry,[3,5] Jacqueline Kowarzyk,[1] and Vincent Messier[1,4]

[1]Département de Biochimie et Médecine Moléculaire, Université de Montréal, Montréal, Québec H3C 3J7, Canada; [2]Department of Structural Biology, Weizmann Institute of Science, Rehovot 76100 Israel; [3]Département de Biologie, Institut de Biologie Intégrative et des Systèmes, PROTEO-Québec Research Network on Protein Function, Structure and Engineering, Université Laval, Québec, Québec G1V 0A6, Canada; [4]Department of Medical Research, University of Toronto, Toronto, Ontario M5S 3E1, Canada

Protein-fragment complementation assays (PCAs) can be used to study protein–protein interactions (PPIs) in any living cell, in vivo or in vitro, in any subcellular compartment or membranes. Here, we present a detailed protocol for performing and analyzing a high-throughput PCA screening to study PPIs in yeast, using dihydrofolate reductase (DHFR) as the reporter protein. The DHFR PCA is a simple survival-selection assay in which *Saccharomyces cerevisiae* DHFR (scDHFR) is inhibited by methotrexate, thus preventing nucleotide synthesis and causing arrest of cell division. Complementation of cells with a methotrexate-insensitive murine DHFR restores nucleotide synthesis, allowing cell proliferation. The methotrexate-resistant DHFR has two mutations (L22F and F31S) and is 10,000 times less sensitive to methotrexate than wild-type scDHFR, but retains full catalytic activity. The DHFR PCA is sensitive enough for PPIs to be detected for open reading frame (ORF)-PCA fragments expressed off of their endogenous promoters.

MATERIALS

It is essential that you consult the appropriate Material Safety Data Sheets and your institution's Environmental Health and Safety Office for proper handling of equipment and hazardous materials used in this protocol.

RECIPES: Please see the end of this protocol for recipes indicated by <R>. Additional recipes can be found online at http://cshprotocols.cshlp.org/site/recipes.

Reagents

Commercial bleach solution (10%) (4% sodium hypochloride)
Dimethylsulfoxide (DMSO)
Glycerol solution (75%)
  Sterilize by autoclaving for 30 min at 121°C.

---

[5]Correspondence: stephen.michnick@umontreal.ca; emmanuel.levy@weizmann.ac.il; christian.landry@bio.ulaval.ca

Copyright © Cold Spring Harbor Laboratory Press; all rights reserved
Cite this protocol as *Cold Spring Harb Protoc*; doi:10.1101/pdb.prot090027

TABLE 1. Plasmid and oligonucleotide list (Tarassov et al. 2008)

Plasmid or oligonucleotide	Characteristics

pAG25-linker-DHFR-F[1,2]
  Antibiotic resistance: nourseothricin
pAG32-linker-DHFR-F[3]
  Antibiotic resistance: hygromycin-B
	For creating integration cassettes of DHFR PCA fragments to study yeast PPIs at endogenous levels of expression. Cassettes are integrated by homologous recombination.

Forward oligonucleotides for creating integration cassettes	61 base pairs long, consisting of a 40-nucleotide sequence homologous to a region located 5′ to the ORF stop codon, followed by a 21-nucleotide sequence that anneals to the linker: 5′-GGCGGTGGCGGATCAGGAGGC-3′
Reverse oligonucleotides for creating integration cassettes	64 base pairs long, consisting of a 40-nucleotide sequence homologous to the sequence of genomic DNA immediately 3′ to the stop codon of each ORF, followed by a 24-nucleotide sequence that anneals to the 3′ end of the TEF terminator region of the antibiotic resistance cassettes: 5′-TTCGACACTGGATGGCGGCGTTAG-3′

Oligonucleotides for amplification of protein-fragment complementation assay (PCA) fragment cassettes

Forward and reverse oligonucleotides for creation of homologous recombination cassettes are available for most genes on request. Should this not be the case, they can be designed following the specifications provided in Table 1.

Polymerase chain reaction (PCR) reagents

Plasmids containing the dihydrofolate reductase protein-fragment complementation assay (DHFR PCA) fragments, followed by a terminator and an antibiotic selection marker (Table 1)
  pAG25-linker-DHFR-F[1,2]
  pAG32-linker-DHFR-F[3]

  We also have DHFR expression plasmids which allow bait-coding sequences to be modified with mutations and screened against prey arrays. These plasmids can be requested from the authors.

PLATE solution <R>

Yeast culture media
  Synthetic complete (SC) medium, minus adenine, plus methotrexate (35 mL in OmniTrays) <R>
  Synthetic complete (SC) medium, minus lysine and methionine, plus antibiotics (35 mL in OmniTrays) (optional; see Step 18) <R>
  YPD for the DHFR PCA (solidified with 2% agar [w/v] in Petri dishes with suitable antibiotic for the competent strain; Table 1) <R>
  YPD for the DHFR PCA (solidified with 3% agar [w/v]; 35 mL in OmniTrays) <R>

  For prey strain selective medium (Step 11), prepare with 100 µg/mL nourseothricin (for MATa recombinant strains) or 250 µg/mL hygromycin B (for MATα recombinant strains). For diploid selective medium (Step 17), prepare with both 100 µg/mL nourseothricin and 250 µg/mL hygromycin B.

  YPD for the DHFR PCA (without agar) <R>

  For bait strain selective medium (Step 8), prepare with 100 µg/mL nourseothricin (for MATa recombinant strains) or 250 µg/mL hygromycin B (for MATα recombinant strains).

Yeast strains (competent, for use in Step 2)
  BY4741 recombinant strains (MATa; his3Δ 1; leu2Δ 0; met15Δ 0; ura3Δ 0) containing ORFs of interest (Open Biosystems) (to be fused to the complementary DHFR F[1,2] PCA fragment in Step 2)
  BY4742 recombinant strains (MATα his3Δ 1; leu2Δ 0; lys2Δ 0; ura3Δ 0), containing ORFs of interest (Open Biosystems) (to be fused to the complementary DHFR F[3] PCA fragment in Step 2)

  Here, we use MATa strains as bait and MATα as prey strains; however, MATa or MATα strains can be used as bait or prey as desired. Competent BY4741 and BY4742 strains can be used to generate new strains with ORFs tagged with DHFR PCA fragments, if strains containing the ORF of interest are not available at Open Biosystems.

Equipment

  Bench centrifuge (Eppendorf 5810R)
  Centrifuge tubes

Glass beads
Incubator at 30°C
OmniTray single-well plates (Nunc CA62409-600)
PCR equipment
Pintools, robotically or manually manipulated (V&P Scientific)
Manually manipulated
   96-pintool with 1.58-mm, 1-μL slot pins (VP 408Sa)
Robotically manipulated
   96-pintool with 0.787-mm flat round-shaped pins (FP3N)
   384-pintool with 0.457-mm flat round-shaped pins (custom FP1N)
   1536-pintool with 0.457-mm flat round-shaped pins (custom FP1N)
Plate imaging equipment
   Camera (at least 4.0 megapixels, e.g., Canon Powershot A520)
   Plate-shooting platform
      We recommend covering the plate fixation platform with a black velvet cover.

   Stationary arm (70-cm mini repro [Industria Fototecnica Firenze, Italy])
Shaking incubator at 30°C
Water bath or heat block at 42°C

METHOD

Fusing PCA Fragments at Chromosomal Loci by Homologous Recombination

1. PCR-amplify the PCA fragment cassettes from the pAG25 (DHFR fragment [1,2]) and pAG32 (DHFR fragment [3]) plasmids using forward and reverse oligonucleotides (Table 1).

2. Transform the appropriate strain of competent cells (BY4741 or BY4742 recombinant strains containing the ORFs of interest, according to experimental design) with each PCR product.

    i. Mix 20 μL of thawed competent cells with 8 μL (~1 μg/μL) of each amplified cassette DNA (encoding the PCA fragments and a resistance marker).

    ii. Add 100 μL of PLATE solution.

    iii. Incubate for 30 min at room temperature.

    iv. Add 15 μL DMSO and heat shock the cells for 15 min at 42°C.

3. Centrifuge the cells at 805g for 3 min at room temperature. Remove the supernatant and resuspend the cells in 100 μL of YPD medium. Incubate with shaking for 4 h at 30°C.

4. Centrifuge the cells at 805g for 3 min at room temperature. Remove the supernatant and resuspend the cells in 200 μL of YPD medium.

5. Plate the entire 200 μL on a Petri dish containing solid YPD medium plus the suitable antibiotic (Table 1).

6. Incubate the plates for 48–72 h at 30°C.
    Colonies can be further verified using colony PCR.

7. Prepare glycerol stocks of the bait/prey strains for storage.

Performing a Large-Scale DHFR PCA Screen

In the following procedure, an array of "prey" strains is generated and mated to individual "bait" strains of the opposite mating type. Resulting diploids are transferred to methotrexate selection plates for the DHFR PCA (Fig. 1).

It is critical to use both positive and negative controls (baits and preys that are known to directly interact with each other, preferably as shown by multiple methods, as well as baits and preys that have never been shown to interact with each other) to assure the quality of the entire printing and selection process, and to allow the detection of grid

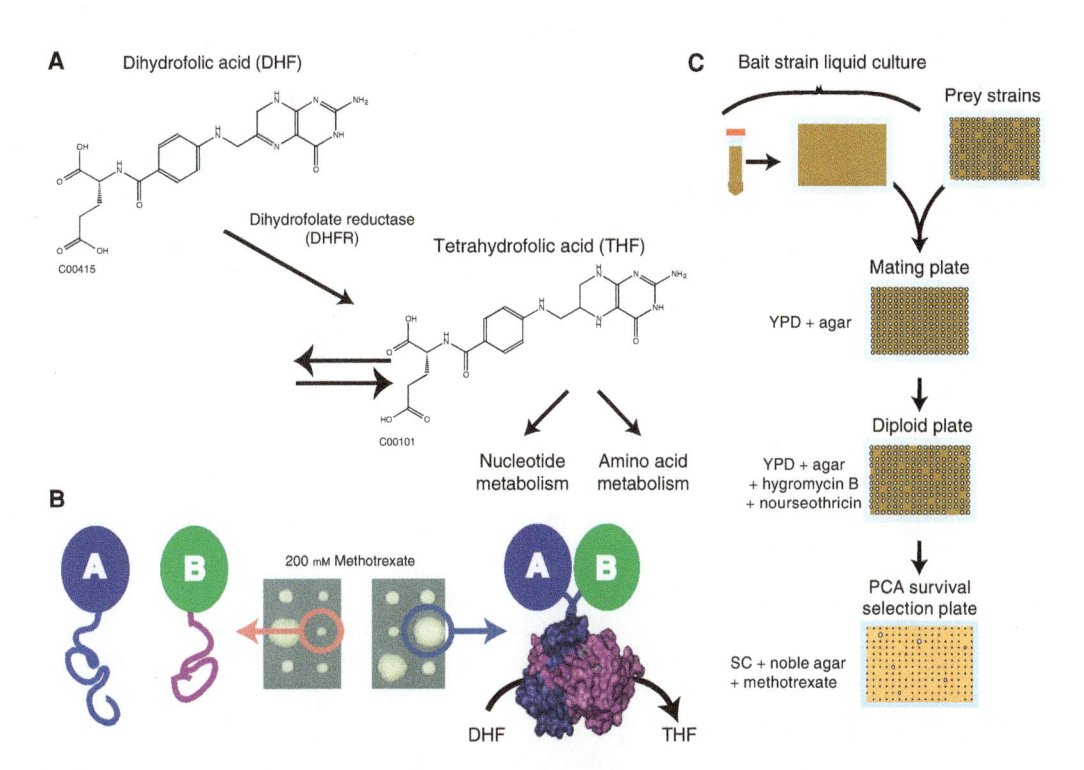

FIGURE 1. The DHFR PCA. (*A*) DHFR catalyzes the reduction of dihydrofolate to tetrahydrofolate, an early step in the synthesis of purine. Methotrexate is a potent and highly specific competitive inhibitor of DHFR. (*B*) Bait and prey proteins are fused to complementary fragments of methotrexate-insensitive *murine* DHFR. No DHFR PCA selection is observed if there is no bait–prey interaction. If there is a bait–prey interaction, DHFR activity is reconstituted, allowing cells to proliferate in the presence of methotrexate. (*C*) Schematic representation of a DHFR PCA screen. The bait strain is incubated in liquid culture and the prey reporter strains are printed on solid medium. A mating plate is created through sequential printing of a bait to the prey strain array grown on agar in nonselection medium. Haploids and diploid mixture strains are transferred to agar plates containing a medium for the selection of diploids. The resulting diploid array is then transferred onto DHFR PCA selection plates containing methotrexate.

mispositioning or batch effects (e.g., differences in media, incubation conditions, or drug concentration) that could change growth rates of colonies on different plates; see Discussion.

Preparing Bait Strains

8. Pick the bait strains from thawed glycerol stocks and add the cells to 100 mL of bait strain selective medium.

9. Allow the resulting culture to reach saturation by shaking at 30°C.
   It is very important to obtain a saturated bait culture to print enough cells for efficient mating on solid phase. See Troubleshooting.

10. Centrifuge the saturated culture at 500*g* for 5 min. Resuspend the cells in 15 mL of YPD medium.

Preparing Prey Strains

11. Pick the prey strains from thawed glycerol stocks and print onto OmniTray plates containing 35 mL of solid prey strain selective medium. Clean the pintool between each cell transfer by soaking the pins twice in a 10% bleach solution containing glass beads, and then washing once in 10% bleach followed by twice in a sterile water bath.
    Either 96-pin manual or robotic pintools can be used to print the prey array, with a total of 384 prints per plate. Step 11 can be repeated from the 384-prints to transfer the prey strain to a maximum of four other 1536-prints per OmniTray, achieving a density of up to 6144 colonies.

12. Incubate the plates for 16 h at 30°C.
    See Troubleshooting.

Performing the PCA Screen

13. Transfer the bait strain suspensions into an empty OmniTray plate.
14. Print the bait strain suspensions from the OmniTray to another OmniTray containing 35 mL of solid YPD medium using an appropriate pintool for the desired colony array density.
15. Transfer the prey strains onto the bait strain using the same density pintool as used for the bait strain.
16. Allow mating to occur by incubating the plate for 16 h at 30°C.
17. Transfer the mixed haploid and diploid colonies to an OmniTray plate containing 35 mL of solid diploid selective medium using an appropriate pintool for the desired colony array density.
18. Incubate the OmniTray for 48 h at 30°C.
    To increase the diploid selection stringency, mixed haploid and diploid colonies can be transferred onto SC medium, minus lysine and methionine, plus antibiotics (both nourseothricin and hygromycin B when using BY4741 MATa and BY4742 MATα). Stringency can also be increased by repeating Step 17.
    See Troubleshooting.
19. Wet the pintool pin tips with 75% glycerol solution.
    Liquid transfer with pintools is sensitive to the retraction speed of the pintool from the liquid, with faster speeds yielding more liquid on the pins. To obtain a homogeneous glycerol microdrop on every pin, the immersion depth and retraction speed may vary with each robotic setup and will need to be optimized, as it depends on several factors, such as the time between the last water wash and the glycerol dip, the room humidity, and the room air-flow.
20. Transfer the diploid selected strains onto an OmniTray plate containing 35 mL of solid SC medium, minus adenine, plus methotrexate, using an appropriate pintool for the desired colony array density.
    SC medium without adenine is used to decrease background growth.
21. Incubate the OmniTray at 30°C. Acquire pictures of the colony array every 24–96 h for up to 2 wk.
    See Troubleshooting.

Analyzing Images of Large-Scale DHFR PCA Screens

22. Measure the number of pixels per colony position on the selection plates.
    Several bioinformatics tools are available to perform colony size measurements from digital images of high-density plates (Collins et al. 2006; Memarian et al. 2007). When densities up to 1536 colonies per plate are used, macros can be implemented in ImageJ (http://rsb.info.nih.gov/ij/) or any other image analysis software. We convert digital images of plates to 8-bit gray scale format and measure colonies by positioning the measurement tool on a colony center and estimating integrated pixel intensity in an area corresponding to a maximal allowed colony size. We iterate this process over all grid positions and then over all of the plates. Colonies should always be found at the same positions in the images. If this is not so, a step can be included to position an analysis grid onto colony positions before measuring the colony sizes. Image analysis should be performed only on the region containing colonies by removing the plate sides from analysis. Images can be corrected for nonuniform illumination with algorithms such as those described here: http://blogs.mathworks.com/steve/2011/09/27/digital-image-processing-using-matlab-reading-image-files/?s_tid=srchtitle. Artifactual objects such as bubbles, gel background, and other anomalies can be removed using the imopen function. When working with high-density plates (e.g., a 1536-position grid or greater), methods must be used to segment adjacent and overlapping colonies (Tarassov et al. 2008).
    Grid positions and intensity values can be saved as text files for processing in your favorite spreadsheet or statistical analysis software.

Performing Statistical Analyses of Colony Data

Both positive and negative controls should be analyzed on every plate; see Discussion.

23. Perform statistical analyses to determine whether there is a significant difference in growth rates among the plates.

    Significant variation in the data requires that it be normalized so that all of the plates have the same average colony size. Alternatively, relative scores of the data can be calculated (e.g., Z-scores) so that data points are transformed into the number of standard deviations separating the data point from the average value of an individual plate. We have combined the Z-score and the raw intensity for best interpretation of our large-scale screens (Tarassov et al. 2008).

24. Convert continuous values into binary values by setting a threshold of intensity above which proteins are inferred to interact. Calculate the positive predictive value (PPV) as a function of threshold of intensities as follows:

$$\text{PPV} = \frac{\text{True positive interactions}}{\text{Total number of inferred positives}},$$

Total number of inferred positives = True positives + Predicted false positives.

For example, at PPV = 95%, 5% of positives are predicted to be false. Thresholds are arbitrary, depending on the stringency one wants to achieve in the analysis. The relative occurrence of positives and negatives in the reference sets must be similar to that of the real positives and negatives to assure that estimated PPV values are accurate (Jansen and Gerstein 2004). For comprehensive genome-wide screens, this fraction is a very low prior probability of finding interactions among all possible pairs of proteins. In small-scale screen of a specific biological process, however, there will be a greater proportion of true positives than in a random screen. Thus, the choice of reference sets must be appropriate to the specific screen, specifically to the size of the potential interactome that will be covered.

25. Establish a confidence score for the selected threshold.

    Confidence values can be applied to PCA interactions by benchmarking observed intensity values to the results of interactions that should be detected in the screen (a set of real positives) compared with interactions that should not be observed under any circumstances (e.g., a set of real negatives representing proteins that are expressed in different cellular compartments and never observed to interact) (Collins et al. 2007). For a given intensity threshold, one can then determine what should be the proportion of true positive and false positive interactions.

26. Perform a statistical analysis to separate interacting pairs from noninteracting pairs.

27. Perform Gene Ontology enrichments and visualize interaction clusters to further assess confidence in the produced data set.

    Choosing true positives and negatives does not guarantee that observed interactions are functionally relevant. Only direct functional assays and evolutionary characterization of interactions are sufficient to determine whether an interaction is functionally relevant to an organism (Levy et al. 2009; Landry et al. 2013).

TROUBLESHOOTING

Problem (Steps 9, 12): Strains are not growing, or prey array growth is incomplete.

Solution: Verify the appropriate culture conditions, because the problem might be caused by erroneous haploid selection. If the problem is caused by low glycerol viability, strains can be streaked on solid agar-selective medium in Petri dishes before inoculation to increase viability. Finally, verify that all pins of the pintool touch the glycerol stocks and the recipient OmniTray plate.

Problem (Step 18): There are few or no colonies on diploid selective plates.

Solution: Verify the mating type of the haploid strains, because the problem might be caused by erroneous haploid strain type. The problem may also be technical, that is, pintool alignment might have changed. No modifications to the pintool positioning should be done between transfers.

Problem (Step 21): There is no colony growth on DHFR PCA survival-selective medium.
Solution: This problem can be caused by erroneous selection conditions. Use the heteromeric complex SspB$_{YGMF}$:SspB$_{LSLA}$ as a positive control to validate DHFR PCA activity (Tarassov et al. 2008). The problem might also be caused by an erroneous DHFR PCA. Perform a strain diagnostic PCR of the complementary PCA fragments and verify DHFR PCA fragment recombinant insertion by genomic sequencing.

Problem (Step 21): All colonies on the DHFR PCA survival-selective medium grow at the same rate.
Solution: Verify DHFR PCA fragment expression by western blot. The problem might be caused by erroneous selection conditions; use DHFR PCA fragment controls alone as negative control. Also verify methotrexate solubility under the conditions used. Stock solutions of methotrexate should not exceed 10 mg/mL in DMSO, and the final concentration in solid agar plates should not exceed 200 µg/mL.

DISCUSSION

In the above protocol, the growth rates of the positive controls provide a rough indication of the intensity threshold that can be expected for strains expressing interacting bait–prey pairs of similar abundance level. Other critical controls are those for spontaneous PCA—cases in which a protein–PCA fragment fusion interacts with the complementary fragment alone (Tarassov et al. 2008). Additional controls can be used to examine the range of dissociation constants for which the DHFR PCA is sensitive (Campbell-Valois et al. 2005) and to test a condition-dependent PPI (Pelletier et al. 1998). A potential source of false positives in a PCA screen is the trapping of nonspecific complexes caused by irreversible folding of the DHFR fragments. However, we have used adenosine 3′,5′-monophosphate-dependent dissociation of the yeast protein kinase A complex (Michnick et al. 2007) to show that the DHFR PCA is fully reversible, and thus the trapping of complexes is unlikely (Tarassov et al. 2008). All of these reagents are available from the authors on request.

In an initial study, Freschi et al. (2013) showed that genetic or environmental perturbations that affect the interaction between the catalytic and regulatory subunits of PKA yield different growth rates when probed with the DHFR PCA (Freschi et al. 2013), indicating that the DHFR PCA is sensitive to the amount of complex formed, with a higher complex concentration leading to a faster growth. We characterized the extent and precision of this relationship in our recent experiment measuring the interactions between (i) a neutral reporter comprising the *venus* yellow fluorescent protein (YFP) fused to DHFR F[1,2] and (ii) the Matα library of ORFs fused to DHFR[3], denoted $P(i)$, with $i = 1,...,N$ (with N corresponding to up to 4808 interactions that can be examined) (Levy et al. 2014). Given that the concentration of YFP is constant, and that YFP only shows weak interactions of comparable K_d with all yeast proteins, the main determinant of complex concentration [YFP · $P(i)$] is the concentration of $P(i)$. Colony size measurements in this experiment enabled us to infer [$P(i)$] with an accuracy comparable with that of mass spectrometry, demonstrating that the DHFR PCA assay is highly quantitative in determining the amount of complex formed. At the same time, this highlights the importance of controlling for protein abundance in a DHFR PCA, as a protein can yield a strong signal because of abundance alone. For example, two proteins present in the cell at nanomolar concentrations and showing a strong (nanomolar) K_a would yield fewer complexes (and therefore lesser growth) than two proteins present at micromolar concentration and showing a weak (e.g., micromolar) K_a.

RECIPES

Amino Acid Mix (10×)

Amino acid or nucleobase	Concentration (10×)
Adenine sulfate	0.4 g/mL
Uracil	0.2 g/mL
L-Tryptophan	0.4 g/mL
L-Histidine hydrochloride	0.2 g/mL
L-Arginine hydrochloride	0.2 g/mL
L-Tyrosine	0.3 g/mL
L-Leucine	0.6 g/mL
L-Lysine hydrochloride	0.3 g/mL
L-Phenylalanine	0.5 g/mL
L-Glutamic acid	1.0 g/mL
L-Asparagine	1.0 g/mL
L-Valine	1.5 g/mL
L-Threonine	2.0 g/mL
L-Serine	3.75 g/mL
L-Methionine	0.2 g/mL

Dissolve the amino acids from the list above (except those to be excluded from any dropout media) in distilled H_2O. Filter-sterilize, store at 4°C, and protect from light.

PLATE Solution

Reagents	Final concentration (1×)
PEG 3350	40% (w/v)
Lithium acetate	100 mM
Tris (pH 7.5)	10 mM
EDTA	0.4 mM

Wear personal protective equipment when handling this reagent. Store at room temperature in a dry and well-ventilated place.

Synthetic Complete Medium (SC), Minus Lysine and Methionine, Plus Antibiotics

Reagents	Final concentration (1×)
Yeast nitrogen base (YNB) medium without amino acids or ammonium sulfate (Bioshop)	1.74 g/L
Noble agar (purified agar; Bioshop)	4% (w/v)
Glucose solution (sterile)	2% (w/v)
Amino acid mix (10×), without lysine or methionine <R>	1×
Nourseothricin (Jena Bioscience)	100 μg/mL
Hygromycin B (Wisent Bioproducts)	250 μg/mL

Synthetic Complete (SC) Medium, Minus Adenine, Plus Methotrexate

Reagents	Final concentration (1×)
Yeast nitrogen base (YNB) medium without amino acids or ammonium sulfate (Bioshop)	1.74 g/L
Noble agar (purified agar; Bioshop)	4% (w/v)
Glucose solution (sterile)	2% (w/v)
Methotrexate	200 μg/mL
Amino acid mix (10×), without adenine <R>	1×

YPD for the DHFR PCA

Reagents	Final concentration (1×)
Yeast extract	1% (w/v)
Peptone	2% (w/v)
Dextrose	2% (w/v)
Agar	2% (w/v)

Store for up to several months at 4°C.

REFERENCES

Campbell-Valois FX, Tarassov K, Michnick SW. 2005. Massive sequence perturbation of a small protein. *Proc Natl Acad Sci* **102:** 14988–14993.

Collins SR, Schuldiner M, Krogan NJ, Weissman JS. 2006. A strategy for extracting and analyzing large-scale quantitative epistatic interaction data. *Genome Biol* **7:** R63.

Collins SR, Kemmeren P, Zhao XC, Greenblatt JF, Spencer F, Holstege FC, Weissman JS, Krogan NJ. 2007. Toward a comprehensive atlas of the physical interactome of *Saccharomyces cerevisiae*. *Mol Cell Proteomics* **6:** 439–450.

Freschi L, Torres-Quiroz F, Dube AK, Landry CR. 2013. qPCA: A scalable assay to measure the perturbation of protein–protein interactions in living cells. *Mol BioSyst* **9:** 36–43.

Jansen R, Gerstein M. 2004. Analyzing protein function on a genomic scale: The importance of gold-standard positives and negatives for network prediction. *Curr Opin Microbiol* **7:** 535–545.

Landry CR, Levy ED, Abd Rabbo D, Tarassov K, Michnick SW. 2013. Extracting insight from noisy cellular networks. *Cell* **155:** 983–989.

Levy ED, Landry CR, Michnick SW. 2009. How perfect can protein interactomes be? *Sci Signal* **2:** pe11.

Levy ED, Kowarzyk J, Michnick SW. 2014. High-resolution mapping of protein concentration reveals principles of proteome architecture and adaptation. *Cell Rep* **7:** 1333–1340.

Memarian N, Jessulat M, Alirezaie J, Mir-Rashed N, Xu J, Zareie M, Smith M, Golshani A. 2007. Colony size measurement of the yeast gene deletion strains for functional genomics. *BMC Bioinformatics* **8:** 117.

Michnick SW, Ear PH, Manderson EN, Remy I, Stefan E. 2007. Universal strategies in research and drug discovery based on protein-fragment complementation assays. *Nat Rev* **6:** 569–582.

Pelletier JN, Campbell-Valois FX, Michnick SW. 1998. Oligomerization domain-directed reassembly of active dihydrofolate reductase from rationally designed fragments. *Proc Natl Acad Sci* **95:** 12141–12146.

Tarassov K, Messier V, Landry CR, Radinovic S, Serna Molina MM, Shames I, Malitskaya Y, Vogel J, Bussey H, Michnick SW. 2008. An in vivo map of the yeast protein interactome. *Science* **320:** 1465–1470.

Protocol 2

Combining the Dihydrofolate Reductase Protein-Fragment Complementation Assay with Gene Deletions to Establish Genotype-to-Phenotype Maps of Protein Complexes and Interaction Networks

Guillaume Diss and Christian R. Landry[1]

Département de Biologie, Institut de Biologie Intégrative et des Systèmes, PROTEO-Québec Research Network on Protein Function, Structure and Engineering, Université Laval, Québec, Québec G1V 0A6, Canada

Systematically measuring the impact of gene deletion on protein–protein interactions is a promising approach to reveal the structural bases of protein interaction networks and to allow a better understanding of how genotypes translate into phenotypes. Genetic and protein-interaction tools in yeast now allow us to explore this third dimension of protein–protein interaction networks. Because it is scalable and quantitative, the protein-fragment complementation assay (PCA) using dihydrofolate reductase (DHFR) as the reporter protein provides an exceptionally powerful tool for such a purpose. Here, we describe a fully automated protocol that combines DHFR PCA for protein–protein interaction measurement and synthetic genetic array (SGA) technology for introducing mutant and other alleles into PCA strains using genetic crosses. In this, PCA strains are crossed with strains carrying a gene deletion and SGA markers, and the recombinant haploid progeny are selected by SGA. The resulting haploid strains, each expressing a DHFR-fragment fusion protein in a gene-specific haploid deletion background, are crossed to measure the interaction between the two recombinant proteins by PCA in a diploid homozygous deletion background. This approach can be used to measure a single protein interaction in a large array of genetic backgrounds or a large number of protein interactions in a small number of genetic backgrounds.

MATERIALS

It is essential that you consult the appropriate Material Safety Data Sheets and your institution's Environmental Health and Safety Office for proper handling of equipment and hazardous material used in this protocol.

RECIPES: Please see the end of this protocol for recipes indicated by <R>. Additional recipes can be found online at http://cshprotocols.cshlp.org/site/recipes.

Reagents

Methotrexate (MTX) medium <R>
Reagents for selective media (see Table 1), prepared in omniplates
  Amino acid mix (10×) <R>
  Canavanine (50 mg/mL stock in water) (Sigma-Aldrich)
  Geneticin/G418 (200 mg/mL stock in water, filter sterilized) (Bioshop)
  Hygromycin B (Hyg; 100 mg/mL stock in water) (Bioshop)

---

[1]Correspondence: christian.landry@bio.ulaval.ca

Copyright © Cold Spring Harbor Laboratory Press; all rights reserved
Cite this protocol as *Cold Spring Harb Protoc*; doi:10.1101/pdb.prot090035

TABLE 1. Selective media for protein-fragment complementation assays (PCAs)

Medium name	Type of medium	Amino acid drop-out	Antibiotic[a]
h1a	SC–MSG	–**His**, –Met, –Arg, –Lys	Geneticin, thialysine, canavanine
h1α	SC–MSG	–**Leu**, –Met, –Arg, –Lys	Geneticin, thialysine, canavanine
h2.1a	SC	–**His**, –Arg, –Lys	Thialysine, canavanine
h2.1α	SC	–**Leu**, –Arg, –Lys	Thialysine, canavanine
h2.3a	SC–MSG	–**His**, –Arg, –Lys	Geneticin, thialysine, canavanine
h2.3α	SC–MSG	–**Leu**, –Arg, –Lys	Geneticin, thialysine, canavanine
h2.4a	SC–MSG	–**His**, –Met, –Arg, –Lys	Geneticin, thialysine, canavanine
h2.4α	SC–MSG	–**Leu**, –Met, –Arg, –Lys	Geneticin, thialysine, canavanine
h2.5a			**Nourseothricin**, geneticin, thialysine, canavanine
h2.5α	YPD		**Hygromycin B**, geneticin, thialysine, canavanine
sel2n1a	YPD		**Nourseothricin**, geneticin
sel2n1α	YPD		**Hygromycin B**, geneticin
sel2n2	YPD		Nourseothricin, hygromycin B, geneticin

[a]Always use the following antibiotic concentrations: geneticin, 200 µg/mL; thialysine, 100 µg/mL; canavanine, 50 µg/mL; nourseothricin (Nat), 100 µg/mL; and hygromycin B, 250 µg/mL.

Nourseothricin (Nat; 100 mg/mL stock in water) (WERNER BioAgents)
Synthetic complete (SC) medium <R>
Synthetic complete monosodium glutamic acid (SC–MSG) medium <R>

> *During the first four steps of haploid selection, synthetic medium is used to be able to select specific mating types with auxotrophic markers. In the last step of haploid selection, we select for the presence of the recombinant protein. We cannot do that in synthetic medium because we found that Hyg selection is weak in this medium. So we do this step on YPD because after four steps of mating type selection, there are no detectable cells of the opposite mating type remaining.*

Thialysine (100 mg/mL stock in water) (Sigma-Aldrich)
YPD for the DHFR PCA <R>
Spo.a and spo.α sporulation media, prepared in omniplates <R>
Yeast strains
    Glycerol stocks of *MAT*a SGAready deletion strains in which open reading frames (ORFs) are replaced by the KanMX marker (Diss et al. 2013)
    Glycerol stocks of *MAT*α SGAready deletion strains in which ORFs are replaced by the KanMX marker (Diss et al. 2013)
    Glycerol stocks of *MAT*a strains in which gene coding sequences are fused to the complementary DHFR F[1,2] fragment (Open Biosystems)
    Glycerol stocks of *MAT*α strains in which gene coding sequences are fused to the complementary DHFR F[3] fragment (Open Biosystems)
YPD medium containing 250 µg/mL hygromycin B (YPD + Hyg), solid (prepared in omniplates) and liquid
YPD medium containing 100 µg/mL nourseothricin (YPD + Nat), solid (prepared in omniplates) and liquid

Equipment

Colony processing robot (BM3-BC; S&P Robotics) equipped with a digital camera (e.g., Canon EOS Rebel T3i)
Ethanol bath
Incubators, set at 22°C and 30°C
Omniplates (86 × 128 mm, sterile plastic) (Nunc)
Pintools with 96, 384, and 1536 pins with diameter sizes of, respectively, 1, 1 and 0.5 mm (S&P Robotics)
Water bath

METHOD

Introgression of Gene Deletions into DHFR PCA Strains

1. Incubate individually the bait and the prey DHFR fragment F[1,2] and F[3] PCA strains picked from glycerol stocks into a 10 mL liquid culture of selection medium (YPD + Nat for the strain expressing the bait recombinant protein, normally *MAT*a, or YPD + Hyg for the strain expressing the prey recombinant protein, normally *MAT*α) and let the culture reach saturation at 30°C.

2. Replicate the two SGA-ready collections from glycerol stocks onto a 35 mL agar solidified omniplate of strain-selective medium (h1a for the *MAT*a collection or h1α for the *MAT*α collection) using four 96 manual or colony processing robotic pintool prints for a total of 384 prints per plate and incubate for 2 d at 30°C.

   Depending on the target screen density, Step 2 can be repeated from the 384 prints to a density of 1536 colonies. The two collections must be assembled the same way—that is, strains with the same deletion must have the same position on equivalent plates to obtain homozygous diploids at the final PCA step when the two arrays are crossed with each other. Use a proper sterilization step (e.g., dipping the pins in a water bath followed by an ethanol bath and dried using an air dryer) for the pin tool to avoid cross-contamination.

3. Pour 10 mL of the saturated preculture from Step 1 onto a 35 mL agar solidified omniplate plate of strain-specific selection medium (YPD + Nat for the strain expressing the bait recombinant protein, and YPD + Hyg for the strain expressing the prey recombinant protein). Let cells adsorb on the agar surface for 5–10 min, remove the excess liquid medium and incubate for 2 d at 30°C.

   Make sure that the entire surface is covered by the saturated culture. No growth or poor growth at some positions could lead to a loss of strains at the corresponding positions in the diploid selection step owing to improper mating between the bait strain and the deletion collection.

4. Print deletion strains from Step 2 onto a 35 mL agar-solidified rich medium (YPD) using the appropriate pintool.

5. Print the PCA strains grown on surface-covered omniplates on top of each plate containing the deletion strains of the opposite mating type (the *MAT*a strain expressing the bait recombinant protein with the *MAT*α deletion collection and the *MAT*α strain expressing the prey recombinant protein against the *MAT*a deletion collection) at the same density as the deletion strain using a pintool appropriate for the desired colony array density. Allow mating to occur and incubate for 2 d at 30°C.

   Steps 4 and 5 can be repeated several times to perform the screen in replicate for more-robust results. We recommend at least four replicates.

6. Transfer the mixed haploid–diploid colonies onto an omniplate containing diploid selection medium [sel2n1a for the cross with the strain expressing the bait recombinant protein or sel2n1α for the cross with the strain expressing the prey recombinant protein (see Table 1)] using the appropriate pintool. Incubate for 2 d at 30°C.

7. Repeat the diploid selection as described in Step 6. Incubate for 2 d at 30°C.

   Images of the plates should be taken at every selection step to be able to track down the causes of missing colonies at the end of the experiment (see Step 18).

Sporulation and Selection of Recombinant Haploid Strains

8. Transfer the diploid colonies onto an omniplate containing solid agar and sporulation medium (spo.a for the cross with the strain expressing the bait recombinant protein or spo.α for the cross with the strain expressing the prey recombinant protein) using the appropriate pintool. Incubate for 7 d at 22°C

   Omniplates should be wrapped in humidified paper towels to avoid drying of the medium.

9. Transfer the sporulated colonies onto an omniplate containing solid agar and a mating-type selection medium (h2.1a for the plates from the cross with the strain expressing the bait recombinant protein, and h2.1α for the plates from the cross with the strain expressing the prey recombinant protein) using the appropriate pintool. Allow germination to occur and incubate the plate for 2 d at 30°C.

Minus His selects for the MATa progeny and –Leu selects for the MATα progeny. Although any of the two mating types can be selected, a different mating type must be selected for colonies expressing either the bait or the prey recombinant protein to allow subsequent crosses for the final PCA step. However, it is advisable to select bait recombinant proteins in MATa cells and prey recombinant protein in MATα cells to avoid growth of the residual parental haploid deletion strains on the haploid selection medium.

10. Repeat the haploid selection as described in Step 9. Incubate for 2 d at 30°C.

11. Transfer the haploid colonies onto an omniplate containing solid agar and a deletion selection medium (h2.3a for the plates from the cross with the strain expressing the bait recombinant protein, and h2.3α for the plates from the cross with the strain expressing the prey recombinant protein) using the appropriate pintool. Incubate the plate for 2 d at 30°C.

12. Transfer the haploid colonies onto an omniplate containing solid agar and a methionine prototrophy selection medium (h2.4a for the plates from the cross with the strain expressing the bait recombinant protein, and h2.4α for the plates from the cross with the strain expressing the prey recombinant protein) using the appropriate pintool. Incubate the plate for 2 d at 30°C.

   This step is optional for the colonies expressing the prey recombinant protein as the parental diploid is homozygous for the presence of the MET15 gene. However, it is advisable to perform the step to avoid confusion between media and steps between the two parallel SGAs and to continue the deletion selection.

13. Transfer the haploid colonies onto an omniplate containing solid agar and a recombinant protein selection medium (h2.5a for the plates from the cross with the strain expressing the bait recombinant protein, and h2.5α for the plates from the cross with the strain expressing the prey recombinant protein) using the appropriate pintool. Incubate the plate for 2 d at 30°C.

Diploid Strain Construction and PCA

14. Transfer the haploid colonies expressing the bait recombinant protein onto an omniplate containing solid agar and a rich medium (YPD) using the appropriate pintool.

15. Transfer the haploid colonies expressing the prey recombinant protein onto the bait strains using the appropriate pintool. Allow mating to occur and incubate for 2 d at 30°C.

    Plates from the two parallel SGAs should be mated in a pair-wise fashion—that is, crossing plates harboring the same deletion mutants against one another to obtain diploid homozygous deletion mutants. Other configurations could be used if combinations such as double-heterozygous strains are to be studied.

16. Transfer the mixed haploid–diploid colonies onto an omniplate containing the diploid selection medium (sel2n2) using the appropriate pintool. Incubate for 2 d at 30°C.

17. Repeat the diploid selection as described in Step 16. Incubate for 2 d at 30°C.

18. Transfer diploid colonies onto an omniplate containing the scoring medium (MTX medium) using the appropriate pintool. Incubate for 4 d at 30°C.

19. Image plates and repeat the selection as described in Step 18. Incubate for 4 d at 30°C while imaging plates every day.

    See Troubleshooting.

Image and Statistical Analysis

20. Proceed to the image-analysis stage, following the method described by Leducq et al. (2012).

21. Conduct a statistical analysis of each SGA–PCA screen by following three major steps that, first, correct the plate-to-plate variation; second, correct for fitness effects, and finally identify regulators of the protein–protein interaction of interest by, respectively, conducting the following actions.

    i. Subtract the median colony size (log transformed) of each plate from each colony size.

    ii. Subtract the colony size of a constitutive interaction measured in the same deletion background to correct for fitness effects and positional biases.

iii. Compute a perturbation score by averaging the net intensity values of the replicates and measuring the difference in colony size between wild-type and mutant backgrounds for each interaction.

iv. Finally, conduct appropriate statistical tests to test the significance of the observed effects.

Troubleshooting

Problem: No colony growth during a selection step.
Solution: Verify that the correct culture conditions have been used and that the mating-type-specific selection medium has not been inverted.

Problem (Step 19): Colonies are missing at the end of the experiment owing to technical issues (improper mating early in the experiment, improper transfers).
Solution: Take images of plates at all selection steps to allow tracking down of missing strains.

Problem (Step 19): Colonies are missing at the end of the experiment owing to biological issues (genetic linkage between the deletion and the fusion gene, in which case the frequency of recombination during sporulation is really low).
Solution: An examination of the genetic distance between *DHFR* fusion genes and deleted genes will allow the identification of any improper strain construction caused by lack of recombination.

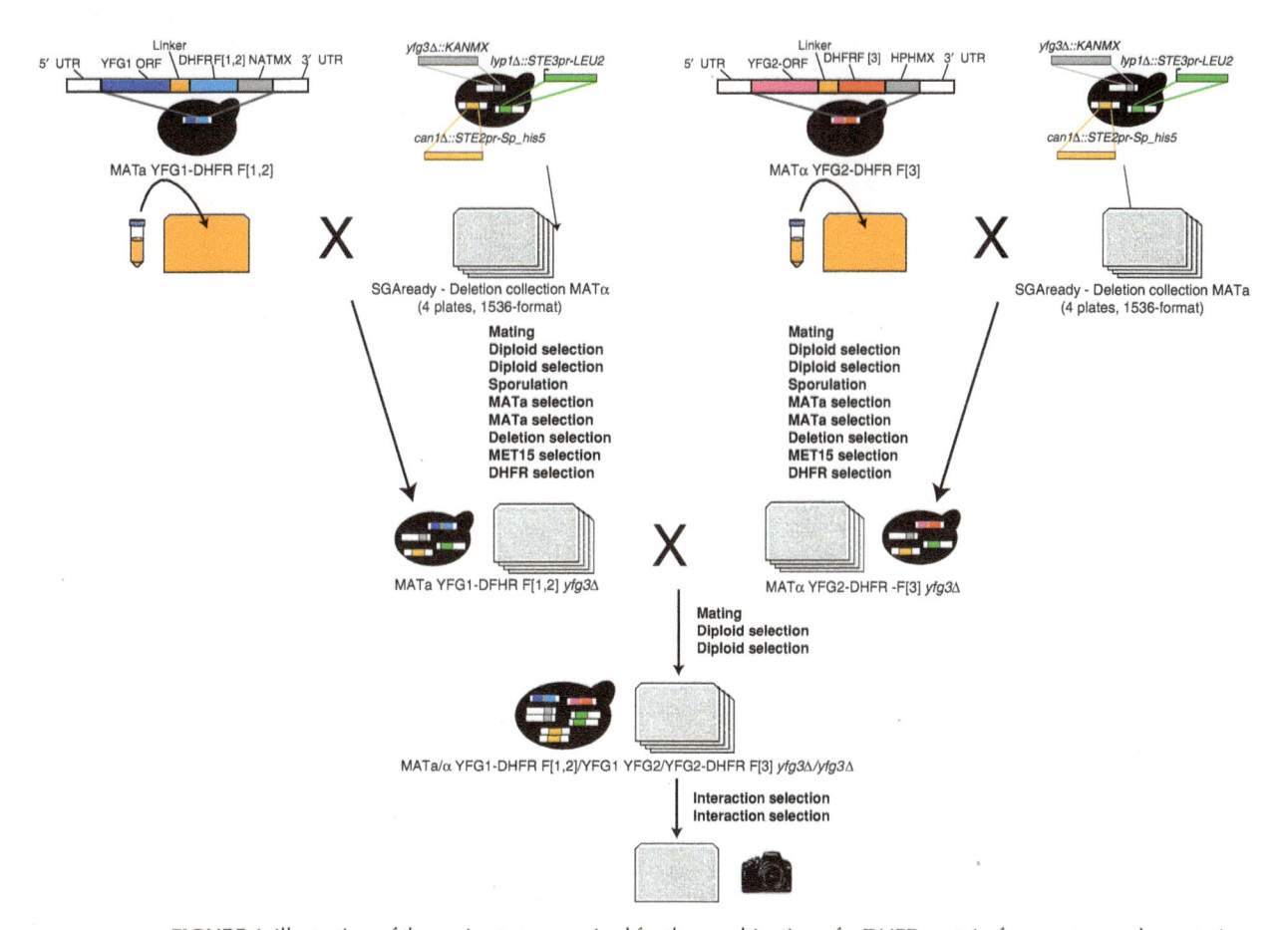

FIGURE 1. Illustration of the main stages required for the combination of a DHFR protein-fragment complementation assay (PCA) and synthetic genetic array (SGA) (Costanzo et al. 2010) to identify regulators of protein–protein interactions. The *MATa* bait and the *MATα* prey strains are mated, respectively, with the *MATα* and *MATa* SGAready collections. The two SGAs are performed in parallel, and the resulting haploid strains, each expressing one of the two recombinant bait or prey proteins in a deletion background, are mated to perform the DHFR PCA assay. DHFR, dihydrofolate reductase.

DISCUSSION

Figure 1 illustrates the screening of one interaction against the deletion collection. Each screen relies on a first step consisting of two parallel SGA procedures (Costanzo et al. 2010) for switching the genetic backgrounds of the DHFR PCA bait and prey strains. The second step consists of the DHFR PCA growth assay. Each step can be repeated to screen more than one interaction. The purpose is to identify genes that, when deleted, enhance or reduce the interaction between two proteins of interest through the measurement of colony size. Colony size variation is the result of two effects: first, the actual deletion effect on the interaction (PCA perturbation score) and, second, the fitness effect of the deletion. The entire screen should be performed in parallel using a constitutive interaction or full-length DHFR enzyme (Leducq et al. 2012; Diss et al. 2013; Lev et al. 2013) in the same genetic backgrounds to control for the fitness effects. In addition, diploids from Step 17 can be transferred in parallel onto a medium similar to the MTX medium but without MTX while still containing the MTX solvent, dimethyl sulfoxide. Colony sizes can then be used to correct for any fitness effect caused by the deletion or the *DHFR* fragment fusions (Rochette et al. 2014). To estimate the variation (both technical and biological) in colony size and the false-discovery rate, a mock deletion (e.g., *HO* or *HIS3*) can be inserted at several positions in the array.

RECIPES

Amino Acid Mix (10×)

Amino acid or nucleobase	Concentration (10×)
Adenine sulfate	0.4 g/mL
Uracil	0.2 g/mL
L-Tryptophan	0.4 g/mL
L-Histidine hydrochloride	0.2 g/mL
L-Arginine hydrochloride	0.2 g/mL
L-Tyrosine	0.3 g/mL
L-Leucine	0.6 g/mL
L-Lysine hydrochloride	0.3 g/mL
L-Phenylalanine	0.5 g/mL
L-Glutamic acid	1.0 g/mL
L-Asparagine	1.0 g/mL
L-Valine	1.5 g/mL
L-Threonine	2.0 g/mL
L-Serine	3.75 g/mL
L-Methionine	0.2 g/mL

Dissolve the amino acids from the list above (except those to be excluded from any drop-out media) in distilled H$_2$O. Filter-sterilize, store at 4°C, and protect from light.

Methotrexate (MTX) Medium

Reagents	Final concentration (1×)
Yeast nitrogen base without amino acids but with ammonium sulfate	6.7 g/L
Noble agar	2.5% (w/v)
Amino acid mix (10×) <R>	1×
Glucose	2% (w/v)
Methotrexate (e.g., Bioshop)	200 µg/mL

Prepare the MTX medium 1–2 d before use and store at room temperature.

Spo.a and Spo.α Sporulation Media

Reagents	Final concentration (1×)
Potassium acetate	10 g/L
Yeast extract	1 g/L
Agar	2% (w/v)
Uracil, histidine, leucine, and lysine mix at a mass ratio of 1:1:5:1	0.1 g/L
Glucose	0.5 g/L
Geneticin	50 µg/mL
Nourseothricin (for spo.a medium only)	25 µg/mL
Hygromycin B (for spo.α medium only)	62.5 µg/mL

Store for up to several months at 4°C.

Synthetic Complete (SC) Medium

Reagents	Final concentration (1×)
Yeast nitrogen base without amino acids and with ammonium sulfate	6.7 g/L
Agar	2% (w/v)
Amino acid mix (10×) <R>	1×
Glucose	2% (w/v)

Store synthetic complete medium for up to several months at 4°C.

Synthetic Complete–Monosodium Glutamic Acid (SC–MSG) Medium

Reagents	Final concentration (1×)
Yeast nitrogen base without amino acids and without ammonium sulfate	6.7 g/L
Agar	2% (w/v)
Amino acid mix (10×) <R>	1×
Glucose	2% (w/v)
Monosodium glutamic acid (MSG)	0.1%

Store SC–MSG medium for up to several months at 4°C.

YPD for the DHFR PCA

Reagents	Final concentration (1×)
Yeast extract	1% (w/v)
Peptone	2% (w/v)
Dextrose	2% (w/v)
Agar	2% (w/v)

Store for up to several months at 4°C.

REFERENCES

Costanzo M, Baryshnikova A, Bellay J, Kim Y, Spear ED, Sevier CS, Ding H, Koh JL, Toufighi K, Mostafavi S, et al. 2010. The genetic landscape of a cell. *Science* **327**: 425–431.

Diss G, Dube AK, Boutin J, Gagnon-Arsenault I, Landry CR. 2013. A systematic approach for the genetic dissection of protein complexes in living cells. *Cell Rep* **3**: 2155–2167.

Leducq JB, Charron G, Diss G, Gagnon-Arsenault I, Dube AK, Landry CR. 2012. Evidence for the robustness of protein complexes to inter-species hybridization. *PLoS Genet* **8**: e1003161.

Lev I, Volpe M, Goor L, Levinton N, Emuna L, Ben-Aroya S. 2013. Reverse PCA, a systematic approach for identifying genes important for the physical interaction between protein pairs. *PLoS Genet* **9**: e1003838.

Rochette S, Gagnon-Arsenault I, Diss G, Landry CR. 2014. Modulation of the yeast protein interactome in response to DNA damage. *J Proteomics* **100**: 25–36.

Protocol 3

Dissecting the Contingent Interactions of Protein Complexes with the Optimized Yeast Cytosine Deaminase Protein-Fragment Complementation Assay

Po Hien Ear,[1,2] Jacqueline Kowarzyk,[1] and Stephen W. Michnick[1,3]

[1]Département de Biochimie et Médecine Moléculaire, Université de Montréal, Montréal, Québec H3C 3J7, Canada; [2]Department of Genetics, Harvard Medical School, Boston, Massachusetts 02115

Here, we present a detailed protocol for studying in yeast cells the contingent interaction between a substrate and its multisubunit enzyme complex by using a death selection technique known as the optimized yeast cytosine deaminase protein-fragment complementation assay (OyCD PCA). In yeast, the enzyme cytosine deaminase (encoded by *FCY1*) is involved in pyrimidine metabolism. The PCA is based on an engineered form of yeast cytosine deaminase optimized by directed evolution for maximum activity (OyCD), which acts as a reporter converting the pro-drug 5-fluorocytosine (5-FC) to 5-fluorouracil (5-FU), a toxic compound that kills the cell. Cells that have OyCD PCA activity convert 5-FC to 5-FU and die. Using this assay, it is possible to assess how regulatory subunits of an enzyme contribute to the overall interaction between the catalytic subunit and the potential substrates. Furthermore, OyCD PCA can be used to dissect different functions of mutant forms of a protein as a mutant can disrupt interaction with one partner, while retaining interaction with others. As it is scalable to a medium- or high-throughput format, OyCD PCA can be used to study hundreds to thousands of pairwise protein–protein interactions in different deletion strains. In addition, OyCD PCA vectors (pAG413GAL1-ccdB-OyCD-F[1] and pAG415GAL1-ccdB-OyCD-F[2]) have been designed to be compatible with the proprietary Gateway technology. It is therefore easy to generate fusion genes with the OyCD reporter fragments. As an example, we will focus on the yeast cyclin-dependent protein kinase 1 (Cdk1, encoded by *CDC28*), its regulatory cyclin subunits, and its substrates or binding partners.

MATERIALS

It is essential that you consult the appropriate Material Safety Data Sheets and your institution's Environmental Health and Safety Office for proper handling of equipment and hazardous material used in this protocol.

RECIPES: Please see the end of this protocol for recipes indicated by <R>. Additional recipes can be found online at http://cshprotocols.cshlp.org/site/recipes.

Reagents

Accuprime Pfx DNA polymerase (Invitrogen)
Antibodies against yCD fragments (Anti-yCD polyclonal; Biogenesis)
"BP Gateway" clonase enzyme mix (Invitrogen)

---

[3]Correspondence: stephen.michnick@umontreal.ca

Copyright © Cold Spring Harbor Laboratory Press; all rights reserved
Cite this protocol as *Cold Spring Harb Protoc*; doi:10.1101/pdb.prot090043

Competent BY4741 MATa (his3Δ 1; leu2Δ 0; met15Δ 0; ura3Δ 0) yeast strains with single gene deletion (*fcy1Δ, cln1Δ, cln2Δ, cln3Δ, clb1Δ, clb2Δ, clb3Δ, clb4Δ, clb5Δ,* and *clb6Δ*) (Giaever et al. 2002), prepared according to Knop et al. (1999)
Dimethylsulfoxide (DMSO)
5-Fluorocytosine (5-FC) (10 mg/mL) <R>
Galactose [20% (w/v) stock solution, sterile]
Geneticin/G418 (200 mg/mL stock in water, filter sterilized) (Bioshop)
"LR Gateway" clonase enzyme mix (Invitrogen)
Nourseothricin/clonNat (WERNER BioAgents)
Plasmids and oligonucleotides (see Table 1)
PLATE (PEG–LiAc–Tris–EDTA) solution <R>
Raffinose [20% (w/v), sterile]
Synthetic complete (SC) agar medium for OyCD PCA <R>
Water (distilled and sterile)
YPD media (liquid and solid) for the OyCD PCA <R>

Equipment

Bench centrifuge (e.g., Centrifuge 5810R, Eppendorf)
Incubator and shaking incubators, preset to 30°C and 37°C
Omniplates (Nunc)
Polymerase chain reaction (PCR) machine
Pintool, robotically manipulated [96 pintool (0.787 mm flat round-shaped pins, #FP3N, V&P Scientific) and 384 pintool (0.457 mm flat round-shaped pins, custom #FP1N, V&P Scientific, Inc.)] or manually manipulated 96 pintool (1.58 mm, 1 µL slot pins, VP 408Sa, V&P Scientific, Inc.)
Plate imager
  Typically required are a ≥4.0 megapixel camera (e.g., Canon Powershot A520), a stationary arm (70 cm mini repro; Industria Fototecnica) and a plate-shooting platform.
Shaker

TABLE 1. Plasmids and oligonucleotides for OyCD PCA

Plasmids and oligonucleotides	Description
BG1805 plasmids with the genes of interest (GE/Dharmacon/Open Biosystems)	Gateway yeast ORF collection expression clones
pDONR221 (Invitrogen)	Gateway donor vector
pAG413GAL1-ccdB-L-OyCD-F[1] and pAG415GAL1-ccdB-L-OyCD-F[2] (Ear et al. 2013)	Gateway destination vectors used to generate fusion genes with the yCD reporter fragments, using the Gateway technology. OyCD-F[1]: amino acid residues 1–77 of yCD with an A23L point mutation. OyCD-F[2]: amino acid residues 57–158 of yCD with V108I, I140L, T95S, and K117E point mutations
pAG25 plasmid (Goldstein and McCusker 1999)	Contains nourseothricin resistance gene, which is used to knockout *FCY1* gene
p415GAL1-Cdk1-L-OyCD-F[2] (Ear et al. 2013)	Used to study PPIs with Cdk1
p415GAL1-L-OyCD-F[2] (Ear et al. 2013)	Serves as negative control when used alone
p413GAL1-Zipp.-L-OyCD-F[1][a] and p415GAL1-Zipp.-L-OyCD-F[2] (Ear et al. 2013) p413GAL1-Ssk1-L-OyCD-F[1] and p415GAL1-Ypd1(W80A)-L-OyCD-F[2] (Ear and Michnick 2009)	Positive control plasmid pairs
Forward-FCY1-KO oligo (5′ to 3′) (Ear et al. 2013)	TGAGAGCCAGCTTAAAGAGTTAAAAATTTCATAGCTAATGgcgccagatctgtttagctt
Reverse-FCY1-KO oligo (5′ to 3′) (Ear et al. 2013)	ATAAAATTAAATACGTAAATACAGCGTGCTGCGTGCTCTAggttaacctggcttatcgaa

[a]Zipp. consists of the GCN4 leucine zipper. The linker (L) comprises a 10-amino-acid flexible polypeptide linker consisting of (Gly.Gly.Gly.Gly.Ser)$_2$ between the protein of interest and the reporter protein fragments.

Abbreviations: Cdk1, cyclin-dependent kinase 1; ORF, open reading frame; OyCD, optimized yeast cytosine deaminase; PPI, protein–protein interaction; yCD, yeast cytosine deaminase.

METHOD

Construction of an Expression Plasmid Containing the Gene of Interest

1. Select genes from the yeast ORF collection (expression clones; see Table 1) to transfer into a Gateway donor vector (pDONR221) to obtain "Entry" clones using BP Gateway clonase enzyme mix (Invitrogen). Perform a diagnostic PCR to determine the presence of the gene of interest or send the plasmids (Entry clones) for sequencing.

   Store plasmids at −20°C.

   See Troubleshooting.

2. Use LR Gateway clonase enzyme mix (Invitrogen) to perform the LR reaction with the Entry clones and the destination vector (pAG413GAL1-ccdB-OyCD-F[1]) to obtain the expression plasmid (pAG413GAL1-GeneX-OyCD-F[1]) (Ear and Michnick 2009). Confirm by diagnostic PCR the presence of your gene of interest ("GeneX") or send the expression plasmids for sequencing.

   Store plasmids at −20°C.

   See Troubleshooting.

Deletion of the *FCY1* Gene in the Cyclin Deletion Strains

In the yeast S. cerevisiae, the FCY1 gene encodes yeast cytosine deaminase (yCD). Therefore, to obtain colony growth only when proteins of interest do not interact, OyCD PCA death selection assays must be performed in an FCY1 deletion strain.

3. Use the oligonucleotides given in Table 1 to amplify the nourseothricin resistance gene from the pAG25 plasmid (Goldstein and McCusker 1999) using the AccuPrime Pfx DNA polymerase.

4. Transform each cyclin deletion strain with the PCR product: mix 50 µL of thawed competent cells with 5 µL (2 µg) of the PCR-amplified cassette DNA encoding the NAT1 resistance marker. Mix and incubate for 30 min at room temperature. Add 500 µL of PLATE solution and 50 µL of DMSO, followed by heat-shock for 20 min at 42°C. Centrifuge at 805*g* for 3 min, discard supernatant, and resuspend cells in 1 mL YPD liquid medium and incubate for 4 h at 30°C with shaking. Centrifuge the cells, remove supernatant and resuspend cells in 200 µL of YPD liquid and then plate on a selection plate (YPD agar medium with 200 µg/mL G418, 100 µg/mL nourseothricin and 100 µg/mL 5-FC).

   Shorter or longer incubation times can result in decreased transformation efficiency. Use fresh competent cells to increase the transformation efficiency. The NAT1 gene encodes a subunit of the protein amino-terminal acetyltransferase, which confers resistance to the antibiotic nourseothricin (trade name: clonNat).

   See Troubleshooting.

5. Incubate the plates for 3–5 d at 30°C. Further verify the colonies by colony PCR methods.

   If time permits, prepare a glycerol stock of the correct strains.

OyCD PCA to Detect Protein–Protein Interactions with Cdk1

Before proceeding with a screen in the various cyclin deletion strains, verify that interactions between the proteins of interest can be detected by the OyCD PCA (Fig. 1). Titrate the amount of 5-FC required for detecting the interactions. We normally try a range between 100 µg/mL and 1 mg/mL of 5-FC. For many cell cycle proteins and Cdk1 interactions, 1 mg/mL of 5-FC is required for detecting OyCD PCA activity. Consider using well-known interacting Cdk1 and Swi6 or the homodimeric GCN4 leucine zippers (Zip–Zip) and noninteracting (Cdk1 and OyCD-F[2] alone) proteins as controls (Table 1).

6. Transform the selected genes fused to the OyCD-F[1] sequence in Gateway expression clones separately into competent BY4741 *fcy1*Δ yeast containing either p415GAL1-Cdk1-L-OyCD-F[2]

Protein-Fragment Complementation Assays in Yeast

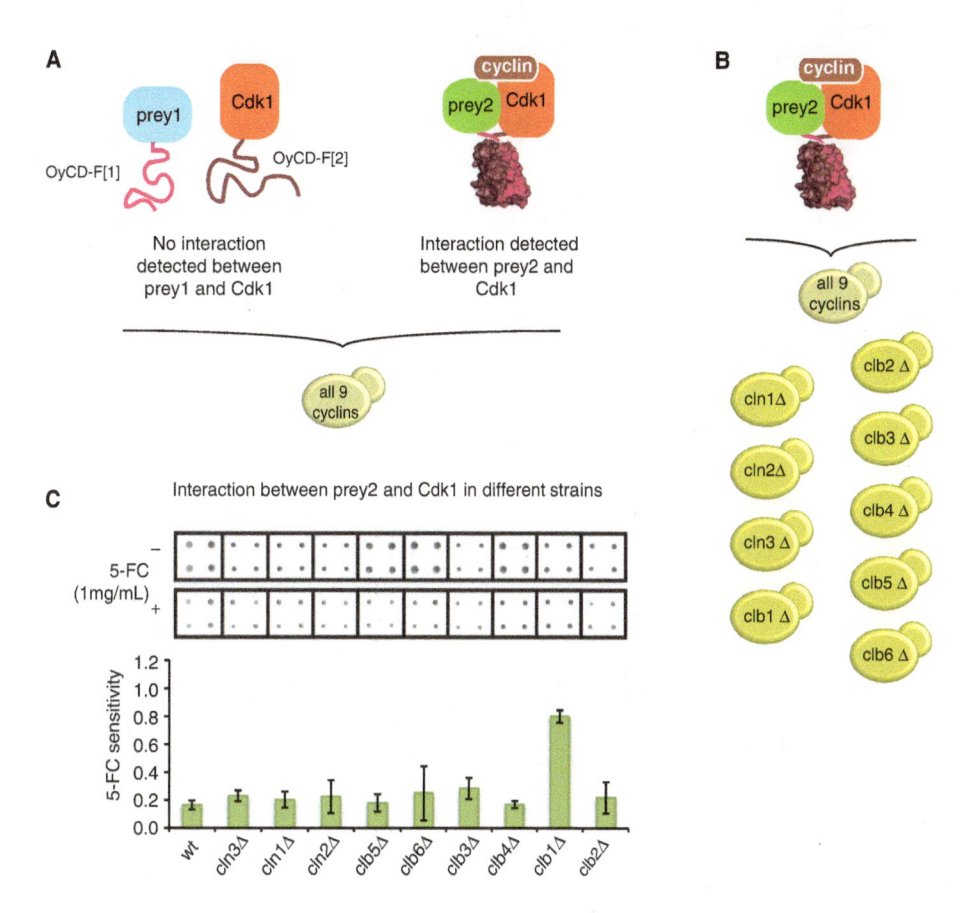

FIGURE 1. Detecting protein–protein interactions between cyclin-dependent kinase 1 (Cdk1) and potential substrates in different yeast strains. (*A*) Cartoon showing the approach for determining the activity of the optimized yeast cytosine deaminase protein-fragment complementation assay (OyCD PCA) (Ear et al. 2013) for interactions between Cdk1 and its prey proteins in a yeast strain harboring all nine cyclin genes. This entails expressing a prey protein in yeast cells in which Cdk1 is fused to one OyCD reporter fragment (F[2]), and the prey protein is fused to the complementary fragment (F[1]). Illustrated is an example in which, if "prey1" does not interact with Cdk1, the OyCD is not reconstituted and folded correctly, and so the cells are resistant to 5-fluorocytosine (5-FC). This contrasts with the example of "prey2" that does interact with Cdk1, meaning that the OyCD fragments can interact and fold correctly, restoring OyCD activity and causing the cells to become sensitive to 5-FC. (*B*) To test for the identity of the acting cyclin(s), the above procedure can be performed in different cyclin deletion strains. (*C*) Determination of the activity of the OyCD PCA for Cdk1 and its interacting partner (prey2) in different cyclin deletion strains and the wild type (wt). If a cyclin is required for maintaining the interaction between Cdk1 and the prey protein (as illustrated for Clb1 here), removal of the cyclin will affect the OyCD PCA activity (resolved as resistance to killing by 5-FC).

or p415GAL1-L-OyCD-F[2]. Mix 10 µL of cells with 200 ng of each yeast expression plasmid, 60 µL of PLATE solution and 8 µL of DMSO. Heat-shock yeast for 20 min at 42°C. Centrifuge at 805*g* for 3 min. Remove the supernatant and resuspend the cells in 50 µL sterile water and plate the cell suspension on SC agar medium plates lacking histidine and leucine and containing 2% (w/v) glucose. Incubate for 3 d at 30°C.

> *Shorter or longer incubation times can result in decreased transformation efficiency. Use fresh competent cells to increase the transformation efficiency.*

7. Inoculate colonies from each transformation in a 96-well plate containing 400 µL of SC liquid medium without histidine and leucine and with 2% (w/v) raffinose and incubate in a shaking incubator for 16 h at 30°C.

   > *If time permits, prepare a glycerol stock of yeast harboring the two plasmids.*

8. Add galactose to each culture at a final concentration of 2% (w/v) and incubate for >1 h at 30°C to induce the expression of the OyCD fusion proteins.

9. Pin the samples onto Nunc omniplates (SC medium lacking histidine and leucine with 2% (w/v) raffinose, 2% (w/v) galactose, 200 µg/mL of G418 and 3% agar with and without 1 mg/mL of 5-FC) using manual or robotic pintools.

10. Incubate plates for 2–3 d at 30°C. Use plate imager assembly to take pictures of the plates at days 2 and 3.

Detecting Protein–Protein Interactions in the Different Cyclin Deletion Strains

11. Screen the interaction between Cdk1 and its potential substrate in yeast strains expressing all nine cyclin genes or lacking one of the nine cyclin genes. Cotransform the potential substrate genes fused to the OyCD-F[1] sequence in Gateway expression vectors with p415GAL1-Cdk1-L-OyCD-F[2] into the *FCY1* deletion (*fcy1Δ*) strain and into the nine double-deletion strains for the single cyclins and *FCY1*. Follow a transformation protocol similar to that in Step 6.

12. Inoculate four clones from each transformation and grow them to saturation in a 96-well plate containing SC medium lacking histidine and leucine with 2% (w/v) glucose and 200 µg/mL (w/v) of G418.

    If time permits, prepare a glycerol stock of these cultures.

13. Pin all samples from the glycerol stock onto omniplates containing the same medium with 3% (w/v) agar and allow to grow for 4 d. To produce more replicates, pin the samples four times onto the omniplates using a 96 pintool and create a 384-colony array.

14. For evaluating the OyCD PCA activity, pin the colonies onto SC medium lacking histidine and leucine with 2% (w/v) raffinose and 2% (w/v) galactose, and 200 µg/mL of G418 and 3% agar plates with and without 1 mg/mL of 5-FC using a robotically manipulated 384 pintool.

15. Acquire pictures after 1, 2, 3, and 4 d of incubation at 30°C.

Analysis of the OyCD PCA Results

16. Quantify cell growth using ImageJ (Abràmoff et al. 2004) by calculating the integrated intensity of each colony. Measure the activity of the OyCD PCA by taking the ratio of integrated intensity of each colony grown on 1 mg/mL of 5-FC over the integrated intensity of colonies grown in the absence of 5-FC.

    Consider the fcy1Δ yeast strain as acting as the control strain in this screen and refer to it as the wild-type (wt) strain as it expresses all nine cyclins.

    See Troubleshooting.

TROUBLESHOOTING

Problem (Steps 1 and 2): No colonies grow after the BP or LR reaction.
Solution: Increase the amount of plasmid used, use new BP or LR enzymes and/or buy high-efficiency DH5α competent cells.

Problem (Step 4): No *fcy1Δ* and *cyclinΔ* double-mutant colonies grow.
Solution: Optimize the PCR reaction to get at least 100 ng of DNA per microliter, use freshly made competent cells and/or incubate cells for 16 h instead of 4 h at 30°C with shaking.

Problem (Step 16): No modulation of OyCD PCA activity occurs in the deletion strain.
Solution: The 5-FC concentration might be too high; use various 5-FC concentrations (from 100 to 1000 µg/mL).

RECIPES

Amino Acid Mix (10×)

Amino acid or nucleobase	Concentration (10×)
Adenine sulfate	0.4 g/mL
Uracil	0.2 g/mL
L-Tryptophan	0.4 g/mL
L-Histidine hydrochloride	0.2 g/mL
L-Arginine hydrochloride	0.2 g/mL
L-Tyrosine	0.3 g/mL
L-Leucine	0.6 g/mL
L-Lysine hydrochloride	0.3 g/mL
L-Phenylalanine	0.5 g/mL
L-Glutamic acid	1.0 g/mL
L-Asparagine	1.0 g/mL
L-Valine	1.5 g/mL
L-Threonine	2.0 g/mL
L-Serine	3.75 g/mL
L-Methionine	0.2 g/mL

Dissolve the amino acids from the list above (except those to be excluded from any drop-out media) in distilled H_2O. Filter sterilize, store at 4°C, and protect from light.

5-Fluorocytosine (5-FC) (10 mg/mL)

Combine 100 mg of 5-FC and 10 mL of distilled water. Vortex. Incubate for 2 h at 37°C, manually mixing the solution every 15 min until the 5-FC dissolves. Filter the solution and use it immediately. Alternatively, aliquot into sterile tubes and store at −20°C.

PLATE Solution

Reagents	Final concentration (1×)
PEG 3350	40% (w/v)
Lithium acetate	100 mM
Tris (pH 7.5)	10 mM
EDTA	0.4 mM

Wear personal protective equipment when handling this reagent. Store at room temperature in a dry and well-ventilated place.

Synthetic Complete (SC) Agar Medium for OyCD PCA

Reagents	Final concentration (1×)
Yeast nitrogen base (YNB) without amino acids and without ammonium sulfate	1.74 g/L
Agar (for solid media only; see below)	2%/3% (w/v)
Amino acid mix (10×), without histidine or leucine <R>	1×
Glucose (if required; see below)	2% (w/v)

Reagent	Concentration
Raffinose (if required; see below)	2% (w/v)
Galactose (if required; see below)	2% (w/v)
G418 (if required; see below)	200 µg/mL (w/v)
5-Fluorocytosine (5-FC) (if required; see below)	1 mg/mL (w/v)

Use 3% (w/v) agar in OmniTray plates or 2% (w/v) for Petri dishes. Prepare the following formulations: (1) solid medium in Petri plates with glucose; (2) solid medium in Omniplates and liquid medium with glucose and G418; (3) liquid medium with raffinose; (4) and solid medium in Omniplates with raffinose, galactose, and G418, with and without 5-FC. For long term storage, keep at 2°C–8°C.

YPD Media (Liquid and Solid) for the OyCD PCA

Reagents	Final concentration (1×)
Yeast extract	1% (w/v)
Peptone	2% (w/v)
Dextrose	2% (w/v)
Agar (for solid medium only)	2% (w/v)
G418	200 µg/mL
Nourseothricin	100 µg/mL
5-Fluorocytosine (5-FC)	100 µg/mL

Prepare in Petri plates. For long term storage, keep at 2°C–8°C.

REFERENCES

Abràmoff MD, Magalhaes PJ, Ram SJ. 2004. Image processing with ImageJ. *Biophotonics Int* 11: 36–42.

Ear PH, Michnick SW. 2009. A general life-death selection strategy for dissecting protein functions. *Nat Methods* 6: 813–816.

Ear PH, Booth MJ, Abd-Rabbo D, Kowarzyk Moreno J, Hall C, Chen D, Vogel J, Michnick SW. 2013. Dissection of Cdk1-cyclin complexes in vivo. *Proc Natl Acad Sci* 110: 15716–15721.

Giaever G, Chu AM, Ni L, Connelly C, Riles L, Veronneau S, Dow S, Lucau-Danila A, Anderson K, Andre B, et al. 2002. Functional profiling of the *Saccharomyces cerevisiae* genome. *Nature* 418: 387–391.

Goldstein AL, McCusker JH. 1999. Three new dominant drug resistance cassettes for gene disruption in *Saccharomyces cerevisiae*. *Yeast* 15: 1541–1553.

Knop M, Siegers K, Pereira G, Zachariae W, Winsor B, Nasmyth K, Schiebel E. 1999. Epitope tagging of yeast genes using a PCR-based strategy: More tags and improved practical routines. *Yeast* 15: 963–972.

Protocol 4

Real-Time Protein-Fragment Complementation Assays for Studying Temporal, Spatial, and Spatiotemporal Dynamics of Protein–Protein Interactions in Living Cells

Mohan Malleshaiah,[1,2] Emmanuelle Tchekanda,[1] and Stephen W. Michnick[1,3]

[1]Département de Biochimie et Médecine Moléculaire, Université de Montréal, Montréal, Québec H3C 3J7, Canada; [2]Department of Systems Biology, Harvard Medical School, Boston, Massachusetts 02115

Here, we present detailed protocols for direct, real-time protein-fragment complementation assays (PCAs) for studying the spatiotemporal dynamics of protein–protein interactions (PPIs). The assays require the use of two fluorescent reporter proteins—the "Venus" version of yellow fluorescent protein (vYFP), and the monomeric infrared fluorescent protein 1.4 (IFP 1.4)—or two luciferase reporter proteins—*Renilla* (Rluc) and *Gaussia* (Gluc). The luciferase PCAs can be used to study the temporal dynamics of PPIs in any cellular compartment and on membranes. The full reversibility of these PCAs assures accurate measurements of the kinetics of PPI assembly/disassembly for processes that occur anywhere in a living cell and over time frames of seconds to hours. vYFP PCA, and all PCAs based on green fluorescent protein and its variants, are irreversible and can be used to trap and visualize rare and transient complexes and follow dynamic relocalization of constitutive complexes. vYFP PCA is limited in that accurate measurements of temporal changes in PPIs are not possible owing to the slow maturation time of vYFP (minutes to hours) and the irreversibility of its PCA that traps the complexes, thereby preventing the dissociation of PPIs that, in some instances, might cause spurious mislocalization of protein complexes. The limitations of vYFP PCA are overcome with IFP PCA, which is fully reversible and thus can be used to study spatiotemporal dynamics of PPIs on the timescale of seconds. All of these PCAs are sensitive enough to detect interactions among proteins expressed at endogenous levels in vivo.

MATERIALS

It is essential that you consult the appropriate Material Safety Data Sheets and your institution's Environmental Health and Safety Office for proper handling of equipment and hazardous material used in this protocol.

RECIPES: Please see the end of this protocol for recipes indicated by <R>. Additional recipes can be found online at http://cshprotocols.cshlp.org/site/recipes.

Reagents

Amino acid mix (10×) <R>
Benzyl-coelenterazine (2 mM in absolute ethanol) (Nanolight)
Biliverdin hydrochloride (Frontier Scientific)
ClonNAT (nourseothricin; Jena Bioscience)
Coelenterazine (prepared following the manufacturer's instructions) (Nanolight)

---

[3]Correspondence: stephen.michnick@umontreal.ca

Copyright © Cold Spring Harbor Laboratory Press; all rights reserved
Cite this protocol as *Cold Spring Harb Protoc*; doi:10.1101/pdb.prot090068

Competent BY4741 *MAT*a (his3Δ 1; leu2Δ 0; met15Δ 0; ura3Δ 0) or BY4742 *MAT*α (his3Δ 1; leu2Δ 0; lys2Δ 0; ura3Δ 0) or BY4743 (diploid) *Saccharomyces cerevisiae* yeast strains prepared according to Knop et al. (1999)

Concanavalin A (Sigma-Aldrich) or poly-L-lysine (mol. wt 30,000–70,000; 1 mg/mL) (Sigma-Aldrich)

Dimethylsulfoxide (DMSO)

Galactose solution [20% (w/v), sterile]

Gene(s) of interest fused to Venus YFP, *R*luc, *G*luc, or IFP PCA fragments in yeast expression vectors (see Table 1)

Glucose [2 M or 20% (w/v), sterile]

Hygromycin B (Wisent Bioproducts)

Low-fluorescence yeast nitrogen base without riboflavin and folic acid medium (LFM) (10×) <R>
  This medium has negligible autofluorescence (within 10% of that of water) (Sheff and Thorn 2004).

PCA fragment cassettes
  Gluc, Rluc, or IFP PCA fragment cassettes containing PCA fragment 1 (PCR amplified from pAG25-linker-F[1]) or fragment 2 (PCR amplified from pAG32-linker-F[2]) to endogenously tag genes of interest (Table 1).

Phosphate-buffered saline (PBS; 10×, pH 7.4) <R>

PLATE solution <R>

Synthetic complete (SC) and synthetic defined (SD) media for PCAs <R>
  Prepare SC as liquid medium with raffinose and with appropriate amino acids or antibiotic according to the chosen expression plasmids or integration cassettes (Table 1). Also prepare SC as solid medium with agar, glucose, and appropriate amino acids or antibiotic (Table 1).

Water (distilled and sterile)

TABLE 1. Plasmids used in real-time PCAs

PCA reporter protein	Plasmids[a]	Drop-out media or antibiotic resistance	Comments
Venus YFP	p413ADH-ProteinA-L-vYFP-F[1]	–His (p413ADH)	Expression plasmids:
	p415ADH-ProteinB-L-vYFP-F[2]	–Leu (p415ADH)	vYFP-F[1]: amino acids 1–158 of vYFP
	Positive controls:		vYFP-F[2]: amino acids 159–239 of vYFP (Malleshaiah et al. 2010)
	p413ADH-Zipp.-L-vYFP-F[1]		
	p415ADH-Zipp.-L-vYFP-F[2]		
*R*luc	p413ADH-ProteinA-L-*R*luc-F[1]	–His (p413ADH)	Expression plasmids:
	p415ADH-ProteinB-L-*R*luc-F[2]	–Leu (p415ADH)	*R*luc-F[1]: amino acids 1–110 of *R*luc
	Positive controls:		*R*luc-F[2]: amino acids 111–312 of *R*luc (Stefan et al. 2007)
	p413ADH- Zipp.- L-*R*luc-F[1]		
	p415ADH- Zipp.- L-*R*luc-F[2]		
	pAG25-L-*R*luc-F[1]	ClonNAT (pAG25)	Used to create integration cassettes of *R*luc PCA fragments to study yeast PPIs at endogenous levels of expression. Cassettes are integrated by homologous recombination.
	pAG32-L-*R*luc-F[2]	Hygromycin B (pAG32)	
*G*luc	**Positive controls:**	Zeocin	Expression plasmids:
	pcDNA3.1/Zeo-Zipp.-L-h*G*luc-F[1][b]		*G*luc-F[1]: amino acids 1–63 of *G*luc
	pcDNA3.1/Zeo-Zipp.-L-h*G*luc-F[2][b]		*G*luc-F[2]: amino acids 64–185 of *G*luc (Remy and Michnick 2006)
IFP	p416GAL1-ProteinA-L-IFP-F[1]	–Ura (p416GAL1)	Expression plasmids:
	p413GAL1-ProteinB-L-IFP-F[2]	–His (p413GAL1)	IFP-F[1]: amino acids 1–132 of IFP
	Positive controls:		IFP-F[2]: amino acids 133–321 of IFP
	p416- Zipp.- L-IFP-F[1]		Zipper residues 250–281 (Tchekanda et al. 2014)
	p413- Zipp.- L-IFP-F[2]		
	pAG25-L-IFP-F[1]	ClonNAT (pAG25)	Used to create integration cassettes of IFP PCA fragments to study yeast PPIs at endogenous levels of expression. Cassettes are integrated by homologous recombination.
	pAG32-L-IFP-F[2]	Hygromycin B (pAG32)	

*G*luc, *Gaussia* luciferase; IFP, infrared fluorescent protein; PCA, protein-fragment complementation assay; PPI, protein–protein interaction; *R*luc, *Renilla* luciferase; vYFP, Venus yellow fluorescent protein.

[a]Proteins A and B are sequences encoding proteins of interest for a particular PPI study. Zipp. consists of the GCN4 leucine zipper. The linker (L) comprises a 10-amino-acid flexible polypeptide linker consisting of (Gly.Gly.Gly.Gly.Ser)$_2$ between the protein of interest and the reporter protein fragments. Fragments (F[1] and F[2]) of the PCA reporter proteins are fused to the carboxyl terminus or amino terminus of the proteins of interest via the linker. Negative controls can be created by replacing proteins A and B by well-characterized noninteracting proteins.

[b]For mammalian expression. Can be used to create compatible plasmids for yeasts.

YPD <R>
> *Prepare without agar for liquid culture, and with agar and with appropriate antibiotics (see Table 1) for selection and growing colonies in Petri dishes.*

Equipment

6-well culture plate or Petri dish
96-well plates (MGB101-1-2-LG, Matrical Bioscience)
96-well special optics plates (Corning COSTAR 3614)
96-well white plates (Molecular Machines)
Bench centrifuge (Centrifuge 5810R, Eppendorf)
Cy5.5 filter (specifications: Scan range 600–800 nm, HQ665/45×, HQ725/50 m Q695LP)
Fluorometer SPECTRAmax GEMINI-XS (Molecular devices)
Incubator shaker (Microtron, INFORS HT)
Luminescence microplate reader (LMax II^{384} Luminometer, Molecular Devices)
MetaMorph software (version 7.5.3, MDS Analytical Technologies)
Microscope equipped for fluorescence and luminescence (Nikon Eclipse TE2000U inverted microscope) with Coolsnap HQ Monochrome CCD camera (Photometrics)
Tubes (sterile) to culture yeast
UV/visible spectrophotometer Ultrospec 2000 (Pharmacia Biotech)
Water bath, set at 42°C

METHOD

Cotransformation of Competent Yeast with PCA Expression Plasmid Pair

1. Thaw competent yeast cells on ice.

2. Mix 10 µL of cells with 1 µL (~250 ng) of each yeast PCA expression plasmid pair (Table 1), 60 µL of PLATE solution and 8 µL of DMSO.
    > *The PCA expression plasmid pair consists of expression plasmids containing "gene of interest A" tagged with fragment 1 (e.g., p413ADH-ProteinA-linker-Rluc-F[1]) and "gene of interest B" tagged with fragment 2 (e.g., p415ADH-ProteinB-linker-Rluc-F[2]). A list of expression plasmids for real-time PCA assays is available in Table 1.*

3. Heat-shock yeast in a water bath for 20 min at 42°C.
    > *Shorter or longer incubation times at higher or lower temperatures can result in decreased efficiency of transformation.*

4. Centrifuge at 805g for 3 min at room temperature. Remove supernatant and resuspend cells in 500 µL SD medium for PCAs without amino acids or glucose.

5. Plate 20 µL of cell suspension per well on SC medium for PCAs with agar and containing 2% glucose and 1× amino acid solution together with the appropriate drop-out mix (Table 1) in a six-well culture plate.

6. Incubate for 48–72 h at 30°C.
    > *See Troubleshooting.*

Fusion of PCA Fragments at Chromosomal Loci by Homologous Recombination

7. PCR amplify PCA fragment cassettes from pAG25 (fragment 1) and pAG32 (fragment 2) plasmids containing the PCA fragments followed by a terminator and an antibiotic selection marker such as ClonNat or hygromycin B resistance (Table 1).

8. Transform the PCR product into suitable competent yeast cells—mix 20 µL of thawed competent cells with 8 µL (~1 µg/µL) of each PCR-amplified cassette DNA encoding the PCA fragments and a resistance marker. Add 100 µL of PLATE solution.

9. Incubate for 30 min at room temperature.

10. Add 15 µL of DMSO and then heat-shock for 15 min at 42°C.

11. Centrifuge at 805g for 3 min at room temperature, remove supernatant, resuspend cells in 100 µL YPD medium, and grow at 30°C in an incubator shaker for 4 h.

12. Centrifuge the cells, remove supernatant and resuspend the cells in 200 µL of YPD.

13. Plate 60 µL per well in a six-well culture plate, or the entire 200 µL on a Petri dish, which contains YPD plus the appropriate antibiotic such as ClonNat or hygromycin B (Table 1).

14. Incubate the plates for 48–72 h at 30°C.

    The colonies observed can be further verified by colony PCR methods.

    See Troubleshooting.

Preparation of Cells for Fluorescence Microscopy Using Venus YFP PCA and for Bioluminescence Assays Using Gluc and Rluc PCA

15. Inoculate a fresh colony for each transformant into 3 mL of SC medium containing glucose and 1× amino acid solution with the appropriate drop-out mix (Table 1), and grow overnight at 30°C with shaking. For cells with fragments fused at chromosomal loci, grow them in SC or YPD medium containing the required antibiotic (Table 1).

16. The following day, measure the OD_{600} of the overnight culture and start a fresh culture of LFM [1× low-fluorescence yeast nitrogen base plus 2% (w/v) glucose and 1× amino acid mix with the appropriate drop-out] or "LFM complete" with suitable antibiotics with enough cells to obtain an OD_{600} of approximately 0.1–0.3 at the time of analysis.

    It is particularly important for the cells to be in the log phase of growth to avoid including dead and unhealthy cells. These cells are highly auto-fluorescent and thus would confound quantitative analysis. Cells in the log phase can be used if they are appropriate to study a particular interaction(s) as long as the condition of the cells is verified by bright-field microscopy.

17. Depending on the reporter used (vYFP, or Gluc and Rluc), complete one of the following.

    - For Venus YFP PCA fluorescence microscopy: coat the wells of a glass-bottom 96-well plate (Matrical Bioscience) with a solution of 1 mg/mL poly-L-lysine, or 50 µg/mL concanavalin A, for 10 min, rinse with distilled water and allow to dry. Transfer 70 µL of cell suspension to each well. Wait 10 min to allow the cells to settle in the wells. Acquire images with a fluorescence microscope equipped with a CCD camera, using a YFP filter cube and an exposure time of ~750 msec.

        It is best to use 60× or 100× objectives to discriminate subcellular structures. Bright-field or phase-contrast images can be acquired for each field of view to compare the morphology of the yeast possessing fluorescent PCA signals. Specific functional assays to further characterize a protein–protein interaction might be performed here.

    - For Gluc and Rluc luminescence assay: transfer 160–180 µL of cell suspension (cells equivalent to 0.1–0.3 OD_{600}) to each well of a 96-well white plate (Molecular Machines). Manually add or inject 20–40 µL of suitable substrate (2 mM benzyl-coelenterazine for Rluc, or coelenterazine, prepared by following the manufacturer's instructions, for Gluc) using the injector of a luminescence microplate reader (Molecular Devices) and initiate the bioluminescence analysis. Optimize the signal integration times depending on the bioluminescence signal strength. For real-time kinetics experiments, add or inject the substrate, immediately initiate the bioluminescence readings with the optimized signal integration time continuously for the desired period. Then, background-correct the bioluminescence signals to obtain a meaningful signal. Afterward, normalize the data to total protein concentration in cell lysates if desired (Bio-Rad protein assay).

        Specific functional assays to further characterize a protein–protein interaction might be performed here. For example, incubation of cells with agents, such as specific enzyme or transport inhibitors, can be performed for various amount of time, before the luminometric analysis.

        See Troubleshooting.

Preparation of Cells for Fluorometric Analysis Using IFP PCA

18. Inoculate a fresh colony for each transformant into 5 mL of SC medium containing raffinose and 1× amino acid solution with the appropriate drop-out mix (Table 1), and then grow overnight (~18 h) in a shaking incubator at 30°C.

    For cells with fragments fused at chromosomal loci, grow them in YPD medium with appropriate antibiotics (Table 1).

19. The next day, start a fresh culture in SC medium containing raffinose and 1× amino acid drop-out solution using enough cells from the overnight culture to reach an OD_{600} of 0.1.

    For cells with fragments fused at chromosomal loci, grow them in YPD medium with appropriate antibiotic (Table 1) and 4 mm biliverdin hydrochloride. Biliverdin hydrochloride is a chromophore required to produce the IFP fluorescent signal.

20. Incubate the cultures for 3–4 h at 30°C with shaking to obtain a log-phase culture.

    For cells with PCA fragment-encoding nucleotides integrated into chromosomal loci and thus lower levels of expression than when proteins are expressed from plasmids, perform controls by measuring the OD600 of the cultures with a UV/visible spectrophotometer using a blank solution containing YPD plus the same concentrations of biliverdin hydrochloride and of the antibiotics that were added to the cultures.

21. Add galactose solution to a final concentration of 2% (w/v) to induce protein expression.

22. Induce the protein expression for 2 h at 30°C with shaking.

23. Add biliverdin hydrochloride to a final concentration of 2–4 mM.

    It is important to completely solubilize the biliverdin hydrochloride using DMSO. However, the solution should be diluted with sterile distilled water to minimize the concentration of DMSO in the culture, which can be toxic to the cells (one volume of the biliverdin hydrochloride–DMSO solution plus nine volumes of sterile distilled water). If biliverdin hydrochloride is not fully solubilized, black spots in bright-field images might be observed.

24. Incubate for 2–3 h at 30°C with shaking, to reach an OD_{600} of 0.4–0.6 at analysis.

25. Collect the cells from 10 mL of cell suspension by centrifuging in a tabletop centrifuge for 5 min at 500g then carefully pour off the medium, retaining the cell pellet in the culture tube, and resuspend the cells in 100 µL of 1× LFM [1× low-fluorescence yeast nitrogen base plus 2% (w/v) galactose plus 1× amino acid drop-out plus 4 mM biliverdin hydrochloride]. Transfer 100 µL of cell suspension to each well of a 96-well special optics plate (Corning COSTAR) and measure using the fluorometer SPECTRAmax the fluorescence signal at excitation: 640 nm, emission: 708 nm; photomultiplier tube (PMT) settings: medium/30 reads.

    For cells with PCA fragment-encoding nucleotides integrated into chromosomal loci, collect the cells from 50 to 100 mL of cell suspension by centrifuging in a tabletop centrifuge for 5 min at 500g then carefully pour off the medium, retaining the cell pellet in the culture tube, and resuspend the cells in 400 µL of 1× LFM (1× low-fluorescence yeast nitrogen base plus 2% (w/v) glucose plus 1× amino acid drop-out plus 4 mm biliverdin hydrochloride).

Preparation of Cells for Fluorescence Microscopy

For preparation of cells for fluorescence microscopy, follow Steps 18–24 and then perform the following additional steps.

26. Coat the wells of a glass-bottom 96-well plate (Matrical Bioscience) with a solution of 1 mg/mL poly-L-lysine, or 50 µg/mL concanavalin A, for 10 min, rinse with distilled water, and allow to dry. Transfer 70 µL of cell suspension to each well. Wait 10 min to allow the cells to settle in the wells.

27. For cells transformed with expression plasmids, acquire images with a fluorescence microscope equipped with a CCD camera, using a Cy5.5 filter and exposure times of 4–6 sec. For cells with PCA fragment-encoding nucleotides integrated into chromosomal loci, acquire with the same microscope and filter and 10–40 sec of exposure time (the exposure time varies with the signal strength of the interaction). For the detection of foci or any other subcellular structures, use 60× objectives, which are better than 100× objectives to achieve a better spatial resolution.

28. Analyze the acquired images using MetaMorph software (version 7.5.3, MDS Analytical Technologies).

TROUBLESHOOTING

Problem (Steps 6 and 14): No colonies grow after transformation.
Solution: Increase quantity of cells and DNA. Increase the number of cells plated on the Petri dish or six-well plate.

Problem (Steps 6 and 14): Too many colonies grow after transformation.
Solution: Dilute cells before plating on the Petri dish or six-well plate.

Problem (Step 14): Fusion protein is not functioning correctly (fragment fusion might interfere with protein expression/function/stability).
Solution: Fuse the PCA fragment to the other end of the protein.

Problem (Step 17): A poor luminescence signal is recorded.
Solution: The signal integration time might be too short: optimize the signal integration times. Increase substrate concentrations and the number of cells used per assay.

Problem (Step 17): No or low signal modulation occurs after stimulus or inhibitor treatment.
Solution: The number of cells and signal integration times might not be optimal: optimize the number of cells and signal integration times. Try different stimulus or inhibitor treatment times and/or concentrations. The peak signal occurs immediately after addition of coelenterazines. Try optimizing the beginning of signal integration after substrate addition. If the signal is very low, find an optimal way to extract the meaningful signal from background. Test appropriate positive and negative controls for the interaction being studied.

DISCUSSION

The fluorescence intensity of the IFP PCA and the reassembled Venus YFP PCA (Fig. 1) varies with the expression levels and the interaction dissociation constants of the protein pairs attached to the PCA fragments (Remy and Michnick 2004; Remy et al. 2004; MacDonald et al. 2006; Manderson et al. 2008; Tchekanda et al. 2014). In the case of one of the simplest positive controls (GCN4 leucine zipper pair fused to the PCA fragments; see Table 1), the reconstituted PCAs represent ~10%–20% of the activity of the full-length Venus YFP and ~50% of the activity of the full-length IFP 1.4 protein. The PCA fusions expressed alone should not result in detectable fluorescence (compared with nontransformed cells) because the individual PCA fragments have no activity. For each study, positive (known interaction), and particularly negative (noninteracting proteins), controls should always be performed in parallel. A PCA response is not expected to be observed if noninteracting proteins are used as PCA partners.

The luminescence intensity of the reassembled *R*luc (Fig. 1) and *G*luc PCAs vary with the strength of interaction between the protein pairs attached to the PCA fragments (Remy and Michnick 2006; Stefan et al. 2007, 2011; Malleshaiah et al. 2010). In the case one of the simplest positive controls (GCN4 leucine zipper pair fused to the PCA fragments; Table 1), the reconstituted PCAs represent ~10%–30% of the activity of the full-length *R*luc or *G*luc enzymes. Again, the PCA fusions expressed alone should not result in detectable luminescence (compared with nontransfected cells) because the individual PCA fragments have no activity. For each study, positive (known interaction), and particularly negative (noninteracting proteins), controls should always be performed in parallel. A PCA response should not be observed if noninteracting proteins are used as PCA partners.

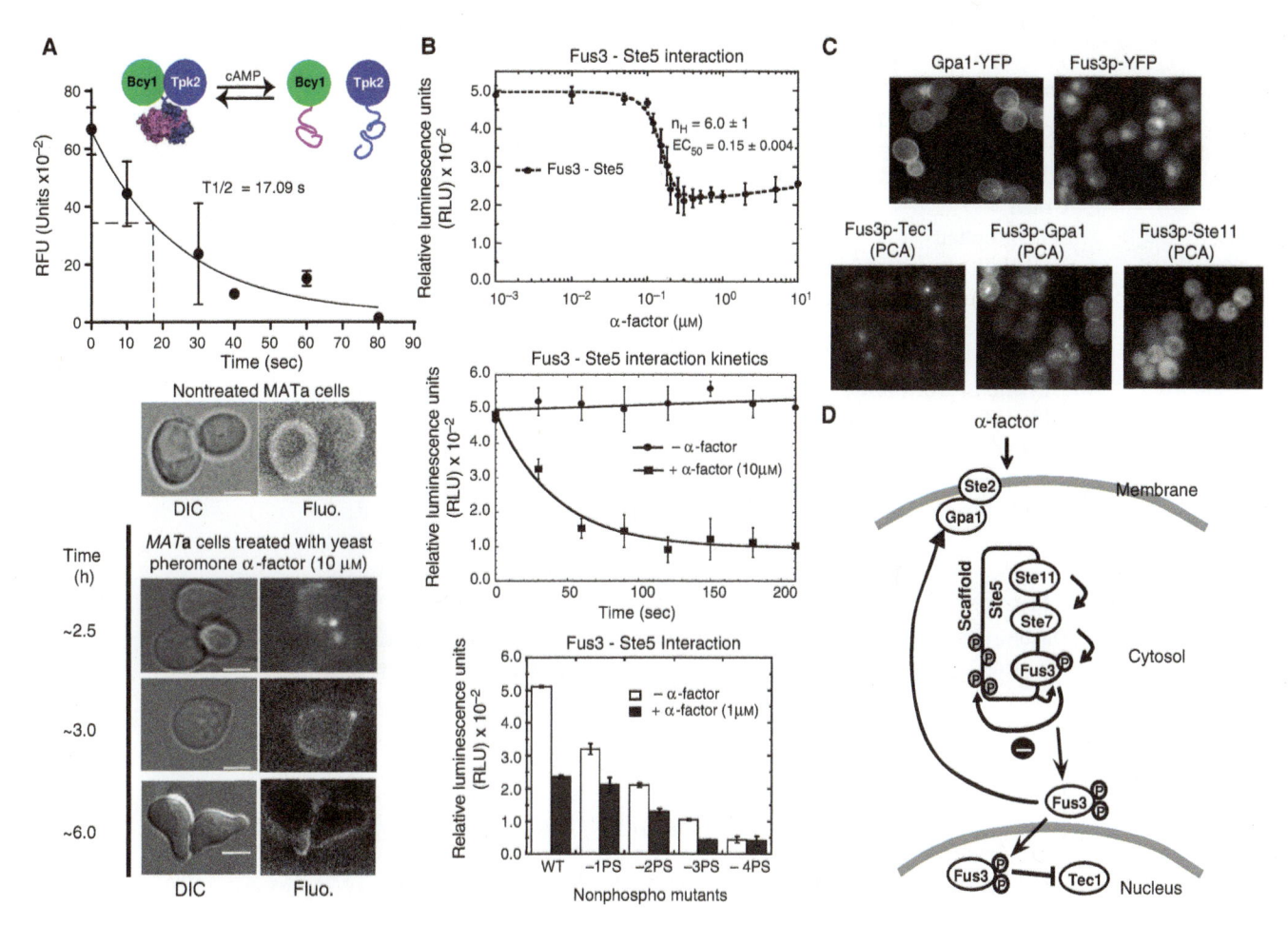

FIGURE 1. Applications of real-time PCAs. (*A*) The infrared fluorescent protein (IFP) PCA detects the dissociation or the assembly of protein complexes in living yeast cells. The dissociation of the endogenous yeast cAMP-dependent protein kinase (PKA) subunits Bcy1 and Tpk2 tagged with IFP fragments F[2] and F[1], respectively, is induced by cAMP (6.5 mM) (error bars indicate s.e.m., $n = 3$). In the absence of α-factor (nontreated MATa cells), a diffuse distribution of IFP-PCA-reported Fus3–Ste5 complexes on the yeast cell cortex is observed [images show cells viewed by differential interference contrast (DIC) and fluorescence (fluo.) microscopy]. Following 2.5 and 3.0 h of treatment with α-factor (10 μM), polarization of IFP-PCA-reported Fus3–Ste5 complexes is evident. Depolarization of the Fus3–Ste5 complexes occurs by 5–6 h of α-factor treatment. Scale bars, 4 μm. (*B*) Example of measuring protein–protein interaction dynamics using *Renilla* luciferase (*R*luc) PCA: the scaffold protein Ste5 interaction with Fus3 (*upper* graph), expressed at their endogenous levels in fusion with PCA fragments, is quantified in response to increasing concentrations of α-factor pheromone. The Hill number (n_H) and EC_{50} are estimated from the curve fit to the standard Hill function (dotted line). *R*luc PCA is effective in measuring both the pheromone dose-dependent dissociation of Fus3 (kinase) from Ste5 and the association of Ptc1 (phosphatase) with Ste5 (data not shown), which continuously compete for the multiple phosphosites on Ste5 (Malleshaiah et al. 2010). The reversibility and faster folding–unfolding nature of *R*luc PCA allows for real-time monitoring of the assembly of protein complexes and their disassembly in vivo. For instance, using *R*luc PCA, the pheromone-dependent real-time dissociation of Fus3 from its complex with Ste5 (*middle* graph) can be monitored in seconds. Exponential decay or constant fits are shown as solid lines. The *R*luc PCA signal reflects the extent of in vivo Fus3 and Ste5 complexes—the signal decreases with the removal of each of the four potential Ste5 phosphosites that, in turn, control the association of Fus3 with Ste5 (*lower* graph). Error bars represent the s.e.m. of triplicates from three independent experiments. (*C*) Venus yellow fluorescent protein (YFP) PCA allows for the detection of the precise location of protein complexes within living cells. As an example (*D*), the yeast pheromone response of simplified mitogen-activated protein kinase (MAPK) pathway protein complexes formed in different regions within living cells is illustrated. The images in *C* show the precise localizations for Fus3p interactions with Gpa1 (Metodiev et al. 2002) at the membrane, Ste11 (Choi et al. 1994) at the cytoplasm and Tec1 (Chou et al. 2004) in the nucleus. As controls for different localizations, Gpa1 fused to full-length Venus YFP protein is seen at the membrane, and similarly Fus3 fused to full-length Venus YFP is found in both the cytoplasm and nucleus. Yeast cells were treated with 1 μM α-factor pheromone for 2–3 h to induce the translocation of Fus3p between plasma membrane, cytosol and nucleus, as observed during pheromone signaling. PS, phosphosite; RFU, relative fluorescence units; RLU, relative luminescence units.

Chapter 20

RECIPES

Amino Acid Mix (10×)

Amino acid or nucleobase	Concentration (10×)
Adenine sulfate	0.4 g/mL
Uracil	0.2 g/mL
L-Tryptophan	0.4 g/mL
L-Histidine hydrochloride	0.2 g/mL
L-Arginine hydrochloride	0.2 g/mL
L-Tyrosine	0.3 g/mL
L-Leucine	0.6 g/mL
L-Lysine hydrochloride	0.3 g/mL
L-Phenylalanine	0.5 g/mL
L-Glutamic acid	1.0 g/mL
L-Asparagine	1.0 g/mL
L-Valine	1.5 g/mL
L-Threonine	2.0 g/mL
L-Serine	3.75 g/mL
L-Methionine	0.2 g/mL

Dissolve the amino acids from the list above (except those to be excluded from any drop-out media) in distilled H_2O. Filter-sterilize, store at 4°C, and protect from light.

Low-Fluorescence Yeast Nitrogen Base without Riboflavin and Folic Acid Medium (LFM) (10×)

Reagents	Final concentration (10×)
Yeast nitrogen base without riboflavin and folic acid	6.7 g/L
$(NH_4)_2SO_4$	5 g/L
KH_2PO_4	1 g/L
$MgSO_4$	0.5 g/L
NaCl	0.1 g/L
Ca_2Cl	0.1 g/L
H_3BO_4	0.5 mg/L
$CuSO_4$	0.04 mg/L
KI	0.1 mg/L
$FeCl_3$	0.2 mg/L
$MnSO_4$	0.4 mg/L
Na_2MoO_4	0.2 mg/L
$ZnSO_4$	0.4 mg/L
Biotin	2 µg/L
Calcium pantothenate	0.4 mg/L
Inositol	2 mg/L
Niacin	0.4 mg/L
Para-aminobenzoic acid (PABA)	0.2 mg/L
Pyridoxine hydrochloride	0.4 mg/L
Thiamine	0.4 mg/L
Galactose or glucose (optional; see below)	2% (w/v)
Amino acid mix (10×) (optional; see below) <R>	1×

Always include sugars and amino acids (as appropriate) for cells that are not to be stressed while performing fluorescence microscopy or while doing the fluorometric or luminescent analyses. Store at 4°C. LFM without amino acids mix added can be stored at room temperature.

Phosphate-Buffered Saline (PBS; 1×, pH 7.4)

Reagent	Quantity	Final concentration
NaCl	8.0 (g)	137.0 mM
KCl	0.20 (g)	2.7 mM
Na_2HPO_4	1.14 (g)	8.0 mM
KH_2PO_4	0.20 (g)	1.5 mM

Combine all ingredients in ~800 mL of H_2O and stir to dissolve. Adjust pH to 7.4 and bring final volume to 1 L. Sterilize by autoclaving. Store at room temperature.

PLATE Solution

Reagents	Final concentration (1×)
PEG 3350	40% (w/v)
Lithium acetate	100 mM
Tris (pH 7.5)	10 mM
EDTA	0.4 mM

Wear personal protective equipment when handling this reagent. Store at room temperature in a dry and well-ventilated place.

Synthetic Complete (SC) and Synthetic Defined (SD) Media for PCAs

Reagents	Final concentration (1×)
Yeast nitrogen base without amino acids and with ammonium sulfate	6.7 g/L
Agar (for solid medium only)	2% (w/v)
Amino acid mix (10×), without histidine and leucine (for *R*Luc and Venus PCA experiments) or without uracil and histidine (for IFP PCA experiments) <R>	1×
Glucose (if required)	2% (w/v)
Raffinose (if required)	2% (w/v)

For SD, omit the amino acids, glucose, and raffinose. For SC, the addition of glucose or raffinose depends on the point in the procedure in which it will be used. Sterilize by autoclaving and store at room temperature.

YPD

Peptone, 20 g
Glucose, 20 g
Yeast extract, 10 g
H_2O to 1000 mL
YPD (YEPD medium) is a complex medium for routine growth of yeast.

To prepare plates, add 20 g of Bacto Agar (2%) before autoclaving. Store at room temperature.

REFERENCES

Choi KY, Satterberg B, Lyons DM, Elion EA. 1994. Ste5 tethers multiple protein kinases in the MAP kinase cascade required for mating in S. cerevisiae. *Cell* **78**: 499–512.

Chou S, Huang L, Liu H. 2004. Fus3-regulated Tec1 degradation through SCFCdc4 determines MAPK signaling specificity during mating in yeast. *Cell* **119**: 981–990.

Knop M, Siegers K, Pereira G, Zachariae W, Winsor B, Nasmyth K, Schiebel E. 1999. Epitope tagging of yeast genes using a PCR-based strategy: More tags and improved practical routines. *Yeast* **15**: 963–972.

MacDonald ML, Lamerdin J, Owens S, Keon BH, Bilter GK, Shang Z, Huang Z, Yu H, Dias J, Minami T, et al. 2006. Identifying off-target effects and hidden phenotypes of drugs in human cells. *Nat Chem Biol* **2**: 329–337.

Malleshaiah MK, Shahrezaei V, Swain PS, Michnick SW. 2010. The scaffold protein Ste5 directly controls a switch-like mating decision in yeast. *Nature* **465**: 101–105.

Manderson EN, Malleshaiah M, Michnick SW. 2008. A novel genetic screen implicates Elm1 in the inactivation of the yeast transcription factor SBF. *PLoS One* **3**: e1500.

Metodiev MV, Matheos D, Rose MD, Stone DE. 2002. Regulation of MAPK function by direct interaction with the mating-specific Gα in yeast. *Science* **296**: 1483–1486.

Remy I, Michnick SW. 2004. A cDNA library functional screening strategy based on fluorescent protein complementation assays to identify novel components of signaling pathways. *Methods* **32**: 381–388.

Remy I, Michnick SW. 2006. A highly sensitive protein–protein interaction assay based on *Gaussia* luciferase. *Nat Methods* **3**: 977–979.

Remy I, Montmarquette A, Michnick SW. 2004. PKB/Akt modulates TGF-β signalling through a direct interaction with Smad3. *Nat Cell Biol* **6**: 358–365.

Sheff MA, Thorn KS. 2004. Optimized cassettes for fluorescent protein tagging in *Saccharomyces cerevisiae*. *Yeast* **21**: 661–670.

Stefan E, Aquin S, Berger N, Landry CR, Nyfeler B, Bouvier M, Michnick SW. 2007. Quantification of dynamic protein complexes using *Renilla* luciferase fragment complementation applied to protein kinase A activities in vivo. *Proc Natl Acad Sci* **104**: 16916–16921.

Stefan E, Malleshaiah MK, Breton B, Ear PH, Bachmann V, Beyermann M, Bouvier M, Michnick SW. 2011. PKA regulatory subunits mediate synergy among conserved G-protein-coupled receptor cascades. *Nat Commun* **2**: 598.

Tchekanda E, Sivanesan D, Michnick SW. 2014. An infrared reporter to detect spatiotemporal dynamics of protein–protein interactions. *Nat Methods* **11**: 641–644.

CHAPTER 21

Protein Complex Purification by Affinity Capture

John LaCava, Javier Fernandez-Martinez, Zhanna Hakhverdyan, and Michael P. Rout[1]

Laboratory of Cellular and Structural Biology, The Rockefeller University, New York, New York 10065

Affinity capture has become a powerful technique for consistently purifying endogenous protein complexes, facilitating biochemical and biophysical assays on otherwise inaccessible biological assemblies, and enabling broader interactomic exploration. For this procedure, cells are broken and their contents separated and extracted into a solvent, permitting access to target macromolecular complexes thus released in solution. The complexes are specifically enriched from the extract onto a solid medium coupled with an affinity reagent—usually an antibody—that recognizes the target either directly or through an appended affinity tag, allowing subsequent characterization of the complex. Here, we discuss approaches and considerations for purifying endogenous yeast protein complexes by affinity capture.

BACKGROUND

Two interacting molecules form the cognate groups of an affinity capture system. These groups may include antibody–antigen interactions and other protein–protein and protein–ligand interactions. Typically, one group is a proteinaceous moiety ("tag") that is fused to a protein of interest via genetic engineering, resulting in the expression of the tagged fusion protein of interest within a model organism. The second group is commonly covalently immobilized on an insoluble resin, gel, or paramagnetic medium and may be any molecule capable of interacting with the tag at high specificity and affinity. When extracts from cells expressing the affinity-tagged protein of interest are exposed to the medium coupled to the affinity capture reagent, the tagged protein becomes immobilized on the medium through interaction with its cognate binding partner, also bringing along its stably associated endogenous interacting proteins (for review, see Urh et al. 2009).

The yeast *Saccharomyces cerevisiae* is readily amenable to homologous recombination–based genomic tagging, resulting in the tagged protein being expressed normally from its endogenous genomic locus. For this reason, together with other useful traits, *S. cerevisiae* has been the leading model organism used for genome-wide tagging and affinity capture (e.g., Ho et al. 2002; Gavin et al. 2006; Krogan et al. 2006). As a result, yeast strains expressing nearly any endogenous protein as a carboxy-terminally tagged fusion protein are commercially available (Ghaemmaghami et al. 2003; Huh et al. 2003; Gavin et al. 2006), and custom strains can be constructed and validated at the bench within ~2 wk. Two commonly used tags are SpA (*Staphylococcus aureus* protein A) and GFP (*Aequorea victoria* green fluorescent protein). SpA interacts with IgG via the constant (Fc) region (Lindmark et al. 1983; Moks et al. 1986) and therefore does not require an antigen-specific antibody for affinity capture, making the affinity medium comparatively inexpensive to produce. Tandem repeats of an artificial domain derived from SpA are included in the TAP-tag (Nilsson et al. 1987; Puig et al. 2001). The use of

---

[1]Correspondence: rout@rockefeller.edu

Copyright © Cold Spring Harbor Laboratory Press; all rights reserved
Cite this introduction as *Cold Spring Harb Protoc*; doi:10.1101/pdb.top077545

GFP as an affinity tag requires a high-quality anti-GFP antibody preparation for producing the affinity capture medium; such preparations are available commercially.

The end point of an affinity capture experiment typically includes mass spectrometric analyses to define the sample composition (e.g., Cristea et al. 2005; Alber et al. 2007; Oeffinger et al. 2007). The copurifying proteins, together with the tagged protein of interest, may constitute one or more functional complexes (or parts thereof) in vivo—providing information about the specific constituents of particular biological machinery. Keep in mind that false positives may vary in affinity capture experiments, depending on the conditions of capture and other handling procedures (Bell et al. 2007; Devos and Russell 2007; Mellacheruvu et al. 2013) (discussed below). When the purified complexes are eluted natively from the affinity medium, they may be further fractionated (e.g., by rate-zonal centrifugation), examined physically (e.g., by electron microscopy to get size and shape information), and assayed functionally (e.g., in vitro enzymatic assays). Affinity-purified protein complexes serve as an important starting material for experimental programs aimed at mapping the interface between the composition, the form, and the function of biological macromolecules (e.g., Alber et al. 2007; Fernandez-Martinez et al. 2012; Lasker et al. 2012).

MAKING AFFINITY CAPTURE WORK

Although affinity capture is a conceptually straightforward approach to protein complex purification, optimizing affinity capture experiments often requires attention to a broad range of variables including pH and buffer type, overall ionic strength, salt type(s) and concentration, detergent type(s) and concentration, and temperature (Fig. 1). These factors, as well as the mechanism of cell breakage and the time required to complete the capture, can have a profound effect on protein interactions (Ugwu and Apte 2004; Oeffinger 2012). A short list of extraction solvent constituents, which may serve as a basic starting point for optimization, is given in Table 1. It is difficult to know a priori the appropriate

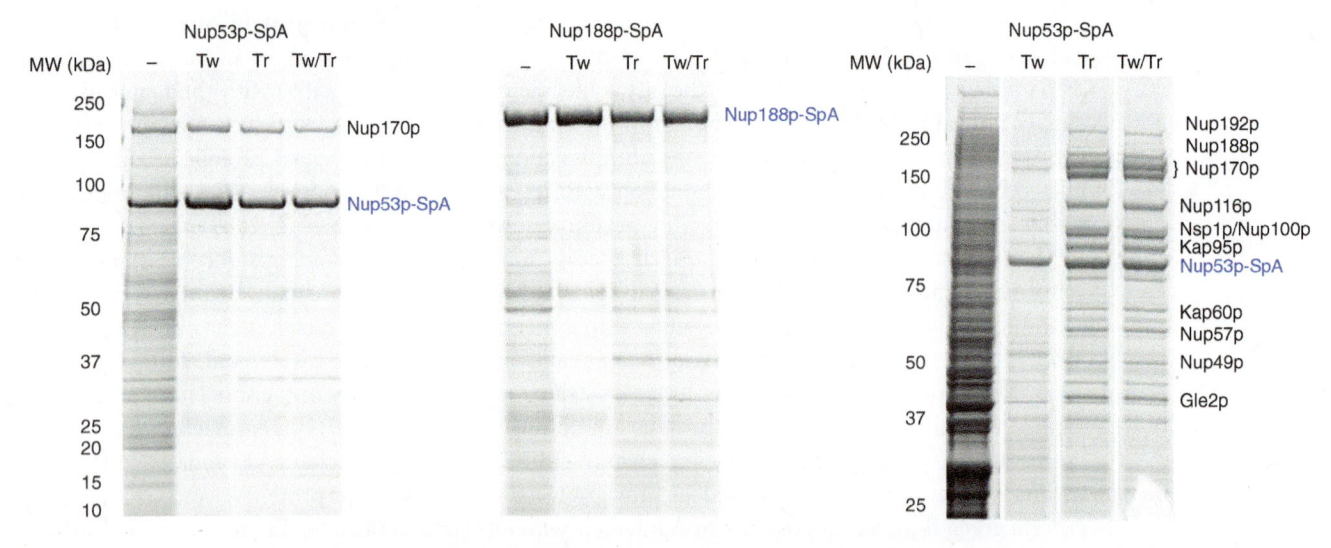

FIGURE 1. Three panels illustrating differences in protein purification patterns depending on the reagents present in the extraction solvent. In the presence of a low trisodium citrate concentration and moderate sodium chloride concentration, Tween 20 provided for comparatively enhanced recovery of the tagged protein while keeping background low: (Left panel) Protein A-tagged Nup53 purified by affinity capture in 40 mM Tris, pH 8.0, 50 mM trisodium citrate, 300 mM NaCl, and either no detergent [−], 0.1% (v/v) Tween 20 [Tw], 1.0% (v/v) Triton X-100 [Tr], or both [Tw/Tr]; and (center panel) an identical experiment is presented for an Nup188-SpA affinity capture experiment. Tween 20 does not work well in conjunction with a high trisodium citrate concentration: (Right panel) Nup53-SpA purified by affinity capture in the same conditions as above, but including 250 mM trisodium citrate. In all three experiments, mixtures of Tween 20 and Triton X-100 showed the pattern associated with Triton X-100 alone, regardless of the relative quantities of trisodium citrate and sodium chloride present.

TABLE 1. A nonexhaustive list of reagents we commonly use in extraction solvents for purifying protein complexes

Reagent	Suggested concentration	Notes
Sodium chloride	0.05–2 M	Higher concentrations improve extraction of total protein and keep background low but may strip away some otherwise stable interactors.
Trisodium citrate	0.05–0.25 M	At alkaline pH (when fully deprotonated in solution), higher concentrations stabilize some protein complexes. Salting-out of total protein can be observed at ~400 mM and above. Can be used alone, or combined with sodium chloride to improve extraction.
Ammonium acetate	0.1–2 M	A salt, consisting of two buffers, that yields a neutral pH solution. Higher concentrations stabilize some protein complexes. Acidic solutions can result from old, improperly stored crystalline stocks on account of ammonia loss. No additional buffer or salt are required in solvents containing ammonium acetate.
Urea	1–3 M	Can strip off background as well as stable core complex components, potentially revealing binary connectivity. Can be used in combination with NaCl.
Tween 20	0.01%–0.1% (v/v)	A nonionic detergent that works well in conjunction with high sodium chloride concentrations for protein extraction (see Fig. 1).
Triton X-100	0.1%–1% (v/v)	A nonionic detergent that works well in conjunction with high sodium citrate concentrations for protein extraction (see Fig 1).
CHAPS	2–5 mM	A zwitterionic detergent. Especially useful for membrane protein complex extraction.
Sarkosyl	0.5–5 mM	An anionic detergent that can strip off stable core complex components, potentially revealing binary connectivity, and keep background low.
DTT	1 mM	A reducing agent effective at alkaline pH. In some cases, this agent may improve protein stability and enzyme activity. High concentrations of DTT and elevated temperature can unlink the chains of the affinity antibody.
EDTA	0.1–1 mM	A chelator of divalent cations typically used to limit the activity of metalloproteases and nucleases. Not necessary in the presence of high concentrations of citrate, which also chelates divalent cations.
Tris–Cl	40 mM (pH 8.0–8.5)	pK_a of 8.8 at 4°C and 8.1 at 25°C.
Na- or K-HEPES	40 mM (pH 7.4–7.6)	pK_a of 7.8 at 4°C and 7.5 at 25°C. NaOH or KOH used for pH equilibration depending on the salt (e.g., NaCl, KCl, or CH_3CO_2K) used in solvent.
Sodium phosphate	40–250 mM (pH 7.0–8.0)	Consult a chart of phosphate buffers for appropriate proportions of mono- and dibasic species required to achieve the desired pH. We often omit salts when working with concentrations greater than 100 mM.

These are suggested reagents and concentrations; conditions outside of these suggestions and reagents not presented in this list may be necessary for success in capturing a particular protein complex. As a general rule, when formulating extraction solvents, we use the minimum number of different additives needed to achieve success in capturing protein complexes—this is determined on an empirical, complex-by-complex basis, and informed by previous successes on related complexes.

CHAPS, 3-[(3-cholamidopropyl)dimethylammonio]-1-propanesulfonate; DTT, dithiothreitol; EDTA, ethylenediaminetetraacetic acid; HEPES, (4-(2-hydroxyethyl)-1-piperazineethanesulfonic acid.

artificial milieu that will best stabilize the constituents of a given protein complex, while minimizing postextraction artifacts at the same time. For this reason, optimized solvents for producing cell extracts typically need to be determined empirically on a complex-by-complex basis. We commonly use Tris-chloride, sodium phosphate, and ammonium acetate as pH buffer systems in our extraction solvents. This may appear to run contrary to the sound logic of Good et al. (1966), who set out several excellent criteria in selecting biological pH buffer systems as alternatives to these, although we commonly use the Good buffer, sodium/potassium-HEPES (see Table 1). We should point out, however, that our overarching objective is the structural preservation of macromolecular protein complexes in an artificial milieu. For this, the reagents listed in Table 1 have all proven effective in our hands, and particular agents such as ammonium acetate, sodium phosphate, and trisodium citrate (not "Good" buffers) have showed effects in protein stabilization distinct from pH buffering (Kunz et al. 2004; Bostrom et al. 2005; Lo Nostro and Ninham 2012). The environment that offers the best in vitro enzymatic activity is not necessarily the same one that best preserves a particular subset of a protein interaction network. All that being said, the criteria set out by Good et al. (1966) in biological buffer selection remain as valid today as ever.

Despite the daunting number of variables, there are some practical bounds to the "optimization space" to be explored for a given affinity capture. It is important to explore within a range of milieux that permit the affinity capture system to function with high specificity and high affinity; and indeed there will be some conditions that directly promote the affinity interaction. For example, the SpA/IgG interaction is promoted by alkaline pH and the presence of, for example, citrate or sulfate (Brown et al. 1998; Schwarz 2000; Ngo and Narinesingh 2008). It is wise to use affinity capture systems that are functionally robust across a wide range of variables and to titrate the quantity of the affinity medium to the minimum needed to extensively deplete the target protein from the cell extracts; excess unbound antibody can contribute to nonspecific experimental noise.

We have also often found the best results, including higher stability of bona fide constituents and fewer false positives, occur when working with concentrated cell extracts of not less than one part wet cell weight of yeast to four parts extraction solvent (w:v). This may be due to the greater stability of protein interactions at high concentration and a more cell-like resulting milieu (Ellis 2001). Concentrated yeast cell extracts tend to run acidic because of the breakage of acidified organelles such as vacuoles; therefore it is important to include the pH-buffering component of the extraction solvent at an appropriately high concentration and check that buffering is achieved. We have found that, for example, 40 mM for Tris at pH 8.0 or HEPES at pH 7.4 possess sufficient strength to buffer a 1:4 (w:v) extract at the expected pH of the buffered extraction solvent.

Affinity capture is typically conducted at 4°C, which retards the disintegration of most protein complexes and helps reduce proteolysis and other enzyme activities within the cell extract. An appropriate cocktail of protease inhibitors should be included at least during protein extraction and binding to the affinity medium. Other kinds of enzyme inhibitors may be important on a case-by-case basis—for example, for preserving posttranslational modifications such as protein phosphorylation or protecting other kinds of macromolecular complex constituents such as nucleic acids. Solvents for extracting protein complexes from cells typically use a near-physiological pH (~7.0–8.0 based on the pH of the cytosol), although there is good reason to vary this parameter for complexes believed to reside in cellular compartments of differing physiological pH.

For a description of a general affinity isolation protocol, including extraction and capture procedures, followed by a denaturing elution in, for example, SDS-PAGE sample buffer for gel-based proteomic analyses, see Protocol 1: Optimized Affinity Capture of Yeast Protein Complexes (LaCava et al. 2015a). For a description of a procedure for nondenaturing elution by specific protease cleavage or competitive displacement from the affinity medium, see Protocol 2: Native Elution of Yeast Protein Complexes Obtained by Affinity Capture (LaCava et al. 2015b). Natively eluted samples may be processed by rate-zonal centrifugation or size-exclusion chromatography for further enrichment of the affinity-purified fraction. Such fractions are typically suitable for electron microscopy studies of protein complex structure. For an approach for rate-zonal centrifugation of natively eluted samples in a sucrose density gradient, see Protocol 3: Density Gradient Ultracentrifugation to Isolate Endogenous Protein Complexes after Affinity Capture (Fernandez-Martinez et al. 2015).

REFERENCES

Alber F, Dokudovskaya S, Veenhoff LM, Zhang W, Kipper J, Devos D, Suprapto A, Karni-Schmidt O, Williams R, Chait BT, et al. 2007. The molecular architecture of the nuclear pore complex. *Nature* 450: 695–701.

Bell AW, Nilsson T, Kearney RE, Bergeron JJM. 2007. The protein microscope: Incorporating mass spectrometry into cell biology. *Nat Methods* 4: 783–784.

Bostrom M, Tavares F, Finet S, Skouri-Panet F, Tardieu A, Ninham B. 2005. Why forces between proteins follow different Hofmeister series for pH above and below pI. *Biophys Chem* 117: 217–224.

Brown NL, Bottomley SP, Scawen MD, Gore MG. 1998. A study of the interactions between an IgG-binding domain based on the B domain of staphylococcal protein A and rabbit IgG. *Mol Biotechnol* 10: 9–16.

Cristea I, Williams R, Chait B, Rout M. 2005. Fluorescent proteins as proteomic probes. *Mol Cell Proteomics* 4: 1933–1941.

Devos D, Russell RB. 2007. A more complete, complexed and structured interactome. *Curr Opin Struct Biol* 17: 370–377.

Ellis RJ. 2001. Macromolecular crowding: Obvious but underappreciated. *Trends Biochem Sci* 26: 597–604.

Fernandez-Martinez J, Phillips J, Sekedat MD, Diaz-Avalos R, Velazquez-Muriel J, Franke JD, Williams R, Stokes DL, Chait BT, Sali A, et al. 2012. Structure–function mapping of a heptameric module in the nuclear pore complex. *J Cell Biol* 196: 419–434.

Fernandez-Martinez J, LaCava J, Rout MP. 2015. Density gradient ultracentrifugation to isolate endogenous protein complexes after affinity capture. *Cold Spring Harb Protoc* doi: 10.1101/pdb.prot087957.

Gavin A-C, Aloy P, Grandi P, Krause R, Boesche M, Marzioch M, Rau C, Jensen LJ, Bastuck S, Dümpelfeld B, et al. 2006. Proteome survey reveals modularity of the yeast cell machinery. *Nature* 440: 631–636.

Ghaemmaghami S, Huh W-K, Bower K, Howson RW, Belle A, Dephoure N, O'Shea EK, Weissman JS. 2003. Global analysis of protein expression in yeast. *Nature* 425: 737–741.

Good NE, Winget GD, Winter W, Connolly TN, Izawa S, Singh RMM. 1966. Hydrogen ion buffers for biological research. *Biochemistry* 5: 467–477.

Ho Y, Gruhler A, Heilbut A, Bader GD, Moore L, Adams S-L, Millar A, Taylor P, Bennett K, Boutilier K, et al. 2002. Systematic identification of protein complexes in *Saccharomyces cerevisiae* by mass spectrometry. *Nature* 415: 180–183.

Huh W-K, Falvo JV, Gerke LC, Carroll AS, Howson RW, Weissman JS, Shea EK. 2003. Global analysis of protein localization in budding yeast. *Nature* 425: 686–691.

Krogan NJ, Cagney G, Yu H, Zhong G, Guo X, Ignatchenko A, Li J, Pu S, Datta N, Tikuisis AP, et al. 2006. Global landscape of protein complexes in the yeast *Saccharomyces cerevisiae*. *Nature* 440: 637–643.

Kunz W, Henle J, Ninham B. 2004. 'Zur Lehre von der Wirkung der Salze' (about the science of the effect of salts): Franz Hofmeister's historical papers. *Curr Opin Colloid Interface Sci* 9: 19–37.

LaCava J, Fernandez-Martinez J, Hakhverdyan Z, Rout MP. 2015a. Optimized affinity capture of yeast protein complexes. *Cold Spring Harb Protoc* doi: 10.1101/pdb.prot087932.

LaCava J, Fernandez-Martinez J, Rout MP. 2015b. Native elution of yeast protein complexes obtained by affinity capture. *Cold Spring Harb Protoc* doi: 10.1101/pdb.prot087940.

Lasker K, Förster F, Bohn S, Walzthoeni T, Villa E, Unverdorben P, Beck F, Aebersold R, Sali A, Baumeister W. 2012. Molecular architecture of the 26S proteasome holocomplex determined by an integrative approach. *Proc Natl Acad Sci* 109: 1380–1387.

Lindmark R, Thorén-Tolling K, Sjöquist J. 1983. Binding of immunoglobulins to protein A and immunoglobulin levels in mammalian sera. *J Immunol Methods* 62: 1–13.

Lo Nostro P, Ninham BW. 2012. Hofmeister phenomena: An update on ion specificity in biology. *Chem Rev* 112: 2286–2322.

Mellacheruvu D, Wright Z, Couzens AL, Lambert JP, St-Denis NA, Li T, Miteva YV, Hauri S, Sardiu ME, Low TY, et al. 2013. The CRAPome: A contaminant repository for affinity purification–mass spectrometry data. *Nat Methods* 10: 730–736.

Moks T, Abrahmsén L, Nilsson B, Hellman U, Sjöquist J, Uhlén M. 1986. Staphylococcal protein A consists of five IgG-binding domains. *Eur J Biochem* 156: 637–643.

Ngo TT, Narinesingh D. 2008. Kosmotropes enhance the yield of antibody purified by affinity chromatography using immobilized bacterial immunoglobulin binding proteins. *J Immunoassay Immunochem* 29: 105–115.

Nilsson B, Moks T, Jansson B, Abrahmsén L, Elmblad A, Holmgren E, Henrichson C, Jones TA, Uhlén M. 1987. A synthetic IgG-binding domain based on staphylococcal protein A. *Protein Eng* 1: 107–113.

Oeffinger M. 2012. Two steps forward—One step back: Advances in affinity purification mass spectrometry of macromolecular complexes. *Proteomics* 12: 1591–1608.

Oeffinger M, Wei KE, Rogers R, DeGrasse JA, Chait BT, Aitchison JD, Rout MP. 2007. Comprehensive analysis of diverse ribonucleoprotein complexes. *Nat Methods* 4: 951–956.

Puig O, Caspary F, Rigaut G, Rutz B, Bouveret E, Bragado-Nilsson E, Wilm M, Séraphin B. 2001. The tandem affinity purification (TAP) method: A general procedure of protein complex purification. *Methods* 24: 218–229.

Schwarz A. 2000. Affinity purification of monoclonal antibodies. *Methods Mol Biol* 147: 49–56.

Ugwu SO, Apte SP. 2004. The effect of buffers on protein conformational stability. *Pharm Technol* 28: 86–108.

Urh M, Simpson D, Zhao K. 2009. Affinity chromatography: General methods. *Methods Enzymol* 463: 417–438.

Protocol 1

Optimized Affinity Capture of Yeast Protein Complexes

John LaCava, Javier Fernandez-Martinez, Zhanna Hakhverdyan, and Michael P. Rout[1]

Laboratory of Cellular and Structural Biology, The Rockefeller University, New York, New York 10065

Here, we describe an affinity isolation protocol. It uses cryomilled yeast cell powder for producing cell extracts and antibody-conjugated paramagnetic beads for affinity capture. Guidelines for determining the optimal extraction solvent composition are provided. Captured proteins are eluted in a denaturing solvent (sodium dodecyl sulfate polyacrylamide gel electrophoresis sample buffer) for gel-based proteomic analyses. Although the procedures can be modified to use other sources of cell extract and other forms of affinity media, to date we have consistently obtained the best results with the method presented.

MATERIALS

It is essential that you consult the appropriate Material Safety Data Sheets and your institution's Environmental Health and Safety Office for proper handling of equipment and hazardous material used in this protocol.

Reagents

Antibody-conjugated paramagnetic beads

There are many choices in paramagnetic media available commercially. We prefer Life Technologies Dynabeads M-270 Epoxy Beads for conjugating to our own choice of antibody in-house. We suggest using 300 mg of antibody-conjugated beads combined with 2 mL of storage buffer or equivalent slurry (i.e., a 15% w:V suspension). For the conjugation procedure, the manufacturer's instructions may be followed with excellent results or see Cristea and Chait (2011).

Extraction solvent (e.g., one of the following):

- 40 mM Tris-Cl (pH 8.0) or 40 mM Na-HEPES (pH 7.4), 0.15–1 M NaCl, and 0.1% (v/v) Tween 20
- 0.25–1.5 M ammonium acetate (pH close to 7.0) and 1% (v/v) Triton X-100
- 250 mM sodium phosphate (pH 7–8) and 1% (v/v) Triton X-100
- 40 mM Tris-Cl (pH 8.0), 50–250 mM sodium citrate, 150 mM sodium chloride, and 1% (v/v) Triton X-100

Determine the optimal extraction solvent composition for each individual protein complex of interest. Use empirical and practical considerations, including prior art gleaned from the literature, to select an appropriate composition. Above we suggest four kinds of formulation, with the ranges of concentrations to be explored for each. They have proven very useful in our hands for a wide variety of yeast complexes. These formulations are reasonable starting points, but they are not universal and may need modifications (e.g., using a different detergent, buffer agent, or pH). For additional discussion on this topic, see Introduction:

---

[1]Correspondence: rout@rockefeller.edu

Copyright © Cold Spring Harbor Laboratory Press; all rights reserved
Cite this protocol as *Cold Spring Harb Protoc*; doi:10.1101/pdb.prot087932

Protein Complex Purification by Affinity Capture [LaCava et al. 2015a]). Use the extraction solvent at room temperature.

Mass-spectrometry-compatible protein stain (for staining proteins in sodium dodecyl sulfate polyacrylamide gel electrophoresis [SDS–PAGE] gels)

We recommend the Blue Silver modified Neuhoff's colloidal Coomassie Blue G-250 stain (Candiano et al. 2004) or Imperial Protein Stain (Thermo Scientific 24615).

Protease inhibitor cocktail

We recommend using an EDTA-free cocktail unless EDTA is specifically required in the optimized extraction buffer. We commonly use product P8340 from Sigma-Aldrich or cOmplete mini, EDTA-free protease inhibitor cocktail tablets from Roche. The protease inhibitors should be reconstituted directly in extraction solvent at the volume recommended by the manufacturer, or in the case of Roche tablets, prepared as a 100× concentrate in water or the extraction solvent and diluted to 1× with the same solvent before use. 100× stocks prepared in water are more versatile to use with multiple different solvents and only impart a 1% dilution on other ingredients.

SDS–PAGE sample buffer (1.1×)

We use 4X NuPAGE LDS Sample Buffer (ThermoFisher Scientific NP0008) and dilute it to 1.1× just before use. The 0.1× "extra" is to account for the reducing agent that is added in Step 12.

SDS–PAGE system

Any SDS–PAGE system can be used for the final separation of protein complex constituents. However, for our routine separations, we prefer the Novex NuPAGE system (Life Technologies) with 1-mm-thick, 4%–12% Bis–Tris gels run in 2-(N-morpholino)ethanesulfonic acid– or 3-(N-morpholino)propanesulfonic acid–SDS buffer. Final elution is achieved by adding SDS (or lithium dodecyl sulfate) containing gel-loading buffer.

Yeast strain of interest

Equipment

Analytical balance (with milligram accuracy)
Heat block (set at 70°C)
Liquid nitrogen
Microcentrifuge (14,000 rpm) (e.g., 5417R from Eppendorf with F-45-30-11 rotor)
Microcentrifuge tubes (1.5- or 2-mL; screw cap recommended)
Micropipettors and tips (e.g., P1000, P200, and P20)
Neodymium magnet (e.g., DynaMag-2 from Life Technologies)
Planetary ball mill (PM100 from Retsch)

Many mechanical milling devices are available, but we use a Retsch PM100 planetary ball mill to produce a micron-scale powder with excellent properties for subsequent affinity capture, providing high-yield, low-noise affinity capture results.

Rotating wheel mixer (or shaker) (set at 4°C)
SDS–PAGE equipment (for one-dimensional gels)
Spatula (small)
Volumetric spoon (e.g., Norpro 3080 Mini Measuring Spoons from Amazon)
Vortex mixer

METHOD

Cryogenic Disruption of Yeast Cells

1. Grow a yeast cell culture to a density of at least 3.0×10^7 cells/mL. Prepare a washed cell pellet that is free of excess liquid and resembles a thick paste (e.g., by centrifugation at ∼3000g for 5 min at 4°C). Press the cell paste through an appropriately sized (e.g., 10-, 20-, or 50-mL) syringe into liquid nitrogen in a clean Dewar flask to generate frozen yeast cell "noodles."

    The precise cell culture density can be changed to suit your biological inquiry. We suggest processing a total yeast pellet of a minimum of ∼5 g wet cell weight, up to tens of grams, total yeast pellet be processed at a time.

2. Prechill all components of the planetary ball mill in liquid nitrogen in a polystyrene container. Add frozen noodles to the milling jar (≤20 mL of noodles/50-mL jar and 20–50 mL of noodles/125-mL jar). Weigh the milling jar with noodles and adjust the counterbalance weight. Add seven to eleven 20-mm stainless steel balls to the 125-mL jar (depending on the volume of noodles) or three balls to the 50-mL jar.

   Ensure no liquid nitrogen is visible within the jar before milling. This is important to avoid venting of cell material during milling or opening of the jar. The milling jar for the PM100 is not airtight, and under high pressure, nitrogen gas will escape. However, although it is unlikely, excess pressure buildup could lead to an explosion. Therefore, if liquid nitrogen enters the jar and is visible, be cautious, and allow it to evaporate before initiating milling.

3. Perform milling in eight cycles. For the 125-mL jar, use the following settings for each cycle: 400 rpm, 3 min, and 1 min reverse rotation with no breaks between rotations. For the 50-mL jar, use the following: 500 rpm, 3 min, and 1 min reverse rotation with no breaks between rotations. Cool the jars in liquid nitrogen between cycles. Do not remove the lid because this may result in cell loss. Keep the lid chilled by carefully pouring liquid nitrogen over the top of the jar. Do not submerge the jar completely because liquid nitrogen may leak into the milling chamber.

   You MUST hear the balls rattling around in the jar. If there is no rattling, add/remove balls until you hear a rattle. It is not considered a milling cycle unless there is rattling.

4. When eight cycles are complete, remove the cryomilled yeast cell powder from the jars with a spatula. Weigh the powder into microcentrifuge tubes using a spatula or volumetric spoon. Use liquid nitrogen to prechill all utensils and tubes, and return the utensils to liquid nitrogen in between uses. Tare an analytical balance with each prechilled tube before adding the cell powder, and then add the cell powder and check the mass added by quickly replacing the tube with powder on the balanced scale. Work quickly to avoid thawing the cell powder. Keep cell powder stocks and aliquots transferred to microcentrifuge tubes in liquid nitrogen.

   Using the measuring spoons suggested in the Equipment list, we have found that one heaped scoop using the spoon labeled "smidgen" gives ~50 mg of yeast cell powder and one level scoop using the spoon labeled "dash" gives ~250 mg of yeast cell powder. However, an initial feel for the "size" of the scoop must be made by the user. We have found that 100 mg of cryomilled yeast cell powder is commonly sufficient to purify a tagged protein complex of a moderate to low abundance at levels detectable by Coomassie blue staining after SDS–PAGE. Depending on the abundance of the complex and the application, the amount of starting material may need to be increased. If >250 mg needs to be used, we recommend weighing out multiple 250-mg samples and pooling the eluates at the end.

   Store the cryomilled yeast cell powder for up to 1 yr at −80°C or proceed immediately to Step 5.

Preparation of Yeast Cell Extracts

5. Remove the tubes containing aliquots of yeast cell powder from liquid nitrogen and let them stand with open (or loosened) caps for ~30 sec (per 100 mg) at room temperature.

   Residual pressure due to evaporating liquid nitrogen held within the cell powder can cause standard microcentrifuge tubes to pop open during warming, potentially resulting in the loss of the cell powder. Snap-cap tubes can become brittle at the temperature of liquid nitrogen, causing the hinge to break. This is why screw caps are preferred at this step. No thawing is observed during this brief incubation at room temperature, but be mindful not to initiate thawing of more material than you can process rapidly in Step 6.

6. Add four volumes of extraction solvent at room temperature supplemented with protease inhibitors to the yeast cell powder (e.g., 400 µL per 100 mg of powder). Briefly vortex (~30 sec) to completely resuspend the cell powder.

   It is important to use extraction solvent at room temperature to expedite the resuspension and avoid ice formation at the powder/liquid interface. Keep resuspended crude extracts on ice until all samples are ready (but try to process all as rapidly as possible). Following this step, the extraction solvent can be held on ice to be used for subsequent washes.

7. Clarify the crude extracts by spinning in a benchtop microcentrifuge at 14,000 rpm for 7 min at 4°C.

Centrifuging in an Eppendorf 5417R with an F-45-30-11 rotor (or similar microcentrifuge) under these conditions is roughly equivalent to centrifuging in a Beckman JA-25.50 rotor at 40,000g for 15 min.

Perform Steps 8 and 9 while the extracts are centrifuging. Proceed immediately to affinity capture.

Equilibration of Antibody-Conjugated Beads

8. While the cell extracts are in the centrifuge, aliquot the magnetic affinity medium into microcentrifuge tubes.

    Typically 5 µL of antibody-conjugated magnetic beads (suspended in a 15% w:V slurry) per 100 mg of frozen powder provides an excellent depletion (≥70%) of the target protein and its accompanying complex(es) from the clarified extract. However, a titration of the quantity of affinity medium needed to efficiently deplete the tagged protein can be performed and monitored by western blot. As the quantity of beads increases relative to extract, the opportunity for off-target protein binding (i.e., experimental noise) increases, potentially polluting the fraction. Thus, a compromise between complete recovery and increasing noise is usually sought; we recommend operating at 70%–95% depletion efficiency as often as possible. You may also want to assess the initial degree of extraction of the tagged protein (i.e., what fraction remains in the pellet versus the supernatant), keeping in mind that greater extraction of the tagged protein into the solvent does not necessarily translate to greater stability of the complex(es) it forms.

9. Pre-equilibrate the beads by washing them three times with 0.5 mL of extraction solvent. To wash the beads, agitate by vortexing to fully resuspend, then put the tube on the magnet briefly to allow the beads to collect on the tube wall, and remove the supernatant. To avoid excessive drying of the affinity medium, remove the final wash as the centrifugation in Step 7 reaches completion.

    In Step 9 as in Step 11, agitating the beads excessively can cause some of the solution to collect away from the test in the case or on the side of the tube. The user should always ensure all the solution is at the bottom of the tube, typically by brief pulse centrifugation in a low-speed bench-top minicentrifuge.

Affinity Capture

10. Without disturbing the pellets, transfer the clarified extracts from Step 7 to the tubes containing the washed and equilibrated affinity medium from Step 9. Save an aliquot from each sample before combining with the affinity medium for analysis by western blot, if necessary. Incubate the mixture on a rotating wheel mixer for 30 min at 4°C (batch binding).

    Although tagged protein binding may increase over several hours, the accumulation of off-target binding often increases disproportionately (Cristea et al. 2005). We advise using the shortest incubation time that gives an excellent depletion with low off-target binding—typically between 15 min and 1 h.

11. After the batch binding is complete, place the tubes on a magnet to collect the beads and remove the supernatant (save an aliquot for analysis by western blot, if necessary). Wash the beads three times with 500 µL of extraction buffer. (This is best accomplished by pipetting the beads up and down or by brief, low-speed vortex mixing. High-speed vortex mixing is not recommended at this stage because, in our experience, some components may be released from the immobilized complexes, presumably because of shearing forces.) During the second wash, transfer the beads to a fresh microcentrifuge tube, and then perform the final wash.

    Changing to a fresh tube helps to reduce the background due to nonspecific adsorption of proteins to the tube walls. The wash volume may be increased up to the nominal volume of the tube for effectively washing larger quantities of affinity medium.

    At this point, the sample can be eluted using denaturing conditions (see Step 12) or using nondenaturing (native) conditions; for the latter, see Protocol 2: Native Elution of Yeast Protein Complexes Obtained by Affinity Capture (LaCava et al. 2015b).

Elution (Denaturing Conditions)

12. Elute the protein complexes by incubating the beads with 20 µL of 1.1× SDS–PAGE sample buffer (without a reducing agent to avoid excessive release of Ig chains into the eluted fraction; 1/10th volume of reducing agent will be added later). Scale the volume of sample buffer according to the quantity of beads and the size of the gel well. (The volume should be sufficient to resuspend the beads and large compared to the residual volume retained by the beads.) To release complexes captured via a protein A tag from magnetic media conjugated to rabbit IgG, incubate for 10 min

at room temperature. To facilitate the release of complexes captured via a GFP tag, depending on the affinity of the anti-GFP antibody, it may be necessary to heat at 70°C. Agitate gently during elution. Place the tubes on the magnet and recover the eluted fraction. Incubate the eluates with a reducing agent (typically, heat for 10 min at 70°C) and perform SDS–PAGE and staining to assess the composition of the purified complex.

RELATED INFORMATION

For further information regarding this procedure, consult Oeffinger et al. (2007); Domanski et al. (2012), and LaCava et al. (2013). The latest versions of our detailed protocols for cryomilling yeast and conjugating antibodies to Dynabeads can be found on our website (http://www.ncdir.org). These protocols evolve with our work and are continuously updated online as improvements are achieved.

REFERENCES

Candiano G, Bruschi M, Musante L, Santucci L, Ghiggeri GM, Carnemolla B, Orecchia P, Zardi L, Righetti PG. 2004. Blue silver: A very sensitive colloidal Coomassie G-250 staining for proteome analysis. *Electrophoresis* 25: 1327–1333.

Cristea IM, Chait BT. 2011. Conjugation of magnetic beads for immunopurification of protein complexes. *Cold Spring Harb Protoc* doi: 10.1101/pdb.prot5610.

Cristea I, Williams R, Chait B, Rout M. 2005. Fluorescent proteins as proteomic probes. *Mol Cell Proteomics* 4: 1933–1941.

Domanski M, Molloy K, Jiang H, Chait BT, Rout MP, Jensen TH, LaCava J. 2012. Improved methodology for the affinity isolation of human protein complexes expressed at near endogenous levels. *Biotechniques* 0: 1–6.

LaCava J, Chandramouli N, Jiang H, Rout MP. 2013. Improved native isolation of endogenous protein A-tagged protein complexes. *Biotechniques* 54: 213–216.

LaCava J, Fernandez-Martinez J, Hakhverdyan Z, Rout MP. 2015a. Protein complex purification by affinity capture. *Cold Spring Harb Protoc* doi: 10.1101/pdb.top077545.

LaCava J, Fernandez-Martinez J, Rout MP. 2015b. Native elution of yeast protein complexes obtained by affinity capture. *Cold Spring Harb Protoc* doi: 10.1101/pdb.prot087940.

Oeffinger M, Wei KE, Rogers R, DeGrasse JA, Chait BT, Aitchison JD, Rout MP. 2007. Comprehensive analysis of diverse ribonucleoprotein complexes. *Nat Methods* 4: 951–956.

| Protocol 2

Native Elution of Yeast Protein Complexes Obtained by Affinity Capture

John LaCava,[1] Javier Fernandez-Martinez,[1] and Michael P. Rout[2]

Laboratory of Cellular and Structural Biology, The Rockefeller University, New York, New York 10065

This protocol describes two options for the native (nondenaturing) elution of protein complexes obtained by affinity capture. The first approach involves the elution of complexes purified through a tag that includes a human rhinovirus 3C protease (PreScission protease) cleavage site sequence between the protein of interest and the tag. Incubation with the protease cleaves immobilized complexes from the affinity medium. The second approach involves the release of protein A–tagged protein complexes using a competitive elution reagent called PEGylOx. The degree of purity of the native assemblies eluted is sample dependent and strongly influenced by the affinity capture. It should be noted that the efficiency of native elution is commonly lower than that of elution by a denaturing agent (e.g., SDS) and the release of the complex will be limited by the activity of the protease or the inhibition constant (K_i) of the competitive release agent. However, an advantage of native release is that some nonspecifically bound materials tend to stay adsorbed to the affinity medium, providing an eluted fraction of higher purity. Finally, keep in mind that the presence of the protease or elution peptide could potentially affect downstream applications; thus, their removal should be considered.

MATERIALS

It is essential that you consult the appropriate Material Safety Data Sheets and your institution's Environmental Health and Safety Office for proper handling of equipment and hazardous material used in this protocol.

Reagents

Desalting spin columns (40 kDa molecular mass cutoff) (optional; see Step 14)

Depending on the volume of the eluted fraction, use either Micro Bio-Spin Columns with Bio-Gel P-30 (Bio-Rad 732-6223) or Zeba Micro Spin Desalting Columns, 75 µL, with 40 kDa molecular mass cutoff (Thermo Scientific 87764). These columns give equivalent results, depleting the peptide >100-fold.

Digestion buffer (for Steps 1–10 only)

Determine the optimal composition of this buffer for each protein complex of interest. A suggested formulation to work from is 20 mM K-HEPES (pH 7.4), 150 mM sodium chloride, 110 mM potassium acetate, 2 mM magnesium chloride, 0.1% Tween 20, and 1 mM DTT.

PEGylOx native elution solution (saturated solution; >2 mM) (for Steps 11–15 only)

PEGylOx is an amino-terminally PEGylated peptide of primary sequence DCAWHLGELVWCT, cyclized by oxidation of the cysteines to cystine. It can be synthesized by standard Fmoc solid-phase synthesis methods. The PEGylOx solution can be prepared using a solvent of your choosing. As long as the saturation concentration of

---

[1]These two authors contributed equally.
[2]Correspondence: rout@rockefeller.edu

Copyright © Cold Spring Harbor Laboratory Press; all rights reserved
Cite this protocol as *Cold Spring Harb Protoc*; doi:10.1101/pdb.prot087940

Chapter 21

peptide is ~2 mM (which will be solvent dependent), effective elution can be expected. We recommend using a 40 mM Tris–Cl buffered elution solution at pH 8.0 and ethanol at 5% (v/v) in any elution solution. Na-HEPES at pH 7.4 can be substituted for Tris-Cl. As a detergent, we use Tween 20 at 0.01% (v/v); a detergent is required for stable high concentration solutions of PEGylOx. The concentration of the peptide solution is monitored by UV spectrophotometry at 280 nm using an estimated molar extinction coefficient ($\varepsilon 280$) of 11,125 M^{-1} cm^{-1}.

PreScission protease (human rhinovirus 3C protease) (GE Healthcare Life Sciences) (for Steps 1–10 only)

PreScission protease inhibitors (optional; see Step 9)

Yeast protein complexes attached to antibody-conjugated Dynabeads as prepared in
Protocol 1: Optimized Affinity Capture of Yeast Protein Complexes (LaCava et al. 2015)

Equipment

Low protein-binding tubes (e.g., Sorenson Biosciences 39640T) (optional; see Step 14)
Microcentrifuge (benchtop; refrigerated)
Microcentrifuge tubes (1.5–2 mL)
Neodymium magnet (e.g., DynaMag-2 from Life Technologies)
Sample mixing block
Vortex mixer

METHOD

Select either Steps 1–10 or Steps 11–15.

Native Elution of Protein Complexes Using a Cleavable Tag

This protocol is suitable for eluting complexes purified through a tag that includes a human rhinovirus 3C protease (PreScission protease) cleavage site sequence (GLEVLFQGPS) between the protein of interest and the tag (Cordingley et al. 1989, 1990; Walker et al. 1994). This approach has also been successful using a cleavage site for tobacco etch virus protease (Dougherty et al. 1989; Polayes et al. 1994).

1. Prepare 1.25 μL of digestion mixture (PreScission protease in digestion buffer at a concentration of 0.08 units/μL) per 5 μL of Dynabeads slurry used in the affinity capture. This solution will be added directly to the beads containing immobilized complexes obtained at the end of Step 11 in Protocol 1: Optimized Affinity Capture of Yeast Protein Complexes (LaCava et al. 2015). Keep the digestion mixture on ice.

   We suggest preparing a small excess of digestion mixture to compensate for pipetting errors. If the protease activity (units) is not known, try 0.08 μg/μL.

2. If the digestion buffer is different from the wash buffer used in Step 11 in Protocol 1: Optimized Affinity Capture of Yeast Protein Complexes (LaCava et al. 2015), wash the slurry of protein complexes attached to beads once with 10 slurry volumes of cold digestion buffer.

3. Add digestion mixture from Step 1 to the beads. Use 1 μL per 5 μL of Dynabeads slurry used in the affinity capture. Tap the tube until the beads are fully resuspended. Do not vortex.

4. Incubate with gentle agitation for 30–45 min at 4°C.

5. Centrifuge the sample at 3500g for 1 min at 4°C to pellet the beads. Place the tube in the magnet, and transfer the supernatant to a clean tube on ice.

6. Add digestion buffer (no enzyme) to the beads. Use 1 μL per 5 μL of Dynabeads slurry used in the affinity capture. Tap the tube until the beads are fully resuspended. Do not vortex.

7. Centrifuge the sample at 3500g for 1 min at 4°C to pellet the beads. Place the tube in the magnet, and transfer the supernatant to the same tube that contains supernatant from Step 5.

8. Centrifuge the combined eluted fraction at 20,000g for 10 min at 4°C to pellet residual magnetic beads and particulates.

9. Transfer the supernatant to a clean tube—this is the natively eluted material.

   The sample should be used as soon as possible to avoid disintegration of the purified complex, but depending on the stability of the complex the sample could be held on ice or at 4°C for up to several hours if further processing by, for example, density gradient centrifugation will be carried out. The sample may be stored at −20°C or lower for longer term, but freeeze–thaw cycles may adversely affect native structures. Consider the addition of a cryoprotectant such as glycerol or ethylene glycol to the sample before storage at −20°C if, for example, enzymatic assays will be conducted on the sample. If the material is not going to be used immediately in subsequent procedures, consider adding PreScission protease inhibitors to the sample. The manufacturer suggests 4 mM Pefabloc or 100 μM chymostatin.

   To prepare a more homogeneous solution of protein complexes (i.e., one without fragments caused by degradation during the purification procedure), subject the eluate to rate-zonal density gradient ultracentrifugation as described in Protocol 3: Density Gradient Ultracentrifugation to Isolate Endogenous Protein Complexes after Affinity Capture (Fernandez-Martinez et al. 2015).

10. Analyze a fraction of this sample by SDS–PAGE as in Step 12 in Protocol 1: Optimized Affinity Capture of Yeast Protein Complexes (LaCava et al. 2015).

    To determine what material remains on the beads after protease cleavage, wash the Dynabeads with SDS–PAGE sample loading buffer.

    See Troubleshooting.

Native Elution of Protein A–Tagged Complexes Using the Competitive Elution Reagent PEGylOx

This procedure uses a synthetic peptide, PEGylOx, whose production, handling and usage is described in LaCava et al. (2013). The peptide competitively releases protein A–tagged complexes from interaction with the IgG constant (Fc) region, and it can be removed from the eluted fraction by spin column gel filtration.

11. Wash protein complexes immobilized on Dynabeads (from the last wash in Step 11 in Protocol 1: Optimized Affinity Capture of Yeast Protein Complexes [LaCava et al. 2015]) with 50 μL of the solution used to dissolve PEGylOx, and then remove the wash solution. Fifty microliters of solution is sufficient for up to 25 μL of Dynabeads slurry used in the affinity capture. Scale up the wash if necessary.

    This step is to equilibrate the beads with the elution solution, while keeping the beads and complexes highly concentrated. It can be omitted if the elution solution and the extraction solvent are the same.

12. Add 26 μL of PEGylOx native elution solution to up to 25 μL of the Dynabeads slurry; scale up as needed. Incubate with gentle shaking (e.g., setting three on a vortex mixer) for 15 min at room temperature.

13. Apply a brief pulse of low-speed centrifugation to ensure all of the slurry is at the bottom of the tube. Place the tube in the magnet, and transfer the supernatant to a clean tube. To ensure no beads have carried over, place the sample on the magnet again, and transfer the supernatant to a fresh tube—this is the natively eluted material.

    The samples should be used as soon as possible to avoid disintegration of the purified complex, but depending on the stability of the complex the sample could be held on ice or 4°C for up to several hours if further processing by, for example, PEGylOx removal and/or density gradient centrifugation will be carried out. The sample may be stored at −20°C or lower for longer term, but freeze–thaw cycles may adversely affect native structures. Consider the addition of a cryoprotectant such as glycerol or ethylene glycol to the sample prior to storage at −20°C if, for example, enzymatic assays will be conducted on the sample. Bear in mind that PEGylox solubility at low temperatures may vary with the composition of the elution solution used—the PEGylOx solution compositions suggested in Materials are stable at the above storage temperatures.

14. (Optional) To deplete the peptide >100-fold, apply the natively eluted samples to gel filtration spin columns with a nominal molecular mass cutoff of 40 kDa. Pre-equilibrate these columns with the desired buffer solution for exchange following the manufacturer's instructions. If necessary, use this step as an opportunity to exchange the solution the sample is in for downstream assays. Consider using low protein-binding tubes to collect the flowthrough.

To prepare a more homogeneous solution of protein complexes (i.e., one without fragments caused by degradation during the purification procedure), subject the eluate to rate-zonal density gradient ultracentrifugation as described in Protocol 3: Density Gradient Ultracentrifugation to Isolate Endogenous Protein Complexes after Affinity Capture (Fernandez-Martinez et al. 2015).

15. Analyze a fraction of the eluate by SDS–PAGE as in Step 12 in Protocol 1: Optimized Affinity Capture of Yeast Protein Complexes (LaCava et al. 2015). To determine what material remains on the beads after PEGylOx treatment, wash the Dynabeads in the pellet with SDS–PAGE sample loading buffer.

TROUBLESHOOTING

Problem (Step 10): Protein complexes are only weakly detected or absent.
Solution: The composition of the digestion buffer may not be optimal. The composition of the digestion buffer is complex dependent and will have to be formulated for each protein complex of interest. Follow these guidelines:

- Do not include protease inhibitors in the buffer.

- To ensure optimal efficiency of the PreScission protease, add DTT at a final concentration of 1 mM.

- As a general rule, prepare the digestion buffer so that its composition is similar to the buffer used for affinity capture (i.e., the extraction solvent used in Protocol 1: Optimized Affinity Capture of Yeast Protein Complexes [LaCava et al. 2015]). Ensure that the ionic strength is enough to elute the complex from the magnetic matrix without causing dissociation of the protein components. Add a low concentration of detergent (e.g., Tween 20) to help release the complex and avoid nonspecific adsorption to the affinity medium.

RELATED INFORMATION

A procedure for eluting affinity-captured protein complexes in a denaturing solvent (SDS–PAGE sample buffer) is included in Protocol 1: Optimized Affinity Capture of Yeast Protein Complexes (LaCava et al. 2015).

REFERENCES

Cordingley MG, Register RB, Callahan PL, Garsky VM, Colonno RJ. 1989. Cleavage of small peptides in vitro by human rhinovirus 14 3C protease expressed in *Escherichia coli*. *J Virol* 63: 5037–5045.

Cordingley MG, Callahan PL, Sardana VV, Garsky VM, Colonno RJ. 1990. Substrate requirements of human rhinovirus 3C protease for peptide cleavage in vitro. *J Biol Chem* 265: 9062–9065.

Dougherty W, Cary S, Parks T. 1989. Molecular genetic analysis of a plant virus polyprotein cleavage site: A model. *Virology* 171: 356–364.

Fernandez-Martinez J, LaCava J, Rout MP. 2015. Density gradient ultracentrifugation to isolate endogenous protein complexes after affinity capture. *Cold Spring Harb Protoc* doi: 10.1101/pdb.prot087957.

LaCava J, Chandramouli N, Jiang H, Rout MP. 2013. Improved native isolation of endogenous protein A-tagged protein complexes. *BioTechniques* 54: 213–216.

LaCava J, Fernandez-Martinez J, Hakhverdyan Z, Rout MP. 2015. Optimized affinity capture of yeast protein complexes. *Cold Spring Harb Protoc* doi: 10.1101/pdb.prot087932.

Polayes D, Goldstein A, Ward G, Hughes AJ. 1994. TEV Protease, recombinant: A site-specific protease for efficient cleavage of affinity tags from expressed proteins. *Focus* 16: 2–5.

Walker PA, Leong L, Ng PWP, Tan S, Waller S, Murphy D, Porter AG. 1994. Efficient and rapid affinity purification of proteins using recombinant fusion proteases. *Biotechnology (NY)* 12: 601–605.

Protocol 3

Density Gradient Ultracentrifugation to Isolate Endogenous Protein Complexes after Affinity Capture

Javier Fernandez-Martinez, John LaCava, and Michael P. Rout[1]

Laboratory of Cellular and Structural Biology, The Rockefeller University, New York, New York 10065

This protocol describes the isolation of native protein complexes by density gradient ultracentrifugation. The outcome of an affinity capture and native elution experiment is generally a mixture of (1) the complex(es) associated with the protein of interest under the specific conditions of capture, (2) fragments of the complex generated by degradation or disassembly during the purification procedure, and (3) the protease or reagent used to natively elute the sample. To separate these components and isolate a homogeneous complex, an additional step of purification is required. Rate-zonal density gradient ultracentrifugation is a reliable and powerful technique for separating particles based on their hydrodynamic volume. The density gradient is generated by mixing low- and high-density solutions of a suitable low-molecular-weight inert solute (e.g., sucrose or glycerol). The gradient is formed in a solvent that could be any of the solvents used for the affinity capture and native elution and should help to preserve the structure and activity of the assembly.

MATERIALS

It is essential that you consult the appropriate Material Safety Data Sheets and your institution's Environmental Health and Safety Office for proper handling of equipment and hazardous material used in this protocol.

Reagents

Buffer for preparing gradients

As a general rule, the buffer used to prepare the gradients should be the same composition as the one used to purify and stabilize the protein complex (i.e., the one used during the native elution during Protocol 2: Native Elution of Yeast Protein Complexes Obtained by Affinity Capture [LaCava et al. 2015]). However, the buffer composition can be modified depending on the downstream sample processing requirements, bearing in mind that the stability and activity of the complex in the new composition should be tested beforehand.

Natively eluted yeast protein complexes prepared by either method described in Protocol 2: Native Elution of Yeast Protein Complexes Obtained by Affinity Capture (LaCava et al. 2015)

Protease inhibitor cocktail (e.g., Sigma-Aldrich P8340)

Sucrose (66%, w/w) or glycerol

Prepare a solution of 66% (w/w) sucrose by dissolving 1710 g of sucrose (molecular biology grade) in 900 mL of high-performance liquid chromatography (HPLC)-grade water. Double-check the concentration using a refractometer and correct if necessary by adding sucrose or water. This solution can be stored at 4°C indefinitely because the low available water in the solution prevents bacterial growth.

As an alternative to sucrose, glycerol is a solute very commonly used to generate density gradients.

---

[1]Correspondence: rout@rockefeller.edu

Copyright © Cold Spring Harbor Laboratory Press; all rights reserved
Cite this protocol as *Cold Spring Harb Protoc*; doi:10.1101/pdb.prot087957

Chapter 21

Equipment

Filter (Millex-GP Syringe Filter Unit [0.22-μm] from EMD Millipore)
Gradient Master (BioComp Instruments)
Piston Gradient Fractionator (BioComp Instruments) (optional; see Step 10)
Polyclear centrifuge tubes (13 × 51-mm) (Seton Scientific)

> When purchased from BioComp Instruments, Inc. (cat. no. 151-513), the tubes are tested for tolerance and compatibility with piston fractionation.

Refractometer (Bausch & Lomb)
SW 55 Ti Rotor (Beckman Coulter)
Ultracentrifuge (Beckman Coulter)

METHOD

Gradient Preparation

Prepare two gradients by following the instructions in the BioComp Gradient Master operator's manual, which are briefly summarized here. The small gradient volumes (~5 mL each) help to maintain the concentration of the purified complex within a reasonable range, as usually the amount of purified native protein assemblies is very limited.

1. Prepare 10 mL each of 5% and 20% (w/w) sucrose solutions in a buffer with the same composition as the one used during the native elution. See Griffith (1986, p. 49) for a chart indicating the appropriate dilutions of a 66% (w/w) sucrose solution to prepare a 5% solution (1/20 dilution) and a 20% solution (1/4.1 dilution). Filter the solutions through a 0.22-μm filter. Verify the final concentration using a refractometer.

   The concentrations of the gradient should be adjusted to the size of the protein complex that needs to be isolated. 5%–20% (w/w) sucrose works well for complexes of 200–800 kDa. For larger complexes, a 10%–40% (w/w) sucrose gradient would be more suitable. If the gradient is prepared using glycerol, the most common gradient used for separation of protein complexes is 10%–30% (v/v).

2. Add protease inhibitor cocktail to each sucrose solution to give a final concentration of 0.001×, and mix carefully by rocking to avoid the formation of air bubbles. Add other kinds of chemical compounds depending on the enzymatic activities that need to be inhibited or preserved.

3. Place two polyclear centrifuge tubes into the SW 50/SW 55 marker block and, using a marker, draw a fine line following the top of the block. Fill each tube with 5% (w/w) sucrose solution until the liquid reaches ~2 mm above the half-full mark. Fill a 10-mL syringe with the 20% (w/w) sucrose solution and attach a blunt-end steel cannula to its tip. Being careful to avoid air bubbles and not to disturb the interface, insert the cannula quickly to the bottom of the tube and slowly layer the 20% solution until the interface formed with the 5% reaches the half-full mark. Quickly withdraw the cannula and cap the tubes using the short rubber caps.

4. Place the tubes in the tube holder and use the Gradient Master to rotate and mix the solutions as outlined in the operator's manual to generate a linear gradient.

5. Take the formed gradients and let them cool down for 45 min to 1 h at 4°C. At this time, ensure that the rotor is also precooled at 4°C.

   The gradients are stable for up to 3 h at 4°C.

Centrifugation

During this series of steps, avoid heating the gradients by working fast or in a cold room.

6. Carefully remove the caps from the gradient tubes, and slowly load between 50 and 200 μL containing 10–100 pmol of natively eluted yeast protein complexes (from Protocol 2: Native Elution of Yeast Protein Complexes Obtained by Affinity Capture [LaCava et al. 2015]) on top of one of the gradients, trying not to disturb it.

   Although these ranges are desirable, lower amounts can be successfully analyzed. We recommend using gel-loading tips for this task.

7. Use the other gradient to balance the rotor; add a volume of water or buffer equal to the volume of sample so as to keep them balanced.

   Alternatively, to analyze several samples, distribute the samples across multiple gradients.

8. Balance the tubes before loading them into the swing-out rotor. Check that the gradients are of comparable weight using a balance, keeping them within ±100 mg of one another by adding solution as needed.

9. Centrifuge at 59,000–237,000g in an SW 55 Ti rotor and Beckman L-80 ultracentrifuge for 6–20 h at 4°C.

   These centrifugation conditions are commonly used, but the time and speed of centrifugation must be determined empirically for each protein assembly because their sedimentation rate depends not only on their size, but also on their shape. Ideally, the complex should move about two-thirds of the way down the gradient to obtain good separation.

10. After centrifugation, fractionate the gradients at 4°C by one of the following methods.

    - Use a BioComp Piston Gradient Fractionator to collect fractions in 2-mm increments (i.e., ~20 fractions of ~225 µL each).
    - Manually fractionate by unloading from the top using a pipette or from the bottom of the tube using a capillary and a peristaltic pump. Collect 12 fractions of 410 µL each or 24 fractions of 205 µL each, depending on how precisely the peak of the complex needs to be determined.

      For assays requiring native complexes (e.g., assays of enzymatic activity or protein complex structure), analyze the fractions immediately. If this is not possible, store for short periods (no more than a few hours) at 4°C, or flash-freeze in liquid nitrogen and store for days to months at –80°C. For assays that do not require native complexes (e.g., sodium dodecyl sulfate-polyacrylamide gel electrophoresis [SDS-PAGE] analysis), store the fractions at –20°C until analysis.

Analysis of Fractions

11. Analyze an aliquot of each fraction by SDS-PAGE protein staining (e.g., Coomassie blue, silver, or SYPRO Ruby) to visualize the components and localize the complex of interest.

    In the event that the final sample concentration in the fractions is insufficient for visualization upon direct loading of an aliquot, the fractions can be effectively precipitated with methanol/chloroform (Wessel and Flügge 1984).

    Data from an experiment in which a native protein complex was purified by affinity capture, released by PreScission protease cleavage, and enriched by density gradient ultracentrifugation is presented in Figure 1.

12. Characterize the purified native complexes as desired.

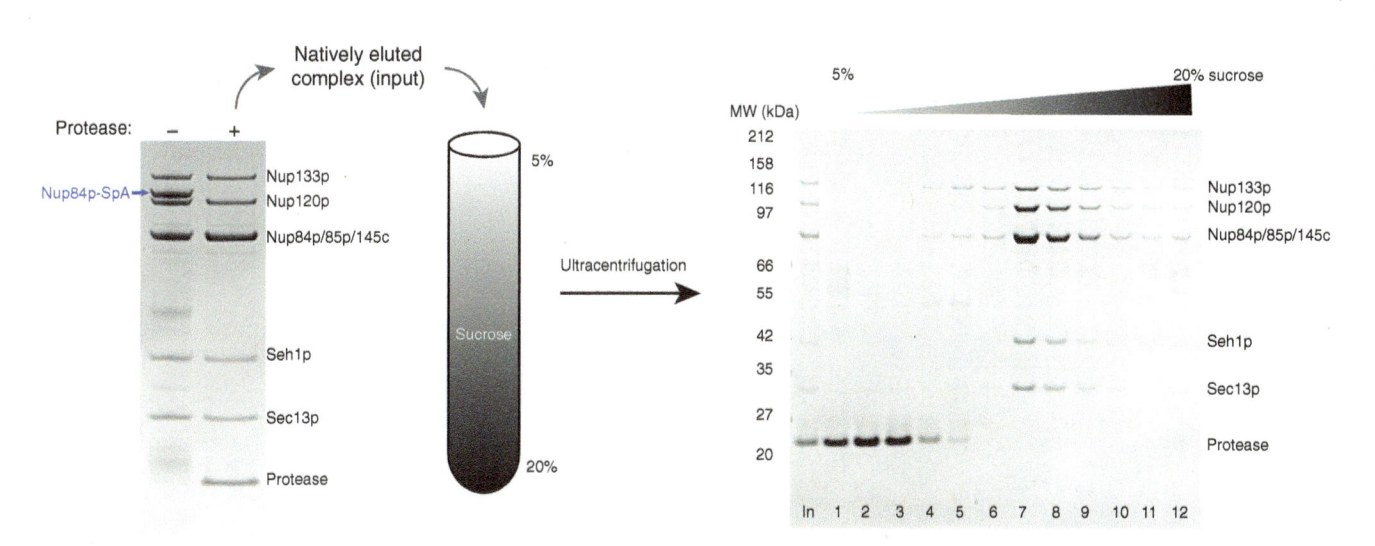

FIGURE 1. The endogenous heptameric Nup84-complex purified by affinity capture, released by PreScission protease cleavage, and enriched by fractionation across a 5%–20% (w/w) sucrose density gradient.

Mass spectrometry is the ideal method to identify the proteins present in each fraction. Depending on the downstream application, the sucrose or glycerol used in the gradient may need to be removed by dialysis, gel filtration, or ultrafiltration. For examples of downstream structural and functional characterization using protein complexes purified with this procedure, see Erickson (2009) and Luhrmann and Stark (2009).

RELATED INFORMATION

For further information regarding the density gradient ultracentrifugation procedure, consult Griffith (1986) and Erickson (2009).

REFERENCES

Erickson HP. 2009. Size and shape of protein molecules at the nanometer level determined by sedimentation, gel filtration, and electron microscopy. *Biol Proced Online* 11: 32–51.

Griffith OM. 1986. *Techniques of preparative, zonal, and continuous flow ultracentrifugation.* Beckman, Palo Alto, CA.

LaCava J, Fernandez-Martinez J, Rout MP. 2015. Native elution of yeast protein complexes obtained by affinity capture. *Cold Spring Harb Protoc* doi: 10.1101/pdb.prot087940.

Luhrmann R, Stark H. 2009. Structural mapping of spliceosomes by electron microscopy. *Curr Opin Struct Biol* 19: 96–102.

Wessel D, Flügge UI. 1984. A method for the quantitative recovery of protein in dilute solution in the presence of detergents and lipids. *Anal Biochem* 138: 141–143.

CHAPTER 22

Proteomic Analysis of Protein Posttranslational Modifications by Mass Spectrometry

Danielle L. Swaney and Judit Villén[1]

Department of Genome Sciences, University of Washington, Seattle, Washington 98195

The addition of posttranslational modifications (PTMs) to proteins is an influential mechanism to temporally control protein function and ultimately regulate entire cellular processes. Most PTMs are present at low stoichiometry and abundance, which limits their detection when analyzing whole cell lysates. PTM purification methods are thus required to comprehensively characterize the presence and dynamics of PTMs using mass spectrometry-based proteomics approaches. Here we describe several of the most influential PTMs and discuss the fundamentals of proteomics experiments and PTM purification methods.

BACKGROUND

Protein posttranslational modifications (PTMs) modulate protein functions and are integral to almost every cellular process. The reversible nature of PTMs enables dynamic regulation of PTM status—functioning as highly versatile switches to control protein structure, activity, subcellular localization, etc. PTMs can function individually or can work in a coordinated fashion with other PTMs. For example, multiple PTMs on the same protein can interact to alter protein function, or individual PTMs on different proteins can interact in series to form interconnected signaling networks and regulate broad cellular processes. To modulate the PTM status of proteins, specific classes of enzymes catalyze the addition of particular modification types to proteins (e.g., kinases and acetyl transferases), whereas alternate classes of enzymes are responsible for the removal of PTMs (e.g., phosphatases and deacetylases). More than 200 different types of PTMs have been described; however, most exist at low stoichiometry on proteins of low abundance (Krishna and Wold 1998; Mann and Jensen 2003). Therefore, to characterize PTMs on a proteome-wide scale, purification methods are necessary to isolate modified proteins and peptides before proteomic analysis. In the associated protocols (Protocol 1: Enrichment of Phosphopeptides via Immobilized Metal Affinity Chromatography [Swaney and Villén 2015a] and Protocol 2: Enrichment of Modified Peptides via Immunoaffinity Precipitation with Modification-Specific Antibodies [Swaney and Villén 2015b]), we describe such purification methods for the proteomic analysis of several of the most widely studied PTMs.

TYPES OF POSTTRANSLATIONAL MODIFICATIONS

Arguably the most well-studied PTM is protein phosphorylation, for which more than 20,000 phosphorylation sites have been detected in yeast, targeting nearly half of the yeast proteome (Stark et al.

---

[1]Correspondence: jvillen@u.washington.edu

Copyright © Cold Spring Harbor Laboratory Press; all rights reserved
Cite this introduction as *Cold Spring Harb Protoc*; doi:10.1101/pdb.top077743

2010). Kinases catalyze the transfer of a phosphoryl group from ATP to a serine, threonine, or tyrosine residue on a protein substrate. The presence of phosphorylation on a specific amino acid can be detected in mass spectrometry–based proteomics experiments by a characteristic mass addition of 79.96633 Da. Phosphorylation is a reversible PTM and can be removed from a substrate by phosphatase enzymes. This reversible nature enables proteins to be dynamically modified to meet cellular needs. For example, cell cycle progression requires a network of phosphorylation events that temporally oscillate to activate protein substrates. Also, protein phosphorylation is the main driving force behind signal transduction.

Another widely studied PTM is ubiquitylation. Ubiquitin is a small (8.6-kDa) protein that is added to substrates via a multistep enzymatic cascade and can be removed by a class of proteins known as deubiquitinating enzymes (DUBs). To add ubiquitin to a substrate, ubiquitin is first activated by an enzyme (E1) and conjugated to a second enzyme (E2). In most cases, a ligase enzyme (E3) recognizes a substrate and complexes with it and the E2 to facilitate transfer of ubiquitin to a lysine residue on the substrate. Unlike phosphorylation and many other PTMs that are monomeric, ubiquitin can be added to a substrate in a polymeric fashion. Here, ubiquitin can serve as both modification and substrate to generate a branched structure propagated through one of its seven lysine residues. This branching adds additional layers of complexity to ubiquitin signaling. Different branching structures have been associated with specific cellular processes. The most abundant branching structure through lysine 48 on ubiquitin (K48) is commonly used to mark proteins for degradation, whereas branching through lysine 63 (K63) is associated with the DNA damage response, signaling, and trafficking of membrane proteins (Spence et al. 1995). Because of the size of the ubiquitin, and its capacity for polymeric addition, proteomics analysis relies on indirect detection methods. Enzymatic proteolysis of lysates with trypsin leaves the two carboxy-terminal glycine residues of ubiquitin attached to lysine. Antibodies with specificity to this diglycyl-lysine residue can then be used to purify sites of lysine ubiquitylation (Emanuele et al. 2011; Kim et al. 2011; Wagner et al. 2011), and detection of a mass addition of 114.04293 Da on lysine residues can confirm the site of ubiquitylation (Peng and Gygi 2001). A caveat of this approach is that other modifications by ubiquitin-like proteins can also produce a diglycyl-lysine residue (e.g., neddylation). In yeast, antibody-based enrichment of diglycyl-lysine has identified more than 5500 ubiquitylation sites on ~1900 proteins, revealing that at least one-third of the proteome is ubiquitylated (Swaney et al. 2013).

The final PTM that has been well characterized in yeast is lysine acetylation. Here, lysine acetyltransferases (KATs; also known as histone acetyltransferases, HATs) catalyze the addition of an acetyl group onto lysine residues. This modification is also reversible and can be removed by lysine deacetylases (KDACs; also known as histone deacetylases, HDACs). Using mass spectrometry detection, the characteristic mass addition of 42.01056 Da on lysine can confirm the presence of acetylation. Proteomic experiments using immunoaffinity purification methods have revealed ~4000 sites of lysine acetylation in yeast (Henriksen et al. 2012). This profiling has helped illustrate roles for lysine acetylation signaling in yeast, such as mitochondrial metabolism, protein synthesis, and chromatin organization (Henriksen et al. 2012).

PROTEOMIC METHODS

Mass spectrometry–based proteomics is a powerful approach to identify proteins modified by a particular PTM. Additionally, characteristic mass shifts in fragmentation spectra (i.e., tandem mass spectrometry [MS/MS]) often permit exact localization of the PTM to an individual amino acid within the protein. Advances in both methodology and technology have dramatically propelled the field of PTM analysis, and over the past decade, the number of PTM sites identified in a proteomics experiment has increased more than 30-fold (Ficarro et al. 2002; Peng et al. 2003; Swaney et al. 2013). In a single proteomics experiment, researchers can now routinely identify thousands of PTMs, as well as quantify their abundance changes in response to stimuli (Olsen et al. 2006; Kim et al. 2011; Henriksen et al. 2012). Such advancements have illustrated the breadth and diversity of modified proteins that regulate cell state, although the challenge of defining the biological function of these PTMs still remains.

Mass spectrometry experiments to characterize PTMs are often separated into single protein or global proteomics approaches. In single protein experiments, mass spectrometry is performed on peptides following isolation of a given protein of interest. This approach is useful for comprehensive identification of the variety of PTMs on a single protein; however, the success of this approach is dependent on effective isolation of the protein of interest. Two common strategies are protein-level immunoprecipitation using antibodies and introduction of a tag to the protein that can be used for purification (e.g., GST, FLAG, His-tag). Such isolation can be highly protein-specific, and thus methods for this approach are not described here.

Alternatively, using a global proteomics approach, one can obtain a cellular wide view of the PTM status of thousands of proteins at once. Here, peptides containing a single PTM type (e.g., phosphorylation) are isolated from a complex mixture of thousands of peptides. Methods to isolate modified peptides largely fall into two categories. First are biochemical-based methods, which exploit differences in chemical properties between modified peptides and nonmodified peptides (e.g., solution charge, metal affinity). This approach is used to isolate phosphopeptides by immobilized metal affinity chromatography (IMAC); see Protocol 1: Enrichment of Phosphopeptides via Immobilized Metal Affinity Chromatography (Swaney and Villén 2015a). The second category consists of immuno-affinity enrichment methods in which antibodies with specificity toward a particular PTM of interest (e.g., acetylation, ubiquitylation, phosphotyrosine) are used to purify PTM containing peptides; see Protocol 2: Enrichment of Modified Peptides via Immunoaffinity Precipitation with Modification-Specific Antibodies (Swaney and Villén 2015b).

Phosphorylation, ubiquitylation, and acetylation represent the PTMs that are most widely studied using global proteomics methods. As mentioned, many additional PTMs exist and regulate important biological processes. However, the study of these PTMs remains a challenge, as sensitive and selective purification methods for most of them have yet to be described.

ACKNOWLEDGMENTS

This work was supported in part by National Institutes of Health/National Cancer Institute (NIH/NCI) grant R00CA140789 and an Ellison Medical Foundation New Scholar Award (to J.V.).

REFERENCES

Emanuele MJ, Elia AEH, Xu Q, Thoma CR, Izhar L, Leng Y, Guo A, Chen Y-N, Rush J, Hsu PW-C, et al. 2011. Global identification of modular cullin-RING ligase substrates. *Cell* 147: 459–474.

Ficarro SB, McCleland ML, Stukenberg PT, Burke DJ, Ross MM, Shabanowitz J, Hunt DF, White FM. 2002. Phosphoproteome analysis by mass spectrometry and its application to *Saccharomyces cerevisiae*. *Nat Biotechnol* 20: 301–305.

Henriksen P, Wagner SA, Weinert BT, Sharma S, Bacinskaja G, Rehman M, Juffer AH, Walther TC, Lisby M, Choudhary C. 2012. Proteome-wide analysis of lysine acetylation suggests its broad regulatory scope in *Saccharomyces cerevisiae*. *Mol Cell Proteomics* 11: 1510–1522.

Kim W, Bennett EJ, Huttlin EL, Guo A, Li J, Possemato A, Sowa ME, Rad R, Rush J, Comb MJ, et al. 2011. Systematic and quantitative assessment of the ubiquitin-modified proteome. *Mol Cell* 44: 325–340.

Krishna RG, Wold F. 1998. Posttranslational modifications. In *Proteins: Analysis and design* (ed. Angeletti RH), pp. 121–207. Academic Press, San Diego.

Mann M, Jensen ON. 2003. Proteomic analysis of post-translational modifications. *Nat Biotechnol*. 21: 255–261.

Olsen JV, Blagoev B, Gnad F, Macek B, Kumar C, Mortensen P, Mann M. 2006. Global, in vivo, and site-specific phosphorylation dynamics in signaling networks. *Cell* 127: 635–648.

Peng J, Gygi SP. 2001. Proteomics: The move to mixtures. *J Mass Spectrom* 36: 1083–1091.

Peng J, Schwartz D, Elias JE, Thoreen CC, Cheng D, Marsischky G, Roelofs J, Finley D, Gygi SP. 2003. A proteomics approach to understanding protein ubiquitination. *Nat Biotechnol* 21: 921–926.

Spence J, Sadis S, Haas AL, Finley D. 1995. A ubiquitin mutant with specific defects in DNA repair and multiubiquitination. *Mol Cell Biol* 15: 1265–1273.

Stark C, Su T-C, Breitkreutz A, Lourenco P, Dahabieh M, Breitkreutz B-J, Tyers M, Sadowski I. 2010. PhosphoGRID: A database of experimentally verified in vivo protein phosphorylation sites from the budding yeast *Saccharomyces cerevisiae*. *Database (Oxford)* 2010: bap026.

Swaney DL, Villén J. 2015a. Enrichment of phosphopeptides via immobilized metal affinity chromatography. *Cold Spring Harb Protoc* doi: 10.1101/pdb.prot088005.

Swaney DL, Villén J. 2015b. Enrichment of modified peptides via immunoaffinity precipitation with modification-specific antibodies. *Cold Spring Harb Protoc* doi: 10.1101/pdb.prot088013.

Swaney DL, Beltrao P, Starita L, Guo A, Rush J, Fields S, Krogan NJ, Villén J. 2013. Global analysis of phosphorylation and ubiquitylation cross-talk in protein degradation. *Nat Methods* 10: 676–682.

Wagner SA, Beli P, Weinert BT, Nielsen ML, Cox J, Mann M, Choudhary C. 2011. A proteome-wide, quantitative survey of in vivo ubiquitylation sites reveals widespread regulatory roles. *Mol Cell Proteomics* 10: M111.013284.

Protocol 1

Enrichment of Phosphopeptides via Immobilized Metal Affinity Chromatography

Danielle L. Swaney and Judit Villén[1]

Department of Genome Sciences, University of Washington, Seattle, Washington 98195

Immobilized metal affinity chromatography (IMAC) is a frequently used method for the enrichment of phosphorylated peptides from complex, cellular lysate-derived peptide mixtures. Here we outline an IMAC protocol that uses iron-chelated magnetic beads to selectively isolate phosphorylated peptides for mass spectrometry–based proteomic analysis. Under acidic conditions, negatively charged phosphoryl modifications preferentially bind to positively charged metal ions (e.g., Fe^{3+}, Ga^{3+}) on the beads. After washing away nonphosphorylated peptides, a pH shift to basic conditions causes the elution of bound phosphopeptides from the metal ion. Under optimal conditions, very high specificity for phosphopeptides can be achieved.

MATERIALS

It is essential that you consult the appropriate Material Safety Data Sheets and your institution's Environmental Health and Safety Office for proper handling of equipment and hazardous materials used in this protocol.

RECIPES: Please see the end of this protocol for recipes indicated by <R>. Additional recipes can be found online at http://cshprotocols.cshlp.org/site/recipes.

Reagents

Empore C18 polymer disks (3M)
Formic acid (10% and 1%)
IMAC resin (Ni-NTA magnetic beads, 5% suspension, QIAGEN)
Metal loading solution: 10 mM iron chloride ($FeCl_3$)
Metal stripping solution <R>
Methanol
Peptide samples (derived from cellular lysates via enzyme digestion)
> *Include phosphatase inhibitors in preparation of peptide samples. Dry and desalt peptide samples. See Villén and Gygi (2008) for a protocol for preparing peptide samples.*

Phosphopeptide binding solution <R>
Phosphopeptide elution solution <R>
Stage tip elution solution <R>
Water (HPLC-grade)
> *Prepare all solutions in HPLC-grade water.*

---

[1]Correspondence: jvillen@u.washington.edu

Copyright © Cold Spring Harbor Laboratory Press; all rights reserved
Cite this protocol as *Cold Spring Harb Protoc*; doi:10.1101/pdb.prot088005

Equipment

Blunt end cannula with a rod (e.g., paper clip) that fits snuggly into the cannula shaft
Magnet
Microcentrifuge
Microcentrifuge tubes

When performing enrichment of several samples simultaneously, 200-µL tube strips or 96-well plates can be used in conjunction with a multichannel pipette or an automated liquid handler to increase throughput of the protocol.

Syringe
Vacuum concentrator
Vortexer or spinning wheel for bead agitation

METHOD

For an overview of this procedure, see Figure 1. The protocol described here for phosphopeptide enrichment using magnetic beads has been adapted from a previous publication (Ficarro et al. 2009). The protocol for phosphopeptide filtration has been adapted from a previous publication detailing microscale peptide desalting (Rappsilber et al. 2003, 2007).

Resin Preparation

1. Take a 1-mL aliquot of the 5% IMAC resin suspension (magnetic beads).
2. Wash the aliquot of beads with 1 mL of water as follows.
   i. Aspirate the liquid with a pipette, using the magnet to retain the beads.
   ii. Remove the magnet, and add 1 mL of water.
   iii. Mix the beads by inverting the tube.
   iv. Repeat Steps 2.i–2.iii two times.

   When working with magnetic particles, special consideration should be taken as to when to apply magnetic force. To ensure that particles are retained in the tube during washing, magnetic force should be applied during washes whenever liquid is being removed from the tube. In contrast, to promote mixing and more effective washing, magnetic force should be removed when adding liquid to magnetic particles.

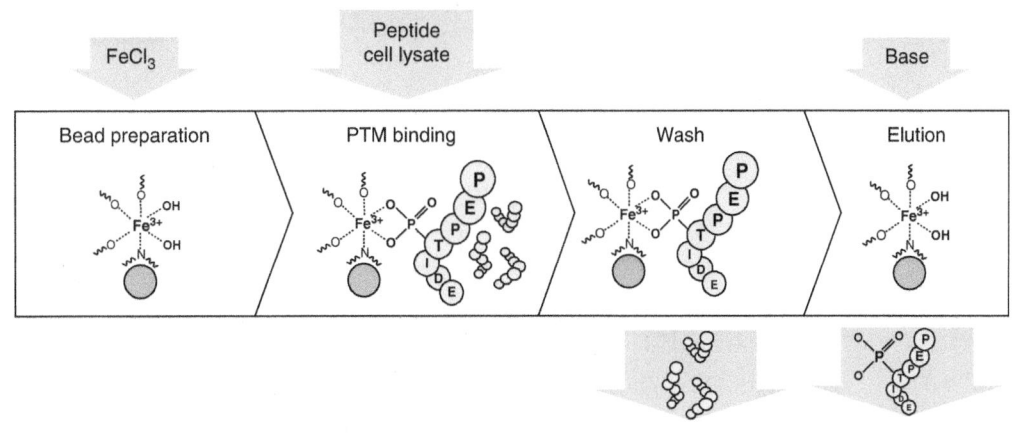

FIGURE 1. Method overview for phosphopeptide enrichment via IMAC. To prepare beads for phosphopeptide enrichment, Fe^{3+} is chelated to the surface of the beads. Next, a peptide cell lysate is added to the beads. Phosphopeptides coordinate with the Fe^{3+} and are retained on the beads. The unbound nonphosphorylated peptides can then be washed away or collected for further analysis. On addition of a basic pH solution, the phosphopeptides elute from the beads and can be collected for mass spectrometry analysis.

Chapter 22

3. Incubate the IMAC resin in 1 mL of metal stripping solution for 30 min with shaking at room temperature.

    This step strips chelated metal ions.

4. Wash the IMAC resin three times with 1 mL of water.

5. Incubate the IMAC resin in 1 mL of metal loading solution for 30 min with shaking at room temperature.

    This step chelates iron to the resin.

6. Wash the IMAC resin three times in phosphopeptide binding solution.

7. Resuspend the IMAC resin in 1 mL of phosphopeptide binding solution to create a 5% suspension.

    This prepared IMAC resin can be stored for several weeks at 4°C until used for phosphopeptide enrichment.

Phosphopeptide Binding

8. Resuspend 50–250 µg of peptide in 100 µL of phosphopeptide binding solution.

9. Add an aliquot of resuspended IMAC resin to the resuspended peptide sample.

    The volume of IMAC added resin should be between 10% and 20% of your peptide weight (i.e., for a 100-µg peptide sample use 10–20 µL of IMAC resin).

10. Incubate the resin with peptides with shaking for 30 min at room temperature.

Preparation of Filtration Tip

11. Prepare the filtration tip by using a blunt-end cannula to cut out two small circular disks of C18 polymer. Using a rod (e.g., paper clip), gently push the C18 disks from the cannula into a 200-µL pipette tip.

    Alternatively, assembled tips can also be purchased from commercial vendors (e.g., NuTip, ZipTip, Stage-Tip, etc.).

12. Equilibrate the C18 disk by the addition of 20 µL of methanol to the tip. Incubate for 10 min at room temperature.

13. Pass the methanol through the C18 disk by centrifugation at 2000 rpm for 3 min at room temperature or by the use of a syringe.

14. Condition the C18 disk by passing 20 µL of stage tip elution solution through it by centrifugation at 2000 rpm for 3 min at room temperature or by the use of a syringe.

15. Add 5 µL of stage tip elution solution to the C18 disk to keep the disk wet while completing the phosphopeptide enrichment.

Resin Wash, Phosphopeptide Elution, and Filtration of Phosphopeptide Sample

16. After phosphopeptide binding to the IMAC resin, use a magnet to retain the resin and remove the solution containing unbound peptides. Save for subsequent analysis if desired.

    This fraction contains nonphosphorylated peptides.

17. Wash the resin three times with 150 µL of phosphopeptide binding solution, using the magnet as in Step 2.

18. Elute phosphopeptides from the resin by adding 100 µL of phosphopeptide elution solution. Incubate for 1 min with agitation at room temperature.

    Take care to perform this step quickly. Extended incubation times (>5 min) in the highly basic solution can cause significant degradation of phosphopeptides through β-elimination.

19. Retaining the resin with a magnet, transfer the phosphopeptide-containing eluate to a prepared filter tip.

20. Collect and acidify the filtered phosphopeptide eluate by using a syringe to pass it through the C18 disk into a microcentrifuge tube containing 30 µL of 10% formic acid.

21. Elute any bound peptides from the C18 disk by passing 20 µL of stage tip elution solution through the C18 disk. Collect this eluate into the same tube as Step 20.

22. Dry the filtered phosphopeptide eluate by vacuum centrifugation at room temperature and resuspend in 10 µL of 1% formic acid for liquid chromatography–tandem mass spectrometry (LC-MS/MS) analysis.

    Drying will take ~40–60 min.
    See Troubleshooting.

TROUBLESHOOTING

Problem: The resin aggregates.
Solution: Never store the IMAC resin in aqueous solution or in dry conditions.

Problem (Step 22): A low yield of phosphopeptides is observed.
Solution: Increase the starting amount of peptide sample. Ensure that adequate magnetic force is being applied, such that magnetic particles are being retained in the tube. Ensure that beads are not lost during washes.

Problem (Step 22): Low specificity (i.e., low percentage of phosphopeptides over total peptides) is observed.
Solution: Adjust the ratio of peptide to IMAC resin by either increasing peptide amount or decreasing resin volume. In addition, add additional washes in Step 17 to increase specificity and take care to remove all solution when performing washes."

RECIPES

Metal Stripping Solution

40 mM EDTA (pH 8.0)

To prepare, add 1.17 g of EDTA to 100 mL of water and stir. Slowly adjust the pH to 8.0 with 2 M NaOH. (Unless using the salt, EDTA often requires extended time and basic pH conditions to completely dissolve.) Store for up to 3 mo at 4°C.

Phosphopeptide Binding Solution

80% acetonitrile
0.1% trifluoroacetic acid (TFA)

Store for up to 1 yr at room temperature.

Phosphopeptide Elution Solution

70% acetonitrile
1% ammonium hydroxide

Store for up to 3 mo at room temperature.

Stage Tip Elution Solution

70% acetonitrile
1% formic acid

Store for up to 1 yr at room temperature.

REFERENCES

Ficarro SB, Adelmant G, Tomar MN, Zhang Y, Cheng VJ, Marto JA. 2009. Magnetic bead processor for rapid evaluation and optimization of parameters for phosphopeptide enrichment. *Anal Chem* **81:** 4566–4575.

Rappsilber J, Ishihama Y, Mann M. 2003. Stop and go extraction tips for matrix-assisted laser desorption/ionization, nanoelectrospray, and LC/MS sample pretreatment in proteomics. *Anal Chem* **75:** 663–670.

Rappsilber J, Mann M, Ishihama Y. 2007. Protocol for micro-purification, enrichment, pre-fractionation and storage of peptides for proteomics using StageTips. *Nat Protoc* **2:** 1896–1906.

Villén J, Gygi SP. 2008. The SCX/IMAC enrichment approach for global phosphorylation analysis by mass spectrometry. *Nat Protoc* **3:** 1630–1638.

Protocol 2

Enrichment of Modified Peptides via Immunoaffinity Precipitation with Modification-Specific Antibodies

Danielle L. Swaney and Judit Villén[1]

Department of Genome Sciences, University of Washington, Seattle, Washington 98195

Immunoaffinity precipitation is an effective method of purifying select protein posttranslational modifications (PTMs) for proteomic analysis via mass spectrometry. Peptides containing a modification of interest are isolated directly from protease-digested cellular protein extracts using an antibody with specificity against the modification, and the modified peptides are analyzed by tandem mass spectrometry. Antibodies now exist with specificity for a variety of individual PTMs, such as phosphotyrosine, acetyl-lysine, methyl-arginine, ubiquitylation (i.e., diglycyl-lysine affinity), etc. Here we outline a generalized protocol for the purification of modified peptides by immunoaffinity precipitation. The main restriction for using this protocol is the availability of an antibody against the modification of interest. To purify modified peptides, antibodies are first conjugated to a solid support, such as agarose beads. The beads are then incubated with a complex peptide mixture, derived from a cellular lysate, under neutral pH to facilitate binding of modified peptides. The incubation time can vary from 30 min to overnight, depending upon the antibody used and the complexity of the peptide sample. Finally, acidic buffer conditions are used to elute the PTM-enriched bound peptides for mass spectrometry analysis.

MATERIALS

It is essential that you consult the appropriate Material Safety Data Sheets and your institution's Environmental Health and Safety Office for proper handling of equipment and hazardous materials used in this protocol.

RECIPES: Please see the end of this protocol for recipes indicated by <R>. Additional recipes can be found online at http://cshprotocols.cshlp.org/site/recipes.

Reagents

Acetonitrile (100%)
Acetonitrile/formic acid (70%/1%; 5%/5%)
Antibody against modification of interest (e.g., phosphotyrosine, acetyl-lysine, diglycyl-lysine, dimethyl-arginine)
Empore C18 polymer disks (3M)
Formic acid (1%)
Glycerol (optional; see Step 8)
IAP buffer <R>
Methanol

---

[1]Correspondence: jvillen@u.washington.edu

Copyright © Cold Spring Harbor Laboratory Press; all rights reserved
Cite this protocol as *Cold Spring Harb Protoc*; doi:10.1101/pdb.prot088013

Chapter 22

Peptide samples (derived from cellular lysates via enzyme digestion)
> Include phosphatase inhibitors in preparation of peptide samples. Dry and desalt peptide samples. See Villén and Gygi (2008) for a protocol for preparing peptide samples.

Phosphate-buffered saline (PBS), without $CaCl_2$ or $MgCl_2$ <R>
Protein A or protein G agarose beads
> The choice of protein A or protein G agarose beads depends on their affinity for the antibody. In general, protein G works well with mouse monoclonal antibodies, and protein A with rabbit polyclonal antibodies, with exceptions. Alternatively, a mixture of protein A and protein G can be used.

Trifluoroacetic acid (TFA; 0.15%)
Water (HPLC-grade)
> Prepare all solutions in HPLC-grade water.

Equipment

Blunt end cannula with a rod (e.g., paper clip) that fits snuggly into the cannula shaft
Insert vial for liquid chromatography–tandem mass spectrometry (LC-MS/MS)
Microcentrifuge
Microcentrifuge tubes
Nutator mixer, spinning wheel, or vortexer
Slip tip syringe
Sonicator (optional; see Step 9)
Vacuum concentrator

METHOD

> For an overview of this procedure, see Figure 1. To study multiple PTMs within a biological sample, multiple orthogonal immunoaffinity precipitations can be performed sequentially, where the flowthrough of one purification is used as input for the next one.

Conjugation of Antibody to Agarose Beads

1. Place a volume of beads appropriate for the amount of antibody to be conjugated (see Step 2) into a microcentrifuge tube. Wash protein A or protein G agarose beads three times by adding 1 mL of PBS and centrifuging 30 sec at 1500g at 4°C. Discard supernatant.

2. Add antibody at 5 mg per mL packed volume of protein A or protein G beads.

3. Incubate beads and antibody overnight at 4°C with rotation.
   > Extended incubation (>24 h) is not advised.

Washes and Storage of Antibody-Conjugated Beads

4. Prepare 50-µL aliquots of packed volume of resin.

5. Centrifuge the beads at 1500g for 2 min at 4°C and discard the supernatant.

6. Wash the resin three times with 4°C PBS. For each wash, proceed as follows.
   i. Add 1 mL of PBS buffer.
   ii. Mix by inverting the tube five times.
   iii. Centrifuge at 1500g for 1 min at 4°C.
   iv. Discard the supernatant.

7. Wash the resin two times as described in Step 6 with 1 mL of IAP buffer.

8. Resuspend the resin (50 µL packed volume) in 50 µL IAP buffer to make a 1:1 slurry, ready for use.
   > To store antibody-conjugated beads for up to 1 wk, resuspend resin in 1 mL of IAP buffer, and store at 4°C. For longer-term storage, resuspend resin in 1 mL of IAP buffer containing 50% glycerol and store at −20°C.

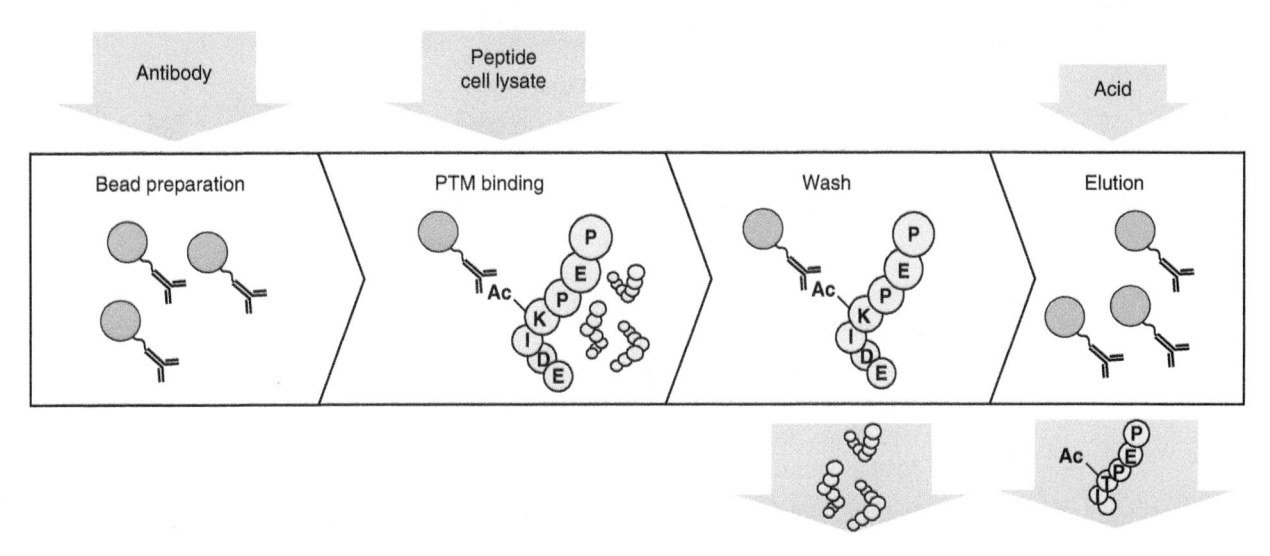

FIGURE 1. Method overview for immunoaffinity purification. To prepare beads for immunoaffinity purification an antibody with specificity to a particular PTM (e.g., lysine acetylation) is conjugated to protein A or protein G beads. Next, a peptide cell lysate is added to the beads and peptides with a specific PTM bind to the antibody. The unbound nonmodified peptides can then be washed away or collected for further analysis. Upon addition of an acidic pH solution the PTM-containing peptides dissociate from the antibody-conjugated beads and can be collected for MS analysis.

Incubation with Peptides

9. Dissolve 10 mg of peptides in 1.0–1.4 mL of IAP buffer.

   Shaking at room temperature and sonication in a room temperature sonicating water bath can help promote dissolving.

10. Centrifuge the solution at 2000g for 5 min at room temperature.

    The pellet of insoluble matter may seem considerable.

11. Transfer the supernatant to a new tube and cool for 10 min on ice.

12. Centrifuge the solution at 2000g for 5 min at 4°C.

13. Transfer the peptide-containing supernatant to the antibody conjugated beads from Step 8.

14. Incubate the antibody/bead mixture for a period of 30 min to overnight at 4°C with rotation.

    With increased incubation time, the amount of modified peptides bound to antibody-conjugated beads increases (i.e., increased sensitivity), but this comes with the cost of increased binding of nonmodified peptides (i.e., decreased specificity).

Washes and Elution of Peptides

15. Centrifuge the mixture at 1500g for 1 min at 4°C. Remove the supernatant and discard, or save for other enrichment strategies.

16. Wash the resin three to five times with IAP buffer followed by two washes with PBS. For each wash, proceed as follows.

    i. Add 1 mL of IAP buffer or PBS.

    ii. Mix by inverting the tube five times.

    iii. Centrifuge at 1500g for 1 min at 4°C.

    iv. Discard the supernatant.

17. Elute the bound peptides from the antibody-conjugated resin twice by adding 100 µL of 0.15% TFA. Tap the bottom of the tube to mix and incubate for 10 min at room temperature.

Chapter 22

18. Centrifuge the solution at 1500g for 1 min at room temperature.
19. Transfer the PTM-enriched supernatant to a new tube.
    PTM-enriched peptides can be kept at room temperature while preparing the stage tips (see steps below).

Concentration and Desalting of Peptides for LC-MS/MS

Modified peptides from Step 19 are desalted on stage tips (Rappsilber et al. 2003, 2007) as described below.

20. Prepare stage tips for each PTM-enriched peptide sample with Empore C18 disks and 200-µL pipette tips. Using a blunt-end cannula, cut out four small circular wedges of C18 polymer and push the polymer into the pipette tip using a rod.
21. Wash the C18 polymer according to the following sequence, using a syringe to pass the liquid through: two times with 20 µL of methanol; once with 20 µL of 100% acetonitrile; once with 20 µL of 70% acetonitrile, 1% formic acid.
22. Equilibrate the polymer by passing 20 µL of 1% formic acid through the microcolumn with a syringe two times.
23. Load the peptide sample from Step 19 on the stage tip microcolumn by passing the peptide solution through the polymer using a syringe.
24. Wash the sample by passing 20 µL of 1% formic acid through the column using a syringe twice.
25. Elute the sample by passing 20 µL of 70% acetonitrile, 1% formic acid through the column with a syringe. Collect the eluate into an insert vial for LC-MS/MS.
26. Dry the sample to completeness in a vacuum concentrator at room temperature.
    Drying typically takes <20 min.
27. Resuspend the sample in 10 µL of 5% acetonitrile, 5% formic acid.
    The sample is ready for LC-MS/MS analysis.
    See Troubleshooting.

TROUBLESHOOTING

Problem (Step 27): Low yield of peptides is observed.
Solution: Take care not to remove any of the resin when performing washes. Increase the starting amount of peptide sample in Step 9. Because of the low abundance of many PTMs, 5–50 mg of peptide lysate may be required. Increase the incubation time in Step 14.

Problem (Step 27): Low specificity (i.e., low percentage of modified peptides over total peptides) is observed.
Solution: Adjust the ratio of peptide to antibody resin by either increasing peptide amount or decreasing resin volume. Decrease the incubation time in Step 14. In Step 16, add additional washes to increase specificity, and take care to remove all solution when performing washes.

DISCUSSION

The protocol described here for general enrichment of PTMs by peptide immunoaffinity precipitation has been adapted from a previous publication (Rush et al. 2005). This protocol was initially developed to enrich for phosphotyrosine-containing peptides. By enriching for a subset of the phosphoproteome, peptide immunoprecipitation yielded significantly greater coverage of the phosphotyrosine proteome than global phosphorylation enrichment strategies. Since then, peptide immunoaffinity precipitation has been successfully applied to enrich for peptides with other phosphorylation motifs

(Matsuoka et al. 2007; Moritz et al. 2010) and peptides with other modifications, such as acetyl-lysine (Choudhary et al. 2009; Hebert et al. 2013), and diglycyl-lysine (remnant from ubiquitin modification after proteolysis with trypsin) (Emanuele et al. 2011; Kim et al. 2011; Wagner et al. 2011).

RECIPES

IAP Buffer (1×)

50 mM MOPS–NaOH (pH 7.2)
10 mM Na_2HPO_4
50 mM NaCl

Prepare IAP buffer at a concentration of 10× and store at room temperature, or prepare at a concentration of 1× and store at 4°C.

Phosphate-Buffered Saline (PBS)

Reagent	Amount to add (for 1× solution)	Final concentration (1×)	Amount to add (for 10× stock)	Final concentration (10×)
NaCl	8 g	137 mM	80 g	1.37 M
KCl	0.2 g	2.7 mM	2 g	27 mM
Na_2HPO_4	1.44 g	10 mM	14.4 g	100 mM
KH_2PO_4	0.24 g	1.8 mM	2.4 g	18 mM
If necessary, PBS may be supplemented with the following:				
$CaCl_2 \cdot 2H_2O$	0.133 g	1 mM	1.33 g	10 mM
$MgCl_2 \cdot 6H_2O$	0.10 g	0.5 mM	1.0 g	5 mM

PBS can be made as a 1× solution or as a 10× stock. To prepare 1 L of either 1× or 10× PBS, dissolve the reagents listed above in 800 mL of H_2O. Adjust the pH to 7.4 (or 7.2, if required) with HCl, and then add H_2O to 1 L. Dispense the solution into aliquots and sterilize them by autoclaving for 20 min at 15 psi (1.05 kg/cm^2) on liquid cycle or by filter sterilization. Store PBS at room temperature.

REFERENCES

Choudhary C, Kumar C, Gnad F, Nielsen ML, Rehman M, Walther TC, Olsen JV, Mann M. 2009. Lysine acetylation targets protein complexes and co-regulates major cellular functions. *Science* 325: 834–840.

Emanuele MJ, Elia AEH, Xu Q, Thoma CR, Izhar L, Leng Y, Guo A, Chen Y-N, Rush J, Hsu PW-C, et al. 2011. Global identification of modular cullin-RING ligase substrates. *Cell* 147: 459–474.

Hebert AS, Dittenhafer-Reed KE, Yu W, Bailey DJ, Selen ES, Boersma MD, Carson JJ, Tonelli M, Balloon AJ, Higbee AJ, et al. 2013. Calorie restriction and SIRT3 trigger global reprogramming of the mitochondrial protein acetylome. *Mol Cell* 49: 186–199.

Kim W, Bennett EJ, Huttlin EL, Guo A, Li J, Possemato A, Sowa ME, Rad R, Rush J, Comb MJ, et al. 2011. Systematic and quantitative assessment of the ubiquitin-modified proteome. *Mol Cell* 44: 325–340.

Matsuoka S, Ballif BA, Smogorzewska A, McDonald ER, Hurov KE, Luo J, Bakalarski CE, Zhao Z, Solimini N, Lerenthal Y, et al. 2007. ATM and ATR substrate analysis reveals extensive protein networks responsive to DNA damage. *Science* 316: 1160–1166.

Moritz A, Li Y, Guo A, Villen J, Wang Y, MacNeill J, Kornhauser J, Sprott K, Zhou J, Possemato A, et al. 2010. Akt-RSK-S6 kinase signaling networks activated by oncogenic receptor tyrosine kinases. *Sci Signal* 3: ra64.

Rappsilber J, Ishihama Y, Mann M. 2003. Stop and go extraction tips for matrix-assisted laser desorption/ionization, nanoelectrospray, and LC/MS sample pretreatment in proteomics. *Anal Chem* 75: 663–670.

Rappsilber J, Mann M, Ishihama Y. 2007. Protocol for micro-purification, enrichment, pre-fractionation and storage of peptides for proteomics using StageTips. *Nat Protoc* 2: 1896–1906.

Rush J, Moritz A, Lee KA, Guo A, Goss VL, Spek EJ, Zhang H, Zha X-M, Polakiewicz RD, Comb MJ. 2005. Immunoaffinity profiling of tyrosine phosphorylation in cancer cells. *Nat Biotechnol* 23: 94–101.

Villén J, Gygi SP. 2008. The SCX/IMAC enrichment approach for global phosphorylation analysis by mass spectrometry. *Nat Protoc* 3: 1630–1638.

Wagner SA, Beli P, Weinert BT, Nielsen ML, Cox J, Mann M, Choudhary C. 2011. A proteome-wide, quantitative survey of in vivo ubiquitylation sites reveals widespread regulatory roles. *Mol Cell Proteomics* 10: M111.013284.

CHAPTER 23

Protein Microarrays: Flexible Tools for Scientific Innovation

Johnathan Neiswinger,[1,2] Ijeoma Uzoma,[1,2] Eric Cox,[1,2,3] HeeSool Rho,[1,2] Guang Song,[1,2] Corry Paul,[1,2] Jun Seop Jeong,[1,2] Kuan-Yi Lu,[4] Chien-Sheng Chen,[4] and Heng Zhu[1,2,5,6]

[1]Department of Pharmacology and Molecular Sciences, Johns Hopkins School of Medicine, Baltimore, Maryland 21287; [2]The Center for High-Throughput Biology, Johns Hopkins School of Medicine, Baltimore, Maryland 21287; [3]Department of Neuroscience, Johns Hopkins School of Medicine, Baltimore, Maryland 21287; [4]Graduate Institute of Systems Biology and Bioinformatics, National Central University, Jhongli 32001, Taiwan; [5]The Sidney Kimmel Comprehensive Cancer Center, Johns Hopkins School of Medicine, Baltimore, Maryland 21287

Protein microarrays have emerged as a powerful tool for the scientific community, and their greatest advantage lies in the fact that thousands of reactions can be performed in a parallel and unbiased manner. The first high-density protein microarray, dubbed the "yeast proteome array," consisted of approximately 5800 full-length yeast proteins and was initially used to identify protein–lipid interactions. Further assays were subsequently developed to allow measurement of protein–DNA, protein–RNA, and protein–protein interactions, as well as four well-known posttranslational modifications: phosphorylation, acetylation, ubiquitylation, and SUMOylation. In this introduction, we describe the advent of high-density protein microarrays, as well as current methods for assessing a wide variety of protein interactions and posttranslational modifications.

PROTEIN MICROARRAYS

In 1983, Tse-Wan Chang spotted a variety of antibodies onto a glass slide in a matrix formation to study "the potential of simultaneous multiple determinations of specific cell surface antigens in one reaction incubation" (Chang 1983). This study represents one of the first examples of what we today refer to as a microarray. Over the next 10 years, this concept was transformed and applied to the generation of DNA microarrays, which allow detection of mRNA expression levels for thousands of genes in parallel. Yet although DNA microarrays provide a useful platform for biological study, they cannot imitate the function of what many consider the pillars of cellular signaling—proteins. As research indicated that mRNA levels do not always associate with protein expression levels, array technology evolved to overcome these limitations through the creation of protein microarrays (Gygi et al. 1999; Zhu et al. 2001; Kopf and Zharhary 2007). In 2001, the first study using high-density protein arrays was presented by the Snyder research group at Yale (Zhu et al. 2001).

The first "proteome" microarray contained approximately 5800 full-length yeast open reading frames (ORFs) individually expressed in yeast. Each ORF was cloned into a yeast high copy expression vector (pEGH-A), which produces proteins that are fused to glutathione S-transferase-polyhistidine (GST-HisX6) at their amino termini, and can be expressed using the inducible *GAL1* promoter. For a detailed protocol describing the protein purification method used to create the array, see the study by Zhu et al. (2001). In brief, yeast cells were grown at 30°C to an OD$_{600}$ of 1.0 and induced for 6 h with

[6]Correspondence: heng.zhu@jhmi.edu

Copyright © Cold Spring Harbor Laboratory Press; all rights reserved
Cite this introduction as *Cold Spring Harb Protoc*; doi:10.1101/pdb.top081471

galactose at a final concentration of 2%. After induction, the yeast were harvested in a 96-deep-well plate and frozen at −80°C. (Cell pellets can be stored in this condition for up to 1 yr.) Lysis buffer containing protease inhibitors and 0.5-mm zirconia beads was added to the pellets, and the yeast cells were sheared by vigorous shaking in a standard hardware store paint shaker. The supernatant of the lysate was collected after centrifugation and transferred to 96-well filter plates (sealed at the bottom with agar), where it was incubated with Sepharose glutathione beads for 1 h to immunoprecipitate the overexpressed protein. After several wash steps, proteins were eluted with buffer containing 40 mM glutathione and collected in a 96-well PCR plate. Eluates were then re-arrayed into a 384-well plate for use with a commercial protein microarrayer. (Because the elution buffer contains 30% glycerol, proteins can be stored in this format for up to 2 yr at −80°C.)

One benefit of using yeast as a means of protein purification for the study was that all proteins were in their native environment. Thus, even the largest and smallest proteins could be purified, including membrane proteins; many of these bound phosphoinositides while printed on the array, signifying that they were properly folded (Zhu et al. 2001). Because the array contained ∼80% of the known yeast proteome, almost all of the protein classes were represented, and >80% of the strains produced detectable amounts of fusion proteins. Lysis buffer containing 0.1% Triton resulted in proteins free of lipids, and the strict washing conditions ensured that other cellular proteins that might noncovalently interact with the purified protein were washed away.

The protein microarrays were fabricated using a commercial robotic system. A standard array indicated that 90% of the spots contained 10–950 fg of protein, an amount detectable using a glutathione S-transferase (GST) antibody. Robotic microarrayers (such as the Bio-Rad VersArray ChipWriter Pro System) use metal pins containing a small capillary on the tip to collect purified proteins from the 384-well plate and disperse them onto a glass surface in a very systematic and precise manner. The glass surface can either be aldehyde-treated, in which case the fusion proteins attach to the surface through primary amines at their amino termini or other residues of the protein (random orientation), or nickel-coated, which orients the fusion proteins uniformly through their HisX6 tag. The latter can produce better signal-to-noise ratios, but the lack of a covalent modification means that reactions performed on the slide cannot withstand harsh washing conditions (e.g., wash buffers containing sodium dodecyl sulfate [SDS]). Arrays fabricated in this manner can be stored for up to 1 yr at −80°C.

At present, there are no commercially available yeast proteome arrays, but some laboratories fabricate them in-house for their own purposes (e.g., the Heng Zhu laboratory at Johns Hopkins), so obtaining these arrays is generally accomplished through collaborative efforts between laboratories. Currently, the only organism with commercially available "proteome" arrays is human. The array available from Life Technologies contains more than 9000 human proteins and costs ∼$1000 per array.

Detecting interactions on a protein array typically involves one of two methods—fluorescent dyes or radiography. Fluorescent dyes are frequently used to detect binding interactions between the proteins printed on the array and a fluorescently labeled binding protein. Fluorescently labeled antibodies can also be used to detect a wide variety of modifications (e.g., acetylation, SUMOylation, and ubiquitylation). Both fluorescence-based methods require a microarray scanner such as the GenePix 4000B for data acquisition. These machines scan standard microarrays by obtaining fluorescent signal at clearly defined wavelengths. Some can be used for multiplex assays by detecting multiple wavelengths for multiple dyes at the same time. Radiography can be used to detect modifications for which antibody specificity is too poor for use on an array (e.g., phosphorylation). This method simply requires standard film and a high-quality tabletop scanner capable of a resolution of 4000 dpi. Once an array has been scanned, microarray analysis software (such as GenePix 7.0) must be used to score and interpret the data. This is the most time-consuming step of the experiment and requires the most precision. Once the data are scored, simple foreground:background ratios can be obtained and a putative hit list can be generated. Because of the immense amount of data obtained, many laboratories collaborate with bioinformaticians to draw more useful conclusions and/or narrow their findings for further validation studies.

APPLICATIONS

The versatility of protein microarrays, coupled with their ability to allow characterization of thousands of proteins in a parallel and high-throughput manner, has resulted in great strides by the scientific community. By assessing interactions between thousands of proteins, many laboratories have been able to identify binding consensus motifs (Zhu et al. 2001; Lu et al. 2012) and phosphorylation motifs (Newman et al. 2013) and assess antibody specificity (Jeong et al. 2012)—all of which would be incredibly difficult using other methods. Arrays offer advantages over traditional yeast two-hybrid screens, which only take place in the nucleus and are thus severely limited in the breadth of interactions that can be observed. Mass spectrometry approaches have been very powerful for the identification of posttranslational modifications but are also time-consuming and not feasible with low-abundance proteins. Another limitation is the lack of direct evidence for the upstream enzyme(s) responsible for catalyzing the addition of the modification. Protein microarrays overcome these hurdles by providing a wider scope of proteins that can be assessed and identifying interactions even with limited protein amounts.

However, microarrays are not without their disadvantages. Array experiments are performed as in vitro assays, so although an interaction or modification is observed, it may not necessarily be present in a cellular context. Also, many methods that require antibodies are limited by antibody specificity, so careful consideration and methodology (e.g., proper controls) must be used when using antibodies for detection. Because of this, further characterization is required for all experiments involving protein microarrays.

Even with these disadvantages, protein microarrays can be used to examine a wide variety of interactions, including protein–protein, protein–DNA, protein–RNA, and protein–lipid interactions. We previously provided a detailed protocol for Characterization of Protein–DNA Interactions Using Protein Microarrays [Hu et al. 2011]). Our current protocols include Protocol 1: Characterization of Protein–Protein Interactions Using Protein Microarrays (Paul et al. 2015), Protocol 2: Characterization of RNA–Binding Proteins Using Protein Microarrays (Song and Zhu 2015), and Protocol 3: Characterization of Lipid–Protein Interactions Using Nonquenched Fluorescent Liposomal Nanovesicles and Yeast Proteome Microarrays (Lu et al. 2015). Although these experiments involve noncovalent binding, functional assays can also be performed on the substrate proteins in the presence of an enzyme. In addition to the aforementioned binding assays, we have also provided a method for detection of phosphorylation, acetylation, ubiquitylation, and SUMOylation that can be adapted for any microarray experiment; see Protocol 4: Posttranslational Modification Assays on Functional Protein Microarrays (Neiswinger et al. 2015).

REFERENCES

Chang TW. 1983. Binding of cells to matrixes of distinct antibodies coated on solid surface. *J Immunol Methods* **65:** 217–223.

Gygi SP, Rochon Y, Franza BR, Aebersold R. 1999. Correlation between protein and mRNA abundance in yeast. *Mol Cell Biol* **19:** 1720–1730.

Hu S, Xie Z, Blackshaw S, Qian J, Zhu H. 2011. Characterization of protein–DNA interactions using protein microarrays. *Cold Spring Harb Protoc* doi: 10.1101/pdb.prot5614.

Jeong JS, Jiang L, Albino E, Marrero J, Rho HS, Hu J, Hu S, Vera C, Bayron-Poueymiroy D, Rivera-Pacheco ZA, et al. 2012. Rapid identification of monospecific monoclonal antibodies using a human proteome microarray. *Mol Cell Proteomics* **11:** O111.016253.

Kopf E, Zharhary D. 2007. Antibody arrays—An emerging tool in cancer proteomics. *Int J Biochem Cell Biol* **39:** 1305–1317.

Lu KY, Tao SC, Yang TC, Ho YH, Lee CH, Lin CC, Juan HF, Huang HC, Yang CY, Chen MS, et al. 2012. Profiling lipid–protein interactions using nonquenched fluorescent liposomal nanovesicles and proteome microarrays. *Mol Cell Proteomics* **11:** 1177–1190.

Lu KY, Chen CS, Neiswinger J, Zhu H. 2015. Characterization of lipid–protein interactions using nonquenched fluorescent liposomal nanovesicles and yeast proteome microarrays. *Cold Spring Harb Protoc* doi: 10.1101/pdb.prot087981.

Neiswinger J, Uzoma I, Cox E, Rho H, Jeong JS, Zhu H. 2015. Posttranslational modification assays on functional protein microarrays. *Cold Spring Harb Protoc* doi: 10.1101/pdb.prot087999.

Newman RH, Hu J, Rho HS, Xie Z, Woodard C, Neiswinger J, Cooper C, Shirley M, Clark HM, Hu S, et al. 2013. Construction of human activity-based phosphorylation networks. *Mol Syst Biol* **9:** 655.

Paul C, Rho H, Neiswinger J, Zhu H. 2015. Characterization of protein–protein interactions using protein microarrays. *Cold Spring Harb Protoc* doi: 10.1101/pdb.prot087965.

Song G, Neiswinger J, Zhu H. 2015. Characterization of RNA-binding proteins using protein microarrays. *Cold Spring Harb Protoc* doi: 10.1101/pdb.prot087973.

Zhu H, Bilgin M, Bangham R, Hall D, Casamayor A, Bertone P, Lan N, Jansen R, Bidlingmaier S, Houfek T, et al. 2001. Global analysis of protein activities using proteome chips. *Science* **293:** 2101–2105.

Protocol 1

Characterization of Protein–Protein Interactions Using Protein Microarrays

Corry Paul,[1,2] HeeSool Rho,[1,2] Johnathan Neiswinger,[1,2] and Heng Zhu[1,2,3,4]

[1]*Department of Pharmacology and Molecular Sciences, Johns Hopkins University School of Medicine, Baltimore, Maryland 21205;* [2]*The Center for High-Throughput Biology, Johns Hopkins University School of Medicine, Baltimore, Maryland 21205;* [3]*The Sidney Kimmel Comprehensive Cancer Center, Johns Hopkins University School of Medicine, Baltimore, Maryland 21287*

Functional protein microarrays allow fast, straightforward, and efficient high-throughput screening of protein–protein interactions. The microarray approach has outpaced other interaction methods, such as yeast two-hybrid screens, in part because of the vast amounts of information that can be obtained during a single assay. This protocol describes how to perform a binding assay for a protein of interest using a proteome microarray composed of thousands of functional, recombinant proteins adhered to a microchip.

MATERIALS

It is essential that you consult the appropriate Material Safety Data Sheets and your institution's Environmental Health and Safety Office for proper handling of equipment and hazardous materials used in this protocol.

RECIPES: Please see the end of this protocol for recipes indicated by <R>. Additional recipes can be found online at http://cshprotocols.cshlp.org/site/recipes.

Reagents

Blocking buffer for protein arrays <R>
   Alternatively, Superblock (Pierce) supplemented with 0.5% BSA and 5% normal goat serum can be used.

Primary antibodies
   Mouse anti-GST (glutathione *S*-transferase) monoclonal (Sigma-Aldrich SAB4200237)
   Primary antibody (Flag-, HA-, or V5-tagged) to protein of interest

Probing buffer <R>
Protein microarrays (from Zhu laboratory; heng.zhu@jhmi.edu for inquiries)
Secondary antibodies
   Goat anti-mouse-Alexa Fluor 555 (Invitrogen A28180)
   Goat anti-tag-Alexa Fluor 647 (according to primary antibody tag)

Tris-buffered saline (TBS; 1×) <R> with 0.1% Tween 20 (TBST)
   Chill TBST to 4°C for Step 17.

---

[4]Correspondence: heng.zhu@jhmi.edu

Copyright © Cold Spring Harbor Laboratory Press; all rights reserved
Cite this protocol as *Cold Spring Harb Protoc*; doi:10.1101/pdb.prot087965

Equipment

Aluminum foil (for arrays without opaque lids)
Benchtop centrifuge (e.g., Heraeus Multifuge 3SR+; Thermo Scientific)
Humidity chamber (homemade)

> To prepare a humidity chamber, place a folded paper towel in the bottom of an empty pipette tip box, add one inch of ddH$_2$O, and replace the tip holder on the box. The arrays will sit on the tip holder and be covered with the lid.

Incubator at 30°C (as needed for Step 4)
LifterSlip glass coverslips, 25 mm × 60 mm (Fisher Scientific 22-035-809)
Microarray chamber trays (6-Sectional-Short, black) (GenHunter)
Microarray scanner (e.g., GenePix 4000B; Molecular Devices)
Micro slide boxes (VWR 48444-004) with laboratory tissues in the bottom
Orbital shaker

METHOD

Protein microarrays can be used for the identification of thousands of interactions in parallel (Fasolo et al. 2011; Li et al. 2012; Shamay et al. 2012). The following method uses fluorescently labeled antibodies for detection of protein–protein binding.

Do not allow the array slides to dry between steps or bubbles to form under the coverslip. This will cause proteins to denature, leading to a loss of binding signal or generation of nonspecific signals.

Blocking the Microarrays

1. Incubate each protein microarray in a microarray chamber tray (with the barcode facing up) with 3 mL of blocking buffer for 1 h at room temperature with gentle shaking.
   > To preserve spot morphology and protein activity, buffer should not be added directly on the array.
   > This step can also be performed for 3 h at 4°C.

2. Following the blocking period, aspirate the blocking buffer while tilting the microarray chamber tray so the buffer runs to a single corner near the barcode.

Probing Microarrays for the Protein of Interest

3. Immediately following aspiration of the blocking buffer, add 200 μL of probing buffer to each microarray and carefully cover with a LifterSlip.

4. Place the arrays in a single layer in a humidity chamber. Incubate the chamber with gentle shaking for 60 min at 30°C.
   > This step can also be performed for 3 h at 4°C.

5. Following the incubation, remove the LifterSlip by sliding it down the length of the microarray.

6. Immediately wash the microarray by quickly dunking it in a beaker filled with TBST.

Incubating Protein Microarrays with Primary Antibody

7. Place each array in a microarray chamber tray. Rinse the arrays three times with 3 mL of probing buffer for 10 min per rinse at room temperature.

8. During the final rinse in Step 7, prepare 3 mL per microarray of blocking buffer containing mouse anti-GST (∼50–100 ng/mL) and the tagged primary antibody of interest (∼50–100 ng/mL). Keep the primary antibody solution on ice until ready to proceed with Step 9.

9. Following the final rinse in Step 7, add 3 mL of the primary antibody solution to each array in the microarray chamber.

10. Incubate for 60 min at room temperature with gentle shaking.
11. Aspirate the primary antibody solution as described in Step 2.
12. Wash each microarray three times with 3 mL of TBST for 5 min per wash at room temperature with gentle shaking.

Incubating Protein Microarrays with Secondary Antibody

The secondary antibody solution is a mix of both secondary antibodies labeled with fluorophores.

13. During the final wash in Step 12, prepare 3 mL per microarray of blocking buffer containing anti-mouse-Alexa Fluor 555 (~200 ng/mL) and anti-tag Alexafluor 647 (~200 ng/mL) antibodies. Keep the secondary antibody solution on ice until ready to proceed with Step 14.
14. Following the final wash in Step 12, add 3 mL of secondary antibody solution to each array in the microarray chamber. If arrays do not have an opaque lid, cover with aluminum foil to prevent light exposure.
15. Incubate for 60 min at room temperature with gentle shaking.
16. Aspirate the secondary antibody as described in Step 2.
17. Wash each array three times with 3 mL of cold (4°C) TBST for 5 min per wash with gentle shaking at room temperature.
18. Rinse the microarrays with distilled H_2O and place each array sideways in a micro slide box containing tissues. Centrifuge the boxes at 1000g for 3 min to dry the arrays.
19. Refer to Characterization of Protein–DNA Interactions Using Protein Microarrays (Hu et al. 2011) for data quantification and analysis using a microarray scanner.

    See Troubleshooting.

TROUBLESHOOTING

Problem (Step 19): Spots display signal saturation during analysis (see Characterization of Protein–DNA Interactions Using Protein Microarrays [Hu et al. 2011]).
Solution: Adjust the laser power and gain after preview. If the problem still persists, use a fluorescently conjugated primary antibody.

RECIPES

Blocking Buffer for Protein Arrays

Phosphate-buffered saline (PBS)
3% Bovine serum albumin (BSA)
0.05% Tween 20

Prepare 3 mL of blocking buffer per array. Store for up to 1 mo at 4°C.

Probing Buffer

1× TBS
5% Bovine serum albumin (BSA)
0.05% Tween 20
1–10 µg/µL purified, tagged protein of interest
1 mM DTT

On the day of the experiment, prepare 200 µL per microarray of probing buffer containing the tagged protein of interest. Add DTT and protein of interest immediately before use.

Tris-Buffered Saline (TBS) (1×)

50 mM Tris-Cl, pH 7.5

150 mM NaCl

To prepare, dissolve 6.05 g Tris and 8.76 g NaCl in 800 mL of H_2O. Adjust pH to 7.5 with 1 M HCl and make volume up to 1 L with H_2O. TBS is stable for 3 mo at 4°C.

REFERENCES

Fasolo J, Sboner A, Sun MG, Yu H, Chen R, Sharon D, Kim PM, Gerstein M, Snyder M. 2011. Diverse protein kinase interactions identified by protein microarrays reveal novel connections between cellular processes. *Genes Dev* 25: 767–778.

Hu S, Xie Z, Blackshaw S, Qian J, Zhu H. 2011. Characterization of protein–DNA interactions using protein microarrays. *Cold Spring Harb Protoc* doi: 10.1101/pdb.prot5614.

Li R, Wang L, Liao G, Guzzo CM, Matunis MJ, Zhu H, Hayward SD. 2012. SUMO binding by the Epstein–Barr virus protein kinase BGLF4 is crucial for BGLF4 function. *J Virol* 86: 5412–5421. doi: 10.1128/JVI.00314-12.

Shamay M, Liu J, Li R, Liao G, Shen L, Greenway M, Hu S, Zhu J, Xie Z, Ambinder RF, et al. 2012. A protein array screen for Kaposi's sarcoma–associated herpesvirus LANA interactors links LANA to TIP60, PP2A activity, and telomere shortening. *J Virol* 86: 5179–5191. doi: 10.1128/JVI.00169-12.

Protocol 2

Characterization of RNA-Binding Proteins Using Protein Microarrays

Guang Song,[1,2] Johnathan Neiswinger,[1,2] and Heng Zhu[1,2,3,4]

[1]Department of Pharmacology and Molecular Sciences, Johns Hopkins University School of Medicine, Baltimore, Maryland 21205; [2]The Center for High-Throughput Biology, Johns Hopkins University School of Medicine, Baltimore, Maryland 21205; [3]The Sidney Kimmel Comprehensive Cancer Center, Johns Hopkins University School of Medicine, Baltimore, Maryland 21287

RNA-binding proteins (RBPs), along with target RNA, play a vital role in the regulation of cellular processing and are especially important in gene transcription and posttranscriptional regulation. Here, we present a high-throughput method for rapid identification of RBPs for a given RNA using protein microarray technology. This protocol includes preparation of Cy5-labeled RNA probes, probe denaturing and refolding, and an RNA-binding assay as performed on a yeast protein microarray.

MATERIALS

It is essential that you consult the appropriate Material Safety Data Sheets and your institution's Environmental Health and Safety Office for proper handling of equipment and hazardous materials used in this protocol.

RECIPES: Please see the end of this protocol for recipes indicated by <R>. Additional recipes can be found online at http://cshprotocols.cshlp.org/site/recipes.

Reagents

Agarose
Blocking buffer for RNA binding <R>
Cy5-UTP (GE Healthcare PA55026)
Dithiothreitol (DTT) (Sigma-Aldrich)
DNA template for in vitro transcription (IVT)

> Three forms of DNA can be used as a template for IVT (Step 3): PCR products, DNA oligonucleotides, and linearized plasmid DNA. PCR products should be gel-purified before beginning this protocol using the QIAquick Gel Extraction kit (QIAGEN 28704) to avoid primer–dimer contamination. Linearized plasmid DNA should be purified using phenol:chloroform as described in Step 1.

> All templates should contain the RNA polymerase promoter, in the correct orientation, upstream of the sequence to be transcribed. For example, we designed the T7 RNA polymerase upstream sequence as 5′-TAATACGACT CACTATAGGG-3′.

Ethanol (Fisher Scientific 16-100-386)

---

[4]Correspondence: heng.zhu@jhmi.edu

Copyright © Cold Spring Harbor Laboratory Press; all rights reserved
Cite this protocol as Cold Spring Harb Protoc; doi:10.1101/pdb.prot087973

Folding buffer (2×) <R>
   Cool to 4°C for Step 10.ii.

Formaldehyde solution (37%) (Sigma-Aldrich 252549)
Glycerol
H₂O, treated with diethyl pyrocarbonate (DEPC) (Sigma-Aldrich D5758)
   Cool to 4°C for Step 10.ii.

Heparin sodium salt from porcine intestinal mucosa (Sigma-Aldrich H3393) (4 mg/µL)
Herring sperm DNA, single-stranded (ssDNA) (10 µg/µL) (Sigma-Aldrich D7290)
HiScribe T7 High Yield RNA Synthesis kit (New England BioLabs E2040S)
Igepal CA-630 (1%) (Sigma-Aldrich)
Protein microarrays (from Zhu laboratory; heng.zhu@jhmi.edu for inquiries)
Reagents for DNA plasmid purification (as needed)
   Chloroform
   Ethanol (70%) at −20°C
   Glycogen (Thermo Scientific R0561)
   Phenol:chloroform:isoamyl alcohol (UltraPure 25:24:1, v/v) (Life Technologies 15593031)
   Sodium acetate (3 M, pH 5.2)

RNA gel buffer (10× MOPS) (Quality Biological 351-059-101)
RNA gel loading dye (Life Technologies R0641)
RNA ladder (RiboRuler High Range; 200–6000 bases) (Thermo Scientific SM1821)
RNase inhibitor, murine (New England BioLabs M0314L)
RNeasy Mini kit (QIAGEN 74104)
TURBO DNase (Life Technologies AM2238) (2 U/µL)
Washing buffer for RNA binding <R>
Yeast tDNA (10 µg/µL)

Equipment

Benchtop centrifuge (e.g., HERAEUS Multifuge 3SR+; Thermo Scientific)
   Cool to 4°C for plasmid DNA purification (Step 1), as needed.

Centrifuge tubes (50-mL) with laboratory tissues in the bottom
Cooling block (Diversified Biotech CHAM-3000)
Equipment for agarose gel electrophoresis (RNase-free)
Heat blocks at 37°C, 70°C, 90°C
Humidity chamber
   To prepare a humidity chamber, place two pieces of laboratory tissue in the bottom of an empty pipette tip box and add 5 mL of RNase-free H₂O. To eliminate RNase contamination, use an RNase decontamination wipe (e.g., RNAseZap) on the inside of the lid and tip rack.

LifterSlip glass coverslips (Fisher 22-035-809)
Microarray analysis software (GenePix Pro 6.0; MDS Analytical Technologies)
Microarray chamber trays (6-Sectional-Short, black) (GenHunter)
Microarray scanner (GenePixPro 4000B; Molecular Devices)
Microcentrifuge tubes (1.5-mL)
NanoDrop 2000 (Thermo Scientific) (optional; see Step 7)
Orbital shaker
UV transilluminator and image acquisition system
Vortex

Chapter 23

METHOD

Protein microarrays have been used to identify both protein and long coding/noncoding RNA (Siprashvili et al. 2012), as well as protein– and oligo–RNA interactions (Zhu et al. 2007). In the following protocol, protein microarrays are used for high-throughput and unbiased identification of RBPs for an RNA of interest.

Preparing the Cy5-Labeled RNA Probe

1. If linearized plasmid DNA will be used as the template for IVT, purify the DNA as follows. If using oligonucleotides or gel-purified PCR product as the IVT template, proceed to Step 2.

    i. Adjust the volume of digested plasmid DNA to 200 µL, and then add an equal volume of phenol:chloroform:isoamylalcohol (25:24:1, v/v).

    ii. Vortex the sample and centrifuge at maximum speed (21,130g) for 10 min at 4°C. Carefully pour the supernatant into a fresh tube.

    iii. Repeat Steps i–ii.

    iv. Extract the sample twice using the same volume of chloroform to remove the residual phenol.

    v. Add 10 µL of 3 M sodium acetate (pH 5.2), 2 µL of glycogen, and 500 µL of ethanol. Vortex and incubate overnight at −20°C.

    vi. Pellet the DNA at maximum speed (21,130g) for 30 min at 4°C. Carefully remove the supernatant.

    vii. Carefully wash the DNA pellet by adding 500 µL of precooled (−20°C) 70% ethanol. Do not disturb the pellet. Centrifuge at maximum speed for 30 min at 4°C. Remove the supernatant carefully and dry the pellet in the fume hood.

    viii. Resuspend the DNA pellet with DEPC-treated H_2O to a concentration of ∼0.5 µg/µL.

2. Prepare an in vitro transcription (IVT) master mix as follows.

    i. Thaw the Cy5-UTP, transcription buffer and NTP components of the HiScribe T7 High Yield RNA Synthesis kit on ice, and pulse spin to collect tube contents.

    ii. Combine the following reagents.

10 × T7 transcription buffer	25 µL
100 mM ATP	12.5 µL
100 mM CTP	12.5 µL
100 mM GTP	12.5 µL
100 mM UTP	11.25 µL
5 mM labeled UTP	20 µL
DEPC-treated H_2O	6.25 µL
Total volume	100 µL

    iii. (Optional) To store the IVT master mix for future use, prepare 25 aliquots containing 4 µL per tube. Store aliquots at −20°C.

3. Assemble the IVT reaction(s) in the following order, mix thoroughly, and pulse spin.

IVT master mix	4 µL
DNA template (0.1–0.5 µg) in DEPC-treated H_2O	4.5 µL
T7 RNA polymerase	1 µL
RNase inhibitor	0.25 µL
0.2 M DTT	0.25 µL
Total volume	10 µL

4. Incubate the IVT reaction(s) for 3 h at 37°C to synthesize RNA.

5. Add 1 µL of TURBO DNase (2 U) per 10-µL IVT reaction and incubate for 15 min at 37°C to eliminate the DNA template.

6. Purify the Cy5-labeled RNA probe using the RNeasy Mini kit components as follows.

   i. Adjust the volume of each DNase-treated IVT sample to 50 µL with RNase-treated H_2O. Add 175 µL (3.5 volumes) of RLT buffer.

   ii. Add 320 µL of 100% ethanol to each sample. Vortex and transfer the full volume (545 µL) to an RNeasy Mini spin column.

   iii. Centrifuge the columns for 15 sec at >8000g. Discard the flowthrough.

   iv. Add 500 µL of RPE buffer and centrifuge for 15 sec at >8000g. Discard the flowthrough.

   v. Add 500 µL of RPE buffer and centrifuge for 2 min at >8000g.

   vi. Transfer each column to a new collection tube and centrifuge once more to remove carryover RPE buffer.

   vii. Transfer each column to a 1.5-mL tube. Add 50 µL of DEPC-treated H_2O and centrifuge for 1 min at >8000g to elute the RNA.

   We recommend using freshly prepared RNA probes for the RNA-binding assay on a protein microarray; however, RNA samples may be stored at −80°C. In this case, the integrity of the RNA probes after storage should be examined by denaturing agarose gel (Step 8) before every assay.

7. (Optional) Use a NanoDrop to detect the RNA concentration and total yield.

   If a NanoDrop is not available, the amount of RNA probe can be quantified by gel electrophoresis in Step 8.

8. Perform RNA denaturing electrophoresis to visualize RNA probe sizes as follows.

   i. To prepare a 1% denaturing RNA gel, dissolve 0.5 g of agarose in 37 mL of DEPC-treated H_2O by heating. Cool on the bench to 60°C, and then add 5 mL of 10× MOPS RNA gel buffer and 8 mL of formaldehyde (37%). Mix and pour the gel solution into the tank.

   Work with formaldehyde should be performed in a fume hood.

   ii. To prepare samples for loading, mix 0.5 µg of each RNA sample (or 2 µL of RNA ladder) with an equal volume of RNA loading dye. Heat the tubes for 10 min to 70°C, and then transfer to a cooling block on ice for 2 min.

   iii. Load and run the gel in 1× MOPS buffer with 5% formaldehyde at 5–6 V/cm until the bromophenol blue band has migrated two-third length of the gel.

   iv. Visualize the RNA probes with a UV transilluminator and image acquisition system.

   See Troubleshooting.

Performing the RNA-Binding Assay on Protein Microarrays

IMPORTANT: Avoid exposure of Cy5-labeled RNA to light during the following procedure.

9. Remove the protein microarrays from the −80°C freezer and place them in microarray chamber trays with 3 mL of blocking buffer per tray. Incubate for 1.5 h at room temperature with gentle shaking.

   Steps 10–12 should be performed during this 1.5-h incubation.

10. Perform denaturing and refolding of the Cy5-labeled RNA probe as follows.

    i. Heat 20 pmol of RNA probe per array for 2 min to 90°C, and then immediately move to a cooling block on ice for 1 min.

    ii. Adjust the RNA probe volume to 50 µL with precooled DEPC-treated H_2O. Add 50 µL of precooled 2× folding buffer to the denatured RNA probe sample.

    iii. Vortex the sample briefly and incubate for 10 min at 37°C to generate the refolded probe.

11. Combine the following reagents to prepare 100 µL of RNA-binding buffer for each microarray.

2× Folding buffer	50 µL
ssDNA (10 µg/µL)	0.2 µL
Yeast tDNA (10 µg/µL)	0.2 µL
Heparin (4 mg/µL)	0.5 µL
DTT (200 mM)	1 µL
Igepal CA-630 (1%)	2 µL
Glycerol	10 µL
RNase inhibitor	1 µL
DEPC-treated H$_2$O	to 100 µL
Total volume	100 µL

12. Mix the refolded Cy5-labeled RNA probe (100 µL) with same volume of binding buffer (100 µL).
    The final probe concentration will be 100 nM.

13. Remove each protein microarray from blocking buffer, and tap the slide on laboratory tissue to remove excess medium.

14. Place the arrays in the humidity chamber. Immediately add 200 µL of Cy5-labeled RNA probe mixture onto each array. Gently cover each array with a LifterSlip coverslip.

15. Incubate the microarrays for 1 h at room temperature with gentle shaking.

16. Gently slide the LifterSlip off of each array and place each array in a microarray chamber tray.

17. Wash each array three times in 10 mL of washing buffer for 10 min per wash with gentle shaking.

18. Wash each array with 10 mL of DEPC-treated H$_2$O for 10 min with gentle shaking.

19. Place the arrays in individual 50-mL tubes with laboratory tissues in the bottom and centrifuge for 3 min at 500g.

20. Scan the microarrays using a GenePix 4000B scanner with 5-µm resolution detection at 635 nm. Set the appropriate photomultiplier tube (PMT) and power value to ensure no spots have saturated signal intensities.

21. Refer to Characterization of Protein–DNA Interactions Using Protein Microarrays (Hu et al. 2011) for data quantification and analysis, as methods are highly variable.
    See Troubleshooting.

TROUBLESHOOTING

Problem (Step 8.iv): The RNA bands on the denaturing gel are unclear; notably, there is a long tail below the main bands.
Solution: The RNA was partially degraded before gel loading. The RNA electrophoresis buffer should be freshly prepared before every experiment. All containers and tools (electrophoresis tank, comb, etc.) should be treated with DEPC or wiped with an RNase decontamination solution such as RNAseZap.

Problem (Step 21): Some RNA probes produce stronger signals than others.
Solution: Some RNA probes may show nonspecific, high-affinity binding to proteins on the array, especially large-size probes. A lower RNA concentration or lower ratio of Cy5-UTP to unlabeled UTP during in vitro transcription will decrease nonspecific binding. Longer washing periods might also be necessary.

Problem (Step 21): Signals are faint or there are only a few strong signal spots on the entire microarray.
Solution: Some RNA probes have a lower affinity binding to target proteins. A higher RNA concentration or higher ratio of Cy5-UTP to labeled UTP during in vitro transcription can increase signals. Also, take proper care to avoid RNA degradation.

RECIPES

Blocking Buffer for RNA Binding

10 mM Tris (pH 7.5, RNase-free; Quality Biological 351-006-101)
137 mM NaCl (RNase-free; Life Technologies AM9760G)
3 mM KCl (RNase-free; Life Technologies AM9640G)
2 mM MgCl$_2$ (RNase-free; Quality Biological 351-033-721)
2% Bovine serum albumin (BSA) (Jackson ImmunoResearch 001-000-162)
10 µg/mL ssDNA (Sigma-Aldrich D7290)
10 µg/mL yeast tRNA

Combine all ingredients except ssDNA and yeast tRNA. This can be stored for up to 1 mo at 4°C. Add ssDNA and yeast tRNA fresh.

Folding Buffer (2×)

20 mM Tris (pH 8.0, RNase-free; Life Technologies AM9856)
200 mM potassium acetate (pH 8.0, RNase-free)
20 mM MgCl$_2$ (RNase-free; Quality Biological 351-033-721)

Store for up to 6 mo at −20°C.

Tris-Buffered Saline (TBS) (1×)

50 mM Tris-Cl, pH 7.5
150 mM NaCl

To prepare, dissolve 6.05 g Tris and 8.76 g NaCl in 800 mL of H$_2$O. Adjust pH to 7.5 with 1 M HCl and make volume up to 1 L with H$_2$O. TBS is stable for 3 mo at 4°C.

Washing Buffer for RNA Binding

1× Tris-buffered saline (TBS) <R>
0.05% Tween 20
2 mM MgCl$_2$ (RNase-free; Quality Biological 351-033-721)

Store for up to 1 mo at room tempature.

REFERENCES

Hu S, Xie Z, Blackshaw S, Qian J, Zhu H. 2011. Characterization of protein–DNA interactions using protein microarrays. *Cold Spring Harb Protoc* doi: 10.1101/pdb.prot5614.

Siprashvili Z, Webster DE, Kretz M, Johnston D, Rinn JL, Chang HY, Khavari PA. 2012. Identification of proteins binding coding and non-coding human RNAs using protein microarrays. *BMC Genomics* 13: 633.

Zhu J, Gopinath K, Murali A, Yi G, Hayward SD, Zhu H, Kao C. 2007. RNA-binding proteins that inhibit RNA virus infection. *Proc Natl Acad Sci* 104: 3129–3134.

Protocol 3

Characterization of Lipid–Protein Interactions Using Nonquenched Fluorescent Liposomal Nanovesicles and Yeast Proteome Microarrays

Kuan-Yi Lu,[1] Chien-Sheng Chen,[1] Johnathan Neiswinger,[2,3] and Heng Zhu[2,3,4]

[1]*Graduate Institute of Systems Biology and Bioinformatics, National Central University, Jhongli 32001, Taiwan;* [2]*The Center for High-Throughput Biology, Johns Hopkins University School of Medicine, Baltimore, Maryland 21205;* [3]*Department of Pharmacology and Molecular Sciences, Johns Hopkins University School of Medicine, Baltimore, Maryland 21205*

Studying lipid–protein interactions is central to understanding lipid signaling, a key regulatory system in cells. To better identify lipid-binding proteins, we developed a nonquenched fluorescent (NQF) liposome that is able to carry both fluorescent molecules and a lipid of interest. By combining the strength of NQF liposomes with protein microarray technology, the method presented here facilitates high-throughput screening of lipid–protein interactions. This protocol describes how to prepare NQF liposomes and apply the fabricated liposomes to yeast proteome microarrays.

MATERIALS

It is essential that you consult the appropriate Material Safety Data Sheets and your institution's Environmental Health and Safety Office for proper handling of equipment and hazardous materials used in this protocol.

RECIPES: Please see the end of this protocol for recipes indicated by <R>. Additional recipes can be found online at http://cshprotocols.cshlp.org/site/recipes.

Reagents

1,2-Dipalmitoyl-*sn*-glycero-3-phospho-(1′-rac-glycerol) (sodium salt) (DPPG) (Avanti Polar Lipids 770455)

1,2-Dipalmitoyl-*sn*-glycero-3-phosphocholine (DPPC) (Avanti Polar Lipids 770355)

1,2-Dipalmitoyl-*sn*-glycero-3-phosphoethanolamine-*N*-(lissamine rhodamine B sulfonyl) (ammonium salt) (LRB-DPPE) (Avanti Polar Lipids 810158) (1 mg/mL in chloroform solution at −80°C)

Blocking buffer for NQF liposomes <R>

Chloroform (Tedia High Purity Solvents CS1332)

Cholesterol (Avanti Polar Lipids 700000P)

Lipid of interest (1 mg/mL in chloroform solution at −80°C)

Methanol (Tedia High Purity Solvents MS1922)

Sulforhodamine B sodium salt (SRB) (Sigma-Aldrich S1402)

---

[4]Correspondence: heng.zhu@jhmi.edu

Copyright © Cold Spring Harbor Laboratory Press; all rights reserved
Cite this protocol as *Cold Spring Harb Protoc*; doi:10.1101/pdb.prot087981

TBS$_{ex}$ (1×) <R>
TBS$_{in}$ (10×) <R>
TBST for protein arrays (10×) <R>
Yeast proteome microarrays (from Zhu laboratory; heng.zhu@jhmi.edu for inquiries)

Equipment

3D micro shaker
Aluminum foil
Auto dehumidify cabinet, relative humidity (RH) ≤30%
Cell culture dishes (four-well) (Thermo Scientific 167063)
Centrifuge, benchtop, with A-4-62 rotor and 15-mL centrifuge tube buckets (Eppendorf 5810 R)
Desalting columns (2-mL, 7K MWCO [molecular weight cut-off]) (Thermo Scientific 89889)
Flask, single-neck, round-bottom (25-mL, 14/20 outer joint with plug)
Funnel
Hot-plate stirrer at 60°C (LMS HTS-1003)
Lid (sliding) from pipette tip box
Microarray analysis software (GenePix Pro 6.0; MDS Analytical Technologies)
Microarray hybrid chambers (Camlab)
  Add 200 µL of H$_2$O to the inside of the chamber to humidify.

Microarray scanner (GenePix 4000B; MDS Analytical Technologies)
Microcentrifuge (e.g., Force Mini; Select BioProducts)
Microcentrifuge tubes, amber (1.5-mL)
Mini-extruder set with polycarbonate syringe filter and accessories (50-nm pore size) (Avanti Polar Lipids 610000)
Nitrogen (N$_2$) tank with spray gun
Orbital shaker
Parafilm
Polypropylene boxes (120 × 90 × 70-mm)
Serum bottles (500-mL)
Tweezers
Vortex
Water bath at 60°C

METHOD

Conduct Steps 2–12 and Steps 18–19 in the dark to avoid light damage to lipids (SRB [sulfate-reducing bacteria] and LRB-DPPE [lissamine rhodamine B-dipalmitoyl phosphatidylethanol]) and nonquenched fluorescent (NQF) liposomes. Tubes and flasks containing lipids should be completely covered in aluminum foil.

Preparing NQF Liposomes

1. Place the structural lipids (cholesterol, DPPC [dipalmitoylphosphatidylcholine], and DPPG [dipalmitoylphosphatidylglycerol]) and SRB in a dehumidify cabinet (RH ≤ 30%) for 15–20 min to expel moisture.

2. Prepare the SRB solution.

   i. Add 58 mg of SRB to 5 mL of 1× TBS$_{in}$ to prepare a 100× SRB stock solution.

   ii. Dilute 10 µL of 100× SRB solution to 1 mL with 1× TBS$_{in}$ for a final concentration of 200 µM SRB. Store the 100× SRB solution at 4°C.

   iii. Prewarm the 200 µM SRB solution in an amber microcentrifuge tube to 60°C.
   Ensure that tubes containing SRB solution are completely covered with aluminum foil.

3. Add the following compounds to a 25-mL round-bottom flask.

   9 µmol cholesterol (45%)
   1 µmol DPPG (5%)
   9.86 µmol DPPC (49.3%)
   0.06 µmol LRB-DPPE (0.3%)
   0.08 µmol lipid of interest (0.4%)

   The lipid of interest and LRB-DPPE are stored in a chloroform solution (1 mg/mL) at −80°C and should be added to the flask by quick pipetting in a fume hood.

   The total amount of lipids can be adjusted to dictate the concentration or amount of liposomes per batch by following the molar ratio above.

   Ensure that the flask is completely sealed with aluminum foil throughout the procedure to avoid light.

4. Add 1 mL of chloroform/methanol (volume ratio = 5:1) to the flask to dissolve the lipids. Mix thoroughly by slightly swirling the flask.

   DO NOT vortex the flask, as lipids will be splashed onto the flask wall. Lipids can be easily dissolved in the organic solvent by swirling the flask.

5. Dip half the flask in a water bath at 60°C and completely dry the organic solvent with N_2 gas.

   A thin, circular lipid film with a pale red color should be observed at the bottom of the flask.
   See Troubleshooting.

6. Add 300 µL of the prewarmed 200 µM SRB solution to the flask. Seal the flask tightly with a plug and Parafilm. Dip half the flask in a water bath for 45 min at 60°C to allow hydration.

   The formed lipid layer should be fully covered by the SRB solution.

7. Prewash two desalting spin columns with 2 mL of 1× TBS_{ex} two times, following manufacturer's instruction. Set aside until Step 11.

8. Vortex the flask until the lipid layer is completely detached from the wall of flask. Seal the flask tightly with Parafilm again to assure a tight seal and incubate the liposome suspension for an additional 30 min at 60°C to allow the spherical vesicles to further mature and fully hydrate.

9. Prepare the mini-extruder.

   i. Prewet a 50-nm-diameter polycarbonate membrane and four membrane supports with 1× TBS_{ex}.

   ii. Assemble the mini-extruder following the manufacturer's instructions. Extrude the TBS_{ex} back and forth through the membrane 30 times to rinse the mini-extruder, avoiding leakage.

   iii. Prewarm the mini-extruder to 60°C on a hot-plate stirrer.

10. Load the liposome suspension into a syringe and carefully place the syringe into one end of the mini-extruder. Insert an empty syringe into the other end of the mini-extruder. Extrude the liposome suspension back and forth through the membrane 30 times at 60°C.

11. Add the liposome suspension to the center of the compacted resin bed of a prewashed desalting spin column. Centrifuge the column at 1000g for 5 min to purify the NQF liposomes. Transfer the collected NQF liposomes to the second prewashed desalting spin column and repeat.

12. Collect the purified NQF liposomes in an amber microcentrifuge tube. Store at 4°C until Step 17.

Applying NQF Liposomes to Yeast Proteome Microarrays

13. Prepare a four-well dish containing 3.6 mL of blocking buffer per well. Add ∼80 mL of 1× TBST to a sliding lid of a pipette tip box.

14. Remove the protein arrays from −80°C and quickly immerse in the TBST solution. DO NOT touch the microarray surface or scratch the surface against the bottom of the sliding lid that contains TBST, as it will cause a severe damage to the microarray. Gently shake for 4 min.

    Residual nonimmobilized proteins will become competitors of the binding, thus reducing binding signals.

15. Dry each array by tapping the edge on laboratory tissues.
16. Place each array in one well of the four-well dish containing blocking buffer and incubate for 1 h at room temperature with gentle shaking.
17. For each array, dilute 0.3 µL of the NQF liposomes (containing the lipid of interest) to 110 µL in 1× TBS$_{ex}$ in an amber microcentrifuge tube.
18. Perform hybridization as follows.

    i. Use tweezers to remove the arrays from the blocking buffer by grabbing the barcode end.
    ii. Drain the blocking buffer by tapping each slide sideways on laboratory tissues.
    iii. Lift slightly one of the short ends of a microarray and add 55 µL of the prepared NQF liposomes to the lifted end; the liposome solution will cover the microarray surface. Repeat with the remaining 55 µL of NQF liposomes by lifting opposite end.
    iv. Place the microarray in a humidified hybrid chamber and incubate for 80 min at room temperature with very gentle shaking using a 3D shaker.

19. Wash each array as follows.

    i. Prepare four serum bottles each containing 500 mL of 1× TBS$_{ex}$.
    ii. Place a microarray in one corner of a polypropylene box and use a funnel to very slowly add 500 mL of 1× TBS$_{ex}$ from the opposite corner to reduce the shear force.
    iii. With gloved fingertips, hold the microarray carefully in the TBS$_{ex}$ solution and gently rock back and forth a few times. DO NOT touch the microarray surface.
    iv. Repeat Steps 19.ii–19.iii three times per array.
    v. Place the microarray in a fume hood to partially air dry.

    A centrifuge should not be used to dry the slide, as this will severely blur the microarray image resulting in an unanalyzable data.

20. Scan the microarray with a GenePix 4000B scanner with 10-µm resolution detection at 532 nm, carefully avoiding saturation. Save the scanned images as TIF files.
21. Scan the microarrays with different photomultiplier tubes (PMTs) from lower to higher for data storage and normalization.

    Multiple slides for comparison must be scanned using the same PMT.

22. Refer to Characterization of Protein–DNA Interactions Using Protein Microarrays (Hu et al. 2011) for data quantification and analysis, as methods are highly variable.

TROUBLESHOOTING

Problem (Step 5): The formed lipid film appears on places other than the bottom of the flask or in patches, which will not be covered by the SRB solution.
Solution: Scattered lipid films can appear because of the strong force exerted by N$_2$ gas. Add another 1 mL of chloroform/methanol to dissolve the lipids completely. Blow-dry the organic solvent by carefully and gently applying N$_2$ gas to the sidewall of the flask.

Problem (Step 5): The organic solvent is not dried by N$_2$ gas after 30 min or longer.
Solution: There is water residue in the flask or the containers that store the organic solvent. Place the flask in an oven or dehumidify cabinet. Once dried, add another 1 mL of chloroform/methanol to dissolve the lipids and repeat Step 5.

RECIPES

Blocking Buffer for NQF Liposomes

25 mM Tris (pH 7.5)
140 mM NaCl
3% bovine serum albumin (BSA) (Sigma-Aldrich A7030)

Store for up to 1 mo at 4°C.

TBS$_{ex}$ (1×)

25 mM Tris (pH 7.5)
140 mM NaCl
51.77 g/L sucrose

Store for up to 1 mo at 4°C.

TBS$_{in}$ (10×)

0.25 M Tris (pH 7.5)
1.4 M NaCl

Store for up to 6 mo at room temparature.

TBST for Protein Arrays (10×)

0.25 M Tris (pH 7.5)
1.4 M NaCl
0.05% Tween 20

Store for up to 6 mo at room temparature.

REFERENCES

Hu S, Xie Z, Blackshaw S, Qian J, Zhu H. 2011. Characterization of protein–DNA interactions using protein microarrays. *Cold Spring Harb Protoc* doi: 10.1101/pdb.prot5614.

Protocol 4

Posttranslational Modification Assays on Functional Protein Microarrays

Johnathan Neiswinger,[1,2] Ijeoma Uzoma,[1,2] Eric Cox,[1,2,3] HeeSool Rho,[1,2] Jun Seop Jeong,[1,2] and Heng Zhu[1,2,4,5]

[1]Department of Pharmacology and Molecular Sciences, Johns Hopkins University School of Medicine, Baltimore, Maryland 21205; [2]The Center for High-Throughput Biology, Johns Hopkins University School of Medicine, Baltimore, Maryland 21205; [3]Department of Neuroscience, Johns Hopkins University School of Medicine, Baltimore, Maryland 21205; [4]The Sidney Kimmel Comprehensive Cancer Center, Johns Hopkins University School of Medicine, Baltimore, Maryland 21287

Protein microarray technology provides a straightforward yet powerful strategy for identifying substrates of posttranslational modifications (PTMs) and studying the specificity of the enzymes that catalyze these reactions. Protein microarray assays can be designed for individual enzymes or a mixture to establish connections between enzymes and substrates. Assays for four well-known PTMs —phosphorylation, acetylation, ubiquitylation, and SUMOylation—have been developed and are described here for use on functional protein microarrays. Phosphorylation and acetylation require a single enzyme and are easily adapted for use on an array. The ubiquitylation and SUMOylation cascades are very similar, and the combination of the E1, E2, and E3 enzymes plus ubiquitin or SUMO protein and ATP is sufficient for in vitro modification of many substrates.

MATERIALS

It is essential that you consult the appropriate Material Safety Data Sheets and your institution's Environmental Health and Safety Office for proper handling of equipment and hazardous materials used in this protocol.

RECIPES: Please see the end of this protocol for recipes indicated by <R>. Additional recipes can be found online at http://cshprotocols.cshlp.org/site/recipes.

Reagents

Eukaryotic proteome microarrays (from Zhu laboratory; heng.zhu@jhmi.edu for inquiries)
Reagents for the acetylation assay

   Acetylation reaction buffer (100 µL) <R>
   Prepare in Step 4.

   Blocking buffer (acetylation assay) <R>
   H_2O (double-distilled) at 37°C
   Wash buffer I (acetylation assay) <R>

Reagents for the phosphorylation assay
   Blocking buffer (phosphorylation assay) <R>

---

[5]Correspondence: heng.zhu@jhmi.edu

Copyright © Cold Spring Harbor Laboratory Press; all rights reserved
Cite this protocol as *Cold Spring Harb Protoc*; doi:10.1101/pdb.prot087999

H₂O (double-distilled) at 37°C
Kinase reaction buffer <R>
> Prepare in Step 4.

Purified kinase (~100 U/µL)
SDS (0.5% in ddH₂O)

Reagents for the SUMOylation assay

Blocking buffer (ubiquitylation/SUMOylation assays) <R>
Mouse anti-GST mAB (Sigma-Aldrich SAB4200237) (primary antibody)
Rabbit anti-SUMO (yeast SMT3) polyclonal (Abcam ab14405) (primary antibody)
SDS (1% in ddH₂O) at 55°C
SUMOylation reaction buffer (200 µL) <R>
> Prepare in Step 4.

Reagents for the ubiquitylation assay

Blocking buffer (ubiquitylation/SUMOylation assays) <R>
Mouse anti-FLAG mAb (Sigma-Aldrich F3165) (primary antibody)
Rabbit anti-GST polyclonal (Millipore AB3282) (primary antibody)
Ubiquitylation reaction buffer <R>
> Prepare in Step 4.

Sodium dodecyl sulfate (SDS) (1% in TBST) (optional, for acetylation and ubiquitylation assays)
Secondary antibodies for ubiquitylation and SUMOylation assays

Goat anti-mouse Alexa Fluor 555 (Life Technologies: A21422)
Goat anti-mouse Alexa Fluor 647 (Life Technologies: A21236)
Goat anti-rabbit Alexa Fluor 555 (Life Technologies: A21429)
Goat anti-rabbit Alexa Fluor 647 (Life Technologies: A21245)

Tris-buffered saline (TBS) <R>
Tris-buffered saline with 0.05% Tween 20 (TBST)

Equipment

Benchtop centrifuge (e.g., Heraeus Multifuge 3SR+; Thermo Scientific)
Equipment for acetylation and phosphorylation assay processing

Double-sided tape
Film cassette and Kodak BioMax MR Film (870 1302)
Image manipulation software (e.g., Photoshop)
Radiation hood/workbench
Tabletop scanner (with film capacity and scanning resolution up to 4800 dpi)

Humidity chamber
> To prepare a humidity chamber, place a folded paper towel in the bottom of an empty pipette tip box, add one inch of ddH₂O, and replace the tip holder on the box. The arrays will sit on the tip holder and be covered with the lid.

Incubator at 30°C (for phosphorylation and SUMOylation assays)
LifterSlip glass coverslips (Fisher 22-035-809)
Microarray analysis software (GenePix Pro 6.0; MDS Analytical Technologies)
Microarray scanner (GenePix 4000B; MDS Analytical Technologies) (for ubiquitylation and SUMOylation assay processing)
Microcentrifuge tubes
Micro slide boxes (VWR 48444-004) with laboratory tissues in the bottom

Orbital shaker
Pipette tip box (empty) (for ubiquitylation and SUMOylation assays)
Rectangular dishes (four-well) (Nunc 267061)
Slide washing rack (400–500 mL volume) (for phosphorylation and acetylation assays)
Syringe (1-mL) (Becton Dickenson 309625) and syringe-driven filter unit (0.22-μm) (Millex SLGV004SL) (for phosphorylation assay)

METHOD

The following method is a general protocol for performing posttranslational modifications (PTMs) on a functional protein microarray. Assay-specific steps are indicated where applicable.

Blocking the Arrays

1. Quickly dunk the proteome microarrays in a beaker containing 300 mL of TBS to rinse.
2. Transfer each rinsed array into one well of a four-well dish containing 3 mL of the appropriate blocking buffer.
3. Incubate the plate in blocking buffer according to assay.
   - For the phosphorylation, acetylation, and ubiquitylation assays, incubate the plate for 1 h at room temperature with gentle shaking.
   - For the SUMOylation assay, incubate the plate overnight at 4°C with gentle shaking.

Performing the PTM Assays

Preparing the Assay Reaction Mix

4. During blocking (Step 3), prepare a reaction buffer for the assay of interest (phosphorylation, acetylation, ubiquitylation, or SUMOylation) as follows.

   Kinase Reaction Buffer

   i. Prepare 100 μL of kinase buffer per array.
   ii. Centrifuge the buffer at top speed for 10 min at 4°C.
   iii. Aspirate the buffer into a 1-mL syringe, carefully remove the needle, and place a 0.22-μm filter onto the bottom of the syringe. CAREFULLY and slowly filter the mix into a fresh microcentrifuge tube. Keep the mixture on ice.
   iv. Immediately before the end of blocking (Step 3), add 20 μL of kinase to 100 μL of reaction buffer.

   Each array should contain around 200 U of kinase. The activity of commercially purchased kinases varies, and is typically measured in U/mL. Kinase should be diluted accordingly in TBS to a volume of 20 μL before it is added to the reaction mix.

   Acetylation Reaction Buffer

   i. Prepare 100 μL per array of acetylation reaction buffer without HAT, ^{14}C-AcCoA, DTT, sodium butyrate, and trichostatin A (TSA). Keep them on ice.
   ii. Prepare an additional 3 mL per array of acetylation reaction buffer without HAT (histone acetyltransferase), ^{14}C-AcCoA, DTT (dithiothreitol), sodium butyrate, and TSA for Step iii.
   iii. Equilibrate the arrays in 3 mL of buffer from Step ii in a four-well dish for 5 min.
   iv. Immediately before the acetylation assay (Step 6), add HAT, ^{14}C-AcCoA, DTT, sodium butyrate, and TSA to the buffer from Step i and mix well.

Ubiquitylation Reaction Buffer

i. Prepare 100 μL per array of ubiquitylation reaction buffer without ATP, DTT, E1, E2, E3, and ubiquitin. Keep on ice.

ii. Prepare an additional 9 mL per array of ubiquitylation reaction buffer without ATP, DTT, E1, E2, E3, and ubiquitin for Steps iii–iv.

iii. Equilibrate the arrays in 3 mL of the buffer from Step ii in a four-well dish for 5 min.

iv. Wash the arrays twice more in 3 mL of buffer from Step ii for 5 min per wash.

v. Immediately before the ubiquitylation assay (Step 6), add ATP, DTT, E1, E2, E3, and ubiquitin to reaction mix prepared in Step i.

SUMOylation Reaction Buffer

i. Prepare 200 μL per array of the SUMOylation reaction buffer without E1, E2, and E3. Keep on ice.

ii. Add E1, E2, and E3 (if using) immediately before the end of blocking (Step 3).

Performing the PTM Reactions

5. Remove the arrays from the blocking buffer and carefully wick off any liquid by tapping the edges on a paper towel. Place the arrays in a humidity chamber.

6. Carefully add the appropriate volume of the prepared reaction buffer to each slide.
   - For the phosphorylation, acetylation, and ubiquitylation assays, add 100 μL of reaction buffer.
   - For the SUMOylation assay, add 200 μL of reaction buffer.

7. Place a LifterSlip on top of each slide, being careful to avoid bubbles. Incubate the arrays at the specified time and temperature.
   - For the phosphorylation assay, incubate for 30 min at 30°C.
   - For the acetylation and ubiquitylation assays, incubate for 1 h at room temperature.
   - For the SUMOylation assay, incubate for 60–90 min at 30°C (depending on enzyme activity).

Washing the Arrays

All wash steps in this section should be performed at room temperature with gentle shaking.

8. When the incubation is complete, quickly dunk the arrays in a 500-mL beaker containing 400 mL of TBST.

   Coverslips should fall off upon dunking.

9. Dunk the arrays in a second beaker containing 400 mL of TBST.

10. Wash the arrays as follows.
    - For the phosphorylation assay, place the arrays in a slide washing rack and wash three times for 10 min per wash, using enough TBST to completely cover the arrays.
    - For the acetylation assay, place the arrays in a slide washing rack and proceed to Step 11.
    - For the ubiquitylation and SUMOylation assays, place the arrays in an empty pipette tip box and wash three times for 10 min per wash, using enough TBST to completely cover the arrays.
      An empty pipette tip box should be used for all washes in the ubiquitylation and SUMOylation assays from here forward.

11. Perform a harsh wash of the arrays as follows.

Phosphorylation Assay Harsh Wash

i. Wash three times in 0.5% SDS in ddH$_2$O for 10 min per wash.

Acetylation Assay Harsh Wash

i. Wash three times with Wash Buffer I for 15 min per wash.

ii. [Optional] Wash three times with 1% SDS (sodium dodecyl sulfate) in TBST for 15 min per wash.

iii. Wash three times with TBST for 15 min per wash.

Ubiquitylation Assay Harsh Wash

i. Wash three times with TBST for 5 min per wash.

ii. [Optional] Wash three times with 1% SDS in TBST for 5 min per wash.

SUMOylation Assay Harsh Wash

i. Wash three times with TBST for 10 min per wash.

ii. Wash three times for 5 min per wash with 1% SDS in ddH$_2$O warmed to 55°C.

iii. Wash once with ddH$_2$O.

Detecting and Processing Assay Results

For detection and processing of phosphorylation and acetylation reactions, proceed with Step 12. For ubiquitylation and SUMOylation, skip to Step 20.

Phosphorylation and Acetylation Assays

12. Rinse the arrays quickly with ddH$_2$O prewarmed to 37°C.

13. Place each array horizontally into a micro slide box with laboratory tissues on the bottom. Centrifuge the box in a benchtop centrifuge for 3 min at 1000g.

14. Arrange the arrays on piece of paper using double-sided tape to adhere at the barcode.

15. Expose the arrays to Kodak MR film for both 1 d and 5 d for γ-^{32}P and at least 2 wk for ^{14}C. Develop film.

    Film should be placed "shiny side up" for best results. Place at −80°C for longer exposures. Longer exposures may be necessary if signal is low.

    See Troubleshooting.

16. After film development, scan the film at 4800 dpi using dimensions of 2.50″ × 1.00″ in 16-bit grayscale.

17. In Photoshop (or a similar program), rotate the image until it is vertical, using positive controls to align.

    This can be performed in the initial scan, but is easier to do once the image is scanned in.

18. Decrease the size of the image to 53.3% of the original size, invert the image, and save as a TIF file.

    The image is now compatible for use with microarray analysis software such as GenePix.

19. Refer to Characterization of Protein–DNA Interactions Using Protein Microarrays (Hu et al. 2011) for analysis.

Ubiquitylation and SUMOylation Assays

20. Mix the appropriate primary antibodies in blocking buffer at a dilution of 1:1000.

- For the ubiquitylation assay, use rabbit anti-GST and mouse anti-FLAG antibodies.
- For the SUMOylation assay, use mouse anti-GST and rabbit anti-SUMO antibodies.

21. Apply 200 µL of the primary antibody mixture to each array and cover with a LifterSlip.
22. Place the arrays in a humidity chamber and incubate for 1 h at room temperature.
23. Wash the slides three times in 1 in. of TBST in a clean pipette tip box for 10 min per wash.
24. Mix the appropriate fluorescent secondary antibodies in blocking buffer at a dilution of 1:5000.
25. Apply 200 µL of the antibody mixture to each array and cover with a LifterSlip.
26. Place the arrays in a humidity chamber and incubate for 1 h at room temperature.
27. Wash the slides as described in Step 23.
28. Wash the slides once in 1 in. of ddH$_2$O for 5 min to remove residual salts from the surface of the chip.
29. Place each slide horizontally into a micro slide box with laboratory tissues on the bottom. Centrifuge the box in a benchtop centrifuge for 3 min at 1000g.
30. Scan the microarray with a GenePix 4000B scanner with 5-µm resolution detection at 547 or 635 nm, depending on the secondary antibody. Ensure that no spots appear as saturated signals (median pixel intensity = 65,536). Save the scanned images as TIF files.

    All slides that will be compared should be scanned using the same photomultiplier tube (PMT). See Troubleshooting.

31. Refer to Characterization of Protein–DNA Interactions Using Protein Microarrays (Hu et al. 2011) for analysis.

TROUBLESHOOTING

Problem (Step 15): In the phosphorylation assay, spots cannot be resolved.
Solution: If the array density is high (i.e., if spots are closer together than 250 µm), using lower energy ATP-γ-^{33}P will ensure that individual spots can be resolved.

Problem (Step 15): In the phosphorylation assay, the number of spots is very low/too high.
Solution: Adjust the exposure time or increase/decrease the kinase amount.

Problem (Step 30): In the SUMOylation assay, the SUMO modification signal is weak or undetectable.
Solution: Using a well-characterized substrate (such as RanGAP), perform SUMOylation reactions with increasing amounts of enzyme (e.g., twofold up to 25-fold). Increasing the ratio of E2 to E1 may also increase the efficiency of your reaction. If the activity of the enzymes has been confirmed, yet the detection of the signal is weak, confirm that the primary antibody recognizes SUMO by western blot.

RECIPES

Acetylation Reaction Buffer (100 µL)

HAT reaction buffer (5×) <R>	20 µL
Sodium butyrate (1 M)	1 µL
Dithiothreitol (DTT) (100 mM)	1 µL
^{14}C-Acetyl CoA (50 µCi/mL) (Perkin Elmer; 333.3 mM)	0.5 µL
Trichostatin A (TSA) (8.3 mM)	1.2 µL
Purified recombinant HAT enzyme	~20 µg

Bring to 100 µL with ddH$_2$O.

Blocking Buffer (Acetylation Assay)

3% bovine serum albumin (BSA) fraction V, heat shock (Roche 03116999001)
0.05% Tween 20

Prepare in phosphate-buffered saline (PBS) (pH 7.4) <R>.

Blocking Buffer (Phosphorylation Assay)

1% bovine serum albumin (BSA) fraction V, heat shock (Roche 03116999001)
Prepare in Tris-buffered saline (TBS) (pH 7.4) <R>.

Blocking Buffer (Ubiquitylation/SUMOylation Assays)

2% Bovine serum albumin (BSA) fraction V, heat shock (Roche 03116999001)
0.05% Tween 20

Prepare in Tris-buffered saline (TBS) (pH 7.4) <R>.

HAT Reaction Buffer (5×)

250 mM Tris–HCl (pH 7.5)
25% glycerol
5 mM EDTA
250 mM KCl
5 mM PMSF

Prepare a stock solution containing the first four ingredients. Store for up to 1 yr at −80°C. Add PMSF immediately before use.

Kinase Reaction Buffer

50 mM Tris–HCl, pH 7.5
100 mM NaCl
10 mM $MgCl_2$
1 mM $MnCl_2$
1 mM dithiothreitol (DTT)
1 mM EGTA
25 mM HEPES–KOH, pH 7.5
1 mM $NaVO_4$
1 mM NaF
0.1% NP-40
0.0000556 mM ATP-γ-^{32}P (Perkin Elmer BLU002250UC) (4.17 µL/array)

Phosphate-Buffered Saline (PBS)

Reagent	Amount to add (for 1× solution)	Final concentration (1×)	Amount to add (for 10× stock)	Final concentration (10×)
NaCl	8 g	137 mM	80 g	1.37 M
KCl	0.2 g	2.7 mM	2 g	27 mM
Na_2HPO_4	1.44 g	10 mM	14.4 g	100 mM
KH_2PO_4	0.24 g	1.8 mM	2.4 g	18 mM

If necessary, PBS may be supplemented with the following:

$CaCl_2$·$2H_2O$	0.133 g	1 mM	1.33 g	10 mM
$MgCl_2$·$6H_2O$	0.10 g	0.5 mM	1.0 g	5 mM

PBS can be made as a 1× solution or as a 10× stock. To prepare 1 L of either 1× or 10× PBS, dissolve the reagents listed above in 800 mL of H_2O. Adjust the pH to 7.4 (or 7.2, if required) with HCl, and then add H_2O to 1 L. Dispense the solution into aliquots and sterilize them by autoclaving for 20 min at 15 psi (1.05 kg/cm^2) on liquid cycle or by filter sterilization. Store PBS at room temperature.

SUMOylation Buffer (2×)

100 mM HEPES (pH 7.4)
200 mM NaCl
20 mM MgCl$_2$
0.2 mM dithiothreitol (DTT)

Prepare a stock solution containing HEPES, NaCl, and MgCl$_2$. Store for up to 1 yr at −80°C. Add DTT immediately before use.

SUMOylation Reaction Buffer (200 µL)

SUMOylation buffer (2×) <R>	100 µL
SUMO E1 enzyme (SEA1/UBA2) (Boston Biochem E-311 [yeast]; E-315 [human])	4 µg
SUMO E2 enzyme (UBE2I/UBC9) (Boston Biochem E2-645)	2 µg
SUMO E3 ligase (optional)	0.5 µg
Mature SUMO1 (Boston Biochem UL-712) and/or mature SUMO2 (Boston Biochem UL-752) (according to assay)	1 µg
ATP (100 mM) (GE Healthcare 27-2056-01)	10 µL

Dilute to 200 µL with ddH$_2$O. Note that certain E1 and E2 enzymes prefer either SUMO1 or SUMO2/3, whereas other enzymes have no preference.

Tris-Buffered Saline (TBS) (1×)

50 mM Tris-Cl, pH 7.5
150 mM NaCl

To prepare, dissolve 6.05 g Tris and 8.76 g NaCl in 800 mL of H$_2$O. Adjust pH to 7.5 with 1 M HCl and make volume up to 1 L with H$_2$O. TBS is stable for 3 mo at 4°C.

Ubiquitylation Reaction Buffer

25 mM Tris–HCl, pH 7.6
50 mM NaCl
10 mM MgCl$_2$
4 mM ATP (GE Healthcare 27-2056-01)
0.5 mM dithiothreitol (DTT)
100 nM ubiquitin E1 enzyme (UBE1) (Boston Biochem E-300)
500 nM ubiquitin E2 enzyme (Boston Biochem; choice dependent on E3)
500 nM ubiquitin E3 enzyme (according to array)
4 µg FLAG-ubiquitin (Boston Biochem U-120)

Wash Buffer I (Acetylation Assay)

50 mM Na$_2$CO$_3$–NaHCO$_3$, pH 9.3

Dissolve 21.2 g of Na$_2$CO$_3$ and 16.8 g of NaHCO$_3$ in 1 L of ddH$_2$O and at pH 9.3.

REFERENCES

Hu S, Xie Z, Blackshaw S, Qian J, Zhu H. 2011. Characterization of protein–DNA interactions using protein microarrays. *Cold Spring Harb Protoc* doi: 10.1101/pdb.prot5614.

CHAPTER 24

Synthetic Genetic Arrays: Automation of Yeast Genetics

Elena Kuzmin,[1,2] Michael Costanzo,[1,2] Brenda Andrews,[1,2,3] and Charles Boone[1,2,3]

[1]*Department of Molecular Genetics, University of Toronto, Ontario M5S 1A8, Canada;* [2]*Donnelly Centre for Cellular and Biomolecular Research, Toronto, Ontario M5S 3E1, Canada*

Genome-sequencing efforts have led to great strides in the annotation of protein-coding genes and other genomic elements. The current challenge is to understand the functional role of each gene and how genes work together to modulate cellular processes. Genetic interactions define phenotypic relationships between genes and reveal the functional organization of a cell. Synthetic genetic array (SGA) methodology automates yeast genetics and enables large-scale and systematic mapping of genetic interaction networks in the budding yeast, *Saccharomyces cerevisiae*. SGA facilitates construction of an output array of double mutants from an input array of single mutants through a series of replica pinning steps. Subsequent analysis of genetic interactions from SGA-derived mutants relies on accurate quantification of colony size, which serves as a proxy for fitness. Since its development, SGA has given rise to a variety of other experimental approaches for functional profiling of the yeast genome and has been applied in a multitude of other contexts, such as genome-wide screens for synthetic dosage lethality and integration with high-content screening for systematic assessment of morphology defects. SGA-like strategies can also be implemented similarly in a number of other cell types and organisms, including *Schizosaccharomyces pombe, Escherichia coli, Caenorhabditis elegans*, and human cancer cell lines. The genetic networks emerging from these studies not only generate functional wiring diagrams but may also play a key role in our understanding of the complex relationship between genotype and phenotype.

INTRODUCTION

Genome-sequencing projects have assembled a massive catalogue of genes and revealed extensive genetic diversity in a variety of organisms. The mapping of genetic interactions on a large scale in model organisms such as the budding yeast, *Saccharomyces cerevisiae*, provides a powerful approach for deciphering the functional organization of a cell and tracing the principles governing genetic buffering and the genotype-to-phenotype relationship (Hartman et al. 2001). A genetic interaction is defined as an unexpected phenotype that cannot be explained given the effects of the individual mutations (Mani et al. 2008; Dixon et al. 2009). Based on this definition, genetic interactions can be divided into two broad categories, referred to as negative and positive interactions. In the case of cell fitness, a negative genetic interaction is detected when a double mutant is less fit than expected. An extreme example of a negative genetic interaction is synthetic lethality, which occurs when two different deletion mutants, each of which fails to alter cellular fitness as a single mutant, combine to lead to a lethal phenotype (Novick and Botstein 1985). Such interactions are of particular importance because they identify genes that impinge on the same essential process. On the other hand, a positive interaction refers to a double mutant that is more fit than expected. There are a number of different types of positive genetic

[3]Correspondence: brenda.andrews@utoronto.ca; charlie.boone@utoronto.ca

Copyright © Cold Spring Harbor Laboratory Press; all rights reserved
Cite this introduction as *Cold Spring Harb Protoc*; doi:10.1101/pdb.top086652

interactions. For example, they can result from genetic suppression, or they may occur within the same nonessential pathway or complex. It is important to note that the vast majority of both positive and negative interactions occur between pathways and protein complexes; thus, they reveal broad phenotypic connections between functional modules and biological processes (Baryshnikova et al. 2010; Costanzo et al. 2010; Bellay et al. 2011).

THE SYNTHETIC GENETIC ARRAY (SGA) SYSTEM

Mutant Strain Collections

High-throughput genetic interaction screens were made possible by the construction of the yeast deletion mutant collection, in which every yeast open-reading frame (ORF) was replaced with an antibiotic-selectable marker flanked by unique 20-bp sequence identifiers, which served as molecular barcodes (Winzeler et al. 1999; Giaever et al. 2002). Each gene was deleted within a diploid strain and tested for viability in the haploid meiotic progeny, resulting in identification of approximately 1000 essential genes and approximately 5000 viable deletion mutants (Giaever et al. 2002). Strain-specific barcodes are particularly useful for the application of microarray- or sequencing-based measurement of the relative abundance of thousands of different mutants grown competitively in heterogeneous pools (Hillenmeyer et al. 2008; Douglas et al. 2012). A number of different libraries have been constructed for detailed phenotypic and SGA analysis of essential gene mutants, including libraries of conditional temperature-sensitive (Ben-Aroya et al. 2008; Li et al. 2011), tetracycline-repressible (Mnaimneh et al. 2004), and hypomorphic DAmP (Schuldiner et al. 2005) alleles.

SGA Analysis

The availability of arrayed mutant libraries has made yeast genetic analysis amenable to automation. To this end, our group developed the SGA system, a high-throughput method for constructing yeast haploid double mutants and identifying genetic interactions (Tong et al. 2001); see Protocol: Synthetic Genetic Array Analysis (Kuzmin et al. 2015). The SGA methodology enables large-scale mating and meiotic recombination via a series of replica pinning steps of high-density ordered arrays of yeast nonessential deletion mutants or conditional alleles of essential genes. The resulting output array of haploid double mutants can be subsequently probed for a multitude of phenotypes, the simplest of which is colony growth, which is a proxy for fitness.

SGA has been applied on a genome-wide scale, enabling quantitative analysis of approximately 5 million double mutants to identify approximately 170,000 genetic interactions (Costanzo et al. 2010). The resulting global genetic network revealed functions for previously unknown genes, determined pathway and complex membership, and provided an initial glimpse into the functional organization of a eukaryotic cell by highlighting broad connections between diverse pathways and functional modules in the cell (Tong et al. 2004; Costanzo et al. 2010). In addition, these studies emphasized a largely orthogonal relationship between protein–protein and genetic interaction networks. SGA analysis has also been applied to several biased subsets of functionally related genes to map interactions and genetic networks underlying specific biological processes; this is referred to as an E-MAP (Schuldiner et al. 2005; Collins et al. 2007; Wilmes et al. 2008; Fiedler et al. 2009; Aguilar et al. 2010; Zheng et al. 2010).

Alternative Methods

Yeast genetic interactions have also been mapped using other methods. For example, synthetic lethality analysis by microarrays (SLAM) (Ooi et al. 2003) and diploid-based SLAM (dSLAM) (Pan et al. 2004) generate mutants by directly disrupting a query gene via integrative transformation of a haploid or heterozygous diploid pool of deletion mutants, respectively. The latter method takes advantage of a version of the heterozygous yeast deletion collection, which has been modified to contain the SGA reporter in each strain, enabling the selection of haploid double-mutant progeny.

The abundance of each mutant in a pool is measured using barcode microarrays, and genetic interactions are identified by comparing the ratio of double to single mutants. Another related approach is genetic interaction mapping (GIM), which introduces a query mutation into a liquid pool of mutants by mating, and subsequently uses SGA-based steps to select haploid double mutants and measure genetic interactions by microarrays (Decourty et al. 2008). SGA-based methodological developments are depicted in Figure 1.

Quantifying Genetic Interactions

Early studies identified genetic interactions qualitatively, by visual inspection of the colony size of double and corresponding single mutants (Tong et al. 2001). However, high-throughput approaches demand automated and quantitative methods to measure the observed single- and double-mutant phenotypes and estimate expected outcomes. For yeast fitness, the expected double-mutant colony size can be modeled as a multiplicative combination of the single mutants, and a genetic interaction is then a deviation from this expectation. Several methods have been developed to quantify genetic interactions from mutant colony sizes derived from large-scale studies, including calculation of the SGA score (Baryshnikova et al. 2010) and the S score (Collins et al. 2006, 2010). The differential S (dS) score is a newly developed strategy for scoring condition-specific genetic interactions (Bean and Ideker 2012). A MATLAB toolbox has also been devised for image analysis of ultra-high-density arrays (6144 vs. standard 1536 colonies/plate), reducing the cost and increasing the throughput of such functional genomic experiments (Bean et al. 2014).

The aforementioned scoring methods were primarily designed for analysis of large-scale data sets and are not easily adapted for smaller-scale studies. As a result, tools have been developed to facilitate analysis of genetic interactions from individual SGA screens. For instance, "SGAtools" provides a web-based interactive interface that integrates all of the steps necessary for genetic interaction analysis, allowing a user to measure mutant colony sizes, correct for systematic biases, and quantify genetic interactions (Wagih et al. 2013; Wagih and Parts 2014). "Balony" operates similarly to SGA tools by incorporating image scanning into the pipeline, but it uses static gridding rather than segmenting each colony separately (Young and Loewen 2013). Other methods such as "ScreenMill" (Dittmar et al. 2010), "Colonyzer" (Lawless et al. 2010), and "GrowthDetector" (Memarian et al. 2007) are also available but are not accessible through a web-based interface.

Versatility of SGA

In its most general form, SGA is a method for combining marked genetic elements in a high-throughput manner and, thus, is amenable for use in a variety of contexts (Fig. 1). Synthetic dosage lethality (SDL) is one such application of SGA (Sopko et al. 2006). An SDL interaction is observed when the overexpression of one gene combined with the deletion of another gene results in a lethal phenotype (Kroll et al. 1996). Genome-wide SDL screens use SGA to introduce a deletion mutation or a conditional allele of an essential gene into an array of yeast strains, each of which carries a different overexpression plasmid (Sopko et al. 2006). This approach has been applied successfully to uncover substrates for diverse regulatory enzymes, such as kinases, phosphatases, histone deacetylases, or ubiquitin ligases, in standard as well as stress conditions (Douglas et al. 2012; Kaluarachchi Duffy et al. 2012; Sharifpoor et al. 2012; Bian et al. 2014).

SGA can also be combined with high-content screening (HCS) to introduce genes encoding fluorescently marked proteins into the deletion array, enabling quantitative genome-wide screens for defects in pathway-specific reporters and subcellular morphology. To date, an integrated SGA-HCS pipeline has been used to quantitatively assess mitotic spindle defects (Vizeacoumar et al. 2010), identify regulators of nucleolar size (Neumuller et al. 2013), and characterize pathways that respond to DNA replication stress (Tkach et al. 2012). In theory, virtually any pathway can be analogously quantitatively examined within the context of genetic and environmental perturbations given an appropriate fluorescent or cytological marker.

Chapter 24

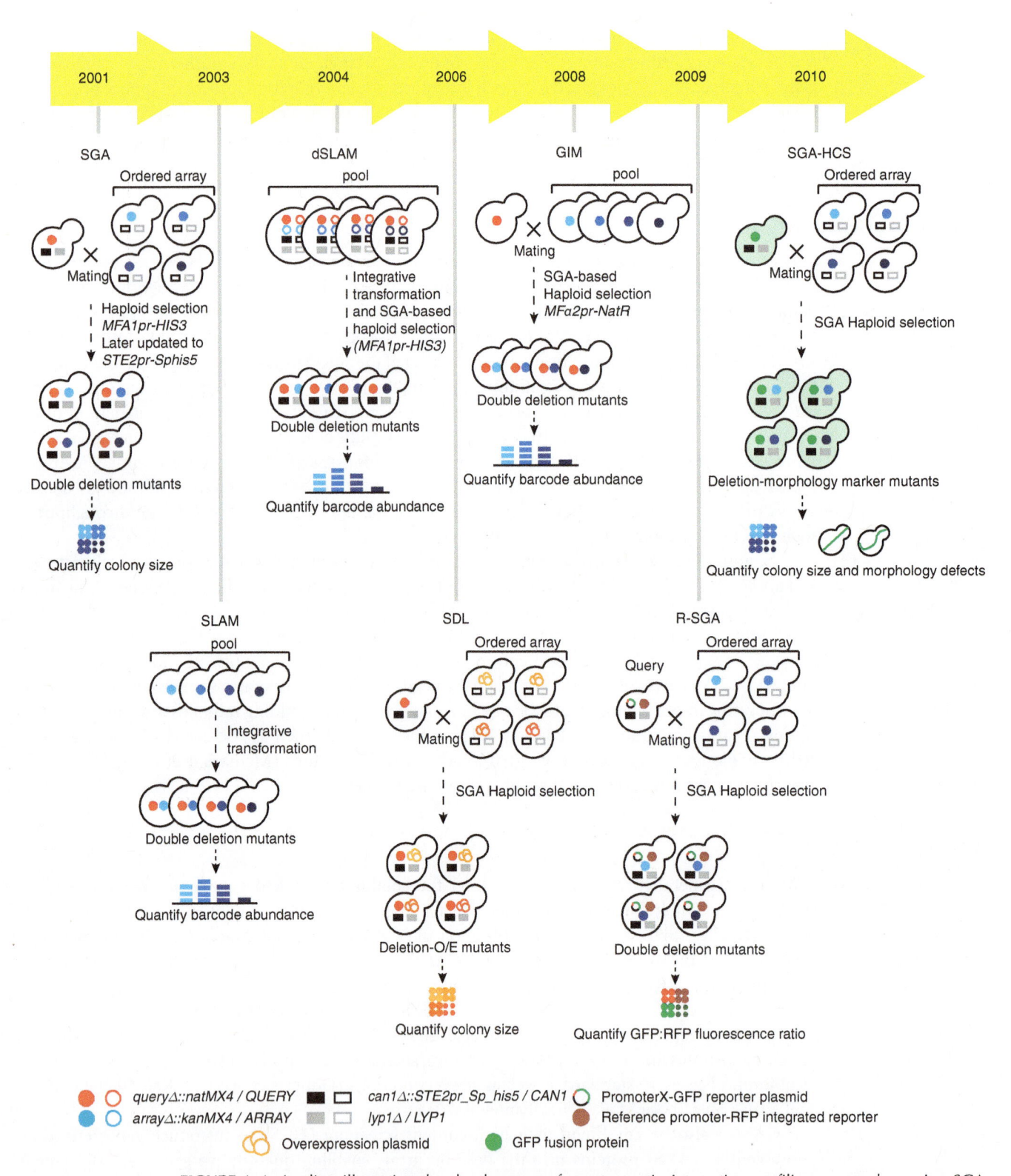

FIGURE 1. A timeline illustrating the development of yeast genetic interaction profiling approaches using SGA and SGA-derived methods. Original articles describing the experiments depicted here are referenced in the main text.

Additional applications of SGA include Reporter-SGA (R-SGA), which assesses the consequence of genomic perturbations on gene expression by using a query strain that carries a GFP reporter under the control of a promoter of interest and a reference promoter driving RFP (Fillingham et al. 2009), and SGA Mapping (SGAM), which enables high-resolution mapping of spontaneous genetic suppressor mutations (Costanzo and Boone 2009) as well as global analysis of meiotic recombination (Baryshnikova et al. 2013). In addition, combining more than two mutant alleles facilitates the interrogation of complex functional redundancies (Tong et al. 2004; Zou et al. 2009; Haber et al. 2013). Finally, assessing genetic interactions of various point mutants can elucidate the functions of specific residues and protein domains (Braberg et al. 2013).

CONCLUSIONS AND PERSPECTIVES

SGA is a powerful technique for large-scale construction of mutants and assessment of phenotypic consequences associated with combinatorial genetic perturbations. This introduction provides an overview of the systematic genetic interaction studies in yeast enabled by SGA (Fig. 1). The SGA method relies on robotic manipulation of high-density arrays of yeast mutants, an easily scored phenotype, such as colony size, and analysis tools to accurately quantify the effects of genetic perturbations. Genetic interaction studies in yeast have inspired the development of similar strategies in other organisms, such as the distantly related yeast species *Schizosaccharomyces pombe* (Dixon et al. 2008; Frost et al. 2012; Ryan et al. 2012), the prokaryotic species *E. coli* (Butland et al. 2008; Typas et al. 2008; Babu et al. 2011; Gagarinova and Emili 2012), the multicellular eukaryote *Caenorhabditis elegans* (Lehner et al. 2006; Tischler et al. 2006; Burga et al. 2011), and human cancer cell lines that allow probing of cancer "vulnerabilities" (Bassik et al. 2013; Laufer et al. 2013; Roguev et al. 2013; Vizeacoumar et al. 2013). In conclusion, global genetic interaction networks generated via the SGA methodology have proven to be instrumental in understanding the functional organization of a model eukaryotic cell and promise to help elucidate the genotype-to-phenotype relationship as well as the genetic mechanisms that underlie complex inheritance patterns of human disease (Zuk et al. 2012).

REFERENCES

Aguilar PS, Frohlich F, Rehman M, Shales M, Ulitsky I, Olivera-Couto A, Braberg H, Shamir R, Walter P, Mann M, et al. 2010. A plasma-membrane E-MAP reveals links of the eisosome with sphingolipid metabolism and endosomal trafficking. *Nat Struct Mol Biol* 17: 901–908.

Babu M, Diaz-Mejia JJ, Vlasblom J, Gagarinova A, Phanse S, Graham C, Yousif F, Ding H, Xiong X, Nazarians-Armavil A, et al. 2011. Genetic interaction maps in *Escherichia coli* reveal functional crosstalk among cell envelope biogenesis pathways. *PLoS Genet* 7: e1002377.

Baryshnikova A, Costanzo M, Kim Y, Ding H, Koh J, Toufighi K, Youn JY, Ou J, San Luis BJ, Bandyopadhyay S, et al. 2010. Quantitative analysis of fitness and genetic interactions in yeast on a genome scale. *Nat Methods* 7: 1017–1024.

Baryshnikova A, VanderSluis B, Costanzo M, Myers CL, Cha RS, Andrews B, Boone C. 2013. Global linkage map connects meiotic centromere function to chromosome size in budding yeast. *G3 (Bethesda)* 3: 1741–1751.

Bassik MC, Kampmann M, Lebbink RJ, Wang S, Hein MY, Poser I, Weibezahn J, Horlbeck MA, Chen S, Mann M, et al. 2013. A systematic mammalian genetic interaction map reveals pathways underlying ricin susceptibility. *Cell* 152: 909–922.

Bean GJ, Ideker T. 2012. Differential analysis of high-throughput quantitative genetic interaction data. *Genome Biol* 13: R123.

Bean GJ, Jaeger PA, Bahr S, Ideker T. 2014. Development of ultra-high-density screening tools for microbial "omics." *PLoS One* 9: e85177.

Bellay J, Atluri G, Sing TL, Toufighi K, Costanzo M, Ribeiro PS, Pandey G, Baller J, VanderSluis B, Michaut M, et al. 2011. Putting genetic interactions in context through a global modular decomposition. *Genome Res* 21: 1375–1387.

Ben-Aroya S, Coombes C, Kwok T, O'Donnell KA, Boeke JD, Hieter P. 2008. Toward a comprehensive temperature-sensitive mutant repository of the essential genes of *Saccharomyces cerevisiae*. *Mol Cell* 30: 248–258.

Bian Y, Kitagawa R, Bansal PK, Fujii Y, Stepanov A, Kitagawa K. 2014. Synthetic genetic array screen identifies PP2A as a therapeutic target in Mad2-overexpressing tumors. *Proc Natl Acad Sci* 111: 1628–1633.

Braberg H, Jin H, Moehle EA, Chan YA, Wang S, Shales M, Benschop JJ, Morris JH, Qiu C, Hu F, et al. 2013. From structure to systems: High-resolution, quantitative genetic analysis of RNA polymerase II. *Cell* 154: 775–788.

Burga A, Casanueva MO, Lehner B. 2011. Predicting mutation outcome from early stochastic variation in genetic interaction partners. *Nature* 480: 250–253.

Butland G, Babu M, Diaz-Mejia JJ, Bohdana F, Phanse S, Gold B, Yang W, Li J, Gagarinova AG, Pogoutse O, et al. 2008. eSGA: *E. coli* synthetic genetic array analysis. *Nat Methods* 5: 789–795.

Collins SR, Schuldiner M, Krogan NJ, Weissman JS. 2006. A strategy for extracting and analyzing large-scale quantitative epistatic interaction data. *Genome Biol* 7: R63.

Collins SR, Miller KM, Maas NL, Roguev A, Fillingham J, Chu CS, Schuldiner M, Gebbia M, Recht J, Shales M, et al. 2007. Functional dissection of protein complexes involved in yeast chromosome biology using a genetic interaction map. *Nature* 446: 806–810.

Chapter 24

Collins SR, Roguev A, Krogan NJ. 2010. Quantitative genetic interaction mapping using the E-MAP approach. *Methods Enzymol* **470**: 205–231.

Costanzo M, Boone C. 2009. SGAM: An array-based approach for high-resolution genetic mapping in *Saccharomyces cerevisiae*. *Methods Mol Biol* **548**: 37–53.

Costanzo M, Baryshnikova A, Bellay J, Kim Y, Spear ED, Sevier CS, Ding H, Koh JL, Toufighi K, Mostafavi S, et al. 2010. The genetic landscape of a cell. *Science* **327**: 425–431.

Decourty L, Saveanu C, Zemam K, Hantraye F, Frachon E, Rousselle JC, Fromont-Racine M, Jacquier A. 2008. Linking functionally related genes by sensitive and quantitative characterization of genetic interaction profiles. *Proc Natl Acad Sci* **105**: 5821–5826.

Dittmar JC, Reid RJ, Rothstein R. 2010. ScreenMill: A freely available software suite for growth measurement, analysis and visualization of high-throughput screen data. *BMC Bioinform* **11**: 353.

Dixon SJ, Fedyshyn Y, Koh JL, Prasad TS, Chahwan C, Chua G, Toufighi K, Baryshnikova A, Hayles J, Hoe KL, et al. 2008. Significant conservation of synthetic lethal genetic interaction networks between distantly related eukaryotes. *Proc Natl Acad Sci* **105**: 16653–16658.

Dixon SJ, Costanzo M, Baryshnikova A, Andrews B, Boone C. 2009. Systematic mapping of genetic interaction networks. *Annu Rev Genet* **43**: 601–625.

Douglas AC, Smith AM, Sharifpoor S, Yan Z, Durbic T, Heisler LE, Lee AY, Ryan O, Gottert H, Surendra A, et al. 2012. Functional analysis with a barcoder yeast gene overexpression system. *G3 (Bethesda)* **2**: 1279–1289.

Fiedler D, Braberg H, Mehta M, Chechik G, Cagney G, Mukherjee P, Silva AC, Shales M, Collins SR, van Wageningen S, et al. 2009. Functional organization of the *S. cerevisiae* phosphorylation network. *Cell* **136**: 952–963.

Fillingham J, Kainth P, Lambert JP, van Bakel H, Tsui K, Pena-Castillo L, Nislow C, Figeys D, Hughes TR, Greenblatt J, et al. 2009. Two-color cell array screen reveals interdependent roles for histone chaperones and a chromatin boundary regulator in histone gene repression. *Mol Cell* **35**: 340–351.

Frost A, Elgort MG, Brandman O, Ives C, Collins SR, Miller-Vedam L, Weibezahn J, Hein MY, Poser I, Mann M, et al. 2012. Functional repurposing revealed by comparing *S. pombe* and *S. cerevisiae* genetic interactions. *Cell* **149**: 1339–1352.

Gagarinova A, Emili A. 2012. Genome-scale genetic manipulation methods for exploring bacterial molecular biology. *Mol Biosyst* **8**: 1626–1638.

Giaever G, Chu AM, Ni L, Connelly C, Riles L, Veronneau S, Dow S, Lucau-Danila A, Anderson K, Andre B, et al. 2002. Functional profiling of the *Saccharomyces cerevisiae* genome. *Nature* **418**: 387–391.

Haber JE, Braberg H, Wu Q, Alexander R, Haase J, Ryan C, Lipkin-Moore Z, Franks-Skiba KE, Johnson T, Shales M, et al. 2013. Systematic triple-mutant analysis uncovers functional connectivity between pathways involved in chromosome regulation. *Cell Rep* **3**: 2168–2178.

Hartman JL 4th, Garvik B, Hartwell L. 2001. Principles for the buffering of genetic variation. *Science* **291**: 1001–1004.

Hillenmeyer ME, Fung E, Wildenhain J, Pierce SE, Hoon S, Lee W, Proctor M, St Onge RP, Tyers M, Koller D, et al. 2008. The chemical genomic portrait of yeast: Uncovering a phenotype for all genes. *Science* **320**: 362–365.

Kaluarachchi Duffy S, Friesen H, Baryshnikova A, Lambert JP, Chong YT, Figeys D, Andrews B. 2012. Exploring the yeast acetylome using functional genomics. *Cell* **149**: 936–948.

Kroll ES, Hyland KM, Hieter P, Li JJ. 1996. Establishing genetic interactions by a synthetic dosage lethality phenotype. *Genetics* **143**: 95–102.

Kuzmin E, Costanzo M, Andrews B, Boone C. 2015. Synthetic genetic array analysis. *Cold Spring Harb Protoc* doi: 10.1101/pdb.prot088807.

Laufer C, Fischer B, Billmann M, Huber W, Boutros M. 2013. Mapping genetic interactions in human cancer cells with RNAi and multiparametric phenotyping. *Nat Methods* **10**: 427–431.

Lawless C, Wilkinson DJ, Young A, Addinall SG, Lydall DA. 2010. Colonyzer: Automated quantification of micro-organism growth characteristics on solid agar. *BMC Bioinform* **11**: 287.

Lehner B, Tischler J, Fraser AG. 2006. RNAi screens in *Caenorhabditis elegans* in a 96-well liquid format and their application to the systematic identification of genetic interactions. *Nat Protocols* **1**: 1617–1620.

Li Z, Vizeacoumar FJ, Bahr S, Li J, Warringer J, Vizeacoumar FS, Min R, Vandersluis B, Bellay J, Devit M, et al. 2011. Systematic exploration of essential yeast gene function with temperature-sensitive mutants. *Nat Biotechnol* **29**: 361–367.

Mani R, St Onge RP, Hartman JL 4th, Giaever G, Roth FP. 2008. Defining genetic interaction. *Proc Natl Acad Sci* **105**: 3461–3466.

Memarian N, Jessulat M, Alirezaie J, Mir-Rashed N, Xu J, Zareie M, Smith M, Golshani A. 2007. Colony size measurement of the yeast gene deletion strains for functional genomics. *BMC Bioinform* **8**: 117.

Mnaimneh S, Davierwala AP, Haynes J, Moffat J, Peng WT, Zhang W, Yang X, Pootoolal J, Chua G, Lopez A, et al. 2004. Exploration of essential gene functions via titratable promoter alleles. *Cell* **118**: 31–44.

Neumuller RA, Gross T, Samsonova AA, Vinayagam A, Buckner M, Founk K, Hu Y, Sharifpoor S, Rosebrock AP, Andrews B, et al. 2013. Conserved regulators of nucleolar size revealed by global phenotypic analyses. *Science Signal* **6**: ra70.

Novick P, Botstein D. 1985. Phenotypic analysis of temperature-sensitive yeast actin mutants. *Cell* **40**: 405–416.

Ooi SL, Shoemaker DD, Boeke JD. 2003. DNA helicase gene interaction network defined using synthetic lethality analyzed by microarray. *Nat Genet* **35**: 277–286.

Pan X, Yuan DS, Xiang D, Wang X, Sookhai-Mahadeo S, Bader JS, Hieter P, Spencer F, Boeke JD. 2004. A robust toolkit for functional profiling of the yeast genome. *Mol Cell* **16**: 487–496.

Roguev A, Talbot D, Negri GL, Shales M, Cagney G, Bandyopadhyay S, Panning B, Krogan NJ. 2013. Quantitative genetic-interaction mapping in mammalian cells. *Nat Methods* **10**: 432–437.

Ryan CJ, Roguev A, Patrick K, Xu J, Jahari H, Tong Z, Beltrao P, Shales M, Qu H, Collins SR, et al. 2012. Hierarchical modularity and the evolution of genetic interactomes across species. *Mol Cell* **46**: 691–704.

Schuldiner M, Collins SR, Thompson NJ, Denic V, Bhamidipati A, Punna T, Ihmels J, Andrews B, Boone C, Greenblatt JF, et al. 2005. Exploration of the function and organization of the yeast early secretory pathway through an epistatic miniarray profile. *Cell* **123**: 507–519.

Sharifpoor S, van Dyk D, Costanzo M, Baryshnikova A, Friesen H, Douglas AC, Youn JY, VanderSluis B, Myers CL, Papp B, et al. 2012. Functional wiring of the yeast kinome revealed by global analysis of genetic network motifs. *Genome Res* **22**: 791–801.

Sopko R, Huang D, Preston N, Chua G, Papp B, Kafadar K, Snyder M, Oliver SG, Cyert M, Hughes TR, et al. 2006. Mapping pathways and phenotypes by systematic gene overexpression. *Mol Cell* **21**: 319–330.

Tischler J, Lehner B, Chen N, Fraser AG. 2006. Combinatorial RNA interference in *Caenorhabditis elegans* reveals that redundancy between gene duplicates can be maintained for more than 80 million years of evolution. *Genome Biol* **7**: R69.

Tkach JM, Yimit A, Lee AY, Riffle M, Costanzo M, Jaschob D, Hendry JA, Ou J, Moffat J, Boone C, et al. 2012. Dissecting DNA damage response pathways by analysing protein localization and abundance changes during DNA replication stress. *Nat Cell Biol* **14**: 966–976.

Tong AH, Evangelista M, Parsons AB, Xu H, Bader GD, Page N, Robinson M, Raghibizadeh S, Hogue CW, Bussey H, et al. 2001. Systematic genetic analysis with ordered arrays of yeast deletion mutants. *Science* **294**: 2364–2368.

Tong AH, Lesage G, Bader GD, Ding H, Xu H, Xin X, Young J, Berriz GF, Brost RL, Chang M, et al. 2004. Global mapping of the yeast genetic interaction network. *Science* **303**: 808–813.

Typas A, Nichols RJ, Siegele DA, Shales M, Collins SR, Lim B, Braberg H, Yamamoto N, Takeuchi R, Wanner BL, et al. 2008. High-throughput, quantitative analyses of genetic interactions in *E. coli*. *Nat Methods* **5**: 781–787.

Vizeacoumar FJ, van Dyk N, F SV, Cheung V, Li J, Sydorskyy Y, Case N, Li Z, Datti A, Nislow C, et al. 2010. Integrating high-throughput genetic interaction mapping and high-content screening to explore yeast spindle morphogenesis. *J Cell Biol* **188**: 69–81.

Vizeacoumar FJ, Arnold R, Vizeacoumar FS, Chandrashekhar M, Buzina A, Young JT, Kwan JH, Sayad A, Mero P, Lawo S, et al. 2013. A negative genetic interaction map in isogenic cancer cell lines reveals cancer cell vulnerabilities. *Mol Syst Biol* **9**: 696.

Wagih O, Parts L. 2014. gitter: A robust and accurate method for quantification of colony sizes from plate images. *G3 (Bethesda)* **4**: 547–552.

Wagih O, Usaj M, Baryshnikova A, VanderSluis B, Kuzmin E, Costanzo M, Myers CL, Andrews BJ, Boone CM, Parts L. 2013. SGAtools: One-stop analysis and visualization of array-based genetic interaction screens. *Nucleic Acids Res* **41:** W591–W596.

Wilmes GM, Bergkessel M, Bandyopadhyay S, Shales M, Braberg H, Cagney G, Collins SR, Whitworth GB, Kress TL, Weissman JS, et al. 2008. A genetic interaction map of RNA-processing factors reveals links between Sem1/Dss1-containing complexes and mRNA export and splicing. *Mol Cell* **32:** 735–746.

Winzeler EA, Shoemaker DD, Astromoff A, Liang H, Anderson K, Andre B, Bangham R, Benito R, Boeke JD, Bussey H, et al. 1999. Functional characterization of the *S. cerevisiae* genome by gene deletion and parallel analysis. *Science* **285:** 901–906.

Young BP, Loewen CJ. 2013. Balony: A software package for analysis of data generated by synthetic genetic array experiments. *BMC Bioinform* **14:** 354.

Zheng J, Benschop JJ, Shales M, Kemmeren P, Greenblatt J, Cagney G, Holstege F, Li H, Krogan NJ. 2010. Epistatic relationships reveal the functional organization of yeast transcription factors. *Mol Syst Biol* **6:** 420.

Zou J, Friesen H, Larson J, Huang D, Cox M, Tatchell K, Andrews B. 2009. Regulation of cell polarity through phosphorylation of Bni4 by Pho85 G1 cyclin-dependent kinases in *Saccharomyces cerevisiae*. *Mol Biol Cell* **20:** 3239–3250.

Zuk O, Hechter E, Sunyaev SR, Lander ES. 2012. The mystery of missing heritability: Genetic interactions create phantom heritability. *Proc Natl Acad Sci* **109:** 1193–1198.

Protocol 1

Synthetic Genetic Array Analysis

Elena Kuzmin,[1,2] Michael Costanzo,[1,2] Brenda Andrews,[1,2,3] and Charles Boone[1,2,3]

[1]Department of Molecular Genetics, University of Toronto, Ontario M5S 1A8, Canada; [2]Donnelly Centre for Cellular and Biomolecular Research, Toronto, Ontario M5S 3E1, Canada

Genetic interaction studies have been used to characterize unknown genes, assign membership in pathway and complex, and build a comprehensive functional map of a eukaryotic cell. Synthetic genetic array (SGA) methodology automates yeast genetic analysis and enables systematic mapping of genetic interactions. In its simplest form, SGA consists of a series of replica pinning steps that enable construction of haploid double mutants through automated mating and meiotic recombination. Using this method, a strain carrying a query mutation, such as a deletion allele of a nonessential gene or a conditional temperature-sensitive allele of an essential gene, can be crossed to an input array of yeast mutants, such as the complete set of approximately 5000 viable deletion mutants. The resulting output array of double mutants can be scored for genetic interactions based on estimates of cellular fitness derived from colony-size measurements. The SGA score method can be used to analyze large-scale data sets, whereas small-scale data sets can be analyzed using SGAtools, a simple web-based interface that includes all the necessary analysis steps for quantifying genetic interactions.

MATERIALS

It is essential that you consult the appropriate Material Safety Data Sheets and your institution's Environmental Health and Safety Office for proper handling of equipment and hazardous material used in this protocol.

RECIPES: Please see the end of this protocol for recipes indicated by <R>. Additional recipes can be found online at http://cshprotocols.cshlp.org/site/recipes.

Reagents

DNA template for temperature-sensitive [*ts*] allele of interest (for query strain construction using two-step polymerase chain reaction (PCR)-mediated integration)
Enriched sporulation medium <R>
This medium is supplemented with G418 to prevent contamination.

Glycerol
MATa haploid selection media
  SD – His,Arg,Lys (+ canavanine,thialysine) selection medium <R>
  SD$_{MSG}$ – His,Arg,Lys (+ canavanine,thialysine,G418 +/− clonNAT) selection medium <R>
  MSG instead of ammonium sulfate is used as a nitrogen source in this medium, because the latter interferes with the activity of the antibiotic. Supplement with 100 µg/mL of clonNAT for Step 42.

[3]Correspondence: brenda.andrews@utoronto.ca; charlie.boone@utoronto.ca

Copyright © Cold Spring Harbor Laboratory Press; all rights reserved
Cite this protocol as *Cold Spring Harb Protoc*; doi:10.1101/pdb.prot088807

*MAT*α haploid selection medium
> SD$_{MSG}$ − Leu,Arg,Lys (+ canavanine,thialysine,clonNAT) selection medium <R>

Oligonucleotide primers for query strain construction (when using PCR-mediated deletion)
> *Refer to Figure 1A for a schematic representation of the primer design.*

> Confirmation primers (forward and reverse) for gene of interest (YFG-F and -R)
>> *Design confirmation primers 400 bp upstream or 400 bp downstream from the gene of interest.*

> Confirmation primers for *natMX4* (Confirmation primer-F and -R; see Table 1)
> Gene-deletion primers for gene of interest
>> *Synthesize the following pair of gene-deletion primers: (1) an upstream primer consisting of the 55 bp immediately upstream (including the start codon) of the gene of interest and 22 bp of the 5' end of the natMX4 cassette (i.e., the MX4-F primer sequence, provided in Table 1); and (2) a downstream primer consisting of the 55 bp immediately downstream (including the stop codon) of the gene of interest and 22 bp of the 3' end of the natMX4 cassette (i.e., the MX4-R primer sequence, provided in Table 1).*

Oligonucleotide primers for query strain construction (when using the marker switching method)
> *Synthesize MX4-F and MX4-R (Table 1) for marker switching. Refer to Figure 1B for a schematic representation of primer location.*

Oligonucleotide primers for query strain construction (when using two-step PCR-mediated integration of a conditional temperature-sensitive [*ts*] allele of interest)
> *Synthesize the following two primer pairs: (1) gene-specific primers to amplify the conditional allele, including 200 bp downstream from its stop codon and an additional 25 bp of sequence complementary to the 5' end of the natMX4 cassette; and (2) primers to amplify the natMX4 cassette such that the reverse 3' primer contains 45 bp of complementary sequence immediately downstream from the gene of interest. Refer to Figure 1C for a schematic diagram of primer design.*

p4339 plasmid DNA (containing the *natMX4* cassette; see Table 1)
Reagents for PCR
> DMSO
> dNTPs (10 mM)
> *Taq* DNA polymerase (5 U/μL) and 10× PCR buffer

Reagents for standard LiAc transformation of yeast cells
Reagents for sterilization of manual pin tools (as needed)
> Bleach (10%)
> Ethanol (95%)

Reagents for sterilization of robotic pin tools (as needed)
> Ethanol (70%)
> H$_2$O from a bottle supply source (sterile, distilled)

Sporulation medium for query strain construction <R>
Yeast extract peptone dextrose (YEPD) (+/− G418 +/− clonNAT) solid medium <R>
> *Supplement YEPD with 200 μg/mL G418 (Steps 9 and 27), 100 μg/mL clonNAT (Steps 2, 8, 14, 15, and 34), or both (Step 38). In addition, prepare (1) YEPD liquid medium without agar (Step 35); and (2) YEPD solid medium in OmniTrays (Nunc) (Step 36).*

Yeast strains (as needed, according to experiment; see Table 1); for example:
> Y7092 (SGA background strain for query strain construction using PCR-mediated gene deletion)
> Y8205 (SGA background strain for query strain construction using the marker switching method)
> Y8835 (wild-type control strain for SGA)
> Yeast deletion collection (*MAT*a deletion strain of interest, for the marker switching method)
> Yeast deletion mutant array for SGA
>> *Construction of a 1536-format deletion mutant array (DMA) is described in Steps 26–32. Frozen glycerol stocks are available through Euroscarf (http://web.uni-frankfurt.de/fb15/mikro/euroscarf/col_index.html).*

Chapter 24

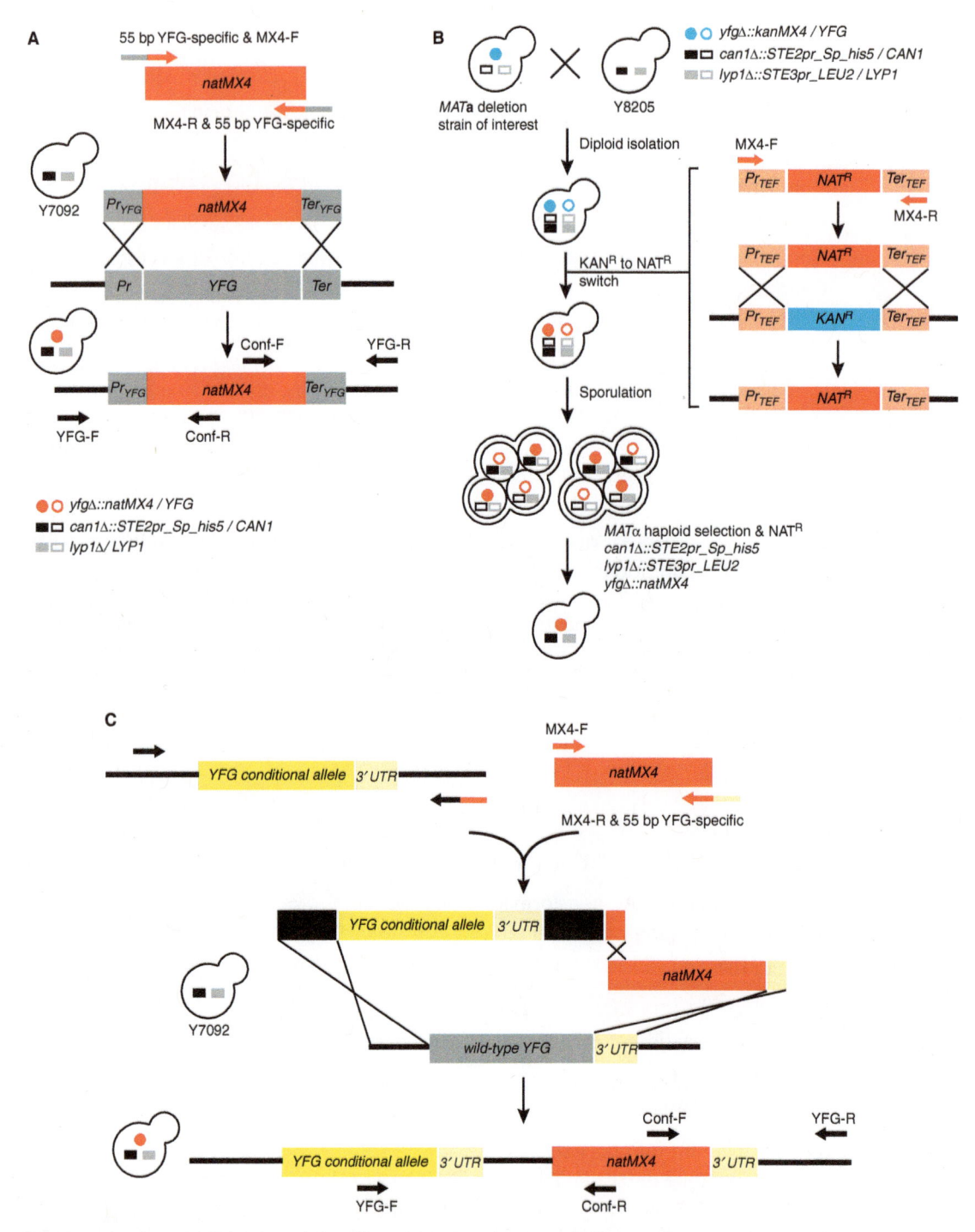

FIGURE 1. SGA query strain construction. (*A*) PCR-mediated deletion of nonessential genes. The *natMX4* cassette is PCR-amplified using a pair of gene-deletion primers that anneal to the cassette as well as carry regions of homology with the gene of interest. Yeast strain Y7092 is then transformed with the resulting PCR product and transformants are selected. The correct integration is confirmed by PCR using primers that anneal within the selectable marker and either upstream or downstream from the deleted gene of interest. (*B*) Marker switching method for nonessential genes. A deletion strain of interest from the *MAT*a deletion collection is crossed to the yeast strain Y8205 and diploid zygotes are isolated. Then, the *kanMX4* marker is switched to *natMX4* by PCR amplification of the *natMX4* cassette, transformation into the diploid strain, and selection for transformants. The resulting diploids, heterozygous for SGA markers, are sporulated, and the desired *MAT*α meiotic progeny resistant to clonNAT are germinated. (*C*) Two-step PCR-mediated integration to introduce a conditional allele of an essential gene into the SGA query strain background. A conditional allele of interest is PCR-amplified using oligonucleotide primers that anneal to the gene of interest and contain a region of homology with the *natMX4* cassette; separately, the *natMX4* cassette is PCR-amplified using primers that carry a region of homology with the 3′ UTR of the gene of interest. The PCR products are cotransformed into Y7092 and successful transformants are selected and confirmed for the *ts* phenotype, and confirmed by PCR. (The illustration was adapted from Li et al. 2011 with permission from Macmillan Publishers Ltd.)

TABLE 1. Genetic information for SGA reagents: yeast strains, plasmids, and primers

Name	Genetic information	Comments
Y7092	MATα can1Δ::STE2pr-Sp_his5 lyp1Δ ura3Δ0 leu2Δ0 his3Δ1 met15Δ0	Background strain for SGA query strains constructed by PCR-mediated gene deletion
SGA query strains (Y7092 background)	MATα yfgΔ::natMX4 can1Δ::STE2pr-Sp_his5 lyp1Δ ura3Δ0 leu2Δ0 his3Δ1 met15Δ0	Genotype of the majority of SGA query strains; yfg is your favorite gene
Y7091	MATa can1Δ::STE2pr-Sp_his5 lyp1Δ ura3Δ0 leu2Δ0 his3Δ1 met15Δ0	MATa version of Y7092
Y8205	MATα can1Δ::STE2pr-Sp_his5 lyp1Δ::STE3pr_LEU2 ura3Δ0 leu2Δ0 his3Δ1 met15Δ0	Background strain for SGA query strains constructed by the marker switching method
SGA query strains (Y8205 background)	MATα yfgΔ::natMX4 can1Δ::STE2pr-Sp_his5 lyp1Δ::STE3pr_LEU2 ura3Δ0 leu2Δ0 his3Δ1 met15Δ0	Genotype of some SGA query strains; yfg is your favorite gene
Y8835	MATα can1Δ::STE2pr-Sp_his5 lyp1Δ ura3Δ0::natMX4 leu2Δ0 his3Δ1 met15Δ0	Wild-type control strain, used for estimating single mutant fitness of MATa kanMX4 deletion array
DMA809	MATa hoΔ::kanMX4 his3Δ1 leu2Δ0 met15Δ0 ura3Δ0	Wild-type control strain, used for estimating single-mutant fitness of MATα natMX4 query strain array
Yeast Deletion Collection	MATa his3Δ1 leu2Δ0 met15Δ0 ura3Δ0	The collection of MATa deletion strains is available for purchase as 96-well agar plates from Invitrogen, American Type Culture Collection, or EUROSCARF, or 96-well agar plates and 96-well plate of frozen stock from Open Biosystems
p4339	pCRII-TOPO::natMX4	Plasmid containing the natMX4 cassette
Primer MX4-F	(Seq 5'-3') ACATGGAGGCCCAGAATACCCT	Forward amplification primer of MX4 series cassettes; anneals to TEF promoter
Primer MX4-R	(Seq 5'-3') CAGTATAGCGACCAGCATTCAC	Reverse amplification primer of MX4 series cassettes; anneals to TEF terminator
Confirmation primer-F	(Seq 5'-3') ATGGGTCCTTCACCACCGACA	Forward primer anneals inside natR gene of natMX4 cassette
Confirmation primer-R	(Seq 5'-3') CGAGTACGAGATGACCACGAAGC	Reverse primer anneals inside natR gene of natMX4 cassette

All yeast strains are in the S288c genetic background.

Equipment

Aluminum sealing tape

Equipment for manual replica pinning (low-throughput) (V&P Scientific)

Extra floating pins (FP) (1.58-mm diameter with chambered tip)

Manual replicator (96- and 384-floating pin E-clip style)

Pin tool cleaning accessories (plastic reservoirs or tip boxes for water and bleach, Pyrex reservoir with lid for alcohol, pin cleaning brush)

Registration accessories (library copier, colony copier)

Equipment for robotic pinning

BioMatrix Colony Arrayer robot (S&P Robotics; www.sprobotics.com) for high-throughput robotic pinning

Medium should be prepared in OmniTrays (Nunc) for all replica pinning steps on a BioMatrix Colony Arrayer robot.

Singer RoToR benchtop robot (Singer Instruments, UK; www.singerinst.co.uk) for medium-throughput robotic pinning

Medium should be prepared in PlusPlates and RePads should be used for all replica pinning steps on a Singer RoToR benchtop robot. Note that RePads can be reused if washed in bleach and sterilized by UV exposure or autoclaved.

High-resolution digital imaging system, e.g., SPImager (S&P Robotics)

Incubators at 22°C, 26°C (as needed), and 30°C

Micromanipulator

Shaking incubators at 22°C (as needed) and 30°C

Chapter 24

METHOD

Constructing the Query Strain

The following describes three different methods for SGA query strain construction (Fig. 1). The first two methods describe approaches to generate query strains involving nonessential genes. PCR-mediated deletion strategy works well for most genes; however, some genes are more difficult to knock out de novo. In these cases, it is desirable to use existing strains available in the deletion collection and so the marker switching strategy is more suitable. The third method is used for generating mutants involving essential genes.

PCR-Mediated Deletion of Nonessential Genes

1. PCR amplify the *natMX4* cassette using gene-deletion primers (Fig. 1A).

    i. Assemble the following reagents in a 100-μL reaction.

H_2O	74.2 μL
10× PCR buffer	10 μL
10 mM dNTPs	2 μL
50 μM forward gene-deletion primer	4 μL
50 μM reverse gene-deletion primer	4 μL
DNA template (p4339 plasmid)	0.1 μg in 0.5 μL of H_2O
DMSO	5 μL
5 U/μL Taq DNA polymerase	0.3 μL
Total reaction volume	100 μL

    ii. Amplify in a thermal cycler using the following program.

	5 min at 95°C
Denaturation	30 sec at 95°C
30 cycles of	30 sec at 55°C
	2 min at 68°C
Final extension	10 min at 68°C
Hold	4°C

    PCR products can be stored at −20°C.

2. Transform 20 μL of PCR product into the SGA background strain Y7092 using standard LiAc transformation. Select for transformants on plates containing YEPD + clonNAT solid medium following a 3- to 4-d incubation at 30°C.

3. Confirm the correct orientation of the *natMX4* cassette in the transformants via PCR as described in Step 1, using *natR* confirmation primers (Confirmation primer-F and -R) with the upstream or downstream gene-specific confirmation primers (YFG-F and -R) (Fig. 1A).

4. Confirm *MATα* mating type and proper marker segregation.

5. Prepare a 20% glycerol stock of the query strain and store at −80°C until ready to proceed with SGA (Step 33).

Marker Switching Method for Nonessential Genes

6. Mate a deletion strain of interest from the *MAT*a deletion collection with the SGA background strain Y8205 and isolate diploid zygotes by micromanipulation (Fig. 1B).

7. To switch the *kanMX4* to the *natMX4* resistance marker, PCR amplify *natMX4* from plasmid p4339 as described in Step 1, using MX4-F and MX4-R primers.

8. Transform 20 μL of the resulting PCR product into the diploid strain using standard LiAc transformation. Select for transformants on plates containing YEPD + clonNAT solid medium following a 3- to 4-d incubation at 30°C (modified from Tong and Boone 2006).

9. Check for a successful marker switch by replica-plating the transformants on YEPD + G418 solid medium to ensure G418 sensitivity.

10. Transfer a clonNAT-resistant colony to sporulation medium and incubate for 7 d at 22°C.

11. Resuspend (using a yellow pipette tip) a small amount of the resulting spores in 1 mL of sterile H$_2$O. Plate 100 µL of the suspension on SD$_{MSG}$ − Leu,Arg,Lys (+ canavanine,thialysine, clonNAT) selective medium. Incubate for 2–3 d at 30°C to select for clonNAT-resistant MATα meiotic progeny.

12. Prepare a 20% glycerol stock of the query strain and store at −80°C until ready to proceed with SGA (Step 33).

Two-Step PCR-Mediated Integration of a Conditional Allele of an Essential Gene

13. PCR-amplify the conditional temperature-sensitive (ts) allele of interest from an appropriate DNA template as described in Step 1, using the gene-specific primer pair for two-step PCR-mediated integration. PCR amplify the natMX4 cassette from plasmid p4339 using the second pair of oligonucleotide primers designed for two-step integration (Fig. 1C).

14. Combine the PCR products and cotransform them into the SGA background strain Y7092 using standard LiAc transformation. Select for transformants on plates containing YEPD + clonNAT solid medium and incubate for 3–5 d at the permissive temperature (∼22°C).

15. Verify the integration of the conditional allele by replica-plating transformants on plates containing YEPD or YEPD + clonNAT solid medium and incubating for 1–2 d at the restrictive temperature.

    Testing on rich medium without clonNAT ensures that the ts phenotype is linked to the gene of interest and not to the natMX4 cassette.

    Conducting confirmation PCR using confirmation primers that anneal inside natMX4 is also desirable at this stage.

16. Prepare a 20% glycerol stock of the query strain and store at −80°C until ready to proceed with SGA (Step 33).

    We recommend storing a diploid version of the resulting ts mutant in addition to the haploid query strain. Isolate a heterozygous diploid by crossing your ts mutant with yeast strain Y7091.

Sterilizing Pin Tools

All SGA pinning steps can be performed using a hand pinner, a Singer RoTor benchtop robot, a BioMatrix colony arrayer robot, or any other desired pinning tool/robotic system along with accompanying software.

Manual Pin Tools

17. Set up five sterile reservoirs containing the following: (1) 30 mL of sterile H$_2$O, (2) 50 mL of sterile H$_2$O, (3) 70 mL of sterile H$_2$O, (4) 40 mL of 10% bleach, and (5) 90 mL of 95% ethanol.

18. Soak the replicator pins for 1 min in the reservoir containing 30 mL of H$_2$O to remove the cells from the pins.

19. Immerse the replicator pins in 10% bleach for ∼20 sec.

20. Transfer the replicator to the reservoir containing 50 mL of H$_2$O and then to 70 mL of H$_2$O to rinse the bleach from the pins.

21. Immerse the replicator pins in 95% ethanol for 30 sec.

22. Shake off the ethanol and flame the replicator pins.

23. Allow the replicator pins to cool before use.

BioMatrix Colony Arrayer (Robotic Pin Tools)

24. Before robot use, clean and sterilize the replicator pins as follows.

    i. Fill the sonicator bath with sterile distilled H$_2$O.

    ii. Clean the replicator pins in the sonicator bath for 5 min.

iii. Replace the H$_2$O with 70% ethanol.

iv. Sterilize the replicator pins in the sonicator bath for 20 sec. Repeat.

v. Dry the replicator over a fan for 22 sec.

25. At the end of each replica pinning step, run the following program.

    i. Automatically fill the waste bath with sterile distilled H$_2$O from a bottle supply source. Manually fill the brush station with sterile distilled H$_2$O and the sonicator with 70% ethanol.

    ii. Soak the replicator pins in the water bath for 10 sec. Repeat four times to remove residual cells from the pins.

    iii. Transfer the replicator to the brush station. Clean the pins for a total of three cycles.

    iv. Sterilize the replicator in the 70% ethanol sonicator bath twice for 20 sec.

    v. Dry the replicator over a fan for 22 sec.

Constructing a Deletion Mutant Array (DMA): From 96- to 1536-Density Format

Any array of interest can be used in the following procedure. In our laboratory, an array of nonessential gene deletion mutants is assembled on 14 agar plates and maintained at a density of 384 colonies/plate for a total of 4294 strains. This excludes 432 slow-growing strains that should be assessed for fitness separately. Also, because colonies on the edges of a plate grow very large as a result of excessive access to nutrients, our array also includes a border control strain (his3Δ1::kanMX4) that is excluded from colony size measurements and genetic interaction analysis. The exact composition of our array is available in the supplementary data file S10 from the study by Costanzo et al. (2010). A schematic of the procedure is provided in Figure 2.

26. Slowly peel off the aluminum sealing tape from each 96-well plate containing the frozen glycerol stock of deletion mutants. Do not cross-contaminate the wells.

27. Thaw the plates and use the Singer RoTor benchtop robot (or selected 96-pin replicator) to mix the stock by stirring. Replicate the array on plates containing YEPD + G418 solid medium.

28. Reseal the stock plates with fresh aluminum sealing tape and immediately return to −80°C.

29. Incubate the cells for 2 d at room temperature.

    [Optional] Prior to this growth step, you can reorganize the strains by randomizing their position with regard to their chromosome coordinates.

30. Condense four 96-format plates to a single 384-format plate of mutant colonies. Incubate the cells for 2 d at room temperature.

31. Replicate each 384-format DMA in quadruplicate onto a single plate to generate a 1536-density array. Incubate the cells for 1 d at room temperature.

32. Duplicate the 1536-format DMA to generate both a working and a master copy.

    A fresh copy of the DMA should be generated for each SGA procedure (Step 37).

Performing the SGA Procedure

In addition to SGA using a query mutant of interest, a second SGA screen using the "wild-type" control query strain Y8835 should be performed as described. This control screen results in an array of single mutants that has been subjected to the same SGA selection conditions as the query mutant of interest.

For temperature-sensitive mutant query or array strains, all the steps described in the SGA procedure should be conducted at 22°C and the final plates should be incubated at a semipermissive temperature, such as 26°C.

A schematic of the SGA procedure is provided in Figure 3.

33. Prepare query strain lawns from the frozen glycerol stocks.

34. Streak out the query strains on plates containing YEPD + clonNAT solid medium. Incubate for 2–3 d at 30°C.

35. Inoculate 5 mL of YEPD liquid medium with a single colony. Incubate for 2 d at 30°C in a shaking incubator (250 rpm).

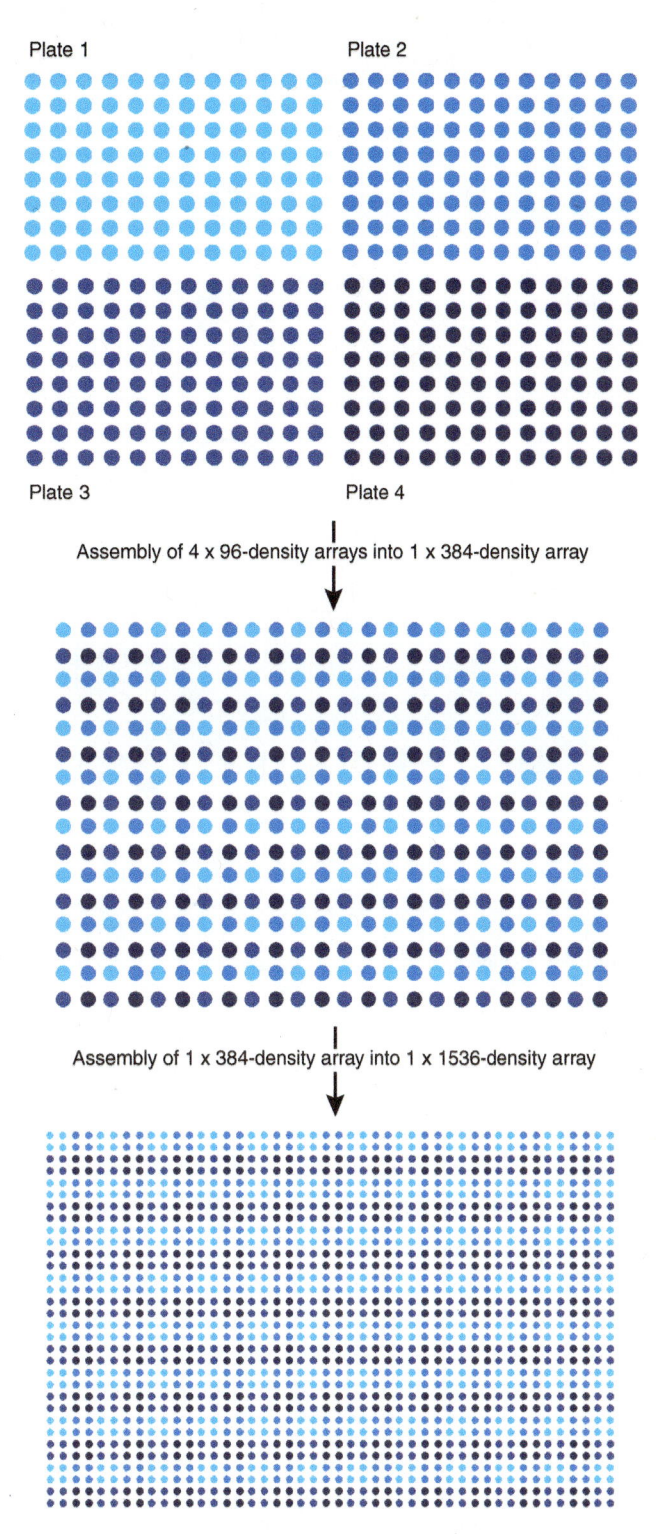

FIGURE 2. Deletion mutant array (DMA) construction in a 1536-density format. The frozen stock of deletion mutants is pinned onto agar plates, and then four 96-format plates are assembled into a single 384-format plate of mutant colonies. After a 2-d incubation period, each 384-format DMA is replicated in quadruplicate onto a single plate to construct a 1536-density array.

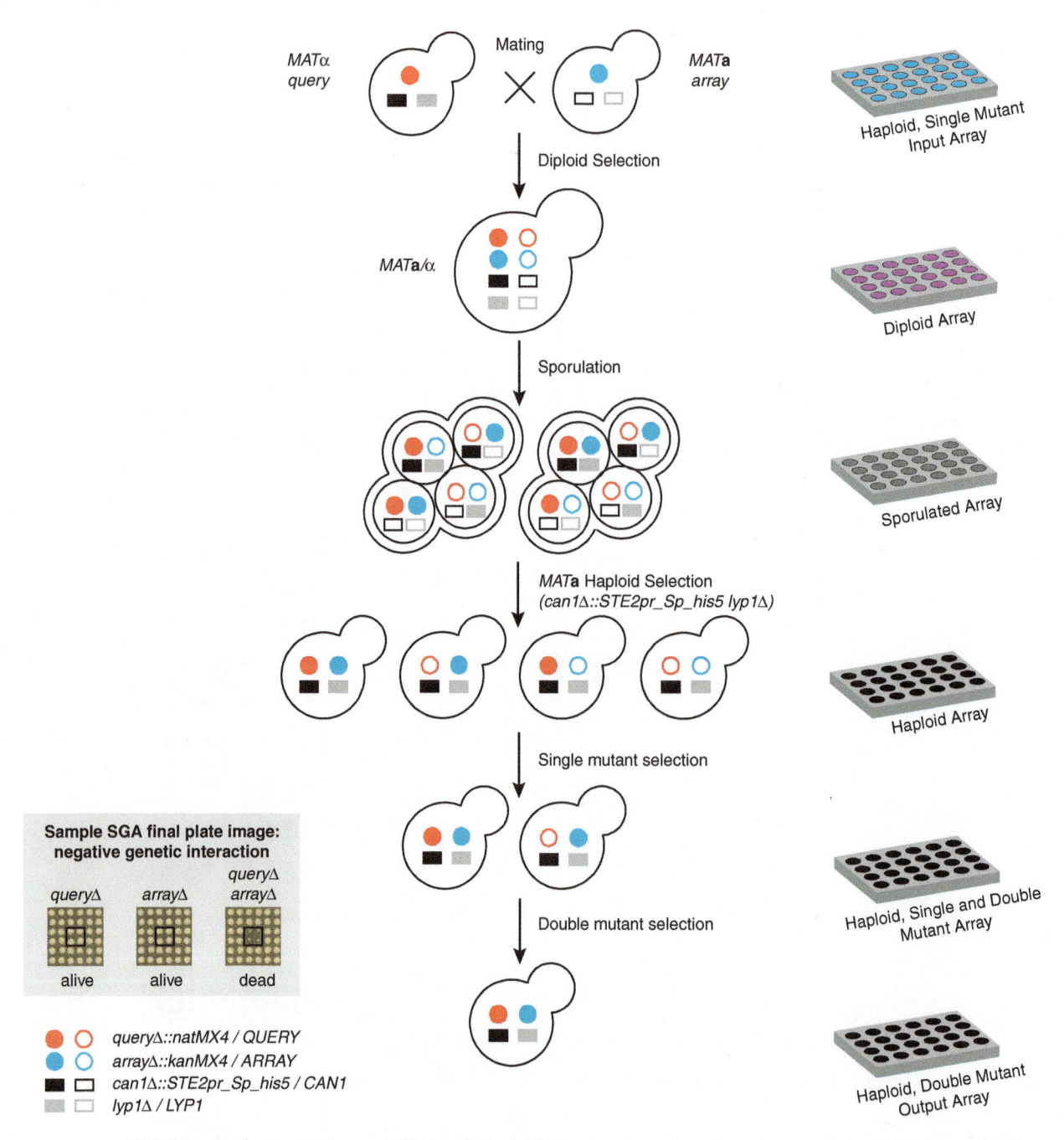

FIGURE 3. Synthetic genetic array (SGA) analysis. A *MAT*α query mutant strain carries a mutation of interest marked by a dominant selectable marker (*natMX4*), which confers resistance to the antibiotic nourseothricin (filled red circle). The query strain also contains deletions of *CAN1* and *LYP1*, whereby the former is replaced by the SGA reporter, *STE2pr_Sp_his5*. The *STE2* promoter (*STE2pr*) is *MAT*a-specific and controls the expression of the *Schizosaccharomyces pombe his5* gene, which complements an *S. cerevisiae his3–1* mutant. In a typical SGA screen, the query strain is crossed to an ordered array of *MAT*a nonessential gene deletion strains ("array" mutants) that are marked by a dominant selectable marker, *kanMX4*, which confers geneticin resistance (filled blue circle). The resulting heterozygous diploids are replica-pinned to a medium containing reduced carbon and nitrogen sources to induce sporulation. The cells are then transferred to synthetic medium lacking histidine but containing canavanine and thialysine to allow for selective germination of *MAT*a haploid meiotic progeny. Deletion of *CAN1* and *LYP1* loci confer sensitivity to canavanine and thialysine, respectively, and are used to select against unsporulated diploids. *MAT*a haploids are then transferred to a medium containing geneticin to select for array mutants and, then, to medium containing both geneticin and nourseothricin to select for double mutants. The *inset* depicts a sample cropped plate image following the final SGA selection step, where the deletion of either a query gene (*query*Δ) or an array gene (*array*Δ) does not result in any observable fitness defects, but deletion of both genes is lethal (i.e., an extreme negative genetic interaction resulting in synthetic lethality). The mutant is represented in quadruplicate and is highlighted with a black box. (Illustration adapted from Baryshnikova et al. 2010a.)

36. Prepare query strain lawns for screening.

    i. Spread 800 μL of the saturated liquid culture on an OmniTray containing YEPD solid medium.

    ii. Repeat to prepare a total of four query strain lawns for one genome-wide screen with 14 DMA plates.

    iii. Allow the lawns to dry, and then incubate for 2 d at 30°C.

37. Mate the query strain with the DMA as follows.

    i. Pin the query strain from the lawn to freshly prepared plates containing YEPD solid medium.

    ii. Pin the freshly prepared DMA on top of the query strain.

    iii. Incubate for 1 d at room temperature.

38. To select for diploids, pin the resulting *MAT*a/α diploid zygotes to plates containing YEPD + G418,clonNAT solid medium. Incubate for 2 d at 30°C.

39. To sporulate the diploids, pin the resulting KAN^R, NAT^R *MAT*a/α diploids onto plates containing enriched sporulation medium. Incubate for 7 d at 22°C.

40. To select for *MAT*a meiotic haploid progeny, pin the spores onto SD − His,Arg,Lys + canavanine, thialysine solid medium. Incubate for 2 d at 30°C.

41. To select for KAN^R *MAT*a meiotic haploid progeny, pin the spores onto SD_{MSG} − His,Arg,Lys + canavanine,thialysine,G418 solid medium. Incubate for 2 d at 30°C.

42. To select for KAN^R, NAT^R *MAT*a meiotic haploid progeny, pin the single mutant haploids onto SD_{MSG}—His,Arg,Lys + canavanine,thialysine,G418,clonNAT solid medium. Incubate for 2 d at 30°C.

    The resulting array consists of a collection of yeast double mutant strains in which each nonessential KAN^R-marked deletion mutant is also deleted for the query gene of interest, which is marked with NAT^R.

    Depending on the downstream application of the array, colony purification and isolation of mutants may be required. Usually, streaking out for single colonies is sufficient, although FACS may be used to confirm ploidy.

43. Image the double-mutant array using a high-resolution digital imaging system.

    Genetic interactions can be assessed qualitatively by comparing double-mutant colony sizes to the colony sizes of the corresponding single mutants derived from the wild-type control SGA screen. Alternatively, genetic interactions can be measured quantitatively using the SGA score method (Baryshnikova et al. 2010b) for large-scale data sets or SGAtools (Wagih et al. 2013; Wagih and Parts 2014) for smaller-scale data sets. See Discussion.

44. Confirm genetic interactions by random spore or tetrad analysis.

DISCUSSION

The method described here should enable the construction of a wide plethora of mutants suitable for numerous purposes. If the mutants are generated for the identification of genetic interactions, then we highly recommend quantitatively scoring mutant colony sizes in addition to visual inspection. A genetic interaction score is computed as the difference between the observed double-mutant fitness (AB) and the expected double-mutant fitness, which is calculated as $(A \times B)$, where A and B denote the fitness value for a mutant in *gene A* and *gene B*, respectively; thus, the genetic interaction score = $AB − (A \times B)$. A genetic interaction is negative when the observed double-mutant colony size is smaller than expected from the combined effects of individual mutants ($AB < A \times B$), and a genetic interaction is positive if the observed double-mutant colony size is larger than expected ($AB > A \times B$). For data sets consisting of a small number of SGA screens, we developed a web-based scoring pipeline called "SGAtools" (http://sgatools.ccbr.utoronto.ca/; Wagih et al. 2013; Wagih and Parts 2014). It contains all the necessary steps for analysis of genetic interactions, including image processing to extract colony size measurements from high-density colony arrays, data normalization to correct for systematic effects, and, finally, measurement of genetic interactions. There are a number of other available scoring systems that are described in our introduction to SGA (see Introduction: Synthetic Genetic Arrays: Automation of Yeast Genetics [Kuzmin et al. 2015]). For large-scale data sets consisting of hundreds/thousands of screens, we developed the SGA score algorithm (Baryshnikova et al. 2010b).

Genetic interaction studies have enhanced our understanding of genetic network architecture and provided insight into gene function. For instance, genes with similar biological roles tend to share many genetic interactions. Therefore, clustering genes according to the similarity of their genetic interaction profiles (all the interactions for a particular gene) using an open source clustering software such as Cluster 3.0 (http://bonsai.hgc.jp/~mdehoon/software/cluster/software.htm) is a powerful tool for predicting the functions of unknown genes. Once the data is clustered, it can be visualized using open source applications, such as Java TreeView (http://jtreeview.sourceforge.net/). Another approach to find genes with similar genetic interaction profiles is to compute pair-wise Pearson correlation coefficients between all genes in the genome and use a network visualization tool, such as Cytoscape (http://cytoscape.org) to organize and lay out the network in an unbiased manner. Applying a force-directed network layout will pull highly correlated genes toward each other and away from less-correlated genes, revealing meaningful biological relationships (Costanzo et al. 2010). All of our published genetic interaction data and visualization tools are accessible from a web database (http://thecellmap.org; M Usaj et al., unpubl.).

RECIPES

Amino Acid Supplement Powder (Drop-Out Mixture)

Reagent	Amount
Adenine	3 g
Uracil	2 g
Inositol	2 g
p-Aminobenzoic acid	0.2 g
Alanine	2 g
Arginine	2 g
Asparagine	2 g
Aspartic acid	2 g
Cysteine	2 g
Glutamic acid	2 g
Glutamine	2 g
Glycine	2 g
Histidine	2 g
Isoleucine	2 g
Leucine	10 g
Lysine	2 g
Methionine	2 g
Phenylalanine	2 g
Proline	2 g
Serine	2 g
Threonine	2 g
Tryptophan	2 g
Tyrosine	2 g
Valine	2 g

To prepare a drop-out (DO) mixture, exclude the desired amino acid from the powder. Mix by inverting end to end for 15 min and store the contents in a dark bottle at room temperature.

Enriched Sporulation Medium

Prepare the following in a 2-L Erlenmeyer flask with a stir bar. Autoclave and cool to ~65°C.

Potassium acetate	10 g
Bacto-Yeast extract	1 g
Glucose (dextrose) (40% w/v in H_2O)	0.5 g
Sporulation amino acid supplement powder <R>	0.1 g
Bacto-Agar	20 g
H_2O	to 1 L

Add 250 μL of G418 (Geneticin; Invitrogen) stock solution (200 mg/mL, prepared in H_2O and filter-sterilized).

Stir for ~15 min to mix, and then pour into plates.

SD—His,Arg,Lys (+Canavanine,Thialysine) Selection Medium

1. In a 250-mL glass bottle, combine the following. Shake to mix, autoclave, and cool to ~65°C.

Yeast nitrogen base without amino acids	6.7 g
Amino acid supplement powder (drop-out mixture; drop out His, Arg, and Lys) <R>	2 g
H_2O	200 mL

2. In a 2-L Erlenmeyer flask with stir bar, combine the following. Stir to mix, autoclave, and cool to ~65°C.

Bacto-Agar	20 g
H_2O	750 mL

3. Combine the solutions from Steps 1 and 2, and then add the following.

Glucose (dextrose) (40% w/v in H_2O)	50 mL
Canavanine (L-canavanine sulfate salt; Sigma-Aldrich) (100 mg/mL, filter-sterilized)	0.5 mL
Thialysine (S-[2-aminoethyl]-L-cysteine hydrochloride; Sigma-Aldrich) (100 mg/mL, filter-sterilized)	0.5 mL

4. Stir the solution for ~15 min to mix, and then pour into plates.

SD$_{MSG}$—His,Arg,Lys (+Canavanine,Thialysine,G418 +/− clonNAT) Selection Medium

1. In a 250-mL glass bottle, combine the following. Shake to mix, autoclave, and cool to ~65°C.

Yeast nitrogen base without amino acids and ammonium sulfate	1.7 g
Monosodium glutamic acid (MSG)	1 g
Amino acid supplement powder (drop-out mixture; drop out His, Arg, and Lys) <R>	2 g
H_2O	200 mL

2. In a 2-L Erlenmeyer flask with stir bar, combine the following. Stir to mix, autoclave, and cool to ~65°C.

Bacto-Agar	20 g
H_2O	750 mL

3. Combine the solutions from Steps 1 and 2, and then add the following.

Glucose (dextrose) (40% in H_2O)	50 mL
Canavanine (L-canavanine sulfate salt; Sigma-Aldrich) (100 mg/mL, filter-sterilized)	0.5 mL

Chapter 24

Thialysine (S-[2-aminoethyl]-L-cysteine hydrochloride; Sigma-Aldrich) (100 mg/mL, filter-sterilized)	0.5 mL
G418 (Geneticin; Invitrogen) (200 mg/mL, filter-sterilized)	1 mL
[Optional] clonNAT (Nourseothricin; Werner BioAgents) (100 mg/mL, filter-sterilized)	1 mL

4. Stir the solution for ~15 min to mix, and then pour into plates.

SD_{MSG}—Leu,Arg,Lys (+Canavanine,Thialysine, clonNAT) Selection Medium

1. In a 250-mL glass bottle, combine the following. Shake to mix, autoclave, and cool to ~65°C.

Yeast nitrogen base without amino acids and ammonium sulfate	1.7 g
Monosodium glutamic acid (MSG)	1 g
Amino acid supplement powder (drop-out mixture; drop out Leu, Arg, and Lys) <R>	2 g
H_2O	200 mL

2. In a 2-L Erlenmeyer flask with stir bar, combine the following. Stir to mix, autoclave, and cool to ~65°C.

Bacto-Agar	20 g
H_2O	750 mL

3. Combine the solutions from Steps 1 and 2, and then add the following.

Glucose (dextrose) (40% in H_2O)	50 mL
Canavanine (L-canavanine sulfate salt; Sigma-Aldrich) (100 mg/mL, filter-sterilized)	0.5 mL
Thialysine (S-[2-aminoethyl]-L-cysteine hydrochloride; Sigma-Aldrich) (100 mg/mL, filter-sterilized)	0.5 mL
clonNAT (Nourseothricin; Werner BioAgents) (100 mg/mL, filter-sterilized)	1 mL

4. Stir the solution for ~15 min to mix, and then pour into plates.

Sporulation Amino Acid Supplement Powder

Reagent	Amount
Histidine	2 g
Leucine	10 g
Uracil	2 g

Mix by inverting end to end for 15 min. Store the contents in a dark bottle at room temperature.

Sporulation Medium for Query Strain Construction

1. Prepare the following in a 2-L Erlenmeyer flask with a stir bar. Autoclave and cool to ~65°C.

Potassium acetate	10 g
Sporulation amino acid supplement powder <R>	0.1 g
Bacto agar	20 g
H_2O	to 1 L

2. Stir for ~15 min to mix, and then pour into plates.

Yeast Extract Peptone Dextrose (YEPD) (+/− G418 +/− clonNAT) Solid Medium

1. In a 2-L Erlenmeyer flask with stir bar, combine the following. Autoclave and cool to ~65°C.

Bacto yeast extract	10 g
Bacto peptone	20 g
Bacto agar	20 g
Adenine	120 mg
H_2O	950 mL

2. Add the following reagents.

Glucose (40% w/v in H_2O)	50 mL
[Optional] G418 (Geneticin; Invitrogen) (200 mg/mL, filter-sterilized)	1 mL
[Optional] clonNAT (Nourseothricin; Werner BioAgents) (100 mg/mL, filter-sterilized)	1 mL

3. Stir the solution for ~15 min to mix, and then pour into plates.

REFERENCES

Baryshnikova A, Costanzo M, Dixon S, Vizeacoumar FJ, Myers CL, Andrews B, Boone C. 2010a. Synthetic genetic array (SGA) analysis in *Saccharomyces cerevisiae* and *Schizosaccharomyces pombe*. Methods Enzymol **470**: 145–179.

Baryshnikova A, Costanzo M, Kim Y, Ding H, Koh J, Toufighi K, Youn JY, Ou J, San Luis BJ, Bandyopadhyay S, et al. 2010b. Quantitative analysis of fitness and genetic interactions in yeast on a genome scale. Nat Methods **7**: 1017–1024.

Costanzo M, Baryshnikova A, Bellay J, Kim Y, Spear ED, Sevier CS, Ding H, Koh JL, Toufighi K, Mostafavi S, et al. 2010. The genetic landscape of a cell. Science **327**: 425–431.

Kuzmin E, Costanzo M, Andrews B, Boone C. 2015. Synthetic genetic arrays: Automation of yeast genetics. Cold Spring Harb Protoc doi: 10.1101/pdb.top086652.

Li Z, Vizeacoumar FJ, Bahr S, Li J, Warringer J, Vizeacoumar FS, Min R, Vandersluis B, Bellay J, Devit M, et al. 2011. Systematic exploration of essential yeast gene function with temperature-sensitive mutants. Nat Biotechnol **29**: 361–367.

Tong AH, Boone C. 2006. Synthetic genetic array analysis in *Saccharomyces cerevisiae*. Methods Mol Biol **313**: 171–192.

Wagih O, Parts L. 2014. gitter: A robust and accurate method for quantification of colony sizes from plate images. G3 (Bethesda) **4**: 547–552.

Wagih O, Usaj M, Baryshnikova A, VanderSluis B, Kuzmin E, Costanzo M, Myers CL, Andrews BJ, Boone CM, Parts L. 2013. SGAtools: One-stop analysis and visualization of array-based genetic interaction screens. Nucleic Acids Res **41**: W591–W596.

CHAPTER 25

Systematic Mapping of Chemical–Genetic Interactions in *Saccharomyces cerevisiae*

Sundari Suresh,[1] Ulrich Schlecht,[1] Weihong Xu,[1] Walter Bray,[2] Molly Miranda,[1] Ronald W. Davis,[1] Corey Nislow,[3] Guri Giaever,[3] R. Scott Lokey,[2] and Robert P. St.Onge[1,4]

[1]Stanford Genome Technology Center, Department of Biochemistry, Stanford University, Palo Alto, California 94304; [2]Department of Chemistry and Biochemistry, University of California Santa Cruz, Santa Cruz, California 95064; [3]Pharmaceutical Sciences, University of British Columbia, Vancouver, British Columbia V6T 1Z3, Canada

Chemical–genetic interactions (CGIs) describe a phenomenon where the effects of a chemical compound (i.e., a small molecule) on cell growth are dependent on a particular gene. CGIs can reveal important functional information about genes and can also be powerful indicators of a compound's mechanism of action. Mapping CGIs can lead to the discovery of new chemical probes, which, in contrast to genetic perturbations, operate at the level of the gene product (or pathway) and can be fast-acting, tunable, and reversible. The simple culture conditions required for yeast and its rapid growth, as well as the availability of a complete set of barcoded gene deletion strains, facilitate systematic mapping of CGIs in this organism. This process involves two basic steps: first, screening chemical libraries to identify bioactive compounds affecting growth and, second, measuring the effects of these compounds on genome-wide collections of mutant strains. Here, we introduce protocols for both steps that have great potential for the discovery and development of new small-molecule tools and medicines.

INTRODUCTION

Chemogenomic profiling in yeast involves measuring the growth of genome-scale sets of genetic variants (most commonly, strains having defined gene copy number changes) in the presence of a chemical inhibitor. These screens can reveal new mechanistic details of known chemical probes and drugs and can also enable the discovery of novel compounds and cognate druggable targets (i.e., proteins that can be selectively targeted with a small molecule). Inhibitors of cell growth, including antifungal, antibiotic, and anticancer drugs, have been used therapeutically for decades. Expanding the repertoire of these inhibitors and their targets will ultimately have a profound impact on the treatment of various human illnesses.

IDENTIFICATION OF USEFUL COMPOUNDS AND NEW TARGETS

Individual chemical–genetic interactions (CGIs) can reveal information regarding the mechanism of action of a growth inhibitor, including identification of the direct cellular target(s) of a compound. A cell's sensitivity to a chemical inhibitor will vary depending on the abundance of the cognate protein target (provided the target is essential for growth); thus, targets can be revealed by measuring strain-

---

[4]Correspondence: bstonge@stanford.edu

Copyright © Cold Spring Harbor Laboratory Press; all rights reserved
Cite this introduction as *Cold Spring Harb Protoc*; doi:10.1101/pdb.top077701

specific sensitivity (Fig. 1). For example, systematic parallel measurement of the drug sensitivity of thousands of heterozygous strains, known as haploinsufficiency profiling (HIP), has accurately identified the cellular targets of many drugs (Giaever et al. 1999, 2004; Lum et al. 2004; Hoepfner et al. 2014; Lee et al. 2014). Similarly, drug targets have been revealed by identifying multicopy and/or overexpression clones that suppress a drug's cytotoxicity (Rine et al. 1983; Luesch et al. 2005; Butcher et al. 2006; Hoon et al. 2008; Ho et al. 2009). CGIs identified when the genetic defect results in the absence of a protein product (haploid deletion strains or diploid homozygous deletion strains) do not generally lead to the direct identification of cellular targets. Rather, they identify gene products that buffer the growth-inhibitory effects of the chemical (Parsons et al. 2004, 2006; Lee et al. 2005; Hillenmeyer et al. 2008). Thus, although the mechanism underlying a CGI can vary depending on the nature of the genetic defect, all CGIs yield information about how a compound affects growth of the organism.

The quantitative nature of CGIs can be used to predict mechanism in other ways. For example, systematic measurement of CGIs across thousands of strains yields a complex signature (i.e., a pattern of CGIs) that constitutes a detailed description of a compound's mechanism of action. By comparing the signatures of different compounds, a "guilt by association" strategy can reveal novel mechanistic insights (Lee et al. 2005; Parsons et al. 2006; Hillenmeyer et al. 2010; Kapitzky et al. 2010). In addition, similar effects are often produced by loss-of-function mutations in a gene and chemical inactivation of the protein product of that gene; therefore, comparing CGI signatures to genetic interaction signatures can link compounds to their cellular targets (Parsons et al. 2004; Costanzo et al. 2010).

The power of chemical genetics is illustrated by the recent identification of inhibitors of yeast Sec14, a conserved phosphatidylinositol transfer protein (PITP) that is required for pathogenesis of certain fungi (Chayakulkeeree et al. 2011). Hoon et al. (2008) screened 188 growth-inhibitory compounds against genome-scale collections of homozygous deletion mutants, heterozygous deletion mutants, and multicopy clones. One compound, a nitrophenyl,[4-(2-methoxyphenyl)piperazin-1-yl]methanone (NPPM), was found to cause specific sensitivity of the SEC14/$sec14\Delta$ heterozygous strain and, moreover, a strain containing additional copies of SEC14 was resistant to the compound (Fig. 2). Subsequent structure-activity characterization in cell-free and cell-based assays identified chemical derivatives with improved properties and firmly established the specificity of these inhibitors for Sec14 (Nile et al. 2013). These compounds represent powerful tools for studying phosphoinositide signaling and underscore how systematic mapping of CGIs can identify not

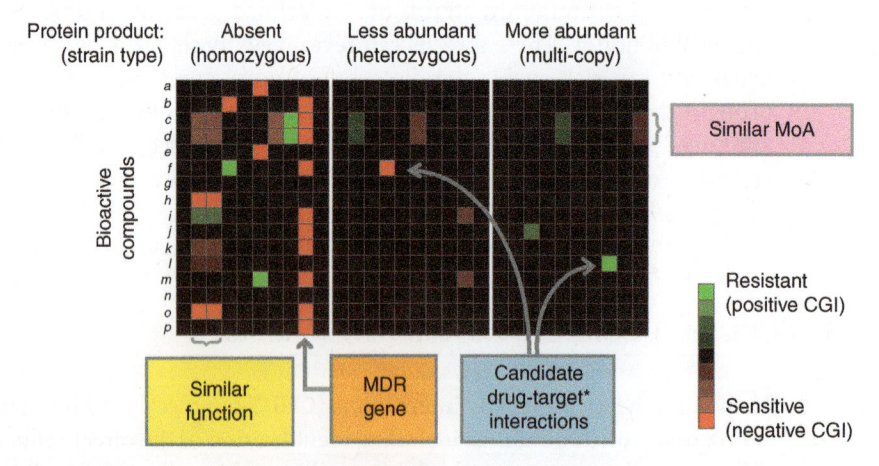

FIGURE 1. Chemical–genetic interactions reveal functional and mechanistic information. Schematic heatmap in which bioactive compounds (a–p) are arranged on the vertical axis and genes (represented by different yeast strains) on the horizontal axis. Chemical–genetic interactions (CGIs) are shown in green and red. Genes showing CGIs with similar compounds are often functionally related (yellow box). Similarly, compounds forming CGIs with similar genes are mechanistically related (pink box). Strains showing sensitivity to many compounds identify multidrug resistance (MDR) genes (orange box). Strains producing less or more of a compound's target will be sensitive or resistant to that compound, respectively (blue box; *indicates that in these scenarios, the target is the product of an essential gene).

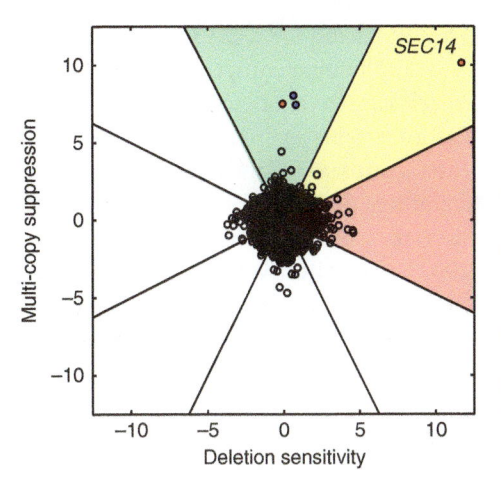

FIGURE 2. Chemical–genetic screening identifies a new chemical inhibitor (4130–1278) of yeast Sec14. (Reproduced with permission from Hoon et al. 2008.) Deletion sensitivity (sensitivity of homozygous or heterozygous deletion strains to 4130–1278) is plotted on the x-axis and multicopy suppression (resistance of genomic library transformants to 4130–1278) is plotted on the y-axis. The green sector identifies genes that are multicopy suppressors, and the pink sector identifies genes where a reduction in copy number results in drug sensitivity. The yellow sector identifies genes that show both characteristics. SEC14 is represented by the red dot in the *upper right* corner of the yellow sector.

only new chemical inhibitors but also new druggable targets that can potentially be exploited for therapeutic purposes.

CHALLENGES IN CGI SCREENING

The above results notwithstanding, robust and efficient identification of CGIs presents several challenges. First, chemical space consists of a mind-boggling number of unique chemical entities. Even when one only considers those compounds that can be synthesized or acquired in sufficient supply for experimental testing, this number is far beyond what can be practically subjected to chemogenomic profiling methods. Studies have shown that the vast majority of compounds do not inhibit yeast growth (Wallace et al. 2011); therefore, efficient methods for screening large numbers of compounds for bioactivity are critical for efficient CGI mapping. The high-throughput (HT) yeast halo assay (Gassner et al. 2007; Woehrmann et al. 2010) that we describe (see Protocol 1: A High-Throughput Yeast Halo Assay for Bioactive Compounds [Bray and Lokey 2015]) allows large libraries of small molecules to be screened in a manner that is cheap and scalable. The HT halo assay is essentially a miniaturized and automated version of the disk diffusion assay pioneered more than half a century ago by Kirby and Bauer (Bauer et al. 1959) and offers advantages over conventional growth assays performed in liquid culture. First, hits in the halo assay can be checked by visual inspection. Symmetric halos centered on sites of compound insertion can be easily detected with the naked eye and distinguished from false positives arising from, for example, dispensing artifacts. Second, a standard curve can be used to calculate IC_{50}s based on halo sizes, allowing potencies to be estimated from the initial screen. This is especially important with libraries containing structurally related homologous series of active compounds and enables structure–activity relationships to be determined from primary screening data.

The yeast cell wall and elaborate chemical defense mechanisms represent a formidable barrier to many compounds. The pleiotropic drug response in yeast is characterized by transcriptional upregulation of multiple drug efflux pumps (including many evolutionarily conserved plasma membrane ATP-binding cassette [ABC] transporters) that actively remove a wide range of potentially toxic compounds from the cell (Bauer et al. 1999; Schlecht et al. 2012). The yeast genes required for this response are known, and as expected, strains lacking these genes show elevated sensitivity to a broader range of chemical compounds compared to wild-type controls (Rogers et al. 2001; Suzuki et al. 2011). Thus, a greater amount of a chemical enters and is retained in these strains, and employing them in primary phenotypic screens can uncover interesting bioactive compounds that also happen to be substrates for drug transporters. Subsequent chemogenomic profiling of such compounds may require drug-sensitized strain collections (Best et al. 2013) or, alternatively, higher compound concentrations if the strain collections used harbor intact drug-transporters.

Comprehensive testing of a bioactive compound for CGIs requires robust phenotyping of thousands of individual genetic mutants while minimizing compound use, as many interesting bioactive compounds are limited in supply or expensive. Here, we describe a simple protocol for the systematic parallel testing of thousands of gene deletion strains against a compound of interest (see Protocol 2: Identification of Chemical–Genetic Interactions via Parallel Analysis of Barcoded Yeast Strains [Suresh et al. 2015]). This protocol, like others described previously (Pierce et al. 2007; Smith et al. 2011), uses the yeast knockout collection, the construction of which was a landmark event for the chemical genetics field (Giaever et al. 2002). Cost-effective DNA synthesis technology permitted two unique molecular barcodes (an "UP tag" and a "DOWN tag") to be included in each strain in the collection. Quantification of these tags allows the abundance of individual strains to be determined following competitive growth of thousands of strains in a single culture. The unbiased and cell-based nature of this assay makes it an extremely powerful approach for investigating the mechanism by which bioactive compounds inhibit cell growth.

ACKNOWLEDGMENTS

We are grateful for funding from the U.S. National Institutes of Health (P01HG000205 to R.W.D. and R01HG003317 to G.G., C.N., and R.W.D.) and the Canadian Institutes for Health Research (C.N. and G.G.).

REFERENCES

Bauer AW, Roberts CE Jr., Kirby WM. 1959. Single disc versus multiple disc and plate dilution techniques for antibiotic sensitivity testing. *Antibiot Annu* 7: 574–580.

Bauer BE, Wolfger H, Kuchler K. 1999. Inventory and function of yeast ABC proteins: About sex, stress, pleiotropic drug and heavy metal resistance. *Biochim Biophys Acta* 1461: 217–236.

Best HA, Matthews JH, Heathcott RW, Hanna R, Leahy DC, Coorey NV, Bellows DS, Atkinson PH, Miller JH. 2013. Laulimalide and peloruside A inhibit mitosis of *Saccharomyces cerevisiae* by preventing microtubule depolymerisation-dependent steps in chromosome separation and nuclear positioning. *Mol Biosyst* 9: 2842–2852.

Bray W, Lokey RS. 2015. A high-throughput yeast halo assay for bioactive compounds. *Cold Spring Harb Protoc* doi: 10.1101/pdb. prot088047.

Butcher RA, Bhullar BS, Perlstein EO, Marsischky G, LaBaer J, Schreiber SL. 2006. Microarray-based method for monitoring yeast overexpression strains reveals small-molecule targets in TOR pathway. *Nat Chem Biol* 2: 103–109.

Chayakulkeeree M, Johnston SA, Oei JB, Lev S, Williamson PR, Wilson CF, Zuo X, Leal AL, Vainstein MH, Meyer W, et al. 2011. SEC14 is a specific requirement for secretion of phospholipase B1 and pathogenicity of *Cryptococcus neoformans*. *Mol Microbiol* 80: 1088–1101.

Costanzo M, Baryshnikova A, Bellay J, Kim Y, Spear ED, Sevier CS, Ding H, Koh JL, Toufighi K, Mostafavi S, et al. 2010. The genetic landscape of a cell. *Science* 327: 425–431.

Gassner NC, Tamble CM, Bock JE, Cotton N, White KN, Tenney K, St.Onge RP, Proctor MJ, Giaever G, Nislow C, et al. 2007. Accelerating the discovery of biologically active small molecules using a high-throughput yeast halo assay. *J Nat Prod* 70: 383–390.

Giaever G, Shoemaker DD, Jones TW, Liang H, Winzeler EA, Astromoff A, Davis RW. 1999. Genomic profiling of drug sensitivities via induced haploinsufficiency. *Nat Genet* 21: 278–283.

Giaever G, Chu AM, Ni L, Connelly C, Riles L, Veronneau S, Dow S, Lucau-Danila A, Anderson K, Andre B, et al. 2002. Functional profiling of the *Saccharomyces cerevisiae* genome. *Nature* 418: 387–391.

Giaever G, Flaherty P, Kumm J, Proctor M, Nislow C, Jaramillo DF, Chu AM, Jordan MI, Arkin AP, Davis RW. 2004. Chemogenomic profiling: Identifying the functional interactions of small molecules in yeast. *Proc Natl Acad Sci* 101: 793–798.

Hillenmeyer ME, Fung E, Wildenhain J, Pierce SE, Hoon S, Lee W, Proctor M, St.Onge RP, Tyers M, Koller D, et al. 2008. The chemical genomic portrait of yeast: Uncovering a phenotype for all genes. *Science* 320: 362–365.

Hillenmeyer ME, Ericson E, Davis RW, Nislow C, Koller D, Giaever G. 2010. Systematic analysis of genome-wide fitness data in yeast reveals novel gene function and drug action. *Genome Biol* 11: R30.

Ho CH, Magtanong L, Barker SL, Gresham D, Nishimura S, Natarajan P, Koh JL, Porter J, Gray CA, Andersen RJ, et al. 2009. A molecular barcoded yeast ORF library enables mode-of-action analysis of bioactive compounds. *Nat Biotechnol* 27: 369–377.

Hoepfner D, Helliwell SB, Sadlish H, Schuierer S, Filipuzzi I, Brachat S, Bhullar B, Plikat U, Abraham Y, Altorfer M, et al. 2014. High-resolution chemical dissection of a model eukaryote reveals targets, pathways and gene functions. *Microbiol Res* 169: 107–120.

Hoon S, Smith AM, Wallace IM, Suresh S, Miranda M, Fung E, Proctor M, Shokat KM, Zhang C, Davis RW, et al. 2008. An integrated platform of genomic assays reveals small-molecule bioactivities. *Nat Chem Biol* 4: 498–506.

Kapitzky L, Beltrao P, Berens TJ, Gassner N, Zhou C, Wuster A, Wu J, Babu MM, Elledge SJ, Toczyski D, et al. 2010. Cross-species chemogenomic profiling reveals evolutionarily conserved drug mode of action. *Mol Syst Biol* 6: 451.

Lee AY, St.Onge RP, Proctor MJ, Wallace IM, Nile AH, Spagnuolo PA, Jitkova Y, Gronda M, Wu Y, Kim MK, et al. 2014. Mapping the cellular response to small molecules using chemogenomic fitness signatures. *Science* 344: 208–211.

Lee W, St.Onge RP, Proctor M, Flaherty P, Jordan MI, Arkin AP, Davis RW, Nislow C, Giaever G. 2005. Genome-wide requirements for resistance to functionally distinct DNA-damaging agents. *PLoS Genet* 1: e24.

Luesch H, Wu TY, Ren P, Gray NS, Schultz PG, Supek F. 2005. A genome-wide overexpression screen in yeast for small-molecule target identification. *Chem Biol* 12: 55–63.

Lum PY, Armour CD, Stepaniants SB, Cavet G, Wolf MK, Butler JS, Hinshaw JC, Garnier P, Prestwich GD, Leonardson A, et al. 2004. Discovering modes of action for therapeutic compounds using a genome-wide screen of yeast heterozygotes. *Cell* 116: 121–137.

Nile AH, Tripathi A, Yuan P, Mousley CJ, Suresh S, Wallace IM, Shah SD, Pohlhaus DT, Temple B, Nislow C, et al. 2014. PITPs as targets for selectively interfering with phosphoinositide signaling in cells. *Nat Chem Biol* 10: 76–84.

Parsons AB, Brost RL, Ding H, Li Z, Zhang C, Sheikh B, Brown GW, Kane PM, Hughes TR, Boone C. 2004. Integration of chemical-genetic and

genetic interaction data links bioactive compounds to cellular target pathways. *Nat Biotechnol* 22: 62–69.
Parsons AB, Lopez A, Givoni IE, Williams DE, Gray CA, Porter J, Chua G, Sopko R, Brost RL, Ho CH, et al. 2006. Exploring the mode-of-action of bioactive compounds by chemical-genetic profiling in yeast. *Cell* 126: 611–625.
Pierce SE, Davis RW, Nislow C, Giaever G. 2007. Genome-wide analysis of barcoded *Saccharomyces cerevisiae* gene-deletion mutants in pooled cultures. *Nat Protoc* 2: 2958–2974.
Rine J, Hansen W, Hardeman E, Davis RW. 1983. Targeted selection of recombinant clones through gene dosage effects. *Proc Natl Acad Sci* 80: 6750–6754.
Rogers B, Decottignies A, Kolaczkowski M, Carvajal E, Balzi E, Goffeau A. 2001. The pleitropic drug ABC transporters from *Saccharomyces cerevisiae*. *J Mol Microbiol Biotechnol* 3: 207–214.
Schlecht U, Miranda M, Suresh S, Davis RW, St.Onge RP. 2012. Multiplex assay for condition-dependent changes in protein–protein interactions. *Proc Natl Acad Sci* 109: 9213–9218.

Smith AM, Durbic T, Oh J, Urbanus M, Proctor M, Heisler LE, Giaever G, Nislow C. 2011. Competitive genomic screens of barcoded yeast libraries. *J Vis Exp* 54: 2864.
Suresh S, Schlecht U, Xu W, Miranda M, Davis RW, Nislow C, Giaever G, St.Onge RP. 2015. Identification of chemical–genetic interactions via parallel analysis of barcoded yeast strains. *Cold Spring Harb Protoc* doi: 10.1101/pdb.prot088054.
Suzuki Y, St.Onge RP, Mani R, King OD, Heilbut A, Labunskyy VM, Chen W, Pham L, Zhang LV, Tong AH, et al. 2011. Knocking out multigene redundancies via cycles of sexual assortment and fluorescence selection. *Nat Methods* 8: 159–164.
Wallace IM, Urbanus ML, Luciani GM, Burns AR, Han MK, Wang H, Arora K, Heisler LE, Proctor M, St.Onge RP, et al. 2011. Compound prioritization methods increase rates of chemical probe discovery in model organisms. *Chem Biol* 18: 1273–1283.
Woehrmann MH, Gassner NC, Bray WM, Stuart JM, Lokey S. 2010. HALO384: A halo-based potency prediction algorithm for high-throughput detection of antimicrobial agents. *J Biomol Screen* 15: 196–205.

Protocol 1

A High-Throughput Yeast Halo Assay for Bioactive Compounds

Walter Bray and R. Scott Lokey[1]

Department of Chemistry and Biochemistry, University of California Santa Cruz, Santa Cruz, California 95064

When a disk of filter paper is impregnated with a cytotoxic or cytostatic drug and added to solid medium seeded with yeast, a visible clear zone forms around the disk whose size depends on the concentration and potency of the drug. This is the traditional "halo" assay and provides a convenient, if low-throughput, read-out of biological activity that has been the mainstay of antifungal and antibiotic testing for decades. Here, we describe a protocol for a high-throughput version of the halo assay, which uses an array of 384 pins to deliver ~200 nL of stock solutions from compound plates onto single-well plates seeded with yeast. Using a plate reader in the absorbance mode, the resulting halos can be quantified and the data archived in the form of flat files that can be connected to compound databases with standard software. This assay has the convenience associated with the visual readout of the traditional halo assay but uses far less material and can be automated to screen thousands of compounds per day.

MATERIALS

It is essential that you consult the appropriate Material Safety Data Sheets and your institution's Environmental Health and Safety Office for proper handling of equipment and hazardous materials used in this protocol.

RECIPES: Please see the end of this protocol for recipes indicated by <R>. Additional recipes can be found online at http://cshprotocols.cshlp.org/site/recipes.

Reagents

Agar (2×) <R>
DMSO (50% v/v)
DMSO stock solutions of compounds formatted in 384-well plates
Ethanol (70% v/v and 95% v/v)
Saccharomyces cerevisiae strain(s) appropriate for experimental goals

Depending on the objectives of the screen, use a strain of wild-type yeast, or for screens designed to identify chemical–genetic interactions, select a strain containing a genetic deletion mutant or a strain with a sensitized genetic background (Nehil et al. 2007).

YPD (1× and 2×) <R>

Equipment

384-notched-pin array (V&P Scientific, VP 384FP3S100)
Air drier manifold

---

[1]Correspondence: slokey@ucsc.edu

Copyright © Cold Spring Harbor Laboratory Press; all rights reserved
Cite this protocol as *Cold Spring Harb Protoc*; doi:10.1101/pdb.prot088047

Cell culture plates (single-well; e.g., Nunc OmniTray 242811)
Circulating, 384-individual-channel wash bath
Culture tube (14-mL; Falcon 352057)
EnVision Multilabel plate reader (PerkinElmer)
Incubator (30°C) with orbital shaker or rotator
Janus MDT Liquid Handling Robot (PerkinElmer)
Lint-free blotting paper (V&P Scientific, VP540D-100)
Microwave oven
Pipeline Pilot software (Accelrys)
Serological pipette (25-mL)
Sonicator (Branson 200)
Water bath

METHOD

Yeast Plate Preparation

1. Inoculate 10 mL of 1× YPD in 14-mL culture tubes with a colony of *S. cerevisiae* and grow overnight at 30°C on a rotator at 70 rpm.

2. Use a microwave oven to melt a stock of 2× agar that is sufficient for the number of plates to be prepared in Step 5. Keep the solution in a 50°C water bath.

3. Dilute the overnight culture into 2× YPD to an OD_{600} of 0.06, or ~10^6 cells per mL.
   Prepare sufficient diluted culture for the number of plates to be prepared in Step 5.

4. Add an equal volume of molten 2× agar to the 2× YPD and yeast mixture.

5. Dispense 20 mL of the resulting YPD, agar, and yeast solution into one-well OmniTrays with a 25-mL serological pipette. Tilt gently to distribute the solution throughout the plate and remove any bubbles with the pipette. Prepare a suitable number of plates for the number of compounds to be screened.

6. Allow the plates to solidify for 90 min at room temperature.

High-Throughput Halo Assay

7. Transfer DMSO stock solutions of compounds formatted in 384-well plates onto the cooled yeast–YPD OmniTrays with a Janus MDT liquid-handling robot using notched pins that each deliver 200 nL (±8%).
   Cycloheximide pinned at 500 μM can be used as a control.

8. Between delivery of compounds, clean pins by dipping into a 70% ethanol bath (three times), sonicating in 50% DMSO (three times), and finally dipping into a 95% ethanol circulating 384-channel bath (three times). Between each wash step blot pins on lint-free blotting paper to absorb excess solvent. At the end of the cleaning cycle, dry pins in an air drier manifold.

9. Incubate the soft agar plates for 18 h at 30°C and then read plates using an EnVision plate reader. Setup the plate reader to scan nine points centered on each drug inoculation point, with four points on each side separated by 0.48 mm. Archive the data for each plate by saving as separate .csv files.

10. Use Pipeline Pilot (Accelrys) or similar software to associate plate reader data to 384-well compound plate maps.

11. Inspect plates visually and score hits based on halo size.

Halos are visible as symmetrical zones of clearing centered on drug inoculation points. The size of a halo may vary depending on the use of a deletion mutant or a sensitized genetic background. Small halos will only show clearing around their own drug inoculation points; medium halos will have edges that touch the four nearest drug inoculation points; large halos will extend beyond the four nearest drug inoculation points. Mark any drug inoculation points as "obscured" if they are underneath halos that extend from neighboring drug inoculation points.

RECIPES

Agar (2×)

15 g agar
1 L water

Sterilize by autoclaving for 20 min at 15 psi (1.05 kg/cm^2) on liquid cycle. Swirl gently to distribute the melted agar evenly throughout the solution. Be careful; the fluid may become superheated and may boil over when swirled. Store for up to 6 mo at 23°C.

YPD (1× and 2×)

	1× YPD	2× YPD
Glucose (20% in water)	100 mL	200 mL
Peptone	20 g	40 g
Yeast extract	10 g	20 g
Water	900 mL	800 mL

Dissolve all components in the water and sterilize by autoclaving for 20 min at 15 psi (1.05 kg/cm^2) on liquid cycle. Store for up to 6 mo at 23°C.

REFERENCE

Nehil MT, Tamble CM, Combs DJ, Kellogg DR, Lokey RS. 2007. Large-scale cytological profiling for functional analysis of bioactive compounds. *Chem Biol Drug Des* **69**: 258–264.

Protocol 2

Identification of Chemical–Genetic Interactions via Parallel Analysis of Barcoded Yeast Strains

Sundari Suresh,[1] Ulrich Schlecht,[1] Weihong Xu,[1] Molly Miranda,[1] Ronald W. Davis,[1] Corey Nislow,[2] Guri Giaever,[2] and Robert P. St.Onge[1,3]

[1]Stanford Genome Technology Center, Department of Biochemistry, Stanford University, Palo Alto, California 94304; [2]Pharmaceutical Sciences, University of British Columbia, Vancouver, British Columbia V6T 1Z3, Canada

The Yeast Knockout Collection is a complete set of gene deletion strains for the budding yeast, *Saccharomyces cerevisiae*. In each strain, one of approximately 6000 open-reading frames is replaced with a dominant selectable marker flanked by two DNA barcodes. These barcodes, which are unique to each gene, allow the growth of thousands of strains to be individually measured from a single pooled culture. The collection, and other resources that followed, has ushered in a new era in chemical biology, enabling unbiased and systematic identification of chemical–genetic interactions (CGIs) with remarkable ease. CGIs link bioactive compounds to biological processes, and hence can reveal the mechanism of action of growth-inhibitory compounds in vivo, including those of antifungal, antibiotic, and anticancer drugs. The chemogenomic profiling method described here measures the sensitivity induced in yeast heterozygous and homozygous deletion strains in the presence of a chemical inhibitor of growth (termed haploinsufficiency profiling and homozygous profiling, respectively, or HIPHOP). The protocol is both scalable and amenable to automation. After competitive growth of yeast knockout collection cultures, with and without chemical inhibitors, CGIs can be identified and quantified using either array- or sequencing-based approaches as described here.

MATERIALS

It is essential that you consult the appropriate Material Safety Data Sheets and your institution's Environmental Health and Safety Office for proper handling of equipment and hazardous materials used in this protocol.

RECIPES: Please see the end of this protocol for recipes indicated by <R>. Additional recipes can be found online at http://cshprotocols.cshlp.org/site/recipes.

Reagents

Agarose gel electrophoresis reagents
Chemical compound(s) under investigation
PCR reagents:

    Deionized water
    dNTPs (10 mM)
    MgCl$_2$ (25 mM)
    Primers (forward and reverse; see Steps 8 and 23)
    Taq polymerase (5 U/µL) with reaction buffer (10×) (NEB M0273L)

---

[3]Correspondence: bstonge@stanford.edu

Copyright © Cold Spring Harbor Laboratory Press; all rights reserved
Cite this protocol as *Cold Spring Harb Protoc*; doi:10.1101/pdb.prot088054

Reagents used only for array-based analysis (Steps 8–22):
- GenFlex Tag 16K Arrays (Affymetrix)
- GenFlex Tag 16K Chip Wash Buffer A <R>
- GenFlex Tag 16K Chip Wash Buffer B <R>
- GenFlex Tag 16K 2× Hybridization Buffer <R>
- GenFlex Tag 16K Labeling Mixture <R>
- GenFlex Tag 16K Tag Hybridization Mixture <R>
- Isopropanol (optional; see Step 17)
- Solvent for chemical compound(s) under investigation
- Tough-spots (ISC BioExpress L-1000–6)
- YeaStar Genomic DNA kit (Zymo Research)
- Yeast Knockout Collection (ThermoScientific)
- YPD (1× and 2×) <R>

Reagents used only for sequencing-based analysis (Steps 23–39):
- Agarose and SYBR safe
- Agencourt AMPure XP beads
- Deionized water
- Ethanol (70% v/v, freshly prepared)
- KAPA Library Quantification kit (Kapa Biosystems)
- MiSeq Reagent kit v2 (50 cycles) (Illumina)
- NaOH (2 N)

Equipment

Absorbance reader
Culture flasks (125 and 500 mL)
Equipment used only for array-based analysis (Steps 8–22):
- Affymetrix Fluidics Station 450
- Affymetrix GeneChip Scanner
- Affymetrix Hybridization Oven
- Boiling water bath
- Kimwipes (optional; see Step 17)
- Microcentrifuge tubes (0.5 mL, Eppendorf Safe-Lock; VWR 20901-505)

Equipment used only for sequencing-based analysis (Steps 23–39):
- 7900HT Fast Real-Time PCR System (Applied Biosystems)
- Bioanalyzer 2100 (Agilent Technologies) (optional; see Step 28)
- Magnetic rack (Invitrogen Dynal)
- MiSeq Benchtop Sequencer (Illumina)
- Qubit fluorometric system (Life Technologies) (optional; see Step 28)
- Vortexer

Freezer (−20°C)
Ice bucket
Microcentrifuge
Microcentrifuge tubes (1.5 mL)
PCR tubes
Temperature-controlled incubator with platform shaker
Thermocycler

METHOD

Unless otherwise noted, all steps are performed at room temperature, and all oligonucleotide sequences are listed 5′ to 3′.

Competitive Growth with and without Chemical Inhibitors

This protocol uses concentrated frozen aliquots of the Yeast Knockout Collection in which each strain is present in roughly equal abundance (Pierce et al. 2007). Note, when combining uniquely barcoded sets of heterozygous and homozygous deletion strains in the same experiment, twice as many cells of each heterozygous strain should be added to compensate for having only one copy of each barcode (homozygous strains contain two). The following steps can be used to assess a pool containing 1165 heterozygous strains (representing essential genes) and 5081 homozygous deletion strains (representing nonessential genes) over nine generations, in triplicate. Longer growth periods may increase assay sensitivity, but they require larger culture volumes (or redilution of smaller cultures) and will also exacerbate depletion of slower growing strains from the cultures. Inclusion of biological replicates increases the total cost of the experiment, but greatly improves detection of CGIs.

1. In a 500 mL flask, dilute an aliquot of pooled deletion strains into 200 mL of YPD to a final concentration of 6.0×10^4 cells/mL.

2. Recover the pool for 4 h at 30°C with modest shaking (200 rpm).

3. Aliquot 25 mL of the recovered culture into each of six 125 mL culture flasks.

4. Add the chemical compound under investigation to three flasks to a final concentration that inhibits growth by roughly 20%, and the compound's solvent (e.g., DMSO) to the remaining three flasks, and continue growth at 30°C with modest shaking (200 rpm).

   Compound concentration is an important variable for robust detection of CGIs, and must be carefully determined before the experiment. Compound and the compound's solvent are each added to three flasks to generate triplicate replicates.

5. Monitor growth of each culture and when they reach a concentration of 3×10^7 cells per mL, remove 1.5 mL (i.e., 4.5×10^7 cells) to a 1.5 mL microcentrifuge tube.

6. Centrifuge the collected cells for 3 min at ~20,000g and discard supernatant. Cell pellets can be stored for 3 mo at −20°C.

7. Extract genomic DNA from cell pellets using the YeaStar Genomic DNA kit, or by similar methods. Genomic DNA can be stored for 3 wk at −20°C.

   For the quantification of each strain within the pooled cultures, proceed to either Step 8 for array-based barcode quantification and data analysis or to Step 23 for sequencing-based barcode quantification and data analysis.

Array-Based Barcode Quantification and Data Analysis

The following steps (Steps 8–22) describe quantification of barcodes using the GenFlex Tag 16k Array from Affymetrix, which was specifically designed for compatibility with the yeast knockout collection (Pierce et al. 2006).

Microarray Hybridization and Scanning

8. Polymerase chain reaction (PCR) amplify UP and DOWN tags from each genomic DNA sample (from Step 7) in separate reactions.

   i. Prepare 50 µL "UP tag" PCRs for each sample as follows.

	Amount	Final concentration
10× reaction buffer	5 µL	1×
25 mM MgCl$_2$	2 µL	1 mM
10 mM dNTPs	0.4 µL	0.08 mM
100 µM UPTAG oligonucleotide (GATGTCCACGAGGTCTCT)	0.5 µL	1 µM
100 µM BUPKANMX4 oligonucleotide (Biotin-GTCGACCTGCAGCGTACG)	0.5 µL	1 µM
5 U/µL *Taq* polymerase	0.5 µL	0.05 U/µL
Purified genomic DNA (from Step 7)	~75 ng	
Deionized water	to 50 µL	

ii. Prepare reactions for the "DOWN tags" in the same way, but use DNTAG oligonucleotide (CGGTGTCGGTCTCGTAG) and BDNKANMX4 oligonucleotide (Biotin-GAAAACGAG CTCGAATTCATCG), in place of UPTAG and BUPKANMX4, respectively.

iii. Perform reactions with the following cycling conditions.

1 cycle	3 min	94°C
30 cycles	30 sec	94°C
	30 sec	55°C
	30 sec	72°C
1 cycle	3 min	72°C
Hold	Forever	4°C

9. Verify that PCRs produce the expected 60 bp product by analyzing 5 µL of each reaction by 2% agarose gel electrophoresis.

    A smaller band of 40 bp is often observed in both the template and no template reactions. This represents common primers forming a primer dimer and, if of lower intensity than the 60 bp band, will not affect the hybridization. If the 40 bp band is of greater intensity than the 60 bp band, PCR should be repeated.

    PCR products can be stored for up to 3 mo at −20°C.

10. Hydrate one GenFlex Tag 16K array for each sample with 75 µL of 1× Hybridization Buffer freshly diluted from 2× Hybridization Buffer, and incubate chips for 10–20 min at 42°C at 20 rpm in an Affymetrix Hybridization Oven.

11. Freshly prepare 90 µL of Tag Hybridization Mixture for each sample in 0.5 mL Safe-Lock microcentrifuge tubes. Add 30 µL of UP tag PCR product and 30 µL of DOWN tag PCR product (from Step 8) for a final volume of 150 µL.

12. Float each tube in boiling water for 2 min, transfer immediately to ice for 2 min, spin down condensate, and replace on ice.

13. Remove 1× Hybridization Buffer from each array, add 90 µL of the chilled PCR product solution prepared in Step 12, and cover both gaskets with Tough-spots.

14. Incubate arrays in an Affymetrix Hybridization oven for 16 h at 42°C, rotating at 20 rpm.

15. Immediately before washing the arrays, prepare Chip wash buffer A, Chip wash buffer B, and the labeling mixture.

16. Wash and label microarrays using an Affymetrix Fluidics Station 450 and the protocol FlexGenflex_Sv3_450 (http://www.affymetrix.com/support/technical/fluidics_scripts.affx) with the following modifications: change the temperature of the wash B step from 40°C to 42°C, and change the temperature of the staining step from 25°C to 42°C.

17. Place Tough-spots over both gaskets, check glass slide for cleanliness and bubbles.

    Glass can be cleaned with a Kimwipe and isopropanol, and bubbles can be removed by slowly pipetting Chip wash buffer A in and out of the array.

18. Scan arrays using an Affymetrix GeneChip Scanner.

Analysis of Microarray Data: Identification of CGIs

The following is a simple analytical method for the detection of CGIs. Additional optimization and/or alternative analyses may also be applied, such as algorithms to identify and remove array hybridization defects (Pierce et al. 2006), alternative methods for array normalization (Smyth et al. 2003; Smyth 2005), and nonparametric approaches for significance testing (Tusher et al. 2001). The performance of UP and DOWN tags from the same strain can often differ greatly and, therefore, CGIs can be identified if either barcode meets desired magnitude and significance thresholds. Ultimately, CGIs can be validated by retesting the chemical sensitivity of individual strains in isogenic cultures (Lee et al. 2005; St.Onge et al. 2007).

19. Extract the raw fluorescence measurements from CEL files and, for each barcode, calculate the average signal of the five replicate probes represented on the array.

    The microarray annotation file relating barcode sequence to strain is available online at cshprotocols.cshlp.org as supplementary material (TAG4_probe_info.txt).

20. Log$_2$-transform and then quantile normalize the total signal between experiments using the normalize.quantiles function of the preprocessCore package in R.

    Because they are amplified in separate PCRs, data for the UP and DOWN tags should be normalized separately.

21. For each barcode, calculate the magnitude of the CGI as the log$_2$-transformed fold change; $\mu_t - \mu_c$, where μ_t is the average of the log$_2$-transformed and normalized treatment replicates, and μ_c is the average of the log$_2$-transformed and normalized control replicates. Chemical sensitivity will thus yield negative scores, and chemical resistance positive scores.

22. For each barcode, determine the significance of the CGI with a moderated T-test using the R package LIMMA (Smyth et al. 2003; Smyth 2005), and a multiple testing correction using the Benjamini and Hochberg False Discovery Rate method (Benjamini and Hochberg 1995).

Sequencing-Based Barcode Quantification and Data Analysis

This alternative protocol for barcode quantification employs Barcode analysis by Sequencing (Bar-Seq), in which next-generation sequencing is used in place of microarrays to quantify barcode abundance following pooled growth (Smith et al. 2009). It represents a modified version of previously described methods (Smith et al. 2010, 2012; Gresham et al. 2011; Robinson et al. 2013) and uses Illumina's MiSeq benchtop sequencer. Processing of six samples in one MiSeq flow cell is described. This protocol is optimized for single-read sequencing and a 50-cycle reagent kit; however, it can be adapted (and further multiplexed) for other Illumina sequencing platforms.

Sequencing

23. PCR amplify UP and DOWN tags from each of the six samples generated in Step 7 in separate reactions using primers containing sequence for the Illumina adapters, short index barcodes for sample multiplexing, and the common priming sites for each tag.

    i. Prepare 50 μL "UP tag" PCRs as follows.

	Amount	Final concentration
10× reaction buffer	5 μL	1×
25 mM MgCl$_2$	2 μL	1 mM
10 mM dNTPs	0.4 μL	0.08 mM
10 μM UP-X oligonucleotide (where X is 1–6; see Table 1)	0.5 μL	0.1 μM
10 μM UP-KanMX oligonucleotide (see Table 1)	0.5 μL	0.1 μM
5 U/μL *Taq* polymerase	0.5 μL	0.05 U/μL
Purified genomic DNA (from Step 7)	~75 ng	
Deionized water	to 50 μL	

    ii. Prepare the reaction mixture for DOWN tags as above, but use DN-X oligonucleotides and DN-KanMX oligonucleotide (see Table 1) in place of UP-X and UP-KanMX, respectively.

    iii. Perform reactions with the following cycling conditions.

1 cycle	3 min	94°C
30 cycles	30 sec	94°C
	30 sec	55°C
	30 sec	72°C
1 cycle	3 min	72°C
Hold	Forever	4°C

24. Verify the PCRs were successful by analyzing a small aliquot of each reaction by 2% agarose gel electrophoresis.

    Each PCR is expected to yield 144 and 147 bp fragments for UP and DN tags, respectively. PCR products can be stored for 3 mo at −20°C.

Chapter 25

TABLE 1. Primer sequences for bar-seq

Oligonucleotide	Sequence (5′ to 3′)
UP-kanMX	CAAGCAGAAGACGGCATACGAGATGTCGACCTGCAGCGTACG
UP-1	AATGATACGGCGACCACCGAGATCTACACTCTTTCCCTACACGACGCTCTTCCGATCT**TGCTAA**GATGTCCACGAGGTCTCT
UP-2	AATGATACGGCGACCACCGAGATCTACACTCTTTCCCTACACGACGCTCTTCCGATCT**AGGTCA**GATGTCCACGAGGTCTCT
UP-3	AATGATACGGCGACCACCGAGATCTACACTCTTTCCCTACACGACGCTCTTCCGATCT**GGATTA**GATGTCCACGAGGTCTCT
UP-4	AATGATACGGCGACCACCGAGATCTACACTCTTTCCCTACACGACGCTCTTCCGATCT**CGTTGA**GATGTCCACGAGGTCTCT
UP-5	AATGATACGGCGACCACCGAGATCTACACTCTTTCCCTACACGACGCTCTTCCGATCT**ATGATC**GATGTCCACGAGGTCTCT
UP-6	AATGATACGGCGACCACCGAGATCTACACTCTTTCCCTACACGACGCTCTTCCGATCT**CTTAAC**GATGTCCACGAGGTCTCT
DN-kanMX	CAAGCAGAAGACGGCATACGAGATGAAAACGAGCTCGAATTCATCG
DN-1	AATGATACGGCGACCACCGAGATCTACACTCTTTCCCTACACGACGCTCTTCCGATCT**TCCCGA**CGGTGTCGGTCTCGTAG
DN-2	AATGATACGGCGACCACCGAGATCTACACTCTTTCCCTACACGACGCTCTTCCGATCT**AAACCT**CGGTGTCGGTCTCGTAG
DN-3	AATGATACGGCGACCACCGAGATCTACACTCTTTCCCTACACGACGCTCTTCCGATCT**CTGACT**CGGTGTCGGTCTCGTAG
DN-4	AATGATACGGCGACCACCGAGATCTACACTCTTTCCCTACACGACGCTCTTCCGATCT**GTCGGA**CGGTGTCGGTCTCGTAG
DN-5	AATGATACGGCGACCACCGAGATCTACACTCTTTCCCTACACGACGCTCTTCCGATCT**GTGTAG**CGGTGTCGGTCTCGTAG
DN-6	AATGATACGGCGACCACCGAGATCTACACTCTTTCCCTACACGACGCTCTTCCGATCT**AGAGGA**CGGTGTCGGTCTCGTAG

Bold sequence indicates 6-base index barcode.

25. Purify the PCR amplicons by adding 50 µL of AMPure XP beads to 30 µL of each PCR, incubate for 5 min at room temperature, and collect the beads using a magnetic rack.

26. Remove the supernatant and wash the beads twice with 200 µL of freshly prepared 70% ethanol while keeping the tubes in the magnetic rack.

27. Aspirate the supernatant, remove tubes from the magnetic rack, and let the beads air-dry for 5 min.

28. Add 30 µL of deionized water to elute the DNA. Pipette the beads up and down a few times until they are resuspended and then leave for 5 min at room temperature. Return the tubes to the magnetic rack to collect the beads. Transfer supernatant to a fresh 1.5-mL tube.

    The size and concentration of the purified PCR amplicons can be analyzed using a highly sensitive electrophoretic system (such as the Bioanalyzer 2100 instrument). Samples can also be quantified with fluorescent DNA-binding dye approaches (such as the Qubit fluorometric system).

29. Dilute a small aliquot of each DNA sample 1:50,000 and 1:100,000 with KAPA Library Quantification kit dilution buffer and then quantify the concentration of every sample using the KAPA Library Quantification kit and a real-time PCR system (such as the Applied Biosystems 7900HT instrument).

30. Mix the 12 DNA samples so that the final concentration of the resulting sequencing library is 1 nM and every sample is represented in equimolar amounts.

31. Thaw the ready-to-use reagent cartridge (containing reagents and buffer HT1) of an Illumina MiSeq Reagent kit for 30 min in deionized water. Keep thawed buffer HT1 on ice.

32. Combine 10 µL of the sequencing library with 10 µL of 0.2 N NaOH (freshly diluted from a 2 N solution), and incubate for 5 min at room temperature.

33. Add 980 µL of prechilled HT1 buffer to the denatured sequencing library, vortex, and place on ice. The concentration of the library is now 10 pM.

34. Dilute the PhiX control stock solution to 2 nM in water, then combine 10 µL of this diluted PhiX control solution with 10 µL of 0.2 N NaOH, and incubate for 5 min at room temperature.

35. Add 980 µL of prechilled HT1 buffer to the denatured PhiX control, vortex, and place on ice. The concentration of the PhiX control is now 20 pM and can be stored for 3 wk at −20°C.

36. Combine 225 µL of the denatured and diluted sequencing library (from Step 33) with 12.5 µL PhiX control (from Step 35) and 762.5 µL HT1 buffer. The final concentrations of the sequencing library and PhiX control are 2.25 and 0.25 pM, respectively.

37. Inject 600 µL of this mixture into the cartridge of the MiSeq Reagent kit and start the sequencing run on the MiSeq instrument.

Data Analysis: Identification of CGIs

38. Extract the sequencing reads from the fastq-file and use the first six bases (representing the sample index) to assign reads to samples, and the bases following the common primer sequences (bases 25–44 for the UP tags and 24–43 for the DOWN tags) to assign reads to each deletion strain. For the UP and DOWN tags, include inexact matches within a Levenshtein distance of 2.

39. Add one to each barcode's count in each sample, and calculate the fold-change ($\mu_t - \mu_c$, where μ_t is the average of the $\log_2$-transformed treatment replicates and μ_c is the average of the $\log_2$-transformed control replicates), and the significance of the CGI using the R package DESeq (Anders and Huber 2010). Significance can be corrected for multiple testing using the Benjamini and Hochberg False Discovery Rate (FDR) method (Benjamini and Hochberg 1995).

    Even though the UP and DOWN barcodes associated with a given strain should in theory produce very similar results, in practice, the baseline signal of tags from the same strain can often differ greatly. Therefore, CGIs can be identified if either barcode meets the desired magnitude and significance thresholds.

RELATED INFORMATION

This protocol is related to previously described methods (Lum et al. 2004; Pierce et al. 2007; Smith et al. 2011) and is both scalable and adaptable for automation (Hillenmeyer et al. 2008; Hoon et al. 2008; Proctor et al. 2011; Hoepfner et al. 2014; Lee et al. 2014). Detailed information on the yeast knockout collection can be found in Winzeler et al. (1999) and Giaever et al. (2002). Related resources that followed are described by Ho et al. (2009).

RECIPES

GenFlex Tag16K Chip Wash Buffer A

Reagent	Amount	Final concentration
SSPE (20×) (Invitrogen 15591-043)	150 mL	6×
Tween 20 (10%)	500 µL	0.01%
Deionized water	349.5 mL	

Prepare 500 mL immediately before use, and filter-sterilize.

GenFlex Tag 16K Chip Wash Buffer B

Reagent	Amount	Final concentration
SSPE (20×) (Invitrogen 15591-043)	75 mL	3×
Tween 20 (10%)	500 µL	0.01%
Deionized water	424.5 mL	

Prepare 500 mL immediately before use, and filter-sterilize.

GenFlex Tag 16K 2× Hybridization Buffer

First prepare ~5 mL of 0.66 M MES hydrate (Sigma-Aldrich M5287) and ~5 mL of 1.78 M MES sodium salt (Sigma-Aldrich M5057). Mix equal volumes of each to create a 12× MES stock solution. Use this to make 50 mL of 2× hybridization buffer as follows.

Reagent	Amount	Final concentration
MES stock (12×)	8.3 mL	2×
NaCl (5 M)	17.7 mL	1.77 M
EDTA (500 mM)	4 mL	40 mM
Tween 20 (10%)	100 µL	0.02%
Deionized water	19.9 mL	

Store buffer for up to 3 mo at 4°C.

GenFlex Tag 16K Labeling Mixture

The following recipe prepares 500 µL. Multiply the recipe according to the total volume required (each GenFlex Tag 16K array requires 500 µL of labeling mixture).

Reagent	Amount	Final concentration
SSPE (20×) (Invitrogen 15591-043)	150 µL	6×
Denhardt's (50×) (Sigma-Aldrich D2532)	10 µL	1×
Tween 20 (1%)	5 µL	0.01%
Streptavidin-phycoerythrin (1 mg/mL) (Invitrogen S866)	0.84 µL	1.68 µg/mL
Deionized water	334.16 µL	

Prepare immediately before use.

GenFlex Tag16K Tag Hybridization Mixture

Reagent	Amount	Final concentration
GenFlex Tag 16K 2× Hybridization Buffer <R>	74.5 µL	1.7×
B213 oligonucleotide (Biotin-CTGAACGGTAGCATCTTGAC) (2 µM)	0.5 µL	11 nM
Blocking oligonucleotides (12.5 µM stocks):		
Uptag (GATGTCCACGAGGTCTCT)	1.5 µL	0.2 µM
Dntag (CGGTGTCGGTCTCGTAG)	1.5 µL	0.2 µM
Uptagkanmx (GTCGACCTGCAGCGTACG)	1.5 µL	0.2 µM
Dntagkanmx (CGAGCTCGAATTCATCG)	1.5 µL	0.2 µM
Uptagcomp (AGAGACCTCGTGGACATC)	1.5 µL	0.2 µM
Dntagcomp (CTACGAGACCGACACCG)	1.5 µL	0.2 µM
Upkancomp (CGTACGCTGCAGGTCGAC)	1.5 µL	0.2 µM
Dnkancomp (CGATGAATTCGAGCTCG)	1.5 µL	0.2 µM
Denhardt's (50×) (Sigma-Aldrich D2532)	3 µL	1.7×

For each sample, combine the above for a total of 90 µL. Prepare immediately before use.

YPD (1× and 2×)

	1× YPD	2× YPD
Glucose (20% in water)	100 mL	200 mL
Peptone	20 g	40 g
Yeast extract	10 g	20 g
Water	900 mL	800 mL

Dissolve all components in the water and sterilize by autoclaving for 20 min at 15 psi (1.05 kg/cm^2) on liquid cycle. Store for up to 1 yr at room temperature.

ACKNOWLEDGMENTS

We are grateful to the U.S. National Institutes of Health (P01HG000205 to R.W.D. and R01HG003317 to G.G., C.N., and R.W.D.), the Canadian Institutes for Health Research and the Canadian Cancer Society Research Institute (C.N. and G.G.) for funding and to Ana Maria Aparicio for helpful discussions.

REFERENCES

Anders S, Huber W. 2010. Differential expression analysis for sequence count data. *Genome Biol* 11: R106.

Benjamini Y, Hochberg Y. 1995. Controlling the false discovery rate: A practical and powerful approach to multiple testing. *J R Stat Soc Series B* 57: 289–300.

Giaever G, Chu AM, Ni L, Connelly C, Riles L, Veronneau S, Dow S, Lucau-Danila A, Anderson K, Andre B, et al. 2002. Functional profiling of the *Saccharomyces cerevisiae* genome. *Nature* 418: 387–391.

Gresham D, Boer VM, Caudy A, Ziv N, Brandt NJ, Storey JD, Botstein D. 2011. System-level analysis of genes and functions affecting survival during nutrient starvation in *Saccharomyces cerevisiae*. *Genetics* 187: 299–317.

Hillenmeyer ME, Fung E, Wildenhain J, Pierce SE, Hoon S, Lee W, Proctor M, St.Onge RP, Tyers M, Koller D, et al. 2008. The chemical genomic portrait of yeast: Uncovering a phenotype for all genes. *Science* 320: 362–365.

Ho CH, Magtanong L, Barker SL, Gresham D, Nishimura S, Natarajan P, Koh JL, Porter J, Gray CA, Andersen RJ, et al. 2009. A molecular barcoded yeast ORF library enables mode-of-action analysis of bioactive compounds. *Nat Biotechnol* 27: 369–377.

Hoepfner D, Helliwell SB, Sadlish H, Schuierer S, Filipuzzi I, Brachat S, Bhullar B, Plikat U, Abraham Y, Altorfer M, et al. 2014. High-resolution chemical dissection of a model eukaryote reveals targets, pathways and gene functions. *Microbiol Res* 169: 107–120.

Hoon S, Smith AM, Wallace IM, Suresh S, Miranda M, Fung E, Proctor M, Shokat KM, Zhang C, Davis RW, et al. 2008. An integrated platform of genomic assays reveals small-molecule bioactivities. *Nat Chem Biol* 4: 498–506.

Lee AY, St.Onge RP, Proctor MJ, Wallace IM, Nile AH, Spagnuolo PA, Jitkova Y, Gronda M, Wu Y, Kim MK, et al. 2014. Mapping the cellular response to small molecules using chemogenomic fitness signatures. *Science* 344: 208–211.

Lee W, St.Onge RP, Proctor M, Flaherty P, Jordan MI, Arkin AP, Davis RW, Nislow C, Giaever G. 2005. Genome-wide requirements for resistance to functionally distinct DNA-damaging agents. *PLoS Genet* 1: e24.

Lum PY, Armour CD, Stepaniants SB, Cavet G, Wolf MK, Butler JS, Hinshaw JC, Garnier P, Prestwich GD, Leonardson A, et al. 2004. Discovering modes of action for therapeutic compounds using a genome-wide screen of yeast heterozygotes. *Cell* 116: 121–137.

Pierce SE, Davis RW, Nislow C, Giaever G. 2007. Genome-wide analysis of barcoded *Saccharomyces cerevisiae* gene-deletion mutants in pooled cultures. *Nat Protoc* 2: 2958–2974.

Pierce SE, Fung EL, Jaramillo DF, Chu AM, Davis RW, Nislow C, Giaever G. 2006. A unique and universal molecular barcode array. *Nat Methods* 3: 601–603.

Proctor M, Urbanus ML, Fung EL, Jaramillo DF, Davis RW, Nislow C, Giaever G. 2011. The automated cell: Compound and environment screening system (ACCESS) for chemogenomic screening. *Methods Mol Biol* 759: 239–269.

Robinson DG, Chen W, Storey JD, Gresham D. 2013. Design and analysis of Bar-seq experiments. *G3 (Bethesda)* 4: 11–18.

Smith AM, Durbic T, Kittanakom S, Giaever G, Nislow C. 2012. Barcode sequencing for understanding drug-gene interactions. *Methods Mol Biol* 910: 55–69.

Smith AM, Durbic T, Oh J, Urbanus M, Proctor M, Heisler LE, Giaever G, Nislow C. 2011. Competitive genomic screens of barcoded yeast libraries. *J Vis Exp* 54: 2864.

Smith AM, Heisler LE, Mellor J, Kaper F, Thompson MJ, Chee M, Roth FP, Giaever G, Nislow C. 2009. Quantitative phenotyping via deep barcode sequencing. *Genome Res* 19: 1836–1842.

Smith AM, Heisler LE, St. Onge RP, Farias-Hesson E, Wallace IM, Bodeau J, Harris AN, Perry KM, Giaever G, Pourmand N, et al. 2010. Highly-multiplexed barcode sequencing: An efficient method for parallel analysis of pooled samples. *Nucleic Acids Res* 38: e142.

Smyth GK. 2005. Limma: Linear models for microarray data. *Bioinformatics and Computational Biology Solutions Using R and Bioconductor* (ed. Gentleman R, Carey V, Huber W, Irizarry R, Dudoit S), pp. 397–420. Springer, New York.

Smyth GK, Yang YH, Speed T. 2003. Statistical issues in cDNA microarray data analysis. *Methods Mol Biol* 224: 111–136.

St.Onge RP, Mani R, Oh J, Proctor M, Fung E, Davis RW, Nislow C, Roth FP, Giaever G. 2007. Systematic pathway analysis using high-resolution fitness profiling of combinatorial gene deletions. *Nat Genet* 39: 199–206.

Tusher VG, Tibshirani R, Chu G. 2001. Significance analysis of microarrays applied to the ionizing radiation response. *Proc Natl Acad Sci* 98: 5116–5121.

Winzeler EA, Shoemaker DD, Astromoff A, Liang H, Anderson K, Andre B, Bangham R, Benito R, Boeke JD, Bussey H, et al. 1999. Functional characterization of the *S. cerevisiae* genome by gene deletion and parallel analysis. *Science* 285: 901–906.

CHAPTER 26

Prions

Dmitry Kryndushkin,[1] Herman K. Edskes,[2] Frank P. Shewmaker,[1] and Reed B. Wickner[2,3]

[1]Department of Pharmacology, Uniformed Services University of Health Sciences, Bethesda, Maryland 20814;
[2]Laboratory of Biochemistry and Genetics, National Institute of Diabetes and Digestive and Kidney Diseases, National Institutes of Health, Bethesda, Maryland 20892-0830

Infectious proteins (prions) are usually self-templating filamentous protein polymers (amyloids). Yeast prions are genes composed of protein and, like the multiple alleles of DNA-based genes, can have an array of "variants," each a distinct self-propagating amyloid conformation. Like the lethal mammalian prions and amyloid diseases, yeast prions may be lethal, or only mildly detrimental, and show an array of phenotypes depending on the protein involved and the prion variant. Yeast prions are models for both rare mammalian prion diseases and for several very common amyloidoses such as Alzheimer's disease, type 2 diabetes, and Parkinson's disease. Here, we describe their detection and characterization using genetic, cell biological, biochemical, and physical methods.

BACKGROUND

[PSI+] and [URE3] were discovered as nonchromosomal genetic elements (Cox 1965; Lacroute 1971), and then found to be prions based on their unusual genetic properties (Wickner 1994, for yeast prion nomenclature see Box 1). Soon thereafter, the [Het-s] prion of *Podospora anserina* (Coustou et al. 1997) and the yeast [PIN+] prion (Derkatch et al. 2001) were found and there are now 10 well-documented prions of *S. cerevisiae* (Table 1; reviewed in Wickner et al. [2013]). Although a prion can be a self-modifying protein, such as the vacuolar protease B that cleaves/activates its own inactive precursor (Roberts and Wickner 2003), most are filamentous ordered polymers of a specific protein, similar to the disease-associated amyloids of mammals. Like the mammalian pathologic amyloids, the yeast and fungal prion amyloids have a cross-β structure, meaning that the β strands run essentially perpendicular to the long axis of the filaments. The [URE3], [PSI+], and [PIN+] prions are based on in-register parallel β sheet amyloid filaments of Ure2p, Sup35p, and Rnq1p, respectively (reviewed in Wickner et al. 2013). Once one of these proteins forms amyloid, it acts catalytically to convert the same protein to the amyloid form. Passage of amyloid filaments to another cell results in that cell, and its progeny, being "infected" with the prion.

GENETIC APPROACHES USED IN STUDYING YEAST PRIONS

Yeast prions began as phenomena in classical genetics, nonchromosomal genetic elements showing 4 prion : 0 segregation in meiosis (Cox 1965; Lacroute 1971). Cytoduction, mating without nuclear fusion followed by separation of the parental nuclei with mixed cytoplasms into the daughter cells (Conde and Fink 1976), is now usually used to show nonchromosomal inheritance, and is an impor-

[3]Correspondence: wickner@helix.nih.gov

Copyright © Cold Spring Harbor Laboratory Press; all rights reserved
Cite this introduction as *Cold Spring Harb Protoc*; doi:10.1101/pdb.top077586

> **BOX 1. YEAST PRION NOMENCLATURE**
>
> Yeast prions are in square brackets like other nonchromosomal genes. Capitals indicate the dominant allele (the prion), and lower case shows the recessive allele (the normal protein). Thus, [URE3] or its absence [ure-o] (as [rho-o] means no mitochondrial DNA and [kil-o] means no M dsRNA—the "killer factor"). The presence of [PSI] is shown as [PSI+] and its absence [psi−].

tant method in most prion studies (see Protocol 1: Genetic Methods for Studying Prions [Wickner et al. 2015]). That the nonchromosomal genes [URE3] and [PSI+] were actually prions was first shown based on their unusual genetic properties (Wickner 1994).

Three Genetic Criteria for a Yeast Prion

Reversible Curing

If a certain treatment of cells carrying a nonchromosomal genetic element results in its elimination from the entire population, one says cells are "cured" and it is known in many such cases that the entire replicon is lost, and not simply mutated. Even if a prion can be cured, it should arise again at some low frequency in the cured strain (Wickner 1994). This is unlike a nucleic acid replicon (such as mitochondrial DNA) that, once cured, does not arise again de novo. The protein capable of forming the prion is still present and can again form amyloid. Reversible curability is frequently misinterpreted to mean just curability by the Hsp104 inhibitor, guanidine (see below). Note that in many strains, guanidine cures or induces mutation of mitDNA (Villa and Juliani 1980), so a trait altered by guanidine cannot be assumed to be caused by a prion (e.g., Halfmann et al. 2012).

Prion Protein Overproduction Increases Prion Generation

The de novo formation of a nonchromosomal genetic element in itself suggests a prion. However, because prions are altered protein forms that have the ability to catalyze their own formation, the increase in the frequency of a prion arising on overproduction of the prion-forming protein is particularly strong evidence that one is dealing with a prion (Wickner 1994).

Phenotype Relationship

The prion protein structural gene is necessary for the propagation of the prion, just as many chromosomal genes are needed for propagation of nucleic acid replicons. Generally, the prion phenotype resembles the phenotype that results from loss of function of the prion protein-coding gene. For example, the prion phenotypes of [PSI+], [URE3], [SWI+], [OCT+], [BETA], [MOT+], and [MOD+] are attributable to deficiency of the normal form of their respective prion proteins. Thus phenotypic similarity between nonchromosomal genes and recessive mutants in the genes that are required for their propagation is evidence that the nonchromosomal gene is a prion (Wickner 1994).

[PIN+] is a prion of Rnq1p that was discovered as a nonchromosomal genetic element necessary for the induction of the [PSI+] prion by overexpression of Sup35p (Derkatch et al. 1997, 2001). The amyloid of Rnq1p in [PIN+] strains occasionally cross-seeds Sup35p amyloid formation (as well as Ure2p amyloid formation) thus facilitating prion formation (Derkatch et al. 2001). Derkatch et al. showed that overproduction of any of several Q/N-rich proteins could seed generation of [PSI+]. Rnq1p is rich in Q and N residues (hence its name) and had been shown capable of self-propagating aggregation (Sondheimer and Lindquist 2000). Derkatch et al. showed that amyloid of Rnq1p was the basis of [PIN+] by applying the above three genetic criteria (Derkatch et al. 2001). Several of the

TABLE 1. Prions of *S. cerevisiae* and *Podospora anserina*

Prion	Affected protein	Normal function	Phenotype	References
[URE3]	Ure2p	Nitrogen catabolite repression	Inappropriate derepression of genes for catabolism of poor nitrogen sources; very slow growth	Wickner 1994; McGlinchey et al. 2011
[PSI+]	Sup35p	Translation termination	Read-through of translation termination codons; lethality; very slow growth	Wickner 1994; McGlinchey et al. 2011
[PIN+]	Rnq1p	Unknown	Increased generation of [PSI+] or [URE3]	Derkatch et al. 1997, 2001
[SWI+]	Swi1p	Chromatin remodeling	Inability to use nonfermentable C sources; slow growth	Du et al. 2008
[OCT+]	Cyc8p	Transcription repression	Derepressed transcription; flocculence	Patel et al. 2009
[MOT+]	Mot3p	Repressor of genes for anaerobic growth	Derepressed anaerobic genes	Alberti et al. 2009
[ISP+]	Sfp1p	Transcription factor	Decreased translation termination codon read-through	Rogoza et al. 2010
[MOD+]	Mod5p	tRNA isopentenyl—transferase	Slow growth; resistance to azole antifungal drugs	Suzuki et al. 2012
[Het-s]	HET-s	Heterokaryon incompatibility	Heterokaryon incompatibility (*Podospora anserina*)	Coustou et al. 1997
[BETA]	Prb1p	Vacuolar protease B	Necessary for normal sporulation, survival in stationary phase.	Roberts and Wickner 2003

other proteins identified in this screen, or similar screens, have also proven to be prions, including Swi1p ([SWI+]), Cyc8p ([OCT+]), and Mod5p ([MOD+]) (Du et al. 2008; Patel et al. 2009; Suzuki et al. 2012). It is expected that this method will be useful in finding prions of other organisms, particularly because the prion domain of Mod5p is not Q/N rich, unlike those of other yeast prions (Suzuki et al. 2012).

Curing Yeast Prions

Although there are no effective treatments for mammalian prion diseases, there are many ways to cure yeast prions. This has been a particularly useful method of detecting genes other than the prion protein structural gene that are involved in prion generation or propagation. Curing of [PSI+] and other prions with millimolar concentrations of guanidine (Tuite et al. 1981) has been an extremely useful tool. Overproduction of the disaggregase Hsp104 cures [PSI+] (Chernoff and Ono 1992), which led to the finding that Hsp104 is also necessary for propagation of [PSI+] (Chernoff et al. 1995) and most other yeast prions. Indeed, it is by inhibition of Hsp104 that guanidine cures prions (Ferreira et al. 2001; Jung and Masison 2001; Jung et al. 2002). Studies of other genes affecting the Hsp104-overproduction curing of [PSI+] have identified many other chaperones and nonchaperone components (Chernoff et al. 1999; Newnam et al. 1999; Allen et al. 2007; Reidy and Masison 2010; Kiktev et al. 2012). Curing of [URE3] by overproduced Ydj1p showed that Hsp40s could be involved in prion propagation (Moriyama et al. 2000). Curing of [URE3] by overproduced nucleotide exchange factors for Hsp70s (Kryndushkin and Wickner 2007) supported Masison's model of nucleotide regulation of the Hsp70 role in prion propagation (reviewed by Sharma and Masison [2009]). A general screen for factors whose overproduction cures [URE3] revealed the somewhat homologous Btn1p and Cur1p, two proteins that, at normal levels, cure most variants of the [URE3] prion arising (Kryndushkin et al. 2008; Wickner et al. 2014).

Transfection

Transfection of yeast cells with amyloid formed from recombinant prion proteins was viewed as a final proof of the prion model, and showed that amyloid was indeed the infectious material (King and Diaz-Avalos 2004; Tanaka et al. 2004; Brachmann et al. 2005). This approach has since become an important method for characterizing prion amyloids (Brachmann et al. 2006); see Protocol 2: Prion Transfection of Yeast (Edskes et al. 2015).

Chapter 26

BIOCHEMICAL METHODS

The protease-resistance of Ure2p in extracts of [URE3]-carrying strains, like that of PrP in scrapie-infected tissues, provided the first biochemical evidence for the yeast prions (Masison and Wickner 1995). The rapid sedimentation of Sup35p in [PSI+] cells was the first evidence for aggregation as the mode of prion formation (Paushkin et al. 1996). Cold sodium dodecyl sulfate (SDS) solubilizes most cellular components, but not amyloid fibers, so agarose gels of cold SDS-treated extracts, the semi-denaturing detergent agarose gel electrophoresis method, have been widely used to characterize different prion variants, and even to screen for prions (Kryndushkin et al. 2003, 2013), and this method is detailed by its developer in an associated protocol (see Protocol 3: Isolation and Analysis of Prion and Amyloid Aggregates from Yeast Cells [Kryndushkin et al. 2015]).

CELL BIOLOGICAL METHODS

Most prions are aggregated forms of normally evenly dispersed soluble proteins. Prion proteins have a special domain that determines the prion properties of the protein, and constitutes the part that forms amyloid. Fusions of the full-length prion protein, or just the prion domain, with green fluorescent protein have been used in yeast to study prion behavior, and to document protein aggregation in vivo (Patino et al. 1996). Fluorescence recovery after photobleaching has been used to measure mobility of prion aggregates in vivo (Wu et al. 2006).

COMPUTATIONAL METHODS

The most successful method for predicting prion-forming ability is the PAPA program developed for Q/N-rich regions such as those that are the prion domains of most yeast prions (Toombs et al. 2012). The method is based solely on the amino acid composition of the domain in question because it is known that shuffling the sequence of the Ure2p or Sup35p prion domains does not adversely affect their ability to become prions (Ross et al. 2005).

PHYSICAL STUDIES OF YEAST PRION AMYLOIDS

Electron Microscopy

The filamentous nature of prion amyloids must be verified by electron microscopy. Their morphological properties are also important as they may differ with prion variants. Amyloid filaments can be uniform or heterogeneous with various widths and lengths. Filaments may remain solitary or laterally bundle and can have a variety of morphologies. Filaments may be of spiral appearance (or not), straight, curved, and helical. Infectious amyloid made from recombinant yeast prion proteins can be examined by making a suspension in water, placing 5 µL on a carbon-coated grid (commercially available) for 2 min, wicking off the excess water, washing once with 5 µL of water, applying 5 µL of 2% uranyl acetate for 1 or 2 min, wicking off the excess, and allowing the grid to air-dry. The details of insertion of the grid into the microscope, focusing the beam, and collection of images are machine-specific.

Measurement of Filament Mass-per-Length

This is important for distinguishing different amyloid architectures (Paravastu et al. 2008). For example, a β helix has ≤ 0.5 molecules per 4.7 Å, while parallel in-register architectures may have 1 (for the yeast prion amyloids of Sup35p, Ure2p, or Rnq1p), 2 (for one kind of Abeta fiber), or 3 (for another Abeta fiber variant). This is generally performed using a scanning transmission electron microscope (Wall and Hainfeld 1986), but can also be done using a transmission electron microscope in tilted beam (dark field) mode (Chen et al. 2009). Using thin-layer graphene as a sample support enables using lower beam intensities to preserve sample integrity and reduces background scattering for more accurate mass measurements (Jeon et al. 2013).

X-Ray Fiber Diffraction

This is important in establishing the cross-β structure (β strands oriented perpendicular to the fiber long axis) of amyloid filaments (Eanes and Glenner 1968; Stubbs 1999). The most difficult part of this method is orienting the amyloid filaments in the capillary tube used for the diffraction experiment. An alternative is electron diffraction using transmission electron microscopy and identifying a field of unstained filaments that are by chance largely bundled and oriented in one direction (Baxa et al. 2005; Shewmaker et al. 2009). The diffraction is measured for the aligned sample to determine repeated atomic spacing. Of course, even unoriented filaments can be used to show that there is a predominant β sheet secondary structure, by either X-ray or electron diffraction.

Hydrogen–Deuterium Exchange

The amide hydrogens of proteins will exchange with deuterium when the protein is placed in D_2O. In the amyloid form, the exchange rate is slowed, compared with the soluble form, by both isolation of a site from the solvent and by hydrogen bonds. Detection of exchange can be done by either nuclear magnetic resonance (NMR) or mass spectrometry. If the material is of homogeneous structure, the signal should show a single exponential decay for each amide hydrogen. As discussed below, heterogeneity of amyloid preparations is a problem in hydrogen–deuterium (HD) exchange experiments as in other physical studies.

Solid-State NMR

Because amyloids do not form crystals and are very large, neither X-ray crystallography nor solution NMR is ideal for examining their structure. However, solid-state NMR (ssNMR) is particularly suited for this problem, as illustrated by Tycko's work on structures of amyloids of Abeta peptide (Alzheimer's disease) and amylin (type II diabetes) (reviewed in Tycko 2006) and the determination of the structure of the prion domain amyloid of HET-s by Meier and coworkers (Ritter et al. 2005; Wasmer et al. 2008).

At least several milligrams of homogeneous amyloid are needed for ssNMR experiments. Proteins are labeled with ^{13}C and ^{15}N by chemical synthesis, or synthesis in *Escherichia coli* with labeled precursors. Chemical synthesis is preferable, but the prion domains of yeast prion proteins are too long for this procedure. Site-specific labeling in *E. coli* is limited to amino acids whose biosynthesis is efficiently repressed by the supplied labeled amino acid and whose metabolism does not result in shuffling of label: Leu, Ile, Val, Met, Tyr, Trp, Lys, and Phe have been used and Ala for short labeling times. Except for HET-s, which has only one prion variant in vivo, and forms a unique structure in vitro (Ritter et al. 2005; Wasmer et al. 2008), amyloid forming spontaneously in vitro from recombinant prion proteins is not generally homogeneous as shown by the spectrum of prion variants resulting from its transfection into yeast cells (e.g., Brachmann et al. 2005). Even filament preparations with a uniform appearance by electron microscopy (EM) may be a mixture of forms. Filaments of Sup35NM (the prion domain of Sup35p), which produce largely homogeneous "strong" or "weak" variants on transfection appear to be heterogeneous based on HD exchange data (Toyama et al. 2007) and may be a mixture of transmission variants (Bateman and Wickner 2013). Favoring unique amyloid structure, are seeding with filaments isolated from cells (King and Diaz-Avalos 2004), and certain filamentation conditions (Helsen and Glover 2012) including low monomer concentration and a high ratio of seeds-to-monomers so that self-seeding is not efficient (e.g., Kryndushkin et al.

2011). Nonetheless, the limited success of these approaches results in two-dimensional NMR experiments showing wide peaks that limit the possibility of complete structural determination.

The chemical shifts distinguish between helical, sheet, and random coil structures. If peaks can be assigned to specific residues, then this chemical shift data can be used to map out the secondary structure of specific amino acid residues.

Experiments to date indicate that the yeast prion amyloids have an in-register parallel folded β sheet architecture (reviewed in Tycko and Wickner [2013] and Wickner et al. [2013]). The evidence for this architecture comes from dipolar recoupling experiments that measure the distance from labeled atoms to the nearest neighbor labeled atom. Labeling specific carbonyl carbons in the prion domain of Sup35p, Ure2p, or Rnq1p with ^{13}C has generally resulted in a distance of ~5 Å, and the nearest neighbor has been shown to be generally on another molecule as expected for this architecture (Shewmaker et al. 2006; Baxa et al. 2007; Wickner et al. 2008). An accessible introduction to NMR is the book *Spin Dynamics* by Malcolm Levitt (2nd edition, 2008), Levitt (2008), and its application to amyloid structure is reviewed by Tycko (2011) and Tycko and Wickner (2013).

ACKNOWLEDGMENTS

This work was supported in part by the Intramural Program of the National Institute of Diabetes and Digestive and Kidney Diseases of the National Institutes of Health.

REFERENCES

Alberti S, Halfmann R, King O, Kapila A, Lindquist S. 2009. A systematic survey identifies prions and illuminates sequence features of prionogenic proteins. *Cell* 137: 146–158.

Allen KD, Chernova TA, Tennant EP, Wilkinson KD, Chernoff YO. 2007. Effects of ubiquitin system alterations on the formation and loss of a yeast prion. *J Biol Chem* 282: 3004–3013.

Bateman D, Wickner RB. 2013. The [*PSI*$^+$] prion exists as a dynamic cloud of variants. *PLoS Genet* 9: e1003257.

Baxa U, Cheng N, Winkler DC, Chiu TK, Davies DR, Sharma D, Inouye H, Kirschner DA, Wickner RB, Steven AC. 2005. Filaments of the Ure2p prion protein have a cross-β core structure. *J Struct Biol* 150: 170–179.

Baxa U, Wickner RB, Steven AC, Anderson D, Marekov L, Yau W-M, Tycko R. 2007. Characterization of β-sheet structure in Ure2p1-89 yeast prion fibrils by solid state nuclear magnetic resonance. *Biochemistry* 46: 13149–13162.

Brachmann A, Baxa U, Wickner RB. 2005. Prion generation in vitro: Amyloid of Ure2p is infectious. *EMBO J* 24: 3082–3092.

Brachmann A, Toombs JA, Ross ED. 2006. Reporter assay systems for [URE3] detection and analysis. *Methods* 39: 35–42.

Chen B, Thurber KR, Shewmaker F, Wickner RB, Tycko R. 2009. Measurement of amyloid fibril mass-per-length by tilted-beam transmission electron microscopy. *Proc Natl Acad Sci* 106: 14339–14344.

Chernoff YO, Ono B-I. 1992. Dosage-dependent modifiers of PSI-dependent omnipotent suppression in yeast. In *Protein synthesis and targeting in yeast* (ed. Brown AJP, Tuite MF, McCarthy JEG), pp. 101–107. Springer, Berlin.

Chernoff YO, Lindquist SL, Ono B-I, Inge-Vechtomov SG, Liebman SW. 1995. Role of the chaperone protein Hsp104 in propagation of the yeast prion-like factor [psi$^+$]. *Science* 268: 880–884.

Chernoff YO, Newnam GP, Kumar J, Allen K, Zink AD. 1999. Evidence for a protein mutator in yeast: Role of the Hsp70-related chaperone Ssb in formation, stability, and toxicity of the [*PSI*] prion. *Mol Cell Biol* 19: 8103–8112.

Conde J, Fink GR. 1976. A mutant of *Saccharomyces cerevisiae* defective for nuclear fusion. *Proc Natl Acad Sci* 73: 3651–3655.

Coustou V, Deleu C, Saupe S, Begueret J. 1997. The protein product of the *het-s* heterokaryon incompatibility gene of the fungus *Podospora anserina* behaves as a prion analog. *Proc Natl Acad Sci* 94: 9773–9778.

Cox BS. 1965. Ψ, a cytoplasmic suppressor of super-suppressor in yeast. *Heredity* 20: 505–521.

Derkatch IL, Bradley ME, Zhou P, Chernoff YO, Liebman SW. 1997. Genetic and environmental factors affecting the de novo appearance of the [*PSI*$^+$] prion in *Saccharomyces cerevisiae*. *Genetics* 147: 507–519.

Derkatch IL, Bradley ME, Hong JY, Liebman SW. 2001. Prions affect the appearance of other prions: The story of [*PIN*]. *Cell* 106: 171–182.

Du Z, Park K-W, Yu H, Fan Q, Li L. 2008. Newly identified prion linked to the chromatin-remodeling factor Swi1 in *Saccharomyces cerevisiae*. *Nat Genet* 40: 460–465.

Eanes ED, Glenner GG. 1968. X-ray diffraction studies on amyloid filaments. *J Histochem Cytochem* 16: 673–677.

Edskes HK, Kryndushkin D, Shewmaker F, Wickner RB. 2015. Prion transfection of yeast. *Cold Spring Harb Protoc* doi: 10.1101/pdb.prot089037.

Ferreira PC, Ness F, Edwards SR, Cox BS, Tuite MF. 2001. The elimination of the yeast [*PSI*$^+$] prion by guanidine hydrochloride is the result of Hsp104 inactivation. *Mol Microbiol* 40: 1357–1369.

Halfmann R, Jarosz DF, Jones SK, Chang A, Lancster AK, Lindquist S. 2012. Prions are a common mechanism for phenotypic inheritance in wild yeasts. *Nature* 482: 363–368.

Helsen CW, Glover JR. 2012. Insight into molecular basis of curing of [*PSI*$^+$] prion by overexpression of 104-kDa heat shock protein (Hsp104). *J Biol Chem* 287: 542–556.

Jeon J, Lodge MS, Dawson BD, Ishigami M, Shewmaker F, Chen B. 2013. Superb resolution and contrast of transmission electron microscopy images of unstained biological samples on graphene-coated grids. *Biochem Biophys Acta* 1830: 3807–3815.

Jung G, Masison DC. 2001. Guanidine hydrochloride inhibits Hsp104 activity in vivo: A possible explanation for its effect in curing yeast prions. *Curr Microbiol* 43: 7–10.

Jung G, Jones G, Masison DC. 2002. Amino acid residue 184 of yeast Hsp104 chaperone is critical for prion-curing by guanidine, prion propagation, and thermotolerance. *Proc Natl Acad Sci* 99: 9936–9941.

Kiktev DA, Patterson JC, Muller S, Bariar B, Pan T, Chernoff YO. 2012. Regulation of the chaperone effects on a yeast prion by the cochaperone Sgt2. *Mol Cell Biol* 32: 4960–4970.

King CY, Diaz-Avalos R. 2004. Protein-only transmission of three yeast prion strains. *Nature* 428: 319–323.

Kryndushkin D, Wickner RB. 2007. Nucleotide exchange factors for Hsp70s are required for [URE3] prion propagation in *Saccharomyces cerevisiae*. *Mol Biol Cell* 18: 2149–2154.

Kryndushkin DS, Alexandrov IM, Ter-Avanesyan MD, Kushnirov VV. 2003. Yeast [*PSI*⁺] prion aggregates are formed by small Sup35 polymers fragmented by Hsp104. *J Biol Chem* 278: 49636–49643.

Kryndushkin D, Shewmaker F, Wickner RB. 2008. Curing of the [URE3] prion by Btn2p, a Batten disease-related protein. *EMBO J* 27: 2725–2735.

Kryndushkin DS, Wickner RB, Tycko R. 2011. The core of Ure2p prion fibrils is formed by the N-terminal segment in a parallel cross-β structure: Evidence from solid-state NMR. *J Mol Biol* 409: 263–277.

Kryndushkin D, Pripuzova N, Burnett B, Shewmaker F. 2013. Non-targeted identification of prions and amyloid-forming proteins from yeast and mammalian cells. *J Biol Chem* 288: 27100–27111.

Kryndushkin D, Pripuzova N, Shewmaker FP. 2015. Isolation and analysis of prion and amyloid aggregates from yeast cells. *Cold Spring Harb Protoc* doi: 10.1101/pdb.prot089045.

Lacroute F. 1971. Non-Mendelian mutation allowing ureidosuccinic acid uptake in yeast. *J Bacteriol* 106: 519–522.

Levitt M. 2008. *Spin dynamics: Basics of nuclear magnetic resonance*, 2nd ed. Wiley. ISBN: 978-0-470-51117-6.

Masison DC, Wickner RB. 1995. Prion-inducing domain of yeast Ure2p and protease resistance of Ure2p in prion-containing cells. *Science* 270: 93–95.

McGlinchey R, Kryndushkin D, Wickner RB. 2011. Suicidal [*PSI*⁺] is a lethal yeast prion. *Proc Natl Acad Sci* 108: 5337–5341.

Moriyama H, Edskes HK, Wickner RB. 2000. [URE3] prion propagation in *Saccharomyces cerevisiae*: Requirement for chaperone Hsp104 and curing by overexpressed chaperone Ydj1p. *Mol Cell Biol* 20: 8916–8922.

Newnam GP, Wegrzyn RD, Lindquist SL, Chernoff YO. 1999. Antagonistic interactions between yeast chaperones Hsp104 and Hsp70 in prion curing. *Mol Cell Biol* 19: 1325–1333.

Paravastu AK, Leapman RD, Yau WM, Tycko R. 2008. Molecular structural basis for polymorphism in Alzheimer's β-amyloid fibrils. *Proc Natl Acad Sci* 105: 18349–18354.

Patel BK, Gavin-Smyth J, Liebman SW. 2009. The yeast global transcriptional co-repressor protein Cyc8 can propagate as a prion. *Nat Cell Biol* 11: 344–349.

Patino MM, Li J-J, Glover JR, Lindquist S. 1996. Support for the prion hypothesis for inheritance of a phenotypic trait in yeast. *Science* 273: 622–626.

Paushkin SV, Kushnirov VV, Smirnov VN, Ter-Avanesyan MD. 1996. Propagation of the yeast prion-like [*psi*⁺] determinant is mediated by oligomerization of the *SUP35*-encoded polypeptide chain release factor. *EMBO J* 15: 3127–3134.

Reidy M, Masison DC. 2010. Sti1 regulation of Hsp70 and Hsp90 is critical for curing of *Saccharomyces cerevisiae* [*PSI*⁺] prions by Hsp104. *Mol Cell Biol* 30: 3542–3552.

Ritter C, Maddelein ML, Siemer AB, Luhrs T, Ernst M, Meier BH, Saupe SJ, Riek R. 2005. Correlation of structural elements and infectivity of the HET-s prion. *Nature* 435: 844–848.

Roberts BT, Wickner RB. 2003. A class of prions that propagate via covalent auto-activation. *Genes Dev* 17: 2083–2087.

Rogoza T, Goginashvili A, Rodionova S, Ivanov M, Viktorovskaya O, Rubel A, Volkov K, Mironova L. 2010. Non-Mendelian determinant [ISP+] in yeast is a nuclear-residing prion form of the global transcriptional regulator Sfp1. *Proc Natl Acad Sci* 107: 10573–10577.

Ross ED, Minton AP, Wickner RB. 2005. Prion domains: Sequences, structures and interactions. *Nat Cell Biol* 7: 1039–1044.

Sharma D, Masison DC. 2009. Hsp70 structure, function, regulation and influence on yeast prions. *Protein Pept Lett* 16: 571–581.

Shewmaker F, Wickner RB, Tycko R. 2006. Amyloid of the prion domain of Sup35p has an in-register parallel β-sheet structure. *Proc Natl Acad Sci* 103: 19754–19759.

Shewmaker F, McGlinchey R, Thurber KR, McPhie P, Dyda F, Tycko R, Wickner RB. 2009. The functional curli amyloid is not based on in-register parallel β-sheet structure. *J Biol Chem* 284: 25065–25076.

Sondheimer N, Lindquist S. 2000. Rnq1: An epigenetic modifier of protein function in yeast. *Mol Cell* 5: 163–172.

Stubbs G. 1999. Developments in fiber diffraction. *Curr Opin Struct Biol* 9: 615–619.

Suzuki G, Shimazu N, Tanaka M. 2012. A yeast prion, Mod5, promotes acquired drug resistance and cell survival under environmental stress. *Science* 336: 355–359.

Tanaka M, Chien P, Naber N, Cooke R, Weissman JS. 2004. Conformational variations in an infectious protein determine prion strain differences. *Nature* 428: 323–328.

Toombs JA, Petri M, Paul KR, Kan GY, Ross ED. 2012. *De novo* design of synthetic prion domains. *Proc Natl Acad Sci* 109: 6519–6524.

Toyama BH, Kelly MJ, Gross JD, Weissman JS. 2007. The structural basis of yeast prion strain variants. *Nature* 449: 233–237.

Tuite MF, Mundy CR, Cox BS. 1981. Agents that cause a high frequency of genetic change from [*psi*+] to [*psi*-] in *Saccharomyces cerevisiae*. *Genetics* 98: 691–711.

Tycko R. 2006. Molecular structure of amyloid fibrils: Insights from solid-state NMR. *Q Rev Biophys* 1: 1–55.

Tycko R. 2011. Solid-state NMR studies of amyloid fibril structure. *Annu Rev Phys Chem* 62: 279–299.

Tycko R, Wickner RB. 2013. Molecular structures of amyloid and prion fibrils: Consensus vs. controversy. *Acc Chem Res* 46: 1487–1496.

Villa LL, Juliani MH. 1980. Mechanism of ρ⁻ induction in *Saccharomyces cerevisiae* by guanidine hydrochloride. *Mutat Res* 7: 147–153.

Wall JS, Hainfeld JF. 1986. Mass mapping with the scanning transmission electron microscope. *Annu Rev Biophys Biophys Chem* 15: 355–376.

Wasmer C, Lange A, Van Melckebeke H, Siemer AB, Riek R, Meier BH. 2008. Amyloid fibrils of the HET-s(218–279) prion form a β solenoid with a triangular hydrophobic core. *Science* 319: 1523–1526.

Wickner RB. 1994. [URE3] as an altered *URE2* protein: Evidence for a prion analog in *S. cerevisiae*. *Science* 264: 566–569.

Wickner RB, Dyda F, Tycko R. 2008. Amyloid of Rnq1p, the basis of the [*PIN*⁺] prion, has a parallel in-register β-sheet structure. *Proc Natl Acad Sci* 105: 2403–2408.

Wickner RB, Edskes HK, Bateman DA, Kelly AC, Gorkovskiy A, Dayani Y, Zhou A. 2013. Amyloids and yeast prion biology. *Biochemistry* 52: 1514–1527.

Wickner RB, Bezsonov E, Bateman DA. 2014. Normal levels of the antiprion proteins Btn2 and Cur1 cure most newly formed [URE3] prion variants. *Proc Natl Acad Sci* 111: E2711–E2720.

Wickner RB, Edskes HK, Kryndushkin D, Shewmaker FP. 2015. Genetic methods for studying prions. *Cold Spring Harb Protoc* doi: 10.1101/pdb.prot089029.

Wu Y-X, Masison DC, Eisenberg E, Greene LE. 2006. Application of photobleaching for measuring diffusion of prion proteins in cytosol of yeast cells. *Methods* 39: 43–49.

Protocol 1

Genetic Methods for Studying Yeast Prions

Reed B. Wickner,[1,3] Herman K. Edskes,[1] Dmitry Kryndushkin,[2] and Frank P. Shewmaker[2]

[1]Laboratory of Biochemistry and Genetics, National Institute of Diabetes and Digestive and Kidney Diseases, National Institutes of Health, Bethesda, Maryland 20892-0830; [2]Department of Pharmacology, Uniformed Services University of Health Sciences, Bethesda, Maryland 20814

The recognition that certain long-known nonchromosomal genetic elements were actually prions was based not on the specific phenotypic manifestations of those elements, but rather on their unusual genetic properties. Here, we outline methods of prion assay, methods for showing the nonchromosomal inheritance, and methods for determining whether a nonchromosomal trait has the unusual characteristics diagnostic of a prion. Finally, we discuss genetic methods often useful in the study of yeast prions.

MATERIALS

It is essential that you consult the appropriate Material Safety Data Sheets and your institution's Environmental Health and Safety Office for proper handling of equipment and hazardous material used in this protocol.

RECIPES: Please see the end of this protocol for recipes indicated by <R>. Additional recipes can be found online at http://cshprotocols.cshlp.org/site/recipes.

Reagents

Media appropriate for assay(s) of interest (see Method)
    Half-strength YPD medium <R>
    Synthetic dextrose (SD) medium <R>
    Yeast extract–peptone–glycerol (YPG) medium <R>
    YPAD medium <R>
        Other media may be required (see Method).

Media supplements or additives for assay(s) of interest (see Method)
    Cycloheximide
        Add cycloheximide to media as needed (e.g., 0.3 mL of 10 mg/mL cycloheximide stock per liter of YPG) after autoclaving and before pouring plates.
    Ethidium bromide
    Guanidine hydrochloride
    Uracil
    Ureidosuccinic acid (USA)
        Dissolve calculated mass of USA (Research Organics 1032C) in water at 20 mg/mL, neutralize with NaOH, and filter sterilize. If stored at 4°C, USA solutions are stable for >2 yr. USA can be top-spread or added to SD media after autoclaving and cooling to 50°C.

---

[3]Correspondence: wickner@helix.nih.gov

Cite this protocol as *Cold Spring Harb Protoc*; doi:10.1101/pdb.prot089029

TABLE 1. Strains of *S. cerevisiae*

Strain	Genotype	References
MA116-8A	MATα his ura2-60 [URE3-1]	Aigle and Lacroute 1975
M328	MATa/MATα [ure-o]	Aigle and Lacroute 1975
BY241	MATa leu2 trp1 ura3 P_{DAL5}:ADE2 P_{DAL5}:CAN1 kar1	Brachmann et al. 2005
74-D694	MATa ade1-14 (UGA) his3 leu2 trp1 ura3	Chernoff et al. 1995
779-6a	MATa ade 2-1 SUQ5 trp1 kar1-1 his3 leu2 ura3 [PSI+]	Jung and Masison 2001
4830ref	MATα ade 2-1 SUQ5 trp1 kar1-1 lys2 leu2 ura3 sup35::kanMX [pin−] [psi−] pH953 (SUP35-reference)	Bateman and Wickner 2012
4830E9	MATα ade 2-1 SUQ5 trp1 kar1-1 lys2 leu2 ura3 sup35::kanMX [pin−] [psi−] pDB89 (SUP35-E9)	Bateman and Wickner 2012

Other supplements or additives may be required (see Method).

Plasmid(s) needed for assay(s) of interest (see Method)
   pKanMX-Rnq1-GFP (for assaying [PIN+] phenotype only; see Steps 14–15) (Nakayashiki et al. 2005)
Strain(s) of *Saccharomyces cerevisiae* for assay(s) of interest (see Table 1 and Method)

Equipment

Fluorescence microscope (for assaying [PIN+] phenotype only; see Steps 14–15)
Incubator set to 30°C
Replica-plating block and velveteen squares

METHOD

Assaying Specific Prion Phenotypes

[PSI+] Phenotype

The [PSI+] assay measures readthrough of the ade2-1 ochre mutation, aided by the weak serine-inserting tRNA suppressor SUQ5 (Cox 1965; Liebman et al. 1975) or by suppression of ade1-14 (opal), which, conveniently, does not require a second tRNA mutation (Inge-Vechtomov et al. 1988). Excess adenine represses the adenine biosynthetic pathway, so color development is best done on half-strength YPD medium with only 5 g/L of yeast extract instead of the usual 10 g/L or on minimal medium with 9 mg/L of adenine sulfate (Sharma et al. 2009). Either ade1 or ade2 mutants accumulate phospho-ribosylaminoimidazole, whose oxidation produces a red pigment, thus forming red Ade− [psi−] colonies and white or pink Ade+ [PSI+] colonies.

1. Streak a strain carrying *ade1-14* or *ade2-1 SUQ5* for single colonies on half-strength YPD solid medium (limiting adenine).

2. Incubate plates at 30°C until colonies appear (usually 2 d).

3. Observe colonies for pink or red color, which is brightened by incubation at room temperature or 4°C for a few days.

   Colony color change is a sensitive indicator of the readthrough efficiency and is often used to distinguish prion variants ("strong [PSI+]" is white, whereas "weak [PSI+]" is pink). Sup35-GFP expressed in [PSI+] strains appears as cytoplasmic dots but is evenly distributed in the cytoplasm of [psi−] cells.

   Alternatively, colonies can be replica-plated to −Ade plates and growth examined.

[URE3] Phenotype

Ure2p is a negative regulator of genes (such as DAL5) encoding enzymes and transporters involved in the assimilation of poor nitrogen sources, turning off these genes when a good nitrogen source, such as ammonia, is present (Cooper 2002; Magasanik and Kaiser 2002). Ureidosuccinate (USA), the product of Ura2p, can be taken up by Dal5p, the allantoate transporter.

Using ura2 Mutation

A ura2 mutant grows without uracil if USA is supplied with a poor nitrogen source (such as proline), but not on ammonia (as in "yeast nitrogen base"). If Ure2p is deficient because of a ure2 mutation or from inactivation of Ure2p by its conversion to the prion form, then ura2 cells can use USA in place of uracil even if ammonia is the nitrogen source. From a ura2 mutant, [URE3] clones are selected on minimal medium with ammonia as the nitrogen source and 30 µg/mL of USA in place of uracil (Lacroute 1971). Isolation of [URE3] clones or testing of clones for the presence of [URE3] is best done on plates containing only the minimum required nutrients, rather than on dropout plates.

4. Grow the strain to be tested carrying a *ura2* mutation with uracil (and other required nutrients) to single colonies or as patches at 30°C.

5. After colonies or patches have grown, replica-plate to SD containing 30 µg/mL USA but without uracil.

6. Grow the replica plates for 1 or 2 d at 30°C.

7. Record growth (indicative of the presence of [URE3]).

Using Uracil Secretion

Here, excess USA (100 µg/mL) will be taken up by the Dal5p permease, converted to uracil and secreted, allowing growth of a ura2/ura2 diploid lawn around the edge of the tested strain. Thus, the presence of [URE3] can be tested in wild strains without special markers.

8. Grow cells in patches on any medium.

9. Seed plates of SD containing 100 µg/mL USA and other required nutrients (except uracil) with ~10^5 cells of a *ura2/ura2* homozygous diploid.

10. Replica-plate the patches grown in Step 8 to the plates with the lawn prepared in Step 9. Incubate these plates at 30°C.

11. After 2 d or longer, record the growth of the lawn around the edges of the patches caused by excess uracil secreted by [URE3]-carrying patches.

Using P_{DAL5}-ADE2

The ADE2 open reading frame fused to the DAL5 promoter can also be used to assay [URE3], with [URE3] clones Ade+ and those lacking it Ade− (Schlumpberger et al. 2001; Brachmann et al. 2005). The Ade2 assay for [URE3] avoids any posttranscriptional regulation of Dal5p as has been shown for amino acid permeases by Kaiser and coworkers (Cain and Kaiser 2011). Checking guanidine curability or transmission by cytoduction ensures that Ade+ clones are not ure2.

12. Grow a strain carrying a P_{DAL5}-ADE2 fusion to single colonies or in patches on half-strength YPD at 30°C.

13. Record the colony color after colonies appear (~2 d).

    As for [PSI+] (Cox 1965), colony color on half-strength YPD reflects prion status: [ure-o] cells are red, "strong" [URE3] is white, and "weak" [URE3] is pink (Schlumpberger et al. 2001; Brachmann et al. 2005).

[PIN+] Phenotype

Rnq1p has no known function, so its deficiency is inapparent. The amyloid of Rnq1p rarely primes [PSI+] formation, elevating the frequency of [PSI+] formation dramatically (Derkatch et al. 1997, 2001; Osherovich and Weissman 2001). A patch of a [PIN+] strain (and the [pin−] control), each carrying a plasmid expressing Sup35p or just Sup35NM from a GAL1 promoter and the ade2-1 (with SUQ5) or ade1-14 mutations to detect [PSI+], are grown on galactose for 2 d, then replica plated to −Ade medium to detect [PSI+] cells. Different [PIN+] variants show different frequencies of [PSI+] formation (Bradley et al. 2002). Alternatively, as described below, expressing Rnq1-GFP from a plasmid shows diffuse cytoplasmic fluorescence in [pin−] cells, and distinct cytoplasmic foci (aggregates) in [PIN+] cells (Derkatch et al. 2001), allowing testing [PIN+] in strains without markers. Substitution of the prion domain of Sup35p with that of Rnq1p allows use of the [PSI+] Ade+/−, red/white system for easier detection of [PIN+] (Sondheimer and Lindquist 2000).

14. Transform the cells to be tested with plasmid pKanMX-Rnq1-GFP expressing *RNQ1-GFP*.

15. Examine transformant clones by fluorescence microscopy.

    [pin−] clones have fluorescence evenly distributed in the cytoplasm. [PIN+] clones have one or several dots of fluorescence in each cell.

Cytoduction

Standard Procedure

The kar1-1 mutation (Conde and Fink 1976) or kar1Δ15 (Biggins and Rose 1994) results in failure of cells to fuse their nuclei following the cytoplasmic fusion that takes place in mating (Conde and Fink 1976). Only one of the mating partners need have this mutation. Transfer of mitochondrial DNA (ρ) is usually used as the indicator of cytoplasmic transfer, and the transfer of another trait is tested. The donor strain is thus ρ^+ and the recipient is made ρ^0 by growing to single colonies on YPAD containing 1 mg/plate of ethidium bromide, then grown up on a separate YPAD plate.

16. Mix a dab of donor 779-6A in 0.3 mL of water at approximately threefold excess over recipient 4830ref ρ^0 or 4380E9 ρ^0, and spot the mixture on a YPAD plate.

17. Incubate plates for ~7 h at 30°C or overnight at room temperature.

    The kar1-1 nuclear fusion defect is not perfect, so if the cell mixture is incubated too long, mostly diploids will be obtained and few cytoductants will be found.

18. Streak the cytoduction mixture for single colonies on plates selecting against the donor (His dropout plate or SD+Ade Trp Lys Leu Ura in this case). Grow for ~3 d at 30°C.

19. Replica-plate His+ colonies to YPG (only ρ^+ clones will grow), SD+Leu Ura Trp Ade (only diploids will grow), and SD+Trp Lys Leu Ura (only [PSI+] clones will grow).

    Clones that grow on YPG and not on SD+Leu Ura Trp Ade are cytoductants. Their growth on SD+Trp Lys Leu Ura tells whether [PSI+] was transmitted. For typical cytoduction results, see Table 2.

Cytoductions Using Cycloheximide Resistance

A Q38 K mutant of Cyh2p (ribosomal 60S subunit protein L28) confers resistance to cycloheximide. Cytoductants are selected on YPG plates containing 3 μg/mL cycloheximide. Both donors and diploids are sensitive to cycloheximide and die, whereas recipient cells that are not cytoductants cannot grow on plates containing glycerol as the sole carbon source. Cytoductants can be directly isolated on this medium by streaking the mating mixture to single colonies. Alternatively, use of this marker allows cytoductions to be performed in larger screens as detailed here.

20. Grid cytoduction donors or recipients (depending on the experiment) on a plate or use ungridded clones growing on a selection plate.

21. Replica-plate the donors onto a velvet.

22. Press a plate containing a freshly grown lawn of ρ^0 cycloheximide-resistant recipient cells on top of the velvet already containing the donors.

23. Press a YPAD plate on top of the velvet, transferring the mating mixtures onto it.

24. Incubate the mating plate for 1 d at 30°C.

25. Replicate the mating plate onto a plate containing YPG and 3 μg/mL cycloheximide, and allow cytoductants to grow for 2 d at 30°C.

TABLE 2. Typical cytoduction results

Donor	Recipient	Ade+ cytoductants	Total cytoductants	% Ade+
779-6aReference Sup35 [PSI+]	4830 reference Sup35 ρ^0	80	80	100
	4830 E9 Sup35 ρ^0	16	86	19
	4830 Δ19 Sup35 ρ^0	13	101	13

In the first cytoduction, all of the cytoductants (clones with the nuclear genotype of the recipient strain and able to grow on glycerol (ρ^+) had become Ade+, indicating their acquisition of the [PSI+] prion. Recipients with a Sup35 protein having a different sequence (E9 or Δ19) often did not acquire [PSI+]. The inefficient prion transfer because of sequence differences in the prion protein is a general feature of prions, called a "transmission barrier" or, in some cases, a "species barrier."

26. Subsequently, identify [PSI+] cytoductants by replicating onto SD medium lacking adenine but containing 10 μg/mL cycloheximide. Grow for 2 d at 30°C, at which point [PSI+]-containing patches should be visible.

Analogous tests for other prions may be done.

Prion Formation Induction by Protein Overproduction

Overproduction of the prion protein dramatically increases the frequency of the prion form arising de novo. This is best done by transient overproduction using a controllable promoter such as the GAL1 promoter. Overexpression of the prion domain alone usually gives a greater frequency of prion induction than does the full length protein (Masison and Wickner 1995; Kochneva-Pervukhova et al. 1998).

27. Grow cells carrying a plasmid expressing the prion domain from a *GAL1* (or *CUP1*) promoter for 1–2 d in 2% galactose-2% raffinose minimal medium (or Cu^{2+}-containing medium).

28. Plate the cells on SD medium (without added copper) selective for the prion phenotype.

Guanidine-Curing of Prions

29. Cure [PSI+], [URE3], or [PIN+] by growing cells to single colonies on rich medium containing 3–5 mM guanidine HCl (Tuite et al. 1981; Wickner 1994; Derkatch et al. 2000).

[PIN+] is cured somewhat slower than [PSI+] (Derkatch et al. 2000) and may require additional passages on guanidine-containing medium. Curing with guanidine requires cell growth (Byrne et al. 2007) and occurs via Hsp104 inhibition (Ferreira et al. 2001; Jung and Masison 2001; Jung et al. 2002).

Other Genetic Strategies

30. Consider other genetic strategies that may be useful in prion studies.

As we have mentioned elsewhere (see Introduction: Prions [Kryndushkin et al. 2015]), isolation of high-copy clones that cure a prion has been useful in defining cellular components that impinge on prion propagation (e.g., Kryndushkin et al. 2008). Likewise, isolation of high-copy clones that have a Pin+ effect—increasing the frequency of [PSI+] generation—has been a useful means of finding new prions (Derkatch et al. 2001; Du et al. 2008; Patel et al. 2009; Suzuki et al. 2012). The prion variant phenomenon is perhaps the most unique feature of prions. A single protein sequence can support the generation of a wide variety of yeast prion variants with different structural and biological features. Among these are lethal prions (McGlinchey et al. 2011), whose detection requires a condition under which the prion is not lethal. Lethal and other extremely toxic variants of [PSI+] based on near complete sequestration of the essential Sup35 protein were detected by expressing from a URA3 plasmid in the same cells a strictly limited amount of Sup35C, the part of the protein essential for translation termination, and selecting for elevated translation suppression. Among these [PSI+] isolates were those which could not survive loss of the plasmid expressing Sup35C (McGlinchey et al. 2011).

RECIPES

Half-Strength YPD Medium

Reagent	Final concentration
Yeast extract	5 g/L
Peptone	20 g/L
Dextrose	20 g/L
Agar (for solid plates)	20 g/L

Autoclave the medium. Store at 4°C (this is stable for at least 6 mo).

Synthetic Dextrose (SD) Medium

Reagent	Final concentration
Yeast nitrogen base without amino acids (Difco)	6.7 g/L
Dextrose	20 g/L
Agar (for solid medium)	20 g/L

Autoclave the medium. Store at 4°C (this is stable for at least 6 mo).

Yeast Extract–Peptone–Glycerol (YPG) Medium

Reagent	Final concentration
Yeast extract	10 g/L
Glycerol	20 mL/L
Peptone	20 g/L
Agar (for solid medium)	20 g/L

Autoclave the medium. Store at 4°C (this is stable for at least 6 mo).

YPAD Medium

Reagent	Final concentration
Yeast extract	10 g/L
Peptone	20 g/L
Dextrose	20 g/L
Adenine sulfate	0.4 g/L
Agar (for solid plates)	20 g/L

Autoclave the medium. Store at 4°C (this is stable for at least 6 mo).

ACKNOWLEDGMENTS

This work was supported in part by the Intramural Program of the National Institute of Diabetes and Digestive and Kidney Diseases, and by the Uniformed Services University of the Health Sciences.

REFERENCES

Aigle M, Lacroute F. 1975. Genetical aspects of [URE3], a non-Mendelian, cytoplasmically inherited mutation on yeast. *Mol Gen Genet* **136:** 327–335.

Bateman DA, Wickner RB. 2012. [*PSI*⁺] Prion transmission barriers protect *Saccharomyces cerevisiae* from infection: Intraspecies "species barriers." *Genetics* **190:** 569–579.

Biggins S, Rose MD. 1994. Direct interaction between yeast spindle pole body components: Kar1p is required for Cdd31p localization to the spindle pole body. *J Cell Biol* **125:** 843–852.

Brachmann A, Baxa U, Wickner RB. 2005. Prion generation in vitro: Amyloid of Ure2p is infectious. *EMBO J* **24:** 3082–3092.

Bradley ME, Edskes HK, Hong JY, Wickner RB, Liebman SW. 2002. Interactions among prions and prion "strains" in yeast. *Proc Natl Acad Sci* **99**(Suppl 4): 16392–16399.

Byrne LJ, Cox BS, Cole DJ, Ridout MS, Morgan BJ, Tuite MF. 2007. Cell division is essential for elimination of the yeast [*PSI*⁺] prion by guanidine hydrochloride. *Proc Natl Acad Sci* **104:** 11688–11693.

Cain NE, Kaiser CA. 2011. Transport activity-dependent intracellular sorting of the yeast general amino acid permease. *Mol Biol Cell* **22:** 1919–1929.

Chernoff YO, Lindquist SL, Ono B-I, Inge-Vechtomov SG, Liebman SW. 1995. Role of the chaperone protein Hsp104 in propagation of the yeast prion-like factor [*PSI*⁺]. *Science* **268:** 880–884.

Conde J, Fink GR. 1976. A mutant of *Saccharomyces cerevisiae* defective for nuclear fusion. *Proc Natl Acad Sci* **73:** 3651–3655.

Cooper TG. 2002. Transmitting the signal of excess nitrogen in *Saccharomyces cerevisiae* from the Tor proteins to the GATA factors: Connecting the dots. *FEMS Microbiol Rev* **26:** 223–238.

Cox BS. 1965. Ψ, a cytoplasmic suppressor of super-suppressor in yeast. *Heredity* **20:** 505–521.

Derkatch IL, Bradley ME, Zhou P, Chernoff YO, Liebman SW. 1997. Genetic and environmental factors affecting the de novo appearance of the [*PSI*⁺] prion in *Saccharomyces cerevisiae*. *Genetics* **147:** 507–519.

Derkatch IL, Bradley ME, Masse SV, Zadorsky SP, Polozkov GV, Inge-Vechtomov SG, Liebman SW. 2000. Dependence and independence of [*PSI*⁺] and [*PIN*⁺]: A two-prion system in yeast? *EMBO J* **19:** 1942–1952.

Derkatch IL, Bradley ME, Hong JY, Liebman SW. 2001. Prions affect the appearance of other prions: The story of [*PIN*]. *Cell* **106:** 171–182.

Du Z, Park K-W, Yu H, Fan Q, Li L. 2008. Newly identified prion linked to the chromatin-remodeling factor Swi1 in *Saccharomyces cerevisiae*. *Nat Genet* **40:** 460–465.

Ferreira PC, Ness F, Edwards SR, Cox BS, Tuite MF. 2001. The elimination of the yeast [*PSI*⁺] prion by guanidine hydrochloride is the result of Hsp104 inactivation. *Mol Microbiol* **40:** 1357–1369.

Inge-Vechtomov SG, Tikhodeev ON, Karpova TS. 1988. Selective systems for obtaining recessive ribosomal suppressors in *Saccharomyces cerevisiae*. *Genetika* **24:** 1159–1165.

Jung G, Masison DC. 2001. Guanidine hydrochloride inhibits Hsp104 activity in vivo: A possible explanation for its effect in curing yeast prions. *Curr Microbiol* **43:** 7–10.

Jung G, Jones G, Masison DC. 2002. Amino acid residue 184 of yeast Hsp104 chaperone is critical for prion-curing by guanidine, prion propagation, and thermotolerance. *Proc Natl Acad Sci* **99:** 9936–9941.

Kochneva-Pervukhova NV, Poznyakovski AI, Smirnov VN, Ter-Avanesyan MD. 1998. C-terminal truncation of the Sup35 protein increases the frequency of de novo generation of a prion-based [*PSI*⁺] determinant in *Saccharomyces cerevisiae*. *Curr Genet* **34:** 146–151.

Kryndushkin D, Shewmaker F, Wickner RB. 2008. Curing of the [URE3] prion by Btn2p, a Batten disease-related protein. *EMBO J* **27:** 2725–2735.

Kryndushkin D, Edskes HK, Shewmaker FP, Wickner RB. 2015. Prions. *Cold Spring Harb Protoc* doi: 10.1101/pdb.top077586.

Lacroute F. 1971. Non-Mendelian mutation allowing ureidosuccinic acid uptake in yeast. *J Bacteriol* **106:** 519–522.

Liebman SW, Stewart JW, Sherman F. 1975. Serine substitutions caused by an ochre suppressor in yeast. *J Mol Biol* **94:** 595–610.

Magasanik B, Kaiser CA. 2002. Nitrogen regulation in *Saccharomyces cerevisiae*. *Gene* **290:** 1–18.

Masison DC, Wickner RB. 1995. Prion-inducing domain of yeast Ure2p and protease resistance of Ure2p in prion-containing cells. *Science* 270: 93–95.

McGlinchey R, Kryndushkin D, Wickner RB. 2011. Suicidal [*PSI*+] is a lethal yeast prion. *Proc Natl Acad Sci* 108: 5337–5341.

Nakayashiki T, Kurtzman CP, Edskes H, Wickner RB. 2005. Yeast prions [URE3] and [*PSI*+] are diseases. *Proc Natl Acad Sci* 102: 10575–10580.

Osherovich LZ, Weissman JS. 2001. Multiple Gln/Asn-rich prion domains confer susceptibility to induction of the yeast [*PSI*+] prion. *Cell* 106: 183–194.

Patel BK, Gavin-Smyth J, Liebman SW. 2009. The yeast global transcriptional co-repressor protein Cyc8 can propagate as a prion. *Nat Cell Biol* 11: 344–349.

Schlumpberger M, Prusiner SB, Herskowitz I. 2001. Induction of distinct [URE3] yeast prion strains. *Mol Cell Biol* 21: 7035–7046.

Sharma D, Martineau CN, Le Dall M-T, Reidy M, Masison DC. 2009. Function of SSA subfamily of Hsp70 within and across species varies widely in complementing *Saccharomyces cerevisiae* cell growth and prion propagation. *PLoS ONE* 4: e6644.

Sondheimer N, Lindquist S. 2000. Rnq1: An epigenetic modifier of protein function in yeast. *Mol Cell* 5: 163–172.

Suzuki G, Shimazu N, Tanaka M. 2012. A yeast prion, Mod5, promotes acquired drug resistance and cell survival under environmental stress. *Science* 336: 355–359.

Tuite MF, Mundy CR, Cox BS. 1981. Agents that cause a high frequency of genetic change from [psi+] to [psi−] in *Saccharomyces cerevisiae*. *Genetics* 98: 691–711.

Wickner RB. 1994. [URE3] as an altered *URE2* protein: Evidence for a prion analog in *S. cerevisiae*. *Science* 264: 566–569.

Protocol 2

Prion Transfection of Yeast

Herman K. Edskes,[1] Dmitry Kryndushkin,[2] Frank P. Shewmaker,[2] and Reed B. Wickner[1,3]

[1]*Laboratory of Biochemistry and Genetics, National Institute of Diabetes and Digestive and Kidney Diseases, National Institutes of Health, Bethesda, Maryland 20892-0830;* [2]*Department of Pharmacology, Uniformed Services University for the Health Sciences, Bethesda, Maryland 20814*

Transfection of yeast with amyloid filaments, made from recombinant protein or prepared from extracts of cells infected with a prion, has become an important method in characterizing yeast prions. Here, we describe a method for transmission of [URE3] with Ure2p amyloid that is based on a previously published protocol for transfection with Sup35p filaments to make cells [PSI+]. This method may be used for other prions by changing just the amyloid source, host strain, and plating medium.

MATERIALS

It is essential that you consult the appropriate Material Safety Data Sheets and your institution's Environmental Health and Safety Office for proper handling of equipment and hazardous material used in this protocol.

RECIPES: Please see the end of this protocol for recipes indicated by <R>. Additional recipes can be found online at http://cshprotocols.cshlp.org/site/recipes.

Reagents

Complete synthetic medium without leucine and containing adenine (CS + A.1-L) <R>
Complete synthetic medium without leucine and containing adenine sulfate (CS + A5-L) <R>
Lyticase solution <R>
Protein extract or amyloid filaments
  Amyloid filaments prepared from bacterially expressed Ure2p (e.g., see Taylor et al. 1999; Brachmann et al. 2005) should be sonicated as described in Step 5 immediately before addition to spheroplasts. Amyloid filaments have been used at concentrations of 0.04–6 µg/µL.

pRS425 plasmid (2 µg/µL; Christianson et al. 1992)
PTC buffer <R>
Single-stranded DNA (10 mg/mL; Worthington-Biochemicals)
SOS medium <R>
ST buffer <R>
STC buffer <R>
Tris-Cl (1 M, pH 7.4)
Yeast strains
  BY241 (*MATa leu2 trp1 ura3* P_{DAL5}:*ADE2* P_{DAL5}:*CAN1 kar1*)
  BY256 (*MATa his3 trp1 leu2* P_{DAL5}:*ADE2* P_{DAL5}:*CAN1 kar1*)
YPAD medium <R>

---

[3]Correspondence: wickner@helix.nih.gov

Cite this protocol as *Cold Spring Harb Protoc*; doi:10.1101/pdb.prot089037

Chapter 26

Equipment

Air supply for blowing cell lysates to new tube after breakage (see Step 2)
Branson 250 Sonifier equipped with a microtip
Conical screw cap microtubes (2 mL)
Glass beads (0.5 mm)
Incubator at 30°C and 50°C
Microcentrifuge
Mini-BeadBeater-8 (Biospec)
Spectrophotometer
Vortex mixer (e.g., Vortex Genie 2)

METHOD

1. Grow [URE3]-containing cells overnight at 30°C in 20 mL of complete synthetic medium without adenine (i.e., CS + A.1-L without any adenine) to select for maintenance of the prion. Collect the cells, resuspend them in 50 mL of YPAD to an OD_{600} of 0.2–0.5, and grow them for 2–3 doublings at 30°C. Collect the cells, and wash the cell pellets twice with water.

   At this point, the pellets can be stored frozen at −80°C.

2. Resuspend pellets of [URE3]-containing cells in 600 µL of water, and transfer them to a 2-mL conical screw cap microtube. Fill the tubes with 0.5-mm glass beads, and break the cells for 3 min in a Mini-BeadBeater-8 at 4°C. Briefly centrifuge the tubes, pierce them at the top and bottom, and blow the cell lysates into a clean microcentrifuge tube using an air hose. Clear the lysates by centrifugation for 5 min at 4°C, and store them at −80°C until recipient yeasts are ready.

3. Grow recipient [ure-o] cells overnight at 30°C in YPAD. Then use 1 mL of culture to inoculate 50 mL of YPAD, and grow the cells for 2–3 doublings at 30°C. Wash the cells once with 20 mL of water and twice with 25 mL of ST buffer, and then resuspend the cells in 5 mL of ST buffer.

4. Convert the cells into protoplasts by incubation for 40 min at 30°C with 4 µL of lyticase solution. Collect the protoplasts by centrifugation at 250g for 3 min, wash them two times with 10 mL of STC buffer, and then resuspend them in 1 mL of STC buffer.

   To reduce shearing, perform all manipulations of protoplasts using pipette tips from which the ends are trimmed to have a wide bore size.

5. Sonicate lysates of [URE3]-containing cells (from Step 2) three times for 45 sec (duty cycle 20%, output 4) using a Branson 250 Sonifier equipped with a microtip. Keep the lysates on ice between sonications.

6. Add the following to 100 µL of protoplasts.

Single-stranded DNA (10 µg/µL)	1 µL
pRS425 plasmid (2.0 µg/µL)	2 µL
Protein extract or amyloid filaments	9 µL

7. Incubate the protoplast/DNA/protein mixture for 10 min at room temperature. Add 900 µL of PTC buffer, and incubate the mixture for 20 min at room temperature.

8. Collect the protoplasts by centrifugation at 400g for 3 min in a microcentrifuge. Add 200 µL of SOS buffer to the pellet, and leave the protoplasts to recover for 30 min at 30°C.

9. Pipette the recovered protoplasts into 10 mL of either CS + A.1-L or CS + A5-L medium kept at 50°C. Mix each solution by inverting the tubes and directly pour them into Petri dishes containing 20 mL of the same solidified medium.

10. Incubate the plates for 6 d at 30°C.

CS + A5-L medium allows all transformants to grow and thus provides a measurement for the transformation efficiency. CS + A.1-L medium is selective for [URE3] cells. Although [URE3] cells do not need adenine-supplemented medium to grow, primary transformants need it to allow phenotypic establishment of the prion (Brachmann et al. 2005).

DISCUSSION

The first yeast proteins identified genetically to form prions, Ure2p and Sup35p (Wickner 1994), are also the first yeast prion forming proteins shown to form amyloid (King et al. 1997; Taylor et al. 1999). The first attempt to infect yeast with a protein extract from a prion containing strain was made by Sparrer et al. (2000). However, the liposome fusion method was inefficient, and it was not clear if the prions detected were formed in the recipient cell following the introduction of prion domain protein—in effect induction of prion formation by overproduction of the prion domain. Using ballistic transformation Maddelein et al. (2002) showed that introduction of HET-s amyloid into the filamentous fungus *Podospora anserine* results in infection of mycelium with the [Het-s] prion; introduction of heat-aggregated HET-s did not result in infection. This method has not been used for *S. cerevisiae*. In 2004 both Tanaka et al. and King et al. created more efficient methods to deliver prion particles into yeast. King et al. used a protoplast fusion system (King and Diaz-Avalos 2004). In this system, protoplasts from two genetically marked strains are incubated with protein extracts. Cell fusion also introduces species present in the medium, including prion particles, into the cells. This method has not found wide application although, in addition to [PSI+], it was shown to work for [PIN+] (Sharma and Liebman 2013). Tanaka et al. created a method that uses plasmid uptake to identify yeast protoplasts that have taken up media contents (Tanaka et al. 2004). In addition to [PSI+], this method has been used to infect cells with prion particles and amyloid of several prion proteins (Ure2p, Rnq1p, Cyc8p, and Swi1p) (Brachmann et al. 2005; Patel and Liebman 2007; Patel et al. 2009; Du et al. 2010), and it is our minor modification (for [URE3]) of this method that we describe here.

Yeast strains used for protein transformation experiments aimed at [URE3] contain an *ADE2* ORF controlled by the *DAL5* promoter. The *DAL5* promoter is tightly controlled by Ure2p activity and is activated in [URE3] cells when soluble Ure2p levels drop because of sequestering in amyloid filaments (Schlumpberger et al. 2001; Brachmann et al. 2005). Cells also contain at least one nutritional marker to allow selection for cells that have obtained a yeast replication plasmid. Plasmid uptake is used to identify cells that have taken up buffer contents including Ure2p amyloid filaments (Brachmann et al. 2005). Dramatic differences in transfection efficiency have been noted among yeast strains examined with BY241 (*MATa leu2 trp1 ura3 P_{DAL5}:ADE2 P_{DAL5}:CAN1 kar1*) and BY256 (*MATa his3 trp1 leu2 P_{DAL5}:ADE2 P_{DAL5}:CAN1 kar1*) providing particularly good results (Brachmann et al. 2005).

RECIPES

Complete Synthetic Medium without Leucine and containing Adenine (CS + A.1-L)

Yeast nitrogen base without amino acids (Difco)	6.7 g/L
Dextrose	20 g/L
Agar (for solid medium)	20 g/L
Sorbitol	1 M
L-Tyrosine	50 mg/L
L-Arginine HCl	50 mg/L
L-Aspartic acid	80 mg/L

L-Histidine HCl	20 mg/L
L-Isoleucine	50 mg/L
L-Lysine HCl	50 mg/L
L-Methionine	20 mg/L
L-Phenylalanine	50 mg/L
L-Threonine	100 mg/L
L-Tryptophan	50 mg/L
Uracil	20 mg/L
L-Valine	140 mg/L
Adenine hemisulphate	0.1 mg/L

Dissolve the amino acid L-tyrosine first. (When it is added according to its alphabetical position it will not go into solution.) Autoclave. For solid medium, add 20 mL to each Petri plate. Store at 4°C (this is stable for at least 6 mo).

Complete Synthetic Medium without Leucine and containing Adenine Sulfate (CS + A5-L)

Yeast nitrogen base without amino acids (Difco)	6.7 g/L
Adenine sulfate	5 mg/L
Dextrose	20 g/L
Agar (for solid medium)	20 g/L
Sorbitol	1 M
L-Tyrosine	50 mg/L
L-Arginine HCl	50 mg/L
L-Aspartic acid	80 mg/L
L-Histidine HCl	20 mg/L
L-Isoleucine	50 mg/L
L-Lysine HCl	50 mg/L
L-Methionine	20 mg/L
L-Phenylalanine	50 mg/L
L-Threonine	100 mg/L
L-Tryptophan	50 mg/L
Uracil	20 mg/L
L-Valine	140 mg/L
Adenine hemisulphate	0.1 mg/L

Dissolve the amino acid L-tyrosine first. (When it is added according to its alphabetical position it will not go into solution.) Autoclave. For solid medium, add 20 mL to each Petri plate. Store at 4°C (this is stable for at least 6 mo).

Lyticase Solution

Reagent	Final concentration
Lyticase (Sigma-Aldrich L2524)	25 units/µL
Glycerol	20%
Tris-Cl (pH 7.4)	10 mM

Store at 4°C (this is stable for at least 6 mo).

PTC Buffer

Reagent	Final concentration
PEG8000	20% (w/v)
Tris-HCl (pH 7.4)	10 mM
$CaCl_2$	10 mM

Store at 4°C (this is stable for at least 6 mo).

SOS Medium

Reagent	Final concentration
Sorbitol	1 M
$CaCl_2$	7 mM
Yeast extract	3.3 g/L
Peptone	6.7 g/L
Glucose	6.7 g/L

Store at 4°C (this is stable for at least 6 mo).

ST Buffer

Reagent	Final concentration
Sorbitol	1 M
Tris-HCl (pH 7.4)	10 mM

Store at 4°C (this is stable for at least 6 mo).

STC Buffer

Reagent	Final concentration
Sorbitol	1 M
Tris-HCl (pH 7.4)	10 mM
$CaCl_2$	10 mM

Store at 4°C (this is stable for at least 6 mo).

YPAD Medium

Reagent	Final concentration
Yeast extract	10 g/L
Peptone	20 g/L
Dextrose	20 g/L
Adenine sulfate	0.4 g/L
Agar (for solid plates)	20 g/L

Store at 4°C (this is stable for at least 6 mo).

ACKNOWLEDGMENTS

This work was supported in part by the Intramural Program of the National Institute of Diabetes and Digestive and Kidney Diseases of the National Institutes of Health.

REFERENCES

Brachmann A, Baxa U, Wickner RB. 2005. Prion generation in vitro: Amyloid of Ure2p is infectious. *EMBO J* 24: 3082–3092.

Christianson TW, Sikorski RS, Dante M, Shero JH, Hieter P. 1992. Multifunctional yeast high-copy-number shuttle vectors. *Gene* 110: 119–122.

Du Z, Crow ET, Kang HS, Li L. 2010. Distinct subregions of Swi1 manifest striking differences in prion transmission and SWI/SNF function. *Mol Cell Biol* 30: 4644–4655.

King CY, Diaz-Avalos R. 2004. Protein-only transmission of three yeast prion strains. *Nature* 428: 319–323.

King C-Y, Tittmann P, Gross H, Gebert R, Aebi M, Wuthrich K. 1997. Prion-inducing domain 2-114 of yeast Sup35 protein transforms in vitro into amyloid-like filaments. *Proc Natl Acad Sci* 94: 6618–6622.

Maddelein ML, Dos Reis S, Duvezin-Caubet S, Coulary-Salin B, Saupe SJ. 2002. Amyloid aggregates of the HET-s prion protein are infectious. *Proc Natl Acad Sci* 99: 7402–7407.

Patel BK, Liebman SW. 2007. "Prion proof" for [PIN⁺]: Infection with in vitro-made amyloid aggregates of Rnq1p-(132-405) induces [PIN⁺]. *J Mol Biol* 365: 773–782.

Patel BK, Gavin-Smyth J, Liebman SW. 2009. The yeast global transcriptional co-repressor protein Cyc8 can propagate as a prion. *Nat Cell Biol* **11:** 344–349.

Schlumpberger M, Prusiner SB, Herskowitz I. 2001. Induction of distinct [URE3] yeast prion strains. *Mol Cell Biol* **21:** 7035–7046.

Sharma J, Liebman SW. 2013. Exploring the basis of [PIN+] variant differences in [PSI+] induction. *J Mol Biol* **425:** 3046–3059.

Sparrer HE, Santoso A, Szoka FC, Weissman JS. 2000. Evidence for the prion hypothesis: Induction of the yeast [PSI+] factor by in vitro-converted Sup35 protein. *Science* **289:** 595–599.

Tanaka M, Chien P, Naber N, Cooke R, Weissman JS. 2004. Conformational variations in an infectious protein determine prion strain differences. *Nature* **428:** 323–328.

Taylor KL, Cheng N, Williams RW, Steven AC, Wickner RB. 1999. Prion domain initiation of amyloid formation in vitro from native Ure2p. *Science* **283:** 1339–1343.

Wickner RB. 1994. [URE3] as an altered *URE2* protein: Evidence for a prion analog in *S. cerevisiae*. *Science* **264:** 566–569.

Protocol 3

Isolation and Analysis of Prion and Amyloid Aggregates from Yeast Cells

Dmitry Kryndushkin, Natalia Pripuzova, and Frank P. Shewmaker[1]

Department of Pharmacology, Uniformed Services University of the Health Sciences, Bethesda, Maryland 20814

Amyloid fibers are large and extremely stable structures that can resist denaturation by strong anionic detergents, such as sodium dodecyl sulfate or sarkosyl. Here, we present two complementary analytical methods that exploit these properties, enabling the isolation and characterization of amyloid/prion aggregates. The first technique, known as semidenaturating detergent agarose gel electrophoresis, is an immunoblotting technique, conceptually similar to conventional western blotting. It enables the targeted identification of large detergent-resistant protein aggregates using antibodies specific to the protein of interest. The second method, called the technique for amyloid purification and identification, is a nontargeted approach that can isolate amyloid aggregates for analysis by tandem mass spectrometry. The latter approach requires no special genetic tools or antibodies, and can identify amyloid-forming proteins, such as prions, as well as proteins tightly associated with amyloid, from a variety of cell sources.

MATERIALS

It is essential that you consult the appropriate Material Safety Data Sheets and your institution's Environmental Health and Safety Office for proper handling of equipment and hazardous material used in this protocol.

RECIPES: Please see the end of this protocol for recipes indicated by <R>. Additional recipes can be found online at http://cshprotocols.cshlp.org/site/recipes.

Reagents

Acrylamide gel (e.g., gradient or 10% gel, Bio-Rad) (for the technique for amyloid purification and identification [TAPI] only)

Gels can be prepared using standard protocols or purchased from commercial sources. A range of acrylamide concentrations or gradients may be used, but low acrylamide concentrations in the stacking well are essential. Use gels with large wells to maximize load volume. Alternatively, preparatory gels can be used for loading larger volumes.

Agarose (for semidenaturating detergent agarose gel electrophoresis [SDD-AGE] only)
Ammonium bicarbonate (50 mM, pH 8.5) (for TAPI only)
Amyloid-prion buffer R (for TAPI only) <R>
Amyloid-prion resuspension buffer <R>
Amyloid-prion transfer buffer (for SDD-AGE only) <R>
Bromophenol blue (1%)
Dithiothreitol (DTT; 20 mM)
Endoproteinase Lys-C, sequencing grade (Promega) (for TAPI only)
Formic acid (for TAPI only)
Glycerol (100%)

[1]Correspondence: frank.shewmaker@gmail.com

Cite this protocol as *Cold Spring Harb Protoc*; doi:10.1101/pdb.prot089045

Chapter 26

    HiPPR detergent-removal resin (Thermo Scientific) (for TAPI, Strategy 1 only)
    I-Block (Life Technologies) (for SDD-AGE only)
    Iodoacetamide (IAA) (for TAPI only)
    RapiGest (Waters) or ProteaseMax (Promega) (optional for TAPI, Strategy 1; see Step 17)
    RNase A
    Sodium dodecyl sulfate (SDS)
    Sucrose (20%–40%, for sucrose pad; see Step 5)
    Tris-buffered saline (TBS; 10×, pH 7.5) <R>
    Tris-borate-EDTA buffer (TBE; 10×) (Thermo Scientific) (for SDD-AGE only)
    Triton X-100
    Trypsin, sequencing grade (Promega) (for TAPI only)
    Urea, ultrapure (9 M, freshly prepared; MP Biomedicals) (for TAPI, Strategy 2 only)
    Yeast cells and appropriate culture medium
    Yeast disruption buffer <R>

Modified radio-immunoprecipitation assay (RIPA) buffer supplemented with nucleases <R> can be used in place of yeast disruption buffer if amyloid aggregates are being analyzed from mammalian cell culture.

Equipment

    Amicon Ultra Centrifugal Filters (Ultracel, 30K; Millipore) (for TAPI, Strategy 2 only)
    Bench centrifuge
    Conical tubes (50 mL) (optional; see Step 3)
    Electro-Eluter (Model 422, Bio-Rad) (optional for TAPI; see Step 15)
    Flasks for culture growth (100 mL)
    Glass beads (0.5 mm)
    Heating block
    Horizontal DNA electrophoresis casting chamber and tank (for SDD-AGE only)
    Incubator
    Mechanical cell disruptor (e.g., Scientific Industries Disruptor Genie) (optional; see Step 3)
    Microcentrifuge
    Microcentrifuge tubes (1.5 and 2 mL)
    Razor blade (for TAPI only)
    SpeedVac concentrator (e.g., Thermo Scientific Savant DNA120) (for TAPI only)
    Ultracentrifuge (e.g., Beckman Coulter L-80 with SW55 rotor) and tubes
    Vacuum blotter, including filter paper, polyethylene mask, and polyvinylidene fluoride (PVDF) membrane (e.g., GE Healthcare VacuGene XL Vacuum Blotting System) (for SDD-AGE only)
    Vertical electrophoretic chamber (for TAPI only)
    Vortex mixer (e.g., Vortex Genie)
    Zeba spin desalting column (7K; Thermo Scientific) (for TAPI, Strategy 1 only)

METHOD

Both SDD-AGE (presented in Steps 8–12 below) and TAPI (Steps 13–31) share a similar set of steps for cell disruption and the initial isolation of amyloid aggregates from cellular lysates by ultracentrifugation (Steps 1–7). After separation of high-molecular-weight aggregates, samples can be analyzed by SDD-AGE. For TAPI, samples undergo further purification (see Fig. 1). Aggregates are first resuspended in SDS or sarkosyl at room temperature and then electrophoretically separated by SDS-PAGE (without boiling). High-molecular-weight aggregates are trapped at the start of the gel, and are subsequently eluted, purified, and cleaved by trypsin in preparation for tandem mass spectrometry. Two strategies (Steps 17–22 and Steps 23–30) are given for trypsin digestion.

Preparation of Cell Lysates and Sedimentation of Amyloid Aggregates

There are many yeast cell-lysis protocols, and various standard protocols can be used for SDD-AGE or TAPI, if aggregates are maintained in the lysate (amyloid can be lost during centrifugation steps). Protocols that use high-

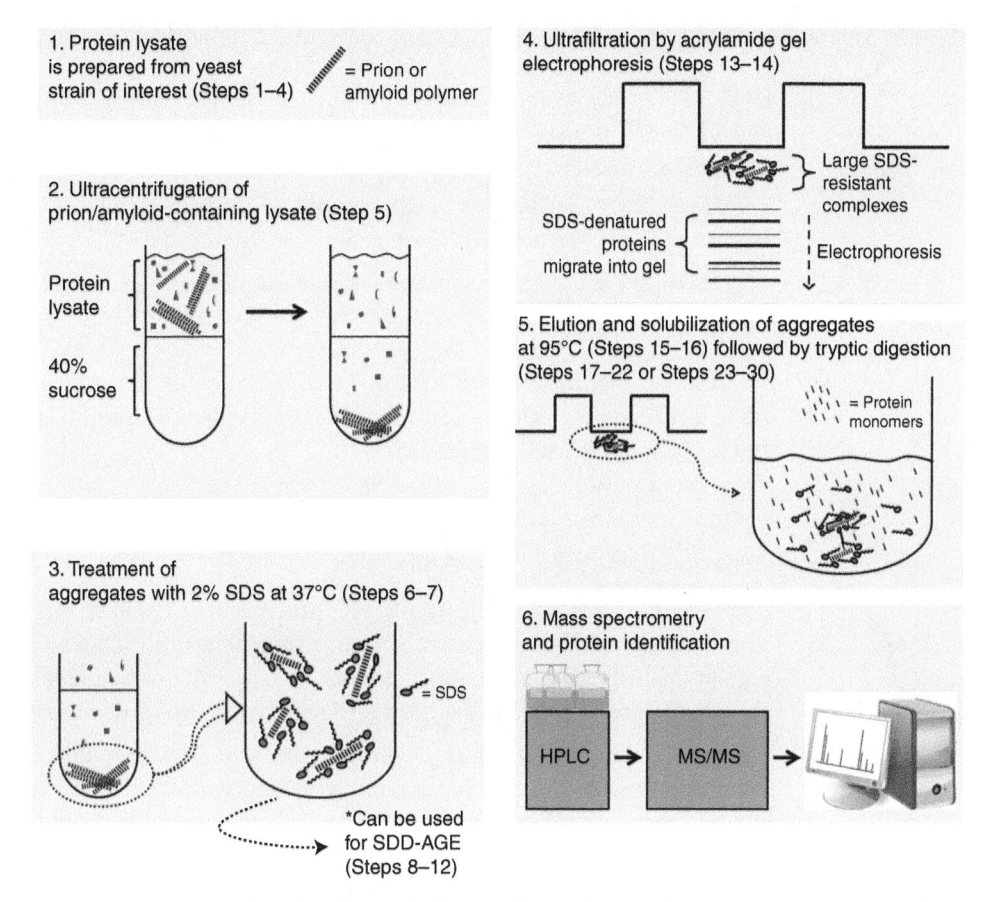

FIGURE 1. The stepwise workflow for the TAPI. The initial steps for sample preparation are similar for TAPI and for SDD-AGE. (Adapted from Kryndushkin et al. 2013.)

molar urea or extended heating in detergent should be avoided if they solubilize the protein aggregates of interest. Here, we provide an effective protocol that preserves amyloid aggregates for downstream biophysical characterization.

Preparation of Yeast Cell Lysate

1. Grow yeast cells in 100-mL flasks to early stationary phase, or ~75% of the maximum cell density for a given medium (either synthetic or rich).

   Usually 40 mL of medium is sufficient for isolation and identification of yeast prions. If more amyloid is required (for structural studies), the procedure may be scaled up.

2. Collect cells by centrifugation ($4000g$ for 5 min). Wash cell pellet with cold $1 \times$ TBS once.

   The cells can be used immediately or stored for several months at $-80°C$.

3. To lyse cells with glass beads, add an equal volume of 0.5-mm beads and two volumes of yeast disruption buffer. Break the cells in the same tube by vortexing at top speed for a total of 5 min, with cooling on ice for 30 sec after each minute. Centrifuge the lysate at $800g$ for 5 min at $4°C$, and transfer the supernatant to a new tube.

   A mechanical cell disruptor can be used instead of hand vortexing. Small cell pellets (0.5 g or less) can be efficiently disrupted in 2-mL microcentrifuge tubes, whereas for larger cell pellets, 50-mL conical tubes are preferred. Mild sonication may be applied for particular cells to achieve complete cell disruption.

4. Add RNase A to the lysate at 0.1 mg/mL, and incubate for 10 min at room temperature. Then add Triton X-100 to 0.5% and incubate for 10 min on ice. Spin the lysate at $2000g$ for 10 min at $4°C$. Keep the supernatant.

Chapter 26

Sedimentation of Amyloid Aggregates

5. Load the yeast lysate containing amyloid (from Step 4) gently onto the top of a 1–2 mL sucrose pad (20%–40%) in an ultracentrifuge tube. Spin the lysates for 1–2 h at 200,000g (SW55 rotor) and carefully discard the supernatant (including the sucrose pad).

6. Resuspend the pellet in a sufficient volume of amyloid-prion resuspension buffer by pipetting 10–20 times, and incubate the suspension for 10 min at 37°C.

   The volume and stringency of resuspension buffer should be adjusted based on the size of the pellet and stability of the amyloid aggregates. Likewise, the temperature may be increased or decreased as necessary. See Troubleshooting.

7. Spin the clear solution at 5000g for 10 min and carefully transfer the supernatant to a new tube; avoid disturbing the pellet. Add 1 μL of 1% bromophenol blue and 5 μL of 100% glycerol for each 100 μL of solution, and mix briefly.

   Proceed to either Step 8 (for SDD-AGE) or Step 13 (for TAPI).

Agarose Gel Electrophoresis of Amyloid and Prion Aggregates

8. Prepare a 1.8% agarose gel using 1 × TBE buffer with 0.1% SDS. First, dissolve the agarose in 1 × TBE by bringing the suspension to boiling; then add 0.1% SDS and gently mix by inverting. Use regular high-melting-point agarose and a standard horizontal DNA electrophoresis casting chamber to cast the gel. Allow the gel to solidify for 1 h at room temperature.

9. Place the gel inside a horizontal electrophoretic tank (designed for DNA electrophoresis), and fill the chamber with 1 × TBE buffer supplemented with 0.1% SDS. Load the solution containing amyloid aggregates (from Step 7) into the wells of the agarose gel and run the electrophoresis at a constant 120 V until the bromophenol blue touches the opposite edge of the gel (~1 h).

10. Assemble the vacuum blotter. Soak several filter papers (0.8 mm thick) and a piece of PVDF membrane in the amyloid-prion transfer buffer for 5–10 min. On the base of the apparatus, place presoaked filter paper, a polyethylene mask with a gel-sized window, another filter paper, and the PVDF membrane. Apply vacuum. Rinse the gel with distilled water and place it on the PVDF membrane while avoiding making bubbles between the gel and the membrane. Connect the gel and a reservoir with the amyloid-prion transfer buffer by a long sheet of filter paper, allowing constant flow of the buffer from the reservoir to the upper surface of the gel under vacuum. Transfer proteins from the gel to the membrane overnight at 30 mbar.

11. Block the membrane with an appropriate volume of I-Block solution (or similar blocking reagent) for at least 1 h at room temperature.

12. Proceed with a standard immunostaining procedure using primary antibodies against the amyloid-forming protein, followed by appropriate secondary antibodies and detection protocols.

    Amyloid aggregates are usually detected as high-molecular-weight smears. Monomers of the amyloid-forming protein migrate further and as more discreet bands, but might not always be detected (Kryndushkin et al. 2003).

Proteomic TAPI

Isolation of Amyloid Proteins

13. Load the solution containing amyloid aggregates (from Step 7) onto an acrylamide gel placed in a standard vertical electrophoretic chamber (designed for protein electrophoresis). If necessary, use multiple lanes on the gel to accommodate the sample volume. Run for 15 min at 70 V and then 40 min at 200 V.

14. Rinse the gel with water. Cut out the top 3 mm of the stacking gel with a razor blade and place it in a sterile 1.5-mL microcentrifuge tube. Combine gel pieces from several wells in one tube. Freeze the tube for 20 min or overnight at −20°C.

15. Add 200 μL of amyloid-prion buffer R to the tube with the gel pieces, resuspend by vortexing, and incubate in the heating block for 15 min at 98°C. Remove the sample from the block, vortex again, spin at 4000g for 1 min, and carefully transfer the supernatant into a new tube. Repeat this elution two more times; combine the supernatants.

See Troubleshooting.

During heating in the presence of SDS in amyloid-prion buffer R, amyloid aggregates fall apart and monomeric proteins diffuse from the gel into solution. As an alternative, after heating, the elution from gel pieces can be done using Electro-Eluter (Model 422, Bio-Rad). However, the efficiency of the electro-elution is comparable with the simple resuspension procedure described here.

16. Reduce the sample volume to 150 μL using a SpeedVac Concentrator.

    Proceed to either Step 17 or Step 23.

Sample Digestion for Identification by Mass Spectrometry

Two different strategies are presented; follow Steps 17–22 or 23–30. The second strategy employs high-molar urea for chaotropic disruption of resistant aggregates.

Strategy 1

17. Remove SDS from the samples by mixing them with an equal volume of HiPPR detergent-removal resin, incubating for 10 min inside Zeba spin desalting columns, and eluting with 50 mM ammonium bicarbonate (pH 8.5) by centrifugation (following the resin manufacturer's instructions). Repeat the procedure once.

    Repetition is unnecessary if the SDS concentration is <2%. If desired, mass spectrometry-compatible surfactants, such as RapiGest or ProteaseMax, can be added to prevent protein aggregation and enhance efficiency of subsequent trypsin digestion.

18. Add 20 mM DTT to the eluate and incubate for 5 min at 95°C. Cool the sample to room temperature, add 50 mM IAA, and incubate for 30 min at room temperature (protect from light).

19. Dissolve Lys-C in 50 mM ammonium bicarbonate and then add it to the sample at 1 part to 50 by protein content. Incubate the sample for 2–3 h at 37°C.

    Protein concentration can be estimated using standard protocols. Typically the final concentration of Lys-C is ~5 μg/mL.

20. Dissolve trypsin in 50 mM ammonium bicarbonate and add it to the sample 1 part to 20 by protein content. Incubate for 16 h at 37°C.

    Typically, the final concentration of trypsin is ~20 μg/mL.

21. Stop the reaction by adding formic acid to a final concentration of 0.5%. If surfactant was used at Step 17, incubate for 30 min at 37°C, spin at 16,000g for 10 min, and discard the pellet (surfactant degrades at acidic pH). Dry the supernatant using a SpeedVac Concentrator.

22. Reconstitute the sample in appropriate buffer for subsequent liquid chromatography and mass spectrometry.

Strategy 2

23. Add 20 mM DTT to the sample from Step 16 and incubate it for 5 min at 95°C. Cool the sample to room temperature, add IAA to a final concentration of 50 mM, and incubate for 30 min at room temperature (protect from light).

24. Mix the sample with 1.2 mL of freshly prepared 9 M urea (dilute the SDS concentration to 0.2% or less). Add the mixture to an Amicon Ultra Centrifugal Filter (30K).

25. Centrifuge at 16,000g for 5–10 min to reduce the sample volume to ~100 μL.

    This step might need to be repeated to accommodate the entire diluted sample.

26. Add 400 μL of 9 M urea to the filter. Carefully mix by inverting and then centrifuge at 16,000g for 5–10 min. Repeat once.

27. Add 400 µL of 25 mM triethylammonium bicarbonate, rinsing the sides of the filter, and then centrifuge at 16,000g for 5–10 min. Repeat this once.

28. Dissolve trypsin in 50 mM ammonium bicarbonate, add 10 µg of trypsin per filter in 300 µL total volume. Carefully mix by inverting. Digest for 3 h and spike with fresh trypsin overnight.

29. Spin at 16,000g for 10 min and collect the flowthrough. Wash twice with 300 µL of 50 mM ammonium bicarbonate, and then spin at 16,000g for 5 min. Combine the flowthrough and dry the mixture using a SpeedVac concentrator.

30. Reconstitute the sample in appropriate buffer for subsequent liquid chromatography and mass spectrometry.

TROUBLESHOOTING

Problem (Step 6): Different types of amyloid aggregates have different detergent and/or temperature resistance. Overdilution in SDS-containing buffer (i.e., high molar ratios of SDS to total protein) or overheating may cause a decreased amount of recovered amyloid-forming protein.

Solution: Some amyloid-like protein aggregates may be sensitive to SDS-treatment, so 1% Sarkosyl may be used in the amyloid-prion resuspension buffer instead of 2% SDS. Or, try a minimal amount of buffer first (e.g., 30 µL) and add more if the pellet is not solubilized completely by pipetting and incubating (the temperature can be varied). Solubilization is monitored visually, aiming for the solution to become clear (not cloudy), without major particulates. This step-by-step solubilization strategy helps to avoid overtreatment with SDS.

Problem (Step 15): Gel pieces tend to block normal pipetting.

Solution: Cut pipette tip ends and avoid drawing up gel pieces. To remove small gel pieces that occasionally persist in solution after elution, spin the sample at 6000g for 1 min, and transfer the supernatant to a new tube.

DISCUSSION

The strategies described here enable the identification of amyloid-forming proteins by immunostaining (SDD-AGE) or mass spectrometry (TAPI; Kryndushkin et al. 2013). The advantage of using mass spectrometry is that it does not require specific antibodies and thus can be used to identify previously unknown amyloid-forming proteins. However, it is important to note that nonamyloid control samples (i.e., isogenic cells that do not harbor the prion or amyloid-forming protein) are necessary for proper qualitative comparison. Amyloid-forming proteins identified by mass spectrometry should be absent from control samples. Of course, mass spectrometry hits should always be confirmed by secondary analysis, such as antibody-based immunostaining methods.

RECIPES

Amyloid-Prion Buffer R

Reagent	Final concentration
Tris-HCl (pH 7.4)	10 mM
Dithiothreitol (DTT)	5 mM
Sodium dodecyl sulfate (SDS)	0.4%

Use buffer immediately, or prepare and keep it without DTT at room temperature for months. Add fresh DTT immediately before use.

Amyloid-Prion Resuspension Buffer

Reagent	Final concentration
Tris-HCl (pH 7.4)	50 mM
NaCl	100 mM
Sodium dodecyl sulfate (SDS)	2%
Dithiothreitol (DTT)	5 mM
Glycerol	5%
Complete protease inhibitors with EDTA (Roche)	1×

Use buffer immediately, or prepare and keep it without DTT at 4°C for several weeks. Add fresh DTT immediately before use.

Amyloid-Prion Transfer Buffer

Reagent	Final concentration
Tris-borate-EDTA (TBE) buffer (Thermo Scientific)	0.5×
Sodium dodecyl sulfate (SDS)	0.02%

Prepare and keep at room temperature for several months.

Modified RIPA Buffer Supplemented with Nucleases

Reagent	Final concentration
Tris-HCl (pH 7.4)	50 mM
NaCl	100 mM
NP-40	1%
Sodium deoxycholate (DOC)	1%
Sarkosyl	0.1%
Dithiothreitol (DTT)	5 mM
$MgCl_2$	5 mM
Glycerol	3%
Benzonase (e.g., Sigma-Aldrich #######)	0.5 µL/mL
RNase A	0.2 mg/mL
Complete protease inhibitors with EDTA (Roche)	1×
Phenylmethylsulfonyl fluoride (PMSF)	5 mM

Use buffer immediately, or prepare and keep it without enzymes and DTT at 4°C for several weeks. Add the enzymes and DTT immediately before use.

Tris-Buffered Saline (TBS; 10×, pH 7.5)

Reagent	Amount	Final concentration
Tris	24.2 g	200 mM
NaCl	87.7 g	1.5 M

Combine ingredients in ~800 mL of H_2O. Adjust pH to 7.5 and bring final volume to 1 L. Sterilize by autoclaving.

Yeast Disruption Buffer

Reagent	Final concentration
Tris-HCl (pH 7.4)	50 mM
NaCl	40 mM
Dithiothreitol (DTT)	5 mM
MgCl$_2$	5 mM
Glycerol	5%
Benzonase (e.g., Sigma-Aldrich #E1014, 250 units/μL)	0.5 μL/mL
RNase A	0.1 mg/mL
Complete protease inhibitors with EDTA (Roche)	1×
Phenylmethylsulfonyl fluoride (PMSF)	5 mM

Use buffer immediately, or prepare and keep it without enzymes and DTT at 4°C for several weeks. Add the enzymes and DTT immediately before use.

REFERENCES

Kryndushkin DS, Alexandrov IM, Ter-Avanesyan MD, Kushnirov VV. 2003. Yeast [*PSI*[+]] prion aggregates are formed by small Sup35 polymers fragmented by Hsp104. *J Biol Chem* **278:** 49636–49643.

Kryndushkin D, Pripuzova N, Burnett B, Shewmaker F. 2013. Non-targeted identification of prions and amyloid-forming proteins from yeast and mammalian cells. *J Biol Chem* **288:** 27100–27111.

CHAPTER 27

Gene-Centered Yeast One-Hybrid Assays

Juan I. Fuxman Bass,[1] John S. Reece-Hoyes, and Albertha J.M. Walhout[1]

Program in Systems Biology, University of Massachusetts Medical School, Worcester, Massachusetts 01605

An important question when studying gene regulation is which transcription factors (TFs) interact with which *cis*-regulatory elements, such as promoters and enhancers. Addressing this issue in complex multicellular organisms is challenging as several hundreds of TFs and thousands of regulatory elements must be considered in the context of different tissues and physiological conditions. Yeast one-hybrid (Y1H) assays provide a powerful "gene-centered" method to identify the TFs that can bind a DNA sequence of interest. In this introduction, we describe the basic principles of the Y1H assay and its advantages and disadvantages and briefly discuss how it is complementary to "TF-centered" methods that identify protein–DNA interactions for a known protein of interest.

INTRODUCTION

Precise spatiotemporal gene expression plays a central role in development, homeostasis, and response to environmental cues. Gene expression is largely controlled by the specific binding of transcription factors (TFs) to DNA regulatory regions such as promoters and enhancers. TFs comprise 5%–10% of the protein-coding genes of most organisms (Reece-Hoyes et al. 2005; Kummerfeld and Teichmann 2006; Vaquerizas et al. 2009). Interactions between TFs and regulatory DNA sequences can be experimentally identified using approaches that are either "TF-centered," which identify the DNA targets of a protein of interest, or "gene-centered," which identify the repertoire of TFs that bind a DNA sequence of interest (Walhout 2006; Arda and Walhout 2009). TF-centered approaches include chromatin immunoprecipitation (ChIP) (Ren et al. 2000) and protein-binding microarrays (Berger et al. 2006), whereas the yeast one-hybrid (Y1H) assay (Deplancke et al. 2004) is a gene-centered approach.

PRINCIPLES OF Y1H ASSAYS

Y1H assays involve two main components (Fig. 1): first, a reporter construct in which a DNA fragment of interest (the DNA "bait") is cloned upstream of a reporter gene(s), and, second, a plasmid that expresses a "prey" hybrid protein (hence the one-hybrid name) comprising a TF of interest fused to the activation domain (AD) of the yeast TF Gal4. Both components are introduced into a budding yeast strain, and the bait is used to "fish" for interacting preys: If the hybrid TF binds to the DNA of interest, the AD will induce the expression of the reporter gene(s). Interactions involving both TF activators and repressors are detectable by this assay, as the activation of reporter expression is driven

[1]Correspondence: juan.fuxmanbass@umassmed.edu; marian.walhout@umassmed.edu

Copyright © Cold Spring Harbor Laboratory Press; all rights reserved
Cite this introduction as *Cold Spring Harb Protoc*; doi:10.1101/pdb.top077669

Chapter 27

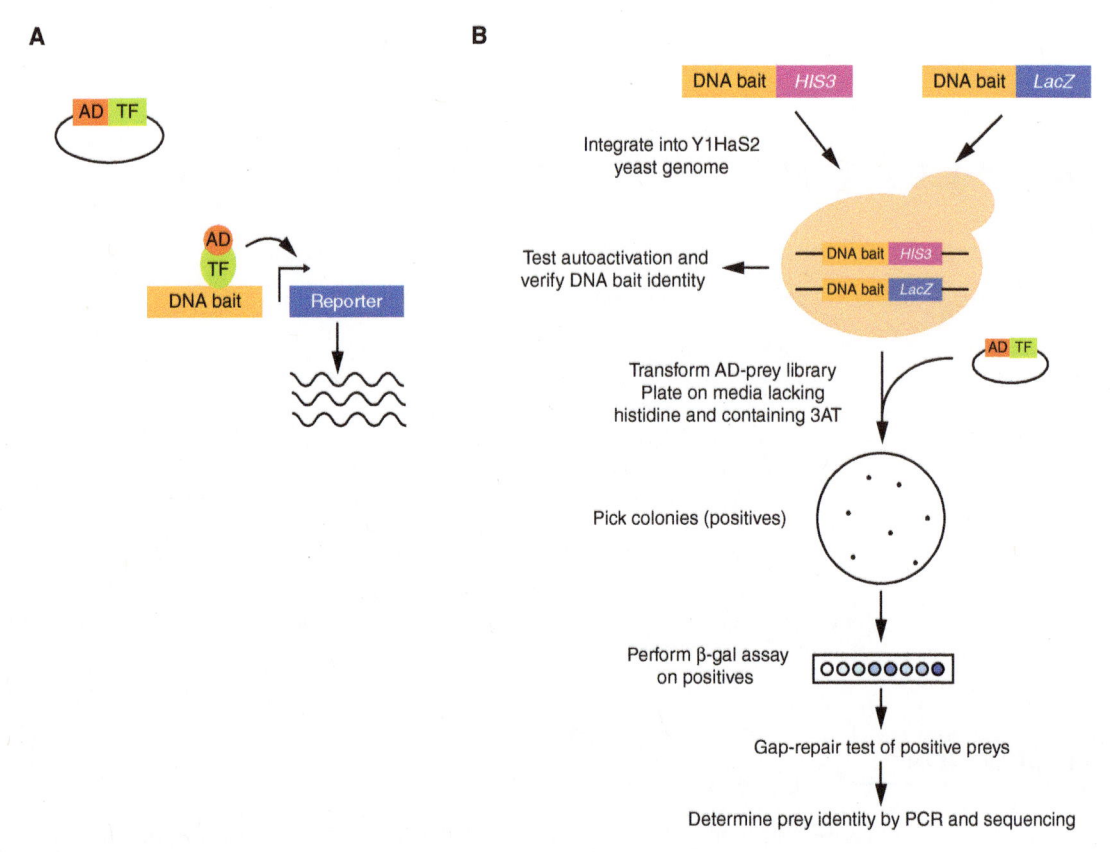

FIGURE 1. The principles underlying the yeast one-hybrid technique and the pipeline for library screening. (*A*) Yeast one-hybrid (Y1H) assays detect protein–DNA interactions between transcription factors (TFs) fused to the Gal4 activation domain (AD–TF "prey") and a DNA fragment of interest (DNA "bait"). The bait is cloned upstream of reporter genes, and each bait::reporter construct is integrated into the yeast genome. If the TF binds to the DNA of interest, the AD moiety will induce expression of the reporter. (*B*) Outline of steps involved in generating a DNA bait yeast strain and performing an AD-cDNA library screen, as described in other protocols. The procedure involves integrating, by homologous recombination into a Y1HaS2 yeast strain, the two reporter cassettes with the bait sequence cloned upstream of both the *HIS3* auxotrophic marker and the colorimetric *LacZ* marker. The screen is performed by transforming the yeast with an AD–prey library and selecting for growth on medium lacking histidine but containing 3-aminotriazole (3AT), a competitive inhibitor of the Hisp3 enzyme, followed by the colorimetric detection of β-galactosidase (β-gal; leading to formation of a blue compound). Interactions are confirmed by conducting gap-repair tests, followed by polymerase chain reaction (PCR) and sequencing to identify the prey.

by the yeast AD. However, to distinguish between activators and repressors, functional assays should be performed in an endogenous system using nonhybrid TFs.

The Y1H system can be used to identify protein–DNA interactions (PDIs) with short *cis*-regulatory elements (as single copy or tandem repeats) (Li and Herskowitz 1993; Deplancke et al. 2006; Reece-Hoyes et al. 2009) as well as with longer and more complex DNA fragments such as promoters or enhancers (Martinez et al. 2008; Arda et al. 2010; Reece-Hoyes et al. 2011a; Fuxman Bass et al. 2014, 2015e). Cloning of the DNA baits to generate reporter constructs can be achieved by traditional restriction enzyme cloning or by recombination-based technologies such as Gateway cloning (Walhout et al. 2000b).

Multiple reporters have been used to detect interactions in Y1H assays, including auxotrophic genes, such as *HIS3*, *URA3*, *TRP1*, and *LEU2*, that enable growth in the absence of histidine, uracil, tryptophan, and leucine, respectively, and *LacZ*, which encodes for the bacterial enzyme β-galactosidase, which is detectable in colorimetric assays. The Y1H protocol that we have developed employs two reporter genes, *HIS3* and *LacZ*, integrated into the yeast genome at two different loci (Deplancke et al. 2004). Integration ensures that every yeast cell has the same number of reporter constructs, thus eliminating confusion between yeast that are positive in the assay because of a PDI and those that appear positive owing to basal expression in colonies harboring higher copy numbers of the reporter plasmids. Furthermore, for an interaction to be detected, it needs to induce reporter expression from two separate loci in the same yeast nucleus—thus the interaction is tested twice, which provides an inherent retest. Activation of the *HIS3* reporter results in yeast growth in the absence of histidine and in the presence of enough 3-aminotriazole (3AT), a competitive inhibitor of the Hisp3 enzyme, to inhibit growth driven by background *HIS3* expression. Activation of the *LacZ* reporter results in higher levels of β-galactosidase expression, which is monitored using a colorimetric assay in which colorless X-gal is modified into a blue compound.

An important aspect of undertaking Y1H assays is the selection of prey source to be screened. The most widely used source of prey molecules is a cDNA library generated by extracting mRNA from a tissue/organism of interest and cloning the reverse-transcribed products into a vector backbone that encodes an adjacent AD moiety (Walhout et al. 2000a,b). Such libraries are commercially available, can be requested from other academic laboratories, or can be generated in house. However, Y1H assays that screen cDNA libraries suffer from two drawbacks: first, it is challenging to know whether all TFs are present in the library and, second, TFs are generally expressed at low levels and therefore are relatively uncommon in any cDNA library (especially nonnormalized ones). The latter drawback can be somewhat overcome by screening large numbers of colonies to ensure that even uncommon transcripts are interrogated. However, the former can only be overcome by screening a different prey source comprising a characterized set of TF clones, such as those available from Gateway-compatible collections of cloned open reading frames ("ORFeomes") (Reboul et al. 2003; Rual et al. 2004). These TF clones can be combined and screened as a "TF minilibrary" (Deplancke et al. 2004), but a more efficient technique is to screen these clones individually as an array to ensure that every TF is interrogated and to remove the need to sequence the clones from the positive yeast because the TF identity at each array position is known (Vermeirssen et al. 2007). Recently, high-throughput Y1H platforms have been developed for four widely studied organisms (*Caenorhabditis elegans*, *Drosophila melanogaster*, *Arabidopsis thaliana*, and human) (Gaudinier et al. 2011; Hens et al. 2011; Reece-Hoyes et al. 2011a,b). These platforms use robotics to manipulate AD–TF arrays in a 1536-colony format. This greatly increases throughput and sensitivity and allows testing all bait–prey combinations four times, providing independent technical replicates that reduce both false-positive and false-negative rates.

In associated protocols, we present detailed methods for conducting gene-centered Y1H assays. First, we outline how to produce bait strains and conduct high-efficiency transformations (see Protocol 1: Generating Bait Strains for Yeast One-Hybrid Assays [Fuxman Bass et al. 2015a]). Next, we detail how to perform a TF screen by transforming the yeast bait strain with an AD–prey library (see Protocol 2: Performing Yeast One-Hybrid Library Screens [Fuxman Bass et al. 2015b]). Finally, various steps in these procedures require the last two protocols that we present (Protocol 3: Colony Lift Colorimetric Assay for β-Galactosidase Activity [Fuxman Bass et al. 2015c] and Protocol 4: Zymolase Treatment and Polymerase Chain Reaction Amplification from Genomic and Plasmid Templates from Yeast [Fuxman Bass et al. 2015d]).

ADVANTAGES AND DISADVANTAGES OF Y1H ASSAYS

The fact that Y1H assays detect PDIs in the milieu of the yeast nucleus rather than in their endogenous biological context provides both advantages and disadvantages (Table 1). As with any methodology, Y1H assays are subject to false-negative and false-positive interactions. There are several explanations for missed PDIs: (1) Not every clone is transformed into yeast and gets the chance to be detected (this is most relevant when using libraries as a TF prey source); (2) the TF is absent from the clone source; (3) the TF might only bind to DNA as a heterodimer (and the current Y1H system expresses only one TF at a time); (4) to bind DNA, the TF needs posttranslational modification(s) not available in yeast; and (5) the TF hybrid protein does not fold correctly in yeast. Regarding false positives, there are two different types to be considered. Technical false positives are interactions that cannot be reproduced when repeating the assay and can be filtered out by testing any detected PDI (at least) one more time. Biological false

TABLE 1. Advantages and disadvantages of the yeast one-hybrid technique

Advantages	Disadvantages
Identifies binding of multiple TFs with a DNA fragment of interest	Might identify PDIs that do not occur in vivo (biological false positives)
Condition-independent: can identify interactions with low-abundance TFs that are difficult to detect directly from tissue	PDIs that require posttranslational modifications of TFs are not detected
	Not (yet) adapted to detect PDIs involving heterodimers

TF, transcription factor; PDI, protein–DNA interaction.

positives are interactions that can be robustly detected in yeast but never occur in vivo. It is challenging to determine whether an interaction is clearly a biological false positive because assays used to validate the interactions in vivo have their own false-negative rate (Walhout 2011). For instance, the effect of knocking down or knocking out a TF can be masked by functionally redundant interactions.

Perhaps the most important advantage of Y1H is that it can determine which TFs from a collection of many hundreds bind and (potentially) regulate a gene of interest. This would be very challenging to achieve through TF-centered approaches, as this would require assaying all TFs individually. Furthermore, PDIs that are rare in vivo (because they occur for only a limited time, in only a few cells, or under a rare environmental condition) are difficult to detect directly from tissue. In Y1H assays, all the TFs are expressed in yeast from the same promoter, and so the ability of the assay to detect a PDI is independent of the in vivo conditions required for the interaction.

Ultimately, the comprehensive detection of the PDIs that drive gene regulation will require a combination of complementary approaches, such as ChIP, functional assays with reporter genes in the endogenous system, and genome editing of the TF binding sites followed by measures of target gene expression.

ACKNOWLEDGMENTS

This work was supported by the National Institutes of Health grants DK068429 and GM082971 to A.J.M.W. J.I.F.B. was partially supported by a postdoctoral fellowship from the Pew Latin American Fellows Program.

REFERENCES

Arda HE, Walhout AJM. 2009. Gene-centered regulatory networks. *Brief Funct Genomics* doi: 10.1093/elp049.

Arda HE, Taubert S, Conine C, Tsuda B, Van Gilst MR, Sequerra R, Doucette-Stam L, Yamamoto KR, Walhout AJM. 2010. Functional modularity of nuclear hormone receptors in a *C. elegans* gene regulatory network. *Mol Syst Biol* 6: 367.

Berger MF, Philippakis AA, Qureshi AM, He FS, Estep PW 3rd, Bulyk ML. 2006. Compact, universal DNA microarrays to comprehensively determine transcription-factor binding site specificities. *Nat Biotechnol* 24: 1429–1435.

Deplancke B, Dupuy D, Vidal M, Walhout AJM. 2004. A Gateway-compatible yeast one-hybrid system. *Genome Res* 14: 2093–2101.

Deplancke B, Mukhopadhyay A, Ao W, Elewa AM, Grove CA, Martinez NJ, Sequerra R, Doucette-Stam L, Reece-Hoyes JS, Hope IA, et al. 2006. A gene-centered *C. elegans* protein–DNA interaction network. *Cell* 125: 1193–1205.

Fuxman Bass JI, Reece-Hoyes JS, Walhout AJM. 2015a. Generating bait strains for yeast one-hybrid assays. *Cold Spring Harb Protoc* doi: 10.1101/pdb.prot088948.

Fuxman Bass JI, Reece-Hoyes JS, Walhout AJM. 2015b. Performing yeast one-hybrid library screens. *Cold Spring Harb Protoc* doi: 10.1101/pdb.prot088955.

Fuxman Bass JI, Reece-Hoyes JS, Walhout AJM. 2015c. Colony lift colorimetric assay for β-galactosidase activity. *Cold Spring Harb Protoc* doi: 10.1101/pdb.prot088963.

Fuxman Bass JI, Reece-Hoyes JS, Walhout AJM. 2015d. Zymolase treatment and polymerase chain reaction amplification from genomic and plasmid templates from yeast. *Cold Spring Harb Protoc* doi: 10.110/pdb.prot088971.

Fuxman Bass JI, Sahni N, Shrestha S, Garcia-Gonzalez A, Mori A, Bhat N, Yi S, Hill DE, Vidal M, Walhout AJ. 2015e. Human gene-centered transcription factor networks for enhancers and disease variants. *Cell* 161: 661–673.

Fuxman Bass JI, Tamburino AM, Mori A, Beittel N, Weirauch MT, Reece-Hoyes JS, Walhout AJ. 2014. Transcription factor binding to *Caenorhabditis elegans* first introns reveals lack of redundancy with gene promoters. *Nucleic Acids Res* 42: 153–162.

Gaudinier A, Zhang L, Reece-Hoyes JS, Taylor-Teeples M, Pu L, Liu Z, Breton G, Pruneda-Paz JL, Kim D, Kay SA, et al. 2011. Enhanced Y1H assays for *Arabidopsis*. *Nat Methods* 8: 1053–1055.

Hens K, Feuz J-D, Iagovitina A, Massouras A, Bryois J, Callaerts P, Celniker S, Deplancke B. 2011. Automated protein–DNA interaction screening of *Drosophila* regulatory elements. *Nat Methods* 8: 1065–1070.

Kummerfeld SK, Teichmann SA. 2006. DBD: A transcription factor prediction database. *Nucleic Acids Res* 34: D74–D81.

Li JJ, Herskowitz I. 1993. Isolation of the ORC6, a component of the yeast origin recognition complex by a one-hybrid system. *Science* **262**: 1870–1874.

Martinez NJ, Ow MC, Barrasa MI, Hammell M, Sequerra R, Doucette-Stamm L, Roth FP, Ambros V, Walhout AJM. 2008. A *C. elegans* genome-scale microRNA network contains composite feedback motifs with high flux capacity. *Genes Dev* **22**: 2535–2549.

Reboul J, Vaglio P, Rual JF, Lamesch P, Martinez M, Armstrong CM, Li S, Jacotot L, Bertin N, Janky R, et al. 2003. *C. elegans* ORFeome version 1.1: Experimental verification of the genome annotation and resource for proteome-scale protein expression. *Nat Genet* **34**: 35–41.

Reece-Hoyes JS, Deplancke B, Shingles J, Grove CA, Hope IA, Walhout AJM. 2005. A compendium of *C. elegans* regulatory transcription factors: A resource for mapping transcription regulatory networks. *Genome Biol* **6**: R110.

Reece-Hoyes JS, Deplancke B, Barrasa MI, Hatzold J, Smit RB, Arda HE, Pope PA, Gaudet J, Conradt B, Walhout AJ. 2009. The *C. elegans* Snail homolog CES-1 can activate gene expression in vivo and share targets with bHLH transcription factors. *Nucleic Acids Res* **37**: 3689–3698.

Reece-Hoyes JS, Barutcu AR, Patton McCord R, Jeong J, Jian L, MacWilliams A, Yang X, Salehi-Ashtiani K, Hill DE, Blackshaw S, et al. 2011a. Yeast one-hybrid assays for high-throughput human gene regulatory network mapping. *Nat Methods* **8**: 1050–1052.

Reece-Hoyes JS, Diallo A, Kent A, Shrestha S, Kadreppa S, Pesyna C, Lajoie B, Dekker J, Myers CL, Walhout AJM. 2011b. Enhanced yeast one-hybrid (eY1H) assays for high-throughput gene-centered regulatory network mapping. *Nat Methods* **8**: 1059–1064.

Ren B, Robert F, Wyrick JJ, Aparicio O, Jennings EG, Simon I, Zeitlinger J, Schreiber J, Hannett N, Kanin E, et al. 2000. Genome-wide location and function of DNA binding proteins. *Science* **290**: 2306–2309.

Rual J-F, Hirozane-Kishikawa T, Hao T, Bertin N, Li S, Dricot A, Li N, Rosenberg J, Lamesch P, Vidalain P-O, et al. 2004. Human ORFeome version 1.1: A platform for reverse proteomics. *Genome Res* **14**: 2128–2135.

Vaquerizas JM, Kummerfeld SK, Teichmann SA, Luscombe NM. 2009. A census of human transcription factors: Function, expression and evolution. *Nat Rev Genet* **10**: 252–263.

Vermeirssen V, Deplancke B, Barrasa MI, Reece-Hoyes JS, Arda HE, Grove CA, Martinez NJ, Sequerra R, Doucette-Stamm L, Brent M, et al. 2007. Matrix and Steiner-triple-system smart pooling assays for high-performance transcription regulatory network mapping. *Nat Methods* **4**: 659–664.

Walhout AJM. 2006. Unraveling transcription regulatory networks by protein–DNA and protein–protein interaction mapping. *Genome Res* **16**: 1445–1454.

Walhout AJM. 2011. What does biologically meaningful mean? A perspective on gene regulatory network validation. *Genome Biol* **12**: 109.

Walhout AJM, Sordella R, Lu X, Hartley JL, Temple GF, Brasch MA, Thierry-Mieg N, Vidal M. 2000a. Protein interaction mapping in *C. elegans* using proteins involved in vulval development. *Science* **287**: 116–122.

Walhout AJM, Temple GF, Brasch MA, Hartley JL, Lorson MA, van den Heuvel S, Vidal M. 2000b. GATEWAY recombinational cloning: Application to the cloning of large numbers of open reading frames or ORFeomes. *Methods Enzymol* **328**: 575–592.

Protocol 1

Generating Bait Strains for Yeast One-Hybrid Assays

Juan I. Fuxman Bass,[1] John S. Reece-Hoyes, and Albertha J.M. Walhout[1]

Program in Systems Biology, University of Massachusetts Medical School, Worcester, Massachusetts 01605

Yeast one-hybrid (Y1H) assays are used to identify which transcription factor (TF) "preys" can bind a DNA fragment of interest that is used as the "bait." Undertaking Y1H assays requires the generation of a yeast "bait strain" for each DNA fragment of interest that features the DNA bait coupled to a reporter(s). Plasmids encoding TFs fused to the Gal4 activation domain (AD) are then introduced into the bait strain, and activation of the reporter(s) indicates that a TF–DNA interaction has occurred. Here, we present a protocol for the first part of the strategy—the generation of a bait strain for Y1H assays. We assume that the DNA bait has already been cloned into two different reporter constructs: One places the fragment of interest upstream of *HIS3*, an auxotrophic growth marker, whereas the other places the DNA bait upstream of *LacZ*, a colorimetric marker that changes colorless X-gal into a blue compound. Briefly, generation of the bait strain involves using homologous recombination to integrate the two reporters into the genome of the yeast strain, screening individual integrants for background reporter expression (i.e., expression in the absence of a TF), and using polymerase chain reaction (PCR) and sequencing to confirm the DNA bait identity from both integrated reporter cassettes.

MATERIALS

It is essential that you consult the appropriate Material Safety Data Sheets and your institution's Environmental Health and Safety Office for proper handling of equipment and hazardous material used in this protocol.

RECIPES: Please see the end of this protocol for recipes indicated by <R>. Additional recipes can be found online at http://cshprotocols.cshlp.org/site/recipes.

Reagents

Agarose gel (1%, w/v, in 0.5× TBE buffer]
Bovine serum albumin (BSA; 1 mg/mL in distilled water)
DNA bait cloned into either pHisi-1 and pLacZi (Clontech) or pMW#2 and pMW#3 (Addgene)

The reporter constructs pHisi-1 and pLacZi from Clontech allow restriction-endonuclease-based cloning, whereas pMW#2 and pMW#3 are from the Addgene repository and allow Gateway recombination-based cloning.

DNA molecular mass markers
Glycerol solution (15%, v/v, in sterile water)
LiAc (10×; 1 M in water)
Nitrocellulose filters (45 µm, 137 mm; Fisher Scientific WP4HY13750)
PEG 3350 (40% w/v, sterile; Fisher Scientific)
Primers

[1]Correspondence: juan.fuxmanbass@umassmed.edu; marian.walhout@umassmed.edu

Copyright © Cold Spring Harbor Laboratory Press; all rights reserved
Cite this protocol as *Cold Spring Harb Protoc*; doi:10.1101/pdb.prot088948

HISRV primer (anneals both pHisi-1 and pMW#2; 5′-GCTTTCTGCTCTGTCATCTTTG-3′)
LACFW primer (anneals both pLacZi and pMW#3; 5′-GTTCGGAGATTACCGAATCAA-3′)
LACRV primer (anneals both pLacZi and pMW#3; 5′-ATCTGCCAGTTTGAGGGGAC-3′)
pHisi-1FW primer (5′-AACAAATAGGGGTTCCGC-3′)
pMW#2FW primer (5′- AGCTATGACGTCGCATGCAC-3′)

Restriction enzymes (various, see Step 1) and associated 10× buffers
The suitable restriction enzymes are AflII, ApaI, BseR1, NcoI, NsiI, StuI, and XhoI.

Salmon sperm DNA (ssDNA; 10 mg/mL; Life Technologies)
Sc-Ura-His plates (150 mm) ± 3AT (see Step 19) <R>
TBE buffer <R>
TE (10×; 100 mM Tris–HCl [pH 8.0] and 10 mM EDTA)
Water (MilliQ, sterile)
YAPD liquid medium <R>
YAPD medium is yeast extract peptone dextrose (YEPD) medium with extra adenine, which is added because the yeast strain used has mutations in the adenine synthesis pathway.

YAPD plates <R>
Y1HaS2 yeast strain (genotype: *MATa, ura3-52, his3-Δ1, ade2-101, ade5, lys2-801, leu2-3,112, trp1-901, tyr1-501, gal4Δ, gal80Δ, ade5::hisG*)

Equipment

Agarose gel electrophoresis equipment
Camera (optional; see Step 20)
Conical tubes (50-mL, sterile)
Freezer (−80°C)
Glass beads (sterile)
Replica-plating apparatus (Cora Styles #4006 for 150-mm plates)
Shaking incubator (30°C)
Toothpicks and/or disposable plastic loops (sterile)
Tubes (1.5-mL, sterile)
Velvets (220 × 220-mm pieces of velveteen [100% cotton velveteen without rayon])
Water baths (37°C and 42°C)

METHOD

Linearization of the Y1H Reporter Constructs

The protocol begins with a DNA bait that has already been cloned into the HIS3 (pHisi-1 or pMW#2) and LacZ (pLacZi or pMW#3) reporter constructs.

1. Set up the following digests with restriction enzymes in 1.5-mL tubes and incubate in a water bath for 3 h at 37°C.

   Tube 1:

pMW#2 (or pHisi-1) construct including DNA bait	1 to 4 µg
Bovine serum albumin (1 mg/mL)	2.5 µL
10× Restriction buffer	2.5 µL
AflII, XhoI, NsiI, or BseR1 (select one)	2 µL (20 units)
Water	to 25 µL

Tube 2:

pMW#3 (or pLacZi) construct including DNA bait	1 to 4 µg
Bovine serum albumin (1 mg/mL)	2.5 µL
10× Restriction buffer	2.5 µL
NcoI, ApaI, or StuI (select one)	2 µL (20 units)
Water	to 25 µL

The restriction enzyme of choice must not cut within the DNA-bait sequence. These digests linearize the Y1H reporter constructs such that regions of homology with the yeast genome occur at both ends. When these linear constructs are transformed into the Y1HaS2 host yeast strain, the pMW#2 (or pHisi-1) construct is integrated into the mutant HIS3 locus (his3-Δ1), whereas the pMW#3 (or pLacZi) construct is integrated into the mutant URA3 locus (ura3-52).

2. Using agarose gel electrophoresis equipment, verify linearization of the constructs by running 1–2 µL of the restriction digest reaction mixture on a 1% (w/v) agarose gel in 0.5× TBE buffer (pH 8.3) next to an equal amount of the undigested constructs as well as DNA molecular mass markers.

High-Efficiency Transformation to Generate Integrated Colonies of Y1HaS2

3. Using a sterile toothpick or loop, thickly spread the Y1HaS2 strain onto a YAPD plate. Incubate overnight at 30°C.

4. Using a sterile toothpick or loop, resuspend the Y1HaS2 strain into 80 mL of liquid YAPD to an OD_{600} of 0.15–0.20.

5. Incubate the culture in a shaking incubator (30°C, 200 rpm) until the OD_{600} reaches 0.4–0.6.
   This usually takes ~4–6 h.

6. Pellet the cells by centrifugation (700g) in conical 50-mL tubes for 5 min at room temperature.

7. Discard the liquid from the pellet formed, and wash the cells with 20 mL of sterile water. Resuspend the cells by pipetting or by shaking the tube. Do not vortex.

8. Centrifuge the yeast suspension as in Step 6, discard the supernatant, and resuspend the cells in 5 mL of a freshly made solution of 1× LiAc in 1× TE in sterile water ("TE–LiAc solution") at room temperature. Do not vortex.

9. Centrifuge the suspension as in Step 6, remove the supernatant carefully by aspiration, and then resuspend the cells in 400 µL of TE–LiAc solution by pipetting up and down.

10. Prepare 50 µL of denatured salmon sperm DNA (ssDNA, 10 mg/mL) by boiling for 5 min, and then keep on ice.

11. Add 40 µL denatured ssDNA to the yeast suspension (from Step 9) and mix by pipetting up and down.

12. In a 1.5-mL tube, transform 20 µL of *both* linearized constructs for the same DNA bait (using the digest reaction mixes from Step 1) into the *same* 200 µL of competent Y1HaS2 yeast cell suspension in TE–LiAc–ssDNA.
    Remember to also prepare a negative control in which no linearized vector DNA is added to the yeast.

13. Add 1250 µL of 40% PEG in TE–LiAc solution to the tube and mix by multiple inversions or by pipetting up and down (do not vortex).

14. Incubate all transformation reactions for 30 min at 30°C.

15. Heat-shock the cells for 20 min at 42°C (in a water bath).

16. Centrifuge the tubes for 1 min at room temperature at 700g and remove the supernatant by aspiration.

17. Resuspend each transformation in 700 μL of sterile water and use sterile glass beads to spread on a 150-mm Sc-Ura-His plate. Incubate the plates for 3–5 d at 30°C until individual colonies appear.

    The Y1H reporter constructs are integrative (YIp) vectors that cannot be replicated within yeast, so any colonies that grow at this stage must be integrants. The minimal promoter in pHisi-1 or pMW#2 drives enough HIS3 expression to enable growth in the absence of histidine. The number of colonies obtained usually varies between 10 and 100. The negative control should give rise to no colonies.

    See Troubleshooting.

Test DNA Bait Strain Autoactivity

Autoactivity is the level of reporter expression in the absence of binding to the AD–prey. Autoactivation is likely to be due to an endogenous yeast TF binding the DNA bait (Deplancke et al. 2004). Different integrant strains arising from the same transformation can show varying levels of autoactivity, which can be because differing numbers of reporter cassettes have been integrated into each strain (and even at each locus within each strain). It is important to test several independent integrants to select the one with the lowest autoactivity for both reporters, so that interactions are easily detected in subsequent Y1H assays.

18. Using a sterile toothpick or loop, transfer six to 24 yeast colonies from the Sc-Ura-His plate in Step 17 to a new 150-mm Sc-Ura-His plate in "96-spot format." Incubate the plates for 1–2 d at 30°C.

    If yeast strains that show low, medium, and high HIS3 and lacZ reporter expression levels are available, add these to the Sc-His-Ura plate.

19. Using sterile velvets and the replica-plating apparatus, replica-plate the yeast from Step 18 onto a fresh 150-mm Sc-Ura-His plate (without 3AT), a set of 150-mm Sc-Ura-His+3AT plates (containing 10, 20, 40, 60, or 80 mM 3AT), and a 150-mm YAPD plate onto which a nitrocellulose (NC) filter has been placed (so that the yeast will grow on the filter). Using sterile velvets and the replica-plating apparatus, replica-clean only the 3AT-containing selective medium plates until no yeast are visible (usually three cleans are needed). Incubate all of the plates at 30°C.

    For replica-plating, a sterile velvet is placed onto the replica-plating block and locked into place using the collar. The yeast plate is placed yeast side down onto the velvet and evenly pressed to transfer the yeast from the plate to the velvet. Then, the yeast plate is removed, and a new plate is placed medium side down onto the velvet and pressed evenly to transfer the yeast from the velvet to the new plate. In this way, multiple (up to five) transfers can be performed from one velvet. To replica-clean the 3AT-containing plates, press the plate (yeast side down) onto a series of sterile velvets (used only once) to remove excess yeast that can interfere with the yeast growth assay.

    The 3-aminotriazole (3AT) acts as a competitive inhibitor of the His3p enzyme.

20. After 1 d at 30°C, use the nitrocellulose filter–YAPD plate from Step 19 in a β-galactosidase colorimetric assay (see Protocol 3: Colony Lift Colorimetric Assay for β-Galactosidase Activity [Fuxman Bass et al. 2015a]). Record how much blue compound was generated by each integrant strain by taking a photograph with a camera and/or noting the color in a qualitative manner (e.g., white, light blue, dark blue, very dark blue).

21. After 5–10 d at 30°C, inspect the Sc-Ura-His+3AT plates from Step 19 and record how much 3AT was required to inhibit the growth of each integrant strain. This information will help determine which yeast bait strain(s) to screen in Y1H assays, as explained in Step 22.

Confirming the Identity of the Integrated DNA Bait

22. Choose up to four integrant strains showing the lowest autoactivity for *both* reporters. For each integrant, verify that the correct DNA bait is upstream of *both* reporters by amplification of the inserts, as described elsewhere (see Protocol 4: Zymolase Treatment and Polymerase Chain Reaction Amplification from Genomic and Plasmid Templates from Yeast (Fuxman Bass et al. 2015b), using the appropriate primers, followed by DNA sequencing of the PCR products.

The primers for pLacZi and pMW#3 (LACFW and LACRV) will add ~640 nucleotides to the original bait sequence, pHisi-1 (pHisi-1FW and HISRV) ~240 nucleotides, and pMW#2 (pMW#2FW and HISRV) ~210 nucleotides. Proceed to Step 23 while waiting for the sequencing results.

Preparing Glycerol Stocks of Yeast Integrants

23. Using a sterile toothpick, transfer the integrants selected in Step 22 from the Sc-Ura-His plate replicated in Step 19 to a new Sc-Ura-His plate and incubate the plate overnight at 30°C.

    Only one integrant strain per DNA bait is needed for library screens, but it is useful to have at least one backup strain.

24. Using a sterile toothpick, transfer a match-head-sized amount of this freshly grown yeast from Step 23 into 200 µL of sterile 15% (v/v) glycerol solution in 1.5-mL tubes. Vortex the yeast–glycerol solution for 5 sec and store the resulting yeast suspension in a freezer at −80°C.

    The frozen yeast stocks can be stored, for up to 2 yr, and then revived by transferring some (~5 µL) of the frozen stock to a YAPD plate and allowing growth for at least 2 d at 30°C. Remember to discard any frozen stocks of yeast that proved incorrect via sequencing in Step 22.

TROUBLESHOOTING

Problem (Step 17): Transformation fails to generate yeast colonies.
Solution: Assuming that the medium was made correctly (e.g., glucose was not forgotten), the next-most-common issue is that one of the reporters failed to integrate. Check this by transforming only the *HIS3* construct and plating on Sc-His, or only the *LacZ* construct and plating on Sc-Ura. It might be necessary to digest the refractory construct with a different enzyme (Step 1), generate a new or more-concentrated miniprep, or even remake the construct. Integration efficiency is also dependent on the "health" of the yeast, so only use Y1HaS2 revived from frozen stocks <7 d previously, and do not mix harshly at any step.

DISCUSSION

Optimal integrant strains generate a little blue compound in the colorimetric assay (driven by the minimal promoter upstream of *LacZ*), and have some growth on 10 mM 3AT plates, but minimal or no growth on 20 mM 3AT plates. If no blue compound at all is observed, or no growth even on 10 mM 3AT plates is observed, this might indicate a problem with either reporter construct, and such integrants should ideally not be selected. The lowest concentration of 3AT that totally prevents growth should be used for library screens (i.e., integrants that grow on 20 mM 3AT but not on 40 mM 3AT should be screened at 40 mM) (see Protocol 2: Performing Yeast One-Hybrid Library Screens [Fuxman Bass et al. 2015c]). Integrants that grow strongly on 80 mM 3AT cannot be used in Y1H library screens because few (if any) protein–DNA interactions can activate the *HIS3* reporter enough to overcome this high background, and because higher concentrations of 3AT are toxic to yeast. However, integrants with low *HIS3* autoactivity and high *LacZ* autoactivity can be screened because interactions are detected by growth assay using the *HIS3* reporter first, and the *LacZ* activity in these "HIS-positive" yeast might be higher than background when observed closely. However, results obtained with highly autoactive baits should be judged with caution. Note that, for some DNA baits (10%–20%), the autoactivity levels for all integrants will be too high, making Y1H screens impossible (Deplancke et al. 2004; Reece-Hoyes et al. 2013). For these highly autoactive DNA baits, it might be desirable to use smaller fragments of the DNA bait that confer lower autoactivity.

RECIPES

Sc-Ura-His Plates (150 mm) ± 3AT

Reagent	Quantity (for 2 L)
Drop-out mix synthetic minus histidine, leucine, tryptophan, and uracil, adenine rich (2 g) without yeast nitrogen base (US Biological, D9540-02)	2.6 g
Yeast nitrogen base (YNB) without amino acids and without ammonium sulfate	3.4 g
Ammonium sulfate	10 g
Agar	35 g
Glucose (40%, w/v, in water, sterile)	100 mL
Leucine (100 mM, filter sterilized)	16 mL
Tryptophan (40 mM, filter sterilized)	16 mL
3-Amino-1,2,4-triazole (3AT) (2 M, filter sterilized) (optional)	10–80 mL

Dissolve the drop-out mix, the YNB, and the ammonium sulfate in 920 mL of water, and pH to 5.9 with NaOH (5 M). Pour into a 2-L flask and add a stir bar. In a second 2-L flask, add the agar to 950 mL of water (do not add a stir bar as it will cause the agar to boil over in the autoclave). Autoclave for 40 min at 15 psi on liquid cycle. Immediately pour the contents of the first flask, including the stir bar, into the agar-containing flask. Add the glucose, mix well on a stir plate, and cool to 55°C. Add the leucine and the tryptophan. Mix well on a stir plate and pour into 150-mm sterile Petri dishes (~80 mL per dish), dry for 3–5 d at room temperature, wrap in plastic bags, and store for up to 6 mo at room temperature. For Sc-Ura-His plates containing 3AT, add 10–80 mL of 3AT together with the leucine and the tryptophan for a final concentration of 10 to 80 mM 3AT, and store them for up to 1 mo at room temperature.

TBE Buffer

Prepare a 5× stock solution in 1 L of H_2O:
  54 g of Tris base
  27.5 g of boric acid
  20 mL of 0.5 M EDTA (pH 8.0)

The 0.5× working solution is 45 mM Tris-borate/1 mM EDTA. TBE is usually made and stored as a 5× or 10× stock solution. The pH of the concentrated stock buffer should be ~8.3. Dilute the concentrated stock buffer just before use and make the gel solution and the electrophoresis buffer from the same concentrated stock solution. Some investigators prefer to use more concentrated stock solutions of TBE (10× as opposed to 5×). However, 5× stock solution is more stable because the solutes do not precipitate during storage. Passing the 5× or 10× buffer stocks through a 0.22-µm filter can prevent or delay formation of precipitates.

YAPD Liquid Medium

Reagent	Quantity (for 1 L)
Peptone	20 g
Yeast extract	10 g
Adenine hemisulfate dehydrate	0.16 g
Glucose (40%, w/v, in water; sterile)	50 mL

Dissolve powders in 950 mL of water in a 1-L bottle. Autoclave for 40 min at 15 psi on liquid cycle. Add the glucose, mix well, and store for up to 6 mo at room temperature.

YAPD Plates

Reagent	Quantity (for 2 L)
Peptone	40 g
Yeast extract	20 g
Adenine hemisulfate dehydrate	0.32 g
Glucose (40%, w/v) in water, sterile	100 mL
Agar	35 g

Dissolve the first three powders in 950 mL of water in a 2-L flask and add a stir bar. In a second 2-L flask, add the agar to 950 mL of water (do not add a stir bar as it will cause the agar to boil over in the autoclave). Autoclave for 40 min at 15 psi on liquid cycle. Immediately pour the contents of the first flask, including the stir bar, into the agar-containing flask. Add the glucose, mix well on a stir plate, and cool to 55°C. Pour into 150-mm sterile Petri dishes (~80 mL per dish), dry for 3–5 d at room temperature, wrap in plastic bags, and store at room temperature for up to 6 mo.

ACKNOWLEDGMENTS

This work was supported by the National Institutes of Health grants DK068429 and GM082971 to A.J.M.W. J.I.F.B was partially supported by a postdoctoral fellowship from the Pew Latin American Fellows Program.

REFERENCES

Deplancke B, Mukhopadhyay A, Ao W, Elewa AM, Grove CA, Martinez NJ, Sequerra R, Doucetta-Stamm L, Reece-Hoyce JS, Hope IA, et al. 2006. A gene-centered *C. elegans* protein-DNA interaction network. *Cell* 125: 1193–1205.

Fuxman Bass JI, Reece-Hoyes JS, Walhout AJM. 2015a. Colony lift colorimetric assay for β-galactosidase activity. *Cold Spring Harb Protoc* doi: 10.1101/pdb.prot088963.

Fuxman Bass JI, Reece-Hoyes JS, Walhout AJM. 2015b. Zymolase treatment and polymerase chain reaction amplification from genomic and plasmid templates from yeast. *Cold Spring Harb Protoc* doi: 10.1101/pdb.prot088971.

Fuxman Bass JI, Reece-Hoyes JS, Walhout AJM. 2015c. Performing yeast one-hybrid library screens. *Cold Spring Harb Protoc* doi: 10.1101/pdb.prot088955.

Reece-Hoyes JS, Pons C, Diallo A, Mori A, Shrestha S, Kadreppa S, Nelson J, Diprima S, Dricot A, Lajoie BR, et al. 2013. Extensive rewiring and complex evolutionary dynamics in a *C. elegans* multiparameter transcription factor network.. *Mol Cell* 51: 116–127.

Protocol 2

Performing Yeast One-Hybrid Library Screens

Juan I. Fuxman Bass,[1] John S. Reece-Hoyes, and Albertha J.M. Walhout[1]

Program in Systems Biology, University of Massachusetts Medical School, Worcester, Massachusetts 01605

Yeast one-hybrid (Y1H) assays are used to identify which transcription factor (TF) "prey" molecules can bind a DNA fragment of interest that is used as "bait". Y1H assays involve introducing plasmids that encode TFs into a yeast "bait strain" in which the DNA fragment of interest is integrated upstream of one or more reporters, and activation of these reporters indicates that a TF–DNA interaction has occurred. These plasmids express each TF as a hybrid protein (hence the "one-hybrid" name) fused to the activation domain (AD) of the yeast TF Gal4. The AD moiety activates reporter expression even if the TF to which it is fused typically functions as a repressor. Here, we describe how to perform a Y1H screen of a library of cDNA fragments cloned into a pPC86 plasmid expressing the protein encoded by the cDNA as an AD fusion. The method assumes availability of either commercially available libraries or libraries generated in house using mRNA extracted from a tissue of interest. We also assume that users have access to a yeast bait strain that possesses the DNA fragment of interest integrated upstream of two different reporters—*HIS3*, an auxotrophic marker, and *LacZ*, a colorimetric marker that changes colorless X-gal into a blue compound. Briefly, the screen involves transforming the AD-cDNA library into the yeast bait strain, identifying colonies that show activation of both reporters, retesting the interaction in a freshly grown bait strain, and sequencing the cDNA insert to identify the interacting TF.

MATERIALS

It is essential that you consult the appropriate Material Safety Data Sheets and your institution's Environmental Health and Safety Office for proper handling of equipment and hazardous material used in this protocol.

RECIPES: Please see the end of this protocol for recipes indicated by <R>. Additional recipes can be found online at http://cshprotocols.cshlp.org/site/recipes.

Reagents

AD-cDNA library DNA (1 µg/µL in TE, pH 7.0)
Agarose gel [1% (w/v) in 0.5× TBE buffer]
Bovine serum albumin (BSA; 1 mg/mL in water)
LiAc (10×; 1 M in water)
pPC86 plasmid (Clontech)
Primers:
    AD primer (5′-CGCGTTTGGAATCACTACAGGG-3′)
    TERM primer (5′-GGAGACTTGACCAAACCTCTGGCG-3′)
Restriction enzymes SalI and BglII and associated 10× buffers

[1]Correspondence: juan.fuxmanbass@umassmed.edu; marian.walhout@umassmed.edu

Copyright © Cold Spring Harbor Laboratory Press; all rights reserved
Cite this protocol as *Cold Spring Harb Protoc*; doi:10.1101/pdb.prot088955

Chapter 27

Salmon sperm DNA (ssDNA; 10 mg/mL; Life Technologies)
This boiled DNA acts as a carrier and increases transformation efficiency.

Sc-Ura-His-Trp plates (150 mm) ± 3AT (without tryptophan) <R>
The concentration of 3AT in the plates should be that at which growth of each strain is minimal, as determined by the HIS3 autoactivation test (see Protocol 1: Generating Bait Strains for Yeast One-Hybrid Assays [Fuxman Bass et al. 2015a]).

TBE buffer <R>
TE (10×; 100 mM Tris–HCl [pH 8.0] and 10 mM EDTA)
YAPD plates <R>
Water (MilliQ, sterile)

Equipment

Agarose gel electrophoresis equipment
Glass beads (sterile)
Incubators (set at 30°C and 37°C)
Nitrocellulose filters (45 µm, 137 mm; Fisher Scientific WP4HY13750)
PCR plates (96-well)
Replica-plating apparatus (Cora Styles #4006 for 150-mm plates)
Thermal cycler
Toothpicks (sterile) and/or sterile disposable plastic loops
Tubes (sterile; 1.5-mL and 15-mL)
Tube shaker
Velvets (sterile, 22-cm × 22-cm pieces of velveteen)

METHOD

Transformation of the AD–Prey Library into the Y1H Bait Strain

1. For each bait strain to be assayed, perform the following high-efficiency yeast transformations, following the procedure described elsewhere (see Reece-Hoyes and Walhout 2012 or follow Steps 3–17 of Protocol 1: Generating Bait Strains for Yeast One-Hybrid Assays [Fuxman Bass et al. 2015a]).

    i. Add 3 µg of AD-cDNA library into ten 1.5-mL tubes, each containing 200 µL of TE–LiAc–ss DNA bait yeast suspension.

    ii. Add 50 ng of empty pPC86 plasmid (i.e., possessing no cDNA insert) into 30 µL of TE–LiAc–ss DNA bait yeast suspension in a 1.5-mL tube.

    To prepare 2 mL of competent bait yeast cells, the volumes described in Protocol 1: Generating Bait Strains for Yeast One-Hybrid Assays (Fuxman Bass et al. 2015a) should be scaled accordingly (e.g., start with a 400 mL culture of the bait yeast strain). The amount of library DNA used in the transformation will depend on the complexity of the library. The amount listed here is for a Caenorhabditis elegans cDNA library of 4×10^7 clones, representing approximately 16,000 of the approximately 20,000 genes, in the pPC86 backbone. pPC86 features a TRP1 expression cassette that enables the Y1H bait strain to grow in the absence of tryptophan. Remember to also prepare a negative control that has no plasmid added to the yeast, which should yield no growth.

2. After the heat-shock stage in the high-efficiency yeast transformation protocol (i.e., Step 15 in Protocol 1: Generating Bait Strains for Yeast One-Hybrid Assays [Fuxman Bass et al. 2015a]), resuspend each aliquot of the cDNA library transformations in 500 µL of sterile water and combine in a 15-mL tube. Resuspend the empty plasmid control yeast in 30 µL of sterile water.

3. To determine the transformation efficiency, dilute 5 µL of the library transformation mix from Step 2 in 495 µL of sterile water in a 1.5-mL tube to create a 1/100 dilution. Take 50 µL from the

1/100 dilution and add 450 μL of sterile water to create a 1/1000 dilution. Use sterile glass beads to spread each dilution onto a 150-mm Sc-Ura-His-Trp plate.

> *An effective library screen needs to have generated at least 1 million clones. For libraries of higher complexity (i.e., more genes and/or isoforms), a higher number of clones might need to be screened. If the yield is low, then the screen might need to be repeated until the required number of colonies is attained.*

4. Using sterile glass beads, plate the remainder of the library transformation across ten 150-mm Sc-Ura-His-Trp+3AT plates.

5. Pipette 5 μL of the empty pPC86 control yeast suspension as a single spot onto a Sc-Ura-His-Trp plate.

> *There is no need to spread for single colonies. The yeast transformed with empty vector will be used in Step 8 as a "negative control" yeast to show reporter expression levels when no interaction is occurring.*

6. Use incubator to maintain all plates from Steps 3 to 5 at 30°C.

7. After 3 d, check for yeast colonies, which should be obvious on the plates from Step 3 and Step 5. If colonies appear on the negative control plate (with no plasmid) prepared in Step 1, the screen cannot be used as there has been a contamination in one of the reagents. Calculate the transformation efficiency using data from the plates from Step 3. Store the plate from Step 5 at room temperature until needed in Step 8.

8. Continue to incubate the plates from Step 4 at 30°C for up to 14 d. Check the plates regularly and, using sterile toothpicks or loops, transfer "HIS-positive" colonies as they become visible on 150-mm Sc-Ura-His-Trp plates in "96-spot format." Also transfer some of the empty pPC86 control yeast from Step 5 to these plates. Incubate the plates for 1–2 d at 30°C.

> *The number and size of colonies obtained varies greatly and depends on the bait used. It is important to select faster- and slower-growing colonies where feasible because different TFs drive different levels of reporter activation, resulting in different yeast phenotypes.*

> *If positive-control yeast strains that express known levels of reporters are available, they should also be transferred to these plates. The number of colonies picked per bait is limited only by how many grow, but we usually do not exceed 96 (which can be plated on a single 150-mm plate and is convenient for 96-well polymerase chain reaction [PCR]). Some bait strain transformations might not yield any "HIS-positive" colonies despite good transformation efficiency. Such baits can be screened again, perhaps using medium with lower concentrations of 3AT or using a different cDNA library (e.g., obtained with mRNA from another tissue or source). However, it is also possible that some bait strains might yield many hundreds of colonies. Such baits might be interacting with many partners or they interact with a highly abundant protein that is overrepresented in the library (this latter issue is particularly relevant for non-normalized cDNA libraries). Baits with many interactors might need to be screened again using medium with higher concentrations of 3AT.*

Identification of "Double-Positive" Yeast

9. Using sterile velvets and replica-plating apparatus, replica-plate the "HIS-positive" (and empty pPC86 control) yeast onto a fresh 150-mm Sc-Ura-His-Trp plate, a 150-mm Sc-Ura-His-Trp plate containing the 3AT concentration used for screening, and a 150-mm YAPD plate onto which a nitrocellulose filter has been placed (so that the yeast will grow on the filter).

10. Using sterile velvets and the replica-plating apparatus, replica-clean the plate containing 3AT (usually three cleanings are needed) immediately after replica-plating and incubate for 5–10 d at 30°C.

> *This plate will confirm the "HIS-positive" result.*

11. After 1–2 d incubation of the nitrocellulose filter and YAPD plate at 30°C, perform a colony-lift colorimetric assay (see Protocol 3: Colony-Lift Colorimetric Assay for β-Galactosidase Activity [Fuxman Bass et al. 2015b]). Combining the results observed in this step and in Step 10, identify which colonies are "double-positive" (i.e., show induction of both reporters above the level observed in the negative control yeast).

> *For baits that display moderate to high autoactivity, it might be necessary to compare potential positive yeast with the negative control very often and early in the readout period to see increased reporter expression.*

Chapter 27

INTERACTION RETESTING BY GAP-REPAIR

It is necessary to retest every interaction identified in Step 11 because some of the "double-positive" yeast phenotypes are not actually due to the AD–prey interacting with the bait. The easiest method to confirm many interactions is to use gap-repair. Gap-repair involves using PCR to amplify the cDNA from the pPC86 clone in each positive yeast colony, and then cotransforming each amplicon with linear pPC86 vector into fresh bait strain yeast. Homologous recombination within the yeast reconstitutes the vector and insert into a new construct, and reporter activation within the resulting transformed yeast can then be reassayed. Without gap repair, every interaction would need to be confirmed by transforming the AD–prey construct into fresh yeast, and, although extracting the plasmid from the double-positive yeast for such a transformation is technically possible, it is challenging to do this for many interactions simultaneously.

12. Using sterile toothpicks or loops (or by replica-plating), transfer the double-positive yeast identified in Step 11 to fresh Sc-Ura-His-Trp plates and incubate overnight at 30°C.

13. Use yeast colony PCR from these freshly grown yeast with the AD primer and TERM primer to amplify the insert from all interacting library clones (see Protocol 4: Zymolase-Treatment and Polynerase Chain Reaction Amplification from Genomic and Plasmid Templates from Yeast [Fuxman Bass et al. 2015c]). Store the PCR products at −20°C.

    The PCR extension time should reflect the average insert size. Use in later steps only PCR reactions that amplify a single band. Yeast lysates that generate multiple bands likely contain multiple AD–prey clones, and will not give clean sequence data to identify the interactor.

14. In a sterile 1.5-mL microcentrifuge tube, set up the following restriction enzyme digest:

pPC86 plasmid	10 µg
Bovine serum albumin (1 mg/mL)	4 µL
10× Restriction buffer	4 µL
SalI	2 µL (20 units)
BglII	2 µL (20 units)
Water (molecular biology grade)	to 40 µL

    Mix the digest with a tube shaker and incubate at 37°C overnight. With agarose gel electrophoresis equipment, verify the digest by running 2 µL of the restriction digest on a 1% (w/v) agarose gel in 0.5× TBE buffer next to an equal amount of the undigested plasmid as well as DNA molecular mass markers.

    This digest excises a small (~20-bp) fragment from the multicloning site of the vector, resulting in a linear fragment with noncompatible ends. If the library of choice is not pPC86-based then restriction enzymes should be chosen for the appropriate vector that results in the same type of linear fragment. There is no need to purify the linear plasmid backbone from the digest mix.

15. For each bait yeast strain, perform a high-efficiency yeast transformation, as described elsewhere (see Reece-Hoyes and Walhout 2012 or follow Steps 3–17 of Protocol 1: Generating Bait Strains for Yeast One-Hybrid Assays [Fuxman Bass et al. 2015a]), that introduces 40 ng of linear pPC86 and 5 µL of prey PCR product (from Step 13 of this protocol) into 20 µL of TE–LiAc–ss DNA yeast suspension. Three negative controls without PCR product should be included: no DNA, 40 ng of linear pPC86 alone, and 40 ng of uncut pPC86.

    This transformation can be performed in a 96-well PCR plate (200 µL well volume), with incubations in a thermal cycler. At either end of these PCR products is ~100 bp of sequence matching the vector sequence on either side of the pPC86 multicloning site. These sequences will facilitate homologous recombination.

16. Resuspend all transformations in 20 µL of sterile water by pipetting up and down and use 5 µL of each resuspension to generate a single spot (in 96-spot format) onto Sc-Ura-His-Trp plates. If positive-control yeast strains that express known levels of reporters are available, also transfer them to these plates. Incubate the plates for 2 d at 30°C.

    The number of transformants should be an order of magnitude higher in gap-repair samples and uncut pPC86 controls compared with the linear pPC86-alone control. No transformants should be present in the no-DNA control.

17. Replica-plate the transformants from Step 16 onto a fresh 150-mm Sc-Ura-His-Trp plate as well as a 150-mm Sc-Ura-His-Trp plate with the 3AT concentration used for screening, and a 150-mm

nitrocellulose-filter-equipped YAPD plate for assays of *LacZ* expression (see Protocol 3: Colony Lift Colorimetric Assay for β-Galactosidase Activity [Fuxman Bass et al. 2015b]).

The first plate is required to maintain the yeast transformants, whereas the other two plates are for retesting the HIS3 and LacZ expression, respectively.

18. Identify double-positive yeast colonies, as outlined in Steps 10 and 11.

Identification of Prey Identity

19. Determine the identity of the double-positive preys that retest successfully by gap repair by sequencing the PCR products generated in Step 13 using the AD primer.

The majority (>95%) of retrieved proteins will be predicted regulatory TFs containing known DNA-binding domain(s). Interactors that are not obvious TFs might possess a novel DNA-binding domain. If an expected interaction is not observed, bear in mind that interactions that require additional protein partners or post-translational modifications not available in yeast will not be recorded using this assay.

RECIPES

Sc-Ura-His-Trp Plates (150 mm) ± 3AT

Reagent	Quantity (for 2 L)
Drop-out mix synthetic minus histidine, leucine, tryptophan, and uracil, adenine rich (2 g) without yeast nitrogen base (US Biological, D9540-02)	2.6 g
Yeast nitrogen base (YNB) without amino acids and without ammonium sulfate	3.4 g
Ammonium sulfate	10 g
Agar	35 g
Glucose (40%, w/v, in water, sterile)	100 mL
Leucine (100 mM, filter sterilized)	16 mL
3-Amino-1,2,4-triazole (3AT) (2 M, filter sterilized) (optional)	10–80 mL

Dissolve the drop-out mix, the YNB, and the ammonium sulfate in 920 mL of water, and pH to 5.9 with NaOH (5 M). Pour into a 2-L flask and add a stir bar. In a second 2-L flask, add the agar to 950 mL of water (do not add a stir bar as it will cause the agar to boil over in the autoclave). Autoclave for 40 min at 15 psi on liquid cycle. Immediately pour the contents of the first flask, including the stir bar, into the agar-containing flask. Add the glucose, mix well on a stir plate, and cool to 55°C. Add the leucine and the tryptophan. Mix well on a stir plate and pour into 150-mm sterile Petri dishes (~80 mL per dish), dry for 3–5 d at room temperature, wrap in plastic bags, and store for up to 6 mo at room temperature. For Sc-Ura-His-Trp plates containing 3AT, add 10–80 mL of 3AT together with the leucine for a final concentration of 10 to 80 mM 3AT, and store them for up to 1 mo at room temperature.

TBE Buffer

Prepare a 5× stock solution in 1 L of H_2O:
  54 g of Tris base
  27.5 g of boric acid
  20 mL of 0.5 M EDTA (pH 8.0)

The 0.5× working solution is 45 mM Tris-borate/1 mM EDTA. TBE is usually made and stored as a 5× or 10× stock solution. The pH of the concentrated stock buffer should be ~8.3. Dilute the concentrated stock buffer just before use and make the gel solution and the electrophoresis buffer from the same concentrated stock solution. Some investigators prefer to use more concentrated stock solutions of TBE (10× as opposed to 5×). However, 5× stock solution is more stable because the solutes do not precipitate during storage. Passing the 5× or 10× buffer stocks through a 0.22-μm filter can prevent or delay formation of precipitates.

YAPD Plates

Reagent	Quantity (for 2 L)
Peptone	40 g
Yeast extract	20 g
Adenine hemisulfate dehydrate	0.32 g
Glucose (40%, w/v) in water, sterile	100 mL
Agar	35 g

Dissolve the first three powders in 950 mL of water in a 2-L flask and add a stir bar. In a second 2-L flask, add the agar to 950 mL of water (do not add a stir bar as it will cause the agar to boil over in the autoclave). Autoclave for 40 min at 15 psi on liquid cycle. Immediately pour the contents of the first flask, including the stir bar, into the agar-containing flask. Add the glucose, mix well on a stir plate, and cool to 55°C. Pour into 150-mm sterile Petri dishes (~80 mL per dish), dry for 3–5 d at room temperature, wrap in plastic bags, and store for up to 6 mo at room temperature.

ACKNOWLEDGMENTS

This work was supported by the National Institutes of Health grants DKO68429 and GM082971 to A.J.M.W. J.I.F.B was partially supported by a postdoctoral fellowship from the Pew Latin American Fellows Program.

REFERENCES

Fuxman Bass JI, Reece-Hoyes JS, Walhout AJM. 2015a. Generating bait strains for yeast one-hybrid assays. *Cold Spring Harb Protoc* doi: 10.1101/pdb.prot088948.

Fuxman Bass JI, Reece-Hoyes JS, Walhout AJM. 2015b. Colony lift colorimetric assay for β-galactosidase activity. *Cold Spring Harb Protoc* doi: 10.1101/pdb.prot088963.

Fuxman Bass JI, Reece-Hoyes JS, Walhout AJM. 2015c. Zymolase-treatment and polymerase chain reaction amplification from genomic and plasmid templates from yeast. *Cold Spring Harb Protoc* doi: 10.1101/pdb.prot088971.

Reece-Hoyes JS, Walhout AJM. 2012. High-efficiency yeast transformation. In *Molecular cloning: A laboratory manual*, 4th ed. (ed. Green MR, Sambrook J), pp. 1799–1802. Cold Spring Harbor Laboratory Press, Cold Spring Harbor, NY.

Protocol 3

Colony Lift Colorimetric Assay for β-Galactosidase Activity

Juan I. Fuxman Bass,[1] John S. Reece-Hoyes, and Albertha J.M. Walhout[1]

Program in Systems Biology, University of Massachusetts Medical School, Worcester, Massachusetts 01605

In this protocol, we present a qualitative assay for monitoring the level of expression of β-galactosidase, an enzyme encoded by the *LacZ* gene, in yeast. This is useful both for determining autoactivity of *LacZ* expression in yeast DNA "bait" strains and for assessing *LacZ* reporter gene activation mediated by a transcription factor "prey" interaction with a DNA bait of interest in yeast one-hybrid (Y1H) assays. In this colorimetric assay, yeast are lysed in liquid nitrogen and then assayed for β-galactosidase expression using the colorless compound X-gal, which turns blue in the presence of this enzyme.

MATERIALS

It is essential that you consult the appropriate Material Safety Data Sheets and your institution's Environmental Health and Safety Office for proper handling of equipment and hazardous material used in this protocol.

RECIPES: Please see the end of this protocol for recipes indicated by <R>. Additional recipes can be found online at http://cshprotocols.cshlp.org/site/recipes.

Reagents

β-mercaptoethanol
Liquid nitrogen
X-gal (4%, w/v, in dimethyl formamide)
YAPD plates (each containing a nitrocellulose filter) <R>
Z-buffer <R>

Equipment

Camera
Forceps
Fume hood
Incubator (set at 37°C)
Liquid nitrogen bath
Petri dishes (150-mm)

---

[1]Correspondence: juan.fuxmanbass@umassmed.edu; marian.walhout@umassmed.edu

Copyright © Cold Spring Harbor Laboratory Press; all rights reserved
Cite this protocol as *Cold Spring Harb Protoc*; doi:10.1101/pdb.prot088963

Tubes (plastic, 15-mL)
Waste bottle (glass)
Whatman filters (125-mm diameter, hardened; Sigma-Aldrich 1452125)

METHOD

1. For each nitrocellulose filter-YAPD plate to be analyzed, place two Whatman filters in an empty 150-mm Petri dish.

2. Move to a fume hood for Steps 3–5.

3. For each plate, in a 15-mL plastic tube, set up a reaction mix at room temperature containing:

Z-buffer	6 mL
β-mercaptoethanol	11 µL
X-gal (4% solution)	100 µL

   Use this entire mix to completely soak the Whatman filters in the Petri dish from Step 1. Remove any air bubbles using forceps to lift the filters and squeeze the bubbles to the sides, and then remove excess liquid into a waste bottle by tipping the plate.

   It is important to work in the fume hood with gloves at all times as β-mercaptoethanol and dimethyl formamide are toxic, irritants, and permeators. Once made, Z-buffer can be stored for up to 1 yr at room temperature.

4. Lift the nitrocellulose filter from the YAPD plate using forceps and place the filter, yeast side up, in a liquid nitrogen bath for 10 sec. Discard the YAPD plate.

5. Use the forceps to place the frozen nitrocellulose filter, with the yeast facing up, onto the wet Whatman filters, and use forceps (or a needle) to remove air bubbles under the nitrocellulose filter quickly as the filter (and yeast lysate) thaws.

6. Use an incubator to maintain each plate at 37°C.

   Check for blue coloring regularly (every hour if necessary) over a maximum 24 h period. Timing is important as yeast expressing high levels of β-galactosidase can produce a saturating intensity of blue within just a few hours. See Troubleshooting.

7. Take pictures with a digital camera to show the amount of blue compound generated by each yeast lysate.

   This is a qualitative assay, and so intensities of blue should be compared only within a plate and within a bait: either between different strains for the same DNA bait in autoactivation assays, or between a yeast DNA bait strain transformed with empty vector versus the same strain transformed with an AD-prey clone. Pictures should be taken at different time points to better capture the differences in blue intensities. For quantitative measures of β-galactosidase production, a liquid assay using ortho-nitrophenyl-β-galactoside can be performed (Deplancke et al. 2004; Pruneda-Paz et al. 2009).

TROUBLESHOOTING

Problem (Step 6): No blue compound is generated.
Solution: If, after 24 h at 37°C, no blue compound has been generated by any yeast lysate, it is possible that one of the reagents has been omitted or incorrectly prepared. The most important considerations regarding ingredients are that the Z-buffer should be pH 7.0 and that the X-gal is at the correct concentration. The easiest way to ensure that the reagents are in good order is to test the assay using yeast control strains that are known to express differing levels of β-galactosidase.

RECIPES

YAPD Plates

Reagent	Quantity (for 2 L)
Peptone	40 g
Yeast extract	20 g
Adenine hemisulfate dehydrate	0.32 g
Glucose (40%, w/v) in water, sterile	100 mL
Agar	35 g

Dissolve the first three powders in 950 mL of water in a 2-L flask and add a stir bar. In a second 2-L flask, add the agar to 950 mL of water (do not add a stir bar as it will cause the agar to boil over in the autoclave). Autoclave for 40 min at 15 psi on liquid cycle. Immediately pour the contents of the first flask, including the stir bar, into the agar-containing flask. Add the glucose, mix well on a stir plate, and cool to 55°C. Pour into 150-mm sterile Petri dishes (~80 mL per dish), dry for 3–5 d at room temperature, wrap in plastic bags, and store for up to 6 mo at room temperature.

Z-Buffer

60 mM Na_2HPO_4 (anhydrous)
60 mM NaH_2PO_4
10 mM KCl
1 mM $MgSO_4$

Adjust the pH to 7.0 with 10 M NaOH.

ACKNOWLEDGMENTS

This work was supported by the National Institutes of Health grants DK068429 and GM082971 to A.J.M.W. J.I.F.B. was partially supported by a postdoctoral fellowship from the Pew Latin American Fellows Program.

REFERENCES

Deplancke B, Dupuy D, Vidal M, Walhout AJ. 2004. A Gateway-compatible yeast one-hybrid system. *Genome Res* **14:** 2093–2101.

Pruneda-Paz JL, Breton G, Para A, Kay SA. 2009. A functional genomics approach reveals CHE as a component of the *Arabidopsis* circadian clock. *Science* **323:** 1481–1485.

Protocol 4

Zymolyase-Treatment and Polymerase Chain Reaction Amplification from Genomic and Plasmid Templates from Yeast

Juan I. Fuxman Bass,[1] John S. Reece-Hoyes, and Albertha J.M. Walhout[1]

Program in Systems Biology, University of Massachusetts Medical School, Worcester, Massachusetts 01605

Here, we present a protocol for amplifying DNA fragments from the genome of, or plasmids transformed into, yeast strains that require the use of the lytic enzyme zymolyase to break open the yeast cells by digesting the cell wall. Yeast strains requiring such treatment include YM4271 and Y1HaS2, whereas other yeast strains (e.g., MaV103) might not require treatment with Zymolyase.

MATERIALS

It is essential that you consult the appropriate Material Safety Data Sheets and your institution's Environmental Health and Safety Office for proper handling of equipment and hazardous material used in this protocol.

RECIPES: Please see the end of this protocol for recipes indicated by <R>. Additional recipes can be found online at http://cshprotocols.cshlp.org/site/recipes.

Reagents

Agarose gel (1%, w/v, in 0.5× TBE buffer)
DNA molecular mass markers
dNTP solution (PCR grade)
Primers (various, specific for desired template)
Taq DNA polymerase (thermostable) and associated 10× PCR buffer

We have found that Taq DNA polymerase from Invitrogen (10342-053) works robustly.

TBE buffer <R>
Water (sterile, PCR grade)
YAPD plates (150-mm) <R> or selective medium

YAPD medium is yeast extract peptone dextrose (YEPD) medium with extra adenine, which is added because the yeast strain used has mutations in the adenine synthesis pathway. Growing on YAPD, rather than on selective medium, increases the efficiency for amplifying the reporter constructs from the yeast genome of integrated yeast one-hybrid (Y1H) DNA-bait strains, but colonies grown on selective medium are suitable for amplifying from plasmid templates.

Zymolyase suspension <R>

Equipment

Incubator (set at 30°C)
PCR plates (96-well) or PCR tubes (0.2-mL)
Thermal cycler

---

[1]Correspondence: juan.fuxmanbass@umassmed.edu; marian.walhout@umassmed.edu

Copyright © Cold Spring Harbor Laboratory Press; all rights reserved
Cite this protocol as *Cold Spring Harb Protoc*; doi:10.1101/pdb.prot088971

Toothpicks and/or pipette tips (sterile)
Tubes (sterile; 1.5-mL)

METHOD

1. Grow yeast on appropriate plates (e.g., solid YAPD or selective medium) overnight at 30°C in an incubator.

2. For each yeast colony, aliquot 15 µL of zymolyase suspension into a sterile polymerase chain reaction (PCR) tube (0.2 mL) or the wells of a 96-well PCR plate.

   Because Zymolyase has low solubility in water, it is important to mix the suspension thoroughly before and periodically (every 30 sec) during distribution into the tubes or wells.

3. Using a sterile toothpick or pipette tip, transfer a small amount (approximately one-eighth of a match head) of each yeast colony grown in Step 1 to each Zymolyase aliquot from Step 2.

   Too much yeast will inhibit the PCR reaction.

4. Transfer the PCR tubes or plate to a thermal cycler and incubate the yeast–enzyme mix for 30 min at 37°C. Then heat-inactivate the enzyme for 10 min at 95°C.

5. Remove the PCR tubes or plate from the thermal cycler and dilute the lysate by adding 85 µL of sterile PCR-grade water.

   The lysate can be stored at −20°C to await the subsequent PCR reactions.

6. For each amplification, in a sterile PCR tube (or well of a PCR plate) prepare the following PCR reaction mix:

Diluted lysate (from Step 5)	5 µL
Forward primer (20 µM)	1 µL
Reverse primer (20 µM)	1 µL
dNTPs (1 mM)	5 µL
10× PCR buffer	5 µL
Taq (or other thermostable) DNA polymerase	2 units
Water	to 50 µL

   Remember to include a negative-control PCR reaction that lacks template.

7. Place the tubes in a thermal cycler and run the following PCR program, as tabulated below.

Cycle number	Denaturation	Annealing	Polymerization
1	2 min at 94°C		
35 cycles	1 min at 94°C	1 min at 56°C	1 min per kb of sequence to be amplified at 72°C
Last cycle			7 min at 72°C

   The conditions of the PCR reaction might need to be optimized.

8. Run 5–10 µL of the PCR reaction from Step 7 on a 1% (w/v) agarose gel in 0.5× TBE buffer (pH 8.3) alongside DNA molecular mass markers.

   When analyzing the PCR products, remember that amplification using vector primers will generate PCR products larger than the insert.

   See Troubleshooting.

TROUBLESHOOTING

Problem (Step 8): Yeast colony PCR fails to generate a product.
Solution: Aside from the generic problems that can affect PCR (e.g., forgotten, degraded, or incorrectly prepared reagents) there are factors specific to yeast colony PCR that require consideration. The

first is that the yeast need to be effectively lysed by the zymolyase to release the template for the PCR. Zymolyase is an enzyme and needs to be treated somewhat carefully by ensuring that melted suspensions are always kept on ice and do not freeze-thaw the same aliquot of suspension more than three times. Also remember that the zymolyase is a suspension that needs to be regularly mixed (every 30 sec) while aliquoting or some samples will receive less enzyme than others. The second factor to consider is that adding too much yeast to the lysis solution will inhibit the PCR. An effective method for determining what is the correct amount of yeast to lyse is to setup a series of Zymolyase treatments with different amounts of yeast and then test which lysate is successful in the PCR.

RECIPES

TBE Buffer

Prepare a 5× stock solution in 1 L of H_2O:
  54 g of Tris base
  27.5 g of boric acid
  20 mL of 0.5 M EDTA (pH 8.0)

The 0.5× working solution is 45 mM Tris-borate/1 mM EDTA. TBE is usually made and stored as a 5× or 10× stock solution. The pH of the concentrated stock buffer should be ~8.3. Dilute the concentrated stock buffer just before use and make the gel solution and the electrophoresis buffer from the same concentrated stock solution. Some investigators prefer to use more concentrated stock solutions of TBE (10× as opposed to 5×). However, 5× stock solution is more stable because the solutes do not precipitate during storage. Passing the 5× or 10× buffer stocks through a 0.22-µm filter can prevent or delay formation of precipitates.

YAPD Plates

Reagent	Quantity (for 2 L)
Peptone	40 g
Yeast extract	20 g
Adenine hemisulfate dehydrate	0.32 g
Glucose (40%, w/v) in water, sterile	100 mL
Agar	35 g

Dissolve the first three powders in 950 mL of water in a 2-L flask and add a stir bar. In a second 2-L flask, add the agar to 950 mL of water (do not add a stir bar as it will cause the agar to boil over in the autoclave). Autoclave for 40 min at 15 psi on liquid cycle. Immediately pour the contents of the first flask, including the stir bar, into the agar-containing flask. Add the glucose, mix well on a stir plate, and cool to 55°C. Pour into 150-mm sterile Petri dishes (~80 mL per dish), dry for 3–5 d at room temperature, wrap in plastic bags, and store for up to 6 mo at room temperature.

Zymolyase Suspension

Mix 200 mg of Zymolase-100T powder (Associates of Cape Cod 120493-1) in 100 mL of sterile 0.1 M sodium phosphate buffer (pH 7.5). (The powder will not dissolve totally; some precipitate will be visible even after 30 min of mixing.) Divide into aliquots of 1 mL and store for up to 12 mo at −20°C.

ACKNOWLEDGMENTS

This work was supported by the National Institutes of Health grants DK068429 and GM082971 to A.J.M.W. J.I.F.B. was partially supported by a postdoctoral fellowship from the Pew Latin American Fellows Program.

CHAPTER 28

Transposon Calling Cards

David Mayhew and Robi D. Mitra[1]

Department of Genetics, Center for Genome Sciences & Systems Biology, Washington University School of Medicine, St. Louis, Missouri 63110

Identifying the genomic targets of transcription factors is an important step in understanding the regulatory networks of gene transcription in yeast. We have developed a method that utilizes what we refer to as transposon "calling cards," in which a transcription factor directs the Ty5 retrotransposase to insert transposons into the genome adjacent to where the transcription factor binds. This method is designed to be multiplexed with many barcoded transcription factors and has the potential to decrease the labor required for the study of large numbers of transcription factors.

CALLING CARD ANALYSIS OVERVIEW

Calling card analysis is an alternative method to chromatin immunoprecipitation (ChIP)-based approaches (Horak and Snyder 2002; Johnson et al. 2007) for identifying the genomic targets of DNA-binding proteins (see Protocol 1: Calling Card Analysis in Budding Yeast [Mayhew and Mitra 2015]). The calling card method utilizes the specificity of the Ty5 retrotransposon (Xie et al. 2001) to identify regions of the genome where a transcription factor (TF) binds, causing the transposon to insert there (Fig. 1). This redirection of insertion preference is achieved by cloning a domain of the Sir4 heterochromatin protein to the TF of interest. This domain of Sir4 (amino acids 951–1200) contains the Ty5 integrase-interacting domain, which attracts Ty5 transposition (Zhu et al. 2003). The TF-Sir4 fusion can be overexpressed from a plasmid or expressed from its native locus. When expression is induced, the Ty5 transposon will insert into the genome in close proximity to where the tagged TF binds and leave behind a permanent mark or "calling card" to record its visit. These insertions are recovered and sequenced in parallel to provide a global map of the binding sites of the tagged TF. Figure 2 shows a sample data set.

The resulting analysis gives a high-resolution map of the DNA sequence bound by a given TF comparable to that obtained by ChIP-seq (Wang et al. 2011). Unlike ChIP-based methods, the recovery of DNA for analysis requires no antibodies. Instead, mapping the Ty5 insertions requires restriction endonuclease digestion, DNA ligation, and polymerase chain reaction (PCR) to prepare the library for Illumina sequencing.

ADVANTAGES AND LIMITATIONS OF THE APPROACH

A calling card experiment can be easily scaled up, as the marginal cost of analyzing an additional TF is minimal. There is a requisite increase in the number of induced yeast cells and plates with the

[1]Correspondence: rmitra@genetics.wustl.edu

Copyright © Cold Spring Harbor Laboratory Press; all rights reserved
Cite this introduction as *Cold Spring Harb Protoc*; doi:10.1101/pdb.top077776

Chapter 28

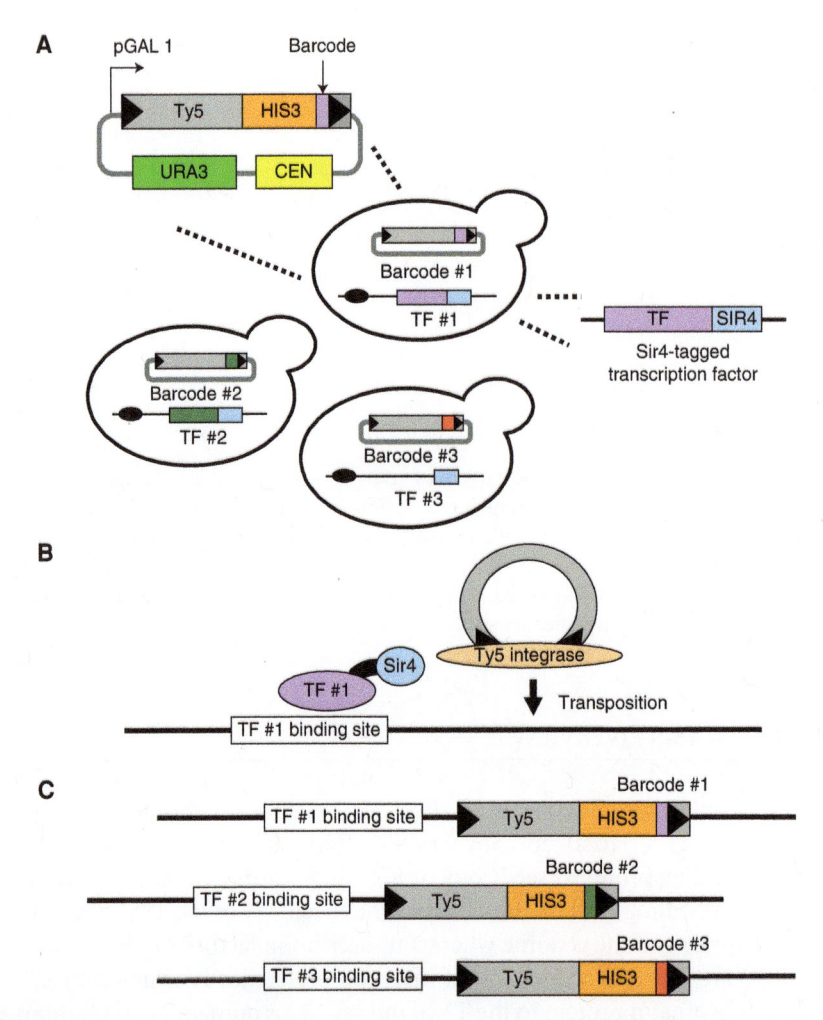

FIGURE 1. Calling card analysis uses the Ty5 retrotransposon to map the binding sites of transcription factors. (*A*) Mapping requires a barcoded Ty5 plasmid transformed into a yeast strain with a TF-Sir4 fusion. (*B*) When expression of Ty5 is induced, the integrase will be targeted to the TF-Sir4 fusion binding site and will insert into nearby genomic DNA. (*C*) HIS3+ cells can be selected for genomic insertions.

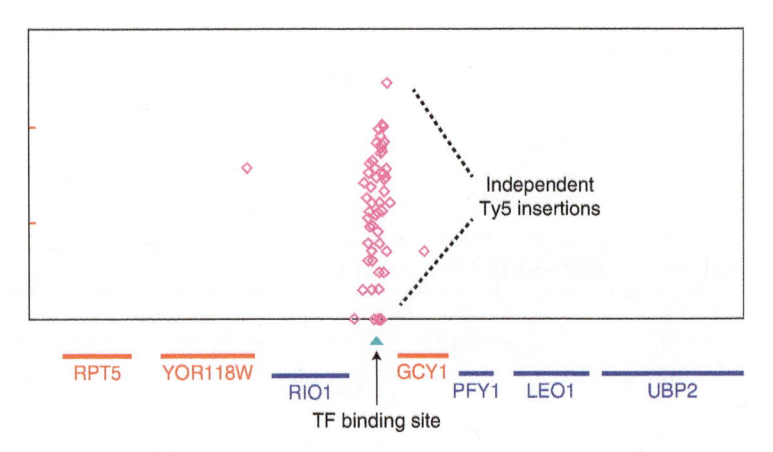

FIGURE 2. Sample calling card data for Gal4-Sir4 directed Ty5 insertions cluster around the Gal4p target gene *GCY1*. Each purple diamond represents a mapped Ty5 insertion. The *x*-axis specifies gene position; the *y*-axis is the number of sequencing reads for each insertion.

examination of additional TFs. However, because all TFs in the experiment are pooled together from the induction step through library construction and sequencing, scaling up the experiment does not adversely affect the time required for subsequent steps in the protocol. Twenty TFs can be easily pooled in a single experiment and sequenced on one lane of an Illumina GA or MiSeq device. With a HiSeq instrument, the limit is even higher.

One disadvantage of the technique is the lack of yeast strains available with Sir4-tagged TFs relative to the number of ChIP-ready tagged strains. Cloning the TF-Sir4 strains may require a moderate investment in time at the start of an experiment. Another disadvantage is that the current system induces Ty5 expression from the *GAL1* promoter, requiring galactose to be present in the experiment. Examining yeast grown in other carbon-source conditions will require a different inducibility system for the retrotransposon.

The ability to scale up the number of TFs examined in an experiment makes several potential large-scale experiments less laborious. Provided the TF-Sir4 fusion strains were first created, every TF in the yeast genome could be barcoded and studied in a single experiment without requiring a separate immunoprecipitation for each TF. This would enable the comprehensive mapping of the yeast regulatory network under a number of different growth conditions. It is also possible to analyze multiple strains in a single calling card experiment. For example, it is possible to analyze the binding of a single TF in various barcoded deletion strains. This would enable the large-scale analysis of cooperative DNA binding. The multiplexed version of the calling card method has been used in yeast to map the binding of 28 transcription factors during the transition from unicellular to filamentous growth (Mayhew and Mitra 2014). Multiplexed calling cards can also be useful to analyze the growth of a small number of transcription factors in different genetic backgrounds, such as different knockout strains, as was done to analyze the role of the tripartite transcription factor complex Mfg1, Flo8, and Mss11 in filamentous growth (Ryan et al. 2012). Although the Ty5 transposon is not active in mammalian cells, the calling card method can be used in mouse and human by utilizing the piggyBac transposon (Wang et al. 2012), indicating that it could find application in a wide variety of systems. In summary, this method provides a useful alternative to ChIP methods to provide high-resolution mapping of TF binding sites in a multiplexed fashion.

REFERENCES

Horak CE, Snyder M. 2002. ChIP-chip: A genomic approach for identifying transcription factor binding sites. *Methods Enzymol* 350: 469–483.

Johnson DS, Mortazavi A, Myers RM, Wold B. 2007. Genome-wide mapping of in vivo protein–DNA interactions. *Science* 316: 1497–1502.

Mayhew D, Mitra RD. 2015. Transcription factor regulation and chromosome dynamics during pseudohyphal growth. *Mol Biol Cell* 25: 2669–2276.

Mayhew D, Mitra RD. 2015. Calling card analysis in budding yeast. *Cold Spring Harb Protoc* doi: 10.1101/pdb.prot086918.

Ryan O, Shapiro RS, Kurat CF, Mayhew D, Baryshnikova A, Chin B, Lin ZY, Cox MJ, Vizeacoumar F, Cheung D, et al. 2012. Global gene deletion analysis exploring yeast filamentous growth. *Science* 337: 1353–1356.

Wang H, Mayhew D, Chen X, Johnston M, Mitra RD. 2011. Calling cards enable multiplexed identification of the genomic targets of DNA-binding proteins. *Genome Res* 5: 748–755.

Wang H, Mayhew D, Chen X, Johnston M, Mitra RD. 2012. "Calling cards" for DNA-binding proteins in mammalian cells. *Genetics* 190: 941–949.

Xie W, Gai X, Zhu Y, Zappulla DC, Sternglanz R, Voytas DF. 2001. Targeting of the yeast Ty5 retrotransposon to silent chromatin is mediated by interactions between integrase and Sir4p. *Mol Cell Biol* 21: 6606–6614.

Zhu Y, Dai J, Fuerst PG, Voytas DF. 2003. Controlling integration specificity of a yeast retrotransposon. *Proc Natl Acad Sci* 100: 5891–5895.

Protocol 1

Calling Card Analysis in Budding Yeast

David Mayhew and Robi D. Mitra[1]

Department of Genetics, Center for Genome Sciences and Systems Biology, Washington University School of Medicine, St. Louis, Missouri 63110

Calling card analysis is a high-throughput method for identifying the genomic binding sites of multiple transcription factors in a single experiment in budding yeast. By tagging a DNA-binding protein with a targeting domain that directs the insertion of the Ty5 retrotransposon, the genomic binding sites for that transcription factor are marked. The transposition locations are then identified en masse by Illumina sequencing. The calling card protocol allows for simultaneous analysis of multiple transcription factors. By cloning barcodes into the Ty5 transposon, it is possible to pair a unique barcode with every transcription factor in the experiment. The method presented here uses expression of transcription factors from their native loci; however, it can also be altered to measure binding sites of transcription factors overexpressed from a plasmid.

MATERIALS

It is essential that you consult the appropriate Material Safety Data Sheets and your institution's Environmental Health and Safety Office for proper handling of equipment and hazardous materials used in this protocol.

RECIPES: Please see the end of this protocol for recipes indicated by <R>. Additional recipes can be found online at http://cshprotocols.cshlp.org/site/recipes.

Reagents

Agarose gels (0.7%) and running buffer
Barcoded Ty5 plasmid (pRM1001-series; 200 ng/µL)
  These plasmids, which are available from the corresponding author, each contain an 8-bp barcode within the transposon that will be matched to each tagged strain.

Betaine (5 M)
BSA (20 mg/mL; molecular biology grade)
Carrier (Sheared Salmon Sperm) DNA Solution (10 mg/mL; Invitrogen)
Chloroform
dH_2O
dNTP mix (10 mM each of dATP, dCTP, dGTP, dTTP)
Ethanol (70%, 100%)
Galactose – Ura agar plates <R>
Genomic DNA from yeast strain yRM1009
  This strain is available from the corresponding author.

---

[1]Correspondence: rmitra@genetics.wustl.edu

Copyright © Cold Spring Harbor Laboratory Press; all rights reserved
Cite this protocol as *Cold Spring Harb Protoc*; doi:10.1101/pdb.prot086918

Glucose – His agar plates <R>
Glucose – His 5-fluoroorotic acid (5-FOA) agar plates <R>
Glucose – Ura medium <R>
Glycerol (20%; optional; see Step 16)
HindPI1 restriction enzyme and appropriate buffer
HpaII restriction enzyme and appropriate buffer
Lithium acetate (1 M)
Phenol:choloroform:isoamyl alcohol (PCA) (25:24:1)
Phusion DNA polymerase
Phusion HF buffer (5×)
Polyethylene glycol 3500 (PEG 3500; 50% w/v)
Primers (all sequences are listed 5′ to 3′)

"Forward" cloning primer: (N)$_{40}$AGAGTGTCGCATAGTGATAC

This primer includes 40 bp upstream of the stop codon of the ORF of the DNA-binding protein of interest (without including the stop codon) plus 20 bp to amplify the tagging domain.

"Reverse" cloning primer: (N)$_{40}$CGCACTTAACTTCGCATCTG

This primer includes the reverse complement of the 40 bp downstream from the ORF of the DNA-binding protein of interest (without including the stop codon) plus 20 bp to amplify the tagging domain.

"Forward" inverse PCR primer: AATGATACGGCGACCACCGAGATCTACACTCTTTCCCTACACGACGCTCTTCCGATCTAATTCACTACGTCAACA

"Reverse" inverse PCR primer: CAAGCAGAAGACGGCATACGAGATCGGTCTCGGCATTCCTGCTGAACCGCTCTTCCGATC

RedTaq DNA polymerase and associated buffer
RNaseA (20 mg/mL)
Sodium acetate (3 M, pH 5.2)
TaqI restriction enzyme and appropriate buffer
T4 DNA ligase and associated buffer
Yeast calling card lysis buffer <R>
Yeast strain of desired background

Strain must be deficient for SIR4, HIS3, and URA3 and sensitive to nourseothricin (NAT) antibiotic.

YPD <R> (liquid medium and agar plates)

YPD agar plates with and without 100 µg/mL NAT are needed.

Equipment

Agarose gel apparatus
Centrifuge
Conical centrifuge tubes (15-mL)
Glass beads (0.2 mm diameter)
Glass culture tubes (5-mL; disposable)
Glass yeast spreader
Heat blocks
Incubator (30°C)
Microcentrifuge
Microcentrifuge tubes (1.5-mL)
Microcentrifuge tube rotator
PCR (polymerase chain reaction) tubes
QIAquick PCR purification kit (Qiagen)
Sequencing machine (Illumina GA, Illumina Miseq, or Illumina HiSeq with paired-end capabilities)

Chapter 28

Spectrophotometer for DNA quantification (Nanodrop [Thermo Scientific] or Qubit fluorometer [Invitrogen])
Thermocycler
Vacuum concentrator (e.g., Speedvac)
Velvet replica plating pads and base
Vortexer
Water bath

METHOD

Cloning Strain

This stage involves cloning the Ty5 integrase-interacting domain of Sir4 to the carboxyl terminus of the desired TF. If the desired tagged strains have already been created, proceed to Step 11. See Figure 1 for an overview of the entire protocol.

1. Amplify the Sir4-NatR sequence from genomic DNA of strain yRM1009. Add the following PCR mix components and mix well:

    1× Phusion HF buffer

    0.5 μM "forward" cloning PCR primer

    0.5 μM "reverse" cloning PCR primer

    0.2 mM of each dNTP

    1 unit Phusion DNA Polymerase

    100 ng genomic DNA of yRM1009

    dH$_2$O to 50 μL

2. Perform the following cycling program:

No. of cycles	Temperature	Time
1	98°C	30 sec
30	98°C	10 sec
	58°C	30 sec
	72°C	2 min
1	72°C	5 min
1	12°C	Hold

3. Run 5 μL of the PCR product on 0.7% agarose gel using standard techniques to confirm success of PCR.

    A single product at 1.8 kb is expected.

4. Inoculate the yeast strain of the desired background into 5 mL of YPD in a disposable 5-mL glass culture tube. Grow to saturation overnight at 30°C with shaking. Dilute 50 μL of the saturated starter culture into 5 mL YPD in a disposable 5-mL glass culture tube. Grow at 30°C with shaking to an OD$_{600}$ between 0.8 and 1.0 (∼6 h).

5. Pellet the cells by centrifuging the culture tube at 3000 rpm for 2 min at room temperature. Remove the supernatant. Resuspend the cells in 1000 μL of dH$_2$O and transfer the cells to a microcentrifuge tube. Pellet the cells by centrifuging the tube at 10,000 rpm for 2 min at room temperature. Remove the supernatant.

6. Add the following components of the transformation mix with the remaining PCR product from Step 3 directly on top of the pelleted cells to the following final concentrations and to a total volume of 360 μL:

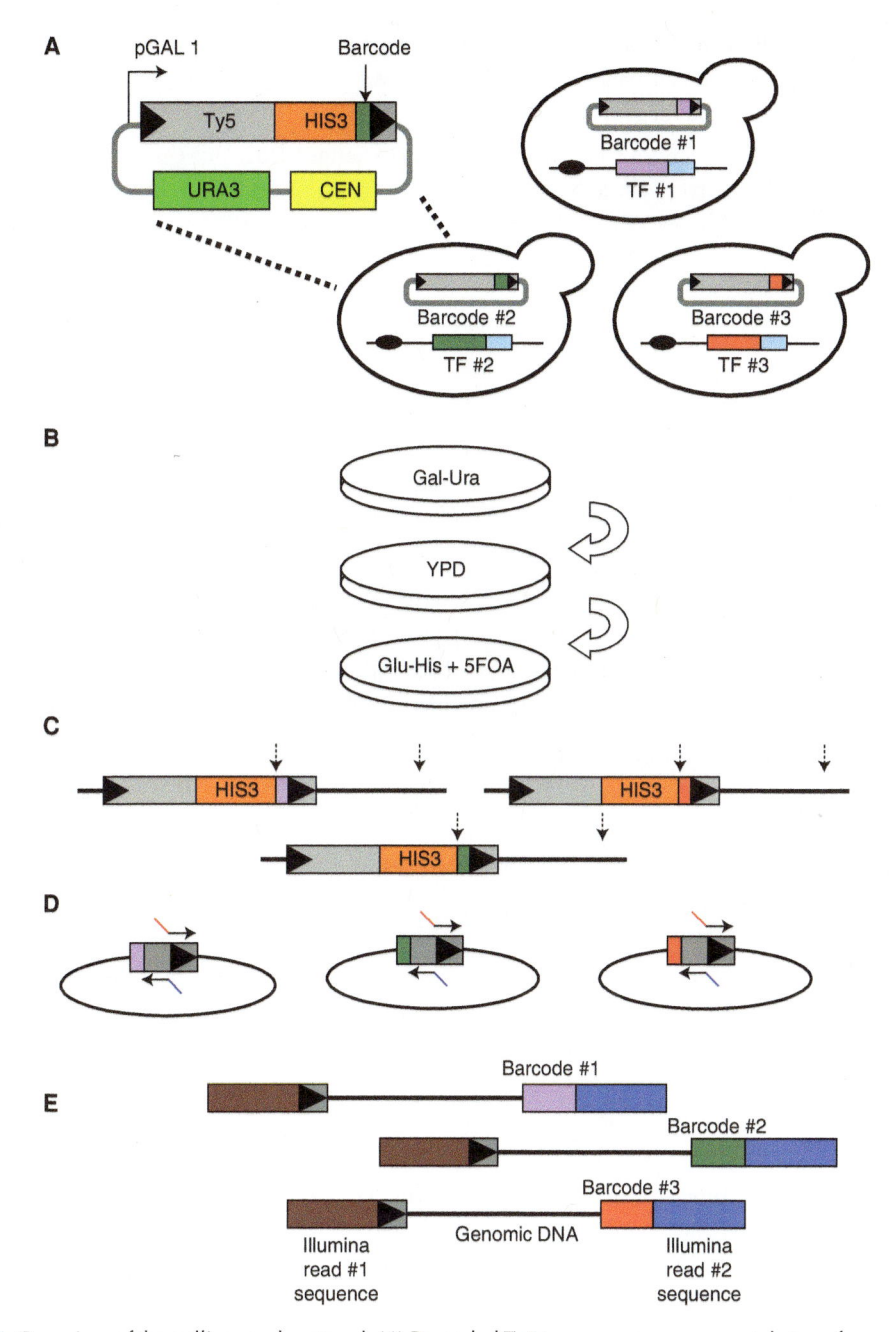

FIGURE 1. Overview of the calling card protocol. (A) Barcoded Ty5 transposons are separately transformed into yeast strains with TFs tagged with Sir4. (B) The strains are pooled, expression of Ty5 is induced, and the subsequent growth medium selects for genomic insertions. (C) Genomic DNA is digested with restriction endonucleases. (D) Digested fragments are self-ligated. (E) Inverse PCR adds Illumina sequencing adaptors.

Component	Amount (per reaction)	Final concentration
PEG 3500 (50% w/v)	240 µL	33.3%
Lithium acetate (1 M)	36 µL	0.1 M
Carrier DNA solution	10 µL	250 ng/µL
PCR product plus dH$_2$O	74 µL	

7. Vortex the mixture and incubate in a water bath for 40 min at 42°C.

8. Centrifuge the tube at 10,000 rpm for 2 min at room temperature and remove supernatant. Resuspend the cell pellet in 1 mL of YPD and incubate with rotation for 60 min at 30°C.

Chapter 28

9. Centrifuge the tube at 10,000 rpm for 2 min at room temperature and remove YPD. Resuspend the cell pellet in 50 µL dH$_2$O and plate on a YPD agar plate containing 100 µg/mL NAT and incubate for 2 d at 30°C.

10. Select NAT resistant colonies and Sanger sequence the DNA bordering the tagged ORF to verify a correct clone.

    Select at least three clones for each cloned strain.

Plasmid Transformation and Induction

This stage requires four different growth steps on separate days with 1 h of hands-on time each day.

11. Select a unique barcoded Ty5 donor plasmid for each transcription factor–tagged strain to be examined.

12. Separately inoculate each tagged strain from step 10 into 5 mL of YPD in a disposable 5-mL glass culture tube. Grow to saturation overnight at 30°C with shaking. Dilute 50 µL of the saturated starter culture into 5 mL YPD in a disposable 5-mL glass culture tube. Grow at 30°C with shaking to an OD$_{600}$ between 0.8 and 1.0 (∼ 6 h).

13. Pellet the cells by centrifuging the culture tube at 3000 rpm for 2 min at room temperature. Remove the supernatant and wash cells as in Step 5.

14. For each tagged strain, add the following components of the transformation mix with a barcoded Ty5 plasmid to the following final concentrations and to a total volume of 360 µL:

Component	Amount (per reaction)	Final concentration
PEG 3500 (50% w/v)	240 µL	33.3%
Lithium acetate (1.0 M)	36 µL	0.1 M
Carrier (Sheared Sperm DNA Solution)	10 µL	250 ng/µL
Ty5 plasmid (200 ng/µL)	1 µL	0.6 ng/µL
dH$_2$O	73 µL	

15. Vortex the mixture and incubate in a water bath for 40 min at 42°C.

16. Centrifuge tube at 10,000 rpm for 2 min at room temperature and remove the transformation mix. Resuspend the cell pellet in 50 µL of dH$_2$O and plate on a Glucose – His plate. Incubate at 30°C for 2 d.

    Transformed yeast can be frozen in 20% glycerol at −80°C at this step to avoid repeating the previous steps in future experiments.

17. Inoculate a single colony from each barcoded-plasmid transformed strain in 5 mL of liquid Glucose – Ura medium in a disposable 5-mL glass culture tube. Grow to saturation overnight at 30°C with shaking. Pellet cells by centrifuging the culture tube at 3000 rpm at room temperature for 2 min. Remove medium.

18. Resuspend the cells in each tube in 1 mL dH$_2$O. Combine all transformed strains into a single tube and mix well. Plate 100 µL of this mixture onto each Galactose – Ura agar plate and incubate at room temperature for 2 d.

    Yeast should form a confluent lawn.

    The Ty5 transposon is induced by galactose, so this is the step at which transcription factor binding is measured. If additional growth conditions are desired in the experiment, these should be supplied at this step. The number of Galactose – Uracil plates should scale with the number transcription factors being studied. As a rule of thumb, use two 10-cm plates for each strain in the experiment (enough for 5 × 10^3 insertions), but DNA-binding proteins with a very large number of targets may require more plates to gather enough insertions to map all of their binding sites.

19. Replica plate cells to YPD agar plates by pressing the yeast-side of the Galactose – Ura agar plate to a clean velvet, removing, and then pressing a new YPD agar plate against the velvet. Use a clean velvet for each pair of plates. Incubate the plates for 1 d at 30°C.

    Yeast should form a confluent lawn.

20. Replica plate cells from YPD plates to Glucose – His 5-FOA agar plates and incubate for 2 d at 30°C.

    This step selects for genomic insertions of Ty5 and counter-selects against cells still containing the plasmid. Each 10-cm plate should have ~2 × 10³ colonies by the second day of growth.

    See Troubleshooting.

DNA Extraction

Requires a total of 2 h.

21. Pipette 1 mL of liquid YPD onto each Glu – His 5-FOA plate and scrape the yeast with a glass yeast spreader to put the cells into solution. Pipette the solution into a 15-mL conical tube. Combine cells from different plates into a single conical tube. Mix by gently inverting the tube.

22. Transfer 500 μL of the liquid mixture to a 1.5-mL microcentrifuge tube. Centrifuge at 13,000 rpm for 10 min at room temperature in a microcentrifuge and pour off the liquid.

23. To the pelleted cells add:

    400 μL Yeast calling card lysis buffer

    400 μL Glass beads (measure using a microcentrifuge tube)

    400 μL Phenol:choloroform:isoamyl alcohol (25:24:1)

24. Vortex for 10 min at room temperature.

25. Centrifuge at 13,000 rpm for 10 min at room temperature.

26. Transfer the aqueous supernatant to a new microcentrifuge tube.

27. Add 400 μL of phenol:choloroform:isoamyl alcohol (25:24:1). Repeat Steps 24–26.

28. Add 400 μL of chloroform. Repeat Steps 24–26.

29. Precipitate DNA by adding 40 μL of 3 M sodium acetate and 1000 μL of ethanol (100%) at room temperature. Vortex and incubate for 60 min at −70°C.

30. Centrifuge at 13,000 rpm for 10 min at room temperature. Carefully remove the supernatant.

31. Carefully wash the pellet with 500 μL of ethanol (70%) at room temperature.

32. Centrifuge at 13,000 rpm for 1 min at room temperature.

33. Remove ethanol supernatant with a micropipette and dry the pellet in a vacuum concentrator until all liquid is removed (~ 10 min). Reconstitute the pellet in 50 μL dH₂O and 2 μL RNaseA and incubate for 30 min at 37°C.

    At this step, genomic DNA can be frozen at −20°C and stored long-term before subsequent steps.

Genomic Digestions

Requires a total of 3.5 h.

34. Measure DNA concentration with a spectrophotometer.

35. Put 4 μg of DNA into each of three PCR tubes. Perform the following three digestions in a total volume of 50 μL each with the final concentrations and conditions as listed:

    i. <u>TaqI digestion</u>

       4 Units TaqI enzyme

       1× TaqI-compliant buffer

       1× BSA

       dH₂O to 50 μL

       Incubate for 3 h at 65°C followed by a heat inactivation for 20 min at 80°C.

Chapter 28

    ii. <u>HindPI1 digestion</u>

        4 Units HindPI1 enzyme

        1× HindPI1-compliant buffer

        dH$_2$O to 50 μL

        Incubate for 3 h at 37°C followed by a heat inactivation for 20 min at 65°C.

    iii. <u>HpaII digestion</u>

        4 units HpaII enzyme

        1× HpaII-compliant buffer

        dH$_2$O to 50 μL

        Incubate for 3 h at 37°C followed by a heat inactivation for 20 min at 65°C.

36. Purify digestions using the QIAquick PCR purification kit, following the manufacturer's protocol. Elute DNA in 50 μL of dH$_2$O.

Self-Ligation

Requires a total of 14 h or overnight.

37. To each of the eluates from Step 36 add the reagents below to the following final concentrations:

    18 units T4 ligase

    1× T4 ligation buffer

    dH$_2$O to 400 μL

38. Incubate for at least 12 h at 14°C.

    Overnight incubation is preferable.

39. Precipitate DNA by following Steps 29–32.

40. Remove ethanol and dry the pellet in a vacuum concentrator. Resuspend the pellet in 50 μL dH$_2$O and vortex.

Inverse PCR

Requires a total of 4 h.

41. Add the following PCR mix components to half (25 μL) of each of the reconstituted DNA samples from Step 40 to yield the indicated concentrations:

    1× RedTaq PCR Buffer

    0.5 μM "forward" inverse PCR primer

    0.5 μM "reverse" inverse PCR primer

    0.2 mM of each dNTP

    0.5 M betaine

    4 units RedTaq Enzyme

    dH$_2$O to 50 μL

42. Perform the following cycling program:

No. of cycles	Temperature	Time
1	93°C	2 min
30	93°C	30 sec
	60°C	6 min
1	12°C	Hold

43. Run 5 μL of each PCR product on a 0.7% agarose gel using standard techniques to confirm the success of PCR.

    Expect products ranging from 200 bp to 2 kb.
    See Troubleshooting.

44. Purify PCR products using QIAquick PCR purification kit, following the manufacturers's protocol. Elute DNA in 50 μL of dH$_2$O and quantify the DNA concentration using Nanodrop/Qubit.

45. Combine equimolar quantities of the 3 PCR products and dilute to 10 nM in dH$_2$O for Illumina sequencing.

    Assume an average PCR product size of 200 bp, which makes 10 nM equal to 1.32 ng/μL. At least 20 μL of the final solution is needed for Illumina sequencing.

46. Sequence the DNA on a paired-end Illumina sequencing run according to the manufacturer's instructions.

47. Analyze data by demultiplexing reads using the 8-bp barcode (the first eight bases on the second read), removing the first 17 bp of the read (corresponds to the Ty5 transposon), and aligning the remaining bases of the read to the reference genome of the background yeast strain.

    See Troubleshooting.

TROUBLESHOOTING

Problem (Step 20 or 47): A high background of yeast on the Glucose − Histidine + 5-FOA selection step or a high percentage of plasmid reads in the sequencing is observed.
Solution: The 5-FOA selection step is not killing all of the yeast still containing the Ty5 plasmid. An additional round of replica plating on Glucose − Histidine + 5-FOA plates will reduce plasmid in the extracted DNA.

Problem (Step 43): A thick band is observed at 100 bp with a smear of genomic DNA above this product.
Solution: Primer–dimers have formed during the inverse PCR. Perform a QIAquick gel extraction (QIAGEN) to separate the larger products from the primer–dimer size fragments before sequencing.

DISCUSSION

Chromatin immunoprecipitation read out by either microarrays (ChIP-chip) (Horak and Snyder 2002) or next-generation sequencing (ChIP-seq) (Johnson et al. 2007) has been an invaluable technique for identifying the binding sites of transcription factors (TFs) in yeast. For studies requiring large numbers of TFs, the nature of ChIP-based methods requires increased time and work for each additional TF in the experiment. Large sequencing capacities of next-generation sequencers allow for ChIP-seq libraries of many TFs to be barcoded and sequenced in parallel (Lefrançois et al. 2009); however, the immunoprecipitation for each TF is performed independently. This limitation makes large-scale ChIP experiments laborious. The ability of the calling card technique to multiplex many TFs through most of the steps in this protocol makes scaling up these experiments more feasible.

The number of TFs that can be included in this protocol is a function of the downstream sequencing capacity. To ensure adequate coverage of all of the insertions with each barcode, $1-2 \times 10^6$ reads are required for each barcode/TF included in the multiplex. Experiments with large numbers of TFs may make the growth and selection steps on solid plates unsuitable. For these larger experiments the protocol can be altered to move these steps (18–20) to liquid cultures. For more information on the data analysis or altering the protocol to overexpress the TF-Sir4 fusion from a plasmid, please see Wang et al. (2011).

A limitation of calling cards is that there are currently fewer existing yeast strains with TFs tagged for the method compared with ChIP-ready tagged strains. The study of some TFs or background strains will necessitate cloning them as described in the first steps of this protocol. The method is easy to use and provides high signal-to-noise mapping of DNA-binding with an accuracy and resolution comparable with ChIP-seq (Wang et al. 2011). This protocol should be adaptable to many experiments for mapping binding sites of DNA-binding proteins, especially when multiple proteins need to be studied.

RECIPES

Drop-Out Mix

Reagent	Amount to add
Adenine	0.5 g
Alanine	2.0 g
Arginine	2.0 g
Asparagine	2.0 g
Aspartic acid	2.0 g
Cysteine	2.0 g
Glutamine	2.0 g
Glutamic acid	2.0 g
Glycine	2.0 g
Histidine	2.0 g
Inositol	2.0 g
Isoleucine	2.0 g
Leucine	10.0 g
Lysine	2.0 g
Methionine	2.0 g
para-Aminobenzoic acid	0.2 g
Phenylalanine	2.0 g
Proline	2.0 g
Serine	2.0 g
Threonine	2.0 g
Tryptophan	2.0 g
Tyrosine	2.0 g
Uracil	2.0 g
Valine	2.0 g

Combine the appropriate ingredients, in 1 L of medium minus the relevant supplements, and mix in a sealed container. Turn the container end-over-end for at least 15 min; add several clean marbles to help mix the solids.

Galactose – Ura Agar Plates

1.7 g	Yeast nitrogen base without amino acids and ammonium sulfate
1 g	Dropout mix (without uracil) <R>
5 g	Ammonium sulfate
20 g	Agar

Add H_2O to 900 mL, stir well, and autoclave. Add 100 mL of 20% galactose (sterile) and stir. Pour into sterile Petri dishes and allow to solidify.

Glucose – His Agar Plates

1.7 g	Yeast nitrogen base without amino acids and ammonium sulfate
1 g	Drop-out mix (without histidine) <R>
5 g	Ammonium sulfate
20 g	Agar

Add H$_2$O to 900 mL, stir well, and autoclave. Add 100 mL of 20% glucose (sterile) and stir. Pour into sterile Petri dishes and allow to solidify.

Glucose – His 5-Fluoroorotic Acid (5-FOA) Agar Plates

1.7 g	Yeast nitrogen base without amino acids and ammonium sulfate
1 g	Drop-out mix (without histidine) <R>
5 g	Ammonium sulfate
20 g	Agar

Add H$_2$O to 900 mL, stir well, and autoclave. Add 100 mL of 20% glucose (sterile) and 1 g 5-FOA, and stir. Pour into sterile Petri dishes and allow to solidify.

Glucose – Ura Medium

1.7 g	Yeast nitrogen base without amino acids and ammonium sulfate
1 g	Drop-out mix (without uracil) <R>
5 g	Ammonium sulfate

Add H$_2$O to 900 mL, stir well, and autoclave. Add 100 mL of 20% glucose (sterile) and stir. Store for up to 3 mo at 4°C.

Yeast Calling Card Lysis Buffer

2% Triton X-100
1% SDS
100 mM NaCl
10 mM Tris–HCl (pH 8.0)
1 mM EDTA (pH 8.0)

Store for up to 12 mo at 4°C.

YPD

Peptone, 20 g
Glucose, 20 g
Yeast extract, 10 g
H$_2$O to 1000 mL
YPD (YEPD medium) is a complex medium for routine growth of yeast.

To prepare plates, add 20 g of Bacto Agar (2%) before autoclaving.

REFERENCES

Horak CE, Snyder M. 2002. ChIP-chip: A genomic approach for identifying transcription factor binding sites. *Methods Enzymol* **350:** 469–483.

Lefrançois P, Euskirchen GM, Auerbach RK, Rozowsky J, Gibson T, Yellman CM, Gerstein M, Snyder M. 2009. Efficient yeast ChIP-Seq using multiplex short-read DNA sequencing. *BMC Genomics* **10:** 37.

Johnson DS, Mortazavi A, Myers RM, Wold B. 2007. Genome-wide mapping of in vivo protein-DNA interactions. *Science* **316:** 1497–1502.

Wang H, Mayhew D, Chen X, Johnston M, Mitra RD. 2011. Calling Cards enable multiplexed identification of the genomic targets of DNA-binding proteins. *Genome Res* **5:** 748–755.

CHAPTER 29

Transcription Factor–DNA Binding Motifs in *Saccharomyces cerevisiae*: Tools and Resources

Joshua L. Schipper and Raluca M. Gordân[1]

Department of Biostatistics and Bioinformatics, Center for Genomic and Computational Biology, Duke University, Durham, North Carolina 27708

The DNA binding specificity of transcription factors (TFs) is typically represented in the form of a position weight matrix (PWM), also known as a DNA motif. A PWM is a matrix that specifies, for each position in the DNA binding site of a TF, the "weight" or contribution of each possible nucleotide. DNA motifs can be derived from various types of TF–DNA binding data, from small collections of known TF binding sites to large data sets generated using high-throughput technologies. One drawback of this simple model of DNA binding specificity is that it makes the implicit assumption that individual base pairs within a TF binding site contribute independently to the TF–DNA binding affinity. Although this assumption does not always hold, PWM models have been shown to provide reasonable approximations to the DNA binding specificity, and they are still widely used in practice. DNA motifs are currently available for more than 150 *Saccharomyces cerevisiae* TFs. Here, we briefly describe how these models are built, we provide information on databases containing DNA motifs for *S. cerevisiae* TFs, and we introduce guidelines on how to interpret the motifs and use them in practice to generate hypotheses about transcriptional regulatory regions.

WHAT IS A TRANSCRIPTION FACTOR–DNA BINDING MOTIF?

A DNA binding motif, or position weight matrix (PWM), is a simple and intuitive model for representing the DNA binding preferences of a transcription factor (TF). There are several accepted definitions of PWMs, depending on what the entries in the matrix represent (Stormo 2000, 2013). Typically, a PWM model M is defined as a matrix with four rows (corresponding to the four nucleotides) and W columns (where W is the width of the TF binding site). Each element $M(i, j)$ in the matrix represents the probability of seeing nucleotide i at position j in the binding site (Fig. 1A). This type of matrix is also known as a position frequency matrix (PFM) or position-specific scoring matrix (PSSM). In other definitions of the PWM, entries in the matrix may correspond to the energy contribution of each nucleotide at each position, the binding affinity for each nucleotide at each position, or the log ratio of the probability of each nucleotide at each position in the TF binding site relative to its background probability. The exact meaning of the entries in a PWM will affect how the PWM is used to score putative TF binding sites, as described below.

SCORING A PUTATIVE TF BINDING SITE ACCORDING TO A DNA MOTIF

Given a putative TF binding site S and a DNA motif M containing the probabilities of each nucleotide at each position, we can compute a "score" for site S as

---

[1]Correspondence: raluca.gordan@duke.edu

Copyright © Cold Spring Harbor Laboratory Press; all rights reserved
Cite this introduction as *Cold Spring Harb Protoc*; doi:10.1101/pdb.top080622

Chapter 29

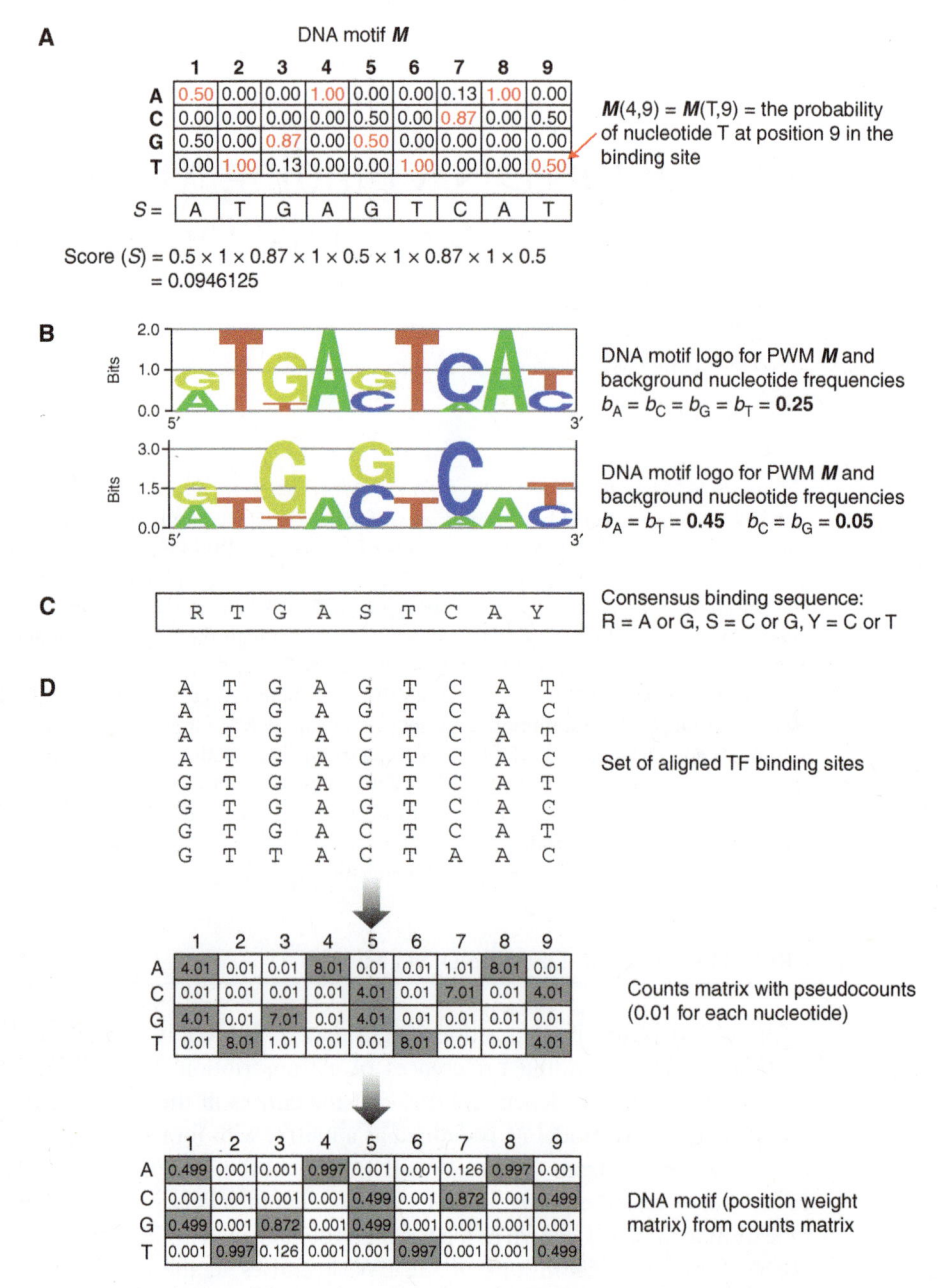

FIGURE 1. TF–DNA binding models. (*A*) Representation of a TF–DNA binding motif as a position weight matrix (PWM). (*B*) DNA motif logos generated using the enoLOGOS web server (Workman et al. 2005) for the PWM in *A* using a uniform background distribution $b_A = b_C = b_G = b_T = 0.25$ (*top* logo) or $b_A = b_T = 0.45$, $b_C = b_G = 0.05$ (*bottom* logo). (*C*) Consensus binding sequence derived from DNA motif in *A*. (*D*) Building a DNA motif from a set of aligned TF binding sites.

$$\text{Score}(S) = P(S|M) = \prod_{i=1}^{W} M(S_i, i), \quad (1)$$

where $M(S_i, i)$ is the probability of seeing nucleotide S_i at position i in the binding site (Fig. 1A). If we consider the PWM as a generative model (i.e., a model that generates binding sites for the TF of interest), then we can interpret Score(S) as the probability of generating binding site S from the PWM model M (i.e., $P(S|M)$). Because the PWM model assumes that the contributions of nucleotides at different positions in the binding site are independent of each other, the probability of generating site

548

S from the PWM is simply the product of generating, independently, each nucleotide S_i at each position i (Fig. 1A). In the case of PWMs where the entries represent energies, the product in Equation 1 should be replaced by a sum, as each nucleotide in a binding site contributes *additively* to the binding energy for that site.

An alternative way of scoring putative TF binding sites using a PWM is to use the log-likelihood ratio (LLR). Let B be a background model representing the nucleotide frequencies in "background" or random DNA sequences (i.e., sequences that are not binding sites for the TF of interest). The LLR score of a putative DNA site S is the natural logarithm of the ratio between the probability of generating S from the PWM model M, and the probability of generating S from the background model B:

$$\text{Score}_{\text{LLR}}(S) = \ln \frac{P(S|M)}{P(S|B)} = \ln \prod_{i=1}^{W} \frac{M(S_i, i)}{b_{S_i}} = \sum_{i=1}^{W} \ln \frac{M(S_i, i)}{b_{S_i}}, \quad (2)$$

where b_j is the frequency of nucleotide j in background DNA sequences. A positive LLR score indicates that site S is more likely to be a TF binding site than to be part of the background DNA. A negative LLR score indicates that site S is more likely to come from background DNA than to be a binding site for a TF with the DNA motif M.

For both the PWM score (Equation 1) and the LLR score (Equation 2), higher scores correspond to sites that are more likely to be bound by the TF. In the case of PWMs where the entries in the matrix represent log ratios of generating each nucleotide from the motif model versus a background model, these entries are equal to $\ln(M(i, j)/b_i)$, and the score according to such a PWM is exactly the log-likelihood ratio score described in Equation 2. This type of matrix is sometimes referred to as a "log-odds" matrix.

VISUALIZING TF–DNA BINDING MOTIFS

TF–DNA binding motifs can be visualized using motif logos (Fig. 1B). A motif logo contains as many positions as the DNA motif. At each position, the relative heights of the letters are proportional to their probabilities, as defined by the PWM. The total height of each column j is the information content (IC) of that column, represented as the number of bits of information and defined as

$$\text{IC}_j = \sum_{i \in \{A, C, G, T\}} M(i, j) \times \log_2 \frac{M(i, j)}{b_i}, \quad (3)$$

where $M(i, j)$ is the probability of nucleotide i at position j in the PWM, and b_i is the background frequency of nucleotide i. Intuitively, high information content positions are very stringent—that is, mutations in TF binding sites that occur at positions with high IC will significantly affect the binding affinity. Low information content corresponds to a more degenerate position, where several nucleotides are accepted in the TF binding site; mutations at positions with low IC have a lesser effect on the binding affinity.

Publicly available software and web tools for generating motif logos include the following:

- enoLOGOS: http://www.benoslab.pitt.edu/cgi-bin/enologos/enologos.cgi (Workman et al. 2005) and

- WebLogos: http://weblogo.berkeley.edu/logo.cgi (Crooks et al. 2004)

Such software tools allow users to build logos for both DNA and protein sequences. Each software tool has several user-defined parameters and can accept DNA motifs in different formats, including position frequency matrices (e.g., the PWM in Fig. 1A), aligned TF binding sites, or PWMs downloaded directly from DNA motif databases (such as TRANSFAC; Matys et al. 2006). Refer to the manual of each tool for more details.

An important parameter that the user should always set when generating DNA motif logos is the background nucleotide distribution. Different background distributions will lead to different motif logos, as shown in Figure 1B. For generating DNA motif logos, the most widely used background nucleotide distribution is the uniform distribution: $b_A = b_C = b_G = b_T = 0.25$. In some DNA motif discovery tools, such as enoLOGOS (Workman et al. 2005), this distribution can be specified by setting the GC content parameter (called "%GC" in enoLOGOS) to 50%. This type of background model will lead to DNA motifs where the information content at each position (i.e., the y-axis in the motif logo) is between 0 (the most degenerate position possible) and 2 (the most stringent position possible). Figure 1B shows examples of motif logos generated using the same PWM (the one in Fig. 1A) but different background nucleotide distributions (a uniform distribution in the top logo, and a highly skewed distribution, with GC content of 10%, in the bottom logo). According to the DNA motif generated using the uniform background distribution, positions 2, 4, 6, and 8 are very stringent (i.e., not degenerate), and A or T nucleotides at these positions seem to be highly preferred. When a background GC content of 10% is used, positions 2, 4, 6, and 8 appear more degenerate, and their height relative to other positions decreases. Intuitively, this is due to the fact that when the background GC content is 10%, the general frequencies of A and T nucleotides are very high (the AT content is 90%), so even DNA sites that are not putative TF binding sites will contain a large number of A and T nucleotides simply by chance. Thus, having an A or T in a putative TF binding site is not as significant as in the case of a uniform background distribution. In general, we recommend using a uniform background nucleotide distribution when generating DNA motif logos.

FROM DNA MOTIFS TO CONSENSUS SEQUENCES

DNA motifs can be used to derive consensus binding sequences, which use the four DNA bases and a set of IUPAC (International Union and Pure and Applied Chemistry) codes to specify the most common nucleotides at each position in a TF binding site (Fig. 1C). Consensus sequences have an intuitive interpretation: The TF has a high affinity for sequences matching the consensus and a low affinity for sequences that deviate from the consensus. However, this separation into binding sites and nonbinding sites is artificial because TFs bind a wide range of DNA sites with a continuum of binding affinities (Siggers and Gordân 2014).

The PWM model is a generalization of the consensus sequence model. The advantage of the PWM is that it provides position-specific penalties for deviations from the consensus; thus, when comparing a putative TF binding site to a PWM (as opposed to a consensus sequence) not all mismatches are treated equally, and some mismatches are more detrimental than others. For example, the DNA site ATGAGTCAT has the highest possible score according to the PWM in Figure 1A, and it matches the consensus sequence shown in Figure 1C (RTGASTCAY). If we mutate the G at position 3 to either an A or a T, then we obtain sites that do not match the consensus and thus would not be considered putative TF binding sites according to the consensus model. However, if we use the PWM model, then the score for ATGAGTCAT will be higher than the score of both mutated sites (ATAAGTCAT and ATTAGTCAT), but the two mutated sites are considered of different affinity, with ATTAGTCAT being preferred over ATAAGTCAT because it has a higher PWM score: $Score_{LLR}(ATGAGTCAT) = 10.12$, $Score_{LLR}(ATTAGTCAT) = 8.05$, and $Score_{LLR}(ATAAGTCAT) = 5.65$.

BUILDING DNA BINDING MOTIFS

Generating TF–DNA binding motifs depends, in part, on the DNA binding data available for the TF of interest. The simplest case is when we are given a set of aligned, known TF binding sites; such data are available from databases such as TRANSFAC (Matys et al. 2006) and JASPAR (Mathelier et al. 2014). Given this set of aligned binding sites, PWM models can be derived by computing the counts for each nucleotide at each position and then transforming the counts into frequencies by normalizing the values in each column to sum to 1 (i.e., by dividing each item in the matrix by the number of sites used to build the PWM) (Fig. 1D). Typically, before normalizing each column, a small number (e.g., 0.01) representing pseudocounts is added to each entry in the matrix to avoid occurrences of zero values in the PWM. Such zero probabilities are unlikely because we cannot be 100% certain that a particular nucleotide will never occur in a binding site for a particular TF. In addition, zero values may confound the results of applications that use PWMs. For example, when computing log-likelihood ratio scores using PWMs with zero entries, we run into the problem of having to compute log(0), which is not defined. Setting the exact pseudocounts to add when deriving a PWM is somewhat ad hoc. In general, small values such as 0.001 or 0.01 are used; however, when the number of aligned binding sites is very large (hundreds or thousands of sites), larger pseudocounts are recommended to avoid having very small values in the PWM. A value close to 0.1% of the total number of sites often yields reasonable results.

When aligned TF binding sites are not available, one can derive PWMs from sets of genomic or artificial sequences bound by the TF of interest (e.g., sequence reported to be bound according to ChIP-chip [Ren et al. 2000] or ChIP-seq [Johnson et al. 2007] data, or promoter sequences of genes believed to be regulated by a common TF). Once a set of "bound" DNA sequences has been identified, one can use motif discovery tools to identify PWMs that are enriched in those sequences. Hundreds of motif discovery methods have been developed to date, but few of them are easily accessible to researchers without a strong computational background. One of the easily accessible and well-supported motif discovery tools is MEME (Bailey and Elkan 1994). MEME is widely used and can be applied to relatively small sets (tens or hundreds) of DNA sequences (http://meme.nbcr.net/meme/), as well as high-throughput data generated using ChIP-seq assays (http://meme.nbcr.net/meme/cgi-bin/meme-chip.cgi). The MEME Suite also provides additional tools for analysis of TF–DNA binding data.

Figure 2 shows an example of running MEME to find overrepresented DNA motifs in ChIP-chip data for the *Saccharomyces cerevisiae* TF Gcn4 (Harbison et al. 2004). A total of 58 sequences with ChIP-chip *p*-value < 0.001 were used in this analysis. When run with the default settings, MEME found three DNA motifs (Fig. 2). For each motif, the tool also reported the number of DNA sites containing matches to that motif, as well as the statistical significance of the motif, represented as an

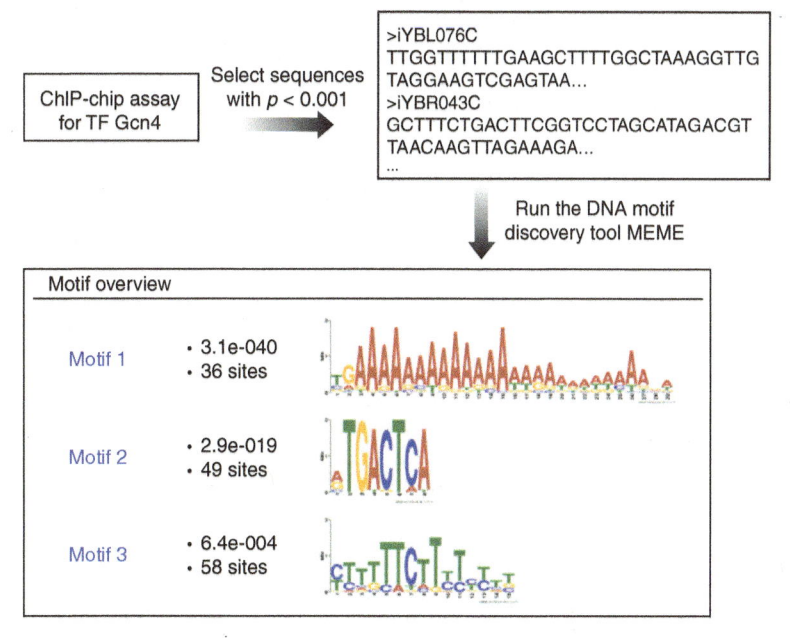

FIGURE 2. Example of running the MEME motif discovery tool on a set of DNA sequences bound by *Saccharomyces cerevisiae* TF Gcn4, according to ChIP-chip (chromatin immunoprecipitation with DNA microarray) data (Harbison et al. 2004).

E-value. Intuitively, the *E*-value can be interpreted as the number of motifs as good as or better than the motif of interest, identified in a set of random DNA sequences (for more details, see the documentation provided with the MEME tool). Interpreting the results of a motif discovery tool is not trivial. In general, motifs with lower *E*-values (i.e., with higher statistical significance) are preferred. However, as can be seen in Figure 2, motifs with very low *E*-values, such as Motif 1, sometimes capture DNA sequences that are common in the set of input DNA sequences but are unlikely to represent TF binding sites. In this example, Motif 2 is the correct motif for TF Gcn4; it has a high significance (*E*-value = 2.9e−019) and a large number of DNA sites (49 sites in the 58 sequences given as input to the algorithm). Motif 3 also has a large number of occurrences, but its significance is relatively low (*E*-value = 6.4e−4), which makes it unlikely to be the DNA motif for the TF tested in the ChIP-chip experiment (i.e., TF Gcn4). Further information about the DNA motifs identified by MEME is available in the results provided by the algorithm.

REPOSITORIES OF DNA BINDING MOTIFS FOR *S. cerevisiae* TFs

DNA motifs for more than 150 *S. cerevisiae* TFs are currently available in online databases and in various publications.

- *Saccharomyces* Genome Database (SGD): http://www.yeastgenome.org (Cherry et al. 2012). For each TF, SGD contains the DNA binding motif reported in the YeTFaSCo database (http://yetfasco.ccbr.utoronto.ca) (de Boer and Hughes 2012), as well as links to other DNA motif databases. Importantly, the YeTFaSCo database also provides expert curated information for the available DNA motifs.

- DNA motifs for *S. cerevisiae* TFs have been derived from large collections of ChIP-chip experiments by Harbison et al. (2004) and MacIsaac et al. (2006).

- TRANSFAC: www.biobase-international.com/product/transcription-factor-binding-sites (Matys et al. 2006). For each TF, the TRANSFAC database contains consensus binding sequences, motif logos, and PWM descriptions (typically as counts instead of frequencies), as well as the exact binding sites used to generate the PWMs, when available. The PWMs reported in TRANSFAC are sometimes derived from small-scale assays (such as electrophoretic mobility shift assays or footprinting) or from small collections of binding sites reported in the literature. Such PWMs may be biased toward the high-affinity sites characterized previously and may not accurately reflect moderate affinity TF binding sites. We recommend using PWMs derived from higher-throughput experimental data such as in vivo ChIP-chip/ChIP-seq assays (Ren et al. 2000; Johnson et al. 2007) or in vitro protein-binding microarray (PBM) (Berger et al. 2006) or mechanically induced trapping of molecular interactions (MITOMI) (Maerkl and Quake 2007) experiments. Access to the latest version of the TRANSFAC database requires an annual subscription. Older versions are freely available.

- JASPAR: http://jaspar.genereg.net (Mathelier et al. 2014). For each TF, the JASPAR database contains motif logos and PWM descriptions (represented as counts), as well as the exact binding sites used to generate the PWMs, when available. The type of data used to generate each PWM is specified. Compared to TRANSFAC, the JASPAR database contains information for fewer TFs; however, the data available in JASPAR is typically curated and restricted to high-quality motifs. In addition, a unique feature of JASPAR PWMs (compared with other sources of *S. cerevisiae* motifs described here) are the links to the TFBSshape database (Yang et al. 2014), which contains information on the DNA structure of binding sites for certain TFs.

- UniPROBE: http://the_brain.bwh.harvard.edu/uniprobe/ (Robasky and Bulyk 2011). The UniPROBE database contains TF–DNA binding data and PWMs derived from in vitro universal PBM experiments (Berger et al. 2006). Compared with other assays used to generate DNA motif models, universal PBM assays are unbiased and comprehensive, as they measure the binding specificity of a TF for all possible ungapped 10-mers and a large number of 10-mers containing gaps (for details on the experimental design, see Berger et al. 2006; Berger and Bulyk 2009). In addition, PBM data are advantageous in that they reflect the direct DNA binding specificity of the tested TF, in contrast to in vivo assays such as ChIP, for which both direct and indirect TF–DNA binding events are represented in the data. Gordân et al. (2009) performed a detailed analysis of direct and indirect TF–DNA binding for S. cerevisiae TFs by combining in vivo ChIP-chip data (Harbison et al. 2004) and PWMs derived from universal PBMs (Badis et al. 2008; Zhu et al. 2009).

- The PWMs available in UniPROBE were derived from the universal PBM data reported in a large-scale study of yeast TFs (Zhu et al. 2009). Later, Gordân et al. (2011) added PBM data for 27 additional yeast TFs and reanalyzed the data reported by Zhu et al. (2009), as well as the PBM data of Badis et al. (2008) and the MITOMI data of Fordyce et al. (2010). In total, in vitro TF–DNA binding data for 150 yeast TFs was curated in Gordân et al. (2011). The quality of each motif was assessed based on the available in vitro and in vivo data and, for each TF, the DNA motif that best explains the in vivo binding data for that factor was reported. These curated DNA motifs are available as Data File S1 in Gordân et al. (2011). In addition, this curated collection reports, for each of the 150 TFs analyzed in the study, the binding specificities for all possible 8-mers, derived from the universal PBM data. For each TF and each 8-mer, an enrichment score (or E-score) is reported, which reflects the enrichment of the 8-mer among the DNA probes bound with high affinity in the PBM assay for that TF. The E-score is a modified form of the Wilcoxon–Mann–Whitney statistic and ranges from -0.5 (least favored sequence) to $+0.5$ (most favored sequence), with values >0.35 corresponding, in general, to sequence-specific DNA binding of the tested TF (Berger et al. 2006).

Importantly, it has recently been shown that about one-third of S. cerevisiae TFs have two distinct modes of binding DNA, described by two PWMs (typically called "primary" and "secondary" motifs; Gordân et al. 2011). Such motifs were derived from PBM data for yeast TFs and reported in Data File S1 in Gordân et al. (2011). TFs with multiple modes of binding have long been reported in the literature, but the large number of such TFs has recently been revealed by large-scale in vitro studies. The distinct modes of DNA binding are believed to reflect slightly different conformations of the TFs bound to DNA (see Gordân et al. 2011 for more details and information on yeast TFs with multiple modes of DNA binding). When using DNA motifs to identify putative binding sites for such TFs, it is important to take into account the primary and the secondary motifs, as both motifs have been shown to be relevant for TF–DNA binding in the cell (Gordân et al. 2011).

USING DNA MOTIFS TO IDENTIFY PUTATIVE TF BINDING SITES

DNA motifs can be used to scan genomic regions of interest (such as gene promoters) to identify putative TF binding sites. Given a TF–DNA binding motif M of width W and a DNA sequence X of length L, we can use the DNA motif to scan the sequence X and score each site of size W in the sequence using, for example, the LLR score. Figure 3A shows an example of scanning part of an intergenic yeast sequence (iYJL089W) using the LLR score with a DNA motif for TF Gcn4 and a uniform background distribution. In general, when scanning a sequence with a DNA motif to identify putative TF binding sites, we will report all sites for which the score represents at least a fraction F of the score for the best possible site according to that DNA motif. The score of the best site can be computed by taking, from each PWM column, the entry containing the highest probability.

Most DNA motif databases mentioned above also provide tools for identifying putative TF binding sites using the reported motifs. For example, the JASPAR web server (Mathelier et al. 2014) contains a simple online tool that can be used to scan a FASTA-formatted DNA sequence using the PWM model (s) selected by the user. Only one parameter is required: the relative score threshold, which represents the ratio between the LLR score of a putative TF binding site versus the LLR score of the best site according to the selected PWM model (i.e., the F fraction mentioned above and in Fig. 3A). Figure 3B

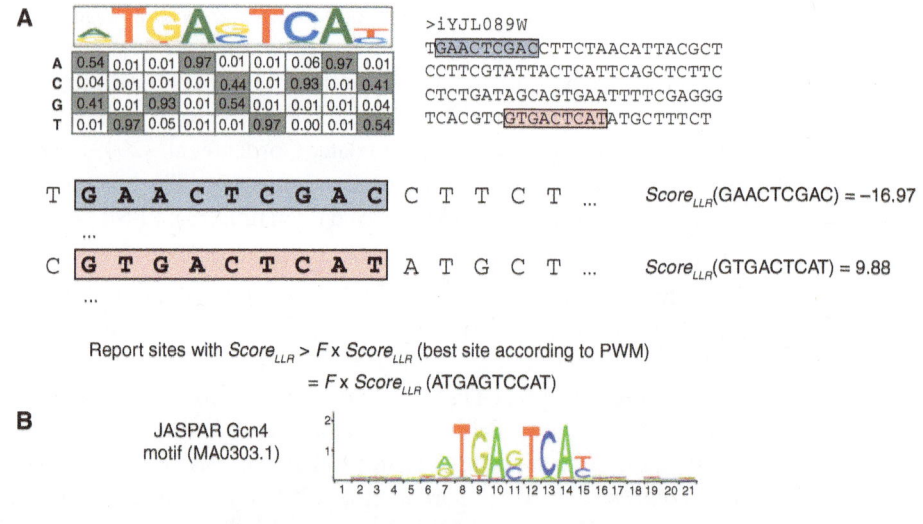

Chapter 29

FIGURE 3. Identifying putative TF binding sites using a DNA motif. (A) A DNA motif for TF Gcn4 is used to identify putative Gcn4 binding sites, defined as sites with LLR score > F × the LLR score of the best DNA site according the motif (here, ATGAGTCAT). (B) Putative Gcn4 binding sites reported by JASPAR (Mathelier et al. 2014) using the Gcn4 motif MA0303.1 and a relative score cutoff $F = 0.8$.

shows the putative Gcn4 binding sites identified using the JASPAR Gcn4 motif and a relative score threshold of 0.8.

Additional online tools for calling putative TFBSs using DNA motifs include the following:

- FIMO: http://meme.nbcr.net/meme/cgi-bin/fimo.cgi (part of the MEME Suite)
- PATSER: http://rsat.ulb.ac.be/patser_form.cgi (part of the RSA-tools suite)

Further information about these tools, their output, and the interpretation of the output is available on the tools' websites.

The 8-mer PBM data reported in Badis et al. (2008) and Zhu et al. (2009) and curated in Gordân et al. (2011) can also be used to identify putative TF binding sites. Briefly, one can scan the sequences of interest to search for 8-mers with enrichment scores (E-scores) >0.35, which likely represent sequence-specific TF–DNA binding sites. For increased confidence, one can search for two consecutive, overlapping 8-mers with E-scores above a certain cutoff (as done, e.g., in Gordân et al. 2013).

COMPUTING THE ENRICHMENT OF A DNA MOTIF IN A SET OF DNA SEQUENCES

Certain motif discovery tools, such as MEME, automatically provide a measure of statistical significance for the identified DNA motifs. However, there are situations when a PWM for the TF of interest is already available, and we are interested in determining whether or not the PWM is *enriched* in a set of DNA sequences (e.g., in promoters of a set of co-regulated genes). The hypergeometric distribution (i.e., Fisher's exact test) can be used for this purpose. To apply this test, we first need to define a "foreground" set containing DNA sequences that are likely to be bound by the TF of interest, and a "background" set containing unbound sequences (e.g., if the foreground set contains promoters of co-regulated genes, the background set can be composed of randomly chosen promoters of other genes in the *S. cerevisiae* genome). Next, we use the DNA motif to scan each foreground and background sequence and determine how many of the sequences contain putative TF binding sites (at a specific relative score threshold F). Let N be the total number of sequences, in both the foreground and background sets, m the number of sequences containing putative TF binding sites, n the number of foreground sequences, and k the number of foreground sequences that contain putative TF binding sites. Then we can use the hypergeometric distribution (Equation 4) and compute the p-value as the cumulative probability $P(X \geq k)$ (a small p-value will indicate that the DNA motif is enriched in the foreground sequences compared to the background sequences):

$$P(X = k) = \frac{\binom{m}{k}\binom{N-m}{n-k}}{\binom{N}{n}}. \tag{4}$$

REFERENCES

Badis G, Chan ET, van Bakel H, Pena-Castillo L, Tillo D, Tsui K, Carlson CD, Gossett AJ, Hasinoff MJ, Warren CL, et al. 2008. A library of yeast transcription factor motifs reveals a widespread function for Rsc3 in targeting nucleosome exclusion at promoters. *Mol Cell* 32: 878–887. doi: 10.1016/j.molcel.2008.11.020.

Bailey TL, Elkan C. 1994. Fitting a mixture model by expectation maximization to discover motifs in biopolymers. *Proc Int Conf Intell Syst Mol Biol* 2: 28–36.

Berger MF, Bulyk ML. 2009. Universal protein-binding microarrays for the comprehensive characterization of the DNA-binding specificities of transcription factors. *Nat Protoc* 4: 393–411. doi: 10.1038/nprot.2008.195.

Berger MF, Philippakis AA, Qureshi AM, He FS, Estep PW 3rd, Bulyk ML. 2006. Compact, universal DNA microarrays to comprehensively determine transcription-factor binding site specificities. *Nat Biotechnol* 24: 1429–1435. doi: 10.1038/nbt1246.

Cherry JM, Hong EL, Amundsen C, Balakrishnan R, Binkley G, Chan ET, Christie KR, Costanzo MC, Dwight SS, Engel SR, et al. 2012. Saccharomyces genome database: The genomics resource of budding yeast. *Nucleic Acids Res* 40: D700–D705. doi: 10.1093/nar/gkr1029.

Crooks GE, Hon G, Chandonia JM, Brenner SE. 2004. WebLogo: A sequence logo generator. *Genome Res* 14: 1188–1190. doi: 10.1101/gr.849004.

de Boer CG, Hughes TR. 2012. YeTFaSCo: A database of evaluated yeast transcription factor sequence specificities. *Nucleic Acids Res* 40: D169–D179. doi: 10.1093/nar/gkr993.

Fordyce PM, Gerber D, Tran D, Zheng J, Li H, DeRisi JL, Quake SR. 2010. De novo identification and biophysical characterization of transcription-factor binding sites with microfluidic affinity analysis. *Nat Biotechnol* 28: 970–975. doi: 10.1038/nbt.1675.

Gordân R, Hartemink AJ, Bulyk ML. 2009. Distinguishing direct versus indirect transcription factor-DNA interactions. *Genome Res* 19: 2090–2100. doi: 10.1101/gr.094144.109.

Gordân R, Murphy KF, McCord RP, Zhu C, Vedenko A, Bulyk ML. 2011. Curated collection of yeast transcription factor DNA binding specificity data reveals novel structural and gene regulatory insights. *Genome Biol* 12: R125. doi: 10.1186/gb-2011-12-12-r125.

Gordân R, Shen N, Dror I, Zhou T, Horton J, Rohs R, Bulyk ML. 2013. Genomic regions flanking E-Box binding sites influence DNA binding specificity of bHLH transcription factors through DNA shape. *Cell Rep* 3: 1093–1104. doi: 10.1016/j.celrep.2013.03.014.

Harbison CT, Gordon DB, Lee TI, Rinaldi NJ, MacIsaac KD, Danford TW, Hannett NM, Tagne JB, Reynolds DB, Yoo J, et al. 2004. Transcriptional regulatory code of a eukaryotic genome. *Nature* 431: 99–104. doi: 10.1038/nature02800.

Johnson DS, Mortazavi A, Myers RM, Wold B. 2007. Genome-wide mapping of in vivo protein-DNA interactions. *Science* 316: 1497–1502. doi: 10.1126/science.1141319.

MacIsaac KD, Wang T, Gordon DB, Gifford DK, Stormo GD, Fraenkel E. 2006. An improved map of conserved regulatory sites for *Saccharomyces cerevisiae*. *BMC bioinformatics* 7: 113. doi: 10.1186/1471-2105-7-113.

Maerkl SJ, Quake SR. 2007. A systems approach to measuring the binding energy landscapes of transcription factors. *Science* 315: 233–237. doi: 10.1126/science.1131007.

Mathelier A, Zhao X, Zhang AW, Parcy F, Worsley-Hunt R, Arenillas DJ, Buchman S, Chen CY, Chou A, Ienasescu H, et al. 2014. JASPAR 2014: An extensively expanded and updated open-access database of transcription factor binding profiles. *Nucleic Acids Res* 42: D142–D147. doi: 10.1093/nar/gkt997.

Matys V, Kel-Margoulis OV, Fricke E, Liebich I, Land S, Barre-Dirrie A, Reuter I, Chekmenev D, Krull M, Hornischer K, et al. 2006. TRANSFAC and its module TRANSCompel: Transcriptional gene regulation in eukaryotes. *Nucleic Acids Res* 34: D108–D110. doi: 10.1093/nar/gkj143.

Ren B, Robert F, Wyrick JJ, Aparicio O, Jennings EG, Simon I, Zeitlinger J, Schreiber J, Hannett N, Kanin E, et al. 2000. Genome-wide location and function of DNA binding proteins. *Science* 290: 2306–2309. doi: 10.1126/science.290.5500.2306.

Robasky K, Bulyk ML. 2011. UniPROBE, update 2011: Expanded content and search tools in the online database of protein-binding microarray data on protein–DNA interactions. *Nucleic Acids Res* 39: D124–D128. doi: 10.1093/nar/gkq992.

Siggers T, Gordân R. 2014. Protein–DNA binding: Complexities and multi-protein codes. *Nucleic Acids Res* 42: 2099–2111. doi: 10.1093/nar/gkt1112.

Stormo GD. 2000. DNA binding sites: Representation and discovery. *Bioinformatics* 16: 16–23.

Stormo GD. 2013. Modeling the specificity of protein–DNA interactions. *Quant Biol* 1: 115–130.

Workman CT, Corcoran DL, Ideker T, Stormo GD, Benos PV. 2005. enoLOGOS: A versatile web tool for energy normalized sequence logos. *Nucleic Acids Res* 33: W389–W392. doi: 10.1093/nar/gki439.

Yang L, Zhou T, Dror I, Mathelier A, Wasserman WW, Gordân R, Rohs R. 2014. TFBSshape: A motif database for DNA shape features of transcription factor binding sites. *Nucleic Acids Res* 42: D148–D155. doi: 10.1093/nar/gkt1087.

Zhu C, Byers KJ, McCord RP, Shi Z, Berger MF, Newburger DE, Saulrieta K, Smith Z, Shah MV, Radhakrishnan M, et al. 2009. High-resolution DNA-binding specificity analysis of yeast transcription factors. *Genome Res* 19: 556–566. doi: 10.1101/gr.090233.108.

CHAPTER 30

The *Saccharomyces* Genome Database: A Tool for Discovery

J. Michael Cherry[1]

Department of Genetics, Stanford University School of Medicine, Stanford, California 94305-5120

The *Saccharomyces* Genome Database (SGD) is the main community repository of information for the budding yeast, *Saccharomyces cerevisiae*. The SGD has collected published results on chromosomal features, including genes and their products, and has become an encyclopedia of information on the biology of the yeast cell. This information includes gene and gene product function, phenotype, interactions, regulation, complexes, and pathways. All information has been integrated into a unique web resource, accessible via http://yeastgenome.org. The website also provides custom tools to allow useful searches and visualization of data. The experimentally defined functions of genes, mutant phenotypes, and sequence homologies archived in the SGD provide a platform for understanding many fields of biological research. The mission of SGD is to provide public access to all published experimental results on yeast to aid life science students, educators, and researchers. As such, the SGD has become an essential tool for the design of experiments and for the analysis of experimental results.

INTRODUCTION

The wisdom of the budding yeast research community is represented by the body of experimental work published in the last five decades. Access to these results is available to anyone who has the opportunity to explore this literature and a lot of time. Reviews on specific areas of yeast biology are also available; however, for the student of yeast biology, it is a challenge to digest the literature. What if a researcher is interested in only a small portion of what is known on a particular topic or gene, the relevant information being spread across many papers? What about bioinformaticists or computer scientists who wish to use computational analyses to explore results that are in the supplemental pages of many papers? The answer to accessing relevant information is an encyclopedic database of integrated experimental results that is annotated by expert biocurators. This database is the *Saccharomyces* Genome Database (SGD; www.yeastgenome.org; Cherry et al. 2012).

Historically, the core of the SGD comprises chromosomal features—defined regions of the chromosome that are associated with a function or product. Information on protein-coding genes and non-protein-coding RNA genes, such as tRNA and rRNA genes, is typically what is available. The basic entry point is a gene name that leads directly to summary information about the locus. A keyword describing a function, phenotype, selective condition, or text from an abstract will also provide entry into SGD. Gene or chromosomal regions can be identified for exploration using a DNA or protein sequence with an integrated BLAST sequence search tool. Unique identifiers from many sources also provide valid entry points, such as protein and DNA sequence accession numbers, PubMed and NCBI identifiers, author names, and Gene Ontology (GO) function terms. The information provided by SGD has been gathered and is maintained by a group of scientists working as biocurators and software developers. This team is devoted to providing users with up-to-date information and connections to

---

[1]Correspondence: cherry@stanford.edu

Copyright © Cold Spring Harbor Laboratory Press; all rights reserved
Cite this introduction as *Cold Spring Harb Protoc*; doi:10.1101/pdb.top083840

Chapter 30

all the major research resources and tools that allow previously published data to be explored and the user's own results to be interpreted. The goal is to aid experimental design and the analysis of results as well as to facilitate education and further study of budding yeast.

SGD MAINTAINS THE *S. cerevisiae* REFERENCE GENOME SEQUENCE

The SGD is the keeper of *S. cerevisiae* nomenclature and reference genome sequence (Engel et al. 2014). It has been responsible for maintaining this important community documentation since 1996. The reference genome has been updated several times since its first release in 1996, typically for a single region only after specific rules were met; for example, both strands from a region in the S288C strain were sequenced to confirm that a change was necessary. As described by Engel et al. (2014), a major update of the reference genome sequence occurred in 2010 to incorporate data generated by next-generation sequencing from a single colony. The sequence of this reference is deemed to be of such high quality that now only gross changes to the sequence will be considered. This is because any single colony is likely to have at least one nucleotide difference compared with its parent, excluding any sequence differences that arise because of technical errors. Thus, it is appropriate to assume that any minor differences observed are the result of allelic variation compared to the reference.

ANNOTATIONS AND ONTOLOGIES IN THE SGD

The SGD provides thousands of annotations for yeast genes and their products. Experimental results are captured as annotations using detailed precise vocabulary, typically in the form of an ontology, and include details of an experiment's evidence and methods and the appropriate literature citation. An ontology is a highly structured form of controlled vocabulary. Each entry in the ontology is commonly called a term. The name of the term is a short descriptive phrase, such as "carnitine dehydratase activity." However, each term also has a definition, similar to the definition of an English word found within a dictionary, which provides the complete usage and detailed explanation of the word/term. It is critical to consult the term's definition because the distinction between terms can be subtle. One important difference between the definition of an ontology term and that of an English word is that the term has only one meaning; however, there can be synonyms. The creation of ontology terms and their definitions often involves debate but the result has been a descriptive language. The use of ontologies has been successful in unifying communication between scientific communities and in providing a standard dictionary for topics such as molecular functions, biological processes, mutant phenotypes, chemical properties, and structures. In addition to terms and definitions, ontologies require a relationship between terms to define the type of connection (relationship). In an ontology, a term can have more than one parent term, the term above it in the ontology, as well as more than one child, terms below it in the ontology. GO is a system used to describe gene function and is used extensively in SGD annotations. Further details on the GO project are available from several sources (Gene Ontology Consortium 2013). GO annotations are often used to illustrate the structure of an ontology, but many other ontologies are also used to construct annotations in SGD, as is the case in all modern biological databases.

SGD USER INTERFACE

The main entry point into SGD is the home page, www.yeastgenome.org. Across the top of all SGD pages is a purple bar with several features collected into groups of similar purpose. For example, the drop-down menus, "Analyze" for computational tools or "Function" for collections of experimental results such as microarray expression or mutant phenotype data, can be selected. Also at the upper right corner of every page is located the "Search" box. Here text can be entered to search the contents of the database. For example, if the word "actin" is slowly typed into the search box a list appears and changes with the addition of each letter. Entry of "actin" should bring down many phrases that contain "actin." The results of a search include all the entries for all types of information found within SGD that matches the input string. Selecting a search result takes the user to the relevant page; for example, selecting the "actin" search result, *ACT1*, brings up the Locus Overview page, which presents the collected information on *ACT1* and its product. The locus overview page for another gene, *SNF1* (AMP-activated serine/threonine protein kinase), is illustrated in Figure 1. All of the information provided by SGD via its web pages or search tools can be downloaded from its dedicated file retrieval site by selecting the "Download" button at the top of every SGD page. This takes the user to http://downloads.yeastgenome.org.

The accompanying protocols provide guidance on navigating SGD. See Protocol 1: The *Saccharomyces* Genome Database: Exploring Biochemical Pathways and Mutant Phenotypes (Cherry 2015a), Protocol 2: The *Saccharomyces* Genome Database: Advanced Searching Methods and Data Mining (Cherry 2015b), Protocol 3: The *Saccharomyces* Genome Database: Gene Product Annotation of Function, Process, and Component (Cherry 2015c), and Protocol 4: The *Saccharomyces* Genome Database: Exploring Genome Features and Their Annotations (Cherry 2015d).

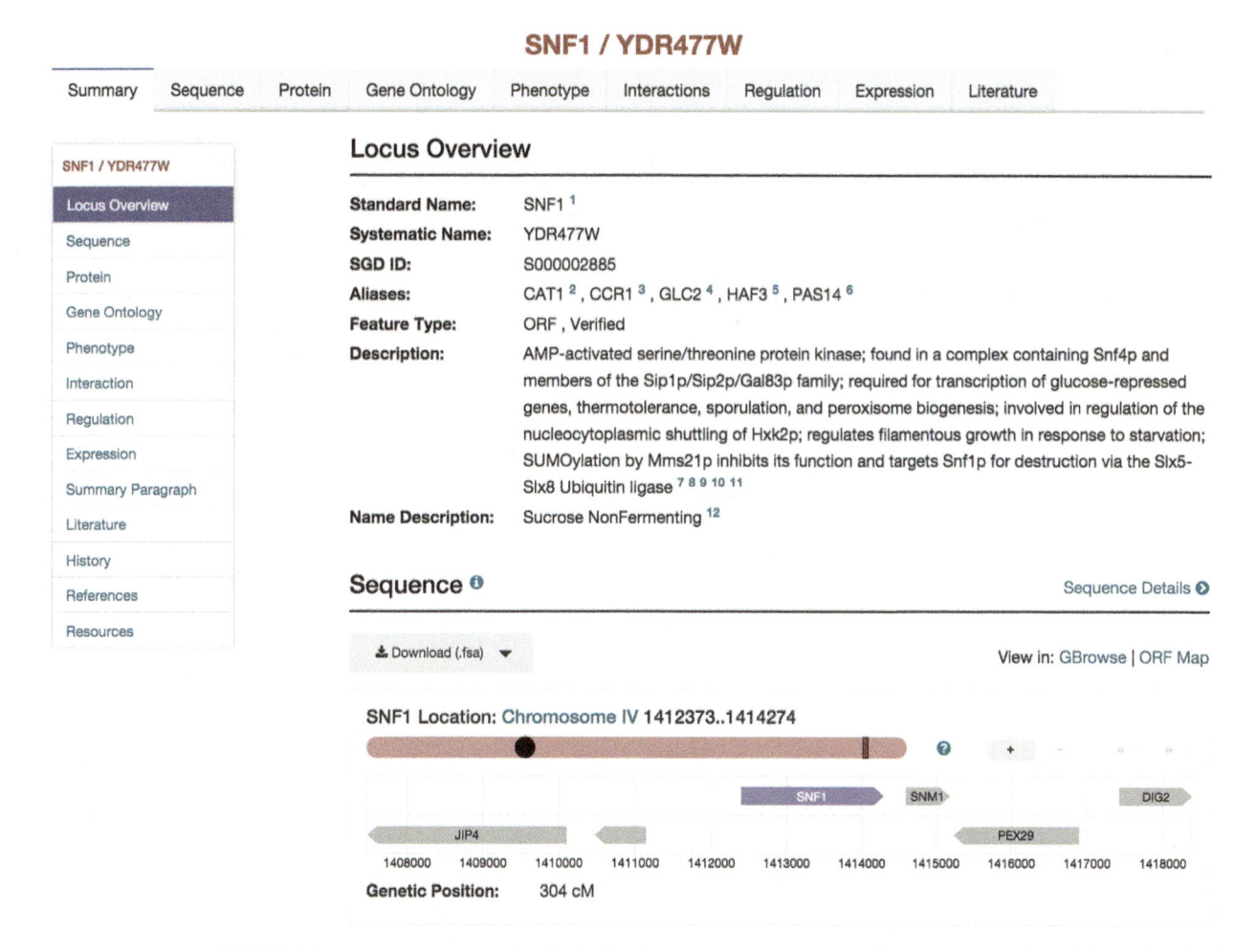

FIGURE 1. Locus summary page for *SNF1*. The locus summary page provides a summary of the information that has been compiled for a gene. Complete information about the gene is available within the topic pages, listed as tabs across the top of the window. Each page also provides quick navigation within the page via the left-hand menu.

ACKNOWLEDGMENTS

I am grateful to all the present and past staff of the *Saccharomyces* Genome Database project for their dedication to accuracy and their service to life science educators and researchers. I also want to thank the yeast research community for their support and suggestions. This work was supported by the National Human Genome Research Institute (grant number U41 HG001315) and Funding for open access charge was provided by the National Institutes of Health. The content is solely the responsibility of the author and does not necessarily represent the official views of the National Human Genome Research Institute or the National Institutes of Health.

REFERENCES

Cherry JM. 2015a. The *Saccharomyces* Genome Database: Exploring biochemical pathways and mutant phenotypes. *Cold Spring Harb Protoc* doi: 10.1101/pdb.prot088898.

Cherry JM. 2015b. The *Saccharomyces* Genome Database: Advanced searching methods and data mining. *Cold Spring Harb Protoc* doi: 10.1101/pdb.prot088906.

Cherry JM. 2015c. The *Saccharomyces* Genome Database: Gene product annotation of function, process, and component. *Cold Spring Harb Protoc* doi: 10.1101/pdb.prot088914.

Cherry JM. 2015d. The *Saccharomyces* Genome Database: Exploring genome features and their annotations. *Cold Spring Harb Protoc* doi: 10.1101/pdb.prot088922.

Cherry JM, Hong EL, Amundsen C, Balakrishnan R, Binkley G, Chan ET, Christie KR, Costanzo MC, Dwight SS, Engel SR, et al. 2012. *Saccharomyces* Genome Database: The genomics resource of budding yeast. *Nucleic Acids Res* 40: D700–D705.

Engel SR, Dietrich FS, Fisk DG, Binkley G, Balakrishnan R, Costanzo MC, Dwight SS, Hitz BC, Karra K, Nash RS, et al. 2014. The reference genome sequence of *Saccharomyces cerevisiae*: Then and now. *G3 (Bethesda)* 4: 389–398.

Gene Ontology Consortium. 2013. Gene Ontology annotations and resources. *Nucleic Acids Res* 41: D530–D535.

Protocol 1

The *Saccharomyces* Genome Database: Exploring Biochemical Pathways and Mutant Phenotypes

J. Michael Cherry[1]

Department of Genetics, Stanford University School of Medicine, Stanford, California 94305-5120

Many biochemical processes, and the proteins and cofactors involved, have been defined for the eukaryote *Saccharomyces cerevisiae*. This understanding has been largely derived through the awesome power of yeast genetics. The proteins responsible for the reactions that build complex molecules and generate energy for the cell have been integrated into web-based tools that provide classical views of pathways. The Yeast Pathways in the *Saccharomyces* Genome Database (SGD) is, however, the only database created from manually curated literature annotations. In this protocol, gene function is explored using phenotype annotations to enable hypotheses to be formulated about a gene's action. A common use of the SGD is to understand more about a gene that was identified via a phenotypic screen or found to interact with a gene/protein of interest. There are still many genes that do not yet have an experimentally defined function and so the information currently available can be used to speculate about their potential function. Typically, computational annotations based on sequence similarity are used to predict gene function. In addition, annotations are sometimes available for phenotypes of mutations in the gene of interest. Integrated results for a few example genes will be explored in this protocol. This will be instructive for the exploration of details that aid the analysis of experimental results and the establishment of connections within the yeast literature.

MATERIALS

Equipment

Internet-connected computer with web browser

METHOD

At any step in this protocol, support is available via the comprehensive Help documents maintained by SGD. These can be accessed via the help button at the top of each page or, for specific features, a small red button with a question mark in its center is provided and will link to the help pages specific for that feature.

The SGD is continually updated; therefore, specific items presented in this protocol may not appear on the SGD website exactly as described.

Exploring Biochemical Pathway Annotation

This series of steps explores the gene CAT2, describing the pathway tools available from SGD and focusing on the Locus pages. Locus pages are at the core of SGD, where information about each chromosomal feature is provided.

---

[1]Correspondence: cherry@stanford.edu

Copyright © Cold Spring Harbor Laboratory Press; all rights reserved
Cite this protocol as *Cold Spring Harb Protoc*; doi:10.1101/pdb.prot088898

Chapter 30

The collected information is divided into several tabs by topic. The default starting tab is the Summary page, which provides an overview of a gene and its product.

1. Open the URL www.yeastgenome.org in any modern web browser (e.g., Chrome, Firefox, or Safari).

2. In the upper right Search box slowly enter "acetylcarnitine," and observe the list of displayed matches that become more specific as additional letters are added. (By the time you have entered "acetylc" you are down to two choices, one representing our entry of choice.) Press return at any time to select the top string to see a full list of all entries available in the database matching that string, or mouse click to select one of the alternative items of interest. To continue with this exercise, select "acetylcarnitine."

    The "Search" box in the upper right of all SGD pages is the starting point to locate things of interest. Most of the textural information included at SGD has been indexed and is provided as choices from the Search box. Often, as text is entered, you will find additional items of interest that you can immediately explore using this search feature.

3. From the search results page click on "CAT2" to display the *CAT2* Locus Summary page. To go directly to the *CAT2* Locus Summary page enter "CAT2" in the Search box and press return.

    When the search text matches a gene name, you are directly taken to the matching Locus Summary page. The Locus Summary page includes basic information about a chromosomal feature, in this case the CAT2 gene. The Description states that Cat2p is carnitine acetyl-CoA transferase and that it is located in both mitochondria and peroxisomes. The Description includes the citation for this information. There are several subsections to the page that provide detailed annotations using ontologies—for example, the Gene Ontology subsection leads to Molecular Function, Biological Process, and Cellular Component ontologies.

4. To explore the "Pathways" in which this enzyme participates, select "carnitine shuttle" under the Pathways subcategory to be taken to the biochemical pathway tool called Yeast Pathways. View the graphical representation for the carnitine shuttle, including a summary of the products and intermediates, that is provided. At the top of the page, select the button "More Detail" once or twice to zoom into the pathway illustration and view the chemical structures of the molecules involved in this reaction (see Fig. 1).

    Cat2p is one of the three enzymes that catalyzes L-carnitine to O-acetylcarnitine using the cofactor acetyl-CoA to donate an acetate group resulting in Coenzyme A. Here you will also learn that Cat2p is the peroxisomal carnitine O-acetyltransferase.

    Yeast Pathways was created using MetaCyc software (Caspi et al. 2014).

5. Explore the pathway by moving the mouse cursor over gene names, compounds, cofactors, enzyme names, and reactions to bring up more information. Also click on the highlighted text. For example, click on Coenzyme A to see a detailed molecular structure, other reactions that it is involved in, and hyperlinks to other web resources. Click on the hyperlink to PubChem, for example, and you are taken to the chemical component database at the NCBI (pubchem.ncbi.nlm.nih.gov).

    The references listed in the References subcategory on the CAT2 Locus Summary page provide the experimental results for these annotations. The Yeast Pathways page presents the gene name as CAT2 (gene nomenclature), but, here, the protein name, Cat2p (protein nomenclature) is used, because the function of the protein is known.

Exploring Mutant Phenotype Annotation

6. Search for *AIM17* via the Search box and go to its summary page.

    The summary page description indicates that Aim17p has been purified from mitochondria and that mutants have an impact on mitochondrial genome loss. The GO annotations subsection states that the Molecular Function is unknown. Thus no experimental results have been published that describe this protein's function in the cell.

7. Select the Phenotype tab, then select the "Phenotype details" button at top right.

    A bar graph indicates there are several annotations of AIM17 from knockouts, that is null mutations, and that more were obtained from large-scale survey experiments than from classical genetic experiments.

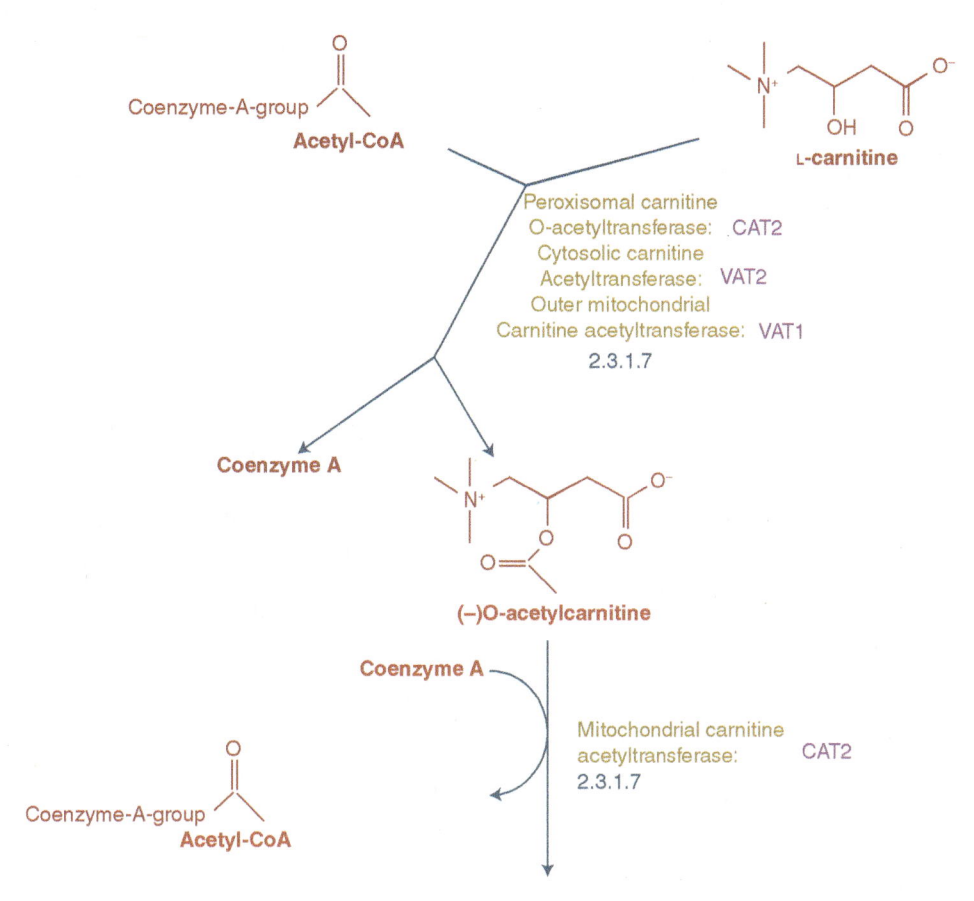

FIGURE 1. Reaction catalyzed by Cat2p. Yeast Pathways provides a detailed illustration (redrawn for clarity) of the reaction, including molecular structures of reactants and products.

8. Scroll to the bottom of the page to view a network visualization of the phenotype observables and the other genes that are annotated to them.

   Controlled vocabularies are used for mutant phenotypes, thus allowing all genes that have the same mutant phenotype to have a consistent annotation. This consistency is important as it guarantees your search for a particular phenotype will find all genes with that specific mutant phenotype. The use of free text would not provide such a guarantee. You will find this type of network visualization for different types of information on other tabs.

9. Experiment by selecting the yellow *AIM17* ball and dragging it around.

   In this way the display changes and you can explore the relationship of AIM17 to the observables and the other genes that have been annotated to the same observables (see Fig. 2).

10. Move the slider at the bottom of the network to change the number of phenotypes shared between *AIM17* and other genes.

    Notice that the dark balls represent other genes that share eight phenotype observables. We know that this gene product is localized to the mitochondria and that the null mutant has similar phenotypes to a variety of characterized genes. This set of phenotypes and genes may lead you to a hypothesis about the function of AIM7.

11. From the Phenotype Annotations table click the phenotype "nickel cation accumulation: increased" button to view all the genes annotated to the same observable phenotype (nickel cation accumulation) and qualifier (increased). Toward the top of the page, select the text "nickel cation accumulation" to see all annotations to this observable phenotype and a diagram of the relationship of this term within the Yeast Phenotype Ontology (Harris et al. 2013).

Chapter 30

Shared Phenotypes

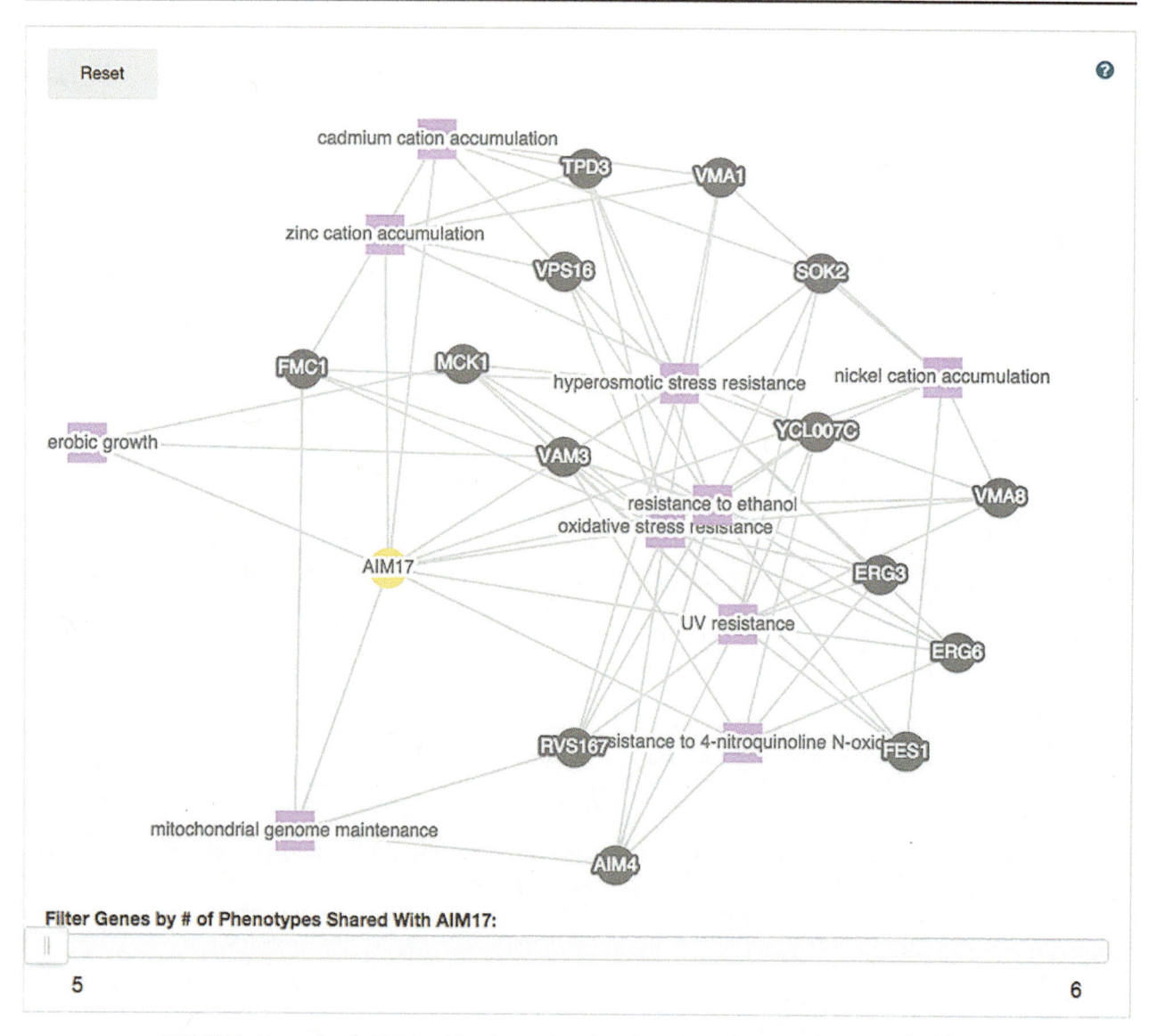

FIGURE 2. Network of *AIM17* with other genes that share five phenotype annotations. *AIM17*, yellow ball, is linked with other genes, dark balls, via phenotype annotations, purple squares. The user can dissect regions that have a complex structure by dragging the balls or squares. The network can also be simplified by increasing the number of shared annotations required for the linkage to be shown by moving the slider at the bottom of the graph.

ACKNOWLEDGMENTS

I am grateful to all the present and past staff of the *Saccharomyces* Genome Database project for their dedication to accuracy and service to life science educators and researchers. I also want to thank the yeast research community for their support and suggestions. This work was supported by the National Human Genome Research Institute (grant number U41 HG001315) and funding for open access charge was provided by the National Institutes of Health. The content is solely the responsibility of the author and does not necessarily represent the official views of the National Human Genome Research Institute or the National Institutes of Health.

REFERENCES

Caspi R, Altman T, Billington R, Dreher K, Foerster H, Fulcher CA, Holland TA, Keseler IM, Kothari A, Kubo A, et al. 2014. The MetaCyc database of metabolic pathways and enzymes and the BioCyc collection of Pathway/Genome Databases. *Nucleic Acids Res* **42:** D459–D471.

Harris MA, Lock A, Bähler J, Oliver SG, Wood V. 2013. FYPO: The fission yeast phenotype ontology. *Bioinformatics* **29:** 1671–1678.

Protocol 2

The *Saccharomyces* Genome Database: Advanced Searching Methods and Data Mining

J. Michael Cherry[1]

Department of Genetics, Stanford University School of Medicine, Stanford, California 94305-5120

At the core of the *Saccharomyces* Genome Database (SGD) are chromosomal features that encode a product. These include protein-coding genes and major noncoding RNA genes, such as tRNA and rRNA genes. The basic entry point into SGD is a gene or open-reading frame name that leads directly to the locus summary information page. A keyword describing function, phenotype, selective condition, or text from abstracts will also provide a door into the SGD. A DNA or protein sequence can be used to identify a gene or a chromosomal region using BLAST. Protein and DNA sequence identifiers, PubMed and NCBI IDs, author names, and function terms are also valid entry points. The information in SGD has been gathered and is maintained by a group of scientific biocurators and software developers who are devoted to providing researchers with up-to-date information from the published literature, connections to all the major research resources, and tools that allow the data to be explored. All the collected information cannot be represented or summarized for every possible question; therefore, it is necessary to be able to search the structured data in the database. This protocol describes the YeastMine tool, which provides an advanced search capability via an interactive tool. The SGD also archives results from microarray expression experiments, and a strategy designed to explore these data using the SPELL (Serial Pattern of Expression Levels Locator) tool is provided.

MATERIALS

Equipment

Internet-connected computer with web browser

METHOD

At any step in this protocol, support is available via the comprehensive Help documents maintained by the SGD. These can be accessed via the help button at the top of each page or, for specific features, a small red button with a question mark in its center is provided and will link to the help pages specific for that feature.

The SGD is continually updated; therefore, specific items presented in this protocol may not appear on the SGD website exactly as described.

Using YeastMine

This series of steps presents a method to identify all uncharacterized yeast genes that are essential for life (lethal when mutated) and that have a human homolog associated with a disease phenotype. The set of yeast genes that fulfill these

---
[1]Correspondence: cherry@stanford.edu

Copyright © Cold Spring Harbor Laboratory Press; all rights reserved
Cite this protocol as *Cold Spring Harb Protoc*; doi:10.1101/pdb.prot088906

criteria can be found using YeastMine (Balakrishnan et al. 2012). YeastMine is a query tool for all the data that are contained in the SGD and downloads site. Video tutorials are available for the basic features of YeastMine and can be accessed from links in the help section, www.yeastgenome.org/help/video-tutorials. These videos provide an effective way to learn the interface and operations that are provided by YeastMine. The procedure described here will use Templates, Lists, and List Operations. There are many situations where an advanced query of the data is the only way to obtain the desired information. In general, this occurs when a very specific answer is sought from a complex set of criteria or when large amounts of data are desired. YeastMine is a specialized data warehouse containing everything that is included within the main SGD database in addition to other large data sets. YeastMine is built using an open source tool called InterMine (Kalderimis et al. 2014).

An effective strategy to address a complex query is to break down the query into smaller questions and to then combine the results of these smaller questions. The answer to a complex question is obtained using basic list operations. There are four basic operations: **union**—the combination of two lists; **intersection**—the elements that are common between two lists; **subtraction**—the elements that are not common between two lists; and **asymmetric difference**—the elements that are not common in one of the two lists. In the procedure below, three simple queries are used to define a list of genes that are essential, a list of all genes that have a human homolog, and a list of genes that are associated with human disease. The union of the first two lists gives all essential yeast genes that have a human homolog, making a fourth list. The fourth list is compared to the third to provide the answer to the question: What are all the essential yeast genes that have a human homolog associated with a disease phenotype?

This procedure describes how to identify the set of uncharacterized essential yeast genes that have a human homolog associated with a disease phenotype annotation defined by OMIM (Online Mendelian Inheritance in Man; www.omim.org; Amberger et al. 2008). This is an easy query for YeastMine and is not something you would typically expect to find precomputed. An open-reading frame (ORF) defined as uncharacterized is believed to be a protein-coding gene; however, it does not yet have a function assigned (Fisk et al. 2006). OMIM is a detailed encyclopedia of human genetic diseases. OMIM includes details on genes and their variants and uses phenotype ontologies to define the characteristics of the disease.

1. Open the URL www.yeastgenome.org in any modern web browser (e.g., Chrome, Firefox, or Safari).

2. Open the "Help" page and under "Video Tutorials" open the "YeastMine" page. Watch the YeastMine video tutorials to become familiar with its basic features.

3. Open YeastMine (Fig. 1) by selecting "Advanced Search" at the top right of any SGD page (just below the Search box) and set up a myMine account so you can save the lists you create and any custom templates you define.

   Custom templates can also be created using instructions from the video tutorials (Step 2).

4. Make a list of all essential yeast genes. From the YeastMine homepage, select the Templates tab and then select the Phenotypes filter, followed by the Phenotype –> Genes template and locate "inviable" from the long list of observables. Start the search by clicking Show Results; there will be over a thousand rows in the resulting table. Now click on "Create/Add to list," select "All 1284 Genes" (listed in columns 1, 2, 3, 4, 5), and name this list "essential genes."

5. Create a list of all yeast genes with a human homolog that have a disease phenotype defined in OMIM. From the YeastMine homepage, select the "Templates" tab and then select the "Homology" filter, followed by the "Yeast gene –> OMIM human homolog(s) –> OMIM Disease Phenotype(s)." Select "constrain to be" and the predefined gene list "Uncharacterized_ORFs," then "Show Results." This results in 562 rows. Create another list of the 197 genes found in columns 1, 2, and 3; the columns used are indicated as you mouse over the options in the "Create New List" menu. Name this list "uncharacterized_with_human_disease_phenotype."

6. Select "Lists" from the purple bar and then "View". Locate the two lists you have just created, "essential genes" and "uncharacterized_with_human_disease_phenotype." Select your two lists and click "Intersect" to find the genes in common. Name this list "yeast-gene-human-phenotype." Click on the list name to begin exploring more information about these genes.

   There are three uncharacterized yeast genes that fit the criteria of having an inviable phenotype annotation, have a human homolog, and the human homolog has a disease phenotype annotation defined by OMIM. The three genes are FMP27/YLR454W, FSF1/YOR271C, and YJR141W. It is possible that fewer genes may qualify for this list if any of these three are reclassified to the Verified set (i.e., genes that have been experimentally shown to be expressed and to have a function in the cell). SGD is continually defining new templates as new data types are being made available via YeastMine. New templates are also added

Chapter 30

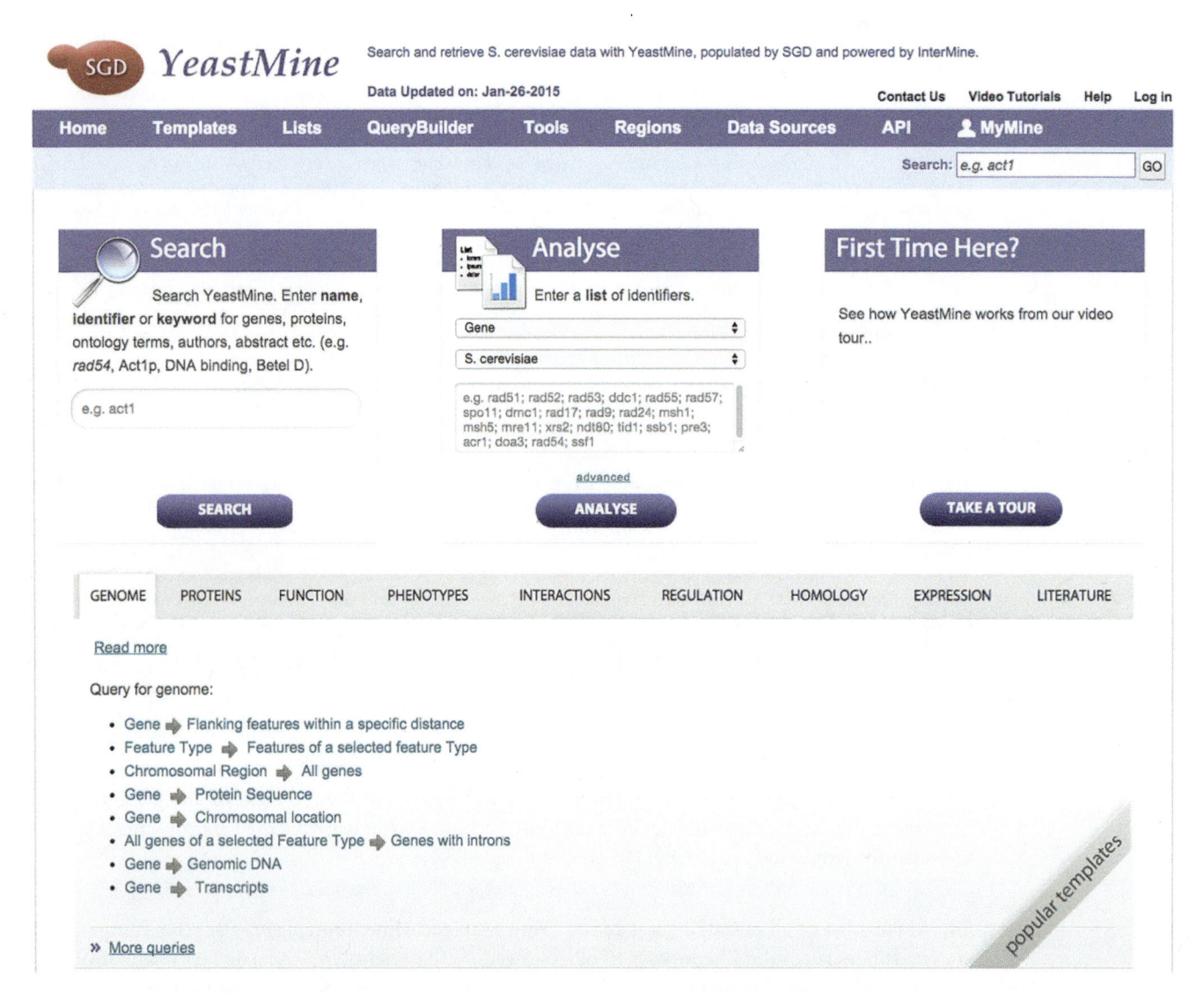

FIGURE 1. Home page for YeastMine. All information available from the SGD database and downloads site is available via a powerful search interface. Predefined templates are provided for many common searches.

at the request of users. In the near future more regulation, protein complex, protein modification, transcription, and cellular pathway details will be provided by YeastMine.

Exploring Microarray Data

The SGD provides access to a complete set of microarray expression data sets via a tool called SPELL (Hibbs et al. 2007). This tool allows a gene or set of genes to be used as a query to more than 250 data sets (studies) that include more than 400 experimental conditions (arrays). There are several links from the SGD Locus pages to the expression array tool.

7. In SGD, search for *AIM17* and open the *AIM17* Locus Overview page. Click on the Expression tab link at the top of the page. Then click on "SPELL" under the Expression Overview graph to go to the site spell.yeastgenome.org.

   This provides standard red/green visualization of the microarray expression ratios. The columns are experimental conditions, and the rows are genes. With a search for AIM17 the rows represent those genes that are most similar to AIM17 based on the Pearson correlation statistic comparing the expression ratios. The second row is TPS2 indicating it has the best-adjusted correlation score across all arrays presented. The rows below these two are genes in decreasing order that are less correlated with the observed expression of AIM17. This tool allows selection of a group of genes (query set) and then observation of the genes that are most similar to the query set.

8. Select the check box for *TPS2* (keeping the *AIM17* box checked) and then the "Update" icon.

   The resulting view changes to provide the correlation of AIM17 and TPS2 with the other genes. The most correlated genes are TPS1 and TSL1. The contribution of each experiment to the correlation is provided along the top of the graphic. Scrolling to the right illustrates that data sets from Orlando et al. (2008) are the top ranking experiments but they only contribute 1.4% of the correlation within the queried data sets.

9. Select the plus sign above the graphic labeled "Options for Filtering Results by Dataset Tags." This reveals topic tags (keywords) that can be queried for each dataset. A complete list of all tags and their definitions are found by clicking on "Dataset Tags." For this example with *AIM17* and *TPS2*, select only respiration and then Update.

   The top two correlated genes remain TPS1 and TSL1. Below the correlation graphic, the GO Term Enrichment results are shown. In this case, the set of genes shown are enriched for "trehalose metabolic process." This is not surprising as TPS1 and TPS2 are components of the alpha, alpha-trehalose-phosphate synthase complex.

   The YeastMine tool explored above (Steps 1–7) also has the numerical values for these expression experiments. You can use templates from the Expression tab to identify expression scores for genes, or list out all genes associated with an expression dataset.

ACKNOWLEDGMENTS

I am grateful to all the present and past staff of the *Saccharomyces* Genome Database project for their dedication to accuracy and service to life science educators and researchers. I also want to thank the yeast research community for their support and suggestions. This work was supported by the National Human Genome Research Institute (grant number U41 HG001315), and Funding for open access charge was provided by the National Institutes of Health. The content is solely the responsibility of the author and does not necessarily represent the official views of the National Human Genome Research Institute or the National Institutes of Health.

REFERENCES

Amberger J, Bocchini CA, Scott AF, Hamosh A. 2008. McKusick's Online Mendelian Inheritance in Man (OMIM). *Nucleic Acids Res* **37**: D793–D796.

Balakrishnan R, Park J, Karra K, Hitz BC, Binkley G, Hong EL, Sullivan J, Micklem G, Cherry JM. 2012. YeastMine—An integrated data warehouse for *Saccharomyces cerevisiae* data as a multipurpose tool-kit. *Database (Oxford)* doi:10.1093/database/bar062.

Fisk DG, Ball CA, Dolinski K, Engel SR, Hong EL, Issel-Tarver L, Schwartz K, Sethuraman A, Botstein D, Cherry JM. 2006. *Saccharomyces* Genome Database Project. *Saccharomyces cerevisiae* S288C genome annotation: A working hypothesis. *Yeast* **23**: 857–865.

Hibbs MA, Hess DC, Myers CL, Huttenhower C, Li K, Troyanskaya OG. 2007. Exploring the functional landscape of gene expression: Directed search of large microarray compendia. *Bioinformatics* **23**: 2692–2699.

Kalderimis A, Lyne R, Butano D, Contrino S, Lyne M, Heimbach J, Hu F, Smith R, Štěpán R, Sullivan J, et al. 2014. InterMine: Extensive web services for modern biology. *Nucleic Acids Res* 2014 **42**: W468–W472.

Orlando DA, Lin CY, Bernard A, Wang JY, Socolar JE, Iversen ES, Hartemink AJ, Haase SB. 2008. Global control of cell-cycle transcription by coupled CDK and network oscillators. *Nature* **453**: 944–947.

Protocol 3

The *Saccharomyces* Genome Database: Gene Product Annotation of Function, Process, and Component

J. Michael Cherry[1]

Department of Genetics, Stanford University School of Medicine, Stanford, California 94305-5120

An ontology is a highly structured form of controlled vocabulary. Each entry in the ontology is commonly called a term. These terms are used when talking about an annotation. However, each term has a definition that, like the definition of a word found within a dictionary, provides the complete usage and detailed explanation of the term. It is critical to consult a term's definition because the distinction between terms can be subtle. The use of ontologies in biology started as a way of unifying communication between scientific communities and to provide a standard dictionary for different topics, including molecular functions, biological processes, mutant phenotypes, chemical properties and structures. The creation of ontology terms and their definitions often requires debate to reach agreement but the result has been a unified descriptive language used to communicate knowledge. In addition to terms and definitions, ontologies require a relationship used to define the type of connection between terms. In an ontology, a term can have more than one parent term, the term above it in an ontology, as well as more than one child, the term below it in the ontology. Many ontologies are used to construct annotations in the *Saccharomyces* Genome Database (SGD), as in all modern biological databases; however, Gene Ontology (GO), a descriptive system used to categorize gene function, is the most extensively used ontology in SGD annotations. Examples included in this protocol illustrate the structure and features of this ontology.

MATERIALS

Equipment

Internet-connected computer with web browser

METHOD

The GO (Gene Ontology Consortium 2013) is the major ontology used to communicate functional characteristics of gene products. This protocol introduces tools available at SGD to use GO annotations. The statistical treatment of a collection of genes with GO annotations uses techniques typically called term enrichment. There are many different algorithms and tools that implement these methods of determining significantly shared annotations between a set of genes. However, there is little agreement on which metric determines the best enrichment tool. This is partly because of the number of components used to determine enrichment, the sets of annotations, version of the ontology, amount of filtering by evidence, and the background frequency used. Huang et al. (2009) provide a discussion on term enrichment methods that are used to determine which annotations are significant. Shah et al. (2013) provide background on the use of gene enrichment with specific case examples. SGD has chosen to integrate the GO TermFinder tool (Boyle et al. 2004) and it is available in the Analysis menu. GO Term Enrichment requires an input list of genes that have been

---

[1] Correspondence: cherry@stanford.edu

Copyright © Cold Spring Harbor Laboratory Press; all rights reserved
Cite this protocol as *Cold Spring Harb Protoc*; doi:10.1101/pdb.prot088914

GO annotated and considers all the genes' annotations to determine which GO terms are represented more than expected by chance.

At any step in this protocol support is available via the comprehensive Help documents maintained by SGD. These can be accessed via the help button at the top of each page or, for specific features, a small red button with a question mark in its center is provided and will link to the help pages specific for that feature.

The SGD is continually updated; therefore, specific items presented in this protocol may not appear on the SGD website exactly as described.

1. Open the URL, www.yeastgenome.org, in any modern web browser (e.g., Chrome, Firefox or Safari).

2. Enter "RTT103" in the Search box and press return to go directly to the *RTT103* Locus Summary page. Click on the Gene Ontology tab.

   Rtt103p is a protein involved in RNA polymerase II transcription termination and is also involved in the regulation of Ty1 transposition. Manual curation includes four annotations to GO biological process terms. Of the four annotations two are to the same term, "mRNA 3'-end processing." The difference between these two annotations is the Evidence code that was defined. For each manually created GO annotation the biocurator reviews the published work and determines the appropriate GO term that describes the experimental result. The creation of the annotation also includes identifying the type of evidence reported for the result (http://www.geneontology.org/page/guide-go-evidence-codes). The code IMP stands for Inferred from Mutant Phenotype and IGI stands for Inferred from Genetic Interaction. In Kim et al. (2004) two forms of results were available and thus both were used to create individual annotations. The IMP annotation is based on an investigation of a RTT103 knockout and that for IGI is based on a synthetic genetic array (SGA) assay where the rtt103 deletion strain was crossed with other deletion strains. The SGA assay showed that RTT103 makes a synthetic lethal with REF2, CTK1 and CTK2. Manual annotations are also presented for Molecular Function and Cellular Component with IDA and IPI (Inferred from Physical Interaction) evidence.

3. Scroll down the *RTT103* Gene Ontology page to view the High-throughput annotations section.

   This section includes results reported in published large-scale assays. In this example there are no HTP annotations.

4. Scroll down the *RTT103* Gene Ontology page to view the Computational predictions annotations section.

   These annotations are not reviewed by curators and are the result of computational analyses, such as protein motif searches, BLAST analysis, and rule systems from several database projects such as UniProt (http://www.uniprot.org).

5. Scroll to the bottom of the *RTT103* Gene Ontology page and view the Shared Biological Processes section.

   This type of visualization is on many of the SGD pages and is a useful alternative to tables of information. In this example you can see the Biological Process annotations (green boxes) annotated to RTT103 (yellow circle). In addition, other genes are included (dark gray circles) that share annotations with RTT103.

6. While not available on the RTT103 visualization, you can typically explore the network by moving the slider at the bottom to increase or decrease the number of shared annotation terms.

7. Manipulate the network by dragging the circles and squares.

   Changing the placement of the nodes can enhance the view.

8. Double click on a node. The circles go to genes and the squares go to GO Term pages. For example, find and click on "mRNA 3″-end processing" to read complete information about this term.

   The term's details include a network display of the parent and child terms. Further down all annotations to this GO term are provided as a table.

9. Select the "GO Term Finder" option from the Analyze pull-down in the purple bar at the top of the page to explore the annotations associated with a list of genes using the GO Term Finder. Enter the following four gene names in "Step 1: Query Set" box: PCF11, TFA1, TFA2, and RTT103. Do not use commas to separate the gene names, rather use space or return. In the "Step 2: Choose Ontology and Set Cutoff" box, select function and then click on Search to use the default settings.

Chapter 30

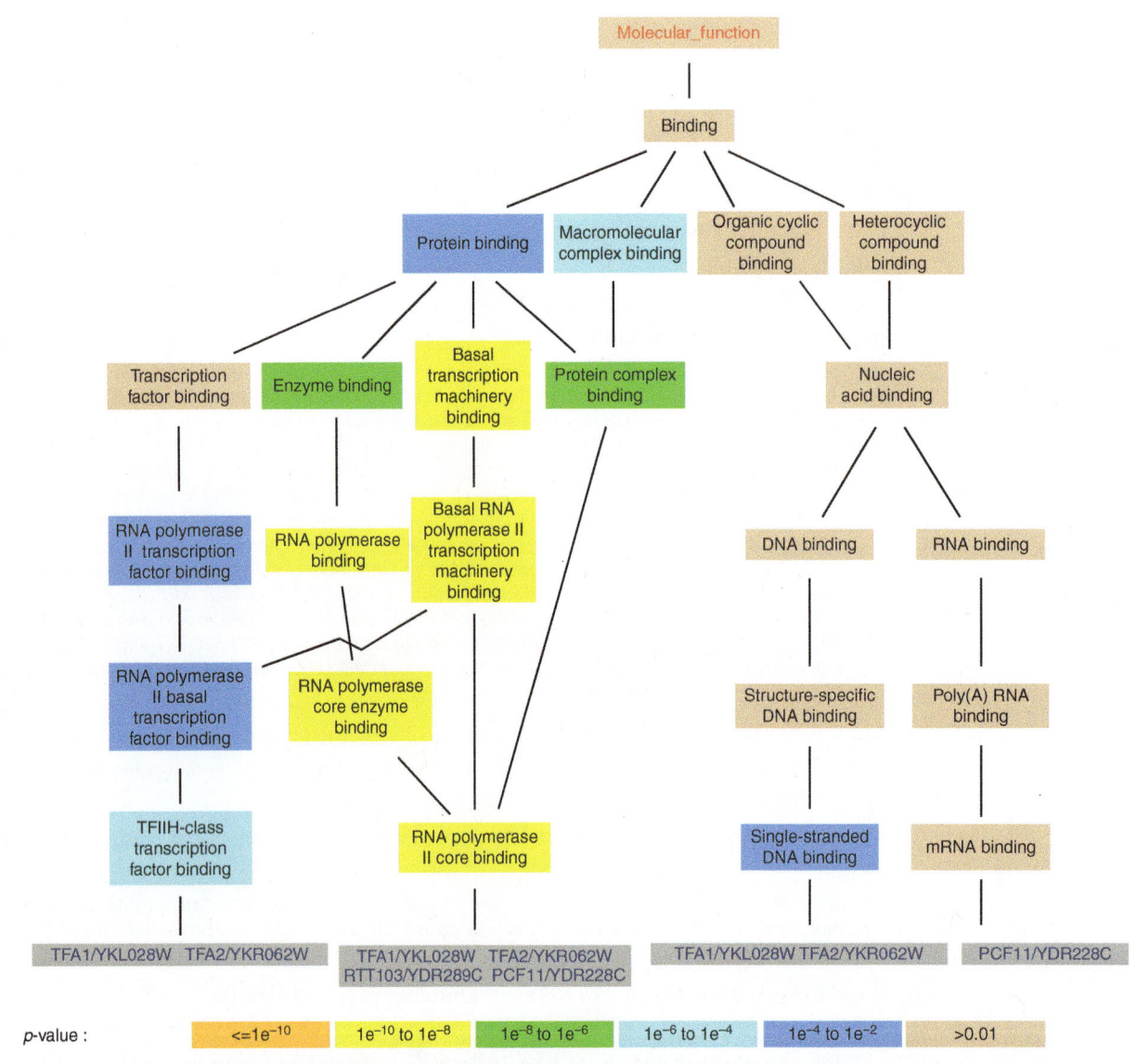

FIGURE 1. Screenshot (redrawn for clarity) of the Gene Ontology Term Finder output graphic. This is the result from performing term enrichment on the annotations of the four genes: PCF11, TFA1, TFA2, and RTT103. The graphic shows the organization of a portion of the Molecular Function graph. Each manually curated annotation for these four genes is shown. The term enrichment results are indicated by color-coded terms. In this graphic the most significant terms are shown in yellow, followed by green, cyan, blue and beige. The significance indicates the likelihood that this set of genes would be annotated to the term by chance.

The results may take a few minutes to return. The graphic (Fig. 1) shows the GO terms that describe the annotations for this set of genes, color-coded by significance. The significance indicates the likelihood that this set of genes would be annotated with this set of GO Terms by chance. The graphic also shows which GO terms have been used to annotate the genes. In this case, using the Molecular Function annotations, the shared terms with greater significance are in yellow such as "RNA polymerase II core binding"; however, parent terms are also highlighted. The table at the bottom of the page includes the p-values for the results.

ACKNOWLEDGMENTS

I am grateful to all the present and past staff of the *Saccharomyces* Genome Database project for their dedication to accuracy and service to life science educators and researchers. I also want to thank the yeast research community for their support and suggestions. This work was supported by the National

Human Genome Research Institute (grant number U41 HG001315), and Funding for open access charge was provided by the National Institutes of Health. The content is solely the responsibility of the author and does not necessarily represent the official views of the National Human Genome Research Institute or the National Institutes of Health.

REFERENCES

Boyle EI, Weng S, Gollub J, Jin H, Botstein D, Cherry JM, Sherlock G. 2004. GO::TermFinder—Open source software for accessing Gene Ontology information and finding significantly enriched Gene Ontology terms associated with a list of genes. *Bioinformatics* **20**: 3710–3715.

Gene Ontology Consortium. 2013. Gene Ontology annotations and resources. *Nucleic Acids Res* **41**: D530–D535.

Huang DW, Sherman BT, Lempicki RA. 2009. Bioinformatics enrichment tools: Paths toward the comprehensive functional analysis of large gene lists. *Nucleic Acids Res* **37**: 1–13.

Kim M, Krogan NJ, Vasiljeva L, Rando OJ, Nedea E, Greenblatt JF, Buratowski S. 2004. The yeast Rat1 exonuclease promotes transcription termination by RNA polymerase II. *Nature* **432**: 517–522.

Shah NH, Cole T, Musen MA. 2013. Chapter 9: Analyses using disease ontologies. *PLoS Comput Biol* **8**: e1002827. doi: 10.1371/journal.pcbi.1002827.

Protocol 4

The *Saccharomyces* Genome Database: Exploring Genome Features and Their Annotations

J. Michael Cherry[1]

Department of Genetics, Stanford University School of Medicine, Stanford, California 94305-5120

Genomic-scale assays result in data that provide information over the entire genome. Such base pair resolution data cannot be summarized easily except via a graphical viewer. A genome browser is a tool that displays genomic data and experimental results as horizontal tracks. Genome browsers allow searches for a chromosomal coordinate or a feature, such as a gene name, but they do not allow searches by function or upstream binding site. Entry into a genome browser requires that you identify the gene name or chromosomal coordinates for a region of interest. A track provides a representation for genomic results and is displayed as a row of data shown as line segments to indicate regions of the chromosome with a feature. Another type of track presents a graph or wiggle plot that indicates the processed signal intensity computed for a particular experiment or set of experiments. Wiggle plots are typical for genomic assays such as the various next-generation sequencing methods (e.g., chromatin immunoprecipitation [ChIP]-seq or RNA-seq), where it represents a peak of DNA binding, histone modification, or the mapping of an RNA sequence. Here we explore the browser that has been built into the *Saccharomyces* Genome Database (SGD).

MATERIALS

Equipment

Internet-connected computer with web browser

METHOD

At any step in this protocol, support is available via the comprehensive Help documents maintained by SGD. These can be accessed via the help button at the top of each page. The drop-down Help menu at the top of the gbrowse page leads you directly to extensive help and to video tutorials created by SGD for the use of gbrowse.

The SGD is continually updated; therefore, specific items presented in this protocol may not appear on the SGD website exactly as described.

1. Open the URL, www.yeastgenome.org, in any modern web browser (e.g., Firefox, Chrome, or Safari).

2. Enter "PIG1" in the Search box and press return to go directly to the *PIG1* Locus Summary page. In the Sequence section, click on "View in: GBrowse". This opens the graphical browser page (http://browse.yeastgenome.org).

[1]Correspondence: cherry@stanford.edu

Copyright © Cold Spring Harbor Laboratory Press; all rights reserved
Cite this protocol as *Cold Spring Harb Protoc*; doi:10.1101/pdb.prot088922

On the left side of the browser page, above the graphic is a search box labeled "Landmark or Region"; it should currently say "chrXII:689,083..691,029," which are the genomic coordinates for the PIG1 gene (Fig. 1). The yeast chromosomes are labeled with roman numerals, and thus PIG1 is on chromosome 12 from base 689,083 to 691,029. The PIG1 open-reading frame (ORF) is on the reverse strand, going right to left, but that is not apparent from the coordinates. The ORF name syntax informs you of this orientation. ORFs ending in W for Watson are on the left to right (forward) strand and those ending in C for Crick are on the other (reverse) strand. Above the red arrow labeled "PIG1, Verified, Putative targeting subunit for . . ." is the PIG1 ORF name, YLR273C, indicating PIG1 is on the reverse strand. In addition, the red arrow points to the left and indicates OFR orientation. More information about the naming conventions of yeast chromosomal features can be found at SGD, http://www.yeastgenome.org/help/community/nomenclature-conventions. The gbrowse view provides a convenient way to see which genes are near each other, including tRNAs and transposon long terminal repeats (LTRs). The feature downstream from PIG1, labeled tV(AAC)L, is the tRNA for valine (V) with the anticodon AAC; the L indicates its location on chromosome 12, L being the 12th letter of the alphabet. There are two LTRs associated with Ty1 downstream from PIG1. These are labeled YLRCdelta16 (delta element number 16 on chromosome 12, right of the centromere on the complement or Crick strand) and YLRWdelta17 (delta element number 17 on chromosome 12, right of the centromere on the Watson strand).

3. Locate the scrolling and zooming buttons at the top of the page and navigate along the chromosome.

   The ">" moves the view a little to the right and the ">>" moves the view a lot to the right. The yellow "+" zooms in and the "−" zooms out. A drop down is provided to quick zoom in or out to a specific window size for the chromosome. The flip button changes the orientation of the view.

4. There are two metered regions on the upper two chromosome tracks that are highlighted in light purple. Drag the solid purple region to scroll from side to side. Click and drag in the shaded region to define a different-sized region, or a new region. These movements only work in the metered region.

   There are several hundred tracks available for display, but it is best to only show a few at a time. By default the "All Annotated Sequence Features" track will be shown; however, gbrowse remembers the last set of tracks viewed.

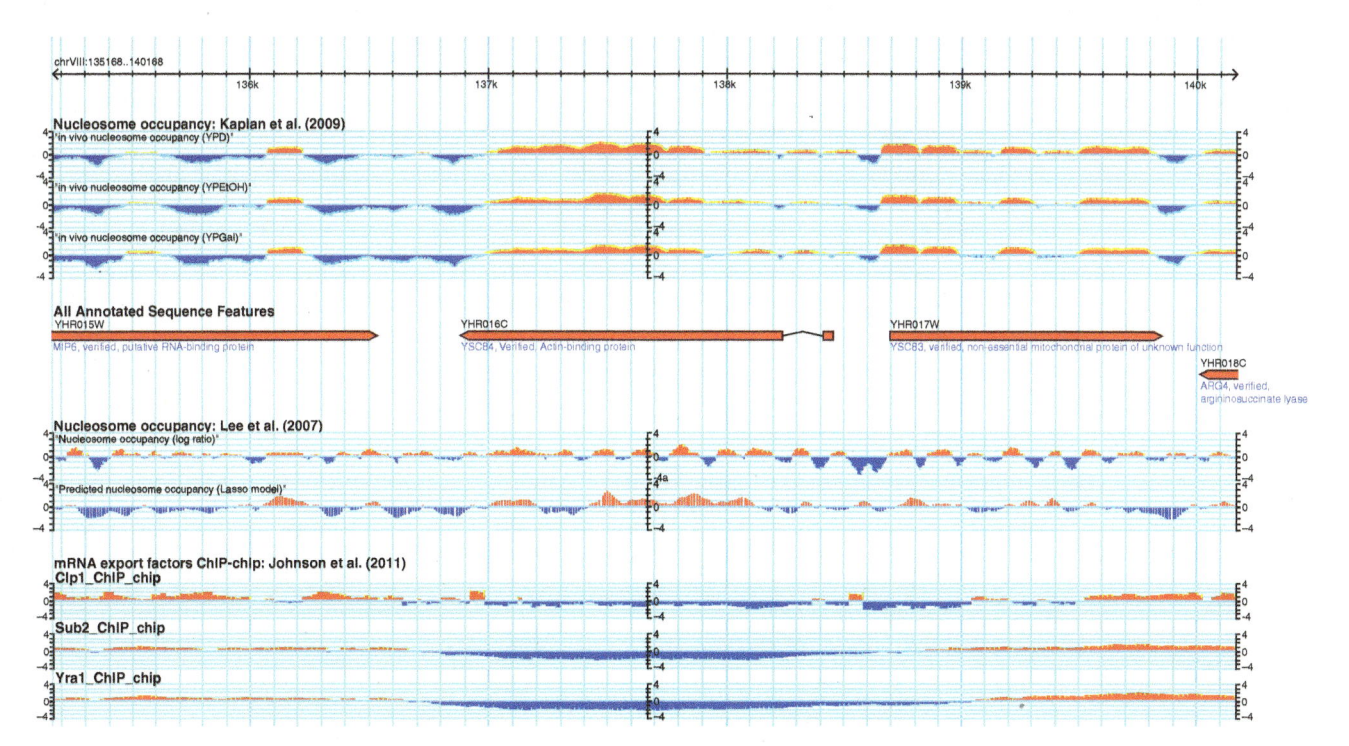

FIGURE 1. View of the genome around the *PIG1* locus. This display is created by the GBrowse software and shows horizontal tracks providing chromosomal feature annotations and results of genomic experiments. The top three tracks show the experimental results for the nucleosome occupancy from Kaplan et al. (2009). The first two tracks below the feature annotations provide nucleosome occupancy results from Lee et al. (2007), and at the bottom tracks show ChIP-chip results from Johnson et al. (2011) for Clp1, Sub2, and Yra1.

5. Scroll to the bottom of the page and click on the "Select Tracks" button. A large list of available tracks from published genomic studies will appear.

6. Click on the "All on" button next to Histone modifications, then the "Back to Browser" button located in the upper left.

    You will see the results from five publications. All of these publications have multiple tracks that correspond to multiple data sets.

7. Mouse over the little buttons to the right of the star to learn about their functions for modifying a track. To move a track up or down in the view click on the publication authors name and drag the group of tracks.

8. Return to the "Select Tracks" page. Click on the "All off" button next to Histone modification. To turn on tracks from a specific study mouse over the citation and click when it turns dark. A check mark will appear to its left to indicate that track is now on.

9. Select a subset of tracks from a study by clicking on the blue text that includes "subtracks selected."

10. Click on a star to denote the track as a favorite (clicking on a star does not turn the track on). View all favorites by selecting "Show Favorites Only" at the top of the "Select Tracks" page.

11. Return to the Browser page and click on an ORF. This takes you to the Locus Summary page for that gene.

12. Click on a track. This takes you to the paper for that study.

13. Click on Preferences. This allows you to control the image created by gbrowse.

    This can be useful for creating a screenshot for a presentation.

14. Pull down the File menu at the top of the gbrowse page, just below the horizontal SGD option bar.

    The File menu is used to export the image in PNG or SVG graphic format, or as GFF (genome feature format) or FASTA (sequence) files.

ACKNOWLEDGMENTS

I am grateful to all the present and past staff of the *Saccharomyces* Genome Database project for their dedication to accuracy and service to life science educators and researchers. I also want to thank the yeast research community for their support and suggestions. This work was supported by the National Human Genome Research Institute (grant number U41 HG001315), and Funding for open access charge was provided by the National Institutes of Health. The content is solely the responsibility of the author and does not necessarily represent the official views of the National Human Genome Research Institute or the National Institutes of Health.

REFERENCES

Kaplan N, Moore IK, Fondufe-Mittendorf Y, Gossett AJ, Tillo D, Field Y, LeProust EM, Hughes TR, Lieb JD, Widom J, et al. 2009. The DNA-encoded nucleosome organization of a eukaryotic genome. *Nature* 458: 362–366.

Lee W, Tillo D, Bray N, Morse RH, Davis RW, Hughes TR, Nislow C. 2007. A high-resolution atlas of nucleosome occupancy in yeast. *Nat Genet* 39: 1235–1244.

Johnson SA, Kim H, Erickson B, Bentley DL. 2011. The export factor Yra1 modulates mRNA 3′ end processing. *Nat Struct Mol Biol* 18: 1164–1171.

CHAPTER 31

BioGRID: A Resource for Studying Biological Interactions in Yeast

Rose Oughtred,[1] Andrew Chatr-aryamontri,[2] Bobby-Joe Breitkreutz,[3] Christie S. Chang,[1] Jennifer M. Rust,[1] Chandra L. Theesfeld,[1] Sven Heinicke,[1] Ashton Breitkreutz,[3] Daici Chen,[2] Jodi Hirschman,[1] Nadine Kolas,[3] Michael S. Livstone,[1] Julie Nixon,[4] Lara O'Donnell,[3] Lindsay Ramage,[4] Andrew Winter,[4] Teresa Reguly,[3] Adnane Sellam,[2] Chris Stark,[3] Lorrie Boucher,[3] Kara Dolinski,[1,5] and Mike Tyers[2,3,4,5]

[1]Lewis-Sigler Institute for Integrative Genomics, Princeton University, Princeton, New Jersey 08544; [2]Institute for Research in Immunology and Cancer, Université de Montréal, Montréal, Québec H3C 3J7, Canada; [3]Samuel Lunenfeld Research Institute, Mount Sinai Hospital, Toronto, Ontario M5G 1X5, Canada; [4]School of Biological Sciences, University of Edinburgh, Edinburgh, EH9 3JR, United Kingdom

The Biological General Repository for Interaction Datasets (BioGRID) is a freely available public database that provides the biological and biomedical research communities with curated protein and genetic interaction data. Structured experimental evidence codes, an intuitive search interface, and visualization tools enable the discovery of individual gene, protein, or biological network function. BioGRID houses interaction data for the major model organism species—including yeast, nematode, fly, zebrafish, mouse, and human—with particular emphasis on the budding yeast *Saccharomyces cerevisiae* and the fission yeast *Schizosaccharomyces pombe* as pioneer eukaryotic models for network biology. BioGRID has achieved comprehensive curation coverage of the entire literature for these two major yeast models, which is actively maintained through monthly curation updates. As of September 2015, BioGRID houses approximately 335,400 biological interactions for budding yeast and approximately 67,800 interactions for fission yeast. BioGRID also supports an integrated post-translational modification (PTM) viewer that incorporates more than 20,100 yeast phosphorylation sites curated through its sister database, the PhosphoGRID.

BACKGROUND

The Biological General Repository for Interaction Datasets (BioGRID; http://www.thebiogrid.org) is an open source database that curates and disseminates collections of protein and genetic interactions from major model organism species from yeast to human (Stark et al. 2006; Chatr-Aryamontri et al. 2013). The BioGRID was originally developed as a budding yeast–specific database to house and visualize protein interaction data from high-throughput proteomic studies (Ho et al. 2002; Breitkreutz et al. 2003a; Stark et al. 2006). Subsequently, comprehensive curation of protein and genetic interactions from the entire budding yeast literature was undertaken to compare emerging high-throughput interaction data to the extensive body of interaction data reported in thousands of focused studies (Reguly et al. 2006). Importantly, the evidence for each interaction in BioGRID is recorded as a structured evidence code derived from the primary experimental data. These evidence codes are concordant and interoperable with high-level stratification of the detailed Proteomics Standards Initiative-Molecular Interaction (PSI-MI) ontology (Hermjakob et al. 2004a; Kerrien et al. 2007). All curated data within BioGRID is fully archived as monthly releases and all records are date-stamped

[5]Correspondence: kara@genomics.princeton.edu; md.tyers@umontreal.ca

Copyright © Cold Spring Harbor Laboratory Press; all rights reserved
Cite this introduction as *Cold Spring Harb Protoc*; doi:10.1101/pdb.top080754

and mapped to individual curators to ensure data integrity. Curation efforts at BioGRID have since been expanded to capture biological interaction data from each of the major model organism species. These data sets serve as a readily accessible resource for interrogation of biological interactions, discovery of gene function, and computational analysis of interaction networks (Dolinski et al. 2013).

BioGRID CURATION STATISTICS

The September 2015 release of BioGRID (version 3.4.128) contains more than 812,000 interactions curated from both high-throughput data sets and low-throughput focused studies found in the literature. These interactions have been distilled from more than 45,000 publications covering 57 different organisms, including the budding yeast *Saccharomyces cerevisiae*, the fission yeast *Schizosaccharomyces pombe*, the yeast *Candida albicans* SC5314, the nematode *Caenorhabditis elegans*, the fruit fly *Drosophila melanogaster*, the mouse *Mus musculus*, the plant *Arabidopsis thaliana*, and *Homo sapiens* (Stark et al. 2011; Chatr-Aryamontri et al. 2013). BioGRID interaction data sets are shared with the respective model organism databases (Cherry et al. 2012; Inglis et al. 2012; Lamesch et al. 2012; Wood et al. 2012; Yook et al. 2012; Marygold et al. 2013), with other interaction databases (Luc and Tempst 2004; Razick et al. 2008; Chautard et al. 2009; Matthews et al. 2009; Cerami et al. 2011; Franceschini et al. 2013), and with metadatabases (Benson et al. 2004; Matthews et al. 2009). Complete coverage of the entire literature for *S. cerevisiae* and *S. pombe*, as well as for the model plant *A. thaliana*, is maintained through continuous monthly updates. As of the latest BioGRID release, approximately 335,400 (225,700 unique) interactions have been curated for *S. cerevisiae* genes/proteins from more than 13,000 publications, and approximately 67,800 (55,400 unique) interactions have been curated for *S. pombe* genes from nearly 2100 publications (Table 1). Of these interactions, 63% of budding yeast and 85% of fission yeast interactions derive from genetic experiments, and for both organisms, some 80% of interactions are derived from high-throughput data sets. Recently, more than 400 interactions have also been curated from nearly 40 papers for the pathogenic yeast model species, *C. albicans*. All yeast genetic interactions include associated phenotypes curated using the structured Ascomycete Phenotype Ontology (APO) developed by the *Saccharomyces* Genome Database (SGD; Engel et al. 2010). In addition, more than 20,100 phosphorylation sites mapped onto nearly 3200 budding yeast proteins are documented in a sister database called PhosphoGRID (http://www.phosphogrid.org; Sadowski et al. 2013) and are available through a new posttranslational modification (PTM) viewer integrated within BioGRID.

USING THE BioGRID DATABASE

The research community can access these extensive interaction data sets using the BioGRID web interface (http://www.thebiogrid.org), which provides users with a tabular interaction summary for each query gene or protein, as well as a link to the abstract for each curated publication and associated PubMed identifier. Details including interaction type, evidence code, and data source are provided in a condensed format on each summary page. Interaction data may also be viewed using an interactive network visualization tool embedded within BioGRID, downloaded in bulk for local analysis, or captured through stand-alone visualization applications, such as Osprey and Cytoscape (Breitkreutz et al. 2003b; Shannon et al. 2003; Cline et al. 2007).

For detailed instructions on how to use the BioGRID website to query genetic or protein interactions for any gene of interest, how to visualize the associated interactions using an embedded interactive network viewer, and how to download data files for either selected interactions or the entire BioGRID interaction data set, see Protocol 1: Use of the BioGRID Database for Analysis of Yeast Protein and Genetic Interactions (Oughtred et al. 2015).

TABLE 1. Summary of current yeast interactions curated in BioGRID

	Total interactions	Curated publications	Protein interactions	Genetic interactions	Unique phenotypes
Saccharomyces cerevisiae	335,427 (225,753)	13,060	125,534 (81,115)	209,893 (151,379)	602
HTP	258,619 (196,923)	345	80,657 (66,045)	177,962 (133,341)	57
LTP	81,739 (44,028)	12,951	45,011 (21,300)	36,728 (26,165)	600
Schizosaccharomyces pombe	67,795 (55,428)	2084	10,062 (7,261)	57,733 (48,922)	318
HTP	55,686 (48,510)	49	3947 (3,759)	51,739 (44,782)	13
LTP	12,144 (7,552)	2074	6147 (3,768)	5997 (4399)	317
Candida albicans	414 (375)	43	146 (113)	268 (263)	11
HTP	254 (254)	2	0 (0)	254 (254)	1
LTP	160 (121)	41	146 (113)	14 (9)	10

BioGRID yeast curation statistics as of September 2015 (BioGRID version 3.4.128). To date, more than 403,600 total interactions have been curated from more than 15,100 publications. These cover 6504 *S. cerevisiae* proteins, 3964 *S. pombe* proteins, and 372 *C. albicans SC5314* proteins. The number of unique interactions is given in parentheses, and the number of interactions derived from high-throughput or low-throughput studies is given for each category. The number of unique phenotypes refers to the number of nonredundant phenotypes curated for genetic interactions using the Ascomycete Phenotype Ontology (APO).

HTP, high-throughput; LTP, low-throughput.

SCOPE OF THE BioGRID

The BioGRID will continue to expand the curation of protein and genetic interactions from the biomedical literature, as well as associated attributes such as PTMs, protein variants, phenotypes, and chemical or drug interactions. In addition to BioGRID, the following members of the International Molecular Exchange (IMEx) consortium (http://www.imexconsortium.org/) (Orchard et al. 2012) also actively curate and freely disseminate yeast protein interaction data:

- DIP (Database of Interacting Proteins: http://dip.doe-mbi.ucla.edu/) (Xenarios et al. 2002; Salwinski et al. 2004) and
- IntAct (IntAct Molecular Interaction Database: http://www.ebi.ac.uk/intact/) (Hermjakob et al. 2004b; Kerrien et al. 2012).

The MPact yeast protein interaction database also supports the PSI-MI standard (Guldener et al. 2006) (http://mips.helmholtz-muenchen.de/genre/proj/mpact) and provides protein interaction data contained in the MIPS Comprehensive Yeast Genome Database (CYGD) (Guldener et al. 2005).

BioGRID is currently unique among interaction databases in that it is the only open access resource that curates both genetic and protein interactions for different yeast species from published high- and low-throughput studies. In particular, BioGRID provides comprehensive interaction curation coverage for the budding and fission yeast literature, which is updated via continuous monthly increments that are fully archived (Reguly et al. 2006). BioGRID is also the only current resource that fully captures phenotypes using the APO for all curated yeast genetic interactions.

The ongoing development of the PhosphoGRID database allows full integration of documented yeast phosphorylation sites into the BioGRID PTM viewer, and thereby provides an integrated display of all curated phosphosites and their corresponding interactions in a single resource. In conjunction with WormBase, BioGRID has also developed a new Genetic Interactions (GI) ontology that will allow consistent curation of genetic interactions across various model organisms and interaction databases, enabling better comparisons of phenotypic information across different species (CA Grove, R Oughtred, A Winter, et al., unpubl.). The GI ontology has recently been incorporated into the PSI-MI ontology, with the intention that the GI ontology will be adopted as a community standard.

The BioGRID will continue to facilitate the use of yeast model systems for understanding the role of biological interaction networks in human biology and disease. Protein and genetic interactions, and entire interaction networks, are often physically and functionally conserved (Bandyopadhyay et al. 2006), such that the detailed interrogation of interactions in genetically tractable systems can prove extremely informative in biomedical contexts. To this end, efforts are now underway at BioGRID for parallel curation of yeast and human interactions that are implicated in human disease. These focused curation drives may be either biological process-centric, such as curation of the ubiquitin–proteasome system (UPS) that controls the stability, localization, and activity of most of the proteome, or disease-centric, such as curation of interaction networks implicated in HIV or other infectious diseases, neurobiological disorders, and metabolic enzymes, all of which are currently in progress. New features planned for BioGRID include the incorporation of ubiquitination sites, which are comprehensively curated as part of the UPS project, into the integrated PTM display. Many expansive interaction networks implicated in prevalent human diseases will be the focus of future curation drives across multiple model organism species, from yeast to human. The integration of these network data sets with other data types, including expression data, quantitative phenotype data, and high-resolution sequence data, should help to enable predictive medicine and future drug discovery efforts.

ACKNOWLEDGMENTS

The authors thank Chris Grove and Paul Sternberg at WormBase for ongoing collaborative development of the Genetic Interaction Ontology. This work was supported by National Institutes of Health (NIH) grants R01OD010929 and R24OD011194 to M.T. and K.D., the Biotechnology and Biological Sciences Research Council (grant number BB/F010486/1 to M.T.), the Canadian Institutes of Health Research (grant number FRN 82940 to M.T.), and a Genome Québec International Recruitment Award and a Canada Research Chair in Systems and Synthetic Biology to M.T.

REFERENCES

Bandyopadhyay S, Sharan R, Ideker T. 2006. Systematic identification of functional orthologs based on protein network comparison. *Genome Res* 16: 428–435.

Benson DA, Karsch-Mizrachi I, Lipman DJ, Ostell J, Wheeler DL. 2004. GenBank: Update. *Nucleic Acids Res* 32: D23–D26.

Breitkreutz BJ, Stark C, Tyers M. 2003a. The GRID: The General Repository for Interaction Datasets. *Genome Biol* 4: R23.

Breitkreutz BJ, Stark C, Tyers M. 2003b. Osprey: A network visualization system. *Genome Biol* 4: R22.

Cerami EG, Gross BE, Demir E, Rodchenkov I, Babur O, Anwar N, Schultz N, Bader GD, Sander C. 2011. Pathway Commons, a web resource for biological pathway data. *Nucleic Acids Res* 39: D685–D690.

Chatr-Aryamontri A, Breitkreutz BJ, Heinicke S, Boucher L, Winter A, Stark C, Nixon J, Ramage L, Kolas N, O'Donnell L, et al. 2013. The BioGRID interaction database: 2013 update. *Nucleic Acids Res* 41: D816–D823.

Chautard E, Ballut L, Thierry-Mieg N, Ricard-Blum S. 2009. MatrixDB, a database focused on extracellular protein–protein and protein–carbohydrate interactions. *Bioinformatics* 25: 690–691.

Cherry JM, Hong EL, Amundsen C, Balakrishnan R, Binkley G, Chan ET, Christie KR, Costanzo MC, Dwight SS, Engel SR, et al. 2012. Saccharomyces Genome Database: The genomics resource of budding yeast. *Nucleic Acids Res* 40: D700–D705.

Cline MS, Smoot M, Cerami E, Kuchinsky A, Landys N, Workman C, Christmas R, Avila-Campilo I, Creech M, Gross B, et al. 2007. Integration of biological networks and gene expression data using Cytoscape. *Nat Protoc* 2: 2366–2382.

Dolinski K, Chatr-Aryamontri A, Tyers M. 2013. Systematic curation of protein and genetic interaction data for computable biology. *BMC Biol* 11: 43.

Engel SR, Balakrishnan R, Binkley G, Christie KR, Costanzo MC, Dwight SS, Fisk DG, Hirschman JE, Hitz BC, Hong EL, et al. 2010. Saccharomyces Genome Database provides mutant phenotype data. *Nucleic Acids Res* 38: D433–D436.

Franceschini A, Szklarczyk D, Frankild S, Kuhn M, Simonovic M, Roth A, Lin J, Minguez P, Bork P, von Mering C, et al. 2013. STRING v9.1: Protein–protein interaction networks, with increased coverage and integration. *Nucleic Acids Res* 41: D808–D815.

Guldener U, Munsterkotter M, Kastenmuller G, Strack N, van Helden J, Lemer C, Richelles J, Wodak SJ, Garcia-Martinez J, Perez-Ortin JE, et al. 2005. CYGD: The Comprehensive Yeast Genome Database. *Nucleic Acids Res* 33: D364–D368.

Guldener U, Munsterkotter M, Oesterheld M, Pagel P, Ruepp A, Mewes HW, Stumpflen V. 2006. MPact: The MIPS protein interaction resource on yeast. *Nucleic Acids Res* 34: D436–D441.

Hermjakob H, Montecchi-Palazzi L, Bader G, Wojcik J, Salwinski L, Ceol A, Moore S, Orchard S, Sarkans U, von Mering C, et al. 2004a. The HUPO PSI's molecular interaction format—A community standard for the representation of protein interaction data. *Nat Biotechnol* 22: 177–183.

Hermjakob H, Montecchi-Palazzi L, Lewington C, Mudali S, Kerrien S, Orchard S, Vingron M, Roechert B, Roepstorff P, Valencia A, et al. 2004b. IntAct: An open source molecular interaction database. *Nucleic Acids Res* 32: D452–D455.

Ho Y, Gruhler A, Heilbut A, Bader GD, Moore L, Adams SL, Millar A, Taylor P, Bennett K, Boutilier K, et al. 2002. Systematic identification of protein complexes in Saccharomyces cerevisiae by mass spectrometry. *Nature* 415: 180–183.

Inglis DO, Arnaud MB, Binkley J, Shah P, Skrzypek MS, Wymore F, Binkley G, Miyasato SR, Simison M, Sherlock G. 2012. The Candida genome database incorporates multiple Candida species: Multispecies search and analysis tools with curated gene and protein information for Candida albicans and Candida glabrata. *Nucleic Acids Res* 40: D667–D674.

Kerrien S, Orchard S, Montecchi-Palazzi L, Aranda B, Quinn AF, Vinod N, Bader GD, Xenarios I, Wojcik J, Sherman D, et al. 2007. Broadening the horizon—Level 2.5 of the HUPO-PSI format for molecular interactions. *BMC Biol* 5: 44.

Kerrien S, Aranda B, Breuza L, Bridge A, Broackes-Carter F, Chen C, Duesbury M, Dumousseau M, Feuermann M, Hinz U, et al. 2012. The IntAct molecular interaction database in 2012. *Nucleic Acids Res* 40: D841–D846.

Lamesch P, Berardini TZ, Li D, Swarbreck D, Wilks C, Sasidharan R, Muller R, Dreher K, Alexander DL, Garcia-Hernandez M, et al. 2012. The Arabidopsis Information Resource (TAIR): Improved gene annotation and new tools. *Nucleic Acids Res* 40: D1202–D1210.

Luc PV, Tempst P. 2004. PINdb: A database of nuclear protein complexes from human and yeast. *Bioinformatics* 20: 1413–1415.

Marygold SJ, Leyland PC, Seal RL, Goodman JL, Thurmond J, Strelets VB, Wilson RJ, FlyBase c. 2013. FlyBase: Improvements to the bibliography. *Nucleic Acids Res* 41: D751–D757.

Matthews L, Gopinath G, Gillespie M, Caudy M, Croft D, de Bono B, Garapati P, Hemish J, Hermjakob H, Jassal B, et al. 2009. Reactome knowledgebase of human biological pathways and processes. *Nucleic Acids Res* 37: D619–D622.

Orchard S, Kerrien S, Abbani S, Aranda B, Bhate J, Bidwell S, Bridge A, Briganti L, Brinkman FS, Cesareni G, et al. 2012. Protein interaction data curation: The International Molecular Exchange (IMEx) consortium. *Nat Methods* **9:** 345–350.

Oughtred R, Chatr-aryamontri A, Breitkreutz B-J, Chang CS, Rust JM, Theesfeld CL, Heinicke S, Breitkreutz A, Chen D, Hirschman J, et al. 2015. Use of the BioGRID database for analysis of yeast protein and genetic interactions. *Cold Spring Harb Protoc* doi: 10.1101/pdb.prot088880.

Razick S, Magklaras G, Donaldson IM. 2008. iRefIndex: A consolidated protein interaction database with provenance. *BMC Bioinformatics* **9:** 405.

Reguly T, Breitkreutz A, Boucher L, Breitkreutz BJ, Hon GC, Myers CL, Parsons A, Friesen H, Oughtred R, Tong A, et al. 2006. Comprehensive curation and analysis of global interaction networks in *Saccharomyces cerevisiae*. *J Biol* **5:** 11.

Sadowski I, Breitkreutz BJ, Stark C, Su TC, Dahabieh M, Raithatha S, Bernhard W, Oughtred R, Dolinski K, Barreto K, et al. 2013. The PhosphoGRID *Saccharomyces cerevisiae* protein phosphorylation site database: Version 2.0 update. *Database (Oxford)* doi: 10.1093/database/bat026.

Salwinski L, Miller CS, Smith AJ, Pettit FK, Bowie JU, Eisenberg D. 2004. The database of interacting proteins: 2004 update. *Nucleic Acids Res* **32:** D449–D451.

Shannon P, Markiel A, Ozier O, Baliga NS, Wang JT, Ramage D, Amin N, Schwikowski B, Ideker T. 2003. Cytoscape: A software environment for integrated models of biomolecular interaction networks. *Genome Res* **13:** 2498–2504.

Stark C, Breitkreutz BJ, Reguly T, Boucher L, Breitkreutz A, Tyers M. 2006. BioGRID: A general repository for interaction datasets. *Nucleic Acids Res* **34:** D535–D539.

Stark C, Breitkreutz BJ, Chatr-Aryamontri A, Boucher L, Oughtred R, Livstone MS, Nixon J, Van Auken K, Wang X, Shi X, et al. 2011. The BioGRID interaction database: 2011 update. *Nucleic Acids Res* **39:** D698–D704.

Wood V, Harris MA, McDowall MD, Rutherford K, Vaughan BW, Staines DM, Aslett M, Lock A, Bahler J, Kersey PJ, et al. 2012. PomBase: A comprehensive online resource for fission yeast. *Nucleic Acids Res* **40:** D695–D699.

Xenarios I, Salwinski L, Duan XJ, Higney P, Kim SM, Eisenberg D. 2002. DIP, the Database of Interacting Proteins: A research tool for studying cellular networks of protein interactions. *Nucleic Acids Res* **30:** 303–305.

Yook K, Harris TW, Bieri T, Cabunoc A, Chan J, Chen WJ, Davis P, de la Cruz N, Duong A, Fang R, et al. 2012. WormBase 2012: More genomes, more data, new website. *Nucleic Acids Res* **40:** D735–D741.

Protocol 1

Use of the BioGRID Database for Analysis of Yeast Protein and Genetic Interactions

Rose Oughtred,[1] Andrew Chatr-aryamontri,[2] Bobby-Joe Breitkreutz,[3] Christie S. Chang,[1] Jennifer M. Rust,[1] Chandra L. Theesfeld,[1] Sven Heinicke,[1] Ashton Breitkreutz,[3] Daici Chen,[2] Jodi Hirschman,[1] Nadine Kolas,[3] Michael S. Livstone,[1] Julie Nixon,[4] Lara O'Donnell,[3] Lindsay Ramage,[4] Andrew Winter,[4] Teresa Reguly,[3] Adnane Sellam,[2] Chris Stark,[3] Lorrie Boucher,[3] Kara Dolinski,[1,5] and Mike Tyers[2,3,4,5]

[1]Lewis-Sigler Institute for Integrative Genomics, Princeton University, Princeton, New Jersey 08544; [2]Institute for Research in Immunology and Cancer, Université de Montréal, Montréal, Québec H3C 3J7, Canada; [3]Samuel Lunenfeld Research Institute, Mount Sinai Hospital, Toronto, Ontario M5G 1X5, Canada; [4]School of Biological Sciences, University of Edinburgh, Edinburgh, EH9 3JR, United Kingdom

The BioGRID database is an extensive repository of curated genetic and protein interactions for the budding yeast *Saccharomyces cerevisiae*, the fission yeast *Schizosaccharomyces pombe*, and the yeast *Candida albicans SC5314*, as well as for several other model organisms and humans. This protocol describes how to use the BioGRID website to query genetic or protein interactions for any gene of interest, how to visualize the associated interactions using an embedded interactive network viewer, and how to download data files for either selected interactions or the entire BioGRID interaction data set.

MATERIALS

Equipment

Computer (Internet-connected with web browser)

BioGRID is compatible with all modern web browsers including Firefox, Chrome, Safari, and Internet Explorer, but is not compatible with Internet Explorer 6 and earlier versions.

METHOD

BioGRID (Chatr-Aryamontri et al. 2013) is a member of the International Molecular Exchange (IMEx) consortium (http://www.imexconsortium.org/) (Orchard et al. 2012), which is a network of public databases that provide curated protein interaction data in a standardized Proteomics Standards Initiative Molecular Interaction (PSI-MI) format (http://www.psidev.info/). BioGRID is also a source of protein and genetic interaction data for the following model organism databases: Saccharomyces Genome Database (SGD) (http://www.yeastgenome.org) (Cherry et al. 2012), FlyBase (http://flybase.org/) (Marygold et al. 2013), PomBase (http://www.pombase.org/) (Wood et al. 2012), TAIR (http://www.arabidopsis.org/) (Lamesch et al. 2012), and WormBase (http://www.wormbase.org) (Yook et al. 2012).

[5]Correspondence: md.tyers@umontreal.ca; kara@genomics.princeton.edu

Copyright © Cold Spring Harbor Laboratory Press; all rights reserved
Cite this protocol as *Cold Spring Harb Protoc*; doi:10.1101/pdb.prot088880

Search Interactions for a Gene or Protein of Interest

This series of steps describes how to search BioGRID for interactions that involve a gene or protein of interest and to view additional details about interactions and associated posttranslational modifications.

1. Use any modern web browser, except Internet Explorer 6, to open the following URL: http://www.thebiogrid.org.

2. Locate the BioGRID search form in the upper right corner (Fig. 1), and search "By gene" or "By publication." For example, to search for the gene *SWE1*, type "SWE1" into the search bar, select "*Saccharomyces cerevisiae*" from the organism pull-down menu, and then press the "SUBMIT GENE SEARCH" button.

   The two search options ("By gene" and "By publication") are available as vertical tabs on the right-hand side of the search form (Fig. 1). When searching for a specific protein or gene, the search can be restricted to an organism of interest, or performed on all 57 organisms simultaneously. Any of the following common types of search terms will be recognized by the gene search field:

   - *Gene symbols (e.g., SWE1), gene accession numbers from NCBI (e.g., 853252), or RefSeq (e.g., NM_001181620);*
   - *Protein identifiers from UniProtKB (e.g., P32944) or NCBI RefSeq (e.g., NP_012348); and*
   - *Systematic identifiers from the corresponding model organism database (e.g., YJL187C from the SGD).*

   When searching by publication, it is possible to search for a PubMed identifier or by keywords found in the title or abstract. Additional instructions are provided in the "Advanced Search" and "Search Tips" links (Fig. 1).

3. View the search results (i.e., the Result Summary page [Fig. 2A], which provides a summary of curated interactions for the gene(s) or protein(s) of interest).

FIGURE 1. Search in BioGRID for interactions of a gene or protein of interest. The BioGRID home page is shown with available search options in the *upper right* corner (arrow). In the *top* menu, links are also provided to the help document, online tools, BioGRID statistics, and download options. In this gene search example, "SWE1" is entered as the search term and "*Saccharomyces cerevisiae*" is selected as the organism.

i. Locate at the top left of the page a brief description of the gene/protein; links to relevant external resources such as SGD, Entrez Gene, and UniProtKB; and an option to download associated interactions.

ii. Locate the curated interaction statistics for the gene or protein of interest in the upper right "Stats & Options" panel.

The total number and kind of interactions are listed and are color-coded throughout the page in green for genetic interactions and in yellow for protein interactions. The interactions may also be filtered via the "Search Filters" option according to whether they are derived from low- versus high-throughput experimental studies. A link to a graphical viewer, which allows for visualization of the gene- and/or protein-specific interaction network, is also available and is described in more detail in Steps 7 and 8.

iii. Further down on the page, locate the full interactors table that displays the list of unique interactors (see Fig. 2A, "Interactors View"). View the total number of independent evidence codes for genetic and protein interactions on the far right of the interactors table. Click on the "details" link to obtain information about each of the listed interaction partners.

This action reveals a detailed list of the different experimental evidence codes associated with each interaction, including the context of the interactor as either a bait or hit and whether the experimental evidence is from a low- or high-throughput study. The different types of interaction data curated by BioGRID are described in the "Experimental Evidence Codes" help document (http://wiki.thebiogrid.org/doku.php/experimental_systems). Pointing to specific icons in the right-hand "Notes" column reveals further applicable details. These include the following:

- *Tag Icon: phenotypes curated by BioGRID using the Ascomycete Phenotype Ontology (APO) developed by SGD (Engel et al. 2010);*
- *Globe icon: posttranslational modifications associated with the in vitro "Biochemical Activity" experiment type, most commonly phosphorylation or ubiquitination;*
- *Histogram icon: quantitative scores published in the original low-throughput study or high-throughput data set, when available; and*
- *Tablet icon: additional notes provided by BioGRID curators to explain experimental details or significance scores.*

4. Use the "Switch View" bar at the top of the full interactors table to navigate from the "Interactors" view to an "Interactions Table" (as shown for *SWE1* in Fig. 2A). Click on the column headers to sort by interactor, role (bait/hit), organism, experimental evidence code, data set publication, throughput (high/low), or interaction scores provided in the original publications.

5. Click on the "PTM Sites" button on the "Switch View" bar to access the Posttranslational Modification (PTM) viewer (Fig. 2B).

  i. View the phosphorylation sites in the protein sequence (in red; e.g., Fig. 2B, arrow b). Click on a modified residue in the protein sequence to display links to publications with supporting evidence.

  ii. Locate the table immediately below the protein sequence, which lists the individual residues and their phosphorylation sites (Fig. 2B, arrow c) along with the interacting proteins and their relationship (e.g., kinase or phosphatase), as applicable.

  Additional notes associated with each phosphosite are provided in the right hand "Notes" column under a tablet icon.

  iii. Scroll down to the separate "Relationship Details" table (Fig. 2B, arrow d), which provides information about the enzymes involved in the PTMs, identifying whether each protein listed in the left column phosphorylates or dephosphorylates the query protein of interest.

  The BioGRID Posttranslational Modification (PTM) viewer integrates more than 20,100 phosphorylation sites on approximately 3200 yeast proteins, curated by PhosphoGRID (version 2.0; www.phosphogrid.org) from both low- and high-throughput S. cerevisiae studies (Sadowski et al. 2013). More than 2500 of these phosphosites are associated with specific kinases and/or phosphatases, and more than 3200 of the sites occur on 1100 proteins under specific environmental conditions. The PTM viewer has been expanded to include sites of ubiquitin modification and will include other modifications in the future.

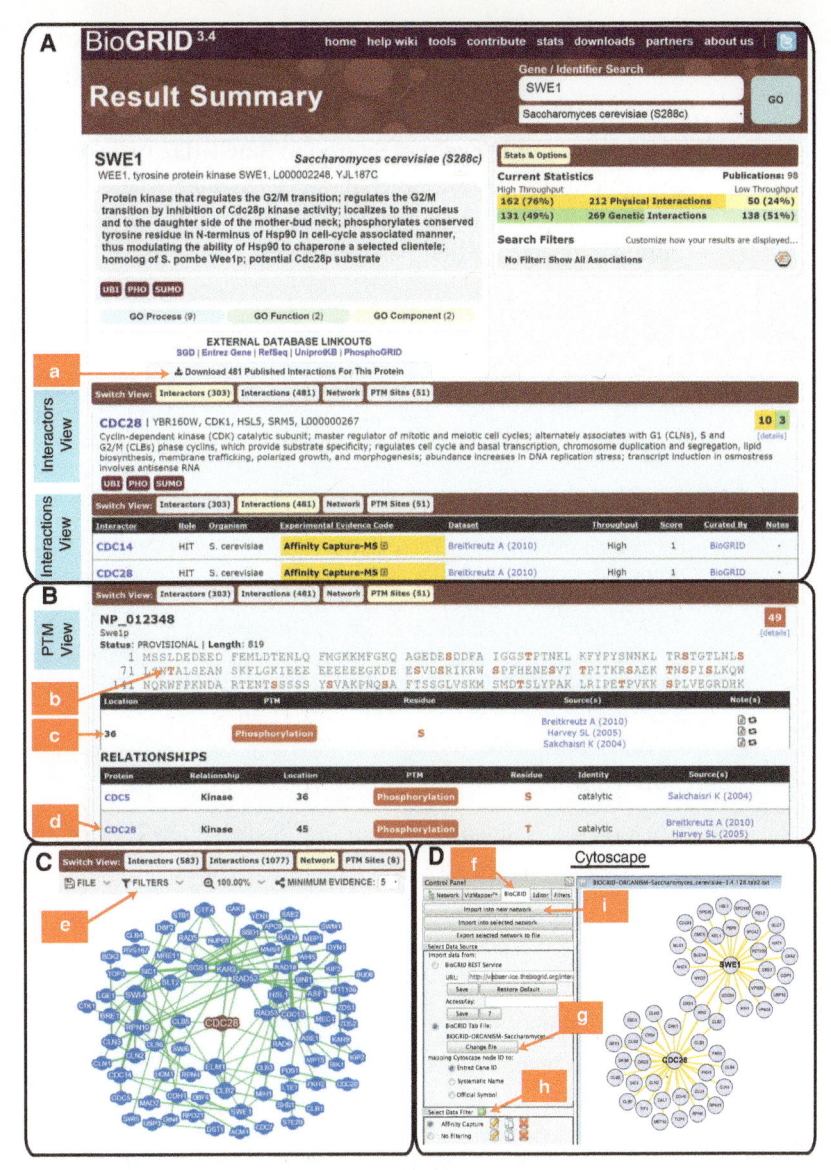

FIGURE 2. (*A*) Result Summary page that displays information for the query gene or protein. Results for the query gene *SWE1* are shown. The Interactors View displays unique interactors for the query gene or protein, sorted from highest to lowest number of curated interactions. The Interactions View allows results to be ordered based on interactor, role (bait/hit), organism, experimental evidence code, data set publication, throughput (high/low), or interaction scores given in the original publications. Genetic and protein interactions are color-coded in green and yellow, respectively, throughout the display. Filtering options in the *top right* "Stats & Options" box may be used to sort the table based on high- or low-throughput studies. Genetic and protein interactions for the query gene can be downloaded in all supported BioGRID download formats by clicking "Download Published Interactions for this Protein" (arrow a). (*B*) Integrated posttranslational modification (PTM) viewer within BioGRID. The Swe1 protein sequence is displayed with documented phosphorylation sites shown in red (arrow b). The table below the sequence shows additional details for each phosphosite, including the corresponding enzyme and its relationship to the query protein (arrow c). The relationship table provides further details, such as whether the interactor in the *left* column phosphorylates or dephosphorylates the main query protein (arrow d). The PTM viewer initially incorporated phosphosites curated by PhosphoGRID, but has also been expanded to include other types of PTMs, such as ubiquitination sites, also as curated by BioGRID. (*C*) Interaction network viewer within BioGRID. A filtered view is shown for interactions of CDC28, which is located at the center of the viewer. Larger-diameter nodes have a greater number of connections, and thicker edges are supported by more experimental evidence. All nodes can be dragged by the user to optimize the layout. Filters are applied by choosing the "FILTERS" option in the top menu of the graphic (arrow e). In this instance, interactions have been filtered to show only genetic interactions. The MINIMUM EVIDENCE option allows edges to be shows or hidden based on the number of curated interactions for that gene/protein pair. The higher the number, the more evidence in support of that interaction. The network also shows secondary interactor interactions, with the same level of minimum evidence. Clicking on each edge in the network reveals experimental evidence related to the interacting pairs. The network may also be exported as a graphic file using the FILE option in the top menu. (*D*) Use of Cytoscape with the BiogridPlugin2.2. After copying the BiogridPlugin2.2.jar into the plugins directory of Cytoscape, select the "BioGRID" tab in the control panel (arrow f). In the "Import data from" window select the "BioGRID Tab File" and choose the tab2 file of interest (arrow g). In the "Select Data Filter" window, click on the green plus symbol (arrow h). After setting up the filter described in Step 15 in the text, click the "Import into new network" button to view the interaction network (arrow i).

Cite this protocol as *Cold Spring Harb Protoc*; doi:10.1101/pdb.prot088880

6. Navigate to another gene of interest in BioGRID by clicking on any gene name link listed in one of the table views. For example, in the "PTM sites" view, click on any interacting gene listed on the left-hand side of the "Relationships" table (e.g., *CDC28*; Fig. 2B, arrow d) to go to the "Result Summary" page for that particular gene.

Interaction Network Visualization

An interactive network viewer is embedded within the BioGRID interface to allow users to immediately visualize interactions of interest. This new Javascript-based BioGRID visualization tool was built using the Cytoscape.js platform (Shannon et al. 2003). The viewer shows the interaction network of the gene of interest based on its direct genetic and/ or protein interaction partners. In addition to BioGRID's integrated network viewer linked from each gene page, selected interaction data may be downloaded in formats compatible with either the Cytoscape (http://www .cytoscape.org/) (Shannon et al. 2003; Cline et al. 2007) or Osprey (Breitkreutz et al. 2003; Stark et al. 2006) network visualization systems. These formats include the BioGRID TAB 2.0 delimited file format for Cytoscape, and an Osprey Custom Network File format for the Osprey visualization system. A BioGRID Cytoscape plugin 2.0 (http://wiki.thebiogrid.org/doku.php/biogridplugin2) is available to facilitate the import of BioGRID interaction data into Cytoscape to visualize interaction networks and integrate interactions with gene expression profiles (Winter et al. 2011). GeneMania (http://www.genemania.org/) is also compatible with BioGRID formats (Montojo et al. 2010; Zuberi et al. 2013). This next series of steps describes how to visualize a local network of interactions in the BioGRID viewer and in Cytoscape.

BioGRID Viewer

7. Click on the "Network" link in the "Switch View" bar of any gene page (e.g., CDC28).

    This action opens up a network view that graphically displays the interaction network around the query gene. The query gene is represented by a node at the center of the layout, and the interacting genes are displayed as associated nodes with each line (edge) between a pair of genes representing interactions (Fig. 2C). Greater node size represents increased connectivity, and thicker edge sizes represent increased evidence supporting the interaction.

8. Display the desired information about the gene of interest as follows.

    i. Filter the network display via options provided in the top "FILTERS" menu option (Fig. 2C, arrow e).

       Pairwise interactions may be restricted to either physical interactions (yellow edges) or genetic interactions (green edges), or both may be displayed. Any interacting pairs associated with both genetic and physical evidence are linked by purple edges.

    ii. Limit the number of displayed interactions via the "MINIMUM EVIDENCE" option in the upper menu.

       Selecting a number allows edges to be shown or hidden based on the number of curated interactions associated with the gene/protein pair. The higher the selected number, the more evidence is associated with that interaction. In the example given in Fig. 2C, the displayed interactions for CDC28 have been restricted to genetic interactions that are associated with at least five unique curated interactions (minimum evidence = 5). This network includes secondary interactor interactions that are also associated with the same level of minimum evidence.

    iii. Click on any edge (e.g., between CLB5 and CDC28) to view the "Association Details" between interacting genes/proteins.

       This opens up a window with "Association Details" between the two interactors including relevant experimental evidence codes and publications.

Cytoscape

9. Download BioGRID interaction data in tab2 format as follows.

    i. Visit the BioGRID download page (http://thebiogrid.org/download.php).

    ii. Click on "Current Release," and then click on the "BIOGRID-ORGANISM-(VERSION NUMBER).tab2.zip" to begin downloading the available interaction data, which will be broken into distinct files by organism.

iii. Open the zip file and locate "BIOGRID-ORGANISM-Saccharomyces_cerevisiae-(VERSION NUMBER).tab2.txt".

This file contains all of the budding yeast interactions from BioGRID and will be used in this example.

10. Download a copy of Cytoscape version 2.8.2 from the following URL: http://www.cytoscape.org/download.html.

    Note that the BioGRID Cytoscape plugin is not currently compatible with version 3.x of Cytoscape but will be updated in the future.

11. After installing Cytoscape, download the BiogridPlugin version 2.2 located in the folder "BiogridPlugin2" within "Cytoscape Plugins" on the BioGRID download page (http://thebiogrid.org/download.php).

12. Copy the downloaded file, "BiogridPlugin2.2.jar," into the plugins directory of Cytoscape, located in the folder in which Cytoscape is installed.

    Typically this will be located in "C:\Program Files\Cytoscape_v2.8.2\plugins" on Windows-based computers, or "/Applications/Cytoscape_v.2.8.2/plugins" on a Mac.

13. Launch Cytoscape and select the "BioGRID" tab in the control panel (Fig. 2D, arrow f) for a list of input variables for the BioGRID plugin.

14. In the "Import data from:" window (Fig. 2D, arrow g) select the "BioGRID Tab File:" option box and change the file to the one that was downloaded in Step 9.iii.

15. Filter interaction data using the plugin. As an example, set up a filter to show only Affinity Capture–MS interactions for the budding yeast proteins CDC28 and SWE1.

    i. Bring up the filter window by clicking on the green and white plus symbol just to the right of the "Select Data Filter" window (Fig. 2D, arrow h).

    ii. Assign the filter a name: In this example use "Affinity Capture-MS Filter."

    iii. In the "Filter by gene" window, add the protein names CDC28 and SWE1, make sure the drop down is set to "Include only," and check the box for "and their primary interactors."

    iv. In the "Filter by evidence code (include checked)" window, select the experimental system "Affinity Capture-MS" (Fig. 2D, arrow h). Remove any genetic evidence codes.

    v. Click "OK" to save this filter.

16. Once the filter is setup, click the "Import in new network" button (Fig. 2D, arrow i), which will produce an interaction network containing the query genes of interest and their interactors.

17. To better visualize the network, select one of the built-in layouts in Cytoscape. Navigate to the "Layout" menu at the top and select "Cytoscape Layouts → Forced-Directed Layout → (unweighted)."

    Cytoscape will generate a network display with each query gene at the center of its set of interactors (Fig. 2D).

Download Options and Formats

Interaction data in BioGRID are updated on a monthly basis and may be downloaded for each query gene or protein, or by publication, in various formats.

18. Obtain interactions for individual genes or publications as follows.

    ### For a Single Query Gene

    i. Click on the "Download Published Interactions for this Protein" button on any gene page above the "Switch View" option (Fig. 2A, arrow a).

    ii. When the "BioGRID Downloads" page has opened, select the desired file format.

    Documentation is available at http://wiki.thebiogrid.org/doku.php/downloads. The BioGRID Tab 2.0 format is the most commonly downloaded tab-delimited text format and is provided as the default option. Other formats include the IMEx-compatible PSI-MI XML or MITAB formats.

iii. Click the "Download Interactions" button to obtain the requested file.

Based on Publication

i. Search BioGRID using the "By Publication" option (see Step 2).
Entering any PubMed ID returns the publication of interest.

ii. Click on the "Download Interactions For This Publication" option below the abstract.

19. Obtain the complete collection of curated interactions as follows.

    i. Click on the "Downloads" menu option at the top right of any BioGRID page to go to the main download page (http://thebiogrid.org/download.php).

    ii. Select the "Current Release" folder to expand a list of downloadable file formats.
    A description of the different formats is available in the right hand panel of the page.

    iii. Download the interaction data based on organism type or experimental system and in the desired format.
    The entire collection is available in various formats, including the BioGRID Tab 2.0 format, and the IMEx-compatible PSI-MI XML or MITAB formats.
    The source code for BioGRID is also available without any restrictions (http://wiki.thebiogrid.org/doku.php/development_overview).

20. Build a specific data set based on a subset of genes, proteins, or publications as follows.

    i. Click on the "custom download generator" link toward the bottom of the main download page to open up the BioGRID Download Dataset Generator page (http://thebiogrid.org/downloadgenerator.php).

    ii. Select the PublicationID(s) or Gene Identifier(s) option.
    The former is for PubMed IDs, and the latter is for official gene symbols, systematic names, and external database IDs. Identifiers that may be used are listed at http://wiki.thebiogrid.org/doku.php/identifiers. Up to five PubMed IDs or ten Gene IDs may be entered per search.

    iii. If desired, refine the search by selecting an organism of interest.

    iv. Choose an output file format.

    v. Click the "Build Download Dataset" button to obtain the specific interaction data set.

21. (Optional) Retrieve interaction data in BioGRID automatically via the BioGRID REST service.
Details regarding the REST service are documented on a wiki page (http://wiki.thebiogrid.org/doku.php/biogridrest) and are also published (Winter et al. 2011). The documentation includes information on how to obtain a unique access key, which is required for all queries to the web service, and how to fetch and filter interactions for a single query gene. Several examples are given along with a table listing the complete set of query parameters that may be used. A formal description of the REST service is also provided in Web Application Description Language (WADL) at http://webservice.thebiogrid.org/application.wadl. BioGRID is also part of the PSICQUIC project (https://code.google.com/p/psicquic/) (del-Toro et al. 2013), which is part of the HUPO Proteomics Standard Initiative (HUPO-PSI) to standardize automated access to molecular interaction databases by providing a standard web service and common Molecular Interactions Query Language (MIQL).

Community and User Feedback

22. Contact the BioGRID team at biogrid.admin@gmail.org as needed.
The BioGRID values feedback from the yeast research community and other users to correct curation errors, extend interaction coverage, request new features, or support and/or contribute published or prepublication interaction data. The BioGRID programming team will assist with extraction of data features, resolution of data discrepancies, and links to BioGRID pages or the REST service. The BioGRID curation team will correct any reported errors in the subsequent monthly release and assist with confidential upload of new data sets to coincide with publication date. Finally, BioGRID welcomes collaborators with an interest in contributing domain expertise or curation capacity. Further details can be found at http://wiki.thebiogrid.org/doku.php/contribute.

RELATED INFORMATION

For background on the BioGRID database, see Introduction: BioGRID: A Resource for Studying Biological Interactions in Yeast (Oughtred et al. 2015).

ACKNOWLEDGMENTS

The authors thank Chris Grove and Paul Sternberg at WormBase for ongoing collaborative development of the Genetic Interaction Ontology. This work was supported by National Institutes of Health (NIH) grants R01OD010929 and R24OD011194 to M.T. and K.D., the Biotechnology and Biological Sciences Research Council (grant number BB/F010486/1 to M.T.), the Canadian Institutes of Health Research (grant number FRN 82940 to M.T.), and a Genome Québec International Recruitment Award and a Canada Research Chair in Systems and Synthetic Biology to M.T.

REFERENCES

Breitkreutz BJ, Stark C, Tyers M. 2003. Osprey: A network visualization system. *Genome Biol* **4:** R22.

Chatr-Aryamontri A, Breitkreutz BJ, Heinicke S, Boucher L, Winter A, Stark C, Nixon J, Ramage L, Kolas N, O'Donnell L, et al. 2013. The BioGRID interaction database: 2013 update. *Nucleic Acids Res* **41:** D816–D823.

Cherry JM, Hong EL, Amundsen C, Balakrishnan R, Binkley G, Chan ET, Christie KR, Costanzo MC, Dwight SS, Engel SR, et al. 2012. Saccharomyces Genome Database: The genomics resource of budding yeast. *Nucleic Acids Res* **40:** D700–D705.

Cline MS, Smoot M, Cerami E, Kuchinsky A, Landys N, Workman C, Christmas R, Avila-Campilo I, Creech M, Gross B, et al. 2007. Integration of biological networks and gene expression data using Cytoscape. *Nat Protoc* **2:** 2366–2382.

del-Toro N, Dumousseau M, Orchard S, Jimenez RC, Galeota E, Launay G, Goll J, Breuer K, Ono K, Salwinski L, et al. 2013. A new reference implementation of the PSICQUIC web service. *Nucleic Acids Res* **41:** W601–W606.

Engel SR, Balakrishnan R, Binkley G, Christie KR, Costanzo MC, Dwight SS, Fisk DG, Hirschman JE, Hitz BC, Hong EL, et al. 2010. Saccharomyces Genome Database provides mutant phenotype data. *Nucleic Acids Res* **38:** D433–D436.

Lamesch P, Berardini TZ, Li D, Swarbreck D, Wilks C, Sasidharan R, Muller R, Dreher K, Alexander DL, Garcia-Hernandez M, et al. 2012. The *Arabidopsis* Information Resource (TAIR): Improved gene annotation and new tools. *Nucleic Acids Res* **40:** D1202–D1210.

Marygold SJ, Leyland PC, Seal RL, Goodman JL, Thurmond J, Strelets VB, Wilson RJ, FlyBase c. 2013. FlyBase: Improvements to the bibliography. *Nucleic Acids Res* **41:** D751–D757.

Montojo J, Zuberi K, Rodriguez H, Kazi F, Wright G, Donaldson SL, Morris Q, Bader GD. 2010. GeneMANIA Cytoscape plugin: Fast gene function predictions on the desktop. *Bioinformatics* **26:** 2927–2928.

Orchard S, Kerrien S, Abbani S, Aranda B, Bhate J, Bidwell S, Bridge A, Briganti L, Brinkman FS, Cesareni G, et al. 2012. Protein interaction data curation: The International Molecular Exchange (IMEx) consortium. *Nat Methods* **9:** 345–350.

Oughtred R, Chatr-aryamontri A, Breitkreutz B-J, Chang CS, Rust JM, Theesfeld CL, Heinicke S, Breitkreutz A, Chen D, Hirschman J, et al. 2015. BioGRID: A resource for studying biological interactions in yeast. *Cold Spring Harb Protoc* doi: 10.1101/pdb.top080754.

Sadowski I, Breitkreutz BJ, Stark C, Su TC, Dahabieh M, Raithatha S, Bernhard W, Oughtred R, Dolinski K, Barreto K, et al. 2013. The PhosphoGRID *Saccharomyces cerevisiae* protein phosphorylation site database: Version 2.0 update. *Database (Oxford)* **2013:** bat026.

Shannon P, Markiel A, Ozier O, Baliga NS, Wang JT, Ramage D, Amin N, Schwikowski B, Ideker T. 2003. Cytoscape: A software environment for integrated models of biomolecular interaction networks. *Genome Res* **13:** 2498–2504.

Stark C, Breitkreutz BJ, Reguly T, Boucher L, Breitkreutz A, Tyers M. 2006. BioGRID: A general repository for interaction datasets. *Nucleic Acids Res* **34:** D535–D539.

Winter AG, Wildenhain J, Tyers M. 2011. BioGRID REST Service, BiogridPlugin2 and BioGRID WebGraph: New tools for access to interaction data at BioGRID. *Bioinformatics* **27:** 1043–1044.

Wood V, Harris MA, McDowall MD, Rutherford K, Vaughan BW, Staines DM, Aslett M, Lock A, Bahler J, Kersey PJ, et al. 2012. PomBase: A comprehensive online resource for fission yeast. *Nucleic Acids Res* **40:** D695–D699.

Yook K, Harris TW, Bieri T, Cabunoc A, Chan J, Chen WJ, Davis P, de la Cruz N, Duong A, Fang R, et al. 2012. WormBase 2012: More genomes, more data, new website. *Nucleic Acids Res* **40:** D735–D741.

Zuberi K, Franz M, Rodriguez H, Montojo J, Lopes CT, Bader GD, Morris Q. 2013. GeneMANIA prediction server 2013 update. *Nucleic Acids Res* **41:** W115–W122.

CHAPTER 32

Exploratory Analysis of Biological Networks through Visualization, Clustering, and Functional Annotation in Cytoscape

Anastasia Baryshnikova[1]

Lewis-Sigler Institute for Integrative Genomics, Princeton University, Princeton, New Jersey 08544

Biological networks define how genes, proteins, and other cellular components interact with one another to carry out specific functions, providing a scaffold for understanding cellular organization. Although in-depth network analysis requires advanced mathematical and computational knowledge, a preliminary visual exploration of biological networks is accessible to anyone with basic computer skills. Visualization of biological networks is used primarily to examine network topology, identify functional modules, and predict gene functions based on gene connectivity within the network. Networks are excellent at providing a bird's-eye view of data sets and have the power of illustrating complex ideas in simple and intuitive terms. In addition, they enable exploratory analysis and generation of new hypotheses, which can then be tested using rigorous statistical and experimental tools. This protocol describes a simple procedure for visualizing a biological network using the genetic interaction similarity network for *Saccharomyces cerevisiae* as an example. The visualization procedure described here relies on the open-source network visualization software Cytoscape and includes detailed instructions on formatting and loading the data, clustering networks, and overlaying functional annotations.

MATERIALS

Equipment

BiNGO (Maere et al. 2005) or other relevant app for network annotation (see Step 17)
Computer system
  System requirements are available at www.cytoscape.org/documentation_users.html.

Cytoscape 3.0 (Smoot et al. 2011) (see Step 1)
Network data to be analyzed or sample data (see Steps 2 and 3)

METHOD

Install Cytoscape and Prepare Data

1. Download Cytoscape 3.0 from www.cytoscape.org/download.php.
    Instructions for installation are available at www.cytoscape.org/documentation_users.html.
    Proceed to either Step 2 (if using sample data) or Step 3.

---

[1]Correspondence: abarysh@princeton.edu

Copyright © Cold Spring Harbor Laboratory Press; all rights reserved
Cite this protocol as *Cold Spring Harb Protoc*; doi:10.1101/pdb.prot077644

2. Download the following sample data from www.baryshnikova-lab.org/publications/4/.

   - Data file 1. A tab-delimited text file containing the genetic interaction similarity data from Costanzo et al. (2010): Costanzo_Science_2010_correlation_network.txt.

     This file contains a list of gene pairs (Columns 1 and 2) whose genetic interaction profile similarity, as measured by Pearson correlation coefficient, is >0.2 (Column 3). Here, a genetic interaction is defined as an unexpected phenotype arising from combining two mutations in the same organism. In Costanzo et al. (2010), genetic interactions were quantitatively measured for 30% of all possible double-mutant combinations in the yeast Saccharomyces cerevisiae. *Genes sharing similar genetic interactions are known to share similar functions. We can therefore build a network of functional relationships between all genes in the data set by computing a quantitative measure of profile similarity, such as Pearson correlation, and selecting the gene pairs with the highest observable correlation.*

   - Data file 2. A tab-delimited text file containing the functional annotation standard used for visualization in Figure 1 in Costanzo et al. (2010): Costanzo_Science_2010_functional_annotation.txt.

     This file contains the list of genes in the Costanzo et al. network and four different node attributes: CommonName (Column 2; the three-letter code for the gene), ORF (Column 3; the systematic name for the gene), FunctionalAnnotation (Column 4; discrete values indicating the association of the gene with one of 13 different functional groups; see Step 18), and Essentiality (Column 5; binary 0–1 values indicating whether the gene is known to be essential or not). Column 1 contains a unique StrainID that matches the node label used in Data file 1.

3. (Optional) If not using sample data, prepare network data for import.

   Network data can be imported into Cytoscape using a variety of file formats (wiki.cytoscape.org/Cytoscape_3/UserManual/Network_Formats/). The simplest format is a delimited text file that defines a network by listing its interactions (Fig. 1A; Data file 1). Each interaction (or "edge") must report the labels of the two connected nodes (i.e., gene or protein names) and, optionally, may also include any number of interaction properties (or "edge attributes"), such as type, strength, and confidence of the interaction. Columns containing node labels and edge attributes must be separated by a special character (e.g., a tab or a comma) that is never encountered within a node label (Fig. 1A; Data file 1).

   The delimited text file is the simplest way to store and exchange network data because it can be generated and edited in many applications on many platforms and is easily interpretable by both humans and computers. The limitation of delimited text files is that they do not carry any information relative to the positioning of nodes in the network and do not specify any node attribute, such as alternative labels, shape, color, or annotation. This information, if available, must be stored in an additional file and loaded separately (see Step 18i—Loading node attributes). More complex file formats, such as XGMML, store all network information in a single file and allow a user to import organized and annotated networks in a single step. XGMML files, however, must be generated using a specialized tool, involve relatively complex syntax (http://cgi5.cs.rpi.edu/research/groups/pb/punin/public_html/XGMML/), and are not easily manipulated. As a result, XGMML files are more appropriate for advanced users to exchange visualizations across different applications or across different networks within the same application.

Load the Network

4. In Cytoscape, go to File → Import → Network → File. Choose the file containing the network data and click Open.

   In the sample data provided, the file is Costanzo_Science_2010_correlation_network.txt.

5. View the Preview section of the window, which displays a sample of the input file and shows how information in the file will be parsed.

6. If parsing appears incorrect (e.g., two node labels are merged in the same column or, conversely, a single node label is split into multiple columns), click on Show Text File Import Options in the Advanced section of the window to specify a different column delimiter character. In addition, indicate whether the input file contains any header rows (Start Import Row) and/or whether the first row should be used to label the columns (Transfer first line as column names).

7. In the Interaction Definition section of the window, specify which columns contain the labels of the interacting nodes (Source Interaction and Target Interaction).

   The corresponding columns in the Preview section will be highlighted in purple and orange, respectively.

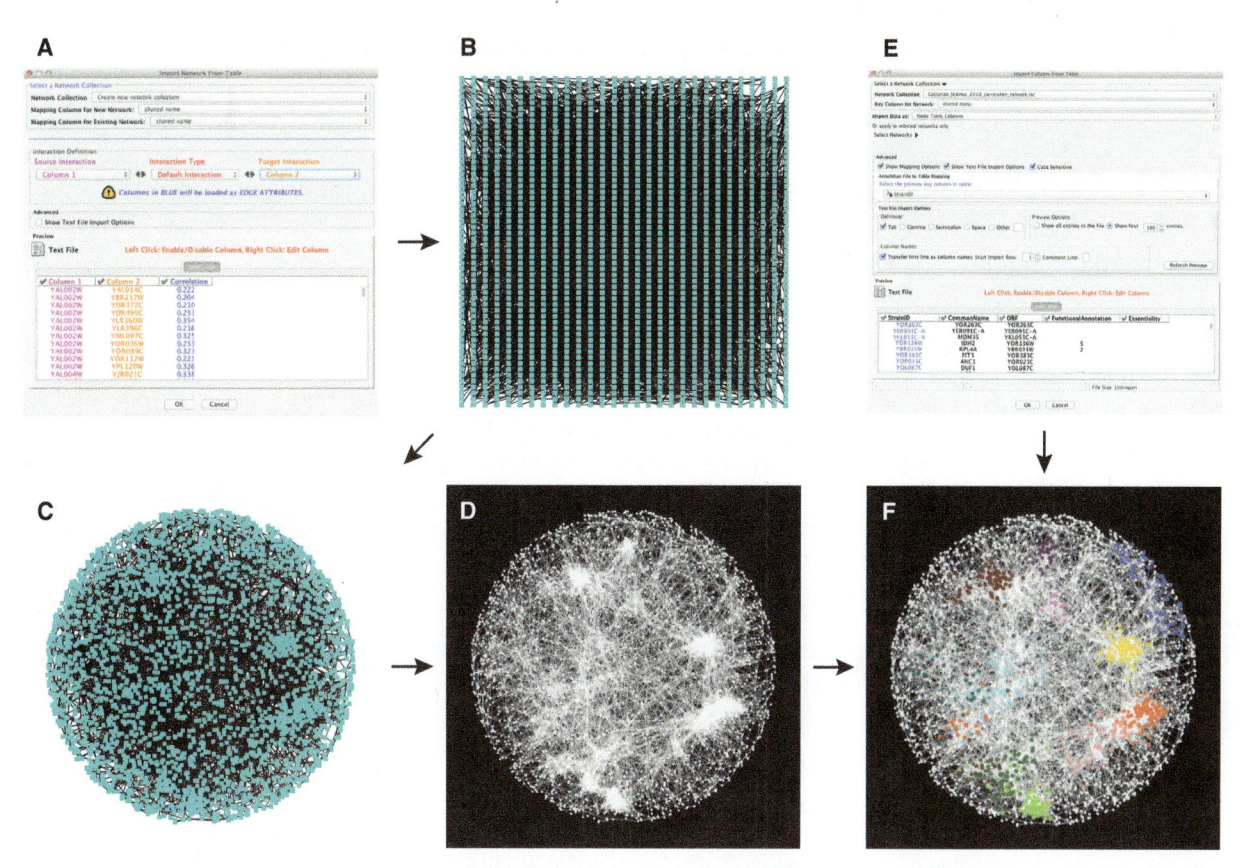

FIGURE 1. Key steps for visualizing a large-scale biological network using Cytoscape 3.0. (*A*) Loading the network data from a tab-delimited text file. (*B*) Creating a preliminary view of the network. (*C*) Organizing the network using the edge-weighted spring-embedded layout. (*D*) Adjusting the visual appearance of the network by changing color, shape, size, and transparency of nodes and edges. (*E*) Loading customized functional annotations as node attributes. (*F*) Adjusting node appearance based on node attributes.

For undirected edges (e.g., protein–protein interactions), it does not matter which node is labeled as Source and which one as Target, unless, for example, it is important to preserve the order in which the interaction was detected (Source and Target may correspond to bait and prey proteins, respectively). For directed edges (e.g., kinase–substrate relationships), Source and Target should indicate the upstream (kinase) and downstream (substrate) nodes, respectively.

In the example provided (undirected edges), Columns 1 and 2 should be labeled as Source and Target (or vice versa) (Fig. 1A).

8. Click on the columns containing edge attributes to select them. Each selected column will be highlighted in blue. To name an attribute and/or to specify whether it contains texts or numerical values, right-click (or Command-click in Mac OS X) on the column and input the relevant information.

    In the example provided, Column 3 contains a measure of similarity (Pearson correlation coefficients) between the genetic interaction profiles of Source and Target genes. This attribute should be named Correlation and assigned to the Floating Point data type (Figure 1A). Alternatively, if, for example, this column contained a discrete score (from 1 to 5) representing our level of confidence in the correlation, we would have named the edge attribute Confidence and assigned it to the Integer data type.

9. Click OK to load the network.

10. If the network does not appear automatically, right-click on the name of the network in the Control Panel on the left and select Create View.

    This step is usually necessary for large networks because they are not automatically visualized to preserve memory.

Chapter 32

> *In the sample data provided, the Control Panel should list a network named Costanzo_Science_2010_correlation_network.txt with 2838 nodes and 10,189 edges (Fig. 1B).*

11. Once the network has been loaded and visualized, save it as a Cytoscape Session file (∗.cys), which can then be reopened at a later time on the same or on a different computer and will maintain all the properties (layout, annotations, and visual style) of the original network.

Organize the Network

For a discussion of possible ways to organize the network see the Discussion section.

12. As an example, we here apply the edge-weighted spring-embedded layout, which mimics the behavior of a system of connected springs and reorganizes the network such that densely connected nodes are positioned close to each other, whereas disconnected nodes are spread farther apart. To apply the layout, select Layout → Edge-weighted Spring Embedded. As an option, an edge attribute can be used to specify the spring force coefficient. In our example, choose the Correlation node attribute to apply stronger forces to highly correlated gene pairs or select (none) to assign equal forces to all gene pairs.

   > *Because of a randomization step in the algorithm, the edge-weighted spring-embedded layout generates a new network configuration at every run. Although these configurations are equivalent in terms of force balance between the edges connecting each node, it is not possible to reproduce the same exact network visualization in two independent Cytoscape sessions.*

Adjust the Visual Style of the Network

The procedure described below reproduces the visual style of the correlation network published as Fig. 1 in Costanzo et al. (2010). It is provided as an example of the visual properties that can be adjusted in Cytoscape to generate informative network visualizations (Fig. 1D). Additional options can be explored by the user within Cytoscape.

13. Set the background color to black.

    i. Go to Control Panel → VizMapper. Click on the image in the Defaults (Click to edit) section.

    ii. Click on Network at the bottom right of the window. Double-click on Network Background Paint and choose the black color.

14. Make the edges white and semitransparent.

    i. In the same window as above, click on Edge at the bottom right of the window.

    ii. In the bottom center, click on Show All.

    iii. Scroll through the properties to find Edge Stroke Color (Unselected), double-click, and choose the white color.

    iv. Scroll through the properties to find Edge Transparency, double-click, and input 50 (values 0 and 255 correspond to 0% and 100% opacity, respectively).

15. Make the nodes white, small, and round.

    i. In the same window as above, click on Node at the bottom right of the window.

    ii. In the bottom center, click on Show All.

    iii. Scroll through the properties to find Node Size, double-click, and input 20.

    iv. Scroll through the properties to find Node Fill Color, double-click, and choose the white color.

    v. Scroll through the properties to find Node Shape, double-click, and choose Ellipse.

16. Click on Apply in the bottom left corner of the window (Fig. 1D).

    > *Depending on the network size, some of the visual changes might not be immediately visible in the Cytoscape main window. To see the changes, it may be necessary to zoom in or out of the current network view ("+" and "−" magnifying glasses in the top toolbar) or export the current network view into a PDF file (see Step 19).*

Annotate the Network

For a discussion of how and why to use network annotation, see the Discussion section.

17. Perform functional enrichment.

    While functional enrichment tests are not directly implemented in the default version of Cytoscape, numerous third-party apps have been developed that provide this capability (http://apps.cytoscape.org/apps/with_tag/enrichmentanalysis). As an example, I describe the app BiNGO (Maere et al. 2005) with most of its default settings. BiNGO is particularly suitable for functional annotation of biological networks because it enables the user to test selected network regions for enrichment against common functional annotation standards, such as Gene Ontology biological process terms, and returns a standard significance p-value.

    i. To download and install BiNGO, choose the menu Apps → App Manager. In the Search window, type BiNGO, select the app from the list, and click on Install.

    ii. To run a functional enrichment analysis, select a subset of nodes (e.g., a cluster of interest) from the network and choose the menu Apps → BiNGO.

    iii. Type in a name for the selected cluster.

    By default, BiNGO extracts node labels directly from the network (Get Cluster from Network, selected by default) and compares them against the specified Gene Ontology file (lower in this window, Select ontology file: GO_Biological_Process, GO_Cellular_Component, GOSlim_Yeast, etc.). Some of the node labels, however, may not match the gene identifiers listed in GO due to the existence of gene name aliases (e.g., SRS2 or HPR5 for YJL092W) and/or custom annotations appended to node labels (e.g., in our example, YAL041W_tsq148 denotes one of the several CDC24 temperature-sensitive alleles present in the data set). In theory, it is possible to provide an additional node attribute containing a standard gene identifier for all nodes in the network (see Step 18i—Loading node attributes). However, BiNGO cannot perform functional enrichment on node attributes. Instead, standard gene identifiers corresponding to the selected nodes can be manually copied from the network window (or any other source) and pasted directly into the BiNGO settings window (following the selection of the Paste Genes from Text option).

    iv. Leave all other options as their defaults (for more details, see Maere et al. (2005)).

    v. Click Start BiNGO.

    vi. View results.

    The results window lists the GO terms whose members are significantly overrepresented/enriched among the genes in the selected cluster. In addition, BiNGO generates a network visualizing the distribution of significantly overrepresented GO terms in the overall hierarchy of GO.

18. Perform visual annotation using functional standards.

    In Cytoscape, node attributes (i.e., numerical values or text labels associated with each node) can be used to associate nodes with unique visual properties, such as size, shape, and color. Using this feature, genes acting in known biological pathways can be easily identified and assessed qualitatively with respect to network topology. For example, we can determine whether members of a biological pathway cluster together in a network module or whether essential genes tend to be highly connected and act as network hubs.

    i. Check the file containing node attribute data: Similarly to network data, the simplest format for importing node attributes is a delimited text file where each row corresponds to a node and each column corresponds to a different node attribute. Columns must be separated by a special character (e.g., a tab or a comma) (Fig. 1E; Data file 2). One of the columns must contain a node identifier: the primary identifier used in the network data file (see Step 3) or a secondary identifier previously loaded as a node attribute (e.g., see Step 18v).

    In our example (Fig. 1E), the node attribute file contains a customized functional annotation: genes are assigned a functional group (numbered from 1 to 13) if they (a) belong to one of the large functionally enriched clusters in the network and (b) are annotated to the biological process most represented in that cluster. These numbers were generated specifically for the genetic interaction similarity network using a multistep procedure involving network clustering, functional enrichment, and manual annotation (Costanzo et al. 2010).

    ii. Go to File → Import → Table → File. Choose the file containing the node attribute data and click Open.

    iii. View the Preview section of the window, which displays a sample of the input file showing how it will be parsed for information.

iv. If parsing appears incorrect, click on Show Text File Import Options in the Advanced section of the window to specify a different column delimiter character. In addition, you can indicate whether or not the first row of the file should be used to label the columns.

v. To select a different node label in the network, click on the Key Column for Network drop-down menu (second from the top of the window) and select the desired node label.

> By default, Cytoscape will match the first column of the node attribute file to the primary node label in the network. Both of these options can be changed.
>
> It may be necessary to load additional node labels before being able to load the functional annotation. For example, the primary node identifier in the network may be the common name of a gene, whereas the functional annotation standard lists genes by their systematic ORF (open reading frame) names. As a result, it is first necessary to create a node attribute containing the systematic ORF name corresponding to each common name label. This can be done using the same exact procedure described in this section (Steps i–v). Once the node attribute containing the systematic ORF name has been created, it can be used as Key Column for Network to load the functional annotation standard.

vi. To select a different column from the node attribute file, click on Show Mapping Options and select the correct column from the Select the primary key column in table drop-down menu. The corresponding column will be highlighted in blue in the Preview section.

vii. By default, all columns of the node attribute file will be loaded as attributes and will be named Column 1, Column 2, Column 3, etc. To avoid loading unnecessary information, click on a column in the Preview section to deselect it. To rename an attribute and/or specify whether it contains text or numerical values, right-click (or Command-click in Mac OS X) on the column and input the relevant information.

viii. To load the node attribute file, click OK.

ix. To assign visual properties to node attributes (Fig. 1F): Go to Control Panel → VizMapper.

x. In the Visual Mapping Browser section, click on Show All. Scroll down all the Node and Edge properties and double-click on Node Fill Color.

xi. Click on Please select a value and choose the node attribute that should be represented as a color.

> In our example, choose the FunctionalAnnotation node attribute.

xii. Click on the cell next to Mapping Type and select Discrete Mapping.

> Discrete Mapping will associate a distinct color to each of the unique values in the FunctionalAnnotation node attribute. As a result, if the network contains 10 nodes annotated to functional group 1 and this functional group is associated with the color red, the network will show 10 red nodes.

xiii. Right-click (or Command-click in Mac OS X) on Node Fill Color, select Mapping Value Generators → Rainbow (or any other option). This will assign random colors to the functional annotation groups.

xiv. To adjust a color associated with a specific functional group, click on the category in Vizmapper, then click on ... and select the appropriate color.

xv. Similarly, node attributes can be associated with different node shape, size, and transparency.

19. Finalize network visualization: It is often convenient to manually adjust the visualization of a network using a vector graphic editor such as Adobe Illustrator. To do this, export the network from Cytoscape into an editable PDF file by choosing File → Export → Network View as Graphics → PDF File (*.pdf). Click on Save a File to specify the name and the location of the PDF file.

DISCUSSION

Network Organization

Cytoscape provides a wide range of automatic network layout options (http://wiki.cytoscape.org/Cytoscape_3/UserManual#Cytoscape_3.2BAC8-UserManual.2BAC8-Navigation_Layout.Automatic_

Layout_Algorithms) that may be useful for different visualization strategies. For example, to examine the frequency of interactions between specific pathways or protein complexes, nodes can be grouped according to one of their attributes (Layout → Group Attributes Layout). The procedure for loading functional annotations as a node attribute is described in Step 18i—Loading Node Attributes. Alternatively, nodes can be sorted by their degree (i.e., number of interactions) to identify network hubs and analyze their common properties (Layout → Degree Sorted Circle Layout).

To explore the network in an unbiased way, it is often useful to arrange nodes based on their interaction strength and density. This approach visually clusters the network, revealing densely connected regions and their positioning with respect to each other, and has the potential to uncover biologically relevant modules. Different network layouts use different algorithms to achieve this effect. In the edge-weighted spring-embedded layout, for example, every edge acts as a spring that pulls nodes together with a strength proportional to an edge attribute (e.g., genetic interaction profile similarity) that is specified by the user (Kamada and Kawai 1989; Fruchterman and Reingold 1991). This and other similar network layouts have been widely adopted to represent biological networks because they often produce clean, symmetrical, and esthetically pleasant visualizations, with a minimal number of crossing edges and overlapping nodes (Fig. 1C) (Freeman et al. 2007; Atkinson et al. 2009; Costanzo et al. 2010; Breuer et al. 2013).

Network Annotation

To evaluate a network and determine how well it recapitulates known biology, it is often useful to visually associate nodes and/or node clusters with specific biological processes. For example, a functional enrichment test may be required to determine whether members of a particular network cluster are overrepresented/enriched for genes annotated to a particular Gene Ontology term. Gene Ontology (GO) is a standardized vocabulary that describes every gene's biological role, molecular function, and cellular localization (Ashburner et al. 2000), and provides a unique resource for evaluating the output of genomic experiments and large scale networks (see Step 17—Functional enrichment). As an alternative to functional enrichment and GO, the user may consider directly labeling the network using a custom set of functional annotations such as, for example, a set of mutants with a phenotype of interest or a hit list from a systematic experiment, and visually assessing the distribution of functional groups throughout the network (see Step 18—Visual annotation using functional standards). This latter approach, while less rigorous, often provides a rapid and useful overview of functional information within the network that may motivate the implementation of more accurate statistical tests.

The level of detail with which a network should be annotated depends on the purpose of the visualization and on the availability of properly defined annotation standards. Although GO is certainly the most widely adopted standard, its hierarchical structure often hinders its direct use in network visualization without preliminary manipulation, such as compression and/or filtering (e.g., Myers et al. 2006). One potential alternative is GO Slim, which has a flat structure, fewer terms, and broader definitions. The list of GO Slim terms for *Saccharomyces cerevisiae* can be downloaded from the Gene Ontology webpage (www.geneontology.org/GO_slims/goslim_yeast.obo), whereas the gene to term mapping can be obtained from the *Saccharomyces* Genome Database (SGD, http://downloads.yeastgenome.org/curation/literature/go_slim_mapping.tab).

Although simpler than GO, GO Slim may still provide a functional categorization that is too detailed for many visualization purposes (e.g., 167 terms in the yeast version). To address this issue, more general annotation standards have been developed. For example, the most recent yeast genetic interaction study grouped 4414 genes into 17 functional categories (Supplementary Data File S6 in Costanzo et al. (2010)).

Similarly, it is often useful to visualize functional annotations that are more specific than GO Slim, such as, for example, individual biological pathways or protein complexes. In addition to a few systematic analyses of co-complex associations (Gavin et al. 2006; Krogan et al. 2006), several groups have curated and integrated the results of multiple experimental data sets to produce a compendium of protein complexes for yeast (http://downloads.yeastgenome.org/curation/literature/go_protein_complex_slim.tab) (Pu et al. 2009; Baryshnikova et al. 2010).

ACKNOWLEDGMENTS

I thank Michael Costanzo for providing critical feedback on the manuscript.

REFERENCES

Ashburner M, Ball CA, Blake JA, Botstein D, Butler H, Cherry JM, Davis AP, Dolinski K, Dwight SS, Eppig JT, et al. 2000. Gene ontology: Tool for the unification of biology. The Gene Ontology Consortium. *Nat Genet* 25: 25–29.

Atkinson HJ, Morris JH, Ferrin TE, Babbitt PC. 2009. Using sequence similarity networks for visualization of relationships across diverse protein superfamilies. *PLoS One* 4: e4345.

Baryshnikova A, Costanzo M, Kim Y, Ding H, Koh J, Toufighi K, Youn JY, Ou J, San Luis BJ, Bandyopadhyay S, et al. 2010. Quantitative analysis of fitness and genetic interactions in yeast on a genome scale. *Nat Methods* 7: 1017–1024.

Breuer K, Foroushani AK, Laird MR, Chen C, Sribnaia A, Lo R, Winsor GL, Hancock RE, Brinkman FS, Lynn DJ. 2013. InnateDB: Systems biology of innate immunity and beyond—Recent updates and continuing curation. *Nucleic Acids Res* 41: D1228–D1233.

Costanzo M, Baryshnikova A, Bellay J, Kim Y, Spear ED, Sevier CS, Ding H, Koh JL, Toufighi K, Mostafavi S, et al. 2010. The genetic landscape of a cell. *Science* 327: 425–431.

Freeman TC, Goldovsky L, Brosch M, van Dongen S, Maziere P, Grocock RJ, Freilich S, Thornton J, Enright AJ. 2007. Construction, visualisation, and clustering of transcription networks from microarray expression data. *PLoS Comput Biol* 3: 2032–2042.

Fruchterman TMJ, Reingold EM. 1991. Graph drawing by force-directed placement. *Software* 21: 1129–1164.

Gavin AC, Aloy P, Grandi P, Krause R, Boesche M, Marzioch M, Rau C, Jensen LJ, Bastuck S, Dumpelfeld B, et al. 2006. Proteome survey reveals modularity of the yeast cell machinery. *Nature* 440: 631–636.

Kamada T, Kawai S. 1989. An algorithm for drawing general unidirected graphs. *Processing Lett* 31: 7–15.

Krogan NJ, Cagney G, Yu H, Zhong G, Guo X, Ignatchenko A, Li J, Pu S, Datta N, Tikuisis AP, et al. 2006. Global landscape of protein complexes in the yeast *Saccharomyces cerevisiae*. *Nature* 440: 637–643.

Maere S, Heymans K, Kuiper M. 2005. BiNGO: A Cytoscape plugin to assess overrepresentation of gene ontology categories in biological networks. *Bioinformatics* 21: 3448–3449.

Myers CL, Barrett DR, Hibbs MA, Huttenhower C, Troyanskaya OG. 2006. Finding function: Evaluation methods for functional genomic data. *BMC Genomics* 7: 187.

Pu S, Wong J, Turner B, Cho E, Wodak SJ. 2009. Up-to-date catalogues of yeast protein complexes. *Nucleic Acids Res* 37: 825–831.

Smoot ME, Ono K, Ruscheinski J, Wang PL, Ideker T. 2011. Cytoscape 2.8: New features for data integration and network visualization. *Bioinformatics* 27: 431–432.

CHAPTER 33

Metabolomics in Yeast

Amy A. Caudy,[1,2] Michael Mülleder,[3] and Markus Ralser[3,4,5]

[1]Donnelly Centre for Cellular and Biomolecular Research, University of Toronto, Toronto, Ontario M5S3E1, Canada; [2]Department of Molecular Genetics, University of Toronto, Toronto, Ontario M5S3E1, Canada; [3]Department of Biochemistry and Cambridge Systems Biology Centre, University of Cambridge, Cambridge CB2 1GA, United Kingdom; [4]The Francis Crick Institute, Mill Hill Laboratory, London NW7 1AA, United Kingdom

Budding yeast has from the beginning been a major eukaryotic model for the study of metabolic network structure and function. This is attributable to both its genetic and biochemical capacities and its role as a workhorse in food production and biotechnology. New inventions in analytical technologies allow accurate, simultaneous detection and quantification of metabolites, and a series of recent findings have placed the metabolic network at center stage in the physiology of the cell. For example, metabolism might have facilitated the origin of life, and in modern organisms it not only provides nutrients to the cell but also serves as a buffer to changes in the cellular environment, a regulator of cellular processes, and a requirement for cell growth. These findings have triggered a rapid and massive renaissance in this important field. Here, we provide an introduction to analysis of metabolomics in yeast.

METABOLOMICS IN YEAST

Research on metabolism and metabolic enzymes dominated the early days of molecular biology. Interest in the topic decreased with the appearance of polymerase chain reaction (PCR) and molecular genetics during the 1980s and 1990s, although the importance of budding yeast in biotechnology ensured continued progress in the understanding of its physiology. Yeast has been used in winemaking, brewing, and baking since ancient times. The economic importance of fermentation triggered the first attempts to breed and manipulate yeast species and led to the purification, crystallization, and characterization of the first enzymes in the early 20th century (with major contributions from Sumner and Northrop in the United States and Warburg in Germany). Yeast experiments were so important in the development of biochemistry that the word "enzyme" is derived from the Greek word for *in leaven* (yeast).

The full set of metabolic reactions in the cell is referred to as the metabolic network. Although complex, the basic structure of the metabolic network is largely conserved among organisms, indicating that it is of a common evolutionary origin (Jeong et al. 2000; Ravasz et al. 2002). Compared with the large number of chemical reactions and possible mechanisms, metabolism operates with only a subset of the reactions and prefers short paths in its functionality (Noor et al. 2010). Overall, this conservation means that metabolomics is a particularly attractive technique, as in principle any analytical method can be applied to a large variety of different species. However, sample preparation methods are species specific, because composition of the cell membrane, presence of a cell wall, cellular resistance to physical parameters, and relative and absolute metabolite content differ from species to species.

[5]Correspondence: mr559@cam.ac.uk

Copyright © Cold Spring Harbor Laboratory Press; all rights reserved
Cite this introduction as *Cold Spring Harb Protoc*; doi:10.1101/pdb.top083576

In principle, two values are important for a metabolic intermediate: its concentration and its turnover rate. Both values provide different information about metabolism. Whereas the concentration values can point to perturbations in the pathway, allow conclusions on the functionality of individual enzymes, and thus serve as a functional parameter for the regulatory cross talk of metabolism, the turnover rate (referred to as metabolic flux on a pathway and network scale) is informative about the activity of a metabolic pathway and its yield and consumption of cofactors and intermediate metabolites (Grüning et al. 2010; Heinemann and Sauer 2010).

Certain parameters are crucial in a yeast metabolomics experiment, and they determine not only the appropriate methods for sample preparation and analysis but also which yeast strains and growth conditions can be applied. These parameters include the following.

- *Nature of application: Targeted versus shotgun approach.* Although it sounds attractive to quantify a large number of metabolites at the same time, quantitative accuracy declines in broad experiments. Different metabolite classes have different stabilities, chemical properties, and turnover rates. There is therefore no universal sample-quenching procedure that works equally well for all metabolites. Our accompanying protocols (see Protocol 1: Metabolite Extraction from *Saccharomyces cerevisiae* for Liquid Chromatography–Mass Spectrometry [Rosebrock and Caudy 2015], Protocol 2: A High-Throughput Method for the Quantitative Determination of Free Amino Acids in *Saccharomyces cerevisiae* by Hydrophilic Interaction Chromatography–Tandem Mass Spectrometry [Mülleder et al. 2015], and Protocol 3: Spectrophotometric Analysis of Ethanol and Glucose Concentrations in Yeast Culture Media [Caudy 2015]) suggest several widely applicable methods, but it is important to define the biological questions of interest in advance to select the optimal experimental method.

- *Turnover rates.* Turnover is fast, and concentration changes of metabolites occur rapidly. For instance, applying hydrogen peroxide to yeast cells to cause oxidative stress leads to an inactivation of the metabolic enzyme GAPDH within 2–15 sec, while equally fast, an increase in the concentration of pentose phosphate pathway metabolites is observed (Ralser et al. 2009). Sampling procedures need to take this speed of metabolism into account. Washing and centrifugation steps in water or chemical buffers quickly cause starvation phenotypes and should be avoided. To preserve the state of metabolism, cells should be quenched as quickly as possible (i.e., by rapid freezing in cold organic solvents [Rabinowitz 2007; de Koning and van Dam 1992]), as described in Protocol 1: Metabolite Extraction from *Saccharomyces cerevisiae* for Liquid Chromatography–Mass Spectrometry (Rosebrock and Caudy 2015).

- *Auxotrophies and media supplements.* Auxotrophic markers are very attractive tools in yeast genetics, but be aware that they represent a major problem in yeast metabolism research. For example, the common laboratory strains BY4741 or W303 possess four to five autotrophies; in practice, this means that highly active biosynthetic pathways are simultaneously perturbed. The physiological effects are only partially complemented by media supplementation, and different nutrients are depleted at different rates during the growth phase. Thus, matching autotrophies between strains within an experiment is essential, and where possible, prototrophic cells are preferred in a metabolomics experiment (Pronk 2002; Mülleder et al. 2012; VanderSluis et al. 2014).

- *Growth.* Yeast metabolism strongly differs between growth phases. Some metabolite concentrations differ between early-, mid-, and late-exponential phases, and stationary phase before or beyond the diauxic shift. In batch cultures, these effects are amplified for auxotrophic strains, because essential supplements are consumed at a different rate. It is important to carefully control growth conditions for cell harvesting.

ANALYSIS OF METABOLOMICS

Metabolites have very different physicochemical properties and can span a concentration range of many orders of magnitude. Therefore, appropriate sample preparation and analytical methods must be used to reliably determine the metabolites of interest and provide a comprehensive snapshot of the entire metabolome. Currently, metabolomics is mostly conducted by nuclear magnetic resonance (NMR) and mass spectrometry (MS).

Nuclear Magnetic Resonance

NMR is a widely used analytical platform for metabolism research and structural elucidation of compounds. Its application in metabolomics is limited by sensitivity (typically micromolar to nanomolar) compared with MS (femtomolar to attomolar) range. However, it has the advantage of allowing in vivo measurements and direct observation of metabolic reactions when hyperpolarization is used to increase sensitivity (Meier et al. 2011; Rodrigues et al. 2014).

Mass Spectrometry

Different platforms and strategies exist for MS, and they have advantages and limitations according to the question to be answered. Mass analyzers with high accuracy, acquisition speed, and resolution are the main requirement for untargeted approaches; this application is currently dominated by time-of-flight (TOF) and Fourier transform–based mass spectrometry (FT-MS) using Orbitrap instruments. In targeted analyses, sensitivity and dynamic range are the important parameters. These analyses are typically performed on (triple) quadrupole platforms or hybrid or tandem platforms combining quadrupoles with high-resolution mass analyzers (i.e., Quadrupole-TOF [qTOF]). Quantification of metabolites is achieved with external or internal standards (stable isotope-labeled standard compounds) or metabolic labeling (extracts from cells grown in isotope-enriched media). Samples can be measured directly by surface-based methods or by flow injection analysis (FIA); however, to increase specificity and quantitative reliability, the complexity of the samples is usually reduced by separation techniques such as gas chromatography (GC), liquid chromatography (LC), capillary electrophoresis (CE), or ion mobility. GC-MS is very useful for volatile, thermostable, and energetically stable molecules and achieves high selectivity and reproducibility. Many polar metabolites, however, are poorly volatile in GC and therefore need prior derivatization to ionize. In contrast, LC-MS allows separation of a broad range of metabolite classes and is therefore the most widely used platform for metabolomics. The dominant LC technique coupled to MS is reversed phase high-performance liquid chromatography (HPLC). Hydrophilic interaction liquid chromatography (HILIC) and ion pairing reagents have proven very useful for the separation of hundreds of polar hydrophilic compounds of primary metabolism (Bajad et al. 2006; Buescher et al. 2010). Although surface-based MS techniques (e.g., matrix-assisted laser desorption/ionization [MALDI]) are limited to abundant metabolites, the ability to measure metabolites at single cell concentrations promises to drive future analysis (Amantonico et al. 2008; Ibáñez et al. 2013).

CONCLUSION

Metabolism is involved in many human diseases, including diabetes, cancer, and neurodegeneration (Hsu and Sabatini 2008; Grüning et al. 2010; Hanahan and Weinberg 2011; Buescher et al. 2012; Keller et al. 2014). Yeast as a model organism continues to be of central importance to the study of metabolism. With the availability of the yeast genome sequence and the wealth of knowledge on the biochemistry of the yeast cell, researchers have generated a reconstruction of its metabolic network. The reconstruction is under continuous refinement and is currently the best eukaryotic model available (Mo et al. 2009). It not only describes the network topology but also allows for simulation and prediction of phenotypes, such as growth, gene knockouts, production of important compounds, and nutrient uptake. Genome-scale models can be combined with transcriptomic, proteomic, and/or metabolomic profiles to better understand an observation, and they have also been successfully applied for metabolic engineering and strain optimization (Oberhardt et al. 2009). Furthermore, they allow for estimation of the metabolic flux under defined steady-state conditions, providing valuable information about pathway activities and aiding in the determination of in vivo flux rates from isotope tracer distributions in metabolic flux analyses (Sauer 2006). It is important to note that because the models are based on our knowledge, they are prone to errors due to incorrect or missing information; therefore, predictions must always be verified experimentally (Österlund et al. 2012).

REFERENCES

Amantonico A, Oh JY, Sobek J, Heinemann M, Zenobi R. 2008. Mass spectrometric method for analyzing metabolites in yeast with single cell sensitivity. *Angew Chem Int Ed Engl* 47: 5382–5385.

Bajad SU, Lu W, Kimball EH, Yuan J, Peterson C, Rabinowitz JD. 2006. Separation and quantitation of water soluble cellular metabolites by hydrophilic interaction chromatography-tandem mass spectrometry. *J Chromatogr A* 1125: 76–88.

Buescher JM, Moco S, Sauer U, Zamboni N. 2010. Ultrahigh performance liquid chromatography-tandem mass spectrometry method for fast and robust quantification of anionic and aromatic metabolites. *Anal Chem* 82: 4403–4412.

Buescher JM, Liebermeister W, Jules M, Uhr M, Muntel J, Botella E, Hessling B, Kleijn RJ, Le Chat L, Lecointe F, et al. 2012. Global network reorganization during dynamic adaptations of *Bacillus subtilis* metabolism. *Science* 335: 1099–1103.

Caudy AA. 2015. Spectrophotometric analysis of ethanol and glucose concentrations in yeast culture media. *Cold Spring Harb Protoc*. doi: 10.1101/pdb.prot089102.

de Koning W, van Dam K. 1992. A method for the determination of changes of glycolytic metabolites in yeast on a subsecond time scale using extraction at neutral pH. *Anal Biochem* 204: 118–123.

Grüning NM, Lehrach H, Ralser M. 2010. Regulatory crosstalk of the metabolic network. *Trends Biochem Sci* 35: 220–227.

Hanahan D, Weinberg RA. 2011. Hallmarks of cancer: The next generation. *Cell* 144: 646–674.

Heinemann M, Sauer U. 2010. Systems biology of microbial metabolism. *Curr Opin Microbiol* 13: 337–343.

Hsu PP, Sabatini DM. 2008. Cancer cell metabolism: Warburg and beyond. *Cell* 134: 703–707.

Ibáñez AJ, Fagerer SR, Schmidt AM, Urban PL, Jefimovs K, Geiger P, Dechant R, Heinemann M, Zenobi R. 2013. Mass spectrometry-based metabolomics of single yeast cells. *Proc Natl Acad Sci* 110: 8790–8794.

Jeong H, Tombor B, Albert R, Oltvai ZN, Barabási AL. 2000. The large-scale organization of metabolic networks. *Nature* 407: 651–654.

Keller MA, Turchyn AV, Ralser M. 2014. Non-enzymatic glycolysis and pentose phosphate pathway-like reactions in a plausible Archean ocean. *Mol Syst Biol* 10: 725.

Meier S, Karlsson M, Jensen PR, Lerche MH, Duus JØ. 2011. Metabolic pathway visualization in living yeast by DNP-NMR. *Mol Biosyst* 7: 2834–2836.

Mo ML, Palsson BO, Herrgård MJ. 2009. Connecting extracellular metabolomic measurements to intracellular flux states in yeast. *BMC Syst Biol* 3: 37.

Mülleder M, Capuano F, Pir P, Christen S, Sauer U, Oliver SG, Ralser M. 2012. A prototrophic deletion mutant collection for yeast metabolomics and systems biology. *Nat Biotechnol* 30: 1176–1178.

Mülleder M, Bluemlein K, Ralser M. 2015. A high-throughput method for the quantitative determination of free amino acids in *Saccharomyces cerevisiae* by hydrophilic interaction chromatography–tandem mass spectrometry. *Cold Spring Harb Protoc*. doi: 10.1101/pdb.prot089094.

Noor E, Eden E, Milo R, Alon U. 2010. Central carbon metabolism as a minimal biochemical walk between precursors for biomass and energy. *Mol Cell* 39: 809–820.

Oberhardt MA, Palsson BØ, Papin JA. 2009. Applications of genome-scale metabolic reconstructions. *Mol Syst Biol* 5: 320.

Österlund T, Nookaew I, Jens N. 2012. Fifteen years of large scale metabolic modeling of yeast: Developments and impacts. *Biotechnol Adv* 30: 979–988.

Pronk JT. 2002. Auxotrophic yeast strains in fundamental and applied research. *Appl Env Microbiol* 68: 2095–2100.

Rabinowitz JD. 2007. Cellular metabolomics of *Escherichia coli*. *Expert Rev Proteomics* 4: 187–198.

Ralser M, Wamelink MM, Latkolik S, Jansen EE, Lehrach H, Jakobs C. 2009. Metabolic reconfiguration precedes transcriptional regulation in the antioxidant response. *Nat Biotechnol* 27: 604–605.

Ravasz E, Somera AL, Mongru DA, Oltvai ZN, Barabasi AL. 2002. Hierarchical organization of modularity in metabolic networks. *Science* 297: 1551–1555.

Rodrigues TB, Serrao EM, Kennedy BWC, Hu D-E, Kettunen MI, Brindle KM. 2014. Magnetic resonance imaging of tumor glycolysis using hyperpolarized 13C-labeled glucose. *Nat Med* 20: 93–97.

Rosebrock AP, Caudy AA. 2015. Metabolite extraction from *Saccharomyces cerevisiae* for liquid chromatography–mass spectrometry. *Cold Spring Harb Protoc*. doi: 10.1101/pdb.prot089086.

Sauer U. 2006. Metabolic networks in motion: 13C-based flux analysis. *Mol Syst Biol* 2: 62.

VanderSluis B, Hess DC, Pesyna C, Krumholz EW, Syed T, Szappanos B, Nislow C, Papp B, Troyanskaya OG, Myers CL, et al. 2014. Broad metabolic sensitivity profiling of a prototrophic yeast deletion collection. *Genome Biol* 15: R64.

Protocol 1

Metabolite Extraction from *Saccharomyces cerevisiae* for Liquid Chromatography–Mass Spectrometry

Adam P. Rosebrock[1,2] and Amy A. Caudy[1,2,3]

[1]Donnelly Centre for Cellular and Biomolecular Research, University of Toronto, Toronto, Ontario M5S3E1, Canada; [2]Department of Molecular Genetics, University of Toronto, Toronto, Ontario M5S3E1, Canada

Prior to mass spectrometric analysis, cellular small molecules must be extracted and separated from interfering components such as salts and culture medium. To ensure minimal perturbation of metabolism, yeast cells grown in liquid culture are rapidly harvested by filtration as described here. Simultaneous quenching of metabolism and extraction is afforded by immediate immersion in low-temperature organic solvent. Samples prepared using this method are suitable for a range of downstream liquid chromatography–mass spectrometry analyses and are stable in solvent for >1 yr at −80°C.

MATERIALS

It is essential that you consult the appropriate Material Safety Data Sheets and your institution's Environmental Health and Safety Office for proper handling of equipment and hazardous materials used in this protocol.

Reagents

Extraction solvent (methanol:acetonitrile:water [MeOH:MeCN:H$_2$O] 40:40:20)

Use high-performance liquid chromatography (HPLC)-grade solvents, including H$_2$O. Solvents marketed for HPLC use are often of sufficient quality and can provide better value than solvents marketed for liquid chromatography–mass spectrometry (LC-MS) use.

Each lot of MeOH or MeCN contains different contaminants; therefore, it is critical to perform a "mock extraction" of a dry filter in extraction solvent for every experiment as described in Step 9. It is also useful to screen several potential suppliers and grades of solvents to select one that has a contaminant profile compatible with your analysis method.

Minimal medium for yeast cell growth

Use of minimal medium (e.g., yeast nitrogen base [YNB] + glucose or other appropriate carbon source) is encouraged wherever feasible; see Discussion. It is critical that cells are grown in the same nutrient source for both the overnight culture (Step 1) and the final growth phase (Step 2). Overnight growth on one medium (e.g., rich medium) followed by a switch into a different medium (e.g., minimal medium) will likely generate results that do not reflect a steady-state measurement of metabolism in the final growth conditions. Instead, profiles will reflect cells in the midst of altering gene expression and protein levels to respond to the new nutrient source. The kinetics of these transcriptional and physiological changes differ as a function of nutrient and genotype.

Yeast cells for starter culture

Samples prepared using the following method are suitable for a range of downstream LC-MS analyses, including hydrophobic interaction (Pluskal et al. 2010) and reverse phase or ion-paired reverse phase chromatography (Ralser et al. 2007; Evans et al. 2009; Lu et al. 2010). The number of cells required is extremely dependent on the

---

[3]Correspondence: amy.caudy@utoronto.ca

Copyright © Cold Spring Harbor Laboratory Press; all rights reserved
Cite this protocol as *Cold Spring Harb Protoc*; doi:10.1101/pdb.prot089086

combination of chromatography and mass spectrometric detection selected. Input from the MS analyst is key, and preliminary experiments are necessary to optimize culture amounts. We have successfully processed up to 50 mL of culture by this method.

Equipment

Centrifuge at 4°C (with swinging-bucket rotor)

Dry ice

Dry ice should be contained in a Styrofoam box suitable for 15-mL tubes. Do not add alcohol or acetone.

Filtration assembly or manifold

A single filter assembly (typically held together with a clamp, e.g., Millipore XX1002530 or VWR 26316-692) is sufficient for processing a small number of samples. For experiments involving time courses or many samples, it is often more convenient to use a multiposition manifold, such as the Hoefer FH 225V 10-position filter holder. In situations where large input culture volumes are required, alternative filter systems can be used.

Forceps, straight with sharp edges (Dumont #1 or #4 preferred)

Labels for sample tubes

Use laser-printed or thermal transfer labels where possible, as spilled extraction solvent will dissolve most markers.

Microcentrifuge tubes (1.5-mL)

As with 15-mL tubes, high-quality consumables are preferred to provide consistent extractable contaminants. Eppendorf microcentrifuge tubes are recommended for their low and consistent solvent extractable profile.

Nylon filters (0.2- to 0.5-µm, supported) (25-mm diameter, or sized for filter apparatus)

Suitable filters include GE Osmonics MAGNA R04SP02500 and Pall Nylaflo 28140-028. Use caution when changing vendors, as background contaminants differ between vendors and part numbers.
Larger-diameter filters may be useful for analyses that require larger cell quantities.

Positive-displacement pipettes

Positive-displacement pipettes are recommended, as the high vapor pressure of the extraction solvent makes volumetric transfer with air-displacement pipettes challenging. A particularly convenient option is a disposable syringe-style repeater pipette, such as the Eppendorf Repeater or Gilson Distriman.

Spectrophotometer

Tubes (polypropylene, 15-mL, conical bottom, or 5-mL flip top)

Select a single vendor for tubes and use them throughout a project; each supplier's plastic has different contaminants. We have found that "premium" name-brand tubes have more consistent contaminant profiles than discounted brands, which may use different manufacturers or plastic feedstocks for a single part number.

Vacuum pump with liquid trap suitable for aqueous filtrates

Vortex mixer

Warm room or steel manifold chimneys (optional)

Yeast culture growth and harvesting can be performed entirely in a warm room or using prewarmed steel manifold chimneys. Alternatively, cultures can be grown and harvested at room temperature. Avoid harvesting cells at a different temperature than that of growth, as changes in temperature rapidly alter metabolite levels.

METHOD

Timing is critical throughout the following protocol. Work quickly and consistently. When cells rest on a filter without growth medium, a starvation response is detectable within seconds (Brauer et al. 2006). Thus, filtration of the sample must be carefully monitored such that filters can be removed and immersed in −20°C-tempered extraction solvent immediately after filtration of each sample is complete (Step 6). Label all tubes and organize the work area in advance to speed sample handling, and handle one sample at a time so that samples are harvested and immediately returned to the freezer. For filtration manifolds, set up multiple positions with filters to allow sampling in rapid succession (not parallel!) without replacing filter membranes. Mock extractions should be performed before processing of critical samples to build familiarity with the protocol.

1. Grow a yeast starter culture overnight in minimal medium (e.g., YNB with glucose or other carbon source as appropriate). Prepare a sufficient amount of culture to allow for the dilutions and desired number of optical density measurements in Step 2.

   Triplicate samples (at a minimum) are typically used for measurements of optical density.

2. Dilute the overnight culture to $OD_{600} = 0.1$ in ≥ 6 mL of medium. Grow the cells for two doublings until $OD_{600} = 0.4$.

   Harvesting at this phase of early- to mid-log growth provides reproducible measurements. Do not confound final density at harvest with the requirement for cells to be actively growing in early log phase. Cells should be permitted to grow for two doublings after dilution. A common misstep is to dilute an overnight culture to the desired final OD_{600}—0.4 in this case—and immediately harvest.

   We find that metabolic profiles are strongly correlated with growth phase. Cells at $OD_{600} = 0.8$ and greater have alterations in metabolism when compared with less dense samples. We observe ~3 h per doubling for cells in glucose minimal medium at room temperature; empirical determination of doubling time is required for other nutrient sources, genotypes, or strain backgrounds.

3. While cells are growing in Step 2, prepare 15-mL conical tubes containing 1.2 mL of extraction solvent per tube. Store the tubes at −20°C until Step 6.

   If using a larger quantity of cells and larger diameter filters, the extraction solvent volume should be scaled up to maintain the same ratio of extraction solvent to filter area and amount of medium inevitably retained during filtration.

   Extraction solvent must be cooled to −20°C; it will freeze solid at −80°C (such as on dry ice or in an ultralow freezer).

4. Immediately before harvest (Step 5), measure and record the OD_{600} of the cells.

   This information will be used to ensure equal loading of cells. For some experiments, direct sizing (e.g., with a Coulter counter) may be more appropriate to enable normalization by total cell volume. We find that two OD_{600} units of cells (e.g., 4 mL of cells at $OD_{600} = 0$), or approximately 6×10^7 cells, is a good starting point. If $OD_{600} < 0.2$, a larger volume of cells should be harvested and noted accordingly.

5. Remove 5 mL of culture and transfer to the filtration apparatus.

   The number of cells harvested may need to be adjusted depending on the number of cells required to detect the analytes of interest. Ensure that both the optical density and harvest volume are recorded for each sample, as these are key parameters for normalizing sample amounts.

6. Apply a vacuum to harvest the cells by filtration onto the filter. Immediately after the liquid is drawn through the filter, use forceps to pick up the filter. Immediately transfer the filter to a 15-mL tube containing cold extraction solvent.

7. Immediately invert the tube twice and pulse for 10–15 sec on a vortex to remove cells from the filter and mix with extraction solvent. (The suspension should become turbid when cells have detached from filter.) Pulse for additional time if the cells have not released. Immediately place the tube on dry ice and proceed to additional samples, as appropriate.

8. Prepare a medium-only control: Filter sterile growth medium (no cells) and process as described in Steps 5–7.

   This critical control allows identification of compounds present in the growth medium. Depending on the experiment and analytes of interest, it may also be desirable to analyze conditioned culture medium that has been cleared of cells by centrifugation or filtration. This will permit the identification of compounds produced from the cells.

9. Prepare a background control by performing a mock extraction without medium or cells: Place a clean, dry filter into extraction solvent in a tube and vortex as described in Step 7.

   This critical control allows determination of background contamination introduced from the plasticwar, and allows comparison to the compounds present in cell medium.

10. After completing extraction of all samples and controls, leave the tubes on dry ice until all the sample solvents have frozen.

11. Thaw the samples by moving the tubes to −20°C.

12. Return all samples to dry ice.

13. Repeat Steps 11–12 for a total of three freeze–thaw cycles. Store the samples at −20°C until ready to proceed with Step 14.

    Freeze–thaws can also be performed by cycling samples between freezers at −80°C and −20°C.

14. Remove the filters from the 15-mL tubes by fishing them out with a clean weigh spatula or forceps. Discard the filters.

15. Centrifuge the 15-mL tubes in a swinging-bucket rotor at ~3000–4000 × g for 10 min at 4°C to pellet the cells.

16. Using a 1000-µL micropipette, transfer 1 mL of supernatant to prelabeled 1.5-mL microcentrifuge tubes. Avoid the pellet and the immediate liquid. (Approximately 200 µL of cells and solvent will remain in the tube.) Use a different pipette tip for each sample.

    The cell pellet can be saved for measurement of DNA content by flow cytometry (see Protocol 3 of Chapter 14: Analysis of the Budding Yeast Cell Cycle by Flow Cytometry [Rosebrock 2015]), western blotting, or RNA extraction.

17. Store the extracts in microcentrifuge tubes at −80°C.

    If shipping to an off-site collaborator, the samples can be shipped on dry ice. In the typical workflow for polar metabolites, samples are dried under a flow of nitrogen or lyophilized and resuspended for LC-MS analysis in aqueous buffer in a volume adjusted for cell number. Resuspension volumes and solvent requirement will vary based on the LC-MS platform used. Consult your MS analyst BEFORE resuspending samples.

18. Clean the filtration assembly by vacuum-aspirating 10 mL of H_2O three times. (For a filtration manifold, wash each position.) Disassemble the manifold and rinse all parts with H_2O. If using a filtration manifold with stainless steel chimneys, wash the chimneys and store them with the flat top side down, separately from the manifold.

    Care should be taken with chimney bottoms, as nicks on the bottom ring can prevent proper sealing of the filter against the base and result in sample loss.

DISCUSSION

For this protocol, it is highly preferable to use prototrophic strains grown in minimal medium, and whole-genome mutant prototrophic libraries are available (Mülleder et al. 2012; Gibney et al. 2013; VanderSluis et al. 2014). Supplementation of growth medium with auxotrophic nutrients perturbs cellular metabolic pathways, prevents quantitation of supplemented compounds due to carryover of growth medium, and can decrease the sensitivity of MS analysis because of in-source suppression. When the use of supplements is unavoidable, all strains in a given experiment should be chosen to have the same auxotrophies, as the metabolic changes resulting from auxotrophies are difficult to deconvolve from other experimental manipulations. The high concentration of ammonium sulfate (5 g/L) in standard minimal medium can cause chromatographic disturbances in some workflows. In addition, sulfate is nonvolatile and may result in increased maintenance requirements for mass spectrometers. We routinely use 0.5 g/L ammonium sulfate with 2.62 g/L ammonium acetate as a nitrogen source in minimal medium prepared using YNB. This provides the equivalent concentration of ammoniacal nitrogen and adequate concentration of the essential nutrient sulfur. Acetate is not used as a carbon source when glucose is present.

Rapid harvesting is essential to reduce metabolic perturbation, as starvation responses are detectable within 30 sec after nutrients are removed (Brauer et al. 2006). A commonly used method for quenching is the direct mixing of liquid cultures with chilled methanol (Villas-Boas et al. 2005). Detailed comparison of methanol quenching versus filtration has shown that the filtration method presented here produces equivalent results, with the advantage of higher signal from nucleotide triphosphates and fewer salts in the sample (Boer et al. 2010). In addition, using a mixture of acetonitrile and methanol as the extraction solvent produces an equal or greater yield of metabolites (Boer et al. 2010).

This protocol includes minimal sample processing between extraction and LC-MS analysis. Sample cleanup can be performed (e.g., using solid phase extraction) as necessary. Because of the diverse chemical nature of cellular metabolites, further processing generally results in the enrichment of a subset of compounds while removing or excluding others.

ACKNOWLEDGMENTS

A.A.C. is the Canada Research Chair in Metabolomics for Enzyme Discovery and is supported by the Ontario Early Researcher Award, the Canadian Institutes for Health Research, the Natural Sciences and Engineering Research Council of Canada, the Canadian Foundation for Innovation and the Ontario Leader's Opportunity Fund. A.P.R. is supported by the Connaught Innovator Award and the Canadian Institutes for Health Research. A.P.R. and A.A.C. collaborate with Agilent Technologies to develop LC-MS methodologies.

REFERENCES

Boer VM, Crutchfield CA, Bradley PH, Botstein D, Rabinowitz JD. 2010. Growth-limiting intracellular metabolites in yeast growing under diverse nutrient limitations. *Mol Biol Cell* 21: 198–211.

Brauer MJ, Yuan J, Bennett BD, Lu W, Kimball E, Botstein D, Rabinowitz JD. 2006. Conservation of the metabolomic response to starvation across two divergent microbes. *Proc Natl Acad Sci* 103: 19302–19307.

Evans AM, DeHaven CD, Barrett T, Mitchell M, Milgram E. 2009. Integrated, nontargeted ultrahigh performance liquid chromatography/electrospray ionization tandem mass spectrometry platform for the identification and relative quantification of the small-molecule complement of biological systems. *Anal Chem* 81: 6656–6667.

Gibney PA, Lu C, Caudy AA, Hess DC, Botstein D. 2013. Yeast metabolic and signaling genes are required for heat-shock survival and have little overlap with the heat-induced genes. *Proc Natl Acad Sci* 110: E4393–E4402.

Lu W, Clasquin MF, Melamud E, Amador-Noguez D, Caudy AA, Rabinowitz JD. 2010. Metabolomic analysis via reversed-phase ion-pairing liquid chromatography coupled to a stand alone orbitrap mass spectrometer. *Anal Chem* 82: 3212–3221.

Mülleder M, Capuano F, Pir P, Christen S, Sauer U, Oliver SG, Ralser M. 2012. A prototrophic deletion mutant collection for yeast metabolomics and systems biology. *Nat Biotechnol* 30: 1176–1178.

Pluskal T, Nakamura T, Villar-Briones A, Yanagida M. 2010. Metabolic profiling of the fission yeast *S. pombe*: Quantification of compounds under different temperatures and genetic perturbation. *Mol Biosyst* 6: 182–198.

Ralser M, Wamelink MM, Kowald A, Gerisch B, Heeren G, Struys EA, Klipp E, Jakobs C, Breitenbach M, Lehrach H, et al. 2007. Dynamic rerouting of the carbohydrate flux is key to counteracting oxidative stress. *J Biol* 6: 10.

Rosebrock AP. 2015. Analysis of the budding yeast cell cycle by flow cytometry. *Cold Spring Harb Protoc* doi: 10.1101/pdb.prot088740.

VanderSluis B, Hess DC, Pesyna C, Krumholz EW, Syed T, Szappanos B, Nislow C, Papp B, Troyanskaya OG, Myers CL, Caudy AA. 2014. Broad metabolic sensitivity profiling of a prototrophic yeast deletion collection. *Genome Biol* 15: R64.

Villas-Boas SG, Hojer-Pedersen J, Akesson M, Smedsgaard J, Nielsen J. 2005. Global metabolite analysis of yeast: Evaluation of sample preparation methods. *Yeast* 22: 1155–1169.

Protocol 2

A High-Throughput Method for the Quantitative Determination of Free Amino Acids in *Saccharomyces cerevisiae* by Hydrophilic Interaction Chromatography–Tandem Mass Spectrometry

Michael Mülleder,[1] Katharina Bluemlein,[1,3] and Markus Ralser[1,2,4]

[1]Department of Biochemistry and Cambridge Systems Biology Centre, University of Cambridge, Cambridge CB2 1GA, United Kingdom; [2]The Francis Crick Institute, Mill Hill Laboratory, the Ridgeway, Mill Hill, London NW7 1AA, United Kingdom

Amino acids are the building blocks for protein synthesis and the precursors for many biomolecules, such as glutathione and *S*-adenosylmethionine. Their intracellular concentrations provide valuable information about the overall metabolic state of the cell, as they are closely connected to carbon and nitrogen metabolism and are tightly regulated to meet cellular demands in ever-changing environments. Here, we describe a fast and simple method enabling metabolic profiling for free amino acids for large numbers of yeast strains. Metabolites are extracted with boiling ethanol and, without further conditioning, analyzed by hydrophilic interaction chromatography–tandem mass spectrometry (HILIC-MS/MS). Several hundred samples can be prepared in a single day with an analytical runtime of 3.25 min. This method is valuable for functional characterization, identification of metabolic regulators and processes, or monitoring of biotechnological processes.

MATERIALS

It is essential that you consult the appropriate Material Safety Data Sheets and your institution's Environmental Health and Safety Office for proper handling of equipment and hazardous materials used in this protocol.

RECIPES: Please see the end of this protocol for recipes indicated by <R>. Additional recipes can be found online at http://cshprotocols.cshlp.org/site/recipes.

Reagents

Amino acid standards (analytical grade)

The levels of free amino acids span three orders of magnitude; hence, standards must be prepared at appropriate concentrations. See Table 1 for the recommended standard mix.

Ethanol (absolute)

Saccharomyces cerevisiae stock culture (cryopreserved)

This method is optimized for yeast culture using 2% glucose as the carbon source and 5 g/L of ammonium sulfate as the nitrogen source. Yeast strains should be prototrophic to eliminate the need for supplementation and additional washing steps when collecting the cells.

Solvent A for HILIC analysis of amino acids <R>

---

[3]Present address: Fraunhofer Institute for Toxicology and Experimental Medicine, 30625 Hannover, Germany
[4]Correspondence: mr559@cam.ac.uk

Copyright © Cold Spring Harbor Laboratory Press; all rights reserved
Cite this protocol as *Cold Spring Harb Protoc*; doi:10.1101/pdb.prot089094

TABLE 1. Mass spectrometric parameters for targeted analysis of amino acids

Compound name	Compound abbreviation	SRM transition	Fragmentor (V)	Collision energy (V)	Retention time (min)	Standard mix concentration (μM)
Phenylalanine	F	166.1 > 120	100	9	1.23	25
Leucine	L	132.1 > 86	80	8	1.24	25
Tryptophane	W	205.1 > 188	85	5	1.29	2.5
Isoleucine	I	132.1 > 86	80	8	1.33	25
Methionine	M	150.1 > 104	40	8	1.47	10
Valine	V	118.1 > 71.9	100	10	1.57	100
Proline	P	116.1 > 70.1	100	13	1.6	50
Tyrosine	Y	182 > 165	90	5	1.64	500
Cysteine	C	122 > 76	60	10	1.74	0.5
Alanine	A	90 > 44.1	50	8	1.87	500
Threonine	T	120.1 > 74	80	9	1.96	250
+ Homoserine	+HS					NA
Glycine	G	76 > 30.1	50	5	1.98	250
Glutamine	Q	147.1 > 84	50	16	2.07	2500
Glutamate	E	148.1 > 84.1	75	10	2.09	2500
Serine	S	106 > 60	40	9	2.1	250
Asparagine	N	133.1 > 74	80	9	2.12	100
Aspartate	D	134.1 > 74	80	10	2.24	500
Histidine	H	156.1 > 110.2	80	12	2.36	250
Arginine	R	175.1 > 70	100	15	2.37	250
Lysine	K	147.1 > 84	50	16	2.41	100

Note that these settings will require adjustments for other liquid chromatography–tandem mass spectrometry instruments.

SRM, selected reaction monitoring.

Solvent B for HILIC analysis of amino acids <R>
Synthetic minimal liquid medium <R>
  Prepare on the day of cultivation.

Synthetic minimal solid medium <R>
  Prepare on the day of cultivation.

EQUIPMENT

Aluminum sealing foil (suitable for −80°C)
Centrifuge
Incubator at 30°C
Liquid chromatography system (Agilent Infinity 1290 HPLC or comparable instrument with operating pressure of up to 1000 bar) and ACQUITY UPLC BEH amide columns (130Å, 1.7 μm, 2.1 × 100 mm)
Mass spectrometer (Agilent Triple Quadrupole 6460 or comparable instrument)
Microcentrifuge
Microcentrifuge tubes
Micropipette (8-channel for 200 μL) and reservoirs
Plate centrifuge
Plate incubator/mixer (Heidolph Titramax 1000) at 30°C
Plates with lids
  96-well (360 μL)
  96-well (conical-well; 230 μL)
  96-well (deep-well; 2.2 mL)
    To enhance mixing during cultivation, place one 4-mm, acid-washed borosilicate bead in each well of the 96-deep-well plates and sterilize by autoclaving for 90 min at 121°C.
  384-well

Chapter 33

Sealing film (pierceable, compatible with high-performance liquid chromatography system)
Singer Rotor or 96-pin manual replicator
Spectrophotometer
Tubes (50-mL)
Ultrasonic bath

Immediately before use (Step 13), add ice to the ultrasonic bath to keep the samples cold and prevent evaporation of the ethanol solution.

Vortex
Water bath at 80°C

METHOD

In the following protocol, cultivation of S. cerevisiae strains is performed in a multiwell plate format to allow the simultaneous preparation of hundreds of samples. Proteogenic amino acids are heat-stable and have relatively high abundance—prerequisites for this high-throughput method, as little biological material is available and the specified metabolite extraction does not preserve heat-labile compounds (e.g., sugar phosphates or citrate cycle intermediates). The boiling ethanol extraction immediately arrests metabolism, lyses the cells, and releases the free amino acids (Entian et al. 1977). The extract is then cleared of debris and can be analyzed without additional conditioning.

Cultivating Samples

1. Thaw the stock cultures and transfer the cells to synthetic minimal solid medium using the Singer Rotor or a sterile manual replicator. Grow the cultures for 2 d at 30°C.

2. Inoculate precultures in synthetic liquid minimal medium in a 96-well plate (200 µL/well) using the Singer Rotor or a sterile manual replicator. Grow overnight at 30°C.

3. Dilute the precultures 1:20 (85 µL to 1.6 mL/well) in fresh synthetic liquid minimal medium in a 96-deep-well plate. Close the plate with a lid and place the culture in a plate incubator at 30°C with mixing (900 rpm, $r = 1.5$ mm). Incubate the strains for exactly 8 h before proceeding to collection and metabolite extraction (Step 5).

   Aliquots for optical density determination can be removed ~30 min after inoculation and again just before collection (Step 5). Care must be taken so that the cultures are well mixed and cells are kept at room temperature only for a minimal time.

   Slow growing strains can be inoculated with a higher initial optical density; however, all cultures should complete at least two doublings before collection.

Collecting Samples and Extracting Metabolites

4. For each 96-well culture plate, prepare a 50-mL tube containing 25 mL of absolute ethanol. Five minutes before cell collection (Step 5), heat the ethanol in the water bath to 80°C.

   Be careful; ethanol is toxic and highly flammable!

5. After 8 h of growth, collect the cells by centrifugation for 3 min at 3000g and discard the supernatant by inverting the plate. Tap the plate on a tissue to remove residual medium.

   Because of capillary force, ~50 µL will remain in each well.

6. Working under a fume hood, pour the ethanol into a reservoir and add 200 µL to each well using a multichannel pipette.

   The approximate final concentration of ethanol in each well is 80%.

   As soon as the cells are deprived of nutrients, the metabolite levels will be affected. Thus, the time from discarding the medium to addition of the boiling ethanol should be minimized. The boiling ethanol should be added as quickly as possible to the cells; the solvent cools down very quickly as it evaporates.

7. Close the plate with the lid, vortex, and incubate 2 min in the water bath at 80°C.

8. Repeat Step 7.

9. Clear the extract from cell debris by centrifugation for 3 min at 3000g.

10. Transfer 150 µL of each supernatant to a 96-conical-well plate.

11. Centrifuge the plate for 5 min at 3000g to precipitate insoluble substances. Transfer a minimum of 30 µL/well of each supernatant to a 384-well plate. Seal the plate with aluminum foil and store at −80°C until further processing.

    Up to four 96-well plates can be accommodated on one 384-well plate and stored for several months. A second aliquot should be prepared in an additional 384-well plate in case of unforeseeable problems.

12. Prepare aliquots of a representative pooled quality control (QC) sample as follows.

    i. Combine the residual metabolite extracts from several plates. Transfer the pooled sample mixture to microcentrifuge tubes.

    ii. Centrifuge the tubes for 5 min at maximum speed to precipitate insoluble compounds.

    iii. Prepare sufficient aliquots to provide at least one aliquot for each day of hydrophilic interaction chromatography–tandem mass spectrometry (HILIC-MS/MS) analysis. Store at −80°C until use.

Conditioning Samples

13. Before HILIC-MS/MS analysis, thaw the extracts and sonicate them in an ultrasonic bath with ice for 10 min to bring all metabolites into solution.

14. Centrifuge the plates for 10 sec at 1000g to collect droplets from the plate seal. Exchange the aluminum foil for a piercable seal.

Performing HILIC-MS/MS

15. Perform metabolite quantification by HILIC-MS/MS using Solvent A and Solvent B.

    i. Separate free amino acids on an ACQUITY UPLC BEH amide column by gradient elution at a constant flow rate of 0.9 mL/min and a column temperature of 25°C.

        Chromatographic conditions for the gradient elution are as follows: Solvent B is kept for 0.7 min at 85% before a steady decrease to 5% B until 2.55 min. Solvent B is then kept at 5% for 0.05 min before returning to the initial conditions of 85% within 0.05 min. This is followed by an equilibration step until 3.25 min before injection of the next sample. The autosampler temperature is maintained at 4°C. As HILIC is susceptible to changes in mobile phase composition, <3 µL of sample (~80% ethanol) should be injected to maintain the separation capacity.

    ii. Direct the sample stream to an Agilent 6460 triple quadrupole mass spectrometer operated in SRM mode.

        The ion source settings on the 6460 are as follows: Cell acceleration voltage set at 7 V, gas flow at 8 L/min and 300°C, sheath gas flow at 11 L/min and 300°C, nebulizer pressure at 50 psi, negative capillary voltage at 3000 V, and nozzle voltage at 500 V. Please note that these settings vary among different triple quadrupole instruments, ion source manufacturers, and the individual instrument configuration and thus need to be tuned on an individual basis.

16. Identify metabolites by matching retention time and fragmentation pattern with commercially obtained standards (Fig. 1). Quantify by external calibration with serial dilutions of a standard mix covering a range of 15× lower and higher than the expected sample concentrations (Table 1).

    Note that homoserine (isothreonine) is not separated from threonine by chromatographic means or with a specific m/z transition. However, homoserine is present at less than one-tenth of the amount of intracellular threonine, so measurements predominantly represent threonine.

DISCUSSION

This protocol builds on a previously reported high-throughput strategy for metabolomics (Ewald et al. 2009). The method has been optimized to not only offer a fast and easy-to-use approach but also to take economic considerations into account with costs of <$1 per sample. Up to 768 samples (two 384-

Chapter 33

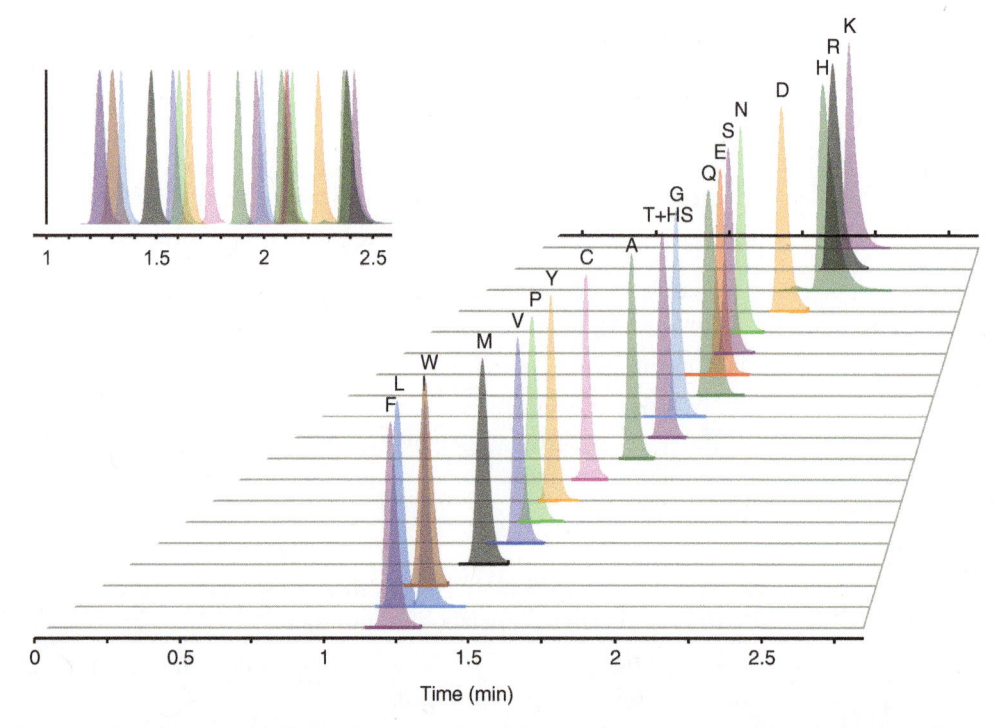

FIGURE 1. Chromatographic separation of an amino acid standard mix. Separation occurs within 1.5 min; total runtime is 3.25 min. Each chromatographic peak is scaled to the same height and the identity of the amino acid is indicated using the single letter code.

well plates) can be measured without a drop in sensitivity or deterioration of chromatographic separation. The presented workflow is best suited for initial screening of high numbers of samples; for instance, we have applied it to measure differences in amino acid concentrations in a global collection of *Schizosaccharomyces pombe* natural isolates (Jeffares et al. 2015). For verification and follow-up experiments, conditions can be more rigorously controlled and/or the replicate number increased. It is critical to carefully supervise the cultivation procedure to assure comparable growth of the strains. Probabilistic quotient normalization and the workflow presented in the MIPHENO R package allow correction for dilution effects, batch-to-batch variation, and elimination of technical outliers (Dieterle et al. 2006; Bell et al. 2012). Strains with significantly changed metabolite profiles are identified using robust multivariate outlier analysis (Rousseeuw and Leroy 2005). For functional characterization, principal component analysis (PCA) or hierarchical clustering analysis (HCA) with an appropriate distance measure (e.g., Mahalanobis distance) is used to group samples with similar metabolic profiles (Fiehn et al. 2000).

RECIPES

Synthetic Minimal Liquid Medium

1. In a 0.5-L bottle or flask, combine 3.4 g of yeast nitrogen base (YNB) without amino acids (Sigma-Aldrich Y0626) and 475 mL of H_2O. Autoclave for 30 min at 15 psi.
2. Add 25 mL of 40% (w/v) glucose (separately sterilized by autoclaving or filtration).

Synthetic Minimal Solid Medium

1. In a 0.5-L bottle, combine 200 mL of H_2O and 10 g of agar. In a 250-mL bottle, combine 3.4 g of yeast nitrogen base (YNB) without amino acids (Sigma-Aldrich Y0626) and 200 mL of H_2O. Autoclave each solution for 30 min at 15 psi and then combine.

It is critical to autoclave agar and YNB separately, as the low pH of YNB will hydrolyze the agar and yield semisolid plates.

2. Add 25 mL of 40% (w/v) glucose (separately sterilized by autoclaving).
3. Pour plates as required for use.

Solvent A for HILIC Analysis of Amino Acids

Using UPLC-grade solvents (including H_2O), prepare a mixture of 10 mM ammonium formate and 0.176% formic acid (analytical grade) in 95/5/5 acetonitrile/methanol/H_2O.

Solvent B for HILIC Analysis of Amino Acids

Using UPLC-grade solvents (including H_2O), prepare a mixture of 10 mM ammonium formate and 0.176% formic acid (analytical grade) in 50/50 acetonitrile/ H_2O.

ACKNOWLEDGMENTS

We thank the Wellcome Trust (RG 093735/Z/10/Z) and the European Research Council (ERC) (Starting grant 260809) for funding. M.R. is a Wellcome Trust Research Career Development and Wellcome-Beit Prize fellow.

REFERENCES

Bell SM, Burgoon LD, Last RL. 2012. MIPHENO: Data normalization for high throughput metabolite analysis. *BMC Bioinformatics* **13**: 10.

Dieterle F, Ross A, Schlotterbeck G, Senn H. 2006. Probabilistic quotient normalization as robust method to account for dilution of complex biological mixtures. Application in 1H NMR metabonomics. *Anal Chem* **78**: 4281–4290.

Entian KD, Zimmermann FK, Scheel I. 1977. A partial defect in carbon catabolite repression in mutants of *Saccharomyces cerevisiae* with reduced hexose phosphyorylation. *Mol Gen Genet* **156**: 99–105.

Ewald JC, Heux S, Zamboni N. 2009. High-throughput quantitative metabolomics: Workflow for cultivation, quenching, and analysis of yeast in a multiwell format. *Anal Chem* **81**: 3623–3629.

Fiehn O, Kopka J, Dörmann P, Altmann T, Trethewey RN, Willmitzer L. 2000. Metabolite profiling for plant functional genomics. *Nat Biotechnol* **18**: 1157–1161.

Jeffares DC, Rallis C, Rieux A, Speed D, Převorovský M, Mourier T, Marsellach FX, Iqbal Z, Lau W, Cheng TM, et al. 2015. The genomic and phenotypic diversity of *Schizosaccharomyces pombe*. *Nat Genet*: **47**: 235–241.

Rousseeuw PJ, Leroy AM. 2005. *Robust regression and outlier detection*, Vol. 589. John Wiley & Sons, Hoboken, NJ.

Protocol 3

Spectrophotometric Analysis of Ethanol and Glucose Concentrations in Yeast Culture Media

Amy A. Caudy[1,2,3]

[1]*Donnelly Centre for Cellular and Biomolecular Research, University of Toronto, Toronto, Ontario M5S3E1, Canada;* [2]*Department of Molecular Genetics, University of Toronto, Toronto, Ontario M5S3E1, Canada*

Fermentative growth on glucose is one of the most widely studied conditions of yeast growth in the laboratory. The production of ethanol from sugars is relevant to the wine, beer, and bread industries and to production of biofuels. Assaying the levels of glucose and ethanol in yeast growth medium allows the experimenter to determine the consumption of the carbon source glucose and the production of ethanol. This protocol describes enzyme-coupled assays for determination of glucose and ethanol concentrations in a sample of cell-free culture medium. Enzymes convert glucose or ethanol into other compounds through chemical reactions that reduce $NAD(P)^+$ to $NAD(P)H$, and the production of $NAD(P)H$ is measured using a spectrophotometer. The methods presented are highly sensitive, with a detection limit of ∼0.4 mg/L of glucose and 50 mg/L of ethanol, and also have the advantage of high specificity. For example, glucose and fructose have identical chemical formulas and thus cannot be distinguished by a mass spectrometer, but the enzyme assay presented here is specific for glucose. The glucose assay can be coupled to other assays to determine the quantity of additional carbohydrates such as fructose, trehalose, and glycogen.

MATERIALS

It is essential that you consult the appropriate Material Safety Data Sheets and your institution's Environmental Health and Safety Office for proper handling of equipment and hazardous materials used in this protocol.

Reagents

Kits for glucose and ethanol determination

These kits are widely used in the food industry for monitoring product quality. It may be possible to source the recommended kits from industrial service vendors in your area at a lower cost than from scientific suppliers. Similar kits are available from a number of other manufacturers; the instructions presented here can be applied to other products. Bring all solutions to room temperature before use.

Ethanol assay kit (R-Biopharm 10 176 290 035) (approximately 60 assays/kit)

All kit contents are stable at 2°C–8°C. The tablets containing NAD (nicotinamide adenine dinucleotide) and aldehyde dehydrogenase should be reconstituted on the day of the assay: Dissolve one tablet in 3 mL of potassium diphosphate buffer (Solution 1) for every two assays to produce the reaction mixture. This solution is stable for up to 1 d at 2°C–8°C.

Glucose assay kit (R-Biopharm 0 716 251) (approximately 240 assays/kit)

[3]Correspondence: amy.caudy@utoronto.ca

Copyright © Cold Spring Harbor Laboratory Press; all rights reserved
Cite this protocol as *Cold Spring Harb Protoc*; doi:10.1101/pdb.prot089102

Add highly pure H₂O (MilliQ or similar) in the quantity directed by the manufacturer to the triethanolamine sample buffer (Bottle 1) containing magnesium sulfate, ATP (adenosine triphosphate), and NADP (nicotinamide adenine dinucleotide phosphate). Cap the bottle and mix by inversion until the powder is completely dissolved. This solution is stable for 4 wk at 2°C–8°C, or for 2 mo at −15°C to −25°C.

The enzyme suspension of hexokinase and glucose-6-phosphate dehydrogenase (Bottle 2) and the control solution of glucose (Bottle 3) are stable at 2°C–8°C. Use an aseptic technique when handling the control solution to avoid contamination with yeast or other microbes.

Yeast cultures in growth medium

Equipment

Centrifuge tubes or filters (0.45-μm or smaller) for preparation of cell-free medium (see Step 1)

Cuvettes and caps (disposable, UV transparent)

Brand disposable UV cuvettes (BrandTech 759210) or Fisherbrand methacrylate disposable cuvettes (Fisher Scientific 14-955-128) are recommended.

Spectrophotometer capable of measuring absorbance at 340 nm

The method presented here describes the use of a spectrophotometer for NAD(P)H determination, but fluorimetry with excitation at 350 nm and emission at 480 nm can also be used. Fluorometric values can be converted to concentrations by comparison with a standard curve of NADH or NAD(P)H as appropriate.

Tubes (15-mL or 50-mL conical polypropylene), for reconstitution of reagents

METHOD

Performing the Glucose Assay

In the following assay, the concentration of glucose in a sample mixture is determined by enzymatic oxidation to D-gluconate-6-phosphate, a reaction that reduces NADP to NADPH. Two enzymes are added together with other substrates in a single reaction. First, glucose is phosphorylated by hexokinase, producing glucose-6-phosphate:

$$\text{D-glucose} + \text{ATP} \xrightarrow{\text{Hexokinase}} \text{Glucose-6-phosphate} + \text{ADP}. \quad (1)$$

This product is then oxidized to D-gluconate-6-phosphate by glucose-6-phosphate dehydrogenase with the concomitant reduction of NADP to NADPH:

$$\text{Glucose-6-phosphate} + \text{NADP}^+ \xrightarrow{\text{Glucose-6-phosphate dehydrogenase}} \text{D-gluconate-6-phosphate} + \text{NADPH} + \text{H}^+. \quad (2)$$

At the start of this procedure, background absorbance at the near-UV range of 340 nm is determined. Then, the required substrates and enzymes are added and the reaction is incubated. The absorbance at 340 nm is measured again following the reaction. After correction for the background absorbance of the sample, the concentration of NADPH produced in the reaction is calculated. The reaction is assembled with enzymes and substrates in excess so that the glucose is quantitatively oxidized, yielding NADPH at a quantity equal to that of glucose.

1. Prepare 1- to 2-mL samples of cell-free growth medium using one of the following methods.

    i. Filter yeast cultures through a 0.45-μm (or smaller) filter into a fresh tube.

    ii. Alternatively, centrifuge yeast cultures at 3000g (or greater) for 5 min and transfer the cleared supernatant to a fresh tube.

    Samples should be stored at −20°C (to prevent growth of contaminating cells) and equilibrated to room temperature before analysis.

2. Dilute the samples with H₂O, as needed, according to culture type in Table 1.

3. Prepare samples in cuvettes using the glucose assay kit according to Table 1.

    i. Prepare a blank reaction by adding sample buffer and H₂O.

    ii. Prepare a control reaction by adding sample buffer, control glucose solution, and H₂O.

    iii. Prepare sample reactions according to culture type.

TABLE 1. Assay setup for glucose determination according to sample type

Assay setup	Experimental blank	Control	Typical growth medium (2% dextrose) from log phase culture	Late log phase culture	Glucose-limited chemostat or starved culture
Sample buffer	500 µL	500 µL	500 µL	500 µL	500 µL
Recommended sample dilution in H$_2$O	–	–	1:5	None	None
Diluted sample	–	–	50 µL	50 µL	1000 µL
Control sample	–	50 µL	–	–	–
H$_2$O	1000 µL	950 µL	950 µL	950 µL	–
Enzyme suspension (added after determination of sample background; Step 6)	20 µL	20 µL	20 µL	20 µL	20 µL

4. Cap the samples and invert five to 10 times to mix. Incubate 3 min at room temperature. Before measurement, examine each cuvette for bubbles interfering with the light path, and tap the cuvettes to dislodge.

5. Measure and record absorbance at 340 nm for all samples.

6. Add 20 µL of enzyme suspension to each sample (Table 1). Cap the cuvettes, invert five to 10 times to mix, and incubate for 15 min at room temperature. Remove any bubbles from the cuvette light path before measurement as in Step 4.

7. Measure and record absorbance at 340 nm.

    If the final absorbance is greater than 1, the sample is too concentrated.
    See Troubleshooting.

8. [Optional] When testing the protocol for the first time or when analyzing a new type of sample, measure the absorbance again after 5 min.

    If absorbance has continued to increase, this is evidence of additional enzyme activities in your sample.
    See Troubleshooting.

9. Determine the change in absorbance, ΔA, for each sample by subtracting the initial absorbance (Step 5) from the final absorbance (Step 7).

    If this value is less than 0.100, the sample is too dilute.
    See Troubleshooting.

10. Calculate the $\Delta A_{corrected}$ by subtracting the ΔA of the experimental blank from the ΔA for each sample.

    This is typically a small adjustment; if the ΔA value of the experimental blank is large, there may be contamination of the experimental blank with glucose.

11. Adjust the $\Delta A_{corrected}$ value for each sample by multiplying by the sample dilution factor and dividing by the sample volume (µL) divided by 50 µL to produce the ΔA_{final}.

    This corrects the sample values to the same range as the 50 µL of undiluted control sample. For example, when using 50 µL of sample diluted 1:5, the $\Delta A_{corrected}$ is multiplied by 5 (dilution factor) and divided by 1 (50 µL/50 µL); when using 1000 µL of undiluted sample, the $\Delta A_{corrected}$ is multiplied by 1 (dilution factor) and divided by 20 (1000 µL/50 µL).

12. Calculate the empirically determined concentration of glucose (expressed in g/L) in all samples by multiplying the ΔA_{final} by 0.8636.

13. Determine the correction factor for the assay.

    i. Locate the actual concentration of glucose in the control solution.

        This is usually 0.500 g/L, listed on the bottle of control solution provided in the kit.

ii. Divide this known value by the empirical result from Step 12 to yield a correction factor to apply to the samples.

> *For example, if the control solution is 0.500 g/L and your empirically determined value is 0.492 g/L, then dividing 0.500 g/L by 0.492 g/L produces a value of 1.0163. Multiplying the empirically determined value of the control by this factor will produce the actual value of 0.500 g/L in the control.*

14. Multiply all sample values by the correction factor to obtain the concentration of glucose in the samples.

Performing the Ethanol Assay

In the following assay, the concentration of ethanol in a sample is determined by oxidizing alcohol to acetaldehyde with alcohol dehydrogenase, which uses NAD^+ as a reductant for the reaction:

$$\text{Ethanol} + NAD^+ \xleftrightarrow{\text{Alcohol dehydrogenase}} \text{Acetaldehyde} + NADH + H^+. \quad (3)$$

The NADH produced is proportional to the ethanol concentration. Although the equilibrium of the bidirectional alcohol dehydrogenase reaction lies toward ethanol production, the use of high pH conditions and the consumption of the acetaldehyde by aldehyde dehydrogenase,

$$\text{Acetaldehyde} + NAD^+ + H_2O \xleftrightarrow{\text{Aldehyde dehydrogenase}} \text{Acetic acid} + NADH, \quad (4)$$

drive the reaction, and the acetaldehyde dehydrogenase step contributes a second molecule of NADH per molecule of ethanol. This is an extremely sensitive reaction.

Methanol does not react in this assay because of a low binding constant for alcohol dehydrogenase. Glycerol is also not a substrate for alcohol dehydrogenase nor are other secondary and tertiary alcohols. Other primary alcohols, including n-propanol and butanol, are oxidized in this reaction, although ethanol is by far the most abundant alcohol produced during fermentation by S. cerevisiae and many other yeasts.

Ethanol is a volatile compound; samples should be tightly capped at all times.

15. Prepare 1- to 2-mL samples of cell-free growth medium using one of the methods described in Step 1.

    > *Samples should be stored at −20°C (to prevent growth of contaminating cells) and equilibrated to room temperature before analysis.*

16. Prepare samples in cuvettes using the ethanol assay kit according to Table 2.
    i. Prepare a blank reaction by adding reaction mixture and H_2O.
    ii. Prepare a control reaction by adding reaction mixture, control ethanol solution, and H_2O.
    iii. Prepare sample reactions by adding reaction mixture and experimental samples as described.

17. Cap the samples and invert five to 10 times to mix. Incubate for 3 min at room temperature. Before measurement, examine each cuvette for bubbles interfering with the light path, and tap the cuvettes to dislodge.

18. Measure and record absorbance at 340 nm for all samples.

19. Add 25 µL of enzyme suspension to each sample (Table 2). Cap the cuvettes, invert five to 10 times to mix, and incubate for 10 min at room temperature. Remove any bubbles from the cuvette light path before measurement as in Step 17.

20. Measure and record absorbance at 340 nm.

    > *If the final absorbance is greater than 1, the sample is too concentrated.*
    > *See Troubleshooting.*

21. [Optional] When testing the protocol for the first time or when analyzing a new type of sample, measure the absorbance again after 5 min.

    > *If absorbance has continued to increase, this is evidence of additional longer-chain alcohols in your sample, or of compounds in the sample that interfere with enzyme activities.*
    > *See Troubleshooting.*

Chapter 33

TABLE 2. Assay setup for ethanol determination according to sample type

Assay setup	Experimental blank	Control	Typical growth medium (2% dextrose) from log phase culture
Reaction mixture (reconstituted from tablet and phosphate sample buffer)	1500 µL	1500 µL	1500 µL
Experimental sample	–	–	250 µL (use 50 µL or dilute sample as appropriate if reading is too high)
Control sample	–	50 µL	–
H$_2$O (µL)	Equal to experimental sample volume	Experimental sample volume minus control sample volume (total volume of control and H$_2$O will equal experimental sample volume)	–
Aldehyde dehydrogenase enzyme suspension (added after determination of sample background; Step 19)	25 µL	25 µL	25 µL

22. Determine the change in absorbance, ΔA, for each sample by subtracting the initial absorbance (Step 18) from the final absorbance (Step 20).

    If this value is less than 0.100, the sample is too dilute.
    See Troubleshooting.

23. Calculate the $\Delta A_{corrected}$ by subtracting the ΔA of the experimental blank from the ΔA for each sample.

    This is typically a small adjustment; if the ΔA of the experimental blank is large, there may be contamination of the experimental blank with ethanol.

24. Adjust the $\Delta A_{corrected}$ value for each sample by multiplying by the sample dilution factor and dividing by the value of the sample volume (µL) divided by 50 µL to produce the ΔA_{final}.

    This corrects the sample values to the same range as the 50 µL of undiluted control sample. For example, when using 250 µL of undiluted sample, the $\Delta A_{corrected}$ is multiplied by 1 (dilution factor) and divided by 5 (250 µL/50 µL).

25. Calculate the empirically determined concentration of ethanol (expressed in g/L) in all samples by multiplying the ΔA_{final} by 0.1152.

26. Determine the correction factor for the assay.

    i. Locate the actual concentration of ethanol in the control solution.

        This is usually 0.500 g/L, listed on the bottle of control solution provided in the kit.

    ii. Divide this known value by the empirical result from Step 25 to yield a correction factor to apply to the samples.

        See the sample calculation provided in Step 13.ii.
        If the correction factor is greater than 1.1, suspect the loss of ethanol during your assay.
        See Troubleshooting.

27. Multiply all sample values by the correction factor to obtain the concentration of ethanol in the samples.

TROUBLESHOOTING

Problem (Steps 7, 20): Sample absorbance values are greater than 1 absorbance unit.
Solution: Use less of the sample or dilute appropriately. For the glucose assay, dilution is recommended. The ethanol assay can be used with 50–250 μL of sample input. (Adjust the blank and control samples accordingly.)

Problem (Steps 8, 21): Sample absorbance increases over time after measurement.
Solution: In the glucose assay, contaminating cellular enzymes can convert other sugars in the solution into glucose. Incubating tightly capped samples in a water bath at 80°C before the glucose assay will usually inactivate such enzymes. In the ethanol assay, increases in absorbance over time can be caused by other longer-chain alcohols, or, more rarely, the sample can contain compounds that inhibit the enzymes. The presence of inhibitors can be tested by mixing a portion of the control sample with the experimental sample; inhibitors will prevent the full detection of the added control sample.

Problem (Steps 9, 22): Sample absorbance values are less than 0.100 absorbance units.
Solution: Increase the sample volume to a maximum of 1000 μL for the glucose assay and 1225 μL for the ethanol assay.

Problem (Step 26): The empirically determined value of the ethanol control is >10% lower than the actual value.
Solution: This usually results from loss of ethanol by evaporation; keep all samples and controls tightly capped at all times.

DISCUSSION

The enzyme assays described here are two of the many commercially available enzyme-coupled assays to determine the concentrations of carbohydrates and alcohols. Assays for other sugars, including sucrose and fructose, are available and are accomplished using enzymes that convert precursor sugars into glucose or glucose-6-phosphate to enter into the reactions described here. Similarly, the concentration of the storage carbohydrates trehalose and glycogen can be determined (typically from cell lysates) by including enzymes that degrade the carbohydrate polymers to glucose. The principles mentioned here should be useful in other enzyme-coupled assays for yeast cell medium analysis.

A range of methods is commercially available for detecting glucose, reflecting the importance of its measurement not only in food products but also in human blood and urine in the context of diabetes. Glucose test strips for determination of glucose in urine are convenient for semiquantitative verification that glucose is present in growth medium. However, test strips for urinalysis are typically limited to detecting glucose of a concentration of more than ~350 mg/L (Penders et al. 2002), which is 1000 times higher than the lower limit of detection of this analysis.

Likewise, a range of methods is available for the measurement of ethanol. Available methods vary in cost, complexity, sensitivity, and accuracy. Regardless of the method selected for analysis, the recommendations described here for sample handling, appropriate controls, comparison with known standards, and troubleshooting are applicable.

ACKNOWLEDGMENTS

This protocol is an adaptation of a method developed by Dr. Maitreya Dunham and Cheryl Christianson from manufacturer protocols. A.A.C. is the Canada Research Chair in Metabolomics for Enzyme

Discovery and is supported by the Ontario Early Researcher Award, the Canadian Institutes for Health Research, the Natural Sciences and Engineering Research Council of Canada, the Canadian Foundation for Innovation, and the Ontario Leader's Opportunity Fund.

REFERENCES

Penders J, Fiers T, Delanghe JR. 2002. Quantitative evaluation of urinalysis test strips. *Clin Chem* **48:** 2236–2241.

CHAPTER 34

High-Throughput Yeast Strain Sequencing

Katja Schwartz and Gavin Sherlock[1]

Department of Genetics, Stanford University Medical School, Stanford, California 94305-5120

The original yeast genome sequencing project was a monumental task, spanning several years, which resulted in the first sequenced eukaryotic genome. The 12 Mbp reference sequence was generated from yeast strain S288c and was of extremely high quality. In the years since it was published, sequencing technology has advanced apace, such that it is within the reach of most labs to sequence yeast strains of interest almost as a matter of standard practice, either via core facilities at their institution or through commercial sequencing services. Because of the availability of the high-quality reference sequence (which itself has received approximately 1500 updates derived from high-throughput sequencing data), reliable identification of differences between a strain of interest and the reference is relatively straightforward, at least for the nonrepetitive regions of the genome. In this introduction, we describe current high-throughput sequencing technology and methods for analysis of the resulting data.

BACKGROUND

Since 2001, DNA sequencing technology has changed dramatically in both its throughput and its cost. Per megabase, costs have fallen by five orders of magnitude in the last dozen years, as researchers have transitioned away from long-read, capillary-based ABI Sanger sequencing to massively parallel technologies that produce short(er) reads with generally higher error rates (e.g., see Mardis 2011). Yeast was the first sequenced eukaryote, and its genome sequence, published in 1996, was the result of an international consortium involving more than 600 scientists working over several years (Goffeau et al. 1996). The project cost several millions of dollars. Using current technologies, it is now possible for a single researcher to sequence many yeast genomes simultaneously in a couple of weeks for a reagent cost of <$100 per genome. This has created many new possibilities for studying our favorite model eukaryote, including straightforward identification of mutations by screening that would have taken months to map, whole-genome genotyping of meiotic segregants (Wilkening et al. 2013), identification of beneficial mutations that have arisen during adaptive evolution (e.g., Araya et al. 2010; Kvitek and Sherlock 2011), identification of the causal single-nucleotide polymorphisms (SNPs) underlying quantitative tract loci (QTLs) (e.g., Parts et al. 2011), and creation of catalogs of variation within species through sequencing of diverse strains.

CURRENT TECHNOLOGIES

There are numerous high-throughput sequencing platforms, each with different characteristics related to throughput, cost, read length, and sequence quality (Table 1). Available platforms include Ion Torrent (Rothberg et al. 2011), Pacific Biosciences (Eid et al. 2009), and Illumina (e.g., Bentley et al.

[1]Correspondence: gsherloc@stanford.edu

Copyright © Cold Spring Harbor Laboratory Press; all rights reserved
Cite this introduction as *Cold Spring Harb Protoc*; doi:10.1101/pdb.top077651

Chapter 34

TABLE 1. Comparison of major sequencing platforms

	Platform			
	Illumina HiSeq	Illumina MiSeq	Ion Torrent Proton	Pacific Biosciences
Approach	Reversible dye-terminator	Reversible dye-terminator	pH change	Single-molecule, zero-mode wave guide
Read length	Up to 2 × 150 bp	Up to 2 × 300 bp	Up to 200 bp	Up to 30 kb
Data per run	Up to 500 Gb/flowcell	Up to 15 Gb	Up to 10 Gb	~375 Mb
Run time	1–11 d	24–65 h	2–4 h	~3 h
Cost per gigabase	~$25	~$500	~$100	~$2000

Note that prices are constantly decreasing and throughput is constantly increasing, whereas read lengths are improving and error rates decreasing, so this should be considered a snapshot at a moment in time (i.e., 2015).

2008). The dominant platform currently for whole-genome resequencing is Illumina HiSeq, which provides sufficiently long reads to permit unique mapping to the vast majority of the genome and has sufficient throughput to generate high coverage. Thus, our protocol describes construction of libraries for sequencing using the HiSeq platform (Protocol 1: Preparation of Yeast DNA Sequencing Libraries [Schwartz and Sherlock 2015]).

The general approach to library construction (Fig. 1) is to first fragment the genome—either randomly by physical means or pseudorandomly using enzymes—and then to repair the DNA ends before ligating adapters onto the double-stranded DNA fragments. These adapters facilitate polymerase chain reaction (PCR)-based amplification of the sequencing library and clustering of the DNA on the Illumina flow cell. In addition, the adapters may contain multiplexing sequences or barcodes—short sequences (typically 6- or 8-mers) that will be read as part of the sequencing run—that "tag" particular samples. This allows the genomes for multiple strains to be simultaneously sequenced within a single lane of an Illumina flow cell. As of 2015, a typical run of the Illumina HiSeq 2500 may yield on the order of 200 million 2 × 100-bp paired-end reads. Given the length of the haploid yeast genome (12 Mbp), this yields ~3300× coverage. For paired-end 2 × 100-bp reads derived from 300-bp fragments, almost 95% of the genome is uniquely mappable (Lee and Schatz 2012). Thus, with sufficient multiplexing barcodes, 96 strains can be sequenced to almost 35× coverage. This is sufficient for most purposes, such as SNP and indel calling and identification of structural variants. When sequencing diploids, additional coverage is useful for confidently calling heterozygous sites.

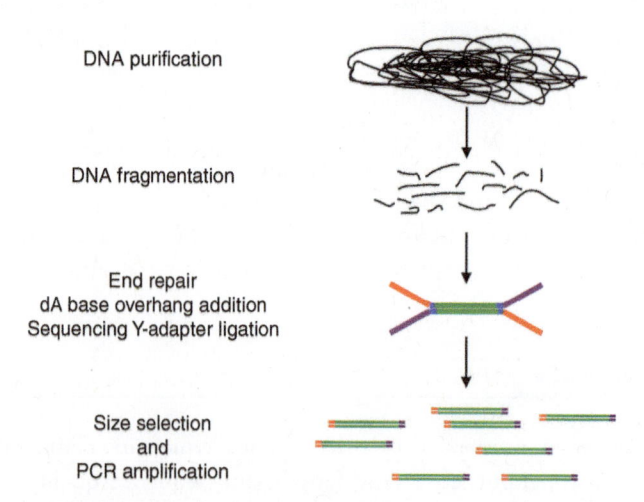

FIGURE 1. General flow of Illumina sequencing. DNA is fragmented, end-polished, and phosphorylated, and a dA is added. The preannealed Y-shaped adaptors are ligated, and PCR and size selection are performed to produce the sequencing library. Green indicates the library insert (with the filling-in and dA addition); blue indicates the ends of the Illumina adapters, which can be ligated to the dA additions (not shown after the PCR step); red and purple indicate the forked part of each Illumina adapter, which allows the creation of different sequences at each end of the insert following PCR.

Once a sample has been sequenced, the resulting data are usually provided to the researcher in the form of a fastq file, a format that contains the reads themselves and the associated quality scores for each base within the reads (Cock et al. 2010). For paired-end sequencing runs, reads are typically provided in two files—one file for the forward reads and one for the reverse. To identify salient features within the data, files are typically processed through an in-house bioinformatics pipeline. The data must first be split based upon multiplexing barcodes (e.g., using the FASTX toolkit) and then can be mapped to the yeast genome. There are many different open source software tools for mapping short-read fastq data to the reference S288C sequence, including bwa (Li and Durbin 2009) and bowtie2 (Langmead and Salzberg 2012). All of these, in conjunction with samtools, can output the mapped data in sorted bam format—a compact format describing the position or positions where each sequence read maps in the genome, if one exists (Li et al. 2009). A sorted bam file can be subsequently used for downstream processing steps for the identification of SNPs and indels, typically performed using the GATK (McKenna et al. 2010; DePristo et al. 2011; Van der Auwera et al. 2013). SNPs and indels are output by the GATK in a tab-delimited vcf (variant call format) file, which describes the location of each variant as well as the reference and variant alleles and the quality of the nucleotide sequences supporting that variant (Danecek et al. 2011). Once SNPs and indels have been identified, they need to be annotated—are they in a gene, or an intergenic region? If in a gene, do they result in a coding change, and if so, what are the predicted consequences of that change? A widely used piece of software for annotating the consequences of mutations is called snpEff (Cingolani et al. 2012), which will accept a vcf file (containing a list of the mutations) and a gff file (indicating the location of all the genes, available from the *Saccharomyces* Genome Database) and use them to generate a report indicating that genes contain mutations and whether their consequences are likely severe or not.

FUTURE PERSPECTIVES

DNA sequencing technology will certainly continue to develop, and it is likely that throughput will increase. This will allow a greater depth of coverage (especially useful when sequencing population samples) and/or allow more strains to be multiplexed, which will lower cost. Illumina has improved their technology by transitioning to regularized arrays of clusters on flow cells, as opposed to randomly placed clusters, although this is yet not widely available. This new technology allows for much higher densities and will make postprocessing of cluster images more straightforward. As chemistry is further improved, it is also likely that error rates will continue to decrease, and as they do, read lengths will also improve. Perhaps most interesting is the prospect of what have been referred to as "synthetic reads," a technology that Illumina acquired when it purchased the company Moleculo. Moleculo technology begins by fragmenting genomic DNA to ~10 kb, and dividing the fragments into small pools such that no given fragment is expected by chance to overlap another in the same pool. Next, the fragments in each pool are clonally amplified, sheared, and marked with a unique barcode for their pool. The pools are then combined and sequenced with Illumina technology. Using the barcodes, the short sequence reads originating from each pool are separated and then assembled into contigs, or synthetic reads that correspond to the 10-kb fragments in each pool. This approach has generated synthetic reads from 6 to 8 kb in length, and it results in an error rate that is orders of magnitude lower than standard Illumina sequencing (Voskoboynik et al. 2013). The reads themselves can be assembled using traditional or possibly new assembly algorithms. Currently, this approach is not supported for smaller genomes (<100 Mbp), but if it is developed to work for genomes as small as yeast, it is likely this technology will make the entire yeast genome mappable as well as possible to assemble, including the repeated and subtelomeric regions, which are often the most different between strains yet hardest to analyze. In addition, in diploid strains, it may allow phasing of large tracts of the genome without generating haploid spores (not always trivial with wild-type strains). Finally, an alternative sequencing protocol, which costs only ~$5/library to prepare, was recently developed for use when sequencing large numbers of strains in groups of 96 (Kryazhimskiy et al. 2014; Baym et al. 2015). As a result, the sequencing of hundreds of strains is now within the reach of most labs.

REFERENCES

Araya CL, Payen C, Dunham MJ, Fields S. 2010. Whole-genome sequencing of a laboratory-evolved yeast strain. *BMC Genomics* **11:** 88.

Baym M, Kryazhimskiy S, Lieberman TD, Chung H, Desai MM, Kishony R. 2015. Inexpensive multiplexed library preparation for megabase-sized genomes. *PLoS ONE* **10:** e0128036.

Bentley DR, Balasubramanian S, Swerdlow HP, Smith GP, Milton J, Brown CG, Hall KP, Evers DJ, Barnes CL, Bignell HR, et al. 2008. Accurate whole human genome sequencing using reversible terminator chemistry. *Nature* **456:** 53–59.

Cingolani P, Platts A, Wang le L, Coon M, Nguyen T, Wang L, Land SJ, Lu X, Ruden DM. 2012. A program for annotating and predicting the effects of single nucleotide polymorphisms, SnpEff: SNPs in the genome of *Drosophila melanogaster* strain w1118; iso-2; iso-3. *Fly (Austin)* **6:** 80–92.

Cock PJ, Fields CJ, Goto N, Heuer ML, Rice PM. 2010. The Sanger FASTQ file format for sequences with quality scores, and the Solexa/Illumina FASTQ variants. *Nucleic Acids Res* **38:** 1767–1771.

Danecek P, Auton A, Abecasis G, Albers CA, Banks E, DePristo MA, Handsaker RE, Lunter G, Marth GT, Sherry ST, et al. 2011. The variant call format and VCFtools. *Bioinformatics* **27:** 2156–2158.

DePristo MA, Banks E, Poplin R, Garimella KV, Maguire JR, Hartl C, Philippakis AA, del Angel G, Rivas MA, Hanna M, et al. 2011. A framework for variation discovery and genotyping using next-generation DNA sequencing data. *Nat Genet* **43:** 491–498.

Eid J, Fehr A, Gray J, Luong K, Lyle J, Otto G, Peluso P, Rank D, Baybayan P, Bettman B, et al. 2009. Real-time DNA sequencing from single polymerase molecules. *Science* **323:** 133–138.

Goffeau A, Barrell BG, Bussey H, Davis RW, Dujon B, Feldmann H, Galibert F, Hoheisel JD, Jacq C, Johnston M, et al. 1996. Life with 6000 genes. *Science* **274:** 546, 563–567.

Kryazhimskiy S, Rice DP, Jerison ER, Desai MM. 2014. Microbial evolution. Global epistasis makes adaptation predictable despite sequence-level stochasticity. *Science* **344:** 1519–1522.

Kvitek DJ, Sherlock G. 2011. Reciprocal sign epistasis between frequently experimentally evolved adaptive mutations causes a rugged fitness landscape. *PLoS Genet* **7:** e1002056.

Langmead B, Salzberg SL. 2012. Fast gapped-read alignment with Bowtie 2. *Nat Methods* **9:** 357–359.

Lee H, Schatz MC. 2012. Genomic dark matter: The reliability of short read mapping illustrated by the genome mappability score. *Bioinformatics* **28:** 2097–2105.

Li H, Durbin R. 2009. Fast and accurate short read alignment with Burrows-Wheeler transform. *Bioinformatics* **25:** 1754–1760.

Li H, Handsaker B, Wysoker A, Fennell T, Ruan J, Homer N, Marth G, Abecasis G, Durbin R. 2009. The Sequence Alignment/Map format and SAMtools. *Bioinformatics* **25:** 2078–2079.

Mardis ER. 2011. A decade's perspective on DNA sequencing technology. *Nature* **470:** 198–203.

McKenna A, Hanna M, Banks E, Sivachenko A, Cibulskis K, Kernytsky A, Garimella K, Altshuler D, Gabriel S, Daly M, et al. 2010. The Genome Analysis Toolkit: A MapReduce framework for analyzing next-generation DNA sequencing data. *Genome Res* **20:** 1297–1303.

Parts L, Cubillos FA, Warringer J, Jain K, Salinas F, Bumpstead SJ, Molin M, Zia A, Simpson JT, Quail MA, et al. 2011. Revealing the genetic structure of a trait by sequencing a population under selection. *Genome Res* **21:** 1131–1138.

Rothberg JM, Hinz W, Rearick TM, Schultz J, Mileski W, Davey M, Leamon JH, Johnson K, Milgrew MJ, Edwards M, et al. 2011. An integrated semiconductor device enabling non-optical genome sequencing. *Nature* **475:** 348–352.

Schwartz K, Sherlock G. 2015. Preparation of yeast DNA sequencing libraries. *Cold Spring Harb Protoc* doi: 10.1101/pdb.prot088930.

Van der Auwera GA, Carneiro MO, Hartl C, Poplin R, Del Angel G, Levy-Moonshine A, Jordan T, Shakir K, Roazen D, Thibault J, et al. 2013. From FastQ data to high-confidence variant calls: The Genome Analysis Toolkit best practices pipeline. *Curr Protoc Bioinform* **43:** 11.10.1–11.10.33.

Voskoboynik A, Neff NF, Sahoo D, Newman AM, Pushkarev D, Koh W, Passarelli B, Fan HC, Mantalas GL, Palmeri KJ, et al. 2013. The genome sequence of the colonial chordate. *Botryllus schlosseri*. *Elife* **2:** e00569.

Wilkening S, Tekkedil MM, Lin G, Fritsch ES, Wei W, Gagneur J, Lazinski DW, Camilli A, Steinmetz LM. 2013. Genotyping 1000 yeast strains by next-generation sequencing. *BMC Genomics* **14:** 90.

Protocol 1

Preparation of Yeast DNA Sequencing Libraries

Katja Schwartz and Gavin Sherlock[1]

Department of Genetics, Stanford University Medical School, Stanford, California 94305-5120

This protocol provides a detailed description of how to prepare a DNA sequencing library from yeast genomic DNA for use with the Illumina sequencing platform. This method does not require purchase of Illumina kits for library preparation but instead employs specific reagents purchased largely from New England BioLabs, which significantly reduces the cost of library preparation. Although we assume here that users intend to generate libraries with ∼400-bp insert sizes for paired-end sequencing, it is relatively straightforward to modify the shearing and size selection steps for longer or shorter inserts.

MATERIALS

It is essential that you consult the appropriate Material Safety Data Sheets and your institution's Environmental Health and Safety Office for proper handling of equipment and hazardous materials used in this protocol.

RECIPES: Please see the end of this protocol for recipes indicated by <R>. Additional recipes can be found online at http://cshprotocols.cshlp.org/site/recipes.

Reagents

DNA polymerase I, large fragment (Klenow) (New England BioLabs M0210L)
DNA purification reagents
    β-mercaptoethanol (Bio-Rad 161-0710)
    Chloroform (Sigma-Aldrich C2432)
    Ethanol (200-proof)
    Phenol, saturated (pH 7.9) (Amresco 0945)
    Potassium acetate (5 M)
    RNase A (Invitrogen 12091-021) (10 mg/mL)
    Sodium acetate (Ambion AM9740) (3 M, pH 5.5)
    Sodium dodecyl sulfate (SDS) (Bio-Rad 161-0301) (10%)
    Sorbitol solution (pH 8.0) <R>
    TE buffer (pH 8) <R>
    Tris/EDTA solution <R>
    Zymolyase-100T (MP Biomedicals 08320931) (30 mg/mL in sorbitol solution)

> We use the above reagents with the yeast DNA preparation protocol described in Step 1 (adapted from Treco 1987). Although time consuming, this approach yields large quantities of high-quality DNA (30–150 µg). Alternatively, Genomic-tip 20/G (QIAGEN 10223) or Yeastar Genomic DNA (Zymo Research D2002) columns can be used for genomic DNA extraction.

---

[1]Correspondence: gsherloc@stanford.edu

Copyright © Cold Spring Harbor Laboratory Press; all rights reserved
Cite this protocol as *Cold Spring Harb Protoc*; doi:10.1101/pdb.prot088930

E-Gel SizeSelect agarose gels (2%) (Invitrogen G661002) or standard 2% agarose gels
Size selection on a 2% Size Select E-gel is recommended (Step 8) since it prevents cross-contamination between samples. Alternatively, samples can be run on a standard 2% agarose gel, excised with a clean scalpel, extracted from the agarose using a gel extraction kit (QIAGEN 28704), and eluted with 30 µL of Buffer EB.

Illumina paired-end adapters and sequencing primers
Adapters and primers can be ordered from Illumina, or synthesized at Integrated DNA Technologies. The 6–8-bp barcodes in the adapter sequences allow multiple samples to be combined in the same lane in the Illumina flow cell.

Klenow fragment (3′ → 5′ exo-) with 10× NEB2 buffer (New England BioLabs M0212L)
MinElute PCR purification kit (QIAGEN 28006)
Nucleotides

dATP (10 mM) (Invitrogen 18252-015)
The dATP reagent should be frozen in aliquots (not repeatedly thawed and frozen).

dNTPs (10 mM) (New England BioLabs N0447S)

PCR purification kit (QIAGEN 28106 or equivalent)
Phusion High-Fidelity DNA polymerase with 5× Phusion HF buffer (Thermo Scientific)
Qubit dsDNA HS assay kit (Invitrogen Q32854) (as needed, for DNA quantitation)
T4 DNA ligase with 10× T4 DNA ligase buffer (New England BioLabs M0202S)
T4 DNA polymerase (New England BioLabs M0203L)
T4 polynucleotide kinase (New England BioLabs M0201L)
Yeast cells for liquid culture

Equipment

Access to Illumina sequencing equipment
Centrifuge
Centrifuge tubes
E-Gel apparatus or other electrophoresis equipment
We use the E-Gel iBase and Safe Imager kit (Invitrogen G6465) in combination with 2% SizeSelect E-Gels.

Glass rod or pipette tip
Incubators at 20°C, 37°C, and 65°C
Microcentrifuge tubes
PCR tubes (200-µL, thin-walled)
Qubit fluorometer (Invitrogen Q32866), or access to a Bioanalyzer
DNA concentration should be measured using a Qubit, Bioanalyzer, or fluorometer but not a spectrophotometer, because yeast DNA preparations may contain significant amounts of ribonucleotides.

Sonicator or other equipment for DNA fragmentation
We prefer to use the Covaris S-2 series with microTUBES (6 × 16-mm [Covaris 520045]).

Thermocycler

METHOD

1. Extract genomic DNA from yeast cells.
    i. Grow yeast in 20–50 mL of liquid culture overnight.
    ii. Centrifuge the overnight culture and wash with 10 mL of sorbitol solution. Resuspend the cells in 5 mL of sorbitol solution.
        Samples can now be frozen at −20°C.

iii. Add 2.5–5.0 µL of Zymolyase-100T (30 mg/mL in sorbitol solution) and 5 µL of β-mercaptoethanol to the cells. Incubate for 30 min at 37°C.

iv. Gently centrifuge (1500 rpm) the spheroplasted cells. Resuspend the cells in 5 mL of Tris/EDTA solution.

v. Add 500 µL of 10% SDS to the cells and invert gently to mix.

   The solution should become very viscous upon cell lysis.

vi. Add 2 mL of 5 M potassium acetate and shake to mix. Incubate for 30 min on ice.

vii. Centrifuge the sample for 10 min at 10,000 rpm. Pour the supernatant into a new tube.

viii. Add 15 mL of 100% ethanol to the supernatant and incubate at room temperature for 5–10 min to precipitate the DNA.

ix. Centrifuge the sample for 10 min at 10,000 rpm. Air dry the DNA pellet and resuspend in 0.5–1 mL of TE (pH 8).

x. Add 50 µL of RNase A (10 mg/mL) to the DNA solution. Incubate for 15 min at 65°C.

xi. Extract the sample at least once using phenol/chloroform. Precipitate with 40 µL of 3 M sodium acetate (pH 5.5) and 1 mL of 100% ethanol, inverting gently. (Do not vortex.)

xii. Spool the DNA with a glass rod or pipette tip and air-dry the sample. Resuspend the DNA in 200–300 µL of TE (pH 8.0).

2. Quantitate the genomic DNA.

   A starting amount of 0.5–2 µg of DNA should be used for library construction; however, when the DNA is very pure, the amount of starting material can be reduced to as low as 50–100 ng.

3. Fragment the extracted genomic DNA by sonication, nebulization, or enzymatic methods (Knierim et al. 2011).

   Using the Covaris S2 series instrument with microTUBES, DNA can be fragmented to ~500 bp by sonicating for 90 sec at Duty = 5%, Intensity = 3, Cycles/Burst = 200.

4. Repair the sheared genomic DNA ends.

   i. Combine the following reagents in a microcentrifuge tube.

Reagent	Amount to add
Genomic DNA (fragmented)	35.5 µL
10× T4 DNA ligase buffer	5 µL
10 mM dNTPs	4 µL
T4 DNA polymerase	2.5 µL
Klenow (large fragment)	0.5 µL
T4 polynucleotide kinase	2.5 µL
Total volume	50 µL

   ii. Incubate the reaction for 30 min at 20°C.

   iii. Purify each reaction using a polymerase chain reaction (PCR) purification column following manufacturer's instructions. Elute with 32 µL of Buffer EB.

5. Attach an A base to DNA ends.

   i. Combine the following reagents in a microcentrifuge tube.

Reagent	Amount to add
End-repaired DNA	32 µL
10× NEB2 buffer	5 µL
10 mM dATP	1 µL
Klenow (3′–5′ exo)	3 µL
H$_2$O	9 µL
Total volume	50 µL

Chapter 34

    ii Incubate the reaction for 30 min at 37°C.

    iii Purify each reaction using a MinElute column following manufacturer's instructions. Elute with 11 µL of Buffer EB.

6. Preanneal the paired-end adapters.

   i. Combine the following reagents in a microcentrifuge tube.

Reagent	Amount to add
10× T4 DNA ligase buffer	5 µL
100 mM adaptor 1	20 µL
100 mM adaptor 2	20 µL
H$_2$O	5 µL
Total volume	50 µL

   ii. Incubate the reaction in a thermocycler as follows.

Time	Temp
5 min	94°C
1 min	70°C
1 min	60°C
1 min	50°C
1 min	40°C
1 min	30°C
1 min	25°C

Preannealed adapters can be stored at −20°C.

7. Ligate the preannealed paired end adapters to the purified DNA.

   i. Combine the following reagents on ice.

Reagent	Amount to add
Purified DNA from Step 5	10 µL
10× T4 DNA ligase buffer	2 µL
T4 DNA ligase	1 µL
40 µM preannealed adaptor mix from Step 6	1 µL
H$_2$O	6 µL
Total volume	20 µL

   ii. Incubate the reaction for 15 min at 20°C.

   iii. Inactivate the ligase by heating the reaction for 10 min to 65°C.

   iv. Place the reaction on ice.

DNA with ligated adapters should not be stored longer than 24–48 h at −20°C.

8. Size-select for 500- to 600-bp DNA fragments with ligated adapters using an E-Gel SizeSelect 2% agarose gel or 2% agarose gel.

If not using E-gels, a separate gel/chamber buffer should be used for each sample to prevent DNA cross-contamination between libraries.

The selected insert size can be decreased to achieve sequencing in both directions if higher fidelity sequencing is required.

9. PCR-amplify the library.

   i. Combine the following reagents in a 200-µL thin-walled PCR amplification tube on ice.

Reagent	Amount to add
Size-selected DNA	2 µL
5× Phusion HF buffer	4 µL

10 mM dNTPs	0.5 µL
100 µM PE 1 primer	0.25 µL
100 µM PE 2 primer	0.25 µL
Phusion DNA polymerase	0.5 µL
H$_2$O	12.5 µL
Total volume	20 µL

ii. Incubate the reaction in a thermocycler as follows.

No. of cycles	Time	Temp
1	5 min	98°C
10–12	1 min	98°C
	1 min	65°C
	1 min	72°C
1 (optional)	5 min	72°C
1	Hold	4°C

When library production is initiated with a higher starting amount of DNA (1–2 µg), this amplification step can be omitted if enough genomic fragments with ligated adapters are generated. The unamplified library concentration can be measured using quantitative PCR technology; see the Illumina qPCR Quantification Protocol Guide.

10. Purify the amplified library using a PCR purification kit following the manufacturer's instructions. Elute with 30 µL of Buffer EB.

11. Quantitate the purified library using the Qubit dsDNA HS assay or a Bioanalyzer.

    Library concentration should be 10–30 ng/µL, and no less than 3.5 ng/µL. The library is now ready for submission for sequencing.

    See Troubleshooting.

TROUBLESHOOTING

Problem (Step 11): Library preparation results in a low yield.
Solution: Low yield usually arises from either poor starting DNA quality or low DNA integrity, so these possibilities should be investigated. Fresh reagents and/or the addition of a couple of amplification cycles may also improve yield.

RECIPES

Sorbitol Solution

Reagent	Final concentration
Sorbitol	0.9 M
Tris–HCl (pH 7.5)	0.1 M
EDTA (disodium salt)	0.1 M

TE Buffer (pH 8)

100 mM Tris-Cl (pH 8.0)
10 mM EDTA (pH 8.0)

Tris/EDTA Solution

Reagent	Final concentration
Tris–HCl (pH 7.5)	50 mM
EDTA (disodium salt)	20 mM

REFERENCES

Knierim E, Lucke B, Schwarz JM, Schuelke M, Seelow D. 2011. Systematic comparison of three methods for fragmentation of long-range PCR products for next generation sequencing. *PLoS ONE* 6: e28240.

Treco DA. 1987. Preparation of yeast DNA. In *Current protocols in molecular biology* (ed. Ausubel FM, et al.), pp. 13.11.11–13.11.12. John Wiley and Sons, New York.

CHAPTER 35

Budding Yeast Strains and Genotype–Phenotype Mapping

Gianni Liti,[1,4] Jonas Warringer,[2,3,4] and Anders Blomberg[2,5]

[1]IRCAN, CNRS UMR 6267, INSERM U998, University of Nice, 06107 Nice, France; [2]Department of Chemistry and Molecular Biology, University of Gothenburg, 40530 Gothenburg, Sweden; [3]Centre for Integrative Genetics (CIGENE), Department of Animal and Aquacultural Sciences, Norwegian University of Life Sciences (UMB), 1432 Ås, Norway

A small number of well-studied laboratory strains of *Saccharomyces cerevisiae*, mostly derived from S288C, are used in yeast research. Although powerful, studies for understanding S288C do not always capture the phenotypic essence or the genetic complexity of *S. cerevisiae* biology. This is particularly problematic for multilocus phenotypes identified in laboratory strains because these loci have never been jointly exposed to natural selection and the corresponding phenotypes do not represent optimization for any particular purpose or environment. Isolation and sequencing of new natural yeast strains also reveal that the total sequence diversity of the *S. cerevisiae* global population is poorly sampled in common laboratory strains. Here we discuss methodologies required for using the natural genetic variation in yeast to complete a genotype–phenotype map.

BACKGROUND

Saccharomyces cerevisiae was the first eukaryotic organism to have its genome sequenced (Goffeau et al. 1996). The sequencing was conducted on the commonly used laboratory strain S288C, and its finalized genome opened the scene for large-scale experimental biology. As a consequence, yeast quickly became the flagship species in functional genomics. A key contribution was the first gene deletion collection in any organism, once again made based on the S288C background (Giaever et al. 2002). Research on *S. cerevisiae* has traditionally focused on a small number of well-studied laboratory strains, S288C being one of the most popular. S288C is an artificial genomic mosaic of predominantly European genetic stock (Mortimer and Johnston 1986; Liti et al. 2009). Repair of some detrimental alleles in S288C, like a missense mutation in *GAL2* impeding galactose uptake and a transposon insertion in *HAP1* impairing the respiratory chain, and the introduction of a variety of auxotrophic markers (Brachmann et al. 1998), gave rise to the BY series, the starting point for most reverse genetics collections. Most other common laboratory genetic backgrounds, such as sigma1278b, YPH, and W303, share the majority of their genomes with S288C, whereas a few others, like SK1 and Y55, primarily stem from West African stock (Liti et al. 2009). None of the commonly used laboratory strains carries substantial genetic material from the Asian or North America populations. Thus, the total sequence diversity of the yeast global population is poorly sampled in common laboratory strains.

---

[4]These authors contributed equally to this work.
[5]Correspondence: anders.blomberg@cmb.gu.se

Copyright © Cold Spring Harbor Laboratory Press; all rights reserved
Cite this introduction as *Cold Spring Harb Protoc*; doi:10.1101/pdb.top077735

Chapter 35

It has become increasingly clear that our efforts to study and understand S288C do not always capture the essence or the complexity of *S. cerevisiae* biology. Many industrial and natural strains have properties completely absent in laboratory strains, such as the ability of several wine strains to ferment xylose (Wenger et al. 2010) and of the Malaysian population to use melibiose (Warringer et al. 2011). Sequencing of globally sampled industrial and natural *S. cerevisiae* revealed up to 1% single-nucleotide polymorphism (SNP) variation as well as a repertoire of novel genes (Liti et al. 2009). The high *S. cerevisiae* variation in genome content, that is, gene gain, loss, and duplication, recently revealed by de novo assemblies (Bergström et al. 2014), may explain the unexpectedly high phenotypic diversity in the species. *Saccharomyces cerevisiae* phenotype variation exceeds that of the closely related species *Saccharomyces paradoxus*, despite many fewer SNPs (Liti et al. 2009; Schacherer et al. 2009; Warringer et al. 2011), and also that of *Schizosaccharomyces pombe* (Brown et al. 2011; Fawcett et al. 2014). That phenotypes, and the contribution of individual genes to phenotypes, are heavily dependent on the genetic background is abundantly clear when comparing effects of gene deletions in two different laboratory strain backgrounds. Even gene essentiality has frequently been shown to be strain-dependent (Dowell et al. 2010). Population structure is very pronounced in *S. cerevisiae* relative to what is observed in higher organisms. Phenotypic variation also tends to follow this population structure, prompting the suggestion that it may largely be caused by genetic drift (Warringer et al. 2011; Zörgö et al. 2012). The use of genetically mosaic laboratory strains like S288C as points of departures for phenotype generalizations is in this context particularly problematic because their multiloci genotypes never have been jointly exposed to natural selection and their corresponding phenotypes therefore do not reflect optimization for any particular purpose. Given that most phenotype variations are caused by combinations of alleles (Lorenz and Cohen 2012), it is to be expected that S288C is a poor representative of natural populations. Indeed, S288C recently emerged as the phenotypically most atypical *S. cerevisiae* strain in a large phenotypic screen over a wide array of conditions of natural and industrial strains (Warringer et al. 2011). This underscores that results obtained in this genetic background may be hard to translate into an understanding of the species as a whole and that the palette of naturally occurring strains analyzed should be extended.

SACCHAROMYCES CEREVISIAE ECOLOGY AND POPULATION GENETICS

The ecology of *S. cerevisiae* remains elusive (Replansky et al. 2008). As long as its ecological information remains scarce, the vast amount of data generated using *S. cerevisiae* will remain difficult to interpret in the context of natural selection. This is not only of relevance for environmental sciences. Understanding yeast's natural ecology also has important implications for yeast molecular biology, as it would allow dissection of gene–function relationships in environmental contexts in which gene function has actually evolved (Peña-Castillo and Hughes 2007). *Saccharomyces cerevisiae* has been encountered in diverse ecological contexts, for example, on the surface of grapes, on oak tree bark, on damaged fruits, in forest soils, on rotten wood and on, or even in, insects and animals. However, it is still unclear if these contexts represent natural habitats in which the species has actually evolved, or places to which yeast has only been accidentally dispersed. No longer than two decades ago, it was argued that all natural lineages of *S. cerevisiae* had gone extinct. Strains found in the wild were thought to be feral escapees from domestic stocks (Replansky et al. 2008). The deep branches of the *S. cerevisiae* population structure, which probably predates the association of yeast with man, show that natural isolates still prevail (Liti et al. 2009). In a recent study from China, thousands of samples were collected from a wide array of biotopes and source material at remote sites. *S. cerevisiae* cultures could be established from >10% of these sources (Wang et al. 2012), indicating that the species frequents a very diverse set of niches that are detached from human activity. This study also highlighted that we have not yet saturated the sampling of the genetic variation within *S. cerevisiae*. The genetic diversity found in natural populations of yeast strains represents a valuable, but so far largely untapped, resource in both molecular and environmental biology, as well as biotechnology. For guidance in isolating new strains of *S. cerevisiae*, see Protocol 1: Isolation and Laboratory Domestication of Natural Yeast Strains (Liti et al. 2015a).

YEAST PHENOMICS

Any observable characteristic (except DNA sequence variation) qualifies as a phenotype, including molecular traits (such as metabolic, proteomic, and transcriptomic properties), organism properties (such as morphology and proliferation), and population properties (such as cell aggregation, colony morphology, and quorum sensing). However, given that gene function has emerged as a consequence of selection on fitness effects and that fitness is a reflection of population net growth, yeast phenomics has largely focused on parameters of the yeast asexual growth curve (Warringer and Blomberg 2014). There are technical advantages to measuring clonal yeast growth on agar surfaces. In particular, enormous throughput can be achieved if state-of-the-art robotics is used. Ideally, the different phases of colony growth, which tend to be differently affected by genetic and environmental factors (Warringer et al. 2008), should be determined. However, because of technical challenges, a composite of colony growth, integrating the various growth phases, is usually considered (Baryshnikova et al. 2010). Measuring growth by microcultivation provides intermediate throughput and, if correctly performed, minimum noise and bias. Even the strongest perturbations of gene function, for example gene deletions, often have marginal phenotypic effects (Thatcher et al. 1998), highlighting the importance of precise measurements (Blomberg 2011). Dense measurements of optical density provide a highly resolved view of changes in population density, and thus independent quantification of the different growth phases, of which the lag, maximum rate, and efficiency of growth are the most important.

YEAST QUANTITATIVE GENETICS OF COMPLEX TRAITS

Yeast is a late arrival to the arena of quantitative genetics, but experimental and intrinsic biological advantages, for example, short sexual generation time, haploid and diploid growth stages, high meiotic recombination rate, small genome, and being easy to maintain in large population size, have rapidly positioned it at the forefront (Liti and Louis 2012; Parts 2014). The heritability of complex traits is explained by many genetic variants, whose individual contributions to trait variation are both small and often depend on the environmental context. Thus, the genetic architecture of traits is most easily studied in laboratory-amenable model organisms, like yeast, for which the environment and genetic context can be kept constant and for which trait measurements can be repeated on a large number of replicates (Swinnen et al. 2012b).

Linkage analysis, in which genetic relatedness is completely known, is now in widespread use in the yeast community (Bloom et al. 2013; Cubillos et al. 2013; Wilkening et al. 2013). Focus has so far been on analyzing the co-inheritance of phenotypes and genotypes in progenies from two-parent crosses (Brem et al. 2002; Cubillos et al. 2011; Warringer et al. 2011; Swinnen et al. 2012a). Overall, the crossover rate between yeast sister chromosomes is about once every 250 kbp (Mancera et al. 2008). If a good number (80–100) of offspring are genotyped, 150 genetic markers evenly distributed over the genome allow mapping of reasonably strong quantitative trait loci (QTLs) to intervals of 10–50 genes (Ehrenreich et al. 2009). See Protocol 2: Mapping Quantitative Trait Loci in Yeast (Liti et al. 2015b) for additional details.

Studies using this design have certainly improved our understanding of complex traits. In a recent study of a two-parent cross between a laboratory strain and a wine strain, roughly 1000 F1 segregants were analyzed for genotypic (Illumina short-read sequencing) and phenotypic (end point growth on agar plates) changes and large-scale linkage of the two was performed. The authors detected an average of 12 QTLs per trait, although the variance around this mean was large. They found that the genetic basis for variation in some traits is almost entirely caused by additive effects, whereas for others approximately half of the heritable component comes from interactions between genetic loci (Bloom et al. 2013). Several rounds of intercrosses will break linkage between nearby sites and improve QTL mapping resolution. In a recent study, a yeast mapping population was obtained from extensive outcrossing (12 generations) of four wild founders representing the main distinct geographic and ecological origins of *S. cerevisiae*. This collection contains 10–100 million segregants with fine-grained mosaic genomes and greatly reduced linkage, and is a well-characterized resource for high-resolution mapping of complex traits (Cubillos et al. 2013).

The confounding influence of the pronounced population structure in *S. cerevisiae*, meaning that many thousands of alleles share a distribution pattern, leads to the problem that genome-wide association studies (GWAS) are challenging (Connelly and Akey 2012). At this stage, only two published studies have relied exclusively on association (Muller et al. 2011; Diao and Chen 2012).

ESTABLISHING CAUSALITY OF GENOTYPE–PHENOTYPE LINKS

Finding the phenotype-causing genes within QTLs is often achieved using engineered reciprocal hemizygotes, that is, two parents are repeatedly crossed to produce diploid hybrids in which either of the two parental alleles of a candidate gene has been deleted (Steinmetz et al. 2002). (For details, see Protocol 3: Yeast Reciprocal Hemizygosity to Confirm the Causality of a Quantitative Trait Loci–Associated Gene [Warringer et al. 2015].) This results in two hybrids that are isogenic, except at the locus of interest, allowing strict assignment of causality. Alternatively, the candidate alleles can be tested by allele-swapping, or reconstruction of individual mutations, in the original strain backgrounds. This approach does not require diploidy, but does change the genetic context in which the effect was originally identified, breaking up epistasis structures. The molecular mechanisms behind the genetics of several trait polymorphisms have been unraveled. Some traits are largely explained by a single gene with large size-effects, for example, copper resistance (*CUP1*, copy number variation) (Warringer et al. 2011), galactose nonutilization (*GAL3*, nonsense mutation) (Warringer et al. 2011), xylose utilization (*XDH1*, gene gain) (Wenger et al. 2010), arsenic tolerance (*ARR* genes, multigene duplication) (Cubillos et al. 2011; Bergström et al. 2014), and sensitivity to the DNA-damaging agent 4-nitroquinolone (*RAD5*, missense mutation) (Demogines et al. 2008). Other polymorphisms, such as heat tolerance (Steinmetz et al. 2002; Parts et al. 2011) and sporulation efficiency (Gerke et al. 2006), are complex and involve substantial numbers of alleles, each contributing only a smaller fraction of the phenotypic variation.

Predicting phenotypes based on the genotype is one central goal of quantitative genetics (Jelier et al. 2011). Despite the vast literature on yeast genetics, we are still quite far from reaching this goal. A more extensive and smart use of the global natural yeast variation will certainly bring us closer to this Holy Grail of genetics.

ACKNOWLEDGMENTS

The support to A.B. from the Swedish Research Council is highly acknowledged. J.W. acknowledges support from Carl Trygger Foundation, CTS 12:521. Research of G.L. is supported by ATIP-Avenir (CNRS/INSERM), FP7-PEOPLE-2012-CIG (grant number 322035) and ANR (ANR-13-BSV6-0006-01-AcrossTrait) and LABEX SIGNALIFE (ANR-11-LABX-0028-01).

REFERENCES

Baryshnikova A, Costanzo M, Kim Y, Ding H, Koh J, Toufighi K, Youn J-Y, Ou J, San Luis B-J, Bandyopadhyay S, et al. 2010. Quantitative analysis of fitness and genetic interactions in yeast on a genome scale. *Nat Methods* **7**: 1017–1024.

Bergström A, Simpson JT, Salinas F, Barré B, Parts L, Zia A, Nguyen Ba AN, Moses AM, Louis EJ, Mustonen V, et al. 2014. A high-definition view of functional genetic variation from natural yeast genomes. *Mol Biol Evol* **31**: 872–888.

Blomberg A. 2011. Measuring growth rate in high-throughput growth phenotyping. *Curr Opin Biotechnol* **22**: 94–102.

Bloom JS, Ehrenreich IM, Loo WT, Lite T-LV, Kruglyak L. 2013. Finding the sources of missing heritability in a yeast cross. *Nature* **494**: 234–237.

Brachmann CB, Davies A, Cost GJ, Caputo E, Li J, Hieter P, Boeke JD. 1998. Designer deletion strains derived from *Saccharomyces cerevisiae* S288C: A useful set of strains and plasmids for PCR-mediated gene disruption and other applications. *Yeast* **14**: 115–132.

Brem RB, Yvert G, Clinton R, Kruglyak L. 2002. Genetic dissection of transcriptional regulation in budding yeast. *Science* **296**: 752–755.

Brown WRA, Liti G, Rosa C, James S, Roberts I, Robert V, Jolly N, Tang W, Baumann P, Green C, et al. 2011. A geographically diverse collection of

Schizosaccharomyces pombe isolates shows limited phenotypic variation but extensive karyotypic diversity. *G3 (Bethesda)* **1:** 615–626.

Connelly CF, Akey JM. 2012. On the prospects of whole-genome association mapping in *Saccharomyces cerevisiae*. *Genetics* **191:** 1345–1353.

Cubillos FA, Billi E, Zörgö E, Parts L, Fargier P, Omholt S, Blomberg A, Warringer J, Louis EJ, Liti G. 2011. Assessing the complex architecture of polygenic traits in diverged yeast populations. *Mol Ecol* **20:** 1401–1413.

Cubillos FA, Parts L, Salinas F, Bergström A, Scovacricchi E, Zia A, Illingworth CJR, Mustonen V, Ibstedt S, Warringer J, et al. 2013. High-resolution mapping of complex traits with a four-parent advanced intercross yeast population. *Genetics* **195:** 1141–1155.

Demogines A, Smith E, Kruglyak L, Alani E. 2008. Identification and dissection of a complex DNA repair sensitivity phenotype in Baker's yeast. *PLoS Genet* **4:** e1000123.

Diao L, Chen KC. 2012. Local ancestry corrects for population structure in *Saccharomyces cerevisiae* genome-wide association studies. *Genetics* **192:** 1503–1511.

Dowell RD, Ryan O, Jansen A, Cheung D, Agarwala S, Danford T, Bernstein DA, Rolfe PA, Heisler LE, Chin B, et al. 2010. Genotype to phenotype: A complex problem. *Science* **328:** 469–469.

Ehrenreich IM, Gerke JP, Kruglyak L. 2009. Genetic dissection of complex traits in yeast: Insights from studies of gene expression and other phenotypes in the BYxRM cross. *Cold Spring Harb Symp Quant Biol* **74:** 145–153.

Fawcett JA, Iida T, Takuno S, Sugino RP, Kado T, Kugou K, Mura S, Kobayashi T, Ohta K, Nakayama J-I, et al. 2014. Population genomics of the fission yeast *Schizosaccharomyces pombe*. *PLoS One* **9:** e104241.

Gerke JP, Chen CTL, Cohen BA. 2006. Natural isolates of *Saccharomyces cerevisiae* display complex genetic variation in sporulation efficiency. *Genetics* **174:** 985–997.

Giaever G, Chu AM, Ni L, Connelly C, Riles L, Véronneau S, Dow S, Lucau-Danila A, Anderson K, André B, et al. 2002. Functional profiling of the *Saccharomyces cerevisiae* genome. *Nature* **418:** 387–391.

Goffeau A, Barrell BG, Bussey H, Davis RW, Dujon B, Feldmann H, Galibert F, Hoheisel JD, Jacq C, Johnston M, et al. 1996. Life with 6000 genes. *Science* **274:** 546, 563–567.

Jelier R, Semple JI, Garcia-Verdugo R, Lehner B. 2011. Predicting phenotypic variation in yeast from individual genome sequences. *Nat Genet* **43:** 1270–1274.

Liti G, Louis EJ. 2012. Advances in quantitative trait analysis in yeast. *PLoS Genet* **8:** e1002912.

Liti G, Carter DM, Moses AM, Warringer J, Parts L, James SA, Davey RP, Roberts IN, Burt A, Koufopanou V, et al. 2009. Population genomics of domestic and wild yeasts. *Nature* **458:** 337–341.

Liti G, Warringer J, Blomberg A. 2015a. Isolation and laboratory domestication of natural yeast strains. *Cold Spring Harb Protoc*. doi: 10.1101/pdb.prot089052.

Liti G, Warringer J, Blomberg A. 2015b. Mapping quantitative trait loci in yeast. *Cold Spring Harb Protoc*. doi: 10.1101/pdb.prot089060.

Lorenz K, Cohen BA. 2012. Small- and large-effect quantitative trait locus interactions underlie variation in yeast sporulation efficiency. *Genetics* **192:** 1123–1132.

Mancera E, Bourgon R, Brozzi A, Huber W, Steinmetz LM. 2008. High-resolution mapping of meiotic crossovers and non-crossovers in yeast. *Nature* **454:** 479–485.

Mortimer RK, Johnston JR. 1986. Genealogy of principal strains of the yeast genetic stock center. *Genetics* **113:** 35–43.

Muller LAH, Lucas JE, Georgianna DR, McCusker JH. 2011. Genome-wide association analysis of clinical vs. nonclinical origin provides insights into *Saccharomyces cerevisiae* pathogenesis. *Mol Ecol* **20:** 4085–4097.

Parts L. 2014. Genome-wide mapping of cellular traits using yeast. *Yeast* **31:** 197–205.

Parts L, Cubillos FA, Warringer J, Jain K, Salinas F, Bumpstead SJ, Molin M, Simpson JT, Quail MA, Moses AM, et al. 2011. Revealing the genetic structure of a trait by sequencing a population under selection. *Genome Res* **21:** 1131–1138.

Peña-Castillo L, Hughes TR. 2007. Why are there still over 1000 uncharacterized yeast genes? *Genetics* **176:** 7–14.

Replansky T, Koufopanou V, Greig D, Bell G. 2008. *Saccharomyces sensu stricto* as a model system for evolution and ecology. *Trends Ecol Evol* **23:** 494–501.

Schacherer J, Shapiro JA, Ruderfer DM, Kruglyak L. 2009. Comprehensive polymorphism survey elucidates population structure of *Saccharomyces cerevisiae*. *Nature* **458:** 342–345.

Steinmetz LM, Sinha H, Richards DR, Spiegelman JI, Oefner PJ, McCusker JH, Davis RW. 2002. Dissecting the architecture of a quantitative trait locus in yeast. *Nature* **416:** 326–330.

Swinnen S, Schaerlaekens K, Pais T, Claesen J, Hubmann G, Yang Y, Demeke M, Foulquié-Moreno MR, Goovaerts A, Souvereyns K, et al. 2012a. Identification of novel causative genes determining the complex trait of high ethanol tolerance in yeast using pooled-segregant whole-genome sequence analysis. *Genome Res* **22:** 975–984.

Swinnen S, Thevelein JM, Nevoigt E. 2012b. Genetic mapping of quantitative phenotypic traits in *Saccharomyces cerevisiae*. *FEMS Yeast Res* **12:** 215–227.

Thatcher JW, Shaw JM, Dickinson WJ. 1998. Marginal fitness contributions of nonessential genes in yeast. *Proc Natl Acad Sci* **95:** 253–257.

Wang Q-M, Liu W-Q, Liti G, Wang S-A, Bai F-Y. 2012. Surprisingly diverged populations of *Saccharomyces cerevisiae* in natural environments remote from human activity. *Mol Ecol* **21:** 5404–5417.

Warringer J, Blomberg A. 2014. *Yeast phenomics—Large-scale mapping of the genetic basis for organismal traits* (ed. Hancock J), pp. 172–206. CRC Press, Boca Raton, FL.

Warringer J, Anevski D, Liu B, Blomberg A. 2008. Chemogenetic fingerprinting by analysis of cellular growth dynamics. *BMC Chem Biol* **8:** 3.

Warringer J, Zörgö E, Cubillos FA, Gjuvsland A, Simpson JT, Forsmark A, Durbin R, Omholt SW, Louis EJ, Liti G, et al. 2011. Trait variation in yeast is defined by population history. *PLoS Genet* **7:** e1002111.

Warringer J, Liti G, Blomberg A. 2015. Yeast reciprocal hemizygosity to confirm the causality of a quantitative trait loci–associated gene. *Cold Spring Harb Protoc*. doi: 10.1101/pdb.prot089078.

Wenger JW, Schwartz K, Sherlock G. 2010. Bulk segregant analysis by high-throughput sequencing reveals a novel xylose utilization gene from *Saccharomyces cerevisiae*. *PLoS Genet* **6:** e1000942.

Wilkening S, Tekkedil MM, Lin G, Fritsch ES, Wei W, Gagneur J, Lazinski DW, Camilli A, Steinmetz LM. 2013. Genotyping 1000 yeast strains by next-generation sequencing. *BMC Genomics* **14:** 90.

Zörgö E, Gjuvsland A, Cubillos FA, Louis EJ, Liti G, Blomberg A, Omholt SW, Warringer J. 2012. Life history shapes trait heredity by accumulation of loss-of-function alleles in yeast. *Mol Biol Evol* **29:** 1781–1789.

Protocol 1

Isolation and Laboratory Domestication of Natural Yeast Strains

Gianni Liti,[1,4] Jonas Warringer,[2,3] and Anders Blomberg[2]

[1]IRCAN, CNRS UMR 6267, INSERM U998, University of Nice, 06107 Nice, France; [2]Department of Chemistry and Molecular Biology, University of Gothenburg, 40530 Gothenburg, Sweden; [3]Centre for Integrative Genetics (CIGENE), Department of Animal and Aquacultural Sciences, Norwegian University of Life Sciences (UMB), 1432 Ås, Norway

The process from yeast isolation to their use in laboratory experiments is lengthy. Historically, *Saccharomyces* strains were easily obtained by sampling alcoholic fermentation processes or other substrates associated with human activity in which *Saccharomyces* was heavily enriched. In contrast, wild *Saccharomyces* yeasts are found in complex microbial communities and small population sizes, making isolation challenging. We have overcome this problem by enriching yeast on media favoring the growth of *Saccharomyces* over other microorganisms. The isolation process is usually followed by molecular characterization that allows the strain identification. Finally, yeast isolated from domestic or wild environments need to be genetically manipulated before they can be used in laboratory experiments.

MATERIALS

It is essential that you consult the appropriate Material Safety Data Sheets and your institution's Environmental Health and Safety Office for proper handling of equipment and hazardous material used in this protocol.

RECIPES: Please see the end of this protocol for recipes indicated by <R>. Additional recipes can be found online at http://cshprotocols.cshlp.org/site/recipes.

Reagents

Potassium acetate medium (liquid or solid) <R>
Reagents for the molecular analysis and genetic manipulation of new isolates (see Steps 9 and 10)
Saccharomyces sensu stricto enrichment medium <R>
Yeast extract–peptone–dextrose (YPD) medium (liquid or solid) <R>

Equipment

Dry incubator (temperature-controlled)
Petri dishes
Sample collection materials (e.g., gloves, sterile cotton balls, scalpels, tweezers, and smart phone)
Shaking incubator (for 30–50 mL tubes; temperature-controlled)
Tubes (sterile; 30–50 mL)

[4]Correspondence: gianni.liti@unice.fr

Copyright © Cold Spring Harbor Laboratory Press; all rights reserved
Cite this protocol as *Cold Spring Harb Protoc*; doi:10.1101/pdb.prot089052

METHOD

Sampling

1. Collect substrates from wild or human-associated environments. Gather a few cubic centimeters of solid substrates (e.g., tree bark, fruit, soil, mushrooms, and insects) or a few milliliters of liquid substrates (e.g., fermenting must, tree sap, and nectar). Alternatively, rub a sterile cotton ball on a surface (e.g., a fermentation vessel).

   Avoid contamination by using gloves and sterilized tools.

2. Meticulously record a description of the sampled item (e.g., take a photograph, note which part of the plant was sampled, record the date, and use a smart phone to precisely record the global positioning system coordinates of the collection site).

   In the future, we hope that such documentation will become compulsory information to report when new strains are isolated.

Enrichment and Isolation

3. Incubate each sample in the *Saccharomyces sensu stricto* enrichment medium at 25°C. Break solid samples into smaller pieces before placing them in the enrichment medium.

   See Troubleshooting.

4. After 7 d, look for signs of fermentation (i.e., an increase in turbidity and CO_2 production). Discard cultures that lack these characteristics.

5. Dilute the actively fermenting cultures, and spread appropriate volumes (~100 µL) on YPD agar plates (or plates with a composition resembling the enrichment medium). Incubate for 2–5 d at 25°C.

6. Check the color and morphology (size and shape) of colonies by microscopy, and discard any obvious nonyeast microbes (e.g., bacteria). Retain those resembling *Saccharomyces* colonies for further analysis.

7. Incubate colonies that resemble yeast in potassium acetate medium at 23°C.

   Potassium acetate medium is a respiratory medium lacking nitrogen that induces Saccharomyces *meiosis.*

8. After 3–7 d of sporulation, look for the presence of ascus structures containing four spores, which is a distinctive trait of *Saccharomyces cerevisiae*-related yeasts. *Saccharomyces sensu stricto* strains can be stored indefinitely in the −80°C laboratory freezer in 25% sterile glycerol solution.

Species Identification

9. Use a molecular method to identify the yeast species.

   The choice of the method depends on the equipment available and personal expertise. A combination of screening procedures may be used. We favor sequencing a small fragment of rDNA (e.g., ITS1 and ITS2; James et al. 1996) and pulsed-field gel electrophoresis karyotyping for an unequivocal identification of different Saccharomyces sensu stricto *species (Liti et al. 2005). The karyotype of* Saccharomyces sensu stricto *yeast species is distinct from that of other yeast species with a larger number of smaller chromosomes. Karyotyping also provides information about chromosomal rearrangements and potential strain hybridization (Liti et al. 2005). Restriction analysis of rDNA and mtDNA can give an indication of the species as can a number of recently developed species-specific primers (Muir et al. 2011). An alternative method is the use of tester strains and genetic crosses for the biological identification of the species (Naumov et al. 2010). This test relies on measuring gamete viability in crosses with species representatives.*

Preparing Strains for Laboratory Work

10. Once the species has been identified, perform further molecular characterization, phenotyping, and/or genetic manipulation in preparation for future work.

    The genetic manipulation is dictated by the nature of the downstream analysis. The majority of wild strains are diploid homothallic, harbor varying levels of heterozygosity, and are mostly, but not always, able to grow in minimal medium. The removal of heterozygosity is generally required. This is performed by isolating monosporic derivatives, which otherwise will segregate in downstream steps. Deletion of the *HO* gene using a dominant resistance marker amplified from the drug-resistance plasmids pFA6 (kanMX4), pAG25 (natMX4), or pAG32 (hphMX4) produces stable haploid strains (Goldstein and McCusker 1999).

The availability of inexpensive restriction site-associated DNA (RAD) sequencing approaches allows a deep analysis of genome phylogeny and structure (Cromie et al. 2013). Molecular barcodes can be added in the polymerase chain reaction construct to uniquely label strains (Cubillos et al. 2009). Additional auxotrophic markers can be introduced by using the marker-less plasmids strategy (Louvel et al. 2014). Marker-less plasmids include pAD1 (*LEU2*), pAD2 (*LYS2*), and pAD4 (*MET15*); for further details, see Brachmann et al. (1998). A standard transformation protocol works well in the majority of strains, although efficiency varies greatly depending on genetic background and conditions may need adjustment (see Troubleshooting).

TROUBLESHOOTING

Problem (Step 3): Growth conditions for recovery of the new strain are unknown.

Solution: Incubation at 30°C tends to favor *S. cerevisiae* strains and closely related species (e.g., *S. paradoxus* and *S. mikatae*). A lower incubation temperature (e.g., 23°C–25°C) will recover a broader range of *Saccharomyces* species (Sampaio and Gonçalves 2008). The *Saccharomyces sensu stricto* enrichment medium preferentially selects *Saccharomyces* over other microorganisms (Naumov et al. 1997; Wang et al. 2012). It may be necessary to make adjustments to the concentration of components of this enrichment medium because different species display quantitative variation in tolerance to these reagents (e.g., to ethanol). A possible solution is to split the sample into aliquots and test different enrichment conditions on these samples, thereby maximizing chances of recovering multiple species and strains.

Problem (Step 10): Some isolates perform very badly in laboratory protocols for sporulation, spore viability, and transformation efficiency.

Solution: The availability of multiple isolates from the same area or substrate might help with the selection of the right strain to work with. There is a considerable variation in these phenotypes, even in closely related strains (e.g., because of the presence of rare loss-of-function alleles; Bergström et al. [2014]). Karyotype analysis can help to identify hybrid *Saccharomyces* strains, which are interesting from an ecological and evolutionary perspective, but are not amenable to genetic manipulation (e.g., as a result of poor spore viability). Protocols may need to be strain optimized. For example, strains respond differently to sporulation conditions (e.g., liquid or solid, potassium acetate medium or minimal potassium acetate medium). Similarly, several steps of the lithium acetate transformation protocol can be adjusted to optimize the transformation process. We suggest testing multiple volumes for re-inoculation after the overnight cultivation. Another critical parameter is the length of the 42°C heat shock. We observed increased transformation efficiency in multiple isolates by extending the incubation time to 30–45 min.

DISCUSSION

Increasing interest in the ecology and evolution in natural habitat of the budding yeast *S. cerevisiae* has prompted field surveys aimed at isolating natural strains (Hittinger 2013; Boynton and Greig 2014). A landmark study that focused on several areas of China allowed isolation of hundreds of *S. cerevisiae* isolates from many environments not previously probed (Wang et al. 2012). These studies documented the presence of *S. cerevisiae* in wild settings but also indicated that the success of isolation is low, perhaps caused by low population size. Isolation based on enrichment media allows specific species to be recovered. Minor differences in environmental factors (e.g., temperature) can favor one species over another (Sampaio and Gonçalves 2008). Future metagenomic studies have the potential to give a better understanding of the microbial communities before selection by enrichment media.

Yeast geneticists are also increasingly exploiting natural variation as a tool to understand the relationship between genotype and phenotype (Nieduszynski and Liti 2011; Fay 2013). For example, see Protocol 2: Mapping Quantitative Trait Loci in Yeast (Liti et al. 2015). The "raw material" for these experiments is naturally diverged yeast strains, such as those isolated using the procedures described above.

RECIPES

COM Drop-Out Powder

Reagent	Quantity
Adenine	800 mg
Arginine	800 mg
Aspartic acid	4000 mg
Histidine	800 mg
Leucine	800 mg
Lysine	1200 mg
Methionine	800 mg
Phenylalanine	2000 mg
Threonine	8000 mg
Tryptophan	800 mg
Tyrosine	1200 mg
Uracil	800 mg

Mix the reagents together, and store the powder at room temperature.

Potassium Acetate Medium (Liquid or Solid)

Reagent	Quantity (for 1 L)
Potassium acetate	16 g
Yeast extract	1.76 g
Dextrose	0.4 g
COM drop-out powder <R>	0.7 g
Agar (for solid medium only)	20 g

Dissolve the potassium acetate, yeast extract, and dextrose in 800 mL of distilled water while stirring continuously. Adjust the pH to 7.0 with 1 M HCl or 2.5 M NaOH. Add the COM drop-out powder and the agar (if used). Adjust the volume to 1 L. Autoclave at 0.5 bar for 20 min at 121°C. Store for several weeks at room temperature.

Saccharomyces Sensu Stricto Enrichment Medium

Reagent	Quantity (for 1 L)
Yeast extract	3 g
Malt extract	3 g
Bacto peptone	5 g
Glucose	10 g
Ethanol	80 mL
Chloramphenicol	200 mg
HCl (1 M)	1 mL

Dissolve the reagents in a final volume of 1 L of distilled water. Autoclave at 0.5 bar for 20 min at 121°C. Store for several weeks at room temperature.

Yeast Extract–Peptone–Dextrose (YPD) Medium (Liquid or Solid)

Reagent	Quantity
Yeast extract	10 g
Peptone	20 g
Dextrose	20 g
Agar (for plates only)	20 g

Dissolve reagents in 1 L of distilled water. Autoclave for 20 min at 121°C and 0.5 bar. Store for several weeks at room temperature.

REFERENCES

Bergström A, Simpson JT, Salinas F, Barré B, Parts L, Zia A, Nguyen Ba AN, Moses AM, Louis EJ, Mustonen V, et al. 2014. A high-definition view of functional genetic variation from natural yeast genomes. *Mol Biol Evol* **31:** 872–888.

Boynton PJ, Greig D. 2014. The ecology and evolution of non-domesticated *Saccharomyces* species. *Yeast* **31:** 449–462.

Brachmann CB, Davies A, Cost GJ, Caputo E, Li J, Hieter P, Boeke JD. 1998. Designer deletion strains derived from *Saccharomyces cerevisiae* S288C: A useful set of strains and plasmids for PCR-mediated gene disruption and other applications. *Yeast* **14:** 115–132.

Cromie GA, Hyma KE, Ludlow CL, Garmendia-Torres C, Gilbert TL, May P, Huang AA, Dudley AM, Fay JC. 2013. Genomic sequence diversity and population structure of *Saccharomyces cerevisiae* assessed by RAD-seq. *G3 (Bethesda)* **3:** 2163–2171.

Cubillos FA, Louis EJ, Liti G. 2009. Generation of a large set of genetically tractable haploid and diploid *Saccharomyces* strains. *FEMS Yeast Res* **9:** 1217–1225.

Fay JC. 2013. The molecular basis of phenotypic variation in yeast. *Curr Opin Genet Dev* **23:** 672–677.

Goldstein AL, McCusker JH. 1999. Three new dominant drug resistance cassettes for gene disruption in *Saccharomyces cerevisiae*. *Yeast* **15:** 1541–1553.

Hittinger CT. 2013. *Saccharomyces* diversity and evolution: A budding model genus. *Trends Genet* **29:** 309–317.

James SA, Collins MD, Roberts IN. 1996. Use of an rRNA internal transcribed spacer region to distinguish phylogenetically closely related species of the genera Zygosaccharomyces and Torulaspora. *Int J Syst Bacteriol* **46:** 189–194.

Liti G, Peruffo A, James SA, Roberts IN, Louis EJ. 2005. Inferences of evolutionary relationships from a population survey of LTR-retrotransposons and telomeric-associated sequences in the *Saccharomyces* sensu stricto complex. *Yeast* **22:** 177–192.

Liti G, Warringer J, Blomberg A. 2015. Mapping quantitative trait loci in yeast. *Cold Spring Harb Protoc* doi: 10.1101/pdb.prot089060.

Louvel H, Gillet-Markowska A, Liti G, Fischer G. 2014. A set of genetically diverged *Saccharomyces cerevisiae* strains with markerless deletions of multiple auxotrophic genes. *Yeast* **31:** 91–101.

Muir A, Harrison E, Wheals A. 2011. A multiplex set of species-specific primers for rapid identification of members of the genus *Saccharomyces*. *FEMS Yeast Res* **11:** 552–563.

Naumov GI, Naumova ES, Sniegowski PD. 1997. Differentiation of European and Far East Asian populations of *Saccharomyces paradoxus* by allozyme analysis. *Int J Syst Bacteriol* **47:** 341–344.

Naumov GI, Naumova ES, Masneuf-Pomarède I. 2010. Genetic identification of new biological species *Saccharomyces arboricolus* Wang et Bai. *Antonie Van Leeuwenhoek* **98:** 1–7.

Nieduszynski CA, Liti G. 2011. From sequence to function: Insights from natural variation in budding yeasts. *Biochim Biophys Acta* **1810:** 959–966.

Sampaio JP, Gonçalves P. 2008. Natural populations of *Saccharomyces kudriavzevii* in Portugal are associated with oak bark and are sympatric with *S. cerevisiae* and *S. paradoxus*. *Appl Environ Microbiol* **74:** 2144–2152.

Wang Q-M, Liu W-Q, Liti G, Wang S-A, Bai F-Y. 2012. Surprisingly diverged populations of *Saccharomyces cerevisiae* in natural environments remote from human activity. *Mol Ecol* **21:** 5404–5417.

Protocol 2

Mapping Quantitative Trait Loci in Yeast

Gianni Liti,[1,4] Jonas Warringer,[2,3] and Anders Blomberg[2]

[1]*IRCAN, CNRS UMR 6267, INSERM U998, University of Nice, 06107 Nice, France;* [2]*Department of Chemistry and Molecular Biology, University of Gothenburg, 40530 Gothenburg, Sweden;* [3]*Centre for Integrative Genetics (CIGENE), Department of Animal and Aquacultural Sciences, Norwegian University of Life Sciences (UMB), 1432 Ås, Norway*

Natural *Saccharomyces* strains isolated from the wild differ quantitatively in molecular and organismal phenotypes. Quantitative trait loci (QTL) mapping is a powerful approach for identifying sequence variants that alter gene function. In yeast, QTL mapping has been used in designed crosses to map functional polymorphisms. This approach, outlined here, is often the first step in understanding the molecular basis of quantitative traits. New large-scale sequencing surveys have the potential to directly associate genotypes with organismal phenotypes, providing a broader catalog of causative genetic variants. Additional analysis of intermediate phenotypes (e.g., RNA, protein, or metabolite levels) can produce a multilayered and integrated view of individual variation, producing a high-resolution view of the genotype–phenotype map.

MATERIALS

It is essential that you consult the appropriate Material Safety Data Sheets and your institution's Environmental Health and Safety Office for proper handling of equipment and hazardous material used in this protocol.

RECIPES: Please see the end of this protocol for recipes indicated by <R>. Additional recipes can be found online at http://cshprotocols.cshlp.org/site/recipes.

Reagents

Double-selective medium
  For example, use synthetically defined (SD) double-selective drop-out medium <R> or prepare YPD medium with, for example, G418 and hygromycin (see Step 1).

Parent *Saccharomyces cerevisiae* strains (or closely related species amenable for genetic crosses)
  See Protocol 1: Isolation and Laboratory Domestication of Natural Yeast Strains (Liti et al. 2015).

Potassium acetate medium (liquid or solid) <R>
Reagents for genotyping (see Step 5)
Yeast extract–peptone–dextrose (YPD) medium (liquid or solid) <R>

[4]Correspondence: gianni.liti@unice.fr

Chapter 35

Equipment

Standard yeast culturing and handling equipment
Statistical package for analyzing quantitative trait loci (QTL) data (e.g., rQTL [Arends et al. 2010])

METHOD

Generating F$_1$ Segregants

1. Cross founder strains of opposite mating types in YPD medium and incubate overnight. Then select diploid hybrids on appropriate double-selective medium.

   For example, delete the HO gene in both parents with distinct dominant drug-resistance markers (e.g., KanMX, HygMX, and NatMX) and select hybrids in medium supplemented with both drugs. Drug-resistance markers relieve the need for auxotrophies that can have dramatic impact on a large variety of phenotypes (Mülleder et al. 2012). However, natural variations in drug tolerance exist, and drug concentrations may require strain optimization.

2. Streak hybrids for single colonies on plates containing double-selective medium.

3. Use a single clone to induce sporulation in potassium acetate medium.

   Sporulation itself is a highly complex trait (Gerke et al. 2009), and the timing and efficiency varies greatly among crosses and might require optimization of, for example, temperature and medium. Spore viability is highly affected by sequence divergence and chromosome rearrangements.

4. Dissect spores on YPD plates using a dissecting microscope. Reduce the risk of confounding aneuploidies by selecting segregants from tetrads with four viable spores that show a correct 2:2 segregation ratio for all segregating markers. Maximize the distribution of recombination events by selecting each spore from a different tetrad (Brem et al. 2002).

   The resulting spore population will be used for mapping. Classical linkage analyses use a sample size of 100–200 individuals and allow mapping of QTLs with high or intermediate effect within larger chromosomal intervals (Steinmetz et al. 2002; Ehrenreich et al. 2009; Romano et al. 2010; Cubillos et al. 2011). The more segregants that are obtained, the lower the QTL effect sizes (QTLs with small phenotypic effect) detected and the smaller the QTL regions will be (Bloom et al. 2013).

   Full tetrad analysis can provide important insights into phenotype segregation within single meioses (Cubillos et al. 2011; Ludlow et al. 2013). Colonies selected from spore dissection can be stored indefinitely in the −80°C laboratory freezer in 25% sterile glycerol solution.

QTL Mapping

5. Obtain genome-wide genotype information on the mapping population.

   Given the total number of recombination events (approximately 50–100 crossovers per meiosis, again number and distribution are strain dependent), low-coverage genotyping of 200 polymorphic markers provides a genetic map sufficient for linkage analysis (Kim and Fay 2007; Gerke et al. 2009; Cubillos et al. 2011). Genome-wide genotyping studies were initially performed by SNP arrays (Steinmetz 2002; Sinha et al. 2008). New sequencing approaches such as low-coverage (Cubillos et al. 2013; Illingworth et al. 2013; Wilkening et al. 2013) or restriction site associated DNA (RAD) sequencing (Cromie et al. 2013) result in a very dense genetic map, providing a more complete view of the recombination landscape compared with single-nucleotide polymorphism (SNP) arrays (especially if inferred from full tetrad analysis to distinguish crossover from noncrossover events. DNA extraction (Parts et al. 2011) and sequencing library preparation (Wilkening et al. 2013) have been adapted to 96-well plate format.

6. Record phenotyping information. Use a high number of replicate measurements and a technically precise method of measurement (e.g., Warringer and Blomberg 2003). Randomize replicates over space (instruments, plates, and plate positions) and time (runs and batches), and normalize to internal controls (Warringer et al. 2003).

   Successful linkage analysis typically requires a large fraction of the phenotypic variability to be explained by additive genetic variation, rather than environmental variation or epistatic effects between gene variants, and high measurement precision and low measurement bias. The noisier and more subject to bias trait measurements are, the lower the power of detection will be.

7. Perform linkage analysis using a statistical package.

    There are many such packages, and rQTL is a popular choice (Arends et al. 2010). Most of these packages are designed for species with diploid genomes, but can be adapted to yeast haploid segregants. Models, permutation tests, and parameters can be changed, affecting significance calls (logarithm of the odds [LOD]). Parameter choices depend on sample size, density, and structure of the genetic map and phenotype distribution.

    See Troubleshooting.

TROUBLESHOOTING

Problem (Step 7): QTL are mapped in large genomic intervals or not detected.

Solution: To improve resolution, segregants can be manually obtained from multigeneration crosses by controlling mating, sporulation, and dissection. However, this procedure is enormously labor intensive. Instead, a simple system in which sexual generations are forced in bulk by using two auxotrophic markers (selectable in drop-out SD double-selective medium) that are engineered in identical chromosomal positions in the two parents can be used (Parts et al. 2011). This method has also been successfully applied to a multiparent mapping population (Cubillos et al. 2013). In most cases, bulk segregant analysis is limited to traits where artificial selection can be applied. However, this approach can be adjusted to traits that are not amenable to artificial selection with individual segregants phenotyped and individuals with extreme phenotype values pooled together for genome sequencing (Swinnen et al. 2012). The problem of detecting QTLs with a small effect size has been partially solved by analyzing segregants as large pools rather than individually. This approach is based on selecting thousands (Ehrenreich et al. 2010) or even millions (Parts et al. 2011) of pooled segregants for a phenotype and then analyzing genome-wide changes in allele frequencies within the population.

DISCUSSION

Although yeast QTL mapping has only recently gained speed (Liti and Louis 2012; Parts 2014), a number of technological advances (e.g., Connelly and Akey 2012) position *S. cerevisiae* at the forefront of quantitative genetics. However, there are some aspects that remain to be solved to optimize linkage analysis efficiency. The choice of parent strains is often largely based on extreme phenotypic characteristics, crossing a very high and a very low performing strain. These phenotype differences appear often to be due to major, but rare, loss-of-function variants. Instead, strain selection would be better based on knowledge of population structure, crossing strains that are genetically different and where different allele constellations account for a trait. Another important aspect related to parent genomes is variability in terms of gene content and gene position. Large variations are observed especially in subtelomeric regions. These regions are mapped and QTLs within them will be determined, because of their linkage with less subtelomeric regions, but their gene content is largely unknown. Therefore, it is not possible to identify the causative gene and understand the mechanistic basis of the trait variation. This is also a problem for the mitochondrial genome, often not investigated, either because only one parent is represented in the genome or because the recombinant configuration is difficult to analyze (Dimitrov et al. 2009) due to incomplete genome information.

RELATED INFORMATION

Once candidate genes are identified based on gene-trait associations, other methods must be used to assess their causality. See, for example, Protocol 3: Yeast Reciprocal Hemizygosity to Confirm the Causality of a Quantitative Trait Loci–Associated Gene (Warringer et al. 2015).

Chapter 35

RECIPES

COM Drop-Out Powder

Reagent	Quantity
Adenine	800 mg
Arginine	800 mg
Aspartic acid	4000 mg
Histidine	800 mg
Leucine	800 mg
Lysine	1200 mg
Methionine	800 mg
Phenylalanine	2000 mg
Threonine	8000 mg
Tryptophan	800 mg
Tyrosine	1200 mg
Uracil	800 mg

Mix the reagents together, and store the powder at room temperature.

Potassium Acetate Medium (Liquid or Solid)

Reagent	Quantity (for 1 L)
Potassium acetate	16 g
Yeast extract	1.76 g
Dextrose	0.4 g
COM drop-out powder <R>	0.7 g
Agar (for solid medium only)	20 g

Dissolve the potassium acetate, yeast extract, and dextrose in 800 mL of distilled water while stirring continuously. Adjust the pH to 7.0 with 1 M HCl or 2.5 M NaOH. Add the COM drop-out powder and the agar (if used). Adjust the volume to 1 L. Autoclave at 0.5 bar for 20 min at 121°C. Store for several weeks at room temperature.

Synthetically Defined (SD) Double-Selective Drop-Out Medium

Reagent	Quantity (for 1 L)
Yeast nitrogen base (without amino acids and nitrogen substrate)	1.7 g
Ammonium sulfate	5 g
Dextrose	20 g
Complete supplement mixture (CSM) of amino acids and nucleotides	Varies
Agar (for solid plates)	20 g

Dissolve the yeast nitrogen base, ammonium sulfate, and dextrose in water. In amounts specified by the supplier, dissolve the CSM (minus those amino acids or nucleotides used for selection) in the solution. Add the agar (if preparing plates). Bring the final volume to 1 L with water. Autoclave for 20 min at 121°C and 0.5 bar. Store for several weeks at room temperature.

Yeast Extract–Peptone–Dextrose (YPD) Medium (Liquid or Solid)

Reagent	Quantity
Yeast extract	10 g
Peptone	20 g
Dextrose	20 g
Agar (for plates only)	20 g

Dissolve reagents in 1 L of distilled water. Autoclave for 20 min at 121°C and 0.5 bar. Store for several weeks at room temperature.

REFERENCES

Arends D, Prins P, Jansen RC, Broman KW. 2010. R/qtl: High-throughput multiple QTL mapping. *Bioinformatics* **26**: 2990–2992.

Bloom JS, Ehrenreich IM, Loo WT, Lite T-LV, Kruglyak L. 2013. Finding the sources of missing heritability in a yeast cross. *Nature* **494**: 234–237.

Brem RB, Yvert G, Clinton R, Kruglyak L. 2002. Genetic dissection of transcriptional regulation in budding yeast. *Science* **296**: 752–755.

Connelly CF, Akey JM. 2012. On the prospects of whole-genome association mapping in *Saccharomyces cerevisiae*. *Genetics* **191**: 1345–1353.

Cromie GA, Hyma KE, Ludlow CL, Garmendia-Torres C, Gilbert TL, May P, Huang AA, Dudley AM, Fay JC. 2013. Genomic sequence diversity and population structure of *Saccharomyces cerevisiae* assessed by RAD-seq. *G3 (Bethesda)* **3**: 2163–2171.

Cubillos FA, Billi E, Zörgö E, Parts L, Fargier P, Omholt S, Blomberg A, Warringer J, Louis EJ, Liti G. 2011. Assessing the complex architecture of polygenic traits in diverged yeast populations. *Mol Ecol* **20**: 1401–1413.

Cubillos FA, Parts L, Salinas F, Bergström A, Scovacricchi E, Zia A, Illingworth CJR, Mustonen V, Ibstedt S, Warringer J, et al. 2013. High-resolution mapping of complex traits with a four-parent advanced intercross yeast population. *Genetics* **195**: 1141–1155.

Dimitrov LN, Brem RB, Kruglyak L, Gottschling DE. 2009. Polymorphisms in multiple genes contribute to the spontaneous mitochondrial genome instability of *Saccharomyces cerevisiae* S288C strains. *Genetics* **183**: 365–383.

Ehrenreich IM, Gerke JP, Kruglyak L. 2009. Genetic dissection of complex traits in yeast: Insights from studies of gene expression and other phenotypes in the BYxRM cross. *Cold Spring Harb Symp Quant Biol* **74**: 145–153.

Ehrenreich IM, Torabi N, Jia Y, Kent J, Martis S, Shapiro JA, Gresham D, Caudy AA, Kruglyak L. 2010. Dissection of genetically complex traits with extremely large pools of yeast segregants. *Nature* **464**: 1039–1042.

Gerke J, Lorenz K, Cohen B. 2009. Genetic interactions between transcription factors cause natural variation in yeast. *Science* **323**: 498–501.

Illingworth CJR, Parts L, Bergström A, Liti G, Mustonen V. 2013. Inferring genome-wide recombination landscapes from advanced intercross lines: Application to yeast crosses. *PLoS One* **8**: e62266.

Kim HS, Fay JC. 2007. Genetic variation in the cysteine biosynthesis pathway causes sensitivity to pharmacological compounds. *Proc Natl Acad Sci* **104**: 19387–19391.

Liti G, Louis EJ. 2012. Advances in quantitative trait analysis in yeast. *PLoS Genet* **8**: e1002912.

Liti G, Warringer J, Blomberg A. 2015. Isolation and laboratory domestication of natural yeast strains. *Cold Spring Harb Protoc* doi: 10.1101/pdb.prot089052.

Ludlow CL, Scott AC, Cromie GA, Jeffery EW, Sirr A, May P, Lin J, Gilbert TL, Hays M, Dudley AM. 2013. High-throughput tetrad analysis. *Nat Methods* **10**: 671–675.

Mülleder M, Capuano F, Pir P, Christen S, Sauer U, Oliver SG, Ralser M. 2012. A prototrophic deletion mutant collection for yeast metabolomics and systems biology. *Nat Biotechnol* **30**: 1176–1178.

Parts L. 2014. Genome-wide mapping of cellular traits using yeast. *Yeast* **31**: 197–205.

Parts L, Cubillos FA, Warringer J, Jain K, Salinas F, Bumpstead SJ, Molin M, Simpson JT, Quail MA, Moses AM, et al. 2011. Revealing the genetic structure of a trait by sequencing a population under selection. *Genome Res* **21**: 1131–1138.

Romano GH, Gurvich Y, Lavi O, Ulitsky I, Shamir R, Kupiec M. 2010. Different sets of QTLs influence fitness variation in yeast. *Mol Syst Biol* **6**: 346.

Sinha H, David L, Pascon RC, Clauder-Münster S, Krishnakumar S, Nguyen M, Shi G, Dean J, Davis RW, Oefner PJ, et al. 2008. Sequential elimination of major-effect contributors identifies additional quantitative trait loci conditioning high-temperature growth in yeast. *Genetics* **180**: 1661–1670.

Steinmetz LM, Sinha H, Richards DR, Spiegelman JI, Oefner PJ, McCusker JH, Davis RW. 2002. Dissecting the architecture of a quantitative trait locus in yeast. *Nature* **416**: 326–330.

Swinnen S, Schaerlaekens K, Pais T, Claesen J, Hubmann G, Yang Y, Demeke M, Foulquié-Moreno MR, Goovaerts A, Souvereyns K, et al. 2012. Identification of novel causative genes determining the complex trait of high ethanol tolerance in yeast using pooled-segregant whole-genome sequence analysis. *Genome Res* **22**: 975–984.

Warringer J, Blomberg A. 2003. Automated screening in environmental arrays allows analysis of quantitative phenotypic profiles in *Saccharomyces cerevisiae*. *Yeast* **20**: 53–67.

Warringer J, Ericson E, Fernandez-Ricaud L, Nerman O, Blomberg A. 2003. High-resolution yeast phenomics resolves different physiological features of the saline response. *Proc Natl Acad Sci* **100**: 15724–15729.

Warringer J, Liti G, Blomberg A. 2015. Yeast reciprocal hemizygosity to confirm the causality of a quantitative trait loci–associated gene. *Cold Spring Harb Protoc* doi: 10.1101/pdb.prot089078.

Wilkening S, Tekkedil MM, Lin G, Fritsch ES, Wei W, Gagneur J, Lazinski DW, Camilli A, Steinmetz LM. 2013. Genotyping 1000 yeast strains by next-generation sequencing. *BMC Genomics* **14**: 90.

Protocol 3

Yeast Reciprocal Hemizygosity to Confirm the Causality of a Quantitative Trait Loci–Associated Gene

Jonas Warringer,[1,2,4] Gianni Liti,[3] and Anders Blomberg[1]

[1]Department of Chemistry and Molecular Biology, University of Gothenburg, 40530 Gothenburg, Sweden; [2]Centre for Integrative Genetics (CIGENE), Department of Animal and Aquacultural Sciences, Norwegian University of Life Sciences (UMB), 1432 Ås, Norway; [3]IRCAN, CNRS UMR 6267, INSERM U998, University of Nice, 06107 Nice, France

Pinpointing causal alleles within a quantitative trait loci region is a key challenge when dissecting the genetic basis of natural variation. In yeast, homing in on culprit genes is often achieved using engineered reciprocal hemizygotes as outlined here. Based on prior information on gene–trait associations, candidate genes are identified. In haploid versions of both founder strains, a candidate gene is then deleted. Gene knockouts are independently mated to a wild-type version of the other strain, such that two diploid hybrid strains are obtained. These strains are identical with regard to the nuclear genome, except for that they are hemizygotic at the locus of interest and contain different alleles of the candidate gene. If correctly measured, a trait difference between these reciprocal hemizygotes can confidently be ascribed to allelic variation at the target locus.

MATERIALS

It is essential that you consult the appropriate Material Safety Data Sheets and your institution's Environmental Health and Safety Office for proper handling of equipment and hazardous material used in this protocol.

RECIPES: Please see the end of this protocol for recipes indicated by <R>. Additional recipes can be found online at http://cshprotocols.cshlp.org/site/recipes.

Reagents

Haploid yeast strains (isogenic MATa and MATα, HO–, one selection marker private to each mating type, e.g., in URA3, HIS3, LEU2, LYS2, and MET17)

Reagents for deleting candidate genes and confirming their removal (see Step 2)

> Gene deletion is preferably achieved using drug-resistance markers. Drug-resistance plasmids pFA6 (kanMX4), pAG25 (natMX4), and pAG32 (hphMX4) confer resistance to G418 (geneticin), nourseothricin, and hygromycin, respectively (see Goldstein and McCusker 1999).

Reagents for genotyping (see Step 4)

Synthetically defined (SD) double-selective drop-out medium <R>

Yeast extract–peptone–dextrose (YPD) medium (liquid or solid) <R>

Equipment

Standard yeast culturing and handling equipment

Statistical package for analyzing data (e.g., the statistical computing freeware R, https://www.r-project.org/)

---

[4]Correspondence: Jonas.warringer@cmb.gu.se

Copyright © Cold Spring Harbor Laboratory Press; all rights reserved
Cite this protocol as Cold Spring Harb Protoc; doi:10.1101/pdb.prot089078

METHOD

For further information regarding identification of genes using engineered reciprocal hemizygotes, see Steinmetz et al. (2002) and Stern (2014). For an illustration of reciprocal hemizygosity, see Figure 1.

1. Identify high-likelihood candidate genes in regions of interest using prior knowledge of gene–function associations.

   To identify quantitative trait loci (QTL) in yeast, see Protocol 2: Mapping Quantitative Trait Loci in Yeast (Liti et al. 2015). Linkage or association regions usually span 10–50 kb and encompass up to a few dozen genes. Gene–phenotype or gene–function associations, determined mostly from experiments in the S288C background, can be downloaded from the Saccharomyces Genome Database (SGD; http://www.yeastgenome.org/). These can be compared with lists of genes with genetic variations likely to impact gene function. Genes containing frameshift, nonsense, or nonsynonymous mutations predicted to be detrimental to function (Kumar et al. 2009) are good candidates, as are genes with copy number variations. Variant lists covering many natural and industrial yeast genomes are available (Liti et al. 2009; Bergström et al. 2014).

2. Delete candidate genes separately, or in bulk (Pais et al. 2013), in haploid versions of natural strains to be crossed. (For essential genes, perform deletions in diploid hybrids.) For convenience, we recommend the following approach.

   i. Amplify KanMX4 deletion cassettes from corresponding deletion strains in the budding yeast (BY) deletion collection using primers and polymerase chain reaction (PCR) protocols as described at http://www-sequence.stanford.edu/group/yeast_deletion_project/PCR_strategy.html and by Giaever et al. (2002). Purify the PCR fragments.

   ii. Transform the strains of interest with the purified PCR fragments.

   In most strains, sequence similarity to the reference genome is sufficient for recombination to guide the PCR fragment to the correct target locus. The transformation procedure may need to be optimized according to the genetic background (Louvel et al. 2014).

   iii. Select transformants using the resistance markers. Determine the optimal drug concentrations required.

   Innate tolerance to drugs commonly used for selection varies depending on genetic background (Warringer et al. 2011).

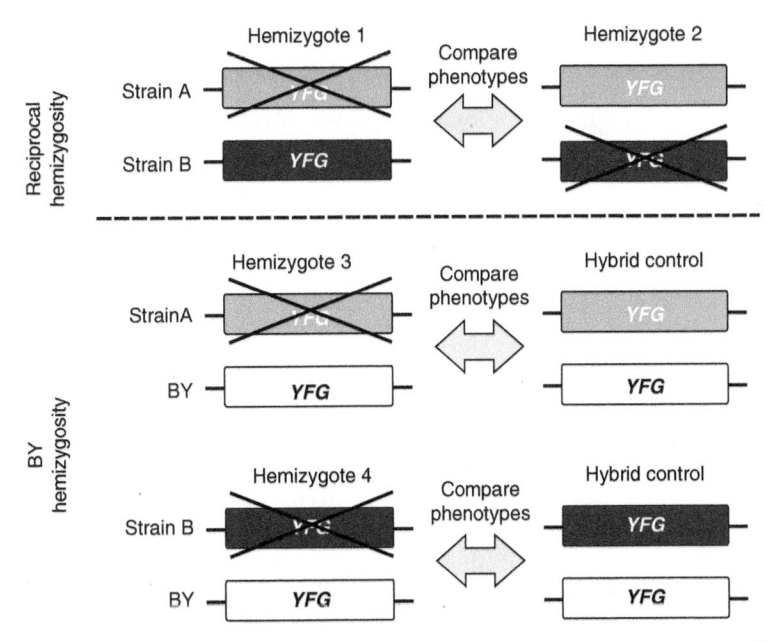

FIGURE 1. Reciprocal hemizygosity. (*Upper* panel) Classical reciprocal hemizygosity. Two diploid hybrids containing single but different alleles at a locus are compared. (*Lower* panel) BY hemizygosity design. Strains are independently mated to a BY single-gene deletion. Hemizygotes are independently compared to their respective cognate diploid hybrid. YFG, your favorite gene.

iv. Use diagnostic PCR and/or sequencing to verify gene deletion.

See Troubleshooting.

3. Construct three diploid hybrids: strain A (WT) × strain B (WT), strain A (WT) × strain B (Δgene), and strain A (Δgene) × strain B (WT). In addition, autodiploidize wild-type (WT) haploids to provide isogenic diploid founder strains. For each mating, drop 10 μL of an OD = 1.0 cell suspension of each strain together onto an identical position on an YPD agar plate. Incubate the plates overnight at 30°C.

Autodiploidization is important because ploidy often has a strong impact on phenotype (Zörgö et al. 2013).

4. Select diploid hybrids using strain-specific selection markers as follows. (Auxotrophies induced by incapacitating mutations in *URA3*, *HIS3*, *LEU2*, *LYS2*, and *MET17* are reliable in a wide variety of genetic backgrounds.)

   i. Transfer a loop full of cells from each mated population to a few mL of SD medium lacking selection nutrients.

   ii. Pipette 10 μL of the cell mixture onto SD agar plates lacking selection nutrients. Spread the cells evenly, and cultivate for 2 d at 30°C.

   iii. Take a single solitary colony of average size, and single-streak this colony on an SD agar plate lacking selection nutrients. Incubate for 2 d at 30°C with selection applied. Verify the hemizygote or homozygote genotype by diagnostic PCR or sequencing directed to the target loci.

   Use independent single streaking of multiple colonies to provide true biological repeats for phenotyping.

   See Troubleshooting.

5. Perform phenotyping using a setup similar to the one used in the linkage or association analysis (see Protocol 2: Mapping Quantitative Trait Loci in Yeast [Liti et al. 2015]), and compare the measured phenotypes (e.g., using a parametric two-group test).

   i. Compare hemizygotes to the WT diploid hybrid to reveal haploinsufficiency and haploproficiency, that is, when a single allele is superior or inferior to two.

   ii. Compare the WT diploid hybrid to autodiploidized parents to reveal heterosis/hybrid superiority.

   This may be of particular interest for breeding efforts to improve commercial traits.

   iii. Compare autodiploidized and haploid parents to reveal ploidy effects on phenotypes.

   If multiple genes are evaluated, adjust the test statistics to correct for the number of hypotheses tested, for example, through calculation of false discovery rates (Benjamini and Hochberg 1995).

   See Troubleshooting.

TROUBLESHOOTING

Problem (Step 2): No colonies emerge after gene deletion, or colonies emerge but diagnostic PCR reveals that the targeted gene is still present.

Solution: Optimize the concentrations of the selection agent used so that no growth occurs on negative control plates (strain without selection marker) and good growth occurs on positive control plates (strain with selection marker in another locus). If controls show expected behavior, verify and possibly change the sequence used for homologous recombination into the target locus. Revisit the information about the genome sequences and consider the possibility that paralogous loci confound either deletion construct targeting or diagnostic PCR. Consult the SGD and verify that the gene to be deleted is nonessential in S288C and that any genes in the immediate vicinity are nonessential.

Problem (Step 4): Diploid hybrids are not obtained.
Solution: Verify that strains to be mated are haploid and of the right mating types, for example, by diagnostic PCR directed at the mating type locus.

Problem (Step 5): No candidates in the QTL region contribute to the phenotype.
Solution: Ensure that phenotyping conditions are as identical as possible to the initial screen. Ensure that parent phenotypes are not vastly different in diploids and haploids. Repeat phenotyping with a higher number of replicates. Repeats should be independent clones. A high number of repeats usually allows detection of small phenotype contributions. Repeats should ideally be randomized over space and time to avoid bias. Extend the number of candidates tested. If necessary, several genes may first be deleted in bulk to narrow down the region. See also Discussion and Figure 1, lower panel, for an alternative high-throughput variant.

DISCUSSION

Mitochondrial genes and genes that have close paralogs in one or both backgrounds do not easily lend themselves to reciprocal hemizygosity. Genes will also need to be exempted because of the requirement for accurately assembled and annotated sequence information. Assembly and annotation is typically achieved by read alignment to the S288C reference genome. Genes in chromosomal regions with unclear correspondence in S288C can therefore not be targeted. This prevents QTL dissection in QTL-rich (Cubillos et al. 2011) subtelomeres.

Intrusive gene manipulation is mutagenic and risks introducing confounding secondary site mutations (Scherens and Goffeau 2004). Reciprocal hemizygosity studies are less prone to be affected by secondary effects than gene deletion studies in haploids, as recessive effects are masked in diploid hybrids.

Logistics is a challenge. Targeting all potentially causative genes in all identified regions is rarely possible. Some QTLs will therefore evade successful genetic dissection. Given that prior knowledge on gene-trait associations is used for candidate gene selection, successfully dissected QTLs rarely disclose novel gene-trait associations. Gene selection and manual deletion can largely be evaded by mating parental haploids to deletion constructs from the BY gene deletion collection (Fig. 1, lower panel). This vastly increases the number of genes that can be targeted. Both natural strains of interest are independently mated to BY single gene deletion mutants corresponding to genes in QTL regions of interest (Gutiérrez et al. 2013). The procedure is easily automated. Causality is called when the phenotype of one, but not the other, of the two hemizygotes at an individual locus deviates from that of the cognate diploid hybrid retaining both alleles. The approach assumes allele contributions to a trait to be independent of whether the carrier strain is crossed to BY or another natural strain. This is not true for all alleles (Cubillos et al. 2011), resulting in both false negatives and false positives. To weed out false positives, confirmation of positive hits using standard reciprocal hemizygosity is required.

RECIPES

Synthetically Defined (SD) Double-Selective Drop-Out Medium

Reagent	Quantity (for 1 L)
Yeast nitrogen base (without amino acids and nitrogen substrate)	1.7 g
Ammonium sulfate	5 g
Dextrose	20 g

Reagent	Quantity
Complete supplement mixture (CSM) of amino acids and nucleotides	Varies
Agar (for solid plates)	20 g

Dissolve the yeast nitrogen base, ammonium sulfate, and dextrose in water. In amounts specified by the supplier, dissolve the CSM (minus those amino acids or nucleotides used for selection) in the solution. Add the agar (if preparing plates). Bring the final volume to 1 L with water. Autoclave for 20 min at 121°C and 0.5 bar. Store for several weeks at room temperature.

Yeast Extract–Peptone–Dextrose (YPD) Medium (Liquid or Solid)

Reagent	Quantity
Yeast extract	10 g
Peptone	20 g
Dextrose	20 g
Agar (for plates only)	20 g

Dissolve reagents in 1 L of distilled water. Autoclave for 20 min at 121°C and 0.5 bar. Store for several weeks at room temperature.

REFERENCES

Benjamini Y, Hochberg Y. 1995. Controlling the false discovery rate: A practical and powerful approach to multiple testing. *J R Stat Soc Series B Methodol* **57**: 289–300.

Bergström A, Simpson JT, Salinas F, Barré B, Parts L, Zia A, Nguyen Ba AN, Moses AM, Louis EJ, Mustonen V, et al. 2014. A high-definition view of functional genetic variation from natural yeast genomes. *Mol Biol Evol* **31**: 872–888.

Cubillos FA, Billi E, Zörgö E, Parts L, Fargier P, Omholt S, Blomberg A, Warringer J, Louis EJ, Liti G. 2011. Assessing the complex architecture of polygenic traits in diverged yeast populations. *Mol Ecol* **20**: 1401–1413.

Giaever G, Chu AM, Ni L, Connelly C, Riles L, Véronneau S, Dow S, Lucau-Danila A, Anderson K, André B, et al. 2002. Functional profiling of the *Saccharomyces cerevisiae* genome. *Nature* **418**: 387–391.

Goldstein AL, McCusker JH. 1999. Three new dominant drug resistance cassettes for gene disruption in *Saccharomyces cerevisiae*. *Yeast* **15**: 1541–1553.

Gutiérrez A, Beltran G, Warringer J, Guillamón JM. 2013. Genetic basis of variations in nitrogen source utilization in four wine commercial yeast strains. *PLoS One* **8**: e67166.

Kumar P, Henikoff S, Ng PC. 2009. Predicting the effects of coding nonsynonymous variants on protein function using the SIFT algorithm. *Nat Protoc* **4**: 1073–1081.

Liti G, Carter DM, Moses AM, Warringer J, Parts L, James SA, Davey RP, Roberts IN, Burt A, Koufopanou V, et al. 2009. Population genomics of domestic and wild yeasts. *Nature* **458**: 337–341.

Liti G, Warringer J, Blomberg A. 2015. Mapping quantitative trait loci in yeast. *Cold Spring Harb Protoc* doi: 10.1101/pdb.prot089060.

Louvel H, Gillet-Markowska A, Liti G, Fischer G. 2014. A set of genetically diverged *Saccharomyces cerevisiae* strains with markerless deletions of multiple auxotrophic genes. *Yeast* **31**: 91–101.

Pais TM, Foulquié-Moreno MR, Hubmann G, Duitama J, Swinnen S, Goovaerts A, Yang Y, Dumortier F, Thevelein JM. 2013. Comparative polygenic analysis of maximal ethanol accumulation capacity and tolerance to high ethanol levels of cell proliferation in yeast. *PLoS Genet* **9**: e1003548.

Scherens B, Goffeau A. 2004. The uses of genome-wide yeast mutant collections. *Genome Biol* **5**: 229.

Steinmetz LM, Sinha H, Richards DR, Spiegelman JI, Oefner PJ, McCusker JH, Davis RW. 2002. Dissecting the architecture of a quantitative trait locus in yeast. *Nature* **416**: 326–330.

Stern DL. 2014. Identification of loci that cause phenotypic variation in diverse species with the reciprocal hemizygosity test. *Trends Genet*. **30**: 547–554.

Warringer J, Zörgö E, Cubillos FA, Gjuvsland A, Simpson JT, Forsmark A, Durbin R, Omholt SW, Louis EJ, Liti G, et al. 2011. Trait variation in yeast is defined by population history. *PLoS Genet* **7**: e1002111.

Zörgö E, Chwialkowska K, Gjuvsland AB, Garré E, Sunnerhagen P, Liti G, Blomberg A, Omholt SW, Warringer J. 2013. Ancient evolutionary trade-offs between yeast ploidy states. *PLoS Genet* **9**: e1003388.

CHAPTER 36

Genetic Analysis of Complex Traits in *Saccharomyces cerevisiae*

Ian M. Ehrenreich[1,3] and Paul M. Magwene[2,3]

[1]*Molecular and Computational Biology Section, University of Southern California, Los Angeles, California 90089-2910;* [2]*Department of Biology and Center for Systems Biology, Duke University, Durham, North Carolina 27708*

Defining the relationship between genotype and phenotype is a central challenge in biology. A powerful approach to this problem is to determine the genetic architecture and molecular basis of phenotypic differences among genetically diverse individuals. *Saccharomyces cerevisiae* is an important model system for such work. Current genetic mapping approaches for this species exploit high-throughput phenotyping and sequencing to facilitate the detection of a large fraction of the genomic loci that underlie trait variation among isolates. Once identified, several methods exist to determine the specific genes and genetic variants that underlie these loci and cause phenotypic variations. In this introduction, we provide a brief overview of research on complex traits in yeast and discuss different genetic mapping approaches applied to yeast studies.

INTRODUCTION

Many traits of agricultural, evolutionary, and medical significance vary among members of the same species. These phenotypes, which are referred to as "complex" or "quantitative" traits, often show a continuous range of variation and are influenced by multiple genetic loci that interact with each other and the environment. Although researchers have sought to understand the genetic basis of complex traits for more than a century (Falconer and Mackay 1996; Lynch and Walsh 1998), recent advances in high-throughput genotyping and sequencing have made it possible to address this problem at unprecedented depth. Indeed, current genetic mapping techniques in *Saccharomyces cerevisiae* permit the detection of many loci that contribute to traits of interest, and studies that use these methods often identify 10 or more loci underlying variations in chemical resistance, growth rate, morphology, and other traits (e.g., Ehrenreich et al. 2010, 2012; Parts et al. 2011; Bloom et al. 2013; Granek et al. 2013). Analysis of how these loci collectively exert their effects can provide valuable insights into the relationship between genotype and phenotype that cannot be obtained from standard molecular genetic approaches (Rockman 2008).

CHALLENGES WHEN DISSECTING THE GENETIC BASIS OF COMPLEX TRAITS

Geneticists studying complex traits face two main challenges: detecting loci that contribute to a trait and resolving these loci to specific genes and genetic variants (Mackay et al. 2009) (Fig. 1). The former problem depends on the statistical power of a given mapping study—that is, the probability that a locus will be detected given that it has a biological effect. Statistical power is influenced by some factors

[3]Correspondence: ian.ehrenreich@usc.edu; paul.magwene@duke.edu

Copyright © Cold Spring Harbor Laboratory Press; all rights reserved
Cite this introduction as *Cold Spring Harb Protoc*; doi:10.1101/pdb.top077602

Chapter 36

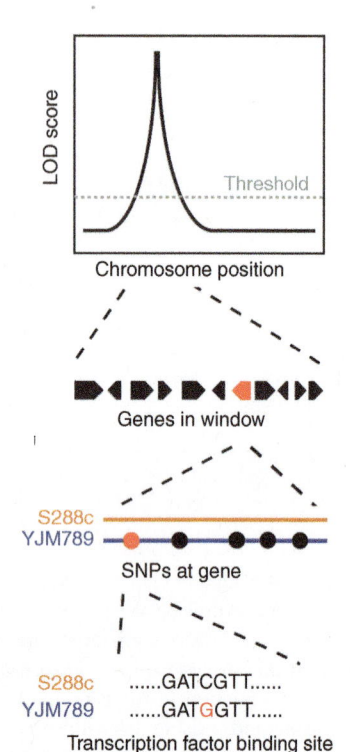

FIGURE 1. Steps in identifying and resolving a causal locus. A significant locus is identified in a cross-based genetic mapping study involving the reference strain S288c and the clinical isolate YJM789. A number of genes underlie the locus, with only one actually influencing the trait (in red). At this gene, YJM789 possesses multiple single-nucleotide polymorphism (SNP) differences relative to S288c (shown as filled circles), most of which do not have effects. Identification of the causal SNP (shown as red circle) results in the discovery of allelic variation in a transcription factor binding site that causes a change in gene expression in *cis*. Note that other types of variants, such as amino acid changes, can also have phenotypic effects.

that are under the control of the investigator, such as sample size and the breeding design of a mapping population, and by some factors that are unknown, such as the genetic architecture of a trait and the effect sizes of causal loci. In terms of the resolution at which loci are detected, the primary factor involved is the amount of recombination that has occurred in a mapping population. For example, panels of wild isolates that have experienced many outcrossings in nature will typically provide better resolution than cross populations generated in the laboratory. Genetic mapping studies in yeast that use F_2 populations often identify causal loci at a resolution of 10 to 20 genes, and even higher resolution can be achieved using mapping populations that have experienced multiple rounds of recombination (e.g., Parts et al. 2011).

CURRENT STRATEGIES FOR DISSECTING COMPLEX TRAITS IN YEAST

The main challenges confronting researchers studying complex traits are common to all organisms; however, optimal strategies to overcome these issues often rely on organism-specific characteristics. Here, we provide an overview of some of the primary genetic mapping methods used by geneticists studying complex traits, as well as the respective advantages and disadvantages of these techniques as applied to yeast (Table 1).

Genome-Wide Association Study

Genome-wide association study (GWAS) involves identifying associations between genetic variants and traits using genetically diverse individuals sampled from a population or species (McCarthy et al. 2008). The advantages of GWAS include higher genetic mapping resolution relative to cross-based approaches and a more extensive sampling of genetic diversity present in a species. GWAS is most effective at detecting genetic variants that are common in a population, and thus causal variants at low frequencies may go unidentified (Manolio et al. 2009). Further, GWAS can have high false-positive rates when population structure is present in a panel of individuals (Lander and Schork 1994; Pritchard and Rosenberg 1999). A variety of methods exist to correct for population structure (e.g.,

TABLE 1. Genetic mapping methods used to study complex traits in yeast

	GWAS	Linkage mapping	BSA
Resolution	High	Modest with F_2 populations	Modest with F_2 populations
Relative cost of genotyping	High	High	Low
Relative phenotyping effort	High	High	Low to high, depending on whether individuals are phenotyped or *en masse* selection is used
Computational/statistical complexity associated with analysis	High	Modest	Low
Advantages	Large amount of genetic diversity assayed; genotyped individuals can be phenotyped for many traits	Genotyped individuals can be phenotyped for many traits; allows estimation of effect sizes and genetic interactions	Amenable to many different phenotyping or selection designs; analysis of similar traits in many different crosses can be performed at relatively low cost
Disadvantages	Complicated by population structure; poor statistical power for alleles that are at low frequency across isolates	Modest resolution despite high genotyping costs when used in F_2 populations	May not allow estimation of effect sizes, dominance, and genetic interactions

GWAS, genome-wide association study; BSA, bulk segregant analysis.

Kang et al. 2008, 2010; Segura et al. 2012), and these approaches enable the detection of large effect loci among yeast isolates (Connelly and Akey 2012; Diao and Chen 2012). More than 100 sequenced yeast isolates are now available (Liti et al. 2009; Schacherer et al. 2009; Bergström et al. 2014; Strope et al. 2015), and ongoing efforts to generate even larger panels of *S. cerevisiae* isolates may make GWAS a more viable strategy for yeast researchers in the future.

Linkage Mapping

Linkage mapping in yeast involves mating two or more haploid parental strains and then phenotyping and genotyping a panel of recombinant offspring from these crosses. Causal loci are identified using statistical tests for association between markers (polymorphic sites) and a trait of interest (Lander and Botstein 1989). When samples on the order of approximately 1000 or more individuals are used, it is possible to detect a large fraction of the loci that have additive effects (Bloom et al. 2013). A major expense associated with linkage mapping is the cost of genotyping, as each cross progeny must be genotyped at a large number of markers throughout the genome. However, a number of low-cost genotyping methods, based on next-generation sequencing, have made it feasible to generate large mapping populations in a cost-effective manner (e.g., Baird et al. 2008; Andolfatto et al. 2011). One of the advantages of linkage mapping, relative to bulk segregant analysis (below), is that the effect sizes and interactions of identified loci can be quantified. Linkage mapping is a particularly good strategy when many phenotypes are to be examined in a single cross.

Bulk Segregant Analysis

As with linkage mapping, bulk segregant analysis (BSA) entails crossing two or more parental strains and phenotyping their recombinant offspring (Michelmore et al. 1991). However, in contrast to linkage mapping, BSA involves selective genotyping of subsets of segregants, usually the extremes of the phenotypic distribution. BSA is a very useful technique in yeast because it is easy to generate large numbers of cross progeny in this organism (Ehrenreich et al. 2010; Parts et al. 2011). These offspring can be phenotyped either individually (e.g., Brauer et al. 2006; Wenger et al. 2010) or *en masse* (e.g., Segre et al. 2006; Ehrenreich et al. 2010, 2012), with offspring of interest then sequenced in bulk (i.e., pooled samples) (Ehrenreich et al. 2010; Wenger et al. 2010; Parts et al. 2011; Granek et al. 2013). Regions of the genome that cause variability in a trait of interest are identified based on significant skew in their allele frequencies (Magwene et al. 2011). Generating bulk samples that contain large numbers of individuals, typically 50 or more, is an important component of BSA because it can minimize random sampling effects on genome-wide allele frequency measurements. Overall, BSA is a cost-effective genetic mapping technique, and it is particularly useful in cases where a number of different cross populations need to be examined, rapid results are desired, or there is little a priori information about the genetic architecture of the trait(s) of interest. For additional details on BSA, see Protocol 1: Genetic Dissection of Heritable Traits in Yeast Using Bulk Segregant Analysis (Ehrenreich and Magwene 2015).

STRATEGIES FOR IDENTIFYING CAUSAL GENES AND GENETIC VARIANTS

One of the distinguishing strengths of yeast as a model system for complex trait genetics is that targeted genetic engineering techniques can be used to identify specific genes and genetic variants that underlie loci causing trait variation. Pinpointing these factors is crucial for moving from a statistical understanding to a mechanistic, molecular characterization of a trait. Several genetic engineering–based methods, including allele replacement (Storici et al. 2001; Reid et al. 2002) and reciprocal hemizygosity (Steinmetz et al. 2002), exist for identifying the specific causal gene from a set of candidates. Typically allele replacement involves first integrating a counter-selectable marker into the gene or site of interest and then knocking an alternate sequence into that site. Allele replacement is one of the greatest advantages of working with yeast, but second-site mutations, errors introduced during polymerase chain reaction (PCR) steps, and genetic background effects can influence results from this technique. The application of CRISPR/Cas9-based genome editing to yeast (e.g., DiCarlo et al. 2013; Mans et al. 2015) is a promising alternative to traditional allelic replacement strategies that is likely to accelerate the rate at which candidate variants can be engineered and tested. In reciprocal hemizygosity, both of the possible hemizygotes of a gene are generated in an otherwise isogenic, hybrid diploid background (Steinmetz et al. 2002). Comparison of these hemizygotes can determine whether the two alleles of a gene possess functional differences. Although reciprocal hemizygosity is a powerful tool, the technique may not work if a gene shows haploinsufficiency or other differences in function between haploids and diploids.

CONCLUSION

Increasingly, genetic analysis of complex traits is being used to explore the molecular basis of phenotypes of interest in *S. cerevisiae*. We have provided a brief summary of work and associated methods on this topic; however, we stress that genetic analysis of complex traits in yeast is a rapidly developing area, with frequent advances in basic understanding and methodology. For researchers interested in applying complex trait genetics, we recommend careful consideration of the different techniques available for genetic mapping in yeast. Attention to experimental design details, such as the amenability of the trait to high-throughput phenotyping, the need for quantitative measurement of allelic effects, and the usefulness of existing resources for studying a trait of interest, can position a study to have the greatest chance for success. Studies that effectively deal with these concerns have the potential to shed light on basic statistical and molecular principles underlying the genotype–phenotype map in yeast.

ACKNOWLEDGMENTS

We thank members of our respective laboratories for helpful comments during the preparation of this contribution. I.M.E. was supported by startup funds from the University of Southern California, as well as National Science Foundation (NSF) grant MCB-1330874, National Institutes of Health (NIH) grants R01GM110255 and R21AI108939, and a grant from the Rose Hills Foundation. P.M.M. was supported by NSF grant DEB-1019753 and NIH grant R01GM098287.

REFERENCES

Andolfatto P, Davison D, Erezyilmaz D, Hu TT, Mast J, Sunayama-Morita T, Stern DL. 2011. Multiplexed shotgun genotyping for rapid and efficient genetic mapping. *Genome Res* 21: 610–617.

Baird NA, Etter PD, Atwood TS, Currey MC, Shiver AL, Lewis ZA, Selker EU, Cresko WA, Johnson EA. 2008. Rapid SNP discovery and genetic mapping using sequenced RAD markers. *PLoS ONE* 3: e3376.

Bergström A, Simpson JT, Salinas F, Barré B, Parts L, Zia A, Nguyen Ba AN, Moses AM, Louis EJ, Mustonen V, et al. 2014. A high-definition view of functional genetic variation from natural yeast genomes. *Mol Biol Evol* 31: 872–888.

Bloom JS, Ehrenreich IM, Loo WT, Lite TL, Kruglyak L. 2013. Finding the sources of missing heritability in a yeast cross. *Nature* 494: 234–237.

Brauer MJ, Christianson CM, Pai DA, Dunham MJ. 2006. Mapping novel traits by array-assisted bulk segregant analysis in *Saccharomyces cerevisiae*. *Genetics* 173: 1813–1816.

Connelly CF, Akey JM. 2012. On the prospects of whole-genome association mapping in *Saccharomyces cerevisiae*. *Genetics* 191: 1345–1353.

Diao L, Chen KC. 2012. Local ancestry corrects for population structure in *Saccharomyces cerevisiae* genome-wide association studies. *Genetics* 192: 1503–1511.

DiCarlo JE, Norville JE, Mali P, Rios X, Aach J, Church GM. 2013. Genome engineering in *Saccharomyces cerevisiae* using CRISPR-Cas systems. *Nucleic Acids Res* 41: 4336–4343.

Ehrenreich IM, Magwene PM. 2015. Genetic dissection of heritable traits in yeast using bulk segregant analysis. *Cold Spring Harb Protoc* doi: 10.1101/pdb.prot088989.

Ehrenreich IM, Torabi N, Jia Y, Kent J, Martis S, Shapiro JA, Gresham D, Caudy AA, Kruglyak L. 2010. Dissection of genetically complex traits with extremely large pools of yeast segregants. *Nature* 464: 1039–1042.

Ehrenreich IM, Bloom J, Torabi N, Wang X, Jia Y, Kruglyak L. 2012. Genetic architecture of highly complex chemical resistance traits across four yeast strains. *PLoS Genet* 8: e1002570.

Falconer DS, Mackay TF. 1996. *Introduction to quantitative genetics*, 4th ed. Pearson Education Limited, Harlow, UK.

Granek JA, Murray D, Kayrkci O, Magwene PM. 2013. The genetic architecture of biofilm formation in a clinical isolate of *Saccharomyces cerevisiae*. *Genetics* 193: 587–600.

Kang HM, Zaitlen NA, Wade CM, Kirby A, Heckerman D, Daly MJ, Eskin E. 2008. Efficient control of population structure in model organism association mapping. *Genetics* 178: 1709–1723.

Kang HM, Sul JH, Service SK, Zaitlen NA, Kong SY, Freimer NB, Sabatti C, Eskin E. 2010. Variance component model to account for sample structure in genome-wide association studies. *Nat Genet* 42: 348–354.

Lander ES, Botstein D. 1989. Mapping mendelian factors underlying quantitative traits using RFLP linkage maps. *Genetics* 121: 185–199.

Lander ES, Schork NJ. 1994. Genetic dissection of complex traits. *Science* 265: 2037–2048.

Liti G, Carter DM, Moses AM, Warringer J, Parts L, James SA, Davey RP, Roberts IN, Burt A, Koufopanou V, et al. 2009. Population genomics of domestic and wild yeasts. *Nature* 458: 337–341.

Lynch M, Walsh B. 1998. *Genetics and analysis of quantitative traits*. Sinauer, Sunderland, MA.

Mackay TF, Stone EA, Ayroles JF. 2009. The genetics of quantitative traits: Challenges and prospects. *Nat Rev Genet* 10: 565–577.

Magwene PM, Willis JH, Kelly JK. 2011. The statistics of bulk segregant analysis using next generation sequencing. *PLoS Comput Biol* 7: e1002255.

Manolio TA, Collins FS, Cox NJ, Goldstein DB, Hindorff LA, Hunter DJ, McCarthy MI, Ramos EM, Cardon LR, Chakravarti A, et al. 2009. Finding the missing heritability of complex diseases. *Nature* 461: 747–753.

Mans R, van Rossum HM, Wijsman M, Backx A, Kuijpers NG, van den Broek M, Daran-Lapujade P, Pronk JT, van Maris AJ, Daran JM. 2015. CRISPR/Cas9: A molecular Swiss army knife for simultaneous introduction of multiple genetic modifications in *Saccharomyces cerevisiae*. *FEMS Yeast Res* 15: fov004.

McCarthy MI, Abecasis GR, Cardon LR, Goldstein DB, Little J, Ioannidis JP, Hirschhorn JN. 2008. Genome-wide association studies for complex traits: Consensus, uncertainty and challenges. *Nat Rev Genet* 9: 356–369.

Michelmore RW, Paran I, Kesseli RV. 1991. Identification of markers linked to disease-resistance genes by bulked segregant analysis: A rapid method to detect markers in specific genomic regions by using segregating populations. *Proc Natl Acad Sci* 88: 9828–9832.

Parts L, Cubillos FA, Warringer J, Jain K, Salinas F, Bumpstead SJ, Molin M, Zia A, Simpson JT, Quail MA, et al. 2011. Revealing the genetic structure of a trait by sequencing a population under selection. *Genome Res* 21: 1131–1138.

Pritchard JK, Rosenberg NA. 1999. Use of unlinked genetic markers to detect population stratification in association studies. *Am J Hum Genet* 65: 220–228.

Reid RJ, Lisby M, Rothstein R. 2002. Cloning-free genome alterations in *Saccharomyces cerevisiae* using adaptamer-mediated PCR. *Methods Enzymol* 350: 258–277.

Rockman MV. 2008. Reverse engineering the genotype–phenotype map with natural genetic variation. *Nature* 456: 738–744.

Schacherer J, Shapiro JA, Ruderfer DM, Kruglyak L. 2009. Comprehensive polymorphism survey elucidates population structure of *Saccharomyces cerevisiae*. *Nature* 458: 342–345.

Segre AV, Murray AW, Leu JY. 2006. High-resolution mutation mapping reveals parallel experimental evolution in yeast. *PLoS Biol* 4: e256.

Segura V, Vilhjalmsson BJ, Platt A, Korte A, Seren U, Long Q, Nordborg M. 2012. An efficient multi-locus mixed-model approach for genome-wide association studies in structured populations. *Nat Genet* 44: 825–830.

Steinmetz LM, Sinha H, Richards DR, Spiegelman JI, Oefner PJ, McCusker JH, Davis RW. 2002. Dissecting the architecture of a quantitative trait locus in yeast. *Nature* 416: 326–330.

Strope PK, Skelly DA, Kozmin SG, Mahadevan G, Stone EA, Magwene PM, Dietrich FS, McCusker JH. 2015. The 100-genomes strains, an *S. cerevisiae* resource that illuminates its natural phenotypic and genotypic variation and emergence as an opportunistic pathogen. *Genome Res* 25: 762–774.

Storici F, Lewis LK, Resnick MA. 2001. In vivo site-directed mutagenesis using oligonucleotides. *Nat Biotechnol* 19: 773–776.

Wenger JW, Schwartz K, Sherlock G. 2010. Bulk segregant analysis by high-throughput sequencing reveals a novel xylose utilization gene from *Saccharomyces cerevisiae*. *PLoS Genet* 6: e1000942.

Protocol 1

Genetic Dissection of Heritable Traits in Yeast Using Bulk Segregant Analysis

Ian M. Ehrenreich[1,3] and Paul M. Magwene[2,3]

[1]Molecular and Computational Biology Section, University of Southern California, Los Angeles, California 90089-2910; [2]Department of Biology and Center for Systems Biology, Duke University, Durham, North Carolina 27708

Bulk segregant analysis (BSA) is commonly used to determine the genetic basis of complex traits in yeast. This technique involves phenotyping progeny from a cross and then selectively genotyping pooled subsets of offspring with extreme phenotypes. Analysis of these genotype data can identify loci that show skewed allele frequencies in a group of phenotypically extreme individuals and that are likely to contribute to a trait. BSA can be applied to diverse strain crosses, including ones involving nonreference isolates. Further, given the high throughput of next-generation sequencing, it is possible to conduct many BSA experiments in parallel. Here, we present a BSA protocol for the generation of recombinant cross progeny. We then describe general BSA strategies for conducting phenotyping, causal loci detection, and candidate gene identification in a statistically powerful manner.

MATERIALS

It is essential that you consult the appropriate Material Safety Data Sheets and your institution's Environmental Health and Safety Office for proper handling of equipment and hazardous materials used in this protocol.

RECIPES: Please see the end of this protocol for recipes indicated by <R>. Additional recipes can be found online at http://cshprotocols.cshlp.org/site/recipes.

Reagents

β-glucuronidase
Canavanine (optional; see Step 7.ii)
Yeast nitrogen base (YNB) agar plates (optional; see Step 7.ii) <R>
Yeast sporulation medium <R>
Yeast strains of interest (see Step 1)
YPD (for liquid medium and agar plates) <R>
Water, sterile

Equipment

Glass beads (425- to 600-μm, sterile, acid-washed)
Hemocytometer

[3]Correspondence: ian.ehrenreich@usc.edu; paul.magwene@duke.edu

Copyright © Cold Spring Harbor Laboratory Press; all rights reserved
Cite this protocol as *Cold Spring Harb Protoc*; doi:10.1101/pdb.prot088989

Ice bucket
Incubator
Light microscope
Microcentrifuge
Microcentrifuge tubes
Micromanipulator
Plates (10-cm)
Shaker
Sonicator (100-W)
Vortex mixer

METHOD

Generation of Segregants

1. Generate diploid progenitors of the mapping population.

   To make the diploid parent of your mapping population, mate two genetically distinct haploid strains that have opposite mating types. Crosses can be performed using standard yeast mating techniques, such as micromanipulation, to recover zygotes or selection for diploids that have inherited a different dominant marker from each parent.

   If a haploid mapping population is desired, ensure that the cross parents are heterothallic. If they are not, delete the homothallic (HO) endonuclease from the parents using polymerase chain reaction (PCR)-mediated transformation with a dominant drug resistance marker (Wach et al. 1994; Goldstein and McCusker 1999). If a diploid mapping population is desired, use cross parents that have a functional version of HO.

2. Sporulate diploids.

   Use standard sporulation approaches to get diploids to undergo meiosis (e.g., Sherman 1991).

3. Observe a subsample of cells using microscopy to ensure that your population has sporulated.

   Depending on the strain backgrounds used in a cross, diploid cultures can show sporulation frequencies that range from <10% to nearly 100%. Elrod et al. (2009) describe media and culture conditions for optimizing sporulation efficiency across diverse Saccharomyces cerevisiae *strain backgrounds.*

Obtain Recombinant Cross Progeny

4. Disrupt the ascii of sporulated diploids by centrifuging 100 µL of the sporulated culture in a microcentrifuge tube at room temperature and 13,000 rpm. Aspirate the supernatant.

5. Resuspend the cell pellet in 100 µL of a 5:1 mixture of water and β-glucuronidase. Incubate the cell suspension for 45 min at 30°C.

   Optimization for this preparation may be needed depending on the specific cross as the ideal the concentration of β-glucuronidase used to digest ascii as well as the length of digestion can differ among strains.

6. Separate spores from tetrads using sonication or vortexing with glass beads.

   - To separate by sonication, place samples on ice and sonicate for ~3 min on full power (based on a 100-W sonicator). Avoid touching the tip of the sonicator on the bottom or sides of the sample tube.

     During sonication, sample tubes should be kept on ice at all times to avoid overheating.

   - To separate by vortexing, add 50 µL of glass beads and vortex for 2 min.

     Unsporulated diploids are not desirable in subsequent steps of this protocol; thus, care should be taken to eliminate them to the maximum extent possible. Multiple strategies exist to kill unsporulated diploids, including using selectable markers (Ehrenreich et al. 2010) or a combination of zymolyase, detergent, and heat (Goddard et al. 2005).

7. Estimate the concentration of spores in the sample using a hemocytometer. Dilute the sample with water to the appropriate plating density depending on whether segregants will be subsequently phenotyped individually or as pools.

- If segregants will be phenotyped individually, plate spores on YPD agar plates at a density of ~100 colonies per 10 cm plate to enable germination.

    For downstream phenotyping experiments conducted on individuals, plating densities should be low enough to prevent undesired matings between different spores. If the cross parents were HO, sporulated cells will undergo mating type switching and self-diploidize.

- If segregants will be phenotyped as pools, plate spores on YPD agar plates at a density of approximately 1 million colonies per 10 cm plate to enable germination. Incubate the plates for ~2 d at 30°C. Scrape the offspring off the plate using a sterile spreader and 10 mL of water, and harvest by pipetting. Then centrifuge cells at 13,000 rpm for 1 min at room temperature. Aspirate the supernatant and resuspend the entire pool of segregants in 1.5 mL of YPD broth. Keep this mixture at 30°C with shaking at 200 rpm for 1 h before phenotyping experiments.

    If downstream phenotyping experiments are to be conducted on pools of haploid segregants, we recommend isolating very large numbers of recombinant haploid offspring of the same mating type. The synthetic genetic array (SGA) marker system, which requires mating haploid parent strains with the genotypes can1Δ::STE2pr-SpHIS5 his3Δ and his3Δ, is one way to generate such populations (Tong et al. 2001; Tong and Boone 2006; Ehrenreich et al. 2010). Plating of progeny from such a cross on YNB plates containing canavanine instead of on YPD plates will select for MATa segregants that are histidine prototrophs (Tong et al. 2001; Tong and Boone 2006). We typically try to generate around a million MATa segregants per 10-cm plate.

    Note that if the SGA markers are used, causal loci in close linkage to CAN1 or MAT may be masked in downstream analyses.

Acquisition of Phenotypically Extreme Segregants

8. Phenotype the segregants.

    Bulk segregant analysis (BSA) can be applied to almost any measurable trait, but its success critically depends on high-quality trait measurement. We discuss phenotyping considerations for both individual- and population-level phenotyping experiments.

    When working with individual segregants, we recommend examining at least 500 individuals, which should be a large enough population to enable identification of moderate and large effect loci. Using even larger populations of segregants, in the order of thousands to tens of thousands of individuals, can provide even higher statistical power.

    Population-level phenotyping is another way to conduct BSA. Pools containing very large numbers of cross progeny can be subjected to phenotypic selection en masse. For phenotypes that can be examined by fluorescence-activated cell sorting (FACS), a specific fraction of the total population can be acquired (e.g., the top 1% of all cells) (Ehrenreich et al. 2010). For other traits that are not amenable to FACS, such as drug response, multiple pieces of information, including knowledge of population density and population-level dose–response curves, can be used to estimate the fraction of all individuals that show a particular level of a trait and to determine an appropriate number of phenotypically extreme individuals required for analysis (Ehrenreich et al. 2012).

9. Form bulks.

    The optimal strategy for forming bulks depends on the design of your experiment and the trait of interest. For phenotypes selected en masse, a one-tail design is typically used in which a pool of phenotypically extreme segregants is compared with a control pool of segregants that did not experience selection. For phenotypes where it is possible to obtain both tails of the phenotype distribution, bulks can be formed from these two tails and compared. In addition, more than two bulks may be useful in some cases. For example, Granek et al. showed that the inclusion of an intermediate bulk in combination with extreme bulks can help to distinguish adjacent loci (2013).

    For experiments on individual segregants, Magwene et al. (2011) showed that bulks containing the top or bottom 10%–15% of individuals provide good statistical power. In the case of pooled phenotyping experiments, it is more difficult to control the exact level of phenotypic selection. With pooled phenotyping, we typically aim to include individuals that are in the top or bottom 1% or less of the phenotype distribution. In both types of phenotyping experiments, pooling of individuals in equimolar fractions is most desirable, but may not always be feasible. For example, with pooled phenotyping, it is not possible to know exactly how many different genotypes and how many copies of each genotype are present in a bulk.

Sequencing of Bulks and Detection of Causal Loci

10. Sequence bulks.

    The next step in BSA is to determine the frequencies of thousands to tens of thousands of genetic markers genome-wide. Sequencing is the easiest and arguably most accurate way to generate such information, although we note that a number of studies have successfully used microarrays in BSA (e.g., Brauer et al. 2006; Segre et al. 2006; Ehrenreich et al. 2010, 2012). Multiple methods exist for generating whole or reduced representation genome sequencing libraries (e.g., Baird et al. 2008; Andolfatto et al. 2011). Any of these approaches can work for BSA. Because BSA depends on an accurate detection of allele frequencies, technical artifacts in allele frequency measurement, such as PCR bias, should be minimized. To achieve an appropriate sequencing coverage for a pool, we recommend a coverage level that is at least equal to the number of segregants that contributed to the pool. Beyond this consideration, higher sequencing coverage will produce more accurate allele frequency measurements, which may improve the resolution at which causal loci are detected.

11. Analyze sequencing data.

    Sequencing reads can be mapped to a reference genome (ideally constructed from a consensus of the genomes of the cross parents) using publicly available software [e.g., BWA (Li and Durbin 2009) or bowtie (Langmead and Salzberg 2012)]. Allele counts at each polymorphic site can be estimated from "pileup" files produced by these programs (Li et al. 2009).

12. Detect causal loci.

    Allele counts can be used in statistical tests to detect loci that show significant allele frequency skew. Linked sites on a chromosome will show correlated frequencies and thus provide similar information; therefore, analysis techniques should be used that take this linkage information into account. Different approaches have been proposed for this type of analysis. For example, Magwene et al. (2011) described a modified contingency-test based on the G statistic. Alternatively, Edwards and Gifford (2012) have described a hidden Markov model (HMM)-based approach that can precisely identify causal loci. Software implementing these analysis techniques is freely available at sites such as https://bitbucket.org/pmagwene/bsaseq and https://github.com/matted/multipool.

Identification of Candidate Genes and Variants at Detected Loci

13. Identify candidate genes and variants at causal loci.

    The identification of candidate genes and variants at each causal locus can often be a difficult task. There is no definitive strategy to approach this, but variants that lead to nonsynonymous amino acid substitutions in protein coding regions, or that disrupt or modify known or predicted transcription factor binding sites, are often good starting candidates. Functional annotation and gene ontology information about genes underlying loci can also be used to prioritize follow-up functional validation studies. Studies to determine the specific causal genes underlying loci can include allele replacements (Storici et al. 2001) or reciprocal hemizygosity tests (Steinmetz et al. 2002).

RECIPES

Yeast Nitrogen Base Agar Plates

Reagent	Quantity	Final concentration
Agar	20 g	2%
Dextrose	20 g	2%
Difco yeast nitrogen base without amino acids	6.7 g	0.67%
Distilled water	1 L	

Combine components and sterilize by autoclaving at 20 psi for 20 min at 121°C. Cool to ~60°C and pour into Petri dishes.

Yeast Sporulation Medium

Reagent	Quantity	Final concentration
Dextrose	0.5 g	0.05%
Distilled water	1 L	
Potassium acetate	10 g	1%
Yeast extract	1 g	0.1%

If the strains that will be sporulated possess auxotrophies, then supplement the sporulation medium with the appropriate amino acids: Addition of 25% of the level of an amino acid that is required for growth in YNB is recommended. Combine all components and sterilize by autoclaving at 20 psi for 20 min at 121°C.

YPD

Peptone, 20 g
Glucose, 20 g
Yeast extract, 10 g
H_2O to 1000 mL

YPD (YEPD medium) is a complex medium for routine growth of yeast.
To prepare plates, add 20 g of Bacto Agar (2%) before autoclaving.

REFERENCES

Andolfatto P, Davison D, Erezyilmaz D, Hu TT, Mast J, Sunayama-Morita T, Stern DL. 2011. Multiplexed shotgun genotyping for rapid and efficient genetic mapping. *Genome Res* 21: 610–617.

Baird NA, Etter PD, Atwood TS, Currey MC, Shiver AL, Lewis ZA, Selker EU, Cresko WA, Johnson EA. 2008. Rapid SNP discovery and genetic mapping using sequenced RAD markers. *PLoS ONE* 3: e3376.

Brauer MJ, Christianson CM, Pai DA, Dunham MJ. 2006. Mapping novel traits by array-assisted bulk segregant analysis in *Saccharomyces cerevisiae*. *Genetics* 173: 1813–1816.

Edwards MD, Gifford DK. 2012. High-resolution genetic mapping with pooled sequencing. *BMC Bioinformatics* 13: S8.

Ehrenreich IM, Torabi N, Jia Y, Kent J, Martis S, Shapiro JA, Gresham D, Caudy AA, Kruglyak L. 2010. Dissection of genetically complex traits with extremely large pools of yeast segregants. *Nature* 464: 1039–1042.

Ehrenreich IM, Bloom J, Torabi N, Wang X, Jia Y, Kruglyak L. 2012. Genetic architecture of highly complex chemical resistance traits across four yeast strains. *PLoS Genet* 8: e1002570.

Elrod SL, Chen SM, Schwartz K, Shuster EO. 2009. Optimizing sporulation conditions for different *Saccharomyces cerevisiae* strain backgrounds. *Methods Mol Biol* 557: 21–26.

Goddard MR, Godfray HC, Burt A. 2005. Sex increases the efficacy of natural selection in experimental yeast populations. *Nature* 434: 636–640.

Goldstein AL, McCusker JH. 1999. Three new dominant drug resistance cassettes for gene disruption in *Saccharomyces cerevisiae*. *Yeast* 15: 1541–1553.

Granek JA, Murray D, Kayrkci O, Magwene PM. 2013. The genetic architecture of biofilm formation in a clinical isolate of *Saccharomyces cerevisiae*. *Genetics* 193: 587–600.

Langmead B, Salzberg SL. 2012. Fast gapped-read alignment with Bowtie 2. *Nat Methods* 9: 357–359.

Li H, Durbin R. 2009. Fast and accurate short read alignment with Burrows–Wheeler transform. *Bioinformatics* 25: 1754–1760.

Li H, Handsaker B, Wysoker A, Fennell T, Ruan J, Homer N, Marth G, Abecasis G, Durbin R. 2009. The sequence Alignment/Map format and SAMtools. *Bioinformatics* 25: 2078–2079.

Magwene PM, Willis JH, Kelly JK. 2011. The statistics of bulk segregant analysis using next generation sequencing. *PLoS Comput Biol* 7: e1002255.

Segre AV, Murray AW, Leu JY. 2006. High-resolution mutation mapping reveals parallel experimental evolution in yeast. *PLoS Biol* 4: e256.

Sherman F. 1991. Guide to Yeast Genetics and Molecular. In *Methods in enzymology* (eds. Guthrie C, Fink GR), pp. 3–21. Elsevier Academic Press, San Diego.

Steinmetz LM, Sinha H, Richards DR, Spiegelman JI, Oefner PJ, McCusker JH, Davis RW. 2002. Dissecting the architecture of a quantitative trait locus in yeast. *Nature* 416: 326–330.

Storici F, Lewis LK, Resnick MA. 2001. In vivo site-directed mutagenesis using oligonucleotides. *Nat Biotechnol* 19: 773–776.

Tong AH, Boone C. 2006. Synthetic genetic array analysis in *Saccharomyces cerevisiae*. *Methods Mol Biol* 313: 171–192.

Tong AH, Evangelista M, Parsons AB, Xu H, Bader GD, Page N, Robinson M, Raghibizadeh S, Hogue CW, Bussey H, et al. 2001. Systematic genetic analysis with ordered arrays of yeast deletion mutants. *Science* 294: 2364–2368.

Wach A, Brachat A, Pohlmann R, Philippsen P. 1994. New heterologous modules for classical or PCR-based gene disruptions in *Saccharomyces cerevisiae*. *Yeast* 10: 1793–1808.

CHAPTER 37

Chemostat Culture for Yeast Physiology and Experimental Evolution

Maitreya J. Dunham,[1] Emily O. Kerr, Aaron W. Miller, and Celia Payen

Department of Genome Sciences, University of Washington, Seattle, Washington 98195

Continuous culture provides many benefits over the classical batch style of growing yeast cells. Steady-state cultures allow for precise control of growth rate and environment. Cultures can be propagated for weeks or months in these controlled environments, which is important for the study of experimental evolution. Despite these advantages, chemostats have not become a highly used system, in large part because of their historical impracticalities, including low throughput, large footprint, systematic complexity, commercial unavailability, high cost, and insufficient protocol availability. However, we have developed methods for building a relatively simple, low-cost, small footprint array of chemostats that can be run in multiples of 32. This "ministat array" can be applied to problems in yeast physiology and experimental evolution.

BATCH CULTURE

The most common method of growing yeast cell cultures is the standard "batch" culture. In this regime, a small number of cells are inoculated into nutrient-rich medium and allowed to divide at maximal growth rate through nutrient exhaustion and into saturation. Sampling of the culture for the experiment of choice is most typically done from the saturated culture or from cells growing in mid-log phase before the diauxic shift. Batch culture provides a number of obvious benefits—it is easy, uses glassware commonly available in any laboratory, and is consistent with a vast literature. However, batch cultures can be problematic for certain applications. When comparing cells of different genotypes, for example, differences in the maximal growth rate are a common phenotype. Many other phenotypes co-vary with growth rate, leading to a nonspecific suite of cell biological, gene expression, and other physiological changes that may not be directly related to the mutation of interest (Regenberg et al. 2006; Castrillo et al. 2007; Brauer et al. 2008).

Batch culture has also been a standard for long-term evolution experiments. Depending on the question being asked, batch culture with serial dilution can be a simple and scalable solution (e.g., Zeyl et al. 2003; Lang et al. 2011). However, such cultures are only rarely propagated in a relatively constant environment and growth rate regime because doing so can require a heroic sampling regimen where back dilution is performed multiple times per day (e.g., Torres et al. 2010). More typically, cultures are allowed to exhaust nutrients before transfer to new medium, leading to a variation in growth rate and nutrient access over the evolutionary time course. These discontinuities in selective pressure can lead to complex subpopulation structures in which different genotypes specialize for the various stages of the growth cycle (e.g., dividing a few extra times or failing to die after saturation, shortening lag phase,

---

[1]Correspondence: maitreya@uw.edu

Copyright © Cold Spring Harbor Laboratory Press; all rights reserved
Cite this introduction as *Cold Spring Harb Protoc*; doi:10.1101/pdb.top077610

or increasing maximal growth rate, all of which have been observed in bacterial serial-dilution-based evolution experiments [Lenski et al. 1998; Rozen and Lenski 2000]).

CONTINUOUS CULTURE

Cell physiology and experimental evolution can be studied in a more controlled manner using continuous cultures. Continuous culture refers to the utilization of a class of growth apparatus that maintains a constant environment in one of several dimensions, including chemostats (constant growth rate), turbidostats (constant turbidity), and other devices that feed back on pH, oxygen tension, or other parameters. For the purposes of this introduction, we will focus on chemostats. In chemostat culture, cells are grown at a set dilution rate to a nutrient-limited steady state, at which all inputs equal all outputs. This steady state is highly desirable for modeling applications (e.g., Knijnenburg et al. 2009; Reznik et al. 2013) and for perturbation experiments (e.g., Ronen 2006; McIsaac et al. 2012). The precise control over growth rate also allows matching of growth rates between wild-type cultures and more slowly growing mutants, or among diverse strain backgrounds, by forcing the cells to grow at a steady state that is below the maximum growth rate of both (e.g., Hayes et al. 2002; Torres et al. 2007; Skelly et al. 2013). Even small differences in competitive fitness can be accurately measured using the chemostat, making it a useful tool for characterizing individual mutants (Baganz et al. 1998) and collections of strains (Delneri et al. 2008).

Steady state is typically achieved within the first several volume replacements after inoculation of the chemostat and onset of media flow. Steady state can be maintained for up to at least 35 generations before being detectably perturbed by the inevitable influx of de novo mutations and subsequent selection, turning a physiology experiment into an evolutionary one. For evolution experiments, the advantages are continuous growth in a stable environment. This assumption is never strictly true, however, as cultures can change in density and media composition as their genotype composition shifts. Despite this complexity, the primary selection pressure is largely maintained by the nutrient limitation, which is selected by the experimenter.

CHEMOSTAT DESIGN

The basic plumbing of all chemostat platforms is generally the same (see Fig. 1). Medium is added at a defined rate while spent medium, including cells, overflows the culture at that same rate. Aeration, if desired, is provided by vigorous stirring and/or bubbling of gas through the culture. The media feed,

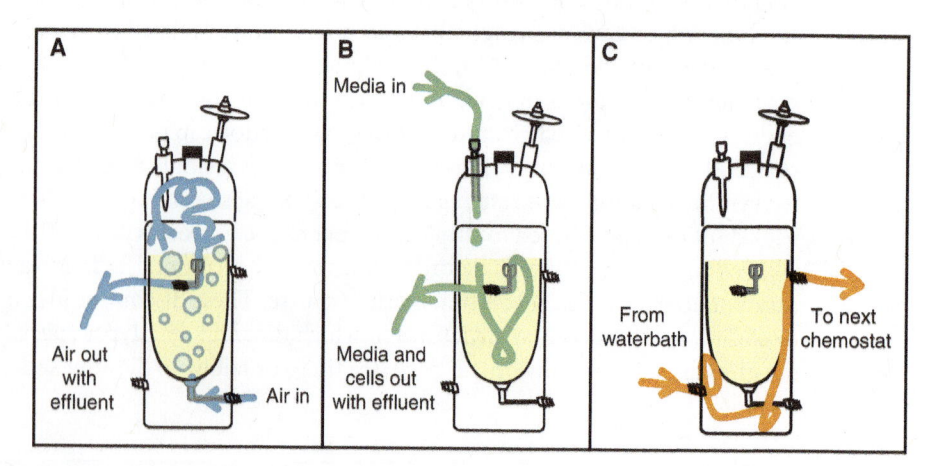

FIGURE 1. Diagram illustrating a basic chemostat. This example of a glass-blown chemostat provides (A) mixing by vigorous aeration, (B) a constant influx of media and overflow of culture effluent, and (C) temperature control via a water jacket.

aeration, and mixing must be optimized to distribute cells and media components uniformly in both space and time. Chemostat platforms can vary in scale over several orders of magnitude with respect to cell population and working volume (a table of vendors and plans can be found in Dunham [2010]). Commercial fermenters that have been modified for chemostat use, such as those from New Brunswick or Infors, typically range from 50 mL to 2 L. Custom glass-blown or otherwise home-built devices typically fall toward the lower end of this range and below (e.g., Klein et al. 2013). Microfluidic implementations weigh in at just a few microliters (Cookson et al. 2005; Groisman et al. 2005; Dénervaud et al. 2013). There are several important considerations to take into account when determining which system scale is most appropriate. Media usage in large fermenters may be prohibitively expensive for long-running experiments and many devices would consume a large laboratory space. However, such a large-scale system might be required to maintain adequate representation of highly complex libraries. At the other extreme, although microfluidic devices help address problems with media consumption and parallelization, population sizes may be insufficient for most evolutionary applications. Run times in microfluidic devices are also most compatible with shorter term physiology and perturbation experiments, although this has been improving (Jakiela et al. 2013).

We have sought a compromise solution to these considerations by using a mid-range 20 mL working volume that can be multiplexed to 32 units in a single array (Miller et al. 2013). For a yeast culture growing at a typical working density, this represents a population size on the order of 10^8–10^9 cells. We most typically grow the cultures at a dilution rate of 0.17 vol/h, a rate chosen to sit comfortably below the critical dilution rate at which most laboratory strains switch from mixed respiro-fermentative growth to primarily fermentation when grown in glucose limitation (van Dijken et al. 2000) and comfortably above the point where synchronized metabolic cycling has most commonly been observed (Tu 2005). This dilution rate is also convenient for collecting 1 mL of culture in 15–20 min for practical passive sampling from the overflow port. One 32-chemostat setup costs approximately $10,000 and fits on a benchtop or in a vertical rack configuration. Although the system is still complex with many parts, it can be operated by appropriately trained and supervised undergraduate researchers.

Our setup does have several limitations in its current implementation. First, all 32 chemostats must be operated off the same peristaltic pump. Running a subset of vessels at different dilution rates can be achieved by changing their working volume, or by using additional pumps that run independently. The off-the-shelf construction also lacks many of the sensors that commercial fermenters are equipped with. New advances in miniaturized sensor technology will hopefully address this limitation. Finally, we have not solved the long-appreciated, platform-independent problems of wall growth and media line colonization. Media lines could conceivably be outfitted with an additional air gap to try to limit this problem, although colonization could still exist at the site of entry. We are currently exploring strain-engineering approaches to abrogate this problem.

Despite these technical issues, the chemostat design presented in Protocol 1: Assembly of a Mini-Chemostat Array (Miller et al. 2015) has been successful in hundreds of experiments lasting an aggregated sum of tens of thousands of generations. We previously showed its equivalence to commercial chemostat platforms with respect to physiology and experimental evolution outcomes (Miller et al. 2013). We hope that our basic chemostat model and detailed protocols for applying the ministat array to problems in yeast physiology (see Protocol 2: Chemostat Culture for Yeast Physiology [Kerr and Dunham 2015]) and experimental evolution (see Protocol 3: Chemostat Culture for Yeast Experimental Evolution [Payen and Dunham 2015]) will encourage additional laboratories to adopt this useful culturing technology for their own applications. These protocols have been optimized to allow utilization of undergraduate researchers working as part of a research team.

ACKNOWLEDGMENTS

This work was supported by grants from the National Institute of General Medical Sciences (P41 GM103533 and R01 GM094306) from the National Institutes of Health and by National Science

Foundation grant 1120425. M.J.D. is a Rita Allen Foundation Scholar and a Senior Fellow in the Genetic Networks program at the Canadian Institute for Advanced Research. A.W.M. was supported in part by National Institutes of Health (T32 HG00035).

REFERENCES

Baganz F, Hayes A, Farquhar R, Butler PR, Gardner DCJ, Oliver SG. 1998. Quantitative analysis of yeast gene function using competition experiments in continuous culture. *Yeast* 14: 1417–1427.

Brauer MJ, Huttenhower C, Airoldi EM, Rosenstein R, Matese JC, Gresham D, Boer VM, Troyanskaya OG, Botstein D. 2008. Coordination of growth rate, cell cycle, stress response, and metabolic activity in yeast. *Mol Biol Cell* 19: 352–367.

Castrillo JI, Zeef LA, Hoyle DC, Zhang N, Hayes A, Gardner DCJ, Cornell MJ, Petty J, Hakes L, Wardleworth L, et al. 2007. Growth control of the eukaryote cell: A systems biology study in yeast. *J Biol* 6: 4.

Cookson S, Ostroff N, Pang WL, Volfson D, Hasty J. 2005. Monitoring dynamics of single-cell gene expression over multiple cell cycles. *Mol Syst Biol* 1: 2005.0024.

Delneri D, Hoyle DC, Gkargkas K, Cross EJM, Rash B, Zeef L, Leong H-S, Davey HM, Hayes A, Kell DB, et al. 2008. Identification and characterization of high-flux-control genes of yeast through competition analyses in continuous cultures. *Nat Genet* 40: 113–117.

Dénervaud N, Becker J, Delgado-Gonzalo R, Damay P, Rajkumar AS, Unser M, Shore D, Naef F, Maerkl SJ. 2013. A chemostat array enables the spatio-temporal analysis of the yeast proteome. *Proc Nat Acad Sci* 110: 15842–15847.

Dunham MJ. 2010. Experimental evolution in yeast: A practical guide. *Methods Enzymol* 470: 487–507.

Groisman A, Lobo C, Cho H, Campbell JK, Dufour YS, Stevens AM, Levchenko A. 2005. A microfluidic chemostat for experiments with bacterial and yeast cells. *Nat Methods* 2: 685–689.

Hayes A, Zhang N, Wu J, Butler PR, Hauser NC, Hoheisel JD, Lim FL, Sharrocks AD, Oliver SG. 2002. Hybridization array technology coupled with chemostat culture: Tools to interrogate gene expression in *Saccharomyces cerevisiae*. *Methods* 26: 281–290.

Jakiela S, Kaminski TS, Cybulski O, Weibel DB, Garstecki P. 2013. Bacterial growth and adaptation in microdroplet chemostats. *Angew Chem Int Ed Engl* 52: 8908–8911.

Kerr EO, Dunham MJ. 2015. Chemostat culture for yeast physiology. *Cold Spring Harb Protoc* doi: 10.1101/pdb.prot089003.

Klein T, Schneider K, Heinzle E. 2013. A system of miniaturized stirred bioreactors for parallel continuous cultivation of yeast with online measurement of dissolved oxygen and off-gas. *Biotechnol Bioeng* 110: 535–542.

Knijnenburg TA, Daran J-MG, van den Broek MA, Daran-Lapujade PA, de Winde JH, Pronk JT, Reinders MJT, Wessels LFA. 2009. Combinatorial effects of environmental parameters on transcriptional regulation in *Saccharomyces cerevisiae*: A quantitative analysis of a compendium of chemostat-based transcriptome data. *BMC Genomics* 10: 53.

Lang GI, Botstein D, Desai MM. 2011. Genetic variation and the fate of beneficial mutations in asexual populations. *Genetics* 188: 647–661.

Lenski RE, Mongold JA, Sniegowski PD, Travisano M, Vasi F, Gerrish PJ, Schmidt TM. 1998. Evolution of competitive fitness in experimental populations of *E. coli*: What makes one genotype a better competitor than another? *Antonie Van Leeuwenhoek* 73: 35–47.

McIsaac RS, Petti AA, Bussemaker HJ, Botstein D. 2012. Perturbation-based analysis and modeling of combinatorial regulation in the yeast sulfur assimilation pathway. *Mol Biol Cell* 23: 2993–3007.

Miller AW, Befort C, Kerr EO, Dunham MJ. 2013. Design and use of multiplexed chemostat arrays. *J Vis Exp* e50262.

Miller AW, Kerr EO, Dunham MJ. 2015. Assembly of a mini-chemostat array. *Cold Spring Harb Protoc* doi: 10.1101/pdb.prot088997.

Payen C, Dunham MJ. 2015. Chemostat culture for yeast experimental evolution. *Cold Spring Harb Protoc* doi: 10.1101/pdb.prot089011.

Regenberg B, Grotkjaer T, Winther O, Fausbøll A, Akesson M, Bro C, Hansen LK, Brunak S, Nielsen J. 2006. Growth-rate regulated genes have profound impact on interpretation of transcriptome profiling in *Saccharomyces cerevisiae*. *Genome Biol* 7: R107.

Reznik E, Mehta P, Segrè D. 2013. Flux imbalance analysis and the sensitivity of cellular growth to changes in metabolite pools. *PLoS Comput Biol* 9: e1003195.

Ronen M. 2006. Transcriptional response of steady-state yeast cultures to transient perturbations in carbon source. *Proc Nat Acad Sci* 103: 389–394.

Rozen D, Lenski R. 2000. Long-term experimental evolution in *Escherichia coli*. VIII. Dynamics of a balanced polymorphism. *Am Nat* 155: 24–35.

Skelly DA, Merrihew GE, Riffle M, Connelly CF, Kerr EO, Johansson M, Jaschob D, Graczyk B, Shulman NJ, Wakefield J, et al. 2013. Integrative phenomics reveals insight into the structure of phenotypic diversity in budding yeast. *Genome Res* 23: 1496–1504.

Torres EM, Sokolsky T, Tucker CM, Chan LY, Boselli M, Dunham MJ, Amon A. 2007. Effects of aneuploidy on cellular physiology and cell division in haploid yeast. *Science* 317: 916–924.

Torres EM, Dephoure N, Panneerselvam A, Tucker CM, Whittaker CA, Gygi SP, Dunham MJ, Amon A. 2010. Identification of aneuploidy-tolerating mutations. *Cell* 143: 71–83.

Tu BP. 2005. Logic of the yeast metabolic cycle: Temporal compartmentalization of cellular processes. *Science* 310: 1152–1158.

van Dijken JP, Bauer J, Brambilla L, Duboc P, Francois JM, Gancedo C, Giuseppin MLF, Heijnen JJ, Hoare M, Lange HC, et al. 2000. An interlaboratory comparison of physiological and genetic properties of four *Saccharomyces cerevisiae* strains. *Enzyme Microb Technol* 26: 706–714.

Zeyl C, Vanderford T, Carter M. 2003. An evolutionary advantage of haploidy in large yeast populations. *Science* 299: 555–558.

Protocol 1

Assembly of a Mini-Chemostat Array

Aaron W. Miller, Emily O. Kerr, and Maitreya J. Dunham[1]

Department of Genome Sciences, University of Washington, Seattle, Washington 98195

Here, we describe instructions for the assembly of an array of miniature (20-mL) chemostats or "ministats" built from relatively inexpensive off-the-shelf parts. In experiments with yeast cultures, we have observed reproducibility in cellular physiology, gene expression patterns, and evolutionary outcomes with different ministats as well as between ministats and commercial large-volume platforms. Growth in continuous culture is a primary means for the characterization of yeast steady-state physiology, competition between strains, and long-term evolution experiments. We hope that these relatively inexpensive and high-throughput devices make the advantages of continuous culture growth more accessible to researchers.

MATERIALS

It is essential that you consult the appropriate Material Safety Data Sheets and your institution's Environmental Health and Safety Office for proper handling of equipment and hazardous material used in this protocol.

RECIPES: Please see the end of this protocol for recipes indicated by <R>. Additional recipes can be found online at http://cshprotocols.cshlp.org/site/recipes.

Reagents

Defined minimal medium appropriate for the experiment
    Glucose-limited chemostat medium <R>
    Nitrogen-limited chemostat medium <R>
    Phosphate-limited chemostat medium <R>
    Sulfate-limited chemostat medium <R>

Select an appropriate nutrient-limited medium, such as one of the above. Alternative sources for each limiting nutrient can be substituted. For example, carbon sources other than glucose can be used in glucose-limited medium. Be sure to confirm that the limiting nutrient is in fact limiting when using any new strains or media formulations. Use high-quality chemicals and water when preparing media. For measuring large volumes, calibrate a graduated cylinder to ensure accurate measurements. Mix the components of the medium in a clean 10-L carboy and then filter them through a 0.2-μm 1-L bottle-top filter into another sterile 10-L carboy (see Steps 27–36).

Ethanol

Equipment

Air filters (0.45-μm; PTFE)
Air pumps (designed for 80-gal aquarium)
Aluminum foil
Autoclave

---

[1]Correspondence: maitreya@uw.edu

Copyright © Cold Spring Harbor Laboratory Press; all rights reserved
Cite this protocol as *Cold Spring Harb Protoc*; doi:10.1101/pdb.prot088997

Block for dry block heater (6 × 25-mm test tubes per block) (VWR 13259-210)
Bottle (glass; 100-mL; 45-mm-wide mouth; with cap)
Bottle top filter (1 L; 0.2-μm pore size; 45-mm diameter) (Corning 431174)
Carboy (vacuum-safe 10-L reservoir bottle with bottom hose outlet)
Connector (1/8-inch internal diameter; barbed Y)
Connector, female luer (1/8-inch barb)
Connector, male luer lock (1/8-inch barb)
Connector, male luer slip (1/8-inch barb)
Connector, reducing 1/4- to 1/8-inch (PVDF)
Culture tubes (55-mL; outer diameter 25 mm; screw cap; Pyrex)
Day pinchcock (metal clamp for tubing)
Dry block heater (VWR 12621-100)
Electrical tape (Scotch #35; green)
Flask (1-L; with sidearm)
Forceps
Ice pick
Inline valved quick-connectors (male and female; 1/4-inch I.D.; polypropylene)
Manifold, four-port (Cole Parmer EW-06464-85)
Micropipette tips (1-mL)
Needles (16-gauge × 5-inch for effluent port; spinal 18-gauge × 6-inch for air port; 20-gauge × 1.5-inch for media port)
Needles, blunt (20-gauge × 1.5-inch)
Needles, stainless steel (blunt; with polypropylene hub; 25-gauge × 1/2-inch)
Peristaltic pump (16-channel cartridge pump) (205S/CA16 from Watson-Marlow)
Pump head extension (16-channel) (205CA from Watson-Marlow)
Pump tubing (orange/green Marprene from Watson-Marlow)
Ring stand (with 10.5-inch three-prong clamp)
Rubber stopper (#2, with two holes)
Serological pipettes (10 mL; e.g., Stripette)
Silicone tubing, medium (1/4-inch × 3/8-inch)
Silicone tubing, small (3/32-inch × 7/32-inch; uses 1/8-inch connectors)
Stopper, pink foam silicone (Nonstandard size 2) (Cole Parmer EW-06298-06)
Stopper, silicone (no. 8 with 3/8-inch hole)
Stopper, yellow foam silicone (Nonstandard size 12) (Cole Parmer EW-06298-22)
Tube rack (for 25-mm-diameter tubes)
Tubing clamps (large; 12-position)
Vacuum pump
Vent filter (EMD Millipore SLFG 050 10)

METHOD

The ministat array consists of carboys supplying media, a peristaltic pump, an aeration system, a set of culture tubes placed in a heat block, and a set of effluent collection vessels (Fig. 1). Using the instructions below, arrays of up to 32 chemostats can be run off a single 32-plexed peristaltic pump. For further details regarding the design and utilization of these arrays, see Miller et al. (2013).

Assembly of the Culture Chambers and Attached Tubing

1. Clean the glass culture tubes, and rinse them thoroughly with water. Mark the exact location of 21, 20, and 19 mL on each tube (e.g., see Fig. 1) because there is a significant variation in the internal diameter of these tubes.

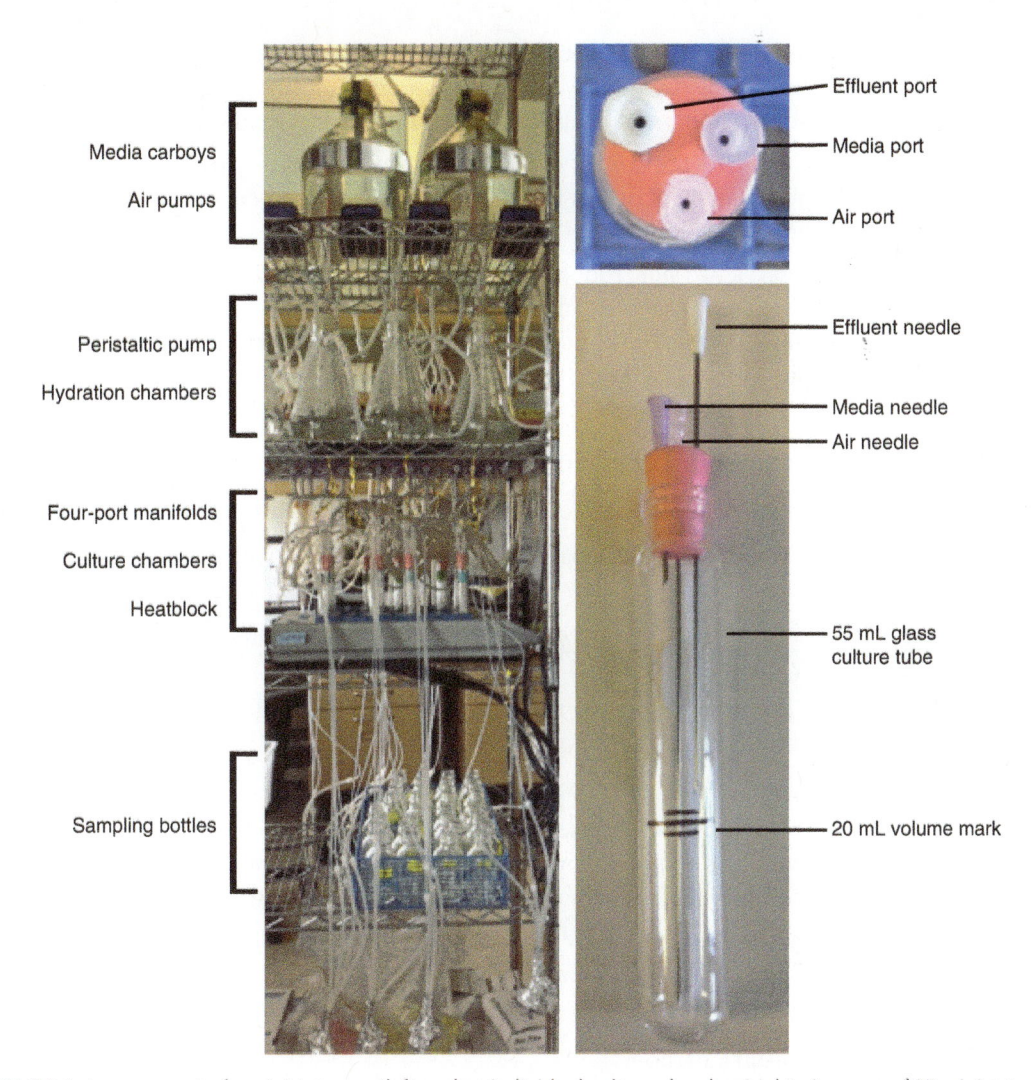

FIGURE 1. Arrangement of a ministat array (*left*) and an individual culture chamber (*right*). An array of 32 ministats and associated parts are shown ready for use in an experiment. The effluent sampling chambers are shown without stoppers, which is acceptable for short-term experiments. Also shown are top and side views of an individual chemostat chamber.

2. Make a cork assembly for each culture tube that comprises a pink size 2 foam silicone stopper containing three needles to deliver air, deliver medium, and remove effluent. Place the needles evenly around the stopper's circumference, and ensure that the needles run roughly parallel to the inside wall of the tube (see Fig. 1).

   Always wear safety goggles when working with exposed needles.

3. Assemble air-line tubing for each chamber by fitting a male luer connector to one end of a piece of small silicone tubing that is of sufficient length to reach from the culture chamber in the heat block to the four-port manifold. Place a 0.45-μm air filter on the other end of this tubing. Attach the luer connector to the longest needle in the cork assembly.

4. Assemble media-line tubing for each chamber by fitting a male luer connector to one end of a piece of small silicone tubing that is long enough to reach from the culture chamber in the heat block to the peristaltic pump. Attach the luer connector to the shortest needle in the cork assembly.

5. Use a female quick-connect to attach a 10-L media carboy with a male quick connect on a piece of medium silicone tubing ~4 inch long. Then, attach a male quick connect, and transition to small silicone tubing of sufficient length to reach from the carboy to the peristaltic pump. Next, use a

Chapter 37

series of small Y connectors separated by ~1-inch lengths of small tubing to branch the media line and divide the media flow from one source to up to 32 ministat culture chambers. After the desired number of branches has been put in place, add a male luer connector to the end of each.

6. Gently insert 1/2-inch stainless steel blunt needles into each end of the orange/green Marprene pump tubing. Use this tubing to join the branched end of the media-line tubing with the media-line tubing leading to the culture chamber.

7. Insert a two-hole rubber stopper into each effluent bottle. Fit the underside of the two holes with different length blunt needles. Using a 1/4-inch connector pushed into one hole of the stopper, connect the effluent bottle to the effluent needle on a culture chamber using a sufficiently long piece of small silicone tubing with a male luer connector.

Pre-/Postexperiment Cleanup and Sterilization

Tubing may be reused many times provided it is cleaned immediately after experiments.

8. Place all tubing in separate trays and rinse excessively with ddH_2O. Empty the tubing using an air pump.

9. Clean the cork assemblies with ddH_2O. Wipe the outside of the needles and cork to remove any residual media or cells.

10. Clean the glass tubing with water and ethanol, and remove physical contaminants with forceps and Kimwipes if necessary.

11. Rinse and dry all parts again before use.

12. Immediately before use, reassemble as described in Steps 3–7 and cover all exposed tubing ends and filters with aluminum foil. Wrap the tops of the sampling bottles in foil and autoclave along with the culture chambers in a separate autoclave tray.

13. Fit the assembled array neatly into one or more autoclave trays and autoclave all parts (using the liquid cycle) for 20 min.

Hydrated Aeration of Ministat Chambers

The ministats are aerated with hydrated air delivered from an aquarium pump and bubbled through sterile water.

14. Fill a 1-L sidearm flask with 700 mL of ddH_2O. Remove the cotton filter from a 10-mL serological pipette, and attach a 4-inch piece of medium tubing to the end that previously contained the cotton filter. Carefully work the pipette through a one-hole #8 silicone cork so that it will dip into the water contained in the sidearm flask; shorten the pipette if necessary. Tightly press the cork into the flask.

    One flask will humidify four ministat chambers.

15. Use a 1/4- to 1/8-inch reducing connector to attach a piece of small silicone tubing from the aquarium pump to the medium tubing at the top of each serological pipette.

16. Add medium tubing of sufficient length to the sidearm of the flask so that it reaches the four-port manifolds fastened above the ministat chambers.

17. Place the dials on the four-port manifold so that air is taken in through one of the two side ports and routed through the four main ports and not the other side port.

18. Place 2-inch pieces of medium tubing on each of the four ports to connect to the air-line filters.

Preparation of Sterile Media Carboys

The 10-L carboys used to house media can be assembled, autoclaved, and stored for months before use.

19. Rinse 10-L carboys thoroughly with ddH_2O.

20. Add two centered holes ~1 in apart in a #12 foam cork using an ice pick.
    Be careful when using the ice pick.

21. Carefully place a 1 mL micropipette tip into each of the two holes such that ~1 inch of the wider end of each pipette tip sticks out of the top of the stopper. Cut the narrow ends off the tips.
    Trimming the narrow ends permits the medium and air to flow more rapidly through the carboy.

22. Place a 4-inch piece of medium silicone tubing onto the wider end of one pipette tip and a 10-inch piece onto the wider end of the other tip.

23. Place the filter adaptor from the bottle-top filter used for media filtration (see Steps 28–30 below) onto the longer piece of tubing. Place a large vent filter onto the other piece of tubing.

24. Insert the cork into the top of the carboy, and give it a firm push. Use green electrical tape to secure the cork. Place at least one piece of tape over the top of the cork (between the two holes) and another around the neck of the carboy to prevent the cork from popping out in the autoclave.

25. Cover the ends of the media-in port (top; with the filter adaptor) and the media-out port (bottom) with foil folded such that it is secure but will be easy to remove.

26. Add 25 mL of water to each carboy and sterilize by autoclaving (using the liquid cycle) for 20 min. Store carboys upright or on their sides.
    They can be stored for months before use.

Filtration of Medium

The medium used in a chemostat is typically a defined minimal medium that is filtered instead of autoclaved to preserve vitamins and metals and to ensure precise nutrient concentrations. Good sterile technique is essential during this procedure.

27. With a flame going nearby, loosen the cap on a sterile 100-mL glass bottle that has threads matched to the bottle-top filter, and carefully screw the bottle-top filter on.

28. Dip forceps in ethanol and shake off any excess. Flame-sterilize the forceps and use them to pull the filter plug from where the adapter and vacuum usually attach.

29. Attach the filter to the filter adapter that was autoclaved on the carboy's media-in port (see Step 23).

30. Clamp the bottle into a ring stand next to the carboy.

31. Attach the vacuum hose to the vent filter on the sterile carboy.

32. Route the clamped outlet tube from the mixing carboy into the top of the filter.

33. Place the sterile carboy in a tray to catch any inadvertent overflow.

34. Turn on the vacuum and unclamp the outlet hose from the mixing carboy. Adjust the large clip to alter the flow of medium out of the mixing carboy.
    The 100-mL bottle will fill first and then overflow into the second carboy. If the vacuum is too strong, the cork can be pulled into the carboy. Filtering should take ~30 min/10 L.

35. Clamp the media line with a metal clamp, and use the same foil that was on the adaptor to cover it once again. Remove the vacuum pump tubing, and turn off the vacuum.

36. Incubate the filter bottle at 30°C to give an early warning of contamination, or clean and autoclave it for future use. Use the filtered medium immediately, or store the carboy for a short period of time before use.

Assembly of Carboy with Chemostats and Collection of Effluent

Assemble the ministat array, culture-sampling bottles, and all nonautoclaved parts on a benchtop or in a metal rack (see Fig. 1).

37. Place the autoclaved culture tubes into the heat blocks.
    > We find that placing 16 ministats staggered in a heat block allows for better observation and troubleshooting with the cultures.

38. Place the effluent bottles below or beside the culture chambers. Direct the effluent tubing into each appropriate bottle.
    > The culture bottles may be organized in wire racks or plastic tubs to aid in sampling organization.

39. Label each of the culture tubes, effluent lines, and sampling bottles to decrease the likelihood of sampling error.

40. Place filtered sterile media above or next to the ministat array.

RELATED INFORMATION

For descriptions of how to use the ministat array to characterize yeast physiology and perform long-term evolution experiments, see Protocol 2: Chemostat Culture for Yeast Physiology (Kerr and Dunham 2015) and Protocol 3: Chemostat Culture for Yeast Experimental Evolution (Payen and Dunham 2015).

RECIPES

Glucose-Limited Chemostat Medium

Reagent	Quantity (for 10 L)
Calcium chloride dihydrate	1 g
Sodium chloride	1 g
Magnesium sulfate heptahydrate	5 g
Potassium phosphate monobasic	10 g
Ammonium sulfate	50 g
Glucose	8 g
Metals (1000×) <R>	10 mL
Vitamins (1000×) <R>	10 mL

Dissolve the salts in just <1 L of glass-distilled water. In a separate beaker, dissolve the glucose in just <1 L of glass-distilled water. Bring each solution to a final volume of 1 L. Combine the salt and glucose solutions with 8 L of water, the metals, and the vitamins in a mixing carboy to make 10 L of medium. Stir for ~5 min or until thoroughly mixed.

Metals (1000×)

Reagent	Quantity (for 1 L)
Boric acid	500 mg
Copper sulfate pentahydrate	40 mg
Potassium iodide	100 mg
Ferric chloride hexahydrate	200 mg
Manganese sulfate monohydrate	400 mg
Sodium molybdate dihydrate	200 mg
Zinc sulfate heptahydrate	400 mg

Dissolve the reagents in just <1 L of glass-distilled water in the indicated order. Bring the total volume to 1 L, and pour into a foil-wrapped bottle, making sure no residue remains in the original vessel. Store at room temperature. Shake well before using.

Nitrogen-Limited Chemostat Medium

Reagent	Quantity (for 10 L)
Calcium chloride dihydrate	1 g
Sodium chloride	1 g
Magnesium sulfate heptahydrate	5 g
Potassium phosphate monobasic	10 g
Ammonium sulfate	400 mg
Glucose	50 g
Metals (1000×) <R>	10 mL
Vitamins (1000×) <R>	10 mL

Dissolve the salts in just <1 L of glass-distilled water. In a separate beaker, dissolve the glucose in just <1 L of glass-distilled water. Bring each solution to a final volume of 1 L. Combine the salt and glucose solutions with 8 L of water, the metals, and the vitamins in a mixing carboy to make 10 L of medium. Stir for ~5 min or until thoroughly mixed.

Phosphate-Limited Chemostat Medium

Reagent	Quantity (for 10 L)
Calcium chloride dihydrate	1 g
Sodium chloride	1 g
Magnesium sulfate heptahydrate	5 g
Ammonium sulfate	50 g
Potassium chloride	10 g
Potassium phosphate monobasic	100 mg
Glucose	50 g
Metals (1000×) <R>	10 mL
Vitamins (1000×) <R>	10 mL

Dissolve the salts in just <1 L of glass-distilled water. In a separate beaker, dissolve the glucose in just <1 L of glass-distilled water. Bring each solution to a final volume of 1 L. Combine the salt and glucose solutions with 8 L of water, the metals, and the vitamins in a mixing carboy to make 10 L of medium. Stir for ~5 min or until thoroughly mixed.

Sulfate-Limited Chemostat Medium

Reagent	Quantity (for 10 L)
Calcium chloride dihydrate	1 g
Sodium chloride	1 g
Magnesium chloride hexahydrate	4.12 g
Ammonium chloride	40.5 g
Potassium phosphate monobasic	10 g
Ammonium sulfate	30 mg
Glucose	50 g
Metals (1000×) <R>	10 mL
Vitamins (1000×) <R>	10 mL

Dissolve the salts in just <1 L of glass-distilled water. In a separate beaker, dissolve the glucose in just <1 L of glass-distilled water. Bring each solution to a final volume of 1 L. Combine the salt and glucose solutions with 8 L of water, the metals, and the vitamins in a mixing carboy to make 10 L of medium. Stir for ~5 min or until thoroughly mixed.

Vitamins (1000×)

Reagent	Quantity (for 1 L)
Biotin	2 mg
Calcium pantothenate	400 mg
Folic acid	2 mg
Inositol (*myo*-inositol)	2000 mg
Niacin (nicotinic acid)	400 mg
p-Aminobenzoic acid	200 mg
Pyridoxine hydrochloride	400 mg
Riboflavin	200 mg
Thiamine hydrochloride	400 mg

Mix the reagents in just <1 L of glass-distilled water. (Some of the materials will not completely dissolve.) Bring the total volume to 1 L. Transfer to a beaker and stir. Maintain stirring while transferring 40-mL aliquots into 50-mL conical tubes. Ensure that no residue is left in the original vessel or graduated cylinder. For long-term storage, freeze the aliquots at −20°C. For short-term storage (<1 mo), store the aliquots at 4°C. Shake well before use.

ACKNOWLEDGMENTS

This work was supported by grants from the National Institute of General Medical Sciences (P41 GM103533) from the National Institutes of Health and by National Science Foundation grant 1120425. M.J.D. is a Rita Allen Foundation Scholar and a Senior Fellow in the Genetic Networks program at the Canadian Institute for Advanced Research. A.W.M. was supported in part by National Institutes of Health (T32 HG00035).

REFERENCES

Kerr EO, Dunham MJ. 2015. Chemostat culture for yeast physiology. *Cold Spring Harb Protoc* doi: 10.1101/pdb.prot089003.

Miller AW, Befort C, Kerr EO, Dunham MJ. 2013. Design and use of multiplexed chemostat arrays. *J Vis Exp* (72): e50262. doi:10.3791/50262.

Payen C, Dunham MJ. 2015. Chemostat culture for yeast experimental evolution. *Cold Spring Harb Protoc* doi: 10.1101/pdb.prot089011.

Protocol 2

Chemostat Culture for Yeast Physiology

Emily O. Kerr and Maitreya J. Dunham[1]

Department of Genome Sciences, University of Washington, Seattle, Washington 98195

The use of chemostat culture facilitates the careful comparison of different yeast strains growing in well-defined conditions. Variations in physiology can be measured by examining gene expression, metabolite levels, protein content, and cell morphology. In this protocol, we show how a combination of sample types can be collected during harvest from a single 20-mL chemostat in a ministat array, with special attention to coordinating the handling of the most time-sensitive sample types.

MATERIALS

It is essential that you consult the appropriate Material Safety Data Sheets and your institution's Environmental Health and Safety Office for proper handling of equipment and hazardous material used in this protocol.

RECIPE: Please see the end of this protocol for recipes indicated by <R>. Additional recipes can be found online at http://cshprotocols.cshlp.org/site/recipes.

Reagents

Ammonium bicarbonate (50 mM, pH 7.8), freshly prepared and maintained on ice (for harvesting samples for protein analyses only)

Chemostat medium
  Examples of media are provided in Protocol 1: Assembly of a Mini-Chemostat Array (Miller et al. 2015).

Ethanolamine (10 mM in 0.1 M potassium phosphate buffer [pH 6.7]), freshly prepared (for harvesting samples for microscopy only)
  Addition of ethanolamine will change the pH of the solution from pH 6.5 to pH 6.7.

Formaldehyde (37%) (for harvesting samples for microscopy only)
Liquid nitrogen (for harvesting samples for RNA and protein analyses only)
Methanol (for harvesting samples for metabolite analyses only)
Potassium phosphate buffer (1 M, pH 6.5) (for harvesting samples for microscopy only) <R>
Yeast strain of interest

Equipment

Chemostat array
  Assemble the apparatus as described in Miller et al. (2013) and Protocol 1: Assembly of a Mini-Chemostat Array (Miller et al. 2015).

---

[1]Correspondence: maitreya@uw.edu

Copyright © Cold Spring Harbor Laboratory Press; all rights reserved
Cite this protocol as *Cold Spring Harb Protoc*; doi:10.1101/pdb.prot089003

Dewar flask (1-L)
Filter holder (25 mm; stainless steel support) (VWR 26316-692)
Forceps
MAGNA nylon membrane filters (0.45-µm pore size; 25-mm diameter)
Ring stand with clamp
Serological pipettes (10-mL)
Silicone tubing, medium (1/4-inch × 3/8-inch)
Silicone tubing, small (3/32-inch × 7/32-inch)
Stopper, black rubber (no. 1 with two holes)
Stopper, silicone (no. 8 with 3/8-inch hole)
Tongs
Tubes, conical (15- and 50-mL)
Tubes, with locking lid (2-mL)
Tubing connector, 1/4- to 1/8-inch
Tubing connector, male luer slip, 1/8-inch barb
Vacuum flask (1-L, with sidearm)
Vacuum tubing

METHOD

This method is adapted from Skelly et al. (2013).

Establishing Steady State in the Chemostats

1. Start the ministat array as follows:
    i. Turn on the pump to fill the ministats, and then turn it off when the medium reaches the 20-mL mark.
        Assess the volume (marked on the tubes) without bubbling.
    ii. Inoculate each ministat with 0.1 mL of overnight culture grown in appropriate chemostat medium.
    iii. Allow the cultures to grow for 30 h at 30°C.
        They should look dense.
    iv. Turn on the pump at a speed of ~6 rpm (if using the Watson-Marlow pump suggested in Protocol 1: Assembly of a Mini-Chemostat Array [Miller et al. 2015]). Ensure that it is pumping medium from the carboy to the ministats and not the other way around.
    v. Adjust the ministat volumes (with the air off) by adjusting the heights of the effluent needles.
    vi. When there is overflow from all of the effluent lines, empty the effluent bottles and note the time.

2. Determine when the cultures reach steady state by collecting and analyzing samples as follows:
    i. Record the time.
    ii. Move the effluent corks from the bottles to sample collection tubes.
    iii. Record the volume of effluent (V_{eff}) from each ministat. Use this and the time elapsed to calculate the dilution rate. Adjust the pump or individual pump cartridges as needed to achieve a dilution rate of 0.17 (±0.01) vol/h.
    iv. Once a sufficient sample volume is collected, replace the effluent corks in the rinsed effluent bottles.

v. Assay the cell density by two measurements using a spectrophotometer, colorimeter, and/or flow cytometer. Dilute the samples as needed to ensure that they are in the linear range for each instrument.

vi. Harvest the cells when the cultures have reached steady state, that is, when the cell density readings are within 5%–10% of each other on two consecutive days.

> Steady state is typically attained 3–4 d after inoculation, at approximately generation 10–20. Samples up to generation 35 are generally free of detectable de novo mutation accumulation.

Harvesting the Samples

Physiological variations can be assessed by collecting samples to examine gene expression (Steps 3–15), cell morphology (Steps 16–26), metabolite levels (Steps 27–33), and protein content (Steps 34–38).

Harvesting for RNA

Because of how quickly gene expression changes, the goal here is to filter and freeze the cells within 30 sec of disturbing the chemostat. Prepare and gather everything before you start, and leave the pump on until all the sampling is finished. Plan to do a test run first, to make sure you are not missing something critical.

3. Label 2-mL locking-lid tubes and 15-mL conical filtrate collection tubes so that there is one for each sample plus one extra conical tube for rinses. Gather the 25-mm nylon filters, 10-mL serological pipettes (and pipette controller), forceps for handling filters, and tongs for retrieving frozen samples from liquid nitrogen. Also set aside an extra chemostat vessel to house the dripping stopper/needle assembly during sampling.

4. Assemble the vacuum apparatus (Fig. 1) as follows, and perform a test run with water to be sure the apparatus is correctly assembled and rinsed.

    i. Arrange a ring stand with clamp to hold the filter apparatus.

    ii. Push a one-hole stopper onto the bottom of the funnel half of the apparatus, followed by a piece of tubing. Ensure that the ring stand clamp holds the stopper so that the apparatus is fairly upright.

    iii. Connect the tubing to a two-hole stopper that fits into the top of the filtrate collection tube, and connect the vacuum trap to tubing from the other hole in that stopper.

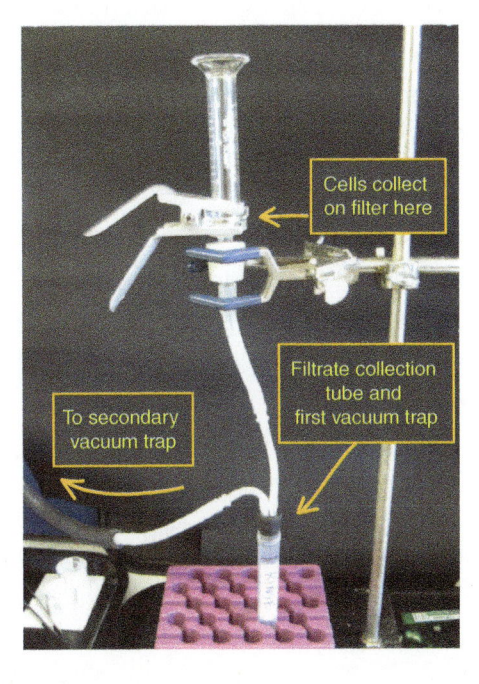

FIGURE 1. Harvesting cultures for RNA assessment. The filter apparatus for RNA sampling.

Alternatively, you could just use a vacuum flask, but this setup allows you to collect and save the filtrate from each sample.

    iv. Fit the stopper in the filtrate collection tube marked "rinse," and turn on the vacuum.

    v. Use forceps to put a filter on the metal screen in the apparatus. Listen for hissing or whistling, which indicate a leak. Reposition or replace whistling filters.

    vi. Clamp the cylinder into place, and pipette 20 mL of ddH$_2$O into the top of the apparatus, checking for leaks. If there are leaks, try adjusting the junction of the two glass pieces or the height of the metal screen. If there are no leaks, reset the apparatus by replacing the filtrate tube and filter.

    vii. Fill a 1-L Dewar flask about half full with liquid nitrogen.

Now is the time to do a test run of the actual harvest as described below. Keep in mind that if you are also going to sample for microscopy, protein, or metabolites, you should have those sample tubes ready and waiting.

5. Make sure the vacuum is on, the filter apparatus is properly assembled with the filtrate collection tube for the first sample, and a fresh filter is in place.

6. Open the 2-mL tube for the first sample and drop it into the liquid nitrogen in the open Dewar flask.

7. Being careful of the exposed needles, take the stopper assembly off the first ministat and place it in an empty/dummy chemostat vessel in a neighboring well of the heat block.

8. Pick up the chemostat and pipette 5 mL of the culture onto the filter.

If you are also taking metabolite or protein samples, pipette these into the specified tubes while the RNA sample is collecting on the filter.

9. As soon as the sample has completely filtered, remove the clamp and top section from the apparatus.

10. Using tongs, dump the liquid nitrogen of the 2-mL tube and put it open in a tube rack.

11. Use the forceps to roll the filter onto itself, so that the cells are to the inside, and put it in the frozen tube.

You may want to use a gloved finger to hold the back of the filter loosely in place while you manipulate it with the forceps.

12. Immediately close the tube as it sits in the rack, and drop it back into the liquid nitrogen.

Now that the samples are frozen, they can be stored at −80°C until further processing.

13. Cap the filtrate sample to preserve the concentration of ethanol or other volatiles of interest, and store at −20°C. Be sure to leave enough headroom in the tube to accommodate sample expansion on freezing.

If also harvesting for microscopy (Steps 16–26), metabolites (Steps 27–33), or protein (Steps 34–38), proceed with that sample handling at this time.

14. Thoroughly rinse the apparatus with ddH$_2$O between samples.

15. Once all samples are taken, put the spent vessel back in place to house the stopper assembly, which will continue to drip media, and move the dummy from position to position.

Harvesting for Microscopy

Prepare fresh solutions before beginning.

16. Pipette 1.2 mL of cells into microcentrifuge tubes containing 0.15 mL of formaldehyde and 0.15 mL of 1 M potassium phosphate buffer (resulting in 3.7% formaldehyde in 0.1 M potassium phosphate solution).

17. Let the samples sit for 10 min at room temperature.

18. Centrifuge the samples in a microcentrifuge at 4500g for 7 min with slow acceleration to pellet the cells.
19. Resuspend the pellets in 1.5 mL of 0.1 M potassium phosphate containing 3.7% formaldehyde.
20. Incubate for 60 min at room temperature.
21. Centrifuge at 4500g for 7 min with slow acceleration to pellet the cells.
22. Resuspend the pellet in 1.5 mL of 0.1 M potassium phosphate containing 10 mM ethanolamine.
23. Incubate for 10 min at room temperature.
    This step deactivates the formaldehyde.
24. Centrifuge at 4500g for 7 min with slow acceleration to pellet the cells.
25. Resuspend the cells in 1 mL of 0.1 M potassium phosphate.
26. Store at 4°C until further processing.

Harvesting for Metabolites

If harvesting for metabolites and protein, the samples can be handled in parallel.

27. Pipette a 10-mL sample into a 50-mL conical tube on ice.
28. Centrifuge at 3000g for 3 min at 4°C and wash the pellet with 10 mL of cold water.
29. Centrifuge again at 3000g for 3 min at 4°C.
30. Immediately resuspend the pellet in 0.5 mL of cold water and quickly add 0.5 mL of cold methanol.
31. Invert to mix, and incubate the samples for 30 min in a dry ice–ethanol bath at approximately −40°C.
32. Thaw the frozen mixture on ice for 10 min, and centrifuge at 3000g for 5 min at 4°C.
33. Freeze the supernatant at −80°C until the time of analysis.

Harvesting for Protein

Note that the full 20-mL culture volume must be used to recover adequate protein for mass spectrometry, leaving none for other sample types. Consider running larger-volume chemostats to allow collection of multiple sample types.

34. Pipette or pour the 20-mL culture into a 50-mL conical tube on ice.
35. Centrifuge the sample at 3000g for 3 min at 4°C.
36. Wash the pellet in cold 50 mM ammonium bicarbonate (pH 7.8).
37. Centrifuge again at 3000g for 3 min at 4°C and freeze the sample using liquid nitrogen.
38. Store at −80°C until the time of analysis.

DISCUSSION

Allowing yeast culture to reach and maintain steady state is a key feature of the chemostat. Before a chemostat culture has stabilized, fluctuations in cell density show that the culture has not fully adjusted to the growth rate imposed by the system. Once the density stabilizes, however, the experimenter can be assured that the doubling time is being regulated by the dilution rate, and that the cells have reached a steady state. In this condition, the cells are coping with the experimental conditions imposed on them, and their physiology can be examined by microscopy, as well as by profiling RNA, metabolites, and protein.

The most time-sensitive part of the protocol is the harvest of cells for RNA analysis. The slightest changes in environment can change gene expression, so great care must be taken to avoid any perturbation of the culture within 24 h before taking the RNA sample. Therefore, it is recommended that only passive sampling via the effluent line be taken before opening the chemostat to take the culture for RNA. Immediately after the cell sample is frozen, and during the initial fixation step for microscopy, cells for protein and metabolites can be handled together, because they both require immediate chilling in an ice bath and the same centrifugation steps at 4°C.

A 20-mL chemostat may not have enough volume to accommodate all of the described sample types from a single vessel. Although the small culture size of these "mini" chemostats may seem to limit the sampling possibilities, the highly adjustable nature of the ministats allows the user to increase the volume with a simple adjustment of the effluent needle. Technological advances will also likely reduce the importance of large sample sizes in the near future.

RECIPE

Potassium Phosphate Buffer (1 M, pH 6.5)

Reagent	Quantity (for 100 mL)
KH_2PO_4	9.5 g
K_2HPO_4	5.25 g

Dissolve the reagents in 80 mL of ddH_2O. Adjust the pH to 6.5. Make the volume up to 100 mL with ddH_2O. Filter sterilize and store at room temperature.

ACKNOWLEDGMENTS

We thank Gennifer Merrihew, Beth Graczyk, Sara Cooper, Eric Muller, and Trisha Davis for assistance with the proteomics, metabolite, and microscopy protocols. This work was supported by a grant from the National Institute of General Medical Sciences (P41 GM103533) from the National Institutes of Health and by National Science Foundation grant 1120425. M.J.D. is a Rita Allen Foundation Scholar and a Senior Fellow in the Genetic Networks program at the Canadian Institute for Advanced Research.

REFERENCES

Miller AW, Befort C, Kerr EO, Dunham MJ. 2013. Design and use of multiplexed chemostat arrays. *J Vis Exp* e50262.

Miller AW, Kerr EO, Dunham MJ. 2015. Assembly of a mini-chemostat array. *Cold Spring Harb Protoc* doi: 10.1101/pdb.prot088997.

Skelly DA, Merrihew GE, Riffle M, Connelly CF, Kerr EO, Johansson M, Jaschob D, Graczyk B, Shulman NJ, Wakefield J, et al. 2013. Integrative phenomics reveals insight into the structure of phenotypic diversity in budding yeast. *Genome Res* 23: 1496–1504.

Protocol 3

Chemostat Culture for Yeast Experimental Evolution

Celia Payen and Maitreya J. Dunham[1]

Department of Genome Sciences, University of Washington, Seattle, Washington 98195

Experimental evolution is one approach used to address a broad range of questions related to evolution and adaptation to strong selection pressures. Experimental evolution of diverse microbial and viral systems has routinely been used to study new traits and behaviors and also to dissect mechanisms of rapid evolution. This protocol describes the practical aspects of experimental evolution with yeast grown in chemostats, including the setup of the experiment and sampling methods as well as best laboratory and record-keeping practices.

MATERIALS

It is essential that you consult the appropriate Material Safety Data Sheets and your institution's Environmental Health and Safety Office for proper handling of equipment and hazardous material used in this protocol.

Reagents

Defined minimal medium appropriate for the experiment
For examples, see Protocol 1: Assembly of a Mini-Chemostat Array (Miller et al. 2015).

Ethanol (95%)
Glycerol (20% and 50%; sterile)
Yeast strain of interest

Equipment

Agar plates (appropriate for chosen strain)
Chemostat array
Assemble the apparatus as described in Miller et al. (2013) and Protocol 1: Assembly of a Mini-Chemostat Array (Miller et al. 2015).

Cryo deep-freeze labels
Cryogenic vials
Culture tubes
Cytometer (BD Accuri C6)
Glass beads, 4 mm (sterile; for plating yeast cells)
Glass cylinder
Kimwipes

---

[1]Correspondence: maitreya@uw.edu

Copyright © Cold Spring Harbor Laboratory Press; all rights reserved
Cite this protocol as *Cold Spring Harb Protoc*; doi:10.1101/pdb.prot089011

Chapter 37

Multiwell plates (96-well)
Sonicator
Spectrophotometer
Tubes, conical (20- and 50-mL)

METHOD

Throughout the experiment, all important parameters and observations should be recorded systematically. For an example of a data collection scheme, see Table 1.

Setting Up the Experiment and Inoculation

Day 1

1. Streak single colonies on an appropriate agar plate and grow at 30°C.

   This is suitable for most Saccharomyces cerevisiae *strains. Optimize the growth temperature for the strain of interest.*

Day 2

2. Assemble the ministats and prepare the medium for the selective conditions desired (see Protocol 1: Assembly of a Mini-Chemostat Array [Miller et al. 2015]). For each evolution culture, inoculate a single colony into 2.5 mL of appropriate medium and let each culture grow to saturation overnight.

   See Troubleshooting.

Day 3

3. Turn the pump on to fill the ministat culture tubes, and then turn it off when the volume reaches the 20-mL mark. Set the heat blocks to the appropriate temperature.

4. Sterilize the tops of the corks with 95% ethanol.

5. Inoculate each chemostat vessel with 0.1 mL from one individual overnight culture using a syringe.

   Do not inoculate multiple vessels from the same overnight culture as this could result in shared mutation content from variants that emerged during the initial outgrowth.

6. Keep a stock of each inoculum by freezing 1 mL of the overnight culture mixed with 0.5 mL 50% glycerol in a cryogenic vial at −80°C.

7. Grow the chemostat cultures to saturation for 30 h before starting the media flow.

Setting Up the Culture Volume and Starting the Continuous Culture

8. Turn the media pump on to a dilution rate of 0.17 vol/h, which corresponds to between 5.75 and 6.5 rpm on the Watson–Marlow pump suggested as part of the ministat setup in Protocol 1: Assembly of a Mini-Chemostat Array (Miller et al. 2015).

TABLE 1. An example of a data collection scheme

Sample	Day	Time	Action	Time elapsed (h)	V_{eff} (mL)	Dilution rate (vol/h)	Generations	Density (cells/µL)	Notes
S1			Inoculation						
S101	1	12 pm	Pump ON 6.25		0				
S102	1	12 pm	Pump ON 6.25		0				
S101	2	9:17 am		21.28	73	0.171	5.3	2000	
S102	2	9:17 am		21.28	72	0.169	5.2	2223	

9. Turn off the air, and adjust the media volume to 20 mL by moving the sampling needle up or down. Turn the air on when done. Wait until the medium starts to exit through the effluent line.
10. Once all the cultures are at 20 mL, empty the effluent bottles. Record the time (this will be time 0), and take samples if desired.

Sampling the Chemostats

The chemostats are ideally sampled every day in a consistent manner and the process may include making a glycerol stock, measuring cell density, sampling for RNA and DNA, etc. The process can take between 30 min (short sampling) and several hours (long sampling). For sampling from overnight effluent collection bottles for DNA extraction, see Steps 19–22. For long-term evolution experiments, samplings can alternate between short and long protocols. Set up all the required materials before you start sampling the culture. Use printer-friendly cryo deep-freeze labels to label the required tubes.

11. Check the whole setup (needle, carboys, and effluent corks) for leaks, contamination, or flocculated cultures.
    See Troubleshooting.
12. Note the time, and start collecting fresh effluent by transferring the sampling corks into labeled 20-mL sterile sampling tubes. Store these tubes on ice while sampling.
13. While the tubes are filling, measure and record the effluent volume (V_{eff}) that has collected in the effluent bottles.
14. Calculate the dilution rate using the time elapsed and the total V_{eff}.
15. Adjust the pump or individual pump cartridges as needed to reach a dilution rate of 0.17 (±0.01) vol/h.
16. Wash the effluent bottles, and replace the sampling corks on the bottles. Note the time.
17. To store frozen aliquots, pipette 1 mL of the fresh sample into 0.5 mL of sterile 50% glycerol in a labeled cryogenic vial. Invert a few times to mix, and store the tubes at −80°C.
18. Measure and record cell density using your favorite method.

    Spectrophotometer
    i. Dilute the sample to reach a linear range.
    ii. Read the optical density at 600 nm using the spectrophotometer and a matched blank.

    Cytometer
    i. Vortex the fresh sample and pipette 20 µL into 80 µL of water in a 96-well plate.
    ii. Seal the plate and sonicate to separate the cells.
    iii. Use the C6 cytometer to count the number of cells and record the data.

    Plating Cells for Viable Counts or Drug Resistance
    i. Vortex a fresh, sonicated sample.
    ii. Make serial dilutions of the culture.
    iii. Plate 250 µL of a 10^{-4} dilution using sterile glass beads onto a labeled agar plate.
        Note that a different dilution may be required depending on your culture density.
    iv. Incubate the plates at the appropriate temperature for 2 d before counting colonies.
    v. Record the data.

Sampling from Overnight Effluent Collection Bottles for DNA Extraction

19. Save 50 mL of the overnight effluent from the effluent bottle. Ideally, sterilize the effluent bottle and effluent tubing before collection begins.

20. Measure and record the remaining volume to calculate total V_{eff}.
21. Centrifuge the samples at 1.5g for 3 min at room temperature.
22. Remove the supernatants and freeze the cell pellets immediately at −20°C for later DNA extraction.

TROUBLESHOOTING

Problem (Step 2): The medium to start the overnight culture for the evolution experiment is not available.
Solution: Overnight cultures can be started in YPD or another nutrient-rich medium and cells can be washed with water before inoculating.

Problem (Step 11): There is colonization of the needle or media line.
Solution: Replace the tubing and the needle with autoclaved pieces, and record the information in your database.

Problem (Step 11): Cells begin to flocculate or clump as the culture evolves.
Solution: Increase sonication time or power until cells are dispersed.

DISCUSSION

Evolution in chemostats can operate for hundreds of generations, but these experiments require daily attention and troubleshooting. The possibility of contamination is the biggest concern and unfortunately this is one common reason for stopping experiments. Observation under a microscope for bacterial and fungal contamination of the fresh sample can be performed every 50 generations. Also, careful attention should be given to sterile technique throughout the entire procedure. Another common reason to stop the experiment is the appearance of clumping and wall growth. These are most frequently seen when using wild strains and certain mutants, but also frequently evolve from laboratory strains given enough time. These traits are also correlated with increased colonization of the media port, which should be monitored daily.

Beyond these issues there is no limit, and depending on your experiment you can run the chemostats for hundreds of generations. With experiments of this scale, you should be aware that a lot of samples are going to be collected and that they will require a lot of space to be stored. Organization is key to track the information and the samples. Data can be recorded using different methods, and samples need to be stored in a logical setup using unique identifiers such as:

F1 9/25/13: Fermenter vessel 1, sample collected on 25 September 2013
S10203: Sulfur, experiment 1, vessel 02, day 03

We have not discussed the experimental design in this protocol, but it is obviously of key importance. Strains, media, population size, experiment length, and other important parameters must be carefully considered before embarking on an experiment. Also, proper controls must be performed to ensure the selective condition is as designed. More discussion of these issues can be found in Dunham (2010).

ACKNOWLEDGMENTS

This work was supported by grants from the National Institute of General Medical Sciences (P41 GM103533 and R01 GM094306) from the National Institutes of Health and by National Science Foundation (grant 1120425). M.J.D. is a Rita Allen Foundation Scholar and a Senior Fellow in the Genetic Networks program at the Canadian Institute for Advanced Research.

REFERENCES

Dunham MJ. 2010. Experimental evolution: A practical guide. *Methods Enzymol* **470:** 487–507.

Miller AW, Befort C, Kerr EO, Dunham MJ. 2013. Design and use of multiplexed chemostat arrays. *J Vis Exp* e50262.

Miller AW, Kerr EO, Dunham MJ. 2015. Assembly of a mini-chemostat array. *Cold Spring Harb Protoc* doi: 10.1101/pdb.prot088997.

CHAPTER 38

Methods to Synthesize Large DNA Fragments for a Synthetic Yeast Genome

Yizhi Cai[1,3] and Junbiao Dai[2,3]

[1]Daniel Rutherford Building G.24, School of Biological Sciences, University of Edinburgh, The King's Buildings, Edinburgh EH9 3BF, United Kingdom; [2]MOE Key Laboratory of Bioinformatics, Center for Synthetic and Systems Biology, School of Life Sciences, Tsinghua University, Beijing 100084, People's Republic of China

De novo DNA synthesis is one of the key enabling technologies for synthetic biology. Methods for large-scale DNA synthesis, in particular, have transformed many facets of life science research, supporting new discoveries in biology through the design of novel synthetic biological systems. This protocol describes in detail the methods currently being used to synthesize and assemble large pieces of DNA for the synthetic yeast genome project. The protocol includes instructions for building block synthesis as well as chunk assembly, each of which can be used as a stand-alone procedure to generate a synthetic DNA of interest.

MATERIALS

It is essential that you consult the appropriate Material Safety Data Sheets and your institution's Environmental Health and Safety Office for proper handling of equipment and hazardous materials used in this protocol.

RECIPES: Please see the end of this protocol for recipes indicated by <R>. Additional recipes can be found online at http://cshprotocols.cshlp.org/site/recipes.

Reagents

Escherichia coli cells, chemically competent
Low-salt LB (Lysogeny Broth) liquid medium <R>
   Prepare low-salt LB liquid medium with kanamycin for Step 19 by adding 1 mL of filtered kanamycin stock solution (40 mg/mL in ddH$_2$O; 1000×) per liter of medium.

Reagents for building block synthesis (Steps 1–24)
   Agarose gel (1%) and reagents for gel electrophoresis
   GoTaq Green Master Mix (Promega)
   High-fidelity polymerase chain reaction (PCR) amplification kit (e.g., Phusion High-Fidelity PCR Kit; New England Biolabs)
   Low-salt LB solid medium containing kanamycin and X-gal <R>
   M13 primers
      M13 forward (M13F; 5′-GTAAAACGACGGCCAG-3′)
      M13 reverse (M13R; 5′-CAGGAAACAGCTATGAC-3′)

[3]Correspondence: yizhi.cai@ed.ac.uk; jbdai@tsinghua.edu.cn

Copyright © Cold Spring Harbor Laboratory Press; all rights reserved
Cite this protocol as *Cold Spring Harb Protoc*; doi:10.1101/pdb.prot080978

Chapter 38

Overlapping oligonucleotides for building block synthesis (300 nM)

Overlapping oligonucleotides of a target synthetic DNA sequence can be designed using the Building Block design module (constant length overlap) of the software GeneDesign (www.genedesign.org) (Richardson et al. 2006, 2010, 2012). A 20-bp overlap is typically used; however, the overlap length can be optimized for different target sequences. Oligonucleotides (normally 40–80 bp in length) can be synthesized in-house or purchased from commercial vendors and must be diluted to a concentration of 300 nM. To avoid the extra work and error-prone procedures of primer dilution, we recommend ordering primers as prediluted liquids from the vendor.

In our initial design, we incorporated building blocks of ~750 bp in length. The purpose of limiting the length of the target sequences was to ensure that one round of Sanger sequencing was adequate to cover the entire sequence, and building blocks were sequenced from both ends to ensure there were no mutations on either strand. If single-strand sequencing is sufficient, building blocks can be as long as 1600–2000 bp. Building block length can also be adjusted to meet various needs, such as the incorporation of restriction enzyme recognition sites.

Zero Blunt TOPO PCR Cloning Kit, with pCR Blunt II-TOPO cloning vector and salt solution (1.2 M NaCl, 0.06 M $MgCl_2$) (Invitrogen)

Reagents for chunk assembly (Steps 25–30)

Backbone vector (e.g., pRS415), linearized by restriction digestion

To minimize background from the undigested backbone, we recommend gel purification of the vector digestion before assembly.

Building blocks for assembly

Each building block should contain at least 40 bp of terminal overlap with its adjacent building blocks. The first building block should contain a 40-bp 5′ overlap with the linearized vector backbone. Building block inserts can be amplified via PCR or digested from the plasmids constructed in the first section of this protocol ("Synthesizing Building Blocks from Oligonucleotides" [Steps 1–24]).

Isothermal reaction (Gibson Assembly) master mix <R>

This protocol uses homemade isothermal reaction master mix for Gibson Assembly; however, a similar product can be purchased from commercial vendors.

Plates containing selective solid growth medium

Choose a selective medium according to the vector backbone. For pRS vectors, use plates containing LB medium with carbenicillin.

Yeast selectable marker (such as *Leu2* or *Ura3*)

PCR-amplify the yeast selectable marker such that it contains a 40-bp 5′ terminal overlap with the last building block and a 40-bp 3′ terminal overlap with the linearized vector backbone.

Equipment

96-well plates (sterile, with lids)
Gel electrophoresis equipment
Glass beads
Heat block or water bath at 42°C
Incubators at 30°C and 37°C
Microcentrifuge tubes
PCR tubes
Thermal cycler

METHOD

Synthesizing Building Blocks from Oligonucleotides

The protocol described here is suitable for high-throughput processing and is ready for automation.

If you are using methods other than GeneDesign for oligonucleotide design, minor adjustments to the PCR programs according to your design parameters may be required for high-quality results.

Preparing Overlapping Oligonucleotides

1. Add 10 μL of each overlapping oligonucleotide to a microcentrifuge tube to prepare a templateless primer mix (TPM). If there are fewer than 20 oligos, add H_2O to bring the final volume to 200 μL. Mix well.

2. Add 25 μL of the first and last oligonucleotides in the sequence to a microcentrifuge tube to prepare an outer oligo mix (OPM). Mix well.

Performing Templateless PCR (TPCR)

3. For each TPCR, assemble the following reaction using a high-fidelity PCR amplification kit. Mix well and keep on ice.

Reagent	Volume
dNTPs (2.5 mM)	2.0 μL
10× reaction buffer	2.5 μL
High-fidelity DNA polymerase	0.25 μL
H_2O	17.75 μL
TPM (~300 nM)	2.5 μL
Total volume	25 μL

4. Perform TPCR using the following program.

Number of cycles	Time	Temperature
1	3 min	94°C
	30 sec	55°C
	1 min	72°C
5	30 sec	94°C
	30 sec	69°C
	1 min	72°C
5	30 sec	94°C
	30 sec	65°C
	1 min	72°C
20	30 sec	94°C
	30 sec	61°C
	1 min/kb	72°C
1	3 min	72°C
	Hold	4°C

Performing Finish PCR (FPCR)

5. Dilute each TPCR sample 1:5 in H_2O.

6. For each FPCR, assemble the following reaction using a high-fidelity PCR amplification kit. Mix well and keep on ice.

Reagent	Volume
dNTPs (2.5 mM)	2.0 μL
10× reaction buffer	2.5 μL
High-fidelity DNA polymerase	0.25 μL
H_2O	15.75 μL
Diluted (1:5) TPM product	2.5 μL
OPM (~300 nM)	2.0 μL
Total volume	25 μL

7. Perform FPCR using the following program.

Number of cycles	Time	Temperature
1	3 min	94°C
	30 sec	55°C
	1 min	72°C
25	30 sec	94°C
	30 sec	55°C
	1 min/kb	72°C
1	3 min	72°C
	Hold	4°C

8. Run 5 µL of the FPCR product on a 1% agarose gel.

   If the PCR was successful, a bright band of the size of the target synthetic DNA should be visible. Successful FPCR products are cloned into the pCR Blunt II-TOPO vector in the next section.

Cloning the Building Blocks

9. Assemble the following ligation reaction using the Zero Blunt cloning kit.

Reagent	Volume
FPCR product	0.5 µL
Salt solution	0.5 µL
pCR Blunt II-TOPO vector DNA	0.5 µL
H$_2$O (sterile)	1.5 µL
Total volume	3 µL

10. Mix the reaction well by pipetting up and down, and incubate for 5 min at room temperature.
11. Place the reaction on ice until ready to proceed with Step 13.

Transforming Bacteria

12. For each transformation, thaw one 25-µL aliquot of competent cells on ice.
13. Add 1 µL of the ligation reaction directly to each vial of cells and mix by stirring gently with a pipette tip.
14. Incubate the vials on ice for 5 min.
15. Heat shock the cells for 30 sec at 42°C. Immediately place the vials on ice for 2 min.
16. Add 125 µL of LB liquid medium to each vial.
17. Using glass beads, spread 125 µL from each vial on plates containing LB solid medium with kanamycin + X-gal.
18. Incubate the plates overnight at 30°C.

Performing Colony Screening PCR (csPCR)

19. Transfer individual white colonies from each transformation plate to sterile 96-well plates containing 100 µL of LB liquid medium with kanamycin per well.

    We recommend picking six clones per transformation plate.

20. Incubate the liquid cultures overnight at 37°C.
21. For each csPCR, assemble the following reaction.

Reagent	Volume
2× GoTaq Green	6.25 µL
Forward primer (M13F)	0.25 µL
Reverse primer (M13R)	0.25 µL
Overnight bacterial culture	1 µL
H$_2$O (sterile)	4.75 µL
Total volume	12.5 µL

22. Perform csPCR using the following program.

Number of cycles	Time	Temperature
1	4 min	94°C
30	30 sec	94°C
	30 sec	55°C
	1 min/kb	72°C
1	3 min	72°C
	Hold	4°C

23. Run 10 µL of csPCR product on a 1% agarose gel without loading dye.

    If the PCR was successful, a bright band of roughly the size of the target sequence length should be visible.

24. Select csPCR-positive clones for sequencing with M13F and M13R.

    Additional primers may be needed if the target sequence is longer than 1000 bp.

Assembling Building Blocks into Chunks

The above method is satisfactory for the synthesis of small building blocks (750 bp). However, to achieve high-efficiency integration and replacement of native chromosomes, much larger chunks of DNA are required. Similarly, many genes and pathways may contain longer DNA sequences. The following describes assembly of multiple building blocks into chunks using Gibson Assembly, the one-step isothermal DNA assembly method by Gibson et al. (2009).

25. Thaw 15 µL of isothermal reaction (Gibson Assembly) master mix on ice.

26. Combine all of the building blocks, vector backbone, and yeast selectable marker in equimolar amounts in a 5-µL volume.

27. Combine the building block mixture with 15 µL of isothermal reaction master mix in a PCR tube. Mix well and keep on ice.

28. Preheat the thermal cycler to 50°C.

29. Pause the thermal cycler at 50°C and transfer the PCR tube to the machine. Incubate the reaction for 30 min at 50°C.

30. Transform 2 µL of the assembly reaction into 50 µL of competent cells as described in Steps 14–18, plating the cells on the appropriate selective solid medium.

    Successfully assembled chunks can be recovered from the bacteria and analyzed by restriction digestion or DNA sequencing.

DISCUSSION

Advances in high-throughput sequencing technologies (i.e., next-generation sequencing or NGS) have greatly enhanced our ability to read genetic information, and we are now able to read more than 15 petabases per year (Schatz and Phillippy 2012). However, our ability to write DNA is much more limited, especially at the genomic level. To date, only a few small genomes (from organisms such as poliovirus, bacteriophage, and mycoplasma) have been synthesized successfully (Cello et al. 2002; Smith et al. 2003; Gibson et al. 2008, 2010). Recently, we reported the synthesis of the first eukaryotic chromosome arms and a full yeast chromosome, beginning the journey toward a complete synthetic

Chapter 38

yeast genome (Dymond et al. 2011; Annaluru et al. 2014). In addition to complete genome synthesis, there is also a great need in the field of metabolic engineering for smaller synthetic genes and genetic pathways. Generally, synthesis of these DNA fragments can be outsourced to commercial vendors; however, despite continual drops in price, commercial gene synthesis remains too costly for many laboratories.

Several methods for gene synthesis have been developed in the past few decades. Currently, all gene synthesis technologies begin with chemical synthesis of short oligonucleotides (30–100 nt). The first-generation method relies on synthesis of oligonucleotides that completely cover both strands of DNA, which are phosphorylated and ligated in vitro. To save on the cost of oligonucleotide synthesis, researchers have also developed PCR-based methods, which require only 1.5× or lower coverage of a given DNA sequence, with no need for phosphorylated oligonucleotides. This is the method we adopted to synthesize the yeast genome, described in detail in this protocol. More recently, Gibson et al. (2009) developed an in vitro isothermal assembly method (now termed Gibson assembly), which allows small DNA fragments to be further assembled into 3 kb or larger fragments. Moreover, using the budding yeast as manufacturer, one can assemble DNA fragments from several kilobases to more than 1 million base pairs using in vivo recombination.

One difficulty of gene synthesis is the reduction of errors in the final products. These errors may come from the oligonucleotides or result from mistakes by DNA polymerase during PCR. To reduce error rates, oligonucleotide sizes must be limited for the current chemistry, or new chemistry must be developed. A few error correction techniques have been widely adopted in gene synthesis, such as the use of enzymes (Carr et al. 2004). To address this issue, we designed this protocol as a hierarchical procedure (Fig. 1). Building blocks (750 bp) are synthesized from overlapping oligonucleotides via high-fidelity PCR and then sequence verified to ensure 100% accuracy. These building blocks are designed to overlap for use with the Gibson Assembly method to form chunks (3–30 kb). Finally, each chunk is swapped into a yeast chromosome using in vivo homologous recombination. The two protocol sections presented here, beginning with overlapping oligonucleotides and covering building block synthesis through assembly of chunks ready for chromosome swap, are essentially independent. They can be used to synthesize DNA sequences of various sizes, and therefore they are not limited to the synthesis of DNA for the yeast genome.

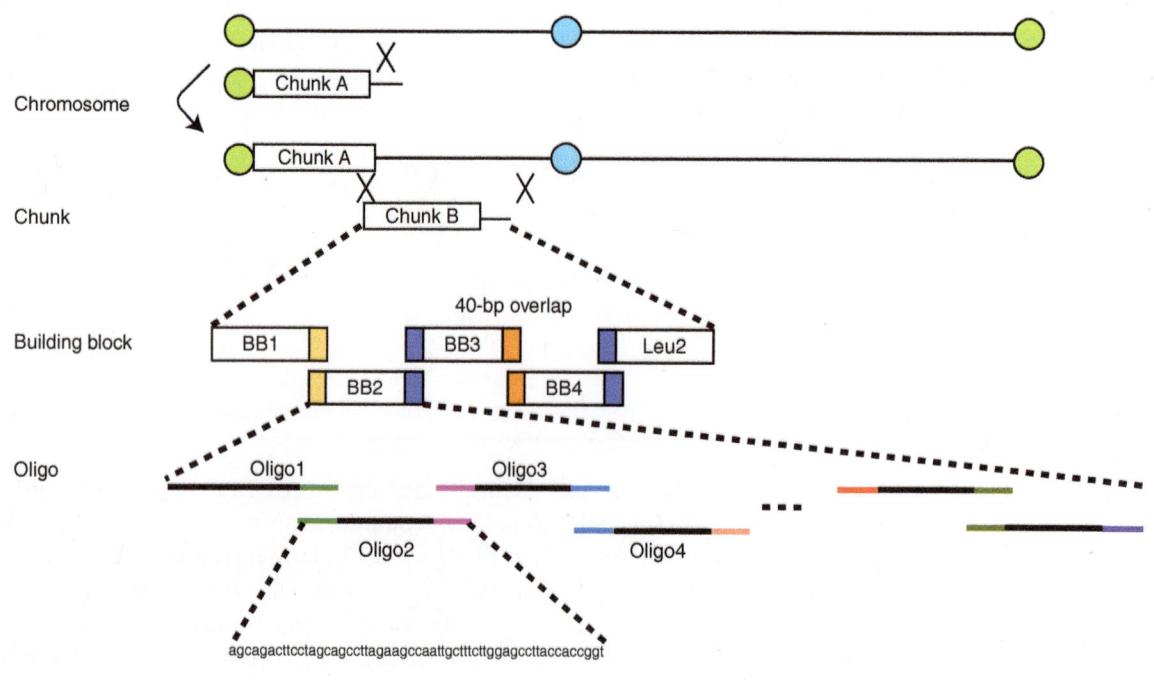

FIGURE 1. Hierarchy of DNA synthesis and assembly of a synthetic yeast chromosome.

RECIPES

ISO Buffer (5×)

1 M Tris–HCl (pH 7.5) (Invitrogen 15567-027)	3000 µL
1 M MgCl$_2$ (Sigma-Aldrich M1028)	300 µL
100 mM dGTP	60 µL
100 mM dTTP	60 µL
100 mM dATP	60 µL
100 mM dCTP	60 µL
1 M DTT (Sigma-Aldrich D9779)	300 µL
PEG-800 (USB Affymetrix 19959)	1.5 g
100 mM NAD (Sigma-Aldrich N1511)	300 µL
H$_2$O	to 6 mL

Combine the above ingredients. Store 40-µL aliquots at −20°C.

Isothermal Reaction (Gibson Assembly) Master Mix

ISO buffer (5×) <R>	40 µL
T5 exonuclease (Epicentre T5E4111K)	1.6 µL
Phusion high-fidelity DNA polymerase (New England BioLabs F-530L)	2.5 µL
Taq ligase (New England BioLabs M0208L)	20 µL
H$_2$O	85.9 µL
Total volume	150 µL

Combine the above ingredients. Store 15-µL aliquots at −20°C. *Do not use a hot start polymerase.*

Low-Salt LB Liquid Medium

10 g tryptone
5 g yeast extract
5 g NaCl

Prepare in H$_2$O. These quantities are for 1 L. Adjust pH to 7.5.

Low-Salt LB Solid Medium Containing Kanamycin and X-gal

10 g Bacto-tryptone
5 g Bacto-yeast extract
5 g NaCl
15 g Agar

Combine the above ingredients in 1 L of ddH$_2$O and autoclave to sterilize. Cool medium to pouring temperature. Add 1 mL of filtered kanamycin stock solution (40 mg/mL in ddH$_2$O; 1000×) and 2 mL of 20% X-gal stock solution (20 mg/mL in dimethyl formamide, 500×).

ACKNOWLEDGMENTS

We thank Dr. Jef Boeke and Dr. Karen Zeller and the Build-a-Genome classes for their collective contribution to the protocol described here. Work in Y.C.'s laboratory is supported by a Chancellor's Fellowship from the University of Edinburgh, a start-up fund from Scottish Universities Life Sciences Alliance, and Biotechnology and Biological Sciences Research Council (BBSRC) grant BB/M005690/1. Work in J.D.'s laboratory is supported by National Science Foundation of China 31471254 and 81171999, Chinese Minister of Science and Technology 2012CB725201, and Tsinghua University Initiative Scientific Research Program 20121087956.

REFERENCES

Annaluru N, Muller H, Mitchell LA, Ramalingam S, Stracquadanio G, Richardson SM, Dymond JS, Kuang Z, Scheifele LZ, Cooper EM, et al. 2014. Total synthesis of a functional designer eukaryotic chromosome. *Science* 344: 55–58.

Carr PA, Park JS, Lee YJ, Yu T, Zhang S, Jacobson JM. 2004. Protein-mediated error correction for de novo DNA synthesis. *Nucleic Acids Res* 32: e162.

Cello J, Paul AV, Wimmer E. 2002. Chemical synthesis of poliovirus cDNA: Generation of infectious virus in the absence of natural template. *Science* 297: 1016–1018.

Dymond JS, Richardson SM, Coombes CE, Babatz T, Muller H, Annaluru N, Blake WJ, Schwerzmann JW, Dai J, Lindstrom DL, et al. 2011. Synthetic chromosome arms function in yeast and generate phenotypic diversity by design. *Nature* 477: 471–476.

Gibson DG, Benders GA, Andrews-Pfannkoch C, Denisova EA, Baden-Tillson H, Zaveri J, Stockwell TB, Brownley A, Thomas DW, Algire MA, et al. 2008. Complete chemical synthesis, assembly, and cloning of a *Mycoplasma genitalium* genome. *Science* 319: 1215–1220.

Gibson DG, Young L, Chuang RY, Venter JC, Hutchison CA 3rd, Smith HO. 2009. Enzymatic assembly of DNA molecules up to several hundred kilobases. *Nat Methods* 6: 343–345.

Gibson DG, Glass JI, Lartigue C, Noskov VN, Chuang RY, Algire MA, Benders GA, Montague MG, Ma L, Moodie MM, et al. 2010. Creation of a bacterial cell controlled by a chemically synthesized genome. *Science* 329: 52–56.

Richardson SM, Wheelan SJ, Yarrington RM, Boeke JD. 2006. GeneDesign: Rapid, automated design of multikilobase synthetic genes. *Genome Res* 16: 550–556.

Richardson SM, Nunley PW, Yarrington RM, Boeke JD, Bader JS. 2010. GeneDesign 3.0 is an updated synthetic biology toolkit. *Nucleic Acids Res* 38: 2603–2606.

Richardson SM, Liu S, Boeke JD, Bader JS. 2012. Design-a-gene with GeneDesign. *Methods Mol Biol* 852: 235–247.

Schatz MC, Phillippy AM. 2012. The rise of a digital immune system. *Gigascience* 1: 4.

Smith HO, Hutchison CA 3rd, Pfannkoch C, Venter JC. 2003. Generating a synthetic genome by whole genome assembly: ϕX174 bacteriophage from synthetic oligonucleotides. *Proc Natl Acad Sci* 100: 15440–15445.

APPENDIX

General Safety and Hazardous Material Information

> This manual should be used by laboratory personnel with experience in laboratory and chemical safety or students under the supervision of such trained personnel. The procedures, chemicals, and equipment referenced in this manual are hazardous and can cause serious injury unless performed, handled, and used with care and in a manner consistent with safe laboratory practices. Students and researchers using the procedures in this manual do so at their own risk. It is essential for your safety that you consult the appropriate Material Safety Data Sheets, the manufacturers' manuals accompanying equipment, and your institution's Environmental Health and Safety Office, as well as the General Safety and Disposal Cautions in this appendix for proper handling of hazardous materials in this manual. Cold Spring Harbor Laboratory makes no representations or warranties with respect to the material set forth in this manual and has no liability in connection with the use of these materials.
>
> All registered trademarks, trade names, and brand names mentioned in this book are the property of the respective owners. Readers should please consult individual manufacturers and other resources for current and specific product information.

Users should always consult individual manufacturers, the manufacturers' safety guidelines and other resources, including local safety offices, for current and specific product information and for guidance regarding the use and disposal of hazardous materials.

PRIMARY SAFETY INFORMATION RESOURCES FOR LABORATORY PERSONNEL

Institutional Safety Office. The best source of toxicity, hazard, storage, and disposal information is your institutional safety office, which maintains and makes available the most current information. Always consult this office for proper use and disposal procedures.

Post the phone numbers for your local safety office, security office, poison control center, and laboratory emergency personnel in an obvious place in your laboratory.

Material Safety Data Sheets (MSDSs). The Occupational Safety and Health Administration (OSHA) requires that MSDSs accompany all hazardous products that are shipped. These data sheets contain detailed safety information. MSDSs should be filed in the laboratory in a central location as a reference guide.

GENERAL SAFETY AND DISPOSAL CAUTIONS

The guidance offered here is intended to be generally applicable. However, proper waste disposal procedures vary among institutions; therefore, always consult your local safety office for specific instructions. All chemically constituted waste must be disposed of in a suitable container clearly labeled with the type of material it contains and the date the waste was initiated.

Appendix

It is essential for laboratory workers to be familiar with the potential hazards of materials used in laboratory experiments and to follow recommended procedures for their use, handling, storage, and disposal.

The following general cautions should always be observed.

- **Before beginning the procedure,** become completely familiar with the properties of substances to be used.

- **The absence of a warning** does not necessarily mean that the material is safe, because information may not always be complete or available.

- **If exposed** to toxic substances, contact your local safety office immediately for instructions.

- **Use proper disposal procedures** for all chemical, biological, and radioactive waste.

- **For specific guidelines on appropriate gloves to use,** consult your local safety office.

- **Handle concentrated acids and bases** with great care. Wear goggles and appropriate gloves. A face shield should be worn when handling large quantities.

    Do not mix strong acids with organic solvents because they may react. Sulfuric acid and nitric acid especially may react highly exothermically and cause fires and explosions.

    Do not mix strong bases with halogenated solvents because they may form reactive carbenes that can lead to explosions.

- **Handle and store pressurized gas containers** with caution because they may contain flammable, toxic, or corrosive gases; asphyxiants; or oxidizers. For proper procedures, consult the Material Safety Data Sheet that is required to be provided by your vendor.

- **Never pipette** solutions using mouth suction. This method is not sterile and can be dangerous. Always use a pipette aid or bulb.

- **Keep halogenated and nonhalogenated** solvents separately (e.g., mixing chloroform and acetone can cause unexpected reactions in the presence of bases). Halogenated solvents are organic solvents such as chloroform, dichloromethane, trichlorotrifluoroethane, and dichloroethane. Nonhalogenated solvents include pentane, heptane, ethanol, methanol, benzene, toluene, *N,N*-dimethylformamide (DMF), dimethylsulfoxide (DMSO), and acetonitrile.

- **Laser radiation,** visible or invisible, can cause severe damage to the eyes and skin. Take proper precautions to prevent exposure to direct and reflected beams. Always follow the manufacturer's safety guidelines and consult your local safety office. See caution below for more detailed information.

- **Flash lamps,** because of their light intensity, can be harmful to the eyes. They also may explode on occasion. Wear appropriate eye protection and follow the manufacturer's guidelines.

- **Photographic fixatives, developers, and photoresists** also contain chemicals that can be harmful. Handle them with care and follow the manufacturer's directions.

- **Power supplies and electrophoresis equipment** pose serious fire hazard and electrical shock hazards if not used properly.

- **Microwave ovens and autoclaves** in the laboratory require certain precautions. Accidents have occurred involving their use (e.g., when melting agar or Bacto Agar stored in bottles or when sterilizing). If the screw top is not completely removed and there is inadequate space for the steam to vent, the bottles can explode and cause severe injury when the containers are removed from the microwave or autoclave. Always completely remove bottle caps before microwaving or autoclaving. An alternative method for routine agarose gels that do not require sterile agar is to weigh out the agar and place the solution in a flask.

- **Ultrasonicators** use high-frequency sound waves (16–100 kHz) for cell disruption and other purposes. This "ultrasound," conducted through air, does not pose a direct hazard to humans, but the associated high volumes of audible sound can cause a variety of effects, including headache, nausea, and tinnitus. Direct contact of the body with high-intensity ultrasound (not medical imaging equipment) should be avoided. Use appropriate ear protection and display signs on the door(s) of laboratories where the units are used.

- **Use extreme caution when handling cutting devices,** such as microtome blades, scalpels, razor blades, or needles. Microtome blades are extremely sharp! Use care when sectioning. If unfamiliar with their use, have an experienced user demonstrate proper procedures. For proper disposal, use the "sharps" disposal container in your laboratory. Discard used needles *unshielded*, with the syringe still attached. This prevents injuries and possible infections when manipulating used needles because many accidents occur while trying to replace the needle shield. Injuries may also be caused by broken Pasteur pipettes, coverslips, or slides.

- **Procedures for the humane treatment of animals** must be observed at all times. Consult your local animal facility for guidelines. Animals, such as rats, are known to induce allergies that can increase in intensity with repeated exposure. Always wear a lab coat and gloves when handling these animals. If allergies to dander or saliva are known, wear a mask.

DISPOSAL OF LABORATORY WASTE

There are specific regulatory requirements for the disposal of all medical waste and biological samples mandated by the U.S. Environmental Protection Agency (see http://www.epa.gov/epawaste/hazard/tsd/index.htm) and regulated by the individual states and territories (see http://www.epa.gov/epawaste/wyl/stateprograms.htm). Medical and biological samples that require special handling and disposal are generally termed Medical Pathological Waste (MPW), and medical, veterinary, and biological facilities will have programs for the collection of MPW and its disposal. Restrictions on how radioactive waste can be disposed of as regulated by the U.S. Nuclear Regulatory Commission can be found in 10 CFR 20.2001, General requirements for waste disposal (see http://www.nrc.gov/reading-rm/doc-collections/cfr/part020/part020-2001.html) or the individual Agreement States. The preferred method for the disposal of radioactively contaminated MPW's is decay-in-storage (see http://www.nrc.gov/reading-rm/doc-collections/cfr/part035/part035-0092.html).

Waste and any materials contaminated with biohazardous materials must be decontaminated and disposed of as regulated medical waste. No harmful substances should be released into the environment in an uncontrolled manner. This includes all tissue samples, needles, syringes, scalpels, etc. Be sure to contact your institution's safety office concerning the proper practices associated with the handling and disposal of biohazardous waste.

Some basic rules are outlined below. For treatment of radioactive and biological waste, see sections on Radioactive Safety Procedures and Biological Safety Procedures.

- In practice, only **neutral aqueous solutions** without heavy metal ions and without organic solvents can be poured down the drain (e.g., most buffers). Acid and basic aqueous solutions need to be neutralized cautiously before their disposal by this method.

- For proper disposal of **strong acids and bases,** dilute them by placing the acid or base onto ice and neutralize them. Do not pour water into them. If the solution does not contain any other toxic compound, the salts can be flushed down the drain.

- For disposal of **other liquid waste,** similar chemicals can be collected and disposed of together, whereas chemically different wastes should be collected separately. This avoids chemical reactions between components of the mixture (see above). Collect at least inorganic aqueous waste, non-halogenated solvents, and halogenated solvents separately.

- Waste **from photo processing and automatic developers** should be collected separately to recycle the silver traces found in it.

Appendix

RADIOACTIVE SAFETY PROCEDURES

In the United States and other countries, the access to radioactive substances is strictly controlled. You may be required to become a registered user (e.g., by attending a mandatory seminar and receiving a personal dosimeter). A convenient calculator to perform routine radioactivity calculations can also be found at http://www.graphpad.com/quickcalcs/ChemMenu.cfm.

If you have never worked with radioactivity before, follow the steps below.

- *Try to avoid it!* Many experiments that are traditionally performed with the help of radioactivity can now be done using alternatives based on fluorescence or chemiluminescence and colorimetric assays, including, for example, DNA sequencing, Southern and northern blots, and protein kinase assays. However, in other cases (e.g., metabolic labeling of cells), use of radioactivity cannot be avoided.

- **Be informed.** While planning an experiment that involves the use of radioactivity, include the physicochemical properties of the isotope (half-life, emission type, and energy), the chemical form of the radioactivity, its radioactive concentration (specific activity), total amount, and its chemical concentration. Order and use only as much as is really needed.

- **Familiarize yourself** with the designated working area. Perform a mental and practical dry run (replacing radioactivity with a colored solution) to make sure that all equipment needed is available and to get used to working behind a shield. Handle your samples as if sterility would be required to avoid contamination.

- **Always wear appropriate gloves**, lab coat, and safety goggles when handling radioactive material.

- **Check the work area** for contamination before, during, and after your experiment (including your lab coat, hands, and shoes).

- **Localize your radioactivity.** Avoid formation of aerosols or contamination of large volumes of buffers.

- **Liquid scintillation cocktails** are often used to quantitate radioactivity. They contain organic solvents and small amounts of organic compounds. Try to avoid contact with the skin. After use, they should be regarded as radioactive waste; the filled vials are usually collected in designated containers, separate from other (aqueous) liquid radioactive waste.

- **Dispose of radioactive waste** only into designated, shielded containers (separated by isotope, physical form [dry/liquid], and chemical form [aqueous/organic solvent phase]). Always consult your safety office for further guidance in the appropriate disposal of radioactive materials.

- Among the experiments requiring **special precautions** are those that use [^{35}S]methionine and ^{125}I, because of the dangers of airborne radioactivity. [^{35}S]methionine decomposes during storage into sulfoxide gases, which are released when the vial is opened. The isotope ^{125}I accumulates in the thyroid and is a potential health hazard. ^{125}I is used for the preparation of Bolton–Hunter reagent to radioiodinate proteins. Consult your local safety office for further guidance in the appropriate use and disposal of these radioactive materials before initiating any experiments. Wear appropriate gloves when handling potentially volatile radioactive substances, and work only in a radioiodine fume hood.

BIOLOGICAL SAFETY PROCEDURES

Biological safety fulfills three purposes: to avoid contamination of your biological sample with other species; to avoid exposure of the researcher to the sample; and to avoid release of living material into the environment. Biological safety begins with the receipt of the living sample; continues with its storage, handling, and propagation; and ends only with the proper disposal of all contaminated materials. A catalog of operations known as "sterile handling" is usually employed in manipulating living matter. However, the actual manner of treatment largely depends on the actual sample, which can be quite diverse: *Escherichia coli* and other bacterial strains, yeasts, tissues of animal or plant origin, cultures of mammalian cells, or even derivatives from human blood are routinely handled in a biological laboratory. Two of these, bacteria and human blood products, are discussed in more detail below.

The Department of Health, Education, and Welfare (HEW) has classified various bacteria into different categories with regard to shipping requirements (see Sanderson and Zeigler 1991). Nonpathogenic strains of *E.coli* (such as K12) and *Bacillus subtilis* are in Class 1 and are considered to present no or minimal hazard under normal shipping conditions. However, *Salmonella*, *Haemophilus*, and certain strains of *Streptomyces* and *Pseudomonas* are in Class 2. Class 2 bacteria are "[a]gents of ordinary potential hazard: agents which produce disease of varying degrees of severity... but which are contained by ordinary laboratory techniques." Contact your institution's safety office concerning shipping biological material.

Human blood, blood products, and tissues may contain occult infectious materials such as hepatitis B virus and human immunodeficiency virus (HIV) that may result in laboratory-acquired infections. Investigators working with lymphoblast cell lines transformed by Epstein–Barr virus (EBV) are also at risk of EBV infection. Any human blood, blood products, or tissues should be considered a biohazard and should be handled accordingly until proved otherwise. Wear appropriate disposable gloves, use mechanical pipetting devices, work in a biological safety cabinet, protect against the possibility of aerosol generation, and disinfect all waste materials before disposal. Autoclave contaminated plasticware before disposal; autoclave contaminated liquids or treat with bleach (10% [v/v] final concentration) for at least 30 minutes before disposal (this is valid also for used bacterial media).

Always consult your local institutional safety officer for specific handling and disposal procedures of your samples. Further information can be found in the Frequently Asked Questions of the ATCC homepage (http://www.atcc.org) and is also available from the National Institute of Environmental Health and Human Services, Biological Safety (http://www.niehs.nih.gov/about/stewardship).

GENERAL PROPERTIES OF COMMON HAZARDOUS CHEMICALS

The hazardous materials list can be summarized in the following categories.

- **Inorganic acids**, such as hydrochloric, sulfuric, nitric, or phosphoric, are colorless liquids with stinging vapors. Avoid spills on skin or clothing. Spills should be diluted with large amounts of water. The concentrated forms of these acids can destroy paper, textiles, and skin and cause serious injury to the eyes.

- **Inorganic bases**, such as sodium hydroxide, are white solids that dissolve in water and under heat development. Concentrated solutions will slowly dissolve skin and even fingernails.

- **Salts of heavy metals** are usually colored, powdered solids that dissolve in water. Many of them are potent enzyme inhibitors and therefore toxic to humans and the environment (e.g., fish and algae).

- Most **organic solvents** are flammable volatile liquids. Avoid breathing the vapors, which can cause nausea or dizziness. Also avoid skin contact.

- **Other organic compounds** including organosulfur compounds, such as mercaptoethanol or organic amines, can have very unpleasant odors. Others are highly reactive and should be handled with appropriate care.

- If improperly handled, **dyes and their solutions** can stain not only your sample but also your skin and clothing. Some are also mutagenic (e.g., ethidium bromide), carcinogenic, and toxic.

Appendix

- **Nearly all names ending with "ase"** (e.g., catalase, β-glucuronidase, or zymolyase) refer to enzymes. There are also other enzymes with nonsystematic names such as pepsin. Many of them are provided by manufacturers in preparations containing buffering substances, etc. Be aware of the individual properties of materials contained in these substances.
- **Toxic compounds** are often used to manipulate cells. They can be dangerous and should be handled appropriately.
- Be aware that several of the compounds listed have not been thoroughly studied with respect to their toxicological properties. Handle each chemical with appropriate respect. Although the toxic effects of a compound can be quantified (e.g., LD_{50} values), this is not possible for carcinogens or mutagens where one single exposure can have an effect. Also realize that dangers related to a given compound may also depend on its physical state (fine powder vs. large crystals/diethyl ether vs. glycerol/dry ice vs. carbon dioxide under pressure in a gas bomb). Anticipate under which circumstances during an experiment exposure is most likely to occur and how best to protect yourself and your environment.

Cold Spring Harbor Laboratory Press (CSHLP) has used its best efforts in collecting and preparing the material contained herein but does not assume, and hereby disclaims, any liability for any loss or damage caused by errors and omissions in the publication, whether such errors and omissions result from negligence, accident, or any other cause. CSHLP does not assume responsibility for the user's failure to consult more complete information regarding the hazardous substances listed in this publication.

REFERENCE

Sanderson KE, Zeigler DR. 1991. Storing, shipping, and maintaining records on bacterial strains. *Methods Enzymol* 204: 248–264.

WWW RESOURCES

ATCC Home page http://www.atcc.org
ATCC, for Sample Handling (in Frequently Asked Questions) http://www.atcc.org/CulturesandProducts/TechnicalSupport/FrequentlyAskedQuestions/tabid/469/Default.aspx
GraphPad Software, Radioactivity Calculations http://www.graphpad.com/quickcalcs/ChemMenu.cfm
National Institute of Environmental Health and Human Services, Biological Safety (NIEHS) http://www.niehs.nih.gov/about/stewardship
U.S. Environmental Protection Agency (EPA), Federal waste disposal regulations, Laboratory http://www.epa.gov/epawaste/hazard/tsd/index.htm
U.S. Environmental Protection Agency (EPA), Individual States and Territories http://www.epa.gov/epawaste/wyl/stateprograms.htm
U.S. Nuclear Regulatory Commission (NRC), Medical Pathological Radioactively Contaminated Waste (Decay-in-Storage) http://www.nrc.gov/reading-rm/doc-collections/cfr/part035/part035-0092.html
U.S. Nuclear Regulatory Commission (NRC), Radioactive Waste Disposal Regulations: General Requirements http://www.nrc.gov/reading-rm/doc-collections/cfr/part020/part020-2001.html

Index

A

Affinity capture
  binding reaction, 391
  density gradient ultracentrifugation of protein complexes
    centrifugation, 398–399
    fraction analysis, 399–400
    gradient preparation, 398
    materials, 397–398
  elution under denaturing conditions
    antibody-conjugated bead equilibration, 391
    cryogenic disruption, 389–390
    extract preparation, 390–391
    materials, 388–389
  native elution
    cleavable tags, 394–395
    competitive elution with PEGylOx, 395–396
    materials, 393–394
    troubleshooting, 396
  optimization, 384–386
  principles, 383–384
Agar medium, 211
α-factor, G_1 synchronization, 243–244
Amyloid-prion buffers, 506–507
Auxotrophic mutants, 12, 15

B

Biofilm
  assays
    culture and photography, 58–59
    materials, 57–58
    overview, 50–51
    recipes, 59
  induction, 50
BioGRID
  curation statistics, 578
  feedback from users, 588
  interaction network visualization
    BioGRID viewer, 586
    Cytoscape, 586–587
    download options and formats, 587–588
  overview, 577–578
  scope, 579–580
  searching for gene or protein of interest, 583–586
  user interface, 578
BRB80, 292, 297

BSA. *See* Bulk segregant analysis
Bulk segregant analysis (BSA)
  candidate gene and variant identification, 659
  materials, 656–657
  overview, 653–654
  phenotypically extreme segregant acquisition, 658–659
  recipes, 659–660
  recombinant cross progeny, 657–658
  segregant generation, 657
  sequencing and causal loci detection, 659

C

Calling card analysis
  advantages and limitations, 533, 535
  cloning strain, 538–540
  DNA extraction, 541
  genomic digestion, 541–542
  inverse polymerase chain reaction, 542–543
  materials, 536–538
  overview, 533–534
  plasmid transformation and induction, 540–541
  recipes, 544–545
  self-ligation, 542
  troubleshooting, 543
CalMorph. *See* High-throughput microscopy
cdc15-2, G_1 synchronization, 245–246
Cell cycle
  drug-induced arrest, 246
  position determination
    flow cytometry
      data analysis, 261
      DNA staining, 260
      enzymatic digestion, 260
      ethanol fixation, 259
      flow cytometry, 260–261
      high-throughput staining, 262
      materials, 258–259
      recipes, 263
      rehydration of fixed cells, 259
      troubleshooting, 262–263
    overview, 240–241
  synchronization
    centrifugal elutriation
      cleanup and preparation for storage, 255
      Coulter counting, 252
      culture pregrowth and inoculation volume calculation, 250
      elutriation setup and sterilization, 251
      exhaustive fractionation, 254–255
      fraction collection and monitoring, 253–254
      G_1 cell collection for sampling, 252–253
      loading, 253
      materials, 248–249
      sample preparation, 252
      troubleshooting, 255–256
    chemical and genetic approaches
      G_1 synchronization using α-factor mating pheromone, 243–244
      M/G_1 synchronization using *cdc15-2*, 245–246
      materials, 242–243
      overview, 239–240
      selection of technique, 256–257
Cell wall
  components, 199–200
  disruption
    disruptors, 201–202
    imaging, 214–215
    materials, 213–214
    recipes, 215
  fluorescent labeling
    materials, 205–206
    staining
      1,3-β-glucan, 206–207
      cell preparation, 206
      chitin, 207–208
      mannoproteins, 207–208
  function, 199
  spore wall integrity testing
    fly feces analysis, 210–211
    materials, 209–210
    prey yeast cells
      Drosophila feeding, 210
      preparation, 210
      recipes, 211–212
  synthesis and assembly, 200–201
Centrifugal elutriation. *See* Cell cycle
CgIs. *See* Chemical–genetic interactions

699

Index

Chemical–genetic interactions (CGIs)
  challenges in screening, 465–466
  halo high-throughput assay
    incubation and analysis, 469–470
    materials, 468–469
    plate preparation, 469
    recipes, 470
  overview, 463–465
  parallel analysis of barcoded yeast strains
    array-based barcode quantification, 473–474
    competitive growth, 473
    data analysis, 474–475, 477
    materials, 471–472
    recipes, 477–478
    sequencing-based barcode quantification, 475–477
Chemogenomics. *See* Chemical–genetic interactions
Chemostat culture
  chemostat design, 662–663
  continuous culture overview, 662
  experimental evolution studies
    culture, 680–681
    inoculation, 680
    materials, 679–680
    sampling
      chemostat, 681
      effluent, 681–682
    troubleshooting, 682
  mini-chemostat array assembly
    carboy assembly with chemostats and effluent collection, 669–670
    cleanup and sterilization, 668
    culture chamber assembly, 666–668
    hydrated aeration, 668
    materials, 665–666
    media
      carboy preparation, 668–669
      filtration, 669
      recipes, 670–672
  physiology studies
    harvesting for analysis
      metabolites, 677
      microscopy, 676–677
      protein, 677
      RNA, 675–676
    materials, 673–674
    recipes, 678
    steady state establishment, 674–675
Chitin, staining in cell wall, 207–208
Chromatin
  conformation studies. *See* Chromosome conformation capture
  organization in yeast, 104–105
Chromosome conformation capture (3C)

chromosome conformation capture carbon copy
  advantages, 125
  ligating fragments, 124
  materials, 121–122
  polymerase chain reaction, 124–125
  probes
    annealing, 123–124
    design, 125–126
    preparation, 122
  troubleshooting, 125
cross-linking chromatin, 110
digestion, 110–111
end-point polymerase chain reaction, 112–113
Hi-C
  biotin removal, 134
  biotinylation of digested ends, 131
  cross-link reversal and ligation product purification, 132–133
  cross-linking chromatin, 129–131
  digestion, 131
  DNA end repair and A-tailing, 134–135
  efficiency estimation, 133–134
  library
    aired-end polymerase chain reaction amplification, 137–138
    fractionation, 135–136p
    sonication, 134
  ligating cross-linked fragments, 131
  ligation product enrichment, 136–137
  materials, 127–129
  recipes, 138–139
  troubleshooting, 138
ligation of cross-linked chromatin fragments, 111
materials, 108–110
overview, 103–106
randomized ligation control
  chromosomal DNA isolation, 116–117
  genomic DNA digestion, 117–118
  ligating digested fragments, 118–119
  materials, 115–116
  recipes, 119–120
recipes, 113
reverse cross-linking, 111–112
Chromosome replication. *See also* Meiosis
overview, 87
single-fiber analysis
  DNA combing
    agarose plug preparation and digestion, 96–97

cell synchronization and bromodeoxyuridine labeling, 95–96
glass surface preparation, 92–93
imaging, 98–99
immunodetection, 97–98
materials, 90–92
plug melting and DNA combing, 97
recipes, 100–101
simple machine preparation, 93–95
troubleshooting, 99
overview, 88
techniques for study, 88
CM medium. *See* Complete minimal medium
COM drop-out powder, 639, 644
Complete minimal (CM) medium, 169–170, 215
Complete synthetic medium, 497–498
Complex traits, genetic dissection
  bulk segregant analysis
    candidate gene and variant identification, 659
    materials, 656–657
    overview, 653–654
    phenotypically extreme segregant acquisition, 658–659
    recipes, 659–660
    recombinant cross progeny, 657–658
    segregant generation, 657
    sequencing and causal loci detection, 659
  causal gene and genetic variant dissection, 654
  challenges, 651–652
  genome-wide association study, 652–653
  linkage mapping, 653
Concanavalin A. *See* High-throughput microscopy
Congenic strain
  conditional effects of mutations, 2–4
  overview, 1–2
Continuous culture. *See* Chemostat culture
Culture
  batch culture, 661–662
  continuous culture. *See* Chemostat culture
  optimal growth conditions, 11–12
  propagating culture, 12–13
  synchronous meiotic cultures
    assessment of efficiency and synchrony, 34
    liquid medium culture, 33–34
    materials, 32–33
    overview, 23
    recipes, 35–36

Culture (*Continued*)
  sample collection
    meiotic recombination analysis, 34–35
    surface spreading of nuclei for immunofluoresence analysis, 35
    western blot analysis, 35

Cytoscape
  BioGRID interaction network visualization, 586–587
  data preparation for import, 592
  installation, 591–592
  networks
    annotation, 595–598
    loading, 592–594
    organization, 594, 596–597
    visualization, 594

Cytosine deaminase protein-fragment complementation assay. *See* Protein-fragment complementation assay

D

DAPI. *See*, 4′,6-Diamidino-2-phenylindole

Deep mutational scanning
  doped synthetic oligonucleotides, 193–194
  enrichment score calculation from DNA sequencing output files
    enrich output file analysis, 196
    materials, 195
    troubleshooting, 197
  functional selection, 192
  high-throughput sequencing, 192–193
  library construction, 192
  principles, 187–189
  sequence-function map analysis, 189

Deletion collections. *See Saccharomyces* Genome Deletion Project

Density gradient ultracentrifugation. *See* Affinity capture; Prions

4′,6-Diamidino-2-phenylindole (DAPI), assessment of efficiency and synchrony of meiotic cultures, 34

Dihydrofolate reductase protein-fragment complementation assay. *See* Protein-fragment complementation assay

Diploids, applications, 13–14

DNA binding motifs. *See* Transcription factor–DNA binding motifs

DNA combing. *See* Chromosome replication

DNA sequencing
  bulk segregant analysis, 659
  high throughput strain sequencing
    platforms, 621–623
    prospects, 623
  library preparation
    fragmentation, 627
    genomic DNA extraction, 626–627
    materials, 625–626
    polymerase chain reaction, 628–629
    recipes, 629
    tailing, 627–628
    troubleshooting, 629

DNA synthesis. *See* Synthetic genome synthesis

Drop-out medium, 169–170, 215, 275, 338, 343, 458, 544

E

Electron microscopy, prion amyloids, 484–485

Electron tomography
  grid preparation, 310
  high-pressure freezing/freeze substitution, 305, 309–310
  materials, 308–309
  prospects for study, 307
  recipes, 311–312
  three-dimensional reconstruction, 305–307, 311
  tilt series acquisition, 310–311
  troubleshooting, 311
  yeast specimen preparation for transmission electron microscopy, 303–305

Evaporative light-scattering detection. *See* Lipids, yeast

F

Fatty acids. *See* Lipids, yeast

Filamentous growth
  assays
    mitogen-activated protein kinase pathway
      materials, 65–66
      mucin secretion profiling, 69–70
      pectinase assay, 69
      recipes, 70–72
      western blot, 67–69
    overview, 50–51
    plate-washing assay
      agar invasion, 54–55
      materials, 53–54
      recipes, 55–56
    single-cell analysis
      culture and microscopy, 62–63
      materials, 61–62
      recipes, 64
  induction, 49–50

Flow cytometry. *See* Cell cycle

5-Fluorocytosine solution, 371
5-Fluoroorotic acid plates, 170

Forward genetics
  mutant identification and selection, 16–17
  overview, 13–14

Freeze-substitution fixative, 311

G

Galactose-Ura plates, 544

Gas chromatography. *See* Lipids, yeast

GenFlex tags, 477–478

Genome synthesis. *See* Synthetic genome synthesis

Genome-wide association study (GWAS), complex trait dissection, 652–653

Genotype–phenotype mapping
  causality confirmation of genotype–phenotype links
    materials, 646
    overview, 634
    recipes, 649–650
    reciprocal hemizygosity, 647–648
    troubleshooting, 648–649
  overview, 631–632
  phenomics, 633
  quantitative trait loci mapping
    F_1 segregant generation, 642
    mapping, 642–643
    materials, 641–642
    overview, 633–634
    recipes, 644–645
    troubleshooting, 643
  Saccharomyces cerevisiae ecology and population genetics, 632
  strain isolation and domestication
    enrichment and isolation, 637
    materials, 636
    recipes, 639–640
    sampling, 637
    species identification, 637
    strain preparation for laboratory work, 637–638
    troubleshooting, 638

Glucose-His plates, 545
Glucose-limited chemostat medium, 670
Glucose-Ura Medium, 545
Glycerophospholipids. *See* Lipids, yeast
GWAS. *See* Genome-wide association study

H

Halo assay. *See* Chemical–genetic interactions

Hi-C. *See* Chromosome conformation capture

Index

High-performance liquid chromatography. *See* Lipids, yeast
High-throughput microscopy
  automated image analysis, 268
  imaging pipelines, 266–267
  morphology studies with CalMorph
    concanavalin A coating of microplates, 279
    fixation, 278
    image acquisition and processing, 279–281
    materials, 277–278
    recipes, 281
    specimen preparation, 279
    staining, 278–279
  overview, 265–266
  synthetic genetic array for fluorescent tagging
    drug treatment and medium switch, 273–274
    imaging, 274
    materials, 271–273
    recipes, 275–276
    subculture preparation, 273
Homologous recombination-based cloning
  applications, 76
  overview, 73–75
  plasmid construction
    competent cell preparation
      Escherichia coli, 83
      yeast, 81–82
    DNA fragment preparation, 81
    genomic DNA preparation, 82
    materials, 78–80
    overview, 80, 83–84
    plasmid recovery from bacteria, 83
    recipes, 84–86
  polymerase chain reaction-free recombination, 75
Hydrogen–deuterium exchange, prion amyloids, 485
Hydrophilic interaction chromatography–tandem mass spectrometry. *See* Metabolomics

I

Immobilized metal affinity chromatography. *See* Proteomics
Immunoaffinity precipitation. *See* Proteomics
Intragenic complementation, 13–14
ISO buffer, 691
Isogenic strain, overview, 1–2, 15
Isothermal reaction master mix, 69

K

Knockout marker cassettes. *See* MX cassettes

L

LB media, 338, 343, 691
LB medium plus ampicillin, 84
LB plates, 338, 343, 591
Lead citrate solution, 312
Linkage mapping, complex trait dissection, 653
Lipids, yeast
  challenges in study, 217–218
  composition by strain, 218
  extraction
    cell growth and harvesting, 224
    materials, 223
    organic extraction, 224–225
    troubleshooting, 225
  fatty acids, 218–219
  gas chromatography
    fatty acid methyl ester derivatization, 232–233
    materials, 231–232
    running conditions, 233–234
    troubleshooting, 234
  glycerophospholipids, 219–220
  high-performance liquid chromatography/evaporative light-scattering detection
    materials, 235–236
    running conditions, 236–237
    troubleshooting, 238
  minor components, 220–221
  overview of analytical techniques, 221
  sphingolipids, 220
  sterols, 220
  thin-layer chromatography
    materials, 227–228
    running and development, 228–229
    troubleshooting, 229
Lipid–protein interactions. *See* Protein microarray
Lyticase solution, 498

M

Mass spectrometry (MS). *See also* Proteomics
  metabolomics
    amino acid analysis with hydrophilic interaction chromatography–tandem mass spectrometry
      conditioning samples, 611
      culture, 610
      materials, 608–610
      recipes, 612–613
      running conditions, 611–612
      sample collection and extraction, 610–611
    cell growth and extraction of metabolites, 604–606
    materials, 603–604
    overview, 601
  technique for amyloid purification and identification, 505–506
Mat formation
  assays
    culture and photography, 58–59
    materials, 57–58
    overview, 50–51
    recipes, 59
  induction, 50
Meiosis
  chromosomes
    segregation, 21
    structure, 22–23
    visualization, 24
  progression regulation, 23
  recombination, 22
  recombination analysis
    chromosome visualization
      fluorescence microscopy, 42
      immunodecoration, 41–42
      surface spreading of nuclei, 40–41
    materials, 38–40
    overview, 24
    physical analysis
      DNA extraction and purification, 42, 44
      restriction enzyme digestion, 44
      Southern blot, 45–46
    recipes, 47–48
  S phase, 22
  spore formation and viability, 24, 26–30
  strain selection for studies, 23
  synchronous cultures
    assessment of efficiency and synchrony, 34
    liquid medium culture, 33–34
    materials, 32–33
    overview, 23
    recipes, 35–36
    sample collection
      meiotic recombination analysis, 34–35
      surface spreading of nuclei for immunofluoresence analysis, 35
      western blot analysis, 35
Membrane yeast two-hybrid system (MYTH)
  bait generation and validation
    integrated MYTH bait generation, 337
    materials, 334–336
    NubGI test for validation, 337–338
    recipes, 338–339
    subcellular localization verification, 338

Index

Membrane yeast two-hybrid system (MYTH) (Continued)
    transitional MYTH bait generation, 336–337
    integrated versus transitional MYTH, 332–333
    overview, 331–332
    screening
        bait-dependency testing, 343
        materials, 340–341
        recipes, 343–345
        secondary screening and prey identification, 342–343
        transformation, 341–342
MES wash buffer, 47
Metabolomics
    amino acid analysis with hydrophilic interaction chromatography–tandem mass spectrometry
        conditioning samples, 611
        culture, 610
        materials, 608–610
        recipes, 612–613
        running conditions, 611–612
        sample collection and extraction, 610–611
    ethanol and glucose analysis in culture media
        ethanol spectrophotometric assay, 617–618
        glucose spectrophotometric assay, 615–617
        materials, 614–615
        troubleshooting, 619
    mass spectrometry
        cell growth and extraction of metabolites, 604–606
        materials, 603–604
        overview, 601
    nuclear magnetic resonance, 601
    overview, 599–601
Metal affinity chromatography. See Proteomics
Metal stripping solution, 407
Methotrexate medium, 364
Microscopy. See High-throughput microscopy; Single-molecule total internal reflection fluorescence microscopy
Mitogen-activated protein kinase. See Filamentous growth
MS. See Mass spectrometry
Mucins, secretion profiling in filamentous growth, 69–70
MX cassettes
    collections for attainment, 144
    gene regulation cassettes, 142–143
    introduction into yeast
        materials, 146–147
        overview, 142
        recipes, 151–152
        transformation, incubation, and amplification, 148–149
        troubleshooting, 149–151
    multiple cassettes and selections, 143
    overview, 141
    polymerase chain reaction amplification, 141–142
    recycling
        confirmation of pop-out, 157
        materials, 153–154
        overview, 143–144
        pop out cassettes flanked by large MX3 or PR direct repeats with counterselection, 154
        without counterselection, 155–156
        pop out cassettes flanked by loxP direct repeats, 156
        recipes, 158–159
        troubleshooting, 157
    types and yeast strain genotypes, 147
MYTH. See Membrane yeast two-hybrid system

N

Nitrogen base agar plates, 659
Nitrogen-limited chemostat medium, 671
NMR. See Nuclear magnetic resonance
Nonquenched fluorescent liposome. See Protein microarray
Nuclear magnetic resonance (NMR)
    metabolomics, 601
    solid-state NMR of prion amyloids, 485–486

O

One-hybrid assay. See Yeast one-hybrid assay

P

PBS. See Phosphate-buffered saline
PCA. See Protein-fragment complementation assay
PCR. See Polymerase chain reaction
Pectinase agar plates, 70–71
Pectinase, filamentous growth assay, 69
PEGylOx. See Affinity capture
Phenol:chloroform, 113, 119–120, 138
Phenomics. See Genotype–phenotype mapping
Phosphate-buffered saline (PBS), 381, 413
Phosphate-limited chemostat medium, 671
Phosphopeptide binding solution, 407
Phosphopeptide elution solution, 407
Phosphopeptides. See Proteomics
Plasmid construction. See Homologous recombination-based cloning
PLATE solution, 357, 371, 381
Polymerase chain reaction (PCR)
    calling card analysis and inverse polymerase chain reaction, 542–543
    chromosome conformation capture and end-point polymerase chain reaction, 112–113
    chromosome conformation capture carbon copy, 124–125
    DNA sequencing, 628–629
    Hi-C, 137–138
    MX cassette amplification, 141–142
    synthetic genome synthesis
        colony screening PCR, 688–689
        finish PCR, 687–688
        templateless PCR, 687
    transposon-insertion libraries, 167–168
    yeast one-hybrid assay genomic and plasmid templates
        amplification, 531
        materials, 530–531
        recipes, 532
        troubleshooting, 531–532
Posttranslational modifications. See Protein microarray; Proteomics
Potassium acetate medium, 639, 644
Potassium phosphate buffer, 678
Potassium phosphate-buffered solution, 71
Presporulation medium, 36
Prions
    approaches for study
        biochemical methods, 484
        cell biology, 484
        computational methods, 484
        genetics, 481–483, 488–492
        physical studies, 484–486
    curing, 482–483, 492
    cytoduction
        cyclohexamide resistance, 491–492
        guanidine curing, 493
        induction by overproduction, 492
        standard cytoduction, 491
    isolation and analysis
        agarose gel electrophoresis, 504
        density gradient sedimentation, 504
        lysate preparation, 502–503
        materials, 501–502
        recipes, 506–508
        technique for amyloid purification and identification
            amyloid protein isolation, 504–505
            digestion, 505–506
            mass spectrometry, 505–506

Prions (*Continued*)
    troubleshooting, 506
    nomenclature, 482
    overproduction and generation, 482, 492
    phenotype assays, 488–491
    phenotype relationship, 482–483
    transfection
        incubation and growth conditions, 496–497
        materials, 495–496
        overview, 483
        recipes, 497–499
    types in yeast, 481, 483
Protein localization. *See* Transposon-insertion libraries
Protein microarray
    applications, 417
    lipid–protein interaction analysis
        liposome applying to microarray, 430–431
        materials, 428–429
        nonquenched fluorescent liposome preparation, 429–430
        recipes, 432
        troubleshooting, 431
    overview, 415–416
    posttranslational modification assays
        blocking, 435
        detection and processing, 437–438
        materials, 433–435
        posttranslational modification reactions, 436
        reaction buffer preparation
            acetylation, 435, 438–439
            phosphorylation, 435, 439
            SUMOylation, 436, 440
            ubiquitylation, 436, 440
        recipes, 438–440
        troubleshooting, 438
        washing, 436–437
    protein–protein interaction analysis
        antibody incubations
            primary antibody, 419–420
            secondary antibody, 420
        blocking, 419
        materials, 418–419
        probing, 419
        recipes, 420–421
        troubleshooting, 420
    RNA-binding protein characterization
        Cy5 labeling of RNA probe, 424–425
        materials, 422–423
        recipes, 427
        RNA-binding assay, 425–426
        troubleshooting, 426
Protein-fragment complementation assay (PCA)
    cytosine deaminase protein-fragment complementation assay
        Cdk1 protein interaction detection, 368–370
        expression plasmid construction, 368
        FCY1 gene deletion, 368
        image analysis, 370
        materials, 366–367
        protein–protein interaction detection in different cyclin deletion strains, 370
        recipes, 371–372
        troubleshooting, 370
    dihydrofolate reductase protein-fragment complementation assay
        homologous recombination of fragments, 352
        large-scale screening
            bait strain preparation, 353
            image analysis, 354
            overview, 352–353
            prey strain preparation, 353–354
            statistical analysis, 355
            tray incubation, 354
        materials, 350–352
        recipes, 357–358
        troubleshooting, 355–356
    general considerations, 348–349
    genotype-to-phenotype mapping of protein complexes and interaction networks
        diploid strain construction, 362
        gene deletion introgression into DHFR PCA strains, 361
        image and statistical analysis, 362–363
        materials, 359–360
        recipes, 364–365
        sporulation and recombinant haploid strain selection, 361–362
        troubleshooting, 363
    principles, 347–348
    real-time assay
        applications, 378–379
        cell preparation
            fluorescence microscopy, 376–378
            fluorometric analysis using infrared fluorescence protein, 377
        homologous recombination of fragments, 375–376
        materials, 373–375
        recipes, 380–381
        transformation of expression plasmid pairs, 375
        troubleshooting, 378
Proteinase K solution, 100, 263
Protein–protein interactions. *See* Affinity capture; BioGRID; Membrane yeast two-hybrid system; Protein-fragment complementation assay; Protein microarray; Proteomics; Yeast two-hybrid system
Proteomics
    immobilized metal affinity chromatography of phosphopeptides
        binding conditions, 406
        filtration tip preparation, 406
        materials, 404–406
        recipes, 407
        resin preparation, 405–406
        troubleshooting, 407
        washing, elution, and filtration, 406–407
    immunoaffinity precipitation of modified peptides
        antibody conjugation to agarose beads, 410
        incubation conditions, 411
        materials, 409–410
        peptide washes and elution, 411–412
        recipes, 413
        sample preparation for liquid chromatography-tandem mass spectrometry, 412
        troubleshooting, 412–413
        washing and storage of beads, 410–411
    posttranslational modification types, 401–402
    protein microarray. *See* Protein microarray
    techniques, 402–403
PTC buffer, 498

Q

Quantitative trait loci. *See* Genotype–phenotype mapping

R

Reciprocal hemizygosity. *See* Genotype–phenotype mapping
Recombination. *See* Homologous recombination-based cloning; Meiosis
RNA-binding proteins. *See* Protein microarray
RNase A, boiled, 363

Index

S

Saccharomyces Genome Database (SGD)
  annotations, 558, 570–572
  biochemical pathway analysis, 561–562
  data mining
    microarray data exploration, 568–569
    YeastMine, 566–568
  genome feature exploration, 574–576
  mutant phenotype analysis, 562–564
  ontology, 558, 570–572
  overview, 557–558
  reference genome sequence, 558
  user interface, 558–559
Saccharomyces Genome Deletion Project
  applications, 176
  collection attainment, 184
  functional profiling of collections
    fitness measurements
      liquid medium culture, 182
      pool construction and growth, 182–183
      solid medium culture, 181–182
    inoculation of collections, 180
    materials, 179–180
    principles, 175–176
    recipe, 184
    troubleshooting, 183
  overview, 173–175
Saccharomyces sensu stricto enrichment medium, 639
Salmon sperm DNA solution, 343
SC medium. *See* Synthetic complete medium
SCE buffer, 100
SDE plates, 151
SDS gel loading buffer, 71
SGA. *See* Synthetic genetic array
SGD. *See* Saccharomyces Genome Database
Single-molecule total internal reflection fluorescence microscopy
  applications, 284–285
  coverslip cleaning and functionalization
    lipid passivation, 290–292
    materials, 287–289
    recipes, 292
    silanization, 289–290
  data analysis, 299–301
  principles, 283–284
  reaction preparation for imaging
    flow chamber assembly, 295–296
    materials, 294–295
    microtubule binding interactions
      dynamic microtubules, 297
      paclitaxel-stabilized microtubules, 296–297
    recipes, 297
Site-directed mutagenesis, 17–18

SLAHD plates, 64
Sodium acetate buffer, 113, 120, 138
Sodium phosphate solution, 281, 338, 344
Sorbitol solution, 629
SOS medium, 499
Southern blot, meiotic recombination analysis, 45–46
Spheroplast fixative solution, 47
Spheroplast lysis buffer, 47
Spheroplast storage buffer, 35
Sphingolipids. *See* Lipids, yeast
SPM plates, 30
SPO agar, 212
Spore
  mutation effects on formation and viability, 24
  sporulation efficiency and viability analysis from tetrad dissection, 27–29
Sporulation medium, 36, 459–460, 660
SSC, 47
ST buffer, 499
STC buffer, 499
Sterols. *See* Lipids, yeast
Strains, *Saccharomyces cerevisiae*
  choice, 6–7
  conditional effects of mutations, 2–4
  congenic versus isogenic strains, 1–2
  expansion, 4–5
  genotype–phenotype mapping. *See* Genotype–phenotype mapping
  high throughput sequencing. *See* DNA sequencing
  isolation and domestication
    enrichment and isolation, 637
    materials, 636
    recipes, 639–640
    sampling, 637
    species identification, 637
    strain preparation for laboratory work, 637–638
    troubleshooting, 638
  lipid composition, 218
  meiosis studies, 23
  prospects, 7
  resources, 4–6
  spore wall integrity testing. *See* Cell wall
  table, 3
  yeast two-hybrid system, 316
Sulfate-limited chemostat medium, 671
Synthetic amino acid dropout medium, 85
Synthetic complete (SC) medium, 169–170, 215, 357, 365, 371–372, 381
Synthetic defined medium, 55–56, 64, 381, 644, 649–650
Synthetic dextrose medium, 492

Synthetic dextrose plates, 151–152
Synthetic drop-out medium, 338–339, 344
Synthetic genetic array (SGA)
  alternative techniques, 442–443
  analysis and imaging, 454, 456–457
  applications, 443, 445, 457–458
  deletion mutant array construction, 454–455
  genetic interaction quantification, 443
  high-throughput microscopy, synthetic genetic arrays for fluorescent tagging
    drug treatment and medium switch, 273–274
    imaging, 274
    materials, 271–273
    recipes, 275–276
    subculture preparation, 273
  materials, 448–451
  mutant strain collections, 442
  pin tool sterilization, 453–454
  principles, 441–442, 444
  query strain construction, 450, 452–453
  recipes, 458–461
Synthetic genome synthesis
  building block synthesis from oligonucleotides
    assembly, 689
    cloning, 688
    overlapping oligonucleotide preparation, 687
    polymerase chain reaction
      colony screening PCR, 688–689
      finish PCR, 687–688
      templateless PCR, 687
    recipes, 691
    transformation, 688
    troubleshooting, 689–690
  materials, 685–686
Synthetic lysine dropout medium, 85
Synthetic minimal medium, 612–613

T

TAPI. *See* Technique for amyloid purification and identification
TB medium. *See* Terrific broth medium
TBE buffer, 139, 519, 525–528, 532
TBS. *See* Tris-buffered saline
TBST, 71, 432
TCA buffer, 71
TE buffer, 48, 113, 120, 139, 171, 629
Technique for amyloid purification and identification (TAPI)
  amyloid protein isolation, 504–505

Index

Technique for amyloid purification and
    identification (TAPI)
    (Continued)
  digestion, 505–506
  mass spectrometry, 505–506
Terrific broth (TB) medium, 86
Tetrad genetics
  crossing over and gene conversion
      analysis in meiosis, 29–30
  overview, 14–15
Thin-layer chromatography. See Lipids,
    yeast
3C. See Chromosome conformation
    capture
TLE buffer, 139
Total internal reflection fluorescence
    microscopy. See Single-
    molecule total internal
    reflection fluorescence
    microscopy
Transcription factor–DNA binding motifs
  consensus sequences, 550
  enrichment computation, 554–555
  generation, 550–552
  overview, 547
  putative binding site identification,
      553–554
  repositories, 552–553
  scoring, 547–549
  visualization, 549
Transmission electron microscopy. See
    Electron tomography
Transposon calling cards. See Calling card
    analysis
Transposon-insertion libraries
  advantages and limitations, 163
  applications, 163–164
  features, 162–163
  overview, 161
  phenotypic screening and protein
      localization
    Cre-lox recombination to
        generate epitope-tagged
        alleles, 168
    insertion site identification with
        inverse polymerase chain
        reaction, 167–168
    materials, 165–166
    recipes, 169–171
    screening of transformants, 167
    transformation, 166–167
  resources, 162

Tris-buffered saline (TBS), 48, 423, 427,
    432, 507
Tween wash buffer, 139
Two-hybrid system. See Membrane yeast
    two-hybrid system; Yeast
    two-hybrid system

V

Vitamin stock solution, 672

W

Western blot
  mitogen-activated protein kinases in
      filamentous growth, 67–69
  synchronous meiotic cultures, 35

X

X-Gal plates, 171
X-ray fiber diffraction, prion amyloids, 485

Y

YAPD medium, 519
YAPD plates, 520, 526, 529, 532
Yeast one-hybrid assay
  advantages and limitations,
      511–512
  bait strain generation
    autoactivity testing, 517
    integrated baits
      glycerol stock preparation, 518
      identity confirmation, 517–518
    materials, 514–515
    recipes, 519–520
    reporter construct linearization,
        515–516
    transformation, 516–517
    troubleshooting, 518
  colony lift β-galactosidase assay
    culture and incubation, 528
    materials, 527–528
    recipes, 529
    troubleshooting, 528
  library screening
    double-positive yeast identification,
        523
    gap-repair for interaction retesting,
        524–525
    materials, 521–522
    recipes, 525–526

transformation, 522–523
    polymerase chain reaction of genomic
        and plasmid templates
      amplification, 531
      materials, 530–531
      recipes, 532
      troubleshooting, 531–532
    principles, 509–511
Yeast two-hybrid system. See also Mem-
    brane yeast two-hybrid system
  array-based screening, 316–317,
      325–327
  bait self-activation testing, 315,
      323–324
  false negatives, 317
  false positives, 317–318
  host strain selection, 316
  library screening
    advantages and disadvantages, 316
    mating, 325
    prey and bait culture preparation,
        324
  materials, 319–322
  overview, 313
  pooled array screening, 317
  rationale, 313–315
  recipes, 328–329
  transformation, 322–323
  troubleshooting, 327–328
  vector choice, 315, 320
YeastMine. See Saccharomyces Genome
    Database
YEP-GAL medium, 72
YEPD medium, 56, 59, 71, 86, 101, 152,
    159, 171, 184, 358, 365, 372,
    381, 470, 478, 492, 499, 545,
    640, 645, 650, 660
YEPD plates, 30, 36, 56, 59, 71, 86, 152,
    212, 329, 372, 461, 640, 645,
    650
YPA agar, 212
YPAD medium, 339, 344–345
YPG medium, 493
YPG plates, 30, 36
YPGal medium, 159

Z

Z-buffer, 529
Zymolase 100-T, 210
Zymolase buffer, 48
Zymolase suspension, 532